FRANK L. PILAR UNIVERSITY OF NEW HAMPSHIRE

CHEMISTRY
THE UNIVERSAL SCIENCE

▲ **ADDISON-WESLEY PUBLISHING COMPANY**
Reading, Massachusetts   Menlo Park, California   London   Amsterdam   Don Mills, Ontario   Sydney

This book is in the
ADDISON-WESLEY SERIES IN CHEMISTRY

Sponsoring Editor: Laura Rich Finney
Production Editor: Emily P. Arulpragasam
Designer: Jean King
Illustrations: Oxford Illustrators
Cover Design: Marshall Henrichs

Library of Congress Cataloging in Publication Data

Pilar, Frank L
   Chemistry, the universal science.

   Includes bibliographies and index.
   1. Chemistry. I. Title.
QD31.2.P55   540   78-55828
ISBN 0-201-05768-9

Copyright © 1979 by Addison-Wesley Publishing Company, Inc. Philippines copyright 1979 by Addison-Wesley Publishing Company, Inc.

All rights reserved. No part of this publication may be reproduced, stored in a retrieval system, or transmitted, in any form or by any means, electronic, mechanical, photocopying, recording, or otherwise, without the prior written permission of the publisher. Printed in the United States of America. Published simultaneously in Canada. Library of Congress Catalog Card No. 78-55828.

ISBN 0-201-05768-9
ABCDEFGHIJ-DO-79

To my first science teacher, Ed Roe, who advised me to become a cartoonist

PREFACE

Why write another general chemistry text when there are so many already on the market—and some excellent ones at that? Obviously, individual writers have their own private reasons; one of mine is that I felt I had something to say—or a way of saying it—that would make a positive contribution to the cause of chemical education. I especially wanted to share with students a personal view of chemical science—a view acquired by years of teaching and reflection—and to convey to all of them something of the beauty of chemistry, a beauty inherent not only in chemistry's simplifying concepts but in its almost infinite diversity and complexity as well. Most of all, perhaps, I wanted to infect students with some of the same excitement I felt—that quickening of the intellectual pulse—when I was a child and my father would explain to me why rain fell, how electricity worked, and why scientists were trying to split the atom. Even today I continue to be fascinated when I learn how some part of the universe operates: how DNA works, what black holes are, and how some new types of compounds behave.

Perhaps no two teachers agree exactly on what should be included—and what should be left out—in a beginning course in chemistry. Furthermore, there is wide diversity in feeling as to what the level of treatment should be, even granted that the intended audience has been agreed upon. I am assuming that students using this text have at least a modest facility with arithmetic and some high school algebra. Although it is a help to have had high school chemistry, my experience is that this is not strictly necessary. Students without such a background usually have to work harder at first in acquiring basic vocabulary and concepts, but by the end of a semester or quarter they usually have caught up with their colleagues and may even surpass them.

Throughout I have attempted to include those topics needed by students who for one reason or another wish to become literate in chemistry. Some students take the course as a foundation for further courses in chemistry or related sciences, some for satisfaction of intellectual curiosity, and others for reasons they themselves may not know. Since I am aware that not all students electing a nonterminal course in chemistry are budding chemists, I have made a deliberate effort to avoid writing at that level of rigor characteristic of texts used for chemistry majors. Yet at the same time I feel strongly that the student is done a disservice unless some minimum level of intellectual rigor is maintained.

Although some topics are covered in greater depth and breadth than others, my guiding philosophy is that expressed long ago by the Chinese philosopher Confucius: "When I have presented one corner of a subject to anyone, and he cannot learn the other three corners from it, I do not repeat my lesson." The point is that an introductory text should be *introductory*: given the proper foundation, students should be able to build upon it as individual needs require. I have tried to avoid telling students more than they need to know. This problem is most acute in an area such as molecular orbital theory, where a great many other texts go far beyond that level which beginning students can possibly appreciate. Indeed, the vast majority of the students will never be exposed to the topic again. Thus, although freshman students should know something about molecular orbitals and how they are used, most of this area is best deferred until the student is able to deal with it on a more sophisticated mathematical level.

Even though this is a textbook of chemistry, I have tried to keep in mind Barry Commoner's statement: "The separation of the laws of nature among the different sciences is a human conceit; nature itself is an integrated whole."* Consequently, where appropriate, I have not hesitated to bring in bits of physics, geology, biology, or any other humanly created subdivision of nature's laws; only in this way can the student acquire a realistic idea of what chemistry is today and—more important perhaps—what it is becoming. So many areas of interest are now being studied at the molecular level that chemistry is virtually the *universal* science. Life scientists—once limited to describing plants and animals in terms of gross anatomy—are now unraveling the complexities of reproduction, survival, and genetics in terms that used to be regarded as the sole province of the chemist. The origin of life—indeed, the origin of the universe itself—is now believed to be at least partly comprehensible in terms of relatively simple chemical laws.

The order of topics in the book is deliberate. Although facts come first and theories follow, it is nevertheless often easier to appreciate facts by studying the theory first. Consequently, I have placed facts and theories in whichever order appeared most pedagogically sound in the particular area studied. Organic chemistry has been introduced early enough so that organic reactions and structures can be used to illustrate important theoretical areas such as rates of reaction and thermodynamics. The placement of nuclear reactions, ostensibly an area of physics, is somewhat arbitrary but was based on the desire to be able to discuss tracer studies in subsequent chapters. However, those teachers who prefer to ignore nuclear reactions or to place them elsewhere can do so without difficulty. Similarly, no great problems would arise if organic chemistry were postponed to a later part of the course, although I personally feel this would be a tactical error. Organic chemistry is such an information-rich field that not to exploit it seems to me an error in judgment. This

---

* *Science and Survival*, Viking, 1967.

is not to denigrate the importance of inorganic chemistry, but choices of emphasis must be made if a book is to remain finite in length.

Many students, friends, and colleagues have helped me write this book. The early organizational work was carried out under a Visiting Professorship at the University of Wisconsin, during the 1970–71 academic year. While I was there, my students and teaching assistants discussed much of the material with me and made many valuable criticisms and suggestions for improvement. I also thank those who reviewed the manuscript in one or more of its versions: Francis Bonner of the State University of New York (Stony Brook), John Burmeister of the University of Delaware, Charles Henrickson of Western Kentucky University, James Huheey of the University of Maryland, Timothy O'Shea of Texas Wesleyan College, Lee Pedersen of the University of North Carolina, Bruce Storhoff of Ball State University, Paul Treichel of the University of Wisconsin, George Bodner of Purdue University, Lawrence Conroy of the University of Minnesota, and Stephen Webber of the University of Texas. Without their substantial help, the present text couldn't have been written. In addition, I am especially grateful to the staff at Addison-Wesley—many of whom worked behind the scenes—for their professional guidance, friendly cooperation, and enthusiastic support.

Finally, above all, I thank my wife, Anita, who typed the entire manuscript (several times!), counseled me as to grammar and style, defended the interests of the struggling student, and put up with an often cranky author.

Durham, New Hampshire F.L.P.
December 1978

A NOTE TO THE STUDENT

Great mountain peaks—like Aconcagua, Denali, and Nanga Parbat—are scaled by two types of people: those who are so strong that they can withstand virtually any physical hardship without visible strain, and those who can force themselves to continue step after interminable step even though climbing is an exhausting effort. Yet, once the summit is attained, both types have the same final accomplishment to their credit. So it is with the mastery of chemistry. Some fortunate few are gifted with photographic memories and quick, efficient reasoning abilities; others must labor hard and long in their deliberate efforts to remember and to learn how to apply. Yet, what matters most is the end result, not how one has attained it. If you are one of those who feel that a knowledge of chemistry is a worthwhile goal and you are reasonably intelligent (if not overwhelmingly gifted), you will find it possible to master chemistry but perhaps not without paying a price of considerable sweat and perhaps even a few tears. It is not enough to look at a page of your lecture notes and to say, "I think I understand all of this." If you do not go considerably beyond this point, a question given to you when you have no access to your notes or text will probably completely confuse you. You need to know the basic structure of the material forwards and backwards and to have obtained some previous practice in applying it in certain typical situations as well as anticipating its application in others. This is the point of "exhaustive climbing" which too many students either do not recognize or refuse to go beyond—yet to stop before this point most certainly means that mastery will not be attained.*

There is far too much in this book to learn all at once. Just as a mountain is scaled by walking over only a very small portion of its surface, so the basic principles of chemistry can be acquired by mastering only certain portions of this book. Your instructor will probably indicate via lectures and assignments just what path to follow—those topics and principles to emphasize. These you should know thoroughly (not merely understand!). Beyond this you need to work problems and to contemplate the significance of what you know and how it is related to other things you have learned before (both in chemistry and in other fields). It is this latter critical cross-referencing in one's mind that leads to real appreciation of knowledge and transforms mere facts into meaningful information.

You should be aware of several features of the book designed to help you to learn and to make studying and problem solving more efficient. On the inside back cover of the book are two tables; the first lists values of the most commonly encountered physical constants. You should be able to locate these quickly since you will

---

* Some excellent advice along this same general line is given by S. Paul Steed, "The EYI Study Technique," in *J. Chem. Educ.* **53**, 746 (1976).

often need these values in solving problems at the ends of the chapters. The second table lists various energy conversions that are frequently encountered. And, above all, don't overlook the appendixes! If you're rusty on the metric system (particularly the new SI units), energy units, temperature scales, exponential notation, logarithms, etc., consult the appropriate appendix.

There is a summary at the end of each chapter; use this to help you form a clear idea of the highlights of the chapter and how these are organized to represent some subarea of chemistry. The list of learning goals identifies those specific things you should know and be able to do after you have studied the chapter and worked many of the problems. Use this list to pace yourself and to uncover gaps in your learning. The list of definitions, terms, and concepts to know may appear like busy work but unless you know the basic vocabulary of the subject area of the chapter, your comprehension is apt to be inefficient and uncertain. Familiarity with the vocabulary will help you to write better exam questions so that you won't need to use the excuse that you understood the question but didn't know how to state the answer clearly. To help you in building this vocabulary, there is a glossary at the end of the text which includes brief descriptions or explanations of all the terms, definitions, and concepts listed at the end of each chapter.

Lastly, after you have familiarized yourself with the contents of a chapter—and not before—work the problems. This is where you *really* learn chemistry.

See you at the top!

CONTENTS

CHAPTER 1. THE FOUNDATIONS OF CHEMISTRY

1.1 The Scientific Method 4
1.2 The Atomic Theory of Matter 7
1.3 Chemical Symbols 11
1.4 The Internal Structure of the Atom 12
1.5 Protons, Neutrons, and Isotopes 19
1.6 The Characterization of Matter 23
1.7 Work and Energy 26
1.8 Unit Conversions 28

CHAPTER 2. CHEMICAL FORMULAS, EQUATIONS, AND PERIODIC PROPERTIES

2.1 Empirical and Molecular Formulas 37
2.2 Structural and Geometric Formulas 40
2.3 Writing and Balancing Chemical Equations 42
2.4 Calculations Involving Balanced Equations 44
2.5 The Periodic Properties of the Elements 47
2.6 Oxidation Numbers of Elements 54

CHAPTER 3. THE GASEOUS STATE

3.1 Boyle's Law 63
3.2 Charles's Law 66
3.3 The General Gas Law 68
3.4 Calculations with the General Gas Law 70
3.5 The Law of Combining Volumes 73
3.6 Dalton's Law of Partial Pressures 73
3.7 The Kinetic Molecular Theory of Gases 74
3.8 Uses of the Kinetic Molecular Equation 77
3.9 Variation of Atmospheric Pressure with Altitude 78
3.10 Critical Phenomena in Real Gases 80
3.11 The van der Waals Equation for Gases 82

CHAPTER 4. THE ELECTRONIC STRUCTURE OF ATOMS

4.1 The Nature of Light and Other Electromagnetic Radiation 94
4.2 The Origins of Quantum Concepts 95
4.3 Wave Mechanics and the Hydrogen Atom 106
4.4 The Quantum Theory of Many-Electron Atoms 112
4.5 Quantum Theory of the Periodic Table 121
4.6 Magnetic Properties of Atoms 127

CHAPTER 5. CHEMICAL BONDING

5.1 Electronegativity 136
5.2 Ionic and Covalent Bonding 138
5.3 Lewis Diagrams of Molecular Structure 143
5.4 Quantum Mechanics of the Covalent Bond 148

CHAPTER 6. MOLECULAR GEOMETRY

6.1 Polarity in Molecules 165
6.2 Lewis Formulas and Molecular Shape 170
6.3 Hybrid Orbitals and Their Use 176
6.4 Intermolecular Forces 185

CHAPTER 7. THE SOLID STATE

7.1 Crystalline and Amorphous Solids 195
7.2 The Crystal Lattice 195
7.3 Varieties of Crystalline Solids 200
7.4 Lattice Energies of Solids 202
7.5 Ionic Radii in Solids 203
7.6 Vibrations in Solids 204
7.7 Defects in Solids 205
7.8 The Specific Heats of Solids 207
7.9 The Electronic Structure of Solids 208

CHAPTER 8. THE LIQUID STATE

8.1 The Surface Tension of Liquids 216
8.2 The Viscosity of Liquids 219
8.3 Diffusion 220
8.4 Liquid Crystals 221

CHAPTER 9. CHANGES OF STATE

9.1 Heating and Cooling Curves of Simple Substances 227

9.2 Vapor Pressures of Solids and Liquids 229
9.3 Melting and Boiling 233
9.4 Energy Changes Accompanying Changes of State 235
9.5 Phase Diagrams of Pure Substances 236
9.6 Le Châtelier's Principle 240

CHAPTER 10. SOLUTIONS

10.1 Solutes and Solvents 247
10.2 Concentration Units 248
10.3 Solubility 250
10.4 Crystallization and Purification of Solids 254
10.5 The Vapor Pressures of Solutions 256
10.6 Fractional Distillation 257
10.7 Constant-Boiling Solutions 260
10.8 Colligative Properties of Solutions 261
10.9 Interionic Interactions in Solutions 266

CHAPTER 11. NUCLEAR TRANSFORMATIONS OF MATTER

11.1 The Discovery of Radioactivity 273
11.1 Natural Radioactive Decay 275
11.3 Artificial Radioactive Decay 277
11.4 The Rate of Radioactive Decay 279
11.5 The Energetics of Nuclear Transformations 283
11.6 Nuclear Fission 284
11.7 Nuclear Fusion 288
11.8 The Hazards of Radioactivity 289
11.9 Radioisotope Dating 293
11.10 Medical Uses of Radioactivity 296

CHAPTER 12. CHEMICAL EQUILIBRIUM

12.1 The Reversibility of Chemical Reactions 303
12.2 Chemical Equilibrium and the Equilibrium Constant 305
12.3 Calculations Using the Equilibrium Constant 309
12.4 Le Châtelier's Principle and Chemical Equilibrium 314
12.5 Equilibria in Multiphase Reactions 316

12.6 Irreversible Reactions 317
12.7 Equilibria in Solutions of Sparsely Soluble Ionic Compounds 318

CHAPTER 13. ACIDS, BASES, AND THEIR REACTIONS

13.1 The Brønsted-Lowry Definition of Acids and Bases 327
13.2 Relative Strengths of Acids and Bases 329
13.3 Acid–Base Equilibrium Constants 331
13.4 Calculations Involving K_a and K_b 334
13.5 The pH Scale 336
13.6 Buffer Solutions 340
13.7 Multiple Acid–Base Equilibria 344
13.8 Acid–Base Titrations 344
13.9 Acid–Base Indicators 347
13.10 The Lewis Definition of Acids and Bases 350

CHAPTER 14. SOME FUNDAMENTAL TYPES OF CHEMICAL REACTIONS

14.1 Some Simple Reaction Types 357
14.2 Oxidation–Reduction Reactions 360
14.3 Balancing Equations by the Oxidation-Number Method 361
14.4 Balancing Equations by the Half-Reaction Method 364
14.5 Disproportionation Reactions 366
14.6 Equivalent Weights 367

CHAPTER 15. HYDROGEN, OXYGEN, AND WATER

15.1 Hydrogen: Some Basic Facts 373
15.2 The Isotopes of Hydrogen 374
15.3 The Preparation of Elemental Hydrogen 376
15.4 Binary Compounds of Hydrogen 379
15.5 Reduction of Metal Oxides and Chlorides with Hydrogen 381
15.6 Atomic Hydrogen 381
15.7 Oxygen: Some Basic Facts 383
15.8 The Isotopes of Oxygen 383

15.9 The Preparation of Elemental Oxygen 384
15.10 Binary Compounds of Oxygen 384
15.11 Ozone 386
15.12 The Oxygen Cycle in Nature 389
15.13 Water: Its Unusual Properties 390
15.14 Hydrates 392
15.15 Heavy Water 395
15.16 The Water Cycle in Nature 395

CHAPTER 16. CARBON AND ITS COMPOUNDS

16.1 Elemental Carbon 407
16.2 The Oxides of Carbon 409
16.3 The Carbon Cycle in Nature 411
16.4 The Carbon-to-Carbon Bond 413
16.5 Isomerism in Organic Compounds 415
16.6 Hydrocarbons 416
16.7 Alkanes 417
16.8 Alkenes 423
16.9 Alkynes and Dienes 427
16.10 Alicyclic Hydrocarbons 429
16.11 Aromatic Hydrocarbons 431
16.12 Classification of Organic Compounds as to Functional Groups 437
　　　Alcohols 438
　　　Ethers 439
　　　Carboxylic acids 440
　　　Aldehydes and ketones 443
　　　Esters 445
　　　Amines 447
　　　Amides 448
　　　Amino acids 449
16.13 Optical Activity of Organic Molecules 450
16.14 Heterocyclic Molecules 454

CHAPTER 17. THE SYNTHESIS OF ORGANIC COMPOUNDS

17.1 Strategy and Tactics in Organic Syntheses 468
17.2 Thirty-Four Syntheses Beginning with Propane 470
17.3 Thirteen Syntheses Beginning with Benzene 478

17.4 Addition Reactions 484
17.5 The Identification of Synthetic Products 485

CHAPTER 18. THE ENERGETICS OF CHEMICAL CHANGE

18.1 The Concept of Energy 493
18.2 Energy Changes in Physical Processes 494
18.3 Energy Changes in Chemical Reactions 496
18.4 The First Law of Thermodynamics 499
18.5 Enthalpy Calculations for Chemical Reactions 499
18.6 Bond Energies 506
18.7 The Second Law of Thermodynamics 508
18.8 Spontaneity and Chemical Equilibrium 510
18.9 Free Energy and Entropy Calculations 513
18.10 Free Energy and Equilibrium 516

CHAPTER 19. ELECTROCHEMISTRY

19.1 Voltaic Cells 526
19.2 Standard Electrode Potentials 528
19.3 Electrochemical Measurement of pH 533
19.4 Storage Batteries 535
19.5 Fuel Cells 536
19.6 Galvanic Corrosion 538

CHAPTER 20. THE DYNAMICS OF CHEMICAL CHANGE

20.1 The Rate of a Reaction 545
20.2 The Effect of Temperature on Chemical Reaction Rates 549
20.3 Activated Chemical Reactions 550
20.4 Mechanisms of Chemical Reactions 553
20.5 The Activation Theory of Reaction Rates 557
20.6 Transition-State Theory of Reaction Rates 561
20.7 Chain Reactions 562
20.8 Catalysis 563
20.9 Enzyme-Catalyzed Reactions 567
20.10 Autocatalysis 570

CHAPTER 21. THE METALS OF GROUPS I AND II

21.1 The Electrolytic Production of Metals 579
21.2 The Preparation of Group IA Metals 582
21.3 The Properties of the Group IA Metals 583
21.4 The Preparation of Group IIA Metals 587
21.5 The Properties of the Group IIA Metals 588
21.6 The Group IB Metals: Copper, Silver, and Gold 591
21.7 The Group IIB Metals: Zinc, Cadmium, and Mercury 594
21.8 Mercury in the Environment 598

CHAPTER 22. THE TRANSITION METALS

22.1 Coordination Complexes of Transition Metals 608
22.2 Iron and Its Compounds 610
22.3 The Biological Role of Iron 611
22.4 Cobalt and Nickel 614
22.5 The Biological Role of Cobalt 615
22.6 The Catalyst Metals: Nickel, Palladium, and Platinum 617
22.7 Chelates 618
22.8 Optical and Magnetic Properties of Transition-Metal Complexes 619
22.9 Quantum Mechanical Models of Bonding in Transition-Metal Complexes 622

CHAPTER 23. NITROGEN, PHOSPHORUS, SULFUR, AND THE HALOGENS

23.1 Nitrogen 633
23.2 Nitrogen Oxides and Smog Production 637
23.3 The Nitrogen Cycle in Nature 638
23.4 Phosphorus 639
23.5 Sulfur 642
23.6 The Aqueous Chemistry of Metal Sulfides 645
23.7 The Halogens 646
23.8 The Hydrogen Halides 649
23.9 The Oxy Acids of the Halogens 650
23.10 Organic Halogen Compounds 651

CHAPTER 24. GIANT MOLECULES

24.1 The Nature of Giant Molecules 662
24.2 Polymers and Monomers 663
24.3 Chain Polymerization 665
24.4 Step-Growth Polymerization 669
24.5 Copolymerization 671
24.6 Steric Control of Polymerization Reactions 673
24.7 Determination of the Molecular Weights of Giant Molecules 675
24.8 The Physical Properties of Synthetic Polymers 679
24.9 The Structure of Proteins 683
24.10 Determination of Protein Structures 688
24.11 Laboratory Synthesis of Proteins 689

CHAPTER 25. THE CHEMISTRY OF LIFE: A BRIEF OVERVIEW

25.1 Photosynthesis and Cell Metabolism 699
25.2 The Chemistry of Reproduction and Survival 701
25.3 Chemical Evolution and the Origin of Life 703
25.4 Chemical Messengers and Regulators 704
25.5 Molecular Pharmacology: The Chemistry of Drugs 705

APPENDIX 1. The Metric System 706

APPENDIX 2. Exponential Notation and Significant Figures 711

APPENDIX 3. Logarithms and their Use 715

APPENDIX 4. The Naming of Simple Inorganic Compounds 723

APPENDIX 5. The Construction and Use of Contour Diagrams 728

Answers to Selected Problems 733

Glossary 744

Index 759

CHEMISTRY THE UNIVERSAL SCIENCE

1

THE FOUNDATIONS OF CHEMISTRY

An alchemist in his laboratory. Exotic objects such as stuffed crocodiles, bat wings, and animal skulls appear to have been standard furnishings. (Courtesy of Brown Brothers.)

So, Franta, you want to become a chemist! Tell me, what does a chemist do these days? I've always heard that chemists spend all their time mixing smelly things, cooking them up to make them stink even worse and sometimes blowing themselves to hell in the process. Wouldn't you be wiser to become a pharmacist instead?

KAREL SLAMŠIDLO, uncle of the author

Chemistry may be identified as the science that deals with the study of matter, its structure, and the changes it undergoes. If one filled a huge barrel with little slips of paper on which were written the names of all the topics of interest and importance to humankind, it would be difficult to draw out one at random to which chemistry could not be applied.

We do not know when humans first began to realize that changes or transformations occur in matter and that these changes can, to some extent, be controlled. Perhaps the discovery and use of fire antedated such a realization. If so, the world's first applied chemists may have been those savages who discovered that cooking certain foodstuffs made them more palatable, easier to chew, and more digestible. And the first conscious attempt to initiate other chemical changes in nature may have been associated with the smelting of metals in the production of tools and weapons. At any rate, the art of extracting copper from its ores was known in Mesopotamia as early as 3500 B.C. and was known to the Egyptians approximately a thousand years later. By the time another thousand years had passed (1400 B.C.) a good grade of glass was being produced at Tell-el-Amarna and natural dyes were in extensive use among various Mediterranean civilizations.

A document discovered in 1828 (the Theban Papyrus) indicates that considerably before A.D. 300, the Egyptians had developed detailed recipes for carrying out chemical operations on common objects to make them resemble rare metals and gems. Although it appears that the practitioners of such arts were under no delusion that the changes they wrought were anything but superficial, it is very likely that out of this legacy grew the medieval practice of **alchemy**, which attempted, among other things, to transform base metals to gold. Alchemy, in turn, led to growing discoveries of technological significance such as new dyes, medicinals, improved apparatus and techniques, and eventually (through the modern science of chemistry) to the production of new elements. Slowly, haltingly—but relentlessly—the chemical knowledge accumulated through the ages has been applied to virtually every area of human interest until today we stand upon the threshold of understanding, and perhaps even controlling, the very chemistry of life itself.

1.1 THE SCIENTIFIC METHOD

Primitive peoples were probably relatively content to accept the crude chemical technology of their time without considering the possibility that it might be ordered around a few general principles. Nevertheless, as more and more knowledge accumulated—whether by accident or by intent—it must have slowly dawned on some contemplative and perceptive individuals that certain underlying patterns were apparent. For example, the winning of metals from their ores involved similar steps regardless of the specific metal processed. It is at such a point that chemistry *as a science* may be said to have had its beginnings. However, it was not until the seventeenth century, when men such as Francis Bacon and Robert Boyle began to speak out for a critical examination of existing knowledge, that an orderly and systematic approach arose and led to what is now usually referred to as the **scientific method**. Under ideal conditions the method may be said to consist of the following steps:

1. The accumulation of observations and their classification into various categories.
2. The summarization or generalization of the observations in the form of succinct statements called **laws**.
3. An attempt to find some overall pattern or design which, if assumed to exist, accounts for the laws in a deductive manner.

Robert Boyle (1627–1691) was an Irish-born natural philosopher who excelled as a linguist and scientific experimenter. He was the first to note that the temperature at which water boils depends on the pressure of the atmosphere, and he published a law of gaseous behavior (see Chapter 3) now known under his name (although actually discovered by his assistant, R. Towneley). Boyle's fame rests most securely on his influence on the development of the experimental method in science.

Francis Bacon (1561–1626) was a British statesman, essayist, and natural philosopher who was convinced that his life was ordained to serve humanity by the production of good discovered by truth. Many of his writings deal with the methods used to produce true conclusions from observations.

In the first step, classification allows a large number of seemingly different observations to be reduced to a few basic types and thus simplifies conceptualization. As understanding based on the original simplified classification grows and becomes more sophisticated, it frequently becomes possible to discover increasingly complex interrelationships and to incorporate these into improved classification schemes.

The process of generalization makes use of **inductive reasoning**, that is, the working from specific situations to general situations. Conclusions arrived at inductively can never be proved absolutely correct; at best their validity can be made to appear more probable as repeated observations are shown to be in accordance with the assumed laws. Thus, as opposed to everyday usage of the word, a scientific or natural *law* is not something which nature must obey; rather, it is simply a description of how nature is thought to behave. For example, the observations of many centuries indicate that the sun rises in the east and sets in the west; this may be regarded as a law of motion of the sun as viewed from earth. Yet, one cannot assert with absolute certainty that the sun will never rise in the west and set in the east. Laboratory observations that a given set of reactants invariably produces a given set of products belong to observations of this same type. Of course, we would be so surprised to observe exceptions in the above examples that it is very difficult to acknowledge the tenuous nature of the laws involved.

In the third step of the scientific method, the overall pattern which appears to underlie a set of observations is usually referred to as a **theory** or **model**. By means of **deductive reasoning**, the theory or model is used to predict details of observations before they are made. Deductive reasoning permits the determination of specific consequences of general situations and thereby predicts the laws a given model will necessarily satisfy. Unlike inductive reasoning, deductive reasoning arrives at conclusions that are rigorously true, *insofar as the general case from which one works is true,* and are not subject to probabilistic uncertainties. For example, if the heliocentric theory of the solar system is accepted, with the planets orbiting about the sun and spinning in a certain sense, then the apparent rising of the sun in the east and setting in the west *must* follow as a logical and necessary consequence. No alternative conclusion exists.*

Perhaps the most important attribute of a good theory or model is that it helps to make some sense out of our observations. A good model helps to establish connections which may have been overlooked before and replaces that which appears erratic and accidental with that which is regular and necessary. A good model not only correlates existing data but also facilitates the classification and interpretation of new data. A model should also suggest new experiments to try or things to look for which might otherwise be overlooked. Of course, the discovery of a single observation which cannot be deduced from the model constitutes a flaw which necessitates modification—or even total abandonment—of the model.

In practice, the growth of a science is rarely, if ever, characterized by a smooth, orderly application of the scientific method. Rather, growth is more apt to be by fits and starts with many halts and temporary reversals. Occasionally, observations may be made carelessly, theories may be formulated prematurely, before the proper observations have accumulated, or models may be clumsily stated. In general, theories are always incomplete in the sense that they cannot account for the laws arising from an ever-growing body of observations. The incompleteness of a theory may be revealed when it is called on to deduce (that is, predict) observations not hitherto recorded. If actual experimentation confirms the predictions, the theory is said to be strengthened; otherwise the theory is incomplete or inadequate

---

* This does not rule out the possibility that a different model also correctly predicts the rising and setting of the sun. In such an event one must invoke additional criteria in order to determine which is the *better* model, that is, which model is in accordance with the greater number of observations.

and must be modified or replaced.* Little by little, a theory is subjected to the close scrutiny of critics who eventually fill in gaps of logic, perfect the terminology, contribute elegance and simplicity, and delineate the limits of applicability. Usually, only then does it become possible for the next step to occur: the construction of a new theory far more comprehensive than the old, yet encompassing the old within it. Each such cycle—which may require decades, generations, or centuries to complete—represents but a single step along the path of our journey toward knowledge of ourselves and our surroundings.

Individual scientists are as human as you or I and tend to be biased observers of nature. The objectivity of science arises only because scientists deliberately expose their data and views to public scrutiny, whereupon the false and unproductive notions become revealed and ultimately are discarded. Human beings, whether scientists or laypeople, are prone to faulty observation and erroneous reasoning, and their viewpoints are easily distorted by emotion. Consequently, the scientific method characterizes the *collective*, not the *individual*, behavior of scientists as a group acting over long periods of time. It is highly unlikely that any individual scientist ever begins a day's work by saying: "Let's see now, what step of the scientific method do I apply next in my experiment?" Many of the major scientific discoveries were made accidentally by people looking for something else (sometimes something rather foolish!) or for nothing at all. Thus the scientific method is usually applied with hindsight when trying to put all the facts together to make sense, to formulate laws, to design models, etc. Ultimately, however, it must be possible to reconstruct science *as if* the growth had occurred via step-by-step application of the scientific method.

It might be argued that all people are artists by nature—artists who choose a variety of widely differing means to give expression to their creative yearnings. Thus the human being as chemist differs only superficially in nature, outlook, goals, and even methodology from the human being as painter, poet, musician, or philosopher.† To be sure, the specific techniques and tools of one type of artist may differ drastically from those of another type, and each chooses a different small subdivision of the universe within which to develop his or her creative instincts. Yet, each type of artist attempts to conceptualize, in essentially abstract form, some interaction (physical, spiritual, or both) with his or her surroundings. Whereas painters do this with color, form, and texture, scientists utilize observations of nature to fashion abstract and stylized descriptions of that for which they have but indirect evidence. Although scientists are obliged to deal with nature as it is, they construct theories using concepts and terms that are not uniquely supplied by nature; bold leaps of the imagination are often needed to take only a bare skeleton of facts and from these construct a general model transcending that which is known. Two different scientific theories may be equally valid in that they lead to the same set of laws, yet one theory may be simple, concise, and replete with appealing imagery while the other is unwieldy, prolix, and colorless. For example, both the heli-

---

* Sometimes very small numerical discrepancies force the drastic modification of a theory. This is one of the reasons some scientists insist on exceedingly accurate measurements.

† J. Bronowski in his book *The Ascent of Man* (Little, Brown, Boston, 1974, p. 412) says: "Man is unique not because he does science, and he is unique not because he does art, but because science and art equally are expressions of his marvellous plasticity of mind."

ocentric and geocentric models of the universe are equally able to account for a variety of astronomical observations; yet the former, in doing this more simply and elegantly, is generally regarded as a more pleasing artistic achievement.

1.2 THE ATOMIC THEORY OF MATTER

In medieval times, alchemists sought in vain for the philosopher's stone—a mythical substance thought to be capable of transforming ordinary, commonly available substances into precious gold. (Had it been found, everything could have been turned into gold—thus making gold valueless!) Though the quest was futile, it nevertheless produced rewards even more valuable than those it sought, because it encouraged the aquisition of a great deal of practical chemical knowledge whose exploitation eventually led to the chemistry of today. Alchemists discovered many new substances, invented new apparatus, perfected standard laboratory techniques such as crystallization and distillation, recorded a variety of observations, and, in many cases, passed on knowledge to others.

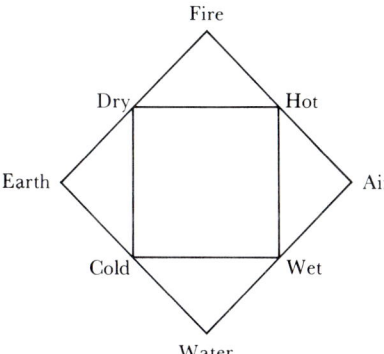

A theory of matter as proposed by the Greek philosopher and statesman Empodocles during the fifth century B. C. Each of the four elements (fire, air, earth, and water) was said to possess two of four basic properties; fire is hot and dry, earth is cold and dry, etc.

Over the course of time it became apparent to perceptive alchemists that matter could be physically separated into different pure substances, each associated with a unique set of properties (color, crystal form, density, etc.). For example, seawater can be separated by distillation into pure water and a solid residue. By use of fractional crystallization, this solid residue can be separated further into sodium chloride (common table salt) and other identifiable substances, many of which are obtainable from other sources. The important observation is that each such pure substance, the water, for example, does not appear to be further separable into components without showing rather abrupt and discontinuous changes in properties. Thus, if an electric current is passed through liquid water, two new substances—both gases—are produced. Some pure substances, iron for example, were found to resist separation into other substances no matter what the means employed.

The inseparable, fundamental—indeed, elemental—substances have come to be known as **elements**; the pure substances that were found to be separable into component substances are called **compounds**. Whenever two or more elements or compounds exist together, such that they can be separated by physical means, the aggregate is termed a **mixture**. Thus seawater is a mixture of many substances: water, sodium chloride, and a variety of other compounds. By the use of **chemical analysis** (the breakdown of substances into their components and the identification of these components) and **chemical synthesis** (the formation of new substances by the union of other substances), chemists have managed to construct a very orderly model of the general composition of all naturally occurring matter in terms of fewer than one hundred different elements.

Early chemists noted that some compounds (for example, those which were gases) were relatively easy to prepare and to study, while others (notably proteins, cellulose, and other substances obtained from living organisms) were difficult to prepare reproducibly and to study quantitatively. Accordingly, most of the fundamental theories of chemistry are based on observations involving the former type of compound. Substances of the latter type are discussed in Chapter 24.

Analyses and syntheses of simple substances reveal several numerical relationships which appear characteristic of them as a class. In the eighteenth century the great French scientist Antoine Lavoisier carried out careful studies which suggested that chemical transformations of one or more substances involve no perceptible change in mass. This observation is now known as the **law of conservation of matter**: matter is neither created nor destroyed during ordinary chemical transformations. This law appears to be universally valid for all types of compounds, simple or complex.

Antoine Lavoisier (1743–1794) is generally regarded as the founder of modern chemistry. His textbook Traité élémentaire de chimie *set the pace for much of the future development of chemistry. Lavoisier provided the first correct explanation of combustion, demonstrated the conservation of matter, and laid the foundation of studies in metabolism. In addition, he contributed extensively to the development of agriculture and to reforms in education and the penal system. On May 8, 1794, he was guillotined by a revolutionary court distinguished more by arrogance and ignorance than by a desire for justice. Lavoisier's statement of the law of conservation of matter is quoted at the beginning of Chapter 2.*

It was also discovered, insofar as simple compounds were concerned, that a given pure compound always contains the same proportion by weight of its constituent elements regardless of how that compound was obtained, whether by purification from natural sources or by synthesis in the laboratory. For example, water is found to consist of 8889 parts by weight of oxygen to 1111 parts by weight of hydrogen, whether it is prepared by distillation of seawater or by burning of wood. This observation is the basis of the **law of definite composition**: a given compound always contains the same elements and these occur in the same relative amounts by weight. This law, it is now known, applies only to gaseous compounds and to some simple compounds which are liquids or solids; compounds such as proteins, synthetic polymers, and certain sulfur-containing ores constitute exceptions to the law of definite composition.

Antoine Lavoisier and his wife. (Courtesy of The Metropolitan Museum of Art; purchase, Mr. and Mrs. Charles Wrightsman Gift, 1977.)

Occasionally, two or more different compounds may contain the same elements but in different proportions. For example, there are two different compounds of phosphorus and oxygen, which lead to the following analyses:

	COMPOUND 1	COMPOUND 2
Unit weights of phosphorus	100	100
Unit weights of oxygen	77	129

The weight ratio of oxygen to phosphorus is 77/100 in compound 1, but 129/100 in compound 2. This means that for a fixed weight of phosphorus, compound 2 has 129/77 or 5/3 as much oxygen as has compound 1. In all similar situations it is found that for a fixed weight of one element, the ratio of the weights of the second (and third, fourth, ... elements, if present) is expressible as small, whole numbers: 5/3 in the example above. This very important observation is referred to as the **law of multiple proportions**.

As another simple example of the law of multiple proportions, there are two compounds of carbon and oxygen that contain 42.9% and 27.3% of carbon by weight, respectively. The two oxygen-to-carbon ratios are 57.1/42.9 = 1.33 and 72.7/27.3 = 2.66. Thus the ratio of the weight of oxygen to that of carbon for a fixed weight of carbon is 1.33/2.66 = 1/2 for the two compounds.

In 1803, an English schoolmaster, John Dalton, proposed a simple model that appeared to account for the three laws: conservation of matter, definite composi-

John Dalton, English schoolmaster who first stated the atomic theory in its modern form. (Courtesy of The Bettmann Archive.)

John Dalton (1766–1844) was an amateur meteorologist, private tutor, and teacher of mathematics and physics at New College, Manchester, England. A devout Quaker, he carried out one of the first studies of color blindness (he himself was afflicted).

tion, and multiple proportions. Stated in modern terms Dalton's theory is:

1. Elements consist of tiny, indivisible particles called **atoms**, which, for a given element, are alike in size, form, and mass. In general, size, form, and mass differ from element to element.
2. Atoms are **immutable**, that is, atoms of one element never change into atoms of another element.
3. Compounds are unions of two or more elements. The unit containing the smallest number of atoms whose union constitutes a compound is called a **molecule**.
4. Atoms combine in simple whole-number ratios.

Dalton's model, now called the **atomic theory**, had been proposed about two thousand years earlier by the Greek philosophers Democritus and Leukippos on the basis of sheer speculation. Since Greek science was not advanced enough to test out the theory, it exerted no influence on the development of the science of chemistry.

Democritus (about 460–370 B.C.) was a Greek philosopher of the so-called atomist school living in Thrace. His teacher, Leukippos, taught him that all things were composed of atoms. Democritus loved to explain how things worked and thus is sometimes called the father of modern (mechanistic) science. He also anticipated the law of conservation of matter.

In spite of the success of Dalton's atomic theory in explaining some of chemistry's most fundamental laws, the theory was not generally accepted by the scientific community until approximately one hundred years later, when Einstein pointed out that Brownian motion could be explained by assuming that the erratically moving particles were being subjected to random collisions by molecules of the suspending medium.* Einstein thus provided rather direct evidence for what were previously regarded as undetectable entities.

1.3 CHEMICAL SYMBOLS

The early chemists originated certain shorthand methods for indicating the elemental composition of compounds. Alchemists employed special mystical symbols to represent elements and certain common substances; John Dalton employed simple symbols for elements which he then combined to designate compounds. In the modern version of Dalton's notation, the 105 or more chemical elements (about 90 occurring naturally, the rest produced artificially) are represented by one- or two-letter symbols derived from the element's name in Latin, Greek, or, occasionally, some other language. Fourteen of the elements have one-letter symbols. For example, H is used to symbolize hydrogen, O oxygen, S sulfur, U uranium, and W tungsten (wolfram). The remainder have two-letter symbols: Na for sodium (natrium), Ar for argon, Cl for chlorine, Mg for magnesium, Fe for iron (ferrum), and Hg for mercury (hydrargyrum). Compounds are indicated by aggregates of atomic symbols called **formulas**. Water and sodium chloride are represented by the formulas H_2O and NaCl, respectively.

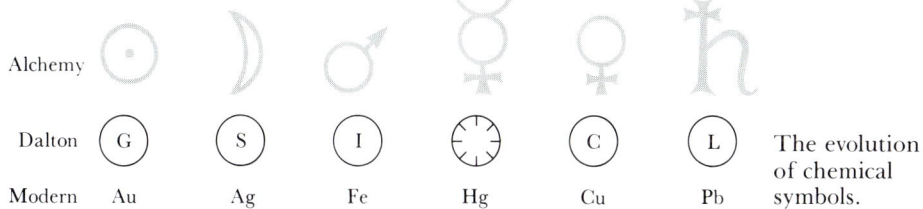

The evolution of chemical symbols.

Symbols such as H, O, Na, and Cl are generally understood to represent either one atom or one **mole** (abbreviated mol) of the element in question. Similarly, the symbol H_2O represents either one molecule or one mole of that compound. The mole is defined more precisely in Section 1.5; for the present it may be regarded as a specific number whose approximate value is 6.02×10^{23}. The mole is sometimes called the "chemist's dozen."† Two atoms or two moles of an element such as hydrogen are designated by writing 2H, and $2H_2O$ represents two molecules or two moles of water molecules.

---

* When a light beam enters a small opening in a darkened room, dust particles are seen to carry out erratic and zigzag motions within the beam. This is called *Brownian motion*, since it was first described by the Scottish physician and botanist Robert Brown (1773–1858).

† To gain some appreciation of the size of a mole, it would take approximately 2×10^{16} yr for a person to count to a mole at a rate of one number per second, night and day. Also, a mole of oranges, each 3 cm in radius, would cover the entire earth (land and oceans) to a depth of over 200 km.

The symbol NaCl for sodium chloride (common table salt) indicates that a mole of this substance contains one mole of sodium ions (Na$^+$) and one mole of chloride ions (Cl$^-$). An **ion** is an atom or group of atoms having a net electrical charge. The ion Na$^+$ has one less electron than the atom Na and thus has a net positive charge. Similarly, the ion Cl$^-$ has one more electron than the atom Cl and thus has a net negative charge. As will be discussed later, some substances are made up of ions and others are not. For example, water (H$_2$O) does not consist of ions but of atoms (two of hydrogen and one of oxygen) held in close proximity by mutual attractions of electrons and nuclei.

Whether a symbol such as 2H means two *atoms* or two *moles* is either apparent from the context, or, as often occurs, is unnecessary to specify. The formula H$_2$O means that a single molecule of water consists of two hydrogen atoms and one oxygen atom; alternatively, the formula also means that every mole of water molecules consists of two moles of hydrogen atoms and one mole of oxygen atoms. In the case of NaCl, no isolable entity such as a molecule exists (except in the gas phase), and the formula merely indicates the 1:1 ratio of sodium to chlorine atoms characteristic of the compound.

It is important, however, to distinguish clearly between the symbols 2H and H$_2$. Although both symbols imply two moles of hydrogen atoms, in the former the hydrogen atoms are in atomic form, that is, uncombined, while in the latter the hydrogen atoms are joined chemically to form a hydrogen *molecule*.

1.4 THE INTERNAL STRUCTURE OF THE ATOM

One of the early objections to Dalton's atomic theory ran something like this: If atoms differ only as to size, form, and mass, how is it possible to account for the different chemical and physical natures of different elements? For example, if Na (sodium) and Au (gold) differ only in the size, form, and mass of their atoms, why does the former react so violently with water while gold is so inert? Why is sodium soft enough to cut with a knife while gold is considerably harder? Some attempts were made to relate such differences to differing *shapes* of atoms, but these were notably unsuccessful; the accumulating evidence pointed to an answer in terms of an atomic *substructure* involving particles smaller than the atom itself.

Michael Faraday (1791–1867) began his career the virtually uneducated son of a blacksmith, to become one of the most important scientists Britain ever produced. As a chemist Faraday discovered benzene and was the first to produce compounds of carbon and chlorine. He also discovered the basic laws of electrolysis (see Section 21.1). Among Faraday's most important achievements were his experiments on electricity and magnetism, which led to his invention of the dynamo and to the development of the theory of electromagnetism. A portrait of Faraday is found in Chapter 21, page 581.

The researches of Michael Faraday (around 1850 and thereafter) on electrical phenomena, along with the work of others, suggested that matter consisted of positively and negatively charged particles present in equal amounts so that the bulk material itself was electrically neutral. Faraday noted that certain substances, NaCl, for example, could conduct electric current (a flow of charge) either in aqueous solution or in molten form. Furthermore, the passage of the electric current through such substances (termed **electrolytes**) produced chemical changes in them. Faraday also made the important discovery that there were some simple numerical relationships between the amount of electric current passing through a solution and the chemical changes produced (see Section 21.1). Earlier, Sir Humphry Davy* had shown that passage of an electric current through aqueous NaCl led to the production of hydrogen gas (H_2), chlorine gas (Cl_2), and a solution of caustic lye (NaOH, sodium hydroxide), whereas if molten NaCl were used, the products were metallic sodium (Na) and chlorine gas (Cl_2) (see Section 21.2).

Around 1880 the Swedish chemist Svante Arrhenius noted that electrolyte solutions (solutions which conduct electric current) exhibited certain anomalous properties. For example, it was known that the addition of certain substances to water depresses its freezing point by an amount very nearly proportional to the number of moles of molecules per unit amount of water (see Section 10.8). Electrolytes, however, depress the freezing point by almost double the expected amount, or by even more. On the basis of these and other observations, Arrhenius postulated that certain substances, when dissolved in water, behave as mobile charged particles now called **ions**. For example, NaCl in water is assumed to consist of positively charged sodium ions (Na^+) and negatively charged chloride ions (Cl^-) (the algebraic signs are assigned on the basis of arbitrary conventions of electrostatics).

Svante Arrhenius (1859–1927) was the first to propose the theory that electrolytes consist of charged particles in solution. Young Arrhenius first presented this theory in his doctoral dissertation. His teachers graded the dissertation as mediocre and granted him his degree with some misgivings. Arrhenius also pioneered in the study of the rates of chemical reactions and the laws governing them (see Section 20.5). He received the Nobel prize in chemistry in 1903. A portrait of Arrhenius is found in Chapter 20, page 559.

In 1890, Stoney† pointed out that the observations of Davy, Faraday, Arrhenius, and others suggested the existence of a fundamental unit of negative electric charge. He proposed to call this unit of negative charge an **electron**.‡

During the last two decades of the nineteenth century, Sir J. J. Thomson of Great Britain and Jean-Baptiste Perrin of France, working independently, followed up some discoveries of Sir William Crookes and obtained strong evidence for the existence not only of electrons but of various positively charged electrical units as

---

* A biographical note on Davy is found in Chapter 19, page 526. Davy's description of his discovery of potassium is quoted at the beginning of that chapter.

† George J. Stoney (1826–1911) was an Irish-born British scientist who carried out research on physical optics and molecular physics.

‡ A comprehensive historical account of the discovery of the electron is given by B. A. Morrow in *J. Chem. Educ.* **46**, 584 (1966).

well. Furthermore, Thomson was able to carry out quantitative measurements of some properties of these charges. By use of a device known as a **cathode ray tube** (see Fig. 1.1), Crookes had earlier demonstrated that the emanations arising from the cathode—the cathode rays, as they were called—possessed certain characteristics:

1. They traveled in straight lines and cast a sharp shadow.
2. Their path could be deflected by magnetic and electric fields; the direction of the deflection indicated that cathode rays were negatively charged.
3. The cathode rays produced were always the same no matter what material the cathode was made of.

Thomson concluded that cathode rays were beams of tiny negatively charged particles and that these were the electrons earlier proposed by Stoney. By balancing the deflections of magnetic and electric fields on the electron beam, Thomson was able to measure the electron charge-to-mass ratio, e/m_e, as -1.759×10^{11} C·kg^{-1}.

In a later experiment, the American physicist Robert Millikan measured the charge on the electron as -1.602×10^{-19} C (see Fig. 1.2).* From this, the mass of the electron can be calculated as follows:

$$m_e = \frac{e}{e/m_e} = \frac{-1.602 \times 10^{-19} \text{ C}}{-1.759 \times 10^{11} \text{ C·kg}^{-1}} = 9.110 \times 10^{-31} \text{ kg}$$

J. J. Thomson (1856–1940), British physicist and director of the Cavendish Laboratory at Cambridge, was winner of the Nobel prize in physics in 1906. Jean-Baptiste Perrin (1870–1942) was a French physicist who studied cathode rays and was one of the first to calculate Avogadro's number. Perrin received the Nobel prize in physics in 1926.

William Crookes (1832–1919), a British scientist, was known as the discoverer of the element thallium and one of the first to study the fundamental properties of cathode rays. Crookes was interested in the occult and conducted many investigations of extrasensory phenomena.

The main observation, originally made by Crookes, that suggested that electrons are particles was that a paddle wheel placed in the path of the cathode beam would turn. It is now known that electrons have masses and velocities too small to produce sufficient momentum to turn the wheel. Rather, the electrons heat the surface of the paddle wheel, this energy is imparted to residual gas molecules in the imperfect vacuum of the tube, and their recoils move the wheel.

---

* The **coulomb** (abbreviated C) is a unit of electrical charge. It is the charge passing through a resistance of one ohm (Ω) during a time interval of one second when the potential difference across the resistance is one volt (V). An electric current of one C·s^{-1} is commonly called an **ampere** (A).

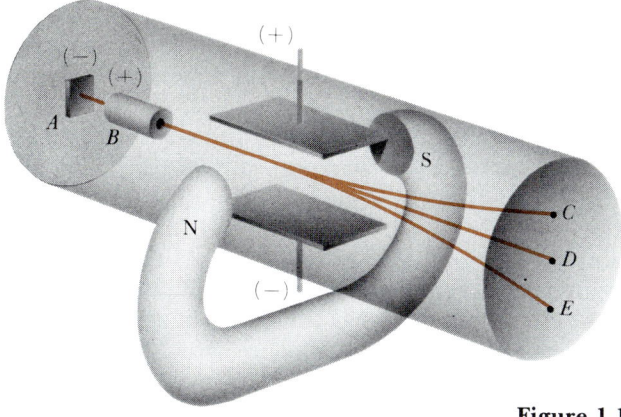

Figure 1.1

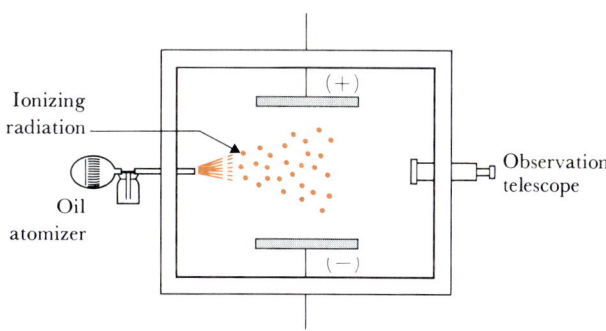

Figure 1.2

Fig. 1.1. Apparatus for measuring the charge-to-mass ratio of the electron. The entire apparatus is housed in a partially evacuated glass tube. Electrons are ejected from the cathode A and accelerated by a high voltage to the hollow plate B. Momentum keeps the electrons going and they travel in a straight line to hit point D at the end of the evacuated glass tube. If an electric field is set up between the horizontal flat plates, the electron beam is deflected upward and hits point D. Similarly, if the large magnet is turned on, the beam of electrons is deflected downward and strikes point E. If the electric field and the magnetic field are turned on simultaneously and adjusted so that their respective deflections cancel, one can calculate the charge-to-mass ratio of the electron from strengths of the electric and magnetic fields at this balance point and from the geometry of the apparatus.

Fig. 1.2. Diagram of Millikan's oil-drop experiment to measure the charge on an electron. Oil droplets are sprayed into the chamber and are ionized by X-ray radiation. The electric field between the parallel horizontal plates at the top and bottom of the box is adjusted so that a drop remains suspended. The field is then turned off and the drop allowed to fall by the action of gravity. Measurement of the rate of fall of the drop allows its mass to be calculated. From this and the strength of the electric field needed to suspend the drop, the charge on the drop may be calculated. Millikan found that the drops always had charges which were whole-number multiples of 1.6×10^{-19} C.

Since the principle of electrical neutrality requires that all negative charges be balanced by positive charges, Thomson looked for traveling positive rays as well. As expected, the positive rays were found traveling in a direction opposite to that of the electrons.* Thomson found that the charge-to-mass ratio of the positive rays varied with the nature of the residual gas in the cathode ray tube. The largest charge-to-mass ratio was obtained when the residual gas was hydrogen (H); the value of e_H/m_H was found to be 9.576×10^7 C·kg^{-1}. When the residual gas was helium (He), e_{He}/m_{He} was found to be 2.394×10^7 C·kg^{-1} and 4.788×10^7 C·kg^{-1}; that is, two different charge-to-mass ratios (one value double that of the other) were obtained. It is now known that the positive rays in the hydrogen-containing tube are hydrogen ions H$^+$ (a hydrogen atom less a unit of negative charge, that is, less an electron) and that the two helium values correspond to the ions He$^+$ and He^{2+}, respectively.

Additional evidence for the existence of charged subunits of the atom was furnished by the phenomenon of radioactive decay (see Chapter 11). Experiments showed that many of the heavier elements emitted electrons (called **β-particles**) and helium ions, He^{2+} (called **α-particles**) when they underwent spontaneous decay and were transformed to different elements.†

Once the idea of atoms assembled from equal numbers of positive and negative particles became accepted as a reliable working hypothesis, scientists began to make conjectures as to the details of the arrangement of the charged particles within the atom. Sir J. J. Thomson personally favored a model in which the atom was a diffuse sphere of positive charge in which were embedded the negative electrons—raisins afloat in a sea of positive pudding. By contrast, the Japanese physicist Nagaoka proposed a model in which a spherical positive charge was surrounded by a ring of swirling electrons—like the gas ring of the planet Saturn. Only one fact about the atom appeared to be certain—its size was on the order of a few angstrom units‡ or less in diameter—an incredibly tiny entity.

Ernest Rutherford (1871–1937), a New Zealand-born British scientist, is considered the father of nuclear physics. During tenures at McGill University in Montreal, the University of Manchester, and the Cavendish Laboratory at Cambridge, Rutherford made fundamental discoveries which laid the foundations of much of modern chemistry and physics. He received the Nobel prize in chemistry in 1908. A fascinating and very readable biography of Rutherford has been written by his colleague, A. S. Eve (Rutherford, New York: Macmillan, 1939).

---

* Thomson found the positive rays by perforating the cathode and looking for them behind this cathode. A second set of magnetic and electric fields in this region allowed the charge-to-mass ratio to be determined.

† As discussed in Chapter 11, the electrons emitted by radioactive decay originate from within the nucleus, whereas the cathode-ray electrons come from the extranuclear part of the atom.

‡ One **angstrom** unit (usually abbreviated by the symbol Å) is 10^{-8} cm or 10^{-10} m. It would take 100 million Å to span 1 cm. SI recommends that the angstrom unit be abandoned and that the nanometer (10^{-9} m) be used instead. Note that 10 Å = 1 nm. See Appendix 1 for a full discussion of these and other metric units.

Ernest Rutherford, originator of the nuclear model of the atom. (Courtesy of Brown Brothers.)

Using an apparatus similar to that diagramed in Fig. 1.3, Rutherford and some of his research students bombarded a foil of gold with helium ions (alpha particles, He^{2+}). As indicated in Fig. 1.4, if the atom's positive charge were diffuse as in the Thomson model, then virtually all of the α-particles would pass through the foil undeflected, or perhaps deflected but slightly, from a straight-line path. Rutherford did find that most of the α-particles were deflected little, if at all. However, there were a few deflections at surprisingly large angles; about one α-particle out of 10,000 was deflected by roughly 180°—straight back toward the source!

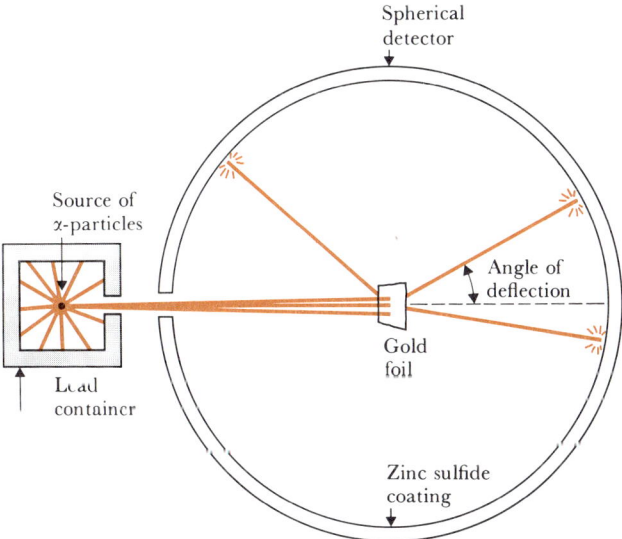

Fig. 1.3. Schematic representation of the Rutherford scattering experiment. The α-particles strike the gold foil, pass through it (usually with only minor deflections), and strike the zinc sulfide coating on the inside of the detector surrounding the foil. Each time an α-particle hits the zinc sulfide screen, a tiny flash of light is seen. Rutherford made a careful count of the number of α-particles and the angles by which they were deflected. He found that most of the α-particles were deflected by very small angles but a very few were deflected by unexpectedly large angles—some by as much as 180°, straight back toward the source.

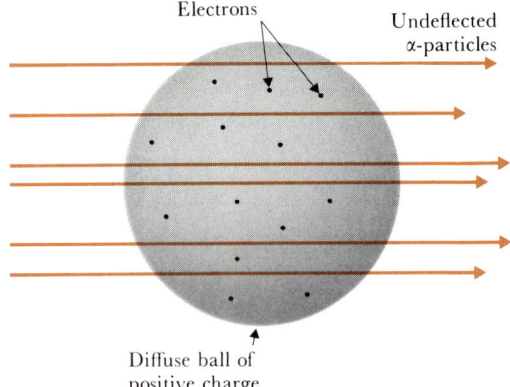

Fig. 1.4. The interaction of α-particles and atoms as expected on the basis of the Thomson model of the atom. The nuclear charge is too thinly spread out to repel the α-particles very much, and the electrons are easy to miss and too light to cause significant deflections upon collision.

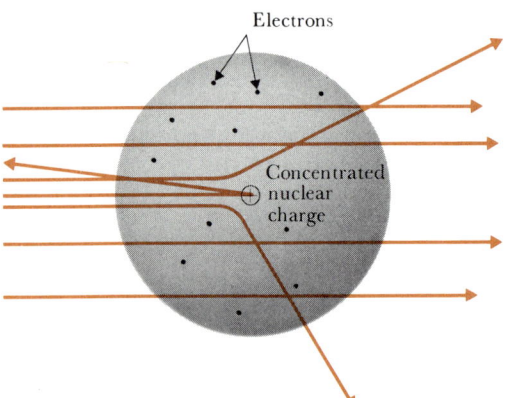

Fig. 1.5. The interaction of α-particles and atoms as expected on the basis of the Rutherford model of the atom. The electrons are as easy to miss as in the Thomson model. The concentrated nuclear charge is easier to miss than in the Thomson model, but if it is hit or even grazed, large deflections are caused.

As indicated in Fig. 1.5, only a head-on or near head-on collision between an α-particle and a dense, compact, positively charged nucleus could account for path deflections as large as 180°. Most of the α-particles would pass far enough from such a nucleus to suffer virtually no deflection at all. According to a well-known law of electrostatics (Coulomb's law) the force of repulsion between an α-particle of charge q_α and a nucleus of charge q_N is given by

$$F_{\alpha N} = K \frac{q_\alpha q_N}{r_{\alpha N}^2}$$

where $r_{\alpha N}$ is the distance between the α-particle and the nucleus and K is a constant whose value is not needed here. Rutherford realized that if this repulsive force was to be large enough to turn back an α-particle by an angle approaching 180°, then the charge on the gold nucleus had to be very, very concentrated—perhaps to a region bounded by only a ten-thousandth of the atom's diameter. A larger, more spread-out nuclear charge could never repel a swiftly moving α-particle strongly enough to hurl it back at the observer.

Reasoning along these lines, Rutherford concluded that the behavior he had observed could be rationalized only by assuming that the positive nucleus occupies a very small portion of the atom's volume (a region about 10^{-14} m or 10 fm in diameter) and that the electrons occupy the largely empty space (about 10^{-10} m or 0.1 nm in diameter) about the nucleus.

1.5 PROTONS, NEUTRONS, AND ISOTOPES

Additional studies by Rutherford and his co-workers led to the determination of nuclear charges, which were found to be multiples of the fundamental unit of charge exhibited by the electron. The simplest model of atomic nuclear structure appeared to be a nucleus of fundamental positively charged particles called **protons**. A proton has a charge of 1.602×10^{-19} C, equal in magnitude to that of the electron but opposite in sign. Since hydrogen is the lightest known element and since its charge-to-mass ratio is the largest found in cathode ray experiments, it is reasonable to assume that a hydrogen atom consists of a single electron and a single proton; that is, a proton is just a hydrogen ion (H^+). Given the measured value of $e_H/m_H = 9.576 \times 10^7$ C·kg^{-1}, the mass of the proton is calculated as

$$m_H = \frac{e_H}{e_H/m_H} = \frac{1.602 \times 10^{-19} \text{ C}}{9.576 \times 10^7 \text{ C·kg}^{-1}} = 1.673 \times 10^{-27} \text{ kg}$$

The ratio of the mass of the proton to the mass of an electron then is

$$\frac{m_H}{m_e} = \frac{1.673 \times 10^{-27} \text{ kg}}{9.110 \times 10^{-31} \text{ kg}} = 1836$$

Clearly, the proton is considerably more massive than the electron.

The charge carried by an α-particle is found to be 3.204×10^{-19} C, just twice that of a proton (hence the symbol He^{2+}). If we use the larger of the two e_{He}/m_{He} values from the cathode ray studies, the mass of the helium nucleus is

$$m_{He} = \frac{e_{He}}{e_{He}/m_{He}} = \frac{3.204 \times 10^{-19} \text{ C}}{4.788 \times 10^7 \text{ C·kg}^{-1}} = 6.692 \times 10^{-27} \text{ kg}$$

The ratio of the helium nuclear mass to that of the proton is

$$\frac{m_{He}}{m_H} = \frac{6.692 \times 10^{-27} \text{ kg}}{1.673 \times 10^{-27} \text{ kg}} = 4.000$$

This result is puzzling in that although the helium nucleus has twice the charge of a proton, its mass is four times that of the proton. Clearly, the helium nucleus contains either charged particles other than protons or, perhaps, neutral particles in addition to protons.

In 1932, Chadwick bombarded a beryllium target with α-particles to produce uncharged particles now called **neutrons** (see Fig. 1.6). Neutrons have almost the same mass as protons and also occur in the nucleus. It is now known that an α-particle (helium nucleus) consists of two protons and two neutrons, thereby accounting for $m_{He}/m_H = 4.000$.

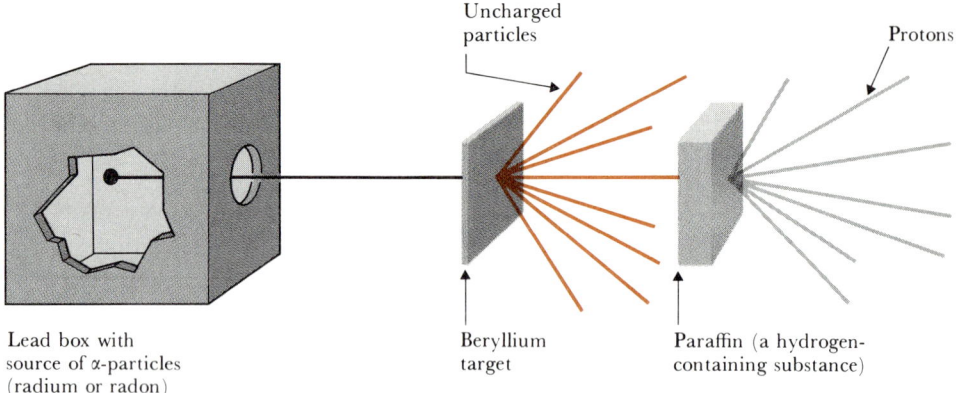

Fig. 1.6. Schematic illustration of the discovery of neutrons. The neutrons reveal themselves indirectly by the protons they produce.

James Chadwick (b. 1891), a British physicist, was a co-worker of Ernest Rutherford at the Cavendish Laboratory. He received the Nobel prize in physics in 1935 for the discovery of the neutron.

Francis W. Aston (1877–1945), the British physicist who invented the mass spectrograph, was a student of Sir J. J. Thomson at the Cavendish Laboratory. Aston was noted as an avid photographer, an enthusiastic traveler, and an excellent musician. He received the Nobel prize in chemistry in 1922.

In 1919, F. W. Aston developed an instrument known as the **mass spectrograph** (or **mass spectrometer**), which is capable of measuring the masses of atoms and molecules with very high accuracy (see Fig. 1.7). Studies of atomic masses show that virtually all the elements have nuclear masses consonant with the presence of both protons and neutrons in the nucleus. Surprisingly, however, it was found that atoms of a given element do not always have the same nuclear composition; some variation in the number of neutrons inevitably occurs. For example, the gaseous element neon (Ne) appears to have three different types of nuclei. Although all neon nuclei contain 10 protons, out of every 10,000 nuclei about 9092 have 10 neutrons, 26 have 11 neutrons, and 882 have 12 neutrons. Collectively, atoms of a given element differing only in the nuclear neutron count are known as **isotopes**. Even the lightest element, hydrogen, has three isotopes containing zero, one, or two neutrons in addition to a single nuclear proton.

Chemists customarily indicate or summarize the composition of elemental nuclei by the use of subscripted and superscripted atomic symbols. For example, $^{20}_{10}\text{Ne}$ represents that isotope of neon which has 10 protons and 10 neutrons. The sum of the numbers of protons and neutrons is called the **mass number** of the isotope and

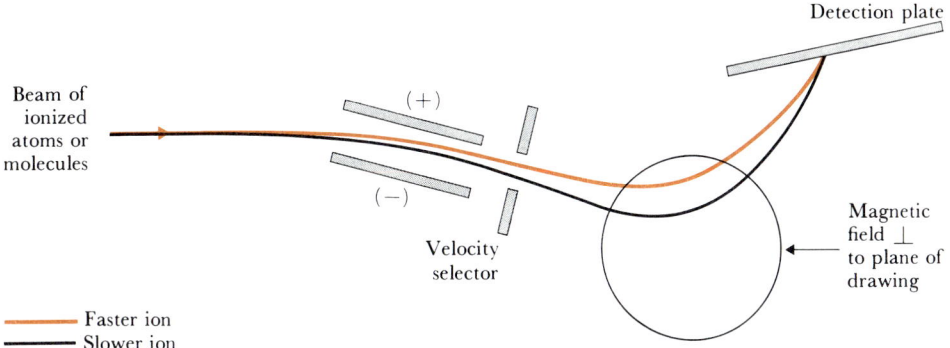

Fig. 1.7. A mass spectrograph. A gaseous sample of atoms or molecules is ionized by bombardment with a beam of electrons. A beam of these ions is then sent between two electrostatically charged plates, where it is deflected (the diagram shows how positively charged ions are deflected downward toward the negative plate). After passing through the velocity selector, which lets through only ions within a particular velocity range, the ions pass between the pole pieces of an electromagnet, where another deflection occurs. A detector plate (for example, a photographic plate) is placed at the focal point of the beam to record the point of impact. The position at which this focal point occurs is a measure of the mass of the ions in the beam; larger masses (for a given charge) are deflected less and are focused more to the right. Ions of identical charge/mass ratios end up at the same spot even if they have different velocities, since the faster ones are deflected less by the electric and magnetic fields and since the two deflections are in opposite directions. The mass spectrograph provides very accurate measurements of the masses of ions and thus leads to accurate masses of individual atoms and molecules.

Henry G. J. Moseley (1887–1915) was a budding British physicist who worked at Oxford and briefly with Rutherford at Manchester. Moseley left Oxford to volunteer for active duty with the Royal Engineers of the British Army in World War I. He was shot and killed while using a field telephone during the Battle of Sulva Bay in Gallipoli. Moseley's lasting contribution to science, gleaned from work with extremely primitive facilities at Oxford, is the establishment of the importance of the atomic numbers of elements.

---

is indicated by the left-hand superscript.* The left-hand subscript always indicates the number of protons and is called the **atomic number**. The atomic number is always the same for all the isotopes of a given element and is numerically equal to the electrical charge on the nucleus in units of 1.602×10^{-19} C. As shown by Moseley, one of Rutherford's co-workers, the atomic number is one of an element's most fundamental properties (see the end of Section 4.5).

In our notation, the three isotopes of neon are written $^{20}_{10}\text{Ne}$, $^{21}_{10}\text{Ne}$, and $^{22}_{10}\text{Ne}$. The three isotopes of hydrogen are $^{1}_{1}\text{H}$, $^{2}_{1}\text{H}$, and $^{3}_{1}\text{H}$. In conversation, $^{20}_{10}\text{Ne}$ becomes "neon-20," $^{21}_{10}\text{Ne}$ becomes "neon-21," and similarly for other isotopes and other elements.

---

* Some chemists and physicists use $_{10}\text{Ne}^{20}$, writing the mass number as a right-hand superscript. We shall reserve the right-hand superscript position for indicating atomic charges, for example as in He^{2+}.

Actually, there is a redundancy in a symbol such as $^{20}_{10}$Ne since only the element neon can have the atomic number 10. Hence, the nuclear symbol ^{20}Ne is sometimes used.

The isotopes of hydrogen have been given special names; $^1_1$H is light hydrogen or **protium**, $^2_1$H is heavy hydrogen or **deuterium**, and $^3_1$H is **tritium**.

Although the masses of isolated protons and neutrons are known with great accuracy, the mass of a nucleus cannot be deduced simply by knowing how many protons and neutrons it contains; invariably the actual mass of the nuclei is slightly less than that of the sum of its constituent **nucleons** (a general term encompassing protons and neutrons). The mass difference, known as the **mass defect**, is related to the energy which binds the nucleons together. (Mass defect is discussed more fully in Chapter 11.) Accordingly, the only way to determine the mass of a nucleus is by actual measurement, for example, with the mass spectrograph.

Masses of atoms and molecules are commonly expressed in terms of **atomic mass units** (abbreviated **amu**). By definition, one amu is exactly one-twelfth the mass of the carbon-12 isotope ($^{12}_6$C). This means that one $^{12}_6$C atom has a mass of exactly 12.0000 . . . amu. This mass scale is chosen in such a way that one mole of $^{12}_6$C atoms weighs exactly 12.000 . . . g. In fact, the precise definition of a mole (see Section 1.3) is the number of atoms in exactly 12.0000 . . . g of $^{12}_6$C. This number cannot be measured with absolute certainty, but the currently accepted approximate value is 6.0229×10^{23}. A less accurate value, 6.02×10^{23}, is often used for ordinary calculations. (The mole is also called **Avogadro's number**.) Thus the mass of a single $^{12}_6$C atom is

$$\frac{12.0000\ldots \text{g}}{6.02 \times 10^{23} \text{ atoms}} = 19.9 \times 10^{-24} \text{ g} \cdot \text{atom}^{-1}$$

Since this mass corresponds to 12.0000 . . . amu, one amu must correspond to

$$\left(\frac{1 \text{ amu}}{12 \text{ amu} \cdot \text{atom}^{-1}}\right)\left(\frac{12 \text{ g}}{6.02 \times 10^{23} \text{ atoms}}\right) = 1.66 \times 10^{-24} \text{ g} \quad \text{or} \quad 1.66 \times 10^{-27} \text{ kg}$$

On this scale the experimentally determined masses of the oxygen isotopes $^{16}_8$O, $^{17}_8$O, and $^{18}_8$O are 15.99491, 16.99914, and 17.99916 amu, respectively. The average of all the various masses (for the naturally occurring isotopes) is known as the **atomic weight** of the element. (Strictly speaking, the term **average atomic mass** is more accurate, but long usage has sanctified the former.) Since naturally occurring oxygen consists of 99.759% (by number) $^{16}_8$O, 0.037% $^{17}_8$O, and 0.204% $^{18}_8$O, the atomic weight of oxygen is*

$$\frac{(99.759)(15.99491) + (0.037)(16.99914) + (0.204)(17.99916)}{100} = 15.9994 \text{ amu}$$

The reader should be aware of the fact that it is easy to confuse the three numerically similar but fundamentally different quantities—*mass number, atomic mass* (in amu), and *atomic weight*. Since protons and neutrons have masses very close to 1

---

* Some scientists prefer to express atomic weight as a dimensionless number with no units of mass or weight attached. This usage can be misleading unless the adjective *relative* is added and, consequently, will not be employed in this text.

amu each, mass numbers and atomic masses will be very close in magnitude. Mass numbers are determined by counting nucleons, atomic masses must be measured experimentally, and atomic weights are averages of atomic masses taken over the naturally occurring isotopes.

1.6 THE CHARACTERIZATION OF MATTER

Since chemistry is the study of **matter**, it is helpful to discuss briefly just what matter is and how it may be described. It is not possible to actually define matter in the strictest sense, yet all of us know what matter is; it comprises all the materials that surround us. Matter, as it reveals itself to our senses, occupies space and possesses other obvious characteristics as well.

One of matter's most important characteristics is its **mass**. The mass of a sample of matter manifests itself by **inertia**, a resistance to a change in the state of motion. A quantity closely related to mass is **weight**, defined as the mathematical product of mass and the local value of the acceleration of gravity. Around sea level the earth's gravity causes a free-falling object to undergo an acceleration of approximately $9.8 \text{ m} \cdot \text{s}^{-2}$. This means that during each second of fall, the velocity of fall increases by 9.8 meters per second. Since Newton's second law of motion defines **force** as the product of mass and acceleration, weight is simply the force involved when matter of mass m is under the influence of a gravitational acceleration.

It should be clear that mass and weight are not synonymous—as casual lay use might suggest—but they are proportional to each other. If two objects of masses m_1 and m_2 are subjected to identical gravitational accelerations, then their weights w_1 and w_2 stand to each other in the ratio $w_1/w_2 = m_1/m_2$, since $w_1 = m_1 g$ and $w_2 = m_2 g$ (g is the acceleration of gravity or gravitational constant). This is why an object's mass can be deduced by weighing it on a balance; both the object weighed and the object used to counterbalance it are subjected to the same gravitational acceleration.* Note that an object would have the same mass on the moon as on the earth but would weigh approximately six times as much on the earth; the moon's gravitational constant is only about one-sixth that of the earth. Thus a given object would be harder to lift on the earth than on the moon. However, the object would be equally hard to roll on the level on either the moon or earth, since resistance to a change in the state of motion would be the same in each case.

Throughout this text it is assumed that all weights have been measured under a common gravitational constant so that the terms mass and weight may be used interchangeably without confusion. Thus, to say that an object weighs 1 kg will be taken to mean that it has a mass of 1 kg. Strictly speaking, the object actually weighs $(1 \text{ kg})(9.8 \text{ m} \cdot \text{s}^{-2})$ or 9.8 newtons (9.8 N).

Matter commonly exists in three forms: **solid**, **liquid**, and **gas**. These forms or states are defined more precisely in later chapters (3, 7, and 8) but, for the moment, it is sufficient to note that solids have a definite shape and volume, liquids have a

---

* This is strictly true only for weighings in a vacuum. In actuality the object weighed—as well as the counterweights used—may be subjected to slightly different buoyancy forces (if their densities differ), and this leads to some error. This same buoyancy effect makes a rock appear lighter under water than in air.

definite volume but assume the shape of their containers, and gases assume both the shapes and volumes of their containers. A somewhat whimsical but nonetheless valid characterization of the three states of matter is depicted in Fig. 1.8.

In any of these states, a sample of matter may be described by its **density**, defined as *mass per unit volume*. This may be expressed in mathematical form as

$$\rho = \frac{m}{V}$$

In all but very refined calculations, the mass m in the above is replaced by weight w—the latter being expressed in mass units. The units of the density ρ are always the units of mass divided by the units of volume. If basic SI units are used, mass will be in kilograms (kg) and volume in cubic meters (m³), so that density is in kg·m⁻³. However, most handbooks and reference books list densities of solids and liquids in g·cm⁻³. For example, the density of a solid such as gold is 19.3 g·cm⁻³ (19.3 × 10³ kg·m⁻³) and that of a liquid such as water is about 1.0 g·cm⁻³ (1 × 10³ kg·m⁻³) (both densities measured at 25 °C). This means that on an equal volume basis, gold is about 19.3 times as heavy (has 19.3 times the mass) as water. Densities of gases depend rather strongly on the pressure (as well as the temperature) and will be discussed in Section 3.4.

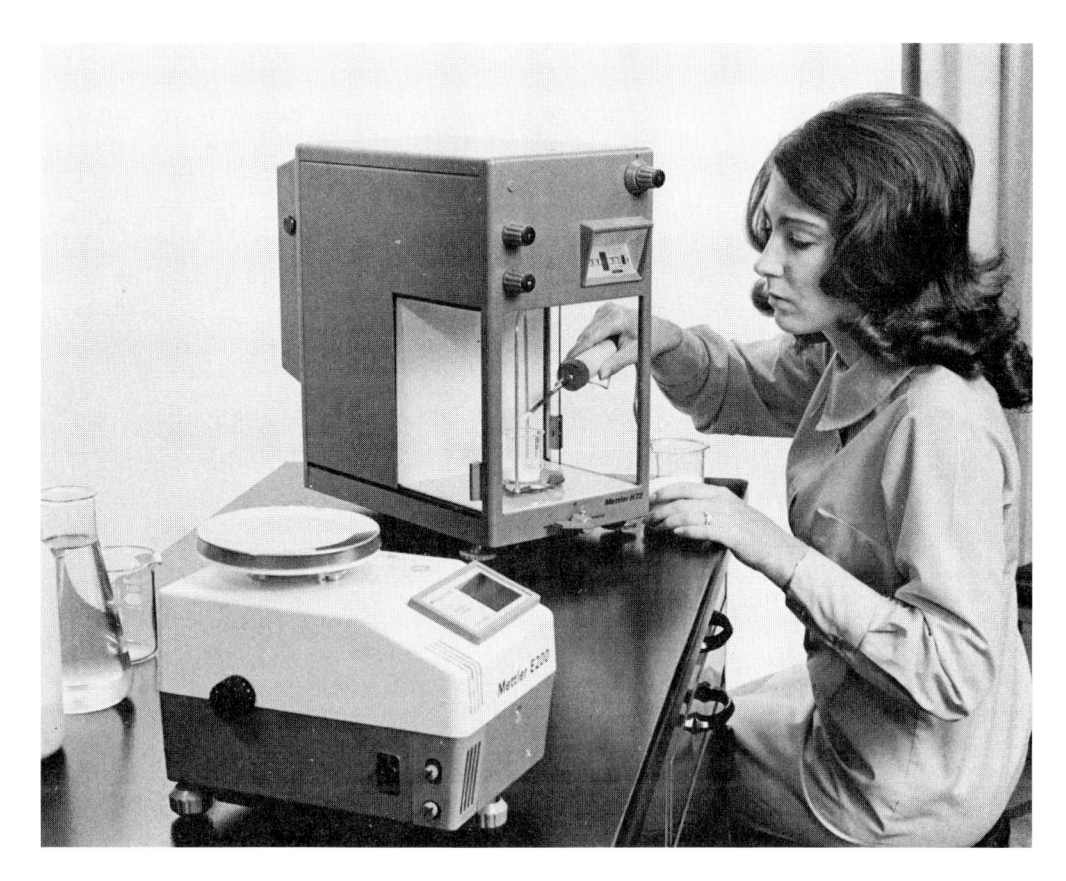

Opposite page: Double-pan balance used during the nineteenth century. (Courtesy of the Bettmann Archive.) Above: Two modern chemical balances: a top-loading balance (left) and an analytical balance (right). (Courtesy of Mettler Instruments Corp., Princeton, N.J.)

Solid:
definite
shape and
volume

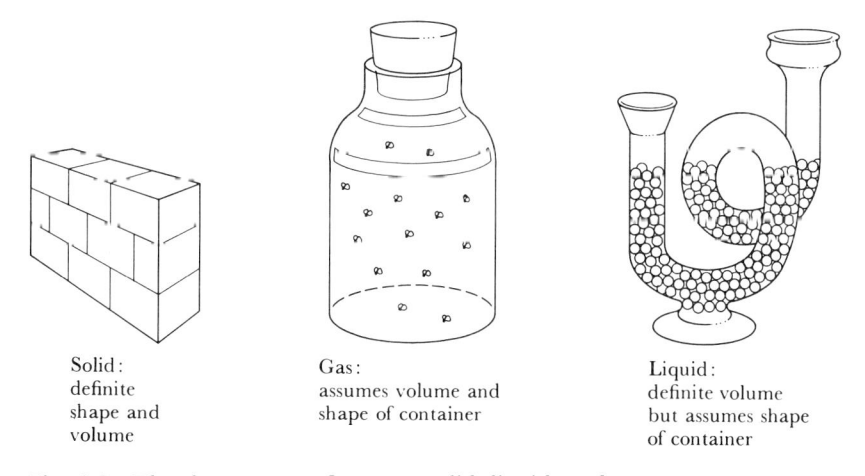

Gas:
assumes volume and
shape of container

Liquid:
definite volume
but assumes shape
of container

Fig. 1.8. The three states of matter: solid, liquid, and gas.

1.6 | THE CHARACTERIZATION OF MATTER 25

Closely related to density is a quantity called the **specific gravity**. The specific gravity of a substance is the ratio of its density to the density of water at the same temperature. Thus specific gravity is always expressed as a dimensionless number. If the densities of gold and water at 25 °C are 19.3 g·cm^{-3} and 1.0 g·cm^{-3}, respectively, their corresponding specific gravities are 19.3 and 1.0.

EXAMPLE 1.1 An object has a volume of 10 cm³ and weighs 23 g. What is its density in (a) g·cm^{-3}, (b) kg·m^{-3}?

SOLUTION

a) $\rho = m/V = 23 \text{ g}/10 \text{ cm}^3 = 2.3 \text{ g·cm}^{-3}$

b) $\rho = m/V = 0.023 \text{ kg}/10 \times 10^{-6} \text{ m}^3 = 23 \times 10^2 \text{ kg·m}^{-3}$

Note that 1 m³ = 10⁶ cm³.

EXAMPLE 1.2 Iron has a density of 7.86 g·cm^{-3} at 25 °C. What is the volume of 50 kg of iron at 25 °C?

SOLUTION Solve for V using $\rho = m/V$. This gives

$$V = \frac{m}{\rho} = \frac{50 \times 10^3 \text{ g}}{7.86 \text{ g·cm}^{-3}} = 6.4 \times 10^3 \text{ cm}^3 \quad \text{or} \quad 6.4 \times 10^{-3} \text{ m}^3$$

EXAMPLE 1.3 The density of water at 25 °C is 0.997 g·cm^{-3}. What is the specific gravity of iron (see Example 1.2) at 25 °C?

SOLUTION Specific gravity of iron at 25 °C = density of iron at 25 °C divided by the density of water at 25 °C:

$$\frac{7.86 \text{ g·cm}^{-3}}{0.997 \text{ g·cm}^{-3}} = 7.88$$

1.7 WORK AND ENERGY

Most people have some notion of what **work** is. We are familiar with the sensation of physical exertion associated with tasks such as walking, swinging a tennis racket, or hefting a mug of beer. When we perform work, we expend **energy**, and when we do work, say, to lift an object, we increase the energy of that object. Consequently, energy is often defined as the ability or capacity to do work.

In many types of problems it is convenient to consider energy of two types: **kinetic** energy and **potential** energy. Kinetic energy is energy associated with motion; a golf ball of mass m and velocity v has a kinetic energy of $\frac{1}{2}mv^2$. Potential energy is the energy a body is potentially capable of expending by virtue of its position or state.

A bowling ball held at waist height has a potential energy due solely to its position; some of this potential energy becomes kinetic energy when the ball is dropped on a toe. Similarly, it takes kinetic energy to compress a spring, thereby increasing its potential energy.

A tankful of gasoline in a stationary car possesses potential energy by virtue of its chemical nature. Burning the gasoline in the engine converts this potential energy to the kinetic energy needed to move the vehicle.

Energy, like mass, is conserved in ordinary physical and chemical processes. This conclusion, based on observation—as is the law of conservation of matter—is known as the **law of conservation of energy**. This law may be expressed in a variety of different but equivalent forms. For example, the total energy (kinetic + potential) of a system never changes, or energy can neither be created nor destroyed.

The basic unit of energy in the SI is the **joule** (symbol J).* When a mass of 1 kg is accelerated by $1 \text{ m} \cdot \text{s}^{-2}$ and moved a distance of 1 m, the work done (energy used) is $1 \text{ kg} \cdot \text{m}^2 \cdot \text{s}^{-2}$ or 1 J. An input of 100 J is required to operate a 100-watt light bulb for one second (a **watt** is a unit of **power,** or rate of doing work, of 1 J per second).

Other units of energy are often used. When a charge of 1 **coulomb** (symbol C) is accelerated by a potential difference of 1 **volt** (symbol V), the energy required is also 1 J. Thus a joule is equivalent to a volt·coulomb (V·C). When a charge of 1.6×10^{-19} C is accelerated by a potential difference of 1 V, the input of energy required is 1.6×10^{-19} J; this amount of energy is also called the **electron volt** (symbol eV) since 1.6×10^{-19} C is the magnitude of the charge on an electron.

In some areas of physics and chemistry the preferred unit of energy is the **calorie** (symbol cal). The calorie was originally defined as the energy required to change the temperature of 1 g of water from 15 °C to 16 °C, but it is now defined as 4.184 J. The current trend is to abandon the calorie (and kilocalorie) in favor of the joule and kilojoule.

The **specific heat** of a substance is defined as the energy required to change the temperature of 1 g of that substance by 1 °C (or 1 K). In the past this has been expressed in units of $\text{cal} \cdot \text{g}^{-1} \cdot \text{deg}^{-1}$ (or $\text{cal} \cdot \text{g}^{-1} \cdot \text{K}^{-1}$). Thus solid tin (Sn) has a specific heat of about $0.05 \text{ cal} \cdot \text{g}^{-1} \cdot \text{K}^{-1}$ at 25 °C. This means that 0.05 cal of energy is needed to change the temperature of 1 g of tin by 1 °C (or 1 K) in the vicinity of 25 °C. Specific heat varies with temperature, usually increasing as the temperature increases.

EXAMPLE 1.4 How much energy is needed to heat 5 g of tin metal from 20 °C to 55 °C? Assume that the average specific heat of tin between 20 °C and 55 °C is just the 25 °C value, that is, $0.05 \text{ cal} \cdot \text{g}^{-1} \cdot \text{K}^{-1}$.

SOLUTION The energy required is

specific heat × mass × change in temperature
$$= (0.05 \text{ cal} \cdot \text{g}^{-1} \cdot \text{K}^{-1})(5 \text{ g})(55 \text{ °C} - 25 \text{ °C}) = 9 \text{ cal}$$

---

* SI units of energy and related quantities are discussed more thoroughly in Appendix 1.

EXAMPLE 1.5 The specific heat of iron is about 0.12 cal·g^{-1}·K^{-1} at 25 °C. What is the temperature of a bar of iron, originally at 25 °C, weighing 15 g to which 10 cal of energy has been supplied by heating?

SOLUTION Assuming that the specific heat does not vary greatly over the temperature range involved, the energy supplied is equal to specific heat × mass × (T − 25 °C), where T is the final temperature of the heated bar. Solving for T one obtains

$$T = \frac{\text{energy supplied}}{\text{specific heat} \times \text{mass}} + 25\,°\text{C}$$

$$= \frac{10\text{ cal}}{(0.12\text{ cal·g}^{-1}\text{·K}^{-1})(15\text{ g})} + 25\,°\text{C}$$

$$= 5.6° + 25\,°\text{C} = 30.6\,°\text{C}$$

1.8 UNIT CONVERSIONS

It is quite common to be given the value of a chemical or physical quantity in units other than those desired. For example, suppose a chemist wishes to know the density of a substance in g·cm^{-3}, but the only available reference lists the density as 0.20 lb·in^{-3}. A quick, efficient way to carry out a conversion is to treat units (g, cm, lb, in., etc.) as algebraic quantities which can be multiplied and divided just as can a, b, and c in expressions such as

$$\left(\frac{a^2b}{a}\right)\left(\frac{c}{b^2}\right) = \frac{ac}{b}$$

A relationship such as "1 in. is equivalent to 2.54 cm" may be written as a quotient:

$$\frac{1\text{ in.}}{2.54\text{ cm}} \quad \text{or} \quad \frac{2.54\text{ cm}}{1\text{ in.}}$$

To convert 13 inches to centimeters, simply multiply 13 inches by that quotient which leads to the answer in centimeters. Since the inches unit must cancel, the correct setup is

$$13\text{ in.}\left(\frac{2.54\text{ cm}}{1\text{ in.}}\right) = 33\text{ cm}$$

Conversion of 0.20 lb·in.$^{-3}$ to g·cm^{-3} can be done in an analogous manner. Suppose the following equivalence relationships are available:

1 kg is equivalent to 2.2 lb

1 in. is equivalent to 2.54 cm

The pounds unit is converted to grams by multiplying by the factor

$$\left(\frac{1 \text{ kg}}{2.2 \text{ lb}}\right)\left(\frac{1000 \text{ g}}{1 \text{ kg}}\right)$$

The conversion from cubic inches to cubic centimeters uses the factor

$$\left(\frac{1 \text{ in.}}{2.54 \text{ cm}}\right)^3$$

Note that this factor comes from the equivalency of 1 in. and 2.54 cm but must be *cubed* in order to obtain proper cancellation of units. The full conversion may be set up in one single expression as follows:

$$0.20 \text{ lb} \cdot \text{in.}^{-3} \left(\frac{1 \text{ kg}}{2.2 \text{ lb}}\right)\left(\frac{1000 \text{ g}}{1 \text{ kg}}\right)\left(\frac{1 \text{ in.}}{2.54 \text{ cm}}\right)^3$$

Inspection shows that the units combine algebraically to produce $g \cdot cm^{-3}$. The answer thus is

$$\frac{(0.20)(1)(1000)(1)^3}{(2.2)(1)(2.54)^3} \text{ g} \cdot \text{cm}^{-3}$$

or

$$5.5 \text{ g} \cdot \text{cm}^{-3}$$

This technique can be applied to a variety of problems. One of its many benefits is that it eliminates confusion as to what the units of answers to problems are. When asked to calculate a volume, for example, you know the answer must come out in a unit of length cubed; the actual unit can always be determined from the units used to do the calculation. The technique can also aid you if you forget a specific formula. As an example, suppose you are to calculate the volume of a sample of matter weighing 10 g and having a density of 2.5 $g \cdot cm^{-3}$. Consideration of units alone tells you the answer is

$$\frac{10 \text{ g}}{2.5 \text{ g} \cdot \text{cm}^{-3}} = 4.0 \text{ cm}^3$$

since no other combination such as

$$(10 \text{ g})(2.5 \text{ g} \cdot \text{cm}^{-3}) \quad \text{or} \quad \frac{2.5 \text{ g} \cdot \text{cm}^{-3}}{10 \text{ g}}$$

leads to the correct units.

From now on, as you work problems, write down units for the quantities involved and check to make certain the answer has the correct units. If not, you have either made an error in the setup or forgotten a conversion. Above all, avoid the bad habit of doing calculations without units and of reporting answers without units. Units are important; you would be justifiably irked if the bank reported your checking account balance as just plain 50 with no indication of whether dollars or cents were intended. Units are *never* taken for granted; they must always be stated.

EXAMPLE 1.6 Use the following relationships (plus any other you are ordinarily familiar with) to convert 15.0 stones per cubic yard to kilograms per cubic meter. The **stone** is a British unit of weight.

1 stone is equivalent to 14 lb

2.2 lb is equivalent to 1 kg

1 m is equivalent to 39.37 in.

SOLUTION

$$15.0 \frac{\text{stone}}{\text{yd}^3} \left(\frac{14 \text{ lb}}{1 \text{ stone}}\right)\left(\frac{1 \text{ kg}}{2.2 \text{ lb}}\right)\left(\frac{1 \text{ yd}}{36 \text{ in.}}\right)^3\left(\frac{39.37 \text{ in.}}{1 \text{ m}}\right)^3$$

$$= \frac{(15.0)(14)(1)^4(39.37)^3}{(1)(2.2)(36)^3(1)^3} \frac{\text{kg}}{\text{m}^3} = 125 \text{ kg} \cdot \text{m}^{-3}$$

In actual practice, you are apt to encounter conversions that involve so many steps that the only way to keep everything straight is to use a systematic procedure such as the above. It is not a good idea to do the conversions one at a time; this leads to extra work and promotes confusion besides. Write out the entire set of conversions, check the unit cancellations, and *then* do the arithmetic.

EXAMPLE 1.7 If there are 0.31 gluk per every zook, how many zooks are there in 15 gluk?

SOLUTION

$$15 \text{ gluk} \left(\frac{1 \text{ zook}}{0.31 \text{ gluk}}\right) = \frac{(15)(1)}{0.31} \text{ zook} = 48 \text{ zook}$$

Note that this problem can be handled with perfect confidence even though you haven't the foggiest notion of what the units actually are.

SUMMARY

Chemistry is the science dealing with the study of matter and the changes it undergoes. Chemistry as a body of knowledge and as an important part of our cultural heritage is based on centuries of observations which can be expressed in terms of laws and theories whose validity is constantly challenged by the scientific method. Laws are statements which attempt to summarize how nature is found to behave; theories serve as models which help us to view observations within systematic and logical frameworks and which aid in the generation of new ideas. In general, a theory is periodically replaced or modified as observations accumulate and our understanding becomes increasingly sophisticated.

Dalton's atomic theory, that matter consists of small, indivisible and immutable particles called atoms, provides one of the major foundations of modern science. Its subsequent modification to encompass the internal structures of atoms in terms of electrons, protons, and neutrons has led to an intimate understanding of how matter is put together and how it behaves under a variety of conditions—including those conditions governing life itself.

Two important characteristics of matter are mass and its ability to occupy space. Mass is a resistance to a change in the state of motion. Weight is a force, the product of mass and the local acceleration of gravity. The ratio of the weights of two objects measured at the same acceleration of gravity equals the ratio of the masses of the two objects. Hence, mass and weight are often used interchangeably. The density of a sample of matter is defined as its mass per unit volume.

Energy is the capacity to do work. Kinetic energy is the energy of motion; it is defined as $\frac{1}{2}mv^2$, where m is the mass of the object and v is its velocity. Potential energy is the energy an object is capable of releasing. The law of conservation of energy states that energy is neither created nor destroyed; that is, the total energy (kinetic plus potential) of a system remains constant. The specific heat of a substance is the amount of energy needed to change the temperature of 1 g of that substance by 1 °C (1 K).

Most of the numbers used in calculations are associated with units. These units should be carried along with the calculations and treated algebraically in order to determine the units associated with the answer. A systematic procedure for handling of units simplifies unit conversions and aids in detecting errors of calculation.

LEARNING GOALS

1. Be familiar with the general features of the scientific method and know the difference between inductive and deductive reasoning.

2. Know what a theory (or model) is, and be able to explain what a law is and how laws arise.

3. Be able to state and explain the law of conservation of matter.

4. Be able to state and explain the laws of definite proportions and multiple proportions.

5. Know the fundamental postulates of Dalton's atomic theory.

6. Know how to use chemical symbols.

7. Know how the mole is defined and what its approximate numerical value is.

8. Know the main components of the atom and be familiar with the basic properties of each.

9. Be able to define an isotope and to give specific examples using proper atomic symbols.

10. Be able to define atomic mass unit, atomic number, mass number, and atomic weight. Know how the latter is calculated from isotopic abundances.

11. Be able to define mass, weight, and density and to do calculations with the density formula. Know what specific gravity is.

12. Be able to define energy and to distinguish between kinetic and potential energy.

13. Be able to state the law of conservation of energy and to apply it to simple problems, for example, to the determination of specific heat.

14. Be familiar with the main units of energy—joule, calorie, and electron volt—and how to convert from one to another.

15. Know how to convert from one set of units to another and how to use units in checking problem setups.

DEFINITIONS, TERMS, AND CONCEPTS TO KNOW

scientific method	molecule	mass number
law	ion	atomic number
theory	ionic compound	atomic weight
model	chemical analysis	protium
inductive and deductive reasoning	chemical synthesis	deuterium
mass	mole	tritium
atomic theory of matter	atomic symbol	atomic mass unit (amu)
law of definite composition	chemical formula	weight
law of multiple proportions	electron	density
law of conservation of matter	cathode rays	work
element	protons	energy
compound	neutrons,	specific heat
mixture	isotopes	law of conservation of energy
atom	mass spectrograph	

QUESTIONS AND PROBLEMS

Note: The use of color for a problem number signifies that the answer to that problem can be found at the back of the book.

Atomic Structure

1. Three different oxides of nitrogen are analyzed to produce the following data: compound A, 46.7% N (by weight); compound B, 63.6% N; compound C, 30.4% N. For a fixed weight of N, determine the ratio of oxygen to nitrogen for the three compounds above. What important law do the above data illustrate?

2. What are the numbers of protons, neutrons, and electrons in each of the following?

a) $^{209}_{83}Bi$ b) $^{66}_{30}Zn^{2+}$ c) $^{59}_{27}Co$

d) $^{84}_{36}Kr$ e) $^{19}_{9}F^-$

3. Complete the following table.

ELEMENT	ATOMIC NUMBER	MASS NUMBER	NUCLEAR CHARGE	NEUTRONS IN NUCLEUS	NUMBER OF ELECTRONS
A	45	103	___	___	___
B	___	59	27	___	___
C	33	___	___	42	___
D	___	61	___	34	___
E	___	___	___	50	53

32 THE FOUNDATIONS OF CHEMISTRY

4. An atom has 8 protons, 9 neutrons, and a net charge of 2⁻. What are the mass number, atomic number, and number of electrons in the atom?

5. Chlorine exists in nature as 75.53% (by number of atoms) ^{35}Cl and 24.47% ^{37}Cl. The masses of ^{35}Cl and ^{37}Cl are 34.969 and 36.966 amu, respectively. Calculate the average atomic weight of chlorine.

6. Show how the data given in Problem 5 can be used to calculate the isotopic composition *by weight*; that is, what fractions of the weight of natural chlorine are due to the individual isotopic species?

7. Complete the following table.

ELE-MENT	ATOMIC NUMBER	MASS NUMBER	PROTONS	NEUTRONS	ELECTRONS
Ca	___	___	20	20	___
___	41	93	___	___	___
Tb	___	151	___	___	65
U	92	___	___	146	___
___	___	31	15	___	___

8. Magnesium occurs in nature in the following abundances: ^{24}Mg (78.70%), ^{25}Mg (10.13%), and ^{26}Mg (11.17%). The corresponding atomic masses (in amu) are: 23.99, 24.99, and 25.98, respectively. Calculate the average atomic weight of magnesium.

9. Is it possible for two or more different elements to have the same mass number? Can two or more different elements ever have the same atomic number? Explain.

10. Iron occurs in nature in the following abundances: ^{54}Fe (5.82%), ^{56}Fe (91.66%), ^{57}Fe (2.19%), and ^{58}Fe (0.33%). The corresponding atomic masses (in amu) are: 53.94, 55.93, 56.94, and 57.93, respectively. Calculate the average atomic weight of iron.

11. Why are the atomic masses of given isotopes not whole numbers? Why are these masses (in amu) very close to the mass numbers?

12. Iron-56 (^{56}Fe) has a mass of 55.9349 amu as determined by the mass spectrograph. Using masses of 1.008665, 1.00785, and 0.00055 amu for a neutron, proton, and electron, respectively, calculate the theoretical mass of an atom of ^{56}Fe. Why do the two numbers not agree?

Density, Mass, and Volume

13. The average density of the earth is about 5.5 g·cm⁻³. What is the mass of the earth if its volume is about 1.1×10^{21} m³?

14. What is the density (in g·cm⁻³ and kg·m⁻³) of a sample of matter having a mass of 0.002 kg and a volume of 0.5 cm³?

15. Magnesium metal has a density of about 1.738 g·cm⁻³ at 25 °C. A mole of magnesium atoms weighs 24.312 g.
   a) What is the density of magnesium in kg·m⁻³?
   b) What is the volume (in cm³) of a mole of magnesium atoms?
   c) What is the apparent volume occupied by one magnesium atom?

16. Silver has a density of about 10.5 g·cm⁻³ at 25 °C. What is the mass of a bar of silver measuring 10 cm × 2 cm × 1 cm?

17. The density of water at 25 °C is 0.997 g·cm⁻³. What is the specific gravity of silver in Problem 16?

18. The density of mercury at 25 °C is 13.6 g·cm⁻³. How much will 0.50 L* of mercury weigh?

19. Iron has a density of 7.86 g·cm⁻³.
   a) What is the weight of a cubic meter of iron when weighed in a vacuum?
   b) A cubic meter of iron weighed under water weighs 6.86×10^3 kg. Explain. [*Hint:* Water exerts a buoyancy effect equal to the mass of the water displaced by the iron.]

20. The density of air at sea level at 25 °C is about 1.19 g·L⁻¹. What is the mass of air in a room of dimensions 20 ft × 30 ft × 7 ft?

21. Is the buoyancy effect of air on two different objects of the same mass always the same? Explain. [*Hint:* The buoyancy effect of air is equal to the mass of air displaced by the object.]

22. Gold has a density of 19.3 g·cm⁻³. The Greek philosopher Archimedes was given a king's crown weighing 5.0 kg to determine if it was pure gold as claimed by the seller, or an alloy. Archimedes immersed the crown in water and found the level rose by 230 cm³. What happened to the seller?

---

* Note that the abbreviation for liter, formerly l, is now L, a change that eliminates longstanding problems in distinguishing the lower-case letter from the numeral 1. For further comments on metric units and their abbreviations, see Appendix 1.

23. An object with a volume of 10 L weighs 25 kg in a vacuum and about 12 g less when weighed in air. What is the density of the air (in $g \cdot L^{-1}$)?

24. Copper and tin have densities of 8.92 and 5.75 $g \cdot cm^{-3}$, respectively. What is the density of an alloy (mixture) of copper and tin containing 90% (by weight) of copper? Assume that the volumes of the two metals in the alloy are strictly additive.

The Mole

25. How many years are there in a mole of seconds? Compare this with the probable age of the universe (approximately 5×10^9 years).

26. Assuming that an average orange has a diameter of 6 cm, calculate the depth to which the earth would be covered with one mole of oranges if these were spread out uniformly over all land and sea areas. The radius of the earth is about 6370 km.

27. How many sunflower seeds are there in $1\frac{1}{2}$ moles of sunflower seeds?

28. The total charge on a mole of electrons has been measured as 96,487 C. Using this and the charge on an electron given in the text, calculate Avogadro's number.

29. If one water molecule weighs 18 amu, how much does one mole of water molecules weigh (in grams)?

30. The mole is now the official SI unit for amount of matter. How many moles of donuts are there in a dozen? Are bakers ever apt to adopt the mole?

Energy and Specific Heat

31. A sample of metal has a mass of 10 g. The metal is heated to 100 °C in a water bath and placed into a container holding 20 g of water at 25 °C. Very shortly the temperature of the system (metal + water) settles down to 28 °C. What is the specific heat of the metal? [*Hint:* Apply the law of conservation of energy in the form: energy lost by the metal = energy gained by the water. The specific heat of water may be taken as 4.184 $J \cdot g^{-1} \cdot K^{-1}$ (1 $cal \cdot g^{-1} \cdot K^{-1}$).]

32. The specific heat of tin is about 0.05 $cal \cdot g^{-1} \cdot K^{-1}$ at 25 °C. What is this in $J \cdot kg^{-1} \cdot K^{-1}$?

33. How much energy (in joules) is needed to accelerate a mole of electrons through a potential difference of 1 V?

34. A well-fed U.S. citizen may consume about 2,500,000 cal (2500 kcal) per day. What is this in joules? In kJ?

35. It takes a minimum of 13.6 eV to pull the proton and electron of a hydrogen atom apart. How much energy (in kJ) would it take to separate a mole of hydrogen atoms into protons and electrons?

36. a) How much energy (in joules) is needed to lift a mass of 10 g to a height of 10 m?
b) What is the increase in potential energy of the object in (a)?
c) What is the kinetic energy of the object after it has dropped 10 m?
d) How do (a) and (c) illustrate the law of conservation of energy?

37. A 2-g sample of nickel is heated to 100 °C and dropped into an insulated container holding 100 g of water at 20 °C. The water temperature increases to 20.2 °C. Estimate the specific heat of nickel.

38. One hundred grams of water at 25 °C is mixed with 25 g of ice at 0 °C. Assuming no heat transfer between the mixture and surroundings, what is the temperature of the water when the ice is all melted? The energy required to melt a gram of ice at 0 °C is about 79 cal (331 J).

39. A compressed spring has enough energy to lift a 150-g mass 5 m against gravity. How much potential energy was in the spring?

40. The potential energy of two electrical charges q_1 and q_2 separated by a distance r is given by Coulomb's law as Kq_1q_2/r, where K is a constant. When the charges are an infinite distance apart, their potential energy is zero. If the particles have charges of the same sign, their potential energy is positive; when the charges have opposite signs, their potential energy is negative. Explain what the sign of the potential energy means physically in each case.

41. If it takes an input of 331 J of energy to melt 1 g of ice, what happens in energy terms when 1 g of liquid water freezes to ice?

42. What is the kinetic energy of a baseball (mass about 0.15 kg) thrown by a major league pitcher with a velocity of about 140 $km \cdot hr^{-1}$? How many grams of ice could you melt (in principle) with this energy?

43. One machine can do 10^5 kJ of work in 10 min and another can do 10^4 kJ of work in 2 s. Which machine has more power?

Unit Conversions

44. Convert 50 BTU·s^{-1}·ft^{-2} to W·m^{-2} (watts per square meter) using the equivalents

   1 BTU is equivalent to 1055 J,
   1 m is equivalent to 39.37 in.,
   1 watt is 1 J·s^{-1}.

45. Coulomb's law for the force F between two electrically charged particles (charges q_1 and q_2) separated by a distance r is $F = Kq_1q_2/r^2$. If F is in newtons (N), charge in coulombs (C), and r in meters (m), what units must the constant K have?

46. The density of compressed bat guano collected from a Peruvian mountain cave is 2.4 gluk·zook^{-1}. The gluk and zook are ancient Quechuan units of measure: 1 gluk = 15 g and 1 zook = 30 cm³. Express the density of the bat guano in modern units.

47. The driver of a foreign automobile notes that his speedometer indicates he is traveling 92 km·hr^{-1} in an area where the speed limit is 55 mph.
   a) Is the driver guilty of speeding?
   b) How many meters per second is the car traveling?
   [*Note:* 1 mile is about 1.61 km.]

48. The specific heat of copper at 25 °C is about 0.10 cal·g^{-1}·K^{-1}. What is this in J·kg^{-1}·K^{-1}?

49. In Example 1.6, why are conversion factors such as 1 yd/36 in. and 39.37 in./1 m cubed?

50. The Appalachian Mountain Club awards 4000-footer patches to anyone who climbs all of the New Hampshire mountains 4000 ft or higher in elevation. What should these patches be called in terms of SI units?

51. Rephrase the following saying in terms of SI units: Give some people an inch and they'll take a mile.

ADDITIONAL READINGS

Matter. Chicago: Time-Life Books, 1965. A relatively nontechnical introduction to the history of chemistry from the alchemists to today.

Stent, Gunther S., "Prematurity and Uniqueness in Scientific Discovery." *Scientific American,* December 1972. A molecular geneticist reflects on the question: Are scientific creations any less unique than artistic creations?

Weeks, Mary Elvira, and Henry M. Leicester, *Discovery of the Elements,* 7th ed. Easton, Pa.: Chemical Education Publishing Company, 1968. A classic account of the discovery of the elements including biographical sketches of some of the world's greatest chemists. Also contains historical accounts of the development of many important theories and concepts of chemistry.

2
CHEMICAL FORMULAS, EQUATIONS, AND PERIODIC PROPERTIES

Some symbols used by the alchemists: upper left, "roasting of silver ore"; lower left, "to compose"; right, "to dissolve a metal in mercury."

We must lay it down as an incontestable axiom, that in all the operations of art and nature, nothing is created; an equal quantity of matter exists both before and after the experiment.
ANTOINE LAVOISIER, *Traité élémentaire de chimie,* 1789

One of the most useful notational devices of chemistry is the **chemical formula**—a combination of atomic symbols which represents considerable information about the molecule. Depending on the type of formula employed, the information may include **molecular weight** (the molecular counterpart of atomic weight), the atomic composition of the molecule, how the atoms are linked together, and their geometrical arrangement.

Formulas, in turn, may be employed in the writing of chemical equations—symbolic representations of chemical changes. By a procedure known as balancing, chemical equations become useful in the quantitative description of chemical changes.

2.1 EMPIRICAL AND MOLECULAR FORMULAS

The simplest type of chemical formula is the **empirical formula** (also called **simplest** formula). An example is CH_3, the empirical formula of a gas called ethane. This formula tells three very precise things about the substance it represents:

1. The elements the substance contains, namely, carbon (C) and hydrogen (H).
2. The ratio of the constituent atoms: one carbon atom for every three hydrogen atoms.
3. The percentage composition (by weight) of the substance:

$$\%C = \frac{12.011 \text{ amu}}{12.011 \text{ amu} + 3(1.008 \text{ amu})}(100) = 79.89$$

$$\%H = \frac{3(1.008 \text{ amu})}{12.011 \text{ amu} + 3(1.008 \text{ amu})}(100) = 20.11$$

where 12.011 amu is the atomic weight of carbon and 1.008 amu (rounded off from 1.0079) is the atomic weight of hydrogen. Any other unit of mass (g, kg, lb, etc.) could also be used in the percentage calculation without affecting the answer.

If the additional information is given that ethane has a molecular weight of 30.07 amu (one mole of ethane has a mass of 30.07 g), then it can be deduced that the **molecular formula** is $(CH_3)_2$ or, according to the preferred style, C_2H_6. This conclusion comes from the fact that one carbon atom and three hydrogen atoms weigh $12.011 + 3(1.008) = 15.035$ amu, so that it takes two carbon atoms and six hydrogen atoms to weigh 30.07 amu (which is just twice 15.035). The molecular formula C_2H_6 means that each molecule of ethane consists of an aggregate of two carbon atoms and six hydrogen atoms. An equivalent statement is that each mole of C_2H_6 consists of two moles of carbon atoms and six moles of hydrogen atoms. It should be realized that six moles of hydrogen atoms would normally occur in the elemental state as three moles of H_2 molecules; their mode of occurrence in the compound C_2H_6 is, as yet, unspecified.

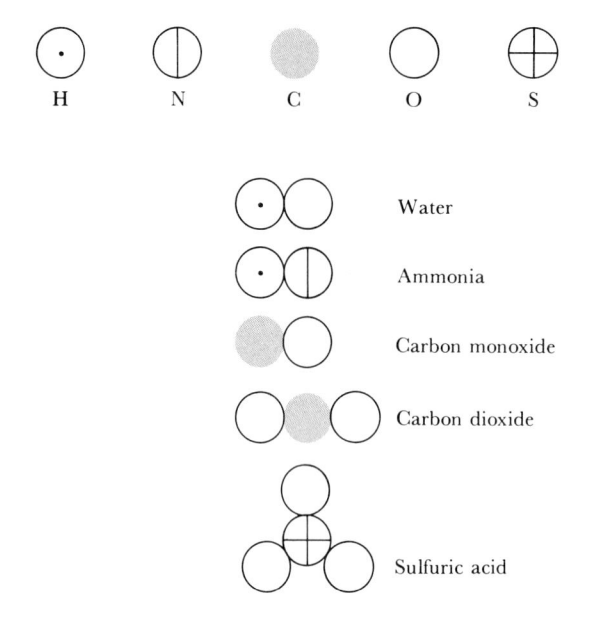

Dalton's formulas for some common molecules. Some of the formulas are obviously at variance with the modern ones; Dalton's water formula (HO in modern symbols) arose partly from his ignorance of the diatomic nature of H_2 and O_2.

Since the percentage composition of elements in a substance can be calculated by knowing the empirical formula, it should be possible to do the reverse: to calculate the empirical formula of a compound from its elemental composition in terms of the relative weights of its constituent atoms. For example, a compound of phosphorus and oxygen is analyzed and found to contain 100 parts by weight of phosphorus for every 229 parts of the substance. The elemental composition is thus $\frac{100}{229}(100) = 43.7\%$ P and, by subtraction, $100\% - 43.7\% = 56.3\%$ O. This means that a 100-g sample of the substance contains 43.7 g of phosphorus and 56.3 g of oxygen. Now 43.7 g of phosphorus contains $(43.7/30.97)(6.02 \times 10^{23})$ phosphorus atoms and 56.3 g of oxygen contains $(56.3/16.00)(6.02 \times 10^{23})$ oxygen atoms (30.97 and 16.00 are the atomic weights of phosphorus and oxygen, respectively). The ratio of phosphorus atoms to oxygen atoms in the compound then is

$$\text{number of P atoms} : \text{number of O atoms} = \frac{43.7}{30.97} : \frac{56.3}{16.00} = 1.41 : 3.51$$

(Note that the number 6.02×10^{23} cancels out in the ratio.) This ratio means that every 1.41 phosphorus atoms is associated with 3.51 oxygen atoms. The empirical formula could be expressed as $P_{1.41}O_{3.51}$, but convention requires that the atom ratios in empirical formulas be reduced to the smallest possible whole numbers. The quickest way to reduce $1.41 : 3.51$ to the smallest whole-number ratio is to first divide both numbers by the smaller of the two. Thus $1.41 : 3.51$ becomes $1 : 2.5$. Multiplying numerator and denominator by 2 gives $2 : 5$, the desired smallest whole-number ratio. Thus the empirical formula of the compound is P_2O_5.

If the compound contains three or more different elements, an analogous procedure is applied using the ratio of three or more numbers. For example, an unknown compound contains 52% carbon, 13% hydrogen, and 35% oxygen. The C:H:O ratio is (using rounded-off values of atomic weights)

number of C atoms : number of H atoms : number of O atoms =

$$\frac{52}{12} : \frac{13}{1} : \frac{35}{16} = 4.33 : 13 : 2.19 = 2 : 6 : 1$$

so that the empirical formula is C_2H_6O. It is important to note that if the analytical data on the composition are not of high accuracy, some judgment must be exercised in converting the decimal ratios to whole numbers. This is especially important for large molecules containing a great number of different elements.

Once the empirical formula is known, the molecular formula may be determined, provided at least an approximate value for the molecular weight of the substance is known. Determination of molecular weights will be discussed in Chapters 3 and 10 and elsewhere. In some cases the mass spectrograph may be used for this purpose. In the case of the compound with the empirical formula P_2O_5 the molecular weight is around 280 amu. If the molecular formula were P_2O_5, the molecular weight would be $2(30.97) + 5(16.00)$ or 141.94 amu. Since this is very close to half of 280, it is concluded that the molecular formula is P_4O_{10} with an exact molecular weight of 2(141.94) or 283.88 amu. Thus an empirical formula plus an approximate molecular weight are sufficient to determine a molecular formula and an exact molecular weight. By *exact* is meant within the limits of accuracy to which the constituent atomic weights are known.

Substances which are ionic (e.g., NaCl) have no molecular formula since unique molecules do not exist. In this case the term **formula weight** is used in lieu of molecular weight. The formula weight of an ionic compound is just the sum of the masses of the atoms in the simplest formula; for NaCl the formula weight is 22.990 amu (for Na) plus 35.453 amu (for Cl), or 58.443 amu.

Closely related to molecular weight (the weight of one molecule) is the **weight of one mole**. These quantities have the same numerical value if the former is in amu and the latter in grams. The weight of one mole of an ionic compound is just the formula weight expressed in grams. Thus the weights of one mole of P_4O_{10} and NaCl are 283.88 and 58.443 g, respectively. The weight of a mole is sometimes called the **gram molecular weight** to distinguish it from the molecular weight. Although not strictly accurate, many chemists also use gram molecular weight when referring to the weight of one mole of an ionic substance. Actually, the correct term is then **gram formula weight**.

2.2 STRUCTURAL AND GEOMETRIC FORMULAS

Suppose it is known that the two carbon atoms in C_2H_6 are adjacent and that three hydrogen atoms are attached to each of the carbon atoms.* These facts may be simply illustrated by the diagram

$$\begin{array}{c} H\ \ H \\ |\ \ \ | \\ H-C-C-H \\ |\ \ \ | \\ H\ \ H \end{array}$$

where dashes symbolize a chemical bond between two atoms. Such a diagram, it is important to note, makes no attempt to show how the atoms are located in a three-dimensional sense. Diagrams such as the above are commonly called **structural formulas**, although **linkage** formula and **network** formula are more accurately descriptive names.

Going one step further, each carbon atom in C_2H_6 occupies the center of a tetrahedron having three hydrogen atoms and the other carbon atom as apexes (see Fig. 2.1). This three-dimensional description of the molecule C_2H_6 can be represented in two dimensions by a variety of **geometric formulas** (see Fig. 2.2). The ball-and-stick model on the right is perhaps the easiest to visualize in a three-dimensional sense, but it is not readily suited for freehand drawing. The line-and-wedge model, on the other hand, is easy to draw (after a little practice) but some imagination must be used in interpreting it in three dimensions. All atoms joined by solid lines are assumed to lie in the plane of the paper, wedges indicate that the attached atom lies above the plane of the paper, and dashed lines indicate that the atom lies below the plane of the paper. The middle diagram is a sort of compromise; it is easier to draw than the ball-and-stick diagram but yet imparts some three-dimensional feeling.

The geometric structure of P_4O_{10} (Fig. 2.3) is somewhat more complicated than that of C_2H_6. The four atoms of phosphorus are located at the apexes of a tetrahedron and each phosphorus atom is at the same time at the center of a tetrahedron with four oxygen atoms serving as apexes.

---

* Early chemists had to determine how atoms were joined on the basis of indirect evidence. Today some more direct methods are available, for example, X-ray studies of solids as discussed in Chapter 7.

Fig. 2.1. Perspective and top views of a regular tetrahedron. There are four faces, each an equilateral triangle. If the carbon atom in C₂H₆ is placed in the center of the tetrahedron, hydrogen atoms are at three of the apexes and the other carbon atom is at the fourth apex.

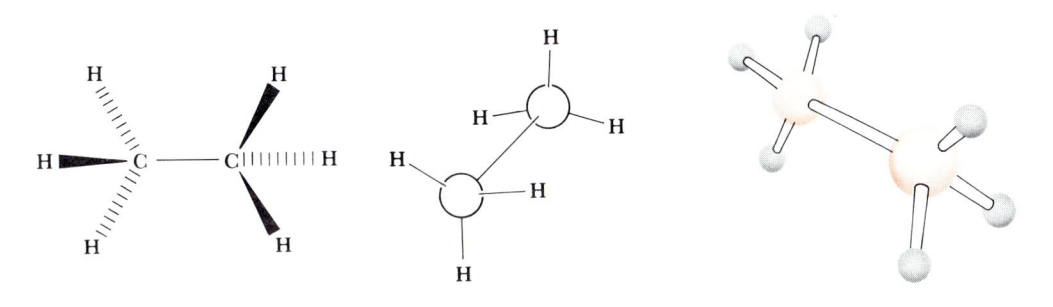

Fig. 2.2. Common representations of the three-dimensional geometrical arrangement of atoms in the ethane (C₂H₆) molecule.

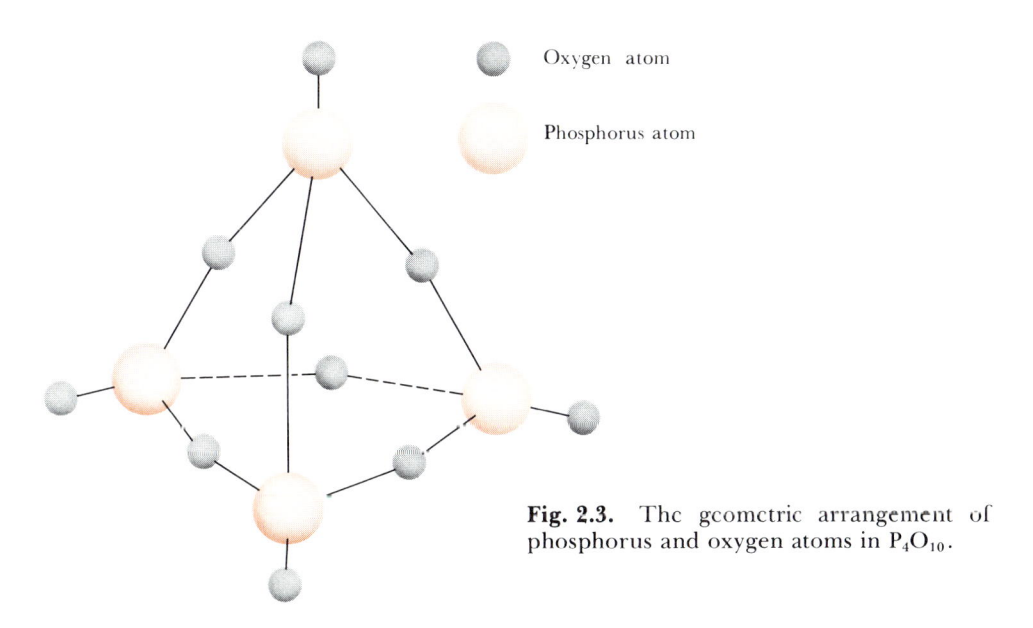

Fig. 2.3. The geometric arrangement of phosphorus and oxygen atoms in P₄O₁₀.

Over the course of years chemists have begun to realize more and more strongly just how important the three-dimensional structure of some molecules can be. In some chemical systems, particularly those involved in living organisms, the three-dimensional shape of a molecule plays as large a role in its chemical properties as do

2.2 | STRUCTURAL AND GEOMETRIC FORMULAS

the atoms making up the molecule.* For this reason, it is wise for students of chemistry to develop the habit of visualizing molecules in their three-dimensional shapes as early as possible. Some simple methods of guessing the geometries of molecules from their molecular formulas will be discussed in Chapter 6.

2.3 WRITING AND BALANCING CHEMICAL EQUATIONS

Chemical transformations are conveniently represented by means of molecular formulas using what is known as a **chemical equation**. For example, suppose hydrogen gas (H_2) is burned in oxygen (O_2) to form water (H_2O). This chemical process may be symbolized by

$$H_2 + O_2 \rightarrow H_2O$$

where the arrow means "goes to" or "produces." Since each molecule of H_2 consists of two atoms of hydrogen and each molecule of O_2 consists of two atoms of oxygen, the above equation—if interpreted literally—implies that one molecule of H_2 and one molecule of O_2 produce one molecule of H_2O. Yet, a molecule of H_2O contains only *one* atom of oxygen; one of the oxygen atoms of O_2 appears to have been lost. Thus, to avoid violating the law of conservation of matter and the immutability of elements, it is customary to **balance** the equation so that the following is true:

number of H atoms on the left = number of H atoms on the right

number of O atoms on the left = number of O atoms on the right

The oxygen atoms can be balanced simply by doubling the number of H_2O molecules, that is, by writing

$$H_2 + O_2 \rightarrow 2H_2O$$

But now the hydrogen atoms are not in balance; there are two hydrogen atoms on the left and four hydrogen atoms on the right. This is corrected by also doubling the number of H_2 molecules on the left. This produces the totally balanced equation

$$2H_2 + O_2 \rightarrow 2H_2O$$

Balancing of the above equation is shown more explicitly by using structural formulas of the molecules involved as shown below.

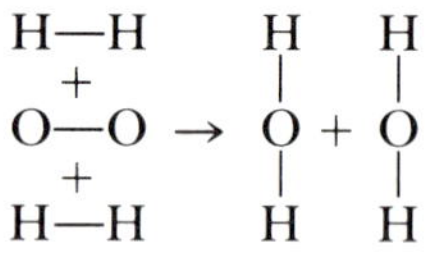

```
H—H      H   H
  +      |   |
O—O  →   O + O
  +      |   |
H—H      H   H
```

---

* One indication of the growing significance of molecular geometry is that the 1975 Nobel prize in chemistry was awarded to Australian-born John W. Cornforth (b. 1917) of the University of Sussex and Yugoslavian-born Vladimir Prelog (b. 1906) of the Swiss Federal Institute of Technology for work on the structures of enzymes, antibiotics, and other organic molecules.

Alternatively, the original equation could have been balanced in one step by halving the number of O_2 molecules on the left. This gives

$$H_2 + \tfrac{1}{2}O_2 \rightarrow H_2O$$

The latter two equations are equivalent; both state that a 2:1 ratio of hydrogen molecules to oxygen molecules is needed to produce water. Furthermore, both equations imply that if x moles of hydrogen is combined with $x/2$ moles of oxygen, then x moles of water is produced.

As another example, suppose glucose ($C_6H_{12}O_6$) is burned in oxygen to produce carbon dioxide (CO_2) and water. The skeleton reaction (unbalanced equation) is

$$C_6H_{12}O_6 + O_2 \rightarrow CO_2 + H_2O$$

The equation is balanced by requiring the following:

number of C atoms on the left = number of C atoms on the right

number of H atoms on the left = number of H atoms on the right

number of O atoms on the left = number of O atoms on the right

The order in which the atoms are balanced is not of fundamental importance, but experience shows that it is sometimes easier to begin with certain atoms. Since the glucose molecule contains six carbon atoms, it will produce six CO_2 molecules. Similarly, the twelve hydrogen atoms in glucose will produce six H_2O molecules. This produces the partially balanced equation

$$C_6H_{12}O_6 + O_2 \rightarrow 6CO_2 + 6H_2O$$

All atoms are balanced except oxygen. Now there are eight oxygen atoms on the left (six in glucose plus two in O_2) but eighteen oxygen atoms (twelve from CO_2 and six from H_2O) on the right. Thus there is a lack of $18 - 8$ or ten oxygen atoms on the left. These ten oxygen atoms can be added to the left simply by changing the one O_2 to six O_2, that is, by adding five O_2 molecules (ten oxygen atoms). The balanced equation then becomes

$$C_6H_{12}O_6 + 6O_2 \rightarrow 6CO_2 + 6H_2O$$

Sometimes the balancing of a chemical equation requires strategy; simply plunging in willy-nilly to balance the first element encountered may lead to confusion. For example, consider the equation

$$C_2H_6 + CoF_3 \rightarrow C_2F_6 + HF + CoF_2$$

An initial perusal of the equation shows that six hydrogen atoms of C_2H_6 are replaced by six fluorine atoms to produce C_2F_6 and that each fluorine atom comes from a conversion of CoF_3 to CoF_2. Thus it is tempting to begin balancing by writing $6CoF_3$ and $6CoF_2$. However, the six hydrogen atoms in C_2H_6 end up as HF so that twelve fluorine atoms (not six) are needed to convert C_2H_6 to C_2F_6. The balanced equation then is

$$C_2H_6 + 12CoF_3 \rightarrow C_2F_6 + 6HF + 12CoF_2$$

After some practice in the balancing of a variety of equations, it becomes easier to recognize what particular strategy is apt to be fruitful. Balancing a chemical equation is one of the easiest things to learn in a chemistry course and, at the same time, it is one of the most important things to be able to do.

Some chemical equations cannot be balanced easily by the method just shown; consequently, a more systematic approach will be discussed in Chapter 14.

EXAMPLE 2.1 Balance the equation

$$Fe_3O_4 + H_2 \rightarrow Fe + H_2O$$

SOLUTION Balance Fe by writing 3Fe on the right: $Fe_3O_4 + H_2 \rightarrow 3Fe + H_2O$. Balance O by putting 4 H_2O on the right: $Fe_3O_4 + H_2 \rightarrow 3Fe + 4H_2O$. Finish by balancing H; put 4 H_2 on the left (for the 8H in $4H_2O$). This gives

$$Fe_3O_4 + 4H_2 \rightarrow 3Fe + 4H_2O$$

2.4 CALCULATIONS INVOLVING BALANCED EQUATIONS

A balanced chemical equation, for example

$$C_6H_{12}O_6 + 6O_2 \rightarrow 6CO_2 + 6H_2O$$

tells us several important facts. This equation states that for every mole of glucose burned, six moles of oxygen is required and six moles of water and six moles of carbon dioxide are formed.* Alternatively, whenever 180 g of glucose (the mass of one mole) is burned, 192 g of oxygen (the mass of six moles) is required and 108 g of water (the mass of six moles) and 264 g of carbon dioxide (the mass of six moles) are formed. Note also that 372 g of reactants (180 g glucose plus 192 g oxygen) produces 372 g of products (108 g water plus 264 g carbon dioxide). If the amount of glucose burned is greater or less than 180 g, the amounts of all other substances are changed proportionately. Also, the weight ratios do not depend on the mass units used; thus 180 lb of glucose requires 192 lb of oxygen for complete combustion and similarly for all other relationships.

Suppose a chemist wants to calculate how many grams of oxygen would be needed to completely burn 100 g of glucose and how many grams of water and carbon dioxide would be produced. A very convenient way of doing the calculation is to determine all amounts in moles and then to convert to ordinary mass units. A systematic way of setting up the calculation is as follows:

a) moles O_2 needed = (MRF)(moles glucose burned)

b) moles H_2O produced = (MRF)(moles glucose burned)

c) moles CO_2 produced = (MRF)(moles glucose burned)

---

* The equation does *not* say that if one mole of glucose is mixed with six moles of oxygen, six moles of water and six moles of carbon dioxide will be produced with no glucose or oxygen remaining. Although this is just about what does happen in this particular case, a balanced equation does not necessarily imply completeness of a reaction. Incomplete reactions are discussed in some detail in Chapter 12.

where MRF stands for **mole-ratio factor**, a quantity which is different for each of the three cases above and which is rapidly determined from the balanced equation. For example, in case (a) the MRF is simply the moles of O_2 appearing in the balanced equation divided by the moles of glucose occurring in the same equation. Thus, for case (a), MRF is 6 mol O_2/1 mol glucose. This means that six moles of oxygen is needed to burn each mole of glucose. Since one mole of glucose has a mass of 180 g, 100 g of glucose is 100/180 mol. Thus case (a) is set up as

moles O_2 needed = (6 mol O_2/1 mol glucose)($\frac{100}{180}$ mol glucose) = 3.33 mol

Note that the units "mol glucose" cancel so that the answer comes out in the units "mol O_2." This fact is important to remember since it will prevent setting up the MRF in inverted form. Always check cancellation of units in a problem setup before doing any arithmetic.

Using 32.0 g as the mass of one mole of O_2 gives

mass of O_2 needed = (3.33 mol)(32.0 g·mol^{-1}) = 107 g

In case (b), MRF is 6 mol H_2O/1 mol glucose, since six moles of H_2O is produced by burning 1 mole of glucose. Thus

moles H_2O produced = (6 mol H_2O/1 mol glucose)($\frac{100}{180}$ mol glucose) = 3.33 mol

Using 18 g as the mass of a mole of H_2O gives

mass of H_2O produced = (3.33 mol)(18 g·mol^{-1}) = 60 g

The reader should verify that the mass of CO_2 produced is 147 g.

It should be noted that the masses of O_2 and H_2O may be calculated directly without an intermediate calculation in terms of moles. The correct setups are:

mass of O_2 needed = 100 g glucose $\left(\dfrac{6 \text{ mol } O_2}{1 \text{ mol glucose}}\right)\left(\dfrac{32 \text{ g } O_2 \cdot \text{mol}^{-1}}{180 \text{ g glucose} \cdot \text{mol}^{-1}}\right)$
= 107 g

mass of H_2O produced = 100 g glucose $\left(\dfrac{6 \text{ mol } H_2O}{1 \text{ mol glucose}}\right)\left(\dfrac{18 \text{ g } H_2O \cdot \text{mol}^{-1}}{180 \text{ g glucose} \cdot \text{mol}^{-1}}\right)$
= 60 g

Note that the units of the answer arise by cancellation of other units in the setup.

EXAMPLE 2.2 Iron(III) oxide is treated with hydrogen as follows:

$$Fe_2O_3 + 3H_2 \rightarrow 2Fe + 3H_2O$$

a) How much hydrogen (in grams) is needed to produce 50 g of iron?
b) How much iron(III) oxide (in grams) is needed to produce 70 g of H_2O?

SOLUTION a) Moles H_2 needed = (MRF)(moles iron produced). The MRF is 3 mol H_2/2 mol Fe (three moles of H_2 is needed to produce two moles of Fe) and 50 g of iron is

50/55.9 mol (one mole of Fe has a mass of 55.9 g). Thus

moles H$_2$ needed = (3 mol H$_2$/2 mol Fe)(50/55.9 mol of Fe) = 1.3 mol

Since a mole of H$_2$ has a mass of 2.0 g, the mass of H$_2$ needed is given by

mass of H$_2$ needed = (1.3 mol)(2.0 g·mol^{-1}) = 2.6 g

b) Moles Fe$_2$O$_3$ needed = (MRF)(moles H$_2$O produced). Here the MRF is 1 mol Fe$_2$O$_3$/3 mol H$_2$O and the moles of H$_2$O produced is 70/18 or 3.9 mol. Thus

moles Fe$_2$O$_3$ needed = (1 mol Fe$_2$O$_3$/3 mol H$_2$O)(3.9 mol of H$_2$O) = 1.3 mol

A mole of Fe$_2$O$_3$ has a mass of 160 g. Thus

mass of Fe$_2$O$_3$ needed = (1.3 mol)(160 g·mol^{-1}) = 21 × 10^1 g

---

An important type of problem arises when one or more reactants are present in excess of that required to react with the remaining reactants. For example, the balanced equation for the combustion of ethane, C$_2$H$_6$, is

$$C_2H_6 + 3\tfrac{1}{2}O_2 \rightarrow 2CO_2 + 3H_2O$$

or, if the balancing is done without fractions,

$$2C_2H_6 + 7O_2 \rightarrow 4CO_2 + 6H_2O$$

Suppose 5.0 moles of ethane is mixed with 5.0 moles of oxygen and the mixture is burned as completely as possible. What are the maximum amounts of CO$_2$ and H$_2$O formed, and is any ethane or oxygen left over? To answer these questions it is convenient to compute the **theoretical mole ratio** of ethane to oxygen and to compare this with the **actual mole ratio** of the problem. The theoretical mole ratio (as determined by the balanced equation) is

$$\frac{\text{moles ethane}}{\text{moles O}_2} = \frac{2}{7} = 0.29$$

Whenever ethane and O$_2$ are mixed in exactly this ratio, both can be completely used up. Using a mixture with a higher ratio indicates ethane is in excess. Oxygen is then said to be the **limiting reagent**, since its amount limits the extent of reaction; when O$_2$ is all gone, no more reaction can proceed even though some ethane still remains. Similarly, using a mixture with a lower ratio than 0.29 means O$_2$ is in **excess**. In this case ethane would be the limiting reagent. To determine which situation obtains, calculate the actual mole ratio

$$\frac{\text{moles ethane}}{\text{moles O}_2} = \frac{5}{5} = 1$$

Since this ratio is larger than 0.29, ethane is in excess and O$_2$ is the limiting reagent. This means that all of the oxygen will be consumed but some ethane will remain.

Specifically,

$$\text{moles } C_2H_6 \text{ burned} = (\text{MRF})(\text{moles } O_2 \text{ consumed})$$
$$= (2 \text{ mol ethane}/7 \text{ mol } O_2)(5.0 \text{ mol } O_2)$$
$$= 1.4 \text{ mol}$$

Thus there will be 5.0 − 1.4 or 3.6 mol ethane left over. The amounts of CO_2 and H_2O produced are given by

$$\text{moles } CO_2 \text{ produced} = (4 \text{ mol } CO_2/7 \text{ mol } O_2)(5.0 \text{ mol } O_2) = 2.9 \text{ mol}$$
$$\text{moles } H_2O \text{ produced} = (6 \text{ mol } H_2O/7 \text{ mol } O_2)(5.0 \text{ mol } O_2) = 4.3 \text{ mol}$$

Note that in each case it is the amount of O_2 (the limiting reagent) that determines the amounts of other substances used up or produced.

EXAMPLE 2.3 Powdered aluminum and iron(III) oxide undergo a vigorous, heat-producing reaction

$$2Al + Fe_2O_3 \rightarrow 2Fe + Al_2O_3$$

Suppose 10 g of Al and 10 g of Fe_2O_3 are mixed and ignited. How much Fe and Al_2O_3 are produced? Is any substance left over and, if so, what is it and in what amount?

SOLUTION The theoretical Al/Fe_2O_3 mole ratio is 2; the actual ratio is

$$\frac{(10 \text{ g Al})/(26.98 \text{ g Al} \cdot \text{mol}^{-1})}{(10 \text{ g Fe}_2O_3)/(159.70 \text{ g Fe}_2O_3 \cdot \text{mol}^{-1})} = \frac{0.37}{0.063} = 5.9$$

Thus Al is in excess and Fe_2O_3 is the limiting reagent (which is used up completely). The amount of iron produced is 2 × 0.063 or 0.126 mol, just twice the number of moles of Fe_2O_3 used up. Similarly, the amount of Al_2O_3 produced is 0.063 mol. All of this information follows at once from the balanced equation. The amount of Al left unreacted is the original amount less the amount reacted or 0.37 − 0.126 = 0.24 mol. This is also (0.24 mol) (26.98 g·mol^{-1}) or 6.6 g. Note how simple the numerical calculations are in terms of moles.

2.5 THE PERIODIC PROPERTIES OF THE ELEMENTS

When scientists are faced with the task of trying to make sense out of a large number of individual facts—for example, the huge number of different chemical reactions observed—they attempt to reduce the apparent hopelessness of the task by looking for *similarities* which will enable them to classify the facts into a manageable number of specific categories, each of which can then be studied in isolation. In this way they may be able to reduce a million seemingly isolated observations into, say, a dozen general categories, each having perhaps no more than 100,000 specific examples. By

Dmitri Mendeleev, Russian chemist who first called attention to periodic relationships in the chemical and physical properties of the elements. (Courtesy of the Library of Congress.)

Dmitri Mendeleev (1843–1907) was a Siberian-born Russian chemist who discovered the periodic law of the elements (with Lothar Meyer of Germany) and the critical temperature phenomenon of liquids and their vapors (with T. Andrews of Ireland). Throughout his life Mendeleev exhibited wide interest in art, education, and economics in addition to prolific work in the systemization of chemistry as a science.

Julius Lothar Meyer (1830–1895), a German chemist who independently discovered the periodic law of the elements, had an abiding interest in education and wrote extensively on this subject.

Mendeleev and Meyer actually ordered the periodic table largely on the basis of atomic weight, since the concept of atomic number was then unknown. Nevertheless, Mendeleev fully realized that some elements (such as cobalt and nickel) fitted properly only if arranged in reverse order of atomic weights. It was only much later that it was realized that the apparent reverse order in atomic weights was the correct order in terms of increasing atomic numbers.

understanding the nature of each individual category—perhaps by the use of models—they then find it easier to learn something about any new observation they make simply by establishing which category it fits into. This process, known as **classification**, is often a necessary first step in the growth of a science. The binomial system of nomenclature for the naming of plants and animals established by the Swedish naturalist Carl Linnaeus (1707–1778) is one of the best-known examples of the use of classification for simplification.

During the nineteenth century many scientists noted that chemical reactions of the elements exhibited certain similarities. Although many classification schemes were proposed (some of them more fanciful than useful), none was truly valuable until 1869, when Dmitri Mendeleev and Lothar Meyer simultaneously and independently proposed a classification of the then known chemical elements according to their **periodic properties**. In Fig. 2.4 is shown one of the many modern versions of the Mendeleev-Meyer classification, known as a **periodic table of the elements**.

Mendeleev and Meyer arranged the table so that the vertical columns contain **groups** of elements whose properties are remarkably similar although at the same time changing progressively as the atomic number increases. Each horizontal section (a **period**) represents a progression of property changes which is repeated approximately by each successive underlying period. This behavior is similar to the way in which hourly variations in temperature establish a general pattern that is repeated, at least approximately, on a daily basis.

Figure 2.5 shows how atomic volume (the atomic weight divided by the density) varies in a periodic fashion with atomic number. The atomic number, whose fundamental importance was first recognized by Moseley (Section 1.5 and end of Section 4.5), besides being equal to the number of protons in the nucleus, hence to the nuclear charge (in units of 1.602×10^{-19} C), also functions as the ordinal number of an element's position in the periodic table. This important fact is often expressed in the form of the **periodic law**, which states that the physical and chemical properties of the elements are periodic functions of their atomic numbers. Although there are exceptions to the orderly progression of properties as implied by the arrangement within the periodic table, the table has nevertheless been of inestimable value to the chemist as an aid in the correlation, interpretation, and prediction of chemical and physical properties.

One of the most obvious features of the periodic table is that it shows a rather sharp separation of the elements into two broad categories: **metallic** and **nonmetallic**. This division is indicated by the heavy zigzag line toward the right-hand side of the periodic table. Elements to the left of this line are metals; elements to the right of this line are nonmetals. The metallic elements are those that tend to have a silvery luster, to be solids at room temperature (except for mercury, a liquid), and to be good thermal and electrical conductors. Nonmetallic elements, on the other hand, often have dull lusters (if solid) and may exist as solids, liquids, or gases at room temperature. With the exception of the graphite form of carbon, the nonmetallic elements are poor conductors of electricity and are, in fact, usually classed as insulators. When the elements occur in ionic compounds, the metallic elements are found as positively charged ions (for example, Na^+ in NaCl) and the nonmetallic elements are often found as negatively charged ions (for example, Cl^- in NaCl).

Elements next to the border between metals and nonmetals tend to exhibit properties of both. For example, carbon is a nonmetal but one of its forms (graphite) conducts electric current much like a metal. The elements boron, silicon, germanium, arsenic, and tellurium—although officially nonmetals—are often called **semimetals** because some of their properties resemble those of metals.

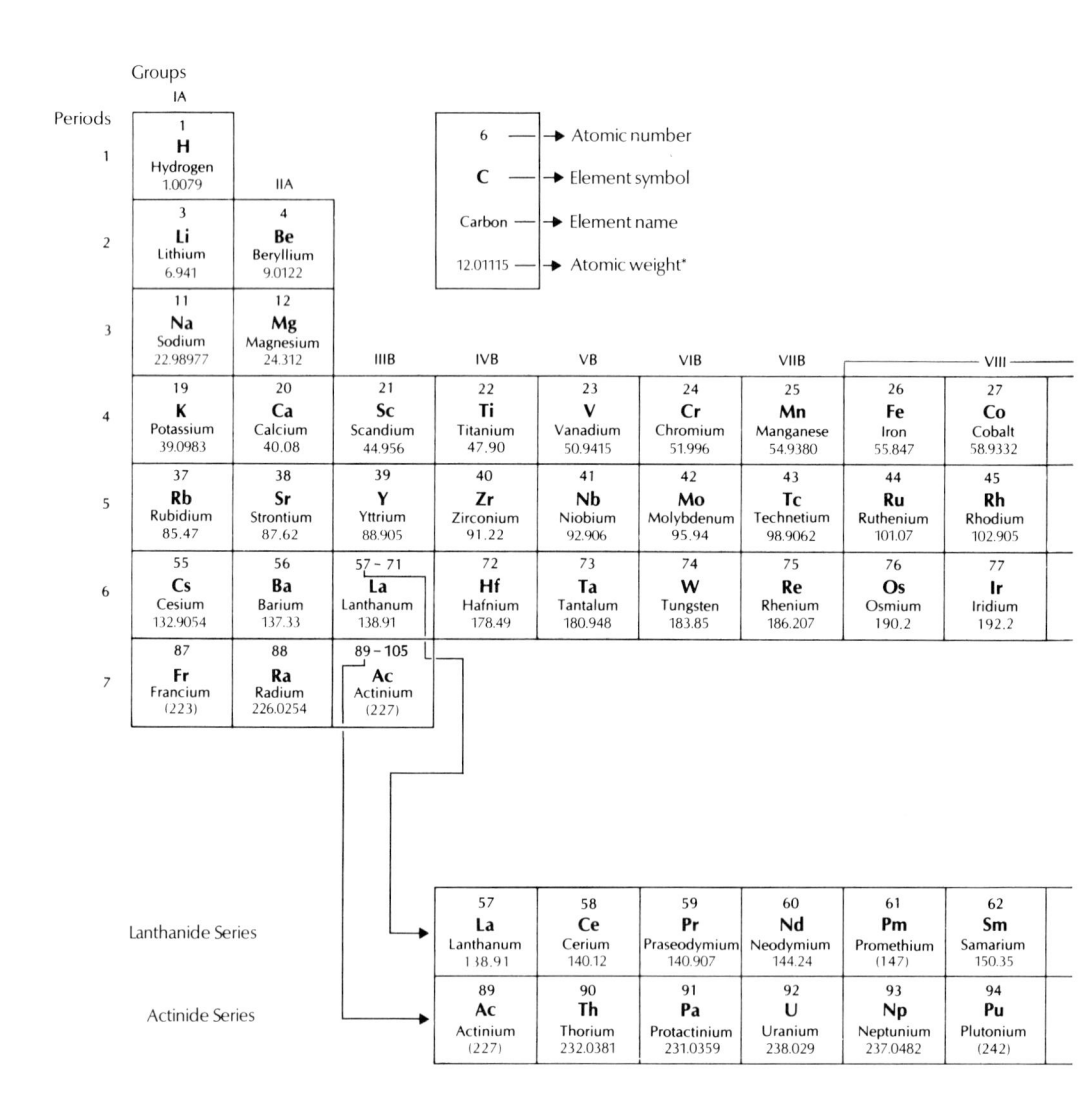

Periodic Table

Periods / Groups

Legend:
- 6 → Atomic number
- C → Element symbol
- Carbon → Element name
- 12.01115 → Atomic weight*

Period	IA	IIA	IIIB	IVB	VB	VIB	VIIB	VIII		
1	1 **H** Hydrogen 1.0079									
2	3 **Li** Lithium 6.941	4 **Be** Beryllium 9.0122								
3	11 **Na** Sodium 22.98977	12 **Mg** Magnesium 24.312								
4	19 **K** Potassium 39.0983	20 **Ca** Calcium 40.08	21 **Sc** Scandium 44.956	22 **Ti** Titanium 47.90	23 **V** Vanadium 50.9415	24 **Cr** Chromium 51.996	25 **Mn** Manganese 54.9380	26 **Fe** Iron 55.847	27 **Co** Cobalt 58.9332	
5	37 **Rb** Rubidium 85.47	38 **Sr** Strontium 87.62	39 **Y** Yttrium 88.905	40 **Zr** Zirconium 91.22	41 **Nb** Niobium 92.906	42 **Mo** Molybdenum 95.94	43 **Tc** Technetium 98.9062	44 **Ru** Ruthenium 101.07	45 **Rh** Rhodium 102.905	
6	55 **Cs** Cesium 132.9054	56 **Ba** Barium 137.33	57–71 **La** Lanthanum 138.91	72 **Hf** Hafnium 178.49	73 **Ta** Tantalum 180.948	74 **W** Tungsten 183.85	75 **Re** Rhenium 186.207	76 **Os** Osmium 190.2	77 **Ir** Iridium 192.2	
7	87 **Fr** Francium (223)	88 **Ra** Radium 226.0254	89–105 **Ac** Actinium (227)							

Lanthanide Series

| 57 **La** Lanthanum 138.91 | 58 **Ce** Cerium 140.12 | 59 **Pr** Praseodymium 140.907 | 60 **Nd** Neodymium 144.24 | 61 **Pm** Promethium (147) | 62 **Sm** Samarium 150.35 |

Actinide Series

| 89 **Ac** Actinium (227) | 90 **Th** Thorium 232.0381 | 91 **Pa** Protactinium 231.0359 | 92 **U** Uranium 238.029 | 93 **Np** Neptunium 237.0482 | 94 **Pu** Plutonium (242) |

			IIIA	IVA	VA	VIA	VIIA	VIIIA
								2 **He** Helium 4.0026
			5 **B** Boron 10.811	6 **C** Carbon 12.01115	7 **N** Nitrogen 14.0067	8 **O** Oxygen 15.9994	9 **F** Fluorine 18.99840	10 **Ne** Neon 20.183
	IB	IIB	13 **Al** Aluminum 26.98154	14 **Si** Silicon 28.0855	15 **P** Phosphorus 30.97376	16 **S** Sulfur 32.064	17 **Cl** Chlorine 35.453	18 **Ar** Argon 39.948
28 **Ni** Nickel 58.70	29 **Cu** Copper 63.546	30 **Zn** Zinc 65.38	31 **Ga** Gallium 69.72	32 **Ge** Germanium 72.59	33 **As** Arsenic 74.9216	34 **Se** Selenium 78.96	35 **Br** Bromine 79.904	36 **Kr** Krypton 83.80
46 **Pd** Palladium 106.4	47 **Ag** Silver 107.868	48 **Cd** Cadmium 112.41	49 **In** Indium 114.82	50 **Sn** Tin 118.69	51 **Sb** Antimony 121.75	52 **Te** Tellurium 127.60	53 **I** Iodine 126.9044	54 **Xe** Xenon 131.30
78 **Pt** Platinum 195.09	79 **Au** Gold 196.967	80 **Hg** Mercury 200.59	81 **Tl** Thallium 204.37	82 **Pb** Lead 207.19	83 **Bi** Bismuth 208.9804	84 **Po** Polonium (210)	85 **At** Astatine (210)	86 **Rn** Radon (222)

*If in parentheses, this is the mass number of the most stable isotope.

†Names not yet officially adopted.

63 **Eu** Europium 151.96	64 **Gd** Gadolinium 157.25	65 **Tb** Terbium 158.924	66 **Dy** Dysprosium 162.50	67 **Ho** Holmium 164.9304	68 **Er** Erbium 167.26	69 **Tm** Thulium 168.934	70 **Yb** Ytterbium 173.04	71 **Lu** Lutecium 174.967		
95 **Am** Americium (243)	96 **Cm** Curium (247)	97 **Bk** Berkelium (247)	98 **Cf** Californium (251)	99 **Es** Einsteinium (254)	100 **Fm** Fermium (253)	101 **Md** Mendelevium (256)	102 **No** Nobelium (254)	103 **Lr** Lawrencium (257)	104 † (261)	105 † (260)

Fig. 2.4. The periodic table of the chemical elements. Atomic weights in parentheses are not known accurately. The heavy solid line toward the right separates metals (right) from nonmetals (left). Note that nonmetals constitute a small minority of the elements. Elements near the border may exhibit some properties of both groups; for example, carbon (C) is a nonmetal but one of its solid forms (graphite) conducts electric current like a metal.

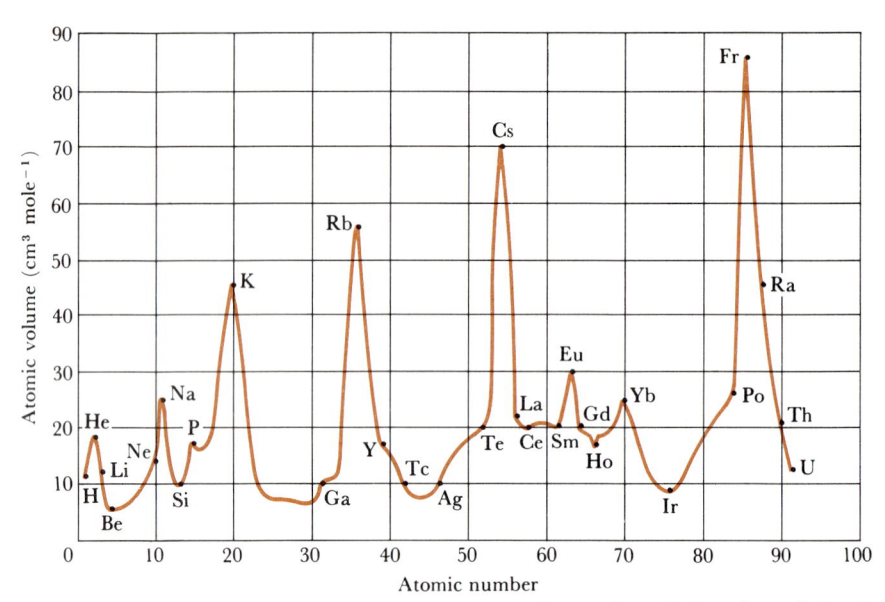

Fig. 2.5. Periodic variation of atomic volume with atomic number. Only a few of the elements are explicitly indicated. The solid line indicates the approximate values for the remaining elements. Note that each of the higher peaks (except for the first) represents one of the alkali metals.

The elements found in the leftmost column of the table (group IA) are called the **alkali metals**. All of these elements—lithium (Li), sodium (Na), potassium (K), rubidium (Rb), and cesium (Cs)—are soft enough to cut with a knife, have a bright, silvery luster when freshly cut, and react readily with water.* The simple oxides all have the empirical formula M_2O and the chlorides MCl.† The oxides react with water to form compounds called **hydroxides** with the empirical formula MOH; NaOH is called sodium hydroxide.‡ Water solutions of the group IA hydroxides consist entirely of the ions M^+ and OH^- except for LiOH, which appears to be only partly Li^+ and OH^-. The reactions of the oxides and the free metals with water are represented by the equations

$$M_2O + H_2O \rightarrow 2MOH$$

$$2M + 2H_2O \rightarrow 2MOH + H_2$$

---

* Francium (Fr) is so rare that its behavior has not been studied in detail. There is perhaps no more than 30 g of francium in the entire earth's crust. Francium may be produced artificially by bombarding bismuth with neutrons.

† With the exception of lithium, the simple oxides tend to form the peroxides M_2O_2 when excess oxygen is present. Other types of oxygen compounds also occur, but their study belongs to a specialized course.

‡ The naming of chemical compounds is discussed in Appendix 4.

The chlorides, MCl, are soluble in water and the solutions are good conductors of electric current. The molten chlorides also conduct electric current. This behavior suggests that the compound MCl is an aggregate of the ions M^+ and Cl^-; that is, MCl compounds are ionic.

The elements of group IIA are called the **alkaline earth metals** and consist of beryllium (Be), magnesium (Mg), calcium (Ca), strontium (Sr), barium (Ba), and radium (Ra). All of these elements form oxides, MO, and chlorides, MCl_2. The free elements are of silvery luster when freshly cut and are somewhat harder than the alkali metals. The elements of group IIA (except for beryllium) react with water less vigorously than do the alkali metals to form hydroxides:

$$M + 2H_2O \rightarrow M(OH)_2 + H_2$$

Magnesium reacts with water only at temperatures considerably above room temperature and beryllium does not appear to react at all. The compounds $Be(OH)_2$ and $Mg(OH)_2$ are not strongly dissociated into M^{2+} and OH^- ions. None of the hydroxides of group IIA is very soluble in water.

In the next to the last column of the table are found the group VIIA elements—fluorine (F), chlorine (Cl), bromine (Br), and iodine (I).* These are known as the **halogens** (Greek for salt formers). At room temperature fluorine and chlorine are gases, bromine is a liquid, and iodine is a solid. All of the halogens exist as diatomic molecules X_2. The reaction of chlorine with an element of group IA or IIA to form a chloride, as described above, is typical of the reaction of a halogen with a group IA or group IIA metal:

group IA: $2M + X_2 \rightarrow 2MX$

group IIA: $M + X_2 \rightarrow MX_2$

All of the halogens react with hydrogen according to the reaction

$$H_2 + X_2 \rightarrow 2HX$$

With the exception of HF, all the hydrogen halides react with water to form strong electrolytes. (An **electrolyte** is a substance which changes chemically while conducting an electric current. The electrolyte is said to be strong if it is a good electrical conductor.)

The rightmost column of the periodic table, headed group VIIIA (also labeled VIII or 0 by some chemists), contains the gases helium (He), neon (Ne), argon (Ar), krypton (Kr), xenon (Xe), and radon (Rn). All of these exist as monatomic gases in their elemental form; all other elements which are gases at room temperature are diatomic, for example, H_2, O_2, N_2, Cl_2, and F_2. At one time it was thought that the elements of group VIIIA were completely inert and could enter into no chemical combinations. For this reason, the group VIIIA elements were called the **inert** or **noble** gases; these names are still in use today even though they are not literally

---

* Astatine (At) is so rare that its behavior has not been studied in detail. It has been estimated that there is no more than 30 g of astatine in the earth's crust. Astatine may be produced artificially by bombarding thorium with protons.

correct. As early as 1930, the American chemist Linus Pauling predicted that xenon should form stable fluorides if a way could be found to prepare such compounds, and it is now known that the so-called inert gases do form some very stable compounds. For example, xenon forms XeF_2, XeF_4, XeF_6, XeO_4, Na_4XeO_6, and other compounds. Krypton and radon form similar compounds. Even helium, the most inert of all elements, has been found in unstable combinations such as HeH and He_2^+.

Chapters 21, 22, and 23 examine in more detail these and certain other groups of the periodic table. From the brief description just presented it should be clear that the periodic table arranges the elements in a very systematic way as to chemical and physical properties. The elements of a given group usually have very similar properties and those in adjacent groups, though still somewhat similar, differ more in their properties. When Mendeleev first proposed the periodic table some elements had not yet been discovered. Nevertheless, Mendeleev was able to predict the properties of these with remarkable accuracy simply on the basis of the periodic law. Table 2.1 compares one of Mendeleev's predictions with the results of subsequent studies.

TABLE 2.1 MENDELEEV'S PREDICTION OF SOME OF THE CHEMICAL AND PHYSICAL PROPERTIES OF GERMANIUM

PROPERTY	PREDICTED (1871)	OBSERVED (1886)
Atomic weight (amu)	72	72.32
Atomic volume (cm³)	13	13.22
Specific gravity	5.5	5.47
Specific heat (cal·g^{-1}·K^{-1})[a]	0.073	0.076
Color	Dark gray	Grayish white
Formula of oxide	GeO_2	GeO_2
Specific gravity of oxide	4.7	4.703
Reaction with H_2O	Slightly reactive with steam	Does not react with H_2O
Reactivity with acids	Slight	Not attacked by hydrochloric acid
Formula of chloride	$GeCl_4$	$GeCl_4$
Specific gravity of chloride	1.9 at 0 °C	1.887 at 18 °C
Boiling point of chloride	Below 100 °C	86 °C

[a] The SI unit of specific heat is J·kg^{-1}·K^{-1}, but very few publications list specific heats in this unit. This may change in the near future.

2.6 OXIDATION NUMBERS OF ELEMENTS

Chemists have found it convenient to assign signed numbers to the atoms existing in chemical compounds. These numbers, known as **oxidation numbers**, are useful in checking the correctness of formulas and, more important, they constitute a useful basis for the classification of elements as they occur in chemical combination. The oxidation number of an element in a compound is assigned on the basis of the following rules:

1. The oxidation number of an element in the elemental state is zero.
2. The oxidation state of hydrogen in all compounds except hydrides is +1; in hydrides (see Section 15.4) it is −1.
3. The oxidation number of oxygen in all compounds except peroxides and OF_2 is −2; in peroxides* it is −1, and in OF_2 it is +2.
4. The algebraic sum of all the oxidation numbers of all the atoms in any neutral molecule is zero; in ions the sum equals the net charge on the ion.

TABLE 2.2 OXIDATION NUMBERS OF SOME OF THE ELEMENTS IN THEIR COMPOUNDS

COMPOUND OR ION	OXIDATION NUMBERS		SUM
NaCl (sodium chloride)	Na = +1	Cl = −1	0
HCl (hydrogen chloride)	H = +1	Cl = −1	0
H_2O (water)	H = +1	O = −2	0
SO_4^{2-} (sulfate ion)	S = +6	O = −2	−2
H_2 (hydrogen)	H = 0		0
NH_3 (ammonia)	H = +1	N = −3	0
NH_4^+ (ammonium ion)	H = +1	N = −3	+1
CO_2 (carbon dioxide)	C = +4	O = −2	0
CO (carbon monoxide)	C = +2	O = −2	0
N_2O_3 (dinitrogen trioxide)	N = +3	O = −2	0
NO_3^- (nitrate ion)	N = +5	O = −2	−1
NaH (sodium hydride)	Na = +1	H = −1	0
H_2O_2 (hydrogen peroxide)	H = +1	O = −1	0

Table 2.2 lists some compounds and ions along with the oxidation numbers of all the component elements. Notice that the elements in a given group tend to have common oxidation numbers. For example, all elements of group IA (sodium in Table 2.2) have +1 oxidation numbers in their compounds and all group IIA elements have +2 oxidation numbers. Some groups are characterized by more than one oxidation number: The group IVA elements have +2 and +4 in common and group VA elements (nitrogen in the table) have +3 and +5 in common. However, nitrogen also has oxidation numbers of −3, −2, −1, +1, +2, and +4 depending on the compound in which it is found. The group VIA elements all have −2 oxidation numbers and all exhibit some positive oxidation numbers as well, for example sulfur

---

* Peroxides are discussed in Chapter 15. There are also oxygen compounds in which the oxidation number of oxygen is −½, but these are rare exceptions. KO_2, potassium superoxide, is an example.

(S) and selenium (Se) exhibit +4 and +6 oxidation numbers, and oxygen exhibits +2 in OF_2. In group VIIA all the elements exhibit the −1 oxidation number; all except fluorine have some positive oxidation numbers as well.

The full significance of oxidation numbers will not become clear to the student until the structures of atoms and molecules are explored in more detail; for the present it is sufficient to be able to use them in certain simple situations. For example, the fact that sulfur may have oxidation numbers of +4 and +6 means that two different oxides, with empirical formulas SO_2 and SO_3, may exist. Similarly, the formula for sodium chloride would never be written mistakenly as $NaCl_2$ or Na_2Cl since neither of these obeys rule 3.

SUMMARY

The chemical composition of compounds is expressed by combining atomic symbols to form chemical formulas. The simplest formula of a compound reveals the elements composing it, the number and weight ratios of the constituent atoms, and the percentage composition of the compound. Molecular formulas indicate, in addition to the above, the actual number of atoms of each kind per molecule. Given the percentage composition of a substance, its simplest formula may be deduced. If the substance's molecular weight is known, its molecular formula may be deduced as well.

Chemical formulas may also be written in such a way that they indicate details of how the atoms are bonded to each other: which are nearest neighbors and where the atoms are located in three-dimensional space.

Chemical reactions are summarized by using the formulas of reactants and products in balanced chemical equations that reflect the law of conservation of matter and the immutability of elements. Such equations form the basis of quantitative calculations that allow the chemist to predict how much of one substance is needed to produce a given amount of another substance.

Many of the chemical and physical properties of the elements are periodic; that is, these properties follow trends as the atomic number increases and these trends recur or repeat every so many atoms. This makes it possible to arrange the elements in tabular form such that each column contains elements of similar properties.

Elements in compounds may be assigned positive and negative numbers known as oxidation numbers. The oxidation number of an ion consisting of a single atom is just its charge. The oxidation numbers of other elements are usually deduced by assigning +1 to hydrogen and −2 to oxygen (with some exceptions) and employing the rule that the sum of the oxidation numbers of a molecule equals the net charge of the molecule. The principal exceptions are −1 for oxygen in peroxides, +2 for oxygen in OF_2, and −1 for hydrogen in hydrides. Oxidation numbers aid in the classification of elements and in writing correct formulas.

LEARNING GOALS

1. Know how to calculate the empirical formula of a compound from percentage composition data.

2. Be able to deduce the molecular formula from the simplest formula using molecular weight.

3. Know the difference between molecular weight and formula weight.

4. Be able to calculate percentage composition from simplest or molecular formulas.

5. Know what structural and geometric formulas are and what information they convey.

6. Know how to balance chemical equations.

7. Be able to use balanced chemical equations to calculate how much of one substance produces a given amount of another substance (or vice versa).

8. Be able to convert weights of a substance to moles and vice versa.

9. Be able to recognize and work a limiting reagent problem.

10. Be able to assign oxidation numbers to the elements in a compound.

11. Using given oxidation numbers for two elements, be able to predict simplest formulas for their compounds.

12. Know what the periodic table is and its general arrangement.

DEFINITIONS, TERMS, AND CONCEPTS TO KNOW

empirical (simplest) formula
molecular formula
structural (linkage) formula
geometric formula
molecular weight
formula weight
weight of a mole
chemical equation

mole-ratio factor
excess reagent
limiting reagent
periodic property
periodic law
alkali metals
alkaline earth metals
groups

periods
metals
nonmetals
halogens
inert gases
oxidation number

QUESTIONS AND PROBLEMS

Simplest and Molecular Formulas

1. Three different compounds of nitrogen and oxygen have the following amounts of nitrogen (in percent by weight): compound A, 46.7%; compound B, 63.6%; compound C, 30.4%. Determine the empirical formulas of these compounds and show how they illustrate the law of multiple proportions (see Problem 1 in Chapter 1).

2. A gaseous compound contains only carbon and hydrogen. When a 1.00-g sample is burned in O_2, 3.00 g CO_2 and 1.64 g H_2O are produced.
   a) Calculate the percentage composition (by weight) of the above compound.
   b) What is the simplest formula of the above compound?
   c) Can the molecular formula of the above compound be deduced from the data given? Explain.

3. Chemical analysis of an unknown compound yielded 66.70% carbon, 7.41% hydrogen, and 25.90% nitrogen. The molecular weight was found to be 106 (±3%) amu.
   a) What is the empirical formula of the compound?
   b) What is the molecular formula of the compound?
   c) What is the precise molecular weight of the compound?

4. A compound of carbon, hydrogen, and oxygen contains 82.6% (by weight) carbon and 9.6% oxygen. The molecular weight is around 330 amu. Determine the molecular formula.

5. A compound of carbon, hydrogen, and chlorine contains 10.04% carbon and 89.12% chlorine. What is the empirical (simplest) formula of the compound?

6. A compound of carbon, hydrogen, and chlorine is analyzed and found to contain 39.7% carbon and 58.7% chlorine. What is its simplest formula?

7. Nicotine has the simplest formula C_5H_7N and a molecular weight of about 160 amu. What is its molecular formula?

8. A compound of carbon and fluorine is found to be 17.4% carbon by weight. Its approximate molecular weight is 135 amu.
   a) What is the simplest formula of the compound?
   b) What is the molecular formula of the compound?

9. A compound of iron, sulfur, and oxygen contains 36.8% iron and 21% sulfur (by weight). What is the empirical formula of the compound?

10. A compound is found to contain 85.69% carbon. The only other element it contains is hydrogen. The molecular weight is estimated to be about 56 amu. What is the molecular formula of the compound?

11. A compound contains 2.094 g of iron and 2.6595 g of chlorine. Determine the empirical formula.

12. Sucrose or table sugar has the molecular formula $C_{12}H_{22}O_{11}$. Calculate the percentage composition by weight in terms of carbon, hydrogen, and oxygen.

13. A compound of xenon and chlorine contains 44.8% chlorine by weight. What is the empirical formula of the compound?

14. Calculate the molecular weight or formula weight of each of the following substances.
   a) C_2H_5OH b) $Ca(NO_3)_2$ c) $(NH_4)_2SO_4$
   d) $Cu(NH_3)_4Cl_2$ e) $C_6H_7NO_2$

15. One form of elemental sulfur has a molecular weight of about 256 amu. What is the formula of such a sulfur molecule?

16. Why are extremely accurate analytical data needed in determining empirical formulas of molecules such as $C_{100}H_{201}NO$?

17. What important chemical law is illustrated by the compounds CH_4, C_2H_6, C_2H_2, and $C_{10}H_8$?

18. A sample of pure water found on the shelf of a laboratory in Oak Ridge, Tennessee, is found to contain 80% oxygen rather than the expected 88.8%. Explain how this can be so and determine the empirical formula of the water. [*Hint:* What is meant by "heavy water"?]

19. The insecticide DDT is a compound of carbon, hydrogen, and chlorine. A sample of DDT weighs 0.1620 g. When burned, this sample produces 0.2815 g of CO_2 and 0.0371 g of H_2O. The residue also produces 0.3276 g of AgCl when dissolved in water and treated with $AgNO_3$. Calculate the molecular formula of DDT if its molecular weight is about 360 amu.

Balancing Chemical Equations

20. Balance the following chemical equations.
   a) $NH_4NO_2 \rightarrow N_2 + H_2O$
   b) $CH_4 + O_2 \rightarrow CO_2 + H_2O$
   c) $O_3 \rightarrow O_2$
   d) $N_2 + H_2 \rightarrow NH_3$
   e) $Cu + AgNO_3 \rightarrow Ag + Cu(NO_3)_2$
   f) $Mg + N_2 \rightarrow Mg_3N_2$
   g) $NH_3 + O_2 \rightarrow NO + H_2O$
   h) $C_{12}H_{22}O_{11} + O_2 \rightarrow CO_2 + H_2O$
   i) $C_6H_{12}O_6 + O_3 \rightarrow CO_2 + H_2O$
   j) $C_6H_4N_2O_4 + O_2 \rightarrow CO_2 + H_2O + NO_2$

21. Write balanced chemical equations for each of the following processes.
   a) Iron reacts with oxygen to form Fe_3O_4.
   b) Ammonia (NH_3) reacts with oxygen to form water and NO.
   c) Ammonia (NH_3) reacts with oxygen to form water and NO_2.
   d) Benzene (C_6H_6) reacts with chlorine to form chlorobenzene (C_6H_5Cl) and hydrogen chloride (HCl).
   e) Anhydrous copper(II) sulfate, $CuSO_4$, reacts with water to form copper(II) sulfate pentahydrate ($CuSO_4 \cdot 5H_2O$).
   f) Diborane (B_2H_6) reacts with oxygen to form water and boron oxide (B_2O_3).

22. Balance the following equations:
   a) $Na_3PO_4 + BaCl_2 \rightarrow Ba_3(PO_4)_2 + NaCl$
   b) $C_3H_8 + O_2 \rightarrow CO_2 + H_2O$

c) $C_6H_{12}O_6 \to C_2H_5OH + CO_2$
d) $Na_2O_2 + H_2O \to NaOH + O_2$
e) $Mg_3N_2 + H_2O \to Mg(OH)_2 + NH_3$

23. Write balanced equations for each of the following (use Appendix 4 to determine formulas).
 a) Calcium carbonate forms calcium oxide and carbon dioxide.
 b) Calcium plus nitrogen produces calcium nitride.
 c) Iron plus sulfur produces iron(III) sulfide.
 d) Nitric acid plus zinc produces zinc nitrate and water.
 e) Aluminum plus fluorine produces aluminum fluoride.

Calculations with Balanced Chemical Equations

24. Soda pop (*tonic* to New Englanders!) fizzes due to spontaneous decomposition of carbonic acid, H_2CO_3. The bubbles given off are carbon dioxide gas, CO_2. Water is the only other product formed.
 a) Write a balanced chemical equation for the fizzing of soda pop.
 b) If a bottle of pop gives off 11.0 g of CO_2 before going flat, what amount of H_2CO_3 (in moles) was originally present? (Assume all the H_2CO_3 decomposes.)
 c) How many grams of water would be formed in part (b)?

25. A convenient method of preparing oxygen in the laboratory makes use of the process $2KClO_3 \to 2KCl + 3O_2$.
 a) How many pounds of O_2 could be prepared from 24.5 lbs of $KClO_3$? (Assume complete reaction.)
 b) How many klotdunks (a lower Slobbovian unit of weight) of $KClO_3$ would be required to prepare 50 klk of O_2?

26. Lithium amide, $LiNH_2$, may be prepared by the following sequences:

$N_2 + H_2 \to NH_3$
$NH_3 + Li \to LiNH_2 + H_2$

 a) Balance the above equations.
 b) What is the maximum weight of $LiNH_2$ obtainable from 28 g of N_2 and 28 g of Li?
 c) How much of which material (N_2 or Li), if any, is left unreacted under the conditions of part (b)?

27. A sample of europium(II) chloride, $EuCl_2$, weighs 1.00 g and is treated with aqueous $AgNO_3$. Solid AgCl is formed according to the reaction

$EuCl_2 + 2AgNO_3 \to 2AgCl + Eu$

The weight of AgCl obtained is 1.28 g. What is the atomic weight of europium?

28. Many years ago, elemental chlorine was prepared industrially by the Deacon's process:

$4HCl + O_2 \to 2H_2O + 2Cl_2$

 a) How many grams of HCl and O_2 would be needed to produce 1 kg of chlorine? (Assume all the HCl and O_2 are used up.)
 b) How many grams of water would be produced in the above?

29. One of the steps in the commercial preparation of nitric acid is

$4NH_3 + 5O_2 \to 4NO + 6H_2O$

 a) When 170 g of NH_3 is completely oxidized, how much NO is produced?
 b) How much oxygen is required in part (a)?
 c) What products are obtained when 100 g of NH_3 is mixed with 100 g of O_2 and allowed to react as completely as possible?

30. Consider the reaction

$Cu_2S + 2O_2 \to 2CuO + SO_2$

 a) What is the maximum weight of Cu_2S (in g) that can be converted to CuO by 0.5 mol of oxygen?
 b) How much SO_2 (in moles) is produced in (a)?

31. Analysis of a sample of pure $Al_2(CO_3)_3$ shows that 1.32 moles of aluminum is present.
 a) How much carbon and oxygen is present in the sample?
 b) What is the weight of the sample prior to analysis?

32. Thirty grams of iron(III) oxide is treated with carbon monoxide to produce carbon dioxide and elemental iron.
 a) Write the balanced equation for the reaction.
 b) How much iron is produced in the reaction?
 c) How much CO (in grams) is required in the reaction?

33. An ore of copper contains an average of 1.9% by weight of copper(I) sulfide. Assuming a process that is 65% efficient, how many tonnes of copper metal can be obtained from 500 t of the ore?

34. The insecticide chloropicrin (CCl_3NO_2) is made from nitromethane (CH_3NO_2) by the process

$$CH_3NO_2 + 3Cl_2 \rightarrow CCl_3NO_2 + 3HCl$$

a) Assuming complete reaction of the limiting reagent, how much chloropicrin is produced from 1 tonne each of nitromethane and chlorine?

b) What is the excess reagent and how much of it remains unused?

35. A train car load of bauxite, an ore of aluminum, contains 86.3% Al_2O_3 by weight. What is the maximum weight of aluminum that can be prepared from 100 t of bauxite?

36. If a sheet of iron weighing 25 g rusts away to iron(III) oxide, how much rust is formed?

37. Consider the equation

$$Fe_2O_3 + 3C \rightarrow 2Fe + 3CO$$

a) How many grams of iron could be obtained from 1 kg of iron(III) oxide?

b) How many pounds of iron could be obtained from 1 lb of iron(III) oxide?

c) How many moles of carbon monoxide would be produced from 10 kg of iron(III) oxide?

38. How many moles of magnesium oxide would be produced by burning 30.0 g of magnesium?

39. Zinc sulfate is produced from the action of sulfuric acid on zinc. How much zinc sulfate could be produced from 30.0 g of zinc?

Oxidation Numbers

40. Determine the oxidation number of each of the elements in each of the following compounds: $RbClO_3$, SnO_2, $K_2Cr_2O_7$, NH_4Br, H_3PO_3, H_3PO_4, $S_2O_3^{2-}$, Fe_2O_3, FeO, Fe_3O_4, $S_4O_6^{2-}$, O_3, HN_3.

41. Five possible oxidation numbers of nitrogen are +1, +2, +3, +4, and +5.

a) Write simplest formulas for all possible oxides of nitrogen.

b) Write formulas for two ions of nitrogen (each with 1− charge) having several oxygen atoms and one nitrogen atom of +3 or +5 oxidation number.

42. The oxidation number of carbon in CH_4 is −4; in CO_2 it is +4. Explain.

General

43. Below is a portion of the periodic table.

A B C D
E F G H
I J K L

a) Which element has the largest atomic volume?

b) Which element has the smallest atomic volume?

44. The density of liquid mercury at room temperature (about 25 °C) is 13.6 g·cm^{-3}. What is the apparent atomic volume of a mercury atom in cm^3?

45. A compound of formula XeF_n (n is unknown) weighs 0.311 g and contains 9.03×10^{20} molecules. What is the value of n?

46. The molecular formula of common table sugar is $C_{12}H_{22}O_{11}$.

a) Calculate the weight of one molecule of sugar in amu.

b) Calculate the weight composition (in percent) of carbon, hydrogen, and oxygen in sugar.

47. a) Calculate the number of molecules in 10 g of pure ozone, O_3.

b) Calculate the number of atoms in the above.

c) Calculate the number of moles of ozone in part (a).

d) Calculate the number of atoms in 1.0 g of platinum.

48. Which oxide of nitrogen (N_2O, NO, N_2O_3, NO_2, or N_2O_5) has the greatest percentage (by weight) of nitrogen?

49. Proteins are often analyzed for nitrogen by the Kjeldahl method, in which nitrogen is converted to NH_3 and measured as such. What is the percentage of nitrogen in a 3.00-g protein sample that yields 0.20 g of NH_3?

ADDITIONAL READINGS

Chemistry: An Experimental Science. San Francisco: Freeman, 1963. This very successful text is widely used in many high schools. Chapters 2, 3, and 6 form good background reading for the material presented so far.

Mahan, Bruce H., *University Chemistry,* 3rd ed. Reading, Mass.: Addison-Wesley, 1975. Chapter 1 provides a somewhat different approach to some of the material presented so far. Chapter 13 deals with the periodic table.

Pauling, Linus, and Roger Hayward, *The Architecture of Molecules.* San Francisco: Freeman, 1964. A collection of attractive color drawings of ball-and-stick models of several dozen important molecules. The accompanying text provides much interesting and important chemical information.

3
THE GASEOUS STATE

The famous Magdeburg experiment. On 8 May 1654 Otto von Guericke, inventor of the vacuum pump, demonstrated that air exerts pressure by showing that sixteen horses could not pull apart two halves of a hollow iron sphere surrounding a vacuum. (Courtesy of Brown Brothers.)

It is evident, that as common air, when reduced to half its wonted volume, obtained near about twice as forcible a spring as it had before; so that this thus comprest air being further thrust into half this narrow room, obtained thereby a spring about as strong again as it last had, and consequently four times as strong as that of common air.
ROBERT BOYLE, *New Experiments Physico-Mechanicall Touching the Spring of the Air*, 2nd Ed., 1662

Whereas solids are characterized by a high degree of geometric orderliness, gases, by contrast, are characterized by the opposite extreme—almost total disorder. In fact, the word "gas" was coined by a Belgian physician, Joannes van Helmont (1579–1644), after the Greek *chaos*. Yet, in some respects the gaseous state is the simplest and the most thoroughly studied of the states of matter. This is partly a legacy of the human race's long and continued interest in gambling and games of chance. This interest encouraged the development of a branch of mathematics known as **statistics and probability theory**, which describes the behavior of random events and, consequently, applies to the behavior of gases.

One of the most obvious characteristics of a laboratory-size sample of a gas is that it readily assumes the size and shape of its container. This is not true of extremely large samples of gas; for example, the air in the earth's atmosphere is prevented from filling the entire universe by the gravitational attraction of the earth, which concentrates gases close to its surface.

Many of the gases commonly encountered at room temperature are monatomic or diatomic. All of the monatomic gases come from group VIIIA of the periodic table, namely, He, Ne, Ar, Kr, Xe, and Rn. The common diatomic gases are H_2, F_2, Cl_2, O_2, N_2, CO, and HCl. Examples of common polyatomic gases are methane (CH_4), ammonia (NH_3), carbon dioxide (CO_2), and sulfur dioxide (SO_2).

3.1 BOYLE'S LAW

Around 1662, Robert Boyle found that for a constant mass of gas at a given temperature, the pressure (p) and volume (V) of many gases appeared to satisfy the relationship

$$p = \frac{k_1}{V} \quad \text{(temperature constant)} \tag{3.1}$$

where k_1 is a constant dependent on the temperature alone. This relationship is known as **Boyle's law**.* The constant k_1 increases as the temperature increases; that

---

* This law was also published in 1676 by the French scientist Edmé Marriotte and is known in many European countries as the Boyle-Marriotte law.

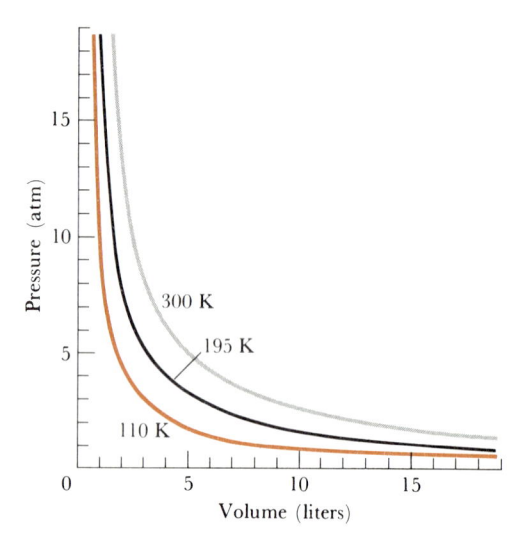

Fig. 3.1. Variation of pressure with volume at several different temperatures—a graphical representation of Boyle's law. Each of the curves is called an *isotherm* (from the Greek *isos*, equal, and *therme*, heat) since each connects points of equal temperatures.

is, the product pV varies directly with temperature. Figure 3.1 illustrates Eq. (3.1) for several different temperatures.

The **pressure** exerted by a gas is of dynamic origin; that is, it results from the impact of swiftly moving atoms or molecules on the walls of the containing vessel. Nevertheless, it is perhaps easier to understand pressure as a general concept if it is examined first in static situations. Pressure is defined as *force per unit area*. A cube of ice with a mass of 1.000 kg resting on a flat surface of the earth exerts a downward force of $(1.000 \text{ kg})(9.80 \text{ m} \cdot \text{s}^{-2})$ or 9.80 N. (Note that 9.80 m·s^{-2} is the value of the gravitational constant—the rate at which the velocity of a freely falling body changes when dropped to earth.) Since the area of contact with the earth's surface is $(0.100 \text{ m})^2$ or 1.00×10^{-2} m^2, the pressure exerted by the ice on the earth's surface is 9.80 N/1.00 $\times 10^{-2}$ m^2 or 9.80×10^2 N·m^{-2}.

If the ice were in the form of a vertical column 1.00 cm^2 (1.00×10^{-4} m^2) in cross section, the force exerted by it on the earth would be the same but the pressure would be 9.80 N/1.00 $\times 10^{-4}$ m^2 or 9.80×10^4 N·m^{-2}, that is, 100 times greater than in the first instance. This is why a pointed object is easier to push into the ground than is a blunt object. The pointed object allows concentration of the force on a smaller area so that a greater pressure is obtained with the same force.

The pressure exerted by a gas is due, as we have said, to the bombardment of the container walls by a large number of swiftly moving particles. Each individual collision between particle and wall produces a force upon the wall; the sum of all such forces divided by the area of the wall is the pressure of the gas.

Gas pressure is measured experimentally by balancing the force per unit area exerted by the gas with a known force per unit area. For example, the gas pressure can be employed to move a diaphragm which in turn rotates an indicator needle against the tension of a spring. By measuring the weights needed to rotate the needle a given amount, the instrument can be calibrated to indicate the pressures exerted by gases. A simple and rapid device for measuring the *pressure difference* between two gases is the **differential manometer**; the principle of operation of this device is illustrated and explained in Fig. 3.2.

Gas pressures are often expressed in terms of a unit called the **atmosphere** (atm), defined as the pressure exerted by a column of mercury at 0 °C which is 76 cm high and has a cross-sectional area of 1 cm². Strictly speaking, the measurement should also be carried out at sea level and at 045° latitude. A pressure of 1 atm is approximately equal to the pressure due to the weight of the earth's atmosphere per square centimeter of surface at sea level. Since the density of mercury at 0 °C is 13.6 g·cm^{-3}, one atm represents a force of (76 cm³)(13.6 g·cm^{-3})(0.001 kg·g^{-1})(9.80 m·s^{-2}) or about 10 N acting upon each square centimeter of surface. The atmosphere pressure unit is subdivided into 760 units, each of which is called a **torr**; 760 torr is equivalent to 1 atm. A torr is also referred to as 1 mm of mercury [often written **mm(Hg)**].

The torr is named in honor of the Italian mathematician and inventor Evangelista Torricelli (1608–1647). Besides making contributions to the physics of falling bodies and flowing liquids, Torricelli invented the mercury barometer and wrote several papers in pure mathematics. Torricelli was Galileo's assistant for a few months before the latter's death and succeeded him as mathematician at the Court of Tuscany.

The official SI unit of pressure is the newton per square meter (N·m^{-2}) and is called the **pascal** (Pa). One atm is equivalent to 1.01325×10^5 Pa. Since it is unlikely that chemists will use the pascal for gas pressures, only atm and torr will be used in this text. However, some weather bureaus (particularly in Canada) are beginning to report barometric pressures in kPa; one atm is approximately 100 kPa.

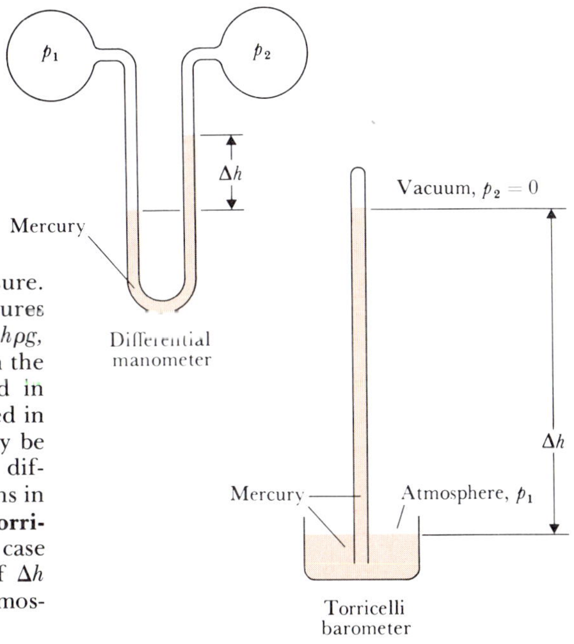

Fig. 3.2. The measurement of gas pressure. In the **differential manometer** the pressures p_1 and p_2 are related by $p_2 = p_1 + \Delta h \rho g$, where ρ is the density of the mercury in the tube. If pressures are to be expressed in torr, then $p_2 = p_1 + \Delta h$ if Δh is expressed in millimeters. If p_1 is known, then p_2 may be measured simply by measuring Δh, the difference in height of the mercury columns in the two legs of the manometer. The **Torricelli barometer** represents a special case where $p_2 = 0$ so that measurement of Δh gives p_1 (the pressure of the earth's atmosphere) directly.

A typical application of Boyle's law is illustrated in the following problem.

EXAMPLE 3.1 At a given temperature a sample of gas has a volume of 10.3 L and a pressure of 0.90 atm.

    a) What is the value of the constant k_1?

    b) If the pressure is changed to 2.0 atm, what is the volume of the gas at the given temperature?

    c) If the volume is changed to 15 L, what is the pressure of the gas at the given temperature?

SOLUTION

    a) $pV = k_1 = (0.90 \text{ atm})(10.3 \text{ L}) = 9.3 \text{ L} \cdot \text{atm}$

    b) $V = \dfrac{k_1}{p} = \dfrac{9.3 \text{ L} \cdot \text{atm}}{2.0 \text{ atm}} = 4.7 \text{ L}$

    c) $p = \dfrac{k_1}{V} = \dfrac{9.3 \text{ L} \cdot \text{atm}}{15 \text{ L}} = 0.62 \text{ atm}$

3.2 CHARLES'S LAW

Although Boyle and others noted that the pV product increases as the temperature increases, attempts to express this observation in a simple mathematical way were unsuccessful until about 125 years later. The reasons for this were the lack of suitable means of expressing temperature quantitatively, and the lack of the concept of **absolute temperature**. The most widely used **relative temperature** scale, the centigrade or Celsius scale,* is defined by assigning the value of 0 °C to the freezing point of pure water and the value of 100 °C to the boiling point of water at 1 atm pressure. The concept of absolute temperature, that is, temperature measured from some unique starting point, is easily introduced by examining how the pV product behaves as the relative temperature is lowered. Experiments show that as the relative temperature t decreases, pV becomes smaller and smaller in roughly linear fashion. As shown in Fig. 3.3, the absolute zero of temperature may be defined as that temperature for which the pV product is zero.† This temperature, which must be obtained by graphical extrapolation of the data, is −273.15 °C; it is assumed that no temperature lower than −273.15 °C can be obtained.

The absolute temperature scale (also called the **Kelvin scale**) is established by replacing the lowest possible temperature, −273.15 °C, with 0 K; the freezing and

---

* Named after the Swedish astronomer, Anders Celsius (1701–1744). Actually, Celsius's original proposal to the Swedish Academy of Sciences was to let 0 ° be the boiling point of water and 100 ° be the freezing point. There is no evidence that Anders practiced yoga by standing on his head!

† The pV product must be evaluated at very low pressures; otherwise the extrapolation to zero pV gives different values for different gases.

boiling points of water then become 273.15 K and 373.15 K, respectively, on the absolute temperature scale. Note that in accord with the SI conventions, no degree symbol is used with K; the symbol K itself means *a degree on the Kelvin scale*. For most ordinary calculations it is permissible to use $-273\,°C$ rather than $-273.15\,°C$ for absolute zero and to use 273 K and 373 K as the freezing and boiling points of water, respectively.

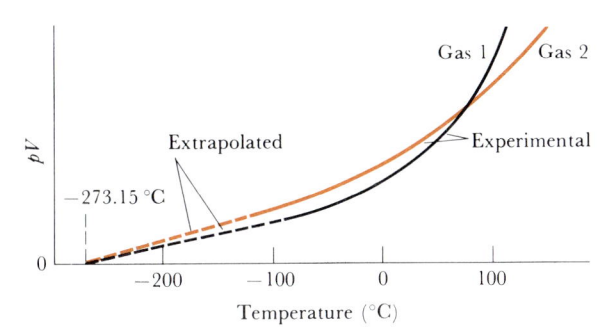

Fig. 3.3. Definition of absolute zero in terms of the extrapolated vanishing of the pressure-volume product as the relative temperature is lowered. If the pV product is evaluated at sufficiently low pressure, the extrapolated temperature is independent of the gas.

The Kelvin scale is named after William Thomson, Baron Kelvin of Largs (1824–1907). An Irish-born British physicist, Thomson made outstanding contributions to the fields of electromagnetism, thermodynamics, telegraphy, and navigation, and he played a crucial role in the laying of the first Atlantic cable.

Jacques Charles (1746–1823) was a French scientist noted for his pioneering work on gases and balloon ascensions. Charles was a talented public lecturer, an inventor of scientific apparatus, and the first man to fly in a hydrogen-filled balloon, eventually attaining an altitude of 3000 meters (9840 ft).

Joseph Louis Gay-Lussac (1778–1850), a younger compatriot of Charles, is noted for work on gas laws and the properties of iodine and cyanogen, $(CN)_2$. Gay-Lussac repeated Charles's experiments on the dependence of gas volume on absolute temperature and published the results. His most important contribution was the discovery that gases combine among themselves by simple integral proportions by volume; this constituted strong support for Dalton's atomic theory, but Dalton rejected it. In 1804 Gay-Lussac ascended to 7000 m (23,000 ft) in a balloon to obtain air samples to analyze.

In 1787 Jacques Charles showed that the volume of a fixed amount of gas at fixed pressure obeyed the relationship

$$V = k_2 T \quad \text{(pressure constant)} \tag{3.2}$$

where k_2 is a constant dependent only upon p, and T is the absolute temperature. This relationship was confirmed in 1802 by Joseph Gay-Lussac and hence is known both as **Charles's law** and as **Gay-Lussac's law**. In Fig. 3.4 is shown a graph of Charles's law for several different pressures.

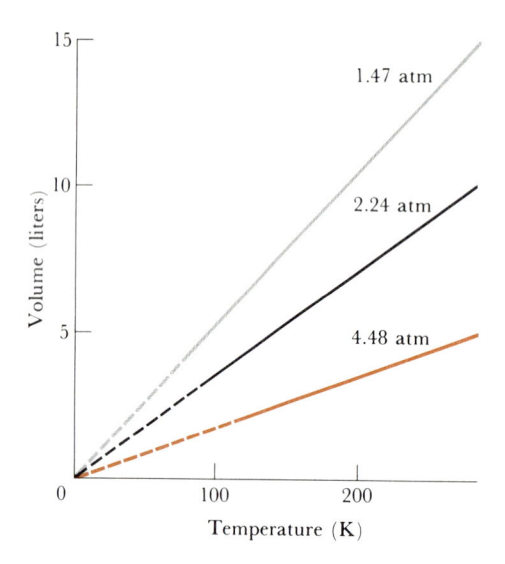

Fig. 3.4. Variation of volume with absolute temperature at several different pressures—a graphical representation of Charles's law. Each of the lines is called an *isobar* (from the Greek *isos*, equal, and *baros*, weight) since each connects points of equal pressures.

A typical application of Charles's law is illustrated in the following example.

EXAMPLE 3.2 At a given pressure a sample of gas has a volume of 25 L at 20 °C.

a) What is the value of the constant k_2?

b) At what temperature will the volume be 15 L if the pressure is unchanged?

c) If the temperature is raised to 50 °C, what is the volume at the same pressure?

SOLUTION

a) $V/T = k_2 = 25 \text{ L}/293 \text{ K} = 0.085 \text{ L} \cdot \text{K}^{-1}$

b) $T = V/k_2 = 15 \text{ L}/0.085 \text{ L} \cdot \text{K}^{-1} = 176 \text{ K } (-97 \text{ °C})$

c) $V = k_2 T = (0.085 \text{ L} \cdot \text{K}^{-1})(323 \text{ K}) = 27 \text{ L}$

Note that the temperature T must be expressed on the Kelvin scale.

3.3 THE GENERAL GAS LAW

The laws of Boyle and Charles may be combined into a single law by the use of a simple algebraic treatment. Consider a sample of gas at temperature T_1, pressure p_1, and volume V_1. Now, keeping the temperature constant, change the pressure from p_1 to some new value p_2. Boyle's law states that the pV products at T_1 obey the relationship

$$p_1 V_1 = p_2 V_x \tag{3.3}$$

where V_x is the volume when the pressure is p_2. Solving for V_x produces

$$V_x = \frac{p_1 V_1}{p_2} \tag{3.4}$$

Now, considering the gas at p_2, V_x, and T_1, keep the pressure constant at p_2 but change the temperature from T_1 to T_2. According to Charles's law the V/T ratios at p_2 obey the relationship

$$V_x/T_1 = V_2/T_2 \tag{3.5}$$

where V_2 is the new volume when the temperature is T_2. Solving for V_x gives

$$V_x = \frac{V_2 T_1}{T_2} \tag{3.6}$$

Since Eqs. (3.4) and (3.6) are alternative expressions for V_x, they may be equated to give

$$\frac{p_1 V_1}{p_2} = \frac{V_2 T_1}{T_2} \tag{3.7}$$

Rearranging so that all like subscripts are on the same side of the equality sign leads to

$$\frac{p_1 V_1}{T_1} = \frac{p_2 V_2}{T_2} \quad \text{or} \quad \frac{pV}{T} = \text{constant} \tag{3.8}$$

Amedeo Avogadro (1776–1856) was an Italian scientist whose name is often used to designate the number now known as a mole. After qualifying in law and practicing for many years, Avogadro turned to physics and chemistry. His 1811 hypothesis, that equal volumes of different gases at the same temperature and pressure contain the same number of molecules, was not recognized as significant until about 1860 (four years after Avogadro died).

In 1811 the Italian scientist Avogadro hypothesized that equal volumes of different gases, all at the same temperature and pressure, contained the same number of molecules. This principle has since been given the name **Avogadro's law**. Now, if the constant in Eq. (3.8) is given the symbol R when V is the volume of one mole of gas, a sample of n moles of gas is described by

$$\frac{pV}{T} = nR \quad \text{or} \quad pV = nRT \tag{3.9}$$

The above relationship is known as the **general gas law** or the **ideal gas equation**. The constant R is often referred to as the **ideal gas constant**. This very important constant appears in many equations of physics and chemistry, not all of which deal explicitly with gases.

Numerical values of R depend on the units in which p and V are expressed. For one mole of gas at 273 K (0 °C), experiment shows that pV for a great many gases appears to be about 22.4 L·atm.* Thus

$$R = \frac{pV}{nT} = \frac{22.4 \text{ L·atm}}{(1 \text{ mole})(273 \text{ K})} = 0.0821 \text{ L·atm·K}^{-1}\text{·mol}^{-1}$$

If other units are used for p or V or both, the value of R changes accordingly.†

Experience has shown that the general gas law is most valid at low pressures or high temperatures or both. Theory shows that the law should be exact only for *hypothetical* gases consisting of strictly noninteracting particles, each of zero volume. The fact that all gaseous substances exhibit intermolecular interactions and have nonzero atomic or molecular volumes means that deviations become most apparent when the pressure is high (closeness of molecules increases interparticle interactions) or when the temperature is low (motions of the particles are too slow to swamp out the effects of interactions). Imaginary or hypothetical gases which obey the general gas law are sometimes called **ideal** or **perfect gases**. No known real gases are ideal, yet all real gases behave somewhat like ideal gases if the pressure is low enough (at a given temperature) or the temperature is high enough (at a given pressure), or both. Gases such as He, N_2, H_2, and O_2 (all with rather low intermolecular forces) obey the ideal gas equation reasonably well for pressures of 1 atm and lower and temperatures of 273 K and higher.

3.4 CALCULATIONS WITH THE GENERAL GAS LAW

Since the general gas law contains four basic quantities (besides R, whose value is fixed and assumed to be known), any one of these (p, V, n, or T) can be calculated, provided that the remaining three are known or can be deduced from available information.

Two caveats should be kept in mind when using the general gas law: First, do not forget to use *absolute* temperatures even though the problem is stated in terms of some other scale; second, the units of p and V must agree with those of R.

---

* This means that one mole of an ideal gas at 0 °C and 1 atm pressure (often called **standard temperature and pressure** or **STP**) occupies 22.4 L. This is almost exactly the volume of three regulation-size basketballs.

† A liter atmosphere (L·atm) has the same dimensions as work or energy (mass × length^2 × time^{-2}). Thus the gas constant R has the units of energy per degree per mole. The value of R in SI units of energy is 8.314 J·K^{-1}·mol^{-1} or 8.314 Pa·m^3·K^{-1}·mol^{-1}.

The following illustrate some typical uses of the general gas law.

EXAMPLE 3.3 A sample of H_2 gas at 25 °C has a pressure of 800 torr and a volume of 2500 cm³. How many moles of H_2 must there be in the sample, assuming ideal gas behavior?

SOLUTION Substitute p, V, T, and R into the equation

$$n = \frac{pV}{RT}$$

If $R = 0.0821$ L·atm·K^{-1}·mol^{-1}, p must be in atm and V in liters. Thus $p = 800$ torr/760 torr·atm^{-1}, or 1.05 atm, and $V = 2.500$ L. Also, $T = 25$ °C $+ 273$, or 298 K. Using these values produces

$$n = \frac{(1.05 \text{ atm})(2.500 \text{ L})}{(0.0821 \text{ L·atm·K}^{-1}\text{·mol}^{-1})(298 \text{ K})} = 0.107 \text{ mol}$$

Note the cancellation of units that occurs when the calculation is set up correctly.

EXAMPLE 3.4 Calculate the pressure exerted by 2.00 moles of N_2 gas at 300 K and confined to a volume of 25.0 L.

SOLUTION Use the equation $p = nRT/V$ and substitute for n, R, T, and V. This leads to

$$p = \frac{(2.00 \text{ mol})(0.0821 \text{ L·atm·K}^{-1}\text{·mol}^{-1})(300 \text{ K})}{25.0 \text{ L}} = 1.97 \text{ atm or } 1.50 \times 10^3 \text{ torr}$$

As the following example shows, the value of R may not be needed in certain types of problems.

EXAMPLE 3.5 A sample of CO at 25 °C and 725 torr occupies a volume of 17.8 L. What will the new volume be if the temperature is changed to 50 °C and the pressure is raised to 995 torr?

SOLUTION Let

$p_1 = 725$ torr $p_2 = 995$ torr
$V_1 = 17.8$ L $V_2 = ?$
$T_1 = 298$ K $T_2 = 323$ K

There are at least two "different" ways of solving for V_2. First, use the ideal gas law to obtain a relationship for V_2:

$$V_2 = \frac{nRT_2}{p_2}$$

Note that n is not given directly and thus must be calculated from other available information. Since p_1, V_1, and T_1 are all known, n is given by

$$n = \frac{p_1 V_1}{RT_1}$$

Rather than evaluating n numerically at this point it is more efficient to substitute its algebraic value into the V_2 equation. This yields

$$V_2 = \left(\frac{p_1 V_1}{RT_1}\right)\frac{RT_2}{p_2} = \frac{p_1 T_2}{p_2 T_1} V_1 = \frac{(725 \text{ torr})(323 \text{ K})}{(995 \text{ torr})(298 \text{ K})}(17.8 \text{ L}) = 14.0 \text{ L}$$

Note that R cancels out in the last step so that its numerical value is not needed.

A second and faster way of solving this problem is to note that the data given allow direct application of Eq. (3.8) to give

$$V_2 = \frac{p_1 V_1}{T_1}\left(\frac{T_2}{p_2}\right)$$

Of course, the two ways are not really different since they are mathematically equivalent.

In certain types of problems some manipulation of the basic equation may be required. This is illustrated in the following two examples.

EXAMPLE 3.6 A volume of 100 cm³ of gas has a mass of 0.10 g at 27 °C and a pressure of 0.82 atm. Calculate the molecular weight of the gas, assuming it behaves ideally.

SOLUTION Begin with $pV = nRT$ but note that n may be replaced with w/M, where w is the mass of the gas (in grams) and M is the weight of one mole. Then

$$pV = \frac{w}{M} RT$$

$$M = \frac{wRT}{pV} = \frac{(0.10 \text{ g})(0.0821 \text{ L} \cdot \text{atm} \cdot \text{K}^{-1} \cdot \text{mol}^{-1})(300 \text{ K})}{(0.82 \text{ atm})(0.100 \text{ L})} = 30 \text{ g} \cdot \text{mol}^{-1}$$

Therefore the molecular weight is 30 amu.

EXAMPLE 3.7 Calculate the density (in $g \cdot L^{-1}$) of methane (CH_4) at 50 °C and 700 torr.

SOLUTION Density is defined in Section 1.6 as $\rho = w/V$. Rearrangement of $pV = wRT/M$ leads to

$$\frac{w}{V} = \rho = \frac{pM}{RT} = \frac{(700 \text{ torr}/760 \text{ torr} \cdot \text{atm}^{-1})(16.03 \text{ g} \cdot \text{mol}^{-1})}{(0.0821 \text{ L} \cdot \text{atm} \cdot \text{K}^{-1} \cdot \text{mol}^{-1})(323 \text{ K})}$$

$$= 5.57 \times 10^{-1} \text{ g} \cdot \text{L}^{-1}$$

3.5 THE LAW OF COMBINING VOLUMES

In 1808 Gay-Lussac announced the observation that gases combine chemically by simple integral proportions by volume, provided all the volumes (reactants and products) are measured at the same temperature and pressure. This observation, now called the **law of combining volumes**, is illustrated by the following.

1 volume hydrogen + 1 volume chlorine → 2 volumes hydrogen chloride

1 volume nitrogen + 1 volume oxygen → 2 volumes nitric oxide

1 volume nitrogen + 3 volumes hydrogen → 2 volumes ammonia

Avogadro's hypothesis of 1811, that equal volumes of different gases at the same temperature and pressure contain the same number of molecules, further implies that

1 molecule of hydrogen + 1 molecule of chlorine → 2 molecules hydrogen chloride

1 molecule of nitrogen + 1 molecule of oxygen → 2 molecules nitric oxide

1 molecule of nitrogen + 3 molecules of hydrogen → 2 molecules of ammonia

The latter may be written symbolically

$$N_x + 3H_y \rightarrow 2N_{x/2}H_{3y/2}$$

Once it is known that the empirical formula of ammonia, based on chemical analysis, is NH_3, it follows that x and y are equal and cannot be less than 2. Similarly, it can be shown that oxygen and chlorine are at least diatomic since the empirical formula of nitric oxide is NO and that of hydrogen chloride is HCl. Accurate molecular weights of NH_3, HCl, and NO clearly establish that hydrogen, nitrogen, chlorine, and oxygen are diatomic. This conclusion is supported by spectroscopic studies showing that each of these gases has but a single vibrational frequency and thus must be diatomic.

3.6 DALTON'S LAW OF PARTIAL PRESSURES

The **partial pressure** of each ideal gas in a mixture of ideal gases may be defined as the pressure each gas would have if it alone occupied the given container at the temperature of the mixture. For example, in a mixture of 0.400 mol of N_2 and 0.100 mol of O_2 in a 10.0-L flask at 300 K, the partial pressures of the component gases N_2 and O_2 are

$$p_{N_2} = \frac{n_{N_2}RT}{V} = \frac{(0.400 \text{ mol})(0.0821 \text{ L} \cdot \text{atm} \cdot \text{K}^{-1} \cdot \text{mol}^{-1})(300 \text{ K})}{10.0 \text{ L}} = 0.985 \text{ atm}$$

$$p_{O_2} = \frac{n_{O_2}RT}{V} = \frac{(0.100 \text{ mol})(0.0821 \text{ L} \cdot \text{atm} \cdot \text{K}^{-1} \cdot \text{mol}^{-1})(300 \text{ K})}{10.0 \text{ L}} = 0.246 \text{ atm}$$

Partial pressures defined in this way have the property that their sum equals the total pressure of the mixture. Thus, if p_A, p_B, and p_C are the partial pressures of gases A,

B, and C, respectively, in a mixture of gases, the total pressure of the mixture is given by

$$p_{tot} = p_A + p_B + p_C$$

Thus the total pressure of the N_2/O_2 mixture is 0.985 + 0.246 or 1.231 atm. Precisely the same pressure results when 0.500 mol of any ideal gas occupies 10.0 L at 300 K. It should be noted that since nitrogen contributes 80% of the total number of molecules in the mixture, its partial pressure (0.985 atm) is 80% of the total pressure, that is, (0.985/1.231)100 = 80%.

The observation that the total pressure of a mixture of gases is a sum of the partial pressures of the component gas is due to John Dalton and is known as **Dalton's law of partial pressures**. The law holds fairly accurately for mixtures of gases that do not interact strongly.

One example of the use of partial pressures is illustrated in the following.

EXAMPLE 3.8 A sample of wet methane (CH_4) at 3.00 atm and 27 °C occupies a 1000-L container. When the gas is passed over a drying agent, 100 g of H_2O vapor is removed. Assuming that CH_4 and H_2O vapor behave as ideal gases, calculate the partial pressure of the methane in the wet mixture.

SOLUTION The problem can be approached by solving for p_{CH_4} in terms of the ideal gas law and then determining n_{CH_4} from the data given. A simpler solution is to note that since

$$p_{CH_4} = n_{CH_4}\frac{RT}{V} \quad \text{and} \quad p_{tot} = (n_{H_2O} + n_{CH_4})\frac{RT}{V}$$

then dividing p_{CH_4} by p_{tot} leads to

$$\frac{p_{CH_4}}{p_{tot}} = \frac{n_{CH_4}}{n_{H_2O} + n_{CH_4}} \quad \text{or} \quad p_{CH_4} = \frac{n_{CH_4}}{n_{H_2O} + n_{CH_4}} p_{tot}$$

and, since n_{H_2O} = 100 g/18.0 g·mol^{-1} = 5.56 mol and

$$n_{tot} = \frac{p_{tot}V}{RT} = \frac{(3.00 \text{ atm})(1000 \text{ L})}{(0.0821 \text{ L·atm·K}^{-1}\text{·mol}^{-1})(300 \text{ K})} = 122 \text{ mol}$$

then n_{CH_4} = 122 − 5.56, or 116 mol. Thus

$$p_{CH_4} = \frac{116}{122}(3.00 \text{ atm}) = 2.85 \text{ atm}$$

3.7 THE KINETIC MOLECULAR THEORY OF GASES

Many of the most important characteristics of ideal gas behavior can be deduced by the use of a model based on the following postulates.

1. The molecules of an ideal gas are treated as mathematical points, that is, as having zero volume (the mass of a molecule is nonzero, however).
2. The molecules neither attract nor repel each other.
3. The molecules are in constant random motion and are continually undergoing elastic collisions among themselves and with the walls of their containers.

By the use of probability theory it is possible to use the above postulates to derive mathematical equations that express the probability that individual molecules will possess given values of physical properties such as velocity and energy. The results are displayed in graphical form in Figs. 3.5 and 3.6. The student should become familiar with the basic characteristics of these two graphs and be able to sketch them from memory.

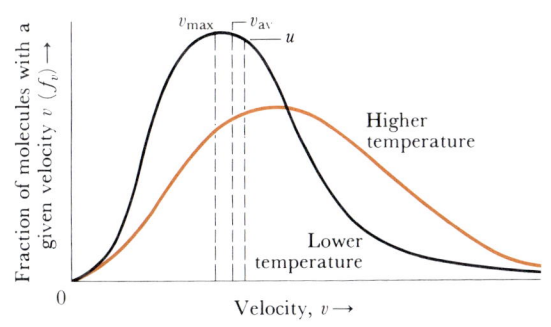

Fig. 3.5. Graph showing how the random velocities of ideal gases are distributed. Note that no molecules have zero velocity ($f_0 = 0$) and no molecules have infinite velocity ($f_\infty = 0$); rather, most molecules have some intermediate velocity. The graph also shows that as the temperature increases, the distribution tends to flatten out. At very high temperatures the velocity distribution tends to become more uniform. The graph also illustrates the relative positions of three important quantities: v_{max} (the velocity for which f_v is the largest), v_{av} (the average velocity), and u (the square root of the average of v^2). The quantity u (called the **root-mean-square velocity**) is needed in order to define the average kinetic energy. The distribution shown in the graph is often called a **Maxwell-Boltzmann distribution**, after the two scientists who first used it in conjunction with the behavior of gases.

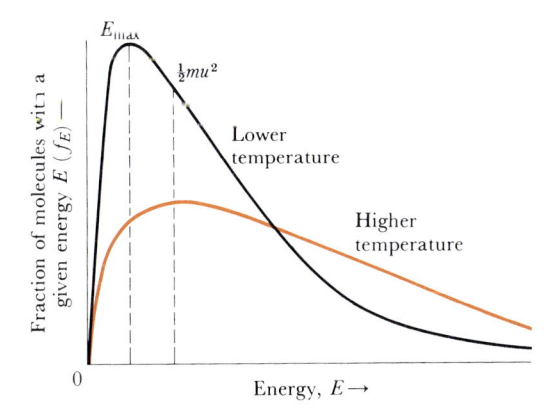

Fig. 3.6. Graph showing how the random energies ($E = \tfrac{1}{2}mu^2$) of ideal gas molecules are distributed. Note that no molecules have zero energy ($f_0 = 0$) and that no molecules have infinite energy ($f_\infty = 0$); rather, most molecules have some intermediate energy. The quantity E_{max} is the value of E for which f_E is the largest; $\tfrac{1}{2}mu^2$ is the average kinetic energy.

The total energy of an isolated sample of matter may be expressed as the sum of two types of energy: **kinetic energy** and **potential energy**. This may be written

$$E_{tot} = E_{kin} + E_{pot} \tag{3.10}$$

The kinetic energy is the energy due to the motions of the individual molecules through space. Since not all the molecules of a gas have identical energies, it is convenient to speak of the *average* kinetic energy of N molecules rather than of the individual molecular kinetic energies. For a collection of N molecules, each of mass m, the average kinetic energy (denoted by $\bar{E}_{kin}$) is

$$\bar{E}_{kin} = \tfrac{1}{2} N m u^2 \tag{3.11}$$

where u^2 is the average of the velocity squared, or $(v^2)_{av}$, of the molecules.* The quantity u, the square root of $(v^2)_{av}$, is called the **root-mean-square velocity**.

Potential energy is the energy due to the relative positions or states of bodies. Since the model assumes no interactions among the molecules, the potential energy is zero. Thus the total energy of N molecules is simply their kinetic energy.

Now examine a specific sample of N molecules of ideal gas, each of mass m, in a cubic container of sides L and volume L^3. As a somewhat crude approximation imagine that at any given time one-third of the molecules are traveling between a pair of opposite faces of the cube at a velocity u. (Recall that the pressure exerted by the gas is a result of collisions of molecules with the faces of the cube.) Elementary mechanics shows that the force of each collision is simply the change in momentum (mass times velocity) occurring upon each collision, divided by the time between successive collisions. Since the momentum of a particle moving with velocity u is mu (and $-mu$ when moving in the opposite direction), the change in momentum per collision is

$$mu - (-mu) = 2mu \tag{3.12}$$

Since $u = 2L/t$, where $2L$ is the distance traveled by a molecule between two successive collisions on a given face and t is the time between two such collisions, then $t = 2L/u$ and the force per collision is $2mu/(2L/u)$ or mu^2/L. Since pressure is force per unit area, the total pressure exerted by the bombardment of one face of area L^2 by one-third of the molecules is

$$p = \frac{N}{3} \frac{mu^2/L}{L^2} = \frac{Nmu^2}{3L^3} = \frac{Nmu^2}{3V} \quad \text{or} \quad pV = \tfrac{1}{3} Nmu^2 \tag{3.13}$$

If Eq. (3.11) is used to bring in the average kinetic energy, the above equation leads to

$$pV = \tfrac{2}{3} \bar{E}_{kin} \tag{3.14}$$

---

* Note that the average of a squared quantity does not usually equal the square of the average of the quantity. For example, the average of the numbers 1, 2, and 3 is 2, and the square of this is 4. But the average of 1^2, 2^2, and 3^2 is $4\tfrac{2}{3}$. Thus, as shown in Fig. 3.5, u is larger than v_{av}.

Since $pV = nRT$ and $n = N/N_0$ (if N_0 denotes Avogadro's number), then

$$\tfrac{2}{3}\bar{E}_{kin} = \frac{N}{N_0} RT \quad \text{or} \quad \bar{E}_{kin} = \frac{3}{2} \frac{N}{N_0} RT$$

When $N = N_0$ (that is, when one mole of gas is used) the average kinetic energy is

$$\bar{E}_{kin} = \tfrac{3}{2}RT \tag{3.15}$$

Equation (3.15) is called the **kinetic molecular equation**; it establishes the very important fact that the absolute temperature is a direct measure of the average kinetic energy of a gas. Thus at absolute zero (0 K) all kinetic energy of motion disappears.* An object feels "cold" to the touch when the average kinetic energy of "hand molecules" is greater than the average kinetic energy of the molecules of the object touched so that there is a net transfer of molecular motion from the hand to the object. Similarly, an opposite net transfer of molecular motion occurs when a "hot" object is touched. This net transfer of motion or kinetic energy is usually termed a **flow of heat**.

One of the attractive features of the kinetic molecular theory is that it provides simple explanations of Boyle's and Charles's laws. Consider a sample of gas at fixed temperature. This means that the average kinetic energy of a gas molecule is also fixed and thus, on the average, all molecules will always hit the walls of the container with the same force. If the volume is made larger, a molecule must move a greater distance between collisions. Consequently, there will be fewer collisions per unit time and these will be distributed over a larger area. This results in a decrease of pressure—precisely the effect described by Boyle's law. Now consider a sample of gas at fixed pressure. This means that the sum of all the forces due to collisions between gas molecules and walls, divided by the area of the walls, is constant. Suppose the temperature is increased. This will increase the pressure since the molecules will have increased kinetic energy and will hit the walls harder and more often. But if the volume is increased, the area of the walls increases. Thus, even though the molecules hit harder and more often, the collisions are spread out over a greater area and, if the volume is increased enough, the pressure can be reduced to its original fixed value. This is precisely the effect described by Charles's law.

3.8 USES OF THE KINETIC MOLECULAR EQUATION

Two important uses of the kinetic molecular equation are illustrated in the following two examples.

EXAMPLE 3.9 Calculate the root-mean-square velocity of a molecule of O_2 at 27 °C.

SOLUTION Since $\tfrac{1}{2}Mu^2 = \bar{E}_{kin} = \tfrac{3}{2}RT$ (where M, whose value is $N_0 m$, is the weight of one mole of O_2), solving for u leads to

$$u = \sqrt{3RT/M}$$

---

* This is true only for **translational motion**, that is, motion from one point in space to another. As discussed in Section 7.5, atoms in a crystal have vibrational motion even at 0 K.

If u is to be expressed in $m \cdot s^{-1}$, an appropriate value of R must be used. To ascertain the appropriate value of R, go back to Eq. (3.9) and express p in $N \cdot m^{-2}$ and V in m^3. This produces $R = 8.3 \, J \cdot K^{-1} \cdot mol^{-1}$ (recall that 1 joule = $1 \, kg \cdot m^2 \cdot s^{-2}$). Thus

$$u = \left(\frac{3 \times 8.3 \, J \cdot K^{-1} \cdot mol^{-1} \times 300 \, K}{32 \, g \cdot mol^{-1} \times 10^{-3} \, kg \cdot g^{-1}} \right)^{1/2} = 4.8 \times 10^2 \, m \cdot s^{-1}$$

This is about 1100 miles per hour. Note that the units cancel to give u in $m \cdot s^{-1}$. Also, it should be noted that u depends only on the temperature and molecular weight, and not on pressure.

EXAMPLE 3.10 Calculate the relative rates of diffusion of two gases with molecular weights M_1 and M_2 at any given temperature T.

SOLUTION The relationship for u developed in Example 3.9 leads to

$$u_1 = \sqrt{3RT/M_1} \quad \text{and} \quad u_2 = \sqrt{3RT/M_2}$$

The velocity ratio then is $u_1/u_2 = \sqrt{M_2/M_1}$.

The result obtained in Example 3.10 shows that the rate of diffusion is inversely proportional to the square root of the molecular weight. This relationship is usually known as **Graham's law** and was originally discovered experimentally. It is of practical importance for processes involving the flow and mixing of gases. This result and that of Example 3.9 can be used to quickly calculate the root-mean-square velocity of any gas besides O_2 at 27 °C. For example, since H_2 has a molecular weight of 2 amu, then the root-mean-square velocity of H_2 at 27 °C is just four times that of O_2, namely, about $2.0 \times 10^3 \, m \cdot s^{-1}$ or about 4400 mph.

Graham's law is named for Thomas Graham (1805–1869), Scottish-born British chemist who served as the first president of the Chemical Society of London. Graham conducted studies on the diffusion of gases, the absorption of gases by charcoal, and the nature of phosphorus and its compounds. Shortly before his death, he discovered the remarkable ability of metals such as palladium to absorb large volumes of hydrogen.

3.9 VARIATION OF ATMOSPHERIC PRESSURE WITH ALTITUDE

When dealing with laboratory-size samples of gases, the fact that the gas molecules are affected by the gravitational field of the earth can be ignored. The effect of the gravitational field is to cause a greater concentration of gas molecules at the bottom of the container with gradually decreasing concentration toward the top. Although this concentration change is too small to detect in an ordinary-size container, it is very important in large samples of gas such as that represented by the earth's atmosphere. The concentration differences in the earth's atmosphere are manifested largely as pressure differences; at a constant temperature T the variation of pressure with altitude of an ideal gas is given by

$$p = p_0 10^{-Mgh/(2.3RT)} \quad \text{or} \quad p_0 e^{-Mgh/(RT)} \tag{3.16}$$

where p_0 is the pressure at some suitable reference point (usually sea level), h is the altitude above the reference point, g is the gravitational constant, and M is the molecular weight of the gas. The alternative but equivalent expressions arise from the fact that e (the base of natural logarithms) and 10 are related by $e^{2.3} = 10$. The form involving 10 instead of e is more suitable for calculations with decimal log tables (when an electronic calculator is not available).

Equation (3.16) is valid for the earth's atmosphere only if air can be treated as an ideal gas, and if the temperature of the atmosphere does not change with altitude. Although neither of these assumptions is strictly valid, the relationship works remarkably well in practice. For example, a barometer can be used to calculate the height of a mountain simply by measuring air pressures at its base and its summit; mountain hikers frequently carry small pocket-size barometers in which pressure readings are directly converted to altitudes and read on a dial. Many automobile and aircraft altimeters are also constructed on the basis of this principle. Figure 3.7 is a plot of Eq. (3.16) showing how pressure is predicted to drop off exponentially with altitude. The graph also shows some typical measured pressures at the same temperature.

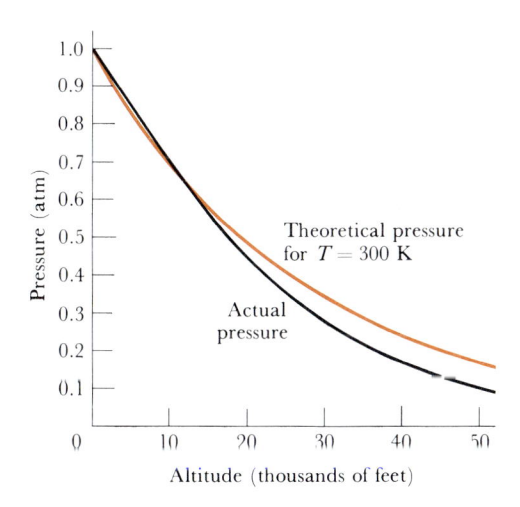

Fig. 3.7. Variation of atmospheric pressure with altitude. The actual pressure is lower than the theoretical pressure at higher altitudes. Part of this difference is due to the fact that the air temperature drops rapidly at higher altitudes; the rest is due to nonideality of atmospheric gases.

An example of the use of Eq. (3.16) is given below.

EXAMPLE 3.11 The architect Frank Lloyd Wright once proposed building a skyscraper one mile high. If the ground floor of such a building were at a pressure of 760 torr, what would be the pressure at the top floor, assuming an average air temperature of 20 °C?

SOLUTION If an electronic calculator is not available, it is convenient to use Eq. (3.16) in the logarithmic form:

$$\log p = \log p_0 - Mgh/(2.3RT)$$

Using p_0 as 1 atm leads to $\log p_0 = 0$. The "molecular weight" of air is taken as an average of the molecular weights of nitrogen and oxygen, assuming air is 79% N_2 and 21% O_2; this produces $M = 29$ g·mol^{-1} or 29×10^{-3} kg·mol^{-1}. A height of one mile is equivalent to 1.609×10^3 m, and $g = 9.80$ m·s^{-2}. Thus

$$\log p = \frac{-Mgh}{2.3RT} = -\frac{(29 \times 10^{-3} \text{ kg·mol}^{-1})(9.80 \text{ m·s}^{-2})(1.609 \times 10^3 \text{ m})}{2.3(8.314 \text{ J·K}^{-1}\text{·mol}^{-1})(293 \text{ K})}$$

$$= -0.0816 \quad \text{or} \quad 9.9184 - 10$$

(See Appendix 3 on the use of logarithms.) The antilog of $9.9184 - 10$ is 0.83; thus the pressure is 0.83 atm or about 630 torr. With an electronic calculator, simply compute Mgh/RT, change sign, and depress the e^x key. Since $p_0 = 1$ atm the result is p (in atm).

3.10 CRITICAL PHENOMENA IN REAL GASES

If real gases obeyed $pV = nRT$ at all pressures and temperatures, then at 0 K the volume of a gas should vanish. In reality, if the pressure is held constant the gas eventually liquefies or solidifies, or both. Beyond this point a further reduction in temperature causes little diminution in volume. A mole of water vapor occupies about 30,000 cm³ at 373 K and 1 atm. At the same pressure the same amount of water condensed to a liquid or solid occupies only about 18 cm³, a volume which does not change very much as the temperature approaches 0 K.

Experiments show that there exists a certain temperature T_c above which a gas cannot be liquefied, no matter how high the pressure. This temperature is known as the **critical temperature** of the gas. The existence of the critical temperature is evidence for intermolecular interactions. The fact that such interactions exist indicates that the potential-energy part of the total energy of a gas is not really zero as assumed in the kinetic molecular model. Rather, the potential energy of a gas varies as the distances between gas molecules vary. This relationship is shown in graph form in Fig. 3.8. In a rather rough sense, a gas may be characterized as having $|E_{\text{kin}}|$ greater than $|E_{\text{pot}}|$—greater kinetic energy of motion than potential energy of intermolecular interaction. This indicates that the "tearing apart" tendency due to kinetic energy of motion predominates over the "sticking together" tendency due to the potential energy of attraction. In the liquid and solid states the situation is expected to be reversed, with $|E_{\text{kin}}|$ less than $|E_{\text{pot}}|$. The higher of the two dashed lines in Fig. 3.8 represents the critical temperature at which $|E_{\text{kin}}|$ and $|E_{\text{pot}}|$ are of about equal magnitudes. At this temperature or higher there is no way that $|E_{\text{pot}}|$ can be made greater than $|E_{\text{kin}}|$ in order to liquefy the gas; an increase in pressure merely decreases $|E_{\text{pot}}|$ (the nuclei begin to get so close that they repel each other strongly) without changing $|E_{\text{kin}}|$. However, at temperatures below T_c (for example, T' in the diagram) an increase in pressure does bring about an increase in $|E_{\text{pot}}|$ so that condensation of the gas may be possible.

Figure 3.9 suggests a method for determining whether a sample of material at very high pressure is above or below its critical temperature.

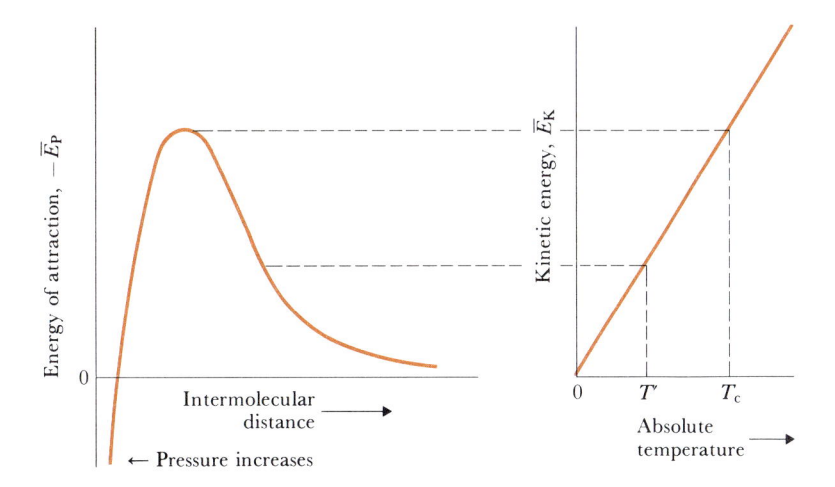

Fig. 3.8. The liquefaction of a gas. At the temperature T_c the potential and kinetic energies are of comparable magnitudes. At this temperature or above there is no way to increase the potential energy/kinetic energy ratio in order to liquefy the gas. At the lower temperature T' an increase in pressure can increase this ratio, making it possible to liquefy the gas.

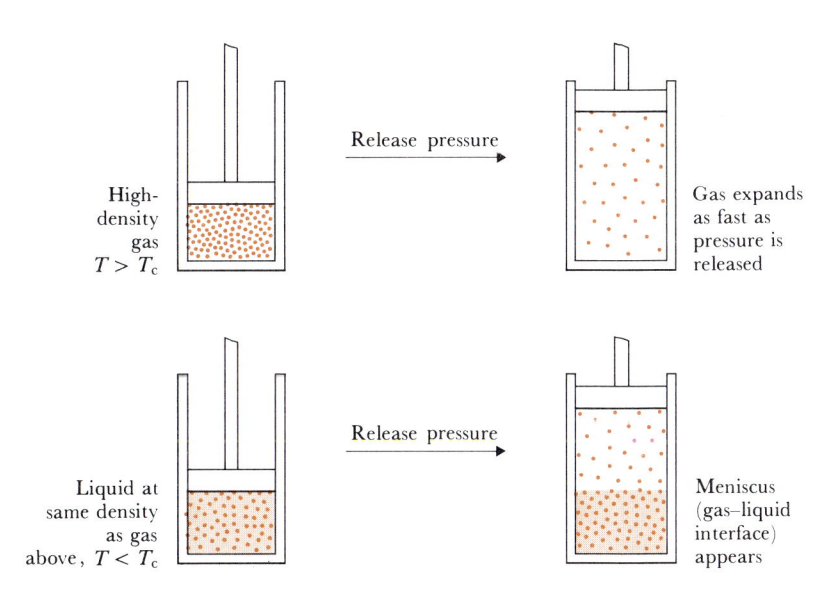

Fig. 3.9. A way of telling whether a substance is above or below its critical temperature. If a goldfish were placed in the high-density gas at $T > T_c$, it would swim about just as in a liquid. But if the pressure were released, the goldfish would fall "clunk" to the bottom of the cylinder.

3.10 | CRITICAL PHENOMENA IN REAL GASES

Another experimental manifestation of intermolecular forces in gases is the so-called **Joule-Thomson effect**. Anyone who has ever removed the valve core from an inflated auto or bicycle tire and felt the out-rushing stream of air with a finger has probably noticed how cold the air feels. Conversely, a heating of the gas occurs when the tire is pumped back up to pressure; this is one of the reasons (besides friction) why a tire pump becomes hot after a few minutes of use. The cooling observed when the air expands is accounted for by an increase in the potential energy at the expense of the kinetic energy. Under the conditions of the expansion, since there is little transfer of energy between the air and its surroundings, the total energy (a sum of the kinetic and potential energies) tends to remain constant. As the air expands and the molecules move farther apart, their attractions decrease; this means that the potential energy increases. Since the sum $E_{kin} + E_{pot}$ is constant (barring energy losses by conduction), the increase in potential energy necessitates a decrease in the kinetic energy and, thereby, a drop in the temperature. When the tire is pumped up again the reverse effect occurs: molecules become closer together, the potential energy decreases, and the kinetic energy increases.

When the Joule-Thomson effect is investigated at room temperature for H_2 or He gas it is found that these substances warm upon expansion and cool upon compression—just the reverse of the effect observed for all other gases; only at very low temperatures do H_2 and He behave "normally." Evidently, at room temperature and high pressure the repulsive part of the potential-energy curves of H_2 and He is increasing. Thus a decrease in pressure decreases the potential energy (attraction increases) and the kinetic energy increases; the expanding gas then feels warm to the hand. Careful studies of H_2 and He indicate that the above interpretation is correct; whereas in most gases the attraction part of the potential energy is dominant at higher temperatures, in H_2 and He repulsive forces dominate except at very low temperatures.

A practical application of the Joule-Thomson effect is in the liquefaction of gases. If a cyclic process is established in which gases are compressed, cooled by expansion, recompressed, and cooled further by expansion, then eventually the temperature of the gases becomes low enough for liquefaction under modest pressure requirements. It should be noted that the heat produced by each compression must be removed by being carried away by the surroundings.

Refrigerators operate on a modification of the Joule-Thomson effect. A gas is compressed to a liquid outside the cold compartment and introduced into the cold compartment, where it picks up heat from the objects to be cooled. This heat vaporizes the liquid, the vapor is pumped outside, and recompression to liquid (which releases the heat to the surroundings) begins the cycle anew.

3.11 THE VAN DER WAALS EQUATION FOR GASES

A convenient way of illustrating the failure of real gases to obey the ideal gas equation is in terms of the **compressibility factor** Z, defined by

$$Z = \frac{pV}{nRT} \tag{3.17}$$

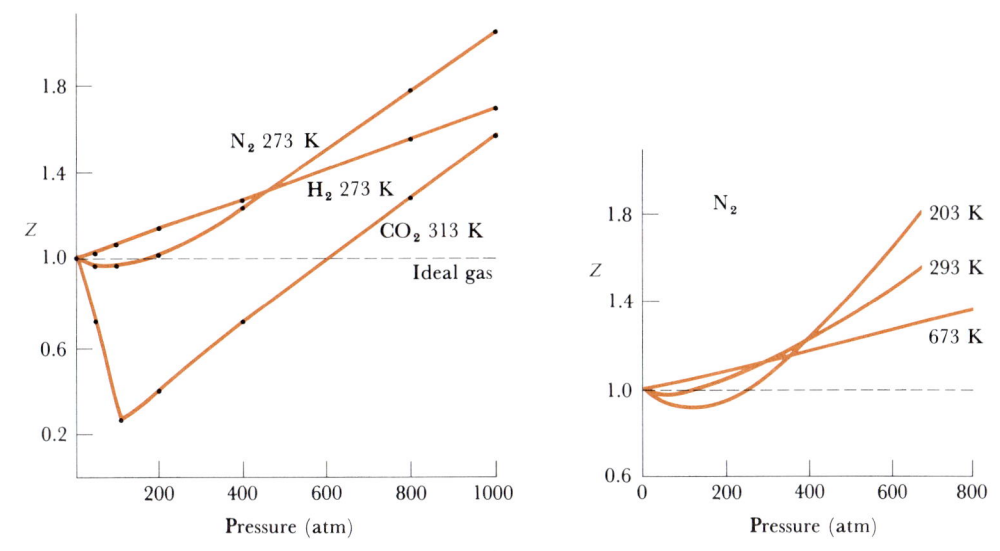

Fig. 3.10 Compressibility factor ($Z = pV/nRT$) of gases as a function of pressure. The smaller graph shows how Z of N_2 tends to straighten out at higher temperatures. This—and the fact that all the curves coincide at $p = 0$—shows that all gases tend to become ideal at low pressures and high temperatures.

For ideal gases, Z is always equal to unity. Figure 3.10 shows how Z varies with pressure for N_2, H_2, and CO_2 at the temperature indicated. Were these gases ideal at the indicated temperatures, Z would be unity at *all* pressures. The illustration also shows how Z of one of the gases (N_2) changes as the temperature changes. Since the real and the ideal gas curves coincide at $p = 0$ and since the former tends to approach $Z = 1$ at higher temperatures, it is apparent that all gases would behave ideally at very low pressures and/or very high temperatures.

Although many attempts have been made to replace the ideal gas equation with a more accurate equation applicable to real gases over a wide range of conditions, no one has been successful in developing an equation that is both accurate and simple to use. Much of the difficulty can be traced to the fact that no one really knows an accurate and simple way to express the potential energy of a real gas. One of the first attempts to improve upon the ideal gas equation was based on the realization that it would be a definite advantage to retain the form of the ideal gas equation insofar as possible. This can be done very simply by introducing the equation

$$p'V' = nRT \tag{3.18}$$

where p' and V' are a *fictitious* pressure and volume, respectively, chosen so that the equation is satisfied for all values of T. However, such an equation is of trivial value unless p' and V' can be related to the experimental values of p and V.

A simple and reasonable relationship between V' and V is

$$V' = V - nb$$

where V is the actual gas volume and b is a constant related to the volume occupied by a mole of closely packed molecules. This is a simple way of correcting for the

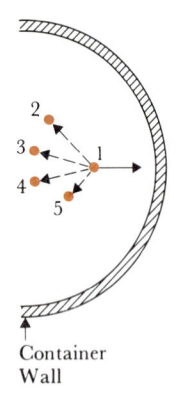

Fig. 3.11. The internal pressure of a real gas. Molecule 1, about to collide with the container wall, is retarded by an attraction to its neighbors (molecules 2, 3, 4, 5, and others). This retardation (which all molecules undergo mutually) acts as an *internal pressure*—a pressure exerted only upon the molecules themselves and not upon the container walls.

unrealistic assumption of zero molecular volume used in the ideal gas equation. The pressure p' may be related to the actual pressure p by

$$p' = p + p_{int}$$

where p_{int} is a correction term accounting for the fact that the molecules attract each other. As illustrated in Fig. 3.11, a molecule about to change its momentum upon collision with the wall of a container is attracted "backward" by its neighbors. Thus p_{int} may be regarded as an "internal" pressure, a pressure that the molecules exert only against each other and at the expense of exerting it against the container wall. Such an internal pressure due to a single molecule should be directly proportional to the density of the neighboring molecules. Thus

$$p_{int} \text{ (per molecule)} \propto \rho$$

where ρ is the density of the bulk gas (the symbol $\propto$ means "is proportional to"). For N molecules the total internal pressure is simply

$$p_{int} \propto N\rho$$

However, N itself is proportional to the density so that the total p_{int} should be proportional to ρ^2. Furthermore, for a fixed mass of gas, ρ is inversely proportional to the volume V so that the final definition of p_{int} is

$$p_{int} = n^2 a / V^2$$

where a is a constant of proportionality and the factor n^2 is introduced to make a independent of the amount of gas.

Equation (3.18) is now written

$$(p + n^2 a / V^2)(V - nb) = nRT \tag{3.19}$$

This is known as the **van der Waals equation**. When V is large (low p, or high T, or both), then the term $p + n^2 a / V^2$ approaches p, $V - nb$ approaches V, and the van der Waals equation becomes equivalent to the ideal gas equation.

One of the disadvantages of the van der Waals equation is that it requires three constants (a, b, and R) instead of the single constant R of the ideal gas equation. Furthermore, the values of a and b vary from gas to gas (in fact, even R varies from gas to gas!)*; and since they vary somewhat with temperature, they cannot really even be considered constants. In addition, the van der Waals equation is *cubic* (of third degree) in the volume and thus somewhat algebraically cumbersome to use.†

Although the van der Waals equation is far from perfect, it is used widely in certain areas of physics and chemistry where the ideal gas equation is too inaccurate to be used. Table 3.1 lists the van der Waals constants of some common gases. Note that He and H_2 have the smallest values of the volume correction term b, suggesting that these molecules are relatively small.

TABLE 3.1 THE VAN DER WAALS CONSTANTS FOR SOME COMMON GASES

NAME	FORMULA	a ($L^2 \cdot atm \cdot mol^{-2}$)	b ($L \cdot mol^{-1}$)
Ammonia	NH_3	4.170	0.03707
Argon	Ar	1.345	0.03219
Carbon dioxide	CO_2	3.592	0.04267
Chlorine	Cl_2	6.493	0.05622
Hydrogen	H_2	0.2444	0.02661
Hydrogen chloride	HCl	3.667	0.04081
Helium	He	0.03412	0.02370
Nitrogen	N_2	1.390	0.03913
Oxygen	O_2	1.360	0.03183
Water	H_2O	5.464	0.03049

If the van der Waals equation is used as a model of real gases, a general expression can be obtained for a quantity called the **Joule-Thomson coefficient**. This quantity states the number of degrees by which the temperature of a gas changes per unit change in pressure arising from expansion of the gas. This relationship is

$$\mu_{JT} = \frac{1}{C_p}\left(\frac{2a}{RT} - b\right) \tag{3.20}$$

where C_p is the **molar heat capacity** of the gas (the amount of thermal energy required to change the temperature of a one-mole sample by one degree) and a and b are the van der Waals constants. For most gases at around room temperature, the

---

* In practice it is generally satisfactory to treat R as a constant and to use the ideal gas value.

† Rearrangement of the van der Waals equation leads to

$$V^3 - n\left(b + \frac{RT}{p}\right)V^2 + \frac{n^2 a}{p}V - \frac{n^3 ab}{p} = 0$$

term in parentheses is positive so that μ_{JT} is positive and the gas cools upon expansion. The data in Table 3.1 show that $2a/RT$ is about 0.11 L·mol^{-1} for O_2 and N_2 at 300 K. Since b is 0.03 L·mol^{-1} for O_2 and 0.04 L·mol^{-1} for N_2, μ_{JT} is definitely positive at this temperature. However, for a gas such as H_2, $2a/RT$ has a value of 0.02 L·mol^{-1} at 300 K and b is about 0.03 L·mol^{-1}, so that μ_{JT} is negative, that is, at 3000 K H_2 becomes warmer on expansion. This also occurs for helium at 300 K. It is apparent that as T decreases, μ_{JT} will eventually become positive even for H_2 and He.

The Joule-Thomson coefficient is but one example of the relationships that can be derived from the van der Waals equation. The equation, even though often unsatisfactory for numerical calculations, frequently leads to a correct qualitative interpretation of various phenomena.

SUMMARY

Gases consist of small molecules moving about in ceaseless, random motion, colliding among themselves and with the walls of their containers. Boyle's law states that the product of the pressure and the volume of a gas is a constant for a given temperature and for a given amount of gas. Charles's law states that the volume-to-absolute-temperature ratio of a gas is also constant for a given pressure and for a given amount of gas. These two relationships may be combined with Avogadro's law—that equal volumes of different gases at the same pressure and temperature contain the same number of molecules—to produce the general gas law: $pV = nRT$. This equation describes the behavior of a hypothetical gas (the *ideal gas*) composed of molecules of zero volume and lacking intermolecular interactions. Mathematical treatment of an ideal gas—the kinetic molecular theory—predicts that the average kinetic energy of gas molecules is directly proportional to the absolute temperature.

Avogadro's law and Gay-Lussac's law of combining volumes (that gases combine chemically by simple integral ratios of volume when reactants and products are at equal pressures and temperatures) provide a basis for the determination of molecular formulas of gaseous compounds.

Real gases have molecules with nonzero volumes and the molecules do interact. As a consequence, gases can be liquefied—provided the temperature is below the critical temperature of the gas in question. Intermolecular attractions account for the Joule-Thomson effect and form the basis of refrigeration.

More accurate calculations on gases result when the ideal gas equation is replaced with the van der Waals equation—an equation that takes into account the actual volumes of gas molecules and the interactions among them.

LEARNING GOALS

1. Know how pressure is defined and how it is measured in the laboratory.

2. Know Boyle's law and when to use it; be able to calculate pressures and volume changes for a given amount of gas at a given temperature.

3. Know how absolute temperature is defined and how absolute zero can be determined experimentally.

4. Know Charles's law and when to use it; be able to calculate volume and temperature changes for a given amount of gas at a given pressure.

5. Know Avogadro's law.

6. Be able to derive the ideal gas equation from the laws of Boyle, Charles, and Avogadro. Know the significance of the ideal gas constant R.

7. Be able to use $pV = nRT$ for calculation of p, V, n, or T when any three of these quantities are given.

8. Be able to calculate molecular weights from gas densities or vice versa.

9. Know Gay-Lussac's law of combining volumes and its significance.

10. Be able to define partial pressure and to calculate it.

11. Know the postulates of the kinetic molecular theory and how these lead to an understanding of what absolute temperature represents.

12. Know the kinetic molecular equation and be able to obtain Graham's law from it.

13. Be able to calculate the root-mean-square velocity of a gas and know how to calculate relative velocities with Graham's law.

14. Know how air pressure varies with altitude and be able to calculate the pressure at a given altitude.

15. Know what is meant by the critical temperature of a gas and what its existence implies about non-ideality of gases.

16. Be able to state what the Joule-Thomson effect is and know its practical significance.

17. Be familiar with the van der Waals equation and how it arises. Know what the constants a and b represent.

DEFINITIONS, TERMS, AND CONCEPTS TO KNOW

Boyle's law
pressure
atmosphere (as a unit of pressure)
torr
differential manometer
Torricelli barometer
absolute temperature
Charles's law

ideal or perfect gas
ideal gas constant
general gas law
partial pressure
Avogadro's law
law of combining volumes
kinetic molecular theory of gases
kinetic energy

potential energy
root-mean-square velocity
kinetic molecular equation
heat
Graham's law
critical temperature of a gas
Joule-Thomson effect
van der Waals equation

QUESTIONS AND PROBLEMS

Boyle's Law

1. In each case below, temperature and amount of gas are held constant. Supply the value of the missing quantity.

	p_1	V_1	p_2	V_2
a)	10 atm	_____	3 atm	30 L
b)	200 torr	10 cm³	_____	100 cm³
c)	_____	25 ft³	10 lb·in.$^{-2}$	50 ft³
d)	15 Q	20 M	30 Q	_____

2. The scientists of Lower Slobbovia use units of klotdunks (k) for pressure, grubniks (G) for volume, and brrs (°b) for temperature. A sample of vodka vapor at -13 °b has a volume of 15 G and a pressure of 10 k. What is the pressure of the vodka vapor when its volume is increased to 150 G at -13 °b?

3. The pressure reading given by a tire gauge is the pressure above and beyond atmospheric pressure.

a) What is the actual air pressure (in atm) in a tire when the gauge reads 30 lb·in.$^{-2}$? (1 atm is 14.7 lb·in.$^{-2}$.)

b) What volume does 10 L of air in a tire at 30 lb·in.$^{-2}$ occupy when it is released to the atmosphere?

4. A skin diver at maximum depth is subjected to about 5 atm pressure. Assuming no temperature change between this depth and the surface, what volume will 1 L of air inhaled at the maximum depth represent at the surface?

5. A steel cylinder of nitrogen gas contains about 0.5 m^3 of gas at a pressure of about 100 atm. What volume would this gas occupy at the same temperature and a pressure of 1 atm?

Charles's Law

6. In each case below, pressure and amount of gas are held constant. Supply the value of the missing quantity.

	V_1	T_1	V_2	T_2
a)	100 L	300 K	———	600 K
b)	10 ft^3	27 °C	20 ft^3	———
c)	———	200 K	50 cm^3	500 K
d)	30 bbl	500 K	———	250 K

7. Absolute zero in Lower Slobbovia is at -100 °b (see Problem 2). The absolute temperature scale is called the *noodnik* scale (named after Alexander Ivanovich Noodnik, the first Lower Slobbovian chemist to freeze to death while washing test tubes in the ice-choked Pollutski river). A sample of vodka vapor has a volume of 10 G at a pressure of 3 k and a temperature of 5 °b. What is the volume of this vapor when the temperature is increased to 110 °b at a pressure of 3 k?

8. An automobile tire is filled to 30 lb·in.$^{-2}$ pressure at 25 °C. After hard driving the tire heats up to 40 °C. What is the pressure in the warm tire?

9. The tight-fitting lid of a pail will pop off when the inner pressure reaches 2.1 atm. If the pail contains air at 1 atm at 25 °C, at what temperature will the lid pop off, assuming the lid tightness is not affected by temperature?

10. A sample of gas has a volume of 15 L at 300 K when the pressure is 1 atm. This means that for any combination of volume and temperature at 1 atm, V/T is always equal to 15 L/300 K or 0.50 L·K^{-1}. Suppose the pressure is increased to 2 atm. Does the ratio V/T decrease, increase, or stay the same?

General Gas Law Calculations

11. What volume would be occupied by 16 g of SO_2 at 25 °C and 720 torr?

12. Calculate the temperature (in °C and K) of 2 moles of ideal gas in a 40-L container with a pressure of 2 atm.

13. How many moles of gas would there be in a sample having a volume of 30,000 cm^3 at 500 °C and 700 torr?

14. A weather balloon holds 10 L of helium gas at 27 °C and 1 atm pressure. How many times could you fill such a balloon from a 100-L tank of compressed helium which is originally at a pressure of 70 atm at 27 °C?

15. A flask can withstand a pressure of 3000 torr without breaking. The flask contains gas at 30 °C and 760 torr. At what temperature will the flask break if the gas is heated?

16. A mixture of gases consists of 14 g of N_2 and 9.6 g of O_2 at 27 °C and has a pressure of 1600 torr.

a) Calculate the partial pressure of O_2 at 27 °C.

b) Calculate the partial pressure of O_2 at 50 °C if the volume is kept constant.

17. Flask A (200 cm^3) contains N_2 at 25 °C and 700 torr, and flask B (800 cm^3) contains O_2 at 25 °C and 1000 torr. Calculate the following when the two flasks are connected and the contents mixed.

a) Partial pressure of N_2 in the mixture.

b) Partial pressure of O_2 in the mixture.

c) Total pressure of the mixture.

18. Two flasks at 27 °C are filled with helium. Flask A has a volume of 100 cm^3 and a pressure of 1 atm; flask B has a volume of 500 cm^3 and a pressure of 76 torr.

a) Calculate the number of moles of helium in flask A.

b) Calculate the number of moles of helium in flask B.

c) If the flasks are connected and the contents mixed, what is the pressure of the resulting mixture?

19. The pressure of H_2O vapor at 25 °C is 23.8 torr. What is the density of this vapor in $g \cdot L^{-1}$?

20. Ten grams of an unknown gas occupies 15 L at 300 K and a pressure of 380 torr. What is the molecular weight of the gas?

21. The density of an ideal gas is 3.54 $g \cdot L^{-1}$ at 25 °C and 1520 torr. What is the density of the same gas at 0 °C and 760 torr?

22. Tests by the Foster D. Snell Laboratories (see *Readers' Digest,* October 1976, pp. 114–118) show that a cigarette smoker may take in 100–640 cm^3 of CO (measured at 37 °C and 1 atm) per day. How many moles of CO is this on a yearly basis?

23. Use the relationship 1 atm is equivalent to 101.3 kPa to show the following.

  a) One mole of ideal gas at 0 °C and a pressure of 100 kPa has a volume of 22.7 L.
  b) The value of the ideal gas constant R is also equal to 8.3 $kPa \cdot L \cdot K^{-1} \cdot mol^{-1}$ or 8.3 $Pa \cdot m^3 \cdot K^{-1} \cdot mol^{-1}$.
  c) Show that the value of R in the above is also equal to 8.3 $J \cdot K^{-1} \cdot mol^{-1}$, that is, a $kPa \cdot L$ is a $Pa \cdot m^3$, which is a $N \cdot m$ or a joule.

24. (Work Problem 23 before doing this one.) Use $R = 8.3\ J \cdot K^{-1} \cdot mol^{-1}$ in each of the following problems.

  a) Calculate the pressure of 2.0 mol of ideal gas having a volume of 10 L at 250 K. Note that if V is in liters, P will come out in kilopascals, but if V is in cubic meters, p will come out in pascals.
  b) Calculate the volume of 5.0 mol of ideal gas having a pressure of 90 kPa at 350 K. Note that if p is in kilopascals, the volume will come out in liters, but if p is in pascals, the volume will come out in cubic meters.

25. The following reaction goes virtually to completion above 400 K:

$$H_2(g) + \tfrac{1}{2}O_2(g) \rightarrow H_2O(g)$$

A sealed 10 L tube at 227 °C contains 5 mol of H_2 and an unknown amount of O_2. The total pressure in the tube before reaction is 100 atm.

  a) What is the partial pressure of H_2 before the reaction?
  b) What is the partial pressure of O_2 before the reaction?
  c) How much O_2 (in moles) is initially present?
  d) After the reaction is complete and the tube cooled to 227 °C, what are the total pressure and the partial pressures of product and excess reactants (if any)?

Pressure and Altitude

26. What pressure would you expect on the summit of Pike's Peak (altitude 14,110 ft or 4301 m) when sea level pressure is 1 atm and the temperature is 15 °C? (The average molecular weight of air may be taken as 29 $g \cdot mol^{-1}$).

27. An intrepid mountain climber carries a barometer to the summit of Mt. Rainier and measures a pressure of 448 torr. Assuming an average temperature of 10 °C and any other data you may need, estimate the altitude of Mt. Rainier in feet above sea level (1 m is equivalent to 3.28 ft).

28. The Swiss-born geographer Arnold Guyot was one of the first to determine accurate elevations in what is now the Great Smoky Mountain National Park in Tennessee and North Carolina. Guyot did this using a mercury barometer which he personally carried, not wishing to trust assistants with this delicate and unwieldy instrument. Approximately what pressure reading would Guyot have measured on the summit of Clingman's Dome (elevation 6642 ft) on a day when the temperature was 25 °C?

Graham's Law and Gas Velocities

29. Calculate the root-mean-square velocity (in $m \cdot s^{-1}$) of gaseous NH_3 (ammonia) at 50 °C.

30. Calculate the relative rates of diffusion of helium and CH_4. How does an increase in temperature change this relative rate?

31. The uranium isotopes ^{238}U and ^{235}U were first separated by making use of their different diffusion rates in the form of the gas UF_6. What is the percentage difference in the velocities of UF_6 (with ^{238}U) and UF_6 (with ^{235}U)?

32. If it takes 10 s for 15 cm^3 of oxygen to pass through a pinhole, how long will it take 15 cm^3 of hydrogen to pass through the same pinhole?

33. If molecules of hydrogen have an average velocity of $1.84 \times 10^3\ m \cdot s^{-1}$ at a given temperature, what is the average velocity of carbon dioxide molecules at the same temperature?

34. To what temperature must nitrogen be heated in order to acquire a root-mean-square molecular velocity of $9.4 \times 10^2\ m \cdot s^{-1}$?

Van der Waals Equation

35. Calculate the pressure exerted by 10 g of O_2 in a 1-L flask at 27 °C using (a) the general gas equation, and (b) the van der Waals equation.

36. Suppose real gas molecules had nonzero volumes but there were no interactions between molecules. Which of the equations below would you expect to best describe such a gas? Explain.

a) $pV = nRT$

b) $\left(p + \dfrac{n^2a}{V^2}\right)(V - nb) = nRT$

c) $p(V - nb) = nRT$

d) $\left(p + \dfrac{n^2a}{V^2}\right)V = nRT$

37. Carry out the algebra which shows that the van der Waals equation is cubic in the volume.

38. a) Below what temperature would you expect H_2 to cool on expansion?

b) Would O_2 and N_2 ever warm up on expansion? Explain.

General

39. Name at least one important scientific contribution of each of the following: Boyle, Charles, Gay-Lussac, Torricelli, Celsius, Kelvin, Avogadro, Graham, Joule, Thomson.

40. The maximum weight a hot-air balloon of volume V can lift is given by

$$w_{T_1} - w_{T_2} - w$$

where w_{T_1} is the mass of air of volume V at the ambient temperature T_1, w_{T_2} is the mass of the same volume of air heated to T_2, and w is the weight of the fabric, guy lines, and other structural features.

a) If a balloon has a volume of 50 m³ and weighs 5 kg, to what temperature must the air be heated if the balloon is to just lift itself at 27 °C and 1 atm?

b) What is the maximum weight the same balloon could lift if filled with helium gas?

41. Why does a balloon filled with H_2 or He float? What factors determine how high a balloon will go? (See Problem 40.)

42. Estimate the mass of O_2 in the entire atmosphere of the earth, assuming an average radius of 6.0×10^6 m and an average atmospheric pressure of 1 atm (assume air is 21% O_2 by volume).

43. On the average, how many molecules of O_2 traveling with velocities of 5×10^2 m·s^{-1} must strike a surface 10 Å by 10 Å each second in order to produce a pressure of 1 atm?

44. The melting point of pure farcium imaginide on the Merdeducheval temperature scale is 0 °M and the boiling point is 200 °M. The corresponding temperatures on the Celsius scale are 35 °C and 115 °C, respectively.

a) Convert 15 °M to °C.
b) Convert 10 °C to °M.
c) What are the melting point and boiling point of H_2O in °M?

45. Fifty cubic feet of propane gas (C_3H_8) measured at 2 atm and 273 K is burned in air (21% O_2 and 79% N_2 by volume).

a) What volume of CO_2 (measured at 2 atm and 273 K) is obtained?

b) What volume of air (measured at 2 atm and 273 K) is needed for the combustion?

46. A quantity of O_2 exerts a pressure of 10 atm. Suppose two-thirds of the mass of O_2 is removed, the volume of the container is doubled, and the absolute temperature is halved. What is the pressure of the remaining O_2?

47. A mole of gas 1 has a molecular weight of 50 amu and a kinetic energy of X units at 200 K. Gas 2 has a molecular weight of 200 amu. What is the kinetic energy of 1 mole of gas 2 at 400 K?

48. The deepest part of the ocean is about 11,000 m. Assuming an average seawater density of 1.03 g·cm^{-3}, what is the water pressure (in atm) at this depth?

49. A skin diver can tolerate a pressure of about 5 atm. How deep is it safe for a skin diver to descend, assuming that mercury has a density about 13.2 times that of seawater?

ADDITIONAL READINGS

Parsonage, N. G., *The Gaseous State.* Elmsford, N.Y.: Pergamon, 1966.

Mahan, Bruce H., *University Chemistry,* 3rd ed. Reading, Mass.: Addison-Wesley, 1975. A somewhat more advanced treatment of gases than in the present text. Calculus is used in the kinetic molecular theory development.

Neville, R. G., "The Discovery of Boyle's Law, 1661–62," *J. Chem. Educ.* **39**, 356 (1962).

4
THE ELECTRONIC STRUCTURE OF ATOMS

The universe is not only queerer than we suppose, but queerer than we can suppose.
J. B. S. HALDANE, *Possible Worlds*

The Rutherford model of the atom, although in accord with many chemical and physical facts, fails to provide insight into some of the most important aspects of atomic and molecular behavior: How do electrons and protons interact so as to produce so many different types of atoms with such diverse natures? Perhaps more puzzling: Why are the physical and chemical properties of atoms periodic, and why do atoms form molecules only in certain combinations?

It requires a profound revolution in our approach to "understanding" nature to provide satisfactory answers to the above questions. Prior to about 1900, scientists were highly successful in explaining almost all natural phenomena in ways that seemed intuitively logical and made common sense. There were even a few scientists who suggested that all of nature could be understood by merely thinking about it in a logical manner; there was no need to do experiments! The nature of the atom soon put this optimistic view to rest.

Almost all of what we humans call intuition and common sense is based on experiences with the visible world about us—experiences we begin to collect almost from the moment of birth. Most of us have a very good feeling for how balls bounce, how Frisbees fly, how cars spin on icy roads, how metals bend, how glass breaks—even though we may not be able to define all of these phenomena in precise terms. But nature did not design the atom as a reflection of human experience; at times it even appears as if she chose patterns completely contrary to it. Consequently, in order to understand how electrons and protons are put together to form atoms and how atoms combine to form molecules, we must be prepared to accept some strange concepts—concepts which often appear to violate common sense and which may appear totally arbitrary and incomprehensible. To attempt a study of the atom on any other basis is to give up all hope of ever acquiring that new level of consciousness that modern science requires. Fortunately, the road to this level of consciousness is not all that difficult to travel, provided one is willing to adopt a flexible response to new and strange ideas. Only those who stubbornly in-

◀ Niels Bohr, Danish theoretical physicist who first used quantum mechanics to explain the structure of atoms. (Courtesy of Brown Brothers.)

sist on reducing everything to the ping-pong ball level will find the path frustrating and, perhaps, impossible.

This chapter presents a simplified—mostly nonmathematical—account of the quantum theory of atomic structure. The student should not feel discouraged if he or she does not immediately feel perfectly at ease with all the new concepts introduced here. Like a new marriage partner, some of these require some living with before they seem real.

4.1 THE NATURE OF LIGHT AND OTHER ELECTROMAGNETIC RADIATION

Modern quantum theory originated in various types of studies of interactions between matter and light (or other electromagnetic radiation). Electromagnetic radiation, of which ordinary visible light is a specific example, consists of electric and magnetic oscillations traveling through space. These oscillations have many properties analogous to the waves seen on disturbed water surfaces and, in fact, may be associated with a **wavelength** λ (Greek letter lambda) defined in a similar way. The velocity at which the waves move is related to the wavelength by the expression

$$c = \lambda \nu \tag{4.1}$$

where ν (Greek letter nu) is the **frequency** of the wave motion. This quantity is just the number of successive waves passing a fixed point per unit time (see Fig. 4.1). The wavelength of the radiation is the distance between two successive comparable points (for example, crests or troughs) of the wave disturbance.

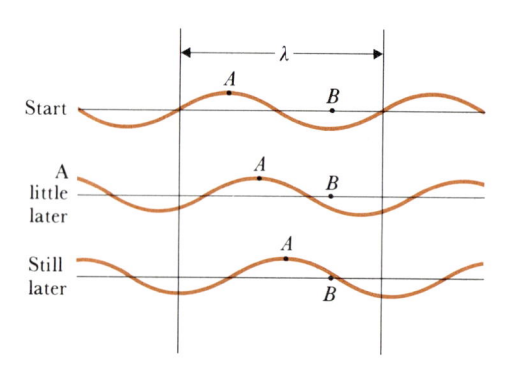

Fig. 4.1. The important characteristics of wave motion. The velocity c is the distance a point such as A moves per unit time. The frequency ν is the number of segments of length λ (the wavelength) moving past a fixed point, such as B, per unit time.

When electromagnetic radiation travels through a vacuum, the velocity c is 3.00×10^8 m·s^{-1} (about 186,000 miles per second). Different types of radiation travel with this same velocity and differ only in wavelength and frequency. For example, X-rays have relatively short wavelengths (on the order of 1 Å or 0.1 nm),* whereas radio waves have relatively long wavelengths (on the order of kilometers).

---

* Recall from Chapter 1 that an angstrom unit (symbol Å) is 1×10^{-10} m; 1 nm (nanometer) is 1×10^{-9} m or 10 Å.

The frequency of an X-ray of wavelength 1.00 Å (1.00 × 10⁻¹⁰ m) is given by

$$\nu(\text{X-ray}) = \frac{c}{\lambda} = \frac{3.00 \times 10^8 \text{ m}\cdot\text{s}^{-1}}{1.00 \times 10^{-10} \text{ m}} = 3.00 \times 10^{18} \text{ s}^{-1}$$

whereas that of a radio wave of wavelength 10.0 km (10.0 × 10³ m) is

$$\nu(\text{radio wave}) = \frac{c}{\lambda} = \frac{3.00 \times 10^8 \text{ m}\cdot\text{s}^{-1}}{10.0 \times 10^3 \text{ m}} = 3.00 \times 10^4 \text{ s}^{-1}$$

Thus a short wavelength implies a high frequency and vice versa. The unit of frequency (reciprocal second, s⁻¹) is sometimes called **cycles per second**; the symbol **Hz** (for **hertz**) is also used to denote s⁻¹.

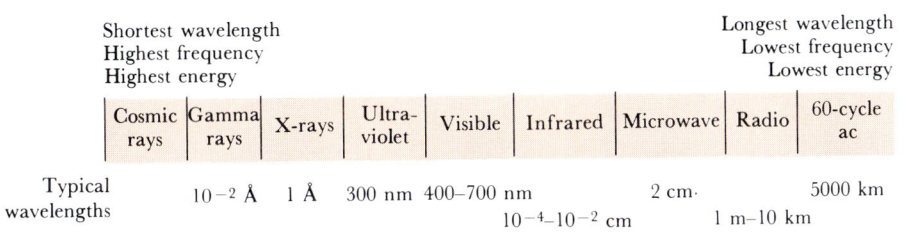

Fig. 4.2. A portion of the electromagnetic spectrum. The sizes of the indicated regions are not to scale.

Figure 4.2 shows schematically some of the more important areas of what is known as the **electromagnetic spectrum**. Note that visible light covers the wavelength range of about 400 to 700 nm; the former is called the *blue* end and the latter the *red* end of the spectrum. Just below the blue end lies the ultraviolet region; the infrared region (heat radiation) lies above the red end.

EXAMPLE 4.1 Calculate the frequencies of light at the extreme ends of the visible spectrum (see Fig. 4.2).

SOLUTION The extreme ends of the visible spectrum are at 400 and 700 nm. The frequency at the blue end is

$$\nu = \frac{c}{\lambda} = \frac{3.00 \times 10^8 \text{ m}\cdot\text{s}^{-1}}{400 \times 10^{-9} \text{ m}} = 7.50 \times 10^{14} \text{ s}^{-1}$$

The frequency at the red end is

$$\nu = \frac{c}{\lambda} = \frac{3.00 \times 10^8 \text{ m}\cdot\text{s}^{-1}}{700 \times 10^{-9} \text{ m}} = 4.29 \times 10^{14} \text{ s}^{-1}$$

4.2 THE ORIGINS OF QUANTUM CONCEPTS

Quantum comes from the Latin word *quantus*, which means "how much." The scientific use of the term is to denote a *discrete unit* or *packet* of a physical quantity such as energy. In a sense, Dalton's atomic theory is a quantum theory since it states that matter, instead of being infinitely subdivisible, consists of discrete units called atoms (quanta of mass). Up to the early 1900s there appeared to be no reason to suspect that energy, as well as matter, was quantized. It soon became apparent, however, that the interaction of matter and radiation could be understood at the atomic level only if some rather radical modifications were introduced into our way of viewing matter and radiation. Perhaps the three most important modifications are the following.

1. **Wave–particle duality.** Light, which ordinary observations suggest is wavelike in nature, also behaves as if it consisted of particles. At the same time particles such as electrons also can behave as if they were waves.

2. **Quantization of energy.** Depending on how hard you hit a tennis ball, it seems perfectly possible to give the ball any precise amount of energy you wish; that is, there seem to be no restrictions (other than your strength) on the energy the ball can have. On the atomic and molecular level, by contrast, there are very definite restrictions on the energy an atom or molecule can have. In general, only certain discrete values of the energy are allowed; all other values are strictly forbidden.

3. **The uncertainty principle.** Assuming that appropriate equipment is available, there appear to be no restrictions *in principle* upon the accuracy with which the trajectory of a moving tennis ball could be described. In the limit of highly refined measurements, it appears possible to be able to tell precisely how fast the ball is moving at a precisely defined position in space. Not so with moving electrons and other small particles on the atomic scale of observation; the more accurately the position is measured, the less precisely it is possible to measure the velocity and vice versa.

The need for wave–particle duality first became apparent when Einstein provided an explanation for the **photoelectric effect**. The photoelectric effect is the name given to the flow of electric current induced when light shines on a metal surface which is part of an electrical circuit (see Fig. 4.3).* Scientists were able to demonstrate that the current was simply a flow of electrons, electrons released by the metal due to energy absorbed from the light. However, prior to Einstein, no one succeeded in explaining why increasing the intensity of the light did not increase the energies of the emitted electrons as ordinary common sense seemed to require. This and other puzzling features were explained by Einstein by assuming light consists of a stream of particles (now called **photons**) whose energies are given by

$$E = h\nu \tag{4.2}$$

where ν is the frequency of the light and h is a proportionality constant now known as **Planck's constant**. This constant has a numerical value of 6.626×10^{-34} J·s. Einstein then postulated that an electron could interact with radiation by absorbing

---

* The "electric eye," once commonly used to open and close supermarket doors, operates by the photoelectric effect. A beam of light across the customer's path keeps a photoelectric cell activated which, in turn, maintains the switch to the door-opening mechanism in the "off" position. Breaking the beam of light turns off the photoelectric cell and allows the door-opening switch to close. Most automatic door openers today are operated by switch mats; stepping on the mat before the door activates a switch to open the door.

Albert Einstein (1879–1955) was a German-born theoretical physicist noted mainly for the theory of relativity. While employed as a Swiss patent clerk, Einstein laid the foundations of relativity theory and, in addition, provided revolutionary insights into the nature of the photoelectric effect, Brownian motion, the specific heats of solids, and the nature of radiation. He received the 1921 Nobel prize in physics for his work on the photoelectric effect. A good recent biography of Einstein is Einstein: The Life and Times, *by Ronald W. Clark (New York: World Publishing Co., 1971).*

The proportionality constant h was introduced about 1900 by the German theoretical physicist Max Planck (1858–1947), generally regarded as the father of quantum theory. While studying the emission of radiation from heated solids, Planck was forced to postulate the existence of quantization, a concept which ultimately led others to the quantum theory of today. He received the Nobel prize in physics in 1918.

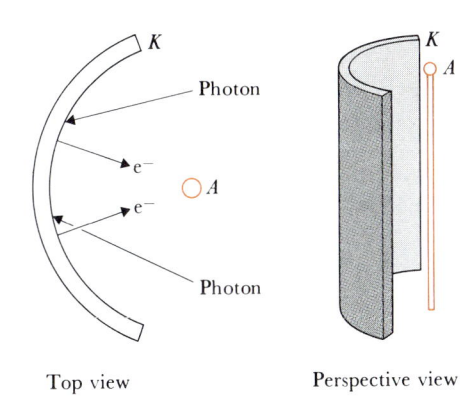

Fig. 4.3. The photoelectric effect. Light falls on the metal cathode K and ejects electrons which travel to the anode A. This flow of electrons constitutes an electric current. The cathode and anode are usually inside an evacuated glass tube. Electrons return to the cathode via an external circuit from anode to cathode.

either a whole photon or none at all. Once this unusual assumption is made, the law of conservation of energy states that

$$h\nu = \tfrac{1}{2}mv^2 + h\nu_0 \tag{4.3}$$

where m is the mass of the electron, v is its velocity, and ν_0 is the **threshold frequency**—the minimum light frequency for which the photoelectric effect is observed. (This frequency varies from metal to metal.) This equation says that the energy of the photon ($h\nu$) is used to move an electron to the surface of the metal (this takes energy $h\nu_0$) and then to eject it with a kinetic energy $\tfrac{1}{2}mv^2$. Thus it is the frequency of the light—and not its intensity—that determines the kinetic energies of the emitted electrons. About ten years later, the U.S. physicist Robert A. Millikan provided a rigorous experimental verification of the Einstein photoelectric equation.

Robert A. Millikan (1868–1953) was the American physicist who made the first measurement of the charge on the electron (see Fig. 1.2). For this achievement he received the Nobel prize in physics in 1923. Millikan also pioneered in cosmic ray research and wrote several important textbooks of physics.

EXAMPLE 4.2 The minimum energy needed to bring an electron to the surface of a metal just short of emission is called its **work function** and is given by $h\nu_0$ in Einstein's photoelectric equation. What is the maximum wavelength a photon can have if it is to eject a photoelectron from cadmium (work function 4.1 eV)?

SOLUTION

$$h\nu_0 = \frac{hc}{\lambda} = 4.1 \text{ eV } (1.6 \times 10^{-19} \text{ J/eV}) = 6.6 \times 10^{-19} \text{ J}$$

$$\lambda = \frac{hc}{6.6 \times 10^{-19} \text{ J}} = \frac{(6.6 \times 10^{-34} \text{ J} \cdot \text{s})(3.00 \times 10^8 \text{ m} \cdot \text{s}^{-1})}{6.6 \times 10^{-19} \text{ J}}$$

$$= 3.0 \times 10^{-7} \text{ m} \quad \text{or} \quad 300 \text{ nm}$$

The concept of the quantization of atomic energy first appeared when Niels Bohr introduced a novel treatment of the hydrogen atom which accounted for the spectral lines of that element. When a glass tube filled with a gas such as hydrogen is subjected to an electrical discharge, the gas glows and emits radiation.* When this radiation is passed through a prism (a device which separates light according to wavelength), only certain lines, corresponding to definite wavelengths, are observed. The light emitted by other gases, or by metal salts heated to incandescence in a gas flame, behaves similarly. By contrast, ordinary sunlight passing through the same prism produces a continuous rainbow of colors. These contrasting phenomena are illustrated in Fig. 4.4. Further studies show that the emitted lines are characteristic of the elements present in the incandescing substance or the glowing gas; in fact, the emitted lines are useful for accurate identification of the elements just as fingerprints are used to identify specific people. Just as with fingerprints, no two different elements ever emit the same set of lines. These lines are called the **spectrum** of the element.

Prior to Bohr, no one was able to account for the origin of the spectral lines of atoms. Bohr first assumed that the electron of the hydrogen atom moved about the proton in a circular orbit much as the moon revolves about the earth and that the coulombic force on the accelerating electron equalled the force due to the acceleration. The mathematical expression for this condition is

$$\frac{mv^2}{r} = \frac{Ze^2}{r^2} K \tag{4.4}$$

where m is the mass of the electron, v is its orbital velocity, e is the unit of electronic

---

* If hydrogen gas is used, the tube originally contains the diatomic molecules H_2, but some of these are dissociated to atoms by the electrical discharge and excited to higher energies.

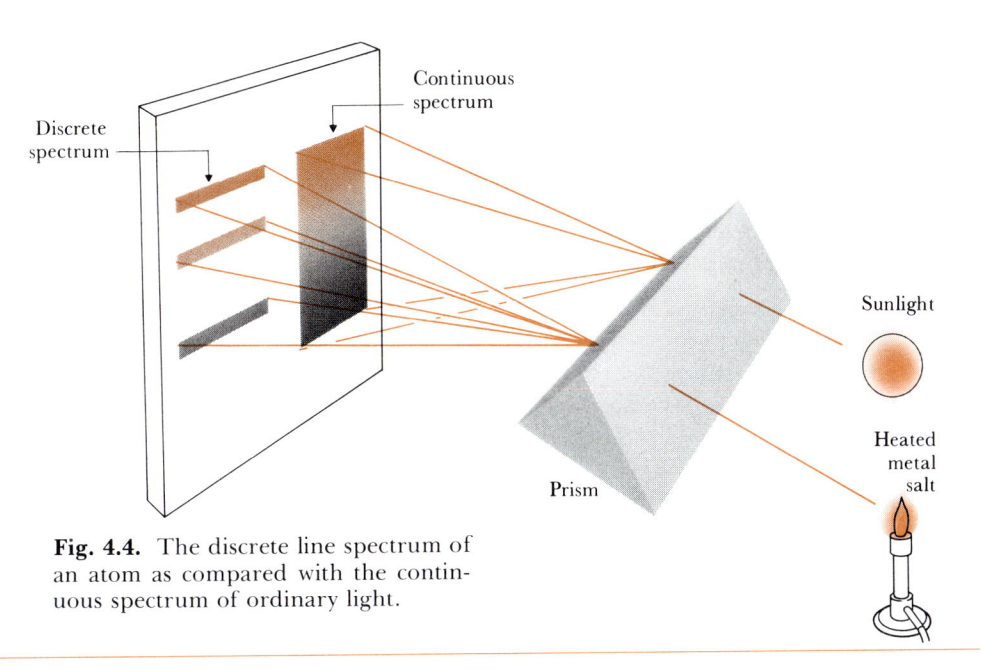

Fig. 4.4. The discrete line spectrum of an atom as compared with the continuous spectrum of ordinary light.

Niels Bohr (1885–1962), a Danish theoretical physicist (and an ex football player of national prominence), worked as a young man with J. J. Thomson at the Cavendish Laboratory and with Ernest Rutherford at the University of Manchester. He later headed the Institute for Theoretical Physics in the University of Copenhagen. Bohr contributed to the philosophical development of quantum theory and aided in the development of nuclear fission for the United States in World War II. During his later years, Bohr was greatly concerned with international control of nuclear energy and was one of the founders of CERN (European Center for Nuclear Research at Geneva, Switzerland). He received the Nobel prize in physics in 1922. Colleagues of Bohr report that he was an addict of U.S.-made western movies.

Actually, the apparently continuous spectrum produced by ordinary sunlight contains some missing frequencies known as **Fraunhofer lines**, *named after their discoverer, the German pioneering scientist Josef von Fraunhofer (1787–1826). The Fraunhofer lines correspond to energy transitions in the elements that are present in the sun's corona (mainly hydrogen and helium) and in the earth's atmosphere. The light emitted from the surface is continuous (has no missing frequencies), but some of the frequencies are absorbed by the cooler gases forming the sun's "atmosphere." A gas such as hydrogen can absorb light of the same frequency as it can emit; the 6563-Å line (Fig. 4.6) which the hydrogen atom emits when excited in a discharge tube corresponds to one of the most prominent Fraunhofer lines. The element helium was first discovered [by Sir Joseph Lockyer (1836–1920), British astrophysicist] on the sun when observation of the sun's corona during an eclipse revealed Fraunhofer lines not attributable to the emission spectrum of any substance then known on earth.*

charge, r is the radius of the orbit, and K is a constant (equal to 9×10^9 J·m·C^{-2}).*
The quantity Z is the atomic number of the nucleus (also the magnitude of its positive charge). This is unity for the hydrogen atom but is included since the Bohr model also applies to one-electron ions such as He$^+$ ($Z = 2$), Li^{2+} ($Z = 3$), etc. Next, Bohr made what was truly an unexpected and novel suggestion: the angular momentum (mvr)† of the orbiting electron is quantized. This may be written

$$mvr = \frac{nh}{2\pi}, \quad n = 1, 2, 3, \ldots \tag{4.5}$$

This means that the angular momentum could have values such as $h/2\pi$, h/π, $3h/2\pi$, etc. but no values in between any of these.

Although Bohr introduced this postulate somewhat arbitrarily, its justification is now known to reside in wave–particle duality. Experiments with beams of electrons reveal that the electrons exhibit some properties ordinarily associated with waves. Suppose the electron in a Bohr orbit behaves as a wave. Since waves can interact with other waves so as to reinforce each other or to weaken each other, the electron wavelength in a Bohr orbit must satisfy the condition

$$n\lambda = 2\pi r \tag{4.6}$$

This means that an integral number of wavelengths ($n\lambda$) must fit into the length of one complete orbit ($2\pi r$); otherwise the wave becomes out of phase with itself after each revolution and destroys itself (see Fig. 4.5).

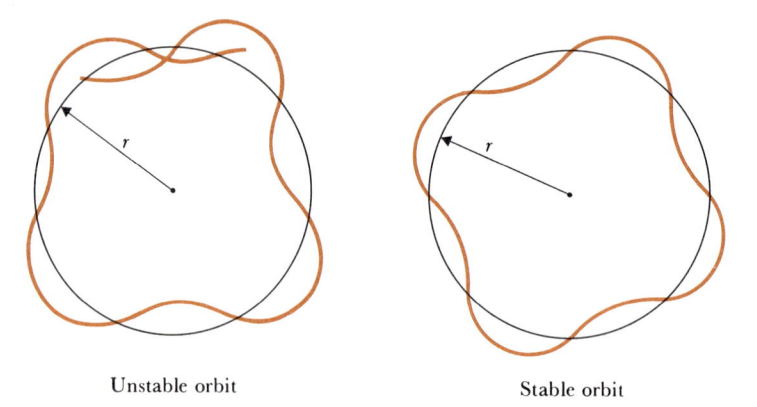

Unstable orbit   Stable orbit

Fig. 4.5. Matter–wave interpretation of a stable electron orbit in an atom. In (a) there are a nonintegral number of wavelengths in an orbit so that the wave destroys itself by phase interference. In (b) there are an integral number of wavelengths in an orbit so that no phase interference occurs and the orbit remains stable indefinitely.

---

* The constant K is needed to have the coulombic force come out in the SI unit of newtons when all other quantities in the equation are also in SI units.

† Angular momentum is a physical quantity useful in describing certain types of motion in which velocity is constantly changing. The angular momentum associated with a spinning top may be viewed as directed parallel to the spin axis. The top tends to stay upright since a torque, or tipping force, must be exerted to change the angular momentum, which, in the absence of friction to slow the top, tends to remain constant.

Louis Victor de Broglie (b. 1892), a French theoretical physicist, received the Nobel prize in physics in 1929. De Broglie presented his ideas on duality in a doctoral dissertation in 1924. His original idea for the dual nature of matter arose largely from certain mystical notions about the symmetry of nature.

The mathematical form of wave–particle duality is given by the de Broglie equation,

$$\lambda = \frac{h}{mv} \quad (4.7)$$

This equation states that a wave of wavelength λ will have a particlelike momentum mv, and that a particle of momentum mv will have a wavelike wavelength λ; in short, nature is simultaneously particlelike and wavelike. Substituting the above for λ in Eq. (4.6) and rearranging leads to the Bohr postulate of Eq. (4.5).

By suitable combination of Eqs. (4.4) and (4.5) and use of the law of conservation of energy, Bohr showed that the energy of the hydrogen atom obeyed the relationship

$$E_n = -\frac{Z^2 e^2}{2n^2 a_0} K \quad (4.8)$$

where a_0, called the **first Bohr radius**, is the value of r for $n = 1$.* This quantity is given by

$$a_0 = \frac{h^2}{4\pi^2 m e^2 K} \quad (4.9)$$

Its value is about 0.529 Å. In general, the radius of an electron orbit is given by

$$r = \frac{n^2 a_0}{Z} \quad (4.10)$$

The integer n is called a **quantum number**. Equation (4.8) shows that the energy of the hydrogen atom is **quantized**, that is, can have only those values arising from integral values of n. Thus the allowed energy values (called **energy states**) of the hydrogen atom may be likened to the rungs of a ladder, and Eq. (4.8) is a formula for calculating the positions of the rungs.

If $-e^2 K/2a_0$ is replaced with the symbol R_H, Eq. (4.8) may be written in the simpler form

$$E_n = \frac{-Z^2 R_H}{n^2} \quad (4.11)$$

---

* Equation (4.8) is derived as follows: the total energy of the electron is the sum of its kinetic energy $\frac{1}{2}mv^2$ and its potential energy $-Ze^2K/r$. Using Eq. (4.4) for mv^2 leads to $E = -Z^2 e^2 K/2r$. Solving Eq. (4.4) for v^2 and Eq. (4.5) for v^2, equating, and solving for r leads to $r = n^2 a_0/Z$ if a_0 is defined as in Eq. (4.9). Using this value of r in E produces Eq. (4.8).

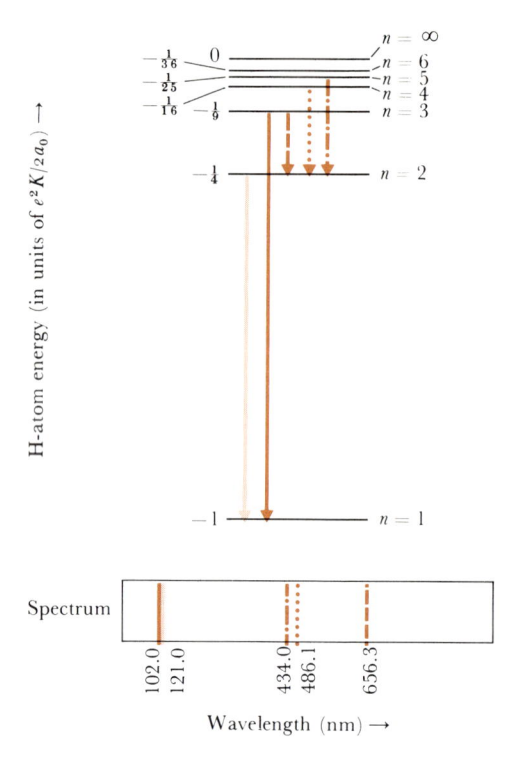

Fig. 4.6. A portion of the energy-level diagram and spectrum of the hydrogen atom. Only a few selected transitions and corresponding spectral lines are shown. The stationary states between $n = 6$ and $n = \infty$ become very crowded and are not shown. The energy difference between the $n = 1$ state and the $n = \infty$ state corresponds to the ionization energy of the atom.

The numerical value of R_H is 2.179×10^{-18} J. Alternatively, using the conversion factor of 1.602×10^{-19} J per eV, R_H is also equal to 13.6 eV. (An electron volt (eV) is the energy required to accelerate a particle of charge $\pm e$ through a potential difference of 1 volt). Thus the lowest two rungs ($n = 1$ and 2) are $13.6/1 - 13.6/4$ or 10.2 eV apart, the next two rungs ($n = 2$ and 3) are $13.6/4 - 13.6/9$ or about 1.9 eV apart, and successive spacings become less and less (see Fig. 4.6). Ultimately the rungs become so close together (as n approaches infinity) that they run into each other. This point represents an ionized hydrogen atom ($E_\infty = 0$). Thus the minimum energy needed to carry out the process

$$H \rightarrow H^+ + e^-$$

when H is in its lowest possible energy state ($n = 1$) is given by

$$E_\infty - E_1 = 0 - \left(\frac{-13.6}{1^2}\right) = 13.6 \text{ eV} \quad \text{or} \quad 2.179 \times 10^{-18} \text{ J} \tag{4.12}$$

This agrees exactly with the experimental value for the ionization energy of the hydrogen atom.

EXAMPLE 4.3 Calculate the ionization energy of the process

$$He^+ \rightarrow He^{2+} + e^-$$

SOLUTION Since He$^+$ has only one electron, its allowed energy levels are given by the Bohr equation: $E_n = -Z^2 R_H/n^2$, where $Z = 2$. The ionization energy of He$^+$ is then given

102   THE ELECTRONIC STRUCTURE OF ATOMS

by
$$-E_1 = 4R_H = 4(2.179 \times 10^{-18} \text{ J}) = 8.7 \times 10^{-18} \text{ J}$$
which is about 54.5 eV.

Bohr now explained the spectral lines of hydrogen as follows: Ordinarily a hydrogen atom can exist in one of its quantized energy states indefinitely, but interaction with radiation can bring about a **transition** from one energy state to another. In one case the atom can absorb energy from the radiation to go to a higher energy state; in another case the atom can drop to a lower state and lose energy by emitting radiation. In either case Bohr postulated that the energy *difference* between the two states had to satisfy the relationship

$$\Delta E = h\nu \tag{4.13}$$

where ν is the frequency of the absorbed radiation or the frequency of the emitted radiation. This means that unless a photon of radiation has exactly the right amount of energy to correspond to the difference of two energy states, it cannot interact with the atom. This is exactly in accord with the experimental observation that atoms absorb only certain wavelengths of radiation and not others. Similarly, the postulate accounts for the line spectrum of the emission process. As a clincher to his arguments, Bohr showed that his model predicted numerical values for the wavelengths of the spectral lines which were in excellent agreement with experiment. For example, the wavelength of the transition between the $n = j$ and $n = i$ levels should be given by the relationship

$$\Delta E = E_j - E_i = h\nu = \frac{hc}{\lambda} \tag{4.14}$$

which may be rearranged to

$$\frac{1}{\lambda} = \frac{E_j - E_i}{hc} \tag{4.15}$$

Using Eq. (4.8) for E_i and E_j produces

$$\frac{1}{\lambda} = \frac{R_H}{hc}\left(\frac{1}{n_i^2} - \frac{1}{n_j^2}\right) \tag{4.16}$$

where n_i is the quantum number of the lower energy level E_i and n_j is the quantum number of the higher energy level E_j. Using $n_i = 1$ and $n_j = 2$ leads to $\lambda = 121$ nm. This is one of the wavelengths actually observed in the emission spectrum of hydrogen. Similarly, using other energy levels leads to values for the other experimentally observed lines.

Equation (4.16) is known as the **Balmer-Rydberg-Ritz formula**; a less general version was first discovered accidentally about 1885 by a German schoolmaster, J. J. Balmer. Bohr's derivation of this hitherto mysterious formula constituted a major intellectual achievement.

EXAMPLE 4.4 Calculate the wavelength a photon must have in order to excite the hydrogen atom from the $n = 3$ state to the $n = 15$ state.

SOLUTION

$$\frac{1}{\lambda} = \frac{R_H}{hc}\left(\frac{1}{n_i^2} - \frac{1}{n_j^2}\right)$$

$$= \frac{2.179 \times 10^{-18} \text{ J}}{(6.6 \times 10^{-34} \text{ J}\cdot\text{s})(3 \times 10^8 \text{ m}\cdot\text{s}^{-1})}\left(\frac{1}{3^2} - \frac{1}{15^2}\right)$$

$$= 1.17 \times 10^6 \text{ m}^{-1}$$

Therefore

$$\lambda = \frac{1}{1.17 \times 10^6 \text{ m}^{-1}} = 8.52 \times 10^{-7} \text{ m} \quad \text{or} \quad 852 \text{ nm}$$

EXAMPLE 4.5 What is the energy (in eV) of the photon emitted when the hydrogen atom drops from the $n = 10$ level to the $n = 5$ level?

SOLUTION

$$\Delta E = h\nu = R_H\left(\frac{1}{n_i^2} - \frac{1}{n_j^2}\right) = 2.179 \times 10^{-18} \text{ J}\left(\frac{1}{5^2} - \frac{1}{10^2}\right)$$

$$= 6.537 \times 10^{-20} \text{ J}$$

or

$$6.537 \times 10^{-20} \text{ J}(1 \text{ eV}/1.6 \times 10^{-19} \text{ J}) = 0.41 \text{ eV}$$

The Bohr theory cannot predict numerical values for the spectral lines of any atoms besides hydrogen and one-electron ions, but the general idea of quantized energy states and the $\Delta E = h\nu$ relationship is applicable to all atoms and even to molecules. Figure 4.7 illustrates those aspects of the Bohr theory as they apply to the emission spectrum of a hypothetical atom. The lowest possible energy state in which the atom can exist is called the **ground state**. The next to the lowest energy state is called the **first excited state**, the next is called the **second excited state**, and so on until the **continuum** (where all states merge) is reached. By convention, the continuum is assigned the value of zero and all the quantum states of the atom are assigned negative values, with the ground state having the largest negative value.

The uncertainty principle was one of the last surprises revealed by quantum theory, and, at the same time, it is perhaps the hallmark of quantum theory. The principle was stated by Werner Heisenberg something like this: Consider a particle of mass m moving with velocity v along the x direction. If the particle's velocity is measured to within an uncertainty Δv, then it is impossible to measure where the particle is (its value of the position x) more accurately than

$$\Delta x = \frac{h}{4\pi m \Delta v} \tag{4.17}$$

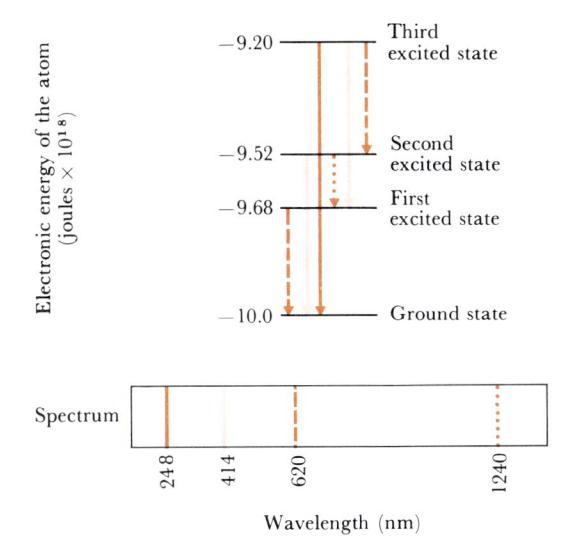

Fig. 4.7. Energy-level diagram and emission spectrum of a hypothetical atom. The Bohr model cannot calculate the energies of individual states in atoms other than hydrogen, but it shows the correct general appearance of the energy states and the spectrum. Note that there are two different transitions leading to the same spectral lines at 414 nm and 620 nm.

Werner Heisenberg (1901–1976) was a German theoretical physicist who received the 1932 Nobel prize in physics for laying the foundation of modern quantum theory. Heisenberg held physics professorships at the Universities of Copenhagen, Leipzig, and Berlin and became director of the Max Planck Institute in Berlin. To develop quantum theory he reinvented (since he didn't know it already existed) a branch of mathematics known as matrix algebra.

Conversely, if the position is measured to within an uncertainty Δx, the velocity at the same time cannot be measured more accurately than

$$\Delta v = \frac{h}{4\pi m \Delta x}$$

Thus v and x are not simultaneously measurable to unlimited accuracy; an increase in the accuracy of one decreases the accuracy of the other.

Strictly speaking, there is no explanation for the uncertainty principle; that's the way we find nature operates and we are forced to accept it, no matter who we are—Nobel prize winners or garbage collectors. However, the principle can be made to appear somewhat more reasonable if it is realized that observation of matter requires that the object to be observed be bombarded with a beam of photons. That's what is happening when we look at a flower in a field (using the sun as a source of photons), or when we observe a colony of bacteria under a microscope. In each case the photons collide with the object to be viewed and give it a jolt or kick whose effect cannot be corrected for. This jolt cannot be avoided since *at least* one photon must be used for the observation. For a large mass such as a baseball (about 0.15 kg) the disturbance due to the jolt is negligible since

$$\Delta x \Delta v = \frac{6.6 \times 10^{-34} \, \text{J} \cdot \text{s}}{4\pi(0.15 \, \text{kg})} \simeq 3 \times 10^{-34} \, \text{m}^2 \cdot \text{s}^{-1}$$

Thus, if the velocity is known to within about $\pm 1.7 \times 10^{-17}$ m·s^{-1}, it is possible to know x to within $\pm 1.7 \times 10^{-17}$ m. For all practical purposes no uncertainties exist. However, in the case of an electron

$$\Delta x \Delta v = \frac{6.6 \times 10^{-34} \, \text{J} \cdot \text{s}}{4\pi(9 \times 10^{-31} \, \text{kg})} \simeq 5 \times 10^{-5} \, \text{m}^2 \cdot \text{s}^{-1}$$

Now if the velocity of the electron is known to within $\pm 7 \times 10^{-3}$ m, the uncertainty in x will be $\pm 7 \times 10^{-3}$ m or ± 0.7 cm. This uncertainty in x is so huge relative to the size of a Bohr orbit (about 0.5×10^{-8} cm) that the whole notion of a definite orbit is meaningless. The electron is so much more sensitive to observational uncertainty because very, very short wavelength (very, very high energy) light is needed to see it. The baseball, on the other hand, can be observed with very, very long wavelength (very, very low energy) light. Thus the measurement disturbs the electron by giving it a large indeterminate "kick" whereas the baseball is affected hardly at all.

4.3 WAVE MECHANICS AND THE HYDROGEN ATOM

The Bohr model of the atom, although providing an excellent description of many properties of the hydrogen atom, does not work well for the remaining one hundred or so elements. One of the most unsatisfactory aspects of the Bohr model is that its assumption of precise electron orbits violates the uncertainty principle. Working independently, Erwin Schrödinger and Werner Heisenberg proposed a new model which incorporates all the known quantum principles and also leads to satisfactory models for the structures of many-electron atoms. This model, known as **wave mechanics**, is a highly mathematical one, so only a qualitative description of it shall be given here.

Instead of assuming definite electron orbits, the new model describes the electron in terms of the *probability* that it is in a given region of space. This is done by describing the **electron distribution** about the nucleus in terms of a quantity called the **electron density**—the amount of negative charge found in a unit volume of space. Such a description is forced upon us by the wave nature of the electron and by our inability to define a definite trajectory for it. Consequently, the electron may be regarded as a cloud of negative charge about the nucleus, a cloud whose thickness or density varies from point to point in space in proportion to the probability of the electron being at that point.

The wave-mechanical treatment of the hydrogen atom predicts the same value for the quantized energy levels as does the Bohr theory, namely

$$E_n = -\frac{Z^2 R_H}{n^2} \qquad (4.18)$$

where n, now called the **principal** quantum number, has the integral values 1, 2, 3, ..., etc. However, the Schrödinger-Heisenberg model introduces two additional

Erwin Schrödinger, Austrian theoretical physicist who improved and generalized Bohr's quantum theory. (Courtesy of Brown Brothers.)

Erwin Schrödinger (1887–1961) was an Austrian physicist who in his later years turned to biology and attempted to explain life on the basis of purely physical concepts. The Britisher Paul A. M. Dirac (b. 1902), who was an electrical engineer before he became a physicist, predicted the existence of the positron on the basis of purely mathematical arguments. Having spent most of his adult life at Britain's Cambridge University, he now resides at Florida State University. In 1933 Schrödinger and Dirac shared the Nobel prize in physics for the development of wave mechanics.

quantum numbers whose symbols are l and m_l. These quantum numbers have the following physical significance.

l is called the **azimuthal** or **orbital angular momentum** quantum number. It is related to the quantization of the angular momentum of the electron and has the values 0, 1, 2, ... up to $n - 1$. Thus if the electron is in the $n = 4$ state, it can have angular momenta of 0, $h/2\pi$, h/π, or $3h/2\pi$, and no other values. This quantum number is related to the *shape* of the electron cloud.

m_l is called the **magnetic** quantum number. Its allowed values are all the integers beginning with $-l$ and running through $+l$. Thus if $l = 2$, the values of m_l can be -2, -1, 0, 1, or 2, and no other. This quantum number describes, among other things, what happens to the angular momentum and energy of the electron when the hydrogen atom interacts with a magnetic field.

A later modification by Dirac introduced a fourth quantum number m_s. This is called the **spin** quantum number and has only two possible values, $-\frac{1}{2}$ and $+\frac{1}{2}$. This quantum number describes a spin angular momentum which the electron must have in addition to its orbital angular momentum. Thus the total angular momentum of the electron is always $(l + m_s)h/2\pi$.

EXAMPLE 4.6 What values may the quantum numbers l and m_l have when $n = 4$?

SOLUTION Since l can have values from 0 up to $n - 1$, the allowed values of l are 0, 1, 2, and 3. Since m_l can go from 0 to $\pm l$, its allowed values are 0 for $l = 0$; 0, ± 1 for $l = 1$; 0, ± 1, ± 2 for $l = 2$; and 0, ± 1, ± 2, ± 3 for $l = 3$.

TABLE 4.1 SUMMARY OF THE MAIN CHARACTERISTICS OF THE FOUR HYDROGEN-ATOM QUANTUM NUMBERS

SYMBOL	NAME	ALLOWED VALUES
n	Principal quantum number	1, 2, 3, ...
l	Orbital angular momentum quantum number	0, 1, 2, ..., $n - 1$
m_l	Magnetic quantum number	0, ± 1, ± 2, ..., $\pm l$
m_s	Spin quantum number	$-\frac{1}{2}$ or $+\frac{1}{2}$

The four quantum numbers of the hydrogen atom and their main characteristics are summarized in Table 4.1. It is customary to replace the l values 0, 1, 2, 3, ... with the letters s, p, d, f, ..., respectively (alphabetical beyond f if needed). The quantum states of the hydrogen atom are symbolized by nl, using a numerical value for n and a letter designation for l. For example, the ground state of hydrogen is symbolized by 1s, and successively higher excited states are written 2s, 2p, 3s, 3p, 3d, 4s, 4p, 4d, 4f, ..., and so on. States having the same n value (such as 2s and 2p) have the same energy since the energy depends only upon n.

Each of the quantum states of hydrogen is associated with a mathematical function called the **wave function** of that state. The wave function by itself has no physical significance, but its square (the wave function multiplied by itself) represents the probability of finding the electron at a given point in space.* As mentioned earlier, this probability may also be viewed as the density of the electron cloud as it varies in the space about the nucleus. An alternative name for these wave functions is **orbital**; an orbital is simply a description of how an electron's probability varies in space.

Figures 4.8 through 4.10 illustrate the electron densities arising from 1s, 2s, and 2p orbitals. Because it is impossible to represent a three-dimensional quantity (the electron density) on two-dimensional paper, each figure shows only how the electron

---

* Imagine a completely dark room occupied by people constrained to stay on a given numbered spot and batting a basketball around the room. Whenever a person is struck by the ball, he hits it to send it someplace else. Now suppose anyone hit by the ball immediately calls out his number and an outside observer records the number of times each spot location is called out. The record of spot-location frequencies would enable one to predict in which portions of the room the ball is most likely to be at any given time but not the path it follows about the room. This is roughly the type of information the square of the wave function provides about the location of an electron in an atom.

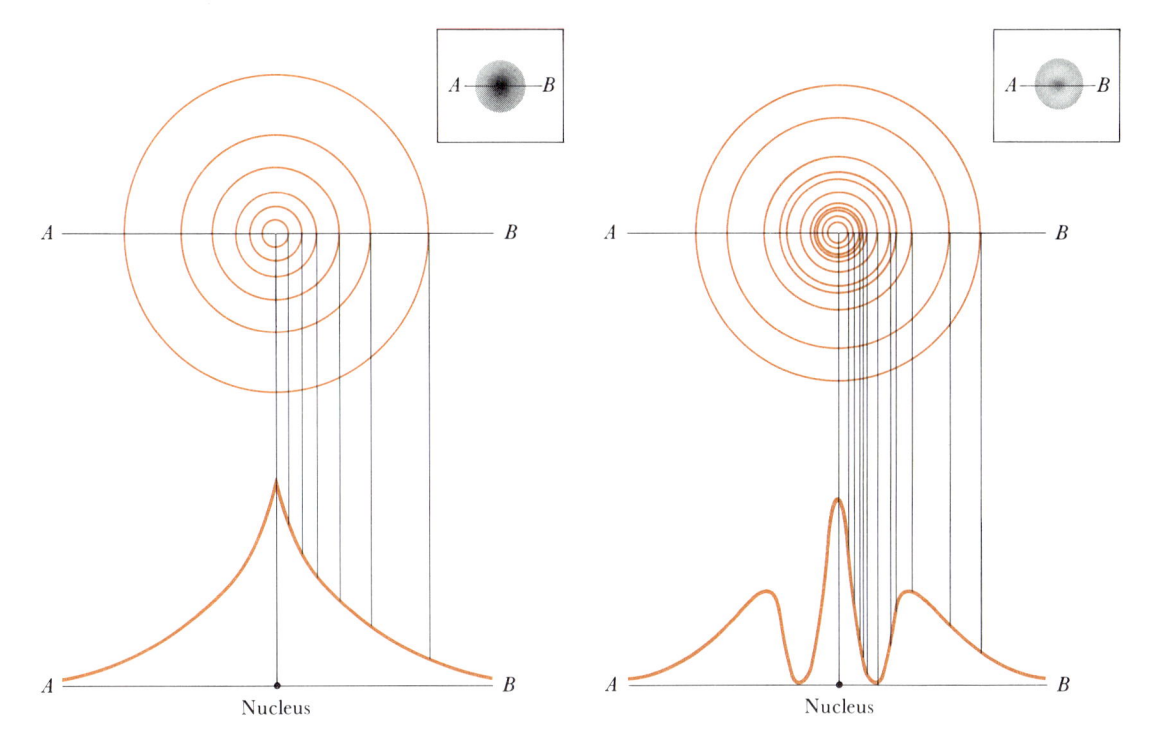

Fig. 4.8 (left). Three different representations of the spatial distribution of an electron in a 1s orbital. The bottom diagram shows how the density varies along a line passing through the nucleus; the contour diagram and the shaded drawing both represent the variation of electron density in a plane containing the nucleus. Note that the 1s electron density is spherically symmetric about the nucleus, with a maximum at the nucleus itself, and decreases rather rapidly in all directions away from the nucleus.

Fig. 4.9 (right). Three different representations of the spatial distribution of an electron in a 2s orbital. The bottom diagram shows how the density varies along a line passing through the nucleus; the contour diagram and the shaded drawing represent the variation of electron density in a plane containing the nucleus. Note that as in the 1s orbital, the 2s electron density is spherically symmetric about the nucleus, with a maximum at the nucleus itself. However, by contrast with the 1s orbital, the 2s electron density drops quickly to zero and then increases again before tailing off at larger distances.

density varies in a plane containing the nucleus.* There are several different ways of conveying this information, *three* of which are employed in each illustration. Toward the bottom of each figure, the height of the line above the base line (A to B) represents the "thickness" or density (charge per unit volume) of the electron cloud along a straight line passing through the nucleus. Immediately above this is a contour diagram (see Appendix 5) that represents electron cloud densities as contour lines, much as map contours represent land elevations.† Finally, the insert in each figure repre-

---

* *Surfaces* of three-dimensional solids are represented quite well on a two-dimensional surface, but cloudlike entities are not; the two-dimensional representation simply can't impart the sense of seeing *into* the cloud.

† Although electron density in a plane is analogous to land elevations, there are some important differences between the two; the electron cloud whose density is being represented is actually confined to the plane, whereas the land whose elevation is being represented rises above the plane represented by the map.

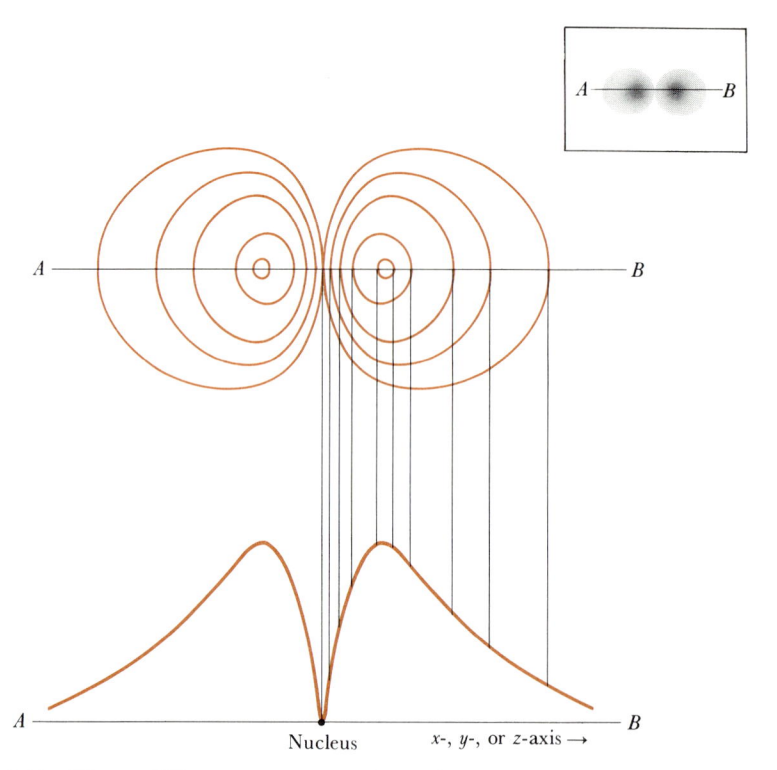

Fig. 4.10. Three different representations of the spatial distribution of an electron in a $2p$ orbital. The bottom diagram shows how the density varies along the x-, y-, or z-axis of the atom; the contour diagram and the shaded drawing represent the variation of electron density in a plane (xy, xz, or yz) containing the nucleus at the origin. The $2p$ orbitals are represented by three such distributions which are called $2p_x$, $2p_y$, and $2p_z$ and which are at right angles to each other. For example, the $2p_x$ distribution looks like that of the illustration when points A and B lie along the x-axis and the plane contains this axis; the $2p_y$ and $2p_z$ distributions are analogous but with respect to the y- and z-axes, respectively. Note that for the characteristic axis, the $2p$ density is zero at the nucleus, rises to a maximum and then decreases with distance. The electron density of a $2p_x$ orbital is zero everywhere in the yz-plane, and similarly for $2p_y$ and $2p_z$, which have zero densities in the xz- and xy-planes, respectively.

sents a cross section of an electron cloud (or probability distribution) and shows directly (by means of degrees of shading) how the density of the electron cloud varies from point to point within a plane containing the nucleus.

The important features of the $1s$, $2s$, and $2p$ electron distributions are as follows: When the electron is in the $1s$ state, the probability of finding it is greatest right at the nucleus and drops off rapidly in all directions from the nucleus. In the $2s$ state the probability is greatest at the nucleus, drops off to zero, rises slightly, then decreases in all directions from the nucleus. In the $2p$ state the probability is zero right at the nucleus and then changes differently depending on what direction is taken in going away from the nucleus. Furthermore, there are actually *three* $2p$ orbitals; all three are equivalent and have the same general appearance but each has its direction of maximum probability at right angles to the other two. These three $2p$ orbitals are labeled $2p_x$, $2p_y$, and $2p_z$ according to the axes (x, y, or z) along which the maximum probabilities lie. When an electron is described by a $2p_x$ orbital, its probability along the x-axis begins at zero, rises to a maximum value, and then drops toward zero.

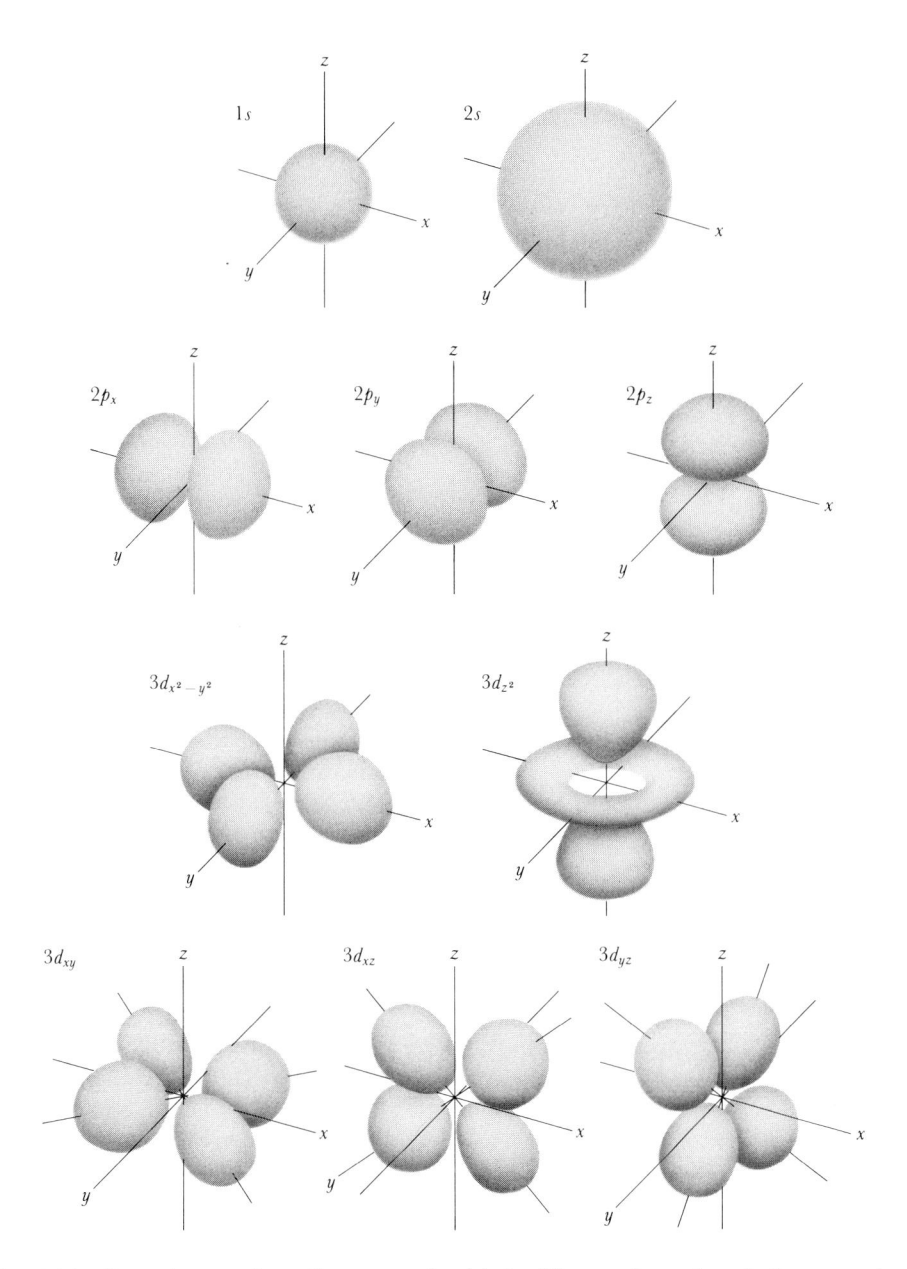

Fig. 4.11. Boundary-surface diagrams of orbitals. The surface of each figure encloses that volume of space within which the electron is most apt to be 95% of the time. This volume also contains 95% of the electron charge cloud. Each point on the surface represents the same electron density.

Along either the y- or z-axis, the $2p_x$ probability is zero. The $2p_y$ and $2p_z$ orbitals are similarly interpreted. Similarly, there are *five* different d orbitals for each value of n. In general, each orbital of quantum number l consists of $2l + 1$ orbitals, differing either in orientation, or shape, or both.

The electron probability becomes very small at fairly short distances from the nucleus (usually within 1 to 2 Å), but theoretically it goes to zero only at infinity. Consequently, chemists frequently represent orbital densities by means of **boundary diagrams**, diagrams in which orbitals are depicted as solid balls, dumbbells, etc. whose *surfaces* represent the boundaries encompassing most (say 95%) of the electron cloud, that is, that volume about the atom where the electron is most likely to be 95% of the time. The boundary surface itself represents a locus of equal electron densities or probabilities. In this representation all s orbitals are shown as spheres and $2p$ orbitals as dumbbells. Figure 4.11 shows boundary diagrams not only for s orbitals, but p and d orbitals as well. In this representation, a $2s$ orbital looks just like a $1s$ orbital but is "larger"; that is, 95% of the electron density of a $2s$ orbital involves a larger volume of space than does the corresponding quantity of a $1s$ orbital. Orbitals such as $2p$ and $3p$ differ in the same way. These latter relationships occur because the volume containing 95% of the electron density increases with the principal quantum number. Thus it may be said that n determines the size of an orbital (in the boundary diagram representation) and that l determines its shape.

4.4 THE QUANTUM THEORY OF MANY-ELECTRON ATOMS

There are formidable mathematical difficulties in applying wave mechanics rigorously to atoms other than hydrogen—even the application to the two-electron helium atom is not simple. These difficulties are due largely to the fact that scientists still have not found a simple way of describing three or more simultaneously interacting particles. In helium, not only are both electrons simultaneously attracted to the nucleus but they repel each other (due to charges of like sign) at the same time. Fortunately, simplified wave-mechanical models have been developed, and these are more than adequate to account for a large number of important atomic and molecular properties. Perhaps the simplest model is one in which electron repulsions are ignored totally at first, and then the model is modified or adjusted slightly to reflect those properties dependent on electron repulsions.

If the electrons did not interact among themselves in any way, then each electron would behave as if all the others were not there. Consequently, each electron would occupy a quantum state very much like one of the states of the lone electron of the hydrogen atom, and the total energy of the atom would be simply the sum of the orbital energies of the electrons. For the moment the **orbital energy** may be regarded as the energy associated with a quantum state of the corresponding one-electron atom. However, experience clearly shows that this simplification cannot be pushed too far. To begin with, it is necessary to limit the occupancy of each hydrogenlike state; otherwise, an atom such as uranium (92 electrons) could be described by putting *all 92 electrons* into the $1s$ state. The occupancy rule—known as the **Pauli exclusion principle**—states that no two electrons may have the same four quantum numbers. There is no simple explanation for this rule; it is another example of an unexpected feature of nature at the subatomic level. An additional rule is that the electrons are placed into the states such that the total energy of the atom—the sum of the orbital energies of the electrons plus an energy term due to repulsions among the electrons—is as low as possible. This **energy minimization principle** is not peculiar to wave mechanics; it plays an important role in many other natural phenomena.

EXAMPLE 4.7 What is the maximum number of electrons that can have $n = 4$ without violating the Pauli exclusion principle?

SOLUTION (See Example 4.6.) For $l = 0$, $m_l = 0$, m_s can be $-\tfrac{1}{2}$ or $+\tfrac{1}{2}$. This accounts for two electrons. For $l = 1$, $m_l = 0, \pm 1$, m_s can be $-\tfrac{1}{2}$ or $+\tfrac{1}{2}$ for each of the three allowed m_l values. This accounts for six more electrons. Similarly, $l = 2$, $m_l = 0, \pm 1, \pm 2$ accounts for ten electrons (twice the number of m_l values) and $l = 3$, $m_l = 0, \pm 1, \pm 2, \pm 3$ accounts for fourteen more electrons. Thus the maximum number of electrons with $n = 4$ is $2 + 6 + 10 + 14 = 32$. As a general rule, each value of n can be used for $2n^2$ electrons and each value of l can be used for $4l + 2$ electrons.

The exclusion principle was proposed by the Austrian theoretical physicist Wolfgang Pauli (1900–1958), who received the Nobel prize in physics in 1945 for discovering the principle. Pauli also contributed to the electronic theory of metals and predicted the existence of a subatomic particle called the neutrino.

To begin with, it is assumed that the electrons of an atom are assigned to orbitals in such a way as to make the sum of the orbital energies as low as possible. This means that $1s$ is used first, then $2s$, followed by $2p$, $3s$, $3p$, etc. Eventually this procedure must be modified slightly to take electron repulsions into account.

Now to apply the Pauli exclusion principle and the energy minimization principle to some elements beyond hydrogen: Helium (with two electrons) is described by placing both electrons into a $1s$ orbital, one with $m_s = +\tfrac{1}{2}$ and the other with $m_s = -\tfrac{1}{2}$. Each electron has $n = 1$, $l = 0$, and $m_l = 0$, but different m_s values in order to satisfy the exclusion principle. This description of the electronic structure of helium is symbolized by writing $1s^2$; this quantity is called the **electron configuration**. The electron configuration indicates what orbitals are being used and how many electrons are in each. Sometimes this is symbolized by a **box diagram** such as the one shown below.

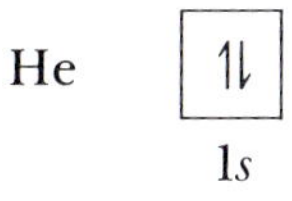

He [↑↓]
     $1s$

The box represents a $1s$ orbital and the arrows represent electrons. The two arrows point in opposite directions to symbolize that they have opposite spin quantum numbers (which is $+\tfrac{1}{2}$ and which is $-\tfrac{1}{2}$ is irrelevant).

For lithium (three electrons) two electrons can go into a $1s$ orbital but the third must go into $2s$. This is symbolized by the electron configuration $1s^2 2s$, or by the box diagram

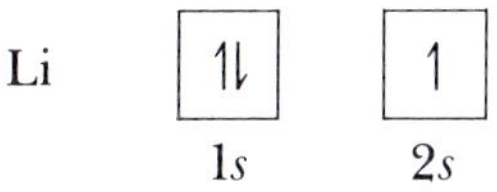

Li [↑↓] [↑]
     $1s$ $2s$

The direction of the arrow in the $2s$ box is immaterial. It should be noted that whereas in the hydrogen atom $2s$ and $2p$ orbitals have the same energy, this is no longer true in many-electron atoms. One of the effects of repulsions between electrons is to make orbital energies slightly dependent on the l quantum number. In general, the higher the l value (for given n), the higher the orbital energy.*

The next atom, beryllium (four electrons) has the electron configuration $1s^22s^2$ and the box diagram

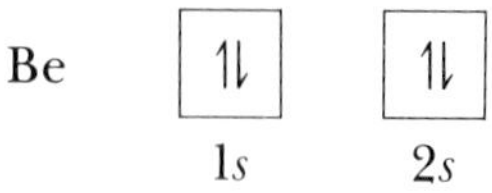

Be [↑↓] [↑↓]
     1s   2s

At this point the reader should verify that all four electrons have a different set of quantum numbers; each electron differs in *at least* one of the four quantum numbers. The set of quantum numbers for beryllium is

n	l	m_l	m_s
1	0	0	$+\frac{1}{2}$
1	0	0	$-\frac{1}{2}$
2	0	0	$+\frac{1}{2}$
2	0	0	$-\frac{1}{2}$

The boron atom (five electrons) has the electron configuration $1s^22s^22p$. The box diagram for this is

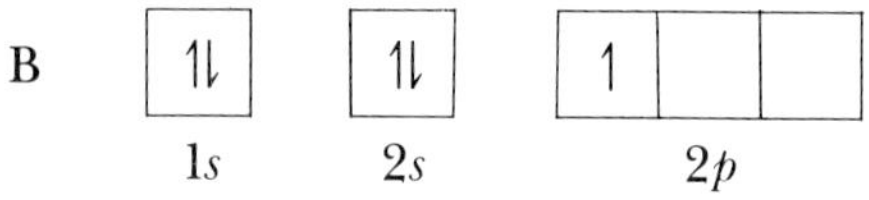

B [↑↓] [↑↓] [↑| |]
   1s   2s   2p

Note that the three $2p$ orbitals are represented as a group of three contiguous boxes. The carbon atom (six electrons) has the electron configuration $1s^22s^22p^2$. The box diagram is

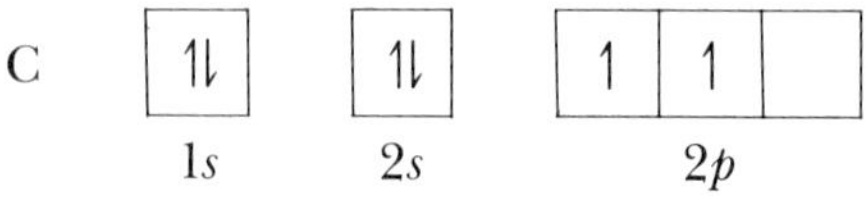

C [↑↓] [↑↓] [↑|↑|]
   1s   2s   2p

Here another unexpected feature comes in; when two or more electrons are placed into orbitals having the same n and l values, their spin quantum numbers must be the same as much as possible (without violating the exclusion principle). This is known as **Hund's rule**.† This rule is a consequence of the fact that two or more electrons with the same n and l quantum numbers repel each other less if they are in different regions of space, that is, in different orbitals. Thus the next atom,

---

* A rough idea of why the $2s$ orbital energy is slightly less than the $2p$ orbital energy can be gleaned from studying Figs. 4.9 and 4.10. An electron in a $2s$ orbital penetrates closer to the nucleus than does an electron in a $2p$ orbital and consequently is more strongly attracted by the positive nucleus. This effect occurs only when the atom has more than one electron.

† Named after Friedrich Hund (b. 1896), another German physicist who pioneered in the development of modern quantum theory.

nitrogen (seven electrons), is $1s^2 2s^2 2p^3$, with the box diagram

N [↑↓] [↑↓] [↑|↑|↑]
   1s   2s   2p

Note that the three electrons in $2p$ have the same n, l, and m_s quantum numbers but different m_l quantum numbers. There is no reason to say which m_l quantum number (-1, 0, or 1) belongs to which electron as long as each electron has a different value.

The next atom, oxygen (eight electrons), is $1s^2 2s^2 2p^4$, with the box diagram

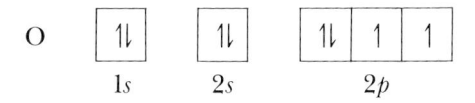

O [↑↓] [↑↓] [↑↓|↑|↑]
   1s   2s   2p

Here the eighth electron is put into a $2p$ orbital already holding one electron. The Pauli exclusion principle is not violated, however, since the two electrons in the same $2p$ orbital have different m_s values.

The two remaining atoms in the first full period are fluorine (nine electrons) and neon (ten electrons). Their electron configurations are $1s^2 2s^2 2p^5$ and $1s^2 2s^2 2p^6$, respectively. The box diagrams are

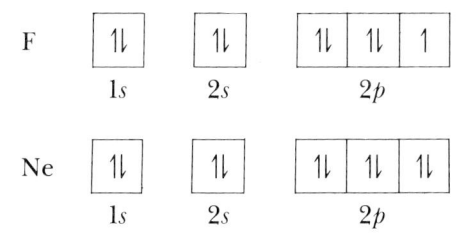

F   [↑↓] [↑↓] [↑↓|↑↓|↑]
    1s   2s   2p

Ne [↑↓] [↑↓] [↑↓|↑↓|↑↓]
    1s   2s   2p

Figure 4.12 shows contour diagrams of the orbital electron densities and the total electron densities of the first six elements. The total electron density is simply a superposition of the individual orbital densities. In other words, orbital densities are but convenient subunits of the total electron distribution of an atom. The electron densities in Fig. 4.12 were calculated with the aid of a high-speed digital computer using the best possible orbital description of each atom. The diagram also lists orbital

---

Fig. 4.12 (following pages). Atomic electronic structure of the first six elements illustrated by contour diagrams of orbital densities. As opposed to topographic maps (see Appendix 5), where contour intervals are constant, each adjacent pair of contour lines in the above represents a tenfold change in electron density (the charges decrease outward from the "center" of each diagram). Note that as the nuclear charge (atomic number) increases, the electron density represented by a given orbital tends to become more concentrated ("pulled in" more closely) about the nucleus. The total electron densities exhibit a similar trend. Whenever two or more $2p$ orbitals are required in the superposition, each has the same general "size" and "shape" but the directional orientations will differ. For example, in carbon the two $2p$ orbitals have their axes at 90° to each other.

Atom	1s	2s	2p	Total orbital superposition	Energies (1s, 2s, 2p, total) (eV)
H					−13.6
He					−25.0 −77.8
Li					−67.5 −5.4 −196.9

116 THE ELECTRONIC STRUCTURE OF ATOMS

−128.7
−8.4
−396.3

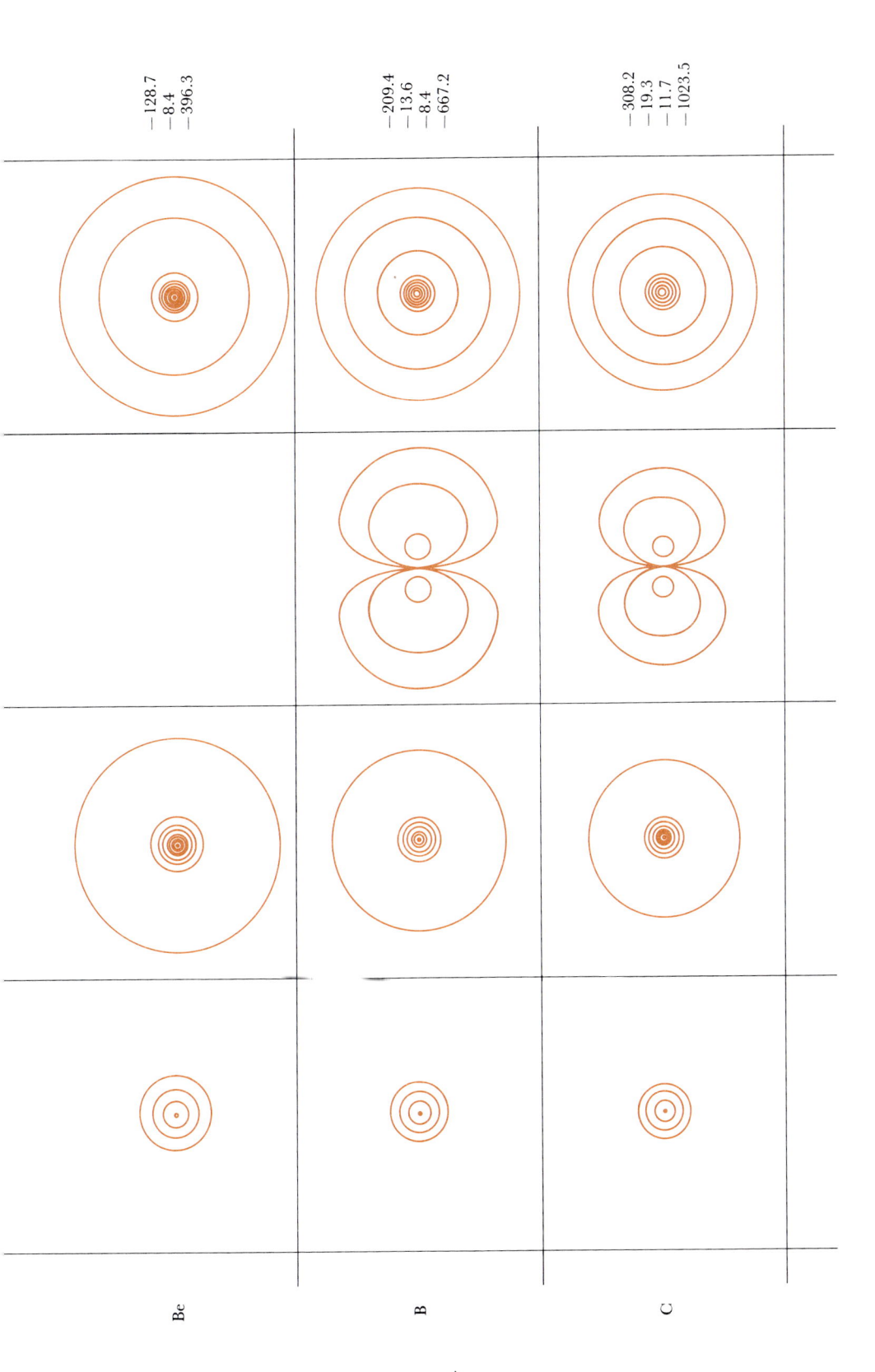

−209.4
−13.6
−8.4
−667.2

−308.2
−19.3
−11.7
−1023.5

Be

B

C

4.4 | THE QUANTUM THEORY OF MANY-ELECTRON ATOMS 117

energies and total atomic energies as calculated by this model. The observant person who has watched this buildup process developing may have noticed that it is *periodic* in nature; in fact, the periodicity of electron configurations closely parallels the experimentally observed periodic properties of atoms as shown in the periodic table. *Thus the periodic table reflects the quantum rules that govern the internal arrangement of electrons in atoms.* Let us now examine the periodicity of electron configurations in more detail.

The first element in the next period, sodium (11 electrons) has the electron configuration $1s^2 2s^2 2p^6 3s$. The box diagram is

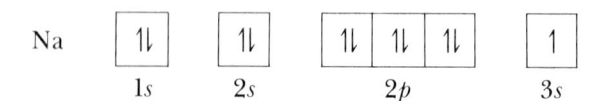

Na [↑↓] [↑↓] [↑↓|↑↓|↑↓] [↑]
    1s   2s   2p   3s

The next seven elements (12 through 18 electrons) have electron configurations and box diagrams as follows.

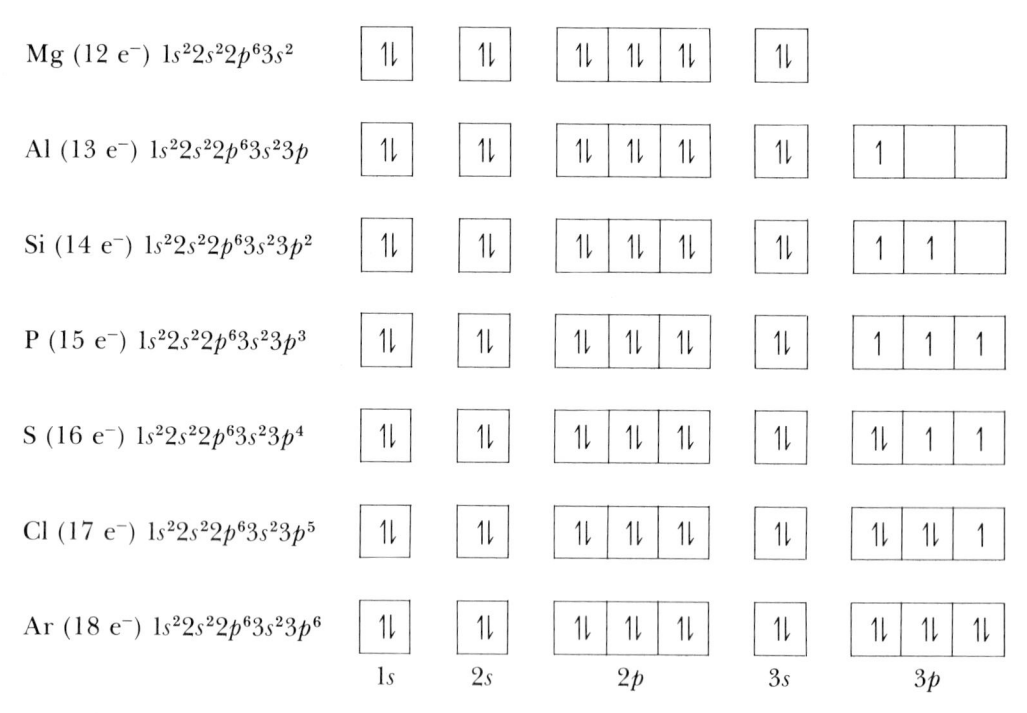

Mg (12 e⁻) $1s^2 2s^2 2p^6 3s^2$ [↑↓] [↑↓] [↑↓|↑↓|↑↓] [↑↓]

Al (13 e⁻) $1s^2 2s^2 2p^6 3s^2 3p$ [↑↓] [↑↓] [↑↓|↑↓|↑↓] [↑↓] [↑| |]

Si (14 e⁻) $1s^2 2s^2 2p^6 3s^2 3p^2$ [↑↓] [↑↓] [↑↓|↑↓|↑↓] [↑↓] [↑|↑|]

P (15 e⁻) $1s^2 2s^2 2p^6 3s^2 3p^3$ [↑↓] [↑↓] [↑↓|↑↓|↑↓] [↑↓] [↑|↑|↑]

S (16 e⁻) $1s^2 2s^2 2p^6 3s^2 3p^4$ [↑↓] [↑↓] [↑↓|↑↓|↑↓] [↑↓] [↑↓|↑|↑]

Cl (17 e⁻) $1s^2 2s^2 2p^6 3s^2 3p^5$ [↑↓] [↑↓] [↑↓|↑↓|↑↓] [↑↓] [↑↓|↑↓|↑]

Ar (18 e⁻) $1s^2 2s^2 2p^6 3s^2 3p^6$ [↑↓] [↑↓] [↑↓|↑↓|↑↓] [↑↓] [↑↓|↑↓|↑↓]
                                          1s   2s   2p   3s   3p

The next element, potassium (19 electrons), might be expected to be $1s^2 2s^2 2p^6 3s^2 3p^6 3d$, since the 3d orbitals come next in energy.* In actuality, as borne out both by calculation and by experiment, the nineteenth electron goes into a 4s

---

* Virtually all textbooks make the incorrect statement that the 3d orbital is not used next since it is higher in energy than is the 4s orbital. Both theory and experiment show this is not true, at least not for elements with atomic number 21 and beyond. See F. L. Pilar, *J. Chem. Educ.*, **55**, 2 (1978).

orbital. Thus potassium is $1s^2 2s^2 2p^6 3s^2 3p^6 4s$ and has the box diagram

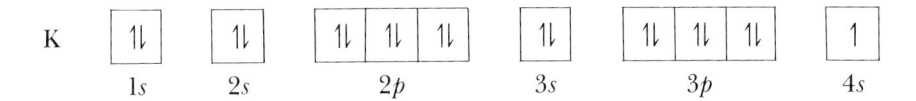

K   [↑↓] [↑↓] [↑↓][↑↓][↑↓] [↑↓] [↑↓][↑↓][↑↓] [↑]
    1s   2s   2p   3s   3p   4s

Note that if this were not so, potassium would not have the same type of electron configuration as the other elements of group IA (hydrogen, lithium, and sodium), all of which have the odd electron in an *s*-type orbital. The next two elements, calcium (20 electrons) and scandium (21 electrons), have the electron configurations $1s^2 2s^2 2p^6 3s^2 3p^6 4s^2$ and $1s^2 2s^2 2p^6 3s^2 3p^6 4s^2 3d$, respectively. The box diagrams are

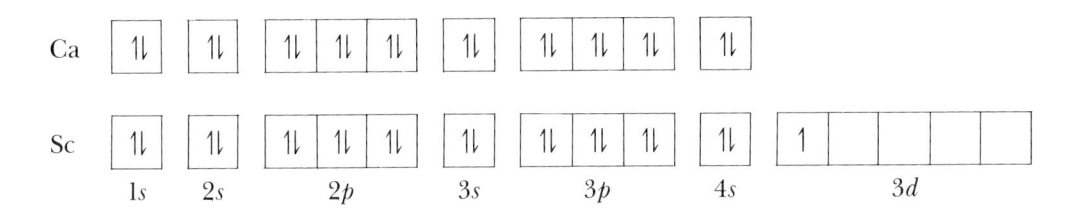

Ca [↑↓] [↑↓] [↑↓][↑↓][↑↓] [↑↓] [↑↓][↑↓][↑↓] [↑↓]

Sc [↑↓] [↑↓] [↑↓][↑↓][↑↓] [↑↓] [↑↓][↑↓][↑↓] [↑↓] [↑][ ][ ][ ][ ]
    1s   2s   2p   3s   3p   4s   3d

The apparent reversal of 3d and 4s has a very simple explanation. When the wave-mechanical treatment of atoms is carried out rigorously, the orbital energy is defined *not* as the energy of a one-electron quantum state, but as the negative of the minimum energy required to remove an electron from an atom. For example, the energy of the 1s orbital of carbon is the energy required to carry out the process

C: $1s^2 2s^2 2p^2 \rightarrow$ C$^+$: $1s 2s^2 2p^2 + e^-$

Similarly, the 2p orbital energy is measured from

C: $1s^2 2s^2 2p^2 \rightarrow$ C$^+$: $1s^2 2s^2 2p + e^-$

Experiments such as these (as well as theoretical calculations) clearly show that the energy of a 4s orbital is *always* higher than the energy of a 3d orbital.* How, then, can the energy minimization process be satisfied if 4s is used before 3d?

The answer resides in the fact that the correct energy of the atom is not simply a sum of the orbital energies of the electrons, as might at first be expected. Such a simple result would be true only if all electron repulsions could be totally ignored. As mentioned earlier, the correct energy of the atom has the form

$$E_{atom} = F - G \tag{4.19}$$

---

* The measurement is carried out by bombarding a sample of the element with high-energy electrons. These electrons remove an electron from the atom (say a 1s electron of carbon); then a 2p electron drops down to fill this "hole" and emits radiation (X-rays) in so doing. The wavelength of the emitted radiation corresponds to the energy difference between the 1s and 2p orbitals. The energy of the 2p orbital is obtained from the first ionization energy of carbon, thereby allowing the energy of the 1s electron to be calculated.

where F is the sum of the orbital energies (always a *negative* quantity) and G is the total electron repulsion energy (always a *positive* quantity). A rigorous examination of the orbital model of atoms reveals the rather unexpected result that orbital energies themselves not only include electron repulsion energies but include them *twice*, that is, *once too often*! This is why the total repulsion energy is *subtracted* from (not added to) the orbital energy sum; in this way the extra repulsion energy is removed. It now becomes easier to see how an atom can have a lower energy with an electron in the higher 4s orbital than in the lower 3d orbital. What happens is that an electron in a 4s orbital penetrates the total electron charge more than does an electron in a 3d orbital. This makes the total electron repulsion greater in the 4s case than in the 3d case—enough greater to more than compensate for the orbital energy difference. Consequently, even though F is less when 3d is used in place of 4s, the opposite is true for $F - G$.*

Some deviations from the simple order occur again in the higher-energy orbitals and for the same reason as above. The order in which orbitals must be filled so as to minimize their total energies and to match the experimental periodicity of atoms is given by the simple diagram below.

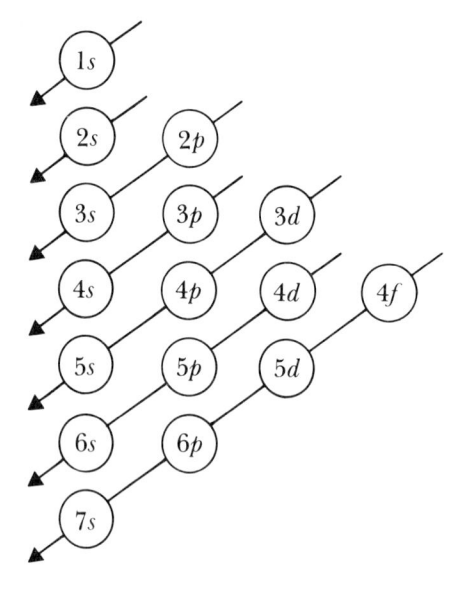

This diagram is based on a very simple pattern which should become apparent on inspection. Students should be able to reconstruct this diagram from memory; with it they have most of the periodic table at their fingertips.

Taken together, the three rules (Pauli exclusion principle, energy minimization, and Hund's rule) that are used to predict electron configurations of atoms constitute what is known as the **aufbau principle**. *Aufbau* is a German word meaning "buildup," or "synthesis."

---

* An analogy would be moving to a more expensive apartment and yet reducing your total living expenses. This might happen if you move to within walking distance of work and no longer have to drive an automobile each morning.

4.5 QUANTUM THEORY OF THE PERIODIC TABLE

Similarities among electronic configurations of different atoms become more obvious if the electrons are divided into two sets.

SET 1: All the electrons assigned to orbitals or groups of orbitals which are totally filled, that is, all those electrons found in filled boxes or filled groups of boxes of a given n quantum number. This part of the electron configuration always corresponds to the electron configuration of one of the inert gases (group VIIIA) such as He: $1s^2$; Ne: $1s^22s^22p^6$; or Ar: $1s^22s^22p^63s^23p^6$. Accordingly, it is often called the **inert-gas core**. For example, in Cl: $1s^22s^22p^63s^23p^5$, there are no more orbitals with n quantum numbers 1 or 2 which can be used, and thus $1s^22s^22p^6$ is the inert-gas core.

SET 2: All the remaining electrons, that is, those in addition to the inert-gas core. This set of electrons is often called the **valence shell**. The $3s^23p^5$ electrons of chlorine belong to this set.

A simplified electron configuration notation arises by replacing each inert-gas core with the corresponding inert-gas symbol placed in parentheses; for example, Li: $1s^22s$ becomes Li: (He)$2s$; C: $1s^22s^22p^2$ becomes C: (He)$2s^22p^2$; and Cl: $1s^22s^22p^63s^23p^5$ becomes Cl: (Ne)$3s^23p^5$. The electron configurations of the first two periods in this new notation are as follows.

Li:	(He)$2s$	Na:	(Ne)$3s$
Be:	(He)$2s^2$	Mg:	(Ne)$3s^2$
B:	(He)$2s^22p$	Al:	(Ne)$3s^23p$
C:	(He)$2s^22p^2$	Si:	(Ne)$3s^23p^2$
N:	(He)$2s^22p^3$	P:	(Ne)$3s^23p^3$
O:	(He)$2s^22p^4$	S:	(Ne)$3s^23p^4$
F:	(He)$2s^22p^5$	Cl:	(Ne)$3s^23p^5$
Ne:	(He)$2s^22p^6$	Ar:	(Ne)$3s^23p^6$

Note that except for an increase in the n value by one unit, the valence shells of the electron configurations parallel each other in the two periods. In other words, electron configurations vary periodically with atomic number just as the chemical and physical properties do. The similar chemical and physical properties of a pair of elements such as lithium and sodium correlate with their similar electron configurations: both have one valence-shell electron in an s orbital. Similarly, the halogens fluorine and chlorine both have seven valence-shell electrons of ns^2np^5 type.

The third period of the periodic table illustrates a slightly different aufbau trend. The electron configurations of elements 19 through 36 are as follows.

K:	(Ar)$4s$	Mn:	(Ar)$4s^23d^5$	Ga:	(Ar)$4s^23d^{10}4p$
Ca:	(Ar)$4s^2$	Fe:	(Ar)$4s^23d^6$	Ge:	(Ar)$4s^23d^{10}4p^2$
Sc:	(Ar)$4s^23d$	Co:	(Ar)$4s^23d^7$	As:	(Ar)$4s^23d^{10}4p^3$
Ti:	(Ar)$4s^23d^2$	Ni:	(Ar)$4s^23d^8$	Se:	(Ar)$4s^23d^{10}4p^4$
V:	(Ar)$4s^23d^3$	Cu:	(Ar)$4s3d^{10}$ (exception)	Br:	(Ar)$4s^23d^{10}4p^5$
Cr:	(Ar)$4s3d^5$ (exception)	Zn:	(Ar)$4s^23d^{10}$	Kr:	(Ar)$4s^23d^{10}4p^6$

Note the exceptions, copper and chromium. Straightforward application of the aufbau order predicts $(Ar)4s^2 3d^4$ for chromium and $(Ar)4s^2 3d^9$ for copper. Theoretical and experimental evidence clearly shows that there is an extra stability imparted by using exactly half-filled $3d$ orbitals ($3d^5$) and exactly filled $3d$ orbitals ($3d^{10}$) along with a single electron in $4s$.*

Similarity to the two previous periods becomes more apparent if the $3d^{10}$ portion of the electron configurations of the elements from gallium through krypton is treated as part of an inert-gas core; the valence shells of the elements potassium, calcium, gallium, germanium, arsenic, selenium, bromine, and krypton are then analogous to those of the preceding two periods. The omitted elements, scandium through zinc, are called **transition elements** and require special study. These are discussed again in Chapter 22.

Below are listed the general electron configurations of groups IA to VIIIA as they apply to the first 36 elements. For convenience the symbol (IGC) is used to designate the usual inert-gas core. (IGC') is used to designate IGC + $3d^{10}$.

GROUP	ELECTRON CONFIGURATION	
IA	(IGC)ns	
IIA	(IGC)ns^2	
IIIA	(IGC)$ns^2 np$	or (IGC')$ns^2 np$
IVA	(IGC)$ns^2 np^2$	or (IGC')$ns^2 np^2$
VA	(IGC)$ns^2 np^3$	or (IGC')$ns^2 np^3$
VIA	(IGC)$ns^2 np^4$	or (IGC')$ns^2 np^4$
VIIA	(IGC)$ns^2 np^5$	or (IGC')$ns^2 np^5$
VIIIA	(IGC)$ns^2 np^6$	or (IGC')$ns^2 np^6$

Note that the entire electron configuration of a group VIIIA element is an inert-gas core.

The valence-shell portion of the electron configuration is the most relevant with respect to the correlation, interpretation, and prediction of the chemical properties of the elements. The maximum oxidation numbers of elements in groups IA to VIIA are equal to the number of valence-shell electrons: +1 for alkali metals, +2 for the alkaline earths, +3 for group IIIA, and so on to +7 for the halogens. Groups IA, IIA, and IIIA generally exhibit only the maximum oxidation numbers, but the remaining groups may exhibit oxidation numbers ranging from the maximum to a minimum value; the latter is generally given by the group number minus eight. For example, the halogens exhibit minimum oxidation numbers of $(7 - 8)$ or -1. All of the oxidation numbers between -1 and $+7$ may actually be encountered, although not always in very stable compounds.

Elements toward the left-hand side of the periodic table (the metals) tend to have positive oxidation states only. Elements toward the right-hand side of the periodic table (the nonmetals) exhibit both negative and positive oxidation numbers. A chemical reaction between a metal such as Na: $(Ne)3s$ and a nonmetal such as Cl: $(Ne)3s^2 3p^5$ may be viewed as the transfer of an electron from sodium to chlorine.

---

* The simple explanations offered for this are incorrect, and all the correct explanations are not simple. The effect arises from details of how an electron correlates its motion with that of all the other electrons.

This may be symbolized by

$$\text{Na: (Ne)}3s + \text{Cl: (Ne)}3s^2 3p^5 \rightarrow \text{Na}^+\text{: (Ne)} + \text{Cl}^-\text{: (Ar)}$$

Note that the reaction product, NaCl, consists of the ions Na$^+$ and Cl$^-$, both of which have inert gas configurations. Many other chemical reactions, particularly those leading to ionic compounds, can be rationalized on a similar basis. One of the basic laws of chemical reactions is that elements often tend to gain or lose electrons in such a way as to attain, insofar as possible, the electron configuration of the nearest group VIIIA element. Thus, through electron configurations, quantum theory provides a model for the law of definite proportions.

Electron configurations also provide a rationale for the variation of **atomic volume** both within a group and within a period. Recall Fig. 2.5, which shows how atomic volume varies periodically with atomic number. Although the volume of a single atom is not a uniquely defined quantity, it may be calculated as follows: The density of aluminum at 25 °C is 2.7 g·cm^{-3}. Thus the *apparent* volume of a single aluminum atom is given by

$$\frac{1}{6.02 \times 10^{23} \text{ atoms·mol}^{-1}} \left(\frac{26.98 \text{ g·mol}^{-1}}{2.7 \text{ g·cm}^{-3}} \right) = 1.7 \times 10^{-23} \text{ cm}^3$$

If it is assumed that the inert-gas electrons form a spherical shield about the nucleus and that repulsions among valence-shell electrons may be ignored, then quantum theory suggests that the apparent radius of an atom ought to be directly proportional to the square of the principal quantum number of the valence-shell electrons and inversely proportional to the effective nuclear charge. This follows at once from the Bohr formula (4.10). The effective nuclear charge is roughly the actual nuclear charge (atomic number) less the number of inert-gas core elctrons which shield the nucleus; that is, it equals the number of valence-shell electrons. This relationship may be written

$$r \propto \frac{n^2}{Z_{\text{eff}}} \tag{4.20}$$

Since n (the principal quantum number of the valence-shell electrons) tends to stay the same from left to right across a period while Z_{eff} increases by one unit from atom to atom, it follows that r decreases and so does the atomic volume (which is roughly $\frac{4}{3}\pi r^3$). Similarly, when going down a group, Z_{eff} remains the same while n increases by one unit from atom to atom. Thus r and the atomic volume increase when proceeding down a group. Figure 2.5 shows that this behavior is in accord with experiment.

The aufbau model also rationalizes the variation of first ionization energies within the periodic table (see Fig. 4.13). Several different energies (corresponding to the negatives of orbital energies) are associated with the removal of unit negative charge (1.602×10^{-19} C) from an atom to form a positive ion of unit charge. This process, called **ionization**, is symbolized by

$$M \rightarrow M^+ + e^-$$

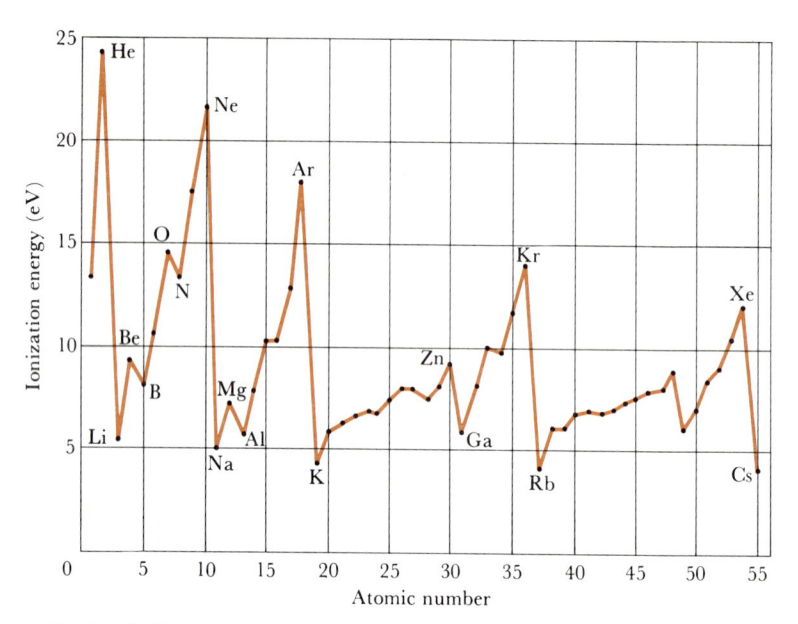

Fig. 4.13. Periodic variation of ionization energies with atomic number. Only the first 55 elements (hydrogen through cesium) are shown. Note that the inert gases form the peaks and the alkali metals form the bottoms of the valleys.

The minimum energy required to remove an electron from the ground state of an atom M is called the **first ionization energy** of the element. The minimum energy required to remove unit negative charge from the ground state of the singly charged ion M$^+$ to form the doubly charged ion M^{2+}, that is,

$$M^+ \rightarrow M^{2+} + e^-$$

is called the **second ionization energy** of the element. Successive ionization energies (third, fourth, and so on) are similarly defined down to formation of the bare nucleus.* Consider the lithium atom as an example; it is found that the process

$$Li \rightarrow Li^+ + e^-$$

requires either 5.39 or 60.11 eV. The lower of these is interpreted as the energy needed to remove an electron from the valence shell, that is, from a 2s orbital, to produce a lithium ion in its lowest quantum state. This is the first ionization energy of lithium. The 60.11 eV value is interpreted as the energy needed to remove an electron from the inert-gas core of lithium, that is, from a 1s orbital. This produces a lithium ion in an excited quantum state. The 1s and 2s orbital energies quoted in Fig. 4.12, −5.4 and −67.5 eV, respectively, correlate well with the experimental ionization energies.

---

* The negative of the sum of all the ionization energies is the total energy of the ground state of the atom.

The above shows that the first ionization energy of an atom is roughly equal to the negative of the orbital energy of the highest-energy orbital of the atom. The Bohr formula (4.8) and the wave-mechanical formula (4.18) both suggest that the energy of the highest occupied orbital should be given approximately by

$$\epsilon \propto \frac{Z_{eff}^2}{n^2}, \qquad (4.21)$$

where, as before, Z_{eff} is the effective nuclear charge and n is the principal quantum number of the highest-energy orbital. In moving from left to right across a period, n tends to stay the same but Z_{eff} increases by one unit with each successive atom. Thus the first ionization energy increases across a period from left to right. In going down a group, Z_{eff} stays the same but n increases by one unit with each successive atom. Thus the first ionization energy decreases in going down a group.

TABLE 4.2 FIRST IONIZATION ENERGIES OF THE ELEMENTS OF THE FIRST FULL PERIOD

ELEMENT	FIRST IONIZATION ENERGY (eV)
Lithium	5.4
Beryllium	9.4
Boron	8.3
Carbon	11.3
Nitrogen	14.1
Oxygen	13.7
Fluorine	17.4
Neon	21.5

In Table 4.2 are listed the first ionization energies of the elements of the first full period. The trend predicted in the previous paragraph is apparent but there are some discrepancies in the details. For example, in going from lithium to beryllium, the first ionization energy increases as predicted but then *drops* by 1.1 eV at boron before increasing again. Similarly, there is a drop of 0.4 eV in going from nitrogen to oxygen. To account for this behavior, an additional rule is often included in the aufbau principle. This rule states that electron configurations in which the valence shells are ns^2, ns^2np^3, ns^2np^6, $ns(n-1)d^5$, $ns^2(n-1)d^5$, $ns(n-1)d^{10}$, or $ns^2(n-1)d^{10}$ (and some other similar configurations) usually are associated with slightly enhanced stabilities. Such valence-shell electron configurations are said to contain **half-filled shells** [np^3 and $(n-1)d^5$] or **filled shells** [ns^2, np^6 or $(n-1)d^{10}$]. Thus it is easier to remove a $2p$ electron from B: $1s^22s^22p$ than to remove a $2s$ electron from Be: $1s^22s^2$, since the latter involves breaking up the filled shell $2s^2$. Similarly, nitrogen has the half-filled shell $2p^3$ whereas oxygen has $2p^4$. This rule also accounts for the sharp drop in ionization energy after each inert gas. This extra stability of half-filled and filled shells also accounts for the electron configuration exceptions of chromium and copper, that is, why $4s3d^5$ and $4s3d^{10}$ are preferred over $4s^23d^4$ and $4s^23d^9$, respectively. In these cases, stability of half-filled and filled $3d$ shells is more important than filled $4s$ shells.

EXAMPLE 4.8 The elements Ca: (Ar)4s² and Zn: (Ar)3d¹⁰4s² have similar electron configurations if the $3d^{10}$ part of zinc is considered part of an inert-gas core. Consequently Z_{eff} should be almost the same (+2) for both elements and both should have almost the same ionization energy. How do you account for the fact that the ionization energies of calcium and zinc are 6.09 and 9.36 eV, respectively?

SOLUTION d electrons are not as effective in shielding the nucleus as are inert-gas core electrons. Consequently, Z_{eff} of zinc is actually greater than +2. Since the ionization energy is directly proportional to Z_{eff}^2, this accounts for the higher zinc value.

The electron configuration of a positive ion (cation) is deduced from that of the neutral atom by removing the electron with the highest orbital energy, that is, the electron with the highest n quantum number and the highest associated l quantum number. Thus, since the carbon atom C is $1s^2 2s^2 2p^2$, the ion C⁺ will be $1s^2 2s^2 2p$. Furthermore, the negative of the energy of the $2p$ orbital provides an approximation to the first ionization energy of carbon. Figure 4.12 shows that the $2p$ orbital of carbon has an energy of -11.7 eV so that the first ionization energy is predicted to be 11.7 eV. The actual measured value is 11.3 eV. Similarly, the ion Sc⁺ should have the electron configuration (Ar)3d4s. To predict the electron configuration of a negative ion (anion) an electron is added to the lowest-energy orbital available. Thus the electron configuration of F⁻ is predicted to be $1s^2 2s^2 2p^6$ [or (Ar)] since that of fluorine is $1s^2 2s^2 2p^5$.

It is important to note that the removal of electrons to form an ion is not the reverse of the buildup of atoms by adding electrons. The former situation involves no change in the nuclear charge as electrons are removed, whereas the latter involves an increase of nuclear charge as electrons are added. Thus there is no reason to expect, as many students do, that the order in which electrons are most easily removed by ionization should be the order of filling the orbitals as given by the aufbau principle.

With the background just presented, it is possible to explain how Moseley (see Sections 1.5 and 2.5) was led to recognition of the atomic number of an atom as one of its most important fundamental properties. Moseley did his work prior to 1913 (before even the Bohr theory was proposed), but we shall rephrase his experiments in terms of modern quantum theory.

When an atom such as sodium is bombarded with a high-energy beam of electrons, the bombarding electrons can knock out an electron from the atom to form a positive ion. One possible process might be

Na: $1s^2 2s^2 2p^6 3s$ → Na⁺: $1s 2s^2 2p^6 3s$

where an electron is removed from the $1s$ orbital of sodium to form a sodium ion in an excited quantum state. The excited state of the sodium ion can emit an X-ray photon and drop to a lower-energy quantum state. The transition to the lowest possible quantum state is symbolized by

Na⁺: $1s 2s^2 2p^6 3s$ → Na⁺: $1s^2 2s^2 2p^6$

where an electron from the $3s$ orbital has replaced the vacancy in the $1s$ orbital. The energy of the emitted X-ray photon is given by

$$h\nu = \epsilon_{3s} - \epsilon_{1s}$$

where ϵ_{3s} and ϵ_{1s} are the energies of the 3s and 1s orbitals of sodium, respectively. What Moseley did was to measure the shortest wavelength X-ray of a series of elements (this corresponds to an electron dropping from the highest occupied orbital to the 1s orbital in each case.) He then showed that a plot of $\sqrt{\nu}$ (the square root of the X-ray frequency) against atomic number produced an almost perfect straight line. A similar plot of $\sqrt{\nu}$ against atomic mass led to a rather jagged line. Hence Moseley concluded—on somewhat meager grounds perhaps—that the atomic number is a more fundamental property than the atomic mass. Modern quantum theory shows that the energy of the X-ray photon ($h\nu$) is very nearly proportional to the square of the atomic number (Z^2), in excellent accord with Moseley's experiments.

4.6 MAGNETIC PROPERTIES OF ATOMS

Hund's rule, mentioned in the preceding section, makes it possible to predict some of the magnetic properties of atoms. Experiment shows that when a sample of a substance is placed between the poles of a magnet, the sample is either repelled by the magnetic field or attracted by it. Substances repelled by the magnetic field are called **diamagnetic**; those attracted by the magnetic field are called **paramagnetic**. These two types of behavior are illustrated in Fig. 4.14.

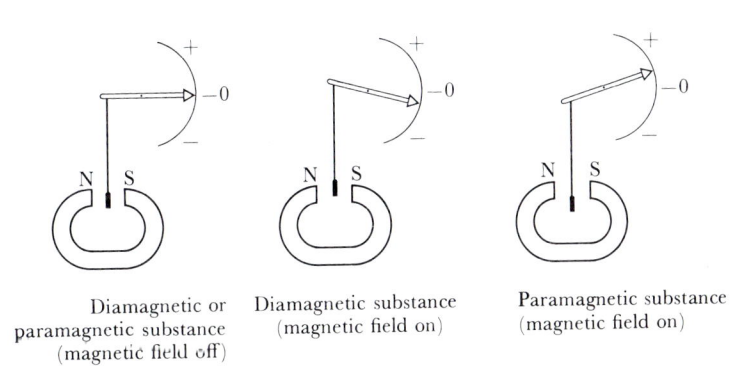

Fig. 4.14. Behavior of paramagnetic and diamagnetic materials in magnetic fields. Compared with its weight when no magnetic field is present (when the magnet is turned off), a diamagnetic substance weighs less in the presence of a magnetic field and a paramagnetic substance weighs more. This occurs because a diamagnetic substance is "pushed" out of the interpole region where the magnetic field is most intense, and the paramagnetic substance is "pulled" into this region.

An electron itself behaves as if it were a tiny magnet; in fact, electrons are paramagnetic. However, when two electrons in the same atom have opposite spin quantum numbers, the magnetic moments cancel and the electron pair behaves diamagnetically. Thus, if an atom has the same number of electrons of one spin (↑)

as it has of the other spin (↓), the element will be diamagnetic. The atom is said to have a *net* electronic spin of zero. If the number of ↑ spins is different from the number of ↓ spins, the element is paramagnetic; its net electronic spin is no longer zero.

It is left to the reader to go back to the electron configurations shown and to predict which elements are diamagnetic and which are paramagnetic. Note that all the group IA elements (hydrogen, lithium, sodium, potassium, etc.) are paramagnetic (there is one more electron of one spin than of the other) and that all the group IIA elements (magnesium, calcium, etc.) are diamagnetic (there are equal numbers of ↑ and ↓ spins). Transition elements having $3d^1$ to $3d^9$ in their electron configurations are also paramagnetic.

EXAMPLE 4.9 Show that iron, cobalt, and nickel are paramagnetic and that all the group IIA metals are diamagnetic.

SOLUTION In looking at the electron configurations of iron, cobalt, and nickel,

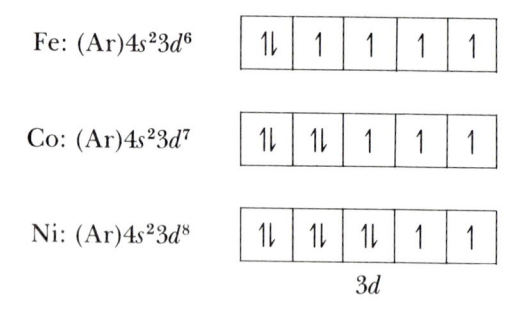

Fe: $(Ar)4s^2 3d^6$ | ↑↓ | ↑ | ↑ | ↑ | ↑ |

Co: $(Ar)4s^2 3d^7$ | ↑↓ | ↑↓ | ↑ | ↑ | ↑ |

Ni: $(Ar)4s^2 3d^8$ | ↑↓ | ↑↓ | ↑↓ | ↑ | ↑ |
$3d$

it is seen that the number of excess electrons with ↑ spin is 4, 3, and 2, respectively. Thus all three atoms are paramagnetic. The group IIA metals have the valence shell configuration ns^2

| ↑↓ |

and have equal numbers of ↑ and ↓ spins; thus they are all diamagnetic.

SUMMARY

The energies of atoms are quantized; that is, an atom can have only certain definite energy states and no others. Transitions between these energy states are accompanied by either the absorption or the emission of radiation. In either case, the difference in energy between the two states must equal Planck's constant h multiplied by the frequency ν of the absorbed or emitted radiation.

Modern quantum theory treats an electron as if it were some composite of a wave and a particle. Consequently, the electrons in an atom are often viewed as clouds of negative charge whose densities vary from point to point in the space about the nucleus. In the case of the hydrogen atom, the density of the electron charge cloud can be predicted quite accurately for each of the allowed quantum states (energy levels) of the atom. The mathematical function describing the density of the charge cloud is called an orbital. Electron configurations of atoms are based on a model in which the electrons are assumed to have spatial distributions similar to those of the energy states of the hydrogen atom. Use of the Pauli exclusion principle and the energy minimization principle leads to electronic structures of the elements which are in accord with their known physical and chemical properties.

The periodic properties of the elements are strikingly revealed in their electron configurations; elements of the same group have very similar electron configurations. Specifically, each atom in a group has the same general type of valence-shell electron configuration. Also, trends in physical properties such as atomic volume and ionization energy are simply rationalized on the basis of electron configurations. Electron configurations, along with Hund's rule, also account for the paramagnetic and diamagnetic natures of the elements.

LEARNING GOALS

1. Know the relationship between wavelength and frequency and be able to use this in calculations. Be able to define wavelength, frequency, and wave velocity in general.

2. Be able to state what wave–particle duality is and how it is expressed mathematically.

3. Know what is meant by quantization of energy.

4. Know what the uncertainty principle is and what it means.

5. Be able to describe the photoelectric effect and to calculate the maximum wavelength of light needed to eject an electron from a given metal.

6. Know what is meant by the spectrum of an element.

7. Know how to derive the Balmer-Rydberg-Ritz formula from Bohr's equation for the quantum states of a one-electron atom.

8. Know what is meant by the ionization energy of an atom and how it is measured and calculated.

9. Be able to define an orbital and tell what type of information it conveys.

10. Know the four quantum numbers used to identify orbitals and know what allowed values each has.

11. Be familiar with the general features of s and p orbitals; know how hydrogen-atom orbitals differ from orbitals of many-electron atoms.

12. Be able to sketch boundary surface diagrams of s and p orbitals.

13. Be able to state the Pauli exclusion principle and to use it to determine how many electrons can have given quantum numbers.

14. Be able to write from memory the full and abbreviated electron configurations of the first 36 elements.

15. Know Hund's rule and be able to use it to draw orbital box diagrams.

16. Be able to predict electron configurations for ions of the elements.

17. Know how atomic volumes and ionization energies vary periodically and why.

18. Be able to look at the periodic table and write the valence–shell electron configuration for any group.

19. Know what paramagnetic and diamagnetic behaviors are and be able to predict these from electron configurations.

20. Be able to predict oxidation numbers of the elements using electron configurations.

DEFINITIONS, TERMS, AND CONCEPTS TO KNOW

quantization
wave–particle duality
uncertainty principle
orbital
Pauli exclusion principle
energy state
wavelength
angstrom unit (Å)
frequency
photon

spectral lines
energy transition
quantum numbers
electron distribution
electron density
atomic volume
aufbau principle
electron configuration
ionization energy
orbital energy

inert-gas core
valence-shell electrons
transition element
half-filled shell
filled shell
Hund's rule
diamagnetic substance
paramagnetic substance

QUESTIONS AND PROBLEMS

Wave Motion

1. An observer is sitting along a lake shore watching water waves as they go by. She notices that in two seconds, 16 successive waves pass the shadow of a eucalyptus tree on the opposite bank and that the distance between successive wave crests is about 3 m. What is the velocity of the wave motion?

2. Calculate the wavelength associated with 60-cycle ac current traveling at the speed of light.

3. A wave disturbance has a velocity of 10 cm·s^{-1}. During a 10-s interval, 500 waves pass a given reference point.
 a) Calculate the wave frequency.
 b) Calculate the wavelength in nm and in Å.

4. When copper salts are heated, green light is emitted. How could you tell whether this light is of a single wavelength or is a mixture of two or more wavelengths?

5. The diameter of the earth is about 12,700 m. What is the frequency of electromagnetic radiation having a wavelength of that value?

Photoelectric Effect

6. Will light at the red end of the visible spectrum (700-nm wavelength) be able to operate an "electric eye" which uses cesium metal (work function 1.9 eV)? Explain.

7. An electron has a velocity of 4×10^5 m·s^{-1} after being ejected from a metal by light of wavelength 500 nm. What is the work function $h\nu_0$ (in eV) of the metal?

8. Sodium metal has a work function of 2.8 eV.
 a) What wavelength must a photon have in order to eject an electron of kinetic energy 1 eV from sodium metal?
 b) What is the longest wavelength a photon can

130 THE ELECTRONIC STRUCTURE OF ATOMS

have in order to be able to eject an electron from sodium metal?

9. Noting that the ionization energy of an element is measured for its gaseous state, how would you expect the ionization energy of an element to compare with its photoelectric work function?

10. The work function of cadmium is 3.7 eV.
   a) What is the longest wavelength of light that is capable of ejecting a photoelectron from cadmium?
   b) Light of a given wavelength ejects cadmium photoelectrons having an average kinetic energy of 1.8 eV. What is the wavelength of this light?
   c) What is the velocity of an electron having a kinetic energy of 1.8 eV?

11. What is the range of work function values (in eV) that metals must have in order to respond photoelectrically to visible light?

Wave–Particle Duality

12. What velocity (in m·s^{-1}) must a neutron have in order to exhibit a wavelength of 1 Å? The mass of a neutron is about 1.7×10^{-27} kg.

13. Calculate the wavelength (in meters) of a man weighing 60 kg walking at a rate of 5 km·hr^{-1}. Would you expect to be able to detect such a wavelength experimentally? Repeat the calculation for an electron (mass about 9×10^{-31} kg) traveling at 2% of the speed of light. Would the wavelength of the electron be experimentally detectable?

14. Calculate the wavelength associated with an α-particle having a kinetic energy of 1.0 MeV.

15. What is the wavelength associated with a hydrogen molecule at 25 °C?

Bohr Theory of the Atom

16. If it requires 13.6 eV of energy to ionize a hydrogen atom in its ground state ($n = 1$), how much energy would be required to ionize it in its $n = 5$ excited state?

17. a) Calculate the energy ΔE (in joules and in eV) for the hydrogen-atom transition from $n = 2$ to $n = 6$.
   b) Calculate the frequency (in s^{-1}) of the radiation associated with this transition.
   c) What wavelength (in nm and in Å) of radiation would a hydrogen atom emit in going from the $n = 6$ to the $n = 2$ state?

18. Calculate the ionization energy of each of the following.
   a) The hydrogen atom in its $n = 10$ state.
   b) The ion Li^{2+} in its ground state.
   c) The ion C^{5+} in its ground state.

19. Two energy states of an atom differ by 9 eV. What is the wavelength (in nm) of the light involved when the atom drops from the higher to the lower state? Is this radiation emitted or absorbed?

20. Verify the wavelength of the 1240-nm line shown in Fig. 4.7. Repeat for the 248-nm line.

21. Below is a partial energy level diagram of a hypothetical atom.

-1×10^{-19} J ——— E_4
-5×10^{-19} J ——— E_3
-15×10^{-19} J ——— E_2
-20×10^{-19} J ——— E_1

   a) What is the value (in joules) of the quantum of energy a photon must have in order to excite the atom from E_1 to E_4?
   b) What *wavelength* of light is needed to excite the atom from the state E_2 to the state E_3?
   c) If the atom is originally in state E_4, what must it do to reach state E_2?
   d) What *frequency* of radiation is involved in part (c)?

22. The ground state energy of a system is labeled E_0 and the excited states (in order of increasing energy) are labeled E_1, E_2, E_3, ...
   a) Draw to scale an energy level diagram for the lowest four states if $E_1 - E_0 = X$ (X is an arbitrary unit of energy), $E_2 - E_1 = 2.5X$, and $E_3 - E_2 = 0.75X$.
   b) If the transition E_0 to E_2 is excited by radiation of wavelength 500 nm, what wavelength will excite the E_1 to E_2 and E_2 to E_3 transitions?
   c) What is the energy (in joules) of the first excited state relative to the ground state?
   d) What is the frequency of the radiation (in s^{-1}) which will excite the E_2 to E_3 transition?

23. a) What is the energy (in joules and eV) associated with radiation of wavelengths 10 Å, 100 Å, and 100,000 Å?
   b) What are the frequencies associated with the above wavelengths?

24. Calculate the energy ΔE associated with radiation of wavelength 500 nm in (a) joules, (b) eV.

25. Derive the expression for the electronic energy of the hydrogen atom (Eq. 4.8). [*Hint:* Use the suggestions given in the footnote on page 101.]

26. Calculate the wavelength of light emitted when the hydrogen atom goes from its $n = 4$ to its $n = 2$ state. What is the energy of the emitted photon?

27. Sodium emits two very intense yellow lines at wavelengths 589.0 and 589.6 nm. What is the difference in energies (in eV) between the two photons emitted to cause these lines?

28. When a sheet of copper is bombarded with electrons, X-rays are given off. The shortest-wavelength X-rays found have a wavelength of 1.380 Å. The minimum ionization energy of copper is 7.68 eV. Estimate the energy of the $1s$ orbital of copper.

29. What is the energy (in eV) of infrared radiation (heat radiation) having a wavelength of 1 μm?

30. How much energy (in kJ) is needed to excite one mole of hydrogen atoms from the ground state to the $n = 4$ state?

31. It takes an energy of about 435 kJ to dissociate one mole of hydrogen molecules to atoms. What wavelength of light has the minimum energy to effect this dissociation? To what region of the electromagnetic spectrum does this radiation belong?

32. It requires at least 170 kJ of energy to break a mole of most chemical bonds. What is the maximum wavelength of radiation that is capable of breaking chemical bonds by photodissociation?

33. Does an α-particle of velocity 1.2×10^4 m·s^{-1} have enough energy to ionize the lithium atom?

Electron Configurations

34. Without referring to the text, write *full* electron configurations for the elements of atomic numbers 8, 14, 18, and 22. Draw box diagrams obeying Hund's rule for numbers 8 and 22.

35. Predict electron configurations for the following.

  a) Br$^-$ c) Ti^{4+} e) O^{2-}
  b) Ti^{2+} d) F$^-$ f) C^{2+}

36. What are the permissible values of l and m_l when $n = 4$? What are the orbitals corresponding to these values?

37. What is the total number of electrons in an atom that may have the following quantum numbers?

  a) $n = 3, l = 1$
  b) $n = 3, m_s = +\frac{1}{2}$
  c) $n = 3, l = 2, m_l = -2$

38. Predict the magnetic natures of the first-row transition elements. Which of these are diamagnetic and which are paramagnetic?

39. Positive ions are generally smaller than the corresponding neutral atoms whereas negative ions are generally larger. Suggest an explanation for this.

40. Write full electron configurations for the second-row transition elements (yttrium through cadmium) and compare with the corresponding first-row elements.

41. Write the full electron configurations of the xenon and gadolinium atoms. An additional rule is needed for the latter; one electron goes into $5d$ before the $4f$ orbitals are used.

42. Which atom in each of the following pairs has the higher ionization energy?

  a) Cs or Au   c) S or P   e) Mg or Al
  b) Rn or At   d) Zn or Cu   f) Rb or Sr

43. Why is it harder to remove an electron from neutral potassium than from the excited state of sodium having the electron configuration $1s^2 2s^2 2p^6 4s$?

44. The molar volumes of zinc and calcium are 10 cm^3 and 28 cm^3, respectively. Why is that of zinc so much smaller?

General

45. Name at least one contribution that each of the following made to quantum theory: Planck, Einstein, Bohr, de Broglie, Heisenberg, Schrödinger, Dirac, and Pauli.

46. Matter may be said to be quantized. In what sense is this true?

47. Elements X, Y, and Z are found in a vertical column of the periodic table with X on top and Z on the bottom. Which element will have (a) the largest atomic radius? (b) the largest ionization energy? (c) the most metallic character?

48. Explain just how the electron density in a 2s orbital differs from that of a 2p orbital.

49. The term "quantum jump" is often used by nonscientists to indicate a large change. If taken literally, does the term really mean what people think it means? What should the term be used for in ordinary conversation?

ADDITIONAL READINGS

Hochstrasser, Robin M., *Behavior of Electrons in Atoms*. New York: Benjamin, 1964. A very complete exposition of the electronic structure of atoms. Available in paperback.

Sebera, Donald K., *Electronic Structure and Chemical Bonding*. New York: Blaisdell, 1964. Chapters 2 and 3 deal fairly thoroughly with atomic electronic structure. Paperback.

Pimentel, George C., and Richard D. Spratley, *Chemical Bonding Clarified Through Quantum Mechanics*. San Francisco: Holden-Day, 1969. Chapters 1 and 2 deal with atoms. Paperback.

5
CHEMICAL BONDING

Linus Pauling, American theoretical chemist and pacifist who pioneered in the application of quantum theory to problems of molecular structure. (Wide World Photos.)

All things by immortal power
  Near or far
  Hiddenly
To each other linked are,
That thou canst not stir a flower
Without troubling of a star.
FRANCIS THOMPSON

Francis Thompson (1859–1907) was a British mystical and religious poet best known for the dreamlike The Hound of Heaven *(1893). After failing to gain recognition as a poet, Thompson became addicted to opium but was saved by friends who helped him to publish his poems. He eventually died of the tuberculosis which plagued him most of his life.*

British poet Francis Thompson probably wasn't thinking about chemistry when he penned the above lines, yet the general idea that the entire universe is one huge interconnected whole does apply to the chemical nature of the matter forming the universe. The universe as a whole—ourselves included—is a huge collection of negatively charged electrons and positively charged nuclei. According to **Coulomb's law**, any pair of charged particles (with charges q_1 and q_2) has associated with it an interparticle force given by

$$F_{12} = K \frac{q_1 q_2}{r^2} \tag{5.1}$$

where r is the distance between the two particles (see Fig. 5.1).* Consequently, only if two charged particles are an infinite distance apart is it possible for them to exert no force on each other. Furthermore, quantum theory shows that the electrons of atoms have some probability of being found at *any* finite distance from the nucleus. Thus, in a sense, the universe is one giant "molecule." In actuality, however, the forces of interaction among the electrons and nuclei of the universe tend to become very concentrated around relatively small groups of nuclei and to become weak and diffuse everywhere else. The concentrated forces lead to what the chemist calls **molecules** and the weak forces account for the existence of solids, liquids, solutions, and nonideal gases. This chapter will discuss some of the more important aspects of how atoms interact to form molecules. The next chapter will discuss how molecules (and unbound atoms) interact among themselves.

---

* If the charges are expressed in coulombs, r is given in meters, and the medium is a vacuum, then $K = 9 \times 10^9$ J·m·C^{-2} and F_{12} is in newtons. A force of 1 N acting through a distance of 1 m is equivalent to an energy of 1 J.

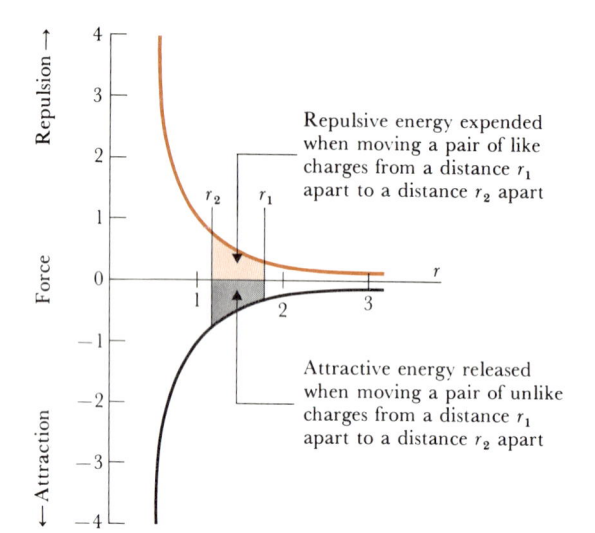

Fig. 5.1. Graphical illustration of Coulomb's law. The shaded area is equal to $Kq_1q_2(1/r_2 - 1/r_1)$ and represents the work expended or obtained when the charges are originally separated by a distance r_1 and are moved until they are separated by a smaller distance r_2.

5.1 ELECTRONEGATIVITY

Whenever two atoms of different elements combine chemically, there is a partial or total transfer of electrons from one atom to another. In order to be able to discuss chemical processes in terms of electron transfers, it is useful to define a property called **electronegativity**. The electronegativity of an atom serves as a measure of that atom's ability to attract electrons from another atom. Mulliken has suggested that the electronegativity of an atom be defined in terms of its *net* tendency to attract electrons. Thus the electronegativity of an atom A may be expressed in mathematical form as

$$\chi_A = k(I_A - E_A) \tag{5.2}$$

where I_A is the **ionization energy** of atom A, E_A is the **electron affinity** of atom A, and k is a constant. The ionization energy of an atom A is the minimum energy required to remove an electron from the ground state of the atom via the process

$$A \rightarrow A^+ + e^-$$

This energy is always positive; that is, it requires an input of energy to remove the electron from the atom. The electron affinity of an atom A is the energy involved in the addition of an electron to the atom via the process

$$A + e^- \rightarrow A^-$$

This energy is usually negative; that is, energy is produced when the atom accepts the electron.

The numerical value of k depends on the units of I_A and E_A and on the scale used to designate the electronegativities. Mulliken's original proposal was to express I_A and E_A in eV and to use $k = \frac{1}{2}$. However, it is now more common to use a scale suggested by Pauling. Pauling developed his electronegativity scale on the basis of an

Robert S. Mulliken (b. 1896), an American physical chemist, pioneered in the development of molecular-orbital theory and the interpretation of molecular spectra. Having retired from the University of Chicago, he now spends winters at the Institute of Molecular Biophysics at Florida State University. Mulliken received the Nobel prize in chemistry in 1966.

Linus Pauling (b. 1901), American chemist and pacifist, has spent most of his life as a teacher and research worker at the California Institute of Technology. He has made fundamental contributions to many areas of chemistry (notably, molecular structure) and molecular biology. He received the Nobel prize in chemistry in 1954 and the Nobel peace prize in 1962 and is the only person to be the sole winner of two Nobel prizes (others have won two Nobel prizes but these have been shared with someone else).

entirely different approach, and he assigned an arbitrary value of 4.0 to fluorine, the most electronegative element. This value ensures that all other elements have positive electronegativity values. Electronegativities on the Mulliken scale can be converted to those on the Pauling scale by a proper choice of the constant k. Since $I_F = 17.4$ eV and $E_F = -4.3$ eV, the value of k turns out to be

$$k = \frac{\chi_F}{I_F - E_F} = \frac{4.0}{17.4 - (-4.3)} = 0.18 \tag{5.3}$$

Electronegativities on the Pauling scale are usually treated as dimensionless numbers.

EXAMPLE 5.1 The ionization energies of hydrogen and chlorine atoms are 13.6 and 13.0 eV, respectively. The corresponding electron affinities are -0.7 and -3.6 eV. Calculate the electronegativities of hydrogen and chlorine on the Pauling scale.

SOLUTION Using the value of k given by Eq. (5.3) in Eq. (5.2) leads to

$$\chi_H = 0.18[13.6 - (-0.7)] = 2.6$$
$$\chi_{Cl} = 0.18[13.0 - (-3.6)] = 3.0$$

This shows that fluorine is more electronegative than chlorine, and that chlorine is more electronegative than hydrogen. Thus, in the molecule HCl, it is expected that there is some tendency for transfer of electrons from H to Cl, that is, from the less electronegative to the more electronegative element.

The procedure of Example 5.1 may be used to calculate electronegativities of all the elements. The results are summarized in Table 5.1. Since electronegativities depend more strongly on the ionization energy than on the electron affinity, they

TABLE 5.1 ELECTRONEGATIVITIES OF SOME SELECTED ELEMENTS IN THEIR MOST COMMON OXIDATION STATES[a]

H 2.1																	He[b] 2.7
Li 1.0	Be 1.5											B 2.0	C 2.5	N 3.0	O 3.5	F 4.0	Ne 4.4
Na 0.9	Mg 1.2											Al 1.5	Si 1.8	P 2.1	S 2.5	Cl 3.0	Ar 3.5
K 0.8	Ca 1.0	Sc 1.3	Ti 1.5	V 1.6	Cr 1.6	Mn 1.5	Fe 1.8	Co 1.8	Ni 1.8	Cu 1.9	Zn 1.6	Ga 1.8	Ge 1.8	As 2.0	Se 2.4	Br 2.8	Kr 3.0
Rb 0.8	Sr 1.0															I 2.5	Xe 2.6
Cs 0.7	Ba 0.9																

[a] Note that electronegativity varies periodically in much the same way as the first ionization energy (Fig. 4.13).
[b] The inert-gas values were calculated by Bing-Man Fung, *J. Phys. Chem.* **69**, 596 (1965). The method used did not employ Mulliken's formula since electron affinities for these gases are not known reliably.

vary periodically just as do the former. Also, since electronegativities may be computed in other ways than illustrated here, there are some variations (usually small) among tabulations in various sources. For example, Table 5.1 shows $\chi_H = 2.1$ rather than 2.6 as calculated in Example 5.1. Variations in electronegativity values for the other elements are usually much smaller than that for hydrogen.

Elements with relatively low electronegativities, for example those of groups IA and IIA, are often said to be **electropositive**. In general, metals are electropositive elements whereas nonmetals are electronegative.

The less electronegative of a pair of bonded atoms generally has the more positive oxidation number. Thus, in HCl, hydrogen has an oxidation number of +1 and chlorine an oxidation number of −1, since hydrogen is less electronegative (2.1) than chlorine (3.0). However, in NaH, hydrogen is −1 and sodium is +1, since hydrogen is more electronegative than sodium (0.9). Oxygen, one of the most electronegative elements (3.5), almost always has the oxidation numbers −1 or −2, but in OF_2 it is +2 since fluorine is even more electronegative than oxygen.

5.2 IONIC AND COVALENT BONDING

Bonding between atoms is usually discussed in terms of two extreme types of bonds called **ionic** (or **electrovalent**) and **covalent**. Actual chemical bonds are generally a composite of these two types of bonding.

Consider two atoms A and B, where A has a low ionization energy and low electron affinity (thus a low electronegativity) and B has a high ionization energy and high electron affinity (thus a high electronegativity). Such a pair of atoms will tend to form a chemical bond in which A exists as a positive ion (**cation**) A^+ and B exists as a negative ion (**anion**) B^-. The attraction between the two ions, which is largely electrostatic in nature, is called an **ionic bond**. The ionic bond can be visualized as forming as follows: Atom A loses an electron to form a cation

$$A \rightarrow A^+ + e^-$$

138   CHEMICAL BONDING

and atom B accepts this electron to form an anion

$$B + e^- \rightarrow B^-$$

The compound has the empirical formula AB and the solid and liquid forms exist as aggregations of the ions A$^+$ and B$^-$. As discussed in more detail in Chapter 7, the solid will generally consist of an orderly array of alternating cations and anions.

Compounds most likely to be ionic are those between metals and nonmetals, particularly the metals of groups IA and IIA and the nonmetals of groups VIA and VIIA. A good example is sodium chloride, NaCl. Here sodium has an electronegativity of 0.9 and chlorine has an electronegativity of 3.0; the difference of electronegativities between the two atoms is 2.1. As a rather rough rule of thumb, whenever two atoms A and B have electronegativity differences around two or more, the bond is apt to be predominantly ionic in nature.

EXAMPLE 5.2 Which of the following compounds would be predicted to have the most pronounced ionic nature: CaO or CaS?

SOLUTION Since $\chi_S = 2.5$, $\chi_O = 3.5$, and $\chi_{Ca} = 1.0$, it is seen that $\chi_S - \chi_{Ca} = 1.5$ and $\chi_O - \chi_{Ca} = 2.5$. Consequently, CaO is predicted to be more ionic in nature than is CaS.

Although solid and liquid forms of ionic substances do not exist as molecules, molecules do exist in the gaseous forms of such compounds. For example, solid table salt consists of the ions Na$^+$ and Cl$^-$ but when vaporized (above 1413 °C) molecules of NaCl exist. The energy required to carry out the process

$$NaCl(g) \rightarrow Na(g) + Cl(g)$$

is called the **binding energy** or **dissociation energy** (symbol D_e) of the molecule. The larger the dissociation energy the more stable is the molecule. A dissociation energy of roughly at least 1 eV is generally needed in order to stabilize a molecule; if the dissociation energy is lower than this the molecule is easily torn apart to atoms by thermal collisions and the like. Sodium chloride has a dissociation energy of 4.25 eV.

The stability of ionic compounds in the solid form depends on the energy holding the ions in their rather rigid locations. This is discussed in Chapters 7 and 18.

A few molecules involve no net transfer of electrons from one atom to another, for example, H$_2$, O$_2$, N$_2$, Cl$_2$, and other homonuclear diatomic molecules. In H$_2$, for example, it is clear that since both atoms have the same electronegativities, there can be no net tendency to transfer electrons from one atom to the other. In this case each nucleus attempts to keep its own electron and to attract the electron of the other atom.* The result is a shared *pair* of electrons known as a **covalent bond**. The total

---

* Although it is convenient to talk about the electrons as if they belonged to a specific nucleus (A or B), in reality all the electrons are indistinguishable and must be regarded as being shared by both nuclei. In fact, it is this mutual sharing of electrons by two or more nuclei that constitutes a major contribution to molecular stability.

energy of the H₂ molecule may be written

$$E_{H_2} = E_{AB} + E_{1,2} - E_{A,1} - E_{A,2} - E_{B,1} - E_{B,2} \tag{5.4}$$

where E_{AB} is the repulsion energy between the two nuclei A and B, $E_{1,2}$ is the repulsion energy between the two electrons 1 and 2, and each of the other terms is an energy of attraction between a nucleus and an electron. Thus E_{H_2} may be written

$$E_{H_2} = E_{\text{repulsion}} - E_{\text{attraction}} \tag{5.5}$$

The H₂ molecule will be stable only if its total energy E_{H_2} is negative, that is, if the attraction of electrons to nuclei surpasses repulsions between nuclei and between electrons.* Thus the existence of a stable molecule depends upon the simultaneous attraction of the electrons to *all* the nuclei. This simultaneous electron–nuclei attraction is the "molecular glue" which holds most matter together.

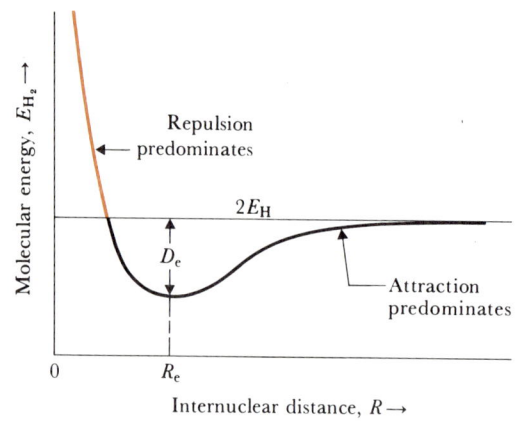

Fig. 5.2. Diagram showing how the molecular energy of a diatomic molecule varies with the intermolecular distance. Note that the electronic energies of the constituent atoms A and B serve as a convenient reference point for the molecular energy. The minimum value of the molecular energy E_{AB} occurs at $R = R_e$. This distance R_e is called the **equilibrium internuclear distance**. The quantity D_e is called the **dissociation energy** of the molecule.

Figure 5.2 illustrates how E_{H_2} varies as two hydrogen atoms are brought closer and closer together. When the two atoms are infinitely far apart, each electron is attracted only to its own nucleus and there is no interaction between the two atoms. As the two atoms are brought closer together, each electron begins to be attracted to both nuclei, leading to a decrease in E_{H_2}. At the same time the two electrons repel each other and so do the two nuclei, but the net effect is an attraction. Finally, at some distance R_e, the repulsions between nuclei and between electrons begin to increase rapidly. Consequently, the nuclei tend to stabilize at the distance R_e, the distance at which the attractions most surpass the repulsions. The distance R_e is often called the **equilibrium bond distance** of the molecule. The difference between the

---

* According to Coulomb's law, repulsions lead to a positive energy, that is, work must be done in order to bring two like charges together. Similarly, attractions lead to a negative energy; two unlike charges coming closer together can perform work (release energy). This means that work must be done to pull the unlike charges farther apart.

energies of the two separated hydrogen atoms and the molecular energy E_{H_2} at its lowest value is the binding energy or dissociation energy of the molecule. This may be written

$$D_e = 2E_H - E_{H_2} \tag{5.6}$$

For the hydrogen molecule, D_e is 4.7 eV at $R_e = 0.74$ Å.

The simultaneous attraction of the electrons by two or more nuclei in a stable molecule leads to a buildup of electron density between the nuclei relative to that which would exist if the atoms did not interact at all. One way of illustrating this is shown in Fig. 5.3. This diagram shows what happens to the electron density along a line joining the nuclei when two hydrogen atoms interact to form the H_2 molecule. It is seen that molecular formation involves a migration of electron charge density from outside the internuclear region into the internuclear region, so that electrons may be regarded as the "glue" holding the nuclei together. This sharing of electrons by a pair of nuclei constitutes what was referred to above as a covalent bond. When the two atoms sharing the electrons are identical, as in H_2, the bond is referred to as a **pure** covalent bond.

The pure covalent bond (as in H_2) and the ionic bond (as in NaCl) are extreme examples of electron transfers accompanying chemical bonding. In the intermediate case, the atoms A and B of a binary compound A_xB_y have different electronegativities—but not different enough to bring about ion formation. In such a case, exemplified by HCl, the more electronegative chlorine attracts electrons more strongly than does hydrogen. This leads to an unequal sharing of electrons and creates a **polar** covalent bond. The term "polar" refers to the fact that the HCl molecule, as a result of the unequal sharing of electrons, becomes polarized; that is, the hydrogen end of the molecule becomes a positive pole (due to loss of negative charge) and the chlorine end of the molecule becomes a negative pole (due to gain of negative charge). In general, chemical bonds between unlike atoms become polarized, with the more electronegative atom becoming the negative pole.

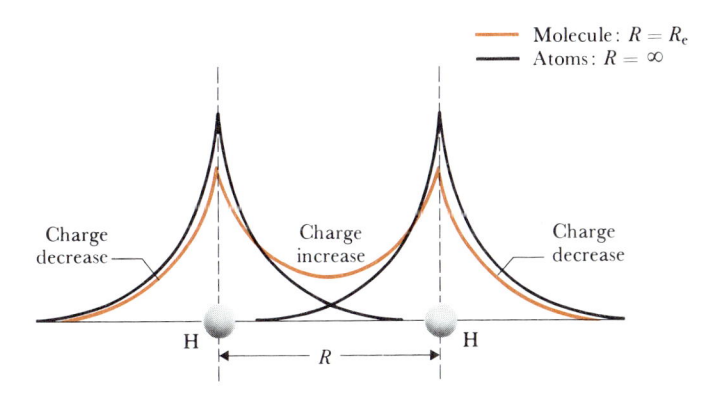

Fig. 5.3. Variation of electron charge density in the hydrogen molecule along the internuclear bond axis. Note that each hydrogen atom attracts the electron of the other so that the electron density increases between the two nuclei and decreases outside the internuclear region.

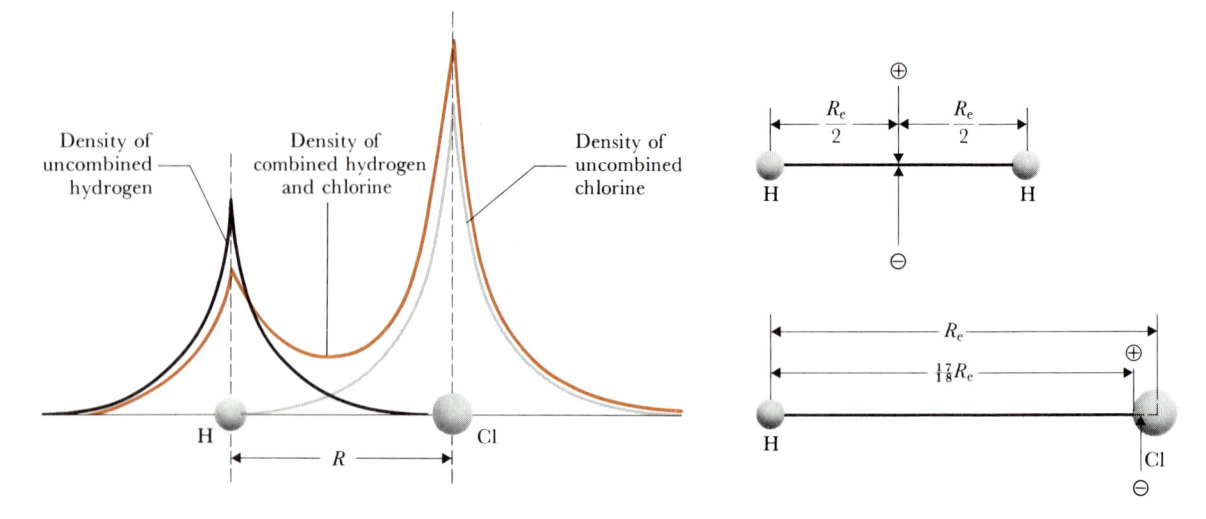

Fig. 5.4 (left). Variation of electron charge density in the hydrogen chloride molecule along the internuclear bond axis. Note that the electron density increases about the chlorine atom at the expense of the hydrogen-atom electron density.

Fig. 5.5 (right). Locations of the centers of positive and negative charges in a nonpolar molecule (H_2) and a polar molecule (HCl).

Figure 5.4 illustrates how atomic electronic densities are shifted in a molecule such as HCl. Note that in contrast to H_2 (a nonpolar molecule), HCl (a polar molecule) shows an unsymmetrical sharing of electron charge. As illustrated in Fig. 5.5, the centers of positive and negative charge coincide for a nonpolar molecule but do not coincide for a polar molecule.* In the case of HCl, the center of positive charge is closer to the chlorine nucleus than to the hydrogen nucleus since the former has a charge of 17+ and the latter a charge of 1+. In fact, this center of positive charge should be located at a distance $\frac{1}{18}R_e$ from the chlorine atom. Now if the hydrogen and chlorine charge distributions had been unaffected by molecular formation, the center of negative charge would be located at the same place as the center of positive charge. In actuality, the center of negative charge is closer to the chlorine nucleus than is the center of positive charge, since the chlorine atom tends to acquire some of the electron charge of the hydrogen atom. It is this separation of positive and negative charge centers that constitutes the polar nature of the molecule; HCl is like a rod charged ⊕ on the hydrogen end and ⊖ on the chlorine end. Note that the bonding in HCl may be regarded as intermediate to that of a pure ionic bond and a pure covalent bond.

The ionic bonding in NaCl can be illustrated in a manner somewhat similar to the covalent bonding in H_2 and HCl. Figure 5.6 shows how electron charge is transferred from the sodium atom to the chlorine atom. The ionic bond is thus seen to be an extreme case of a polar covalent bond.

---

* The center of a charge distribution (analogous to the center of gravity of a solid object) is that point such that the charges on opposite sides of any plane passing through the point are equal.

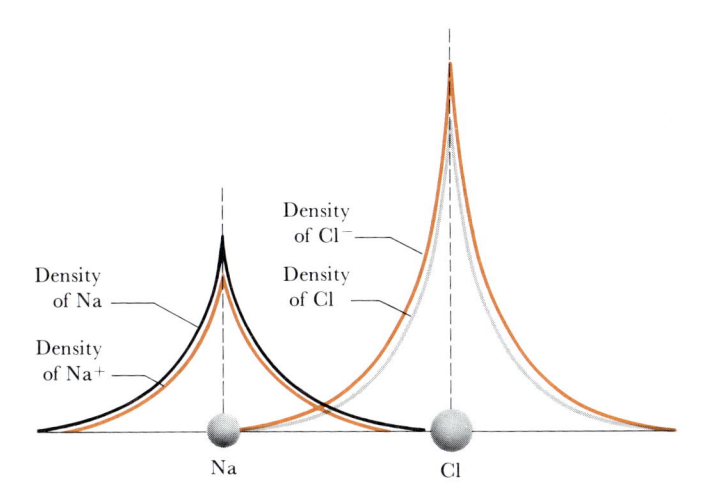

Fig. 5.6. A qualitative representation of the transfer of electron charge from the sodium atom to the chlorine atom in the Na$^+$, Cl$^-$ ion pair.

Many cases of ionic and covalent bonding can be rationalized on the basis of atoms gaining or losing electrons so as to acquire, as nearly as possible, electron configurations of the group VIIIA gases. In the case of an ionic compound such as NaCl, Na: (Ne)$3s$ becomes Na$^+$: (Ar) and Cl: (Ne)$3s^23p^5$ becomes Cl$^-$: (Ne)$3s^23p^6$ or Cl$^-$: (Ar). In the case of a covalent molecule such as H$_2$, no complete transfer of electrons can occur, but by sharing each other's electrons both hydrogen atoms come as close as possible to the H$^-$: ($1s^2$) or H$^-$: (He) configuration.

5.3 LEWIS DIAGRAMS OF MOLECULAR STRUCTURE

Gilbert N. Lewis (1875–1946), was an American physical chemist who made fundamental contributions to thermodynamics (see Chapter 18) and pioneered in the electronic theory of chemical valence. Lewis was also responsible for a general concept of acids and bases (see Section 13.10). Most of his professional life was spent at the University of California at Berkeley.

During the early 1920s the American chemist G. N. Lewis devised a symbolic representation of how atoms are bound in molecules. Although Lewis's representation predated modern quantum mechanics, it correlates surprisingly well with present-day models of chemical bonding. The rules for drawing a Lewis formula of a molecule are as follows.

1. The atoms forming the molecule are arranged, insofar as a two-dimensional representation permits, in the order in which they are actually bonded in the molecule. In H₂O the order is HOH and in CH₄ it is

   $$\begin{array}{c} H \\ H\ C\ H \\ H \end{array}$$

   No attempt is made to convey the correct bond angles or any other details of the actual three-dimensional geometry.

2. The total number of valence-shell electrons supplied by the atoms is determined. For H₂O this is eight (one from each hydrogen and six from oxygen), and for CH₄ it is also eight (one from each hydrogen and four from carbon).

3. The valence-shell electrons, represented by dots or crosses, are placed around the nuclei such that each atom has eight electrons (called a **Lewis octet**) in its immediate neighborhood (except that hydrogen atoms have only two such neighbor electrons). This is called the **octet rule** or **rule of eight**.

As a consequence of the octet rule, Lewis formulas are often called **octet formulas** (also **dot** formulas). The water molecule, H₂O, has the Lewis formula

$$H:\ddot{\underset{..}{O}}:H$$

The electron pairs between adjacent atoms symbolize a covalent bond. Since hydrogen and oxygen have different electronegativities, each O:H bond is polar, with the oxygen being polarized negatively and hydrogen polarized positively. It is convenient to denote the electron pair of a covalent bond by a dash. Thus the Lewis formula for H₂O is often written

$$H-\ddot{\underset{..}{O}}-H$$

The remaining electrons on oxygen are called **nonbonding** electrons or **lone-pair** electrons. They are often omitted in diagrams symbolizing the chemical bonding in molecules.

The Lewis octet formula for CH₄ is

$$\begin{array}{c} H \\ H:\ddot{\underset{..}{C}}:H \\ H \end{array} \quad \text{or} \quad \begin{array}{c} H \\ | \\ H-C-H \\ | \\ H \end{array}$$

Since carbon is more electronegative than hydrogen, the carbon end of the C—H bond is assumed to be negative.

Lewis octet formulas may also be written for ions. That for the simple chloride ion is

$$\left[:\ddot{\underset{..}{Cl}}:\right]^{1-}$$

It is important to note that one of the octet electrons has been obtained from some other atom. Thus the reaction of a sodium atom with a chlorine atom to form NaCl may be written

$$\text{Na}\cdot + \cdot\ddot{\underset{..}{\text{Cl}}}: \rightarrow \text{Na}^+ + \left[:\ddot{\underset{..}{\text{Cl}}}:\right]^{1-}$$

This should be compared with the quantum theory description:

$$\text{Na: (Ne)}3s + \text{Cl: (Ne)}2s^2 2p^5 \rightarrow \text{Na}^+\text{: (Ne)} + \text{Cl}^-\text{: (Ar)}$$

This shows that the Lewis octet rule is related to an atom's tendency to acquire the nearest inert-gas configuration when it enters into chemical combination.

The sulfate ion, SO_4^{2-}, has a total of 32 valence-shell electrons (six from sulfur, six from each oxygen, and two extra from the ion charge). The Lewis formula is

$$\left[\begin{array}{c} :\ddot{O}: \\ :\ddot{O}:S:\ddot{O}: \\ :\ddot{O}: \end{array}\right]^{2-} \quad \text{or} \quad \left[\begin{array}{c} :\ddot{O}: \\ | \\ :\ddot{O}-S-\ddot{O}: \\ | \\ :\ddot{O}: \end{array}\right]^{2-}$$

The ammonia molecule, NH_3, has eight valence-shell electrons. Thus the Lewis structure is

$$\begin{array}{c} H \\ :\ddot{N}:H \\ H \end{array} \quad \text{or} \quad \begin{array}{c} H \\ | \\ :N-H \\ | \\ H \end{array}$$

Note that there are three covalent bonds (nitrogen to hydrogen) and one lone pair. When H^+ (a bare proton, or hydrogen cation) is added to NH_3, the result is the ammonium ion, NH_4^+, symbolized by

$$\left[\begin{array}{c} H \\ H:\ddot{N}:H \\ H \end{array}\right]^{1+} \quad \text{or} \quad \left[\begin{array}{c} H \\ | \\ H-N-H \\ | \\ H \end{array}\right]^{1+}$$

Lewis structures of the homonuclear diatomic molecules H_2, O_2, N_2, and F_2 are

$$\begin{array}{cccc} H:H & \ddot{O}::\ddot{O} & :N::N: & :\ddot{F}:\ddot{F}: \\ H-H & \ddot{O}=\ddot{O} & :N\equiv N: & :\ddot{F}-\ddot{F}: \end{array}$$

From this it can be predicted that H_2 and F_2 are bound by single covalent bonds, that O_2 has a double covalent bond, and that N_2 has a triple covalent bond. The dissociation energies of these molecules are as follows: H_2(4.7 eV), O_2 (5.0 eV), N_2 (9.9 eV), and F_2 (2.8 eV). It is seen that the number of covalent bonds correlates roughly with the dissociation energies.

Lewis showed that if it is possible to construct an octet formula for a given assemblage of atoms, then a molecule corresponding to the formula very likely exists in nature. The converse is often not true; there are many stable molecules for which Lewis octet formulas cannot be written. For example, octet formulas cannot be written for PCl_5 or SF_6, yet these are stable molecules. Consequently, present-day chemists often employ modified Lewis dot formulas that allow up to twelve electrons surrounding an atom. As a rough rule, elements up to about atomic number 10 usually obey the octet rule; elements beyond this often require more than eight electrons. For example, PCl₅ and SF₆ have the dot formulas

$$
\begin{array}{c}
\ddot{\text{:}\ddot{\text{Cl}}\text{:}} \\
\text{:}\ddot{\text{Cl}}\text{·} \quad \text{·}\ddot{\text{Cl}}\text{:} \\
\text{P} \\
\ddot{\text{Cl}} \quad \ddot{\text{Cl}}
\end{array}
\quad \text{and} \quad
\begin{array}{c}
\text{:}\ddot{\text{F}}\text{:} \\
\text{:}\ddot{\text{F}}\text{·} \quad \text{·}\ddot{\text{F}}\text{:} \\
\text{S} \\
\text{:}\ddot{\text{F}}\text{·} \quad \text{·}\ddot{\text{F}}\text{:} \\
\text{:}\ddot{\text{F}}\text{:}
\end{array}
$$

Note there are ten electrons about phosphorus in PCl₅ and twelve electrons about sulfur in SF₆.

Sulfur dioxide, SO₂, has the atom order OSO. Two different octet formulas are possible for this atom order:

$$:\ddot{\text{O}}:\ddot{\text{S}}::\ddot{\text{O}} \leftrightarrow \ddot{\text{O}}::\ddot{\text{S}}:\ddot{\text{O}}:$$

These are called **resonance structures**; the double-headed arrow signifies that the bonding in SO₂ is expected to be a composite of the two different resonance structures. It has been observed that in many cases a molecule or ion for which two or more octet formulas can be written is unusually stable. Also, in the case of SO₂, the fact that the two different octet formulas reverse the relative position of S-to-O single and double bonds has been used to rationalize why the two S-to-O bonds are equivalent and why each behaves as a composite of a single and a double bond.

A useful general strategy for writing Lewis octet formulas is to begin placing eight electrons about all the atoms one by one until all the electrons are used up. For example, in SO₂ this leads to

$$:\ddot{\text{O}}:\ddot{\text{S}}:\ddot{\text{O}}$$

Note that the right-hand oxygen is left without an octet. This is easily remedied by shifting a lone pair of electrons from sulfur to the sulfur-oxygen interatom region. This gives

$$:\ddot{\text{O}}:\ddot{\text{S}}::\ddot{\text{O}}$$

It is interesting to note that a Lewis octet formula can be written for the atom order SOO in SO₂, namely:

$$:\ddot{\text{S}}::\ddot{\text{O}}:\ddot{\text{O}}:$$

Nevertheless, no molecule corresponding to this atom order has ever been isolated.

The nitrate ion, NO₃⁻, can be represented by three different octet formulas or resonance structures:

$$
\left[\begin{array}{c} \ddot{\text{O}}::\text{N}:\ddot{\text{O}}: \\ :\ddot{\text{O}}: \end{array} \right]^{1-}
\leftrightarrow
\left[\begin{array}{c} :\ddot{\text{O}}:\text{N}::\ddot{\text{O}} \\ :\ddot{\text{O}}: \end{array} \right]^{1-}
\leftrightarrow
\left[\begin{array}{c} :\ddot{\text{O}}:\text{N}:\ddot{\text{O}}: \\ :\ddot{\text{O}}: \end{array} \right]^{1-}
$$

Each N-to-O bond is regarded as a superposition of two single bonds and one double bond. Since each N-to-O bond is thus equivalent to the other two, there should be but one N-to-O bond distance in NO_3^-. In the same fashion, SO_2 should have only one S-to-O bond distance. All of these predictions are verified experimentally.

It is possible to write three resonance structures, each obeying the octet rule, for boron trifluoride, BF_3:

$$\begin{array}{ccc}
:\ddot{\ddot{F}}: & :\ddot{\ddot{F}}: & :\ddot{\ddot{F}}: \\
\ddot{B}:\ddot{\ddot{F}}: \leftrightarrow & \ddot{B}::\ddot{F} \leftrightarrow & \ddot{B}:\ddot{F}: \\
:\ddot{\ddot{F}}: & :\ddot{\ddot{F}}: & :\ddot{\ddot{F}}:
\end{array}$$

However, these octet formulas are generally rejected for the following reasons.

1. Each octet formula implies that a pair of electrons is transferred from a more electronegative atom (F: 4.0) to a less electronegative atom (B: 2.0). This is opposite to what would be expected.

2. The three resonance structures imply some double bond character for each B—F bond. There is no experimental evidence to support this. Furthermore, quantum mechanical calculations of electron density in the B—F bond also show no double bond character.

3. BF_3 combines chemically with molecules such as NH_3 to form a single molecule. The simplest explanation of this fact requires a Lewis formula for BF_3 having only *six* electrons about boron.

$$\begin{array}{ccc}
:\ddot{\ddot{F}}: & H & :\ddot{\ddot{F}}:H \\
:\ddot{\ddot{F}}:B \; + & :\ddot{N}:H \; \rightarrow & :\ddot{\ddot{F}}:B:\ddot{N}:H \\
:\ddot{\ddot{F}}: & H & :\ddot{\ddot{F}}:H
\end{array}$$

EXAMPLE 5.3 Draw the Lewis structure of formaldehyde, CH_2O.

SOLUTION It is often informative to write down dot formulas for the individual atoms. These are

$$H \cdot \qquad H \cdot \qquad :\dot{C}: \qquad :\ddot{O}:$$

Some contemplation and a little juggling shows that the only likely way to achieve octets is to group the $2H \cdot$ and $:\ddot{O}:$ about the $:\dot{C}:$ as follows:

$$\begin{array}{c}
H:C:H \\
\ddot{:} \\
:\ddot{O}:
\end{array}$$

This formula suggests that carbon and hydrogen each contribute one electron per C—H bond and that carbon and oxygen each contribute two electrons to the C=O bond.

EXAMPLE 5.4 Benzene, C_6H_6, has six carbon atoms arranged in a circle with a hydrogen atom attached peripherally to each carbon atom. Draw two resonance formulas for benzene.

SOLUTION One fairly obvious possibility is

where C::C alternates with C:C. Switching these double and single bonds produces a second resonance formula.

5.4 QUANTUM MECHANICS OF THE COVALENT BOND

The approximate energy of a molecule can be calculated by the use of quantum mechanics. The procedure is to guess how the probability of finding the electrons varies within the space surrounding the nuclei and to use this probability to average electron–nuclei attractions and electron–electron repulsions over this entire space. In most cases quantum mechanics predicts that the electron probabilities are highest between those atoms which the Lewis structures show as covalently bonded. Thus the two dots symbolizing a covalent bond correlate with a region where electrons are most likely to be as a result of being attracted by more than one nucleus.

The most popular method for guessing the probability distributions of electrons in molecules is known as the **molecular orbital method**. Just as electrons in atoms are assumed to occupy **atomic orbitals**, electrons in molecules are assumed to be assignable to analogous probability distribution functions known as **molecular orbitals**. These molecular orbitals (MO's) are assumed to be represented by an *overlapping* of the atomic orbitals (AO's) of the atoms forming the molecule. This section will show how to set up and use the molecular orbitals of a homonuclear diatomic molecule such as H_2 or N_2. Details of molecular orbital theory and its extension to polyatomic molecules are omitted since few of these can be readily appreciated by beginning chemistry students. Furthermore, the vast majority of students will never be exposed to these again.

Two important types of molecular orbitals are labeled σ and π, the Greek-letter counterparts of s and p atomic orbitals. The labels arise from the fact that electron distributions in σ and π molecular orbitals resemble those in s and p atomic orbitals when the former are viewed by sighting down the bond axis of the molecule. The σ molecular orbitals are formed by overlapping s atomic orbitals of the bonded atoms or by overlapping p, d, f, ... atomic orbitals of the correct type. For example, overlapping p atomic orbitals which coincide with the molecular bond axis also leads to σ molecular orbitals. The lowest-energy σ molecular orbital is called 1σ; it is defined by

$$1\sigma = 1s_A + 1s_B \tag{5.7}$$

where $1s_A$ and $1s_B$ are $1s$ atomic orbitals of atoms A and B of the homonuclear diatomic molecule. For purely mathematical reasons, it is also necessary to use the combination

$$1\sigma^* = 1s_A - 1s_B \tag{5.8}$$

which produces another σ molecular orbital, one higher in energy than 1σ. In this way two atomic orbitals always produce two molecular orbitals and there is no gain or loss of orbitals.

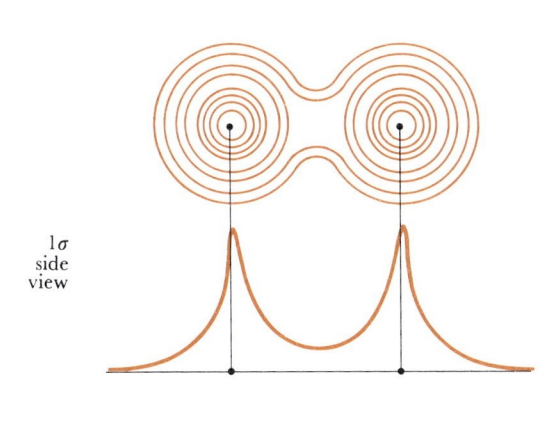

Fig. 5.7. Electron densities arising from 1σ and $1\sigma^*$ molecular orbitals for homonuclear diatomic molecules. The side-view contour diagrams are for a plane containing the nuclei, and the diagrams below each of these illustrate the electron density along the internuclear bond axis. The end-view contour diagram is for a plane containing one nucleus and perpendicular to the bond axis. Note that the $1\sigma^*$ molecular orbital leads to an electron density which is zero in a plane perpendicular to the internuclear axis and passing midway between the nuclei. Note also that both the 1σ and $1\sigma^*$ molecular orbitals resemble a $1s$ atomic orbital in the end view.

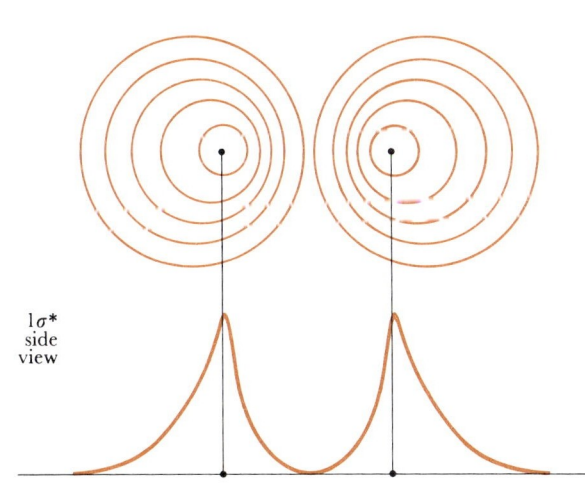

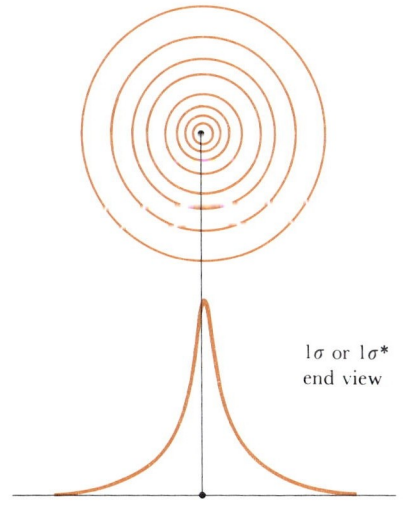

Both the 1σ and $1\sigma^*$ molecular orbitals lead to electron densities which are cylindrically symmetric about the bond axis. When the electron distribution arising from 1σ or $1\sigma^*$ is viewed by sighting down the bond axis (as if atoms A and B were the front and rear lenses of a telescope) the distribution resembles that of a $1s$ atomic orbital. Figure 5.7 illustrates the general characteristics of the 1σ and $1\sigma^*$ molecular orbitals by use of contour diagrams and bond-axis densities. Note that the 1σ molecular orbital shows a buildup of electron density between the two nuclei, whereas the $1\sigma^*$ molecular orbital shows an opposite effect: a buildup of electron charge outside the internuclear region. For this reason 1σ is called a **bonding** molecular orbital and $1\sigma^*$ is called an **antibonding** molecular orbital.

Figure 5.8 illustrates the 1σ and $1\sigma^*$ molecular orbitals by means of boundary surface diagrams; just as in atomic orbitals (Chapter 4), the surface of the figure represents a locus of equal charge densities and denotes that region of space about the two nuclei which has a 95% probability of containing an electron.

Four other σ molecular orbitals are:

$$2\sigma = 2s_A + 2s_B \qquad 3\sigma = 2p_A^{\parallel} + 2p_B^{\parallel}$$
$$2\sigma^* = 2s_A - 2s_B \qquad 3\sigma^* = 2p_A^{\parallel} - 2p_B^{\parallel}$$
(5.9)

where $2p_A^{\parallel}$ signifies a $2p$ orbital of atom A which is *parallel* (superscript $\parallel$) to the A—B bond axis. These orbitals differ in some details from the 1σ and $1\sigma^*$ orbitals but the major features of all σ orbitals are similar—as are the major features of all σ^* orbitals.

The π molecular orbitals (see Fig. 5.9) are formed as follows:

$$1\pi = 2p_A^{\perp} + 2p_B^{\perp} \qquad 1\pi^* = 2p_A^{\perp} - 2p_B^{\perp}$$
(5.10)

where $2p_A^{\perp}$ signifies a $2p$ orbital of atom A which is *perpendicular* (superscript $\perp$) to the A—B bond axis. Since there are two sets of $2p$ orbitals perpendicular to the A—B bond axis ($2p_x$ and $2p_y$ for each atom if the bond axis is along the z-direction), there are actually two 1π and two $1\pi^*$ molecular orbitals. Both 1π molecular orbitals have the same energy, and both $1\pi^*$ molecular orbitals have the same energy, but the 1π molecular orbitals are lower in energy than the $1\pi^*$ molecular orbitals. Similarly, 2π and $2\pi^*$ molecular orbitals are formed by linear combination of $3p$ atomic orbitals.

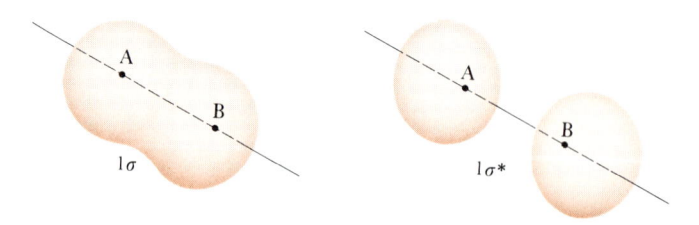

Fig. 5.8. Boundary-surface illustration of 1σ and $1\sigma^*$ molecular orbitals for a diatomic molecule AB. The surface has the same electron probability (cloud density) everywhere and encloses 95% of the electron population. The 1σ molecular orbital (bonding) resembles two spherical drops of liquid beginning to fuse. The $1\sigma^*$ molecular orbital (antibonding) resembles two spherical drops of liquid repelling each other.

When π or π^* molecular orbitals are viewed by sighting along the molecular bond axis, the electron distributions resemble those of atomic p orbitals. This is illustrated in Fig. 5.9.

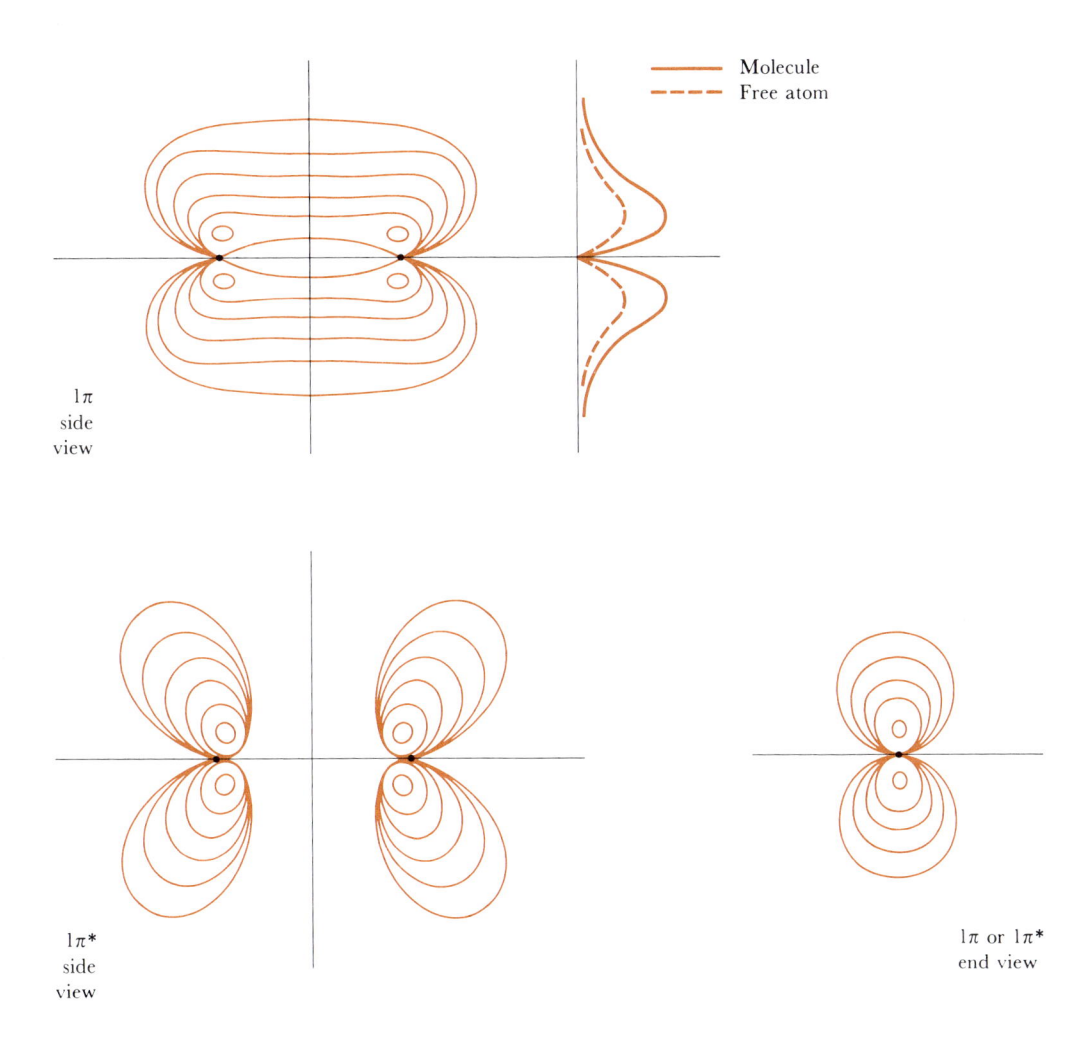

Fig. 5.9. Electron densities arising from 1π and $1\pi^*$ molecular orbitals for homonuclear diatomic molecules. The side-view contour diagrams are for a plane containing the nuclei. The diagram to the right of the 1π side view shows how the electron density varies along a line perpendicular to the internuclear bond axis and passing midway between the nuclei. Both the 1π and $1\pi^*$ molecular orbitals lead to zero electron densities along the internuclear axis; the $1\pi^*$ molecular orbital also leads to zero electron density in a plane perpendicular to the internuclear axis and passing midway between the two nuclei. Note that both the 1π and $1\pi^*$ molecular orbitals resemble a $2p$ atomic orbital in the end view. The dashed-line curve to the right of the 1π contour diagram shows what the electron density arising from an uncombined atom would be at the mid-bond region. When two such atoms combine, this mid-bond electron density is increased.

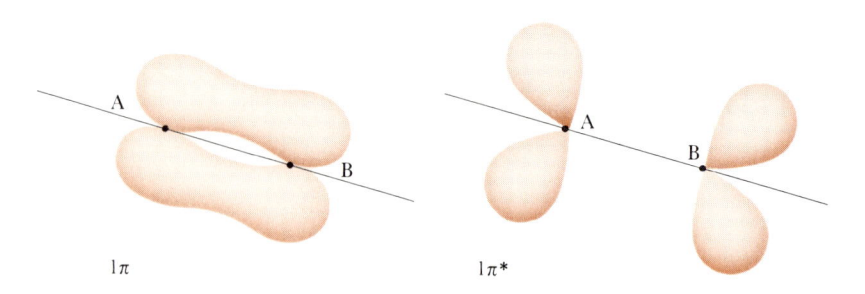

Fig. 5.10. Boundary-surface illustration of 1π and $1\pi^*$ molecular orbitals for a diatomic molecule AB. The surface has the same electron probability (cloud density) everywhere and encloses 95% of the electron population. The 1π molecular orbital (bonding) resembles two parallel vertical dumbbells that have fused, whereas the $1\pi^*$ molecular orbital (antibonding) resembles a repulsion between these two dumbbells. Actually, both the 1π and $1\pi^*$ molecular orbitals are cylindrically symmetric about the bond axis. This can be visualized by imagining that each figure is rotating constantly about the A—B bond axis.

Figure 5.10 shows boundary surface representations of the 1π and $1\pi^*$ molecular orbitals.

Figures 5.7 and 5.9 show that molecular orbitals without asterisks (the "plus" combinations) lead to a buildup of electron charge in the internuclear region (compare with Fig. 5.3) whereas those with asterisks (the "minus" combinations) lead to the opposite effect, that is, to a buildup of electron charge outside the internuclear region at the expense of the internuclear region. Just as with 1σ and $1\sigma^*$ molecular orbitals, 1π molecular orbitals are *bonding* molecular orbitals, and the $1\pi^*$ molecular orbitals are *antibonding* molecular orbitals. It should be noted that σ and σ^* molecular orbitals affect electron densities the most along a line joining the nuclear centers, whereas π and π^* molecular orbitals affect electron densities along an annular ring about the bond axis. Note also that π and π^* molecular orbitals have zero electron densities along a line joining the two nuclear centers.

The aufbau principle for homonuclear diatomic molecules requires that the molecular orbitals be used in the order (first to last): 1σ, $1\sigma^*$, 2σ, $2\sigma^*$, 1π, 3σ, $1\pi^*$, $3\sigma^*$, and so on (this is as far as is needed in this text). As in atoms, molecular orbital energies represent negatives of ionization energies. Each molecular orbital can hold a maximum of two electrons (with opposite spins). Since there are two equivalent π orbitals (and also two equivalent π^* orbitals), four electrons can be placed into the 1π orbitals (two per orbital) and, also, four electrons can be placed into the $1\pi^*$ orbitals. Hund's rule for maximum alignment of spins holds as in atoms. Thus, in using an orbital such as 1π, the first two electrons go into different 1π orbitals and with the same spin.

The molecular orbital electron configuration of the hydrogen molecule H_2, the simplest homonuclear diatomic molecule, is written $(1\sigma)^2$. These two electrons are assigned opposite spins. The He_2 electron configuration is $(1\sigma)^2(1\sigma^*)^2$ and that of Li_2 is $(1\sigma)^2(1\sigma^*)^2(2\sigma)^2$. The stability of a molecule A_2 relative to the dissociation $A_2 \rightarrow 2A$ can be correlated with a quantity called the **bond order**. The order of the A—A bond is given by

$$P_{AA} = \tfrac{1}{2}(N - N^*) \tag{5.11}$$

where N is the number of electrons in bonding molecular orbitals and N^* is the number of electrons in antibonding molecular orbitals. Whenever the bond order P_{AA} is greater than zero, the molecule is predicted to exist long enough to be identified experimentally. For H_2: $(1\sigma)^2$, $P_{HH} = \frac{1}{2}(2 - 0) = 1$; this is known to be a very stable molecule. A bond order of *one* means that the atoms are joined by a single covalent bond, which, by definition, is a pair of shared valence-shell electrons.

The molecule He_2: $(1\sigma)^2(1\sigma^*)^2$ has $P_{HeHe} = \frac{1}{2}(2 - 2) = 0$ and, consequently, should not exist.† Note that the bond order of zero arises due to a cancellation of $(1\sigma)^2$ bonding by $(1\sigma^*)^2$ antibonding.

The molecule Li_2: $(1\sigma)^2(1\sigma^*)^2(2\sigma)^2$ has $P_{LiLi} = \frac{1}{2}(4 - 2) = 1$. Experiment shows this is not a particularly stable molecule, but the vapor of lithium at its boiling point (1317 °C) does contain about 1% Li_2. Note that the single covalent bond of Li_2 is due to the $(2\sigma)^2$ part of the electron configuration, since the bonding due to $(1\sigma)^2$ is cancelled by the antibonding nature of $(1\sigma^*)^2$. The Li—Li bond may be regarded as a sharing of a pair of $2s$ electrons by the two nuclei.

The next diatomic molecule, Be_2: $(1\sigma)^2(1\sigma^*)^2(2\sigma)^2(2\sigma^*)^2$, has $P_{BeBe} = \frac{1}{2}(4 - 4) = 0$ and is not known to exist.

The electron configurations and bond orders of the molecules B_2 through Ne_2 are listed below. For simplicity the $(1\sigma)^2(1\sigma^*)^2(2\sigma)^2(2\sigma^*)^2$ portion common to each electron configuration has been abbreviated as (ISNB); this represents eight **inner-shell nonbonding** electrons similar to those of the inert-gas core in atoms.

MOLECULE	ELECTRON CONFIGURATION	BOND ORDER
B_2	$(ISNB)(1\pi)^2$	1
C_2	$(ISNB)(1\pi)^4$	2
N_2	$(ISNB)(1\pi)^4(3\sigma)^2$	3
O_2	$(ISNB)(1\pi)^4(3\sigma)^2(1\pi^*)^2$	2
F_2	$(ISNB)(1\pi)^4(3\sigma)^2(1\pi^*)^4$	1
Ne_2	$(ISNB)(1\pi)^4(3\sigma)^2(1\pi^*)^4(3\sigma^*)^2$	0

Hund's rule that electrons are to be assigned to atomic orbitals with the same spin as much as possible applies to molecular orbitals as well. Thus B_2 may be written

↑↓	↑↓	↑↓	↑↓	↑	↑
1σ	$1\sigma^*$	2σ	$2\sigma^*$	\multicolumn{2}{c}{1π}

Similarly, O_2 is

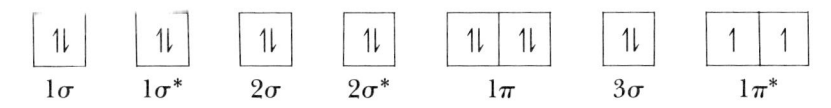

↑↓	↑↓	↑↓	↑↓	↑↓	↑↓	↑↓	↑	↑
1σ	$1\sigma^*$	2σ	$2\sigma^*$	\multicolumn{2}{c}{1π}	3σ	\multicolumn{2}{c}{$1\pi^*$}

Both of these molecules are paramagnetic as predicted by the molecular orbital box

---

† The molecule He_2 has *excited* electronic states which are stable. One such state, approximated by the configuration $(1\sigma)^2(1\sigma^*)(2\sigma)$, is paramagnetic and has a dissociation energy of about 2.6 eV and $R_e = 1.05$ Å. Experimental studies of $^3He_2$ and $^4He_2$ have been discussed by P. Gloersen and G. H. Dieke, *J. Mol. Spectrosc.* **16**, 191 (1965).

diagrams. Molecules such as H_2, Li_2, C_2, N_2, and F_2 are diamagnetic since there are equal numbers of ↑ and ↓ spins.

The C_2 molecule is predicted to be diamagnetic. This is in accord with experiments on carbon vapor. The bond order of 2 indicates that C_2 has a double covalent bond; that is, two pairs of electrons are shared in the bonding. Similarly, it can be verified that N_2 has a triple bond and is diamagnetic and that O_2 has a double bond and is paramagnetic. The O_2 double bond is due to $(1\pi)^2$ and $(3\sigma)^2$ (two of the 1π electrons are cancelled by a pair of $1\pi^*$ electrons). The O_2 and N_2 bond orders are in accord with the Lewis octet formulas but that of C_2 is not; the Lewis octet formula for C_2 predicts a *quadruple* bond! The molecular orbital prediction is the correct one.

A covalent bond resulting from a σ molecular orbital is called a **sigma bond**; that resulting from a π molecular orbital is called a **pi bond**. Molecules such as H_2 and F_2 have one sigma bond each; B_2 has a single pi bond. Oxygen's double bond is said to consist of a sigma bond and a pi bond; nitrogen's triple bond consists of a sigma bond and two pi bonds. The double bond of C_2 consists of two pi bonds.

All other factors being equal, the greater the bond order the greater is the dissociation energy. Table 5.2, which lists dissociation energies and other properties of diatomic molecules, shows that there is such a trend. For example, the dissociation energies of F_2, O_2, and N_2 are very nearly proportional to the corresponding bond orders.

TABLE 5.2 EXPERIMENTAL DISSOCIATION ENERGIES AND OTHER PROPERTIES OF SOME HOMONUCLEAR AND HETERONUCLEAR DIATOMIC MOLECULES AND IONS

MOLECULE	VALENCE-SHELL ELECTRON CONFIGURATION	BOND ORDER	D_e(eV)	R_e(Å)	MAGNETIC TYPE[b]
H_2	$(1\sigma)^2$	1	4.74	0.74	D
Li_2	$(2\sigma)^2$	1	1.03	2.67	D
B_2	$(1\pi)^2$	1	3.6	1.59	P
C_2	$(1\pi)^4$	2	3.6	1.24	D
N_2	$(3\sigma)^2(1\pi)^4$	3	9.9	1.09	D
N_2^+	$(3\sigma)^2(1\pi)^3$	2.5	8.86	1.12	P
O_2	$(3\sigma)^2(1\pi)^4(1\pi^*)^2$	2	5.03	1.21	P
O_2^+	$(3\sigma)^2(1\pi)^4 1\pi^*$	2.5	6.48	1.12	P
F_2	$(3\sigma)^2(1\pi)^4(1\pi^*)^4$	1	2.75	1.44	D
CN	$(1\pi)^4 3\sigma$	2.5	7.6 to 8.2[a]	1.17	P
NO	$(1\pi)^4(3\sigma)^2 1\pi^*$	2.5	5.3 or 6.5[a]	1.15	P

[a] It is not known which is the correct value since different types of experiments lead to conflicting data.
[b] D = diamagnetic, P = paramagnetic

There is also some correlation between bond order and bond length; it would be expected that bond lengths are inversely proportional to bond orders. For example, O_2^+ has the electron configuration $(ISNB)(1\pi)^4(3\sigma)^2(1\pi^*)$, with a bond order of 2.5. From this it can be predicted that O_2^+ will have a shorter internuclear distance (and a larger dissociation energy) than O_2. Table 5.2 shows that both of these predictions are correct.

Fig. 5.11 (following pages). Molecular orbital electron densities for some homonuclear diatomic molecules. As opposed to topographic maps (see Appendix 5), where contour intervals are constant, each adjacent pair of contour lines in this figure represents a twofold change in electron density. [Reproduced by permission of the author and publisher from "Molecular Orbital Densities: Pictorial Studies," by A. C. Wahl, *Science* **151** (25 February 1966), 961–967, copyright © 1966 by the American Association for the Advancement of Science.]

Figure 5.11 shows orbital density diagrams for all the stable homonuclear diatomic molecules from H_2 to F_2. Note that as the nuclear charges on the constituent atoms increase, the electron density represented by a given molecular orbital tends to become more concentrated ("pulled in" more closely) about the nuclei. Figure 5.12 shows molecular orbital energy level diagrams for C_2, N_2, O_2, and F_2. Note that in N_2 and O_2 the 1π and 3σ orbitals have opposite relative energies compared with C_2 and F_2. There is no simple explanation for this fact; it is simply a mathematical consequence of the model.

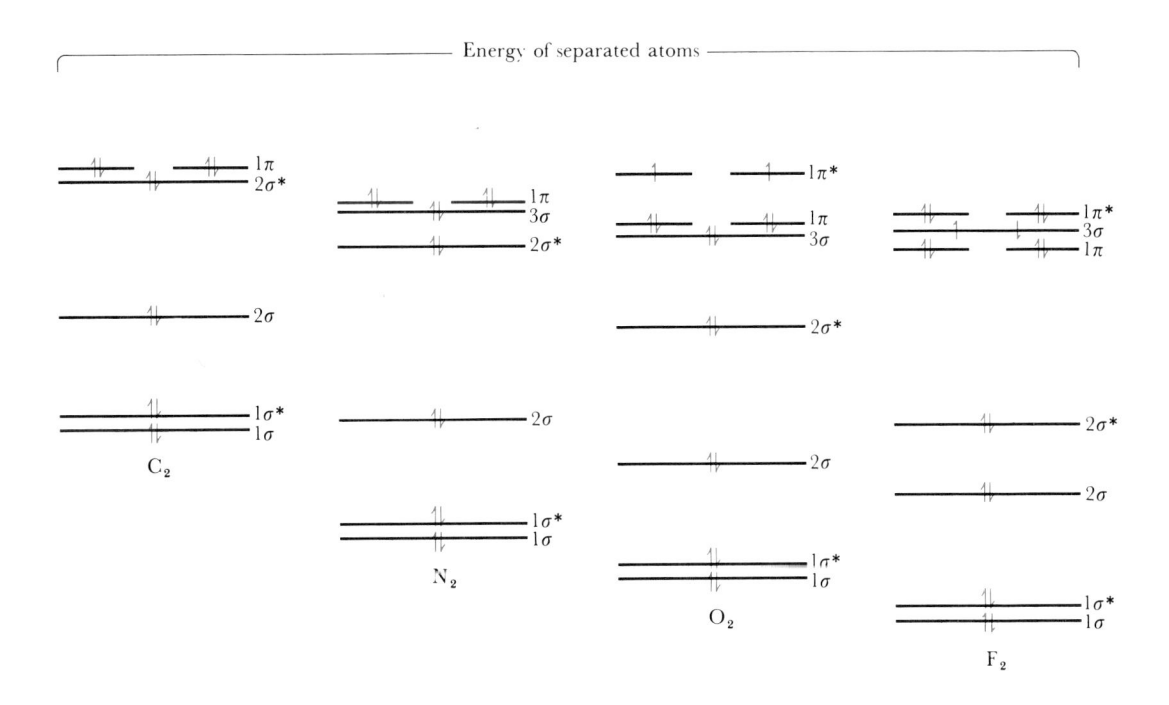

Fig. 5.12. Orbital-energy-level diagrams of the C_2, N_2, O_2, and F_2 molecules. Except for the 1σ and $1\sigma^*$ molecular orbitals, all of the levels are in an approximately correct relative position. In the scale used, the $1\sigma^*$ molecular orbital level should be almost on top of the 1σ molecular orbital level, and the separation between the $1\sigma^*$ and 2σ orbital levels should be much greater. The positions of the highest occupied molecular orbitals predict that the first ionization energies increase in the order C_2, O_2, N_2, and F_2. This agrees with the experimental values 12, 12.1, 15.6, and 15.7 eV, respectively.

Total

$1\pi^*$

1π

3σ

$2\sigma^*$

2σ

$1\sigma^*$

1σ

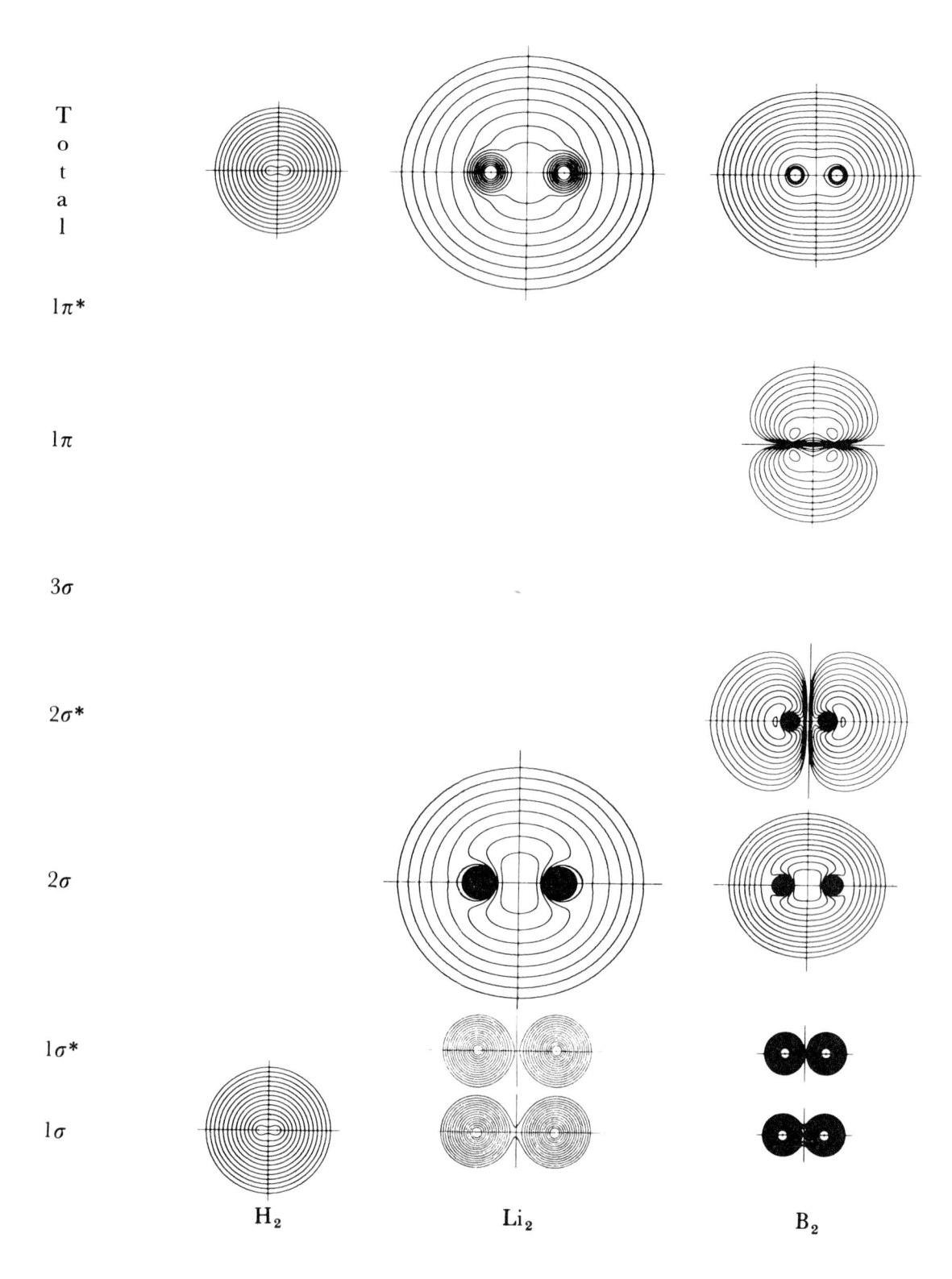

H₂   Li₂   B₂

156 CHEMICAL BONDING

5.4 | QUANTUM MECHANICS OF THE COVALENT BOND

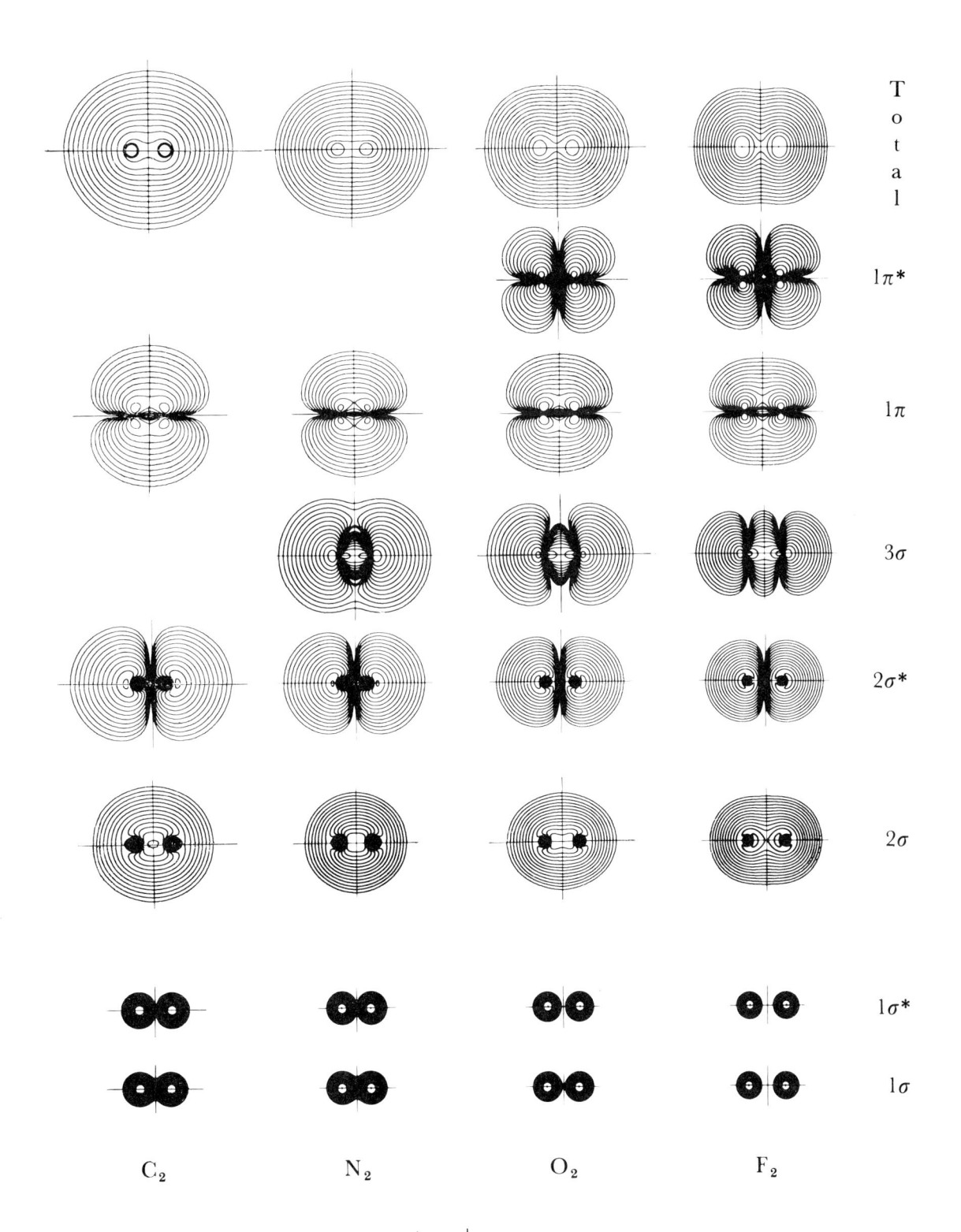

EXAMPLE 5.5 Predict the electron configuration, bond order, and expected stability of the hydrogen-molecule ion, H_2^+.

SOLUTION Since H_2 is $(1\sigma)^2$, H_2^+ should be 1σ. Thus $P = \frac{1}{2}(1 - 0) = \frac{1}{2}$. This ion should be stable enough to be detectable experimentally. In fact, the ion is formed whenever H_2 is subjected to a high-voltage discharge or bombardment with an electron beam. Although the ion quickly picks up an electron to revert to neutral H_2, it does exist long enough to have its mass measured in a mass spectrometer (see Fig. 1.7).

EXAMPLE 5.6 Which would have the longer N-to-N bond distance, N_2 or its ion N_2^+? Explain.

SOLUTION The valence-shell portions of the electron configurations are N_2: $(3\sigma)^2(1\pi)^4$ and N_2^+: $(3\sigma)^2(1\pi)^3$. Thus the bond order of N_2 is $\frac{1}{2}(6 - 0) = 3$ and that of N_2^+ is $\frac{1}{2}(5 - 0) = 2.5$. Since bond length and bond order are inversely proportional, the N_2 molecule has the shorter N-to-N distance. This is verified by the data in Table 5.2.

EXAMPLE 5.7 Which would be predicted to have the stronger O-to-O bond, O_2 or its ion O_2^+? Explain.

SOLUTION The valence-shell electron configurations are O_2: $(3\sigma)^2(1\pi)^4(1\pi^*)^2$ and O_2^+: $(3\sigma)^2(1\pi)^4(1\pi^*)$. Thus the bond order of O_2 is $\frac{1}{2}(6 - 2) = 2$ and that of O_2^+ is $\frac{1}{2}(6 - 1) = 2.5$. Note that removal of an electron *increases* the bond order since the electron is removed from an antibonding molecular orbital. Since bond order and dissociation energies are directly proportional, O_2^+ should have the stronger O-to-O bond. This is verified by the dissociation energy values given in Table 5.2.

The application of molecular orbital theory to heteronuclear diatomic molecules is similar to that for homonuclear diatomic molecules. In the special case that both atoms A and B are from the first full period, the order of superposition of molecular orbitals, as given by the aufbau principle, is very nearly the same as for homonuclear diatomic molecules. The molecular orbitals themselves, however, reflect the polarization of electron density characteristic of two bonded atoms having different electronegativities. For example, the 1σ molecular orbital becomes

$$1\sigma = 1s_A + K 1s_B \tag{5.12}$$

where K is a constant related to the relative electronegativities of atoms A and B. When atoms A and B are the same, K becomes equal to 1.

As an example, NO has the molecular orbital electron configuration $(ISNB)(1\pi)^4(3\sigma)^2(1\pi^*)$, with a bond order of 2.5. Since this molecule has an odd number of electrons, it must be paramagnetic. Paramagnetic molecules are often much more chemically reactive than diamagnetic molecules. NO, a colorless gas at room temperature, reacts very rapidly with atmospheric O_2 (paramagnetic) to form a red-brown gas, NO_2. The molecule $CN:(ISNB)(1\pi)^4(3\sigma)$, with a bond order of 2.5, is also paramagnetic and probably reacts with itself to form $(CN)_2$, a highly toxic gas known as cyanogen.

EXAMPLE 5.8 Would you expect the ion HeH^+ to exist? Explain.

SOLUTION HeH^+ would have two electrons. Thus its electron configuration is $(1\sigma)^2$, with a bond order of 1. This ion is known to exist in gas-discharge tubes containing He and H_2.

SUMMARY

When two different atoms combine chemically, there is a partial or total transfer of electrons from the less electronegative to the more electronegative atom. When there is a total transfer of one or more electrons, the bond between the two nuclei is called *ionic*; when the two share electrons the bond is called *covalent*. A chemical bond is most likely to be ionic if the two atoms have an electronegativity difference of two or more units. A covalent chemical bond is characterized by an increase in electron density between a pair of atoms. This increase in electron density results when one or more electrons are simultaneously attracted by all the nuclei. The molecule is stable toward dissociation to atoms if the electron–nucleus attractions predominate over the nucleus–nucleus and electron–electron repulsions by a sufficient amount.

Bonding of atoms to form molecules may be represented by Lewis structures (dot formulas). Many of the simpler stable molecules found in nature have Lewis formulas in which there are eight electrons (the *Lewis octet*) about each atom (except for hydrogen, which has only two electrons about it). Covalent bonds are represented by pairs of electrons shared by adjacent atoms.

Quantum theory describes electron distributions in molecules by means of *molecular orbitals*, each of which represents an electron density (or probability) as it is distributed about the nuclei of the molecule. These molecular orbitals are constructed by overlapping atomic orbitals of the atoms forming the molecule. Molecular orbitals are used to provide electron configurations of molecules, just as atomic orbitals are used for electron configurations in atoms. The electron configurations of molecules may be used to determine bond orders, which, in turn, provide a measure of molecular stability and may also be used to predict relative bond lengths and dissociation energies.

LEARNING GOALS

1. Be able to define electronegativity and to calculate it on the Pauling scale from ionization energies and electron affinities.

2. Be able to use electronegativities to predict which binary compounds are likely to be ionic.

3. Know how to use Coulomb's law to calculate forces between charged particles and to calculate the energy required to change the distance between charged particles.

4. Know how to predict polarities in chemical bonds (which atom is positive and which is negative).

5. Be able to write Lewis structures for small, common molecules and be familiar with the important exceptions to the octet rule.

6. Be able to write molecular orbital electron configurations for homonuclear and heteronuclear diatomic molecules and their ions for the first full period (up to neon).

7. Be able to determine bond orders for each of these and to interpret these in terms of binding energies and equilibrium bond distances.

8. Be able to predict diamagnetic and paramagnetic molecules and ions from molecular orbital electron configurations.

DEFINITIONS, TERMS, AND CONCEPTS TO KNOW

molecule
Coulomb's law
electronegativity
ionization energy
electron affinity
electropositive element
ionic bond
covalent bond
binding energy (dissociation energy) of a molecule

equilibrium bond distance
polarity
polar bond
the octet rule
nonbonding (lone-pair) electrons
resonance structure
molecular orbital (MO)
σ and σ^* MO's
π and π^* MO's

bonding MO
antibonding MO
bond order
innershell nonbonding electrons
sigma bond
pi bond
diamagnetic substance
paramagnetic substance

QUESTIONS AND PROBLEMS

Electronegativity

1. Complete the following table.

ELEMENT	IONIZATION ENERGY (eV)	ELECTRON AFFINITY (eV)	ELECTRO-NEGATIVITY (PAULING SCALE)
F	17.42	−4.27	_____
Br	11.80	−3.80	_____
I	10.44	−3.43	_____
Li	5.39	≃0	_____
Cs	3.89	≃0	_____

Compare the results with those given in Table 5.1.

2. Using Table 5.1 as a basis, rationalize the electronegativity values given for the noble gases. Many students expect these gases to have zero electronegativity values. Why is this expectation not correct?

3. Shown is a portion of the periodic table.

A	B	C	D
E	F	G	H
I	J	K	L
M			N

160   CHEMICAL BONDING

a) Which element is the most electronegative?
b) Which element is the most electropositive?

4. Criticize the statement: The greater the difference in electronegativity between two elements, the greater is the strength of the bond between them. [*Hint:* Consider NaCl and H_2 as two specific examples.]

5. Arrange the following compounds in decreasing order of ionic character: HI, CsCl, F_2, LiI, and HF.

6. Which of the following substances is more apt to be a solid at 25 °C: AlF_3 or CF_4? Explain.

Ionic and Covalent Bonding

7. Gaseous NaCl has an equilibrium bond distance (R_e) of 2.361 Å and a dissociation energy (D_e) of 4.25 eV. Use Coulomb's law to calculate the energy of attraction of Na^+ and Cl^- ($q_1 = q_2 = 1.6 \times 10^{-19}$ C). Compare the result with D_e. Use $K = 9 \times 10^9$ J·m·C^{-2} in Coloumb's energy formula given in Fig. 5.1. Why is the electrostatic energy higher than the dissociation energy?

8. Which of the two compounds, Na_2S or Cl_2S, would you expect to be ionic? Explain.

9. Which would you expect to be more polar, a C—H bond or a C—F bond? Why?

10. What does Table 5.1 predict about the polarity of the N—H bond in NH_3? Which is the positive end of the bond? What about the P—H bond in PH_3?

11. Atoms A, B, C, D, and E are in the same period and have 1, 2, 3, 4, and 7 valence-shell electrons, respectively.
a) Which atom has the greater electronegativity, B or D?
b) Is the compound between C and D likely to be ionic or covalent?
c) Which atom has the lowest ionization energy?
d) What is a probable formula for the compound between B and E?
e) Which is the positive end of a bond between D and E?

12. Cesium and rubidium do not react with each other, not even at very high temperatures. Explain.

13. Nonmetals generally combine with both metals and other nonmetals but metals generally combine only with nonmetals. Explain.

14. What ions between atomic numbers 7 through 13, inclusive, have the neon electron configuration?

15. NaH (sodium hydride), SO_2 (sulfur dioxide), and SiC (silicon carbide) are examples of compounds in which both elements are in the same group. Discuss each molecule in terms of the tendency of the constituent atoms to acquire inert gas configurations.

16. Which is a better name for BrCl, bromine chloride or chlorine bromide? Explain.

Lewis Structures

17. Draw Lewis formulas for the following molecules and ions.
a) HCN   b) HNNH   c) C_2H_6 d) C_2H_4
e) C_2H_2 f) OCl^- g) CN^- h) PCl_3
i) H_2O_2 j) PO_4^{3-}

18. Show that O_3 has two resonance structures.

19. Show that Lewis octet formulas for N_2O may be written for either NON or NNO atom arrangements (only the latter molecule is known experimentally). Show that NNO has three resonance structures.

20. Draw two resonance structures of the nitrite ion, NO_2^-.

21. Write a Lewis structure for HN_3 (HNNN).

22. Would you expect the substance H_3Sb to exist? Antimony (Sb) has an electronegativity of 1.9. Would SbH_3 be a better way to write the formula of this substance?

23. Using Lewis structures, discuss the various types of bonding in a substance such as Na_2SO_4.

24. Show by use of Lewis structures the most probable structure of the molecule S_2Cl_2.

25. The compound NOCl is called nitrosyl chloride. Which of the Lewis structures below is more likely to represent the actual structure of the molecule?

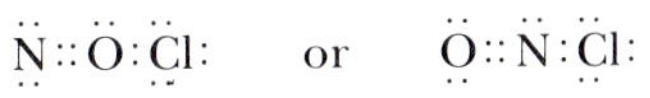

$$\ddot{N}::\ddot{O}:\ddot{Cl}: \quad \text{or} \quad \ddot{O}::\ddot{N}:\ddot{Cl}:$$
$$\text{I} \qquad\qquad\qquad \text{II}$$

[*Hint:* In I, oxygen contributes three electrons to the N=O bond; in II it contributes two electrons to this bond.]

26. Which ion should have the shorter N-to-O distance, NO_2^- or NO_3^-? [*Hint:* Look at the average bond orders implied by the resonance structures.]

27. Arrange the C-to-O bond lengths in the following molecules or ions in decreasing order (longest to shortest):

$$CO_2, \quad HCO^-, \quad CO, \quad CO_3^{2-}, \quad H_3COH$$
$$\quad\quad\quad \| \quad\quad\quad\quad\quad\quad\quad\quad\quad\quad\quad$$
$$\quad\quad\quad O \quad\quad\quad\quad\quad\quad\quad\quad\quad\quad\quad$$

[*Hint:* Draw the Lewis structures (including resonance structures) and consider the implied bond orders.]

28. Write Lewis formulas (including resonance structures) of the oxalate ion $C_2O_4^{2-}$ ($^-O_2CCO_2^-$).

Molecular Orbitals

29. Predict the molecular orbital electron configuration of carbon monoxide, CO.
 a) What homonuclear diatomic molecule has a similar electron configuration?
 b) Which is predicted to have the longer C—O equilibrium distance, CO or CO^+? Explain.

30. Discuss the Lewis structure and molecular orbital electron configuration of the cyanide ion, CN^-.

31. Discuss the expected stability of the ions H_2^+, He_2^+, C_2^{2+}, and NO^+ on the basis of molecular orbital theory.

32. Predict the O—O bond distances of the superoxide ion (O_2^-) and the peroxide ion (O_2^{2-}) relative to those of O_2^+ and O_2 given in Table 5.2. Would you expect these ions to be stable? Explain.

33. Use molecular orbital theory to predict the relative bond lengths and magnetic types for OF^+, OF, and OF^-.

34. Without referring to the text, write down the molecular orbital electron configurations of all the diatomic molecules from H_2 through F_2 and calculate the corresponding bond orders.

35. The dissociation energy of N_2^+ is less than that of N_2, but the dissociation energy of O_2^+ is greater than that of O_2. How is this possible?

36. What molecular orbital electron configuration is implied by the Lewis formula

$$C\!:\!:\!:\!C$$

for C_2? Why is this unlikely?

37. What is the molecular orbital electron configuration of diatomic sodium vapor, Na_2? What is the bond order of this molecule?

38. How could you tell experimentally that F_2 has $1\pi^*$ as the highest energy molecular orbital rather than 3σ? [*Hint:* Look at relative bond lengths and bond energies for F_2 and F_2^+ in the two cases.]

General

39. Name at least one important scientific contribution of each of the following: Coulomb, Pauling, Mulliken, and Lewis.

40. In answering the following questions, recall that the fundamental unit of charge is 1.6×10^{-19} C. Use $K = 9 \times 10^9$ J·m·C^{-2} in Coulomb's law. (See Fig. 5.1.)
 a) What is the electrostatic energy of interaction (in joules) between an α-particle (helium nucleus, $_2^4He^{2+}$) and an electron if the two are separated by a distance of 10 Å?
 b) At what distance will the energy of interaction become ten times greater?
 c) At what distance will the force of interaction become ten times greater?

41. It requires 418 kJ (100 kcal) to decompose a mole of a given compound into its constituent atoms. If this were done by high-energy radiation, what wavelength would need to be used?

42. Given that 1 eV is approximately 1.6×10^{-19} J and that 1 J = 4.184 cal, calculate the energy (in kJ and kcal) of a mole of particles each having an energy of 1 eV.

43. After observing that O_2 reacts with PtF_6 to form $(O_2^+)(PtF_6^-)$, N. Bartlett reasoned that an analogous reaction should occur with xenon, since xenon and O_2 have almost identical ionization energies. Consequently, in 1962, Bartlett produced the first noble gas compound, $Xe(PtF_6)_n$, (where n is between 1 and 2). Compounds of krypton and radon have also been formed since 1962 but not those of the other noble gases. How would you explain this latter fact on the basis of the periodic table?

ADDITIONAL READINGS

Companion, Audrey L., *Chemical Bonding.* New York: McGraw-Hill, 1964. A complete and elementary account of the modern principles of atomic and molecular electronic structure.

Sebera, Donald K., *Electronic Structure and Chemical Bonding.* New York: Blaisdell, 1964.

Wahl, A. C., "Chemistry By Computer," *Scientific American*, April 1970. Discusses how computers are used to produce contour diagrams of electron densities in atoms and molecules.

6
MOLECULAR GEOMETRY

It is from laboratory data such as these that chemists deduce the geometrical structures of molecules. (Photo by J. William Moncrief.)

No one really understands the behaviour of a molecule until he knows its structure—that is to say: its size, and shape, and the nature of its bonds.

C. A. COULSON, *The Shape and Structure of Molecules*, 1973.

Charles Coulson (1910–1974) was Rouse Ball Professor of Mathematics in Oxford University and a Fellow of the Royal Society of London. Beginning in the 1930s, Professor Coulson made numerous pioneering contributions to the application of quantum theory to chemistry. He was an articulate and charming speaker who presented lectures all over the world on various topics of science. He was also known in Great Britain as a lay preacher of the Methodist Church.

The chemical and physical properties of a molecule depend not only upon the atoms forming it but also upon the three-dimensional arrangement of these atoms. Current chemical research, particularly that in biologically and medically oriented areas, reveals with increasing emphasis that molecular geometry may sometimes be just as important as are the actual atoms forming the molecule. This chapter discusses some of the ways chemists determine or predict molecular geometries, and how they describe covalent bonding in polyatomic molecules. The chapter also discusses how molecules bond weakly to each other, thus accounting for the existence of solids, liquids, solutions, and nonideal gases.

6.1 POLARITY IN MOLECULES

Just as a chemical bond A—B between two atoms is polar whenever the atoms A and B have different electronegativity, so can an entire molecule be polar. As will be shown later, the molecular polarity may be related conceptually to the polarities of individual bonds.

Polar and nonpolar molecules are distinguished experimentally by the measurement of a quantity called the **dielectric constant**. The dielectric constant of a sample of matter is defined as the ratio of the **capacitance** of a pair of parallel plates separated by the sample to the capacitance of the same plates separated by a vacuum. Capacitance* is a measure of the ability of a substance to store electrical

---

* The capacitance of a substance is defined as the charge required to raise its potential by one unit. A condenser which can store 1 coulomb of charge per volt of potential is said to have a capacitance of 1 **farad**.

energy; substances with low capacitance are called **insulators**, and substances with high capacitance are called **dielectrics**.

Nonpolar molecules have much lower dielectric constants than do polar molecules; the latter exhibit a large capacity to store electrical energy when an electric field gradient is imposed across the capacitor plates. As illustrated in Fig. 6.1, polar molecules tend to align themselves with the electric field across the capacitor.

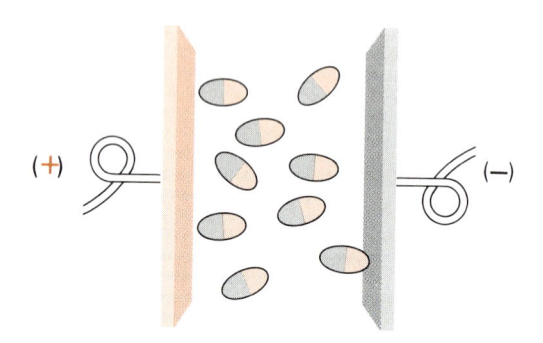

Fig. 6.1. Alignment of polar molecules in an electric field. In actuality, thermal random motions of the molecules tend to disorder the alignment, and very few molecules (perhaps only 1 per 1000) are perpendicular to the charged plates at any given time at room temperature.

The polar nature of a substance may be described quantitatively by a quantity called the **dipole moment**. The arrangement of a charge q^+ and a charge q^- separated by a distance r is called a **dipole**; the dipole moment is given by

$$\mu = qr \tag{6.1}$$

where q is the magnitude of the positive or negative charge (both charges are equal in magnitude but opposite in sign). Dipole moments are expressed in **debye units** (symbol D); one debye unit is 3.34×10^{-30} C·m. Thus, if equal positive and negative charges of 1.6×10^{-19} C are separated by 0.6×10^{-10} m (0.6 Å), the dipole moment is

$$\frac{(1.6 \times 10^{-19} \text{ C})(0.6 \times 10^{-10} \text{ m})}{3.34 \times 10^{-30} \text{ C·m·D}^{-1}} = 2.9 \text{ D}$$

It is found experimentally that substances with zero dipole moments have dielectric constants that do not change with temperature. Substances with nonzero dipole moments, on the other hand, have dielectric constants that change with temperature; in fact, the magnitude of this change in dielectric constant may be used to compute the dipole moment.

A diatomic molecule AB will be polar whenever the atoms A and B have different electronegativities; in general, A and B will have different electronegativities whenever they are different elements. Homonuclear diatomic molecules such as H_2, O_2, N_2, and Cl_2 are nonpolar and consequently have zero dipole moments. The heteronuclear diatomic molecule HCl has a dipole moment of 1.03 D.

The polarity of a polyatomic molecule (a molecule with more than two atoms) is

The debye unit is named after the Dutch physicist and chemist, Peter Debye (1884–1966). Debye's outstanding contributions were the development (along with E. Hückel and L. Onsager) of a modern theory of electrolytes and the measurement of dipole moments and their use in determining molecular structure. He received the Nobel prize in chemistry in 1938.

conveniently discussed by assigning a *hypothetical* dipole moment to each of its bonds and viewing the dipole moment of the molecule as a **resultant** of the dipole moments of the individual bonds. Each pair of bonded atoms in a molecule is treated as if it constituted a diatomic molecule, and a hypothetical dipole moment (called the **bond moment** of that pair) is assigned to this pair. Each such bond moment is represented by a **vector*** drawn as ↔; the left end is the positive end of the dipole and the right end is the negative end of the dipole. As discussed in Section 5.1, the negative and positive ends of the dipole are assigned on the basis of the relative electronegativities of the atoms of the bond. For mathematical calculations the length of the vector must be proportional to the magnitude of the bond moment and the direction of the vector must be that of the A—B bond axis. However, only the direction of the bond moment is needed in order to predict whether a given molecule is polar or nonpolar.

As specific examples, consider the molecules CO_2, H_2O, and CH_4. Carbon dioxide, CO_2, is a linear molecule with the Lewis structure

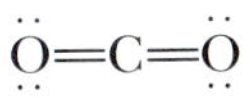

$$\ddot{\underset{..}{O}}=C=\ddot{\underset{..}{O}}$$

Since $\chi_C = 2.5$ and $\chi_O = 3.5$, the bond moment of the left-hand O=C bond may be written ← (the more electronegative atom is at the left or negative end of the bond moment). Similarly, the bond moment of C=O is written → (the lengths of the two bond moments are equal since the two C=O bonds are equivalent). The dipole moment of the molecule is just the vector sum of the two bond moments. Since both bond moments are of equal magnitude but opposite directions, their vector sum is zero. Consequently, CO_2 is nonpolar since its dipole moment is zero.

In actual practice the geometrical shape of a molecule is not known, and the measured dipole moment is used to help determine its shape—not the other way around as the CO_2 example might imply. For example, H_2O could have either the linear or bent structures below.

H—O—H   O
             / \
            H   H

However, experiments show that the dielectric constant of water changes with temperature, thereby indicating that the molecule is polar, that is, has a nonzero dipole moment. This is conclusive proof that the bent structure is the correct one. Similarly, measurement shows that CH_4 has no dipole moment; consequently, its structure must

---

* A **vector** is a mathematical quantity having a definite direction as well as a magnitude. **Force**, since it is always exerted in a definite direction, is a simple example of a vector quantity. Quantities which have magnitude only (volume, for example) are called **scalars**.

be a symmetrical one such as

H—C(—H)(—H)(H)   or   C(H)(H)(H)(H)

planar   tetrahedral

Some other structures are also possible but the two above are the most plausible. It takes additional data to show that the tetrahedral structure* is the correct one, but the lack of a dipole moment at least narrows down the possibilities.

The rule for obtaining the resultant of two vectors, $\mathbf{V}_1$ and $\mathbf{V}_2$, is to connect them tail to head; their resultant (or sum) is a vector drawn from the tail of one vector to the head of the other. Alternatively, the two vectors can be put tail to tail and a parallelogram drawn with these vectors as sides; their sum is now a vector joining the tails of the two vectors and the opposite side of the parallelogram. In either case, the directions of the two vectors being added must not be changed from the original directions (see Fig. 6.2).

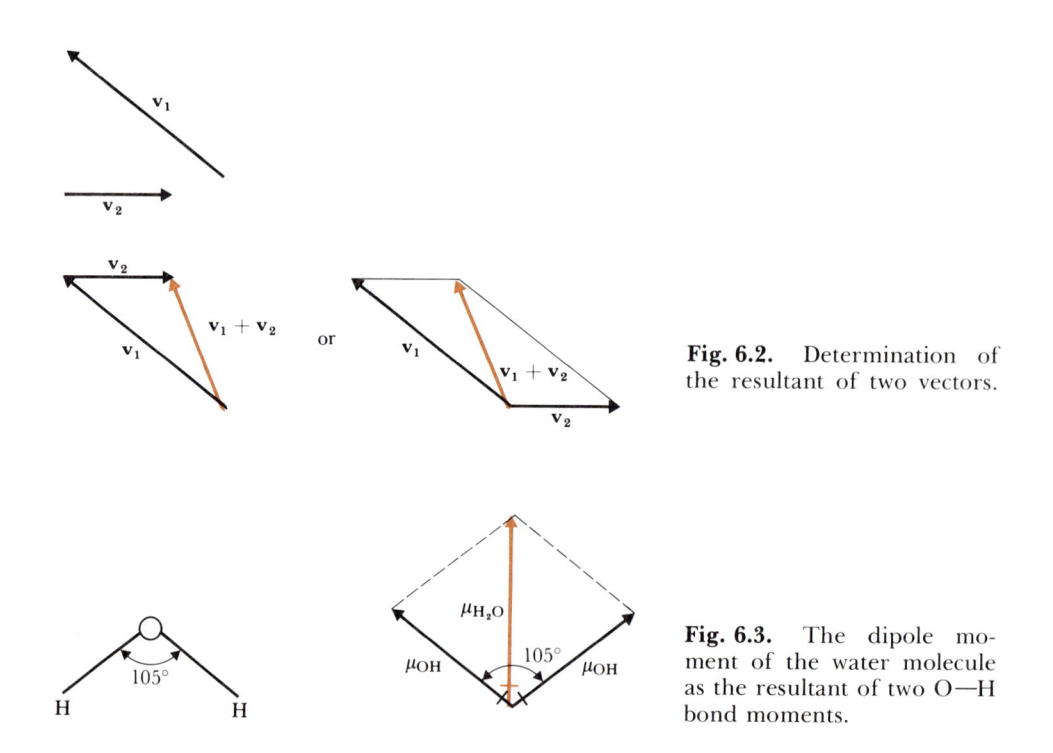

Fig. 6.2. Determination of the resultant of two vectors.

Fig. 6.3. The dipole moment of the water molecule as the resultant of two O—H bond moments.

---

* The precise definition of a tetrahedral structure is given in Section 6.2.

168   MOLECULAR GEOMETRY

Fig. 6.4. Cancellation of the C—H bond moments in CH₄.

Resultant of these is zero.

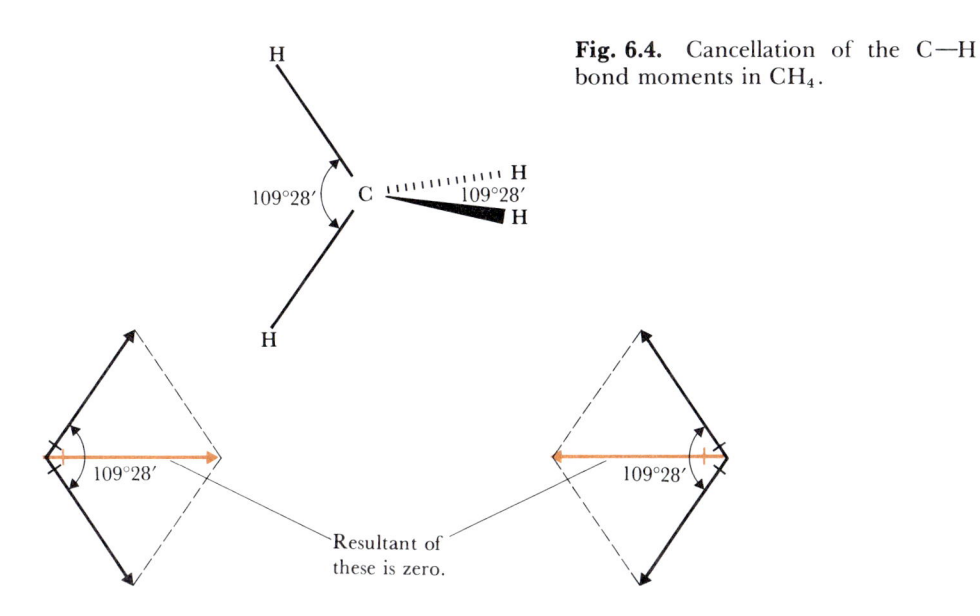

Figures 6.3 and 6.4 show how the bond moments of O—H and C—H bonds add vectorially to produce molecular dipole moments of H₂O and CH₄, respectively. The bond angles shown are those obtained experimentally. Table 6.1 lists the experimental dipole moments of some common molecules. In the case of many diatomic molecules there is a good correlation between the magnitude of the dipole moment and the magnitude of the electronegativity difference between the two atoms.

TABLE 6.1 DIPOLE MOMENTS OF SOME SIMPLE DIATOMIC AND POLYATOMIC MOLECULES

MOLECULAR FORMULA	NAME	DIPOLE MOMENT (D)
HF	Hydrogen fluoride	1.82
HCl	Hydrogen chloride	1.08
HBr	Hydrogen bromide	0.82
HI	Hydrogen iodide	0.44
H_2O	Water	1.85
H_2S	Hydrogen sulfide	0.97
NH_3	Ammonia	1.47
N_2H_4	Hydrazine	1.75
NF_3	Nitrogen fluoride	0.235
CH_4	Methane	0
CCl_4	Carbon tetrachloride	0
CH_3Cl	Methyl chloride	1.87
CO_2	Carbon dioxide	0
CO	Carbon monoxide	0.112
SO_2	Sulfur dioxide	1.63
NO_2	Nitrogen dioxide	0.316
C_2H_6	Ethane	0
C_6H_6	Benzene	0

EXAMPLE 6.1 Explain why CH_3Cl has a large dipole moment whereas both CCl_4 and CH_4 have zero dipole moments.

SOLUTION CH_4 and CCl_4 have zero dipole moments because the C—H and C—Cl bond moments in both cases cancel each other, as shown in Fig. 6.4. However, in CH_3Cl there are three equal C—H moments whose resultant is in the same direction as the C—Cl moment. The resultant of the three C—H moments and the C—Cl moment then add to produce a dipole moment even larger than the C—Cl moment alone.

EXAMPLE 6.2 What can you say about the probable structure of NH_3 based on its nonzero dipole moment?

SOLUTION The nonzero dipole moment of NH_3 rules out a planar structure such as

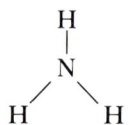

```
       H
       |
       N
      / \
     H   H
```

where each HNH angle is 120° (the three μ_{NH} would cancel in such a structure). The only likely alternative is a trigonal pyramidal shape; that is, the NH_3 molecule is like a tripod with the nitrogen atom supported by legs having hydrogen atoms for feet.

6.2 LEWIS FORMULAS AND MOLECULAR SHAPE

Although the shapes of some simple molecules can be deduced quite conclusively from dipole moments, in general this is not true, and recourse must be had to other experimental measurements. Nevertheless, experience has shown that it is often possible to guess the exact or approximate shape of a molecule using its Lewis formula as a basis. The method for doing this is called the **Valence-Shell Electron-Pair Repulsion** method (VSEPR for short).* The VSEPR method determines the geometric arrangement of atoms about a given atom by first counting the *pairs* of valence-shell electrons shown about that atom in the Lewis structure of the molecule in question. However, if two or more pairs of valence-shell electrons appear in multiple bonds, they are counted as one pair only. The pairs of valence-shell electrons surrounding the atom are then imagined to be placed on the surface of a sphere having that atom as its center. Next—the most crucial feature of the method—it is assumed that each pair of electrons repels all other pairs of electrons and, hence, the pairs arrange themselves on the spherical surface in such a way as to be as far apart as possible. This *hypothetical localization* of the electron pairs defines the apexes or corners of geometrical figures which in turn determine the places where other atoms can group themselves about the central atom.

For example, the carbon atom in CH_4 has an octet of electrons (as evident from the Lewis dot formula). An octet of electrons is four pairs, and four pairs of electrons

---

* Many chemists pronounce VSEPR as if it were spelled "vesper."

on the surface of a sphere maximize their distances apart if they define the corners of a regular **tetrahedron*** with the carbon atom in the center of the tetrahedron. Each hydrogen atom is then placed at the corner of the tetrahedron where the electron pair corresponding to a H—C covalent bond is located. Figure 6.5 illustrates three common ways of depicting the three-dimensional geometry of CH_4 on a two-dimensional surface. The HCH bond angles are all equal to 109° 28′; this angle is characteristic of a regular tetrahedron. Figure 6.6 depicts the shape of CH_4 somewhat more vividly.

As a second example, consider the ammonia molecule, NH_3. Just as in CH_4, the nitrogen atom is surrounded by a regular tetrahedron whose apexes or corners are electron pairs of the Lewis octet. However, there are only three hydrogen atoms

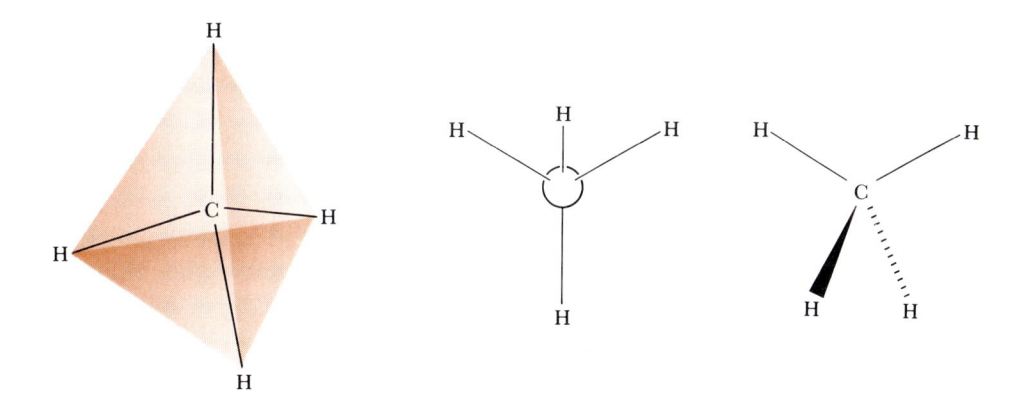

Fig. 6.5. Three common ways of depicting the three-dimensional geometry of methane.

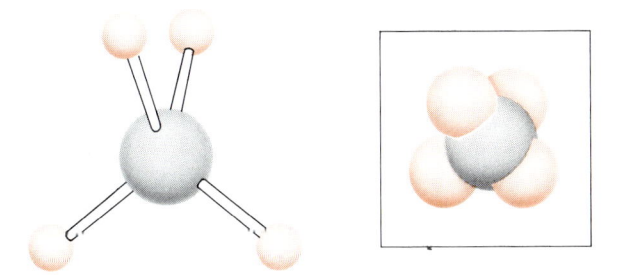

Fig. 6.6. Ball-and-stick model of methane, CH_4. Such models are intended to represent only the correct relative geometrical positions of atoms within a molecule. If the "balls" were used to represent the atom and, say, 95% of its density, then the "sticks" would disappear. The hydrogen atoms would appear partially embedded in the carbon atom, much as shown in the inset above. Models showing the latter features are widely used by chemists when they wish to obtain true-to-scale representations of the "space" requirements of atoms in molecules. Such models represent closely the volumes the molecules occupy in the solid state.

---

* A **tetrahedron** is a four-sided figure with triangular faces. When each face is an equilateral triangle, the tetrahedron is said to be **regular**.

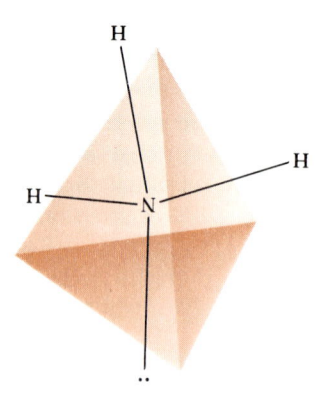

Fig. 6.7. The three-dimensional structure of ammonia as predicted by the VSEPR method. The shape is called a trigonal pyramid.

bonded to nitrogen, so that molecular NH₃ does not have the same geometry as CH₄. Thus NH₃ is represented as in Fig. 6.7. This shape is called a **trigonal pyramid**.*

Figure 6.7 suggests that each HNH angle is 109° 28′ as in CH₄. However, experiment shows that each HNH angle is a few degrees smaller than this—about 106° 47′. This fact is rationalized qualitatively by adopting an auxiliary rule: Lone-pair electrons take up more room than does a covalent-bond pair of electrons.† This means that the three HNH angles are squeezed somewhat closer together than 109° 28′. Consequently, the geometry of NH₃ resembles that of a tripod—a nitrogen atom standing on three legs, each having a hydrogen atom for a foot. This may be indicated as

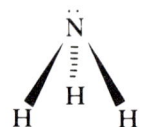

where the pair of dots symbolizes the lone-pair electrons.

The arguments of the preceding paragraph suggest that the geometry of the ammonium ion, NH₄⁺, should be tetrahedral with HNH bond angles equal to 109° 28′ as in CH₄. This is a consequence of the fact that when NH₃ bonds to H⁺ to form NH₄⁺, the lone pair of NH₃ becomes used in an N—H bond. All four N—H bonds are equivalent and a regular tetrahedral geometry results.

In the H₂O molecule there are also four pairs of octet electrons about the oxygen atom, and there are only two hydrogen atoms to form covalent bonds to oxygen. Thus the H₂O geometry is like that of CH₄ with two missing hydrogen atoms. The oxygen atom in the molecule also has two lone pairs of electrons. The geometry may

---

* A **trigonal pyramid** is actually a tetrahedron. Note, however, that the carbon in CH₄ is at the center of a regular tetrahedron whereas the nitrogen in NH₃ is at the apex of a nonregular tetrahedron.

† This rule is readily rationalized as follows: A pair of electrons that bond two atoms are attracted coulombically by two positive nuclei and thus find it difficult to move out of the bond axis region. Lone pairs, on the other hand, are strongly attracted by only one positive nucleus and can spread out in all directions much more easily.

be indicated as

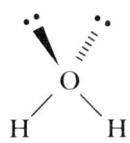

$$\begin{array}{c} \text{:O:} \\ \diagup \quad \diagdown \\ \text{H} \qquad \text{H} \end{array}$$

This is a **planar** molecule with the three atoms forming an isosceles triangle. The geometry is usually called **bent**. The lone-pair electrons do not actually occupy a fixed position in space; quantum theory requires that they behave more like a cloud of varying density. However, the maximum density of the lone-pair clouds does occur roughly in the vicinity of the tetrahedron corners. Similarly, the covalent-bond electron pairs have cloudlike distributions in space with maximum densities along the H—O bond axis.

In accord with the rule concerning the volume effect of lone-pair electrons, the HOH bond angle is not 109° 28' as in a regular tetrahedron but has the somewhat smaller value of 105° 3'. Note that the HOH bond angle is smaller than the HNH bond angle, a fact which is in agreement with the general model, since two lone pairs would be expected to require a greater volume than one lone pair.

The structure of formaldehyde, CH_2O, is slightly more difficult to deduce. The Lewis structure is

$$\begin{array}{c} \text{H:C:H} \\ \text{::} \\ \text{:O:} \end{array} \quad \text{or} \quad \begin{array}{c} \text{H—C—H} \\ \| \\ \text{:O:} \end{array}$$

In accord with the rule that multiple-bond electron pairs count only as one pair, there are three pairs of valence electrons about the carbon atom. These are placed on the surface of a sphere so that they form an equilateral triangle with the carbon atom at the center. This may be depicted as

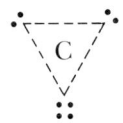

The hydrogen atoms now go to the single pairs and oxygen to the double pairs (as required by the Lewis formula). This produces the structure

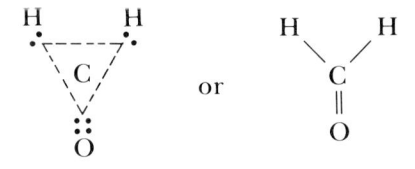

Experiment shows that all four atoms do lie in a plane as predicted. Furthermore, the HCH and HCO bond angles are close to 120° each. Because of the lone pairs of oxygen, the two HCO angles are slightly larger than the HCH angle.

The Lewis formula for CO_2 is

$$\ddot{\text{O}}::\text{C}::\ddot{\text{O}} \quad \text{or} \quad \ddot{\text{O}}=\text{C}=\ddot{\text{O}}$$

There are two pairs of electrons about carbon if each multiple bond is counted as one pair. Two pairs of electrons on the surface of a sphere maximize their distance apart by being on opposite sides of the carbon atom; thus CO_2 is a linear molecule.

The VSEPR model also works for Lewis formulas not obeying the octet rule. For example, BF_3 has the Lewis formula

$$:\ddot{\text{F}}:$$
$$\ddot{\text{B}}:\ddot{\text{F}}:$$
$$:\ddot{\text{F}}:$$

There are three pairs of electrons about boron. These maximize their distance apart if placed to form an equilateral triangle with boron at the center. Thus BF_3 has the planar structure

```
   F     F
    \   /
     B
     |
     F
```

The geometry about boron is sometimes said to be **trigonal**. The experimental FBF bond angles are exactly 120°.

Because of the similarity of formulas, BF_3 and NH_3 might be expected to have the same molecular shapes. However, the difference in the number of electron pairs about the central atom leads to quite different shapes for the two molecules.

The stable molecule PCl_5 has ten electrons about the phosphorus atom. Five pairs of electrons maximize their distance apart if they define a **trigonal bipyramid**, a six-sided figure resembling two trigonal pyramids placed base to base. Thus PCl_5 has the geometry shown in Fig. 6.8.*

Similarly, the molecule SF_6 has six pairs of electrons about the sulfur atom. These six pairs define a regular **octahedron**.† This is illustrated in Fig. 6.9. Note that all FSF bond angles are 90°.

EXAMPLE 6.3 Predict the shape of the molecule S_2Cl_2.

SOLUTION The Lewis formula of this molecule is

$$:\ddot{\text{Cl}}:\ddot{\text{S}}:\ddot{\text{S}}:\ddot{\text{Cl}}:$$

If either sulfur atom is used as a starting point, there are four electron pairs about sulfur, defining a tetrahedron with sulfur as its center. Thus the geometry about

---

* Note that six of the ClPCl bond angles are 90° and the other three are 120°.

† An **octahedron** has eight triangular faces; it is like two square-based pyramids base to base. A **regular** octahedron has an equilateral triangle for each face.

each sulfur atom must be

$$\begin{array}{c} \text{S} \\ \diagup \;\; \diagdown \\ \text{Cl} \quad\; \text{S} \end{array} \quad \text{and} \quad \begin{array}{c} \text{S} \\ \diagup \;\; \diagdown \\ \text{S} \quad\; \text{Cl} \end{array}$$

where the ClSS angle should be somewhat less than 109° 28′ due to lone-pair repulsions on sulfur. Consequently—putting the two together—the entire molecule is most likely to have a geometry between the two possibilities

$$\begin{array}{c} \text{S—S} \\ \diagup \quad\quad \diagdown \\ \text{Cl} \quad\quad\quad \text{Cl} \end{array} \quad \text{and} \quad \begin{array}{c} \quad\quad\quad\;\; \text{Cl} \\ \quad\quad\quad \diagup \\ \text{S—S} \\ \diagup \\ \text{Cl} \end{array}$$

The VSEPR method cannot make a more definite prediction than this.

EXAMPLE 6.4 Predict the shape of the sulfate ion SO_4^{2-}.

SOLUTION The Lewis formula is

$$\left[\; \begin{array}{c} :\ddot{\text{O}}: \\ :\ddot{\text{O}}:\text{S}:\ddot{\text{O}}: \\ :\ddot{\text{O}}: \end{array} \;\right]^{2-}$$

Since there are four pairs of electrons around sulfur and four oxygen atoms bonded to sulfur, the ion is a regular tetrahedron with sulfur at its center and the oxygen atoms at the apexes.

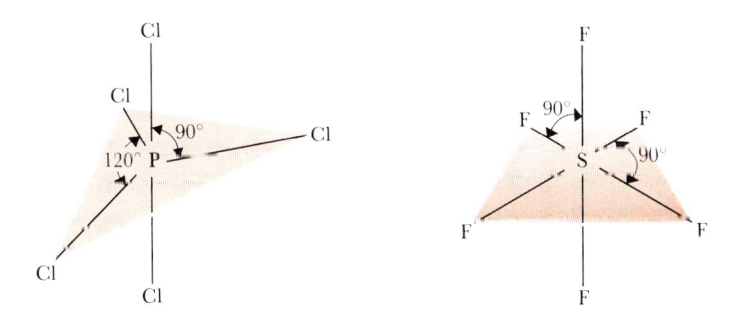

Fig. 6.8 (left). The three-dimensional structure of PCl_5 as predicted by the VSEPR method. The shape is called a trigonal bipyramid. For clarity the lines connecting apexes are omitted.

Fig. 6.9 (right). The three-dimensional structure of SF_6 as predicted by the VSEPR method. The shape is called octahedral. For clarity the lines connecting apexes are omitted.

EXAMPLE 6.5 Predict the shape of the molecule ClF$_3$.

SOLUTION A Lewis octet formula for this molecule (28 valence-shell electrons) cannot be drawn. Hence, assume that there are five pairs of electrons about chlorine and four pairs about each fluorine. This leads to a preliminary trigonal bipyramid about chlorine (see Fig. 6.10). As shown in Fig. 6.8, the five possible atom positions are not equivalent. The three positions at the apexes of the triangle are called **equatorial**; the other two positions are called **axial**. Axial-equatorial positions form an angle of 90° with the chlorine atom as apex; equatorial-equatorial positions form an angle of 120° with the chlorine atom as apex. Due to the large volume requirements of lone pairs they are less crowded at the equatorial positions than at axial positions. Consequently, the two lone pairs are assumed to have equatorial positions and the fluorine atoms are placed at the remaining equatorial position and the two axial positions. From this one would predict a T-shaped molecule with FClF bond angles of 90°. However, the lone pairs "squeeze" these angles somewhat to produce the distorted "T" shown in Fig. 6.10. Since the side Cl—F bonds are not equivalent to the middle Cl—F bond, the two types of bonds should differ in length. Experiment shows an FClF bond angle of 87.5° and bond lengths of 1.7 Å (side Cl—F's) and 1.6 Å (middle Cl—F).

This example illustrates another rule of the VSEPR method: Always put lone pairs in equatorial positions of a trigonal bipyramid.

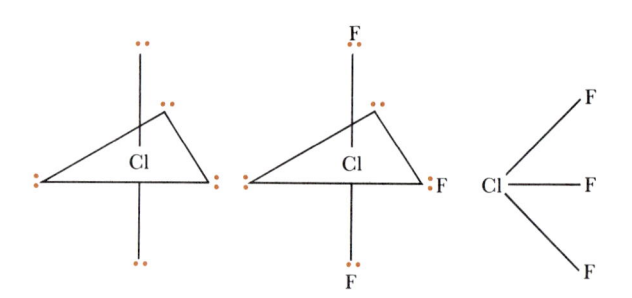

Fig. 6.10. Use of the VSEPR model to predict the three-dimensional structure of ClF$_3$.

6.3 HYBRID ORBITALS AND THEIR USE

Quantum mechanics describes bonding between atoms in a molecule in terms of the *overlapping* of the atomic orbitals of one atom with the atomic orbitals of another atom. This overlapping (or intermingling) of atomic orbitals represents a region of space where the bonding electrons are most apt to be due to their simultaneous attraction to two positive nuclei. However, the ordinary *s*, *p*, *d*, ... atomic orbitals of atoms do not always possess the geometric characteristics needed to bring about overlapping in the right places, that is, between pairs of bonded atoms. For example, it is not possible to overlap the 1*s* atomic orbitals of four hydrogen atoms with the 2*s* and 2*p* atomic orbitals of a carbon atom in such a way as to produce four C—H bonds with an angle of 109° 28' between each pair of bonds. Overlap of 1s_H and 2p_C's leads to three C—H bonds 90° apart while overlap of 1s_H and 2s_C produces no unique directional character whatsoever. Consequently, chemists have *invented* new

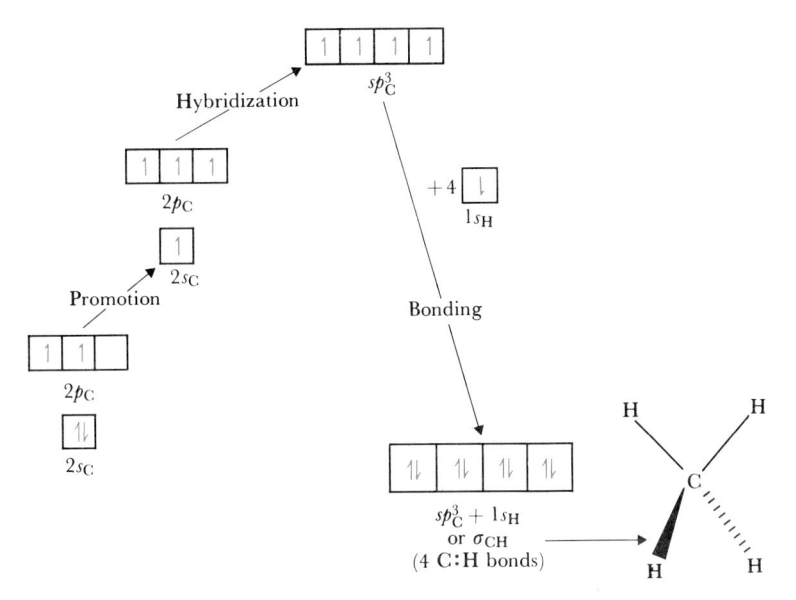

Fig. 6.11. Schematic illustration of the formation of CH₄ using sp^3 hybridization. The positions of orbital "boxes" are meant to suggest relative energies. For example, promotion and hybridization both require an input of energy whereas bonding involves an output of energy.

types of atomic orbitals called **hybrid orbitals**, which are much more flexible than ordinary atomic orbitals in describing a wide range of molecular geometries.

The description of bonding in CH₄ by means of hybrid orbitals is illustrated in Fig. 6.11. The bonding of four hydrogen atoms and one carbon atom is *imagined* to occur via a *hypothetical* process described as follows: A carbon atom in its (He)$2s^2 2p^2$ ground state is *promoted* to a state described by the electron configuration (He)$2s 2p^3$, where all four electrons have the same spin. The four atomic orbitals of carbon ($2s$, $2p_x$, $2p_y$, and $2p_z$) are then **hybridized** (mixed) to form four new orbitals called sp^3, or **tetrahedral**, hybrid orbitals. These hybrid orbitals may be imagined to arise in the following manner: One $2s$ atomic orbital and three $2p$ atomic orbitals ($2p_x$, $2p_y$, and $2p_z$) are placed into an orbital "blender" and thoroughly mixed. The resulting homogeneous "blend" is then divided into four equal parts. Each such part is one part (25%) $2s$ and three parts (75%) $2p$; hence the designation sp^3. The four sp^3 orbitals are equivalent in every way (like four peas in a pod) except in one way—they "point" in different directions; that is, their electron distributions are maximized along axes which are 109° 28′ apart. Hence the alternative name: tetrahedral hybrid orbitals.

Once the four sp^3 orbitals are formed, each can be overlapped with a $1s$ atomic orbital of hydrogen to form a C—H bond orbital. Each C—H bond orbital has the general form

$$\sigma_{CH} = sp^3_C + 1s_H \tag{6.2}$$

A **bond orbital** is simply an overlapping of the atomic orbitals of two bonded atoms. The designation σ_{CH} implies that the electron density in the C:H bond is like that of a σ molecular orbital of a diatomic molecule (see Section 5.4); that is, the increase in electron density due to bonding is greatest along the bond axis itself. The electron configuration of CH_4 may be written in the form

$$CH_4: (1s_C)^2(\sigma_{CH})^8$$

This assigns two electrons to the helium core, $(1s_C)^2$, of carbon and two electrons (with spins ↑↓) to each of the four σ_{CH} bond orbitals.

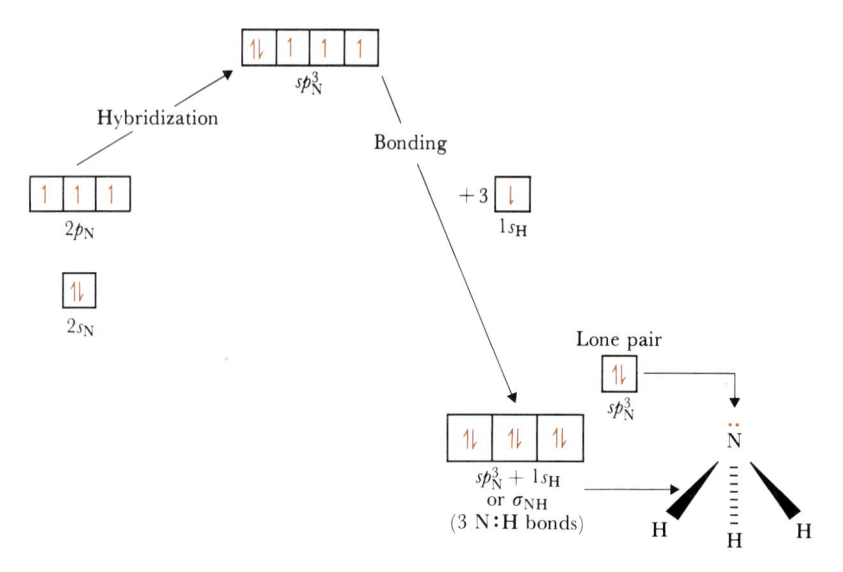

Fig. 6.12. Schematic illustration of the formation of NH_3 using sp^3 hybridization.

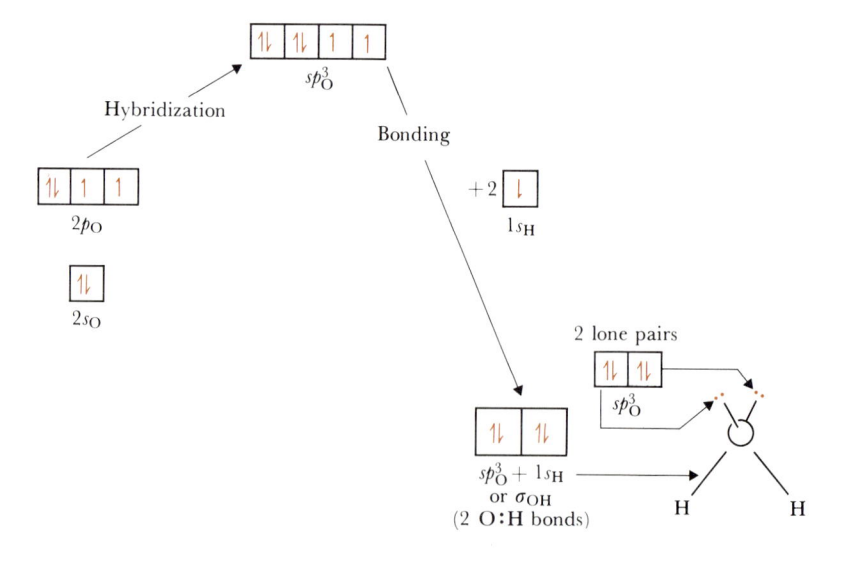

Fig. 6.13. Schematic illustration of the formation of H_2O using sp^3 hybridization.

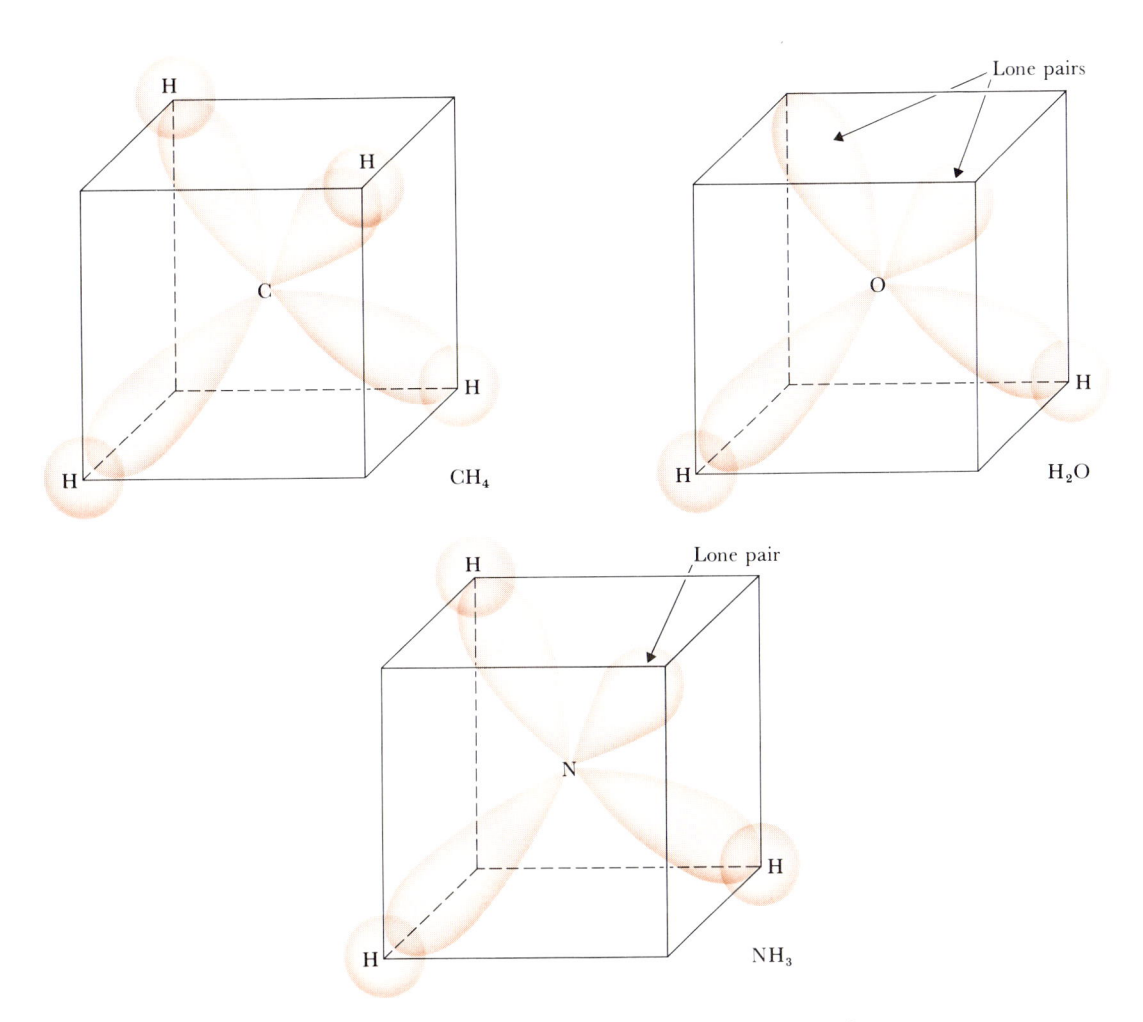

Fig. 6.14. Bonding in methane, ammonia, and water in terms of sp^3 hybrid orbitals and overlap with 1s hydrogen orbitals. The overlap region represents the increase of interatomic electron density constituting the C:H, N:H, and O:H covalent bonds. Each sp^3 orbital has a region of electron density on the opposite side of the bonding region (see Fig. 6.15) which is not shown in the diagram.

Figures 6.12 and 6.13 illustrate analogous procedures for describing bonding in NH_3 and in H_2O by use of sp^3 hybrid orbitals. Actually, it is not strictly correct to use sp^3 hybrid orbitals in these latter molecules since they do not have bond angles of 109° 28′. However, these discrepancies are unimportant for qualitative use of the orbitals. The electron configuration of NH_3 is

$$NH_3: (1s_N)^2(\sigma_{NH})^6(sp_N^3)^2$$

which assigns two electrons to the helium core of nitrogen, six electrons to N:H bonding orbitals, and two electrons (the lone pair) to a nonbonding sp^3 hybrid orbital of nitrogen. Similarly, the H_2O electron configuration is

$$H_2O: (1s_O)^2(\sigma_{OH})^4(sp_O^3)^4$$

Figure 6.14 illustrates the use of sp^3 hybrid orbitals for the description of bonding in CH$_4$, NH$_3$, and H$_2$O in terms of boundary surface diagrams. Figure 6.15 shows a contour diagram of an *sp*^{3} hybrid orbital in a plane containing the hybridized atom (carbon, nitrogen, or oxygen). The electron density in an *sp*^{3} hybrid orbital resembles that in a *2p* atomic orbital (see Fig. 4.10) except for the unequal sizes of the lobes. Overlap of 1s$_H$ is with the larger *sp*^{3} lobe; the smaller lobe is usually omitted in illustrations such as Fig. 6.14.

There are two very important things to keep in mind concerning hybridization: First, it is *not* a physical phenomenon or actual process the atom goes through, but rather a convenient way of thinking about chemical bonding, and, second, hybridization is used not to predict molecular geometry but to describe that geometry once it is known by other means (for example, from experiment or use of the VSEPR model).

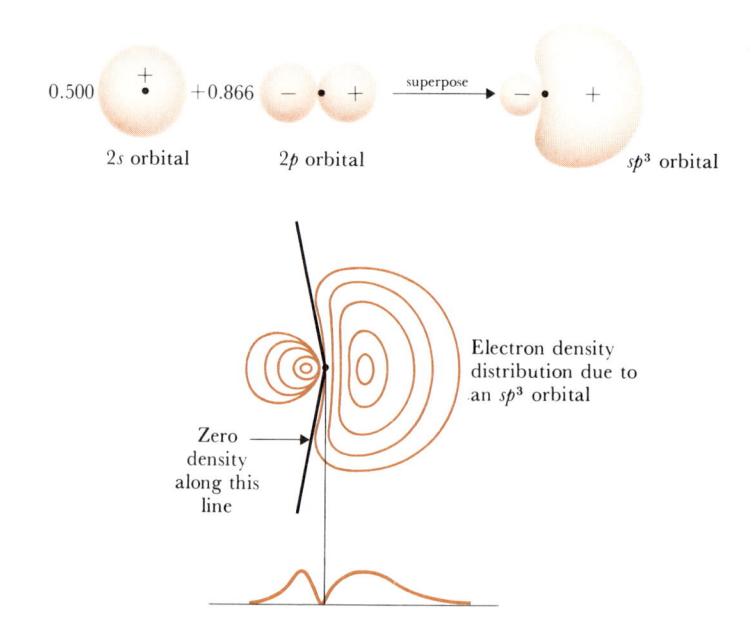

Fig. 6.15. Formation of an *sp*^{3} hybrid orbital from a *2s* and a *2p* orbital. The (+) and (−) signs attached to the orbitals are part of their strictly mathematical properties. When two (+) areas are overlapped (superposed), the result is a larger (+) area; when a (+) area and a (−) area are overlapped, the result is a diminished area. Note that the *sp*^{3} electron density resembles that of a *2p* orbital (Fig. 4.10) but is skewed to one side. The numbers 0.500 and 0.866 mean that the resulting electronic density is $(0.500)^2 \times 100 = 25\%$ of *2s* origin and $(0.866)^2 \times 100 = 75\%$ of *2p* origin.

To describe bonding in a trigonal molecule (such as BF$_3$), a different type of hybrid orbital is needed; the three B—F bonds lie in a plane 120° apart, so three equivalent hybrid orbitals also lying in a plane are needed for boron. Such hybrid orbitals, called ***sp*^{2}** or **trigonal** hybrid orbitals, are made by mixing one part *2s* with

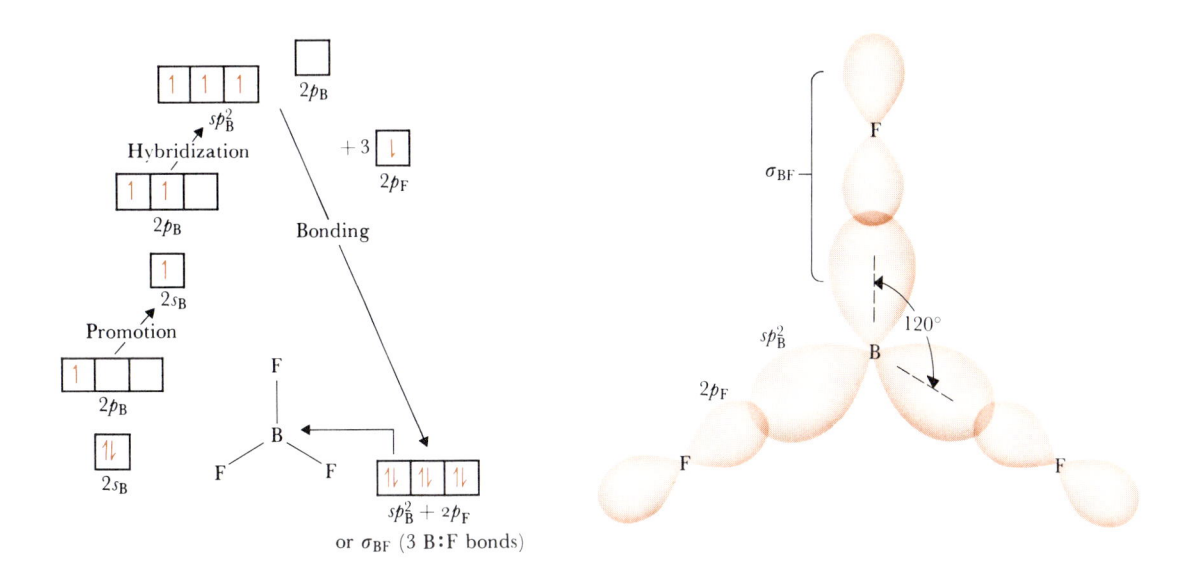

Fig. 6.16 (left). Schematic illustration of the formation of BF$_3$ using sp^2 hybridization. Note that one $2p$ atomic orbital of boron does not become included in the sp^2 hybrid orbitals.

Fig. 6.17 (right). Simplified boundary surface diagram of the bonding in BF$_3$. Each B:F bond is represented by an overlap of a boron sp^2 hybrid orbital and a fluorine $2p$ orbital. Orbitals for fluorine lone pairs are not shown.

two parts $2p$; this procedure is illustrated in Fig. 6.16. Each lobe of an sp^2 hybrid orbital resembles the large lobe of the sp^3 hybrid orbital depicted in Fig. 6.15; sp^2 orbitals also have a small lobe which is usually not shown. Figure 6.17 shows how the bonding in BF$_3$ is usually illustrated with simplified boundary surface diagrams. The electron configuration of BF$_3$ may be written

$$BF_3: (1s_B)^2(1s_F^2 2s_F^2 2p_F^4)^3(\sigma_{BF})^6$$

Some chemists prefer to depict the fluorine atom in terms of sp^3 hybrid orbitals, forming the σ_{BF} bonding orbital by overlap of sp_B^2 and sp_F^3. This puts the lone pairs of fluorine into sp_F^3 hybrid orbitals rather than into $2s_F$ and $2p_F$ as shown. Such an electron configuration might be written

$$BF_3: (1s_B)^2(1s_F)^6(\sigma_{BF})^6(sp_F^3)^{18}$$

The first electron configuration is adequate for most purposes even though the second probably describes the overall electron distributions somewhat better.

Chemical bonding in the ethylene molecule is also described by use of sp^2 hybrid orbitals. This is illustrated in Fig. 6.18. Each C:H bond is represented by the overlap of a carbon sp^2 hybrid orbital and a hydrogen $1s$ orbital. The double bond (C::C) is represented by an overlap of two carbon sp^2 hybrid orbitals (the σ_{CC} bond) and by an overlap of two parallel carbon $2p$ orbitals (the π_{CC} bond). This latter bond utilizes the $2p$ atomic orbitals left over on the carbon atoms upon forming the sp^2 hybrid orbitals.

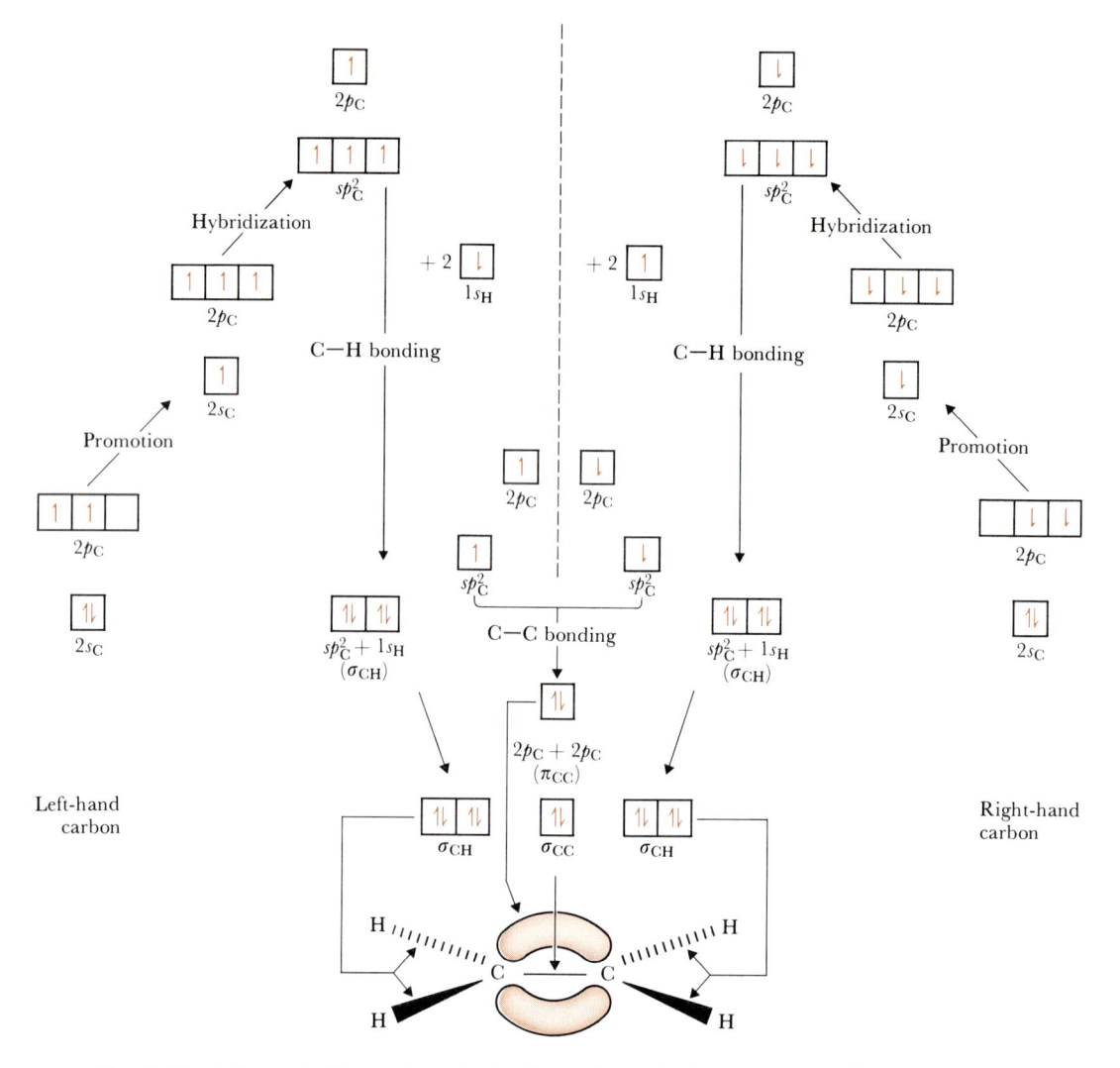

Fig. 6.18. Schematic illustration of the formation of ethylene using sp^2 hybridization. Each carbon atom has one $2p$ atomic orbital which is not included in the sp^2 hybrids; the π_{CC} bond is formed by overlap of the $2p$ atomic orbitals of adjacent carbon atoms.

The electron configuration of ethylene may be written

$$C_2H_4: (1s_{C_1})^2(1s_{C_2})^2(\sigma_{CH})^8(\sigma_{CC})^2(\pi_{CC})^2$$

Figure 6.19 shows contour diagrams for the σ_{CH}, σ_{CC}, and π_{CC} bond orbitals of ethylene.

Bonding in acetylene is described by **sp** or **digonal** hybrid orbitals; these consist of one part $2s$ and one part $2p$. A pair of digonal orbitals on the same carbon atom make an angle of 180° with each other (see Fig. 6.20). The C:H bonds are formed by overlap of $1s$ of hydrogen and sp of carbon, the C∷C triple bond consists of a σ_{CC} (overlap of two carbon sp's) and two π_{CC} (each an overlap of two carbon sp's—those $2p$'s not used to make sp). The two π_{CC} bond orbitals are not shown in Fig. 6.20. The electron configuration of acetylene is

$$C_2H_2: (1s_{C_1})^2(1s_{C_2})^2(\sigma_{CH})^4(\sigma_{CC})^2(\pi_{CC})^4$$

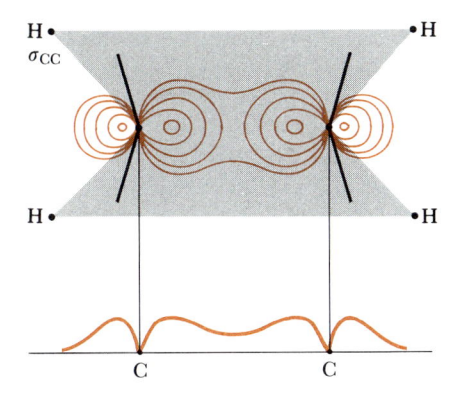

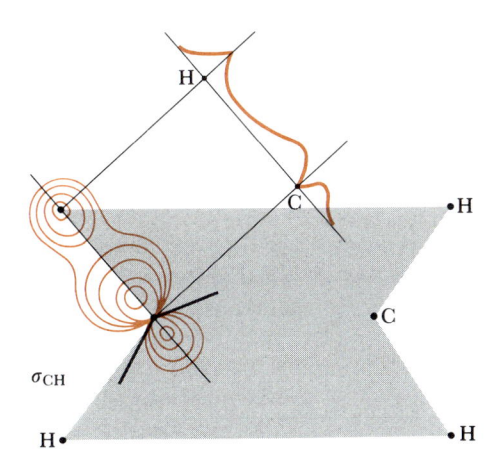

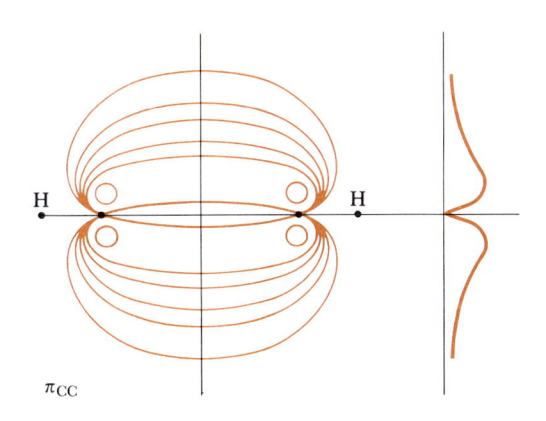

Fig. 6.19. Bond electron densities in the ethylene molecule based on the configuration $(1s_C)^4(\sigma_{CH})^8(\sigma_{CC})^2(\pi_{CC})^2$. The large black dots represent the location of the six atoms of ethylene. In the π_{CC} diagram, only four atoms show since one is viewing the molecule edgewise to the plane of the atoms.

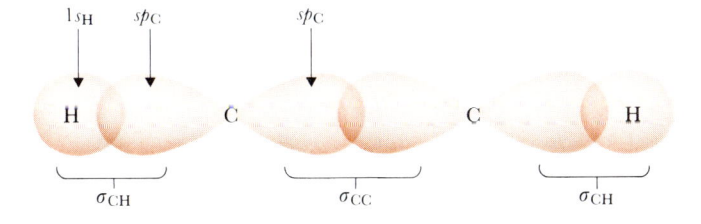

Fig. 6.20. Simplified boundary surface diagram of the bonding in acetylene. The two π_{CC} bonds are not shown; these account for a cylindrical region of increased electron probability about the C—C bond axis.

Since acetylene is a linear molecule, the two π_{CC} bonds cannot be regarded as fixed in space, but rather as if they were spinning about the bond axis. This means that the region of space possessing an increased electron density due to the π_{CC} bonds resembles a hollow cylinder about the C—C bond axis.

6.3 | HYBRID ORBITALS AND THEIR USE   **183**

EXAMPLE 6.6 Discuss the bonding in formaldehyde in terms of the appropriate hybrid orbitals.

SOLUTION Since formaldehyde has the planar structure

$$\begin{array}{c} H \quad\quad H \\ \diagdown \quad \diagup \\ C \\ \| \\ O \end{array}$$

it is convenient to describe each C—H bond as an overlap of a $1s$ atomic orbital of hydrogen and an sp^2 hybrid atomic orbital of carbon. The C=O bonds are a σ bond (overlap of sp^2 of carbon with sp^2 of oxygen) and a π bond (overlap of $2p$ of carbon with $2p$ of oxygen). The maximum electron density of the π bond lies above and below the plane of the molecule. The lone pairs of oxygen are assumed to be in sp^2 atomic orbitals of oxygen. Note that formaldehyde and ethylene have very similar electron distributions; the oxygen atom in conjunction with its two lone pairs is *electronically* similar to the CH_2 part of ethylene.

EXAMPLE 6.7 What type of hybrid orbitals (sp, sp^2, or sp^3) would you need to describe the bonding in each of the numbered atoms in the molecule below?

Note that it is customary to omit lone pairs in chemical formulas. Each nitrogen atom above has one lone pair and each oxygen atom has two lone pairs.

SOLUTION The student must be able to count to four and to remember the following rules in order to assign the hybrid orbitals correctly.

1. The presence of two pairs of electrons about an atom implies linear geometry and, hence, sp hybrids.
2. The presence of three pairs of electrons about an atom implies trigonal geometry and, hence, sp^2 hybrids.
3. The presence of four pairs of electrons about an atom implies tetrahedral geometry and, hence, sp^3 hybrids.

Counting of electron pairs about each numbered atom leads to:

1. sp^3 (four pairs)   4. sp^3 (four pairs)   7. sp^3 (four pairs)
2. sp^2 (three pairs) 5. sp^2 (three pairs)
3. sp^2 (three pairs) 6. sp (two pairs)

Hybrid orbitals to describe various other geometries have been invented. For example, the trigonal bipyramidal geometry of PCl_5 is described by using phosphorus dsp^3 hybrids. Each dsp^3 hybrid is one part $3d$, two parts $3s$, and three parts $3p$. Similarly, octahedral hybrids, d^2sp^3, may be used to describe the bonding in a molecule such as SF_6. Some of these latter hybrid orbitals are discussed in Chapter 22.

In all of the preceding discussions of chemical bonding, it has been implied that chemical bonds are electron distributions confined exclusively to pairs of neighbor atoms. In actuality, the entire molecular framework should be viewed as swimming in a "sea" of electrons. In a stable molecule, the attractions between this electron sea and the nuclei are greater than the nuclear repulsions and the electron repulsions. Such a picture implies that every atom should be considered as bonded to every other atom in the molecule. Thus in ethylene there would be fifteen bonds involving the six atoms two at a time (see Fig. 6.21). However, only five of the above bonds (four single and one double) are associated with large enough internuclear electron densities to merit designation as chemical bonds. Hence, as a very good approximation, the other bonds are neglected.

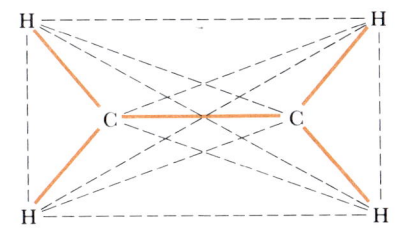

Fig. 6.21. Atom-to-atom bonding in ethylene. The solid lines indicate bonds associated with rather large increases in internuclear electron probabilities. Dotted lines represent much weaker bonding which is usually ignored in ordinary discussions.

6.4 INTERMOLECULAR FORCES

The forces that operate between molecules differ only in degree but not in kind from the forces that bind atoms together to form molecules. Each electron in the universe is attracted simultaneously by every nucleus in the universe. At the same time, all the electrons in the universe repel each other and all the nuclei in the universe also repel each other. At any given instant the balance of forces is such that the net effect is very strong localized attractions in some parts of the universe, less strong attractions elsewhere, and even localized net repulsions in some areas. As stated earlier, those aggregates of atoms called molecules are simply localized regions of very strong attractions which are convenient to isolate—both physically and conceptually—as an aid toward a classification of matter. The very fact that so many compounds exist as liquids and solids is due to the existence of interatomic and intermolecular attractions. If there were no such forces, all substances would exist as ideal gases. Phenomena such as adhesion and friction also depend on intermolecular forces.

The energies required to separate the atoms of simple molecules such as H_2, HCl, H_2O, CH_4, N_2, and CO_2 are usually on the order of 1 to 11 eV; most molecules regarded as stable at 25 °C usually require at least 4 eV. The energies required to separate molecules from each other (such as in a liquid) are usually on the order of 0.2 to 2 eV per pair of molecules.

The forces acting between molecules are conveniently classified into the following four categories.

1. Forces due to a dipole–dipole interaction between two polar molecules. The interactions are due to electrostatic attractions between oppositely charged ends of the dipoles. These are called **dipole–dipole forces**.
2. Forces in which a polar molecule induces a dipole moment in a nonpolar molecule and the two dipoles then interact. These are called **dipole-induced dipole forces**.
3. Forces due to attraction between the electrons of one molecule and the nuclei of another molecule. These forces are referred to as **London dispersion forces**.
4. Forces between molecules in which hydrogen atoms are directly bonded to either oxygen, nitrogen, or fluorine atoms. These forces are called **hydrogen-bonding forces**.

The intermolecular attractions between a pair of molecules may involve one, two, three, or all of the above types of forces.

Dipole–dipole forces account for part of the intermolecular attractions between a pair of polar molecules such as HCl, H_2O, or NH_3. London dispersion forces are also operative but are of negligible magnitude relative to the dipole–dipole forces. Hydrogen-bonding forces are the most important in H_2O and NH_3.

A molecule such as O_2 dissolved in a polar solvent such as H_2O would be subject to dipole-induced dipole intermolecular forces. Specifically, the polar H_2O molecule distorts the electron charge distribution of O_2 so that this latter molecule acquires an *induced* dipole moment. The two dipoles, one permanent and the other induced, interact just as do two permanent dipoles. Again, London dispersion forces would be operative but relatively weak.

Nonpolar molecules such as H_2, CH_4, and CO_2 have intermolecular interactions due only to London dispersion forces. In general, London dispersion forces increase with increasing molecular weight, since the greater the molecular weight the greater is the number of electrons which can be attracted to all the nuclei in question. Thus H_2, F_2, and Cl_2 are gases at room temperature, Br_2 is a liquid, and I_2 is a solid. Figure 6.22 illustrates the interaction of two H_2 molecules due to London dispersion forces.

The first three types of intermolecular forces are often collectively designated as **van der Waals forces**.

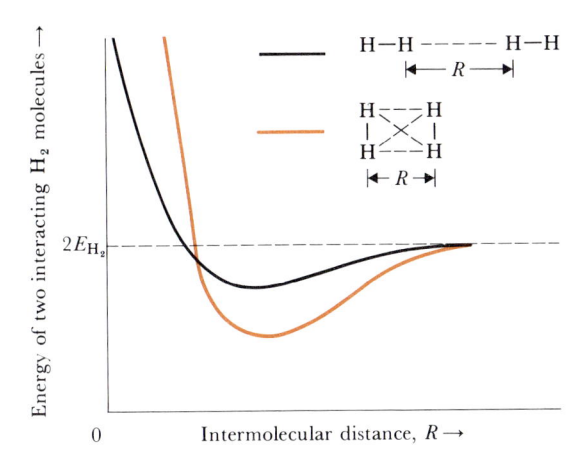

Fig. 6.22. Interaction between two hydrogen molecules as a function of intermolecular distance. Although intermolecular interactions can occur for all possible relative orientations of the two molecules, only two simple possible orientations are shown. The dissociation energy of H_4 (to $2H_2$) would be expected to be very, very small. Compare the above illustration with Fig. 5.2 for a diatomic molecule.

The van der Waals forces are so named after Johannes D. van der Waals (1837–1923), a Dutch physicist best known for his development of an equation describing gases (see Section 3.11). Van der Waals received the Nobel prize in physics in 1910 for his work on relationships among the phases of matter.

Hydrogen-bonding forces are sometimes very large and resemble chemical bonding forces in many respects. In H_2O, for example, the hydrogen atom may be viewed as participating in a quasi-chemical bond with the oxygen atom in another H_2O molecule.

In a sense the hydrogen atom acts as if it doesn't know which oxygen atom it belongs to and, consequently, becomes indentured to both.

It is easy to demonstrate that hydrogen bonding involves much more than just a dipole–dipole interaction. For example, H_2O and H_2S have dipole moments of 1.84 D and 0.97 D, respectively—not different enough to account for the fact that H_2O is a liquid at 25 °C whereas H_2S is a gas.* Similarly, HF and NH_3 are far easier to liquefy than would be expected on the basis of dipole–dipole interactions alone.

In Table 6.2 are listed the dipole moments and boiling points (at 1 atm pressure) of some common molecules. Many of the trends mentioned are apparent from an examination of this table. For example, the boiling points of CH_4, C_2H_6, and C_3H_8

---

* For example, the dipole moment of HBr is about twice that of HI yet the two molecules have similar boiling points (see Table 6.2).

TABLE 6.2 DIPOLE MOMENTS AND BOILING POINTS OF SOME COMMON MOLECULES

MOLECULE	DIPOLE MOMENT (D)	BOILING POINT (°C)
F_2	0	−188
Cl_2	0	−34.6
Br_2	0	58.8
I_2	0	184.4
HF	1.82	19.5
HCl	1.03	−84.9
HBr	0.82	−57
HI	0.44	−35.4
H_2O	1.84	100.0
H_2S	0.97	−60.7
NH_3	1.47	−33.4
PH_3	0.58	−87.4
CH_4	0	−164
C_2H_6	0	−88.6
C_3H_8	0.08	−42.1

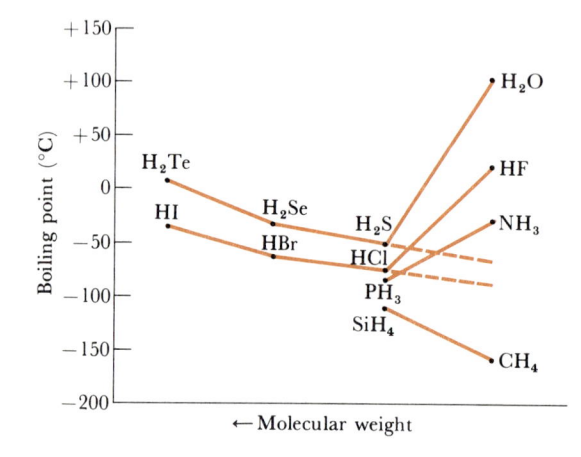

Fig. 6.23. The effect of hydrogen bonding on the boiling points of certain hydrides. Note that except for H_2O, HF, and NH_3, boiling points tend to decrease as the molecular weight decreases.

increase as the molecular weight increases. The intermolecular forces in these molecules are predominantly London dispersion forces. Such forces are roughly proportional to the number of electrons in the molecule. Figure 6.23 shows the abnormally high boiling points associated with H_2O, NH_3, and HF compared with the boiling points of similar molecules in which hydrogen bonding is absent.

The energy required to break a hydrogen bond varies from about 0.2 eV in alcohols (carbon-hydrogen compounds containing the —O—H group) to almost 2 eV in HF_2^-, but it takes less than 0.5 eV to break most hydrogen bonds.

A general representation of the hydrogen bond is

—X_1 ···· H ···· X_2—

where X_1 and X_2 are nitrogen, fluorine, or oxygen atoms with X_1 and X_2 in different molecules. Hydrogen bonding, and other intermolecular interactions as well, play an important role in determining the physical and chemical properties of water, in the

188 MOLECULAR GEOMETRY

structure of the genetic material of living cells, in the structure of proteins, and in many other areas of biology and chemistry.

In general, hydrogen-bonding forces are the strongest of the intermolecular forces, with dipole–dipole, dipole-induced dipole, and London dispersion forces coming next in that order.

SUMMARY

Whenever the centers of negative and positive charges of a molecule do not coincide, the molecule is polar and is said to have a dipole moment. The dipole moment of a molecule is the vector resultant of its bond moments. Whether or not a given molecule has a dipole moment is often a valuable clue to the molecule's shape.

The three-dimensional shape of a molecule often can be predicted on the basis of its Lewis formula. The valence-shell electron pairs of an interior atom of a molecule are imagined to be as far apart as possible on the surface of a sphere about that atom, and the other surrounding atoms are located at these electron-pair positions.

Quantum mechanics shows that there is a large probability of finding electrons between those atoms predicted to be bonded in Lewis structures. In order to be able to discuss the three-dimensional arrangement of bonds in a molecule by molecular orbitals, it is convenient to invent hybrid orbitals to replace the usual hydrogenlike atomic orbitals of atoms. Hybrid orbitals are mixtures of these usual orbitals but have the advantage of providing electron distributions similar to those predicted by Lewis formulas and the VSEPR method. To determine what hybrid orbitals to use for a given atom, count the number of electron pairs about it; for two pairs use sp, for three pairs use sp^2, and for four pairs use sp^3.

There are forces of attraction between molecules and between nonbonded atoms. Such attractions arise from dipole–dipole interactions, from dipole-induced dipole interactions, and from the interactions that always exist among all electrons and all nuclei of matter. A very important type of intermolecular attraction, the hydrogen bond, arises when two very electronegative atoms of two different molecules share a hydrogen atom bonded to only one of the molecules.

LEARNING GOALS

1. Given the three-dimensional shape of a molecule, be able to predict whether or not the molecule is polar.

2. Given the dipole moment of a molecule, be able to predict possible three-dimensional shapes for the molecule.

3. Given the Lewis formula of a molecule or ion, be able to predict its approximate three-dimensional shape.

4. Know what effect lone-pair electrons have on the shape of a molecule.

5. Given the formula of a molecule or ion, be able to describe each bond in terms of the appropriate hybrid orbitals.

6. Know the principal characteristics of sigma and pi bonds in molecules.

7. Know the principal types of intermolecular forces, how they arise, and their relative importance in various types of molecules.

8. Know what a hydrogen bond is, in what types of molecules it is found, and how it determines the nature of a substance such as water.

DEFINITIONS, TERMS, AND CONCEPTS TO KNOW

dielectric constant
capacitance
polar bonds
polar molecules
bond moment
dipole moment
VSEPR model

trigonal geometry
trigonal bipyramid
tetrahedron
octahedron
hybrid orbitals
sp
sp^2

sp^3
bond orbitals
dipole–dipole forces
dipole-induced dipole forces
London dispersion forces
van der Waals forces
hydrogen bonding

QUESTIONS AND PROBLEMS

Polarity and Dipole Moments

1. Would you expect the molecule with the Lewis formula shown below to have (a) zero dipole moment, (b) a large dipole moment, or (c) a small dipole moment? Explain.

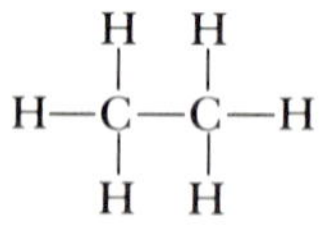

$$\text{H–C(H)(H)–C(H)(H)–H}$$

2. A molecule with the molecular formula $C_2H_2Cl_2$ is known to be planar. Three possible structures are

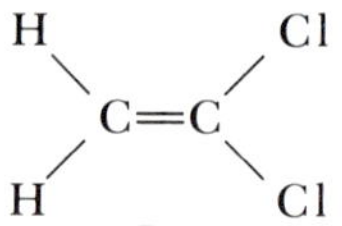

(I) cis H,H on one C, Cl,Cl on other... (II) H,Cl / Cl,H (III) H,H / Cl,Cl

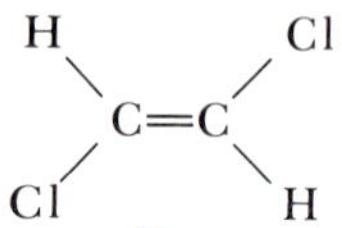

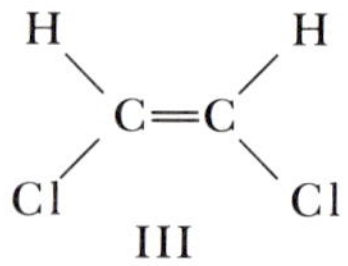

a) If $\mu = 0$, which is the structure?
b) What can you conclude if $\mu \neq 0$?
c) In what directions are the dipole moments of I and III?

3. Explain, using specific examples, how a polyatomic molecule can have polar bonds and yet have no dipole moment.

4. There are three different compounds with the empirical formula $C_6H_4Cl_2$. All three are planar and have the geometries indicated below. (Only one resonance form of each is shown.)

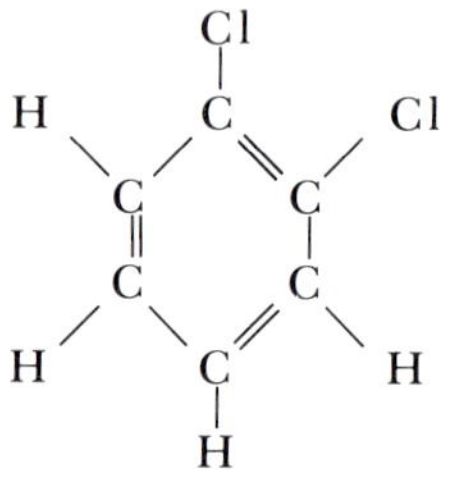

(I, 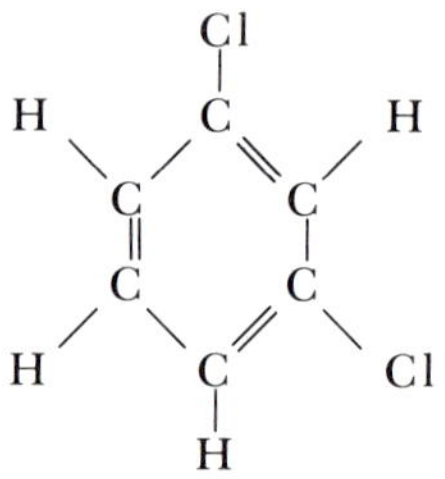 II, 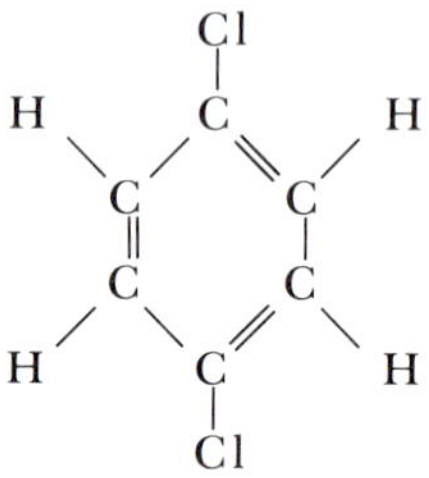 III — dichlorobenzene isomers)

These molecules have dipole moments of 0, 1.72 D, and 2.50 D (not in that order). Match each structure with the expected dipole moment.

5. Gaseous NaCl has an R_e of 2.4 Å, a D_e of 4.25 eV, and a dipole moment of 9.0 D. What would be the dipole moment of NaCl if $q^+ = 1.6 \times 10^{-19}$ C and $q^- = -1.6 \times 10^{-19}$ C (charge on an electron) separated by R_e? Can you suggest at least one reason for the discrepancy?

190   MOLECULAR GEOMETRY

6. The molecule NF₃ has a very small dipole moment (around 0.24 D) although electronegativity differences suggest that the N—F bond should be highly polar. Suggest how the nitrogen lone-pair electrons can account for this apparent discrepancy. Would you expect NH₃ to have a large dipole moment?

7. Discuss the expected dipole moment of formaldehyde based on the geometry given in the text.

8. Show that there is a good correlation between electronegativity differences of constituent atoms and molecular dipole moments in the series HF, HCl, HBr, and HI.

9. The molecule C₂H₄Cl₂ (ClH₂C—CH₂Cl) has virtually no dipole moment. Explain.

10. For which of the following substances would the dielectric constant be expected to change as the temperature changes: NH₃, CH₄, CO₂, CO, CH₂Cl₂, SF₆, CCl₄?

11. Discuss and compare bond polarities in the molecules CH₄ and CF₄. Does either molecule have a dipole moment?

12. Arrange the following molecules in increasing order of dipole moments: CH₂Cl₂, CF₄, CHCl₃, CH₃Cl, CCl₄.

13. The dipole moment of water is 1.84 D. Given that the HOH bond angle is 105°, estimate each O—H bond moment. This can be done graphically by using a protractor and drawing moments to scale.

14. Explain why the absence of a dipole moment proves the structure of CO₂ (linear) but does not do likewise for CH₄.

The VSEPR Method

15. Write Lewis formulas for each of the following. Predict the geometries and whether or not a dipole moment is expected.

a) MgCl₂
b) N₂O (NNO)
c) Cl₂O
d) POCl₃ (OP—Cl with two other Cl)
e) SiH₄
f) SOCl₂
g) NOCl (ONCl)
h) Cl₂O₇
i) S₂Cl₂ (ClSSCl)
j) HgCl₂
k) CS₂
l) AsF₅
m) H₂NOH
n) SO₂
o) SO₃

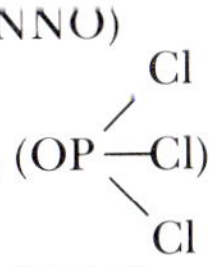

16. Formic acid has the Lewis dot formula

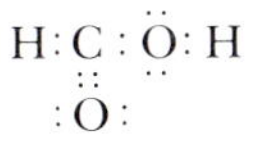

H:C:Ö:H
 ::
 :Ö:

a) Predict the geometrical structure of this molecule. How is the hydrogen atom attached to oxygen located relative to the rest of the molecule?

b) Which of the following is most likely to be true about the HC=O bond angle? Explain. (i) Equal to 120°. (ii) Slightly greater than 120°. (iii) Slightly less than 120°.

17. Two different compounds, ethanol and methyl ether, have the molecular formula C₂H₆O. Ethanol has the carbon and oxygen atoms in the order CCO whereas methyl ether has COC. Discuss the geometries of these two molecules.

18. Use the VSEPR model to discuss the expected geometries of the following molecules.

a) OCN⁻ (cyanate ion)
b) NCCN (cyanogen)
c) H₂NNH₂ (hydrazine)
d) O₃ (ozone)
e) SO₃²⁻ (sulfite ion)
f) PO₄³⁻ (phosphate ion)
g) NO₂⁻ (nitrite ion)
h) CO₃²⁻ (carbonate ion)
i) H₃O⁺ (hydronium ion)
j) PCl₃ (phosphorus trichloride)
k) N₂O (NNO) (nitrous oxide)
l) XeF₄ (xenon tetrafluoride)
m) C₂H₄ (ethylene)
n) 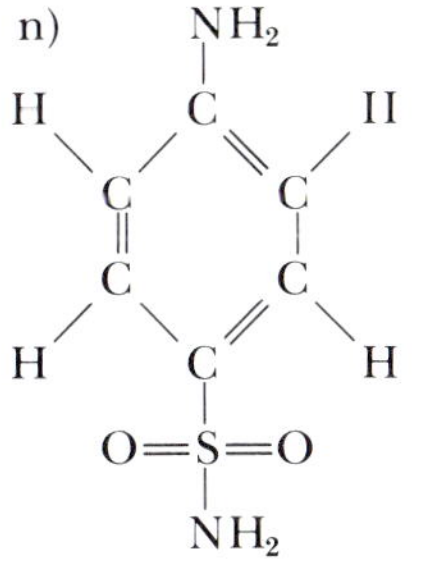 (sulfanilamide structure shown)

19. The C—H bond distance in CH₄ is about 1.07 Å. How far apart are any two hydrogen atoms? [*Note:* If a, b, and c are the sides of a triangle and A is the angle opposite a (between b and c), the following relationship holds: $\cos A = (b^2 + c^2 - a^2)/2bc$.]

20. Experiment shows that SF_4 has the geometry

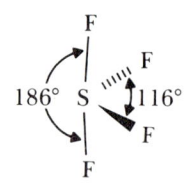

$$186° \quad S \quad 116°$$

with four F atoms attached to S.

Show how the VSEPR model accounts for this geometry.

21. Predict molecular geometries for each of the following.
   a) ICl_4^+ b) BrF_3 c) $SnCl_6^{2-}$

22. Compare geometries of the following molecules: SO_2, SO_3, SO_3^{2-}, and SO_4^{2-}.

23. Suggest a likely structure and geometry for the N_2O_5 molecule.

24. Compare geometries of the molecules HOCl, $HClO_2$, $HClO_3$, and $HClO_4$ (hydrogen is bonded to oxygen in each case.)

25. Suggest likely structures and geometries for Cl_2O and ClO_2.

26. Predict the structure and geometry of the molecule O_2F.

27. Xenon dioxide, XeO_2, is represented by the octet formula

$$:\ddot{O}:Xe:\ddot{O}:$$

where xenon contributes four pairs of electrons. What is the geometric shape of this molecule?

28. Discuss the possible existence and probable geometries of the molecule S_2O_2.

Hybrid Orbitals

29. Discuss the geometry of BF_3 (boron trifluoride) in terms of sp^2 hybrid orbitals about boron and sp^3 hybrid orbitals about fluorine. Would sp^2 or sp orbitals about fluorine work just as well? Explain.

30. Use appropriate orbitals to discuss the bonding and geometry of H_2O_2 (HOOH, hydrogen peroxide). Would you expect this molecule to be planar? Explain.

31. Discuss the bonding in formic acid (See Problem 16.)

32. Discuss the bonding in carbon dioxide, using appropriate hybrid orbitals for both carbon and oxygen.

33. Show that the bonding in molecules such as O_2 and N_2 may also be described in terms of hybrid orbitals and show what the expected bonds would be in terms of hybrid orbitals.

34. One possible Lewis structure for the HN_3 molecule is

$$H:\ddot{N}\diagup\overset{\ddot{N}}{\underset{\ddot{N}}{\|}}$$

Show, on the basis of hybrid orbitals, that this geometry is unlikely. What is a more probable geometry?

35. Assuming that BF_3 is described by boron sp^2 hybrids and unhybridized $2p$ atomic orbitals of fluorine, show why one would not expect a double bond involving overlap of the unused $2p$ of boron and a $2p$ of fluorine parallel to it. How is this related to the fact that the Lewis octet formulas for BF_3 are not taken seriously?

36. Discuss bonding in the molecule CH_3Cl in terms of appropriate hybrid orbitals.

37. Use hybrid orbitals to rationalize the following: Hydrazine (H_2NNH_2) is nonplanar, but the dipositive ion of this molecule is planar. [*Hint:* Consider sp^2 hybridization in the case of the ion.]

38. Guanine, a molecule which plays a role in the genetic code, has the structure

State what type of hybrid orbitals (sp, sp^2, or sp^3) best describe the electron distribution about each carbon and nitrogen atom. Each nitrogen atom has a lone pair of electrons not shown in the above.

39. Discuss the electronic structure of ethane, H_3C—CH_3, in terms of hybrid orbitals.

192 MOLECULAR GEOMETRY

40. Diborane, B_2H_6, has the rather unusual structure shown below

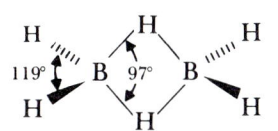

[Structure: H atoms bonded to two B atoms with bridging H's; angles 119° and 97° indicated]

where each boron atom is roughly tetrahedral. Show that if it is assumed that the geometry about each boron atom is exactly tetrahedral then each BHB bond may be described in terms of an overlap of two sp^3 hybrids of boron (one from each boron atom) and a $1s$ orbital of hydrogen.

41. The As—H bond angles in arsine, AsH_3, are very close to 90°. Which is a better description of the As—H bond, overlap of sp^3 of arsenic with $1s$ of hydrogen, or overlap of $4p$ of arsenic with $1s$ of hydrogen? Explain.

Intermolecular Forces

42. Which would be expected to have the higher boiling point, ethanol or methyl ether? (See Problem 17.)

43. Arrange the following in order of increasing boiling points: silane (SiH_4), methane (CH_4), plumbane (PbH_4), and germane (GeH_4).

44. Name all the types of intermolecular forces operative in the following.

a) CF_4
b) H_2NNH_2
c) CH_3OH
d) Ar
e) Between NH_3 and N_2
f) HBr
g) SO_2
h) C_6H_6

45. What is the *predominant* intermolecular force in each case in Problem 44?

46. Which would you expect to have the higher boiling point, CH_4 or CF_4? Explain.

47. Rationalize the trend in boiling points shown for the compounds below.

MOLECULE	BOILING POINT (°C)
CH_4	-161.5
CH_3Cl	-24
CH_2Cl_2	40
$CHCl_3$	61.3
CCl_4	76

48. Compounds with the general formula C_nH_{2n+2} are called alkanes. For $n = 1$ the compound is methane; its formula is CH_4 and it is a gas at 25 °C and 1 atm. For $n = 10$ the compound is decane; its formula is $C_{10}H_{22}$ and it is a liquid at 25 °C and 1 atm. When n is between 20 and 30 (approximately) the compounds are solids and are known collectively as paraffin wax. Explain the reason for the above trend.

49. How does dust collect on ceilings? What holds it there?

ADDITIONAL READINGS

Gillespie, R. J., "The Electron-Pair Repulsion Model for Molecular Geometry." *J. Chem. Educ.* **47**, 18 (1970). Discusses a sophisticated version of the VSEPR model.

Price, Charles C., *Geometry of Molecules*. New York: McGraw-Hill, 1971. This small paperback provides a complete discussion of the factors affecting the three-dimensional structures of molecules. Should serve as a source of inspiration to the mature student.

Ryschkewitsch, G. E., *Chemical Bonding and the Geometry of Molecules*. New York: Reinhold, 1963.

7
THE SOLID STATE

Crystals: An example of matter in the solid state. (Courtesy of Runle/Schoenberger, Grant Heilman.)

> Textbooks & Heaven only are Ideal;
> Solidity is an imperfect state.
> Within the cracked and dislocated Real
> *Nonstoichiometric crystals* dominate.
> Stray Atoms sully and precipitate;
> Strange holes, *excitons,* wander loose; because
> Of Dangling Bonds, a chemical Substrate
> Corrodes and catalyzes—surface Flaws
> Help Epitaxial Growth to fix adsorptive claws.
> JOHN UPDIKE*

Solids are in many ways the opposite of gases. Whereas gases have random, disordered structures, solids are characterized by a high degree of orderliness. Gases are also totally lacking in structural strength, whereas solids may have great rigidity and resistance to deformation.

Solids also have a greater variety of forms than do gases. Whereas the general gas law serves as a single model for all gases (at low pressures and/or high temperatures), there is no corresponding universal model for solids.

7.1 CRYSTALLINE AND AMORPHOUS SOLIDS

The solid state of matter is characterized by a definite volume and by a relative amount of rigidity or structural strength. Liquids also possess a definite volume but generally lack structural strength. Gases assume the volumes of their containers and lack structural strength entirely.

Solids may be classified into two basic types: One type is characterized by an orderly, fixed arrangement of units (usually atoms, ions, or molecules) according to some definite geometric pattern which, like that of wallpaper or yard goods, is repeated throughout the material. Such solids are called **crystalline solids** or, sometimes, **true solids**. The other type of solid shows no orderly arrangement of units. The latter are called **amorphous solids** or, preferably, **glasses**. In some respects glasses resemble highly viscous liquids more than they do crystalline solids.† The discussions in this text are limited almost exclusively to crystalline solids.

7.2 THE CRYSTAL LATTICE

The geometric patterns of crystalline solids may be determined by observing how beams of matter or radiation interact with them. For example, when an atom is subjected to X-ray radiation there is an interaction in which the X-ray photons are

---

* Copyright © 1968 by John Updike. Reprinted from *Midpoint and Other Poems*, by John Updike, by permission of Alfred A. Knopf, Inc.

† For a discussion of how glasses differ from true liquids, see David Dingledy, *J. Chem. Educ.* **39**, 84 (1962).

absorbed by the atom's electronic surroundings and reemitted in random directions. This phenomenon, which also occurs for radiation of other types, is known as **scattering**.

If the scattering atoms are in a plane, such as in a solid, there is a tendency for the reemitted X-rays to be more intense in a given direction. The effect is as if the plane of atoms reflected the X-rays much as a diffraction grating appears to reflect visible light.* If the incident X-rays have a wavelength on the order of the distances between the atoms, then X-rays "reflected" from different parallel planes will either reinforce each other (if they are in phase) or weaken each other (if they are out of phase). If the angle between the plane of the atoms and the direction of the incident X-rays is varied, the emergent X-rays have maximum intensities only at certain critical values of this angle.

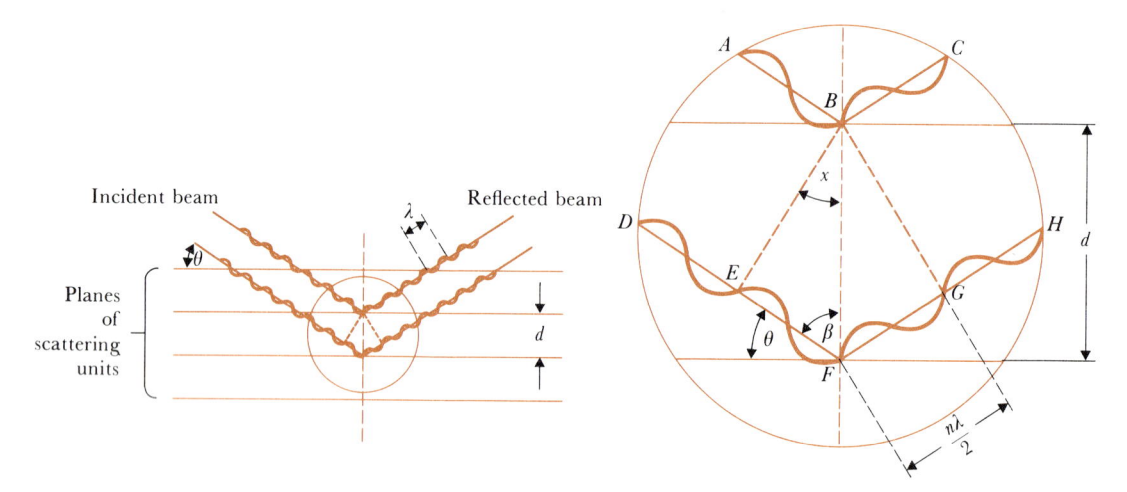

Fig. 7.1. Illustration of the use of X-ray scattering to determine the crystalline structure of solids. The angle θ is adjusted so that the extra distance EFG traveled by beam DFH equals an integral number of wavelengths λ; otherwise the emerging beams BC and GH will be out of phase. Since $\theta + \beta = 90°$ and $\beta + x = 90°$, it follows that $x = \theta$. Thus $\sin \theta = n\lambda/2d$ and $n\lambda = 2d \sin \theta$.

As illustrated in Fig. 7.1, the angle θ at which reinforcement of X-rays of wavelength λ is maximized is related to the distance d between the planes of atoms. The mathematical expression relating θ and λ is

$$n\lambda = 2d \sin \theta \qquad (7.1)$$

and is called the **Bragg equation**. The quantity n is an integer (1, 2, 3, . . .) known as the **order** of the "reflection." Its value can be deduced by observing how differ-

---

* A diffraction grating consists of many fine parallel lines scratched on a glass plate. White light passing through the grating is broken up into colors. You can get the same effect by sighting along a phonograph record against a strong source of light.

ent values of θ lead to successive maximizations of the intensity. For example, for a given crystal the smallest θ for which reinforcement is obtained for X-rays of wavelength 1.754 Å is 18° 7′. As θ is increased, the X-ray intensity drops and then again increases to a maximum at an angle of 38° 28′. Assuming $n = 1$ for the smaller angle, the distance between atomic planes is calculated to be

$$d = \frac{n\lambda}{2 \sin \theta} = \frac{1(1.754 \text{ Å})}{2(\sin 18° 7′)} = \frac{1(1.754 \text{ Å})}{2(0.311)} = 2.82 \text{ Å}$$

If the assumption of $n = 1$ is correct for $\theta = 18° 7′$, then using $n = 2$ for $\theta = 38° 28′$ should lead to the same value of d. The calculation leads to

$$d = \frac{2(1.754 \text{ Å})}{2(\sin 38° 28′)} = \frac{2(1.754 \text{ Å})}{2(0.622)} = 2.82 \text{ Å}$$

This establishes the correctness of the assignments of n values and of the calculated atomic plane spacing.

The Bragg equation is named for Sir William Henry Bragg (1862–1942), the British physicist who founded the field of X-ray crystallography. Sir William and his son, William Lawrence (b. 1890), were awarded the Nobel prize in physics in 1915.

By subjecting the crystalline solid to X-rays under a variety of orientations, all the possible planes of atoms, ions, or other scattering units can be deduced. This information establishes the relative positions of the scattering units within the crystal but does not identify the scattering units themselves. The latter may be identified by making use of the fact that the intensity of the scattered radiation depends not only on the angle θ, but also on the number of electrons around the scattering units. Mathematical procedures are then used to convert the scattering data into electron-density maps of the crystal; these maps often reveal the location and identity of the scattering units. For example, if X-ray studies are carried out on crystalline naphthalene, $C_{10}H_8$, it is possible to determine not only how the $C_{10}H_8$ molecular units are arranged in a repeating pattern but also how the atoms are arranged within each molecule.

Figure 7.2 represents an electron density map of naphthalene based on X-ray studies of solid naphthalene. The electron-density map clearly indicates the hexagonal array characteristic of the carbon atoms since each is associated with many electrons, but it does not clearly reveal the small electron densities associated with the hydrogen atoms. If neutron beams are used instead of X-rays, the positions of the hydrogen atoms become more clearly indicated.

The geometric pattern which denotes the locations of the repeating units in a crystalline solid is known as the **crystal lattice**. Studies of a variety of crystalline solids show that there are only 32 different types of crystal lattices based on a consideration of symmetry properties. These 32 types, or point groups as they are called, fall into

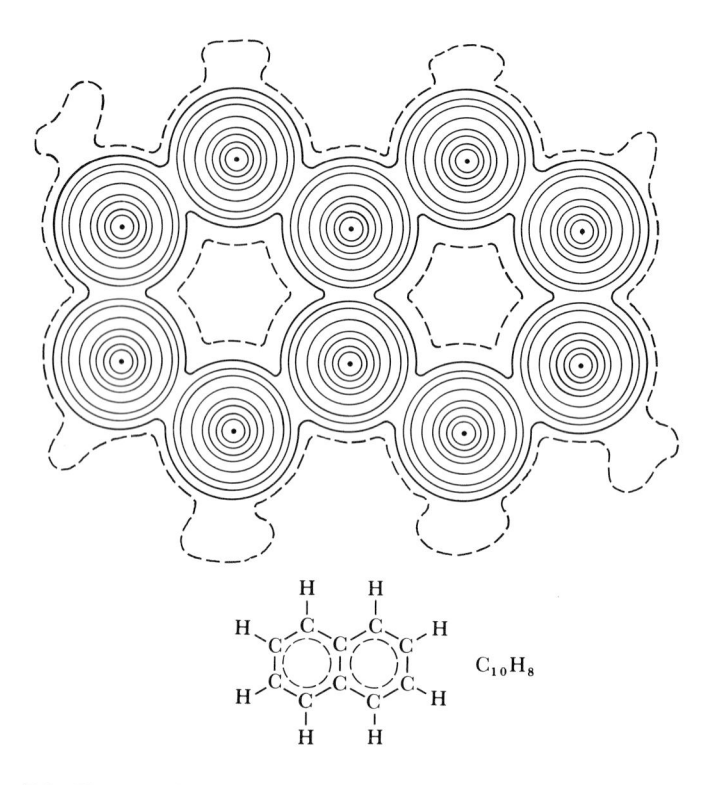

Fig. 7.2. Electron-density map of the naphthalene crystal as constructed from X-ray scattering intensities. The structural formula of naphthalene, $C_{10}H_8$, is shown below the electron-density map. Structures of some very complex molecules have been deduced in a similar way.

seven fundamental crystallographic systems based on axial angles and ratios. The seven systems are illustrated in Fig. 7.3.

Crystal lattice distances may be used, along with other data, to determine **Avogadro's number**, the number of things in a mole. For example, X-ray studies show NaCl forms a face-centered cubic lattice (Fig. 7.4) with a Na^+-to-Cl^- distance of 2.81 Å. Thus a cube of NaCl 2.81 Å on an edge will contain, on the average, one-half of a Na^+ ion and one-half of a Cl^- ion, since each corner of a cube is shared by eight cubes in all (four Na^+ shared by eight cubes is equivalent to $\frac{4}{8}$ or $\frac{1}{2}$ Na^+ ion and similarly for the four Cl^-). The density of NaCl is 2.165 g·cm^{-3}. A mole of NaCl, 58.443 g, thus occupies a volume of

$$V = \frac{58.443 \text{ g}}{2.165 \text{ g} \cdot \text{cm}^{-3}} = 26.99 \text{ cm}^3$$

This will contain 26.99 cm^3/(2.81 × 10^{-8} cm)^3 cubes each with $\frac{1}{2}Na^+$ ion and $\frac{1}{2}Cl^-$ ion. Thus the number of Na^+ ions or Cl^- ions in 58.443 g of NaCl is

$$\frac{26.99 \text{ cm}^3}{(2.81 \times 10^{-8} \text{ cm})^3} \times \frac{1}{2} = 6.08 \times 10^{23}$$

The accepted value of Avogadro's number is 6.02×10^{23}.

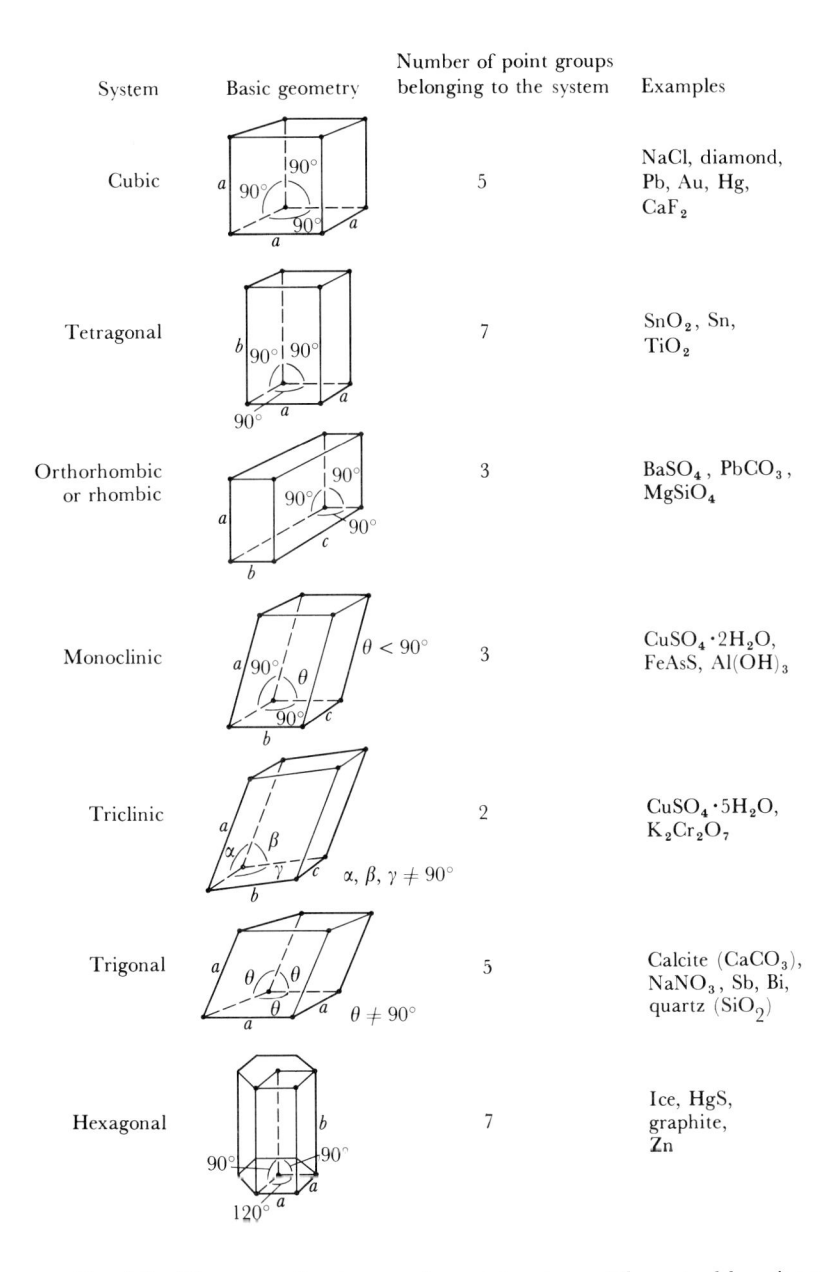

System	Basic geometry	Number of point groups belonging to the system	Examples
Cubic		5	NaCl, diamond, Pb, Au, Hg, CaF_2
Tetragonal		7	SnO_2, Sn, TiO_2
Orthorhombic or rhombic		3	$BaSO_4$, $PbCO_3$, $MgSiO_4$
Monoclinic	$\theta < 90°$	3	$CuSO_4 \cdot 2H_2O$, FeAsS, $Al(OH)_3$
Triclinic	$\alpha, \beta, \gamma \neq 90°$	2	$CuSO_4 \cdot 5H_2O$, $K_2Cr_2O_7$
Trigonal	$\theta \neq 90°$	5	Calcite ($CaCO_3$), $NaNO_3$, Sb, Bi, quartz (SiO_2)
Hexagonal		7	Ice, HgS, graphite, Zn

Fig. 7.3. The seven fundamental crystal systems. The actual locations of atoms, ions, or molecules may differ for a given system from those shown. For example, in the cubic system there are three basic variations called primitive (illustrated), body-centered, and face-centered.

Cl⁻

Na⁺

F⁻

Ca²⁺

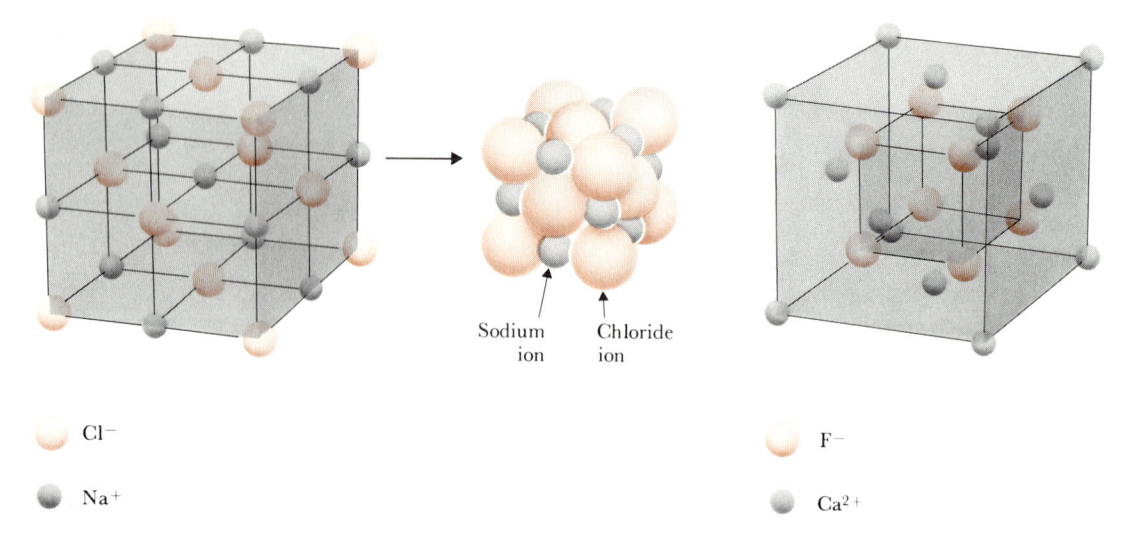

Fig. 7.4. The crystal lattices of NaCl (left) and CaF$_2$ (right). These substances represent two different variations of the cubic system. The NaCl crystal forms a face-centered cubic lattice. The CaF$_2$ crystal consists of a primitive cube of fluoride ions centered in a face-centered cube of calcium ions. The center drawing is a more accurate representation of the NaCl crystal, showing that there is very little empty space in a crystal and that chloride ions occupy much more space than do sodium ions.

7.3 VARIETIES OF CRYSTALLINE SOLIDS

The repeating unit in the crystal lattice may be an atom, an ion, a molecule, or, in some cases, the entire crystal may be viewed as a single giant molecule.

Atomic solids, which consist of repeated atom units, may be either **nonmetallic** (solid group VIIIA elements) or **metallic** (sodium, copper, iron). The former are poor conductors of heat and electricity, whereas the latter are excellent conductors. The nonmetallic solids are held together by relatively weak London dispersion forces and, consequently, have very low melting points. Metallic solids are also held together by London dispersion forces, but these are due mostly to rather loosely held valence electrons (recall that metals generally have low ionization energies). These "mobile" electrons provide a "sea" of negative charge surrounding the positive nuclei (and their inert-gas cores) to constitute moderately strong binding forces. Atomic metallic solids are generally quite ductile and malleable since the atoms are held together by moderately strong forces and are readily able to accommodate deformations due to stretching and compression.

Substances such as NaCl form **ionic crystals**, in which ions constitute the repeating units. NaCl has a cubic structure (see Fig. 7.4) in which Na⁺ and Cl⁻ alternate in occupying the corners of a cube. In CaF$_2$ (also illustrated in Fig. 7.4) the structure is also cubic but the ions Ca²⁺ and F⁻ are distributed differently than in NaCl. The CaF$_2$ crystal may be described as a cube of F⁻ ions located within a cube (of another type) of Ca²⁺ ions. Ionic crystals generally are brittle and have high melting points. Each ion must occupy a rather definite position in the crystal, where it is bound by electrostatic forces, and thus the crystal does not accommodate too readily to deformations such as bending.

Substances such as N_2, CO_2, H_2O, and C_6H_6(benzene) form **molecular crystals**, that is, crystals in which the repeating units are entire molecules. Usually the molecules are packed as efficiently as possible, although some exceptions occur. For example, the H_2O units in ordinary ice are packed in a much looser fashion than might be expected; in fact, the H_2O molecules in ice are farther apart on the average than those in liquid water, thereby accounting for the lower density of ice (see Fig. 15.7).

Some substances have different crystalline forms under different conditions of temperature or pressure, or both. For example, N_2 forms hexagonal crystals stable above $-238\,°C$ and cubic crystals stable below $-238\,°C$. Two or more different crystalline forms of a substance are known as **allotropes**. Sulfur has two well-known allotropic modifications: orthorhombic crystals stable below $95.5\,°C$ and monoclinic crystals stable above this temperature. Perhaps the best-known allotropes are those of carbon: graphite and diamond. The crystalline forms of these are discussed in Chapter 16.

Quartz (empirical formula SiO_2) is an example of an **extended lattice** crystal. The silicon and oxygen atoms occur in the ratio 1:2 but since no aggregate is uniquely definable as a molecule, the entire crystal is sometimes regarded as a single molecule. Figure 7.5 illustrates the relative arrangements of silicon and oxygen atoms as they occur in quartz and similar minerals. Note that each silicon atom is surrounded tetrahedrally by four oxygen atoms. The relative orientation of the different tetrahedra determines the type of mineral formed. Three different orientations found in

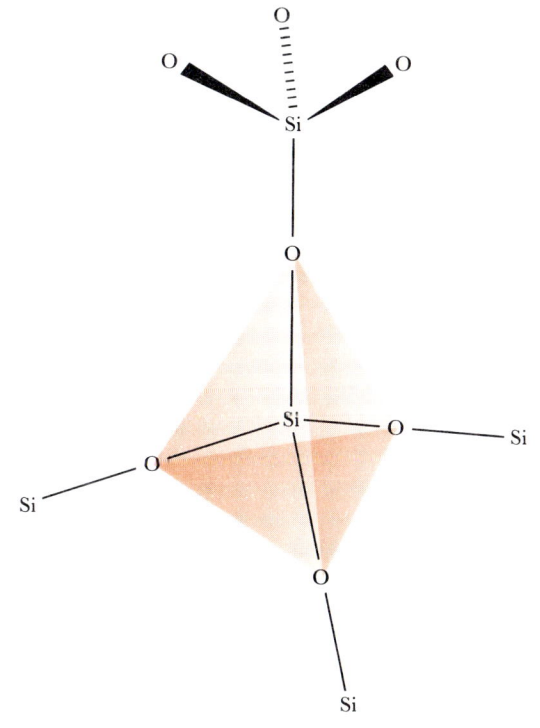

Fig. 7.5. The three-dimensional structure of $(SiO_2)_n$ minerals. The structure is an endless succession of tetrahedra, each consisting of a silicon atom surrounded by four oxygen atoms. For clarity, only two such tetrahedra are shown. Different minerals result depending on just how the tetrahedra are oriented relative to each other.

nature produce the minerals cristobalite, tridymite, and quartz. When some of the silicon atoms are replaced by aluminum, minerals known as **feldspars** are formed. These have the simplest formula $MAlSi_3O_8$, where M is a group IA cation such as Na^+ or K^+. The M^+ and Al^{3+} replace the 4^+ charge of silicon but only the aluminum occupies the silicon lattice site.

Ionic crystals are also a type of extended lattice crystal. Whereas the Si-to-O bonds are primarily covalent, the Na^+-to-Cl^- bonds are primarily electrostatic (ionic). Substances with extended lattice structures are sometimes called **macromolecules** (from the Greek *makros,* long). The term macromolecule is also commonly applied to other types of large molecules (see Chapter 24).

7.4 LATTICE ENERGIES OF SOLIDS

The energy required to break down a crystalline solid into its repeating units is called the **lattice energy** of that solid. If the crystal consists of n repeating units X (atoms or molecules) the breakdown process may be represented by

$$\underset{\text{solid}}{X_n(r = r_0)} \rightarrow \underset{\text{gas}}{nX(r = \infty)}$$

where r_0 is the distance between X units in a stable crystal. It is assumed that the separated X units are an infinite distance apart, that is, so far apart that they no longer interact. The energy required to carry out the above process divided by n is the lattice energy per repeating unit; it is often simply called the lattice energy. For example, the lattice energy of solid N_2 is 0.070 eV; this is the average energy required to remove an N_2 molecule from the solid, where r_0 is about 4 Å, to an infinite distance from all other N_2 molecules. The lattice energy is sometimes expressed in terms of $kJ \cdot mol^{-1}$, that is, the energy in kJ needed to separate a *mole* of X units from each other. For N_2 this quantity is

$$(0.070 \text{ eV per } N_2)(6.02 \times 10^{23} \text{ } N_2 \text{ mol}^{-1})(1.60 \times 10^{-22} \text{ kJ} \cdot \text{eV}^{-1}) \text{ eq } 6.74 \text{ kJ} \cdot \text{mol}^{-1}$$

For an ionic substance MX the lattice breakdown may be symbolized by

$$\underset{\text{solid}}{(M^+, X^-)_n(r = r_0)} \rightarrow \underset{\text{gas}}{(nM^+ + nX^-)(r = \infty)}$$

For example, it takes 8.0 eV to separate a Na^+, Cl^- pair in solid sodium chloride (where r_0 is about 2.8 Å) to the ions Na^+ and Cl^- an infinite distance apart. This lattice energy of 8.0 eV per ion pair is equivalent to 771 $kJ \cdot mol^{-1}$.

Table 7.1 lists the lattice energies of some molecular and ionic crystals. Note that polar molecules such as H_2O, NH_3, and HCl have somewhat higher lattice energies than nonpolar molecules such as N_2, O_2, and CH_4. Also, note that ionic substances have very high lattice energies. Lattice energies of ionic crystals are about as large as dissociation energies of stable diatomic molecules. A large part of the lattice energy of an ionic crystal may be accounted for on the basis of coulombic attraction between charges of opposite sign.

TABLE 7.1 LATTICE ENERGIES OF SOME MOLECULAR AND IONIC CRYSTALS

	MOLECULAR CRYSTALS		IONIC CRYSTALS		
	LATTICE ENERGY IN			LATTICE ENERGY IN	
SUBSTANCE	eV	kJ·mol^{-1}	SUBSTANCE	eV	kJ·mol^{-1}
N$_2$	0.07	6.74	NaCl	8.0	771
O$_2$	0.09	8.67	LiF	10.4	1002
CH$_4$	0.12	11.6	KF	8.3	799
H$_2$O	0.49	47.2	NaF	9.3	896
NH$_3$	0.30	28.9	CsF	7.5	722
HCl	0.22	21.2			

7.5 IONIC RADII IN SOLIDS

Although ions, like atoms, have no definite volumes, it is possible (as well as useful) to define **ionic radii**. To illustrate how this is done, consider the ionic solid NaF. The sum of the radii of the ions Na$^+$ and F$^-$ ($r_{Na^+} + r_{F^-}$) is just the distance between centers of adjacent ions in the crystal lattice; X-ray diffraction shows this distance is 2.31 Å. On the basis of Eq. (4.20), the individual radii are assumed to be inversely proportional to their effective nuclear charges. Thus there is obtained the relationship.

$$\frac{r_{Na^+}}{r_{F^-}} = \frac{Z_{eff}^{F^-}}{Z_{eff}^{Na^+}} = \frac{4.5}{6.5}$$

Although the Z_{eff} values used in Chapter 4 were adequate to predict periodic trends in ionization energies and atomic radii, a more refined method is necessary to obtain the values used above. The use of elementary algebra on the two equations (the sum of the radii and their ratio) leads to r_{Na^+} = 0.95 Å and r_{F^-} = 1.36 Å.

As might be expected, ionic radii exhibit periodic trends just as do neutral atomic radii (or volumes). This is illustrated in Fig. 7.6. The dotted line connects ions of the same group (IA in this case.) Note that as the atomic number increases, the ionic radius increases. Just as in neutral atoms, this result is predicted qualitatively on the basis that Z_{eff} is essentially constant within the group; but the principal quantum number of the valence shell increases in going down the group.

Positive ions are generally smaller than the corresponding neutral atoms since the nucleus has one more unit of charge than the electrons and, hence, can draw these in more tightly. Similarly, anions are generally larger than the corresponding neutral atoms since the extra electron leads to additional repulsions that "swell" the electron cloud.

The solid lines in Fig. 7.6 connect **isoelectronic ions**, that is, ions having equal numbers of electrons. As would be expected, as the charge on an ion in an isoelectronic series becomes more positive, the electrons are drawn in more closely and the radius decreases.

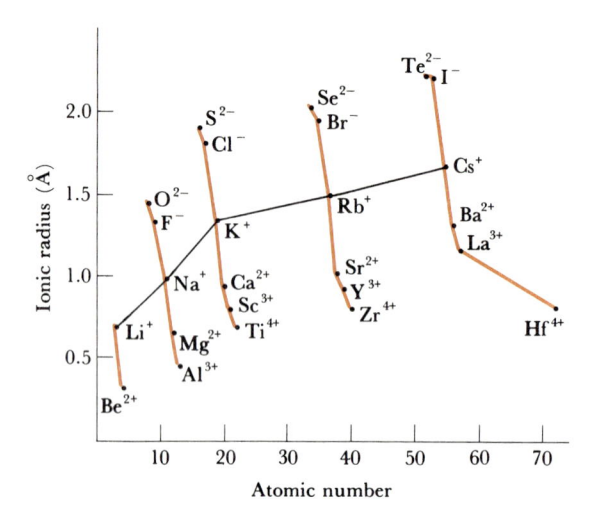

Fig. 7.6. The ionic radii of some cations and anions as a function of atomic number.

EXAMPLE 7.1 X-ray studies show that the interionic distance in solid NaCl is 2.81 Å. Use this information to determine a value for the radius of the chloride ion.

SOLUTION Since $r_{Na^+} + r_{Cl^-} = 2.81$ Å, it follows that $r_{Cl^-} = 2.81 - r_{Na^+}$. Assuming that r_{Na^+} is the same as in NaF (0.95 Å) leads to

$$r_{Cl^-} = 2.81 - 0.95 = 1.86 \text{ Å}$$

A continuation of this process leads to values of radii of other ions. Values obtained in this way vary slightly with the data source; Table 23.7 gives r_{Cl^-} as 1.81 Å.

7.6 VIBRATIONS IN SOLIDS

Although it has been tacitly implied that the repeating units of a solid occupy rigidly fixed positions in the crystal lattice, this is not strictly true. In actuality, the atoms of a solid are perpetually in motion, shivering or quivering about some point in the lattice.

A diatomic molecule AB is, in some respects, the simplest example of a crystal. According to quantum theory the two atoms A and B are not restricted to a fixed distance apart; rather, they move closer together and further apart in a perpetual periodic motion which is called **vibrational motion**. This vibrational motion, like the electronic energies of atoms and molecules, is quantized; that is, it can have only certain definite allowed energy values. The allowed vibrational energies are given by the equation

$$E_v = (v + \tfrac{1}{2})h\nu_0 \tag{7.2}$$

where ν_0 is the fundamental vibrational frequency (depending on the masses of A and B and the nature of the A—B chemical bond) and v is the vibrational quantum number having the values 0, 1, 2, ..., that is, zero or a positive integer. Thus the vibrational energy of the molecule AB may be represented as evenly spaced rungs of a ladder (see Fig. 7.7).

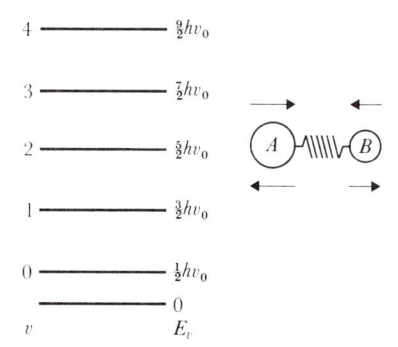

Fig. 7.7. Vibrational energy levels of a vibrating diatomic molecule. Note that a zero value of the vibrational energy is not allowed; the lowest possible vibrational energy is $\frac{1}{2}h\nu_0$. Only the lowest five energy levels are shown. The molecule AB is conveniently viewed as two balls connected by a stiff spring. Electronic energy levels (see Fig. 3.5 for the hydrogen atom), unlike vibrational energy levels, generally are not evenly spaced.

Equation (7.2) shows that the lowest possible vibrational energy (when $v = 0$) is $\frac{1}{2}h\nu_0$. This is called the **zero-point energy**. It is clear that the atoms A and B can never be at rest.

A similar situation holds in crystals. Even at absolute zero (discussed in Chapter 3), where all other types of motion cease entirely, vibrational motions of atoms in crystals persist. Consequently, what are depicted as fixed lattice points in diagrams of crystals are but average positions of vibrating atoms, ions, or groups of atoms.

7.7 DEFECTS IN SOLIDS

Real crystals, whether found in nature or made under carefully controlled laboratory conditions, are almost always flawed by **imperfections** or **defects**. Moreover, such defects account for important chemical and physical properties of many solids. Defects in solids may be ultimately due either to a missing repeating unit (atom, ion, or molecule) somewhere in the lattice or else to the presence of an impurity at some point of the lattice. In either event, defects can lead to a discontinuity in the way subsequent units are incorporated into the structure and may lead to what is often called a **screw dislocation** (see Fig. 7.8). Defects of the latter type are called **substitutional** defects, in contrast to **interstitial** defects, in which foreign atoms occupy otherwise empty spaces in a crystal.

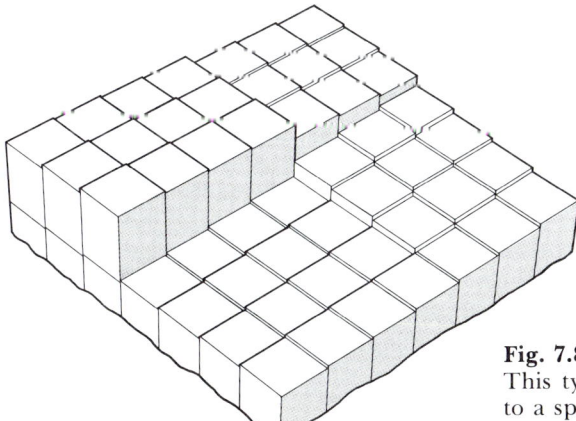

Fig. 7.8. A screw dislocation in a crystal. This type of dislocation may be likened to a spiral ramp. Each cube represents a repeating unit of the crystal.

The octahedral "hole"

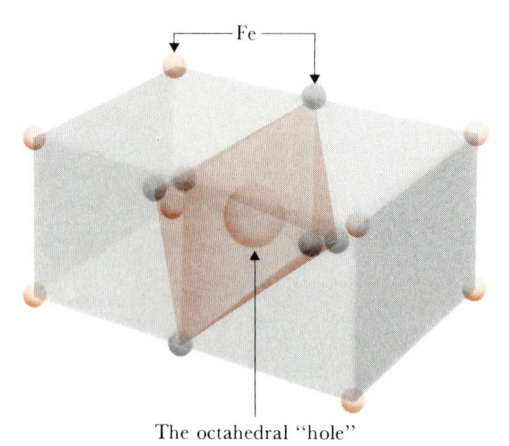

Fig. 7.9. The structure of carbon steel. The above shows two adjacent body-centered cubes. Each cube has an iron atom at each corner and an iron atom at its center. The octahedral hole, at the center of the two adjacent faces, is shown as a large sphere; the six iron atoms closest to this hole are colored black for emphasis.

Steel is an example of an interstitial defect structure; some of the spaces between the iron atoms in steel are filled with small atoms such as carbon, boron, or nitrogen. The iron atoms in steel are arranged as body-centered cubes (see Fig. 7.9). In high-carbon steel, carbon atoms are found occupying the centers of the faces of the cubes. This position, empty in pure iron, is called an **octahedral hole** since it is surrounded by eight iron atoms. Actually, the carbon probably exists as Fe_3C (a compound called **cementite**) and it is this unit which accounts for steel's unusually high ridigity. When steel is heated to redness and allowed to cool slowly, large Fe_3C crystals form and the steel becomes soft. This is called **annealing**. When the cooling takes place rapidly, the Fe_3C stays scattered throughout the lattice instead of becoming localized, and the steel becomes very hard. This process is called **tempering**. When the octahedral hole is occupied by a large atom such as sulfur the steel becomes excessively brittle. It may be that such an interstitial defect produces a dislocation because other atoms cannot pack around it properly.

Some of the impurities present in solids may have a drastic effect on the chemical properties of the substance. For instance, the corrosion of aluminum containing very small traces of iron and other metals may be so severe that boat hulls constructed from this material will not last a single season if used in salt water (see Section 19.6 on corrosion). On the other hand, properly purified aluminum will last indefinitely under the same conditions of use. The rate at which iron rusts is also highly dependent on the impurities present. An object made of iron will usually acquire a highly visible layer of rust very quickly—even overnight if left in the rain. Yet in New Delhi, India, there stands a pillar of iron (the Ashoka Pillar, sixty feet high, inscribed with six lines of Sanskrit) that has stood for 1500 years or more with little trace of rust. No chemical analysis of the pillar has ever been reported, but some people claim it is ultrapure iron; others maintain it is made of a rustless iron alloy. In any event, it is not always easy to decide whether the properties usually attributed to some of the elements are intrinsic to the substance or dependent on traces of impurities.

Pierre Louis Dulong (1785–1838) was professor of physics at the École Polytechnique in Paris and eventually became its director. He lost an eye while preparing NCl₃ in the laboratory, a compound of his own discovery. Alexis-Thérèse Petit (1791–1820) taught physics at the Lycée Bonaparte in Paris and held a titular professorship at the École Polytechnique, where he collaborated with Dulong.

7.8 THE SPECIFIC HEATS OF SOLIDS

In 1819 Dulong and Petit proposed that the molar heat capacities (specific heat times the mass of one mole) of the solid elements were all equal to approximately the same value, 6 cal·K^{-1}·mol^{-1} (about 25 J·K^{-1}·mol^{-1}).* This crude relationship, known as the **law of atomic heats**, found much use in determining approximate atomic weights of newly discovered elements. Once the specific heat of a new element has been measured experimentally, its approximate molar mass can be calculated by the formula

$$\text{mass of one mole} = \frac{6 \text{ cal} \cdot \text{K}^{-1} \cdot \text{mol}^{-1}}{\text{specific heat}}$$

For example, the specific heat of tin at 18 °C is measured to be 0.0542 cal·K^{-1}·g^{-1}. Thus the approximate mass of one mole of tin is

$$\frac{6 \text{ cal} \cdot \text{K}^{-1} \cdot \text{mol}^{-1}}{0.0542 \text{ cal} \cdot \text{K}^{-1} \cdot \text{g}^{-1}} = 111 \text{ g} \cdot \text{mol}^{-1}$$

This approximate molar mass can then be used to determine a more accurate value as follows: Suppose a compound of tin and oxygen is analyzed and found to be 88.12% tin. If the atomic weight of oxygen is known accurately, then the approximate molar mass of tin (111 g·mol^{-1}) leads to SnO as the simplest formula for the compound. This, in turn, implies that the exact weight of tin bound to one mole (16.00 g) of oxygen is given by

$$\frac{88.12 \text{ g Sn}}{11.88 \text{ g O}} \times 16.00 \text{ g} \cdot \text{mol}^{-1} = 118.7 \text{ g} \cdot \text{mol}^{-1}$$

This gives the atomic weight of tin as 118.7 amu. The accepted modern value is 118.70 amu.

In 1905 Einstein applied quantum theory to the specific heats of solids. Einstein assumed that a solid was a collection of vibrating atoms (oscillators) and that the energy of each oscillator was given by

$$E = nh\nu_0 \tag{7.3}$$

where ν_0 is the fundamental vibrational frequency of an oscillator and $n = 1, 2, 3, \ldots$ is a quantum number. This equation is similar (except for the $\frac{1}{2}$ factor) to Eq. (7.2).

---

* Most standard sources of specific heat data use the non-SI units of cal·K^{-1}·g^{-1}. Recall that 1 cal = 4.184 J.

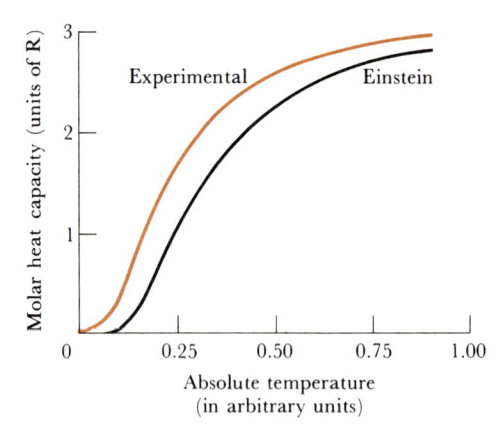

Fig. 7.10. Variation of molar heat capacity with temperature for a monatomic solid. At a high enough temperature (different for different solids), the molar heat capacity approaches $3R$ (about 6 cal·K^{-1}·mol^{-1}).

Absorption of heat by a solid means that the energies of the oscillators increase. Einstein was able to show mathematically that such a model led to a molar heat capacity which increases with temperature and approaches a limiting value at high temperatures. For monatomic solids this limiting value is $3R$ (see Fig. 7.10), where R is the ideal gas constant. Since R is very nearly 2 cal·K^{-1}·mol^{-1} (8.3 J·K^{-1}·mol^{-1}), it is clear why the law of atomic heats works; many monatomic solids have molar heat capacities close to the limiting value at room temperature.

7.9 THE ELECTRONIC STRUCTURE OF SOLIDS

When the atoms of a solid are separated by very large distances, there is little interaction between atoms, and each atom's electrons behave as if the atom were completely isolated. But as the atoms are brought closer together, the electrons of the atoms (particularly the valence electrons) begin to interact. Forgetting that they belong to only one nucleus, they tend to become strongly attracted by all the other nuclei in the solid. The orbital energy levels of the individual atoms now become replaced with a series of closely spaced levels called **bands**. In many respects, the band is simply a series of closely spaced molecular orbitals for an extremely large molecule; each such molecular orbital is a many-center superposition of a large number of atomic orbitals. In some solids there are **gaps** or **forbidden areas** in the energy bands, a certain range of energy values which no electrons in the metal may have. These band gaps provide an explanation of why some solids are electrical insulators and others are electrical conductors. The model utilizing electronic bands is usually termed the **band theory of solids**.

Electrical insulators are characterized by a completely filled band of electrons (the valence band) above which there is a fairly large band gap (usually over 1 eV in width). Above the band gap is a band of empty levels called the **conduction band** (see Fig. 7.11). In order for a solid to conduct an electrical current, electrons must be excited to the conduction band to act as charge carriers. Relatively small voltages are ordinarily inadequate to excite electrons through a band gap of around 1 eV, so the substance behaves as an insulator. However, very high voltages do induce such electron excitations; this is known as a **dielectric breakdown** of the insulator.

Everyday experience shows that even though insulators are poor conductors of electric current, they do conduct some electrical current even at low voltages, and, more important, they become somewhat better conductors as the temperature is increased. Band theory provides a simple explanation of this behavior: Only at absolute zero is the conduction band of an insulator completely empty; at any other temperature a few electrons get into this band by thermal excitation. The number of

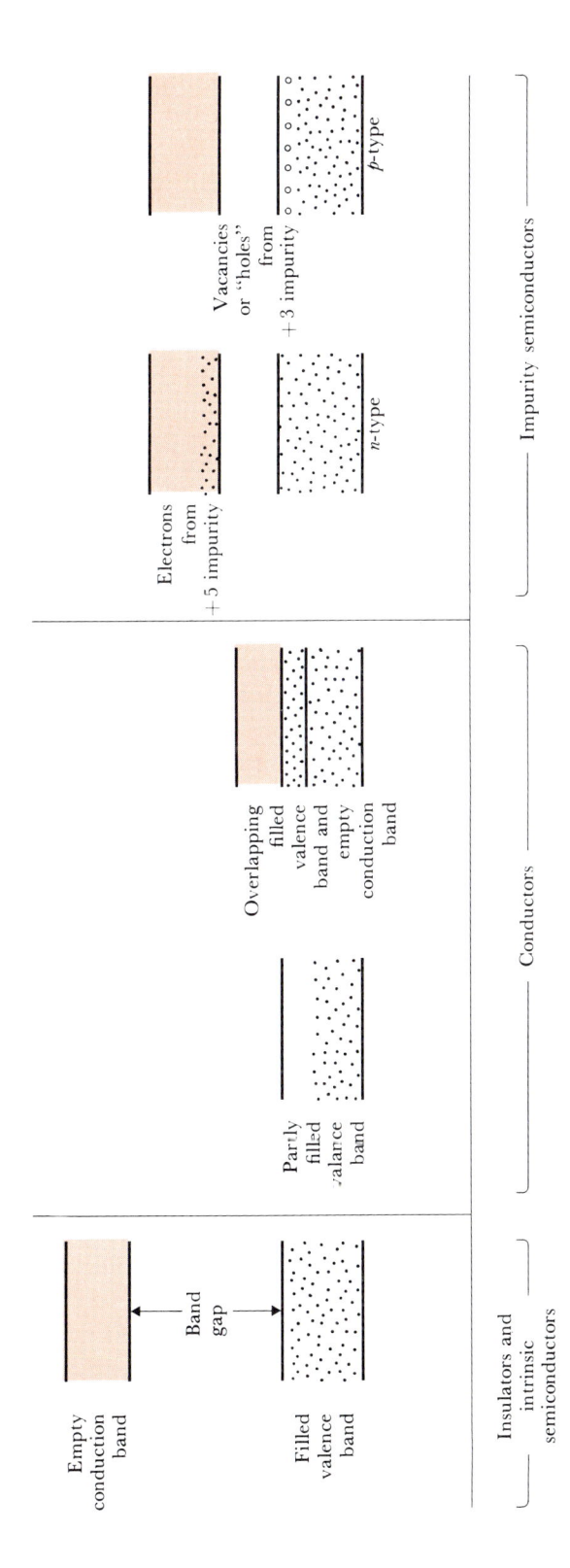

Fig. 7.11. Schematic diagram of the electronic structures of insulators, conductors, and semiconductors, based on the band theory of solids. The bands are actually sets of very closely spaced molecular orbital energy levels.

electrons in the conduction band is inversely proportional to the width of the band gap and directly proportional to the temperature.

If the band gap is relatively small, there may be a small electron population in the conduction band of an insulator at room temperature (around 25 °C) and the substance is known as an **intrinsic semiconductor**. Good insulators normally have conductivities* of about 10^{-17} Ω^{-1} cm^{-1} at room temperature, conductors have conductivities up to 10^7 Ω^{-1} cm^{-1}, and semiconductors have conductivities in an intermediate range from about 10^{-9} to 10^3 Ω^{-1} cm^{-1}.

Materials which are electrical conductors are characterized either by a partially filled valence band or by an overlap between the valence band and the conduction band. In the latter case there is no band gap. Whereas electrons in completely filled bands have no mobility to act as current carriers, electrons in partially filled bands behave in many respects as if they were free. The electrons in a completely filled valence band which overlaps an empty conduction band behave like electrons in a partially filled band.

Experiments show that the conductivity of a conductor typically decreases as the temperature rises. This is explained by assuming that as the temperature increases, the vibrational motions of the atoms in the lattice increase, thereby interfering with electron mobility.

Certain materials that are normally insulators can be made into semiconductors by the addition of small amounts of impurities. Such materials are called **doped** or **impurity semiconductors** to distinguish them from intrinsic semiconductors. Silicon and germanium are typical substances used to make doped semiconductors. The electron configurations of silicon and germanium, $(Ne)3s^23p^2$ and $(Ar)3d^{10}4s^24p^2$, respectively, indicate that both atoms have four valence-shell electrons. Addition of small amounts of elements such as phosphorus, tin, or arsenic (which have five valence-shell electrons per atom) to a silicon or germanium crystal is equivalent to adding a few electrons to the conduction band of the solid. This results in an ***n*-type** (*n* for **negative carrier**) semiconductor. Similarly, adding small amounts of elements such as boron, aluminum, or indium (which have three valence-shell electrons per atom) produces missing electrons or "holes" in the valence band of the crystal. This results in a ***p*-type** (*p* for **positive carrier**) semiconductor. In *n*-type semiconductors the current carriers are the electrons in the conduction band; in the *p*-type semiconductors the current carriers are the "holes" in the valence band.

An electron-hole pair of a semiconductor or insulator in an excited state is often called an **exciton**. Alternatively, the excited state itself is often regarded as an exciton, an excited state that can move from lattice cell to lattice cell.

Doped semiconductors have many uses in electronic devices; they can be used to make rectifiers (devices which change ac current to dc current), to serve as amplifiers, and in general to perform many of the functions formerly carried out by vacuum tubes and other devices. The main advantages of semiconducting electronic devices are their very low power requirements and their very small size. Their smallness has made possible the microminiaturization characteristic of space-age gadgetry, as in the ubiquitous pocket-size transistor radio and electronic calculator.

---

* The symbol Ω stands for **ohm**, a unit of electrical resistance. Since an ohm is the resistance that will allow a current of one ampere (A) to flow under a potential difference of 1 volt (V), a conductivity of 1 $\Omega^{-1} \cdot$cm^{-1} means that when the potential gradient is 1 V$\cdot$cm^{-1}, a current of 1 A (1 C$\cdot$s^{-1}) will flow through each 1-cm^2 cross-sectional area of the conductor.

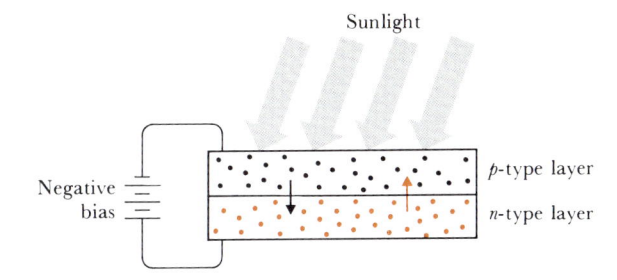

Fig. 7.12. Schematic diagram of a silicon photovoltaic cell. In the absence of light, a small negative bias prevents electrons and holes from moving across the p-n junction. Sunlight produces excess electron-hole pairs which do flow across the p-n junction and thus set up an electric current which may be used to charge a storage battery.

One of the most interesting practical devices based on semiconductors is the **photovoltaic cell**, a device that can transform the radiant energy of the sun directly to electrical energy. As shown in Fig. 7.12, the photovoltaic cell consists of a sandwich formed from a p-type wafer and an n-type wafer in which a small negative bias prevents the flow of electrons and holes across the junction. When sunlight hits the sandwich, some of the atoms are ionized to produce additional holes and electrons. Since the negative bias across the junction is too small to prevent the flow of the excess holes and electrons, an electric current flows across the junction. This current can be used to charge batteries which are then used as sources of energy—and also to maintain the constant polarity across the p-n junction. Many of the weather satellites in orbit are powered by photovoltaic cells. Widespread use of such devices has been hindered by the high cost of producing the p-n junctions, their relatively short lives, and the high cost of storage batteries. Recently developed methods for the mass production of p-n junctions appear to have solved the first problem; the others are receiving considerable research attention.

SUMMARY

Crystalline solids (also called true solids) consist of orderly arrays of ions, atoms, or molecules according to a definite repetitive geometrical pattern. All of the crystals found in nature belong to one of seven basic patterns: cubic, hexagonal, tetragonal, triclinic, trigonal, monoclinic, and orthorhombic. The exact geometric structure of a given crystalline substance may be determined by X-ray diffraction methods.

So-called amorphous solids have no regular arrangements of molecules and resemble liquids in this respect. These solids are also known as glasses.

The physical and chemical properties of solids often depend on defects and impurities in the crystal lattice. For example, steel is formed when certain otherwise vacant lattice sites of iron are occupied by carbon.

The law of atomic heats states that the mass of one mole of a monatomic solid element times the specific heat of the substance is very often equal to approximately 6 cal·K^{-1}·mol^{-1}. This relationship was once used to determine atomic weights of elements.

The electrical properties of solids may be explained on the basis of *band theory*. In this theory the electrons of solids occupy sets of closely spaced "molecular" orbitals called bands; these bands are separated by band gaps. The relative numbers of electrons occupying the valence and conduction bands and the width of the band gap determine whether a given solid is a conductor or an insulator.

Insulators with small band gaps may act as intrinsic semiconductors due to a small population of electrons in the conduction band. Doped semiconductors are made by adding impurities to insulators so as to place a small number of electrons into the conduction band (n-type) or to remove a few electrons from the valence band (p-type).

Photovoltaic cells are devices based on semiconductors and capable of utilizing the energy of sunlight to produce an electric current.

LEARNING GOALS

1. Know the difference between amorphous and crystalline solids.

2. Be able to define what is meant by a crystal lattice and know how this is determined by X-rays.

3. Know how Avogadro's number may be determined from X-ray data on crystals.

4. Be able to define nonmetallic and metallic solids, ionic crystals, molecular crystals, and extended lattice crystals and give an example of each.

5. Be able to define the lattice energy of a solid.

6. Know what is meant by the zero-point energy of a solid and what it implies.

7. Be able to discuss the role of defects in solids and to distinguish between substitutional and interstitial defects.

8. Know what is meant by a screw dislocation.

9. Know the difference between conductors, insulators, and semiconductors and be able to explain these in terms of band theory.

10. Know how a photovoltaic cell works.

DEFINITIONS, TERMS, AND CONCEPTS TO KNOW

X-ray scattering
glasses
atomic solids (metallic and nonmetallic)
ionic crystals
molecular crystals
extended lattice crystals
lattice energy of a solid

isoelectronic ions
crystal defect
screw dislocation
substitutional and interstitial defects
band theory
band gap
conductor

insulator
semiconductor
  (intrinsic and doped)
n-type semiconductor
p-type semiconductor
photovoltaic cell
dielectric breakdown

QUESTIONS AND PROBLEMS

X-Ray Structure Determination

1. A sample of graphite (an allotropic form of carbon) is studied with X-rays of wavelength 0.166 nm. First-order reflection occurs at an angle of 14° 18′. This is interpreted as the distance between layers of carbon atoms (see Chapter 16 and Fig. 16.2).

  a) What is the interlayer spacing in graphite?
  b) At what angle will second-order reflections occur?

2. First-order reflection of X-rays of wavelength 1.54 Å from adjacent layers of ions in a NaCl crystal occurs at an angle of 15° 54′. What is the spacing between the layers?

3. If atoms almost completely fill the crystal lattice (as shown in the inset in Fig. 7.4), how do X-rays penetrate a crystal which may be billions of billions of layers thick?

4. An ionic compound AB has a face-centered cubic structure (like NaCl). The distance between adjacent ions A^+ and B^- is measured to be 3.00 Å. X-ray studies also show there are layers of ions that are separated by a distance of $3.00/\sqrt{2}$ or 2.12 Å. Show that this corresponds to alternating layers, one of A^+ and the other of B^-. What is the distance between any two closest anions or cations?

5. When potassium chloride is studied with X-rays of wavelength 0.154 nm, second order reflection occurs at an angle of 29.4°.

  a) Calculate the distance between nearest K^+ and Cl^- ions.
  b) Using the ionic radius of Cl^- given in Example 7.1, estimate the ionic radius of K^+. Compare with the value given in Table 21.1.

Law of Atomic Heats

6. The specific heat of copper at 20 °C is 0.0921 cal·K^{-1}·g^{-1}. An oxide of copper is analyzed and found to contain 88.82% copper. Using the known atomic weight of oxygen (16.00 amu), determine an accurate value for the atomic weight of copper.

7. Predict the approximate specific heat of gold at 25 °C from its atomic weight.

8. An unknown element has a specific heat of 0.138 J·g^{-1}·K^{-1} at 25 °C. When 1.548 g of the element's oxide is analyzed it is found to contain 1.380 g of the element. Calculate the exact atomic weight of the element. What do you think the element is?

9. The specific heat of carbon (as graphite) is about 0.17 cal·g^{-1}·K^{-1} around 25 °C. Does this follow the law of Dulong and Petit?

Lattice Energies

10. The lattice energy of solid chlorine is 31.1 kJ·mol^{-1}. Suggest why this is so high compared with oxygen and nitrogen (see Table 7.1).

11. The lattice energy of RbCl is 7.1 eV. What is this in kJ·mol^{-1}?

12. Why do ionic crystals generally have such large lattice energies compared with molecular crystals (see Table 7.1)?

13. The lattice energies (in kJ·mol^{-1}) of KF, KCl, KBr, and KI are 896, 690, 665, and 632, respectively. Does the trend follow the expected pattern?

General

14. A sample of the mineral pyrrhotite is found to have the empirical formula $Fe_{11}S_{12}$. Assuming that the mineral is mostly iron(II) sulfide, with some Fe^{3+} ions as impurities, calculate the ratio of Fe^{3+} to Fe^{2+} in the solid.

15. The charge on a mole of electrons is about 96,500 C. How many electrons pass through each 1-cm^2 cross-sectional area of a conductor each second if the conductivity is 1 Ω^{-1}·cm^{-1} and the potential gradient is 2 V·cm^{-1}? (See footnote on page 210).

16. Potassium chloride, KCl, forms a face-centered cubic crystal with a density of 1.984 g·cm^{-3}.

   a) Calculate the distance between adjacent K^+ and Cl^- ions in the crystal.

   b) Using $r_{Cl^-} = 1.86$ Å, calculate the radius of the potassium ion.

17. Vanadium metal forms body-centered cubic crystals with a density of 5.96 g·cm^{-3}. A cube 3.011 Å on an edge contains nine vanadium atoms, one at each corner and one in the center of the cube. Use these data (and any other you may need) to obtain a value for Avogadro's number.

18. Cesium chloride has a cubic lattice in which the corners of a unit cell are occupied by Cl^- ions with a Cs^+ ion in the center. The density of the solid is 3.97 g·cm^{-3}.

   a) Calculate the distance between two adjacent Cl^-. This is the length of the edge of a unit cell.

   b) Calculate the Cs^+-to-Cl^- distance.

   c) Using $r_{Cl^-} = 1.81$ Å, calculate r_{Cs^+}. [*Hint:* The diagonal of a cube having an edge length a is $\sqrt{3}a$.]

19. Solid chromium has a body-centered cubic structure similar to that of vanadium. The edge of a unit cell is 2.884 Å. Calculate the atomic radius of a chromium atom. [*Hint:* The diagonal of a cube of length a is given by $\sqrt{3}a$.]

20. How many atoms, on the average, are there in each unit of the diamond structure shown in Fig. 16.1?

ADDITIONAL READINGS

Bragg, Sir Lawrence, "X-Ray Crystallography." *Scientific American,* July 1968. A winner of a Nobel prize (shared with his father in 1915) discusses various uses of X-rays to probe the structure of matter.

Wannier, Gregory H., "The Nature of Solids." *Scientific American,* December 1952. An excellent summary of the structures of solids by one of the pioneers in the field.

Cuff, Frank B., Jr., and L. McD. Schetky, "Dislocations in Metals." *Scientific American,* July 1955. Discusses how the theory of imperfections is helping to transform metallurgy from an art to a science.

September 1966 issue of *Scientific American*. Thirteen articles written on the general topic of materials. These include "The Solid State" by Sir Nevill Mott and "The Nature of Glasses" by R. J. Charles.

Chalmers, Bruce, "The Photovoltaic Generation of Electricity." *Scientific American,* October 1976. Discusses how the cost of solar-generated electricity may be decreased. Contains an excellent elementary discussion of the photovoltaic effect.

8
THE LIQUID STATE

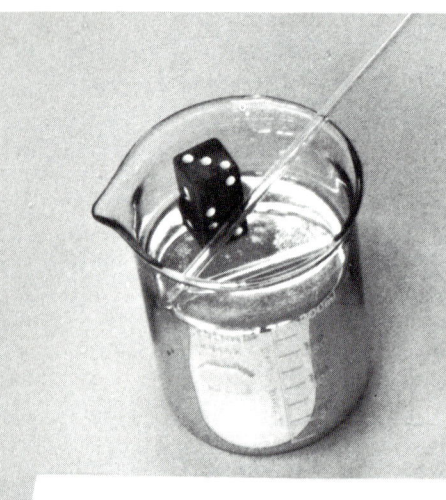

MERCURY
DENSITY: 13.6 g·cm⁻³
OPAQUE
SILVERY

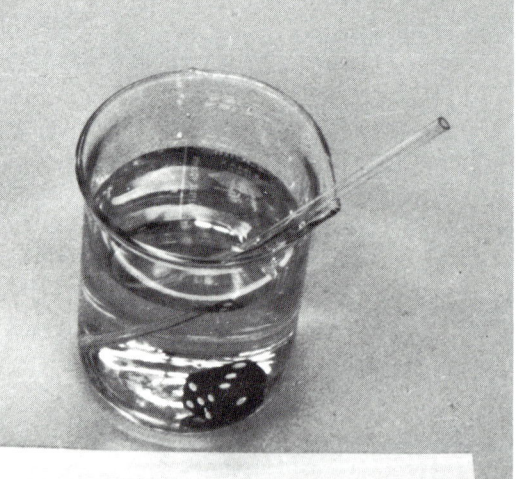

WATER
DENSITY: 1.0 g·cm⁻³
TRANSPARENT
COLORLESS

If a vessel full of water, closed on all sides, has two openings, one a hundred times larger than the other, with a piston carefully fitted to each, a man pressing the small piston will match the strength of a hundred men pressing the piston in the hundredfold greater opening, and will overmaster ninety-nine.

BLAISE PASCAL, *Traité de l'équilibre des liqueurs*, 1663

Blaise Pascal (1623–1662) was a French mathematician, scientist, and philosopher. Pascal, after whom the SI unit of pressure is named, refined the science of hydrostatics, made important discoveries in geometry, and wrote on religious and philosophical matters. The Traité de l'équilibre des liqueurs *appeared one year after Pascal's death.*

Liquids are fluid like gases but incompressible like solids. As Pascal first showed, these characteristics make it possible to transmit forces through liquids—a fact which forms the basis of the hydraulic brake and many other similar devices.

The particles of a solid are characterized by relatively fixed positions in space, that is, a passive orderliness which may be described in precise geometric terms. At the opposite extreme the particles of a gas have random positions which are constantly changing—an active disorder best described by means of statistics and probability. The particles of a liquid are close together as in solids and exhibit some short-range order, but the overall arrangement is random, as in gases; hence the particles have considerable freedom to move about. This in-between nature of liquids means that neither geometry nor statistics adequately describes them. Consequently, there is no really satisfactory general model for the liquid state.

X-ray diffraction studies clearly reveal the intermediate nature of liquids. Whereas X-ray diffraction studies of solids show rather sharp patterns of strong and weak scattering, those of gases show no uniform scattering pattern whatsoever. The diffraction patterns of liquids, however, are somewhat like those of solids but much more diffuse; that is, the difference between strong and weak scattering is less pronounced in liquids than in solids. This suggests that there is some tendency for the liquid molecules to aggregate in a regular geometric fashion, but this tendency is confined to relatively small regions and is not characteristic of an entire liquid sample.

Both gases and liquids belong to a class of substances called **fluids**. A fluid is any substance with a low resistance to flow and the tendency to assume the shape

◀ Examples of two very different liquids.
(Photo by Charles W. Owens, UNH Chemistry Department.)

THE SURFACE TENSION OF LIQUIDS   **215**

of its container. In many simple chemical applications it is useful to think of liquids as very dense fluids, differing from gases mostly in degree rather than in kind.

8.1 THE SURFACE TENSION OF LIQUIDS

When a small-bore glass tube is dipped into a liquid such as water, the liquid flows upward into the tube to a certain height h (see Fig. 8.1). The basis of this phenomenon, known as **capillary flow** (a **capillary** is a fine-bore tube) resides in the intermolecular forces between water molecules and between water and glass. First of all, water molecules are attracted more to glass than they are to themselves. This is why water is said to **wet** glass; wetting of one substance by another means that a substance's molecules attract the molecules of another substance more than they attract each other. By contrast, water does not wet paraffin, since water molecules attract each other more than they attract paraffin molecules. Note that water is a polar substance and thus it wets glass (another polar substance) but not paraffin (a very weakly polar substance). The attraction of water molecules for the glass causes a bunching up of surface water molecules along the capillary wall in such a way as to form a concave surface (Fig. 8.2). If the capillary has a very small radius r, then the concave surface is almost hemispherical with the same radius r. This hemispherical surface is called a **meniscus**.

Almost two centuries ago the French mathematician Pierre Simon Laplace (1749–1827) proved that the two sides of the curved surface of a liquid were at different pressures and that the pressure difference was given by

$$\Delta p = \frac{2\gamma}{r} \tag{8.1}$$

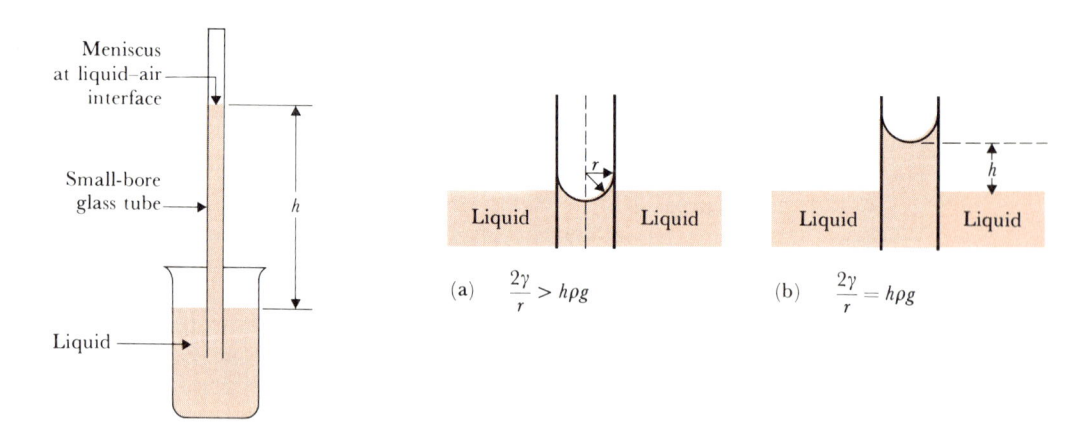

Fig. 8.1 (left). Rise of liquid up a capillary tube.

Fig. 8.2 (right). The balance of forces in a capillary tube. (a) The liquid wets the glass tube and forms a concave curved surface. (b) The liquid rises in the tube until the pressure difference across the meniscus ($2\gamma/r$) equals the pressure due to the column of liquid ($h\rho g$).

216   THE LIQUID STATE

where r is the radius of the curved surface (assumed to be hemispherical) and γ is the **surface tension** of the liquid. The surface tension of a liquid is defined as the force required to increase the perimeter of a bounded liquid surface by unit length, or, equivalently, as the energy required to increase the surface area by unit area.* Thus γ is a measure of the intermolecular forces among the surface molecules of the liquid; the higher γ is, the greater are the intermolecular forces. Why the surface tension is related to forces among surface molecules rather than to forces among interior molecules is most easily seen by examining Fig. 8.3. Molecules in the bulk of the liquid share attractive forces with neighbors on all sides; those on the surface share attractive forces with only those molecules alongside and below. Thus surface molecules are attracted to each other more strongly than are bulk molecules. Consequently, liquids of high surface tension act "as if" their surfaces are covered with a thin membrane.

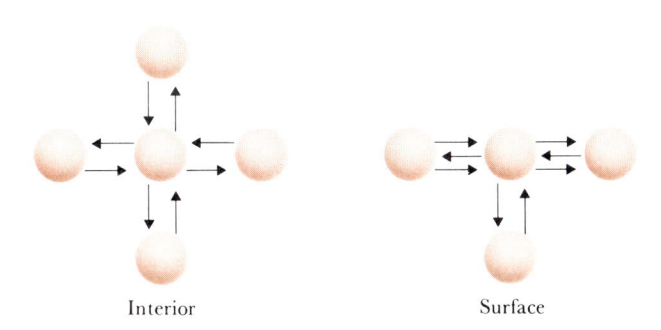

Fig. 8.3. Distribution of intermolecular attraction forces among neighboring molecules (a) in the interior of a liquid, and (b) at the surface of a liquid.

Interior   Surface

The height to which the water rises in the capillary is determined by a balance of opposing forces; the pressure difference across the meniscus must equal the pressure due to the column of water that is contained by the capillary. This balance of forces is expressed mathematically by

$$\frac{2\gamma}{r} = h\rho g \tag{8.2}$$

where ρ is the density of the liquid and g is the gravitational constant ($h\rho$ is just the mass of water per unit cross-sectional area of the capillary; thus $h\rho g$ is the pressure exerted by the water column). The above relationship provides an experimental means of measuring the surface tensions of liquids. Solving for γ leads to

$$\gamma = \tfrac{1}{2} r h \rho g \tag{8.3}$$

The only experimental data needed are: density of the liquid, radius of the capillary, and height of the meniscus above the level of the liquid. As a good approximation, valid when the capillary bore is very small, h may be measured from the bottom of the meniscus.

---

* This equivalence comes about because energy has the same basic units as force times length. Thus force per length and energy per area also have the same basic units.

EXAMPLE 8.1 A capillary tube of diameter 0.20 mm is placed into water at 25 °C. The water rises to a height of 147 mm. Calculate the surface tension of water at 25 °C.

SOLUTION If all units are SI units, γ will come out in newtons per meter ($N \cdot m^{-1}$) or joules per square meter ($J \cdot m^{-2}$).

$r = 0.10$ mm or 1.0×10^{-4} m

$h = 147$ mm or 0.147 m

$\rho = 1.0$ g·cm^{-3} or 0.0010 kg/10^{-6} m^3 = 1.0×10^3 kg·m^{-3}

$g = 9.80$ m·s^{-2}

$\gamma = \frac{1}{2}(1.0 \times 10^{-4}$ m$)(0.147$ m$)(1.0 \times 10^3$ kg·m$^{-3})(9.80$ m·s$^{-2})$

$\quad = 72 \times 10^{-3}$ N·m^{-1} (or J·m^{-2})

Most handbooks list the non-SI value of 72 dyne·cm^{-1} (erg·cm^{-2}). Since a dyne is 10^{-5} N, a dyne·cm^{-1} is just a millinewton per meter (mN·m^{-1}) or a millijoule per square meter (mJ·m^{-2}). Thus

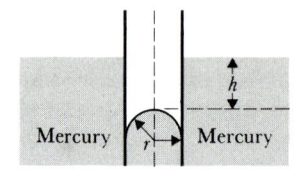

$\gamma = 72$ mN·m^{-1} (mJ·m^{-2})

Fig. 8.4. Inverted meniscus characteristic of a liquid that does not wet glass.

Some liquids, mercury for example, do not wet glass. In this case the level of mercury is depressed in the capillary below the normal liquid level (see Fig. 8.4), the result being an inverted meniscus.

Table 8.1 lists the surface tensions of some common liquids and of some molten metals. In general, the surface tension of a given liquid decreases as the temperature is raised, since increased kinetic energies of the molecules tend to override the intermolecular attractions.

Surface tension plays a role in the ability of the fisher spider and an insect called the water strider to walk on water. Water does not wet the fine feathery hairs on the legs of these creatures, and since they are very light, they do not exert enough pressure on the water surface to push apart water molecules and thereby sink. A steel needle can be floated on the surface of a teacup in the same way if it has first been greased lightly (rubbing with oily fingers will usually suffice) to make it water repellent. Try this, using a fork to lower the needle carefully onto the water surface. Ducks float on water by a somewhat different mechanism. Air trapped in the duck's

TABLE 8.1 SURFACE TENSIONS OF SOME COMMON LIQUIDS AND MOLTEN METALS

SUBSTANCE	TEMPERATURE (°C)	SURFACE TENSION (mN·m⁻¹)
Ethyl ether	20	16.9
Ethanol	20	22.3
Acetone	20	23.7
Benzene	20	28.9
Water	20	72.8
Na(l)	98	191
Pb(l)	327	480
Hg(l)	25	485
Cu(l)	1100	1300

oily feathers provides buoyancy for the duck. Thus when a detergent is added to the water (this lowers its surface tension and makes it a better wetting agent), the water wets the feathers and displaces the trapped air, and the duck sinks. Water striders, fisher spiders, and floating needles can be sunk by the same tactic.

8.2 THE VISCOSITY OF LIQUIDS

Fish swim through water and birds fly through air. In both cases the animal has to push aside surrounding matter in order to make progress. Our own experience with swimming and running tells us that water offers a much greater resistance to progress than does air. This resistance to motion is called **viscosity**.* Thus water is more viscous than air—in fact, air is usually not considered viscous at all.

The viscosity of a fluid (liquid or gas) depends on two main factors: the density of the fluid and the intermolecular attractions among its molecules. The greater the density is, the more energy is required to push aside the fluid. Similarly, the more the fluid molecules attract each other, the harder it is to penetrate the fluid. Thus gases, having generally lower densities than liquids and also lower intermolecular attractions, are much less viscous than liquids.

Not all liquids have the same viscosity. Number 40 motor oil is much more viscous (thicker) than is number 10 motor oil. Viscosity also depends on the size and shape of the liquid's molecules. Oil owes some of its viscous properties to the fact that the molecules in it are very long with high molecular weights. These long molecules are hard to push aside and, furthermore, become entangled with neighboring molecules and drag them along while being themselves pushed aside. A simple way of comparing the viscosities of two different substances is to drop identical steel balls through columns of equal length containing the liquids. The ball drops faster in the less viscous medium. The fact that a feather drops faster in a column containing a vacuum than in the same column containing air demonstrates that air has viscosity.

---

* The word "viscosity" is derived from the Latin name for mistletoe (*viscum*), a plant used to make a sticky substance called birdlime. Hence, viscosity is "stickiness."

8.3 DIFFUSION

The following experiment is easily carried out in the home kitchen (or well-equipped dormitory room). Fill a quart jar about two-thirds full of water and let it stand until the liquid is quiescent. Then carefully place a solid pellet of Easter egg dye on the bottom of the jar (any bright color will do). Over the next several hours the dye dissolves and slowly spreads outward through the water (see Fig. 8.5). Ultimately, the pellet dissolves completely and the dye becomes uniformly distributed throughout the water in the jar. The spontaneous movement of the dye through the surrounding medium is called **diffusion**.

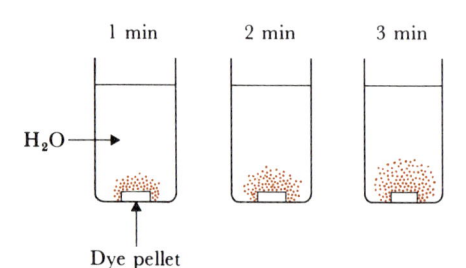

Fig. 8.5. Diffusion of a dye in water.

Careful experiments show there is a definite relationship governing the rate at which molecules diffuse. The distance a molecule travels is directly proportional to the square root of the time of travel. Thus, if a molecule travels 1 cm during the first 10 min, it is likely to travel $\sqrt{2}$ or 1.414 cm after 20 min, $\sqrt{3}$ or 1.732 cm after 30 min, and so on.* This relationship was first explained by Albert Einstein as follows: Imagine that a diffusing molecule behaves like a drunken man who starts out from a lamppost, staggers N_1 steps in one unpredictable direction, N_2 steps in another unpredictable direction, etc. Einstein showed mathematically that the most likely distance of the drunk from the lamppost after $N_1 + N_2 + \ldots$ steps was proportional to the square root of the time elapsed since leaving the lamppost. In the case of the diffusing molecule, the erratic path is due not to alcohol consumption but to repeated collisions and rebounds with solvent molecules. In both cases (the drunkard and the diffusing molecule) it is possible to predict only the *distance* of travel—not the *direction* of travel.

The rate at which a given molecule diffuses through a given medium depends on various factors: the size and shape of the diffusing molecules and the medium molecules, the intermolecular attractions among all the molecules, and the viscosity of the medium. An increase in temperature generally increases the rate of diffusion.

Chemical reactions carried out in liquid solvents are affected by diffusion. For example, if the reaction is between two molecules dissolved in an inert solvent, the molecules must first reach each other by diffusion before they can react. If the molecules react instantaneously on collision, then the rate of reaction is essentially the rate of diffusion. Thus changing to a solvent of different viscosity can alter the rate of reaction.

---

* In practice a single molecule cannot be tracked. What is actually measured is the density of diffusing molecules at varying distances from the source.

8.4 LIQUID CRYSTALS

There exists a class of liquids having some properties so similar to those of solids that they deserve a special study—especially when viewed in the light of their growing importance in technology, medicine, and biology. In 1888, F. Reinitzer noted that solid cholesteryl benzoate melted at 146.6 °C to form a cloudy liquid which turned clear when heated to 180.6 °C. Examination of the cloudy intermediate liquid phase, now called a **mesophase**, showed that it has a crystal-like molecular structure. Because of such crystal-like structure, this and similar materials are now known as **liquid crystals**.

Liquid crystals are classified into two broad types: **thermotropic** and **lyotropic**. In the former the mesophase appears when the temperature changes; in the latter the mesophase appears as the relative concentrations of two or more components change. Although lyotropic liquid crystals appear to be the more important in living systems, they have been little studied so far; hence only thermotropic liquid crystals are discussed in this chapter.

In 1922, G. Friedel showed that thermotropic liquid crystals could be classified into three basic types which he termed **nematic**, **smectic**, and **cholesteric**. Nematic (from the Greek *nema*, thread) liquids consist of long, rodlike molecules which pack together in the liquid much like toothpicks in a box. Such molecules have considerable freedom of movement but tend to remain parallel even when agitated. Smectic (from the Greek *smektikos,* to wash out, hence, soapy) liquids consist of rodlike molecules arranged in parallel sheets or layers. The layered structure of smectic liquids permits the layers of molecules to glide over each other, giving the liquid its characteristic "soapy" feeling when rubbed between the fingers. A soap bubble is a common example of a smectic liquid. The film constituting a soap bubble consists of parallel layers of rodlike soap molecules (their long axes perpendicular to the film surface) separated by a dilute solution of unordered soap molecules. As the soap bubble expands, the film grows by taking in soap molecules from its interior. A schematic representation of nematic and smectic liquids is given in Fig. 8.6.

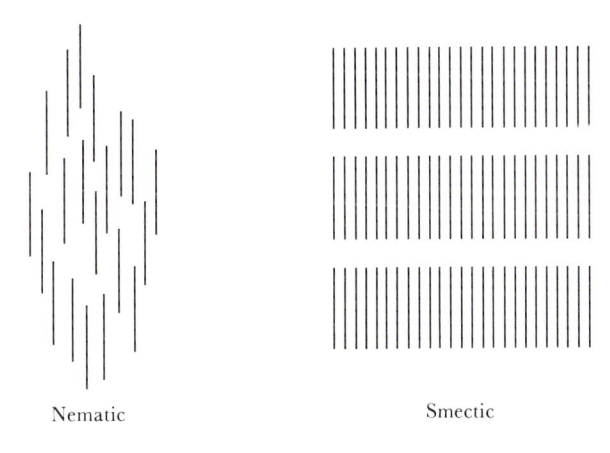

Fig. 8.6. Cross-sectional structure of nematic and smectic liquid crystals. The arrangement of molecules within each layer of smectic liquid may be either regular or disordered.

A cholesteric liquid consists of layers of molecules arranged in a nematic fashion within each layer; that is, the long axes of the molecules are parallel to the layers. This is illustrated in Fig. 8.7. The name cholesteric originated from the observation that many such materials contain the structural unit of **cholesterol*** within their molecular structure. Some chemists prefer to replace the term cholesteric with **twisted nematic** and to group this type with nematic liquid crystals.

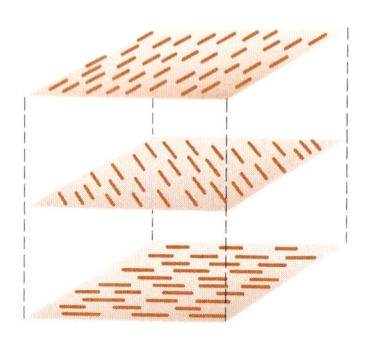

Fig. 8.7. The structure of cholesteric or twisted nematic liquid crystals. The diagram shows three typical layers, each layer having a nematic orientation of molecules. The angle between parallel molecules in adjacent layers is called the *pitch*. The pitch in the above illustration is about 45° per interlayer distance.

Cholesterol itself is not a liquid crystal. However, cholesterol-based deposits in the arteries do appear to be mesophasic around body temperatures. Consequently, medical researchers hope that studies of liquid crystals may eventually shed some light on the mechanism of atherosclerosis (clogging of the arteries by the deposition of fatty substances from the bloodstream) and thus perhaps make it possible to reverse the effects.

One of the most characteristic properties of liquid crystals is that of **optical birefringence**, or the ability of a substance to split ordinary unpolarized light into two polarized components.† A test for optical birefringence is one of the simplest ways of identifying a material as a liquid crystal. Many solids also possess this property;‡ hence the generalization that liquid crystals have the optical properties of solids and the mechanical properties of liquids.

Many liquid crystals have the ability to register minute fluctuations in temperature, mechanical stress, electromagnetic radiation, and chemical environment by

---

* Cholesterol is a white soapy crystalline substance which is a universal tissue constituent and a precursor of vitamin D. The name comes from the Greek words *kholera* (from *kholē*, bile or gall) and *stereos* (hard or solid). Cholesterol deposits in the arteries cause hardening of these vessels and are a major cause of hypertension.

† Ordinary (unpolarized) light consists of a moving wave disturbance accompanied by electric and magnetic fields which fluctuate periodically at right angles to the direction of wave motion. When such light is passed through certain materials, all of the fluctuations but those in a single plane are removed and the light is said to be **plane polarized**. In optical birefringence, *two* plane-polarized beams are produced instead of one.

‡ The double image seen when viewing printed material through a calcite crystal is an example of optical birefringence in a solid. The two polarized beams have different refractive indexes due to optical anisotropy of the crystal.

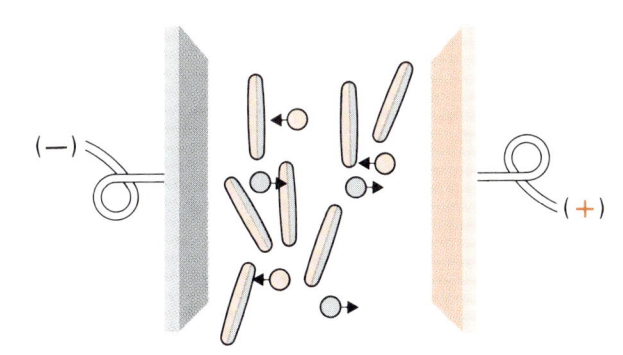

Fig. 8.8. The dynamic scattering of light by a nematic liquid crystal. Here the dipole moment of the rodlike molecules is perpendicular to the long axis. If the dipole moment were parallel to the long axis, the impurity ions would move between the nematic molecules and no turbulence would result.

undergoing pronounced color changes.* The behavior of light as it is reflected by cholesteric material depends very strongly on just how far apart the layers are and the pitch of the layers (see Fig. 8.7 and the accompanying legend). A change in these factors brought about by changes in temperature or mechanical stress is indicated by changes in the color of the relfected light. Even the disturbance brought about by intermolecular interaction with the vapors of other chemical substances is sufficient to bring about color changes in the reflected light; this property has been considered for the basis of construction of highly sensitive odor sensors.

Cholesteric materials show great promise in the thermal mapping of the human body as an aid to medical diagnosis. Veins, arteries, and various body organs generally radiate heat at a different rate than their surrounding tissues. Thus an application of cholesteric material to a large skin surface can be used to locate such organs by virtue of color variations. Such techniques may also be used to detect tumors and varicose veins and to examine the severity of burns. Cholesteric tapes can also be used for temperature monitoring of sleeping infants. A nurse need only glance at the color of a cholesteric-containing tape fastened to a baby and thereby note its temperature without disturbing the baby.

If a nematic liquid containing a small amount of dissolved ionic material is placed between the plates of a condenser, the substance may turn from transparent to cloudy as the electric field is turned on. It appears that whenever the dipole moment of the nematic molecules is directed at right angles to the long axes of the molecules, the molecules tend to become aligned parallel to the condenser plates so as to form barriers to the flow of the positive and negative ions in solution (see Fig. 8.8). Collision between the moving ions and the nematic molecules tends to disrupt the relatively orderly structure and causes optical turbulence. When the electric field is turned off the material immediately becomes clear again. This phenomenon is known as **dynamic scattering**.

If a small amount of cholesteric material is added to a nematic material which exhibits dynamic scattering, the turbulence persists for some time after the electric field is turned off. However, the turbulence can be erased immediately by the application of an ac field of high frequency. This phenomenon promises to make liquid crystals useful for display devices, a function now served by devices such as cathode ray tubes. Some alphanumeric display devices of this type are now in commercial use.

---

* There are liquid crystal thermometers now on the market that indicate temperature by digital readout. A number such as 76 is formed by a pattern of liquid crystal regions that are colorless except at 76 °F. Other numbers respond to other temperatures.

Many solids have the property of being able to rotate the plane of polarized light (see Section 16.13). For example, some varieties of quartz, $(SiO_2)_x$, can rotate this plane by 20° during passage of the light through a layer 1 mm thick. There are some cholesteric liquid crystals that cause a rotation of about 18,000° when passing through the same distance. This means that the plane of the light goes through 50 complete rotations of 360° each, within a distance of 1 mm.

SUMMARY

The properties of liquids are often intermediate to those of gases and solids. Liquids possess some of the fluid properties of gases and also the definite volumes of solids. Liquids also exhibit some short-range order similar to the long-range order of solids.

Two important characteristics of liquids are surface tension and viscosity; both are an indication of the magnitude of the intermolecular forces maintaining the liquid state. Surface tension plays an important role in the phenomenon of capillary flow. Viscosity also depends on the size and shape of the molecules of a liquid.

The diffusion of a given molecule through a liquid medium depends on the viscosity of the medium, the size and shape of the diffusing molecule, and the intermolecular attractions between the diffusing molecule and the medium. The distance a diffusing molecule travels is directly proportional to the square root of the time. This relationship is explained by analogy to the drunkard's walk.

Liquid crystals are a special class of liquids with some properties similar to those of crystalline solids. The three basic types of liquid crystals are nematic, smectic, and cholesteric (twisted nematic). Many liquid crystals have the ability to register minute fluctuations in temperature, mechanical stress, electromagnetic radiation, and chemical environment by undergoing color changes. It is often said that liquid crystals have the optical properties of solids and the mechanical properties of liquids.

LEARNING GOALS

1. Know the general characteristics of liquids and how they differ from gases and solids.

2. Know what surface tension is, how it arises, and how it is measured by capillarity.

3. Know what viscosity is and what factors affect it.

4. Be familiar with the process of diffusion and the factors affecting the rate at which diffusion occurs.

5. Be able to name and give the main characteristics of the three types of thermotropic liquid crystals.

6. Be able to describe the phenomenon of dynamic scattering.

DEFINITIONS, TERMS, AND CONCEPTS TO KNOW

capillary
capillary flow
wetting
surface tension
meniscus

viscosity
diffusion
liquid crystals and their main types
lyotropic, thermotropic
mesophase

optical birefringence
polarized light
dynamic scattering

QUESTIONS AND PROBLEMS

Surface Tension

1. To what height will a column of ethanol (see Table 8.1 for γ) rise in a capillary of radius 0.20 mm at 20 °C? The density of ethanol at 20 °C is 0.79 g·cm^{-3}.

2. A glass tube of 0.50 mm diameter is placed into a container of mercury at 20 °C. How far is the mercury level depressed within the tube? The density of mercury at 20 °C is 13.6 g·cm^{-3}.

3. A liquid has a density of 0.90 g·cm^{-3}. When a capillary tube of 0.10 mm diameter is placed into the liquid, the liquid slowly rises to a height of 11.3 cm. What is the surface tension of the liquid?

4. How high would water (at 20 °C) rise in a glass tube of radius 0.010 mm? In a tube of radius 1.0 mm?

5. Water containing a tiny amount of detergent will not rise as high in a given capillary tube as does pure water. Neglecting small density changes due to the added detergent, what effect does the detergent have on the water?

6. If liquid A has twice the surface tension of liquid B, does it necessarily follow that liquid A will cause twice the capillary rise of liquid B in a capillary of given radius? Explain.

7. For a given capillary tube, where would water rise higher, on the earth or on the moon? What would happen to water in a capillary tube while in a spaceship under zero gravity conditions?

Diffusion and Viscosity

8. A molecule diffuses 13 mm in 20 min in a given medium. How far would it be expected to diffuse in 40 min? In 100 min?

9. Substance X is placed into two different solvents, A and B. X diffuses 6 mm in 31 min in A and 15 mm in 124 min in B. Which solvent, A or B, has the lower viscosity?

10. Two identical columns 100 cm high are each filled with different liquids, A and B. A steel ball, placed at the surface of liquid A and dropped, hits the bottom of the column after a fall of 4.5 s. The same ball has a fall time of 6.0 s in liquid B. Which liquid has the greater viscosity?

11. Why is diffusion generally a faster process in gases than in liquids?

12. Why does one generally use lower-weight (lower-viscosity) motor oil in winter than in summer?

Liquid Crystals

13. How do you explain the appearance of the mesophase in the case of lyotropic liquid crystals?

14. a) Why do liquid crystals invariably consist of long, rodlike molecules?
b) Describe the mesophase, relative to the "surrounding" phases, in terms of intermolecular forces.
c) Why does the mesophase exist only at a given temperature range?

General

15. Why are liquids much less compressible than gases?

16. Which gas would you expect to have a higher critical temperature, NH_3 or CH_4? Explain.

17. Liquid water has a density of about 1.0 g·cm^{-3} whereas the density of liquid mercury is about 13.6 g·cm^{-3}.
a) What is the apparent molecular volume of water?
b) What is the apparent atomic volume of mercury?

ADDITIONAL READINGS

Fergason, James L., "Liquid Crystals." *Scientific American,* August 1964. A general, relatively comprehensive treatment of liquid crystals at an elementary descriptive level.

Bernal, J. D., "The Structure of Liquids." *Scientific American,* August 1960. A discussion in terms of the short-range order observed in many liquids and its use toward a general theory of the liquid state.

Verbit, Lawrence, "Liquid Crystals—Synthesis and Properties." *J. Chem. Educ.* **49**, 36 (1972). Describes some simple laboratory experiments using liquid crystals.

9
CHANGES OF STATE

Three familiar forms of water: vapor, liquid, and ice. (Courtesy of Marshall Henrichs.)

> "The time has come," the Walrus said,
>   "To talk of many things:
> Of shoes—and ships—and sealing wax—
>   Of cabbages—and kings—
> And why the sea is boiling hot—
>   And whether pigs have wings."
>
> LEWIS CARROLL, "The Walrus and the Carpenter"

This chapter talks of many things: not of shoes nor ships nor of most of the things so important to the walrus, but boiling-hot seas do get discussed—in a way. That is, the author knows of no boiling-hot seas, but if there were any, this chapter should help in understanding such a phenomenon.

Previous chapters discussed the three states of matter—gases, liquids, and solids—on an individual basis. The present chapter deals with the transformations from each of these states to the others. Such transformations are called **changes of state**. Changes of state are **physical** processes as contrasted with **chemical** processes. Unlike chemical processes, where one or more substances are transformed to one or more entirely new substances, physical processes merely change the state, form, or appearance of a given substance.

When a change of state occurs, intermolecular forces between atoms, ions, or molecules are disrupted. For example, to melt ice about 0.1 eV must be supplied per H_2O molecule in order to disrupt the forces that hold the crystal lattice together; to boil water requires about 0.4 eV per H_2O molecule to disrupt the forces that keep the molecules in the liquid state. Generally, energies of 0.5 eV or less are sufficient to overcome the intermolecular forces responsible for solid and liquid states. The conversion of a solid from one allotropic form to another is also a physical transformation. This type of physical change typically requires very little energy since the intermolecular forces do not generally differ significantly between two allotropic forms.*

Chemical changes involve the making and breaking of chemical bonds. When hydrogen and oxygen are combined to form water, H—H and O—O bonds are broken and O—H bonds are formed. Each such bond breakage or formation involves around 4 eV—considerably more than that required in physical changes.

9.1 HEATING AND COOLING CURVES OF SIMPLE SUBSTANCES

Observations show that many simple and common substances exist in all three of the physical states of matter: solid, gaseous, and liquid. The various interrelationships among these states are best understood by examining in some detail the process of transforming a solid to a liquid and thence to a gas.

---

* Two allotropic forms of carbon, diamond and graphite (discussed in Section 16.1), are dramatic exceptions.

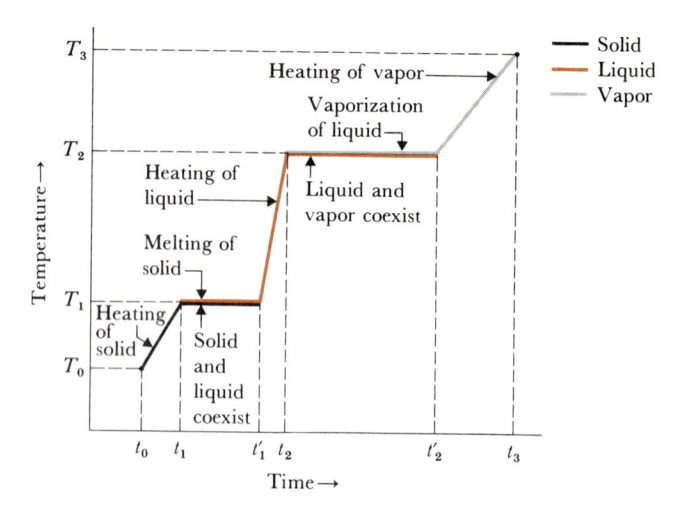

Fig. 9.1. Idealized heating curve of a pure substance, assuming thermal energy is supplied at a uniform rate.

Consider a sample of crystalline solid (ice, for example) at some temperature T_0 and observe what happens as thermal energy is supplied to it at a uniform rate. The overall process is illustrated graphically in Fig. 9.1. Between the times t_0 and t_1 the temperature of the solid increases steadily from T_0 to T_1. During this time, as more and more energy is put into the solid, the atoms or ions within it vibrate with ever increasing magnitude until, finally, the vibrations become strong enough to overcome the forces maintaining the crystal lattice. At temperature T_1 and time t_1 the surface of the solid begins to glisten and some liquid substance begins to appear. This is called **melting** of the solid. Finally, at t_1' the last trace of solid disappears and only liquid is present. During the time t_1 to t_1', while melting occurs, the temperature remains constant at T_1. As heating continues between the times t_1' to t_2 the temperature of the liquid increases steadily from T_1 to T_2. At temperature T_2 and time t_2 the surface of the liquid begins to become highly agitated and to "roll." Vapor begins to form at a fairly constant rate and the amount of liquid decreases; this process is termed **boiling** and occurs while the temperature remains constant at T_2. Finally, at t_2' the last trace of liquid disappears and only vapor is present. The formerly crystalline material is now entirely in the gaseous state* after having passed through the intermediate liquid state. Continued heating beyond time t_2' leads only to a steady increase in the temperature of the vapor.

Not all substances behave in the simple fashion just described. For example, glasses do not show a sharp melting point but rather a range of temperature within which they "soften." Also, crystalline solids that contain impurities exhibit a range of temperature over which melting occurs. In fact, the presence of a sharp, well-defined melting point is frequently used as a criterion of the purity of a crystalline substance. Some substances decompose chemically rather than melting or boiling.

---

* The usual convention is to call a substance a **vapor** when the temperature is below the critical temperature and a **gas** when the temperature is above the critical temperature. The distinction is not important in most of the discussions in this text.

Compounds of very high molecular weight, particularly those found in biological systems, behave in this way.* Such substances are said to be **heat labile** (unstable to heat). Certain solids may undergo crystalline modifications during which the temperature changes nonuniformly. Nevertheless, many of the common substances do behave in a manner very close to that described in Fig. 9.1. For example, all of the elements and most of their compounds of low molecular weight are stable in all three states of matter.

9.2 VAPOR PRESSURES OF SOLIDS AND LIQUIDS

The preceding discussion ignored the fact that at all the temperatures in question some of the crystalline solid was also being transformed directly to the gaseous or vapor state. The vapor arises because a certain number of atoms or molecules in the solid have enough energy to overcome the potential energy of attraction keeping them in the crystal lattice and thus can escape from the lattice surface. The same is true when the substance is a liquid; some of the molecules are energetic enough to vaporize.

If a sample of pure material (solid or liquid) is placed in a closed container, the amount of vapor increases with time but eventually reaches a constant value. When the solid or liquid is first placed in the container, the dominant process is

$$\text{solid or liquid} \rightarrow \text{vapor}$$

in which the more energetic molecules are escaping from the surface of the solid or liquid. At the same time, some of the vapor molecules accidentally strike the solid or liquid surface and these either rebound or, if they lose enough energy by inelastic collision, may "stick" to become part of the solid or liquid once again. The latter process may be symbolized by

$$\text{solid or liquid} \leftarrow \text{vapor}$$

which is just the reverse of the previous process. Eventually, the two opposing processes achieve a balance such that their respective rates are equal and no net transfer between the two physical states occurs. The two continually occurring opposing processes (**condensation** and **evaporation**) are symbolized by double arrows in the single equation

$$\text{solid or liquid} \rightleftharpoons \text{vapor}$$

Any process in which two or more opposing tendencies are in balance (that is, cancel out each other's effects) is said to be in **equilibrium**. Equilibrium can be either static, as in the case of a balanced see-saw, or dynamic, as in the condensation and evaporation example given.

The pressure exerted by the vapor which is in equilibrium with a solid or liquid is called the **vapor pressure** of the substance. Since—just as in a gas—the average kinetic energy of the particles of a solid or liquid increases as the temperature

---

* Large molecules in a solid have large areas of contact and, consequently, are held together by rather large intermolecular forces. A temperature high enough to overcome the intermolecular forces and cause melting is apt to be high enough to break chemical bonds as well.

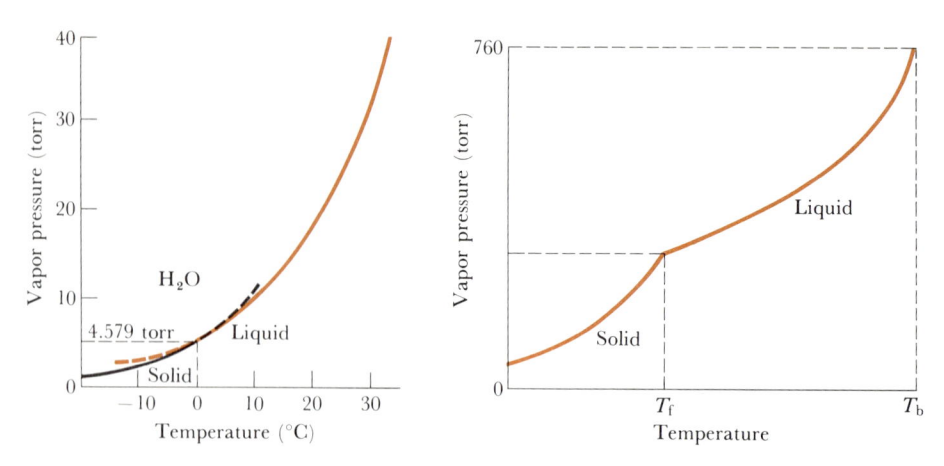

Fig. 9.2. The vapor pressures of ice from −15 to 0 °C and liquid water from 0 to 35 °C (left). The dashed lines at 0 °C indicate that the ice and liquid vapor pressure curves do not match up. The right-hand graph shows in exaggerated fashion the appearance of the entire curve up to the normal boiling point. T_f is the melting point of ice (0 °C) and T_b is the normal boiling point of water (100 °C). At the melting point, the vapor pressure of both ice and liquid water is 4.579 torr.

increases, the vapor pressure of a substance increases with temperature. Figure 9.2 shows how the vapor pressure of ice and liquid water varies with temperature.

Vapor pressures of different substances vary greatly depending on the relative volatility of the substance at the temperature in question. For example, at 25 °C the vapor pressure of moth balls (*p*-dichlorobenzene) is about 1 torr. At the same temperature water has a vapor pressure of 23.8 torr, and ethyl ether has a vapor pressure of 520 torr.

As illustrated in Fig. 9.3, there is a simple way of comparing the vapor pressures of different liquids at a given temperature. A glass tube is sealed at one end, filled with mercury, and inverted over a small reservoir of mercury in such a way that no air gets into the tube. If a small sample of liquid is introduced into the tube, the liquid vaporizes at the top of the tube and depresses the level of the mercury. The amount by which the mercury level is depressed is proportional to the vapor pressure of the liquid.

EXAMPLE 9.1 The maximum recommended level of mercury vapor in air is about 0.1 g per 10^6 L of air. Above this level, mercury vapor constitutes a serious health hazard. Is the vapor pressure of liquid mercury at 25 °C (about 2×10^{-3} torr) high enough to pose a hazard to health?

SOLUTION Assume that mercury vapor obeys the ideal gas law. Then the number of moles of mercury vapor in 10^6 L of mercury-saturated air is given by

$$n = \frac{pV}{RT} = \frac{(2 \times 10^{-3}\ \text{torr}/760\ \text{torr} \cdot \text{atm}^{-1})(10^6\ \text{L})}{(0.0821\ \text{L} \cdot \text{atm} \cdot \text{K}^{-1} \cdot \text{mol}^{-1})(298\ \text{K})} = 1 \times 10^{-1}\ \text{mol}$$

This corresponds to $(1 \times 10^{-1}\ \text{mol})(200.6\ \text{g} \cdot \text{mol}^{-1})$ or about 20 g—an amount that is 200 times the recommended maximum level. It is clear that great care should be taken in the handling and storing of mercury in areas open to personnel.

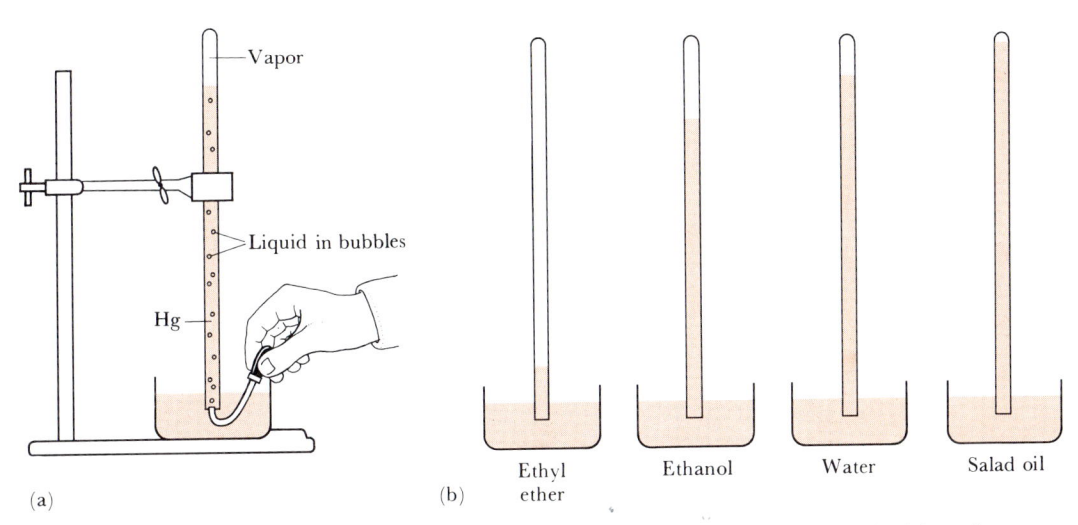

Fig. 9.3. A simple method for comparing the vapor pressures of different liquids at the same temperature. (a) A sample of liquid is introduced into an inverted column of mercury by means of a medicine dropper with a curved tip. (b) The vapors of different liquids depress the column of mercury to different levels in proportion to their vapor pressures.

TABLE 9.1 VAPOR PRESSURES AND NORMAL BOILING POINTS OF SOME COMMON LIQUIDS

LIQUID	VAPOR PRESSURE (torr) AT 25 °C	T_b AT 760 torr (°C)
Hg	0.002	356.6
H_2O	23.8	100
C_2H_6O[a], ethanol	80	78.5
C_6H_6, benzene	87	80
$C_4H_{10}O$, ethyl ether	520	34.5
C_2H_6O[a], methyl ether	4150[b]	−23

[a]Ethanol and methyl ether have the same empirical formulas, but the atoms are linked differently (see pages 438 and 440 for the structural formulas).
[b]This means that at 25 °C, a pressure of 4150 torr would be needed to prevent methyl ether from boiling.

In Table 9.1 are listed the vapor pressures of some common liquids at 25 °C. Generally, the higher the vapor pressure at a given temperature, the lower will be the normal boiling point of the liquid. The data given in Table 9.1 support this generalization.

The vapor pressures of liquids and solids are often tabulated in terms of the equation

$$\log p = -\frac{A}{2.3RT} + B \quad (9.1)$$

where A and B are constants characteristic of the substance. The constant A is generally very close to the energy required to convert one mole of solid or liquid to vapor. For solids A is called the **molar heat of sublimation**; for liquids it is called the **molar heat of vaporization**. The constant B need not be given, provided the heat of vaporization is known along with a vapor pressure value at some given temperature. Suppose A is given along with a vapor pressure p_1 at the temperature T_1. Then it is possible to write

$$\log p_1 = -\frac{A}{2.3RT_1} + B \quad \text{and} \quad \log p_2 = -\frac{A}{2.3RT_2} + B$$

Any other vapor pressure p_2 at a temperature T_2 may be calculated by subtracting the above two equations. This eliminates the constant B and produces

$$\log p_2 = \log p_1 + \frac{A}{2.3R}\left(\frac{1}{T_2} - \frac{1}{T_1}\right)$$

The tabulated values of A and B are valid only for specifically indicated temperature ranges; their use in Eq. (9.1) provides an easy way of determining vapor pressures within the applicable range of temperature.*

EXAMPLE 9.2 What is the vapor pressure of acetic acid ($C_2H_4O_2$) at 50 °C?

SOLUTION Page D-147 of the 51st edition of the *Handbook of Chemistry and Physics* (Cleveland: Chemical Rubber Co., 1970–1971) lists $A = 9486.6$ cal·mol^{-1} and $B = 8.142405$ for the temperature range -17.2 °C to 312.5 °C. Using these values along with $T = 323$ K and $R = 1.987$ cal·K^{-1}·mol^{-1} leads to log $p = 1.724$, where p is in torr. A table of logarithms then produces $p = 53$ torr as the vapor pressure of acetic acid at 50 °C. Note that the above reference includes the numerical value of $1/(2.3R)$, that is, 0.2185 K·mol·cal^{-1}, in the vapor pressure equation.

The actual amount of water vapor present in the earth's atmosphere is normally less than that expected on the basis of the vapor pressure. The relative degree of saturation of the air with water vapor is expressed in terms of a quantity called the **relative humidity**. When expressed in parts per 100 (percent), the relative humidity may be defined by

$$RH \text{ (in \%)} = 100 p_{H_2O}/p_{H_2O}^{eq} \tag{9.2}$$

where p_{H_2O} is the partial pressure of water vapor in air at the given temperature and $p_{H_2O}^{eq}$ is the vapor pressure of water at the same temperature.

EXAMPLE 9.3 A chemist analyzes a 10.0-L sample of outside air and finds it contains 0.17 g of water vapor.
   a) What is the relative humidity that day if the temperature is 25 °C?
   b) What is the relative humidity that day if the temperature is 30 °C?

---

* Tabulation of the constants A and B is found in the *Handbook of Chemistry and Physics*, Cleveland: Chemical Rubber Co., and in various other reference works.

SOLUTION

a) Assuming that the water vapor behaves as an ideal gas, the partial pressure of the 0.17 g of H$_2$O in the 10.0-L air sample is

$$p_{H_2O} = \frac{wRT}{MV} = \frac{(0.17 \text{ g})(0.0821 \text{ L·atm·K}^{-1}\text{·mol}^{-1})(298 \text{ K})}{(18.0 \text{ g·mol}^{-1})(10.0 \text{ L})} = 0.023 \text{ atm or } 18 \text{ torr}$$

The vapor pressure of water at 25 °C is 23.8 torr. Thus the relative humidity is 100(18/23.8) or 76%.

b) The partial pressure due to 0.17 g of H$_2$O vapor at 30 °C is 18(303/298) or 18.3 torr. The vapor pressure of water at 30 °C is 31.8 torr. Thus the relative humidity at 30 °C is 100(18.3/31.8) or 58%.

The above example demonstrates that relative humidity is a measure of the amount of water vapor in the air relative to the maximum amount of water vapor the air can contain at the given temperature.

9.3 MELTING AND BOILING

The melting point of a solid is that temperature at which the solid and liquid phases coexist. This is the temperature plateau labeled T_1 in Fig. 9.1. As shown in Fig. 9.2, the vapor pressure of the solid is equal to the vapor pressure of the liquid at the melting point (T_f in the diagram). The pressure of the atmosphere on the solid has some effect on the melting point but this effect is normally neglected. The melting point of ice, which defines zero on the centigrade or Celsius scale, is based on air-saturated water under an atmospheric pressure of 760 torr. A small amount of air is dissolved in water and this has a small effect on the melting point. At 0 °C the vapor pressure of both solid and liquid water (which is saturated with air at 760 torr) is 4.58 torr.

The melting point of a solid is that temperature at which the average kinetic energy of particles in the crystal lattice becomes sufficient to overcome the potential energy of attraction holding the particles in the lattice. The thermal energy supplied to the solid increases the vibrational energies of the particles; eventually the particles vibrate so strongly that they break loose from the lattice. Some of the liberated particles still attract other particles strongly enough to maintain a loose, sliding contact with them. These loosely bound particles constitute the liquid state. A few of the particles acquire more than the average amount of vibrational energy and these escape the bulk of the substance to form vapor.

The **boiling point** of a liquid (the plateau labeled T_2 in Fig. 9.1) is defined as *that temperature at which the vapor pressure of the liquid equals the atmospheric pressure*. In the special case that the external pressure is 760 torr, the boiling point is called the **normal** boiling point (T_b in Fig. 9.2). The normal boiling point of water defines the 100-degree point on the centigrade or Celsius scale.

It is easy to see that a particle can leave the liquid state (that is, it can vaporize) whenever it has enough kinetic energy to overcome its attraction to its neighbors. But why does the temperature remain constant while the boiling process occurs?

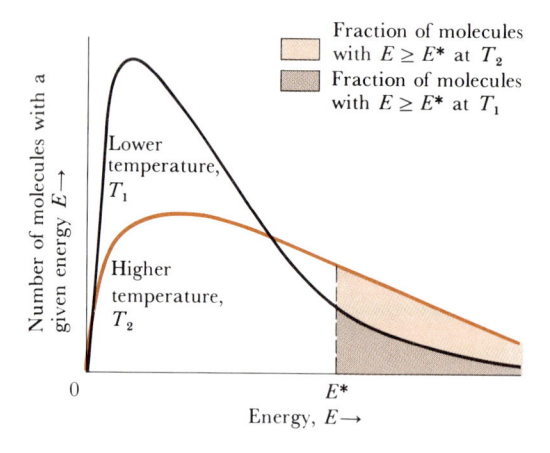

Fig. 9.4. Distribution of the kinetic energies of molecules in a liquid at two different temperatures. E^* is the minimum kinetic energy a molecule must have in order to leave the liquid surface. An increase in temperature increases the fraction of molecules having an energy E^* or greater, and thereby increases the vapor pressure.

Why does the liquid temperature not continue rising while heat is continually added? To answer the question, consider the Maxwell-Boltzmann distribution of kinetic energies within a liquid as a function of temperature (see Fig. 9.4). The shape of this curve is qualitatively the same as that for an ideal gas (compare Fig. 3.6). Let E^* represent the minimum kinetic energy a molecule in the liquid phase needs in order to escape from the liquid surface to the vapor phase. As thermal energy is supplied to the liquid, the fraction of liquid molecules with kinetic energy equal to or greater than E^* increases since the temperature increases. This also means that the vapor pressure is increasing. During this time, dissolved air is coming out of solution and appears as bubbles accumulating along the sides of the containing vessel.† Molecules with energies equal to or greater than E^* can leave the liquid to enter the vapor phase provided they are located at or very near to the liquid surface. The loss of these high-energy molecules slows down the rate at which the liquid temperature increases. Eventually the vapor pressure becomes large enough so that internal bubbles of vapor can form. This occurs when the liquid's vapor pressure just exceeds the surrounding pressure. Evaporation now takes place throughout the bulk of the liquid, accounting for the characteristic "rolling" motion of boiling. At this point *all* molecules with energy equal to E^* or more (not just those near the surface) can enter the vapor phase as fast as this energy is acquired from the heating process. The average kinetic energy of the molecules with energies less than E^* defines the temperature of the liquid phase; this temperature remains fixed until all of the liquid has been evaporated. The fact that different liquids usually have different boiling points is due to the different intermolecular forces they have; E^* will be generally different for different liquids.

In actual practice, boiling does not always occur as simply as implied above. Often the internal bubbles fail to form until the temperature is considerably above the boiling point, whereupon the abrupt formation of bubbles causes the liquid to "bump" and to splatter (sometimes quite violently) out of the container. This undesirable behavior is avoided by adding **boiling chips** to the liquid prior to heating. Boiling chips are simply pieces of porous solid which release entrapped air bubbles that can function as nuclei to begin internal evaporation so that smooth boiling can occur. The phenomenon of the heating of a liquid above its boiling point without the occurrence of boiling is known as **superheating**.

---

† This occurs because gases are usually less soluble in liquids at high temperature than at low temperature. This phenomenon is discussed in Section 10.3.

A similar phenomenon, known as **supercooling**, occurs when a liquid is cooled below its freezing point without solidifying. If a small crystal of the substance in question is added to the supercooled liquid *or* if the liquid is stirred vigorously, instant crystallization usually occurs. Apparently, crystallization cannot occur from the randomly oriented liquid molecules unless a few specifically oriented molecules are able to furnish a pattern or template for the others to follow.† At low temperatures the mobility of liquid molecules may be so low that the chance formation of a template is unlikely. The function of the added crystal is to provide a template to initiate crystallization. Stirring probably agitates the immobile molecules enough so that a few manage to align themselves to form a template. Both superheating and supercooling appear to be more common with highly purified liquids; impurities seem to serve either as boiling chips or as crystallization templates.

9.4 ENERGY CHANGES ACCOMPANYING CHANGES OF STATE

Energy must be supplied to a substance to change its temperature. The amount of energy needed to change the temperature of one gram of a substance by 1 °C (without causing a change of state) is called the **specific heat** of that substance.‡ The amount of energy needed to change the temperatuure of one mole of a substance by 1 °C is called the **molar heat capacity** of that substance. Molar heat capacity is simply the specific heat multiplied by the weight of one mole of the substance.

Although the SI unit of energy is the joule, thermal energy (popularly called *heat energy* or *heat*) is still often expressed in terms of a unit called the **calorie**. A calorie is defined as equivalent to exactly 4.184 joule. This is very nearly the specific heat of water around 15 °C. Nutritionists customarily employ a unit called the **Calorie** (capital C) which is actually 1000 cal or a kilocalorie (kcal). Since carelessness in use of the Calorie (omitting the capital C, for example) can lead to confusion, it is recommended that the unambiguous term kcal be used instead.

Since the gas constant R is equal to $8.314 \text{ J} \cdot \text{K}^{-1} \cdot \text{mol}^{-1}$, it can also be written as $8.314/4.184$ or $1.987 \text{ cal} \cdot \text{K}^{-1} \cdot \text{mol}^{-1}$. A value of $2 \text{ cal} \cdot \text{K}^{-1} \cdot \text{mol}^{-1}$ is often used in approximate calculations.

Energy is also required to bring about a change of state of a substance. The energy needed to convert one gram of solid to one gram of liquid is called the **specific heat of fusion** of the solid. Similarly, the energy required to convert one gram of liquid to one gram of vapor is called the **specific heat of vaporization** of the liquid. Specific heats of fusion and vaporization depend slightly on the temperature at which these processes occur, but this distinction is often ignored. For most practical purposes the values obtained at the melting point and normal boiling point are assumed to be valid at any other temperature. For example, the specific heat of vaporization of water at 100 °C is about $540 \text{ cal} \cdot \text{g}^{-1}$ ($2259 \text{ J} \cdot \text{g}^{-1}$). Vaporization of liquid water at any other temperature is assumed to require the same amount of

---

† This is consistent with the observation that supercooling is most common for those substances having complex molecular shapes. Such molecules must line up just right or a properly oriented crystal cannot form.

‡ The specific heat of a substance changes with temperature but is often approximately constant over a considerable temperature range. It is usually assumed that the specific heat of liquid water is $1 \text{ cal} \cdot \text{g}^{-1} \cdot \text{K}^{-1}$ ($4.18 \text{ J} \cdot \text{g}^{-1} \cdot \text{K}^{-1}$) over the entire liquid range.

energy as at the normal boiling point. The specific heat of fusion of ice at 0 °C is about 79 cal·g^{-1} (331 J·g^{-1}).

Energies of changes of state are often expressed on a molar basis. Thus the **molar heat of fusion** of water is about (79 cal·g^{-1})(18 g·mol^{-1}) or 1.4×10^3 cal·mol^{-1} (5950 J·mol^{-1}). Similarly, the **molar heat of vaporization** of water is 540 cal·g^{-1}(18 g·mol^{-1}) or 9.7×10^3 cal·mol^{-1} (4.1×10^4 J·mol^{-1}). In Table 9.2 are listed the molar heats of fusion and vaporization of some common substances along with their melting points (T_f) and normal boiling points (T_b). Note that the ionic substance NaCl has very high heats of fusion and vaporization and also very high melting and boiling points.

TABLE 9.2 HEATS OF FUSION AND VAPORIZATION, AND MELTING AND NORMAL BOILING POINTS OF SOME COMMON SUBSTANCES

SUBSTANCE	HEAT OF FUSION IN J·g^{-1}	kJ·mol^{-1}	HEAT OF VAPORIZATION IN J·g^{-1}	kJ·mol^{-1}	T_f(°C)	T_b(°C)
H$_2$O	331	5.96	2259	40.6	0	100
C$_6$H$_6$, benzene	131	10.2	435	34.1	5.5	80
NaCl	515	30.2	12550	744	800	1413
Hg	11.3	2.3	301	60.7[a]	−38.9	356.6

[a] This value was obtained by use of Eq. (9.1) and the experimental vapor pressures of mercury at 150 °C and 250 °C: 2.81 torr and 74.38 torr, respectively.

The process that is the reverse of fusion is called **crystallization** or **solidification**. The energy released when a liquid crystallizes to a solid is equal in magnitude to the heat of fusion. Similarly, the reverse of vaporization is called **condensation**; the heat of condensation is equal in magnitude to the heat of vaporization. When 1 g of ice is melted to liquid, 331 J of energy must be supplied. To transfer the liquid water back to ice requires removal of 331 J of energy for each gram of water.

9.5 PHASE DIAGRAMS OF PURE SUBSTANCES

All of the preceding discussions, although presented in terms of changes of state of a single substance, are strictly applicable only to two-component systems: the substance whose changes of state are of interest, and the surrounding air.* The present section considers changes of state as they occur in strictly one-component systems, that is, pure substances isolated from all other substances, including the air.

The behavior of a one-component system is conveniently summarized by the use of a **phase diagram**, a diagram that shows what phases can exist under different combinations of variables such as pressure and temperature. A **phase** is defined as a homogeneous, mechanically separable, and physically distinct portion of a heterogeneous system. Since gases are miscible in all proportions (excluding chemical re-

---

* Air itself is at least a two-component system (nitrogen and oxygen) but it may be treated as a one-component system in the present context, provided neither N$_2$ nor O$_2$ interact strongly with the substance whose phase changes are under study.

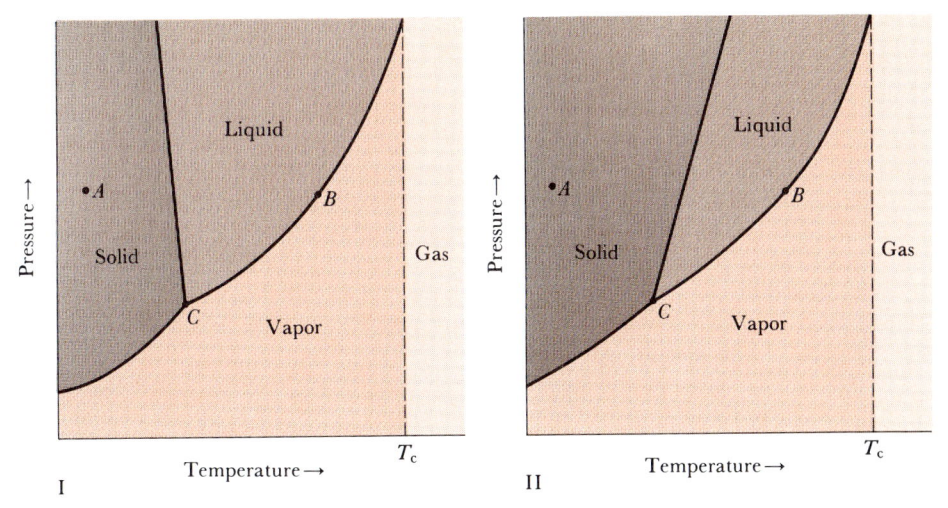

Fig. 9.5. Typical phase diagrams for one-component systems. At point A, only solid can exist; at point B, liquid and vapor can exist in equilibrium; at point C, the triple point, all three phases can coexist in equilibrium.

actions among them), a mixture can have no more than one gas phase. However, in a mixture of solids and/or liquids there can be at least as many phases (solid, liquid, or both) as there are substances. Only the solid, gaseous, and liquid phases of a single substance are considered here.

In Fig. 9.5 are illustrated two common types of simple phase diagrams. The main features of one-component phase diagrams are as follows: Each phase diagram is divided into three areas (the distinction between gas and vapor is not of importance here) which represent combinations of pressure and temperature for which the substance exists as a single phase only. Each of the lines separating the areas represents the combinations of pressures and temperatures for which two phases can coexist in equilibrium. The point at which the three lines intersect represents the single pressure and temperature combination for which all three phases can coexist in equilibrium. Thus the line between solid and liquid represents the pressures and temperatures for which the equilibrium

$$\text{solid} \rightleftharpoons \text{liquid}$$

exists, and similarly for the other lines. The point labeled C represents the three-fold equilibrium

$$\text{liquid} \rightleftharpoons \text{solid} \rightleftharpoons \text{gas}$$

Point C is called the **triple point** of the substance. This is the temperature at which the vapor pressures of all three phases are the same. The practical significance of a phase diagram is that it provides a compact summary of the possible physical states of a substance for a wide range of pressure and temperature combinations—information which would otherwise require voluminous tables of data.

Diagrams I and II in Fig. 9.5 are very similar and differ in only one small but important detail. In diagram I the line representing the solid $\rightleftharpoons$ liquid equilibrium slants toward the left as it leaves the triple point, whereas in diagram II this line slants toward the right. The leftward or rightward slant of the solid $\rightleftharpoons$ liquid line is related to the *sign* of the volume change accompanying the transformation of solid to liquid. If this volume change is defined as $\Delta V = V_{liq} - V_{sol}$, then the volume change ΔV is positive when V_{liq} is greater than V_{sol} and diagram II applies. When V_{sol} is larger than V_{liq}, then ΔV is negative and diagram I applies. Note that V_{liq} and V_{sol} must refer to a fixed amount of substance. Practically all common substances, with the notable exception of water, behave as in diagram II. Since ice floats in liquid water, the density of the solid is less than that of the liquid and V_{liq} is less than V_{sol}. In most other substances the solid does not float in the liquid but sinks to the bottom.

Figure 9.6 is a portion of the phase diagram of water. Omitted are some of the crystalline modifications of ice which exist at high pressure. The triple point of water is at 0.0099 °C and 4.579 torr. This is very close to the melting point of ice (the **ice point**, 0 °C) but should not be confused with it. The ice point refers to air-saturated water (which is not a one-component system) and is lower than the triple point for two reasons. First, the presence of dissolved air depresses the freezing point of water by 0.0024° (such phenomena are discussed in the next chapter, Section 10.8). Next, the ice point is measured when the total external pressure on the water is 760 torr, whereas at the triple point the total pressure is 4.579 torr (due only to water vapor). This change in total pressure, from 4.579 to 760 torr, lowers the freezing point by 0.0075 °C. Thus the total freezing-point depression is 0.0024° + 0.0075° or 0.0099 °C.

The phase diagram of Fig. 9.6, although strictly applicable to a one-component system only, nevertheless provides some information applicable to a two-component system such as air-saturated water. The main effect of the dissolved air is to eliminate the triple point so that solid, liquid, and vapor can coexist at more than one combination of pressure and temperature. The slant of the solid $\rightleftharpoons$ liquid line in Fig. 9.6 shows that an increase in pressure will cause the solid to melt without a change in temperature. It takes a pressure increase of about 1.3×10^7 Pa (around 130 atm) to decrease the melting point of ice by 1 K. It is sometimes said that ice skating is made possible by the skate pressure melting the ice just beneath the blade and forming a lubricating surface. It is true that a skater weighing about 60 kg could exert around 100 atm pressure through a skate blade, but this would melt ice only in the temperature range of 0 to -1 K, and skating is possible well below this range. It is more likely that the glide of the skate over ice is due to low friction enhanced by the plastic response of the ice surface to the blade. When the ice temperature becomes very low, its surface becomes less yielding, frictional drag increases, and skating becomes less easy.

Figures 9.7 and 9.8 show portions of the phase diagrams of carbon dioxide and mercury. The shaded areas represent that range of pressure and temperature in which these substances are ordinarily encountered. For example, carbon dioxide (CO_2) is seen either as a solid (dry ice) or as a gas; few people ever see liquid CO_2. Yet the CO_2 in a fire extinguisher—provided it is at a pressure of 5 atm or more—may be in the liquid state. Pick up a CO_2 extinguisher sometime and shake it to sense the movement of liquid inside. Similarly, the phase diagram of mercury shows why it is usually observed only as a liquid (and the vapor in equilibrium with it) and not

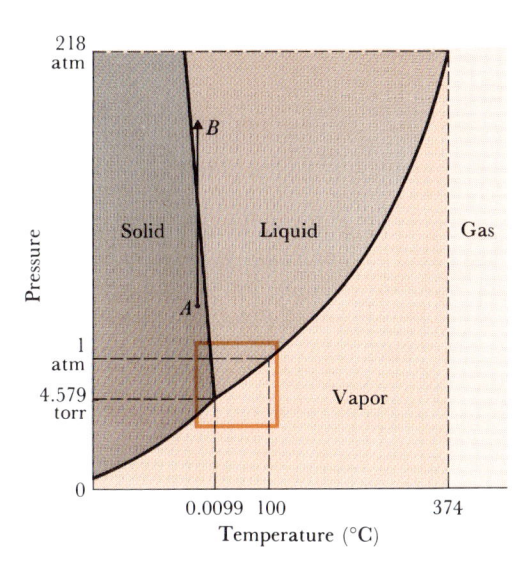

Fig. 9.6. A portion of the phase diagram of pure water (*not to scale*). In order to be able to illustrate certain features, it has been necessary to distort the pressure scale and to exaggerate the extent of slant of the solid ⇌ liquid line. In actuality, this line is extremely close to vertical. The line $A \to B$ illustrates the fact that ice can be melted without changing the temperature by increasing the pressure sufficiently. The boxed area indicates roughly the pressure and temperature ranges under which water is normally observed. At high pressures there exist several different crystalline modifications of ice not shown in this diagram.

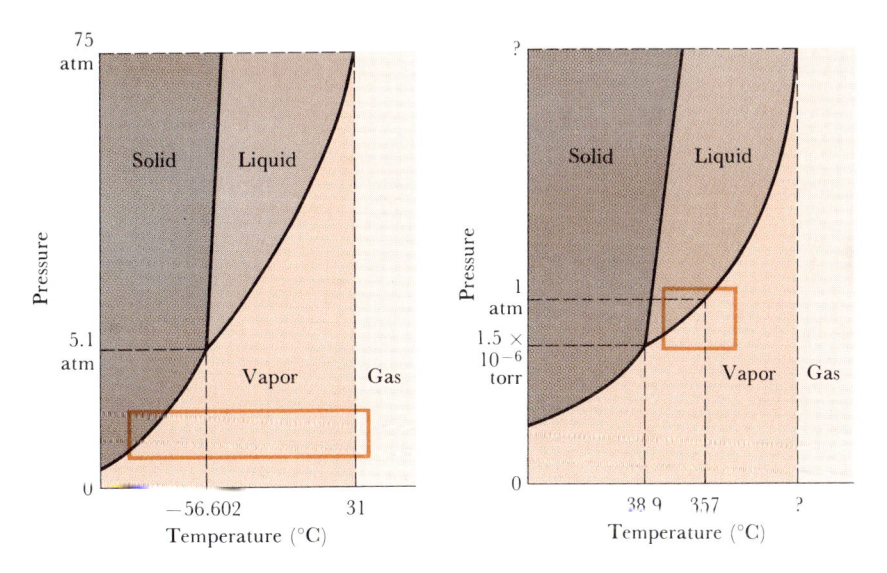

Fig. 9.7 (left). Schematic phase diagram of carbon dioxide (*not to scale*). The boxed area indicates roughly the pressure and temperature ranges under which carbon dioxide is normally observed.

Fig. 9.8 (right). Schematic phase diagram of mercury (*not to scale*). The boxed area indicates roughly the pressure and temperature range under which mercury is normally observed. The critical pressure and critical temperature of mercury are not known accurately, and estimates of these vary widely—from 732 atm to 1150 atm for p_c and from 700 °C to 2600 °C for T_c.

as a solid. Below −39 °C, a temperature few of us ever encounter, mercury becomes a solid.*

The process solid ⇌ vapor, in which solid is transformed directly to vapor (and vice versa), is called **sublimation**.† For example, solid CO_2 (dry ice) sublimes to gaseous CO_2 at room temperature. The element iodine, I_2, also behaves in this way. Under ordinary conditions I_2 is an almost black solid which sublimes readily to form a violet vapor. Most solids of low molecular weight sublime to some extent, but the process is noticeable only if the vapor pressure is appreciable. Of course, the odors of solid substances depend on the sublimed vapors. Many years ago, the eminent physical chemist Irving Langmuir estimated that the metal tungsten, one of the least volatile substances known, had a vapor pressure at 25 °C of one atom per universe. Yet at the temperature of a glowing light-bulb filament, the vapor pressure of tungsten is high enough that gray deposits of the metal slowly accumulate on the relatively cool glass walls. Eventually the filament becomes so thin that a section melts out and breaks the electrical circuit; the bulb "burns out."

Irving Langmuir (1881–1957) was an American physical chemist who pioneered in the field of surface chemistry, for which he received the Nobel prize in 1932. After brief service as a teacher at the Stevens Institute of Technology in Hoboken, N.J., Langmuir worked for the next 41 years at the research laboratories of the General Electric Company at Schenectady, N.Y. Langmuir also pioneered in a variety of other fields, including the production of artificial rain by the seeding of clouds.

Le Châtelier's principle is named after Henri Louis Le Châtelier (1850–1936), French chemist who did pioneering work on the setting of cements, the structures of alloys, and certain areas of thermodynamics. During his later years he thought and wrote extensively on the problems of labor relations and industrial efficiency.

9.6 LE CHÂTELIER'S PRINCIPLE

A system in equilibrium will remain in equilibrium indefinitely as long as there are no external influences acting on the system. However, if the system is subjected to a disturbance, a readjustment occurs and a new state of equilibrium is eventually attained. One of the most useful principles of chemistry, the **principle of Le Châtelier**, states the general nature of the readjustment:

> *Whenever a system in equilibrium is subjected to a stress (disturbance), the system responds in such a way as to undo or lessen the effect of this stress insofar as possible.*

---

* It is rumored that students at the University of Alaska are forbidden to take their Torricelli barometers outdoors during the winter months.

† *Sublimation* also refers to the "round-trip" process of allowing a solid to go directly to vapor and then condensing the vapor back to the solid. Some solids are purified in this way.

Let us apply Le Châtelier's principle to the equilibrium solid ⇌ liquid and consider an increase in pressure as a stress upon the system. Le Châtelier's principle asserts that the system will respond in such a way as to minimize the increase in pressure; a decrease in the volume of the system would tend to have just this effect. In the case of water, the liquid occupies a smaller volume than the solid so that an increase in pressure would favor the melting of additional ice. Equilibrium is thus reestablished, but only after some solid is transformed to liquid. For most other solids, just the opposite effect would occur; an increase in pressure would cause some liquid to become solid.

Similarly, in the equilibrium liquid ⇌ vapor, Le Châtelier's principle readily shows that a vapor may be liquefied by increasing the pressure, or, conversely, a liquid may be vaporized by decreasing the pressure. This process is illustrated in Fig. 9.9. Note that at a given temperature a substance occupies a larger volume as a vapor than as a liquid.

The above examples do not say anything not already evident from the phase diagram of the substance involved. However, Le Châtelier's principle has broad applicability to equilibria of any kind; it will be employed frequently throughout the text.

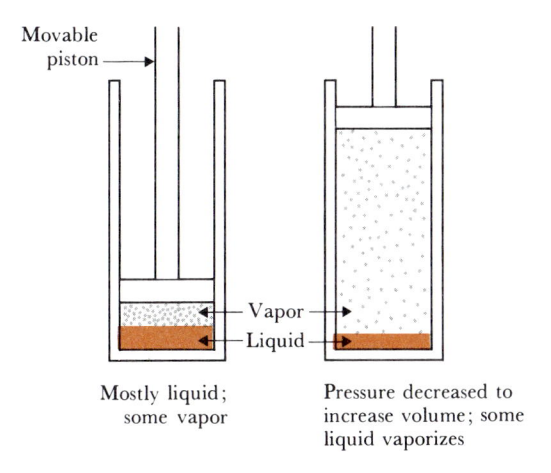

Fig. 9.9. Effect of changing pressure on the liquid ⇌ vapor equilibrium. A sample of liquid is placed in a cylinder containing a movable piston. As the piston is raised, additional liquid vaporizes. The above behavior is in accord with the principle of Le Châtelier; since a given amount of substance occupies a larger volume as a vapor than as a gas, a decrease in pressure (which brings about an increase in volume) can be offset only by vaporization of liquid. If the volume is made large enough (not shown in the illustration), all of the liquid will vaporize.

SUMMARY

Many simple substances can exist in the three states: solid, liquid, and gas. Transformations from one state to another can occur and require definite amounts of energy for a given amount of substance. At a given pressure, transformations from solid to liquid or liquid to gas occur at definite fixed temperatures (the melting point and the boiling point, respectively), generally differing for each specific substance.

The molecules of a solid or liquid exhibit a tendency to enter the vapor phase. The pressure exerted by the vapor of a substance in equilibrium with the solid or liquid phase is called the vapor pressure of that substance.

A given pure substance cannot always exist in all three phases under all conditions. A diagram summarizing the combinations of pressure and temperature for which a substance can exist in one or more phases is called a phase diagram.

Systems in equilibrium always respond to a disturbance of the equilibrium by favoring some process that tends, insofar as possible, to oppose or cancel out the disturbance. This principle, known as Le Châtelier's principle, shows how the relative amounts of a substance in two different phases may be changed by changes of pressure and temperature.

LEARNING GOALS

1. Be able to construct a typical heating or cooling curve for a simple substance and be able to interpret each part of this curve.

2. Be able to describe how equilibrium is reached between a liquid and its vapor. Know how vapor pressure varies with temperature.

3. Know what is meant by relative humidity and how it is calculated.

4. Be able to use the vapor pressure equation, $\log p = -A/(2.3RT) + B$, to calculate vapor pressures using tabulated values of A and B.

5. Be able to define the melting point and boiling point of a substance.

6. Be able to describe superheating and supercooling.

7. Know how the calorie is defined and be able to convert between calories and the SI unit of joules.

8. Be able to define the heats of fusion and vaporization of a substance.

9. Be familiar with a typical phase diagram for a pure substance and be able to interpret what it means. In particular, know what areas, lines, and intersections of lines signify physically.

10. Know Le Châtelier's principle and how it is applied to phase changes.

DEFINITIONS, TERMS, AND CONCEPTS TO KNOW

physical vs. chemical transformations
equilibrium
vapor pressure of a liquid or solid
relative humidity
melting point of a solid
boiling point of a liquid
superheating

supercooling
sublimation
calorie
Calorie
heat of fusion
heat of vaporization
crystallization

condensation
phase
phase diagram
triple point
Le Châtelier's principle

QUESTIONS AND PROBLEMS

Vapor Pressure

1. A sample of liquid was placed into a closed container kept at a fixed temperature. The pressure of the vapor varied with time as follows.

TIME (min)	PRESSURE (torr)
2	15
7	25
12	30
17	33
20	34

a) Is the vapor pressure of the substance at the given temperature equal to 34 torr, less than 34 torr, or at least 34 torr?

b) How could you determine the exact vapor pressure of the substance?

2. Solid calcium chloride ($CaCl_2$) absorbs water strongly and is used extensively as a drying agent. For example, 5 L of air was passed several times through a tube filled with 21.96 g of $CaCl_2$. The contents of the tube were later found to weigh 22.05 g. If the air sample was originally at 30 °C, what was its relative humidity? (The vapor pressure of water at 30 °C is 31.8 torr.)

3. A 1-L sample of air contains 0.020 g of H_2O vapor at a given temperature. When the air is saturated with water vapor at the same temperature it is found to contain 0.025 g of H_2O. What is the relative humidity?

4. Given that A = 10,421 cal·mol^{-1} and B = 8.9373 between −15 °C and 250 °C, estimate the vapor pressure of propanol, C_3H_7OH, at 25 °C (the A and B values above lead to pressure in torr).

5. Assume that 20 L of dry air is passed through a water sample at 25 °C until it is saturated with water vapor. If the water sample decreases in weight by 0.461 g, what is the vapor pressure of water at 25 °C?

6. How much water vapor (in moles) is there in 10^6 L of air at 25 °C if the vapor pressure of water is 23.8 torr at that temperature?

7. Three-minute eggs prepared on top of Long's Peak (elevation 14,255 ft) may be raw. Explain.

8. Given that the vapor pressures of liquid mercury at 150 and 250 °C are 2.81 and 74.38 torr, respectively, show how to estimate the heat of vaporization of mercury.

9. Estimate the temperature at which water would boil on the summit of Mt. Everest (elevation: 29,028 ft). Assume that it requires 40.67 kJ to vaporize one mole of water. [*Hint:* Calculate the atmospheric pressure at the summit, then determine the temperature at which water has that vapor pressure.]

10. The vapor pressure of water at 22.2 °C is 20 torr. The partial pressure of water in a room at that same temperature is 15 torr. What is the relative humidity in the room? Suppose the amount of humidity remains the same but the room is warmed to 25 °C. Does the relative humidity increase, decrease, or stay the same? Explain.

11. Why do people feel uncomfortable under the following conditions?

a) On a very hot day when the relative humidity is high.

b) In a heated house in winter when the relative humidity is very low.

12. The vapor pressures of ethylbenzene at 25.9 and 52.8 °C are 10 and 40 torr, respectively. Estimate the heat of vaporization of one mole of ethylbenzene.

13. A weather report announces that the relative humidity that day is 80%. If the temperature is 77 °F and the barometric pressure is 750 torr, what is the mole fraction of water vapor in the atmosphere? Assume dry air is 20% O_2 and 80% N_2. The vapor pressure of water at 77 °F is 23.76 torr.

14. Hydrogen gas is prepared from zinc and hydrochloric acid and collected over water in an inverted flask as shown in Fig. 15.2. The volume collected is 200 cm^3 at a total pressure of 735 torr and a temperature of 23 °C. The vapor pressure of water at 23 °C is 21.07 torr. Calculate the moles of hydrogen gas produced.

15. An open jar containing 5.0 L of water is placed in an airtight room 8 m × 5 m × 3 m which is thermostatted at 25 °C. The vapor pressure of water at 25 °C is 23.76 torr.

a) Will the room eventually reach 100% relative humidity?

b) If yes, how much liquid water will there be in the jar?

16. A sealed 10-L container at 30 °C contains 0.02 mol of substance A and 0.05 mol of substance B. The vapor pressures of A and B at 30 °C are 60 torr and 75 torr, respectively. Calculate the total pressure inside the sealed container, assuming that the condensed phases of A and B are not mixed.

Energy Calculations

17. Calculate the energy required to transform 10 g of ice at -10 °C to steam at 120 °C and 1 atm. The specific heat of ice and of steam may be taken as 0.5 cal·g^{-1}·K^{-1} (2.1 J·g^{-1}·K^{-1}). The heats of fusion and vaporization of water are 331 and 2260 J·g^{-1}, respectively.

18. How much energy (in joules) is needed to change the temperature of 10 g of aluminum (specific heat = 0.212 cal·g^{-1}·K^{-1}) from 25 °C to 30 °C?

19. The specific heat of iron is 0.11 cal·K^{-1}·g^{-1}. If 10 g of iron at 17 °C absorbs 46 J of energy, what is its final temperature?

20. The heat of fusion of water is 331 J·g^{-1}. How much ice can be melted with 2700 joules?

21. The power output of the sun, as measured at the earth's surface, is about 3.5×10^2 W·m^{-2}. Assuming no loss of heat due to reflection, how long would the sun have to shine on a collector surface 10 m × 10 m to provide enough energy to melt a block of ice 1.0 m^3 in volume? How long would it take to provide the energy to boil 1.0 m^3 of liquid water at 100 °C?

22. Heat is applied to a 10 g sample of water at the rate of 42 J·min^{-1}. Sketch as accurately as possible the heating curve of the sample between -5 °C and 110 °C. The specific heat of ice and steam may be taken to be half that of liquid water; the latter is about 4.2 J·g^{-1}·K^{-1}. The heats of fusion and vaporization of water are 331 and 2260 J·g^{-1}, respectively.

23. The heat of fusion of a substance is typically much less than its heat of vaporization. Explain why this is so.

24. The heat of sublimation is often very close to the sum of the heats of fusion and vaporization. Explain why this might be expected to be so.

Phase Diagrams

25. Use the following data to sketch the phase diagram of carbon dioxide.

Triple point: 216 K and 5.1 atm

Critical temperature: 304 K

Critical pressure: 73 atm

Melting point at the critical pressure: 218 K

a) Will an increase in pressure increase or decrease the melting point of solid carbon dioxide?
b) Approximately what pressure is needed to change the melting point of dry ice by 1 K?
c) What is the physical state of carbon dioxide at 5 atm and -40 °C?

26. What is the physical significance of a steep solid $\rightleftharpoons$ liquid line in a phase diagram? What would a more nearly horizontal line signify?

27. Explain on the basis of a phase diagram why a gas cannot be liquefied by pressure when the temperature is above the critical temperature of that gas. Use H_2O as an example (Fig. 9.6).

28. A certain substance has a triple point at 15 °C and 100 torr. Its boiling point at 760 torr is 120 °C. The substance melts at 20 °C when the pressure is 750 torr.

a) Sketch the phase diagram of the above substance, labeling all lines, areas, etc.
b) Which is greater for the above substance at a given temperature, the density of the solid or the density of the liquid?
c) What happens to the substance if the temperature is held constant at 17 °C and the pressure is changed from 50 torr to 770 torr?
d) What happens in part (c) if the temperature is held constant at 50 °C?

29. Does a substance such as liquid carbon dioxide have a normal boiling point? Explain.

30. Consider a substance whose solid exists as two different allotropic forms. How might this be reflected in the phase diagram of the substance?

31. The information given in a phase diagram could also be presented in tables stating the behavior of the substance at various pressures and temperatures. What, then, is the value of the phase diagram?

32. The solid $\rightleftharpoons$ liquid line in a phase diagram can slant to either the right or the left of vertical depending on the specific substance involved. Can the liquid ⇌ gas line ever slant to the left of vertical? What about the solid ⇌ vapor line? Explain.

Le Châtelier's Principle

33. When a rubber band is stretched, it becomes warmer. Use Le Châtelier's principle to predict what happens when a rubber band is warmed slightly; that is, will it stretch or shrink in length?

34. A system in which solid and liquid water are in equilibrium is heated slightly. Explain what happens in terms of Le Châtelier's principle.

35. If a piece of thin wire with weights on both ends is looped over a block of ice at 0 °C, the wire will eventually pass through the ice, leaving the block in one piece. Explain.

36. A cylinder filled with a movable piston contains a liquid in equilibrium with its vapor. Assume that the contents are thermally insulated from the surroundings. What happens to each of the following when the piston is lowered to halve the volume?
   a) Temperature of the vapor.
   b) Pressure of the vapor.
   c) Relative amounts of liquid and vapor.
Repeat, assuming now that the contents are not insulated but are allowed to acquire the temperature of the surroundings.

General

37. The vapor pressure of DDT is 1.5×10^{-7} torr at 20 °C. Assuming that the effective height of the atmosphere is 11 km and that there is no gravitational effect on pressure, estimate the maximum amount of DDT that could be present in the atmosphere as a vapor. The total cumulative amount of DDT produced in the U.S. up to about 1975 is estimated to be around 1.4×10^9 kg. The molecular weight of DDT is 364.5 amu. The diameter of the earth is 12,740 km.

38. Liquid HCl has a normal boiling point of −84 °C. At the same temperature the vapor pressure of liquid SO_2 is 10 torr.
   a) Which substance has the greater intermolecular forces? Explain.
   b) Which substance has the higher critical temperature? Explain.

39. Why is solid CO_2 (dry ice) formed around the nozzle when a CO_2 fire extinguisher is operated?

40. Warm chloroform (normal boiling point 61.7 °C) feels cold when poured on the hands. Explain.

41. Explain why NaCl has such high values of the heat of fusion, the heat of vaporization, the melting point, and the boiling point.

42. Explain how intermolecular forces are involved in each of the following physical changes: forging of steel, breaking of glass, melting of lead, drying of clothes, cutting of bread, and making a snowball.

43. The accompanying figure shows what happens when a typical pure liquid is cooled by removing heat at a constant rate. What is the significance of the "bump" occurring at the beginning of the solid-formation line? What would the heating curve of this substance look like over the same temperature range?

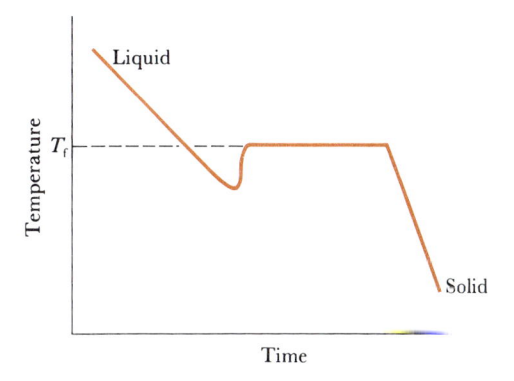

ADDITIONAL READINGS

Chalmers, Bruce, "How Water Freezes." *Scientific American,* February 1959. Explains the strange diversity of forms in which ice appears.

Westwater, J. W., "The Boiling of Liquids." *Scientific American,* June 1954. Discusses the three different ways in which liquids can boil.

10
SOLUTIONS

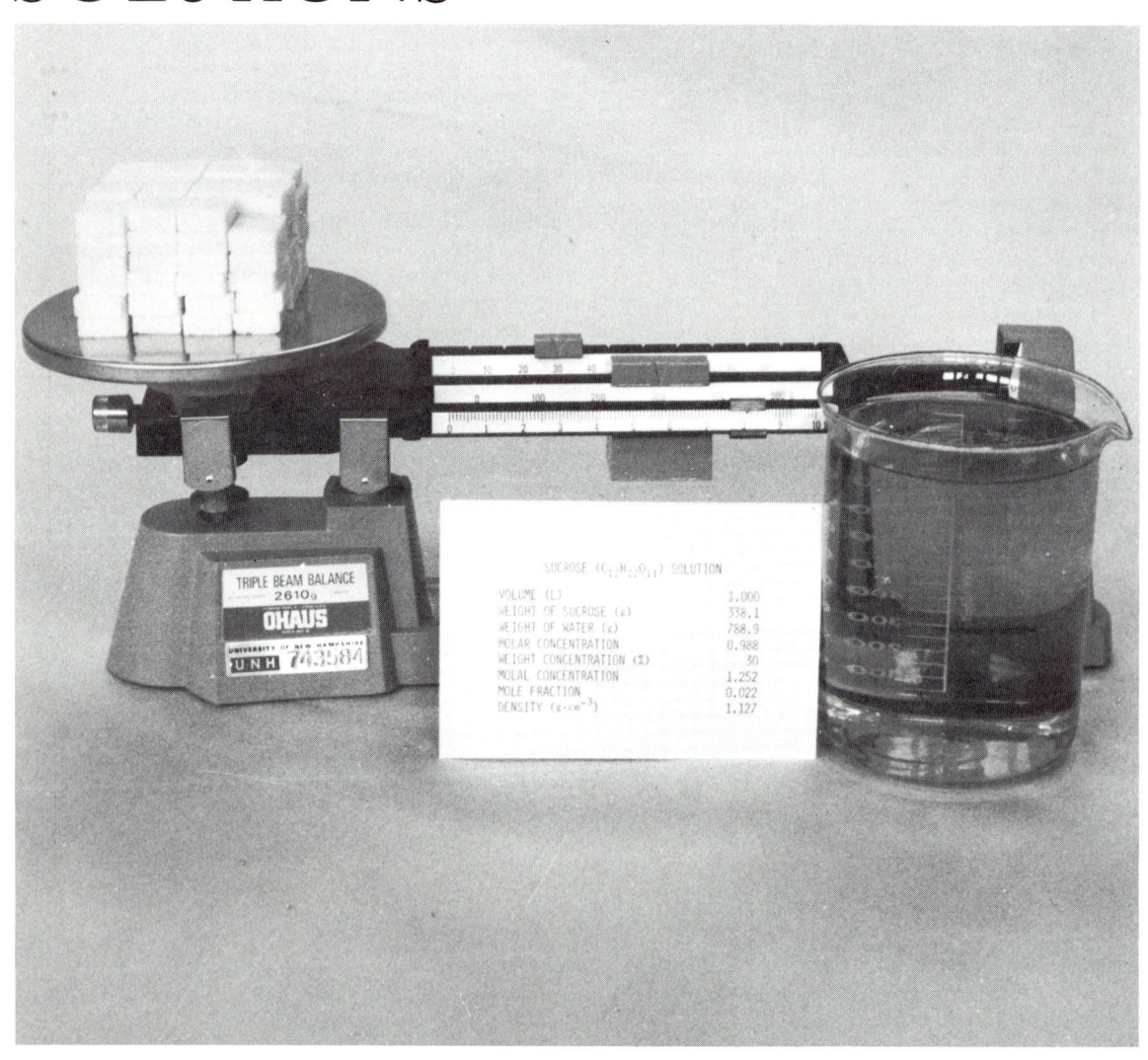

A solution is a mixture of too or more homogenic compounds like if you drink it you can taste them both at once.

FROM A STUDENT EXAM

A **solution** is defined as a homogeneous mixture of two or more substances, that is, a multicomponent single-phase system. Most of the solutions employed in laboratory work contain two components; these solutions are called **binary solutions**. For simplicity the discussion is limited to binary solutions. In addition, it is assumed that there are no chemical reactions between the components.

There are six different combinations of solid, gas, and liquid phases that lead to binary solutions: solid–solid, gas–gas, liquid–liquid, solid–gas, solid–liquid, and gas–liquid. The most important of these are the combinations in which one of the components is a liquid.

10.1 SOLUTES AND SOLVENTS

It is customary to refer to one of the components of a binary solution as the **solute** and the other as the **solvent**. This terminology was originally intended to imply that one of the components of the solution (the solute) is being dissolved by the other (the solvent), that is, solute and solvent are the "dissolvee" and "dissolver," respectively. Today it is realized that the interaction of the two components is completely mutual so that an active role cannot be assigned to one component and a passive role to the other; each component dissolves the other and, in turn, is dissolved by the other. Accordingly, the terms solute and solvent are largely arbitrary, and their use is based on convenience only. In very dilute solutions the component present in the smaller amount is often called the solute. Also, if the solution is used for a purpose in which one of the components plays a secondary role (for example, an inert vehicle for a chemically reacting species), this less important component is called the solvent. Another common usage is to call the solvent that component whose phase does not change upon mixing. In general, it is better to adopt a rather flexible response to these terms.

◂ A solution of sucrose (table sugar) and water. The 1-L solution on the right contains that amount of sugar shown on the balance pan. (Photo by Charles W. Owens, UNH Chemistry Department.)

10.2 CONCENTRATION UNITS

The composition of solutions is expressed in terms of various concentration units, the particular choice of units depending on the specific situation in which the solution is to be used. Basically, all of the concentration units express one important characteristic of the solution: how much of a given component there is in a given amount of solution or solvent.

One of the simplest ways of expressing concentration is to state the **percentage** (or fractional) **composition** on a weight basis. Thus a mixture of 10 g of water and 2.5 g of glucose ($C_6H_{12}O_6$) is said to have a percentage composition (by weight) of

$$\frac{2.5 \text{ g}}{10 \text{ g} + 2.5 \text{ g}} \times 100 = 20\% \text{ glucose (by weight)}$$

and

$$\frac{10 \text{ g}}{10 \text{ g} + 2.5 \text{ g}} \times 100 = 80\% \text{ water (by weight)}$$

Occasionally the composition is expressed in terms of relative volumes (measured before mixing) rather than relative weights. Liquid–liquid solution concentrations are sometimes given in this way.

A common practice is to express composition in terms of the relative number of moles of the components. For example, the above solution contains (10 g)/(18 g·mol^{-1}) or 0.55 mol of water and (2.5 g)/(180 g·mol^{-1}) or 0.014 mol of glucose. The **mole-fraction composition** of the solution is (using the symbol X to represent mole fraction)

$$X_{\text{glucose}} = \frac{n_{\text{glucose}}}{n_{\text{glucose}} + n_{\text{water}}} = \frac{0.014 \text{ mol}}{(0.014 + 0.55) \text{ mol}} = 0.025$$

Since $X_{\text{glucose}} + X_{H_2O}$ must have a value of 1, $X_{H_2O} = 1 - 0.025$ or 0.975. Chemists employ mole-fraction concentrations whenever both components are to be considered on an equivalent basis, that is, when neither component is singled out for emphasis.

When one of the components is more important in a particular context than the other, and usually (but not necessarily) is present in a relatively small amount, the **molal concentration** unit (symbol m) is often employed. If the component of greater importance is called the solute, the molal concentration of the solute is given as the number of moles of solute per kilogram of solvent. In the case of the glucose–water solution used in the previous examples, there is 0.014 mol of glucose dissolved in 0.010 kg of water. The molal concentration of the glucose is

$$\frac{0.014 \text{ mol glucose}}{0.010 \text{ kg water}} = 1.4 \text{ } m$$

The main characteristic of the molal unit is that it expresses the ratio of the number of solute molecules to a fixed mass of solvent. Section 10.8 deals with situations in which this characteristic is important.

In situations where one of the binary solution components acts only as a relatively inert vehicle for the other, it is convenient to use **molar concentration** (symbol M), defined as moles of solute (the more important component) per liter of solution. To calculate the molar concentration of glucose in the above example one must know the volume occupied by a mixture of 10 g of water and 2.5 g of glucose. At 25 °C the density of such a solution is about 1.08 g·cm^{-3}. Thus the volume is

$$V = \frac{\text{mass}}{\text{density}} = \frac{12.5 \text{ g}}{1.08 \text{ g·cm}^{-3}} = 11.5 \text{ cm}^3 \quad \text{or} \quad 0.0115 \text{ L}$$

The molar concentration of the glucose then is

$$\frac{0.014 \text{ mol glucose}}{0.0115 \text{ L solution}} = 1.2 \text{ } M$$

The SI recommendation is that the liter (defined as exactly 1000 cm^3) be replaced by the cubic decimeter (dm^3). Thus molar concentration should be expressed as mol·dm^{-3} rather than mol·L^{-1}. However, only the latter will be used in this book.

The main advantage of the molar concentration unit is that it facilitates the preparation of solute samples containing a desired amount of solute; a given volume of solution is identified with a known mass of solute. This is due to the simple relationship

$$V \text{ (in liters)} \times M = \text{moles}$$

Volumes of liquids are easy to measure rapidly and accurately by a variety of devices: pipets, burets, and graduated cylinders. For example, if it is desired to carry out the reaction

$$\text{HCl(aq)} + \text{NaOH(aq)} \rightarrow \text{NaCl(aq)} + \text{H}_2\text{O (l)}$$

using solutions of 0.1-M HCl and 0.5-M NaOH, then five volumes of HCl solution will react with one volume of NaOH solution to provide a 1:1 mole ratio of HCl to NaOH.* For example: 5 L of 0.1-M HCl contains 0.5 mol of HCl, and 1 L of 0.5-M NaOH contains 0.5 mol of NaOH. Similarly, if 0.03 mol of HCl is needed for some purpose, a volume of (0.03 mol)/(0.10 M) or 0.30 L of 0.1-M solution will provide the desired amount of HCl.

EXAMPLE 10.1 What volume of 18-M H$_2$SO$_4$ (sulfuric acid) is needed to prepare 10 L of 3.0-M H$_2$SO$_4$? Enough water is added to this amount to bring the volume to 10 L.

SOLUTION Let x = the volume (in liters) of 18-M H$_2$SO$_4$ required. Since there is no change in the number of moles of H$_2$SO$_4$ when water is added, the following relationship must hold:

$$(18 \text{ } M)x = (3.0 \text{ } M)(10 \text{ L})$$

$$x = \frac{(3.0 \text{ } M)(10 \text{ L})}{(18 \text{ } M)} = 1.7 \text{ L}$$

---

* The symbol aq, as in HCl(aq), stands for "aqueous" and means the substance is dissolved in water.

EXAMPLE 10.2 Given the reaction

$$2HCl(aq) + Na_2CO_3(aq) \rightarrow 2NaCl(aq) + H_2O(l) + CO_2(g)$$

a) What volume of 2.0-M HCl is needed to react with 250 cm³ of 3.0-M Na$_2$CO$_3$?
b) How many moles of CO$_2$ are produced?

SOLUTION

a) 250 cm³ (0.250 L) of 3.0-M Na$_2$CO$_3$ contains (0.250 L) (3.0 mol·L^{-1}) or 0.75 mol of Na$_2$CO$_3$. The number of moles of HCl needed is given by

$$0.75 \text{ mol Na}_2\text{CO}_3 \times (\text{MRF}) = 0.75 \text{ mol Na}_2\text{CO}_3 \left(\frac{2 \text{ mol HCl}}{1 \text{ mol Na}_2\text{CO}_3}\right)$$

$$= 1.5 \text{ mol HCl}$$

If x = the number of liters of 2.0-M HCl needed,

$$(2.0\text{-}M \text{ HCl})x = 1.5 \text{ mol HCl}$$

$$x = \frac{1.5 \text{ mol}}{2.0 \text{ mol}\cdot\text{L}^{-1}} = 0.75 \text{ L}$$

b) The number of moles of CO$_2$ produced is equal to the number of moles of Na$_2$CO$_3$ used up, namely, 0.75 mol.

10.3 SOLUBILITY

Although some substances mix in all proportions (for example, all gas–gas mixtures, water–ethanol mixtures, and water–ethylene glycol mixtures), it often happens that there is a limit to the amount of one substance which can be mixed with a fixed amount of another substance and yet have stable homogeneity maintained. If enough solute is added to a solvent so that some solute remains undissolved, the solution is said to be **saturated**. The main characteristic of the saturated solution is that dissolved solute and undissolved solute are in equilibrium; at any given time just as much solute is entering the solution phase as is leaving the solution phase. The amount of dissolved solute present in a saturated solution represents the **solubility** of the solute in the given solvent. Solubility may be expressed in any of the concentration units already mentioned, but units of grams of solute per 100 cm³ or 100 g of solvent are commonly used.

The solubility of a substance depends on the solvent and the temperature. For example, at 0 °C, 100 g of water will dissolve 35.7 g of NaCl, whereas 100 g of ethanol will dissolve only about 5 g of NaCl. At the higher temperature of 100 °C, 100 g of water will dissolve 39.12 g of NaCl. As a rough general rule, polar and ionic substances dissolve best in polar liquids, and nonpolar substances dissolve best in nonpolar liquids—a rule often abbreviated to **like dissolves like**. Polar liquids are associated with high dielectric constants (see Section 6.1) and nonpolar liquids have low dielectric constants. For example, water and ethanol (C$_2$H$_5$OH) are both polar (dielectric constants 78.5 and 24.3, respectively, at 25 °C) and form solutions in all proportions; NaCl and KCl (both ionic) also dissolve quite well in water. Similarly, nonpolar substances such as benzene, C_6H_6, and carbon tetrachloride, CCl_4 (dielectric constants 2.3 and 2.2, respectively, at 25 °C), are only slightly soluble in water but are much more soluble in less polar liquids such as alcohols, ethers, and hydrocarbons. The well-known failure of water and oil to mix is another example of the rule. Hydrocarbon fuels such as gasoline, on the other hand, dissolve many oils quite readily. Oils and gasolines have dielectric constants around 2 or less.

The like-dissolves-like rule is most simply rationalized on the basis of relative intermolecular forces. It seems reasonable to expect that a stable solution is characterized by a minimum in its potential energy, that is, by a maximum amount of interparticle attraction. If one substance A has very strong intermolecular attractions (for example, of dipole–dipole type) and another substance B has very weak intermolecular attractions (for example, of London dispersion type), then overall interparticle attractions are lessened by weak A-to-B attractions since these can occur only at the expense of strong A-to-A attractions. In other words, it is more favorable for the substances A and B to attract molecules of their own kind than to attract each other's molecules. Such a preference becomes less important if A and B have intermolecular forces of similar magnitudes. When exceptions to the like-dissolves-like rule occur, it is virtually certain that chemical reactions are taking place between the components. For example, Cl_2 has a solubility in water of about 1.5 g per 100 g of H_2O at 25 °C. This is somewhat higher than would be expected of a nonpolar gas; by contrast, another nonpolar gas, O_2, has a vanishingly low solubility (less than 1% that of Cl_2) at the same temperature. This is explained by the fact that Cl_2 and H_2O react chemically according to the equation

$$Cl_2 + H_2O \rightarrow HOCl + HCl$$

Oxygen and water do not react chemically at 0 °C.

Ionic substances usually dissolve quite readily in polar solvents due to ion–dipole interactions. For example, when sodium chloride dissolves in H_2O, Na^+ ions are attracted by the negative ends of H_2O dipoles and Cl^- ions are attracted by the positive ends of H_2O dipoles. Each ion thus acquires an "overcoat" of water molecules; such ions are said to be **hydrated** and are written $Na^+(aq)$ and $Cl^-(aq)$. The general term for the interaction between an ion and the solvent is **solvation**; hydration is but a specific example of this.

The mixing of components to form a solution is generally accompanied by changes in thermal energy, that is, by heat effects. The **heat of solution** is defined as the energy evolved or absorbed when a given amount of solute is mixed with a given amount of solvent. The heat of solution for a given amount of solute depends on the amount of solvent but tends to approach a fixed value as the amount of solvent becomes very large. Thus heats of solution are often given for a mole of solute dissolved in an infinite amount of solvent. When the solution process evolves energy (heat is produced by the mixing, that is, the solution becomes warmer), the solution process is called **exothermic** and the algebraic sign of the heat of solution is negative. The negative sign means that heat is given up or lost by the system (solute + solvent) and transferred to the surroundings. When the solution process absorbs heat (the solution becomes cooler), the solution process is called **endothermic** and the algebraic sign of the heat of solution is positive. The positive sign means that heat is absorbed or gained by the system (solute + solvent) from the surroundings. For example, the solution of one mole of NaCl in a large amount of water at 25 °C requires the input of about 3880 J; this is an endothermic process and the heat of solution of NaCl in H$_2$O is +3880 J·mol^{-1}. A simple way of viewing such an endothermic process is to recognize that mixing salt and water at 25 °C produces a temperature drop; thus 3880 J of energy must be added for each mole of NaCl in order to keep the temperature constant at 25 °C.

The solubility of substances in a given solvent either decreases or increases as the temperature increases, depending on the sign of the heat of solution. The direction of the solubility change is readily deduced by an application of Le Châtelier's principle. A saturated solution may be described as an equilibrium process between dissolved and undissolved solute, namely

$$\text{undissolved solute in contact with a saturated solution} \rightleftharpoons \text{dissolved solute in a saturated solution}$$

As a specific example, suppose the solution process indicated by the above equilibrium is endothermic. This means that the reverse process (the crystallization of a solid solute, for example) must be exothermic. If the equilibrium system is subjected to the stress of an increased temperature (this is tantamount to adding thermal energy) the system can undo the stress only by carrying out some process which is endothermic, that is, which can use up the added energy. Hence, more solute is dissolved to use up the added energy; in other words, the solubility of any substance with a positive heat of solution will increase as the temperature increases. Thus NaCl, which has an endothermic heat of solution, is more soluble at high temperatures than at low temperatures.

The effect of temperature on solubility may be complicated by the fact that some substances form **hydrates** which may not be stable over the entire temperature range considered. For example, below 32.4 °C the stable solid form of sodium sulfate is the hydrate Na$_2$SO$_4$·10H$_2$O.* The heat of solution of this substance is endothermic and, consequently, the solubility of the substance *below 32.4 °C* increases as the temperature increases. However, above 32.4 °C the anhydrous (without water) salt Na$_2$SO$_4$ becomes the more stable solid form and its heat of solution is exothermic. Consequently, *above 32.4 °C* the solubility decreases as the temperature increases. This example shows that it is important to take into account the possible formation of a hydrate and its range of stability before attempting to use Le Châtelier's principle to predict how solubility is affected by temperature.

Calcium hydroxide, Ca(OH)$_2$, is an example of a substance with an exothermic heat of solution. The solubilities (in grams per 100 g of H$_2$O) at 0, 50, and 100 °C are 0.185, 0.128, and 0.077, respectively. Boiler scale is produced when substances such as this, found in many natural waters, come out of solution as the water is heated.

The heat of solution of a solid in water may be regarded as the sum of two quantities: the lattice energy of the solid and the hydration energy. For example, the solution of NaCl in H$_2$O may be imagined to take place as follows: First, the crystal lattice of NaCl (consisting of rigidly ordered Na$^+$ and Cl$^-$ ions) is broken down into

---

* The H$_2$O's in the formula of a hydrate are incorporated in the crystal lattice of the substance. Na$_2$SO$_4$·10H$_2$O has ten moles of H$_2$O per mole of Na$_2$SO$_4$. See Section 15.14 for a fuller discussion.

Henry's law is named after its discoverer, William Henry (1774–1836), British physician and chemist. Inspired by his friend John Dalton, Henry began research in chemistry and eventually discovered how the solubility of gases in water varies with temperature and pressure. A childhood accident left Henry with chronic ill health and deformed stature and probably led to his eventual suicide.

gaseous Na^+ and Cl^- ions, and second, the Na^+ and Cl^- ions interact with H_2O via ionic-dipole forces to form hydrated ions, that is, ions with some water molecules closely clustered about them like "overcoats." The energy needed for the first process is the lattice energy (see Section 7.4). It is always endothermic (positive) since energy must be put in to break apart the rigid lattice. For NaCl the lattice energy is 769,860 $J \cdot mol^{-1}$. The energy needed to form hydrated ions is called the **hydration energy**; it is always exothermic (negative). The hydration energy of NaCl is $-765,980$ $J \cdot mol^{-1}$. The sum of the energies of the two steps, $769,860 + (-765,980)$, or 3880 $J \cdot mol^{-1}$, is the heat of solution of NaCl in water. If the magnitude of the hydration energy is greater than the magnitude of the lattice energy, then the solution process will be exothermic.

The process of dissolving a gas in a liquid or solid may be viewed in terms of the two steps: liquefaction of the gas (exothermic) and mixing of the liquefied gas with the solvent. The latter process is often almost **athermic**; that is, it involves very little energy change. Thus the solution process is invariably exothermic, and gases are normally less soluble at high temperatures than at low temperatures.* This behavior is exemplified by dissolved air coming out as bubbles when water is heated or by CO_2 fizzing out of warmed carbonated beverages.

The solubility of many gases in liquids or solids has been found to be approximated by the relationship

$$c = kp \tag{10.1}$$

which is known as **Henry's law**. Here c is the concentration of the dissolved gas (in appropriate units), p is the partial pressure of the gas in equilibrium with the solution, and k is a constant whose value depends on the temperature and on the solute and solvent. Table 10.1 lists the Henry's law constants for several gases in water at various

TABLE 10.1 HENRY'S LAW CONSTANTS FOR THE SOLUBILITY OF VARIOUS GASES IN WATER AT SEVERAL TEMPERATURES

GAS	HENRY'S LAW CONSTANT k ($mol \cdot L^{-1} \cdot atm^{-1}$) AT		
	0 °C	20 °C	50 °C
N_2	0.00103	0.00073	0.00051
H_2	0.00096	0.00081	0.00073
O_2	0.0022	0.0014	0.00094
He	0.00042	0.00039	0.00040

* Exceptions to this general rule are discussed by K. J. Mysels in *J. Chem. Educ.* **32**, 399 (1955).

temperatures. Note that k generally decreases with increasing temperature (except for helium, where it remains virtually constant). Henry's law explains why a bottle of carbonated beverage fizzes when the cap is removed: The small amount of CO_2 trapped above the liquid surface and in equilibrium with the dissolved CO_2 escapes, the partial pressure of CO_2 is reduced, and more CO_2 leaves the solution in an attempt to reestablish equilibrium. This is another example of the applicability of Le Châtelier's principle to the disturbance of an equilibrium system.

Henry's law is most accurate quantitatively for gases which interact only weakly with the solvent; it works poorly or not at all for gases such as NH_3 and Cl_2, which react with water.

The following example illustrates a typical application of Henry's law.

EXAMPLE 10.3 Estimate the weight of air dissolved in 1 kg of air-saturated water at 20 °C and 1 atm.

SOLUTION Assume that $p_{O_2} = 0.20$ atm and $p_{N_2} = 0.80$ atm. Then, from the data in Table 10.1, the molar concentrations of O_2 and N_2 are

$$M_{O_2} = (0.0014 \text{ mol} \cdot L^{-1} \cdot \text{atm}^{-1})(0.20 \text{ atm}) = 2.8 \times 10^{-4} M$$

$$M_{N_2} = (0.00073 \text{ mol} \cdot L^{-1} \cdot \text{atm}^{-1})(0.80 \text{ atm}) = 5.8 \times 10^{-4} M$$

A 2.8×10^{-4}-M solution of O_2 in H_2O will contain about 2.8×10^{-4} mol of O_2 per kg of H_2O since 1 L of the solution contains virtually 1 kg of H_2O. Thus the weights of dissolved gases are given by

$$w_{O_2} = (2.8 \times 10^{-4} M)(32 \text{ g} \cdot \text{mol}^{-1}) = 90 \times 10^{-4} \text{ g}$$

$$w_{N_2} = (5.8 \times 10^{-4} M)(28 \text{ g} \cdot \text{mol}^{-1}) = 16 \times 10^{-3} \text{ g}$$

$$w_{O_2} + w_{N_2} = (90 \times 10^{-4} + 16 \times 10^{-3}) \text{g} = 25 \times 10^{-3} \text{ g}$$

No simple models of wide applicability have been found for other types of solutions, for example, liquid–liquid solutions.

10.4 CRYSTALLIZATION AND PURIFICATION OF SOLIDS

If a saturated solution of a solid in a liquid is cooled (assuming a positive heat of solution), crystals of the solid will begin to form. This process, the reverse of dissolution, forms the basis of several purification techniques. One of these, **fractional crystallization**, proceeds as follows: A solid to be purified is dissolved to saturation in a solvent and the solution is cooled below the saturation temperature. The material that crystallizes out is usually purer than the original material, since impurities tend to remain in solution. The solid is removed by filtration, washed with pure, cold solvent, and redissolved to saturation. The process is repeated as often as needed to attain the desired purity. Several different substances can also be separated and purified by this technique by making use of their different solubilities in various solvents.

The faster a saturated solution is cooled, the smaller the crystals of solute formed. Only by very, very slow cooling is it possible to produce large crystals. This behavior

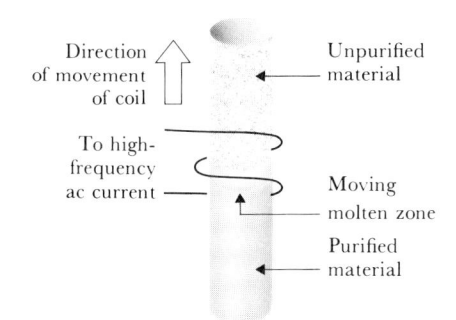

Fig. 10.1. Schematic representation of zone refining. The melting is accomplished by means of a movable narrow coil of wire encircling the vertical bar to be purified. High-frequency ac current flows through the coil. This heating method is called **radio frequency heating**.

accounts for the annealing and tempering of steel discussed in Section 7.6. Molten steel is a solution of Fe$_3$C (cementite) in iron. Rapid cooling causes small Fe$_3$C crystals to be regularly dispersed throughout the octahedral holes of the body-centered lattice (see Fig. 7.9). This leads to a hardened or tempered steel. Slow cooling, on the other hand, allows large Fe$_3$C crystals to form. These, being surrounded by large regions of pure iron, do not strengthen the steel and the steel is annealed.

Another process, useful in producing substances of extraordinarily high purity, is **zone refining**. A bar of a solid substance, purified as much as possible by ordinary means, is suspended vertically and a narrow horizontal band of it is melted (see Fig. 10.1). Then, either the bar is slowly lowered or else the heating device is moved upward so that the molten band moves from the bottom to the top of the vertical bar. Small amounts of impurities present in the substance tend to stay with the moving molten zone and thus to become "swept" toward the top of the bar.* After several such passes of the molten zone, the top end, containing most of the impurities, is cut off and discarded. Materials for semiconductors (see Section 7.9) are made in this way. Generally, semiconductors require about 1 part of impurity per 10^9 parts of silicon or germanium. The most accurate way of attaining this impurity level is to prepare ultrapure germanium or silicon and then add impurities in controlled amounts—a process known as **doping**.

Crystallization does not always occur when a saturated solution of a solid is cooled. Some solid solutes have a pronounced tendency to remain in solution above and beyond their solubility at the temperature in question. This phenomenon is known as **supersaturation**; it bears some resemblance to the supercooling of molten solids discussed in Section 9.3. Supersaturation may be disrupted and crystallization induced either by adding a tiny crystal of the solid solute to the supersaturated solution or else by stirring it vigorously. In either case, the excess solute usually crystallizes out rapidly. As indicated earlier, apparently there must exist a proper orientation of atoms or molecules which forms a pattern or template for the initiation of crystallization; some atoms or molecules must first provide this pattern before others can follow suit.† Such a "nucleus" for crystal growth is provided by the so

---

* In some materials the impurities travel against the molten zone. An understanding of this phenomenon requires a study of solubility and phase relationships involving the material to be purified and its dissolved impurities.

† According to a theory of F. C. Frank, direct growth of a perfect crystal is almost impossible except at enormous supersaturations. If a crystal has smooth sides, it is hard to get atoms to adhere to it long enough for the crystal to grow. Frank believes that all crystal growth takes place along the edges of screw dislocations (Fig. 7.8), the edges of the spiral ramp being able to retain atoms long enough for crystals to form. It is fairly certain that crystals growing from the vapor phase use this mechanism almost exclusively.

called **seed crystal** or by violent agitation—the latter perhaps making it easier for sluggish atoms or molecules to line up properly for crystallization.

Supersaturation is the basis of a very efficient way of growing large, well-formed crystals. A simple way to produce alum crystals is as follows: The alum* is dissolved in hot water to saturation and the solution is purified by filtration. The solution is allowed to cool to room temperature and the cloud of small crystals which appears is removed by filtration. A few of these small crystals are selected as seed and glued to a glass or stainless steel rod, or enclosed in a loop of fine wire. The seed crystals are suspended in the center of the solution while it is kept at constant temperature and allowed to sit undisturbed. As the solvent slowly evaporates, there exists a small but constant amount of supersaturation, and the excess solute slowly crystallizes onto the seed crystal, which gradually grows larger. If the seed crystal is removed, crystal growth stops. Crystals of ordinary table sugar may be produced in an analogous fashion to make what is popularly called "rock" candy. Similarly, in the making of fudge, there is an initial supersaturation of sugar whose crystallization is induced, if necessary, by rapid and vigorous stirring.

10.5 THE VAPOR PRESSURES OF SOLUTIONS

The vapor pressure of a solution depends on the vapor pressures of the components, their relative amounts in the solution, and the extent of their intermolecular interaction. If one component is a liquid and the other is a liquid or solid—and intermolecular interactions are relatively small—the vapor pressure of the solution may often be approximated by the simple relationship

$$p = X_A p_A^0 + X_B p_B^0 \tag{10.2}$$

where p_A^0 and p_B^0 are the vapor pressures of the pure components A and B at the temperature in question. This relationship is known as **Raoult's law**.

Raoult's law forms the basis of several phenomena of interest to chemists, engineers, and biologists. Some of the more important applications are discussed in the following sections.

EXAMPLE 10.4 A solution consists of 50 g of heptane (C_7H_{16}) and 100 g of ethylbenzene (C_8H_{10}) at 25 °C. The vapor pressures of these two substances at 25 °C are 40 and 10 torr, respectively. Calculate the vapor pressure of the solution.

SOLUTION The moles of heptane and of ethylbenzene are given by

$$n_H = \frac{50 \text{ g}}{100 \text{ g} \cdot \text{mol}^{-1}} = 0.50 \text{ mol} \quad \text{and} \quad n_E = \frac{100 \text{ g}}{106 \text{ g} \cdot \text{mol}^{-1}} = 0.94 \text{ mol}$$

---

* **Alum** is a name given to hydrated compounds formed by the combination of trivalent metal sulfates and alkali metal (or ammonium) sulfates. Examples are $KAl(SO_4)_2 \cdot 12H_2O$ (potassium alum) and $NH_4Fe(SO_4)_2 \cdot 12H_2O$ (ammonium iron alum). Note that when the alum contains aluminum, the name of this element is omitted from the name.

The mole fraction of heptane is

$$X_H = \frac{n_H}{n_H + n_E} = \frac{0.50}{0.50 + 0.94} = 0.35$$

The mole fraction of ethylbenzene is

$$X_E = 1 - X_H = 1 - 0.35 = 0.65$$

The partial pressure of heptane in the vapor is

$$X_H p_H^0 = (0.35)(40 \text{ torr}) = 14 \text{ torr}$$

The partial pressure of ethylbenzene in the vapor is

$$X_E p_E^0 = (0.65)(10 \text{ torr}) = 6.5 \text{ torr}$$

Thus the vapor pressure of the solution is

$$p = X_H p_H^0 + X_E p_E^0 = 14 + 6.5 = 20.5 \text{ torr}$$

10.6 FRACTIONAL DISTILLATION

Chemists and chemical engineers often find it necessary to separate the components of a solution and to purify them. Such a process is necessary, for example, in the petroleum industry when crude oil—a mixture of hundreds and even thousands of different compounds—is separated into its components. Although the actual methods of separation and purification are very complex, the basic principle can be understood by considering the application of Raoult's law to a binary solution of liquids.

Consider a solution of two liquids, A and B, assumed to obey Raoult's law. Let A represent the more volatile of the two liquids (this means p_A^0 is greater than p_B^0 at the temperature in question). Figure 10.2 shows a diagram of the solution vapor pressure as a function of composition. The diagram also shows the composition of the vapor in equilibrium with the solution. The key to the separation of A and B is that the vapor in equilibrium with the solution is richer in the more volatile component than is the solution itself; condensation of this vapor produces a new solution enriched in the more volatile component. At the same time the original solution is enriched in the less volatile component. Repeated vaporization and condensation eventually produces a vapor of almost pure A (the more volatile component) and a residue of almost pure B (the less volatile component).

François Marie Raoult (1830–1901) was a French physical chemist who established some of the foundations of the theory of solutions. Raoult's early attempts to find a simple way to determine the alcoholic content of wines led to some of his most important discoveries. (This does not necessarily imply that he got his good ideas while sipping wines.)

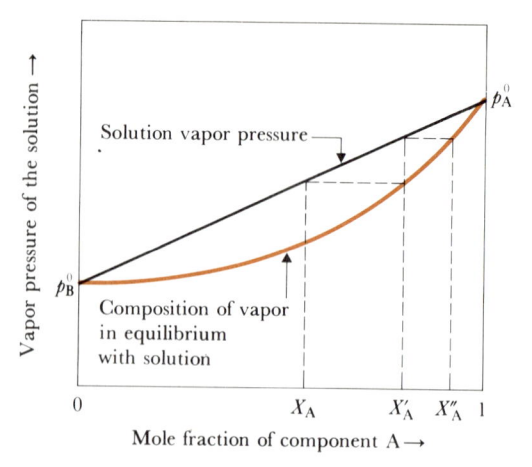

Fig. 10.2. The vapor pressure of a binary liquid–liquid solution as a function of composition, assuming Raoult's law is obeyed. The curved line represents the composition of the vapor which is in equilibrium with the solution. The graph is a plot of the straight line $p = (p_A^0 - p_B^0)X_A + p_B^0$, where $(p_A^0 - p_B^0)$ is the slope and p_B^0 is the intercept. X_A is the amount of A in a given solution and X_A' is the amount of A in the vapor which is in equilibrium with this solution. Note that this means that the vapor is richer in the more volatile component than is the solution. If the vapor of composition X_A' is condensed, the vapor in equilibrium with it has the composition X_A'' and is further enriched in A. Continuation of the process is the basis of the separation and purification of A and B.

The crucial fact that the vapor becomes enriched in the more volatile component can be proved by simple algebra. By Dalton's law of partial pressures (Section 3.6) the total vapor pressure of the solution is given by

$$p = p_A + p_B \tag{10.3}$$

where p_A and p_B are the partial pressures of liquids A and B in the vapor in equilibrium with the solution. If X_A' and X_B' represent the composition of the vapor, then Dalton's law also requires (if the vapors of A and B are ideal) that

$$p_A = X_A' p \quad \text{and} \quad p_B = X_B' p \tag{10.4}$$

Raoult's law provides additional relationships which p_A and p_B satisfy, namely

$$p_A = X_A p_A^0 \quad \text{and} \quad p_B = X_B p_B^0 \tag{10.5}$$

From Eq. (10.4) there is obtained

$$X_A' = p_A/p \tag{10.6}$$

Equation (10.5) gives

$$X_A = p_A/p_A^0 \tag{10.7}$$

Dividing Eq. (10.6) by Eq. (10.5) leads to

$$X_A'/X_A = p_A^0/p \tag{10.8}$$

Since p lies between p_B^0 and p_A^0, and p_A^0 is greater than p_B^0, it follows that X_A'/X_A is greater than unity, so that X_A' is greater than X_A. In other words, the vapor is richer in A, the more volatile component, than is the solution.

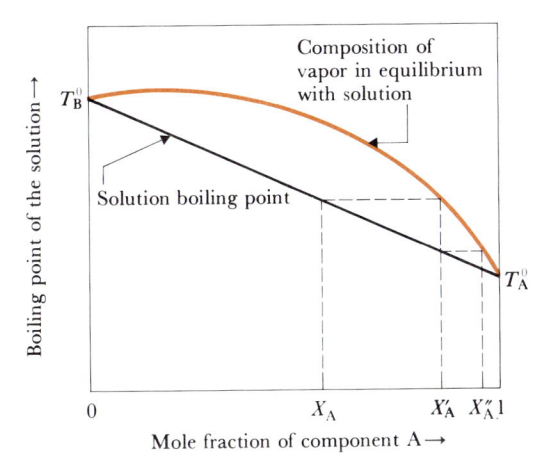

Fig. 10.3. Boiling point of a binary liquid–liquid solution as a function of composition, assuming Raoult's law is obeyed. The curved line represents the composition of vapor which is in equilibrium with the solution. X_A is the amount of A in a given solution and X'_A is the amount of A in the vapor which is in equilibrium with this solution. Note that this means that the vapor is richer in the more volatile component than is the solution. If the vapor of composition X'_A is condensed, the vapor in equilibrium with it has the composition X''_A and is further enriched in A. Continuation of the process will lead to a condensate progressively richer in A and a residue progressively richer in B.

The procedure described above—a succession of condensations and vaporizations which leads to condensates progressively richer in A and residues progressively richer in B—is known as **fractional distillation**. The procedure requires that the temperature be kept constant while the external pressure is adjusted to that of the vapor pressure of the solution (so that boiling can be maintained). In practice it is more convenient to carry out distillations at constant pressure.

EXAMPLE 10.5 What are the mole fractions of heptane and ethylbenzene in the vapor phase in Example 10.4?

SOLUTION The vapor pressure of the solution is 20.5 torr and the partial pressures of heptane and ethylbenzene are 14 and 6.5 torr, respectively. The vapor phase composition, using Dalton's law, is

$$X'_H = \frac{p_H}{p} = \frac{14}{20.5} = 0.68$$

$$X'_E = \frac{p_E}{p} = \frac{6.5}{20.5} = 0.32$$

Note that the vapor is enriched in the more volatile component, heptane.

Figure 10.3 illustrates how the boiling point of a binary liquid solution varies with composition when the vapor pressure is held constant. As the solution boils, the vapor coming off is condensed and reboiled. This new sample has a lower boiling point than the original solution since it is richer in the more volatile component A. The residue, on the other hand, is richer in the less volatile component B and its boiling point increases with successive fractions. Eventually, the residue fraction boils almost at T_B^0, the boiling point of pure B, and the condensate boils almost at T_A^0, the boiling point of pure A. Strict adherence to the above procedure is messy and awkward since it results in the subdivision of the original solution into a large number of small fractions, with gradations of composition running from almost pure A to almost pure B. In practice this annoyance is circumvented by doing the distillation continuously rather than batchwise. Devices for doing continuous distillations are known as **distillation columns**; those used in the petroleum industry may be several stories high.

10.7 CONSTANT-BOILING SOLUTIONS

In actuality, few, if any, solutions obey Raoult's law precisely. Nevertheless, Raoult's law is useful in a qualitative sense even for solutions which depart greatly from it. For example, in some solutions the vapor pressure at some intermediate concentrations is either greater than p_A^0 or else less than p_B^0 (assuming p_A^0 is greater than p_B^0). Each of these situations is illustrated in Fig. 10.4. The fractional distillation of such solutions can be understood on the basis of Raoult's law by dividing each diagram in the figure into two parts at the point where the maximum or minimum vapor pressure (or boiling temperature) occurs and treating each part separately. The main complication entering in is that a complete separation of the components cannot be accomplished by distillation. Eventually, there is obtained a composition for which $X_A = X_A'$ and $X_B = X_B'$ so that the solution behaves as a single component and resists further separation. The material with this composition is called a **constant-boiling mixture** or an **azeotrope**. An important example is ethanol and water. If a mixture of, say, 50% ethanol and 50% water is distilled, the condensate becomes richer in the more volatile alcohol until a composition of 95.6% (by volume) of alcohol is reached at a temperature of 78.2 °C (if the external pressure is held at 1 atm). The normal boiling points of pure ethanol and pure water are 78.5 °C and 100 °C, respectively. This composition is an azeotropic mixture and no further separation by distillation is possible.*

A rather spectacular example of an azeotropic mixture is a solution containing 53 cm³ of methyl formate (normal boiling point 32 °C) and 76 cm³ of n-propanol (normal boiling point 36 °C). If these two are mixed at 25 °C, boiling occurs spontaneously since the azeotrope has a boiling point of 22 °C!

Large deviations from Raoult's law, especially those leading to azeotropic solutions, are indications of rather strong interactions between the components of the solution. In fact, the components of an azeotropic mixture may be regarded as reacting in a quasi-chemical manner; that is, the two molecules act as if they have joined to make a single molecule—the azeotrope. The ethanol–water system may be regarded as a composite of two binary solutions, one a mixture of the azeotrope and alcohol and the other a mixture of the azeotrope and water.

---

* Pure ethanol (called *absolute alcohol*) is obtained by distilling a three-component azeotrope of alcohol, water, and benzene. If a sample of the alcohol–water azeotrope is mixed with benzene and distilled, all of the water, all of the benzene, and only part of the alcohol distills off. Slight traces of moisture in the alcohol may be removed by treating the hot alcohol with magnesium turnings, which remove the H_2O as insoluble $Mg(OH)_2$.

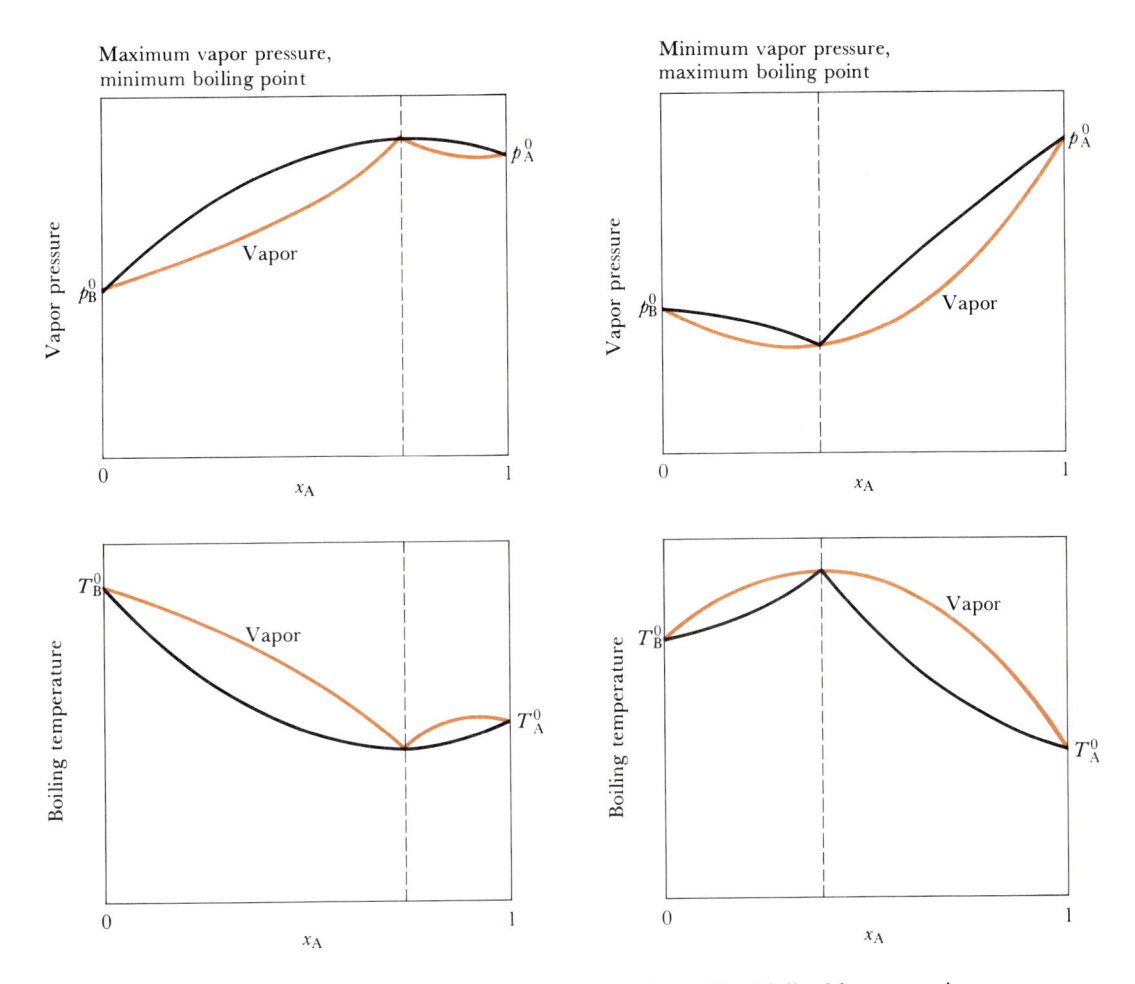

Fig. 10.4. Vapor pressures and boiling points of binary liquid–liquid azeotropic, or constant-boiling, solutions. An example of an azeotrope having maximum vapor pressure and minimum boiling point is nitric acid (HNO_3) and water; an example of an azeotrope having minimum vapor pressure and maximum boiling point is ethanol and water. Note that when $X_A = X'_A$ (indicated by the vertical dotted line) the solution behaves like a one-component system and A and B cannot be separated by distillation.

10.8 COLLIGATIVE PROPERTIES OF SOLUTIONS

A **colligative** property is any property that depends only on the number of solute particles relative to the number of solvent particles, and not on the specific nature of the solute particles. The three most important colligative properties of solutions are: freezing-point depression, boiling-point elevation, and osmotic pressure.

If one of the components of a binary solution is a nonionic solid or a relatively nonvolatile liquid, then dilute solutions of this substance have vapor pressures given approximately by

$$p = X_1 p_1^0, \tag{10.9}$$

where the subscript 1 indicates the solvent. Since X_1 is equal to $1 - X_2$ (where X_2 is the mole fraction of the nonvolatile solute), the preceding equation may be rewritten as

$$p = (1 - X_2)p_1^0$$

and rearranged to

$$\Delta p = p - p_1^0 = -X_2 p_1^0 \tag{10.10}$$

This equation predicts that the solute depresses the vapor pressure of the solvent in proportion to its concentration X_2. If n stands for moles, w for weight (or mass) in g, and M for molecular weight, the solute concentration X_2 is given by

$$X_2 = \frac{n_2}{n_1 + n_2} = \frac{n_2}{w_1/M_1 + w_2/M_2} \tag{10.11}$$

If the solution is dilute (that is, X_2 is much smaller than 1), n_2 is much smaller than n_1 and may be neglected in the denominator of Eq. (10.11). Then X_2 is given approximately by

$$X_2 = \frac{n_2}{w_1/M_1} = n_2 M_1/w_1 \tag{10.12}$$

Replacing w_1 (weight of the solvent in grams) with $1000\, W_1$ (W_1 is the weight of solvent in kg) leads to

$$X_2 = \frac{M_1}{1000} \frac{n_2}{W_1} = \frac{M_1}{1000} m \tag{10.13}$$

where n_2/W_1 or m is the molal concentration of the solute. Combining Eqs. (10.10) and (10.13) leads to

$$\Delta p = -\frac{M_1 p_1^0}{1000} m = km \tag{10.14}$$

where k, which equals $-M_1 p_1^0/1000$, is a constant whose value depends on only the solvent for the temperature in question. Equation (10.14) expresses a very important relationship, namely, that the lowering of the vapor pressure of a solvent by a nonvolatile solute is proportional to the molal concentration of the solute, provided the solution is dilute. Figure 10.5 shows Eq. (10.14) graphically for a temperature range including the freezing point and boiling point of the solid.

Figure 10.5 shows that the effect of lowering the vapor pressure is to *elevate the boiling point* of the solution over that of the pure solvent and to *depress the freezing point* of the solution. If ΔT_b and ΔT_f represent the **boiling-point elevation** and the **freezing-point depression**, respectively, then it is found that for dilute solutions

$$\Delta T_b = K_b m \quad \text{and} \quad \Delta T_f = K_f m \tag{10.15}$$

Figure 10.6 and the accompanying legend show how Eqs. (10.15) may be derived

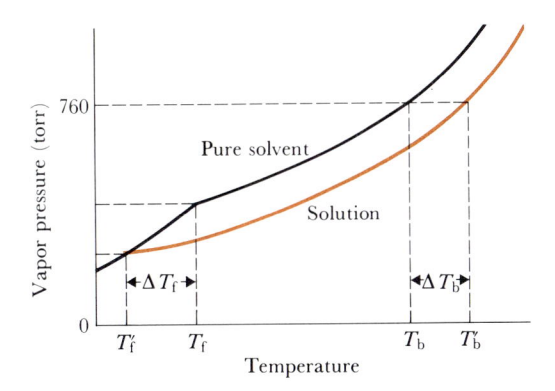

Fig. 10.5. Vapor pressure of a dilute solution of a nonvolatile solute in a liquid solvent as a function of temperature. T_f and T_b are the freezing point and normal boiling point of the pure solvent, respectively. Analogous primed quantities refer to the solution.

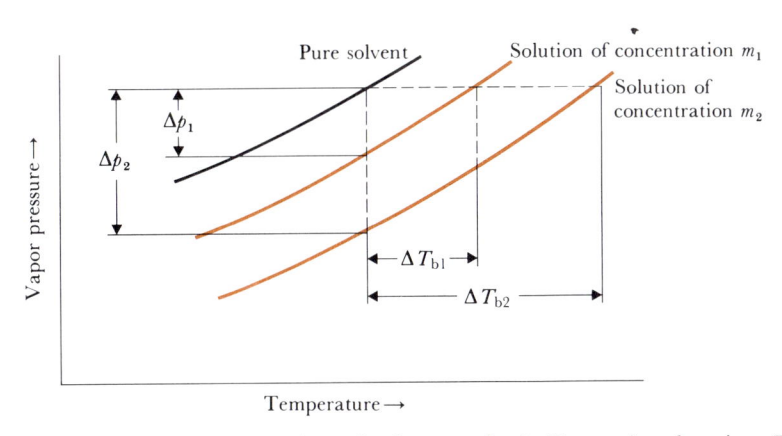

Fig. 10.6. Effect of concentration of solute on the boiling-point elevation. For small changes in vapor pressure (dilute solutions), the vapor pressure lines may be regarded as straight lines. Then by similar triangles, $\Delta p_1 : \Delta T_{b1} = \Delta p_2 : \Delta T_{b2}$. Since Δp is proportional to m (Eq. 10.14), $m_1 : \Delta T_{b1} = m_2 : \Delta T_{b2}$ and $\Delta T_b = K_b m$. A similar treatment leads to $\Delta T_f = K_f m$.

approximately from Eq. (10.14). The quantities K_b and K_f are constants depending only on the solvent. For water, $K_b = 0.52° \, m^{-1}$ (m^{-1} indicates "per molal concentration unit," that is, kg solvent·mol^{-1} solute) and $K_f = 1.86° \, m^{-1}$. When benzene is the solvent, $K_b = 2.53° \, m^{-1}$ and $K_f = 5.12° \, m^{-1}$. K_b and K_f are sometimes called the **ebulliometric** and **cryoscopic constants**, respectively. Values of these constants appear in various handbooks.

Equations (10.15) are often useful in the determination of molecular weights and molecular formulas of compounds, as illustrated in the examples below.

EXAMPLE 10.6 A 10-g quantity of a nonvolatile, nonionic compound is dissolved in 100 g of water. When the solution is cooled, ice crystals begin to appear at -2.1 °C. What is the approximate molecular weight of the solute?

SOLUTION The molal concentration of the solute is

$$m = \Delta T_f/K_f = 2.1°/1.86° \, m^{-1} = 1.1$$

Since the concentration of 10 g of solute in 100 g of H_2O is equivalent to that of 100 g of solute in 1000 g (1 kg) of H_2O, it is clear that the quantity of 100 g of solute is 1.1 mol. One mole of solute then weighs 100 g/1.1 mol or 91 g·mol^{-1}. This is the approximate molecular weight of the solute.

EXAMPLE 10.7 A compound is known to have the empirical formula C_2H_4O. When 5.3 g of this compound is dissolved in 114 cm³ of benzene, the solution begins to freeze (solid benzene first appears) at 3.4 °C. What is the molecular formula for the compound?

SOLUTION The freezing point of pure benzene is 5.5 °C; thus the freezing-point depression is 5.5 − 3.4 or 2.1°. The cryoscopic constant of benzene is 5.12° m^{-1}. The molality of the solute then is

$$m = 2.1°/5.12° \, m^{-1} = 0.41$$

Since the density of benzene is 0.880 g·cm^{-3}, the weight of solvent is (0.880 g·cm^{-3})(114 cm³) or 100 g. Since the concentration of 5.3 g of solute in 100 g of benzene is equivalent to that of 53 g of solute in 1 kg of benzene, 0.41 mol of solute weighs 53 g. Thus the approximate molecular weight of the solute is 53 g/0.41 mol or 129 g·mol^{-1}. The molecular weight associated with a molecular formula C_2H_4O is 44. Since 129 is about 3 × 44 or 132, it follows that the exact molecular weight is 132 g·mol^{-1} and that the molecular formula is $(C_2H_4O)_3$ or $C_6H_{12}O_3$.

The addition of an antifreeze to the radiator of an automobile is a practical application of the freezing-point depression. Here a simple relationship such as Eq. (10.15) does not hold very accurately, but the general principle remains valid.

Another important colligative property of a solution—also related to vapor-pressure lowering—is the osmotic pressure. If two solutions containing a given solvent and different concentrations of solute are separated by a membrane which allows solvent particles to pass through but which is impermeable to the solute particles, there is a net flow of solvent from the less concentrated solution to the more concentrated solution. This process of solvent flow is known as **osmosis** (from Greek *osmos*, "push"). Osmosis is one of the ways in which biological systems regulate fluid flow through cell membranes. For example, regulation of the NaCl concentration on opposite sides of a cell membrane determines in which direction water is able to flow. The **osmotic pressure** of a solution is defined as the pressure that must be exerted on the solution in order to just prevent solvent from flowing through the membrane when only pure solvent is present on the less concentrated side. A simple apparatus for the demonstration of osmotic pressure is shown in Fig. 10.7.

The osmotic pressure of a very dilute water solution is given approximately by the relationship

$$\pi V = n_2 R T \tag{10.16}$$

where π is the osmotic pressure, n_2 is the number of moles of solute, and V is the volume of the solution. The osmotic pressure appears to be just the pressure the

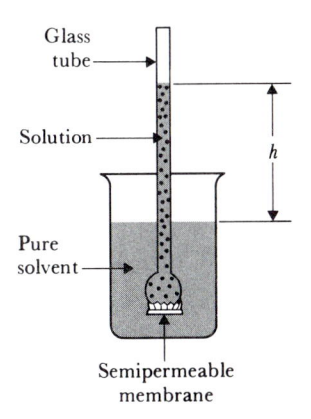

Fig. 10.7. A simple experimental way of demonstrating the osmotic pressure of a solution. Osmosis stops when the hydrostatic head due to the solution in the tube equals the osmotic pressure. The numerical value of the osmotic pressure is given by the weight of the column of solution of height h divided by the cross-sectional area of the tube.

solute would exert if it were an ideal gas. Appearances are deceiving, however; a correct explanation of osmotic pressure does not make use of such a simple model. Actually, solute–water interactions decrease the potential energy of the water so that the flow of water across the membrane is governed by a movement from higher potential energy to lower potential energy.

Two sample calculations illustrate the applicability of Eq. (10.16).

EXAMPLE 10.8 A 1-L water solution contains 34.2 g of sucrose ($C_{12}H_{22}O_{11}$). Calculate the osmotic pressure of the solution at 30 °C.

SOLUTION Since the molecular weight of sucrose is 342 g·mol^{-1}, the molar concentration of sucrose, n_2/V, is (34.2 g/342 g·mol^{-1})/1 L, or 0.1 M. Thus

$$\pi = \frac{n_2}{V} RT = (0.1 \text{ mol} \cdot \text{L}^{-1})(0.082 \text{ L} \cdot \text{atm} \cdot \text{K}^{-1} \cdot \text{mol}^{-1})(303 \text{ K})$$

$$= 2.48 \text{ atm}$$

The experimental value is 2.47 atm.

EXAMPLE 10.9 Calculate the osmotic pressure of a 5.0-M solution of sucrose in water at 30 °C.

SOLUTION Since the temperature is the same as in Example 10.6, the osmotic pressure is just 5.0/0.1 or 50 times that in Example 10.4, that is, 124.2 atm. The experimental value is 187.3 atm. This shows that Eq. (10.16) does not apply as accurately to more concentrated solutions.

Since a colligative property is any property that depends only on the number of solute particles relative to the number of solvent particles, and not upon the specific nature of the solute particles, all 1-m solutions of nonvolatile, nonionic solutes in a given solvent should have the same freezing points, boiling points, and osmotic pressures. In practice this is only approximately true and then only for very dilute solutions.

10.9 INTERIONIC INTERACTIONS IN SOLUTIONS

When the nonvolatile solute is ionic in nature, the colligative properties of water solutions are abnormal; for example, freezing-point depressions are higher than that predicted by Eq. (10.15). This should not be too surprising since a 1-m solution of NaCl, for instance, is actually 2 m in terms of particles (1 m in Na$^+$ and 1 m in Cl$^-$). Similarly, a 1-m solution of CaCl$_2$ is 1 m in Ca^{2+} and 2 m in Cl$^-$, or 3 m in all. However, experiments show that things are not quite as simple as the above might suggest. Very dilute solutions of ionic salts do behave in accordance with the number of particles; for example, a 0.001-m NaCl solution does lead to almost the same colligative properties as a 0.002-m nonionic solution. But a 1-m NaCl solution behaves more like a 1.8-m nonionic solution than like a 2.0-m nonionic solution. A modern explanation of this phenomenon is provided by the **interionic interaction theory** developed by Debye, Hückel, and others. This theory is highly mathematical in structure but some appreciation of it can be imparted in descriptive terms.

Essentially, the theory states that the positive and negative ions of an electrolyte (a solution which conducts an electric current) interfere with each other so that they cannot behave as single particles except in very dilute solutions, where they are far apart and shielded from each other by solvent molecules. A crude analogy to this behavior is as follows: Let us regard positive ions as boys, negative ions as girls, and solvent particles as chaperons and let us imagine that equal numbers of boys and girls are in a room with the assigned task of, say, sorting cards for filing. With just one boy and one girl in the room with many chaperons (dilute solution) it will be difficult for many boy–girl interactions to take place and it is likely that the labor output will be that of two persons. However, as the ratio of boy–girl pairs to chaperons increases (more concentrated solution), heterosexual interaction becomes more likely, the chaperons become less able to discourage such interactions, and the total labor output per boy–girl pair is likely to be less than that of two persons.

SUMMARY

Solutions are homogeneous mixtures of two or more substances. Especially important are binary solutions in which at least one of the components is a liquid.

The composition of a binary solution can be expressed in terms of several different concentration units; fractional composition (by weight), mole fraction, molality, and molarity are commonly employed.

The solubility of one substance in another often obeys the rule that like dissolves like; that is, polar solvents tend to dissolve polar and ionic solutes, and nonpolar solvents tend to dissolve nonpolar solutes. The solubility of a solid in a liquid increases as the temperature increases whenever the heat of solution is endothermic and decreases as the temperature increases whenever the heat of solution is exothermic. This behavior is in accord with Le Châtelier's principle. The sign of the heat of solution itself depends on which is of greater magnitude, the endothermic breakdown of the crystal lattice or the exothermic hydration of the solute. For gases dissolved in a liquid, the solubility normally decreases as the temperature goes up since the heat of solution is almost always exothermic, a combination of an exothermic liquefaction of the gas and an athermal mixing. According to Henry's law, the solubility of a gas at a given temperature is proportional to the partial pressure of the gas in equilibrium with the solution.

Solids may be purified from their solutions by recrystallization or by zone refining.

The vapor pressure of many solutions obeys Raoult's law: The total vapor pressure is a sum of the vapor pressures of the components, each such vapor pressure being equal to the mole fraction of that component multiplied by its vapor pressure as a pure substance. The relative vapor pressures of the components of a liquid–liquid solution form the basis of a method of separating the two components by fractional distillation.

A colligative property is one depending only on the number of solute particles relative to the number of solvent particles and not on their specific natures. The three most important colligative properties of solutions are depression of the freezing point, elevation of the boiling point, and osmotic pressure. All of these may be used in the determination of molecular weights of solutes.

Solutions of ionic solutes exhibit abnormal colligative properties; for example, they depress the freezing point in excess of that expected on the basis of their molal concentrations. Yet, because of interionic interaction, the ions of the solute interfere with each other—especially at higher concentrations—so that the effect on colligative properties is less than that expected of an equal number of nonionic particles.

LEARNING GOALS

1. Know how to express concentrations of solutions in each of the following units: weight fraction, mole fraction, molality, and molarity.

2. Given the concentration of a solution in terms of molality, be able to calculate the amount of solute in a given mass of solvent or the mass of solvent containing a given amount of solute.

3. Given the concentration of a solution in terms of molarity, be able to calculate the amount of solute in a given volume of solution or the volume of solution containing a given amount of solute.

4. Be able to calculate mole-fraction, molal, and molar concentrations from the density and percent composition of the solution.

5. Know what factors affect the mutual solubilities of a pair of substances.

6. Be able to calculate the heat of solution of a solid in a liquid from the lattice energy and the hydration energy of the solid.

7. Given the heat of solution, be able to predict how the solubility of a substance in a given solvent varies with temperature.

8. Know Henry's law and how to apply it to solutions of gases in liquids or solids. Know in general what types of gas–solvent systems are likely to obey Henry's law.

9. Be able to describe: fractional crystallization, zone refining, supersaturation, and growing crystals from solution.

10. Know Raoult's law and be able to use it to describe fractional distillation.

11. Know what an azeotrope is and how it behaves.

12. Know the relationships for freezing-point depression and boiling-point elevation of solutions and be able to use these to calculate approximate molecular weights of solutes.

13. Be able to determine molecular formulas from simplest formulas using approximate molecular weights.

14. Know what osmotic pressure is, how it is measured, and how it can be calculated for dilute solutions.

15. Be able to describe the Debye-Hückel interionic interaction theory for dilute and concentrated solutions.

DEFINITIONS, TERMS, AND CONCEPTS TO KNOW

solution
binary solution
solute and solvent
composition of a solution
   (percentage by weight, mole fraction, molal, and molar)
solubility
solvation
saturated solution

unsaturated solution
supersaturated solution
heat of solution
hydration energy
Henry's law
seed crystal
fractional crystallization
zone refining
doping

Raoult's law
fractional distillation
   (constant temperature and constant pressure)
azeotrope
osmotic pressure
colligative property
freezing-point depression
boiling-point elevation
interionic interaction theory

QUESTIONS AND PROBLEMS

Concentration

1. How many grams of NaCl are present in 250 cm³ of a 5-m solution having a density of 1.021 g·cm⁻³?

2. An aqueous solution of household ammonia contains 30% NH_3 by weight and has a density of 0.892 g·cm⁻³. Calculate (a) the mole fraction of NH_3, (b) the molal concentration of NH_3, (c) the molar concentration of H_2O.

3. A solution contains 20 g of benzene (C_6H_6) and 10 g of toluene (C_7H_8). Calculate the mole fraction of each compound.

4. A solution contains 49 g of H_3PO_4 and 360 g of H_2O. The density is 1.2 g·cm⁻³. Calculate (a) the volume of the solution, (b) the molar concentration of H_3PO_4, (c) the molal concentration of H_3PO_4, (d) the mole fraction of water.

5. Pure ethanol (C_2H_5OH) is labeled 200 proof. Ethanol labeled 100 proof is 50% ethanol and 50% water (by volume). If this is 42.5% ethanol by weight, what is the density of the 100-proof ethanol?

6. Concentrated nitric acid, HNO_3, is 69% by weight HNO_3 and has a density of 1.41 g·cm⁻³ at 20 °C. What volume and what weight of concentrated nitric acid are needed to prepare 100 cm³ of 6-M HNO_3?

7. Given the following chemical reaction in solution:

$CaCl_2(aq) + 2AgNO_3(aq) \rightarrow 2AgCl(s) + Ca(NO_3)_2(aq)$

a) How much 0.010-M $AgNO_3$ (in cm³) is needed to react completely with 0.10 mol of $CaCl_2$?
b) How much AgCl (in moles) is formed in (a)?
c) If the 0.10 mol of $CaCl_2$ is originally in a solution of volume 200 cm³, what is its molar concentration?

8. Describe exactly how you would make 1.0 L of 2-M HCl from a stock solution of 6-M HCl. How much of the 6-M HCl would you need to use?

9. The density of concentrated H_2SO_4 at 25 °C is 1.82 g·cm⁻³. If this contains 98% by weight of H_2SO_4, what is the molar concentration of the acid?

10. Explain how you would make a solution of H_2SO_4 containing 10% H_2SO_4 and having a density of 1.068 g·cm⁻³ from concentrated H_2SO_4 (see Problem 9).

11. A handbook lists the solubility of sodium chloride in water at 30 °C as 36.3 g of NaCl per 100 g of H_2O. What is the molal concentration of a saturated solution of NaCl in H_2O at 30 °C?

12. Show that a 0.001-M solution in water is also very nearly a 0.001-m solution.

13. Solid silver chloride is formed when a solution of a chloride and a solution of a soluble silver salt are mixed. The reaction is

$Ag^+(aq) + Cl^-(aq) \rightarrow AgCl(s)$

What volume of 1.0-M $AgNO_3$ must be added to 500 cm³ of 0.2-M NaCl solution to remove all the chloride ion as AgCl?

268 SOLUTIONS

14. The density of 1.099-M formic acid (HCO$_2$H) is 1.012 g·cm^{-3}. What is the weight percent of formic acid in the solution? What is the mole fraction of formic acid?

15. A water solution of baking soda (NaHCO$_3$) contains 8.76% by weight of soda.
   a) Calculate the molality of the soda solution.
   b) What is the mole fraction of water in the solution?

16. A solution of ethanol (C$_2$H$_5$OH) in water has a mole fraction of alcohol equal to 0.30.
   a) What is the molal concentration of the alcohol?
   b) What is the molal concentration of the water?

17. A water solution contains 10% by weight of glucose (C$_6$H$_{12}$O$_6$). What is the molal concentration and mole fraction of the glucose?

Solubility

18. The solubility of a certain compound in water (in grams of solute per 100 g of H$_2$O) is 15, 30, and 40 at 20 °C, 25 °C, and 30 °C, respectively. A given solution contains 350 g of the solute in 1 L of water. For each of the temperatures below, state whether the solution is saturated, unsaturated, or supersaturated.
   a) 45 °C   b) 35 °C   c) 15 °C

19. A substance has a positive heat of solution. If a small seed crystal is added to a supersaturated solution, is the resulting process exothermic or endothermic? Explain.

20. A 1.0-g quantity of a powdered solid is placed into 30 g of boiling water until dissolution is maximized. The undissolved solid is removed by filtration, dried, and found to weigh 0.90 g
   a) What is the solubility of the above substance (in grams per 100 g of H$_2$O) at 100 °C?
   b) The saturated solution above is cooled to 25 °C and 0.012 g of solid settles out. What is the solubility (in grams per 100 g of H$_2$O) at 25 °C?

21. The lattice energy of sodium uptite is 500 kJ·mol^{-1}. The solvation energy of the compound in Tibetan yak milk is −510 kJ·mol^{-1}.
   a) What is the molar heat of solution of sodium uptite in Tibetan yak milk?
   b) Is sodium uptite more soluble in Tibetan yak milk at 20 °C or at 30 °C? Explain.

22. Show that the melting of a solid is similar to its dissolution in a solvent.

23. How does putting salt on an icy sidewalk in winter cause ice to melt? Does this process warm the sidewalk?

24. Two immiscible liquids are placed into a capped bottle and shaken vigorously. Upon standing, two layers form. Would you expect each layer to be a strictly pure substance? If not, how would you characterize each layer?

25. The solubility of NaNO$_3$ in water at 10 and 90 °C is 80 and 160 grams per 100 g of H$_2$O, respectively. A solution at 90 °C contains 300 g of H$_2$O and 450 g of NaNO$_3$. How much NaNO$_3$ will precipitate out if the solution is cooled to 10 °C?

26. The solubility of KNO$_3$ in water at 80 °C is 170 g per 100 g of H$_2$O. A sample of KNO$_3$ solution at 80 °C weighs 0.500 kg and contains 375 g of KNO$_3$. How much KNO$_3$ will crystallize out upon addition of a seed crystal to the solution?

27. Which is more apt to form supersaturated solutions in water, sodium chloride (NaCl) or sodium acetate (CH$_3$CO$_2$Na)? Explain.

28. Which of the following is predicted to be more soluble in carbon tetrachloride (CCl$_4$) at 25 °C?
   a) Benzene (C$_6$H$_6$) or water.
   b) Sodium chloride or propane (C$_3$H$_8$).
   c) Helium (boiling point −269 °C) or argon (boiling point −186 °C).
Repeat these predictions using methanol (CH$_3$OH) as a solvent.

29. The higher the boiling point of a gas, the more likely it is to be soluble in a given solvent. Explain why this is true.

Henry's Law

30. The partial pressure of oxygen in a high altitude yak herder's yurt is 50 torr. The solubility of oxygen in Tibetan yak milk at that altitude at 5 °C (it's cold in a yak herder's yurt!) is 0.001 mol·L^{-1}.
   a) What is the value of the Henry's law constant for oxygen in Tibetan yak milk at 5 °C?
   b) What is the solubility of oxygen in Tibetan yak milk at 5 °C near sea level where the partial pressure of O$_2$ is 150 torr?
   c) Would the Henry's law constant increase or decrease if the yak herder built a fire of dried yak dung and heated his yurt to 10 °C?

31. At 30 °C and 1 atm pressure, air-saturated water is about 5.3×10^{-4} M in nitrogen. What is the Henry's law constant for N_2 at this temperature? Is this value consistent with the data given in Table 10.1?

32. The Henry's law constants given in Table 10.1 are also valid for molal concentrations of the gases. Explain why this is true.

33. A saturated solution of methane (CH_4) in hexane (C_6H_{14}) at 20 °C and 1 atm pressure of methane has a concentration of 0.024 M (CH_4 as solute). Calculate the solubility of methane in hexane at the same temperature when the partial pressure of methane is 10 atm.

34. The Henry's law constant for a gas dissolved in water is 3.45×10^{-5} g·L^{-1}·torr^{-1} at 30 °C. What gas pressure is needed to produce a solution containing 0.070 g of the gas in 2.00 L of water?

Raoult's Law

35. A solution contains 50 g of benzene (C_6H_6) and 50 g of toluene (C_7H_8) at 20 °C. The vapor pressures of pure benzene and pure toluene at 20 °C are 65 torr and 25 torr, respectively. Assume the solution is ideal, that is, that it obeys Raoult's law.
   a) What is the total vapor pressure of the solution?
   b) What are the partial pressures of benzene and toluene in the vapor?
   c) What is the mole fraction of benzene in the vapor?

36. The vapor pressures of ethanol (C_2H_5OH) and methanol (CH_3OH) at 20 °C are 44.5 and 88.7 torr, respectively. Solutions of these two substances obey Raoult's law fairly well. Assuming a solution is made by mixing 10 g of ethanol and 15 g of methanol, calculate (a) the vapor pressure of the solution at 20 °C, (b) the mole fraction of ethanol in the vapor.

37. Two organic liquids, A and B, have vapor pressures at 20 °C of 100 torr and 300 torr, respectively. Assuming a solution of A and B obeys Raoult's law, draw a graph showing (a) the partial pressures of A and B as a function of the mole fraction of A, (b) the total vapor pressure of the solution as a function of the mole fraction of A.

38. The vapor pressure of pure benzene (C_6H_6) at 25 °C is 97.0 torr. What is the vapor pressure of benzene above a solution containing 100 g of benzene and 10.0 g of naphthalene ($C_{10}H_8$)?

39. A solution at 25 °C contains 10 g of sodium chloride in 250 g of water. Assuming that sodium chloride is 100% ionic and neglecting interionic effects, estimate the vapor pressure of the solution. The vapor pressure of pure water at 25 °C is 23.76 torr.

Colligative Properties

40. A solution contains 3.42 g of solute in 100 g of H_2O and freezes at -0.2 °C. The same amount in 100 g of the solvent dissolvium neverite ($T_f = 10$ °C) freezes at 9.7 °C.
   a) What is the approximate molecular weight of the solute?
   b) What is K_f of dissolvium neverite?

41. A compound has the empirical formula A_2B (the atomic weights of A and B are 5 and 15 amu, respectively). Given that a 3.0-g sample of this compound in 100 g of H_2O freezes at -1.07 °C, calculate (a) the approximate molecular weight of the compound, (b) the exact molecular weight of the compound, (c) the molecular formula of the compound.

42. Assume that 0.211 g of a nonvolatile covalent compound is dissolved in 100 g of the solvent dissolvall. When the solution is cooled, dissolvall crystals begin to form at -10.3 °C (the freezing point of pure dissolvall is -10 °C). When a 2-m solution of sugar in dissolvall is cooled, dissolvall crystals begin to form at -15 °C.
   a) What is the freezing-point constant of dissolvall?
   b) What is the molal concentration of the first compound?
   c) What is the molecular weight of the first compound?

43. Assume that 2.5 g of an unknown compound is dissolved in 50 g of benzene. The solution freezes 2.56° below the freezing point of pure benzene. If K_f of benzene is 5.12 °m^{-1}, what is the molecular weight of the unknown compound?

44. Assume that 0.75 g of sulfur is dissolved in 15.0 g of naphthalene ($K_f = 6.8$ °m^{-1}). Pure naphthalene freezes at 80.55 °C and the sulfur–naphthalene solution freezes at 79.27 °C. What is the molecular weight of sulfur? What is the most likely molecular formula of the sulfur?

45. What osmotic pressure (in torr) would you predict at 25 °C for a water solution containing 0.1 g of NaCl per liter? What height of a column of water is equivalent to this osmotic pressure?

46. Estimate the volume (in liters) of methanol (CH$_3$OH) that must be added to a gallon (about 3.8 L) of water to obtain a solution freezing below -15 °C. The density of methanol around 0 °C is about 0.8 g·cm^{-3}.

47. A substance has the empirical formula AB$_2$. Show how it is possible to tell if the molecule is covalent with formula AB$_2$, ionic A^{2+}, B$_2^{2-}$, or ionic A^{2+}, 2B$^-$. Is it possible to distinguish between AB$_2$ and (A$^+$, AB$_4^-$)?

48. Arrange the following in decreasing order of freezing points: pure water, 1-m sugar solution, 0.5-m alcohol solution, 1-m CaCl$_2$ solution, 0.5-m NaCl solution, and 2-m KCl solution.

49. The freezing-point depression due to an ionic solute in water obeys the relationship

$$\Delta T_f = iK_f m$$

where i (the van't Hoff factor) is related to the number of moles of ions per mole of ionic solute. For nonionic solutes $i = 1$. Due to interionic interactions, i generally decreases as m increases.

a) Sketch the expected appearance of a graph of ΔT_f vs. m for a substance such as CaCl$_2$.
b) What is the value of i at $m = 0$ for CaCl$_2$?

50. A solution contains 0.40 g of a proteinlike substance in 1.0 L of water at 27 °C. The osmotic pressure of the solution, as measured with a water-filled manometer, is 43.5 mm. Estimate the molecular weight of the substance.

51. A 0.010-m solution of ammonia freezes at -0.0193 °C. Ammonia and water react to a small extent according to the equation

$$NH_3(aq) + H_2O(l) \rightarrow NH_4^+(aq) + OH^-(aq)$$

What fraction of the dissolved ammonia has participated in the above reaction?

52. Most modern antifreezes contain ethylene glycol (C$_2$H$_6$O$_2$). How many liters of ethylene glycol (density 1.12 g·cm^{-3}) must be added to 20 L of water to protect an automobile radiator down to -30 °C?

53. A solution of table sugar (C$_{12}$H$_{22}$O$_{11}$) in water freezes at -1.00 °C.

a) What is the weight of sugar dissolved in 1 L of water?
b) What is the boiling point of the above solution?

54. A solution is made up by dissolving 5.75 g of a nonelectrolyte in 185 g of water. The solution boils at 100.14 °C. What is the molecular weight of the solute?

ADDITIONAL READINGS

Gilman, J. J., ed., *The Art and Science of Growing Crystals*. New York: Wiley, 1963. Provides some interesting material for those who would like to begin a fascinating hobby.

Fullman, Robert L., "The Growth of Crystals." *Scientific American*, March 1955. Discusses how crystal growth from the vapor phase begins along a dislocation.

Pfann, William G., "Zone Refining." *Scientific American*, December 1967. A very comprehensive account by the method's inventor, written at a comprehensible level.

11

NUCLEAR TRANSFORM-ATIONS OF MATTER

Marie Curie, Polish-born French physicist who pioneered in the isolation and identification of radioactive elements. (Courtesy of The Bettmann Archive.)

To split the mighty atom
All mankind was intent.
Now any day the atom may
Return the compliment.
ANONYMOUS

Nuclear transformations involve a change in the number of neutrons or protons (or both) in the nucleus of an atom. Such transformations, although ostensibly a subject area of physics, are of great importance to many areas of chemistry. Perhaps one of the most important applications is the use of radioactive elements as tracers; from this chemists have obtained a far more intimate understanding of chemical reactions—including those in biological systems—than was dreamed possible less than half a century ago. Also, with the growing prevalence of radioactive materials in our environment, it is becoming increasingly important for scientists and medical personnel to understand just what chemical processes occur when radiation from radioactive materials interacts with living matter.

11.1 THE DISCOVERY OF RADIOACTIVITY

In 1896 Henri Becquerel was investigating the sunlight-induced fluorescence* of various minerals when he noticed that potassium uranyl sulfate, $K_2UO_2(SO_4)_2 \cdot 2H_2O$, emitted a penetrating radiation which was not related to fluorescence and was independent of previous exposure to light. This new type of radiation passed through paper, glass, and other substances and blackened photographic plates. The phenomenon, found to be characteristic of ores of uranium and some other heavy elements, was named **radioactivity** by Marie and Pierre Curie. After long and arduous chemical treatment of a radioactive ore known as pitchblende, Mme. Curie isolated two new elements having pronounced radioactive natures. One of these she named **polonium** (Po) in honor of her native country, Poland; the other she named **radium** (Ra) after its ability to emit intense radiation. Later, Ernest Rutherford† studied the radiation emanating from a radium prepa-

---

* When certain substances are irradiated with light of one wavelength, they emit light of a longer wavelength. This phenomenon is called **fluorescence**. In sunlight-induced fluorescence, a mineral is exposed to sunlight, then taken into a dark room. The mineral will glow (fluoresce) for a short period of time.

† For a short biographical sketch of Rutherford, see Chapter 1.

Antoine Henri Becquerel (1852–1908) was a French physicist who, in addition to discovering radioactivity, studied the optical properties of solids and the magnetic properties of gases. Becquerel shared the 1903 Nobel prize in physics with the Curies for the discovery and characterization of radioactivity.

ration by passing a narrow beam of this radiation between the pole pieces of a powerful electromagnet (see Fig. 11.1). The beam was found to comprise three different types of radiation named alpha, beta, and gamma (α, β, and γ) rays. The α-rays were deflected in a direction expected of positively charged matter; subsequent work proved that these rays consisted of doubly charged helium ions, that is, what we now call **α-particles**. The β-rays were deflected in a direction opposite to that of the α-particles and thus were thought to consist of negatively charged particles. These **β-particles** were ultimately shown to be ordinary electrons. To distinguish between electrons arising from radioactivity and electrons from other sources, only the former are called β-particles. The **γ-rays** were undeflected by the magnetic field and are now known to be very-short-wavelength electromagnetic radiation similar to X-rays.

By 1903, Rutherford and Soddy were able to demonstrate that α-, β-, and γ-radiation are by-products of spontaneous subatomic changes within atoms. Development of the nuclear model of the atom showed that these subatomic changes are changes in the nuclear composition of the atom. Ingenious and painstaking work by Rutherford and a host of colleagues showed that many radioactive elements decay to other radioactive elements which, in turn, decay further to yet other radioactive elements until, eventually, a nonradioactive element is obtained as a final product.

Marie Sklodovska Curie (1867–1934) and Pierre Curie (1859–1906), a wife-and-husband team, pioneered in the study of radioactivity. Before his marriage to Marie Sklodovska, Pierre Curie was well known for his discoveries of certain fundamental electrical properties of crystals. Later he joined in his wife's research and ultimately shared a Nobel prize in physics with her and with Henri Becquerel in 1903. In 1906 Pierre was run over and killed by a horse-drawn wagon. His widow continued her researches, assumed her husband's chair of physics at the Sorbonne, and received a second Nobel prize (this time in chemistry) for the preparation of pure radium metal. Marie eventually succumbed to leukemia, presumably a result of prolonged exposure to radiation. A sensitive biography of the Curies has been written by their younger daughter, Eve (Madame Curie, New York: Doubleday, 1938). *Marie Curie also wrote a biography of her husband,* Pierre Curie *(New York: The Macmillan Co., 1926; also available as a paperback from Dover Publications, New York, 1963).*

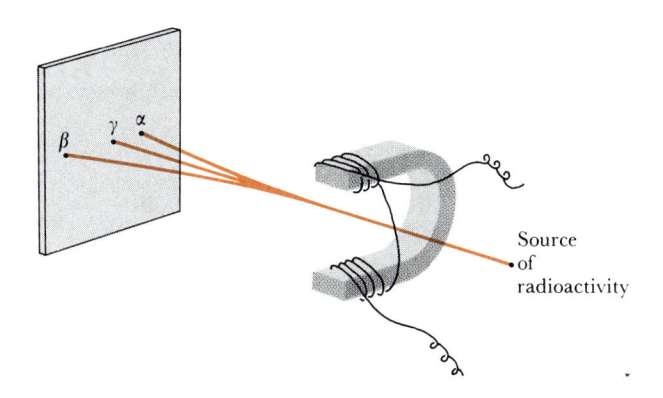

Fig. 11.1. Illustration of Rutherford's identification of α-particles, β-particles, and γ-radiation as the by-products of natural radioactive decay. The source of radioactivity is a radium sample enclosed in a lead box with a tiny hole through which a beam of radiation emerges.

11.2 NATURAL RADIOACTIVE DECAY

Nuclear decay is a process which involves changes in the composition of atomic nuclei and thus may result in the transformation of one element to another. Many nuclear transformations occur spontaneously in nature; others can be induced artificially in the laboratory. Each of these is discussed in turn.

One of the most important naturally occurring nuclear processes is the series of reactions by which radium is transmuted to an isotope of lead:

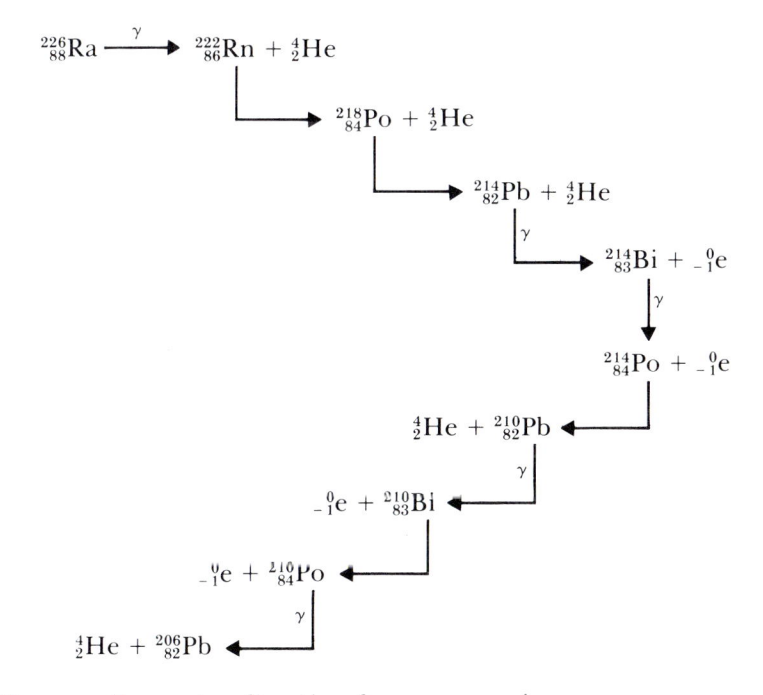

$$^{226}_{88}\text{Ra} \xrightarrow{\gamma} {}^{222}_{86}\text{Rn} + {}^{4}_{2}\text{He}$$
$$\rightarrow {}^{218}_{84}\text{Po} + {}^{4}_{2}\text{He}$$
$$\rightarrow {}^{214}_{82}\text{Pb} + {}^{4}_{2}\text{He}$$
$$\xrightarrow{\gamma} {}^{214}_{83}\text{Bi} + {}^{0}_{-1}\text{e}$$
$$\xrightarrow{\gamma} {}^{214}_{84}\text{Po} + {}^{0}_{-1}\text{e}$$
$$\rightarrow {}^{4}_{2}\text{He} + {}^{210}_{82}\text{Pb}$$
$$\xrightarrow{\gamma} {}^{0}_{-1}\text{e} + {}^{210}_{83}\text{Bi}$$
$$\rightarrow {}^{0}_{-1}\text{e} + {}^{210}_{84}\text{Po}$$
$$\xrightarrow{\gamma} {}^{4}_{2}\text{He} + {}^{206}_{82}\text{Pb}$$

The overall or net radioactive decay process is

$$^{226}_{88}\text{Ra} \xrightarrow{\gamma} {}^{206}_{82}\text{Pb} + 5{}^{4}_{2}\text{He} + 4{}^{0}_{-1}\text{e}$$

Note that a radioactive decay equation is balanced in such a way that total mass numbers and total atomic numbers are the same on both sides of the arrow. Thus

Frederick Soddy (1877–1965) was a British chemist who collaborated with Rutherford in some of the basic studies of the nature of radioactive decay. In 1913 Soddy originated the concept of isotopes and in 1921 he received the Nobel prize in chemistry. During his later years, Soddy—always a controversial person—drifted from science to unpopular political and economic ideas, resigned from academic life, and ended his days ignored and frustrated.

226 = 206 + 5(4) + 4(0) and 88 = 82 + 5(2) + 4(−1). This is true for each of the individual reactions of the above process.

The symbol γ indicates steps that are characterized by the emission of γ-rays. When $^{226}_{88}$Ra decays to $^{222}_{86}$Rn by the loss of an α-particle, 94.5% of the emitted α-particles have energies of 4.781×10^6 eV each, while the remaining 5.5% have energies of 4.598×10^6 eV each; the accompanying γ-radiation has a wavelength of 0.0678 Å. Thus the energy associated with a γ-ray photon is

$$\Delta E = \frac{hc}{\lambda} = \frac{(6.626 \times 10^{-34} \text{ J} \cdot \text{s})(3.00 \times 10^8 \text{ m} \cdot \text{s}^{-1})}{6.78 \times 10^{-12} \text{ m}} = 2.93 \text{ J}$$

Since 1.602×10^{-19} J is 1 eV, this is an energy of 1.83×10^5 eV—precisely the energy difference between the two types of emitted α-particles. As illustrated in Fig. 11.2, the decay process is interpreted as follows: When the radium atom decays, it is transformed to a radon atom whose nucleus can exist in two different quantum states separated by an energy of 1.83×10^5 eV. Specifically, loss of an α-particle of energy 4.598×10^6 eV transforms the radium atom to an excited state of radon, whereas loss of an α-particle of energy 4.781×10^6 eV produces radon in its ground state. Thus the α-particle energy difference, $(4.781 - 4.598) \times 10^6$ eV, or 1.83×10^5 eV, is the energy separation between the two quantum states of the radon nucleus. The excited radon nucleus emits γ-radiation of wavelength 0.0678 Å as it undergoes a spontaneous transition to the ground state.

In many cases a product nucleus in an excited quantum state goes to the ground state or to a lower excited state not by emitting γ-radiation but by one of several alternative processes, such as:

1. Emission of an electron–antineutrino pair.
2. Emission of a positron–neutrino pair.
3. Emission of mesons which decay to other particles and emit γ-radiation.

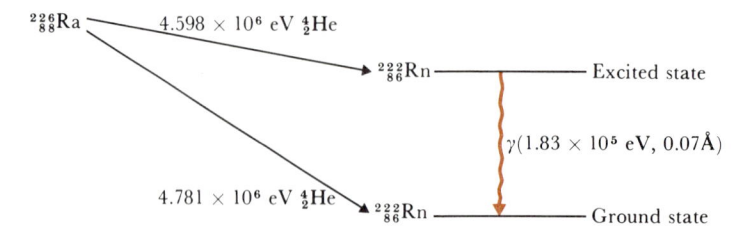

Fig. 11.2. The decay of radium-226 to two different quantum states of the radon-222 nucleus.

Carl D. Anderson (b. 1905), an American physicist, shared the 1936 Nobel prize in physics with Victor Hess of Austria for the discovery of the positron. Along with Seth Neddermeyer, he also discovered the meson (then thought to be a single unique particle) in 1937.

A **positron** (symbol ^0_1e) is a positively charged "electron." It represents the so-called **antimatter** counterpart of an ordinary electron, or **negatron**. Similarly, the ordinary proton (^1_1H) has a negatively charged antimatter counterpart called the **antiproton** ($_{-1}^{1}H$). One of the most important characteristics of antimatter is that it tends to annihilate ordinary matter upon contact. Thus when a positron and an electron collide, both disappear to produce a burst of γ-radiation. The existence of antimatter was predicted in 1929 by Dirac* and was verified experimentally by Anderson in 1933.

Mesons are a class of particles (occurring in about five varieties) with masses from 200 to 1000 times that of an electron and charges of 0 or ± 1. They play important roles in the structure of nuclei.

The **antineutrino** is a small, virtually massless and electrically neutral particle which carries away part of the energy emitted by a β-decay process. The neutrino is the ordinary-matter counterpart of the antineutrino.

Note that loss of an α-particle (^4_2He) decreases the mass number by *four* and the atomic number by *two*, whereas loss of a β-particle causes no change in the mass number but increases the atomic number by *one*. It is also important to note that since an atomic nucleus contains only protons and neutrons, β-particles must be "created" during the decay process. Thus β-emission is equivalent to the transformation of a neutron to a proton via the process

$$^1_0n \rightarrow {}^1_1H + {}^{0}_{-1}e$$

Similarly, positron emission is equivalent to the transformation of a proton to neutron:

$$^1_1H \rightarrow {}^1_0n + {}^0_1e$$

11.3 ARTIFICIAL RADIOACTIVE DECAY

Artificial transmutations are of the general type

    atomic nucleus + projectile $\rightarrow$ products

where the projectile is a high-energy α-particle, β-particle, neutron, proton, deuteron (^2_1H), triton (^3_1H), or some other particle. The first artificial transmutation was carried out in 1919 by Rutherford by bombarding nitrogen atoms with α-particles to produce an isotope of oxygen and a proton. The process is

$$^{14}_{7}N + {}^4_2He \rightarrow {}^{17}_{8}O + {}^1_1H$$

---

* For a short biographical sketch of Dirac, see Chapter 4.

In some cases one or more of the product elements may be radioactive (this is not true of $^{17}_{8}O$). For example, bombardment of $^{238}_{92}U$, which is only slightly radioactive, with neutrons produces highly radioactive $^{239}_{92}U$:

$$^{238}_{92}U + ^{1}_{0}n \rightarrow ^{239}_{92}U \rightarrow ^{239}_{93}Np + ^{0}_{-1}e$$

Virtually all of the radioactive atoms used in tracer and biomedical applications are produced artificially.

In order to bring about an artificial transmutation, the projectile must have an extremely high kinetic energy when it strikes the target nucleus.* If the projectile is a charged particle, it can be accelerated to very high velocities by means of various devices. One of the earliest particle accelerators is the **linear accelerator** diagramed in Fig. 11.3. In Fig. 11.4 is shown a diagram of a **cyclotron**, a particle accelerator which is the prototype of most modern accelerators. The legends of these latter two illustrations provide a brief description of how the accelerators operate. At present it is possible to produce charged-particle projectiles with energies of billions of electron volts. One billion electron volts per projectile corresponds to 23 billion kcal (9.6×10^{10} kJ) per mole of projectiles—quite a high energy indeed!

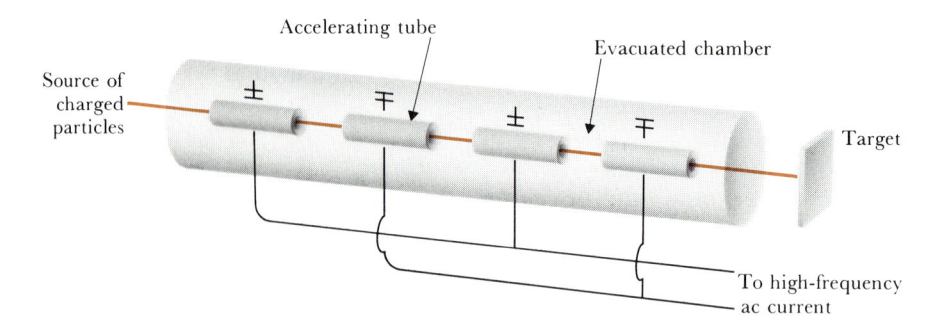

Fig. 11.3. Diagram of a linear accelerator. The frequency of the ac current is adjusted so that the particle always has an accelerating tube of opposite charge in front of it, causing it to go faster and faster.

Since neutrons carry no electrical charge, it is not possible to accelerate them directly. Consequently, an indirect procedure is employed. When beryllium-9 is bombarded with α-particles, neutrons are produced as follows:

$$^{9}_{4}Be + ^{4}_{2}He \rightarrow ^{12}_{6}C + ^{1}_{0}n$$

By varying the energies of the α-particles, some control of the neutron energies is achieved. Since neutrons are uncharged, they are not repelled electrostatically by target nuclei as positively charged projectiles are. Consequently, neutrons have relatively high penetrating power, even at low energies.

---

* The high energies are needed in order to overcome the strongly repulsive electrostatic forces which exist between projectiles such as $^{4}_{2}He^{2+}$ and $^{1}_{1}H^{+}$ and positively charged nuclei. Neutrons do not suffer from this disadvantage.

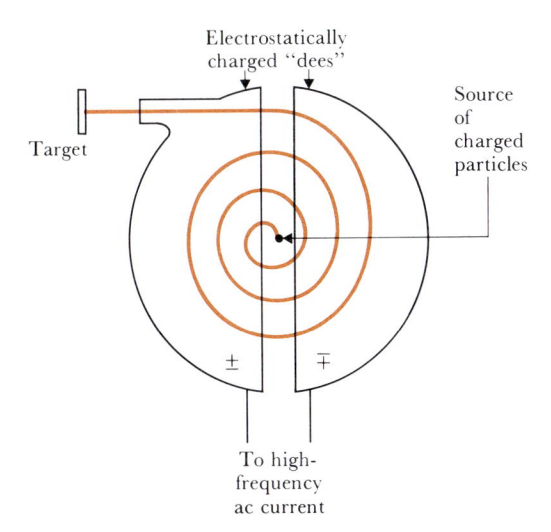

Fig. 11.4. Diagram of a cyclotron accelerator. Not shown is the magnetic field perpendicular to the two "dees," the charged electrodes which accelerate the particles. The magnetic field confines the particles to a curved path while a "dee" is kept oppositely charged in front of the advancing particles. When the particles reach a desired velocity, the magnetic field is turned off and the particles follow a straight line to the target.

11.4 THE RATE OF RADIOACTIVE DECAY

Experiments carried out around the beginning of the twentieth century suggested that the rates at which radioactive nuclei decay follow a general pattern: In each case the rate of decay (the mass of one element being transformed to the mass of another element per unit time) seems to be very nearly proportional to the mass of the decaying element. For example, a 2-g sample of a given radioactive element will have an instantaneous decay rate which is just twice that of a 1-g sample of the same element. Of course, as the decay proceeds, the mass of the decaying element decreases and so does the corresponding decay rate. Variables such as pressure and temperature appear to have no effect on the decay rate.* The decay rate (mass decaying per unit time) may be expressed in the mathematical form

$$\text{rate of decay} = km \tag{11.1}$$

where k is a constant (the **radioactive decay constant**) and m is the mass of the decaying element at any given time. This relationship implies that radioactive decay is a random process in which individual decay steps are independent. Recent careful experiments suggest that Eq. (11.1) is not strictly correct, thereby throwing doubt on the random, independent-event model of decay.† Nevertheless, since Eq. (11.1) is at least approximately correct, it will be used in the present chapter.

---

* Actually, variables such as temperature, pressure, chemical form, and applied electromagnetic fields do have some effect on the decay rate. See Wayne C. Wolsey, *J. Chem. Educ.* **55**, 303 (1978).

† The experiments which throw doubt on the random, independent-event model are discussed in *Ann. Rev. Nucl. Sci.* 165 (1972) and *J. Phys. Chem.* **76**, 3603 (1972); **77**, 3114 (1973). See also *Chemical and Engineering News,* 7 April 1975, p. 2.

In general, every different radioactive element has a different, characteristic radioactive decay constant. By means of the calculus it can be shown that if a radioactive element originally of mass m_0 is allowed to decay for an interval of time t, then the mass remaining is given by

$$m = m_0 10^{-kt/2.303} \quad \text{or} \quad m_0 e^{-kt} \tag{11.2}$$

This expression may also be stated in the logarithmic form*

$$\log \frac{m}{m_0} = \frac{-kt}{2.303} \quad \text{or} \quad \ln \frac{m}{m_0} = -kt \tag{11.3}$$

A decay process obeying such a relationship is said to be an **exponential decay**; the larger the value of k, the faster is the rate of decay for equal masses of different elements.

Although the decay constant k characterizes the decay rate of a given element, it is customary to compare the rates of decay of different elements in terms of a related quantity called the **half-life**. The half-life of a radioactive element is the time it takes for any initial mass m_0 to decay to one-half its value, that is, to $\tfrac{1}{2} m_0$. If $t_{1/2}$ designates the half-life, Eq. (11.3) shows that $m/m_0 = \tfrac{1}{2}$ when $t = t_{1/2}$, so that

$$\log \tfrac{1}{2} = -\frac{kt_{1/2}}{2.303} \quad \text{or} \quad \ln \tfrac{1}{2} = -kt_{1/2} \tag{11.4}$$

Solving for $t_{1/2}$ and using log 2 = 0.301 leads to

$$t_{1/2} = \frac{2.303 \log 2}{k} = \frac{0.693}{k} \quad \text{or} \quad t_{1/2} = \frac{\ln 2}{k} = \frac{0.693}{k} \tag{11.5}$$

As illustrated in Fig. 11.5, half-life is independent of the amount of the decaying element; it takes one half-life for 10,000 kg of a given element to decay to 5000 kg or for 0.1 g of that same element to decay to 0.05 g.†

The half-life of $^{226}_{88}\text{Ra}$ is 1620 yr; that of its decay product $^{222}_{86}\text{Rn}$ is 3.82 da. The $^{238}_{92}\text{U}$ isotope has a half-life of about 4.5×10^9 yr and is considered very weakly radioactive. By contrast, $^{214}_{84}\text{Po}$ has a very short half-life of 1.5×10^{-4} s. Beginning with a 10-g sample of Po-214, there would be 5 g left after 1.5×10^{-4} s, 2.5 g left after 3.0×10^{-4} s, and only about 0.01 g after 1.5×10^{-3} s (ten half-lives later). In each case the Po-214 has been transformed to Pb-210. The presence of the lead is readily determined by chemical analysis.

Rates of radioactive decay are usually measured with a **Geiger counter**.‡ This

---

* Students with pocket electronic calculators will find the second form in Eq. (11.2) and (11.3) more convenient. The first form is better suited for calculations using tables of decimal logarithms.

† Warning! The matter that decays does not vanish; it is transformed to another element. Chemical analysis is required to determine just how much of the element has decayed (been transformed) to another element.

‡ Proportional counters are now used in lieu of Geiger counters. The former have some different operating characteristics.

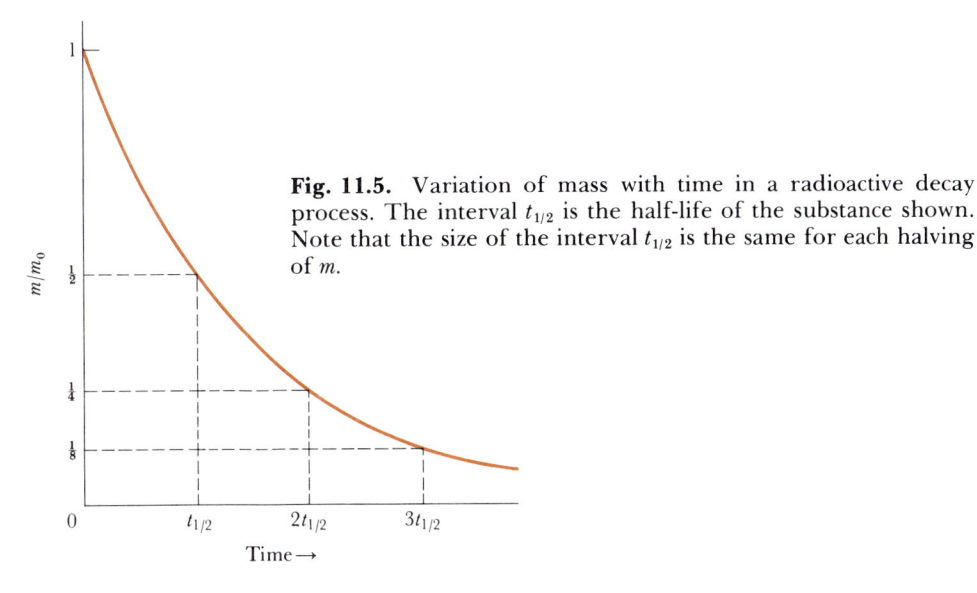

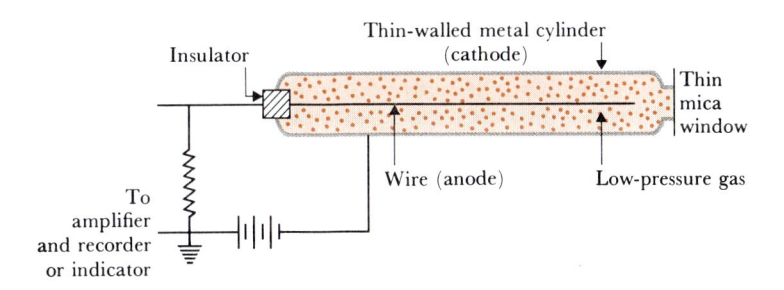

Fig. 11.5. Variation of mass with time in a radioactive decay process. The interval $t_{1/2}$ is the half-life of the substance shown. Note that the size of the interval $t_{1/2}$ is the same for each halving of m.

Fig. 11.6. Diagram of a Geiger counter. The wire and the surrounding metal cylinder act as electrodes and are maintained at a potential just under the breakdown potential of the residual gas separating them. Penetration of the tube by α-, β-, or γ-rays causes ionization of the residual gas and allows a tiny current to flow between the electrodes. This current may be amplified and used to flash a light, operate a clicker, or rotate counting wheels. The thin mica window enables α-particles to be counted, as these do not readily penetrate the metal tube.

device is illustrated in Fig. 11.6 and its operation described in the legend. Since some radioactive material is found everywhere in nature—in rocks, in the bricks of buildings, etc.—all measurements of the radioactivity of elements must be corrected for extraneous radioactivity from such sources. This extraneous radiation—varying from place to place and from day to day—is known as **background radiation**. Typical background radiation for a counter of radius 1 cm and length 10 cm is 50 cpm (counts per minute).

Some representative calculations involving half-life are given in the following examples.

EXAMPLE 11.1. The rate of decay of a radioactive sample was measured with a Geiger counter. At 1015 the counter indicated a count rate of 54,040 cpm; at 1025 (10 min later) only 34,040 cpm were recorded. The average background count (taken with the sample removed) was found to be 40 cpm. Determine the half-life of the element.

SOLUTION Although neither m_0 nor m is known, the ratio m/m_0 is just the ratio of the Geiger-counter readings corrected for background radiation. Thus

$$\frac{m}{m_0} = \frac{30{,}040 - 40}{50{,}040 - 40} = \frac{30{,}000}{50{,}000} = 0.60000$$

$$\log 0.60000 = -\frac{k(10.0 \text{ min})}{2.303}$$

$$k = 5.11 \times 10^{-2} \text{ min}^{-1}$$

$$t_{1/2} = \frac{0.693}{5.11 \times 10^{-2} \text{ min}^{-1}} = 13.6 \text{ min}$$

EXAMPLE 11.2 On 28 August 1973 a sample of radioactive material weighed 5 g. Chemical analysis of the sample on 28 February 1975 showed that there was only 2.5 g of the radioactive element remaining; the rest of the sample consisted of the decay product. What is the half-life of the element?

SOLUTION The time span from 28 August 1973 to 28 February 1975 is 18 mo or 1.5 yr. Since the mass of the radioactive element decreased to one-half during this interval, the half-life of the element is 18 mo or 1.5 yr (the time unit for half-lives is immaterial as long as it is stated).

EXAMPLE 11.3 What is the radioactive decay constant of the sample in Example 11.2?

SOLUTION From Eq. (11.5),

$$k = \frac{0.693}{t_{1/2}} = \frac{0.693}{18 \text{ mo}} = 3.8 \times 10^{-2} \text{ mo}^{-1} \quad \text{or} \quad \frac{0.693}{1.5 \text{ yr}} = 4.6 \times 10^{-1} \text{ yr}^{-1}$$

EXAMPLE 11.4 Starting with a 12.5-g sample of the element in Example 11.2, how long will it take for the element to decay to 1 g?

SOLUTION Solving Eq. (11.3) for t gives

$$t = \frac{2.303 \log m_0/m}{k} = \frac{(2.303)(\log 12.5/1)}{4.6 \times 10^{-1} \text{ yr}^{-1}} = \frac{(2.303)(1.09691)}{4.6 \times 10^{-1} \text{ yr}^{-1}} = 5.5 \text{ yr}$$

11.5 THE ENERGETICS OF NUCLEAR TRANSFORMATIONS

The sum of the masses of the protons, neutrons, and electrons making up a given atom is slightly larger than the experimentally measured mass of that atom. This mass difference, called the **mass defect**, is related to the **nuclear binding energy**—that energy which binds the nuclear particles together. For example, the $^4_2$He atom consists of two protons, two neutrons, and two electrons. The sum of the masses of these particles is computed as follows.

mass of 2 protons	= 2 × 1.007825 or	2.015650 amu
mass of 2 neutrons	= 2 × 1.008665 or	2.017330 amu
mass of 2 electrons	= 2 × 0.00055 or	0.00110 amu
Total mass =		4.03408 amu

The measured mass of a $^4_2$He atom based on mass spectrograph studies is 4.00260 amu, or 0.03148 amu less than the computed value. According to Einstein's theory of relativity, the mass defect Δm is related to the energy ΔE that binds the nuclear particles together by the relationship*

$$\frac{\Delta E}{\Delta m} = c^2 \qquad (11.6)$$

where c is the speed of light in a vacuum (3.00×10^8 m·s^{-1}). Since 1 amu is equivalent to 1.66×10^{-27} kg, the nuclear binding energy associated with 1 amu of mass defect is

$$(1.66 \times 10^{-27} \text{ kg})(3.00 \times 10^8 \text{ m·s}^{-1})^2 = 1.49 \times 10^{-11} \text{ J}$$

Since 1.60×10^{-19} J is equivalent to 1 eV, the nuclear binding energy per amu may also be expressed as 931×10^6 eV. Nuclear scientists frequently employ the energy unit **megaelectron volt**, abbreviated MeV (1 MeV is 1,000,000 eV); the nuclear binding energy per amu is then 931 MeV. The nuclear binding energy of $^4_2$He in MeV is

$$(931 \text{ MeV amu}^{-1})(0.03148 \text{ amu}) = 29.3 \text{ MeV}$$

The binding energy per nucleon (nuclear particle) is

$$\frac{29.3}{4} = 7.33 \text{ MeV}$$

A binding energy of 29.3 MeV per $^4_2$He atom corresponds to about 674 billion cal (2.8×10^9 kJ) per mole of atoms (1 MeV is about 2.3×10^{10} cal or 9.5×10^{10} J per mole).

Figure 11.7 illustrates how the nuclear binding energy per nucleon varies with mass number. Note that the maximum binding energy per nucleon occurs around

---

* This relationship says that the energy and mass of a system must always change in this ratio.

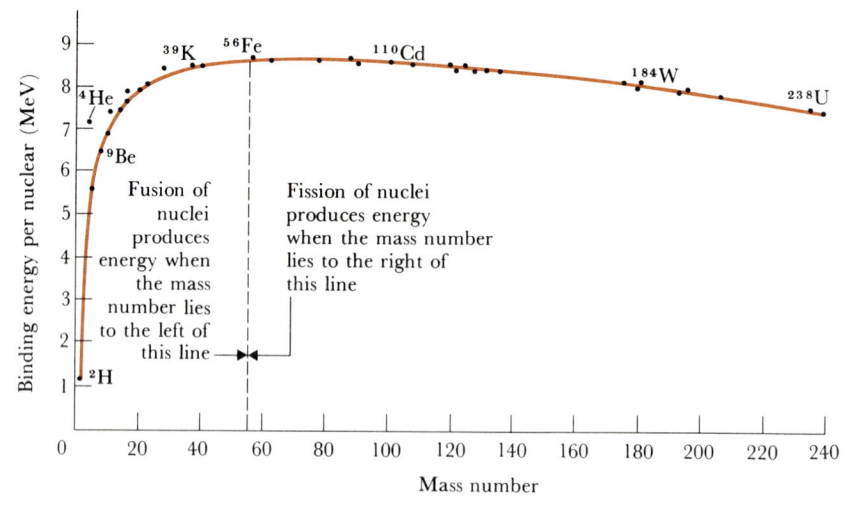

Fig. 11.7. Variation of nuclear binding energy per nucleon with mass number.

mass number 56 (the most common isotope of iron is $^{56}_{26}$Fe). The graph shows that if two or more elements to the left of the maximum combine to form a heavier element, energy should be released. Such a process is called **nuclear fusion**. Similarly, if an element to the right of the maximum splits into two roughly equal masses, energy will also be released. Such a process is called **nuclear fission**.

11.6 NUCLEAR FISSION

In 1934 Enrico Fermi and co-workers bombarded $^{235}_{92}$U with low-energy neutrons.* Since some β-emission was observed, they mistakenly assumed a new element of atomic number 93 had been formed. In 1938 the experiment was repeated by Hahn and Strassmann of Germany, who used ingenious chemical techniques to identify barium as one of the products. The first step in the Fermi reaction is now known to be

$$^{235}_{92}\text{U} + ^{1}_{0}\text{n} \rightarrow ^{236}_{92}\text{U}$$

The unstable $^{236}_{92}$U can fission in several different ways. One way is

$$^{236}_{92}\text{U} \rightarrow ^{142}_{56}\text{Ba} + ^{91}_{36}\text{Kr} + 3^{1}_{0}\text{n}$$

It is this reaction in which Hahn and Strassmann identified barium. Another mode by which $^{236}_{92}$U fissions is

$$^{236}_{92}\text{U} \rightarrow ^{139}_{57}\text{La} + ^{95}_{42}\text{Mo} + 2^{1}_{0}\text{n}$$

Given the atomic masses (in amu) of 235.0439 for U-235, 138.9061 for La-139, 94.9057 for Mo-95, and the neutron mass quoted earlier, the products are calculated to have a mass that is 0.2234 amu less than that of the reactants. The energy released

---

* These workers did not use a sample of pure U-235 but a natural sample of uranium ore containing less than 1% of U-235.

284 NUCLEAR TRANSFORMATIONS OF MATTER

Enrico Fermi (1901–1954) was the Italian physicist responsible for the first controlled nuclear chain reaction. He received the Nobel prize in physics in 1938 for his work in quantum statistics. During World War II, Fermi worked in the United States on the development of nuclear weapons. His wife Laura has written an excellent biography entitled Atoms in the Family *(Chicago: University of Chicago Press, 1954).*

Otto Hahn (b. 1879), a German organic chemist, was one of the pioneers in radiochemistry and received the Nobel prize in chemistry in 1944. He has written a very entertaining account of his work in "The Discovery of Fission," Scientific American, *February 1958.*

is thus (931 MeV·amu^{-1})(0.2234 amu) or 208 MeV—about 4.8 trillion cal (19 × 10^{12} J) per mole of uranium.

In 1942, Fermi and co-workers at the University of Chicago* first demonstrated how energy could be derived continuously from a controlled nuclear fission process using naturally occurring uranium (99.3% U-238 and 0.7% U-235). Experiments showed that U-238 absorbed only fast neutrons—those with energies of 1 MeV or more. U-235, it was found, fissions best with slow neutrons—those with energies far smaller than 1 MeV. Consequently, Fermi and his colleagues constructed an **atomic pile** by building up layers of uranium oxide and uranium metal separated by graphite bricks. Graphite (and certain other light elements) has the ability to slow down neutrons; such substances are called **moderators**.† The function of the moderator is to increase the probability that the neutrons lead to fission by U-235 rather than absorption by U-238. The latter process has a quenching effect upon the energy production and hence must be minimized.

To control the rate at which energy is produced by U-235 fission, the atomic pile was penetrated with long rods of a **control substance**—specifically, cadmium metal—which could be moved in or out at will. Cadmium has the ability to absorb neutrons; thus adjustment of the cadmium rods made it possible to control the neutron flux and, consequently, the rate of U-235 fission.

The atomic pile begins to operate when U-235 is bombarded with neutrons from an outside source. Each U-235 fission then produces enough new neutrons to continue the process. This continuous fission process, using neutrons from one fission to initiate a new fission and so on, is an example of a **chain reaction**.

Modern nuclear reactors may be regarded as modifications of the Fermi atomic pile. In a reactor, the heat produced by the controlled fission reaction is used to

---

* The first controlled nuclear fission occurred on 2 December 1942 in a squash court under the stands of the Amos Alonzo Stagg football field in Chicago. The site is now marked by a bronze sculpture, "Nuclear Energy," by Henry Moore.

† Water is also an effective moderator, especially if it is enriched in deuterium.

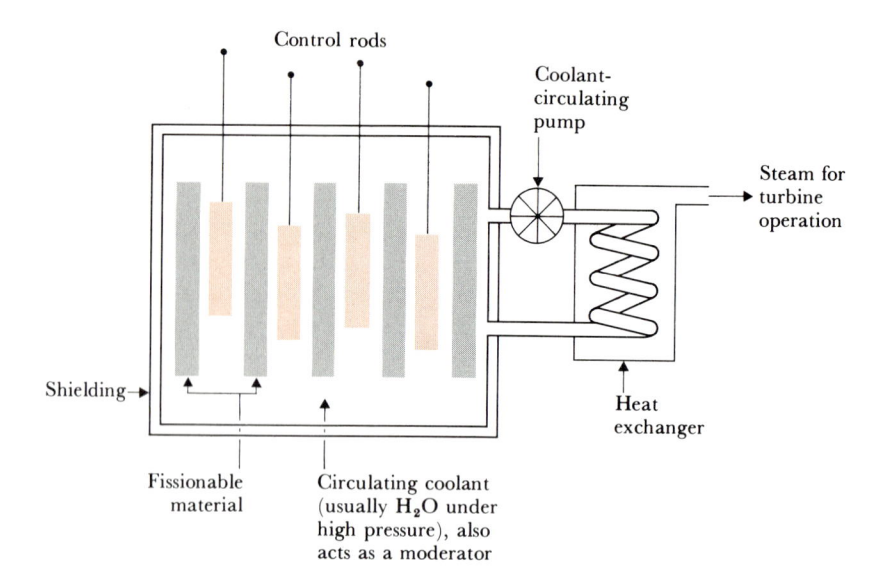

Fig. 11.8. Diagram of the essential features of a typical nuclear reactor. The nuclear fuel is manufactured from ore containing 0.1 to 1% U_3O_8. This is converted to UF_6 and enriched to about 3% in U-235 by means of gaseous diffusion. The enriched UF_6 is converted back to U_3O_8, pressed into small pellets, and packed into stainless steel or zircalloy tubes several meters long.

operate a steam-powered device such as a turbine which, in turn, operates an electrical dynamo (see Fig. 11.8).

The most recent trend in the development of nuclear reactors is to construct so-called **breeder reactors**, that is, reactors which produce more fissionable fuel than they consume. Although Fermi and his colleagues were originally unaware of the fact, the fate of U-238 when it absorbs neutrons is to produce plutonium, an element which resembles U-235 in its ability to carry out a sustained fission reaction. The U-238 reaction is

$$^{238}_{92}U + ^{1}_{0}n \rightarrow ^{239}_{92}U$$

U-239 is unstable and decays to neptunium, Np-239, by β-emission:

$$^{239}_{92}U \rightarrow ^{239}_{93}Np + ^{0}_{-1}e$$

Np-239, in turn, decays to fissionable plutonium, Pu-239, by the process

$$^{239}_{93}Np \rightarrow ^{239}_{94}Pu + ^{0}_{-1}e$$

A breeder reactor is operated by deliberately reducing the amount of moderators so that some fast neutrons are produced to convert U-238 to Pu-239. In this way the more plentiful component of natural uranium becomes usable for energy production. Some of the latest breeder reactors, called **fast breeders**, use virtually no moderators so that a maximum of Pu-239 is produced. The Pu-239 is recovered by shutting down a breeder reactor from time to time and reprocessing the unspent fuel. The Pu-239 may then be used to produce fuel cores for other reactors.

A modern nuclear power plant, characterized by its giant cooling towers. (Courtesy of Grant Heilman.)

Two of the most dramatic uses of nuclear power are exemplified by the surface ship *N. S. Savannah* and the U.S. Navy submarine *Triton*. The nuclear power plant of the *Savannah* consists of 110 lb of U-235 and is capable of sending this vessel on a 300,000-mi trip at a speed of 20 knots. In 1960 the submarine *Triton* traveled around the world completely submerged except for occasional use of the periscope, covering 41,500 mi in 84 days. In 1967 nuclear-powered rocket engines were successfully tested and practical development is now underway.

Since bombardment of U-235 with neutrons produces additional neutrons in excess of those used to produce the fission, there exists the possibility of a very rapid chain reaction. A single U-235 nucleus may fission to produce two neutrons, each of which may fission two U-235 nuclei to produce four neutrons which, in turn, may fission four U-235 nuclei to produce eight neutrons, and so on. If pure U-235 (not the oxide normally used in atomic piles) is prepared in a mass large enough so that each neutron produced by fission produces an average of at least one new neutron before escaping from the mass, an incredibly rapid chain reaction occurs with super-explosive violence. In approximately one one-millionth of a second, solid U-235 becomes a multimillion-degree blast of hot gas. This is the principle of the **nuclear bomb** (commonly misnamed an **atomic bomb**). The minimum mass required for this to happen is called the **critical mass**. The critical mass of U-235 has never been officially released but it is probably a mass about the size of a baseball. If less than a critical mass of U-235 is assembled, many of the neutrons produced by fission escape without reproducing themselves; by making the U-235 mass large enough, the probability that each neutron reproduces itself (on the average) eventually becomes 100% and spontaneous fission of the entire mass occurs. In the nuclear bomb, utilizing either pure U-235 or pure Pu-239, two subcritical masses are hurled together very rapidly by the force of an ordinary explosive charge to produce a supercritical mass. Alternatively, the supercritical mass may be made into a hollow sphere which is then imploded by properly designed explosive charges. In either case the fission reaction may be initiated by a pellet of polonium and beryllium. The polonium spontaneously emits α-particles which, in turn, strike beryllium nuclei to produce neutrons.

11.7 NUCLEAR FUSION

Scientists have long pondered the question: Where does the sun get its energy? Calculations show that chemical reactions produce far too little energy to fuel a sun which we know must have been "burning" for at least several billion years. At present the sun produces about 10^{26} J·s^{-1} (10^{26} watts); the best chemical fuels known can account for only a few thousand years of energy production at this rate, assuming the sun were composed entirely of such fuels. Similarly, other possible energy sources (for example, contraction of the sun due to its powerful gravitational forces) fall far short of accounting for such a prodigious output of energy. Today it is believed that the energy of the sun and that of the stars is due largely to one or both of the following nuclear reaction cycles.

The Carbon–Nitrogen Cycle:
$$^{12}_{6}C + ^{1}_{1}H \rightarrow ^{13}_{7}N$$
$$^{13}_{7}N \rightarrow ^{13}_{6}C + ^{0}_{1}e$$
$$^{13}_{6}C + ^{1}_{1}H \rightarrow ^{14}_{7}N$$
$$^{14}_{7}N + ^{1}_{1}H \rightarrow ^{15}_{8}O$$
$$^{15}_{8}O \rightarrow ^{15}_{7}N + ^{0}_{1}e$$
$$^{15}_{7}N + ^{1}_{1}H \rightarrow ^{12}_{6}C + ^{4}_{2}He$$

The Proton–Proton Chain:
$$2\,^{1}_{1}H \rightarrow ^{2}_{1}H + ^{0}_{1}e$$
$$^{2}_{1}H + ^{1}_{1}H \rightarrow ^{3}_{2}He$$
$$2\,^{3}_{2}He \rightarrow ^{4}_{2}He + 2\,^{1}_{1}H$$

In the carbon–nitrogen cycle the C-12 used up in the first step is resynthesized in the last step. The overall net process is the fusion of four protons to produce helium. Exactly the same net process occurs in the proton–proton chain. Thus, in both cases, the energy-producing process is essentially the fusion reaction

$$4\,^{1}_{1}H \rightarrow ^{4}_{2}He + 2\,^{0}_{1}e$$

It is believed that the proton–proton chain is the more important of the two in the sun and in stars similar to the sun in mass; more massive stars are believed to utilize the carbon–nitrogen cycle to a greater degree.

Since the total binding energy of $^4_2$He is 29.3 MeV and that of a proton is zero, the energy released in the formation of $^4_2$He from protons is 29.3 MeV per atom or about 674 billion cal (2.8×10^{12} J) per mole. By contrast, burning four grams of $^1_1$H (as H$_2$) in oxygen produces about 115,600 cal (4.8×10^5 J); thus the fusion reaction produces about six million times more energy per unit mass of hydrogen than does the oxidation reaction.

The first artificially induced fusion reaction occurred as a test of the so-called hydrogen bomb. In one version of this bomb, a nuclear fission explosion first creates a temperature of several million degrees. This high temperature causes Li-6 (present as $^6_3$Li$^2_1$H, lithium deuteride) to produce tritium by the process

$$^6_3\text{Li} + ^1_0\text{n} \rightarrow ^3_1\text{H} + ^4_2\text{He}$$

The tritium then fuses with deuterium:

$$^3_1\text{H} + ^2_1\text{H} \rightarrow ^4_2\text{He} + ^1_0\text{n}$$

The high temperature is needed to give large enough kinetic energies to the deuterium and tritium nuclei so that they can get close enough to fuse in spite of their very strong coulombic repulsions. Since coulombic repulsion becomes very high at very small internuclear distances, kinetic energies corresponding to over 100 million degrees are needed to initiate fusion. The fusion reaction produces about 17.5 MeV per atom of helium produced.

One of the most exciting applied science problems of our age is to carry out controlled fusion reactions for the generation of energy for domestic and industrial uses. At present it has not been found possible to contain nuclei at temperatures above 100 million degrees long enough to generate more than brief bursts of energy. One approach attempts to contain the nuclei in empty space by the use of strong magnetic and electric fields. According to the so-called Lawson criterion, the product of containment time and density of nuclei must exceed 10^{14} s·cm^{-3} in order for fusion to occur. So far the confinement has not been stabilized for more than a few milliseconds.

In addition to the deuterium–tritium fusion occurring in the lithium deuteride bomb, the deuterium–deuterium fusion

$$^2_1\text{H} + ^2_1\text{H} \rightarrow ^3_2\text{He} + ^1_0\text{n}$$

has been considered as a source of fusion energy. The latter reaction produces only 3.1 MeV per atom of helium produced (compared with 17.5 MeV for the deuterium–tritium fusion) but it has the advantage of being based on relatively abundant deuterium (about 1 atom per 7000 atoms of hydrogen). Tritium is so rare in nature that it must be produced from $^6_3$Li by neutron bombardment. Furthermore, $^6_3$Li constitutes only 7.42% of naturally occurring lithium.

11.8 THE HAZARDS OF RADIOACTIVITY

Many of the early investigators of radioactive phenomena were unaware of the damage caused to living tissue by α-, β-, and γ-radiation, and some suffered severe lesions and even death as a result of exposure. The destructive effect of radiation on living organisms is due mainly to the ionization of water (the main component of most tissues) and the subsequent chemical reactions. When pure water is irradiated with any type of high-energy radiation the primary processes appear to be

$$H_2O \rightarrow H_2O^+ + e^-$$

$$H_2O^+ + H_2O \rightarrow H_3O^+ + OH$$

The OH radicals (not to be confused with the hydroxide ion OH^-) are the most potent oxidizing species known. This radical is capable of reacting chemically with virtually any substance found in a living cell. If the damage occurs in gonads, (male or female) the adverse results may show up genetically in the offspring. If the damage occurs in ordinary body cells, various organs of the carrier may become impaired in function; malignant tumors can result from prolonged exposure to even low levels of radiation.

In rather general terms, the effect of radiation upon a living organism depends on various factors such as the following.

1. Total dosage of the ionizing radiation to which the organism has been subjected.
2. The dosage rate, that is, the period of time over which exposure to a given amount of radiation occurs.
3. The specific part of the organism which has been exposed.

A quantitative measure of radiation exposure is the **gray** (symbol Gy), defined as the amount of radiation that subjects a kilogram of matter to one joule of energy.* Thus $1 \text{ Gy} = 1 \text{ J} \cdot \text{kg}^{-1}$. If 1.0 J of energy is applied to 1.0 kg of water (specific heat of $1 \text{ cal} \cdot \text{g}^{-1} \cdot \text{K}^{-1}$ or about $4.2 \times 10^3 \text{ J} \cdot \text{K}^{-1} \cdot \text{kg}^{-1}$) the temperature rise in the water is

$$\Delta T = \frac{\Delta E}{C_V} = \frac{1.0 \text{ J} \cdot \text{kg}^{-1}}{4.2 \times 10^3 \text{ J} \cdot \text{K}^{-1} \cdot \text{kg}^{-1}} = 2.4 \times 10^{-4} \text{ K}$$

where C_V is the specific heat. This tiny amount of energy—which causes only a negligible heating effect—is capable of producing easily observable biological effects. Radiation is truly an agent of unusual potency.

The three main products of radioactive decay (α-particles, β-particles, and γ-radiation) have different penetrating powers and different abilities to cause ionization. Whereas α-particles are stopped by about 0.006 cm of aluminum (household aluminum foil is approximately of this thickness) or a sheet of ordinary writing paper, β-particles can penetrate about 0.1 cm of aluminum, and γ-radiation can penetrate 10 to 100 times the latter thickness. Ionizing abilities, or the number of ions formed per unit path length, are in inverse proportion to the penetration abilities. The relative ionizing powers of α-, β-, and γ-radiation are roughly $10^4:10^2:1$.

To account for the fact that different types of radiation have different ionizing abilities, a quantity called the **rem** (for "radiation equivalent man") is introduced in calculating radiation exposure:

dosage in rems = dosage in grays $\times$ RBE $\times 10^2$

---

* An older unit still much used is the **rad**. A rad is 10^{-2} Gy.

TABLE 11.1 THE RADIATION BIOLOGICAL EFFECTIVENESS OF VARIOUS RADIOACTIVE SOURCES

RADIATION SOURCE	RBE
X, γ	1
β	1
$_0^1$n (thermal)	2
$_0^1$n (fast)	10
α	10
$_1^1$H	10
Heavy ions	20

TABLE 11.2 RADIATION DOSAGES OF THE AVERAGE AMERICAN DURING 1970 FROM VARIOUS SOURCES*

SOURCE	AVERAGE PER CAPITA DOSE (MILLIREMS PER YEAR)
Natural background	130
Medical diagnostic X-ray	90
Weapons testing fallout	5.1
Nuclear power reactors	<0.01
Total	<225.1

* Data supplied by the Environmental Protection Agency and reported by J. A. Lieberman in *Physics Today*, November 1971, p. 32.

where **RBE** is the **radiation biological effectiveness** of the radiation. The RBE's of various types of radiation are listed in Table 11.1. Notice that β- and γ-radiation have an RBE of 1 whereas α-particles have an RBE of 10. This means that α-particles cause 10 times as much disruption per energy unit per unit weight of target material as does β- or γ-radiation.

The effect of radiation upon living organisms is often expressed as the lethal dose required to kill x percent of a test population; this quantity is symbolized by LD_x. Thus eggs of *Drosophila melanogaster* (common fruit fly) have an LD_{50} of 160 rems; that is, 160 rems of radiation will kill 50% of fruit fly eggs. By contrast, an adult population of fruit flies has an LD_{50} of 10^4 rems.

Radiation exposure comes from a variety of sources. Waste products from mining and processing of nuclear fuel, materials used in research and hospitals, fallout-contaminated foodstuffs, and wastes from nuclear reactors are typical. Table 11.2 shows the average per capita dosage received by U.S. citizens during 1970 from various sources. For the average individual, by far the greatest percentage of radiation exposure comes from natural sources such as cosmic radiation originating in outer space and small amounts of radioactive elements found everywhere in nature. At present, the National Council on Radiation Protection and Measurements places three distinct but interrelated limits on radiation exposure of the general public. These are:

1. No individual should be exposed to more than 500 millirems per year on a whole-body basis.
2. When a group of individuals is subjected to radiation exposure and not all the individual exposures can be determined, the whole-body exposure must not exceed 170 millirems per year.
3. The per capita dosage limitation for male and female reproductive organs is 5 rems in 30 years.

Most radioactive wastes arise from the periodic reprocessing of the primary fuel to remove newly produced fuel such as plutonium or to remove undesirable substances. As fission proceeds, the fuel rods begin to accumulate a great variety of radioactive substances which act as neutron scavengers and thereby reduce the reactor efficiency. Substances such as iodine-131 are not too hazardous as they have rather short half-lives (about 8 days for iodine-131), but longer-lived elements such as strontium-90 (28 yr), cobalt-60 (5.3 yr), cesium-137 (30.2 yr), and krypton-85 (10 yr) present serious disposal problems. Materials such as strontium-90 and cesium-137 enter the food chain and eventually become concentrated in animals. Strontium-90 travels from grass to cows to milk to humans and accumulates in bone marrow to produce leukemia and bone cancer. Cesium-137 enters vegetables and thence is carried into humans to accumulate in soft tissues such as the gonads and the liver. It is obvious that ways must be found to isolate such wastes for long periods of time in order to protect not only the present population but a great many future generations as well. At present, most of these wastes are stored in huge steel tanks at three different places in the United States, namely, the AEC Hanford Works in Washington, the Savannah River Plant in Georgia, and the Idaho Reactor Testing Station. The storage requires a high degree of continuous surveillance and is susceptible to various uncertainties such as the long-term integrity of the tanks and possibility of natural disasters and war, not to mention long-term legal and financial responsibility. An alternative disposal scheme—encapsulation of wastes in inert materials and subsequent deep burial—has been considered.

Some of the radioactive wastes produced by nuclear reactors are difficult to isolate. The gases krypton-85 and tritium are examples. At present both of these are released to the environment. Krypton is essentially chemically inert and does not become caught in food chains, but this is not true for tritium (half-life 12.4 years).

Fusion reactors, if they ever become technically feasible, also have serious radioactivity hazards associated with them. The high neutron fluxes associated with fusion reactors will strike the materials used to construct the reactor and eventually make them dangerously radioactive. The lengthy exposures concomitant with continued reactor operations will lead to a major replacement and disposal problem.

All factors considered, the potential danger from operation of nuclear reactors appears to be far more serious than that from any other human activity. This is why many environmentally concerned citizens call for restraint in developing new nuclear power stations—especially breeder reactors—until it appears reasonably certain that they can be operated safely and their huge amounts of radioactive wastes safely disposed of.

11.9 RADIOISOTOPE DATING

Around 7000 B.C. a band of primitive hunters assembled in the vicinity of what is now Sage Creek, Wyoming, and stampeded a large herd of bison over a high cliff to their deaths in the gorge below. The hunters then proceeded to prepare one of the biggest bison barbecues of the B.C.'s. The date of this event is known today not because of any written records left by the banqueters, nor because of any legends handed down by word of mouth over the ages, but because the carbon supplying the barbecue fires was "bugged," that is, it acted as a recording device.

Carbon's ability to act as a recording device derives from processes occurring in the upper atmosphere of the earth. The earth's atmosphere is constantly bombarded by various particles originating in the sun and other parts of the universe. Among these particles are neutrons, some of which collide with ordinary nitrogen atoms $^{14}_{7}N$, and bring about the transmutation

$$^{14}_{7}N + ^{1}_{0}n \rightarrow ^{14}_{6}C + ^{1}_{1}H$$

The carbon-14 is radioactive and decays by the process

$$^{14}_{6}C \rightarrow ^{14}_{7}N + ^{0}_{-1}e$$

with a half-life of 5730 years. This carbon-14 contributes a small but constant amount to the earth's total carbon pool and becomes incorporated into living organisms, plant and animal alike. As long as an organism is alive, its content of carbon-14 tends to remain constant since the organism continues to ingest materials from the earth's carbon pool. However, when the organism dies, it no longer assimilates carbon compounds, and its carbon-14 content begins steadily to decrease at the rate of the natural decay.

Willard Libby (b. 1908), an American chemist, won the 1960 Nobel prize in chemistry for the development of radiocarbon dating while at the University of Chicago. A past chairman of the Atomic Energy Commission, Libby is now Director of the Institute of Geophysics and Planetary Physics at the University of California at Los Angeles.

In the 1940's, Willard Libby realized that if production of carbon-14 has been constant over the past (or at least during the last 50,000 years or so), then a measurement of the carbon-14 content of a once-living carbon-containing object could be compared with the carbon-14 content of a contemporarily living carbon-containing object and this information used to estimate the time when the once-living object died. When archaeologists discovered the huge piles of bison bones near Sage Creek, they also found traces of charcoal from the barbecue fires. A sample of the charcoal was burned in oxygen, and the resulting CO_2 was collected, purified, and made to react with hot magnesium metal to produce elemental carbon:

$$CO_2 + 2Mg \rightarrow C + 2MgO$$

Subsequent use of a Geiger counter showed that the sample of carbon was undergoing about 231 radioactive disintegrations per hour per gram of carbon. Since it is known that the carbon in living plants contains one atom of carbon-14 per 10^{12} atoms of carbon-12, the number of disintegrations per hour per gram of carbon in a living plant should be

$$km = \frac{0.693}{(5730 \text{ yr})(8760 \text{ hr} \cdot \text{yr}^{-1})} (5.02 \times 10^{10} \text{ atoms } {}^{14}\text{C per g C})$$

$$= 693 \text{ counts} \cdot \text{hr}^{-1} \cdot \text{g}^{-1}$$

Since the count for the barbecue-fire charcoal was only one-third of this, it is deduced that two-thirds of the original carbon-14 must have decayed. Thus $m/m_0 = 1/3$, and the time when the bison feast was held is given by

$$t = \frac{2.303 \log m/m_0}{-k} = \frac{2.303 \log (1/3)}{-(0.693/5730 \text{ yr})} = \frac{2.3030 \log 3}{.693} (5730 \text{ yr}) = 9085 \text{ yr ago}$$

Due to the difficulty of accurately measuring very small count rates (or very small differences in two different count rates) the radiocarbon dating method is not very reliable for times greater than about 45,000 years or less than about 1000 years.

It is now known that Libby's original assumption of a long-term constant rate of carbon-14 production is not quite true. Recently, it has been shown that past fluctuations of carbon-14 production may be corrected by comparing the age as determined by growth rings of the bristlecone pine (*Pinus aristata*), one of the world's most long-lived trees, with the corresponding age as measured by the carbon-14 count. For example, a bristlecone pine sample whose rings show an age of 2500 years has a carbon-14 content which leads to an age of only 2000 years.

The heaviest isotope of hydrogen, tritium (${}_1^3\text{H}$), is used in certain types of dating problems, such as determining the age of wines and subsurface water. Tritium is formed in the upper atmosphere by the process

$${}_0^1\text{n} + {}_7^{14}\text{N} \rightarrow {}_1^3\text{H} + {}_6^{12}\text{C}$$

and decays as

$${}_1^3\text{H} \rightarrow {}_2^3\text{He} + {}_{-1}^0\text{e}$$

with a half-life of 12.4 yr. Assuming that there is one atom of tritium per 10^{17} atoms of hydrogen, a count rate of 4.26 $\text{hr}^{-1} \cdot \text{g}^{-1}$ or 4260 $\text{counts} \cdot \text{hr}^{-1} \cdot \text{L}^{-1}$ should be obtained from the water of a recent rainfall.*

The first use of radioactive decay processes for dating was carried out in 1907 by Yale University physicist Bertram Boltwood. Boltwood noted that a measurement of the amount of certain isotopes of lead arising from natural radioactive decay could be used to date the rocks in which the lead occurred. For example, lead-206 arises from a long decay series beginning with uranium-238, which has a half-life of 4.5

---

* Most of the tritium now on earth was produced by fusion bomb explosions; a fusion bomb produces about 2 kg of tritium per megaton. Prior to bomb testing, the earth had about 2.5 kg of tritium; now it has about 300 kg. This bomb-produced tritium must be taken into account when using the tritium dating method.

A bristlecone pine (*Pinus aristata*), one of the earth's longest lived trees. A study of its growth rings enables scientists to correct errors in the radiocarbon dating method. (Courtesy of Rapho, Photo Researchers, Inc.)

billion years. Lead-208 arises in similar fashion from thorium-232, with a half-life of 13.9 billion years. Basing his calculations on data such as these, Boltwood was able to show that certain rocks had ages approaching at least 500 million years. This finding threw some doubt on the contention of Archbishop Ussher that the earth was created on 26 October 4004 B.C. at 9 o'clock in the morning!*

The element $^{40}_{19}K$ has a half-life of about 1.3 billion years and decays to $^{40}_{20}Ca$ and $^{40}_{18}Ar$.† By heating potassium-containing rocks to a temperature of about 1200 °C and collecting the evolved argon gas, an estimate of the rocks' ages can be made. It was by this means that G. H. Curtis and J. Evernden of the University of California (Berkeley) estimated the ages of two skulls found by Mary and Louis Leakey in Tanzania's Olduvai Gorge. The skulls were found encased in volcanic ash, suggesting that their owners perished in a volcanic eruption. Dating of the ash gave an age of 1.25 to 1.75 million years. Since one of the skulls was of *Homo habilis* and the other of *Australopithecus bosei* (two species of prehistoric humanoids), it appears that representatives of two different branches of the prehuman family tree existed during the same time span at Olduvai Gorge.‡

---

* In 1654 Archbishop James Ussher of Ireland claimed to have determined this as the true date of creation from a study of Hebrew scriptures. Someone later inserted Ussher's date into the margin of an authorized version of the Bible and made it an article of religious dogma.

† The decay of potassium-40 to argon-40 occurs by a process (known as *K*-capture) in which the nucleus "absorbs" one of the atom's electrons and thereby lowers its atomic number by one unit, just as in positron emission.

‡ More recently, Mary Leakey has found other fossil remains at Laetolil, a semiarid region south of Olduvai Gorge, that appear to be about 3.75 million years old. These fossils are thought to belong to the genus *Homo*, the scientific classification which includes *Homo sapiens*, modern humans.

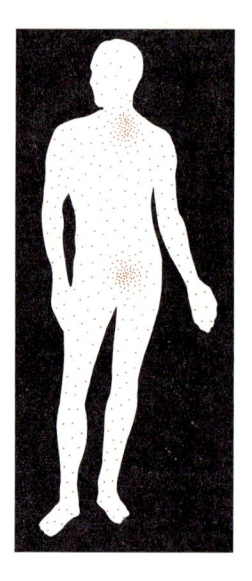

Fig. 11.9. Body scan of cancer victim. The patient is given an intravenous injection of a short-lived radioactive isotope such as gallium-67 (half-life three days) which shows great selectivity between normal and tumor tissues. The output of a scanning sensor is converted to a body map in which high gallium-67 concentrations show as dark areas. The dark areas in the groin and lower neck indicate that the patient depicted has tumor tissue.

11.10 MEDICAL USES OF RADIOACTIVITY

When our ancestors took a drug to relieve localized pain, they had no way of knowing just how the drug worked in detail. Did the drug travel through the body to the site of the pain, or did it act upon the central nervous system? Today it is possible to follow or trace the course of many drugs in the body of a human or other animal by incorporating a radioactive element into the drug. The course of the drug through the body may then be monitored periodically by checking various parts of the body with a radiation detector. Radioactive elements employed in this way are called **tracers**. Much of the intimate knowledge we now possess about physiology is based on tracer studies. Not only drug action but processes such as metabolism may be studied in the same way.

Radioactive isotopes are also useful in the detection of various tumors well in advance of detection by more conventional techniques such as X-rays (see Fig. 11.9). Of great importance is the fact that many radioactive isotopes are very selective toward specific types of malignancies. For example, iodine-131 tends to become concentrated in the thyroid gland, where it can serve both as a detector and as a treatment of cancer. Fluorine-18 plays a similar role in the detection of bone cancers. Tumors of the brain can be detected by the use of technetium-99. Recently, attention has focused on the isotopes indium-111 and gallium-67, since these elements have the ability to detect a broad spectrum of malignancies. This makes it possible to carry out a fairly complete diagnosis on the basis of a single radioisotope injection (usually given intravenously). Gallium-67 has been shown to have a pronounced selectivity for tumor tissues which shows up about 48 hours after injection.

Another medical application of radioactivity is in the destruction of malignant growths by direct radiation. The γ-radiation arising from decay of cobalt-60 ($t_{1/2}$ = 5.26 yr) and cesium-137 ($t_{1/2}$ = 30.2 yr) can be used to destroy cancerous tissue without harming normal tissue. X-rays have long been used in a similar fashion but their use requires bulky apparatus which makes it difficult to reach selected areas of the body. By contrast, radiation treatment with radioactive isotopes is often very simple. For example, a small pellet of radioactive material may be implanted in the malignancy to be destroyed. This is similar to the way an injection of iodine-131, which tends to become localized in the thyroid gland, destroys thyroid cancer.

Both X-rays and γ-rays are used to sterilize food and other objects. For example, the Apollo space ships and their equipment were sterilized by γ-radiation before departing for the moon in order to minimize the probability of contaminating the lunar surface with viral and bacterial organisms.

SUMMARY

Nuclear transformations involve a change in the number of neutrons and/or protons in the nucleus of an element. In this way one element is changed into another element or into a different isotope of the same element. Radioactive elements are found in nature and may also be produced artificially. The rate at which a radioactive element decays is proportional to the mass of the element present. The time required for one-half of a sample of radioactive element to become transformed to another element is called the half-life of that element. The half-life of an element is independent of the mass of the decaying element.

In nuclear fission, a heavy element is split into two or more lighter elements; in nuclear fusion, two light elements combine to form a heavier element. Both of these nuclear transformations are accompanied by the production of energy. This energy production can be carried out at a controlled rate to provide energy for domestic and industrial purposes or it can be carried out extremely rapidly to produce an explosive device.

The α-, β-, and γ-rays associated with radioactive decay constitute serious hazards to life and health. Biological systems are affected by radiation largely via the OH radicals produced when water is irradiated. The OH radical is perhaps the most potent oxidizing species known.

Radioactivity has many uses beneficial to humans, for example, as tracers and in the destruction of malignant tissue. Another important application is in radioisotope dating. By means of such dating techniques, archaeologists have been able to assign relatively firm dates to many events taking place in the distant past—especially those occurring during prehistoric eras.

LEARNING GOALS

1. Be able to give the main physical characteristics of α-, β-, and γ-radiation.

2. Be able to balance a nuclear transmutation equation.

3. Be able to describe the nuclear process which leads to γ-ray emission.

4. Be able to describe how a linear accelerator and a cyclotron operate.

5. Know the basic mathematical relationship for the rate of radioactive decay and how to use it to calculate half-lives.

6. Be able to calculate the half-life from the radioactive decay constant and vice versa.

7. Be able to describe how a Geiger counter operates.

8. Know what the mass defect is and how to calculate it from the Einstein mass/energy ratio.

9. Be able to calculate the binding energy per nucleon for an atomic nucleus.

10. Know how nuclear fission and nuclear fusion arise and be able to calculate the energies released by these processes.

11. Be able to describe what is necessary to set up a chain reaction.

12. Be able to explain the basic operation of an atomic pile.

13. Know what breeder reactors are and how they operate.

14. Be familiar with the relative penetrating powers and ionizing powers of α-, β-, and γ-radiation.

15. Know how radiation exposure is expressed.

16. Be familiar with radioisotope dating and be able to calculate the ages of objects containing carbon-14 and tritium.

17. Know some of the general medical uses of radioactive isotopes.

18. Be able to name at least one important scientific contribution of each of the following: Becquerel, Marie and Pierre Curie, Soddy, Rutherford, Dirac, Anderson, Einstein, Hahn, Fermi, and Libby.

DEFINITIONS, TERMS, AND CONCEPTS TO KNOW

radioactivity
α-particle
β-particle
γ-ray
positron
negatron
antimatter
linear accelerator
cyclotron
Geiger counter

radioactive decay constant
half-life
exponential decay
background radiation
mass defect
nuclear binding energy
nuclear fission
nuclear fusion
atomic pile
moderator

control substance
critical mass
chain reaction
breeder reactor
rad and gray
rem
RBE
radioisotope dating
tracer

QUESTIONS AND PROBLEMS

Nuclear Reactions

1. It is believed that a sample of matter contains about 10^6 radioactive barium-142 atoms (β-emitters) per mole of other substances. Show that by adding a salt of naturally occurring barium, say $BaCl_2$, dissolving the sample in acid, and removing all barium as $BaSO_4$ (a water-insoluble salt), one could demonstrate the presence of the radioactive isotope. [*Note:* Hahn and Strassmann used a similar technique to show that neutrons fission U-235 to produce two elements, rather than transmuting it to Np-235 as Fermi originally believed.]

2. Supply the missing elements in each of the following nuclear reactions.

a) $^{39}_{19}K$ + _____ → $^{36}_{18}Ar$ + $^{4}_{2}He$

b) $^{35}_{17}Cl$ + _____ → $^{35}_{16}S$ + $^{1}_{1}H$

c) $^{23}_{11}Na$ + $^{2}_{1}H$ → _____ + $^{4}_{2}He$

d) $^{62}_{29}Cu$ → $^{0}_{1}e$ + _____

e) _____ → $^{4}_{2}He$ + $^{225}_{88}Ra$

3. Write the nuclear reaction describing each of the following processes.

a) Technetium-101 decays by β-emission.

b) Lanthanum-126 decays by positron emission.

c) Gadolinium-148 decays by α-emission.

4. Suggest possible sources of the β-decay which led Fermi to misinterpret the product of the bombardment of U-235 with neutrons.

5. Nuclear decay reactions are often symbolized as follows: $^9_4Be(\alpha, n)^{12}_6C$. This means that ^9_4Be is bombarded with an α-particle to produce $^{12}_6C$ and a neutron. Interpret the following to write full balanced equations for each process. In each case X represents an element which you are to identify.

a) $^{238}_{92}U(\alpha, 2n)X$
b) $^{63}_{29}Cu(p, n)X$ $(p = ^1_1H)$
c) $^{27}_{13}Al(n, p)X$
d) $^{24}_{12}Mg(d, p)X$ $(d = ^2_1H)$

6. Write balanced nuclear equations for (a) the loss of a β-particle by a $^{12}_5B$ nucleus, (b) the loss of an α-particle by $^{214}_{83}Bi$.

7. Artificial transmutation can be brought about by bombardment with high-energy protons and α-particles. Why are neutrons of very low energies also able to promote nuclear transmutations?

8. A radioactive series begins with $^{241}_{94}Pu$ and ends with $^{209}_{83}Bi$. How many α- and β-particles are involved in the intermediate steps?

Rate of Decay and Half-Life

9. A 0.96-g sample of a radioactive substance decays to 0.015 g in 10 years.
a) What is the half-life of the substance?
b) Calculate the decay constant of the substance.
c) How long will it take for a 5.0-g sample of this substance to decay to 0.63 g?

10. A radioactive substance decays as follows.

MASS (g)	TIME (min)
100	0
70	35
40	92
30	120
20	161

a) What is the half-life of the element?
b) What is the radioactive decay constant of the element?

11. The number of people getting killed per unit time while continuously engaged in a given activity is proportional to the total number of people engaged in that activity. The proportionality constant (the fatality rate) is analogous to a radioactive decay constant. The accompanying table lists estimated fatality rates for various human activities (compiled from data given by Walter Jordan in *Physics Today*, May 1970, p. 72).

ACTIVITY OR EXPOSURE	FATALITY RATE $(hr^{-1} \times 10^6)$ BASED ON 10^6 hrs EXPOSURE
Mountain climbing	40
Motorcycling	6.6
Flying scheduled airlines	2.4
Smoking cigarettes	1.2
Disease and old age	1.0
Private car travel (U.S.)	0.95
Railroad and bus travel	0.08
Radiation (at 5 rems·yr^{-1})	0.05

a) Determine the half-life (in years) of a person engaged in each of the above activities.
b) What other factors need to be considered in order to make the above comparisons realistic?

12. The half-life of $^{72}_{33}As$ is 26 hr.
a) Calculate the radioactive decay constant of $^{72}_{33}As$.
b) Beginning with an original sample weighing 0.25 g, how much $^{72}_{33}As$ would be left 1 week later?

13. At a given time a sample of thorium-234 weighed 1.78 g. Forty-eight days later chemical analysis showed that there was only 0.445 g of thorium remaining; the rest of the sample consisted of the decay product $^{234}_{91}Pa$.
a) What is the half-life of thorium-234?
b) How much $^{234}_{91}Pa$ was present in the above sample?
c) Write the balanced nuclear equation for the decay of thorium-234.

14. A Geiger counter showed that a radioactive sample gave 5690 cpm at 1030. At 2140 the next day the reading had dropped to 930 cpm. Assuming an average background count of 45 cpm, what is the half-life of the element?

15. A radioactive element has a half-life of 30 da. How long will it take for 75% of the element to decay to another element?

16. A sample of radioactive element contains 8.2×10^{24} atoms at a given instant of time. If the element's half-life is 5 da, how many atoms of this element remain 30 da later? What is the radioactive decay constant of the element?

17. The half-life of $^{226}_{88}Ra$ is 1620 yr. How long will it take for 99% of a given sample to decay? For 99.9% to decay? For 100% to decay?

Mass Defect and Binding Energy

18. The experimentally measured atomic mass of $^3_2$He is 3.01693 amu. Calculate the following.
 a) The mass defect of ^{3}He.
 b) The binding energy per nucleon of ^{3}He.

19. Theoretical physicists have estimated that two colliding deuterons must have kinetic energies of around 10^5 eV in order to undergo fusion. To what temperature does this kinetic energy correspond?

20. Calculate the energy liberated in the reaction $^2_1$H + $^3_1$H → $^4_2$He + $^1_0$n. This is thought to be another one of the sun's fusion reactions. Atomic masses are 2.01471 amu for $^2_1$H and 3.01695 amu for $^3_1$H.

21. Astrophysicists believe that the sun will eventually die when its core contracts to bring about a temperature high enough for three He-4 atoms to fuse to a C-12 atom. Calculate the energy released by such a fusion reaction. The atomic mass of C-12 is 12.00386 amu.

22. An atom of $^{37}_{17}$Cl has a mass of 34.96885 amu. Calculate the binding energy per nucleon of this isotope.

23. The total power output of the sun is estimated to be about 4×10^{26} W. What amount of mass (in kg) is lost by the sun each second?

24. Einstein's equation, $\Delta E/\Delta m = c^2$, states literally that every energy change is accompanied by a mass change and that the ratio of the two is always equal to c^2. What is the mass change involving the vaporization of one mole of water? The heat of vaporization of water is 2260 J·g^{-1}. Is such a mass change detectable?

25. Calculate the energy involved in the fusion reaction

$$^7_3\text{Li} + ^1_1\text{H} \rightarrow 2\,^4_2\text{He}$$

The masses of $^7_3$Li and $^4_2$He nuclei are 7.0160 and 4.0026 amu, respectively.

Radioisotope Dating

26. A 3.2-g portion of an Egyptian mummy wrapping is burned to produce 1.3 g of carbon. In 40 min (after correcting for background radiation), 301 counts are obtained.

 a) When was the mummy interred?
 b) Is there any assurance that the mummy was not wrapped in a 1000-yr-old shroud at the time of burial?

27. In mid-1965, two one-magnum bottles of sherry were "liberated" from a cellar in Jerez de la Frontera in Spain. One of the bottles produced 880 counts during 50 min of counting (after correcting for background radiation), and the other produced a mild hangover the next morning. What is the vintage of the wine? [*Note:* A magnum is approximately 2 L.]

28. A 1.5-g sample of carbon obtained from a maizelike cereal grain found along with broken pottery in a New Mexico archaeological excavation produced 383 counts in 30 min (after correcting for background radiation). Estimate the age of the cereal grain.

29. Suppose you had a sample of a carbon-containing fossil from another planet. What uncertainties would you face in using carbon-14 dating to determine the age of the sample?

30. Suppose, while you are sightseeing around the Dead Sea in Israel, that a Bedouin shepherd boy sells you what he claims is one of the Dead Sea scrolls, thought to have been written around 100 B.C. Suppose further you have a sample of the scroll's parchment analyzed to find that it contains 96% of the carbon-14 content of a living plant. Do you now own a genuine Dead Sea scroll?

31. Uranium-238 has a half-life of 4.5×10^9 yr. A sample of uranium ore is found to contain 2.499 g of uranium-238 and 2.163 g of lead-207. What is the age of the ore?

32. An archaeologist collects a carbon sample thought to be from a prehistoric campfire. The carbon is burned to CO_2 and the gas passed through a solution of $Ca(OH)_2$ to produce 1.46 g of $CaCO_3$. A Geiger counter indicates that this $CaCO_3$ sample has a decay rate of 48 counts·hr^{-1} (corrected for background). Estimate the age of the carbon sample.

General

33. The U-235 used to construct the first nuclear bomb was separated from U-238 by making use of the relative rates of diffusion of UF_6 (as a gas) containing the two isotopes. What is this relative rate of diffusion? Would you consider this to be a fast, efficient way of separating the isotope?

34. From the standpoint of radioactive decay, would radioactive sodium behave the same or differently in the forms Na (atomic), Na$^+$ (ionic), and in a compound such as Na$_2$SO$_4$? Explain.

35. In 1911 Rutherford and Boltwood found that 1 g of radium produced 3.4×10^{11} α-particles per second. The same sample also produced 0.039 cm^3 of helium gas in a year's time. Use this data to estimate a value for Avogadro's number. The half-life of the radium is 1620 yr.

ADDITIONAL READINGS

Choppin, Gregory, *Nuclei and Radioactivity*. New York: Benjamin, 1964. A comprehensive account written for use by well-prepared students of general chemistry.

Hammond, Allen L., "Fission: The Pro's and Con's of Nuclear Power." *Science* **178**, 147 (1972).

Gough, William C., and Bernard J. Eastlund, "The Prospects of Fusion Power." *Scientific American,* February 1971.

Seaborg, Glenn T., and Justin L. Bloom, "Fast Breeder Reactors." *Scientific American,* November 1970. The 1951 Nobel prize winner in chemistry and one of his colleagues argue for the development of breeder reactors to meet our growing demands for energy.

Landis, John W., "Fusion Power: Hallmark of the 21st Century." *J. Chem. Educ.* **50**, 658 (1973). The president of the Gulf General Atomic Company presents arguments for the development of fusion power to meet the energy demands of the coming century.

Post, Richard F., "Fusion Power." *Scientific American*, December 1957. Discusses the basic problems of controlling fusion and its potential as an energy source.

Hollaender, Alexander, and George E. Stapleton, "Ionizing Radiation and the Living Cell." *Scientific American,* September 1959. Discusses how radiation damages cells.

Michael, Henry N., and Elizabeth K. Ralph, eds., *Dating Techniques for the Archaeologist*. Cambridge, Mass.: MIT Press, 1972. A concise set of summaries of a large number of dating methods used by professional archaeologists.

Rogers, F. E., "Chemistry in Art: Radiochemistry and Forgery." *J. Chem. Educ.* **49**, 418 (1972). An interesting account of how radiochemistry was used to prove conclusively that the alleged Vermeer painting "Christ and His Disciples at Emmaus" was really painted by a contemporary artist.

Renfrew, Colin, "Carbon 14 and the Prehistory of Europe." *Scientific American,* October 1971. The author uses carbon-14 dates corrected by the bristle-cone pine technique to suggest that cultural advances did not diffuse into Europe from the east as long supposed.

Zimmerman, Joan, "Answering the Question When." *Chemistry* **43** (July 1970): 22–27. An elementary discussion of various methods used for dating.

Kamen, Martin D., "Tracers." *Scientific American,* February 1949. Discusses some of the earliest uses of tracers in biological research.

Biddulph, Susann, and Orlin Biddulph, "The Circulatory System of Plants." *Scientific American,* February 1959. Shows how radioactive tracers have been used to follow the circulation of nutrient materials in plants.

12
CHEMICAL EQUILIBRIUM

The state of chemical equilibrium is analogous to the balancing of these spheres. (Courtesy of Marshall Henrichs.)

In opposing the body A to the compound BC, the combination AC can never take place [completely], but the body C will be divided between the bodies A and B in proportion to the affinity and quantity of each, or in the ratio of their masses.
CLAUDE LOUIS BERTHOLLET, *1801*

Claude Louis Berthollet (1748–1822) was a pioneer in the study of chemical reactions and chemical composition. After receiving an M. D. from the University of Turin in Italy, Berthollet joined the staff of the École Polytechnique in Paris. He established the composition of ammonia, clarified the chemistry of chlorine, and developed early concepts of chemical equilibrium. Unfortunately, some of Berthollet's ideas were erroneous, and chemical equilibrium did not become clearly understood until the concept of mass action was introduced.

The majority of the material in the preceding eleven chapters deals primarily with physical and nuclear transformations of matter. The present chapter now turns to **chemical transformations**, a subject that occupies the rest of the book. In typical chemical reactions, two or more elements may combine to form a compound, two or more compounds may combine to form one or more different compounds, one or more compounds may combine with one or more elements to form one or more different compounds, or a single compound may break up into two or more different elements or compounds. In each case, there are rearrangements in the atom groupings and changes in the electron densities about the various nuclei. Just as many physical transformations—notably changes in state—may be discussed in terms of a dynamic equilibrium, so may chemical reactions. This chapter examines what chemical equilibrium is, how it is characterized quantitatively, and how its concepts are used in practice.

12.1 THE REVERSIBILITY OF CHEMICAL REACTIONS

One of the simplest and most studied chemical reactions is the union of hydrogen and iodine to form hydrogen iodide. The balanced chemical equation for this process as it occurs in the gaseous phase is

$$H_2(g) + I_2(g) \rightarrow 2HI(g)$$

If a given number of H_2 molecules and I_2 molecules is placed into a reaction vessel,

Cato Guldberg (1836–1902) was a Norwegian mathematician who studied the relationship between vapor-pressure lowering and the freezing point of solutions, movements of atmospheric air masses, and rates of chemical processes. He developed the law of mass action in collaboration with his brother-in-law, Peter Waage (1833–1900), a physical chemist and one-time president of the Norwegian equivalent of the YMCA.

the two different molecules will begin to combine to form HI molecules. The rate at which this occurs may be expressed as the number of HI molecules formed per unit time per unit volume of reaction space. In 1864 two Norwegian scientists, Cato Guldberg and Peter Waage, noted that the rate of chemical combination was proportional to the concentrations of the reactants. Thus the number of HI molecules formed per unit time will depend on the numbers of H_2 and I_2 molecules present per unit volume of the reaction vessel. Since the formation of HI molecules uses up H_2 and I_2 (and thus decreases their concentrations), it follows that the rate of HI formation must decrease as time progresses. Figure 12.1 shows how the rate of formation of HI varies with time and also the concomitant changes in molar concentrations of HI, H_2, and I_2. This effect of reactant concentrations on the rate of formation of a product is known as the **law of mass action**. The law is explained on the basis of the assumption that reactant molecules must collide with each other in order to form the product. The greater the number of reactant molecules per unit volume, the greater should be the number of such collisions (occurring in a random fashion).

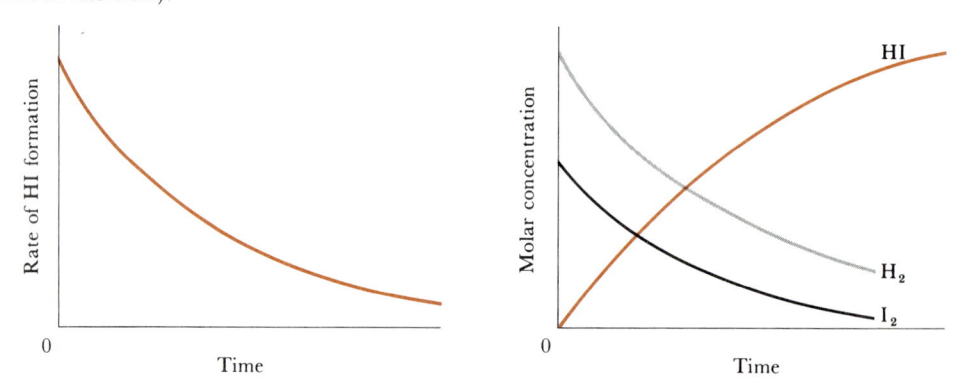

Fig. 12.1. The left-hand graph shows how the rate of formation of HI (molecules per unit volume per unit time) decreases with time. This decrease in rate results from the fact that reactants H_2 and I_2 are being used up (right-hand graph) and thus are less available for HI formation. The right-hand graph also illustrates the increase in HI concentration as the reaction proceeds.

Observation shows that some of the HI molecules break up and reform H_2 and I_2 molecules:

$$2HI(g) \rightarrow H_2(g) + I_2(g)$$

The number of HI molecules breaking up per unit time also follows the law of mass action; initially there are few HI molecules and, consequently, few break up per unit time. Eventually, considerable numbers of H_2 and I_2 molecules have combined to form HI and more and more of these break up. Figure 12.2 illustrates how the rate of HI breakup changes with time. The fact that the reaction proceeds in two directions at once is indicated as follows:

$$H_2(g) + I_2(g) \rightleftharpoons 2HI(g)$$

The double arrows denote what is known as a **reversible reaction**. Although it is customary to call the substances on the left the **reactants** and those on the right the **products**, in strict point of fact all substances in the reaction are simultaneously reactants and products.

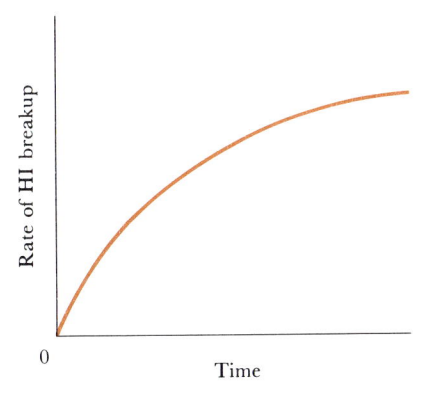

Fig. 12.2. As the concentration of HI increases (right-hand graph in Fig. 12.1) the HI molecules begin to break up into H_2 and I_2 at a faster and faster rate.

12.2 CHEMICAL EQUILIBRIUM AND THE EQUILIBRIUM CONSTANT

Ultimately there comes a time when the number of HI molecules produced per unit time just equals the number of HI molecules breaking up per unit time (see Fig. 12.3). At the same time the concentrations of H_2, I_2, and HI no longer change; that is, they remain at some constant value. This condition of constant composition is called **chemical equilibrium**.

The constancy of chemical composition characteristic of chemical equilibrium suggests that the reaction is static, but quite the opposite is the case. If some of the I_2 of the equilibrium mixture is replaced with radioactive I_2, chemical analysis shows that as time goes on, more and more radioactive iodine begins to show up in HI. This can be explained only by assuming that even at equilibrium, both forward and reverse reactions continue, but at equal rates. Consequently, the net result is a constant number of HI, H_2, and I_2 molecules at equilibrium. Chemical equilibrium, like vapor-pressure equilibrium, is a *dynamic condition*.

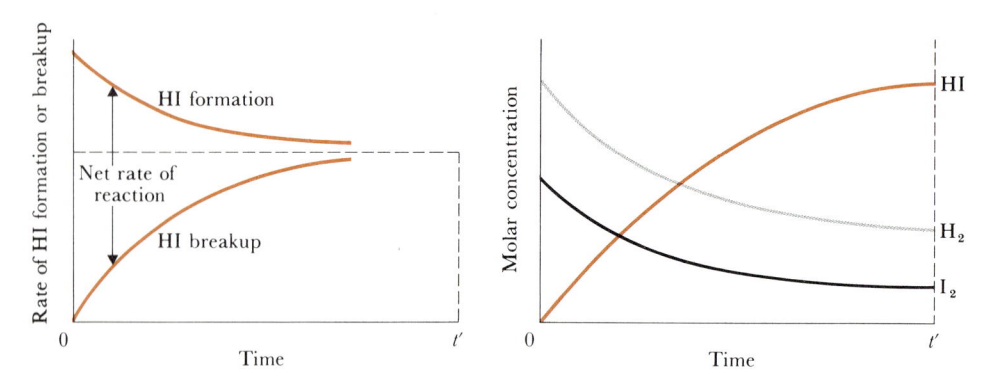

Fig. 12.3. The left-hand graph shows that the two opposing rates (HI formation and HI breakup) tend to become equal as the reaction proceeds. At equilibrium (denoted by t') the rate of HI formation equals the rate of HI breakup and the net rate of reaction is zero. The right-hand graph shows that the concentrations of reactants and products approach constant values as equilibrium is approached.

As a consequence of the reversibility of the H_2, I_2, HI reaction, it is not possible to simply mix H_2 and I_2 and obtain complete conversion to HI; the reverse reaction will always maintain some concentrations of H_2 and I_2. Similarly, an originally pure sample of HI will form H_2 and I_2; if placed in a closed container at a given temperature, an equilibrium mixture of H_2, I_2, and HI will result.

Chemical equilibrium is attained when there is a balance between two opposing tendencies—the tendency to form products from reactants (left-to-right reaction) and the tendency for products to break up into reactants (right-to-left reaction). The point at which the two tendencies are in balance depends on the law of mass action; for any reaction mixture there will be some concentrations of reactants and products such that a net reaction rate of zero is finally attained. In the H_2, I_2, HI system as an example, the concentrations at which equilibrium occurs obey the mathematical relationship

$$\frac{[HI]^2}{[H_2][I_2]} = \text{constant} \tag{12.1}$$

where the numerical value of the constant depends only on the temperature. The symbol [ ] means the molar concentration of the enclosed substance *measured at equilibrium*. Experiments show that, as a fairly good approximation, the same numerical value of the constant is obtained no matter what the starting amounts of H_2, I_2, and HI are—provided the temperature is the same in all cases. This means that the values of $[H_2]$, $[I_2]$, and $[HI]$ adjust themselves so that their mathematical combination given in Eq. (12.1) always leads to the same numerical value.*

The constant in Eq. (12.1) is called the **equilibrium constant**; it is usually given

---

* This statement is not strictly true. The constant is strictly independent of concentrations of products and reactants only if expressed in terms of *effective* (rather than *actual*) concentrations called **activities**. Under certain conditions, low concentrations, for example, activities and molar concentrations become approximately equal.

the symbol K_c.* The equilibrium constant of the H_2, I_2, HI reaction is calculated from experimental data as follows: 0.11337 mol H_2 and 0.07510 mol I_2 are mixed and allowed to come to equilibrium at 425.4 °C in a 10-L flask. Chemical analysis shows that there is 0.1354 mol HI present at equilibrium. Note that it takes one mole of H_2 to produce two moles of HI. Thus, if 0.1354 mol of HI was formed, it took half this number of moles of H_2 (and I_2) to do so. Therefore at equilibrium

$$[H_2] = \frac{\text{original amount}}{\text{volume of solution}} - \left(\frac{1}{2}\right) \frac{\text{moles HI formed}}{\text{volume of solution}}$$

$$= \frac{0.11337 \text{ mol}}{10 \text{ L}} - \left(\frac{1}{2}\right) \frac{0.1354 \text{ mol}}{10 \text{ L}} = 0.004567 \, M$$

Similarly,

$$[I_2] = \frac{0.07510 \text{ mol}}{10 \text{ L}} - \left(\frac{1}{2}\right) \frac{0.1354 \text{ mol}}{10 \text{ L}} = 0.00074 \, M$$

$$[HI] = \frac{0.1354 \text{ mol}}{10 \text{ L}} = 0.01354 \, M$$

The numerical value of K_c at 425.4 °C is

$$K_c = \frac{[HI]^2}{[H_2][I_2]} = \frac{(0.01354)^2}{(0.004567)(0.00074)} = 54.2$$

If a general chemical reaction is represented symbolically by

$$aA + bB \rightleftharpoons lL + mM$$

where A, B, L, and M are elements or compounds and a, b, l, and m are the numbers used to balance the equation, the equilibrium constant may be defined more generally as

$$K_c = \frac{[L]^l[M]^m}{[A]^a[B]^b} \tag{12.2}$$

Note that the numerator is a mathematical product of molar concentrations of chemical products, each raised to the power indicated by the number used to balance it in the chemical equation. Similarly, the denominator is a like quantity involving the reactants.

Two different reactions will generally have different equilibrium constants. For example, the H_2, I_2, HI reaction

$$H_2(g) + I_2(g) \rightleftharpoons 2HI(g)$$

---

* There are several different but interrelated types of equilibrium constants and the symbol K_{eq} is often used to represent any of these. The symbol K_c indicates a particular equilibrium constant in which the amounts of materials are expressed in molar concentration units.

has the equilibrium constant value

$$K_c = \frac{[HI]^2}{[H_2][I_2]} = 54.2 \quad \text{at } 425.4\,°C$$

At the same temperature, a similar reaction involving H_2, Cl_2, and HCl

$$H_2(g) + Cl_2(g) \rightleftharpoons 2HCl(g)$$

has the equilibrium constant value

$$K_c = \frac{[HCl]^2}{[H_2][Cl_2]} = 2 \times 10^7 \quad \text{at } 425.4\,°C$$

The fact that in the latter reaction K_c has a much larger numerical value than it does in the former means that, at 425.4 °C, H_2 and Cl_2 are more completely converted to HCl than H_2 and I_2 are to HI. Thus the equilibrium constant may be used as a measure of the extent to which a given reaction progresses. All other factors considered equal, a large value of K_c denotes a high degree of conversion of reactants to products and a small value of K_c denotes a low degree of conversion.

Since chemical reactions may be written in a variety of different but equivalent ways, a quoted value of an equilibrium constant must always be accompanied by the specific equation to which it refers. For example, if the ammonia synthesis reaction is written

$$N_2(g) + 3H_2(g) \rightleftharpoons 2NH_3(g)$$

then the equilibrium constant is

$$K_c = \frac{[NH_3]^2}{[N_2][H_2]^3} \tag{12.3}$$

This is found to have the numerical value of 9.5×10^{-4} at 500 °C. Suppose the reaction is written in the opposite direction:

$$2NH_3(g) \rightleftharpoons N_2(g) + 3H_2(g)$$

Consistency is maintained and ambiguity avoided if K_c is now written as

$$K_c = \frac{[N_2][H_2]^3}{[NH_3]^2} \tag{12.4}$$

It should be apparent that K_c in Eq. (12.4) is just the reciprocal of K_c in Eq. (12.3). Thus

$$K_c\,[\text{Eq. (12.4)}] = \frac{1}{K_c\,[\text{Eq. (12.3)}]} = \frac{1}{9.5 \times 10^{-4}} = 1.1 \times 10^3$$

Similarly, the ammonia reaction could be written

$$\tfrac{1}{2}N_2(g) + \tfrac{3}{2}H_2(g) \rightleftharpoons NH_3(g)$$

with the equilibrium constant

$$K_c = \frac{[NH_3]}{[N_2]^{1/2}[H_2]^{3/2}} \tag{12.5}$$

In this case

$$\begin{aligned} K_c \text{ [Eq. (12.5)]} &= \sqrt{K_c \text{ [Eq. (12.3)]}} \\ &= \sqrt{9.2 \times 10^{-4}} \\ &= 3.1 \times 10^{-2} \end{aligned}$$

Table 12.1 lists some data to illustrate the reversible nature of the ammonia synthesis reaction. When placed in a reaction vessel, either 1 mol of N_2 and 3 mol of H_2 or 2 mol of NH_3 will produce the same equilibrium composition for a given temperature and total pressure.

It should be pointed out that the actual attainment of chemical equilibrium is largely a laboratory curiosity; processes occurring in nature seldom remain undisturbed long enough to attain equilibrium. For example, in biological reactions the products of a reaction are often removed before they can come to equilibrium with the reactants. The result is that reactants can be converted continuously to products in an essentially one-way or irreversible manner.

TABLE 12.1 THE REVERSIBLE NATURE OF THE AMMONIA REACTION:

$N_2(g) + 3H_2(g) \rightleftharpoons 2NH_3(g)$

AT SEVERAL TEMPERATURES AND PRESSURES

		STARTING AMOUNTS (MOLES)		AMOUNTS AT END OF REACTION (MOLES)		STARTING AMOUNTS (MOLES)
500 °C 200 atm	N_2	1	→	0.824	←	0
	H_2	3	→	2.472	←	0
	NH_3	0	→	0.352	←	2
500 °C 100 atm	N_2	1	→	0.896	←	0
	H_2	3	→	2.688	←	0
	NH_3	0	→	0.208	←	2
400 °C 200 atm	N_2	1	→	0.637	←	0
	H_2	3	→	1.911	←	0
	NH_3	0	→	0.726	←	2

12.3 CALCULATIONS USING THE EQUILIBRIUM CONSTANT

This section considers some typical calculations based on equilibrium constants. One of the most important uses of an equilibrium constant is that once its value is determined for a given reaction at a given temperature, it is possible to predict equilibrium concentrations for any arbitrary mixture of starting materials at this same temperature.

EXAMPLE 12.1 The reaction $A(g) + B(g) \rightleftharpoons C(g)$ is carried out at 250 °C in a 2.00-L flask by mixing 0.060 mol of A and 0.100 mol of B. After equilibrium is attained, the contents of the flask are chilled and analyzed. The mixture contains 0.044 mol of C. Calculate the equilibrium constant of the reaction.

SOLUTION The expression for the equilibrium constant is

$$K_c = \frac{[C]}{[A][B]}$$

The balanced equation tells us that the 0.044 mol of C had to come from 0.044 mol of A and 0.044 mol of B. Thus at equilibrium

$$[A] = \frac{(0.60 - 0.044) \text{ mol}}{2.00 \text{ L}} = 0.0080 \ M$$

$$[B] = \frac{(0.100 - 0.044) \text{ mol}}{2.00 \text{ L}} = 0.028 \ M$$

$$[C] = \frac{0.044 \text{ mol}}{2.00 \text{ L}} = 0.022 \ M$$

Substituting these equilibrium concentrations into the K_c expression produces

$$K_c = \frac{[C]}{[A][B]} = \frac{0.022}{(0.0080)(0.028)} = 98$$

EXAMPLE 12.2 Using the reaction of Example 12.1, 0.020 mol of A and 0.010 mol of B are mixed in a 3.0-L container at 250 °C. Calculate the equilibrium concentrations of A, B, and C.

SOLUTION Let x represent the molar concentration of C at equilibrium. The equilibrium concentration of each substance in terms of x is

$$[A] = \frac{0.020 \text{ mol}}{3.0 \text{ L}} - x \quad \text{or} \quad (0.0067 - x) \ M$$

$$[B] = \frac{0.010 \text{ mol}}{3.0 \text{ L}} - x \quad \text{or} \quad (0.0033 - x) \ M$$

$$[C] = x \text{ (in units of } M)$$

The equilibrium constant in terms of the above is

$$\frac{x}{(0.0067 - x)(0.0033 - x)} = 98$$

Multiplying out the terms and rearranging leads to the quadratic equation

$$98x^2 - 1.98x + 0.002167 = 0$$

This has the two solutions $x = 0.019\ M$ and $0.0012\ M$. The first solution is discarded since x cannot be larger than $0.0067\ M$, the starting concentration of A. Thus

[C] = 0.0012 M
[A] = (0.0067 − 0.0012) M = 0.0055 M
[B] = (0.0033 − 0.0012) M = 0.0021 M

As a check, reevaluate K_c using the above equilibrium values:

$$K_c = \frac{0.0012}{(0.0055)(0.0021)} = 10 \times 10^1$$

The discrepancy with the correct value, 98, is due to round-off error. The only way to avoid this is to carry along many more figures than are significant and then round off at the end.

EXAMPLE 12.3 One mole of nitrogen and two moles of hydrogen are mixed in a 2-L container at 500 °C and allowed to come to equilibrium. Calculate the weight of ammonia present in the equilibrium mixture.

SOLUTION The balanced equation for the reaction is

$$N_2(g) + 3H_2(g) \rightleftharpoons 2NH_3(g)$$

and the equilibrium constant is

$$K_c = \frac{[NH_3]^2}{[N_2][H_2]^3}$$

As given previously, K_c of this reaction is 9.5×10^{-4} at 500 °C. Now let $2x$ represent the molar concentration of ammonia at equilibrium. Since it takes x moles of N_2 to produce $2x$ moles of NH_3, the equilibrium concentration of N_2 may be written

$$[N_2] = \frac{1\ \text{mol}}{2\ \text{L}} - x$$

or

$$(0.5 - x)\ M$$

Similarly,

$$[H_2] = \frac{2 \text{ mol}}{2 \text{ L}} - 3x \quad \text{or} \quad (1 - 3x)\, M$$

$[NH_3] = 2x$ (in units of M)

The equilibrium constant in terms of x is

$$\frac{(2x)^2}{(0.5 - x)(1 - 3x)^3} = 9.5 \times 10^{-4}$$

Direct solution for x is somewhat tedious since it involves x to the fourth power. However, an approximate solution to the above equation is very easy to obtain. Since K_c is very, very small (compared with unity), it is expected that x, the amount of N_2 used up per liter, is also very small. If it can be assumed that x is much smaller than 0.5, then x can be neglected in the denominator of the K_c expression. This means that the quantities $0.5 - x$ and $1 - 3x$ are replaced with 0.5 and 1, respectively, so that x can now be calculated (approximately) from the simpler expression

$$\frac{(2x)^2}{(0.5)(1)^3} = 9.5 \times 10^{-4}$$

This leads to

$x^2 = 1.2 \times 10^{-4}$

$x = 1.1 \times 10^{-2} M$

Checking the original assumption shows that $x/0.5 = 1.1 \times 10^{-2}/0.5 = 0.022$. Thus x is only 2.2% of 0.5. An often used rule of thumb is that x may be neglected in an expression, such as $a - x$, appearing in a denominator if x is no larger than 5% of a.*

Since the concentration of ammonia at equilibrium is $2x$ or $2.2 \times 10^{-2}\,M$, the amount of NH_3 produced is $(2.2 \times 10^{-2} \text{ mol} \cdot \text{L}^{-1})(2\text{ L})$ or 4.4×10^{-2} mol (about 0.75 g).

In some applications it is useful to define the equilibrium constant of a gas phase reaction not in terms of molar concentrations but in terms of the partial pressures of the equilibrium components. In the case of the ammonia reaction $N_2(g) + 3H_2(g) \rightleftharpoons 2NH_3(g)$, the **partial pressure equilibrium constant** is

$$K_p = \frac{p_{NH_3}^2}{p_{N_2} p_{H_2}^3} \tag{12.6}$$

where p_{NH_3} is the partial pressure of NH_3 measured at equilibrium, and similarly for

---

* Had x been 5% of 0.5 or larger, it would have been necessary to solve for x using some other technique, perhaps even solving the full equation involving x to the fourth power.

p_{N_2} and p_{H_2}.* The molar concentration of each component is directly proportional to its partial pressure. If all the component gases are ideal, the relationship is

$$c_i = \frac{n_i}{V} = \frac{p_i}{RT} \tag{12.7}$$

where c_i is the molar concentration of substance i when its partial pressure is p_i. Thus the K_c and K_p equilibrium constants of the ammonia reaction are related by

$$K_c = \frac{[NH_3]^2}{[N_2][H_2]^3} = \frac{p_{NH_3}^2 (1/RT)^2}{p_{N_2} p_{H_2}^3 (1/RT)^4} = K_p (RT)^2$$

In general, for the gas phase reaction $aA + bB \rightleftharpoons lL + mM$,

$$K_c = K_p (RT)^{a+b-l-m} \tag{12.8}$$

For any reaction in which $a + b = l + m$, that is, there is no change in the number of moles, K_c and K_p become numerically equal.

EXAMPLE 12.4 Equal molar amounts of PCl_3 and Cl_2 are mixed at 250 °C in a 2.00-L flask. The reaction is $PCl_3(g) + Cl_2(g) \rightleftharpoons PCl_5(g)$. At equilibrium the total pressure of the mixture is 2.00 atm, and 0.0173 mol of PCl_5 is present. Calculate K_p of the reaction at 250 °C.

SOLUTION Assuming ideal gas behavior, the partial pressure of PCl_5 at equilibrium is given by

$$p_{PCl_5} = n_{PCl_5} \frac{RT}{V} = \frac{(0.0173 \text{ mol})(0.082 \text{ L} \cdot \text{atm} \cdot \text{K}^{-1} \cdot \text{mol}^{-1})(523 \text{ K})}{2.00 \text{ L}} = 0.37 \text{ atm}$$

Since $p_{tot} = p_{PCl_3} + p_{Cl_2} + p_{PCl_5} = 2.00$ atm (Dalton's law) and $p_{PCl_3} = p_{Cl_2}$, then

$$p_{PCl_3} = p_{Cl_2} = \frac{2.00 - 0.37}{2.00} = 0.815 \text{ atm}$$

Thus

$$K_p = \frac{p_{PCl_5}}{p_{PCl_3} p_{Cl_2}} = \frac{(0.37)}{(0.815)^2} = 0.56$$

EXAMPLE 12.5 A sample of pure PCl_5 is placed in a 5.0-L flask and allowed to come to equilibrium at 250 °C. The total pressure of the equilibrium mixture is 1 atm. Calculate the number of moles of PCl_3 and Cl_2 present at equilibrium.

---

* In practice, the partial pressures are divided by a reference pressure of 1 atm (converted to the units in which the partial pressures are expressed) so that K_p is *dimensionless*, that is, a pure number. The same thing is accomplished by always using pressures in atmospheres in K_p expressions. K_c's have units involving $\text{mol} \cdot \text{L}^{-1}$ but it is convenient to omit these.

SOLUTION The total pressure at equilibrium is given by

$$p_{PCl_3} + p_{Cl_2} + p_{PCl_5} = 1 \text{ atm}$$

If $x = p_{PCl_3} = p_{Cl_2}$, then $2x + p_{PCl_5} = 1$ and $p_{PCl_5} = 1 - 2x$. Given the K_p value from Example 12.4,

$$\frac{p_{PCl_5}}{p_{PCl_3} p_{Cl_2}} = 0.56 = \frac{1 - 2x}{x^2}$$

Solving for x gives

$$x = p_{PCl_3} = p_{Cl_2} = 0.44 \text{ atm}$$

Thus the number of moles of each is given by

$$n = \frac{p_{PCl_3} V}{RT} = \frac{p_{Cl_2} V}{RT} = \frac{xV}{RT} = \frac{(0.44 \text{ atm})(5.0 \text{ L})}{(0.082 \text{ L}\cdot\text{atm}\cdot\text{K}^{-1}\cdot\text{mol}^{-1})(523 \text{ K})} = 0.051 \text{ mol}$$

EXAMPLE 12.6 Calculate K_c for the $PCl_5(g) \rightleftharpoons PCl_3(g) + Cl_2(g)$ reaction at 250 °C.

SOLUTION Since K_p for $PCl_3(g) + Cl_2(g) \rightleftharpoons PCl_5(g)$ is 0.56, K_p for the reverse reaction, $PCl_5 \rightleftharpoons PCl_3 + Cl_2$, is 1/0.56 or 1.8. By Eq. (12.8)

$$K_c = K_p(RT)^{1-1-1} = K_p(RT)^{-1} = \frac{1.8}{RT} = \frac{1.8}{(0.082 \text{ L}\cdot\text{atm}\cdot\text{K}^{-1}\cdot\text{mol}^{-1})(523 \text{ K})} = 0.042$$

12.4 LE CHÂTELIER'S PRINCIPLE AND CHEMICAL EQUILIBRIUM

Chapter 9 discusses how Le Châtelier's principle is used to predict the response of phase equilibria to external disturbances. Chemical equilibria may be treated in an analogous manner.

Of particular importance is the use of Le Châtelier's principle to predict how the equilibrium constant changes as the temperature changes. This can be done by knowing only whether the reaction is exothermic or endothermic. The heat of the ammonia reaction, $N_2(g) + 3H_2(g) \rightleftharpoons 2NH_3(g)$ is -91.1 kJ; that is, 91.1 kJ of energy is produced when one mole of N_2 combines with three moles of H_2 to produce two moles of NH_3. The negative sign indicates that the reaction is exothermic.* The reverse process, decomposition of two moles of NH_3 to produce one mole of N_2 and three moles of H_2, is endothermic with a heat of reaction of $+91.1$ kJ. If an increase in temperature is regarded as a stress (addition of thermal energy), the system can respond to minimize or undo this stress by doing something endothermic (to use up the added energy). The simplest endothermic response is to decompose some NH_3

---

* The evolution of energy (exothermic process) may be likened to an expenditure of money (a decrease in wealth, thus negative), whereas the absorption of energy (endothermic process) is like getting wages (an increase in wealth, thus positive).

to N_2 and H_2. Since K_c is the ratio of $[NH_3]^2$ to the product $[N_2][H_2]^3$, K_c must decrease when $[NH_3]$ decreases and $[N_2]$ and $[H_2]$ increase. Thus K_c of the reaction $N_2(g) + 3H_2(g) \rightleftharpoons 2NH_3(g)$ decreases as the temperature increases. This means that the conversion of N_2 and H_2 to NH_3 is favored by low temperatures.

TABLE 12.2 APPLICATION OF LE CHÂTELIER'S PRINCIPLE TO THE REACTION
$N_2(g) + 3H_2(g) \rightleftharpoons 2NH_3(g)$ (Exothermic)

STRESS	EXPECTED RESPONSE
Increase the temperature	Since the reaction is exothermic, some NH_3 will decompose to produce more N_2 and H_2 (equilibrium shifts to the left, K_c decreases)
Add N_2	Use up additional H_2 and produce more NH_3 (K_c remains constant)
Remove NH_3	Use up more N_2 and H_2 to replace some of the removed NH_3 (K_c remains constant)
Add NH_3	Form more N_2 and H_2 to use up some of the added NH_3 (K_c remains constant)
Increase total pressure[a]	Form more NH_3 (K_c remains constant)

[a] The pressure is increased by decreasing the volume of the reaction vessel. The pressure could also be increased by pumping in an inert gas, but this would have no effect on K_c or on the concentrations of N_2, H_2, and NH_3.

In Table 12.2 there are listed some other possible stresses on the ammonia equilibrium and the responses expected. The last effect shown in Table 12.2—the formation of additional NH_3 in response to an increase in total pressure—can be understood by noting that a response capable of lessening the effect of an increase in pressure is a net decrease in the number of moles (thereby a decrease in volume) in the gaseous mixture. Since it requires four moles of reactants to produce two moles of product (a net decrease of two moles), the stress of increased pressure is relieved by the production of more NH_3 at the expense of the N_2 and H_2. Thus ammonia production is favored by low temperature and high pressure. Figure 12.4 summarizes the effect of temperature and pressure on the ammonia equilibrium in terms of the amount of ammonia present in the equilibrium mixture.

To look back for a moment, Table 12.1 also shows the effect of temperature and pressure upon the equilibrium composition. Note that changing the total pressure at a given temperature does change the equilibrium concentrations but not the equilibrium constant. The reason for this is easy to see. If V is the volume of the reaction mixture, K_c may be written

$$K_c = \frac{[NH_3]^2}{[N_2][H_2]^3} = \frac{(n_{NH_3}/V)^2}{(n_{N_2}/V)(n_{H_2}/V)^3} = \frac{n_{NH_3}^2}{n_{N_2} n_{H_2}^3} V^2 \tag{12.9}$$

An increase in pressure will increase the ratio of the number of moles of NH_3 relative to N_2 and H_2, but this will also decrease V. The net result is that the product of the two still equals K_c.

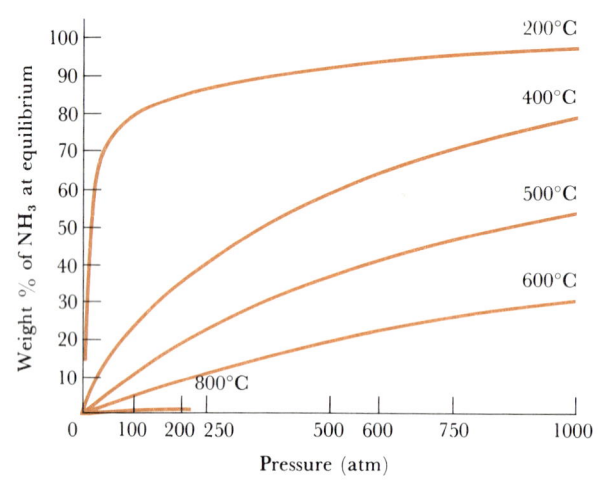

Fig. 12.4. Effect of pressure and temperature on the conversion of a 1:3 molar mixture of nitrogen and hydrogen to ammonia.

Application of Le Châtelier's principle enables the industrial chemist to design a plant that will produce ammonia with optimum efficiency. Not only is it advisable to employ relatively low temperatures and high pressures, but it is also necessary to remove the NH$_3$ as it is produced so that N$_2$ and H$_2$ react continuously in an attempt to reestablish equilibrium. This is the chemist's version of hanging a carrot in front of a donkey's nose to keep it moving.

It should be noted that chemical reactions in which no net change in the number of moles occurs, for example,

$$H_2(g) + I_2(g) \rightleftharpoons 2HI(g)$$

are unaffected by increases in the total pressure.

12.5 EQUILIBRIA IN MULTIPHASE REACTIONS

All the reactions discussed up to now involved a single phase; all reactants and products were gases. In the event that some of the reactants or products are solid or liquid, the equilibrium constant expression must be modified. For example, consider the reaction

$$CaCO_3(s) \rightleftharpoons CaO(s) + CO_2(g)$$

Experiment shows that as long as some solid CaCO$_3$ and CaO are present, then the equilibrium concentration of CO$_2$ is unaffected by their amounts. This may be explained by realizing that the concentration of a solid is just its density and that this density is affected only by temperature. The equilibrium constant for the reaction may be written in the usual way,

$$K_c = \frac{[CaO][CO_2]}{[CaCO_3]}$$

but since [CaO] and [CaCO$_3$] are constant, they may be factored out and grouped with K_c:

$$K_c \frac{[CaCO_3]}{[CaO]} = [CO_2] = K_c'$$

Ordinarily, the prime in K_c' is omitted and the above is written

$$K_c = [CO_2]$$

where the inclusion of [CaO] and [CaCO$_3$] as constants is understood.

In terms of partial pressures the corresponding equilibrium constant for the preceding reaction is

$$K_p = p_{CO_2}$$

where p_{CO_2} is the pressure of CO$_2$ in equilibrium with the CaCO$_3$ and CaO. Experimentally, if a piece of pure CaCO$_3$ is placed into a closed container at some fixed temperature, CO$_2$ is produced until its partial pressure attains the K_p value. At that point the reactants and products are in equilibrium.

12.6 IRREVERSIBLE REACTIONS

Some reactions go so nearly to completion that they are not ordinarily regarded as reversible. Explosions and combustion reactions such as

$$C_6H_{12}O_6(s) + 6O_2(g) \rightarrow 6CO_2(g) + 6H_2O(l)$$

are of this type. Double arrows to signify equilibrium are almost never used for reactions of this type. If it were possible to determine K_c for such a reaction, a very, very large value would be obtained.

Many reactions which would ordinarily be irreversible can be carried out in reversible fashion. For example, in the reaction

$$Mg(s) + 2HCl(aq) \rightarrow MgCl_2(aq) + H_2(g)$$

[recall that the notation (aq) means the substance is in aqueous solution], most of the H$_2$ escapes if the reaction is not carried out in a closed container. Thus H$_2$ cannot participate in the reverse reaction. However, in a closed container the reverse reaction does occur to some extent, especially if the H$_2$ pressure is very large.

The biological reactions occurring in living organisms are usually not treated as equilibrium reactions. Some of these go so completely in one direction that reverse reactions are negligible, but, more important, most of these reactions occur in environments where equilibrium is constantly being disturbed. For example, reactant and product concentrations are continually changing in response to such factors as feeding and physical activity.

12.7 EQUILIBRIA IN SOLUTIONS OF SPARSELY SOLUBLE IONIC COMPOUNDS

Suppose A_xB_y is an ionic compound (its crystal lattice consists of the ions A^{z+} and B^{z-}) which dissolves in water according to the relationship

$$A_xB_y(s) \rightleftharpoons xA^{z+}(aq) + yB^{z-}(aq) \qquad [\textit{Note: } z+ \neq -z- \text{ unless } x = y.]$$

Equilibrium is attained when the solution is saturated. This equation symbolizes the action of the solvent in tearing down the crystal lattice and producing mobile, hydrated ions. The cations A^{z+} are attracted by the negative ends of H_2O dipoles and become hydrated; that is, there results a somewhat loose association between A^{z+} and several water molecules. This association is implied by the (aq) symbol after the ion. Similarly, the anion B^{z-} is attracted to the positive end of the H_2O dipoles. Equilibrium between hydrated ions and the crystalline solid results when there are just as many A^{z+} and B^{z-} ions leaving the crystal lattice per unit time as there are A^{z+} and B^{z-} ions rejoining the crystal lattice per unit time. As in the case of the decomposition of $CaCO_3$ discussed in the preceding section, the equilibrium concentrations of hydrated ions do not depend on the amount of solid A_xB_y present as long as at least some of it is present as a solid. The equilibrium constant for the dissolution reaction is given a special symbol, K_{sp}, and is written

$$K_{sp} = [A^{z+}(aq)]^x [B^{z-}(aq)]^y$$

The equilibrium constant K_{sp} is called the **solubility-product constant** (generally spoken kay-ess-pee). Note that the ions $A^{z+}(aq)$ and $B^{z-}(aq)$ are analogous to atoms or molecules of a gas, with the solvent providing a relatively inert atmosphere.

Practical uses of K_{sp}'s are limited to compounds of very low solubility. For example, the compound AgCl dissolves in water as follows:

$$AgCl(s) \rightleftharpoons Ag^+(aq) + Cl^-(aq)$$

A maximum of 1.79 mg of AgCl dissolves in a liter of water at 25 °C. Since a mole of Ag^+ and Cl^- weighs 143.34 g, the molar concentration of dissolved AgCl is 1.79×10^{-3} g·L^{-1}/143.34 g·mol^{-1} or 1.25×10^{-5} M. Thus

$$[Ag^+(aq)] = [Cl^-(aq)] = 1.25 \times 10^{-5} \, M$$

$$K_{sp} = [Ag^+(aq)][Cl^-(aq)] = (1.25 \times 10^{-5})^2 = 1.56 \times 10^{-10}$$

The following examples indicate some typical uses of K_{sp} data.

EXAMPLE 12.7 The presence of chloride in a water sample may be demonstrated by adding a few drops of the soluble salt AgNO$_3$. The Ag$^+$ in AgNO$_3$ combines with Cl$^-$ to form a milky-white suspension of AgCl. The AgCl eventually settles to the bottom of the sample and may be removed by filtration. What is the minimum amount of chloride that could be detected in this way in a 100-cm^3 sample at 25 °C by adding one drop of 0.2-M AgNO$_3$? Assume one drop is 0.05 cm^3.

SOLUTION One drop of 0.2-M AgNO$_3$ contains (0.2 mol·L^{-1})(0.001 L·cm^{-3})(0.05 cm^3) or 1 × 10^{-5} mol of Ag$^+$(aq). If this is added to a 100-cm^3 solution, the concentration of Ag$^+$(aq) is 1 × 10^{-4} M. Since the product of [Ag$^+$(aq)] and [Cl$^-$(aq)] cannot exceed the K_{sp} value of 1.56 × 10^{-10}, the maximum amount of Cl$^-$(aq) that can exist in equilibrium with 1 × 10^{-4}-M Ag$^+$(aq) is given by

$$[\text{Cl}^-(\text{aq})] = \frac{K_{sp}}{[\text{Ag}^+(\text{aq})]} = \frac{1.56 \times 10^{10}}{1 \times 10^{-4}} = 1.56 \times 10^{-6} \, M$$

This concentration of Cl$^-$(aq) is equivalent to (1.56 × 10^{-6} M)(35.5 g·mol^{-1}) or 5.5 × 10^{-5} g of Cl$^-$(aq) per liter of water. This may also be expressed as equivalent to 55 parts of Cl$^-$(aq) per billion parts of H$_2$O (by weight). Thus water would have to contain at least 55 ppb (parts per billion) of Cl$^-$(aq) in order to be detected by one drop of 0.2-M AgNO$_3$ in a 100-cm^3 sample. Virtually all our public drinking water and natural waters contain considerably more than 55 ppb of Cl$^-$(aq), as a simple AgNO$_3$ test quickly reveals.

TABLE 12.3 SOLUBILITY-PRODUCT CONSTANTS OF SOME SPARSELY SOLUBLE IONIC COMPOUNDS AT 25 °C

COMPOUND	K_{sp}
AgCl	1.56 × 10^{-10}
AgOH	1.5 × 10^{-8}
AgI	1.5 × 10^{-16}
BaF$_2$	1.73 × 10^{-6}
CaF$_2$	3.95 × 10^{-11}
Hg$_2$Cl$_2$	2 × 10^{-18}
Mg(OH)$_2$	1.2 × 10^{-11}
PbF$_2$	3.7 × 10^{-8}

EXAMPLE 12.8 Table 12.3 lists the K_{sp} of PbF$_2$ at 25 °C as 3.7 × 10^{-8}. What is the solubility (in g·L^{-1}) of PbF$_2$ at 25 °C?

SOLUTION In a saturated solution of PbF$_2$, there exists the equilibrium

$$\text{PbF}_2(s) \rightleftharpoons \text{Pb}^{2+}(\text{aq}) + 2\text{F}^-(\text{aq})$$

Let s be the solubility of PbF_2 in $mol \cdot L^{-1}$. At equilibrium

$$[Pb^{2+}(aq)] = s \quad \text{and} \quad [F^-(aq)] = 2s$$

The K_{sp} of PbF_2 is given by

$$K_{sp} = [Pb^{2+}(aq)][F^-(aq)]^2 = (s)(2s)^2 = 3.7 \times 10^{-8}$$

Solving for s leads to

$$s = \left(\frac{3.7 \times 10^{-8}}{4}\right)^{1/3} = 2.1 \times 10^{-3}\, M$$

Since a mole of PbF_2 weighs 245.21 g, this is about $0.5\, g \cdot L^{-1}$.

12.8 THE COMMON-ION EFFECT

The solubility of silver chloride in pure water at 25 °C is $1.79\, mg \cdot L^{-1}$. However, the solubility in water already containing silver or chloride ions from another source is less than the above value. Why this should be so is made evident by the application of Le Châtelier's principle to the equilibrium

$$AgCl(s) \rightleftharpoons Ag^+(aq) + Cl^-(aq)$$

Suppose it is desired to dissolve AgCl in a 0.1-M NaCl solution. The presence of $[Cl^-(aq)] = 0.1\, M$ due to the NaCl has the same effect as adding extra product to an equilibrium system, that is, to shift the equilibrium point to the left. This means that less AgCl will dissolve than if the $Cl^-(aq)$ from NaCl were absent. A similar situation results if the water contains extra $Ag^+(aq)$ due to a compound such as $AgNO_3$.

The above phenomenon is called the **common-ion effect**; the solubility of a sparsely soluble salt decreases if the solvent already contains an ion in common with the salt. Another way of looking at this phenomenon is to note that the product of $[Ag^+(aq)]$ and $[Cl^-(aq)]$ cannot exceed the K_{sp} value. Thus, if $[Cl^-(aq)]$ is increased by addition of $Cl^-(aq)$ from another source, $[Ag^+(aq)]$ must decrease correspondingly. The only way $[Ag^+(aq)]$ can decrease is by having less AgCl dissolve.*

EXAMPLE 12.9 Calculate the solubility (in $mg \cdot L^{-1}$) of AgCl in a 0.1-M NaCl solution at 25 °C.

SOLUTION Since $[Ag^+(aq)][Cl^-(aq)] = 1.56 \times 10^{-10}$ for any solution in equilibrium with AgCl(s) at 25 °C, it follows that the silver ion concentration cannot exceed

$$[Ag^+(aq)] = \frac{1.56 \times 10^{-10}}{[Cl^-(aq)]}$$

---

* However, if $[Cl^-]$ becomes too large, the soluble ion $AgCl_2^-$ forms and AgCl begins to dissolve according to the reaction $AgCl(s) + Cl^-(aq) \rightarrow AgCl_2^-(aq)$.

The value of [Cl⁻(aq)] is due to the 0.1-M NaCl plus that due to dissolved AgCl(s), but the latter is small enough to be neglected. Thus, if [Cl⁻(aq)] = 0.1 M, the maximum [Ag⁺(aq)] possible is

$$[\text{Ag}^+(\text{aq})] = \frac{1.56 \times 10^{-10}}{0.1} = 1.56 \times 10^{-9}\ M$$

This means that only 1.56×10^{-9} mol of AgCl will dissolve per liter of solution. Thus the solubility of AgCl(s) is

$$s = (1.56 \times 10^{-9}\ \text{mol}\cdot\text{L}^{-1})(143.34\ \text{g}\cdot\text{mol}^{-1})(1.00 \times 10^{3}\ \text{mg}\cdot\text{g}^{-1})$$
$$= 2.24 \times 10^{-4}\ \text{mg}\cdot\text{L}^{-1}$$

SUMMARY

The law of mass action states that the rate of a chemical reaction is proportional to the concentrations of reacting substances. Thus, as reactants become depleted during the course of a reaction, the rate of product formation decreases. At the same time, the rate of product breakup (back to reactants) increases as more and more product is formed. Eventually, the two opposing rates become equal and a state of chemical equilibrium results.

Chemical equilibrium is characterized on the macroscopic level by a constant composition of products and reactants. At the microscopic level, the rate at which reactants are forming products is the same as the rate at which products are forming reactants, so that the net rate of reaction is zero.

The mathematical quantity relating the concentrations of products and reactants at equilibrium is called the equilibrium constant. Ideally, the equilibrium constant of a reaction depends only upon temperature and not upon the concentrations of starting materials. Consequently, once an equilibrium constant is determined for a given reaction at a given temperature, its value can be used to predict equilibrium concentrations for any arbitrary amounts of starting materials at the same temperature.

Application of Le Châtelier's principle to chemical equilibria shows that the equilibrium constant of a reaction that is endothermic in the left-to-right direction increases as the temperature increases. Exothermic reactions behave in the opposite fashion—the equilibrium constant decreases as the temperature is increased. The principle also shows how removal of products leads to an increased conversion of reactants to products and how pressure may be used to affect the yield of a reversible reaction.

The principles of chemical equilibrium may be used to describe the dissolution of a sparsely soluble salt in a solvent. The equilibrium constant for such a process, the solubility-product constant, may be used to calculate solubilities and to design procedures for the separation of different substances. The principles of chemical equilibrium and Le Châtelier's principle also provide an explanation for the common-ion effect.

LEARNING GOALS

1. Given a balanced chemical reaction, be able to write its equilibrium constant expression in terms of molar concentrations or partial pressures.

2. Given the appropriate data on equilibrium concentrations of reactants and/or products, be able to calculate numerical values for equilibrium constants.

3. Given the numerical value of an equilibrium constant, be able to calculate equilibrium concentrations of reactants and products from given amounts of starting materials.

4. Be able to convert a K_c value to K_p, and vice versa.

5. Know how to apply Le Châtelier's principle to chemical equilibria with respect to effects of temperature, pressure, and changes in amounts of reactants or products.

6. Be able to write equilibrium constants for multiphase reactions.

7. Be able to write solubility-product constant expressions for the dissolution of sparsely soluble substances.

8. Given the solubility of a sparsely soluble substance, be able to calculate a numerical value for its K_{sp}.

9. Given the K_{sp} of a sparsely soluble substance, be able to predict its solubility.

10. Be able to calculate the effect a common ion has on the solubility of a sparsely soluble substance.

DEFINITIONS, TERMS, AND CONCEPTS TO KNOW

law of mass action
reversible reaction
chemical equilibrium

equilibrium constant
difference between K_c and K_p
solubility-product constant

common-ion effect

QUESTIONS AND PROBLEMS

K_c, K_p, and Equilibrium Concentrations

1. When 0.5 mol of A and 0.25 mol of B are mixed in a 2-L container at 500 °C, the reaction $A(g) + 3B(g) \rightleftharpoons C(g) + 2D(g)$ has $K_c = 10$. Calculate K_c for this reaction given that 1 mol of C, 0.17 mol of D, and 0.03 mol of A are mixed at 500 °C in a 10-L flask.

2. The value of K_c for the reaction $A(g) + B(g) \rightleftharpoons C(g)$ is 100 at 150 °C. Calculate the equilibrium concentration of C if 0.01 mol of A and 0.02 mol of B are mixed in a 2-L container at 150 °C.

3. Use the data in Table 12.1 to calculate the equilibrium constant for the ammonia reaction $N_2(g) + 3H_2(g) \rightleftharpoons 2NH_3(g)$ at 500 °C and 200 atm. Equation (12.9) is useful here. Repeat at 500 °C and 100 atm. Are the two values approximately the same? Explain. [*Note:* The value of 9.2×10^{-4} quoted in the text is based on activities—not on actual molar concentrations. At the high pressures (high concentrations) involved, activities must be used in order to calculate reliable values for equilibrium constants.]

4. In the reaction $2A(g) + B(g) \rightleftharpoons C(g) + D(g)$, 0.01 mol of A and 0.02 mol of B are mixed in a 1-L flask at 50 °C. It is found that $[C] = 0.001\ M$ at equilibrium.
   a) Calculate [A], [B], and [D] at equilibrium.
   b) Calculate K_c for the reaction at 50 °C.

5. In the reaction $2NO(g) + 2H_2(g) \rightleftharpoons N_2(g) + 2H_2O(g)$, 0.1 mol of NO, 0.05 mol of H_2, and 0.1

mol of H_2O are mixed in a 1-L flask. At equilibrium $[NO] = 0.007\ M$.
   a) Calculate $[H_2]$, $[N_2]$, and $[H_2O]$ at equilibrium.
   b) Calculate K_c.

6. The value of K_c for the reaction $H_2(g) + I_2(g) \rightleftharpoons 2HI(g)$ at 448 °C is 50.
   a) If 50 g of I_2 and 1.0 g of H_2 are mixed in a 25-L container and allowed to come to equilibrium, what is the equilibrium concentration of HI at 448 °C?
   b) What is K_p of this reaction at 448 °C?
   c) What is the partial pressure of each substance at equilibrium?
   d) What is the total pressure of the equilibrium mixture?

7. The value of K_c of the reaction $2A(g) + B(g) \rightleftharpoons C(g)$ at 100 °C is 1.5. Calculate K_p of the reaction.

8. Express the equilibrium constant of the reaction $aA + bB \rightleftharpoons lL + mM$ in terms of mole-fraction concentrations (call this K_x). Show how K_x is related to K_c and K_p. [*Note:* Recall Dalton's law $p_A = X_A p_{tot}$, etc. (where X refers to mole-fraction concentration).]

9. The value of K_c for the reaction $PCl_5(g) \rightleftharpoons PCl_3(g) + Cl_2(g)$ is 0.0417 at 250 °C. What is the equilibrium concentration of Cl_2 if 1 mole of PCl_5 is heated at 250 °C in a 10-L container?

10. Complete Example 12.3 by calculating the weights of N_2 and H_2 present at equilibrium.

11. Write the chemical reaction (in symbolic form) that Berthollet is referring to in the quotation at the beginning of this chapter. Also write the equilibrium constant expression for this reaction.

12. Write the K_c expression for the reaction
$$4HCl(g) + O_2(g) \rightleftharpoons 2Cl_2(g) + 2H_2O(g)$$

13. Write K_c expressions for the following reactions.
   a) $CH_3OH(aq) + CH_3CO_2H(aq) \rightleftharpoons CH_3CO_2CH_3(aq) + H_2O(l)$
   b) $2NO(g) + O_2(g) \rightleftharpoons 2NO_2(g)$
   c) $2SO_2(g) + O_2(g) \rightleftharpoons 2SO_3(g)$
   d) $4NH_3(g) + 5O_2(g) \rightleftharpoons 4NO(g) + 6H_2O(g)$
   e) $CH_2{=}CH_2(g) + H_2(g) \rightleftharpoons CH_3CH_3(g)$
   f) $CH_4(g) + 2H_2S(g) \rightleftharpoons CS_2(g) + 4H_2(g)$

14. One mole of N_2O_4 is placed in a 10-L flask at 55 °C. This substance dissociates according to the reaction
$$N_2O_4(g) \rightleftharpoons 2NO_2(g)$$
Analysis of the equilibrium mixture shows that 30% of the N_2O_4 is dissociated.
   a) Calculate $[N_2O_4]$ and $[NO_2]$ at equilibrium.
   b) Calculate K_c and K_p of the reaction at 55 °C.
   c) What is the density of the equilibrium mixture?
   d) What is the value of $[NO_2]$ if the volume of the flask is decreased to 2.5 L? How does this affect K_c and K_p?

15. If the equilibrium constant of a given reaction depends only on temperature, why does K_c (or K_p) change value with the way an equation is balanced or with the direction in which it is written?

Multiphase Equilibria

16. Write equilibrium constant expressions for each of the following chemical reactions.
   a) $3O_2(g) \rightleftharpoons 2O_3(g)$
   b) $HCl(g) + DBr(g) \rightleftharpoons DCl(g) + HBr(g)$
      $(D = {}_1^2H)$
   c) $BaF_2(s) \rightleftharpoons Ba^{2+}(aq) + 2F^-(aq)$
   d) $Fe_2O_3(s) + 6H_2(g) \rightleftharpoons 2Fe(s) + 3H_2O(g)$
   e) $4HCl(g) + O_2(g) \rightleftharpoons 2H_2O(l) + 2Cl_2(g)$

17. At high temperatures $MnCO_3$ decomposes as follows:
$$MnCO_3(s) \rightleftharpoons MnO(s) + CO_2(g)$$
At 298 K the equilibrium pressure of CO_2 is 4×10^{-11} atm.
   a) Calculate K_p and K_c for this reaction.
   b) What is K_c for the following reaction and what does this value signify?
$$MnO(s) + CO_2(g) \rightleftharpoons MnCO_3(s)$$

18. Write equilibrium constant expressions for the following reactions.
   a) $Pt(s) + Cl_2(g) \rightleftharpoons PtCl_2(g)$
   b) $Br_2(l) \rightleftharpoons Br_2(g)$
   c) $H_2O(s) \rightleftharpoons H_2O(g)$
   d) $Ag_2O(s) \rightleftharpoons 2Ag(s) + \frac{1}{2}O_2(g)$
   e) $2HgO(g) \rightleftharpoons 2Hg(l) + O_2(g)$

19. How would you measure the equilibrium constant for the following process, the vaporization of water at some given temperature?

$$H_2O(l) \rightleftharpoons H_2O(g)$$

Le Châtelier's Principle

20. Consider the reaction $2A(g) + B(g) \rightleftharpoons C(g)$. After equilibrium at 25 °C is reached, analysis shows $[A] = 0.2\ M$, $[B] = 1.0\ M$, and $[C] = 2.0\ M$.
  a) Write the general expression for K_c of the above reaction.
  b) Calculate K_c at 25 °C.
  c) What effect would each of the following have on the reaction: increase in total pressure, increase in temperature (assume the reaction is exothermic in the forward direction), removal of some B?

21. Consider the following reactions:

$$4HCl(g) + O_2(g) \rightleftharpoons 2H_2O(g) + 2Cl_2(g) \quad \text{(exothermic)}$$

$$PCl_5(g) \rightleftharpoons PCl_3(g) + Cl_2(g) \quad \text{(endothermic)}$$

State what effect each of the following would have on the amount of products formed at equilibrium.
  a) Decrease in total pressure.
  b) Removal of a reactant.
  c) Increase in temperature.
  d) Addition of a product.

22. Consider the following reactions:

$$PCl_5(g) \rightleftharpoons PCl_3(g) + Cl_2(g)$$

$$H_2(g) + I_2(g) \rightleftharpoons 2HI(g)$$

What effect would each of the following have upon the above equilibria?
  a) Increasing the total pressure by putting in helium gas (volume and temperature of the reaction held constant).
  b) Increasing the total pressure by decreasing the volume (at constant temperature).

23. The reaction $A(g) \rightleftharpoons 2B(g)$ is carried out at 1 atm and 300 K. At equilibrium $p_A = 0.46$ atm and $p_B = 0.54$ atm.
  a) Calculate K_p at 300 K.
  b) What are the partial pressures of A and B at equilibrium if the total pressure is increased to 2 atm?
  c) Calculate the mole fractions of A and B at equilibrium at 1 atm and 2 atm and compare. How do these results compare with the predictions of Le Châtelier's principle?

24. The value of K_p of the gas phase reaction $A + 2B \rightleftharpoons C + D$ is 1.1×10^{-1} at 100 °C and 2.0×10^2 at 1000 °C. Is the reaction endothermic or exothermic?

25. The equilibrium pressure of carbon dioxide in the reaction

$$CaCO_3(s) \rightleftharpoons CaO(s) + CO_2(g)$$

increases as the temperature is raised. Is the reaction endothermic or exothermic? Explain.

26. At elevated temperatures phosgene, $COCl_2$, dissociates according to the reaction

$$COCl_2(g) \rightleftharpoons CO(g) + Cl_2(g)$$

What effect would each of the following have on the equilibrium concentration of Cl_2?
  a) The addition of CO.
  b) The addition of He.
  c) Removal of some Cl_2.
  d) Tripling the volume of the container.

27. The reaction $Ag_2O(s) \rightleftharpoons 2Ag(s) + \frac{1}{2}O_2(g)$ is endothermic. What effect would each of the following have on the equilibrium concentration of O_2?
  a) The removal of some Ag_2O.
  b) The addition of some $Ag(s)$.
  c) The removal of some O_2.
  d) A decrease in temperature.
  e) A doubling of the volume of the reaction vessel.

28. What effect will an increase in pressure have on the concentration of each substance in each of the following reactions? Assume that the temperature is held constant.
  a) $N_2O_4(g) \rightleftharpoons 2NO_2(g)$
  b) $2H_2(g) + O_2(g) \rightleftharpoons 2H_2O(g)$
  c) $2H_2O_2(g) \rightleftharpoons 2H_2O(g) + O_2(g)$
  d) $2NOBr(g) \rightleftharpoons 2NO(g) + Br_2(g)$
  e) $CH_4(g) + Cl_2(g) \rightleftharpoons CH_3Cl(g) + HCl(g)$

29. A large block of ice floats in a tank of water, both maintained at 0 °C and thermally insulated from the surroundings. Describe what happens in each of the two cases below.
  a) A small piece of ice, with a temperature of −10 °C, is added to the tank.
  b) A small quantity of hot water is added to the tank.

30. As shown in Fig. 7.12, a photovoltaic cell con-

324 CHEMICAL EQUILIBRIUM

sists of a p-type semiconductor in contact with an n-type semiconductor. Electrons tend to flow spontaneously across the p-n junction in the n-to-p direction; holes move in the opposite direction. Eventually the following equilibria are established

$$e^-_{(n)} \rightleftharpoons e^-_{(p)} \quad o_{(p)} \rightleftharpoons o_{(n)}$$

where $e^-_{(n)}$ is an electron in the n-layer, $o_{(p)}$ is a hole in the p-layer, etc. When light shines on the p-n junction electrons flow from the n-layer to the p-layer via an external circuit (holes flow in the opposite direction). Is the flow of electrons across the junction (in the n-to-p direction) exothermic or endothermic?

K_{sp} and Solubility

31. The K_{sp} of PbI_2 is 7.5×10^{-9} at 15 °C and 1.4×10^{-8} at 25 °C.
  a) Write the equation for the dissolution of lead iodide in water.
  b) Write the K_{sp} expression for the above reaction.
  c) What is the solubility of lead iodide (in $g \cdot L^{-1}$) at 25 °C?
  d) If 1 L of saturated lead iodide solution at 25 °C is cooled to 15 °C, how much lead iodide settles out if no supersaturation occurs?
  e) Is the dissolution of lead iodide exothermic or endothermic? Explain.

32. The solubility of $CaSO_4$ in water at 25 °C is about $1.9 \text{ g} \cdot L^{-1}$. Calculate K_{sp} of $CaSO_4$ (the ions are Ca^{2+} and SO_4^{2-}).

33. The K_{sp}'s of AgCl and AgBr at 25 °C are 1.56×10^{-10} and 7.7×10^{-13}, respectively. How many drops of 0.2-M $AgNO_3$ must be added to a 100-L solution which is 0.001 M in NaCl and 0.001 M in NaBr in order to precipitate just one compound (AgCl or AgBr)? Which compound precipitates first? (A drop is about 0.05 cm³.)

34. What is the solubility at 25 °C (in $mg \cdot L^{-1}$) of CaF_2 in (a) pure water? (b) a 0.0001-M solution of KF? (c) a 0.0001-M solution of $CaCl_2$?

35. In which would barium sulfate, $BaSO_4$, have the higher solubility, in a 1.0-M KNO_3 solution or a 1.0-M K_2SO_4 solution? Explain.

36. Using Table 12.3, calculate the solubility of magnesium hydroxide in grams per 100 cm³ of solution.

37. The solubility of arsenic sulfide, As_2S_3, is $2.0 \times 10^{-4} \text{ g} \cdot L^{-1}$ at 25 °C. What is the K_{sp} of this substance?

38. Will a precipitate form when 10 cm³ of 0.01-M NaOH is added to 10 cm³ of 0.001-M $MgCl_2$? Explain.

39. A solution is originally 0.001 M in Ag_2CrO_4, a sparsely soluble substance with $K_{sp} = 1.0 \times 10^{-12}$ at 25 °C. Sufficient chromate ion is added to bring the chromate ion concentration to $3 \times 10^{-3} M$.
  a) What is the equilibrium concentration of Ag^+ in the solution?
  b) What fraction of the Ag^+ originally present has been removed from solution?

40. Use the facts that only Ca^{2+} has an insoluble carbonate, only Ni^{2+} has an insoluble sulfide, and only Ag^+ has an insoluble chloride to outline a systematic procedure for analyzing solutions which may contain one or more of the following ions: Ca^{2+}, Ni^{2+}, Ag^+.

ADDITIONAL READINGS

Bard, Allen J., *Chemical Equilibrium*. New York: Harper & Row, 1966. A comprehensive treatment of chemical equilibrium at an elementary level.

Donoghue, John T., *Equilibrium Calculations: Chemistry Problems*. Tarrytown-on-Hudson, N.Y.: Bogden and Quigley, 1971. Contains a variety of illustrative worked-out problems involving many aspects of equilibrium.

Rosenberg, Jerome L., *College Chemistry*, 5th ed. (part of Schaum's Outline Series). New York: McGraw-Hill, 1966. Reviews basic theories in all areas of chemistry and includes many solved equilibrium problems as examples.

13

ACIDS, BASES, AND THEIR REACTIONS

Well, Franta, it's like this: When you mix baking soda with vinegar, the acid atoms fight with the alkali atoms and a lot of smoke [fizz] is given off.
JINDŘICH PILAŘ *TO HIS SON, CA. 1938*

The concept of acids and bases, although a very old one, has changed markedly as the science of chemistry developed. In the 18th century, Antoine Lavoisier expressed all of acid–base theory in the succinct relationships

acid = radical + oxygen

base = metal + oxygen

salt = base + acid

One of the fundamental features of this so-called **duality theory** was that it required oxygen as an integral part of an acid. At that time CO_2 was called **carbonic acid**, since its water solutions behaved as an acid. Today, we call the union of CO_2 and H_2O by the same name and ascribe the acid nature to hydrogen rather than to oxygen. The compound we now write HCl was long thought to contain oxygen since its water solution was an acid (then called **muriatic acid**; modern name, **hydrochloric acid**). Today, there is still active interest in designing new acid–base definitions of ever-increasing generality; this chapter will discuss those most useful for everyday purposes.

13.1 THE BRØNSTED-LOWRY DEFINITION OF ACIDS AND BASES

The first modern definition of acids and bases was proposed in 1887 by the Swedish chemist Svante Arrhenius.* According to Arrhenius' definition, an **acid** is a substance which provides protons (hydrogen ions H^+) in water solution, and a **base** is a substance which furnishes hydroxide ions (OH^-) in water solution. This very restricted definition has now been almost completely supplanted by a much

◀ Vinegar (acetic acid) reacts with baking soda (sodium bicarbonate) to produce bubbles of carbon dioxide gas. (Courtesy of Marshall Henrichs.)

---

* For a biographical sketch of Arrhenius see Chapter 1.

Johannes N. Brønsted (1879–1947) was a Danish physical chemist noted for his thermodynamic investigations and his work on acids and bases. Some of his experimental measurements of solubility provided strong support for the Debye-Hückel theory (see Section 10.9) of interionic interaction. Thomas M. Lowry (1874-1936), a British chemist, was noted for his studies on optical rotation and on concepts of acids and bases.

broader definition proposed in 1923 by J. N. Brønsted in Denmark and T. M. Lowry in Great Britain, both working independently. The **Brønsted-Lowry** definition of an **acid** is any substance which is capable of donating a proton to another substance called a **base**. Conversely, a base is defined as any substance capable of accepting a proton from another substance called an acid. In a nutshell: acids are **proton donors** and bases are **proton acceptors**. The most essential advance of the Brønsted-Lowry definition over that of Arrhenius is that it recognizes the active role of the solvent and makes it possible to discuss acid–base reactions in nonaqueous solutions.

The general form of a Brønsted-Lowry acid–base reaction is

$$acid_1 + base_2 \rightleftharpoons base_1 + acid_2$$

where $acid_1$ and $base_1$ are said to constitute a **conjugate acid-base pair**, and similarly for $acid_2$ and $base_2$. The acid and base making up a conjugate pair are related as follows:

$$base_1 + H^+ = acid_1$$
$$base_2 + H^+ = acid_2$$

That is, the addition of a proton to a base produces the conjugate acid of that base, and, conversely, the loss of a proton from an acid produces the conjugate base of that acid. The proton lost by $acid_1$ when it forms its conjugate $base_1$ is transferred to $base_2$, which then forms its conjugate $acid_2$. The following are some specific examples of Brønsted-Lowry acid–base reactions.

1. $HCl + H_2O \rightleftharpoons H_3O^+ + Cl^-$
2. $NH_3 + H_2O \rightleftharpoons NH_4^+ + OH^-$
3. $NH_2^- + H_2O \rightleftharpoons NH_3 + OH^-$
4. $H_2O + H_2O \rightleftharpoons H_3O^+ + OH^-$

In the first example, HCl and Cl^- form a conjugate acid–base pair, as do H_3O^+ and H_2O. HCl is a Brønsted-Lowry acid which loses a proton to form the conjugate base Cl^-. Similarly, H_3O^+ is an acid which becomes the conjugate base H_2O by losing a proton. Certain substances can be either acids or bases, depending on the specific reaction in question. For example, NH_3 is a base in reaction 2 (since it accepts a proton from H_2O) and an acid in reaction 3 (since it donates a proton to

OH$^-$). Similarly, H$_2$O is an acid in reactions 2, 3, and 4 and a base in reactions 1 and 4. In reaction 4 (the self-ionization of water) H$_2$O functions as both an acid and a base.

The symbol H$_3$O$^+$ represents the **hydronium ion** (also called the **oxonium ion**); the symbol is not to be interpreted literally. At one time it was thought that bare protons, H$^+$, existed as such in solution and were the characteristic acid species. That the existence of H$^+$ in solution is unlikely is based on the fact that the process

$$H(g) \rightarrow H^+(g) + e^-$$

requires 13.6 eV—a very high energy. However, the hydration process

$$H^+(g) + xH_2O(l) \rightarrow H^+(aq)$$

is exothermic and liberates about 11.6 eV. Thus it is likely that H$^+$ exists as a hydrated ion—a proton embedded in a cluster of water molecules. Most likely, the cluster formation depends on electrostatic attraction between the positively charged proton and the negative end of the H$_2$O dipole. Just how many water molecules (the value of x in the above equation) are clustered about a proton on the average is not known, but it is probably more than just one, as the formula H$_3$O$^+$ implies. Some authorities suggest that H$^+$(aq) is (H$_2$O)$_4$H$^+$. Similarly, OH$^-$ are not found free in water; attractions between the negatively charged hydroxide ion and the positive end of the H$_2$O dipole produce clusters believed to be (H$_2$O)$_3$OH$^-$. In any event, it is common practice to symbolize these by H$_3$O$^+$ and OH$^-$ respectively.*
The noncomittal symbols H$^+$(aq) and OH$^-$(aq) are also used to denote the hydrated ions.

13.2 RELATIVE STRENGTHS OF ACIDS AND BASES

Acid A is said to be **stronger** than acid B if it can transfer a proton to some reference base, normally the solvent, more readily than can acid B. Similarly, the strength of a Brønsted-Lowry base refers to its relative ability to accept a proton from a reference acid, normally the solvent. If a large majority of the molecules in a sample of acid donate their protons to a reference base, the acid is said to be strong. Similarly, a strong base is one in which a large majority of the molecules accept protons from a reference acid. This means that if a Brønsted-Lowry acid is strong, its conjugate base must be weak, and vice versa.

In Table 13.1 are listed some important conjugate acid base pairs according to their relative strengths. This table is very useful in predicting the predominant direction of Brønsted-Lowry acid–base reactions. For example, the reaction

$$HCl + CN^- \rightleftharpoons HCN + Cl^-$$

(all in aqueous solution) is predicted to go predominantly to the right since HCl is

---

* According to one interpretation, the proton is visualized as traveling from one H$_2$O cluster to another, probably having a residence time of no longer than 10^{-13} s on any particular cluster.

TABLE 13.1 RELATIVE STRENGTHS OF SOME IMPORTANT BRØNSTED-LOWRY CONJUGATE ACID–BASE PAIRS

CONJUGATE ACID	CONJUGATE BASE
$HClO_4$, perchloric acid	ClO_4^-, perchlorate ion
H_2OAc^+, protonated acetic acid[a]	HOAc, acetic acid[a]
HCl, hydrochloric acid	Cl^-, chloride ion
H_2SO_4, sulfuric acid	HSO_4^-, hydrogen sulfate ion
HNO_3, nitric acid	NO_3^-, nitrate ion
H_3O^+, hydronium ion	H_2O, water
HSO_4^-, hydrogen sulfate ion	SO_4^{2-}, sulfate ion
HOAc, acetic acid[a]	OAc^-, acetate ion[a]
H_2CO_3, carbonic acid	HCO_3^-, hydrogen carbonate ion
H_2S, hydrogen sulfide[b]	HS^-, hydrogen sulfide ion
NH_4^+, ammonium ion	NH_3, ammonia
HCN, hydrogen cyanide[c]	CN^-, cyanide ion
HCO_3^-, hydrogen carbonate ion	CO_3^{2-}, carbonate ion
H_2O, water }[d]	OH^-, hydroxide ion
HS^-, hydrogen sulfide ion	S^{2-}, sulfide ion
NH_3, ammonia	NH_2^-, amide ion
H_2, hydrogen	H^-, hydride ion

← Increasing acid strength

Increasing base strength →

[a] The symbol Ac stands for the acetyl group

```
      H
      |
  H—C—C—
      |  ||
      H  O
```

so that HOAc is a convenient shorthand notation for acetic acid.
[b] Also called **hydrosulfuric acid**.
[c] Also called **hydrocyanic acid**.
[d] H_2O and HS^- are so close together that it is hard to tell which is stronger.

a much better proton donor than is HCN (HCl is nine places above HCN in Table 13.1). The same conclusion is reached by noting that CN^- is a much better proton acceptor than is Cl^-.

EXAMPLE 13.1 Predict the predominant direction of the reaction

$$H_2O + OAc^- \rightleftharpoons HOAc + OH^-$$

SOLUTION H_2O acts as a proton donor (acid) in the left-to-right reaction, and HOAc acts as a proton donor in the right-to-left reaction. Since HOAc is above H_2O in the acid column of Table 13.1, the right-to-left reaction is predominant. Also, it can be seen that any acid above and to the left of any base will tend to transfer H^+ to that base; hence, since HOAc is above OH^- and H_2O is below OAc^-, the stronger tendency is for HOAc to transfer H^+ to OH^-.

Again, from Table 13.1, it is seen that water, the most common solvent on earth, will react strongly with any acid stronger than H_3O^+ and with any base stronger than OH^-. This means that the strongest acid that can exist to any appreciable extent in water solution is H_3O^+; acids stronger than H_3O^+ will lose most of their protons to H_2O to form H_3O^+. For this reason, acids such as $HClO_4$, HCl, H_2SO_4, and HNO_3 will appear to be of equal strength in water solution, whereas HOAc will appear to be a weak acid. If glacial acetic acid (water-free acetic acid) is used as a solvent, $HClO_4$ will appear to be a strong acid and all the other acids will appear weak. This is because $HClO_4$ can react with HOAc to form appreciable amounts of H_2OAc^+, the strongest acid that can exist when HOAc is the solvent. The acids HCl, H_2SO_4, and HNO_3 furnish only a small fraction of their protons to HOAc and thus very little H_2OAc^+ is formed. This effect of a solvent to fail to differentiate the relative strengths of all acids stronger than its conjugate acid is known as the **solvent leveling effect**. Note that HOAc would appear to be a strong acid in liquid ammonia since it is a stronger acid than NH_4^+ (the strongest acid that can exist in liquid NH_3).

A simple analogy helps to make the solvent leveling effect more understandable. Suppose a small boy and his father wish to compare their relative strengths. Both attempt to do so by demonstrating their ability to lift a very small rock. Since both can do this with ease, the test is inconclusive—both appear to be of equal strength. But when both try to lift a 50-lb cinder block, the father lifts it easily whereas the boy can hardly move it. Similarly, both $HClO_4$ and HCl donate protons with ease to H_2O but only $HClO_4$ can donate protons readily to glacial HOAc.

13.3 ACID–BASE EQUILIBRIUM CONSTANTS

The relative strengths of acids and bases with respect to a given solvent may also be expressed in terms of the equilibrium constant for the reaction of the acid or base with the given solvent. All of the following examples use water as the solvent. The reaction of acetic acid with water is

$$HOAc + H_2O \rightleftharpoons H_3O^+ + OAc^-$$

The equilibrium constant for the reaction is

$$K_c = \frac{[H_3O^+][OAc^-]}{[HOAc][H_2O]} \tag{13.1}$$

For relatively dilute solutions, $[H_2O]$ is essentially equal to the preequilibrium value (very little water reacts) and may be treated as a constant. The $[H_2O]$ term then can be factored out and placed on the left-hand side with K_c:

$$K_c[H_2O] = \frac{[H_3O^+][OAc^-]}{[HOAc]} = K_a \tag{13.2}$$

Note that $K_c[H_2O]$ (itself the product of two constants) defines a new constant, K_a, called the **ionization constant** of the acid. For acetic acid at 25 °C, $K_a = 1.8 \times 10^{-5}$. All acids with K_a greater than 1 will be stronger than H_3O^+; acids weaker than H_3O^+ will have K_a less than 1. This means that K_a values for water solutions are expressed relative to the equilibrium

$$H_3O^+ + H_2O \rightleftharpoons H_2O + H_3O^+$$

The K_a of H_3O^+ in water solutions is obviously equal to unity. Acids such as $HClO_4$ and HCl have large K_a values and their reactions with H_2O are generally considered to go to completion.

The ionization constant for the reaction

$$H_2O + H_2O \rightleftharpoons H_3O^+ + OH^-$$

is given the special symbol K_w and is given by

$$K_w = K_c[H_2O]^2 = [H_3O^+][OH^-] \tag{13.3}$$

The ionization constant K_w is called the **ion-product constant** or the **self-ionization constant** of water. At 25 °C it has a value of about 1.0×10^{-14}.

The ionization constants of bases (symbol K_b) are defined analogously to K_a of acids. Thus, for ammonia reacting with H_2O,

$$NH_3 + H_2O \rightleftharpoons NH_4^+ + OH^-$$

the ionization constant of the base is

$$K_b = K_c[H_2O] = \frac{[NH_4^+][OH^-]}{[NH_3]} \tag{13.4}$$

The value of K_b for NH_3 at 25 °C is about 1.8×10^{-5} (that this is the same value as for K_a of acetic acid is merely coincidental). All bases weaker than OH^- will have K_b less than 1; bases stronger than OH^- will have K_b greater than 1. Thus K_b values for water solutions are relative to

$$OH^- + H_2O \rightleftharpoons H_2O + OH^-$$

The K_b of OH^- in water solution is obviously unity.

TABLE 13.2 IONIZATION CONSTANTS OF SOME ACIDS AND BASES IN H_2O SOLUTION AT 25 °C

ACID	K_a	BASE	K_b
HNO_3	22	S^{2-}	8.3
HSO_4^-	1.2×10^{-2}	CN^-	1.4×10^{-5}
HOAc	1.8×10^{-5}	NH_3	1.8×10^{-5}
H_2CO_3	4.3×10^{-7}	HS^-	1.8×10^{-7}
H_2S	5.6×10^{-8}	HCO_3^-	2.3×10^{-8}
NH_4^+	5.6×10^{-8}	OAc^-	5.5×10^{-10}
HCN	7.2×10^{-10}	H_2O	1.0×10^{-14}
H_2O	1.0×10^{-14}	NO_3^-	5.0×10^{-16}
HS^-	1.2×10^{-15}	Cl^-	≈ 0
NH_3	≈ 0		
H_2	≈ 0		

Table 13.2 lists the ionization constants of some important acids and bases at 25 °C. In many cases the ionization constants are listed for both members of a conjugate acid–base pair but, in general, either of these may be calculated from the other. As an example, K_b of the base CN^- is calculated from the K_a of its conjugate acid HCN as follows: The reaction of CN^- as a base with H_2O is

$$CN^- + H_2O \rightleftharpoons HCN + OH^-$$

and the corresponding ionization constant is

$$K_b = \frac{[HCN][OH^-]}{[CN^-]} \tag{13.5}$$

Multiplying numerator and denominator by $[H_3O^+]$ and rearranging leads to

$$K_b = \frac{[HCN][OH^-]}{[CN^-]} \cdot \frac{[H_3O^+]}{[H_3O^+]} = \frac{[HCN]}{[H_3O^+][CN^-]} \cdot [H_3O^+][OH^-]$$

$$= \frac{K_w}{K_a} = \frac{1.0 \times 10^{-14}}{7.2 \times 10^{-10}} = 1.4 \times 10^{-5} \tag{13.6}$$

Similarly, K_a of an acid may be calculated from the K_b of its conjugate base. The above treatment may be generalized to show that for any conjugate acid–base pair in water solution

$$K_a K_b = K_w \tag{13.7}$$

where K_a refers to the acid (HA) and K_b to its conjugate base (A^-). The same relationship holds for a base and its conjugate acid.

Equation (13.7) shows that in pure water

$$K_a(H_2O) = K_b(H_2O) = K_w \tag{13.8}$$

It follows that

$$\left. \begin{array}{l} K_a(H_3O^+) = \dfrac{K_w}{K_b(H_2O)} = 1 \\[2ex] K_b(OH^-) = \dfrac{K_w}{K_a(H_2O)} = 1 \end{array} \right\} \tag{13.9}$$

EXAMPLE 13.2 Calculate the equilibrium constant, K_c for the reaction of Example 13.1:

$$H_2O + OAc^- \rightleftharpoons HOAc + OH^-$$

SOLUTION The above reaction is the sum of the two reactions below:

$$2H_2O \rightleftharpoons H_3O^+ + OH^- \qquad K_w = 10^{-14}$$

$$H_3O^+ + OAc^- \rightleftharpoons HOAc + H_2O \qquad K_a^{-1} = 5.6 \times 10^4$$

13.3 | ACID-BASE EQUILIBRIUM CONSTANTS

Consequently, K_c of the total reaction is the product of the individual equilibrium constants, namely,

$$K_c = K_w K_a^{-1} = (10^{-14})(5.6 \times 10^4) = 5.6 \times 10^{-10}$$

Thus, since $K_c < 1$, the reaction goes predominantly to the left as shown in Example 13.1.

13.4 CALCULATIONS INVOLVING K_a AND K_b

If one out of every four HA molecules dissolved in water donates a proton to H_2O to form H_3O^+ and A^- at equilibrium, the acid HA is said to have a degree of dissociation of $\frac{1}{4}$ or 0.25. Similarly, a base B is said to have a degree of dissociation of 0.25 when, on the average, one out of four B molecules has accepted a proton from H_2O to form BH^+ and OH^- at equilibrium. For the general acid HA as an example, the dissociation reaction is

$$HA + H_2O \rightleftharpoons H_3O^+ + A^-$$

with the ionization constant

$$K_a = \frac{[H_3O^+][A^-]}{[HA]} \tag{13.10}$$

If c represents the moles of HA mixed with H_2O to form each liter of solution and α represents the degree of dissociation of HA at equilibrium, then the equilibrium concentrations may be expressed in terms of α as follows:

$[H_3O^+] = [A^-] = \alpha c$

$[HA] = c - \alpha c = c(1 - \alpha)$

If these expressions are substituted into Eq. (13.10), the result is

$$K_a = \frac{(\alpha c)(\alpha c)}{c(1 - \alpha)} = \frac{\alpha^2 c}{1 - \alpha} \tag{13.11}$$

This expression allows calculation of the degree of dissociation of HA from the ionization constant K_a and the molar concentration c. The calculation proceeds somewhat differently depending on whether HA is a strong acid or a weak acid.

If HA is a very strong acid ($K_a \gg 1$), then it is often a very good approximation to assume that the denominator in Eq. (13.11) is almost zero. Thus α is approximately equal to 1; that is, the acid is completely dissociated. Acids such as HCl and $HClO_4$ belong in this category; their K_a values are so large that they are not measurable with any reliability. If K_a of the acid is known (as for HNO_3, for example), Eq. (13.11) may be solved for α by use of the quadratic formula:

$$\alpha = \frac{-K_a \pm \sqrt{K_a^2 + 4K_a c}}{2c} \tag{13.12}$$

If HA is a very weak acid ($K_a \ll 1$), Eq. (13.12) may also be solved directly for α; however, a simpler, approximate solution is usually quite satisfactory. Weak acids characteristically have values of α much less than 1; thus the assumption is often made that $1 - \alpha$ in the denominator of Eq. (13.11) is approximately equal to 1 so that the equation simplifies to

$$K_a = \alpha^2 c \tag{13.13}$$

The solution is

$$\alpha = \sqrt{\frac{K_a}{c}} \tag{13.14}$$

[*Note:* It is unwise to memorize the above equation. Instead, the student should practice deriving it from the basic equilibrium equation, Eq. (13.10). With a little practice this can be done very quickly.] If Eq. (13.14) leads to an α value which is indeed small compared with 1 (say no greater than about 5% of 1) then the approximation $1 - \alpha \simeq 1$ is justified; otherwise it may be necessary to redo the calculation using the more accurate Eq. (13.12).

EXAMPLE 13.3 Calculate the degree of dissociation of 1.0-M acetic acid at 25 °C.

SOLUTION Since $K_a = 1.8 \times 10^{-5}$, the approximation $1 - \alpha \simeq 1$ is expected to be valid. Thus

$$\alpha = \sqrt{\frac{K_a}{c}} = \sqrt{\frac{1.8 \times 10^{-5}}{1.0}} = 0.004 \quad (0.4\% \text{ dissociated})$$

Since 0.004 is much smaller than 1, the approximation $1 - \alpha \simeq 1$ is valid. Lingering doubts may be allayed by using Eq. (13.12) instead.

EXAMPLE 13.4 Calculate the degree of dissociation of 0.001-M acetic acid at 25 °C.

SOLUTION The problem proceeds as in Example 13.3, but c is now 0.001 M instead of 1.0 M. Assuming $\alpha \ll 1$ leads to

$$\alpha = \sqrt{\frac{K_a}{c}} = \sqrt{\frac{1.8 \times 10^{-5}}{0.001}} = 0.13 \quad (13\% \text{ dissociated})$$

Since α is 13% of 1, the calculation should be repeated using the quadratic formula, Eq. (13.12). This gives $\alpha = 0.125$, which rounds off to 0.13, the same value obtained by the simplified treatment.

The two examples above illustrate the fact that the degree of dissociation increases as the solution becomes more dilute. This general result could also have been predicted on the basis of Le Châtelier's principle by considering the addition of water (dilution) as a stress imposed upon an equilibrium system.

A completely analogous procedure is used to calculate the degree of dissociation of bases.

13.5 THE pH SCALE

Many chemical reactions which occur in aqueous solution, including many reactions occurring in living systems, depend on the concentration of H_3O^+ present in the solution. In 1909 Søren Sørensen proposed a scale of measurement of $[H_3O^+]$ called the **pH scale**. This scale, now in extensive use by chemists, biochemists, biologists, and medical technologists, defines the pH of an aqueous solution as

$$pH = -\log[H_3O^+]_{tot} \tag{13.15}$$

In pure water at 25 °C, $[H_3O^+]_{tot} = [OH^-]_{tot} = 10^{-7}\ M$ [see Eq. (13.3)] so that the pH is $-\log 10^{-7}$ or 7. This point on the pH scale defines neutrality at 25 °C. As the temperature increases, K_w for water increases slightly (it is 1.0×10^{-14} at 25 °C and 1.0×10^{-13} at 60 °C) so that neutrality shifts to a lower pH; at 60 °C neutrality occurs at a pH of 6.5. For most applications the effect of temperature on the pH scale is ignored.

Whenever $[H_3O^+]_{tot} > [OH^-]_{tot}$ (acid solution), then pH < 7, and whenever $[OH^-]_{tot} > [H_3O^+]_{tot}$ (basic solution), then pH > 7. Mathematically, pH may have any value between $-\infty$ and $+\infty$; practically, the scale is seldom used outside the range of approximately 1 to 13.

Since the pH scale is logarithmic, care must be exercised in interpreting the relative acidities of two solutions of different pH; if two solutions differ in pH by one unit, the H_3O^+ concentrations differ by a factor of 10. Also, *low pH* implies *high acidity* and vice versa.

Calculations of pH are carried out differently depending on the source of H_3O^+ and the concentration of the acid or base solute. In general, the total H_3O^+ concentration may be expressed as a sum of two terms—the H_3O^+ arising from the self-ionization of water and the H_3O^+ arising from the acid solute. This may be written

$$[H_3O^+]_{tot} = [H_3O^+]_{H_2O} + [H_3O^+]_{solute}$$

In the most commonly encountered situation, the H_3O^+ arising from the solute is so much greater than that arising from self-ionization of H_2O that the latter may be neglected. In such a case the following approximation is made:

$$[H_3O^+]_{tot} \simeq [H_3O^+]_{solute}$$

The term $[H_3O^+]_{solute}$ may be calculated from the ionization constant and concentration of the solute. For a general acid HA the ionization constant is given by Eq. (13.10). Letting x represent the concentration of $[H_3O^+]$ (the subscript "solute" may

---

Søren Peter Lauritz Sørensen (1868–1939) was a Danish biochemist. After service as consulting chemist to the Royal Danish Navy, Sørensen became Director of the renowned Carlsberg Laboratory and ultimately served as Chairman of the Royal Danish Academy. Sørensen made extensive studies of proteins and the nature of enzyme reactions. He is best known, however, for his invention of the pH scale. The original significance of pH was "potential of hydrogen."

be dropped for convenience) leads to $[H_3O^+] = [A^-] = x$ and $[HA] = c - x$, where c is the molar concentration of HA. Thus Eq. (13.10) may be written

$$K_a = \frac{x^2}{c - x} \tag{13.16}$$

If HA is a strong acid ($K_a \gg 1$) then $c - x$ is approximately zero; that is, c is approximately equal to x. In this case the pH is given by

$$\text{pH} = -\log c \tag{13.17}$$

If HA is a weak acid ($K_a \ll 1$) then x is *much* smaller than c, and $c - x$ in the denominator may be replaced by c. Thus

$$K_a = \frac{x^2}{c}, \quad x = \sqrt{K_a c}, \quad \text{and} \quad \text{pH} = -\log \sqrt{K_a c} \tag{13.18}$$

In case x (that is, $[H_3O^+]$ is *not* much smaller than c, Eq. (13.16) must be solved exactly for x. Use of the quadratic formula gives

$$x = \frac{-K_a \pm \sqrt{K_a^2 + K_a c}}{2} \tag{13.19}$$

The significance of these various approaches should become much clearer upon study of the following examples. Students should *not* memorize Eqs. (13.17) and (13.18), the two key equations for pH calculations. Rather, they should be able to start with Eq. (13.10) and develop Eqs. (13.17) and (13.18) as needed. With a little practice this can be done amazingly quickly.

EXAMPLE 13.5 Calculate the pH of a 0.5-M solution of HCl in water at 25 °C.

SOLUTION Since HCl is a strong acid in water solution ($K_a \gg 1$), it is essentially all in the form of H_3O^+ and Cl^-. Thus $[H_3O^+] = c = 0.5\ M$ and Eq. (13.17) applies. The pH is given by

$$\text{pH} = -\log 0.5 = -(9.699 - 10) = 0.311$$

Since it is seldom meaningful to express pH beyond tenths of units, the above should be rounded off to 0.3.

Note that if the above solution is diluted by a factor of 10, then $[H_3O^+] = 0.05\ M$ and the pH is increased by one unit:

$$\text{pH} = -\log 0.05 = -(8.699 - 10) = 1.3$$

EXAMPLE 13.6 Calculate the pH of 1.0-M acetic acid at 25 °C.

SOLUTION
$$\text{HOAc} + H_2O \rightleftharpoons H_3O^+ + \text{OAc}^-$$

$$K_a = \frac{[H_3O^+][\text{OAc}^-]}{[\text{HOAc}]} = 1.8 \times 10^{-5} \quad \text{at 25 °C}$$

Since acetic acid is a weak acid, Eq. (13.18) must be used to calculate the pH (the student should go through the derivation of this equation before using it):

$$x = [H_3O^+] = \sqrt{K_a c} = \sqrt{1.8 \times 10^{-5} \times 1.0} = 4.24 \times 10^{-3}\ M$$

$$\text{pH} = -\log \sqrt{K_a c} = -\log 4.24 \times 10^{-3} = -0.63 + 3.0 = 2.4$$

Note that a 1.0-M solution of a strong acid would have a pH of 0.0.

If the acetic acid solution is diluted by a factor of 10, then $c = 0.1\ M$, $[H_3O^+] = \sqrt{1.8 \times 10^{-5} \times 0.1} = 1.4 \times 10^{-3}\ M$, and pH $= -\log 1.4 \times 10^{-3} = -0.146 + 3.0 = 2.9$. This is a change of only 0.5 pH unit; dilution of a strong acid by the same factor would have increased the pH by one unit.

EXAMPLE 13.7 Calculate the pH of a 1.2-M solution of NaOH in water at 25 °C.

SOLUTION NaOH in water is assumed to consist entirely of the ions Na$^+$ and OH$^-$. If the [OH$^-$] due to self-ionization of water is neglected, then [OH$^-$] is 1.2 M. Since the ion-product $[H_3O^+][OH^-]$ of *any* water solution must equal 1.0×10^{-14} at 25 °C, $[H_3O^+]$ may be calculated as

$$[H_3O^+] = \frac{1.0 \times 10^{-14}}{1.2} = 8.3 \times 10^{-15}\ M$$

The pH is then given by

$$\text{pH} = -\log[H_3O^+] = -\log 8.3 \times 10^{-15} = -0.916 + 15 = 14.1$$

Problems that require the calculation of [OH$^-$] followed by a calculation of $[H_3O^+]$ may be simplified by defining the quantity pOH $= -\log[OH^-]_{tot}$. This makes it possible to write Eq. (13.3) as pH + pOH = 14. Thus Example 13.7 is treated as follows:

$$\text{pOH} = -\log[OH^-] = -\log 1.2 = -0.079$$

$$\text{pH} = 14 - \text{pOH} = 14 - (-0.079) = 14.1$$

EXAMPLE 13.8 Calculate the pH of a 0.01-M solution of NH$_3$ in water at 25 °C.

SOLUTION Since NH$_3$ reacts with H$_2$O as follows:

$$NH_3 + H_2O \rightleftharpoons NH_4^+ + OH^-$$

then $[NH_4^+] = [OH^-]$. Letting $x = [OH^-]$ and c the concentration of NH$_3$ before dissociation leads to $[NH_3] = c - x$. Thus

$$K_b = \frac{[NH_4^+][OH^-]}{[NH_3]} = \frac{x^2}{c - x} = 1.8 \times 10^{-5}$$

If $x \ll c$, the above equation may be approximated by

$$\frac{x^2}{c} = 1.8 \times 10^{-5} \quad \text{or} \quad x = [OH^-] = \sqrt{1.8 \times 10^{-5} c}$$

For $c = 0.01\ M$ this gives $[OH^-] = 4.24 \times 10^{-4}\ M$. Thus

$$pOH = -\log 4.24 \times 10^{-4}\ M = -0.627 + 4 = 3.4$$
$$pH = 14 - pOH = 14 - 3.4 = 10.6$$

Note that a 0.01-M solution of a strong base would have a pOH of 2 and a pH of 12.

EXAMPLE 13.9 Calculate the pH of a 0.001-M solution of $NaNH_2$ in water at 25 °C.

SOLUTION $NaNH_2$ is an ionic substance consisting of the ions Na^+ and NH_2^-. Table 13.1 shows that NH_2^- is a stronger base than OH^-; thus K_b of NH_2^- is greater than 1. The reaction

$$NH_2^- + H_2O \rightleftharpoons NH_3 + OH^-$$

is assumed to go essentially to completion. Thus $[OH^-] = 0.001\ M$ and

$$pOH = -\log 0.001 = 3.0$$
$$pH = 14 - pOH = 14 - 3.0 = 11.0$$

EXAMPLE 13.10 Calculate the pH of a 0.01-M solution of sodium acetate in water at 25 °C.

SOLUTION Sodium acetate, NaOAc, is an ionic compound consisting of Na^+ and OAc^-. The ion OAc^- is the conjugate base of acetic acid HOAc. Its base reaction with H_2O is

$$OAc^- + H_2O \rightleftharpoons HOAc + OH^-$$

with

$$K_b = \frac{[HOAc][OH^-]}{[OAc^-]}$$

If K_a of HOAc is taken as 1.8×10^{-5}, K_b of the conjugate base OAc^- is calculated as

$$K_b(OAc^-) = \frac{K_w}{K_a(HOAc)} = \frac{10^{-14}}{1.8 \times 10^{-5}} = 5.5 \times 10^{-10}$$

Since this shows that OAc^- is a very weak base, it is permissible to use the approximation $K_b = x^2/c$, where $x = [HOAc] = [OH^-]$. Thus

$$x = [OH^-] = \sqrt{K_b c} = \sqrt{5.5 \times 10^{-10} \times 0.01} = 2.35 \times 10^{-6}\ M$$
$$pOH = -\log 2.35 \times 10^{-6} = -0.37 + 6 = 5.6$$
$$pH = 14 - pOH = 14 - 5.6 = 8.4$$

In extremely dilute solutions it may not be permissible to ignore the $[H_3O^+]$ arising from self-ionization of water. Thus the pH of a 10^{-8}-M HCl solution is not 8.0, as a blind application of Eq. (13.17) implies. A simple solution to this problem proceeds as follows: The total H_3O^+ concentration must satisfy the two relationships

$$[H_3O^+]_{tot} = \frac{K_w}{[OH^-]_{tot}}$$

$$[H_3O^+]_{tot} = [OH^-]_{tot} + [Cl^-]$$

The second relationship simply states that the number of positive ions in the solution must equal the number of negative ions. If it is assumed that $[Cl^-] = c$ (the molar concentration of HCl), then each equation may be solved for $[OH^-]_{tot}$ to yield

$$[OH^-]_{tot} = \frac{K_w}{[H_3O^+]_{tot}}$$

$$[OH^-]_{tot} = [H_3O^+]_{tot} - c$$

Setting the right-hand sides of both equations equal to each other and solving for $[H_3O^+]_{tot}$ produces

$$[H_3O^+]_{tot} = \frac{c + \sqrt{c^2 + 4K_w}}{2} \tag{13.20}$$

It is clear that only when c^2 is much larger than $4K_w$ can the H_3O^+ coming from self-ionization of water be ignored. In such a case, $c^2 + 4K_w \simeq c^2$, and $[H_3O^+]_{tot}$ becomes equal to c just as in Eq. (13.17). It is left to the student to show that the pH of the 10^{-8}-M HCl solution is 6.98—just slightly acid. Common sense tells one as much.

Chemists, biochemists, biologists, and medical technologists frequently need to determine the pH of a solution. To speed their work, especially when there is need to measure pH values of large numbers of samples on a routine basis, there are instruments known as **pH meters** which measure pH very rapidly; the basis of their operation is discussed in Chapter 19.

13.6 BUFFER SOLUTIONS

A solution of a weak acid in which the amount of conjugate base has been augmented has the interesting and important property of resisting changes in pH due to the addition of small amounts of strong acids or strong bases. A solution of a weak base augmented in its conjugate acid behaves in a similar way. Such solutions are called **buffer solutions** or simply **buffers** (a buffer is anything serving to deaden a shock or bear the brunt of a collision). An important example of a buffer solution is the blood of animals. The blood plasma of humans must be maintained at a pH of 7.4 ± 0.1 or death may result. Other parts of the blood, for example, the erythrocytes (red blood cells) have a pH of about 7.2 to 7.3. Since there are many situations such as illnesses and other disorders which can dump considerable amounts of acidic and basic materials into the blood stream, the blood must have some mechanism for keeping the pH of the plasma within 0.1 pH unit of 7.4. Although the blood buffer system is complex and involves more than one buffer, the basic principles can be understood reasonably well by examining the details of a single, simple buffer solution.*

Consider, for example, the HOAc/OAc⁻ buffer system—a solution of acetic acid containing extra conjugate base, acetate ion. The extra acetate ion is added in the form of an ionic salt such as sodium acetate, NaOAc. The basic equilibrium involved between HOAc and its conjugate base is the familiar Brønsted-Lowry equilibrium

$$HOAc + H_2O \rightleftharpoons H_3O^+ + OAc^-$$

The basic principle of buffer action resides in the simple fact that the relatively large amounts of HOAc and OAc⁻ are able to shuffle hydrogen ions between themselves and thus to keep [H_3O^+] from varying greatly. For example, suppose a strong acid is dumped into the buffer solution; that is, excess H_3O^+ is added. The conjugate base OAc⁻ tends to remove this H_3O^+ by converting it to HOAc by the process

$$OAc^- + H_3O^+ \rightarrow HOAc + H_2O$$

This is a simple consequence of Le Châtelier's principle; OAc⁻ absorbs the shock of added H_3O^+ by converting it to HOAc. If excess strong base such as OH⁻ is added, it is neutralized by H_3O^+ to form water, that is

$$H_3O^+ + OH^- \rightarrow 2H_2O$$

However, HOAc acts as a storehouse of H_3O^+ and dissociates to replace the lost H_3O^+ by the process

$$HOAc + H_2O \rightarrow H_3O^+ + OAc^-$$

This, also, is a simple consequence of Le Châtelier's principle; HOAc absorbs the shock of lost H_3O^+ by resupplying part of it.

Calculation of the pH of a buffer solution is quite straightforward. The equilibrium constant of the buffering system is just K_a (the ionization constant of the weak acid), but now [H_3O^+] is no longer equal to [OAc⁻] since extra OAc⁻ has been added. Thus, solving the K_a expression for [H_3O^+] yields

$$[H_3O^+] = K_a \frac{[HOAc]}{[OAc^-]} \tag{13.21}$$

Since both [HOAc] and [OAc⁻] are based on the same volume of solution, the ratio of their concentrations may be replaced with a ratio of their amounts in moles. If n_a represents the moles of acid HOAc at equilibrium and n_b the moles of conjugate base OAc⁻ at equilibrium, Eq. (13.21) becomes

$$[H_3O^+] = K_a \frac{n_a}{n_b} \tag{13.22}$$

---

* Buffer systems present in the blood include CO_3^{2-}/HCO_3^- (carbonate/hydrogen carbonate system), $HPO_4^{2-}/H_2PO_4^-$ (monohydrogen phosphate/dihydrogen phosphate system) along with proteins and various organic acids.

This very important equation shows that the H$_3$O$^+$ concentration of the buffer (and thus its pH) depends on the molar ratio of the weak acid to its conjugate base. Biochemists frequently use a logarithmic form of this equation,

$$\text{pH} = \text{p}K_a + \log \frac{n_b}{n_a} \tag{13.23}$$

(where p$K_a = -\log K_a$) which they call the **Henderson-Hasselbalch equation**.

To illustrate the use of Eq. (13.23), consider an HOAc/OAc$^-$ buffer solution in which [HOAc] = 0.3 M and [OAc$^-$] = 0.2 M. If p$K_a = -\log 1.8 \times 10^{-5} = 4.74$, the pH of the buffer solution is

$$\text{pH} = 4.74 + \log \frac{0.2}{0.3} = 4.74 - 0.18 = 4.56.$$

Now suppose x moles of strong acid is added to 100 cm^3 of this buffer solution. What will be the effect on the pH? Let n_a^0 and n_b^0 represent the moles of HOAc and OAc$^-$ originally present in the 100-cm^3 sample; these values are $n_a^0 = 0.3$ mol·L^{-1} × 0.1 L or 0.03 mol and $n_b^0 = 0.2$ mol·L^{-1} × 0.1 L or 0.02 mol. Using Le Châtelier's principle along with the buffer equilibrium equation shows that the effect of the strong acid (assumed to be all H$_3$O$^+$) is to react with OAc$^-$ to produce HOAc. Assuming that the added H$_3$O$^+$ is removed completely (which is not exactly true), the moles of HOAc and OAc$^-$ present after the new equilibrium is reached are given approximately by

$$n_a = n_a^0 + x \quad \text{and} \quad n_b = n_b^0 - x$$

Thus the resultant pH is given by

$$\text{pH} = 4.74 + \log \frac{n_b^0 - x}{n_a^0 + x} \tag{13.24}$$

The same reasoning shows that addition of y moles of strong base to 100 cm^3 of buffer solution neutralizes y moles of H$_3$O$^+$ and thus uses up y moles of HOAc to form y additional moles of OAc$^-$. The resulting pH is given by

$$\text{pH} = 4.74 + \log \frac{n_b^0 + y}{n_a^0 - y} \tag{13.25}$$

Note that if x (or y) is small compared with n_a^0 and n_b^0, then the ratio

$$(n_b^0 - x)/(n_a^0 + x) \quad [\text{or } (n_b^0 + y)/(n_a^0 - y)]$$

remains almost constant so that the pH also remains almost constant. A numerical example illustrates this more dramatically. Suppose 10 cm^3 of 0.1-M HCl is added to 100 cm^3 of the buffer. This means that $x = 0.1$ mol·L^{-1} × 0.01 L or 0.001 mol. The pH is given by

$$\text{pH} = 4.74 + \log \frac{0.020 - 0.001}{0.030 + 0.001} = 4.74 + \log \frac{0.019}{0.031} = 4.74 - 0.21 = 4.53$$

This is a change in pH of only 0.03 units (from 4.56 to 4.53). By contrast, it is instructive to see what would happen to 100 cm³ of an *unbuffered solution* of pH = 4.56 when 10 cm³ of 0.1-M HCl is added. This solution would have the H_3O^+ concentration

$$[H_3O^+] = \frac{(2.8 \times 10^{-5} + 0.001) \text{ mol } H_3O^+}{0.110 \text{ L}} = 0.01 \ M$$

The pH of this solution is 2; *this is a decrease of almost 2.6 pH units!*

A similar calculation shows that adding 10 cm³ of 0.1-M NaOH would increase the pH by 0.3 units from 4.56 to 4.59; the same amount of strong base added to an unbuffered solution of pH = 4.56 would increase the pH by 2.6 units.

An example of a buffer system made up of a weak base and its conjugate acid is the NH_3/NH_4^+ system. The basic equilibrium is

$$NH_3 + H_2O \rightleftharpoons NH_4^+ + OH^-$$

Here NH_3 and NH_4^+ use up or resupply OH^- as needed. The pH of such a solution is given by

$$pH = 14 - pOH$$

where

$$pOH = pK_b + \log \frac{n_a}{n_b}$$

An equimolar mixture of NH_3 and NH_4^+ (from an ionic compound such as NH_4Cl) would have

$$pOH = pK_b = -\log(1.8 \times 10^{-5}) = 4.74$$
$$pH = 14 - 4.74 = 9.36$$

The pH of the world's oceans is maintained at about 8.15. There are probably two mechanisms by which this pH is controlled; one mechanism provides for slow response and the other for a rapid one. The slow-response system appears to utilize Na^+ and H_3O^+ ions adsorbed on the surface of claylike bottom sediments. Large increases of $[H_3O^+]$ in seawater appears to cause H_2O^+ from water to replace surface Na^+ and thereby keep the water pH unchanged. The fast-response system utilizes the HCO_3^-/CO_3^{2-} buffer. Excess acid is eliminated by the reaction

$$H_3O^+ + CO_3^{2-} \rightarrow HCO_3^- + H_2O$$

and excess base is eliminated by

$$HCO_3^- + OH^- \rightarrow CO_3^{2-} + H_2O$$

This buffer system is also operative in the blood of mammals and other animals.

13.7 MULTIPLE ACID–BASE EQUILIBRIA

Some Brønsted-Lowry acids are capable of donating more than one proton to the solvent. Such acids are said to be **polyprotic**. For example, sulfuric acid (H_2SO_4) reacts with H_2O as follows:

$$H_2SO_4 + H_2O \rightleftharpoons H_3O^+ + HSO_4^-$$

This reaction goes essentially to completion; that is, H_2SO_4 is a strong acid. However, the conjugate base HSO_4^- can also function as an acid. The reaction is

$$HSO_4^- + H_2O \rightleftharpoons H_3O^+ + SO_4^{2-}$$

The K_a value of the HSO_4^- reaction is 1.2×10^{-2} at 25 °C, showing that HSO_4^- (hydrogen sulfate ion) is a weak acid. A solution of H_2SO_4 in water will contain a mixture of H_3O^+, HSO_4^-, and SO_4^{2-} (very little H_2SO_4 exists). Similarly, phosphoric acid (H_3PO_4) is associated with three successive ionizations, all involving weak acids (H_3PO_4, $H_2PO_4^-$, and HPO_4^{2-}). The respective ionization constants at 25 °C are: $K_{a1} = 7.5 \times 10^{-3}$, $K_{a2} = 6.2 \times 10^{-8}$, and $K_{a3} = 1.0 \times 10^{-12}$. Mathematical treatment of such equilibrium systems becomes algebraically cumbersome and will not be dealt with here.*

13.8 ACID–BASE TITRATIONS

Chemists, biochemists, and biologists frequently need to determine the total acid or base content of samples of material. One simple method of acid–base analysis is known as **titration**; an acid of accurately known concentration is added to a sample containing an unknown amount of base (or vice versa) until the two are present in **equivalent** amounts. Then, from a knowledge of the amount of acid added, the amount of base present may be calculated.

To understand what it means to say that an acid and base are present in equivalent amounts, recall that acids are proton donors and bases are proton acceptors. Thus, if a given amount of acid HA is capable of donating x mol of protons, then that amount of base B capable of accepting x mol of protons is said to be equivalent to the acid; it is also said that the respective amounts of the acid and base are equivalent.

The description of acid–base reactions is facilitated by introducing a special concentration term called the **normality** (symbol N). *The normality of an acid or base solution is defined as the moles of replaceable or acceptable protons per liter of solution.* That weight of acid or base containing one mole of replaceable or acceptable protons is often called an *equivalent weight* or, simply, an *equivalent*. Most acids and bases (for example HCl and NH_3) can donate or accept just one mole of protons per mole of acid or base, so that the normality and molar concentration are numerically identical. In substances such as H_2SO_4 and $Ba(OH)_2$, however, each mole can provide or accept *two* moles of protons, so that the normality is twice the molar concentration. In general, normality and molarity are related by

normality (N) = molarity (M) × (moles of protons donated or accepted per mole)

---

* The textbook by Bruce H. Mahan, *University Chemistry*, 3rd ed. (Addison-Wesley, 1975), illustrates some [H_3O^+] calculations for multiple ionizations.

For example, a 1.0-M H_2SO_4 solution is 2.0 N (if used in a reaction in which both protons are given up), but a 1.0-M HCl solution is 1.0 N.

When solutions of an acid HA and a base B react, the process may be described as follows: Before mixing acid and base, each solution is described by the equilibria

$$HA + H_2O \rightleftharpoons H_3O^+ + A^-$$
$$B + H_2O \rightleftharpoons BH^+ + OH^-$$

The chemical reaction between acid and base is the reaction between H_3O^+ (the conjugate acid of water) and OH^- (the conjugate base of water), that is,

$$H_3O^+ + OH^- \rightarrow 2H_2O$$

This is called the **neutralization** reaction; the acid and base are said to **neutralize** each other. The ion pair BH^+ and A^- represents a compound called a **salt** of the acid HA and the base B. The net reaction between the acid HA and the base B is simply

$$HA + B \rightarrow BH^+ + A^-$$

If the base B is a soluble hydroxide, for example, NaOH (or Na^+, OH^-), the net reaction may be written

$$Na^+ + OH^- + HA \rightarrow H_2O + Na^+ + A^-$$

or, if all substances appearing on both sides are cancelled,

$$OH^- + HA \rightarrow H_2O + A^-$$

Several specific examples of acid–base neutralizations are listed below.

1. HCl and NH_3:

    $$HCl + H_2O \rightleftharpoons H_3O^+ + Cl^-$$
    $$NH_3 + H_2O \rightleftharpoons NH_4^+ + OH^-$$

    Neutralization is

    $$H_3O^+ + OH^- \rightarrow 2H_2O$$

    The salt formed is NH_4Cl (ammonium chloride, an ionic compound). The net reaction is

    $$HCl + NH_3 \rightarrow NH_4Cl$$

2. HOAc and NaOH:

    $$HOAc + H_2O \rightleftharpoons H_3O^+ + OAc^-$$

    Neutralization is

    $$H_3O^+ + OH^- \rightarrow 2H_2O$$

The salt formed is NaOAc (sodium acetate, an ionic compound). The net reaction is

$$HOAc + OH^- \rightarrow H_2O + OAc^-$$

3. H_2SO_4 and NaOH:

$$H_2SO_4 + H_2O \rightleftharpoons H_3O^+ + HSO_4^-$$
$$HSO_4^- + H_2O \rightleftharpoons H_3O^+ + SO_4^{2-}$$

Neutralization is

$$H_3O^+ + OH^- \rightarrow 2H_2O$$

Two different salts can be isolated by evaporating the solution to dryness: $NaHSO_4$ (sodium hydrogen sulfate) and Na_2SO_4 (sodium sulfate). Both are ionic salts. The net reactions are

$$H_2SO_4 + OH^- \rightarrow H_2O + HSO_4^-$$
$$H_2SO_4 + 2OH^- \rightarrow 2H_2O + SO_4^{2-}$$

If n_A represents the moles of replaceable protons of the acid and V_A the volume of the acid solution in liters, then the normal concentration (normality) of the solution is given by

$$N_A = \frac{n_A}{V_A} \tag{13.26}$$

Similarly, the normality of a base solution is given by

$$N_B = \frac{n_B}{V_B} \tag{13.27}$$

where n_B is the moles of protons the base solution can accept and V_B is the volume of the base solution in liters. The acid and the base are neutralized when $n_A = n_B$; their respective amounts are *equivalent* at this point. If one of the two substances (acid or base) is being determined by titration, the point of equivalence is called the **endpoint** (or **equivalence point**)* of the titration. In substituting from Eqs. (13.26) and (13.27) for n_A and n_B, the endpoint of the titration must satisfy the mathematical condition

$$N_A V_A = N_B V_B \tag{13.28}$$

This is the fundamental equation of acid–base titration. If, for example, N_A is accurately known and V_A and V_B are measured, then N_B may be calculated as

$$N_B = \frac{N_A V_A}{V_B} \tag{13.29}$$

---

* Strictly speaking, the *equivalence point* is the point at which the two substances are present in equivalent amounts and the *endpoint* is the *apparent* equivalence point as indicated by the titration. The difference between the two is related to the *titration error*.

In this way the concentration (and thus the amount) of the unknown base is deduced. The material whose concentration is known (the acid in the above example) is called the **standard** substance (standard acid in this particular case). Standard acids and standard bases are prepared with great care so that their concentrations are known very accurately.

EXAMPLE 13.11 A quantity of 50 cm³ of an unknown acid is titrated with standard sodium hydroxide having a concentration of 0.153 N. It is found that 35.4 cm³ of standard base is required to neutralize the acid. Calculate the concentration of the unknown acid.

SOLUTION

$$N_A = \frac{N_B V_B}{V_A} = \frac{(0.153\ N)(35.4\ cm^3)}{50.0\ cm^3} = 0.108\ N$$

Note that as long as V_A and V_B are expressed in the same units, Eq. (13.28) is valid for any volume units.

EXAMPLE 13.12 How many moles of replaceable protons are present in the acid sample in Example 13.11?

SOLUTION $n_A = N_A V_A = (0.108\ mol \cdot L^{-1})(0.050\ L) = 5.42 \times 10^{-3}\ mol$
In this case, V_A must be expressed in liters.

13.9 ACID–BASE INDICATORS

If titration is to be a useful method of acid–base analysis, there has to be some way of determining when the endpoint has been reached. There are various physical methods for determining endpoints; for example, there is often an abrupt change in the electrical conductivity of the solution at or near the endpoint. One of the commonly used chemical methods for detecting the endpoint of a titration involves the addition of a small quantity of a substance called an **indicator**. An indicator is itself a weak acid or a weak base which has the property of changing color within a narrow pH range. If HIn represents a general indicator molecule, the indicator participates in the pH-dependent reaction

$$HIn + H_2O \rightleftharpoons H_3O^+ + In^-$$

Le Châtelier's principle tells us that in acid solution (high $[H_3O^+]$) the color of the solution will be that of the predominant species HIn, while in basic solution (low $[H_3O^+]$) the color will be that of the predominant species In^-. The proper indicator for a given titration is chosen by knowing approximately what the pH of the solution at the endpoint will be and finding an indicator that will change color in a narrow range which includes the endpoint pH. In titrations of strong acids with strong bases, the pH at the endpoint will be 7, and hence an indicator changing color between

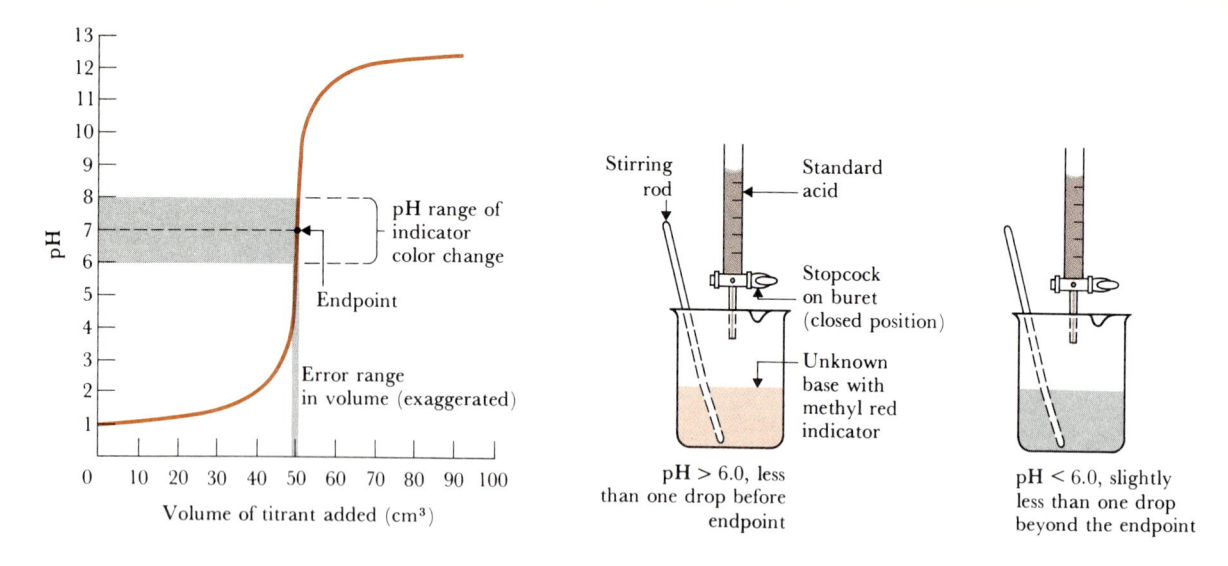

Fig. 13.1 (left). Titration of 50 cm³ of 0.1-N HCl with 0.1-N NaOH. In the titration of a base with an acid, the curve would begin at high pH and end at low pH. When weak acids and bases are titrated, the pH curves change somewhat differently. The endpoints do not occur at pH = 7, and the pH change around the endpoint is less abrupt.

Fig. 13.2 (right). Titration of an unknown weak base with a standard acid using methyl red indicator. Just prior to the endpoint the pH is slightly above 6.0 and the indicator has its basic color (yellow). Somewhere at pH below 6.0 the indicator changes to its acid color (red). If the endpoint of the reaction occurs in the pH region 4.8 to 6.0 (the range in which the indicator changes its color), then the indicator color change may be used to signal the equivalence point of the acid and the base.

pH = 6 and pH = 8 is normally satisfactory. As shown in Fig. 13.1, the pH changes rather rapidly with added titrant (the standard material being added) so that a pH range of 2 units represents a very small spread in titrant volume.

In titrations of a weak base with a strong acid, the pH of the endpoint will be less than 7 since at the endpoint only the conjugate acid of the weak base is present. For example, titrating a solution containing NH_3 (a weak base) with standard HCl leads to the production of NH_4^+ (as NH_4Cl), the conjugate acid of NH_3. The pH of the endpoint can be estimated by doing a rough preliminary titration using almost any indicator and, from this, determining an approximate value of $[NH_4^+]$ at the endpoint. This value may be used, as in Example 13.10, to estimate the endpoint pH.

Similarly, in titrations of a weak acid with a strong base, the pH at the endpoint will be above 7 since only the conjugate base of the weak acid is present at the endpoint. For example, titration of a solution of HOAc with standard NaOH leads to the formation of OAc^-, the conjugate base of HOAc. A rough preliminary titration to estimate the endpoint $[OAc^-]$ makes it possible to estimate the endpoint pH via a calculation such as in Example 13.10.

In Fig. 13.2 is illustrated the titration of an unknown weak base with a standard strong acid using methyl red as an indicator. Table 13.3 lists some commonly used indicators along with their colors and the pH ranges in which the color changes occur.

TABLE 13.3 SOME COMMONLY USED INDICATORS, THEIR COLORS, AND THE pH RANGES IN WHICH THEIR COLOR CHANGES OCCUR

INDICATOR	LOW pH COLOR	HIGH pH COLOR	pH RANGE IN WHICH COLOR CHANGES
Congo red	Blue	Red	3.0–5.0
Methyl orange	Red	Yellow	3.2–4.4
Methyl red	Red	Yellow	4.8–6.0
Bromothymol blue	Yellow	Blue	6.0–7.6
Phenol red	Yellow	Red	6.6–8.0
Phenolphthalein	Colorless	Pink	8.2–10.0
Thymolphthalein	Colorless	Blue	9.3–10.5

EXAMPLE 13.13 The concentration of acid in 50.0 cm³ of spoiled apple cider (essentially acetic acid) is to be determined by titration with 0.115-N NaOH. A preliminary rough titration shows that the HOAc concentration is roughly 1 N. What is the approximate pH at the endpoint and what indicator (see Table 13.3) should be used for accurate titration?

SOLUTION If the HOAc solution is exactly 1.00 N the volume of 0.115-N NaOH required to neutralize it is

$$V_B = \frac{(1.00\ N)(50.0\ \text{cm}^3)}{0.115\ N} = 435\ \text{cm}^3$$

The final solution would have a volume of roughly 50.0 + 435 or 485 cm³ and would contain 50.0/1000 or 0.0500 moles of NaOAc if no reaction with H_2O occurred. Thus $[OAc^-] = 0.0500\ \text{mol}/0.485\ L$ or 0.103 N *if no reaction with H_2O occurs*. Using K_b of OAc^- as 5.5×10^{-10} (Example 13.10) and setting $x = [HOAc] = [OH^-]$ and $[OAc^-] = 0.103 - x$ leads (as in Example 13.10) to

$$\frac{x^2}{0.103 - x} = 5.5 \times 10^{-10}; \qquad x = 7.5 \times 10^{-6}\ M$$

Thus pOH = $-\log x$ = 5.1 and pH = 8.9. Of the indicators listed in Table 13.3, phenolphthalein, which changes color in the pH range 8.2 to 10.0, would be most suitable.

To avoid having to add such a huge volume of standard base to effect neutralization, it would be wiser to use a smaller sample of the apple cider, for example, 5 cm³ instead of 50 cm³. Also, the titration should be repeated at least twice and the results averaged.

When polyprotic acids such as H_2SO_4 and H_3PO_4 are titrated with standard base, more than one endpoint is obtained. Each proton-donating species (for example, H_3PO_4, $H_2PO_4^-$, and HPO_4^{2-}) has its own endpoint.

13.10 THE LEWIS DEFINITION OF ACIDS AND BASES

A more general definition of acids and bases than that of Brønsted and Lowry is the **Lewis** definition.* This definition extends the concept of acids and bases beyond substances which act as proton donors and acceptors by focusing attention on the making and breaking of covalent bonds. A **Lewis acid** is defined as any substance capable of accepting a pair of electrons to form a covalent bond (as depicted by a Lewis dot formula). A **Lewis base** is defined as any substance capable of furnishing an electron pair to form a covalent bond. In short, Lewis acids are **electron-pair acceptors**, and Lewis bases are **electron-pair donors**. Since the Lewis definition is more general than the Brønsted-Lowry concept, all substances which are acids and bases in the Brønsted-Lowry concept are also acids and bases in the Lewis concept. However, there are substances considered acids and bases in the Lewis concept which are not acids or bases in the Brønsted-Lowry concept. For example, BF_3 and $AlCl_3$ are Lewis acids since they lack a pair of electrons about the central atom to make up a Lewis octet.

$$\begin{array}{cc} :\!\ddot{F}\!: & :\!\ddot{Cl}\!: \\ \ddot{B}\!:\!\ddot{F}\!: & \ddot{Al}\!:\!\ddot{Cl}\!: \\ :\!\ddot{F}\!: & :\!\ddot{Cl}\!: \end{array}$$

Since neither of these molecules contains hydrogen, neither is a Brønsted-Lowry acid. An example of an acid–base reaction involving BF_3 is the reaction with NH_3 (a Lewis base as well as a Brønsted-Lowry base):

$$\begin{array}{ccccc} :\!\ddot{F}\!: & & H & & :\!\ddot{F}\!:\!H \\ :\!\ddot{F}\!:\!\ddot{B} & + & :\!\ddot{N}\!:\!H & \rightarrow & :\!\ddot{F}\!:\!\ddot{B}\!:\!\ddot{N}\!:\!H \\ :\!\ddot{F}\!: & & H & & :\!\ddot{F}\!:\!H \end{array}$$

Cations of transition metal elements frequently behave as Lewis acids. If M^{x+} represents such a cation, a typical Lewis acid–base reaction has the form

$$M^{x+} + nB^{y-} \rightarrow (MB_n)^{(x+)+(y-)}$$

where B^{y-} is a Lewis base and n is often (but not always) equal to $2x^+$, twice the charge on the cation. The charge on the Lewis base, y^-, is zero or negative. Typical Lewis bases are H_2O, NH_3, CO, NO, Cl^-, CN^-, OH^-, and NO_2^-. Some specific reactions are:

$$Zn^{2+} + 4NH_3 \rightarrow [Zn(NH_3)_4]^{2+}$$

$$Co^{3+} + 6CN^- \rightarrow [Co(CN)_6]^{3-} \tag{13.30}$$

The structures and properties of such Lewis acid–base compounds are discussed in Chapter 22.

Some Lewis acid–base compounds such as those in Eq. 13.30 can act as proton donors, that is, as Brønsted-Lowry acids. This is most pronounced for compounds of

---

* Named after G. N. Lewis. For a short biographical sketch of Lewis, see Chapter 5.

the type

$$M(H_2O)_n^{x+}$$

that is, hydrated metal ions. The bond between the metal ion and a water molecule is due to sharing of a pair of electrons on oxygen; this drawing away of electron density from the oxygen atom makes the hydrogen electron-deficient and, hence, easier to remove as a proton. The reaction of the hydrated ion with water is

$$M(H_2O)_n^{x+} + H_2O \rightleftharpoons [M(H_2O)_{n-1}(OH^-)]^{(x-1)+} + H_3O^+$$

The greater the ability of the ion M^{x+} to attract the oxygen lone-pair electrons, the greater the strength of the metal-water complex as a Brønsted-Lowry acid. Generally, ions of large positive charge and small size (thus a large charge density) form the strongest Brønsted-Lowry acids. Thus, $[Al(H_2O)_6]^{3+}$ would be expected to be a stronger Brønsted-Lowry acid than $[Mg(H_2O)_4]^{2+}$. The two ions are of comparable size (ionic radii about 0.55 Å) but Al^{3+} has the greater charge, hence the greater charge density (charge per unit volume).

SUMMARY

According to the Brønsted-Lowry concept, acids are proton donors and bases are proton acceptors. Thus acid–base reactions involve the transfer of a proton from an acid to a base; the acid is transformed to its conjugate base and the base is transformed to its conjugate acid.

Acids and bases may be classified as to strength based on their relative abilities to donate or accept protons to or from a reference solvent. The equilibrium constant for the proton-donating or proton-accepting reaction of an acid or base, respectively, with respect to the reference solvent constitutes a quantitative index of acid–base strengths. Acid–base equilibrium constants may be used to calculate degrees of dissociation of acids and bases in solution.

The concentration of H_3O^+ in a solution may be expressed logarithmically in terms of the pH. Mixtures of weak acids and their conjugate bases or weak bases and their conjugate acids have the property of maintaining almost constant pH upon addition of small amounts of strong acids or bases. Such mixtures, called buffers, play important roles in controlling the pH of many natural processes.

There are only three basic types of pH calculations: those for strong acids or bases, those for weak acids or bases, and those for buffer solutions. When the H_3O^+ due to self-ionization of water is negligible, each type of calculation requires a simple, well-defined procedure.

The amount of acid in a sample may be determined by adding to it an equivalent amount of a base of known composition. This process, called titration, is useful in chemical analysis. The point at which the acid and base are equivalent is determined by substances called indicators; indicators are themselves weak acids or bases whose conjugate forms have different colors.

A more generalized concept of acids and bases is provided by the Lewis definition: Acids are substances capable of accepting a pair of electrons to form a covalent bond, and bases are substances capable of furnishing a pair of electrons to form a covalent bond. The class of Lewis acids and bases includes all Brønsted-Lowry acids and bases plus other substances as well.

LEARNING GOALS

1. Know the general form of a Brønsted-Lowry acid–base reaction and be able to identify the conjugate acid–base pairs.

2. Be able to use tables of acid and base strengths to predict the dominant direction of an acid–base reaction. Know how to calculate an equilibrium constant for such reactions.

3. Be able to write ionization constant expressions for weak acids and bases.

4. Be able to calculate K_a of an acid from K_b of the conjugate base, or vice versa.

5. Given the ionization constant of an acid or base, be able to calculate the degree of dissociation.

6. Know how pH is defined and how it is calculated for each of the following situations: (a) a strong acid, (b) a strong base, (c) a weak acid, (d) a weak base, (e) a buffer solution.

7. Know what approximations are normally made in calculating the pH of a solution.

8. Be able to explain how a buffer solution works.

9. Know the basic acid–base titration equation and how to use it in calculations.

10. Be able to determine what indicator to use in a given acid–base titration. Know the theory of how indicators work.

11. Know what Lewis acids and bases are and be able to give examples of each.

DEFINITIONS, TERMS, AND CONCEPTS TO KNOW

Brønsted-Lowry acid
Brønsted-Lowry base
conjugate acid–base pair
solvent leveling effect
acid and base ionization constants
ion-product constant of water
pH
strong acids and bases
weak acids and bases
buffer solution
polyprotic acid
acid–base titration
indicator
endpoint of a titration
equivalent weight
normality
Lewis acid
Lewis base

QUESTIONS AND PROBLEMS

General

1. Complete the table below by supplying the missing acid or base.

CONJUGATE ACID	CONJUGATE BASE
$HClO_2$	_____
_____	H_2O
NH_3	_____
HCO_3^-	_____
_____	HCO_3^-
$Zn(OH)_2$	_____

2. Complete each of the following reactions and predict in which direction the reaction predominates.

a) $HOAc + NO_3^- \rightleftharpoons$

b) $HCN + OAc^- \rightleftharpoons$

c) $HSO_4^- + HS^- \rightleftharpoons$

d) $H_2S + HS^- \rightleftharpoons$

e) $HCO_3^- + HCO_3^- \rightleftharpoons$

3. Use Table 13.1 to predict whether it would be safe to mix $NaHSO_4$ and $NaCN$ in H_2O solution. (HCN is a highly poisonous gas which is not very

352 ACIDS, BASES, AND THEIR REACTIONS

soluble in H_2O.) What is the equilibrium constant for this reaction?

4. Oxyhemoglobin (hemoglobin which is transporting O_2) undergoes the following reaction in body tissues:

oxyhemoglobin + H_3O^+ ⇌ hemoglobin + H_2O + O_2

a) Is oxyhemoglobin acting as a Brønsted-Lowry acid or as a base?
b) What in the above corresponds to a conjugate acid–base pair?

5. H_2S reacts with H_2O at 25 °C as follows:

$H_2S + H_2O \rightleftharpoons H_3O^+ + HS^-$ $K_{a_1} = 1.0 \times 10^{-7}$

$HS^- + H_2O \rightleftharpoons H_3O^+ + S^{2-}$ $K_{a_2} = 1.2 \times 10^{-13}$

Which ion or molecule (other than H_2O) has: (a) the largest concentration in an H_2S solution? (b) the smallest concentration in an H_2S solution?

6. Use the relationship $K_a K_b = K_w$ to show why K_a of NH_3 and K_b of Cl^- in H_2O solution may be taken as zero in most practical situations.

7. If K_a of H_3O^+ is 1, to what reaction does this refer? Repeat for the situation in which K_b of OH^- is 1.

8. The equilibrium constant (comparable to K_w of H_2O) for the reaction $NH_3 + NH_3 \rightleftharpoons NH_4^+ + NH_2^-$ is about 1.0×10^{-30} at -50 °C. An acid HA reacts with NH_3 as

$HA + NH_3 \rightleftharpoons NH_4^+ + A^-$

with a K_a of 2.5×10^{-8} at -50 °C. What is K_b of A^- in this solvent system?

9. Calculate $[NH_4^+]$ in a solution made by mixing 50 cm³ of 0.04-M NH_3 with 50 cm³ of 0.02-M KOH.

10. Calculate the equilibrium constant of the reaction

$HCN + OAc^- \rightleftharpoons HOAc + CN^-$

11. Show by balanced equations the reaction of each of the following with water, and state whether each substance is a base or an acid.

a) H_2S
b) CH_3NH_2
c) NH_4^+
d) $[Cu(H_2O)_4]^{2+}$
e) Cl^-
f) HCO_3^-
g) HS^-
h) PO_4^{3-}

12. The acids in each of the following groups are listed in the order weakest to strongest. What trend do you note and why would you expect such a trend?

a) HClO, $HClO_2$, $HClO_3$, $HClO_4$
b) HNO_2, HNO_3
c) H_2S, H_2SO_3, H_2SO_4

13. Charge densities (charge per unit volume) of the ions Na^+, Mg^{2+}, and Cr^{3+} increase in that order. At a given concentration which would form the most acidic solution, NaCl(aq), $MgCl_2$(aq), or $CrCl_3$(aq)? Write the reaction for the proton donation reaction of each acid species, assuming each ion of charge x^+ forms bonds with $2x^+$ molecules of water.

14. Two salts, NaA_1 and NaA_2, are dissolved in separate containers, each containing 1 L of water. The pH's of the two solutions are 8.0 and 11.0, respectively. Which is the stronger acid, HA_1 or HA_2? Explain.

15. Write net ionic reactions for the following pairs of substances.

a) HCl and NaOAc
b) $HClO_4$ and NH_3
c) HBr and $Ca(OH)_2$

16. Which would you expect to be the stronger acid, H_2SO_4 or H_2SeO_4? [*Hint*: The sulfur atom is smaller than the selenium atom.]

Degree of Dissociation

17. Calculate the fraction of NH_3 dissociated at 25 °C in a 0.01-M NH_3 solution.

18. Calculate the molar concentration of HOAc which is 3% dissociated at 25 °C.

19. The K_a of HNO_3 at 25 °C is 22. What fraction of 1.0-M HNO_3 is dissociated at 25 °C?

20. Assume that α is the degree of dissociation of the weak acid HA.

a) Show that $i = 1 + \alpha$, where i is the van't Hoff factor in $\Delta T_f = iK_f m$ (see Problem 49, Chapter 10).

b) Show that measurement of ΔT_f may be used to calculate K_a or K_b of a weak acid or base, respectively.

21. A 1.00-m solution of HF freezes at -1.91 °C. Calculate the fraction of HF dissociated at equilibrium.

22. Show that the degree of dissociation of a weak acid is given by

$$\alpha = \frac{-K_a \pm \sqrt{K_a^2 + 4K_a c}}{2c}$$

where c is the molar concentration of the acid. Also show that α approaches unity as K_a becomes larger. (Note: According to the binomial theorem of algebra, $\sqrt{x^2 + 4xy}$ approaches $x + 2y$ as x becomes large.)

pH Calculations

23. Calculate the pH of each of the following aqueous solutions at 25 °C.
   a) 0.01-M NaCl
   b) 0.001-M HCl
   c) 0.001-M KOH
   d) 0.2-M HCN
   e) A solution of HCN and NaCN in which [HCN] = 0.5 M and [CN$^-$] = 1.0 M
   f) 0.2-M CH$_3$NH$_2$ ($K_b = 4 \times 10^{-4}$ at 25 °C)

24. Although there are no mathematical limits to the value pH may have from $-\infty$ to $+\infty$, there is at least one physical limit which would restrict pH values to a much smaller finite range. What is the nature of this physical limit?

25. What concentration of NaCN in H$_2$O is needed to make up a solution with pH = 11?

26. The value of K_w of water at 60 °C is 1.0×10^{-13}.
   a) What is the pH of 0.01-M NaOH at 60 °C?
   b) What is the pH of 0.01-M HCl at 60 °C?

27. If a solution contained one HCl molecule in all the water on the earth (approximately 10^{21} L), then [H$_3$O$^+$] = $\{[1/6 \times 10^{23}] \text{ mol}\}/10^{21}$ L = 1.7×10^{-45} M and the pH would be about $+45$. Similarly, a solution with one NaOH in 10^{21} L of water would have [OH$^-$] = 1.7×10^{-45} M and a pOH of $+45$; the pH would be $14 - 45 = -31$. What is the flaw in the above calculations?

28. Calculate [OH$^-$] in (a) 0.01-M HCl, (b) 0.01-M HOAc.

29. The value of K_w of heavy water (water in which the hydrogen is deuterium, $^2_1$H) at 25 °C is 7.4×10^{-14}.
   a) If K_a of $^2_1$HOAc is 2.0×10^{-5} at 25 °C, what is K_b of OAc$^-$ in this solvent?
   b) What is the pH of a neutral solution in heavy water?

30. The value of K_a of HOCN at 25 °C is 2.2×10^{-4}.
   a) Calculate the K_b of OCN$^-$ at 25 °C.
   b) Calculate the pH of a 1.0-M KOCN solution at 25 °C.

31. Calculate the pH of solutions having the following [H$_3$O$^+$] values.
   a) 0.02 M
   b) 1.1 M
   c) 0.55 M
   d) 3×10^{-4} M
   e) 9.2×10^{-2} M
   f) 1.6×10^{-8} M

32. Calculate [H$_3$O$^+$] for solutions having the following pH values.
   a) 6.3
   b) 9.5
   c) 11.2
   d) 1.2
   e) -1.2
   f) 15.1

33. A solution is made by mixing 20 cm^3 of 0.2-M HCl with 60 cm^3 of 0.1-M Ca(OH)$_2$. What is the pH of the solution?

Buffer Solutions

34. Show by means of equations how a solution of KCN and HCN can act as a buffer.

35. Show by means of equations how a solution of C$_2$H$_5$NH$_2$ and C$_2$H$_5$NH$_3$Cl can act as a buffer.

36. What ratio of NH$_3$ concentration to NH$_4$Cl concentration would be needed to make up a buffered solution with a pH of 8.0?

37. What ratio of sodium formate to formic acid would be needed to make up a buffered solution with a pH of 6.0? K_a of formic acid is 1.76×10^{-4}.

38. A solution of a weak acid HA contains both HA and A⁻ and thus should act as a buffer. Why would this make a poor buffer; that is, why is additional A⁻ introduced by adding a soluble salt of HA?

39. Which of the following would be the best buffer? Explain.
   a) A solution of HOAc with pH = 4.6.
   b) A solution with [HOAc] = 0.100 M and [OAc⁻] = 0.075 M.
   c) A solution with [HOAc] = 1.00 M and [OAc⁻] = 0.75 M.
What is the pH of the solutions in (b) and (c)?

Acid–Base Titrations

40. A solution is colored yellow by methyl orange, yellow by methyl red, red by phenol red, and colorless by thymolphthalein. Use Table 13.3 to predict which of the following is the most probable pH of the solution: 4, 7, 9, 1, or 12.

41. The endpoint of an acid–base titration is characterized by [HIn] = [In⁻].
   a) Show that the acid ionization constant of the indicator, K_{ind}, is equal to [H₃O⁺] at this point and thus pK_{ind} = pH (at the endpoint).
   b) The phenolphthalein endpoint occurs at a pH of about 9.1. What is the value of K_{ind} of phenolphthalein?

42. How much 18-N H₂SO₄ (in cm³) would be required to make up 10 L of 3-N H₂SO₄?

43. a) How many grams of NaOH would be required to neutralize 100 cm³ of 5-N HCl?
   b) If this NaOH is dissolved in 500 cm³ of solution, what is its normality?

44. A quantity of 25.1 cm³ of an unknown acid is found to be neutralized with 36.1 cm³ of 0.298-N KOH. Calculate the normality of the acid. How many equivalents of acid are in the sample?

45. Which indicator from Table 13.3 would you choose for each of the following titrations?
   a) HBr with KOH
   b) KOAc with HCl
   c) HOAc with KOH
   d) NH₃ with HCl
   e) NH₃ with HOAc

46. What volume of 0.35-M HCl will be required to neutralize 80 cm³ of 0.55-M Ba(OH)₂?

47. What volume of 0.300-M HCl will be required to neutralize 200 cm³ of 0.400-M Ca(OH)₂? How much CaCl₂ (in moles) is formed?

Lewis Acid–Base Concept

48. Identify the Lewis acids and bases in the following reactions.
   a) BF₃ + F⁻ → BF₄⁻
   b) O=C=O + OH⁻ → O=C(O⁻)(OH)
   c) Ag⁺ + 2NH₃ → [Ag(NH₃)₂]⁺

49. B₂H₆ and CO react according to the equation

   B₂H₆ + 2CO → 2H₃BCO

to form **borine carbonyl**. Draw a Lewis dot formula for borine carbonyl, assuming B and C share a pair of electrons. Would you consider CO a Lewis acid or a Lewis base? Explain.

50. Write Lewis acid–base reactions for each of the following.
   a) Co³⁺ and H₂O
   b) SO₂ and H₂O
   c) Ag⁺(aq) and Cl⁻(aq)

51. Describe the formation of the ion AlCl₄⁻ as a Lewis acid–base reaction. What is the acid–base pair needed to produce this ion?

ADDITIONAL READINGS

Drago, Russell S., and Nicholas A. Matwiyoff, *Acids and Bases.* Lexington, Mass.: D. C. Heath, 1968.

14
SOME FUNDAMENTAL TYPES OF CHEMICAL REACTIONS

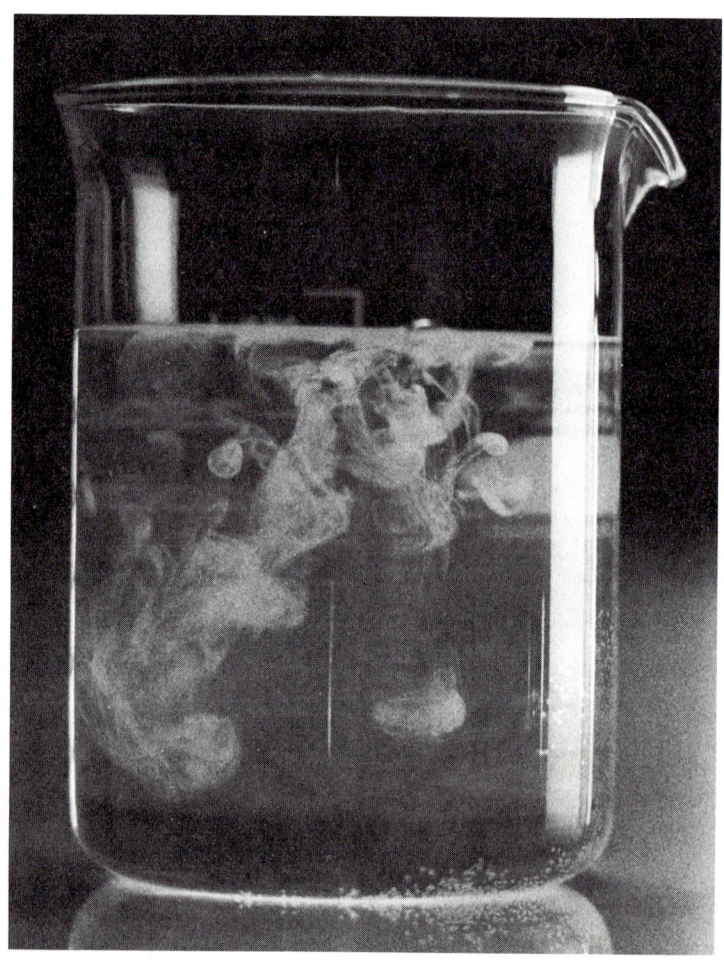

A drop of silver nitrate solution added to a dilute solution of table salt produces insoluble silver chloride. (Courtesy of Marshall Henrichs.)

Examiner: *When a metal is burned it releases phlogiston to produce calx. Why, then, is air necessary for the combustion, and why does the calx weigh more than the metal?*

Student: *Air is necessary to absorb the evolved phlogiston. The calx weighs more than the metal since phlogiston has negative weight.*

Examiner: *Excellent! You will make a superb chemist.*

IMAGINARY ORAL TEST IN CHEMISTRY, *17TH CENTURY*

Today the answer given in the quote above would merit a flat zero. We now know that when a metal is burned it combines with oxygen to produce an oxide (calx). Thus air supplies the oxygen, and there is no need to postulate an unusual entity such as phlogiston. The present chapter will examine briefly a variety of chemical reactions, none of which involves phlogiston.

There are two broad types of chemical reactions: In one type there are no changes in the oxidation numbers (see Section 2.6) of any of the elements involved; in the other type one or more of the elements involved changes oxidation number.

Most of the acid–base reactions discussed in Chapter 13 are of the first type (the reaction of the base H^- with H_2O is an exception). Typical acid–base reactions involve only the transfer of a proton from one substance to another.

The ammonia synthesis reaction

$$N_2(g) + 3H_2(g) \rightleftharpoons 2NH_3(g)$$

is an example of the second type of reaction; the oxidation number of nitrogen decreases from 0 to -3 and that of hydrogen increases from 0 to $+1$.

The present chapter deals briefly with various types of chemical reactions which are classified on a much different basis than that given above. In addition, systematic methods are given for the balancing of those reactions involving oxidation and reduction.

14.1 SOME SIMPLE REACTION TYPES

Although it is not possible to provide a simple, tidy classification of all chemical reactions, some types of reactions are encountered frequently and merit individual attention. Three such types of reaction are: ion-transfer reactions, addition reactions, and substitution reactions.

The most important example of an **ion-transfer reaction** is a Brønsted-Lowry acid–base reaction such as

$$HA + H_2O \rightleftharpoons H_3O^+ + A^-$$

where a hydrogen ion (H⁺) is transferred from HA to H₂O. Such reactions were discussed in some detail in Chapter 13. Similar reactions occur in solvents other than water.

Addition reactions generally involve the combination of two or more substances to form a single substance. One well-known example is the reaction of ethylene and hydrogen to form ethane:

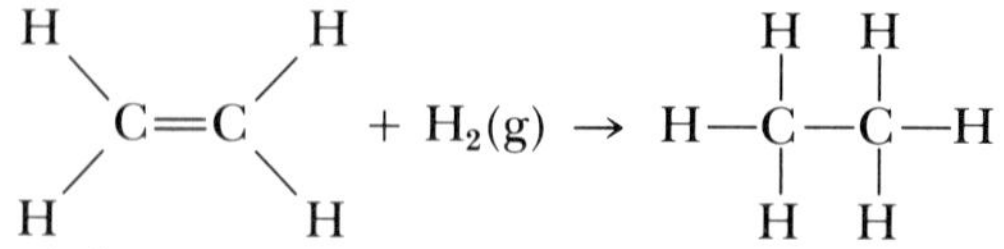

$$\begin{array}{c}H\\ \\H\end{array}\!\!\!\!\!\!\diagdown\!\!\!\!\!\!\!C\!\!=\!\!C\!\!\!\!\!\!\diagup\!\!\!\!\!\!\!\begin{array}{c}H\\ \\H\end{array} + H_2(g) \rightarrow H\!-\!\!\underset{\underset{H}{|}}{\overset{\overset{H}{|}}{C}}\!\!-\!\!\underset{\underset{H}{|}}{\overset{\overset{H}{|}}{C}}\!\!-\!H$$

ethylene   ethane

Such reactions are also called **combination reactions**. Two other examples are

$$2CO(g) + O_2(g) \rightarrow 2CO_2(g)$$

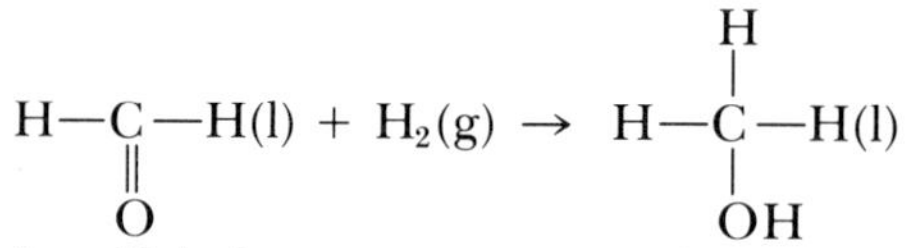

$$H\!-\!\underset{\underset{O}{\|}}{C}\!-\!H(l) + H_2(g) \rightarrow H\!-\!\underset{\underset{OH}{|}}{\overset{\overset{H}{|}}{C}}\!-\!H(l)$$

formaldehyde   methanol

If a reaction of the above type proceeds predominantly in the reverse direction, it is termed a **decomposition reaction**. A simple example is the thermal decomposition of mercury(II) oxide:

$$2HgO(s) \xrightarrow{heat} 2Hg(l) + O_2(g)$$

Decomposition reactions in which at least one of the products is a compound (as opposed to an element) are often called **elimination reactions**. Three examples are:

$$B(OH)_3(s) \xrightarrow{heat} HBO_2(s) + H_2O(g)$$
boric acid   metaboric acid

$$CaCO_3(s) \xrightarrow{heat} CaO(s) + CO_2(g)$$
calcium carbonate   calcium oxide

$$2KClO_3(s) \xrightarrow{heat} 2KCl(s) + O_2(g)$$
potassium chlorate   potassium chloride

Some addition reactions are much more subtle in nature than the above might suggest. For example, if an aqueous solution of silver nitrate (AgNO₃) is mixed with an aqueous solution of sodium chloride (NaCl), a white flocculent substance forms and slowly settles to the bottom of the reaction container. Analysis reveals that this substance is silver chloride (AgCl), a substance with a very low solubility in water (K_{sp} of AgCl in water is 1.56×10^{-10} at 25 °C). If the AgCl is removed by filtration and the clear remaining solution evaporated, crystals of sodium nitrate (NaNO$_3$) are obtained. The apparent reaction is

$$AgNO_3(aq) + NaCl(aq) \rightarrow AgCl(s) + NaNO_3(aq) \tag{14.1}$$

However, AgNO$_3$, NaCl, AgCl, and NaNO$_3$ are all ionic compounds. Thus AgNO$_3$(aq) is Ag$^+$(aq) and NO$_3^-$(aq); NaCl(aq) is Na$^+$(aq) and Cl$^-$(aq); and NaNO$_3$(aq) is Na$^+$(aq) and NO$_3^-$(aq). [AgCl(s) is Ag$^+$ and Cl$^-$, but the ions prefer each other's company rather than becoming hydrated.] Thus the chemical reactions should be written

$$Ag^+(aq) + NO_3^-(aq) + Na^+(aq) + Cl^-(aq) \rightarrow AgCl(s) + Na^+(aq) + NO_3^-(aq)$$

Cancelling out hydrated ions appearing on both sides of the reaction leads to the **net reaction**

$$Ag^+(aq) + Cl^-(aq) \rightarrow AgCl(s) \tag{14.2}$$

(An equilibrium exists between the hydrated ions and the solid, but the reverse reaction is ignored in the present context).

When the reaction is written as in Eq. (14.1), it appears as if Ag$^+$ has traded its partner to Na$^+$ in exchange for Cl$^-$. Hence, reactions of this type were once called **double displacement reactions**.* Appearances notwithstanding, the net reaction represented by Eq. (14.2) is the *correct* description of the chemical reaction. For the preparation of AgCl or NaNO$_3$ from AgNO$_3$ and NaCl, the representation given in Eq. (14.1) is more useful in determining the relative amounts of reactants to use, but Eq. (14.2) is the more instructive insofar as the actual chemistry is concerned. Even Eq. (14.2) is somewhat misleading in that it does not accurately reveal what is happening to the water molecules attached to the hydrated ions. A better equation for the process is

$$Ag^+(H_2O)_x + Cl^-(H_2O)_y \rightarrow AgCl(s) + (x+y)H_2O$$

A simple example of a **substitution reaction** is

$$\underset{\text{toluene}}{C_7H_8(l)} + Cl_2(g) \rightarrow \underset{\text{chlorotoluene}}{C_7H_7Cl(l)} + HCl(g)$$

In this reaction a hydrogen atom on toluene has been substituted by a chlorine atom. Many reactions of this type occur in organic chemistry (Chapters 16 and 17).

A different type of substitution takes place in the following:

$$Zn(s) + 2H^+(aq) \rightarrow Zn^{2+}(aq) + H_2(g)$$

Here Zn^{2+}(aq) replaces 2H$^+$(aq) and H$_2$(g) replaces Zn(s). Such reactions are also called **displacement reactions** since they involve displacement of hydrogen by a metal.

---

* Another name also used is **metathesis** or **metathetical reaction**.

14.2 OXIDATION–REDUCTION REACTIONS

Whenever one atom in a chemical reaction increases its oxidation number (also called its **oxidation state**), another atom (which may be the same element) must *decrease* its oxidation number. For example, in the ammonia synthesis reaction, the oxidation number of hydrogen increases from 0 to +1 and that of nitrogen decreases from 0 to −3.* The net change in oxidation numbers is the algebraic sum of all the individual atomic changes. Its value is always zero. For example, a change in two nitrogen atoms by −3 units each is a change of −6 units; this is exactly balanced by a change in six hydrogen atoms by +1 unit each. In the reaction

$$6H^+ + 5Br^- + BrO_3^- \rightleftharpoons 3Br_2 + 3H_2O$$

five of the bromine atoms increase their oxidation numbers from −1 to 0 and one bromine atom decreases its oxidation number from +5 to 0. Again, the net change in oxidation numbers is zero.

Changes in oxidation number are usually described by the use of two terms, oxidation and reduction, which are defined as follows.

Oxidation is the process whereby an element *increases* its oxidation number. An element whose oxidation number has been increased is said to have been **oxidized** (by some element whose oxidation number has decreased). The ion or compound containing the element whose oxidation number has been decreased is called an **oxidizing agent**.

Reduction is the process whereby an element *decreases* its oxidation number. An element whose oxidation number has been decreased is said to have been **reduced** (by some element whose oxidation number has increased). The ion or compound containing the element whose oxidation number has been increased is called a **reducing agent**.

In the reaction $N_2 + 3H_2 \rightleftharpoons 2NH_3$, hydrogen is oxidized by nitrogen and, conversely, nitrogen is reduced by hydrogen. Oxidation and reduction always appear together; one cannot occur without the other.

There are two main types of oxidation–reduction reactions: those involving nonionic reactants and products and those involving at least one ionic reactant or product. Ionic oxidation–reduction reactions are often called **redox** reactions, but the term is sometimes applied to oxidation–reduction reactions in general.

The ammonia synthesis reaction is of the nonionic type, as is the reaction (unbalanced)

$$HCl(g) + O_2(g) \rightarrow H_2O(g) + Cl_2(g)$$

Nonionic oxidation–reduction reactions can always be balanced by inspection using the rule given in Section 2.3, although it is sometimes less frustrating to use a more organized approach. In the example given, if oxygen is balanced first by writing $2H_2O$, then hydrogen by writing $4HCl$, the remainder ($2Cl_2$) follows at once. The balanced equation is

$$4HCl(g) + O_2(g) \rightarrow 2H_2O(g) + 2Cl_2(g)$$

---

* Increases and decreases in oxidation numbers are defined in the algebraic sense; that is, −1 to +2, −3 to −1, +2 to +3 are all increases and +2 to −1, −1 to −3 and +3 to +2 are all decreases.

360 SOME FUNDAMENTAL TYPES OF CHEMICAL REACTIONS

In nonionic oxidation–reduction reactions, changes in oxidation numbers reflect shifts in electron density due to the different electronegativities of the elements. Generally, it is the more electronegative element that oxidizes the less electronegative element; that is, the more electronegative element draws the center of the electron charge closer to itself.

Balancing of ionic oxidation–reduction reactions requires an additional rule. For example, the reaction

$$Sn^{2+} + Fe^{3+} \rightarrow Sn^{4+} + Fe^{2+}$$

is balanced insofar as atoms are concerned but is not balanced as to ionic charges; there are five positive charges on the left and six positive charges on the right. To balance the reaction as to charge, it is helpful to note that in ionic processes such as the one above there is an actual transfer of electrons between elements. Specifically, each Sn^{2+} loses two electrons and each Fe^{3+} gains an electron. Thus the loss of two electrons by one Sn^{2+} permits *two* Fe^{3+} to gain an electron each. The completely balanced equation then is

$$Sn^{2+} + 2Fe^{3+} \rightarrow Sn^{4+} + 2Fe^{2+}$$

Note that *both* atoms and charges are balanced. Electron transfer in ionic oxidation–reduction reactions is analogous to proton transfer in acid–base reactions.

As with nonionic oxidation–reduction reactions, it is possible to balance ionic oxidation–reduction equations by inspection, but in general it is more efficient to devise a more organized approach. There are several systematic methods of balancing oxidation–reduction reactions. Only two of these, the **oxidation-number method** and the **half-reaction method**, are considered here.

The oxidation-number method is very rapid but it is useful only when it is possible to quickly assign unambiguous oxidation numbers to all the elements in the reactants and products. The half-reaction method is slower but far more general in its application. This method makes it unnecessary to know oxidation numbers; all that is required is identification of principal reactants and products.

14.3 BALANCING EQUATIONS BY THE OXIDATION-NUMBER METHOD

Although the nonionic oxidation–reduction reaction

$$HCl + O_2 \rightarrow H_2O + Cl_2$$

[the (g) designations have been dropped for convenience] is easily balanced by inspection, it also provides a simple illustration of the oxidation-number method. Inspection shows that chlorine goes from -1 to 0 (oxidized) and oxygen goes from 0 to -2 (reduced). This may be indicated as follows:

$$\overset{(0)-(-1)=+1}{\overbrace{HCl + O_2 \rightarrow H_2O + Cl_2}}$$
$$\underset{(-2)-(0)=-2}{\underbrace{}}$$

The arithmetic accompanying each line indicates the *change* in oxidation number of chlorine and oxygen; an increase of 1 for chlorine and a decrease of 2 for oxygen. In order to have a net change of zero in the oxidation numbers, two chlorine atoms must be present for every oxygen atom. Thus a change in two chlorine atoms by +1 each is a +2 change, and a change in one oxygen atom by −2 is a −2 change, so that the net change is +2 − 2 = 0. This means that the balanced equation is

$$4HCl + O_2 \rightarrow 2H_2O + 2Cl_2$$

Ionic oxidation–reduction equations present more complexity. As an example, suppose a compound of Fe(II) is mixed with an acidified permanganate (MnO_4^-) solution. The main products of the reaction are Fe(III) (verified by adding KSCN and observing the red color of the ion $[Fe(SCN)_6]^{3-}$) and Mn(II) (verified by precipitation of pink MnS upon addition of H_2S). These observations suggest that the skeleton reaction is

$$Fe^{2+} + MnO_4^- \rightarrow Fe^{3+} + Mn^{2+}$$

It is important to remember that the first step in the balancing of any equation is to determine what the reactants and the products are. Then, changes in oxidation number are determined and indicated as shown below.

$$\underset{(+2)-(+7)=-5}{\overset{(+3)-(+2)=+1}{Fe^{2+} + MnO_4^- \rightarrow Fe^{3-} + Mn^{2+}}}$$

Since each iron atom increases its oxidation number by 1 while each manganese atom decreases its oxidation number by 5, there must be five iron atoms for each manganese atom. The equation is balanced partially as follows:

$$5Fe^{2+} + MnO_4^- \rightarrow 5Fe^{3+} + Mn^{3+}$$

All atoms are in balance except oxygen. In similar reactions hydrogen atoms may not be balanced at this point. The rules for balancing hydrogen and oxygen for reactions in acid solution are conveniently listed in tabular form as follows:

FOR EACH EXCESS ATOM ON ONE SIDE	ADD TO SAME SIDE	OTHER SIDE
H	nothing	H^+
O	$2H^+$	H_2O

In the present illustration, since the left-hand side has four excess oxygen atoms (that is, four oxygen atoms on the left not balanced by oxygen atoms on the right), add $8H^+$ to the left-hand side of the reaction *and* $4H_2O$ to the right-hand side of the equation. This produces the totally balanced ionic equation

$$5Fe^{2+} + MnO_4^- + 8H^+ \rightarrow 5Fe^{3+} + Mn^{2+} + 4H_2O$$

Note that the net charges on the left are 5(2+) + (1−) + 8(1+) = 17+ and those on the right are 5(3+) + (2+) = 17+.

Ionic reactions such as the latter are usually extremely rapid and often go essentially to completion; that is, they have large K_c values. Nonionic reactions (such as the $4HCl + O_2 \rightleftharpoons 2H_2O + 2Cl_2$ reaction), on the other hand, tend to be much slower and often do not go to completion; that is, they have small K_c values.

The $8H^+$ should actually be written $8H_3O^+$ according to our earlier convention for aqueous protons. However, it has become customary to use H^+ as equivalent to H_3O^+ or $H^+(aq)$ in writing oxidation–reduction equations. Similarly, the other ions above should be written $Fe^{2+}(aq)$, $MnO_4^-(aq)$, $Fe^{3+}(aq)$, and $Mn^{2+}(aq)$. In actuality, free ions such as Fe^{2+} or Fe^{3+} probably do not exist in aqueous solution except, perhaps, in very acid solutions.

If the same reactants [Fe(II) and MnO_4^-] are mixed in basic or neutral solution, different products are obtained. The ion Fe^{2+} in basic solution becomes $Fe(OH)_2$ (a dark-colored sludge), the ion Fe^{3+} becomes $Fe_2O_3 \cdot nH_2O$ (a red-brown precipitate) and instead of Mn^{2+} the Mn(IV) compound MnO_2 (a dark brown flocculent precipitate) is obtained. Thus the skeleton equation is

$$Fe(OH)_2 + MnO_4^- \rightarrow Fe_2O_3 \cdot nH_2O + MnO_2$$

If $n = 3$ is selected as a representative value, the oxidation-number changes are determined as follows.

$$\overset{(+3)-(+2)=+1}{Fe(OH)_2 + \underset{(+4)-(+7)=-3}{MnO_4^- \rightarrow Fe_2O_3 \cdot 3H_2O} + MnO_2}$$

This shows there must be three iron atoms for each manganese atom. The partially balanced equation is

$$6Fe(OH)_2 + 2MnO_4^- \rightarrow 3Fe_2O_3 \cdot 3H_2O + 2MnO_2$$

There are 20 oxygen atoms on the left and 22 oxygen atoms on the right, that is, an excess of 2 oxygen atoms on the right. Thus, add $4H^+$ to the right and $2H_2O$ to the left. Similarly, there are 12 hydrogen atoms on the left and 18 hydrogen atoms on the right, that is, an excess of 6 hydrogen atoms on the right. Thus, add $6H^+$ to the left. The net result is the addition of $2H^+$ and $2H_2O$ to the left side of the equation. The equation now is

$$6Fe(OH)_2 + 2MnO_4^- + 2H^+ + 2H_2O \rightarrow 3Fe_2O_3 \cdot 3H_2O + 2MnO_2$$

However, H^+ has a vanishingly small concentration in basic solution and is not an important reactant. If each such H^+ is replaced with H_2O and an OH^- added to the other side, the reaction becomes properly balanced for basic or neutral environments. The final balanced equation is

$$6Fe(OH)_2 + 2MnO_4^- + 4H_2O \rightleftharpoons 3Fe_2O_3 \cdot 3H_2O + 2MnO_2 + 2OH^-$$

It is a good idea to check for possible errors by running over all atoms and charges to see if they balance:

		LEFT	RIGHT
Atom balance:	Fe	6	6
	Mn	2	2
	O	24	24
	H	20	20
Charge balance:		2−	2−

The most difficult part is determining the reactants and products. Once this is done, balancing is merely a question of patience. At this point students are not expected to know enough descriptive chemistry to correctly identify reactants and products, but if this information is given, they should be able to balance the corresponding chemical equation.

14.4 BALANCING EQUATIONS BY THE HALF-REACTION METHOD

It is not always easy to assign oxidation numbers to all the elements in a reactant or product. For example, the compound aniline ($C_6H_5NH_2$) is oxidized in acidified $K_2Cr_2O_7$ solution to nitrobenzene ($C_6H_5NO_2$). The skeleton equation is

$$C_6H_5NH_2 + Cr_2O_7^{2-} \rightarrow C_6H_5NO_2 + Cr^{3+}$$

Although it is clear that chromium goes from +6 to +3 (reduced) and that nitrogen becomes oxidized, it is not easy for the beginning student to assign oxidation states to nitrogen. The half-reaction method makes it possible to balance this and all other oxidation–reduction equations without knowing any oxidation numbers at all. In this method the skeleton equation is written as two separate half-reactions, one involving the reduced element and the other involving the oxidized element (it is not necessary to know which is which).

$$C_6H_5NH_2 \rightarrow C_6H_5NO_2$$
$$Cr_2O_7^{2-} \rightarrow Cr^{3+}$$

First, balance all atoms except hydrogen and oxygen (unless hydrogen and oxygen are themselves changing oxidation numbers, in which case these are balanced right now):

$$C_6H_5NH_2 \rightarrow C_6H_5NO_2 \text{ (no change from above)}$$
$$Cr_2O_7^{2-} \rightarrow 2Cr^{3+}$$

Now balance hydrogen and oxygen in the same way as in the oxidation-number method:

$$C_6H_5NH_2 + 2H_2O \rightarrow C_6H_5NO_2 + 6H^+$$
$$Cr_2O_7^{2-} + 14H^+ \rightarrow 2Cr^{3+} + 7H_2O$$

Balance the charges in each half-reaction by *adding* the necessary number of electrons to the appropriate side:

$$C_6H_5NH_2 + 2H_2O \rightarrow C_6H_5NO_2 + 6H^+ + 6e^-$$
$$Cr_2O_7^{2-} + 14H^+ + 6e^- \rightarrow 2Cr^{3+} + 7H_2O$$

Next, multiply each half-equation throughout by numerical factors such that each contains the same number of electrons. Both of the above half-equations have the same number of electrons so this step is trivial. Now add the two half-reactions to obtain the final balanced equation:

$$C_6H_5NH_2 + Cr_2O_7^{2-} + 8H^+ \rightarrow C_6H_5NO_2 + 2Cr^{3+} + 5H_2O$$

Examination of the last set of half-reactions shows that aniline is oxidized to nitrobenzene since the electrons appear on the right (oxidation involves *loss* of electrons), and that $Cr_2O_7^{2-}$ is reduced to Cr^{3+} since the electrons appear on the left (reduction involves *gain* of electrons). Thus the half-reaction method identifies what is being oxidized and what is being reduced even though this may not have been clear at the start. For a reaction in basic or neutral solution, the $8H^+$ must be replaced with $8H_2O$, and $8OH^-$ must be added to the other side.

The half-reaction method works for most of the common oxidation–reduction reactions. For example, it can be used to balance the reaction

$$HCl + O_2 \rightarrow H_2O + Cl_2$$

The half-reactions are

$$HCl \rightarrow Cl_2$$
$$O_2 \rightarrow H_2O$$

First, balance chlorine and oxygen (the elements being oxidized and reduced):

$$2HCl \rightarrow Cl_2$$
$$O_2 \rightarrow 2H_2O$$

Next, balance hydrogen atoms (the oxygen atoms are already balanced since oxygen is changing oxidation number):

$$2HCl \rightarrow Cl_2 + 2H^+$$
$$4H^+ + O_2 \rightarrow 2H_2O$$

Next, balance charges by adding electrons:

$$2HCl \rightarrow Cl_2 + 2H^+ + 2e^-$$
$$4H^+ + O_2 + 4e^- \rightarrow 2H_2O$$

Multiply the first half-reaction by two so that the electrons will cancel when the half-reactions are added:

$$4HCl \rightarrow 2Cl_2 + 4H^+ + 4e^-$$
$$4H^+ + O_2 + 4e^- \rightarrow 2H_2O$$

Add the two half-reactions:

$$4HCl + O_2 \rightarrow 2H_2O + Cl_2$$

Note that the $4H^+$ also cancel. Of course, the H^+ are "artifacts" of the method and do not play any role in the actual reaction.

A summary of the half-reaction method is:

1. Determine the half-reactions.
2. Balance the elements undergoing oxidation number changes.
3. Balance hydrogen and oxygen atoms by the rules given.
4. Balance charges in each half-reaction by adding electrons to the appropriate side (different side for each half-reaction).
5. Equalize the electrons in the two half-reactions, add half-reactions, and simplify.
6. If the reaction is carried out in basic or neutral solution, replace any H^+ appearing in it with H_2O and add an OH^- to the other side.

14.5 DISPROPORTIONATION REACTIONS

A **disproportionation reaction** is one in which only a single element changes oxidation number. The reaction of Br^- and BrO_3^- to produce Br_2 is an example. The skeleton reaction is

$$Br^- + BrO_3^- \rightarrow Br_2$$

This equation is balanced by the same methods used for any ionic oxidation–reduction reaction once it is recognized that the Br^- is being oxidized to Br_2 while BrO_3^- is being reduced to Br_2. The oxidation-number method leads to

$$\overset{0 - (-1) = +1}{\overline{Br^- + BrO_3^- \rightarrow Br_2}}$$
$$\underset{0 - (+5) = -5}{}$$

Thus there must be five Br^- for each BrO_3^-. The partially balanced equation is

$$5Br^- + BrO_3^- \rightarrow 3Br_2$$

Balancing the oxygen atoms by adding $6H^+$ to the left and $3H_2O$ to the right gives

$$6H^+ + 5Br^- + BrO_3^- \rightarrow 3Br_2 + 3H_2O$$

In the half-reaction method the two half-reactions are

$$Br^- \rightarrow Br_2$$
$$BrO_3^- \rightarrow Br_2$$

From there the balancing is straightforward.

An important disproportionation reaction is that of NO_2 with water to form nitric acid and NO. The skeleton reaction is

$$NO_2 + H_2O \rightleftharpoons NO_3^- + NO$$

The half-reactions are

$$NO_2 \rightarrow NO_3^-$$
$$NO_2 \rightarrow NO$$

Both of these are balanced as they stand with respect to nitrogen. Balancing oxygens leads to

$$NO_2 + H_2O \rightarrow NO_3^- + 2H^+$$
$$NO_2 + 2H^+ \rightarrow NO + H_2O$$

Balancing charges produces

$$NO_2 + H_2O \rightarrow NO_3^- + 2H^+ + e^-$$
$$NO_2 + 2H^+ + 2e^- \rightarrow NO + H_2O$$

Multiplying the first half-reaction by 2, then adding and simplifying leads to

$$3NO_2 + H_2O \rightleftharpoons 2H^+ + 2NO_3^- + NO$$

14.6 EQUIVALENT WEIGHTS

When dealing with oxidation–reduction reactions it is convenient to express amounts of reactants in terms of a quantity called the **equivalent weight**. *The equivalent weight of a substance in an oxidation–reduction reaction is that weight which either gains or loses one mole of electrons in the reaction.* Equivalent weights defined in this way have a very important property; x equivalent weights of one reactant in a given reaction will react with precisely x equivalent weights of the other reactant.

Equivalent weights are used to define a concentration unit called **normality** (N). The normality of a solution containing an oxidizing agent or a reducing agent is the number of equivalent weights per liter of solution. This concentration term is also used for acids and bases (see Section 13.8). In this case an equivalent weight of acid or base is that weight which donates or accepts one mole of protons in an acid–base reaction.

As the following examples show, equivalent weights (and hence normality) depend on the specific reaction in which the substance is used.

EXAMPLE 14.1 Determine the equivalent weight of potassium permanganate, $KMnO_4$, when used in the reaction

$$2MnO_4^- + 10Cl^- + 16H^+ \rightarrow 2Mn^{2+} + 5Cl_2 + 8H_2O$$

SOLUTION The formula weight of $KMnO_4$ is 158.1 g·mol^{-1}. When MnO_4^- is reduced to Mn^{2+}, five electrons are gained per manganese ion. Thus the equivalent weight of $KMnO_4$ in this reaction is

$$\frac{158.1 \text{ g}}{5} = 31.62 \text{ g}$$

A solution containing one mole of $KMnO_4$ per liter of solution would be 1.00 *M* and also 5.00 *N* (when used in the above reaction).

EXAMPLE 14.2 Determine the equivalent weight of $KMnO_4$ when used in a reaction where MnO_4^- is reduced to MnO_2.

SOLUTION When MnO_4^- is reduced to MnO_2, three moles of electrons are gained per mole of MnO_4^-. Thus the equivalent weight of $KMnO_4$ in this reaction is

$$\frac{158.1 \text{ g}}{3} = 52.70 \text{ g}$$

Thus a 1.00-*M* $KMnO_4$ solution, if used in this latter reaction, would be 3.00 *N*.

SUMMARY

There are a variety of ways of classifying types of chemical reactions. One simple way is to define two broad classes: those that involve oxidation–reduction and those that do not. Another classification leads to types such as ion-transfer, addition, and substitution. These, in turn, contain subclasses such as decomposition and elimination reactions. Disproportionation reactions form a subclass of oxidation–reduction reactions.

Two types of chemical equations are usually easy to balance by simple inspection: those not involving oxidation–reduction and those involving nonionic oxidation–reduction. Oxidation–reduction reactions involving ionic substances are generally easier to balance by a systematic technique such as the oxidation-number method or the half-reaction method.

The half-reaction method is more general and does not require the use of oxidation numbers. This method is also useful in determining what the oxidizing and reducing agents are, especially in those cases when these facts are not clear at the outset.

LEARNING GOALS

1. Be able to identify ion-transfer reactions, addition reactions, and substitution reactions and to give at least one example of each.

2. Know what is meant by decomposition and elimination reactions and be able to give at least one example of each.

3. Know what an oxidation–reduction reaction is and be able to identify what is being oxidized and reduced.

4. Be able to balance oxidation–reduction equations by both the oxidation-number and half-reaction methods.

5. Be able to recognize a disproportionation reaction and to balance it.

6. Know how equivalent weights of reactants are defined and how to calculate these both for oxidation-reduction reactions and acid–base reactions.

DEFINITIONS, TERMS, AND CONCEPTS TO KNOW

addition reaction
substitution reaction
decomposition reaction
ion-transfer reaction
elimination reaction

combination reaction
oxidation
reduction
disproportionation reaction
oxidizing agent

reducing agent
equivalent weight
normality

QUESTIONS AND PROBLEMS

Types of Reactions

1. Classify each of the following reactions as to type, and state whether oxidation and reduction is involved.

a) $NH_2^- + H_2O \rightleftharpoons NH_3 + OH^-$

b) 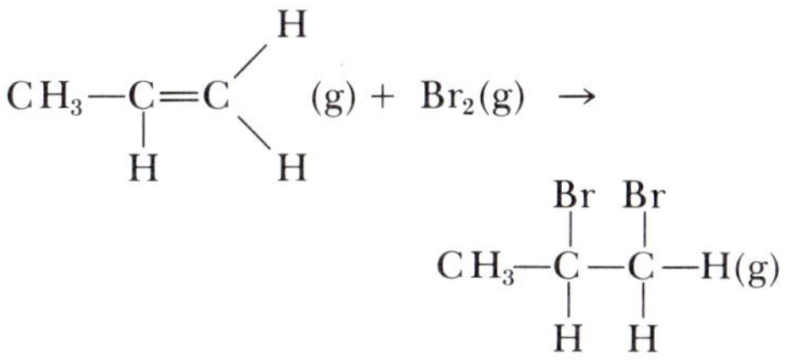 $CH_3-C=C\begin{smallmatrix}H\\H\end{smallmatrix}$ (g) + Br_2(g) $\rightarrow$

$CH_3-\underset{H}{\overset{Br}{C}}-\underset{H}{\overset{Br}{C}}-H$(g)

c) $CH_3-\underset{CN}{\overset{CH_3}{C}}-N=N-\underset{CH_3}{\overset{CH_3}{C}}-CH_3$(aq) $\xrightarrow{heat}$

$CH_3-\underset{CN}{\overset{CH_3}{C}}---\underset{CN}{\overset{CH_3}{C}}-CH_3$(aq) + N_2(g)

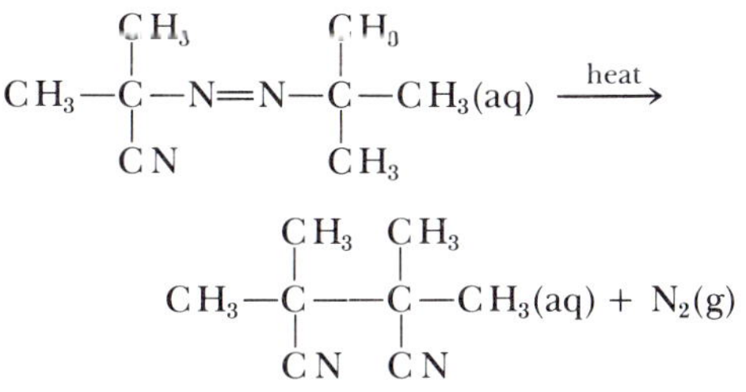

d) $Hg_2(NO_3)_2$(aq) + $2KCl$(aq) $\rightarrow$ Hg_2Cl_2(s) + $2KNO_3$(aq)

e) P_4(s) + OH^-(aq) $\rightarrow$ PH_3(g) + $H_2PO_2^-$(aq)

f) Mg(s) + $2HCl$(aq) $\rightarrow$ $MgCl_2$(aq) + H_2(g)

g) H^+(aq) + Br^-(aq) + $C_2H_5OH \rightarrow$ C_2H_5Br(aq) + H_2O

h) H^-(aq) + H_2O(l) $\rightarrow$ H_2(g) + OH^-(aq)

i) 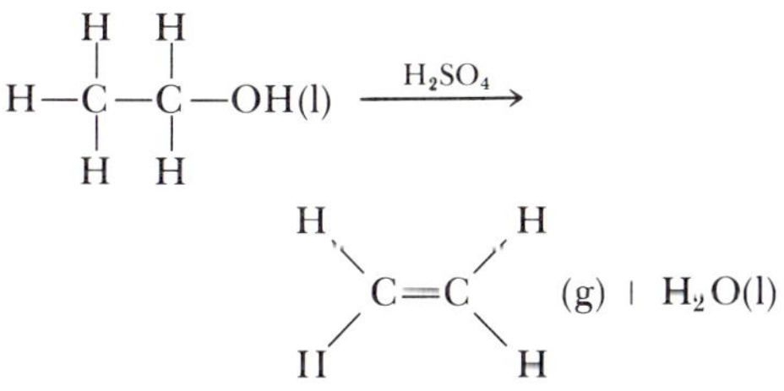 $H-\underset{H}{\overset{H}{C}}-\underset{H}{\overset{H}{C}}-OH$(l) $\xrightarrow{H_2SO_4}$

$\begin{smallmatrix}H\\H\end{smallmatrix}C=C\begin{smallmatrix}H\\H\end{smallmatrix}$ (g) + H_2O(l)

2. The reaction of ethanol (C_2H_5OH) and acetic acid (HOAc) may be written

$C_2H_5OH + HOAc \rightarrow C_2H_5OAc + H_2O$

What two types of reaction are involved in the above?

3. Show that the reaction of the hydride ion with water is not only a Brønsted-Lowry acid-base reaction but an oxidation-reduction reaction as well.

4. What is the main characteristic of each of the following types of reactions?
   a) Elimination reaction.
   b) Substitution reaction.
   c) Addition reaction.
   d) Ion-transfer reaction.
   e) Decomposition reaction.
   f) Disproportionation reaction.

5. What reaction type(s) is (are) represented in the following reaction?

$$CH_3ONa + CH_3Cl \rightarrow CH_3OCH_3 + NaCl$$

Could this be termed a double displacement reaction?

6. Give arguments for and against calling an ionization process such as $Na \rightarrow Na^+ + e^-$ a chemical reaction.

Balancing Redox Equations

7. Write balanced chemical equations for each of the following processes.

	REACTANTS	PRODUCTS	
a)	C_7H_8, MnO_4^-	$C_7H_6O_2$, Mn^{2+}	acid solution
b)	FeS_2, O_2	Fe_2O_3, SO_2	
c)	H_2O_2, MnO_4^-	O_2, Mn^{2+}	acid solution
d)	I^-, NO_2^-	I_2, NO	acid solution
e)	ClO_2^-, O_2^{2-}	ClO_2, H_2O	basic solution
f)	P_4	PH_3, $H_2PO_2^-$	basic solution

Use both methods for each of the above. In each case identify the oxidized and reduced elements.

8. Balance the reaction

$$NO_2 + H_2S \rightarrow H_2O + SO_2 + NO$$

(a) by inspection, (b) by one of the oxidation-reduction equation methods. Which method is less frustrating?

9. The oxidizing agent ClO_4^- is used to produce SO_2 from S^-. What weight of ClO_4^- as $KClO_4$ (potassium perchlorate) would be needed to produce 100 g of SO_2? The other product is Cl^-.

10. The compound butanone, C_4H_8O, may be prepared by oxidation of 2-butanol, $C_4H_{10}O$, by potassium dichromate, $K_2Cr_2O_7$, in acid solution. How much $K_2Cr_2O_7$ (in grams) is needed to prepare one mole of butanone from 2-butanol, assuming complete reaction?

11. Balance the following oxidation–reduction equations.
   a) $PbO_2 + Sb + NaOH \rightarrow PbO + NaSbO_2 + H_2O$
   b) $CrO_2^- + HO_2^- \rightarrow CrO_4^{2-}$ (basic solution)
   c) $ClO_2 \rightarrow ClO_2^- + ClO_3^-$ (basic solution)
   d) $NO_2 \rightarrow NO_3^- + NO$ (acid solution)
   e) $MnO_4^- + S_2O_3^{2-} \rightarrow S_4O_6^{2-} + Mn^{2+}$ (acid solution)
   f) $Ag^+ + AsH_3 \rightarrow H_3AsO_4 + Ag$ (acid solution)
   g) $Zn + NO_3^- \rightarrow [Zn(OH)_4]^{2-} + NH_3$ (basic solution)
   h) $S + NO_3^- \rightarrow SO_4^{2-} + NO_2$ (acid solution)
   i) $FNO_3 \rightarrow O_2 + F^- + NO_3^-$ (basic solution)
   j) $Cu + HNO_3 \rightarrow Cu(NO_3)_2 + NO + H_2O$

12. How many grams of $NaBiO_3$ would be needed to react with 2.0 g of $Mn(NO_3)_2$ to produce $NaMnO_4$ and $Bi(NO_2)_2$? What are the equivalent weights of $NaBiO_3$ and $Mn(NO_3)_2$ in this reaction?

13. Calculate the normality of a solution containing 50.0 g of $K_2Cr_2O_7$ in 800 L of water when used in a reaction that produces Cr^{3+}.

14. What is the equivalent weight of HIO_3 when (a) reduced to I_2, (b) reduced to I^-, (c) used to neutralize NaOH?

15. What volume of 0.500-N $K_2Cr_2O_7$ must be used to oxidize each of the following?
   a) 450 cm³ of 0.300-N HI (to I_2)
   b) 150 cm³ of 0.725-N $H_2C_2O_4$ (to CO_2)

16. An observant Tibetan yak herder noticed that it took a mole of oxygen to consume 8.0 g of the yak dung used to heat his yurt. What is the equivalent weight of yak dung?

General

17. What volumes of 0.1-M $AgNO_3$ and 0.3-M NaCl solution must be mixed in order to prepare 2.0 g of AgCl?

18. Mercury(I) nitrate, $Hg_2(NO_3)_2$, is a water-soluble ionic compound as is potassium chloride, KCl. Mercury(I) chloride, Hg_2Cl_2, is an ionic compound with a K_{sp} of 1.5×10^{-18} at 25 °C [the mercury(I) ion is Hg_2^{2+}].
   a) Write the net reaction that occurs when solutions of $Hg_2(NO_3)_2$ and KCl are mixed.
   b) If the insoluble substance formed in the re-

action is removed by filtration and the remaining solution evaporated to dryness, what substance is obtained?

19. The net result of photosynthesis (a sunlight-produced reaction taking place in the leaves of green plants) may be written as

$$n CO_2 + n H_2O \rightarrow (CH_2O)_n + n O_2$$

where n is typically 6. Identify the oxidizing agent and the reducing agent in the reaction. [*Hint:* Balancing by the half-reaction method may be of help here.]

20. Hydrogen peroxide (H_2O_2) can function both as an oxidizing agent and as a reducing agent. Explain.

ADDITIONAL READINGS

Mahan, Bruce H., *University Chemistry*, 3rd edition. Reading, Mass.: Addison-Wesley, 1975. Discusses the half-reaction method of balancing oxidation–reduction equations.

Theory and Problems of College Chemistry, Schaum's Outline Series, 5th edition. New York: McGraw-Hill, 1966. Chapter 10 discusses the half-reaction method (called the ion-electron partial method) and the oxidation-number method. Many solved examples are shown.

15
HYDROGEN, OXYGEN, AND WATER

Examples of snowflakes showing their characteristic hexagonal symmetries.

I am the daughter of earth and water,
   And the nursling of the sky;
I pass through the pores of the ocean and shores;
   I change, but I cannot die.
For after the rain when, with never a stain,
   The pavilion of heaven is bare,
And the winds and sunbeams with their convex gleams
   Build up the blue dome of air,
I silently laugh at my own cenotaph,
   And out of the caverns of rain,
Like a child from the womb, like a ghost from the tomb,
   I arise and unbuild it again.
PERCY BYSSHE SHELLEY, "The Cloud," 1820

Hydrogen, oxygen, and their most common compound, water, are three of the most important chemical substances known. Hydrogen is the fuel that powers the universe, oxygen plays a vital role in the respiratory processes of life, and water bathes our planet as liquid, vapor, and solid to regulate climate, to transport chemicals, and to fulfill hundreds of other roles in the myriad enterprises of nature. This chapter examines some of the physical and chemical properties of these three simple substances and attempts to show why these properties make the world around us operate as it does.

15.1 HYDROGEN: SOME BASIC FACTS

If several chemists were asked to prepare a list of the elements in the order of their importance, most would probably list hydrogen as number one. For one thing, hydrogen forms a greater number of *different* compounds than any other element; only the so-called inert gases fail to form hydrogen compounds. More important, hydrogen reacts with oxygen to form water, perhaps the most important chemical compound known.

   Elemental hydrogen ordinarily exists as the diatomic molecule H_2. The electron configuration of the ground state is $(1\sigma)^2$, indicating a singlet state (diamagnetic). Hydrogen is a colorless gas at room temperature and is the lightest of all the elements. It has a critical temperature of 33 K and a critical pressure of 12.8 atm.* The bond dissociation energy of H_2 is 4.48 eV; this relatively high value accounts for hydrogen's low level of chemical reactivity. A mixture of H_2 and O_2 at room temperature is quite stable unless subjected to a spark or very high-energy radiation, in which case an explosion occurs. Mixtures of H_2 and some of the halogens (Cl_2 and Br_2) behave similarly. However, H_2 and F_2 combine directly and vigorously—even at low temperatures.

   Hydrogen is by far the most abundant element in the universe; about 92 out

---

* As discussed in Section 3.10, this means that hydrogen cannot be liquefied above 33 K, and that at 33 K it requires at least 12.8 atm pressure to liquefy it.

of every 100 atoms in the universe are hydrogen atoms. However, it is relatively scarce as a component of the earth except on the surface, and elemental hydrogen is virtually absent from the earth's lower atmosphere.* The root-mean-square velocity of H_2 at 300 K (see the discussion following Example 3.10) is about 2×10^3 m·s^{-1} or about one-fifth of the earth's **escape velocity**, the minimum initial velocity a body at sea level must have in order to escape the earth's gravitational field.† This means that the fraction of H_2 molecules having velocities equal to or greater than the earth's escape velocity (see Fig. 3.5) is high enough that most of the H_2 ever present in the earth's atmosphere has escaped during ages past.

Most of the hydrogen found on earth exists in the form of water. Much of the rest is found in combination with carbon in, for example, animals, plants, and fossil fuel deposits. Probably about 10% of the mass of living things consists of hydrogen.

The planet Jupiter appears to have a large outer envelope of liquid *molecular* hydrogen (H_2) and a large inner core of liquid *atomic* hydrogen (H) about an iron-silicate core. Liquid atomic hydrogen has metallic properties similar to that of the group IA metals (lithium, sodium, potassium, etc.). It does not exist on earth since it requires pressures of around 3×10^6 atm and temperatures around 11,000 K.‡

It was pointed out in Section 11.7 that the major source of energy of the sun and many other stars appears to be a series of nuclear fusion reactions whose net effect is

$$4{}^{1}_{1}H \rightarrow {}^{4}_{2}He + 2{}^{0}_{1}e$$

This fusion reaction liberates 2.82×10^9 kJ per mole of helium formed. The radiant energy produced by this process on the sun and received on earth is utilized by plants in the synthesis of various carbon hydrogen compounds. This process, called **photosynthesis**, involves the construction of a carbon skeleton or framework for the support of hydrogen atoms. When the photosynthetically produced compounds are ingested by animal organisms, a biological oxidation of the hydrogen atoms occurs in the cell mitochondria and energy is released. Some of the sun's radiant energy is also used to heat the land, water, and atmosphere of the earth and thus to determine temperature and weather patterns. It is clear that the world and all its organisms operate principally on the energy of the sun's hydrogen fusion reaction.

15.2 THE ISOTOPES OF HYDROGEN

Naturally occurring hydrogen (in either elemental or combined form) consists of three isotopes: ${}^{1}_{1}H$ (ordinary or light hydrogen, also called **protium**), and two heavy

---

* An ultraviolet-sensitive camera left on the moon by the Apollo 16 mission shows a massive cloud of atomic hydrogen enveloping the earth. This is thought to come from the dissociation of water vapor by solar ultraviolet radiation.

† The escape velocity of the earth at sea level is about 25,000 mph, roughly 1.1×10^4 m·s^{-1}.

‡ Russian scientists claimed to have made metallic hydrogen in the laboratory in 1976. The method used was to put solid H_2 between two anvils at 4.2 K and then to apply high pressure. The electrical resistance of the sample dropped sharply, indicating a metallic-like state had been achieved.

hydrogen isotopes, ^2_1H (**deuterium**) and ^3_1H (**tritium**). Deuterium forms about 0.0156% by number of natural hydrogen; that is, about one out of every 6410 atoms of hydrogen is deuterium. It is estimated that only about 1 atom in 10^{17} atoms of hydrogen is tritium.

Although in many respects the chemistry of the three isotopes is the same, there are differences among the isotopes which show up in reaction rates (discussed in Chapter 20) and in equilibrium constant values. For example, in the reaction $N_2(g) + 3D_2(g) \rightleftharpoons 2ND_3(g)$ (where D is the commonly used symbol for ^2_1H), the equilibrium constant is greater at a given temperature than in the corresponding reaction involving light hydrogen. Also, under comparable conditions, H_2 reacts with Cl_2 about 13 times faster than does D_2. Such effects on rates and equilibrium constants are referred to as the **isotope effect**. Isotope effects also occur with other elements, but only in hydrogen—where the isotopic mass differences are relatively large—do they lead to rather large differences in rates and equilibrium constants.

The isotope effect arises as follows: If a light hydrogen atom is attached to some other atom A of mass m_A, the frequency of vibration of the two atoms is given by

$$\nu_{HA} = \frac{1}{2\pi}\sqrt{\frac{k}{\mu_{HA}}}$$

where μ_{HA} is the **reduced mass** of the two vibrating atoms. This quantity is defined by

$$\mu_{HA} = \frac{m_H m_A}{m_H + m_A}$$

The quantity k is called the **force constant** of the H—A bond; it is a measure of the "stiffness" of the electronic bond. Now if deuterium replaces ordinary hydrogen in the H—A bond, the force constant stays approximately the same since it depends mainly on the electron bonding, but the reduced mass becomes

$$\mu_{DA} = \frac{m_D m_A}{m_D + m_A}$$

Since $\mu_{DA} > \mu_{HA}$ it follows that $\nu_{HA} > \nu_{DA}$; that is, the vibrational frequency of the H—A bond is greater than the vibrational frequency of the D—A bond. Chemical reaction rates depend on the rate at which chemical bonds can be broken; the rate at which a chemical bond can be broken depends directly on the vibrational frequency of the bond. As a specific example, if the atom A attached to hydrogen is chlorine, then the reduced mass of HCl is 0.97 amu and that of DCl is 1.89 (using ^{35}Cl in both), and the two vibrational frequencies have the ratio

$$\frac{\nu_{HCl}}{\nu_{DCl}} = \sqrt{\frac{\mu_{DCl}}{\mu_{HCl}}} = \sqrt{\frac{1.89}{0.97}} = 1.4$$

This means that the H—Cl bond vibrates about 40% faster than does the D—Cl bond and, consequently, more H—Cl bonds can be broken per unit time than can D—Cl bonds.

Deuterium and tritium find wide use as tracers (see Section 11.10). Since deuterium is not radioactive, its presence in compounds is detected by the mass spectrograph (see Fig. 1.7). Tritium, as mentioned in Section 11.9, decays by β-emission and has a half-life of 12.4 yr.

15.3 THE PREPARATION OF ELEMENTAL HYDROGEN

A convenient method of preparing hydrogen gas in the laboratory is by **electrolysis** (from *electro* plus Greek *lysis*, loosening; hence loosening or decomposition by means of electricity). The electrolysis (see Fig. 15.1) is carried out by passing low-voltage dc current through a dilute water solution of a strong Brønsted-Lowry acid or base. The function of the acid or base is to make the solution a good electrical conductor; pure water is a very poor electrical conductor. If the water solution is acidic (pH < 7) the processes occurring at the electrodes are:*

cathode (reduction of H): $2H_3O^+ + 2e^- \rightarrow H_2(g) + 2H_2O(l)$

anode (oxidation of O): $6H_2O(l) \rightarrow O_2(g) + 4H_3O^+ + 4e^-$

The net reaction is

$2H_2O(l) \rightarrow 2H_2(g) + O_2(g)$

As shown in Fig. 15.1, the gases H_2 and O_2 are virtually insoluble in H_2O and collect at the tops of the tubes immediately above the cathode and anode, respectively.

The presence of hydrogen and oxygen in the collector tubes of the electrolysis apparatus may be verified by very simple tests. To test for hydrogen, fill a small test tube with the gas to be tested and hold the mouth of the test tube next to a flame. If hydrogen gas is present there will be a mild explosion which gives off a distinct, sharp pop. To test for oxygen, collect a test-tube sample of gas and insert into it a glowing splint (a piece of wood whose flame has been blown out but which still burns like a coal). If oxygen is present, the glowing splint will readily burst into flame again.

If 99% of a water sample is electrolyzed, the remaining 1% is enriched in D_2O (and in HDO, T_2O, THO, etc., where T represents ^3_1H). By electrolyzing large volumes of water, combining residues, reelectrolyzing the combined residues, and repeating the cycle often enough, a residue which is virtually pure D_2O (called **heavy water**) is eventually obtained. This is an example of a practical application of the isotope effect; all other factors being equal, bonds involving hydrogen will be broken at a greater rate than bonds involving deuterium. Electrolysis of heavy water leads to the production of D_2 (and small amounts of DT, T_2, HT, etc.). Fractional distillation of water can also be used to prepare D_2O; a D_2O molecule has more mass than an H_2O molecule and is less likely to be vaporized than is an H_2O molecule. Thus the residue tends to become enriched in D_2O. Relatively pure D_2O is now available in ton quantities.

The alkali metals (group IA) and some of the alkaline earth metals (group IIA)

---

* The electrode at which electrons enter an electrolyte is called the **cathode**. Positive ions (cations) pick up electrons at this electrode and become reduced. The electrode at which negative ions (anions) lose electrons is called the **anode**. Anions are oxidized at the anode.

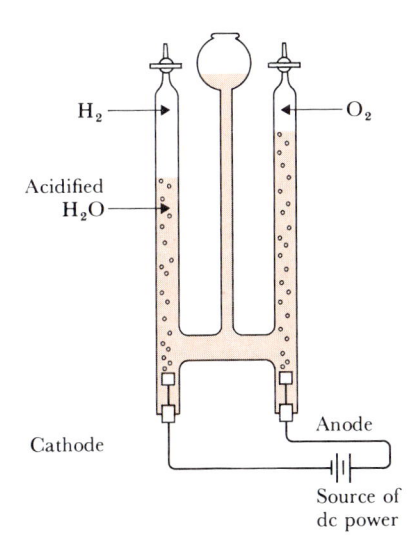

Fig. 15.1. Apparatus for the production of elemental hydrogen and oxygen by the electrolysis of water. The process produces two volumes of H_2 for each volume of O_2. If more water is added as H_2O is used up and H_2 and O_2 are removed, the unelectrolyzed water becomes enriched in heavy water, D_2O.

react with water at room temperature to produce hydrogen gas. For example,

$$2Na(s) + 2H_2O(l) \rightarrow 2NaOH(aq) + H_2(g)$$

$$2Rb(s) + 2H_2O(l) \rightarrow 2RbOH(aq) + H_2(g)$$

$$Ba(s) + 2H_2O(l) \rightarrow Ba(OH)_2(aq) + H_2(g)$$

[Recall that compounds such as NaOH are ionic so that NaOH(aq) represents $Na^+(aq)$ and $OH^-(aq)$.] The reaction with sodium is so violent that splattering occurs and so exothermic that the evolved H_2 is often set afire. Magnesium and calcium will react at an appreciable rate only with hot water:

$$Mg(s) + 2H_2O(l) \xrightarrow{heat} Mg(OH)_2(s) + H_2(g)$$
$$Ca(s) + 2H_2O(l) \xrightarrow{heat} Ca(OH)_2(aq) + H_2(g)$$

[the compound $Mg(OH)_2$ is virtually insoluble in water].

Metals such as iron and zinc react very slowly with water, even at high temperatures, but displace hydrogen quite rapidly from solutions containing relatively large amounts of H_3O^+, that is, solutions of strong Brønsted-Lowry acids such as HCl or H_2SO_4. The reactions are

$$Zn(s) + 2H_3O^+ \rightarrow Zn^{2+}(aq) + 2H_2O(l) + H_2(g)$$

$$Fe(s) + 2H_3O^+ \rightarrow Fe^{2+}(aq) + 2H_2O(l) + H_2(g)$$

A simple, conventional apparatus for the collection of evolved hydrogen gas is illustrated in Fig. 15.2.

Some metals, copper, for example, will not react with H_3O^+ to form H_2. If HCl is used as the source of H_3O^+, no reaction occurs. If HNO_3 or H_2SO_4 is used as a H_3O^+ source, oxidation of copper by NO_3^- or SO_4^{2-} occurs. The reactions are

$$Cu(s) + 2NO_3^-(aq) + 4H_3O^+ \rightarrow Cu^{2+}(aq) + 2NO_2(g) + 6H_2O(l)$$

$$Cu(s) + SO_4^{2-}(aq) + 4H_3O^+ \rightarrow Cu^{2+}(aq) + SO_2(g) + 6H_2O(l)$$

15.3 | THE PREPARATION OF ELEMENTAL HYDROGEN

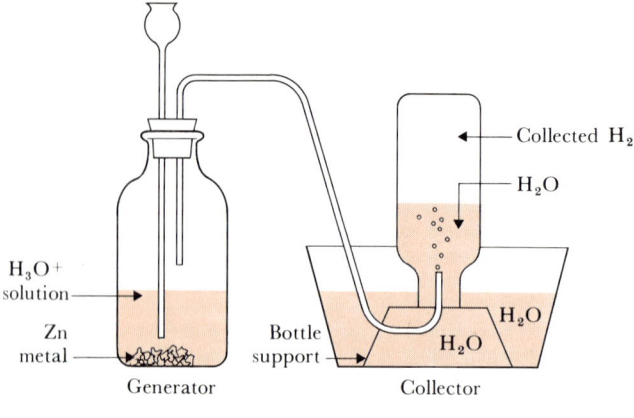

$$Zn(s) + 2H_3O^+ \rightarrow Zn^{2+}(aq) + H_2(g) + 2H_2O(l)$$

Fig. 15.2. Zinc metal and H_3O^+ react in the generator to produce hydrogen gas. The gas is passed into an inverted collector bottle originally filled with water. The hydrogen gas displaces the water and eventually fills the collector bottle. The collected hydrogen gas is contaminated with water vapor at a partial pressure equal to the vapor pressure of water at the temperature of the water in the collector.

TABLE 15.1 THE METAL ACTIVITY SERIES[a]

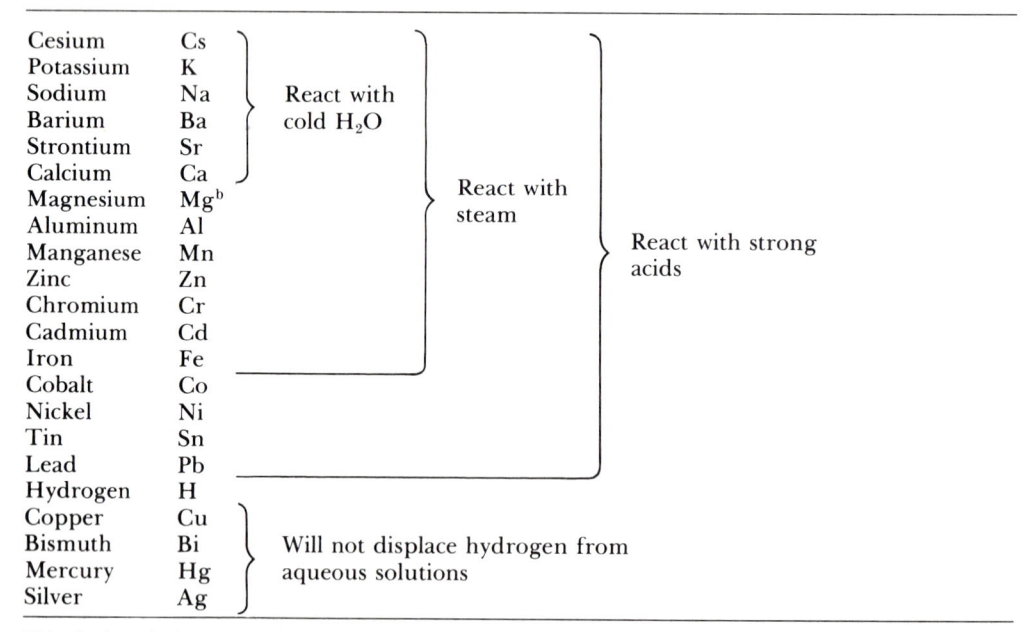

Cesium	Cs	⎫		⎫
Potassium	K	⎪ React with		⎪
Sodium	Na	⎬ cold H_2O		⎪
Barium	Ba	⎪		⎪
Strontium	Sr	⎪		⎪
Calcium	Ca	⎭	⎫	⎪
Magnesium	Mg[b]		⎪ React with	⎪
Aluminum	Al		⎬ steam	⎬ React with strong
Manganese	Mn		⎪	⎪ acids
Zinc	Zn		⎪	⎪
Chromium	Cr		⎭	⎪
Cadmium	Cd			⎪
Iron	Fe			⎪
Cobalt	Co			⎪
Nickel	Ni			⎪
Tin	Sn			⎪
Lead	Pb			⎪
Hydrogen	H			⎭
Copper	Cu	⎫		
Bismuth	Bi	⎬ Will not displace hydrogen from		
Mercury	Hg	⎪ aqueous solutions		
Silver	Ag	⎭		

[a] Metals above hydrogen can displace hydrogen from aqueous solution; those below hydrogen cannot. The farther a metal is above hydrogen, the weaker the Brønsted-Lowry acid required for the reaction.
[b] Magnesium will react with hot water.

(The Cu/NO$_3^-$ reaction can also produce NO or N$_2$O, depending on the concentration of NO$_3^-$). In these reactions the ions NO$_3^-$ and SO$_4^{2-}$ behave as oxidizing agents. The role of H$_3$O$^+$ is to combine with the oxygen atoms that nitrogen and sulfur must give up when they are reduced by copper.

The relative abilities of metals to react with Brønsted-Lowry acids to produce H$_2$ is often tabulated in the form of an **activity series** such as that shown in Table 15.1. Metals located above hydrogen in this series will, under proper conditions, produce H$_2$ from aqueous solution; metals located below hydrogen will not. Actually, a reaction such as

$$Cu(s) + 2H_3O^+ \rightleftharpoons Cu^{2+}(aq) + 2H_2O(l) + H_2(g)$$

does occur but the equilibrium constant is virtually zero. The reverse reaction, as would be expected, has a large equilibrium constant and has been studied quite extensively.

The large-scale industrial preparation of hydrogen is based largely on a petroleum-pyrolysis process known as **cracking**. A typical cracking reaction is

$$n\text{-butane}(g) \xrightarrow[\text{Al}_2\text{O}_3/\text{SiO}_2]{\text{heat}} \text{1,3-butadiene}(g) + 2H_2(g)$$

The primary goal of cracking is not the production of hydrogen (refineries burn some of this as waste) but to convert one petroleum product to another. Another important commercial process in which H$_2$ is a by-product is the electrolysis of salt brine (NaCl in water) to produce chlorine gas and caustic lye (sodium hydroxide). This process is discussed in Chapter 21.

15.4 BINARY COMPOUNDS OF HYDROGEN

A binary compound is one containing two different elements, for example, HCl, H$_2$O, or C$_2$H$_6$. The binary compounds of hydrogen may be classified into two broad types: those in which the oxidation state of hydrogen is +1 and those in which the oxidation state of hydrogen is −1. A more useful classification is as follows.

1. saline or ionic hydrides,
2. covalent hydrides,
3. interstitial hydrides.

The **saline** or **ionic hydrides** are examples of compounds having a −1 oxidation number. They occur with the very electropositive elements, specifically with the alkali metals (group IA) and all of the group IIA metals except beryllium and magnesium. As their name implies, these compounds are ionic; for example, sodium hydride, NaH, consists of the ions Na$^+$ and H$^-$. The hydride ion is a strong Brønsted-Lowry base (see Table 13.1). Thus solution of a saline hydride in water leads to the reaction

$$H^-(s) + H_2O \rightarrow H_2(g) + OH^-(aq)$$

Hydrides in which hydrogen is bonded to an element whose electronegativity is not different enough to produce ionic character are called **covalent hydrides**. The covalent hydrides are examples of both the -1 and $+1$ oxidation states.* Elements of groups IVA, VA, VIA, and VIIA—in addition to the elements beryllium, magnesium, boron, aluminum, and gallium—form covalent hydrides. Many of the covalent hydrides can be prepared by direct union of the elements. For example:

$$H_2 + X_2 \rightarrow 2HX \quad (X = F, Cl, Br, \text{ or } I)$$

$$2H_2 + O_2 \rightarrow 2H_2O$$

$$N_2 + 3H_2 \rightarrow 2NH_3$$

The oxidation state of hydrogen is $+1$ in all of these hydrides since all of the other elements have electronegativities greater than that of hydrogen. The reactions of H_2 and X_2 are very rapid (explosive, in fact) when X is chlorine or bromine and the reaction mixture is sparked or otherwise activated. With fluorine the reaction is spontaneous and vigorous even at very low temperatures. When X is iodine, the reaction is relatively slow and does not go to completion except at very high temperatures. This latter reaction

$$H_2(g) + I_2(g) \rightleftharpoons 2HI(g)$$

is endothermic with a heat of reaction of $25.9 \text{ kJ} \cdot \text{mol}^{-1}$ of HI produced. The reaction of O_2 and H_2 also proceeds with explosive violence once initiated. By contrast, the reaction with nitrogen is very slow under normal temperatures and pressures. The factors affecting the rate of this reaction are discussed in Chapter 20.

Some covalent hydrides (for example, those of the more electropositive elements) are best prepared by indirect routes. To illustrate one case: Direct union of aluminum and H_2 to form AlH_3 is not practicable. An indirect route is to first prepare LiH and to treat this with $AlCl_3$ in ether solution. The reaction is

$$4LiH + AlCl_3 \rightarrow 3LiCl + LiAlH_4$$

The compound $LiAlH_4$ (lithium aluminum hydride) may be treated with excess $AlCl_3$ to produce AlH_3 as follows:

$$3LiAlH_4 + AlCl_3 \rightarrow 3LiCl + 4AlH_3$$

Note that the oxidation number of hydrogen in AlH_3 is -1 since $\chi_{Al} = 1.5$ and $\chi_H = 2.1$.

Although carbon and hydrogen form a very large number of different compounds, direct union of these elements is hard to accomplish. Nature accomplishes

---

* According to the rough rule introduced in Section 5.2, a hydride M_xH_y would be expected to be covalent whenever $|\chi_M - \chi_H|$ is less than about two electronegativity units (Pauling scale). Also, when $\chi_M > \chi_H$, the oxidation state of hydrogen is $+1$; otherwise it is -1.

this union via a variety of complex indirect pathways. For example, the green leaves of plants use the energy of sunlight to carry out the reaction

$$6CO_2 + 6H_2O \rightarrow C_6H_{12}O_6 + 6O_2$$

to produce glucose ($C_6H_{12}O_6$) and similar compounds. When the dead plant becomes buried in the mud of a stagnant pond, **anaerobic** bacteria (bacteria that do not require oxygen for their metabolism) convert $C_6H_{12}O_6$ and similar compounds to substances such as methane, CH_4.* The oxidation number of hydrogen in CH_4 is taken to be +1 since $\chi_C = 2.5$ and $\chi_H = 2.1$.

It is of interest to note that the four hydrides HF, H_2O, NH_3, and CH_4 all have ten electrons but greatly differing properties. Relative to water, HF is acidic, H_2O is acidic and basic, NH_3 is basic, and CH_4 is neutral (no tendency to either gain or lose protons).

The **interstitial hydrides** are usually formed by the absorption of hydrogen gas by transition metals (groups IIIB to VIIIB). The resulting compounds are often nonstoichiometric, that is, they do not obey the law of definite composition (see Section 1.2). This suggests that the H_2 molecules occupy interstitial positions in the metal lattice, that is, the H_2 molecules fit into "empty" space of the crystal. Some transition metals, platinum, palladium, and nickel in particular, have the ability to absorb large volumes of H_2.

15.5 REDUCTION OF METAL OXIDES AND CHLORIDES WITH HYDROGEN

Reaction of metal oxides with H_2, which usually takes place at elevated temperatures, leads to production of the pure metal:

$$CuO(s) + H_2(g) \rightarrow Cu(s) + H_2O(g)$$

$$WO_3(s) + 3H_2(g) \rightarrow W(s) + 3H_2O(g)$$

Such reactions are normally too expensive to use for the commercial production of metals, but are useful in laboratory preparations and in conjunction with various techniques of chemical analysis. The heated chlorides of many metals are also reduced by hydrogen to the free metal, for example:

$$2AgCl(s) + H_2(g) \rightarrow 2Ag(s) + 2HCl(g)$$

15.6 ATOMIC HYDROGEN

Although H_2 is relatively inert, hydrogen atoms are very reactive. Hydrogen atoms may be prepared in a variety of ways: thermal decomposition of H_2 at high temperatures, subjection of H_2 to ultraviolet radiation, and in high-voltage discharge tubes. The hydrogen atoms so formed tend to recombine rather rapidly to form H_2 and

---

* These same bacteria convert mercury to dimethyl mercury, $(CH_3)_2Hg$, a compound which is largely responsible for entry of mercury into the aquatic food chain. See Section 21.8 for a discussion of this topic.

Liquid oxygen (Lox) is used to burn the fuel which powers this space-bound rocket. (Copyright © Dr. Georg Gerster, courtesy of Rapho/Photo Researchers, Inc.)

have a half-life of only a few tenths of a second. One of the most important uses of hydrogen atoms is in the **atomic hydrogen torch**, a device in which H_2 is split into hydrogen atoms and a stream of the gas allowed to impinge on a surface to be heated. The recombination reaction

$$2H(g) \rightarrow H_2(g)$$

which occurs on the surface releases over 418 kJ of energy per mole of H_2 formed. In addition, the H_2 reacts with the O_2 of the air to produce an additional 251 $kJ \cdot mol^{-1}$ of H_2 consumed. The very high temperature attained in the torch is used to weld metals, particularly metals sensitive to oxidation.

Atomic hydrogen can also be used for direct reactions with substances which react slowly or not at all with molecular hydrogen.

15.7 OXYGEN: SOME BASIC FACTS

Oxygen is the most abundant element on our planet. About 486 out of every 1000 atoms on earth are oxygen atoms; this constitutes 27.71% of the earth's mass. Most of the earth's oxygen is found in carbonate rocks such as limestone and in silicates (compounds of silicon and oxygen). About 89% of the mass of water is oxygen. Air contains about 21% oxygen by volume.* It is believed that most of the oxygen in the atmosphere originated in the photosynthetic processes of plants.† On the average, about 65% of the weight of all living things consists of oxygen.

Elemental oxygen exists as a diatomic molecule under ordinary conditions. It is a colorless gas with a critical temperature of 154 K and a critical pressure of 49.7 atm. The bond dissociation energy of about 5 eV is consistent with the relative inertness of O_2 unless activated. Oxygen is found in combination with all other elements, including the inert gases xenon, krypton, and radon. Although most diatomic molecules are diamagnetic, O_2 is paramagnetic (see Section 5.4). The ground state electron configuration, $(ISNB)(1\pi)^4(3\sigma)^2(1\pi^*)^2$, along with Hund's rule, shows O_2 is paramagnetic. However, there is a low-lying diamagnetic state which is believed to play a role in some reactions. This diamagnetic O_2 may be generated in the laboratory by the reaction

$$H_2O_2 + OCl^- \rightarrow Cl^- + O_2 + H_2O$$

The diamagnetic O_2 decays to the paramagnetic ground state and emits red-orange radiation around 634 nm in wavelength.

15.8 THE ISOTOPES OF OXYGEN

The earth's storehouse of oxygen consists of three isotopes: ^{16}O, ^{17}O, and ^{18}O, with abundances of 99.759%, 0.037%, and 0.204%, respectively. The isotopic composition of the moon's oxygen appears not to differ from that of earth. This suggests that

---

* It has been estimated that if the O_2 content of air were increased from 21% to 26%, then fires, once started, would rage out of control.

† Evidence for this is not conclusive. An ultraviolet-sensitive camera placed on the moon by the Apollo 16 mission shows that most of the O_2 presently in the atmosphere comes from the dissociation of water vapor by solar ultraviolet radiation.

both bodies originated in the same general region of the solar system. Meteorites, on the other hand, show a diversity of isotopic ratios, indicating their origins in different parts of the solar system.

Oxygen-17 and oxygen-18 find use as tracers. Since neither is radioactive, both must be detected by the mass spectrograph. It is now possible to purchase water in which the ^{17}O enrichment is 4% or in which the ^{18}O enrichment is 97%. The enrichment is usually carried out by fractional distillation.

Isotope effects are observed in the rates of reaction and equilibria of oxygen-containing compounds, but since the three isotopes do not differ greatly in mass, these effects are not as pronounced as in the case of hydrogen.

15.9 THE PREPARATION OF ELEMENTAL OXYGEN

Oxygen may be prepared in the laboratory by the electrolysis of water (Section 15.3) or by the thermal decomposition of certain oxygen-containing compounds such as $KClO_3$ (potassium chlorate) and mercury(II) oxide (HgO). The latter reactions are

$$2KClO_3(s) \xrightarrow{heat} 2KCl(s) + 3O_2(g)$$

$$2HgO(s) \xrightarrow{heat} 2Hg(l) + O_2(g)$$

The first reaction is very slow unless a material such as MnO_2 powder is mixed with the $KClO_3$, whereupon oxygen production proceeds rapidly (the basis of such effects is discussed in Section 20.8). The second reaction occurs upon relatively gentle heating. In either case the generated O_2 may be collected as shown for H_2 in Fig. 15.2.

The usual method for the commercial preparation of oxygen is distillation of liquid air. Liquid air itself may be prepared by a process utilizing the Joule-Thomson effect (see Section 3.11). Since the normal boiling points of liquid O_2 and liquid N_2 are 90 K and 77 K, respectively, the N_2 distills off in greater proportion (see Section 10.6) leaving the liquid residue enriched in O_2.

Nature produces elemental oxygen via the process of photosynthesis. Carbon dioxide and water are combined in chlorophyl-containing cells of plants to produce oxygen compounds called **carbohydrates**. The reaction in simplified form is

$$xCO_2(g) + xH_2O(l) \rightarrow (CH_2O)_x + xO_2(g)$$

where $(CH_2O)_x$ represents a general carbohydrate. When x is 6, the carbohydrate, $(CH_2O)_6$ or $C_6H_{12}O_6$, is glucose, a sugar. Studies employing oxygen-18 tracers demonstrate that the photosynthetic O_2 comes from the water in the reaction. The consequences of the finding are discussed in Chapter 25.

15.10 BINARY COMPOUNDS OF OXYGEN

The most common oxidation states of oxygen are -1 and -2. Binary compounds of oxygen in the -2 oxidation state are called **oxides**. Oxides can be grouped into two classes known as basic oxides and acidic oxides.

The **basic oxides** are those of the most electropositive metals; the oxygen in such compounds exists in the solid state as the oxide ion O^{2-}. The formation of oxide ions from molecular oxygen

$$\tfrac{1}{2}O_2(g) + 2e^- \rightarrow O^{2-}(g)$$

is endothermic, requiring about 9.4 eV of energy. Thus the ionic solids must have very high lattice energies in order to account for the high stability of basic oxides. The oxide ion is a very strong Brønsted-Lowry base; it reacts with H_2O to form OH^- as follows:

$$O^{2-}(s) + H_2O(l) \rightarrow 2OH^-(aq)$$

The reaction goes virtually to completion. Removal of water from a solution of a basic oxide leads to the production of solid hydroxides. Dissolving Na_2O, for example, in water and evaporating to dryness produces solid NaOH. Basic oxides are sometimes called **basic anhydrides**. Anhydride means "without water." Basic anhydrides are like instant bases; just add water and base (OH^-) is produced.

The **acidic oxides** are covalent compounds of the nonmetals. Many of these form Brønsted-Lowry acids. For example,

$$SO_2(g) + 2H_2O(l) \rightleftharpoons H_3O^+ + HSO_3^-(aq)$$
$$CO_2(g) + 2H_2O(l) \rightleftharpoons H_3O^+ + HCO_3^-(aq)$$

Such oxides are sometimes called **acid anhydrides**. Some oxides such as N_2O and MnO_2 are relatively inert and function as neither basic nor acid anhydrides. Other oxides, notably Al_2O_3 and ZnO, act as both basic and acidic oxides. These are said to be **amphoteric** oxides (from Greek *amphoteros*, both). For example, ZnO reacts with strong acids as

$$ZnO(s) + 2H_3O^+ \rightarrow Zn^{2+}(aq) + 3H_2O(l)$$

With base such as OH^- the reaction is

$$ZnO(s) + 2OH^-(aq) + H_2O \rightarrow [Zn(OH)_4]^{2-}(aq)$$

When certain basic oxides are treated with excess oxygen, **peroxide** compounds are formed:

$$2Na_2O(s) + O_2(g) \rightarrow 2Na_2O_2(s)$$

The ionic peroxides contain the peroxide ion O_2^{2-}, in which the oxidation state of oxygen is -1. The peroxide ion reacts with water as a Brønsted-Lowry base:

$$O_2^{2-}(s) + H_2O(l) \rightarrow HO_2^-(aq) + OH^-(aq)$$

The hydrogen peroxide ion HO_2^- decomposes slowly as follows:

$$2HO_2^-(aq) \rightarrow 2OH^-(aq) + O_2(g)$$

Hydrogen peroxide (H_2O_2) is a covalent compound. It acts as a strong oxidizing agent in either acid or basic solutions. A weak acid, it reacts with water as

$$H_2O_2(aq) + H_2O(l) \rightleftharpoons H_3O^+ + HO_2^-(aq)$$

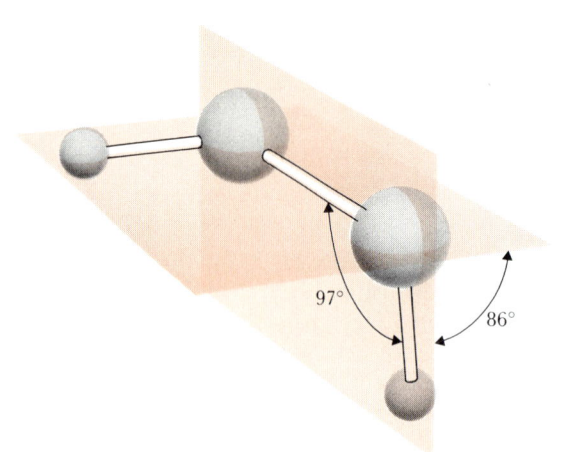

Fig. 15.3. The geometric structure of hydrogen peroxide, H_2O_2. The O—H bond length is 0.97 Å and the O—O length is 1.49 Å.

The ionization constant K_a is 2.4×10^{-12} at 25 °C. Pure H_2O_2 is a pale blue, viscous liquid which has physical properties similar to many of those of H_2O. For example, H_2O_2 freezes at -0.9 °C and has a normal boiling point of 152 °C. Aqueous solutions of H_2O_2 are rendered unstable by traces of heavy-metal-ion impurities and decompose with the evolution of oxygen:

$$2H_2O_2(aq) \rightarrow 2H_2O(l) + O_2(g)$$

The geometry of H_2O_2 is illustrated in Fig. 15.3. The skewed nature of this molecule would not have been predicted on the basis of the VSEPR model (Section 6.2).

Another interesting group of oxygen compounds of metals is the **superoxides**, containing the ion O_2^-. KO_2 is prepared by the action of O_2 on potassium at around 1 atm pressure. The superoxides are very powerful oxidizing agents. Space-science researchers have shown some interest in the reaction

$$4O_2^-(s) + 2CO_2(g) \rightarrow 2CO_3^{2-}(s) + 3O_2(g)$$

for possible use in spaceships to remove CO_2 and to regenerate O_2.

Water, perhaps the most prevalent oxide, is formed whenever hydrogen-containing compounds are burned in oxygen. Some typical reactions involving hydrocarbon fuels are

$$C_3H_6(g) + 5O_2(g) \rightarrow 3CO_2(g) + 4H_2O(l)$$
propane

$$C_2H_6O(l) + 3O_2(g) \rightarrow 2CO_2(g) + 3H_2O(l)$$
ethanol

$$C_6H_6(l) + 7\tfrac{1}{2}O_2(g) \rightarrow 6CO_2(g) + 3H_2O(l)$$
benzene

If the oxygen supply is restricted, some carbon monoxide, CO, is also formed.

15.11 OZONE

Oxygen has an allotropic modification called **ozone**, which has the molecular formula O_3. Ozone is formed in the atmosphere by the high-voltage discharges of electrical storms and by the action of ultraviolet radiation on O_2. The gas has a definite blue

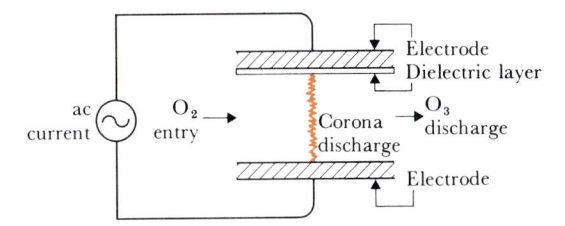

Fig. 15.4. Design of an apparatus for the laboratory production of ozone. The apparatus is kept at as low a temperature as practicable in order to minimize the thermal decomposition of ozone to diatomic oxygen.

color and a pungent odor that is readily detectable after a severe thunderstorm. Formation of O_3 from O_2 is endothermic, requiring about 1.5 eV per O_3 molecule. Ozone is most easily produced in the laboratory by subjecting a stream of O_2 gas to a silent electric discharge (corona) as illustrated in Fig. 15.4.

Two possible Lewis formulas for O_3 are

$$\ddot{O}::\ddot{O}:\ddot{O}: \quad \text{and} \quad :\ddot{O}:\ddot{O}::\ddot{O}$$

This suggests that the molecule has a bent structure (three oxygen atoms forming a "vee") and that each oxygen-to-oxygen bond is a composite of a single bond and a double bond. Since the double-bond length in O_2 is 1.21 Å and the single-bond length in H_2O_2 is 1.49 Å, the bond length in O_3 ought to be between these two values. The experimental value is found to be 1.28 Å, in accord with prediction. The bond angle is about 117°.

Ozone is a powerful oxidizing agent capable of reacting readily with many substances not easily oxidized by O_2. Some water-treatment plants are now using ozone instead of chlorine to remove pathogenic substances and undesirable organic contaminants. Unlike chlorine, ozone does not impart an unpleasant taste or odor to water; ozone is also capable of destroying substances such as viruses, detergents, and pesticide residues which chlorine cannot. Ozone is also useful in the treatment of sewage. Especially desirable in this context is the fact that ozone treatment leaves water saturated with oxygen and thus of optimum usefulness to aquatic life.

Most of the ozone in the earth's atmosphere exists at an altitude of around 25 to 50 km. About 10 ppm (parts per million by weight) of the atmosphere at this height is O_3, compared with roughly 0.04 ppm at altitudes below 10 km. It is believed that the O_3 of this so-called **ozone layer** forms in two steps:

$$O_2(g) + h\nu \rightarrow O(g) + O(g)$$
$$O_2(g) + O(g) + M \rightarrow O_3(g) + M$$

In the first step, ultraviolet radiation of wavelengths below 2400 Å forms atomic oxygen. Then an oxygen atom collides with an oxygen molecule in the presence of a third body M to form ozone. This second reaction is exothermic, producing about 100 kJ per mole of O_3 produced. The third body M, an atmospheric component such as N_2 or another O_2 molecule, is necessary to carry away excess chemical energy

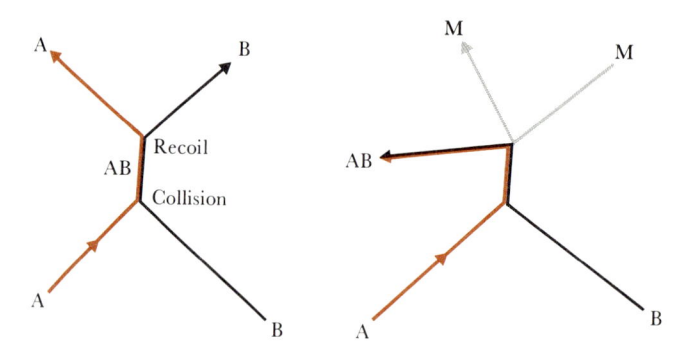

Fig. 15.5. Illustration of a three-body chemical reaction. In the left-hand diagram A and B collide and briefly form a molecule AB. The molecule AB imparts the energy produced by the combination to a recoil of the atoms A and B and thus decomposes. In the right-hand diagram A and B collide as before to form AB but, before recoil can occur, AB collides with a third body M (which may be A, B, AB, some other molecule, or the wall of the container). The third body M may carry away the excess energy of AB so that its atoms remain bound instead of recoiling.

released by the reaction. Without this third body the O_3 molecule would utilize this chemical energy to dissociate back into the reactants (see Fig. 15.5).*

One of the important functions of the ozone layer is to absorb harmful ultraviolet radiation from the sun before it reaches the earth. Ultraviolet radiation of wavelengths around 2600–2700 Å is capable of producing skin cancer and of damaging tissue. Ozone absorbs virtually all radiation below 3000 Å and thus forms a protective shield over the earth. The ultraviolet light is absorbed by the process

$$O_3(g) + h\nu \rightarrow O_2(g) + O(g)$$

It appears that a 1% decrease in the $[O_3]$ of the ozone layer would cause about a 2% increase in the intensity of harmful ultraviolet radiation reaching the earth. This, in turn, is thought to be capable of increasing the incidence of skin cancer by about 2%.

Substances such as nitrogen oxides from supersonic transport aircraft (SST's)(which operate at the ozone-layer level) are capable of removing ozone from the ozone layer. The process appears to be

$$NO(g) + O_3(g) \rightarrow NO_2(g) + O_2(g)$$
$$NO_2(g) + O(g) \rightarrow NO(g) + O_2(g)$$

The net result of this cyclic process in which NO is continually regenerated is the removal of ozone:

$$O_3(g) + O(g) \rightarrow 2O_2(g)$$

---

* This is an example of a general type of process called a **three-body reaction**: Two gaseous substances cannot combine to form a single product without the presence of a third body to carry away the excess energy of reaction. This excess energy ordinarily manifests itself as vibrational energy of the product molecule; the associated vibrations may break the newly-made bonds to reform the original reactants. Solid surfaces such as the walls of reaction vessels often serve as third bodies. In the atomic hydrogen torch (Section 15.6) the object being welded is the third body.

Chlorofluorocarbons, the main components of many refrigerants and of many aerosol propellants, eventually reach the ozone layer where they, too, remove ozone. The steps in the process probably are (using a typical chlorofluorocarbon)

$$CF_2Cl_2(g) + h\nu \rightarrow CF_2Cl(g) + Cl(g)$$
$$Cl(g) + O_3(g) \rightarrow ClO(g) + O_2(g)$$
$$ClO(g) + O(g) \rightarrow Cl(g) + O_2(g)$$

Here, chlorine atoms are recycled to be used over and over again to remove ozone.

15.12 THE OXYGEN CYCLE IN NATURE

Oxygen is found in so many compounds—both in living and nonliving matter—that it is difficult to give a simple and yet accurate account of just how it circulates within our planet. Nevertheless, some broad patterns can be identified; a simple version of some of these is shown in Fig. 15.6.

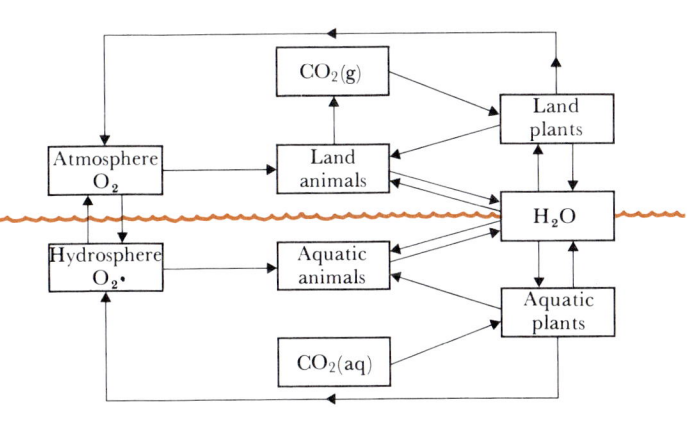

Fig. 15.6. A simplified version of the oxygen cycle in nature. The atmospheric pathway and the hydrospheric pathway are similar in broad outline but differ in many details. Note that the two pathways are linked by exchange of O_2 between water and air and by mutual chemical dependence on water. They are also linked in other ways, for example through interchange of CO_2, consumption of aquatic plants and animals by terrestrial animals, etc.

Most of the world's oxygen is tied up in carbonate and silicate rocks; the role of this oxygen in natural cycles may be ignored except on a long-term basis. The most important active members of the cycle are water, the oxides of carbon, and molecular oxygen. The chemically bound oxygen of water is being removed constantly by photosynthesis and eventually returned via respiration in a process which, on the average, completes one cycle in approximately two million years.

Some CO_2 is removed from the atmosphere to become dissolved in water. Some of the dissolved CO_2 becomes CO_3^{2-} which, in the presence of $Ca^{2+}(aq)$, forms insoluble $CaCO_3$ (limestone). If the limestone is eventually exposed by a geological uplift and subjected to a rainfall of slightly acid water, CO_2 is returned to the atmosphere by

$$CO_3^{2-}(aq) + 2H_3O^+ \rightarrow 3H_2O(l) + CO_2(g)$$

Some oxygen becomes tied up in ions such as $NO_3^-(aq)$ and $SO_4^{2-}(aq)$. Certain organisms (for example, bacteria living in bottom sludges) utilize $NO_3^-(aq)$ and $SO_4^{2-}(aq)$ as oxidizing agents in lieu of molecular oxygen and reduce these to NH_3 (or N_2) and to H_2S, respectively. The cycle is closed when NH_3 (or N_2) and H_2S are oxidized by O_2 to $NO_3^-(aq)$ and $SO_4^{2-}(aq)$, respectively.

It has been conjectured that primitive life probably developed anaerobically (in the absence of oxygen). Indeed, high levels of O_2 would probably have been inimical to molecular evolution. To obtain energy, primitive cells may have burned their fuel by an inefficient process known as fermentation. Somehow, the oxygen content of the primeval atmosphere may have increased (perhaps some primitive photosynthetic organisms arose by mutation), some life forms may have evolved a tolerance for high O_2 levels, and some may have been able to utilize O_2 in metabolic processes of high-energy outputs. In any event, the oxygen-tolerant organisms eventually became predominant.

15.13 WATER: ITS UNUSUAL PROPERTIES

Water is the most abundant compound on the surface of the earth, and it is certainly one of the most familiar. Not only is three-fourths of the earth's surface covered by the seas and oceans, but water is present virtually everywhere as lakes, ponds, rivers, springs, ice fields, snow, rain, and fog and in all plant and animal tissues. We take water so much for granted that we seldom take time to think about the unique physical and chemical properties that make it one of our most valuable treasures.

Water is the only common substance that expands upon freezing. This is what causes not only bursting of automobile radiators not protected by antifreeze, but also the destruction of living cells when ice crystals expand to break membranes and tissues. At the same time this property of water has its benefits as well. Since expansion on freezing produces a decrease in density, ice floats on the surface of water. In natural waters this surface layer acts as insulation to prevent most lakes and ponds from freezing solidly to the bottom, destroying aquatic life in the process. The repeated freezings and thawings of water in the crevices and pores of rocks is also an important step in the production of soil.

The water molecules in ice are arranged in a hexagonal crystal lattice. This hexagonal symmetry is reflected in the six-sided patterns characteristic of snowflakes. More specifically, the oxygen atoms occupy the centers and alternating apexes of cubes in such a way that each oxygen atom is surrounded tetrahedrally by four other oxygen atoms (see Fig. 15.7). The hydrogen atoms are located between pairs of neighbor atoms. The ice crystal turns out to be a somewhat open structure in that the water molecules are inefficiently packed. When ice melts, the open structure closes up somewhat so that each oxygen atom now has five or more nearest neighbors

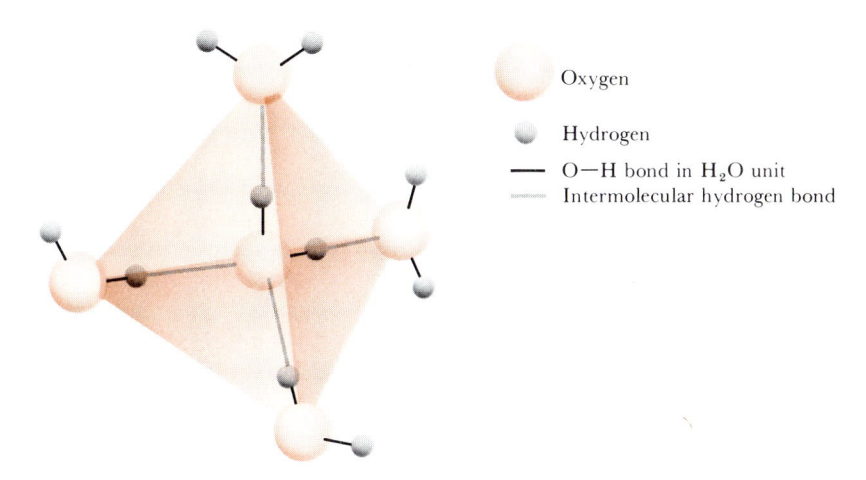

Fig. 15.7. The crystal structure of ice. Each oxygen atom is surrounded tetrahedrally by four other oxygen atoms and each pair of oxygen atoms is hydrogen-bonded. With respect to the entire crystal, the hydrogen atoms are distributed at random; there is no regularity in their locations.

rather than only four. X-ray studies show that the O—O distance in ice is about 2.7 Å whereas that in the liquid is around 2.9 Å. Nevertheless, the more efficient packing in the liquid (more nearest neighbors) results in an increased density in spite of the larger O—O distances.

In Section 9.6 Le Châtelier's principle was used to explain how ice can be melted by the application of pressure. Given the structures of ice and liquid water just discussed, it is easy to see how this happens. External pressure upon the crystal lattice collapses it and forces the water molecules to pack more efficiently; that is, each oxygen atom tends to acquire more nearest neighbors. Hydrogen bonding then becomes weakened, since only a few hydrogen atoms can stay located between two oxygen atoms. Consequently, the molecules can roll past each other in a manner characteristic of a liquid.

If a sample of liquid water at 0 °C is warmed, its density increases slightly and reaches a maximum value at 3.98 °C. Above this temperature the density slowly decreases. Thus, when the ice in a pond melts in the spring, the cold water at the surface sinks to the bottom of the pond. At the same time, water warmer than about 4 °C tends to flow nearer the surface. The net result is a pattern of flow which tends to stir up nutrient-laden bottom sediments and circulate these throughout the body of water. In this way the aquatic life systems of the pond become switched on for another season of activity.

The surface tension (see Section 8.1) of water, 72 mN·m^{-1} at 25 °C, is higher than that of any other common liquid. Thus water will rise higher in a capillary tube of given radius than will any other liquid. The transport of water in plants depends to some extent on capillary flow. Microscopic examination of the stems of plants show that they contain thousands of very small-diameter channels which serve as capillaries. About 0.15% of the water circulated in a typical land plant is used in the photosynthetic production of carbohydrate and oxygen; the remaining 99.85% is used in transpiration, for transport of minerals and other nutrients, and for temperature control. A large tree may use several barrels of water on a hot summer day.

Water also has a higher specific heat, about 4.2 J·K^{-1}·g^{-1}, than any other common liquid. This property enables large bodies of water—the oceans in particular—to act as giant thermostats tending to prevent rapid temperature fluctuations on the earth's surface. Water's high heat of vaporization (2260 J·g^{-1}, again higher than that of any other common liquid) also contributes to its temperature-regulating ability. The unusually high specific heat of water is thought to be due to the ability of individual atoms to become vibrationally responsive to a thermal energy input. In substances with a more rigid arrangement of bonds, several atoms tend to move as one unit in response to thermal excitation and, consequently, absorb less energy per unit rise in temperature.

Water has an unusually high thermal conductivity.* This property, in conjunction with its high specific heat, makes it an excellent coolant. A given mass of water can absorb more heat per degree rise in temperature—and do this more quickly—than a like mass of any other common substance. Of all the relatively common liquids, only mercury, a metal, has a higher thermal conductivity.

Water has a very high dielectric constant; at 25 °C its value is about 80 whereas that of benzene at the same temperature is about 2.3. Even ethanol, a somewhat polar substance, has a dielectric constant of only 24 at the same temperature. Its high dielectric constant makes water not only a good insulator but also a good solvent for polar and ionic substances as well.

Chemically pure water is a very poor conductor of electrical current, a fact which appears to contradict the popular conception of wet hands as a shock hazard around electrical appliances. The explanation is that water is such a good solvent that it very rapidly becomes contaminated with impurities which make it a good conductor. These impurities come from the air (CO_2, for example) and from the containers in which water is stored. Human skin contains a layer of ionic substances (NaCl from sweat, for example) which rapidly turns water into an excellent conductor. Highly purified water must be stored in quartz containers away from air if it is to retain its low electrical conductivity.

15.14 HYDRATES

Many naturally occurring cyrstalline solids appear to contain water molecules incorporated into the lattice structure in definite stoichiometric amounts. For example, when a solution of copper(II) sulfate ($CuSO_4$) is allowed to lose water by evaporation, bright blue triclinic crystals begin to form. The empirical formula of the crystalline material is $CuSO_4 \cdot 5H_2O$; that is, there are five molecules of H_2O associated with each Cu^{2+} and SO_4^{2-} ion pair. When this solid is heated, the H_2O comes off and a white, amorphous powder of $CuSO_4$ remains. Naturally occurring gypsum consists of monoclinic crystals of formula $CaSO_4 \cdot 2H_2O$. Another mineral, mirabilite, consists of

---

* The thermal conductivity of a substance is the amount of thermal energy transported per unit time per unit area of surface per unit temperature gradient. Thus, if 2.0 cal of thermal energy can be passed through a rod 2 cm long with a cross-sectional area of 4.0 cm^2 in 4.0 s when one end of the bar is at 27 °C and the other is at 30 °C, the thermal gradient is (30 − 27) K/2 cm or 1.5 K·cm^{-1} and the thermal conductivity is

$$\frac{2.0 \text{ cal}}{(4.0 \text{ s})(4.0 \text{ cm}^2)(1.5 \text{ K·cm}^{-1})} = 0.083 \text{ cal·s}^{-1} \cdot \text{cm}^{-1} \cdot \text{K}^{-1}$$

This is 35 J·s^{-1}·m^{-1}·K^{-1} in SI units.

monoclinic crystals of composition $Na_2SO_4 \cdot 10H_2O$. The peroxides of the alkali metals form well-defined crystalline compounds containing water; the hexagonal crystals of $Na_2O_2 \cdot 8H_2O$ are an example. Compounds of the above type are called **hydrates**. The substance $CuSO_4 \cdot 5H_2O$ is called copper(II) sulfate pentahydrate, and the other compounds are named in an analogous fashion. All of the hydrates lose water on heating, some more readily than others.

There exists a class of natural and artificial complex silicates containing aluminum (as well as one or more of the elements potassium, sodium, or calcium) called **zeolites** which lose water upon heating but retain, unchanged, their crystalline form. The resulting very porous material contains a multitude of channels which permit the passage of small molecules. Zeolites are sometimes called **molecular sieves** since they can be used to filter molecules, allowing small ones to pass through and keeping larger ones back. By varying the type of ion used in synthesizing the zeolite, the pore sizes can be controlled. Zeolites find many applications such as separating undesirable low-octane components from hydrocarbons used in gasoline blends. They may also be used as water softeners via the reaction

$$2NaAlSiO_4(s) + M^{2+}(aq) \rightarrow M(AlSiO_4)_2(s) + 2Na^+(aq)$$
   a zeolite

where M^{2+} is a metal ion such as Ca^{2+} or Mg^{2+}, two typical components of hard water. The action of the zeolite is to remove an ion such as M^{2+} and replace it with Na^+. The zeolite is recharged by forcing NaCl brine through it; this causes the above reaction to proceed in the reverse direction.

Water forms some unusual hydrates with some highly insoluble molecules. Such hydrates generally form at temperatures above the freezing point of pure water and may resemble ice. For example, an icy slush is frequently found in natural gas lines when the temperature is as high as 20 °C. This hydrate consists of a cluster of H_2O molecules forming a cage about a hydrocarbon molecule such as CH_4. Since CH_4 and H_2O interact very weakly, the surrounded CH_4 decreases the internal pressure of the cluster and allows it to crystallize at temperatures above the normal ice point. Similar hydrates also form in winter wheat, but so slowly that the accompanying expansion does not cause cell damage. These hydrates then prevent damage to the sprouts when severe freezing weather does occur. Similar hydrates also form in corn, but rapidly enough that corn can show signs of frost damage at temperatures as high as 4.5 °C.

Many substances are capable of absorbing water from a moist atmosphere; such substances are said to be **hygroscopic**. Hygroscopic substances are widely used as **desiccants**, that is, to remove water from the atmosphere. For example, dry $CaCl_2$ may be placed into a closed container (called a **desiccator**), and an object such as a damp wrist watch can be dried by allowing it to stand in the desiccator for a few hours. The liquid water in the watch evaporates and reacts with $CaCl_2$ to form a hydrate:

$$CaCl_2(s) + 2H_2O(g) \rightarrow CaCl_2 \cdot 2H_2O(s)$$

Some substances will absorb water from a moist atmosphere to form concentrated solutions; $CaCl_2 \cdot 2H_2O$ itself continues to absorb water (if enough is present) to eventually form such a solution. This phenomenon is called **deliquescence**. When

A portion of the great ice fall of the Muldrow glacier on Mt. McKinley. Most of the earth's water not found in oceans, lakes, and rivers is found in ice fields similar to this. (Courtesy of Bradford Washburn, Boston Museum of Science.)

table salt (NaCl) contains small amounts of MgCl$_2$ as an impurity, the salt becomes damp due to deliquescence of MgCl$_2$.

Under certain conditions hydrates may lose water spontaneously. For example, CuSO$_4 \cdot$5H$_2$O and H$_2$O constitute the equilibrium system

$$\text{CuSO}_4 \cdot 5\text{H}_2\text{O(s)} \rightleftharpoons \text{CuSO}_4\text{(s)} + 5\text{H}_2\text{O(g)}$$

The equilibrium water vapor pressure is 7.8 torr at 25 °C. Whenever CuSO$_4 \cdot$5H$_2$O is placed into an environment (at 25 °C) in which the water vapor pressure is less than 7.8 torr, the hydrate spontaneously loses water. This process is called **efflorescence**.

15.15 HEAVY WATER

Heavy water or deuterium oxide (D$_2$O) has a density about 10% higher than that of ordinary water. Whereas the self-ionization constant K_w of ordinary water is 1.0×10^{-14} at 25 °C, that of heavy water is 1.4×10^{-15} at this same temperature. D$_2$O has a freezing point of 3.8 °C and a normal boiling point of 101.4 °C. The solubility of many substances in heavy water differs slightly from that in ordinary water since D$_2$O has a slightly smaller dielectric constant than ordinary water.

A tonne of ordinary water contains about 28 mg of D$_2$O. Water with greater amounts of D$_2$O may be prepared by electrolysis or fractional distillation (see Section 15.3).

Laboratory experiments show that mice can live almost indefinitely on water which is 30% D$_2$O but they become sterile. When the D$_2$O level exceeds 30% the mice die. Green algae are remarkably tolerant to D$_2$O and eventually adapt to virtually pure D$_2$O. Upon a first exposure to D$_2$O, algae stop growing and the culture takes on a sickly appearance. Eventually, growth resumes again at about one-half the ordinary-water rate. Seeds watered only with D$_2$O fail to sprout.

15.16 THE WATER CYCLE IN NATURE

The entire earth, including its atmosphere, is estimated to contain the equivalent of 1.3×10^{21} L of liquid water. About 97% of this is in the seas, oceans, lakes, rivers, and other surface waters. The remaining 3% is distributed as follows: 77.16% in ice caps and glaciers, 22.8% in underground areas, and a scant 0.03% in the atmosphere.

Ice caps play an important role in determining and regulating global weather patterns. The Antarctic ice cap, for example, cools the surrounding water, which sinks and spreads over the ocean floor. This sets up a gravity-powered circulation pattern which extends to the oceans of the world, causing them to cool and thereby cooling in turn the air masses in contact with them.

Not immediately apparent to the casual observer is the fact that all the earth's water is constantly engaged in intricate circulation patterns collectively termed the **hydrologic cycle** (see Fig. 15.8). This seemingly perpetual motion is powered almost entirely by the sun's radiation, in turn due largely to energy evolved by the hydrogen-to-helium fusion reaction. A very small amount of power is also supplied by the

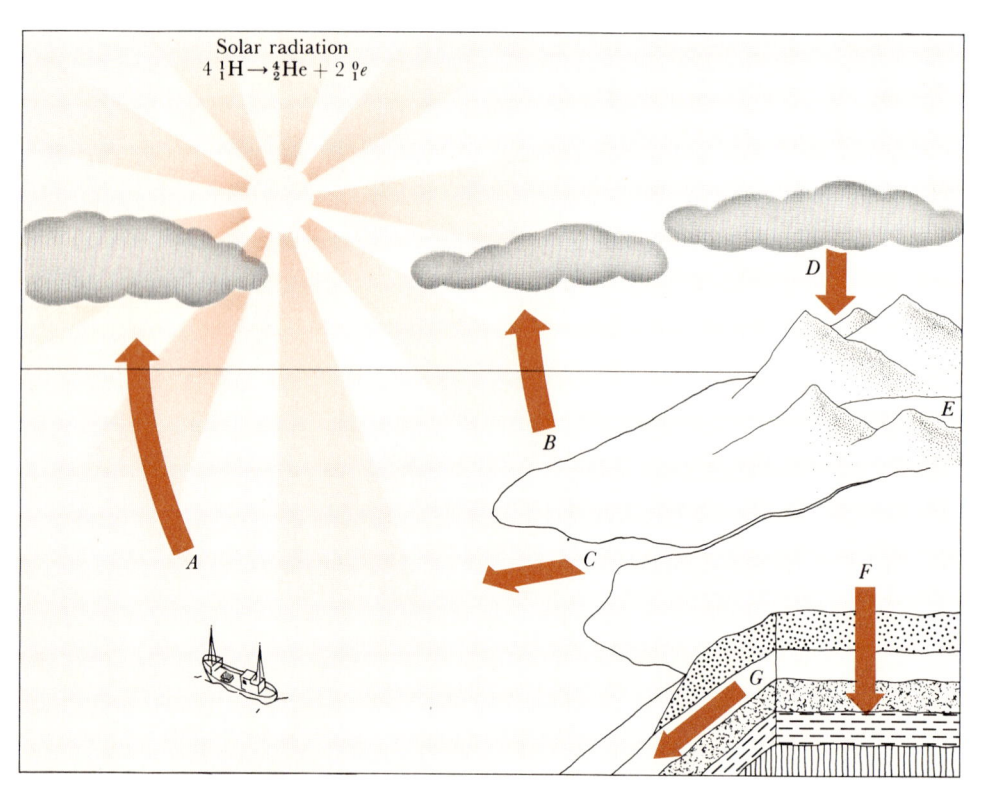

Fig. 15.8. The hydrologic cycle: *A*, evaporation from ocean areas; *B*, evaporation from land areas; *C*, run-off of rainwater to the ocean via rivers and streams; *D*, precipitation as rain or snow; *E*, accumulation as icecaps and glaciers; *F*, seepage to underground; *G*, horizontal flow of water in the water table.

earth's own natural heat, itself the result of radioactive decay processes deep beneath the crust. Basically, the hydrologic cycle consists of two parts: **evaporation** of water from land and sea areas and **precipitation** as rain or snow. Each year about 3.2×10^{17} L of water is evaporated from the oceans by the heat of the sun with the aid of winds, and about 6×10^{16} L is evaporated from the land. Much of the moisture entering the atmosphere from land areas arises from transpiration in plants. A small amount also arises from the oxidation of hydrogen-containing substances, decay of organic material and combustion of petroleum fuels, for example. All of the evaporated water eventually returns to the earth as rain, snow, sleet, hail, or condensed fog; about 75% falls directly into the oceans, about 9% runs off the land areas and quickly reaches the seas, and 16% soaks into the soil. The latter fraction travels downward until it reaches a layer of impermeable rock whence it tends to spread out horizontally to form what is known as the **water table**. It is this underground water which forms the source of much of our potable water. By digging deeply enough water can be found virtually anyplace in the world, including the middle of the driest desert!

The waters which return to the oceans by way of rivers and streams are typically charged with particulate sediment and dissolved salts; perhaps about 4×10^8 tonnes of matter enters the seas each year via this process, which plays a large role in the shaping of land forms. Some water also finds its way to the oceans by means of underground flow.

When water flows underground over limestone (mostly $CaCO_3$) it may carve out huge caverns and form underground streams and lakes. The Carlsbad Caverns in New Mexico, Mammoth Cave in Kentucky, and Wind Cave in the Black Hills of South Dakota were all formed in this way. Although $CaCO_3$ is only slightly soluble in pure water, it dissolves considerably more readily in acidified water. When water falls as rain it dissolves gaseous carbon dioxide

$$CO_2(g) \rightleftharpoons CO_2(aq)$$

The dissolved CO_2 forms carbonic acid, H_2CO_3, via the process

$$CO_2(aq) + H_2O(l) \rightleftharpoons H_2CO_3(aq)$$

Carbonic acid reacts with the CO_3^{2-} in limestone to form soluble acid carbonate HCO_3^-. The process is

$$H_2CO_3(aq) + CO_3^{2-}(s) \rightleftharpoons 2HCO_3^-(aq)$$

The annual rainfall in the United States is about 6.5×10^{15} L. About 6.1×10^{14} L of this is used rather directly by humans; 40% goes to agricultural uses, mainly irrigation, 52% is used by various industries, and 8% is used by public water utilities. Of the industrial portion about 80% is used to produce steam power for generation of electricity. The public water utilities portion represents about 570 L per day per U.S. citizen. The total per capita use of water in the United States is close to 7600 $L \cdot da^{-1}$, compared with about 38 $L \cdot da^{-1}$ typical of many underdeveloped areas of the world.

SUMMARY

Hydrogen, the lightest of the elements, is the most abundant element in the universe. The fusion of hydrogen to helium in the sun produces most of the energy which powers the earth. Elemental hydrogen may be prepared in the laboratory by passing an electric current through acidified water; heavy water becomes concentrated in the undecomposed residue. The relative ability of various metals to displace hydrogen from water or acid solutions may be used as a basis of assigning relative activities to these metals. The binary compounds of hydrogen comprise hydrogen oxidation states of both +1 and −1; the latter occurs for compounds with elements much less electronegative than hydrogen. Hydrogen is a strong reducing agent and can be used to prepare metallic elements from their oxides and chlorides. Atomic hydrogen is a reducing agent of unusual potency.

Oxygen is the most abundant element on earth. The elemental substance may be prepared, as is hydrogen, by electrolysis of water, by thermal decomposition of certain compounds such as $KClO_3$ or HgO, or by fractional distillation of liquid air. Plants produce oxygen by photosynthesis; the sun's energy is used to combine water and carbon dioxide to produce carbohydrates and molecular oxygen. Oxides of the most electropositive metals form bases (hydroxides) when dissolved in water; oxides of nonmetals tend to form acids with water. Compounds of oxygen in the −1 oxidation state are called peroxides. Elemental oxygen also exists in the triatomic allotropic form called ozone. Ozone is a very powerful oxidizing agent. Its presence in the earth's stratosphere acts to shield the earth and its inhabitants from hazardous ultraviolet radiation from the sun.

Water, perhaps the most commonly encountered chemical substance, possesses a variety of unique physical and chemical properties which enable it to play crucial roles in the processes of life, in the control of climate, and in many other operations occurring on or near the surface of the earth.

LEARNING GOALS

1. Be able to state at least three important physical properties of elemental hydrogen and elemental oxygen.

2. Know the names and nuclear compositions of the isotopes of hydrogen and have at least a rough idea of their relative abundance.

3. Be able to describe at least two methods of preparing elemental hydrogen and elemental oxygen.

4. Be familiar with simple tests for identifying hydrogen and oxygen.

5. Know what the activity series is and how to use it.

6. Be familiar with the different types of hydrides and their respective properties. Be able to give specific examples of each type of hydride.

7. Be able to write balanced chemical equations for the reaction of hydrogen with metal halides and metal oxides.

8. Know the nuclear compositions of the isotopes of oxygen and have at least a rough idea of their relative abundance.

9. Be able to write formulas for oxides, peroxides, and superoxides.

10. Be able to write balanced chemical equations

for the reaction of common elements and their compounds with oxygen.

11. Know what basic and acidic oxides are.

12. Know the structure and basic properties of ozone and how it is produced.

13. Be able to discuss the role of ozone in the upper atmosphere of the earth.

14. Be able to discuss how ozone is produced in the upper atmosphere and how it can be removed by fluorinated hydrocarbons and oxides of nitrogen.

15. Be able to give a general outline of the oxygen cycle in nature.

16. Be able to state the most important unique properties of water.

17. Know the structures of liquid and solid water.

18. Know what hydrates are and be able to state some of their most important properties.

19. Be familiar with zeolites and their uses.

20. Be able to give a general outline of the water cycle in nature.

21. Be able to describe the process of cavern formation in limestone, using chemical equations.

DEFINITIONS, TERMS, AND CONCEPTS TO KNOW

escape velocity
photosynthesis
protium, deuterium, and tritium
isotope effect
activity series
saline or ionic hydrides
covalent hydrides
interstitial hydrides

hydrogen bonding
basic oxides
acidic oxides
superoxides
peroxides
three-body reaction
anaerobic process
hydrates

deliquescence
efflorescence
desiccant
desiccator
zeolites
hydrologic cycle
water table

QUESTIONS AND PROBLEMS

Hydrogen

1. Write balanced chemical equations for the following reactions.
 a) Hydrogen with bromine.
 b) Reduction of SnO_2 with hydrogen.
 c) Zinc with acetic acid.
 d) Aluminum with dilute sulfuric acid.
 e) Sodium hydride with water.

2. Hydrogen gas is produced in the laboratory by the reaction of zinc metal and hydrochloric acid. The gas is collected over water at 25 °C.
 a) How many moles of zinc are needed to produce 0.33 mole of H_2?
 b) What weight (in grams) of zinc does this represent?

c) In one experiment at 25 °C, the volume of gas collected at 760 torr atmospheric pressure is 200 cm³. If the vapor pressure of water at 25 °C is 23.8 torr, what is the partial pressure of the H_2 in the collector bottle?
d) How many moles of H_2 are produced in (c)?
e) How many moles of zinc are needed to produce the H_2 in (c)?

3. In each reaction below, identify the oxidizing agent and the reducing agent.
a) $2Na + H_2 \rightarrow 2NaH$
b) $Cl_2 + H_2 \rightarrow 2HCl$

4. Why may the proportions of ^1H, ^2H, and ^3H have varied over the ages? Would this be as likely for ^{16}O, ^{17}O, and ^{18}O?

5. Discuss the operation of the atomic hydrogen torch in terms of a three-body reaction.

6. Give an example of each of the following.
a) Reaction of hydrogen with a metal
b) Reaction of hydrogen with a nonmetal

7. Hydrogen gas is sometimes prepared by the **Bosch process**, in which superheated steam is passed over heated coal or coke and the products mixed with more steam and passed over a metal oxide catalyst. The equations for the two-step process are

$$C(s) + H_2O(g) \rightarrow CO(g) + H_2(g)$$
$$CO(g) + H_2O(g) \xrightarrow[\text{oxide}]{\text{metal}} CO_2(g) + H_2(g)$$

a) What is the net overall reaction corresponding to the above process?
b) How many moles of steam would be required to produce one mole of hydrogen gas?
c) The products of the first step are called **water gas** and are sometimes used as fuel. Write balanced chemical reactions for the combustion of water gas in air.
d) Suggest a way of separating the final products of the reaction using high pressure and cold water.
e) What volumes (measured at 25 °C and 1 atm) of CO_2 and H_2 can be obtained from 100 g of water?

8. What volume of hydrogen at 25 °C and 1 atm would be required to lift a balloon if the balloon and its load weigh 100 kg?

Oxygen

9. Elaborate on the following statement: Since the process $\frac{1}{2}O_2(g) + 2e^- \rightarrow O^{2-}(g)$ is endothermic (9.4 eV), then the lattice energy of solid oxides must be very large to account for their high stabilities.

10. An astronaut produces about 16 L of CO_2 (measured at standard temperature and pressure) per hour when not exercising. What weight of KO_2 would a spaceship have to carry to convert an astronaut's daily production of CO_2 to O_2? How much O_2 (measured at standard temperature and pressure) would be produced per day?

11. a) What is the oxidation number of oxygen in KO_2?
b) Is the ion O_2^- paramagnetic or diamagnetic?
c) What is a likely electron configuration for the superoxide ion?
d) Predict the relative O—O bond lengths in O_2, O_2^{2-} (peroxide ion), and O_2^- (superoxide ion) by use of bond orders.

12. When oxygen in a diamagnetic excited state decays to the paramagnetic ground state, it emits a photon of wavelength 6340 Å. This process has been shown to involve two molecules of O_2 simultaneously emitting a single photon. Estimate the energy difference (in eV) between the two states.

13. Calculate the energy (in kcal·mol⁻¹ and kJ·mol⁻¹) of light of wavelength 3000 Å. Since it normally requires an energy of 100 kcal·mol⁻¹ (418 kJ·mol⁻¹) or more to cause chemical bonds to rupture, why do you think the ozone shield is of importance to life on earth?

14. The process $O_2 + O + M \rightleftharpoons O_3 + M$ is exothermic and produces about 100 kJ of energy for each mole of ozone formed.
a) What effect does the concentration of M have on the equilibrium?
b) What will be the effect of increased total pressure on this reaction?
c) What will be the effect of increased temperature on this reaction?

15. The amount of ozone in air may be measured by reaction with KI solution. The reaction, which goes essentially to completion, is $O_3 + 2KI + H_2O \rightarrow I_2 + 2KOH + O_2$. Measurement of the iodine produced enables one to determine the amount of ozone in the sample. When 2.00 L of air (measured at standard temperature and pressure) is passed through KI solution, 1.05 g of I_2 is produced. What is the volume percent of O_3 in the air sample?

16. Show how the following reaction can be made the basis of a cyclic method for the production of oxygen from the air. Explain thoroughly.

$$2\text{BaO}(s) + O_2(g) \underset{700\,°C}{\overset{500\,°C}{\rightleftharpoons}} 2\text{BaO}_2(s)$$

The temperatures shown are those for which the reaction goes largely to completion in the direction indicated. Is the above reaction endothermic or exothermic?

17. What volume of oxygen, measured at 25 °C and 1 atm, can be prepared from 1.00 kg of potassium chlorate?

18. Oxygen can be prepared from sodium peroxide using the reaction

$$2\text{Na}_2\text{O}_2 + 2\text{H}_2\text{O} \xrightarrow{\text{heat}} 4\text{NaOH} + O_2$$

a) What weight of sodium peroxide is needed to produce 10 L of oxygen at 30 °C and 740 torr?
b) What weight of sodium hydroxide is produced in a)?

19. Complete and balance the following chemical reactions.
a) $CH_4 + O_3 \rightarrow$?
b) $NH_3 + O_2 \rightarrow NO + $?
c) $FeO + O_2 \rightarrow$?
d) $CH_3CH_2OH + O_2 \rightarrow CO_2 + $?
e) $H_2O + O_3 \rightarrow$?
f) $H_2S + O_2 \rightarrow$?   (two possibilities)

20. PbO_2 is called lead peroxide but MnO_2 is called manganese dioxide. Explain.

Water

21. Why is water such a good solvent for many ionic substances?

22. When a basic solution of water is electrolyzed, the skeleton cathode and anode half-reactions are $H_2O \rightarrow H_2$ and $OH^- \rightarrow O_2$, respectively. Write balanced cathode and anode reactions and the net reaction (see Section 14.4).

23. Taking into account the existence of the isotopes ^1H, ^2H, ^3H, ^{16}O, ^{17}O, and ^{18}O and ignoring clusters of water molecules, answer the following.
a) How many different kinds of H_2O molecules can exist?
b) How many different kinds of H_3O^+ can exist?
c) How many different kinds of OH^- can exist?

d) How many different substances are present in water?

24. The heat of vaporization of water is about 2260 $J \cdot g^{-1}$. Given that 1 eV is about 1.6×10^{-19} J, calculate the energy (in eV) needed on the average to remove one H_2O molecule from the liquid.

25. Water from a deep well is analyzed for tritium content and is found to produce about 2000 $counts \cdot L^{-1} \cdot hr^{-1}$. How long ago, on the average, was it since this water fell as rain? (See Section 11.9).

26. If all the water vapor normally present in the atmosphere (on the average) were condensed to liquid and spread uniformly over the earth's land area, how deep would the water be? Assume that the earth's diameter is 12,700 km and that 30% of its total area is land.

27. Justify the following statement: Were it not for hydrogen bonding, all the water in our seas, lakes, and rivers would boil away.

28. Show that the reaction of $CoCl_3$ with water to form $[Co(H_2O)_6]Cl_3$ is a Lewis acid–base reaction between Co^{3+} and H_2O and identify the Lewis acid and the Lewis base.

29. A mixture of 2 L each of hydrogen and oxygen at 25 °C and 1 atm was exploded with a spark. What volume of which gas remained unreacted (if measured at 25 °C and 1 atm) and what weight of water was formed?

30. The boiling points of two similar compounds generally vary directly as their molecular weights, with the higher molecular weight compound having the higher boiling point. Yet, H_2O has a normal boiling point of 100 °C and H_2S has a normal boiling point of −61.8 °C. Explain.

Hydrates

31. A hydrate has the empirical formula $CaCl_2 \cdot xH_2O$, where x is not known. When a 10.0-g sample of the hydrate is strongly heated, its weight decreases to 7.55 g. What is the value of x?

32. What is the percentage of water (by weight) in the hydrate $Na_2SO_4 \cdot 10H_2O$?

33. The equilibrium vapor pressure of water in the reaction

$$CuSO_4 \cdot 5H_2O(s) \rightleftharpoons CuSO_4(s) + 5H_2O(g)$$

is 7.8 torr at 25 °C.

a) Write the K_p expression for the above reaction.
b) What is the numerical value of K_p for the above reaction?
c) What is the numerical value of K_c for the above reaction?
d) The equilibrium vapor pressure of water in the above reaction increases at higher temperature. Is the reaction exothermic or endothermic?

34. Hydrated sodium tungstate (or wolframate), $Na_2WO_4 \cdot nH_2O$, contains 10.9% water by weight. What is the value of n?

35. A hydrate of $Al(ClO_3)_3$ has a molecular weight of 386 amu. What is the formula of the hydrate?

36. Washing soda, $Na_2CO_3 \cdot 10H_2O$, can be purchased in 12-lb bags. How many pounds of water are there in a bag?

37. What is meant by the vapor pressure of a hydrate such as $CuSO_4 \cdot 5H_2O$? Which would be more stable at a given temperature, a hydrate with a low vapor pressure or one with a high vapor pressure? Explain.

General

38. Predict the products of each of the following reactions and balance the equations.

a) $SnO(s) + H_2(g) \rightarrow$
b) $PbCl_2(s) + H_2(g) \rightarrow$
c) $Ca(s) + HCl(aq) \rightarrow$
d) $Na(s) + H_2(g) \rightarrow$
e) $H_2(g) + S(s) \rightarrow$
f) $MgH_2(s) + H_2O(l) \rightarrow$
g) $CaO(s) + H_2O(l) \rightarrow$
h) $P_4O_{10}(s) + H_2O(l) \rightarrow$
i) $H_2O_2(aq) + Fe^{2+}(aq) \rightarrow$
j) $Fe(s) + O_3(g) \rightarrow$
k) $CuSO_4(s) + H_2O(l) \rightarrow$
l) $Cs(s) + H_2O(l) \rightarrow$
m) $Sn(s) + HCl(aq) \rightarrow$
n) $C_6H_6(l) + O_2(g) \rightarrow$
o) $C_{12}H_{22}O_{11}(s) + O_2(g) \rightarrow$
p) $KO_2(s) + CO(g) \rightarrow$
q) $Fe_2O_3(s) + H_2(g) \rightarrow$
r) $H_2O_2(aq) \xrightarrow{\text{heat}}$
s) $B_2H_6(g) + H_2O(l) \rightarrow$
 $[B(OH)_3(aq)$ is one product]
t) $Ag(s) + HNO_3(aq) \rightarrow$
 [assume $NO_2(g)$ as one product]

39. Some ocean-bottom dwelling bacteria use nitrate ions instead of O_2 to oxidize foodstuffs. A typical oxidation reaction carried out during bacterial metabolism is

$$NO_3^- + C_6H_{12}O_6 \rightarrow N_2 + CO_2$$

Balance this equation in acid solution.

40. At the present time rivers carry about 4×10^8 metric tons (tonnes) of salts into the oceans each year. The oceans now contain about 1.5×10^{21} L of water containing about 35 g of salts per liter. How long would it take to attain the present salt level of the oceans at the current rate of river input?

41. Dual water samples are taken at various depths in a shallow freshwater pond. One of the bottles is covered with aluminum foil (the *dark* bottle) and the other is clear plastic (the *light* bottle). The O_2 content of the water in each bottle is measured at the time of sampling and after 24 hours. During this time the bottles are kept suspended in the water at the depth from which the samples were drawn. The change in O_2 content during the 24-hour period is labeled ΔM_{O_2}. The following data were obtained in a 4-m-deep pond.

DEPTH (m) BELOW THE SURFACE	ΔM_{O_2} (g·m^{-3}) LIGHT BOTTLE	DARK BOTTLE
1	3	-1
2	2	-1
3	0	-1
Bottom	-3	-3

(A positive ΔM_{O_2} signifies an increase in O_2; a negative ΔM_{O_2} signifies a decrease in O_2.) Suggest reasons for (a) the differences between the results for the dark bottle and those for the light bottle, (b) the effect of depth.

ADDITIONAL READINGS

Buswell, Arthur M., and Worth H. Rodebush, "Water." *Scientific American,* April 1956. Discusses many of the unusual properties of water.

Runnels, L. K., "Ice." *Scientific American,* December 1966. Discusses the crystal structure of ice and the role played by defects.

Cloud, Preston, and Aharon Gibor, "The Oxygen Cycle"; Penman, H. L., "The Water Cycle." *Scientific American,* September 1970. Two articles in an issue devoted to the biosphere.

Gabianelli, Vincent J., "Water—The Fluid of Life." *Sea Frontiers* **16**, 5 (September–October 1970): 258. A popular-level account of water and its properties.

The Barnstead Basic Book on Water. Boston: The Barnstead Company, 1971. A compilation of facts concerning water. Available for $1.00 from the Barnstead Company, 225 Rivermoor Street, Boston, Mass., 02132. Worth every penny!

Ham, John O., *Study and Interpretation of the Chemical Characteristics of Natural Water* (Geological Survey Water-supply Paper 1473). Washington: U.S. Government Printing Office, 1970. A review of the chemical, geological, and hydrologic principles and processes that control the composition of natural water, with methods for studying and interpreting chemical analyses. Useful as a reference work. Available for $2.25 from the Superintendent of Documents.

Leopold, Luna B., and Kenneth S. Davis, *Water.* New York: Time-Life Books, 1970. A Life Science Library volume containing superb illustrations and much interesting information concerning water.

Eisenberg, David, and Walter Kauzmann, *The Structure and Properties of Water.* New York: Oxford University Press, 1969. Most of the material in this excellent and comprehensive treatment is at an advanced level, but anyone with a modest chemistry background can read parts of it with profit.

Peixoto, José P., and M. Ali Kettani, "The Control of the Water Cycle." *Scientific American,* April 1973. The authors suggest how large-scale properties of the water cycle could be used to design novel control methods.

Leh, F., "Ozone: Properties, Toxicity and Applications." *J. Chem. Educ.* **50**, 404 (1973).

Katz, Joseph J., "The Biology of Heavy Water." *Scientific American,* July 1960. Tells what happens when experimental organisms are raised on heavy water.

Bamberger, C. E., and J. Braunstein, "Hydrogen: A Versatile Element," *American Scientist* **63** (July–August 1975: 438). Discusses how hydrogen generated from water may prove valuable in extending the world's dwindling hydrocarbon supply.

16
CARBON AND ITS COMPOUNDS

Oil-drilling rig on a desert-like plain in Iran. Vast petroleum deposits are often found to underlie rather inhospitable areas of the earth. (Copyright © Paolo Koch, courtesy of Rapho/Photo Researchers, Inc.)

Again the atoms gamboled before my eyes—winding and turning like snakes. And look, what was that? One snake grabbed its own tail and mockingly the shape whirled before my eyes.
FRIEDRICH AUGUST KEKULÉ

Friedrich August Kekulé (1829–1896) was a German chemist of remote Czech ancestry. The original name was Kekule ze Stradonic, which was later Germanized (and made somewhat quasi-French) to Kekulé von Stradonitz. Kekulé abandoned studies in architecture to become a chemist. In 1859, while dozing on a London omnibus, he had a vision of carbon atoms linked in chains; later, while dozing in his study before the fireplace, his vision recurred, this time to suggest the ring structure of benzene.

Early chemists once recognized two major types of chemical compounds; those arising from inanimate matter were called **inorganic** compounds and those arising from animate matter were called **organic** compounds. Since the large majority of organic compounds contain carbon as an important component, the modern definition of **organic chemistry** is the chemistry of carbon and its compounds, even though many of these do not necessarily arise from plants or animals.

This chapter and the next deal with some of the more important aspects of modern organic chemistry: the molecular architecture of organic compounds and their nomenclature, the chemical and physical properties of important classes of compounds, and something of the strategy and tactics chemists use in synthesizing organic molecules in the laboratory. The field of organic chemistry is so rich in information that it is difficult to present a brief overview without greatly underplaying its true scope. Yet, at the same time, even a brief overview tends to strike the student as a bewildering succession of facts, names, structures, and reactions. Nevertheless, the basic framework of organic chemistry is reasonably systematic and relatively easy to discover if the student keeps an eye open for it from the very beginning. Instead of attempting to commit every detail to memory, the student should observe the patterns of structure, the orderly buildup and use of names, the basic classes of molecules, and the general types of reactions involved.

A large chunk of graphite, a naturally occurring allotropic form of carbon. Value less than one cent per gram. (Courtesy of Grant Heilman.)

A diamond, another allotropic form of carbon, in the form of a cut, polished, and mounted gem. Value: $25,000 or more per gram. (Courtesy of Marshall Henrichs.)

16.1 ELEMENTAL CARBON

Although carbon is only the fourteenth most abundant element (in terms of number of atoms) on the earth, it forms a greater number of different compounds than any other element except hydrogen.* Out of every 10,000 atoms making up the bulk of the earth, about 9 are carbon atoms. This amounts to about 4 g of carbon per 10,000 g of matter. The earth's carbon consists of 98.89% of ^{12}C and 1.11% of ^{13}C; about 1 atom out of every 10^{12} atoms of carbon is the radioactive isotope ^{14}C. The carbon-12 isotope serves as the universal standard of the atomic mass scale (see Section 1.5).

Elemental carbon exists in two allotropic modifications: diamond and graphite. The former is one of the hardest substances known, being surpassed in hardness only by certain synthetic materials developed during the 1960s.†

Diamond has a cubic crystalline structure in which each atom of carbon is surrounded tetrahedrally by four other carbon atoms (see Fig. 16.1). The density of diamond is 3.51 g·cm^{-3} at 25 °C while that of graphite is considerably less, 2.22 g·cm^{-3}. This suggests a much more open arrangement of atoms in graphite. X-ray studies show that graphite has a layered structure such as that illustrated in Fig. 16.2. The carbon atoms in each layer are arranged in a hexagonal pattern re-

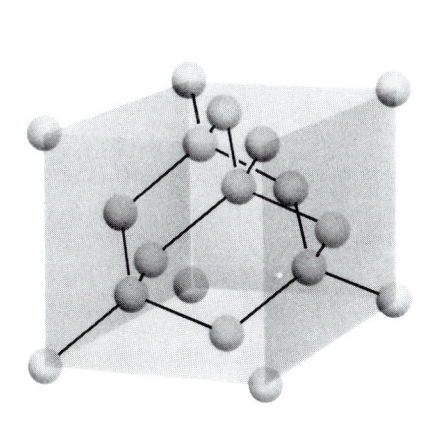

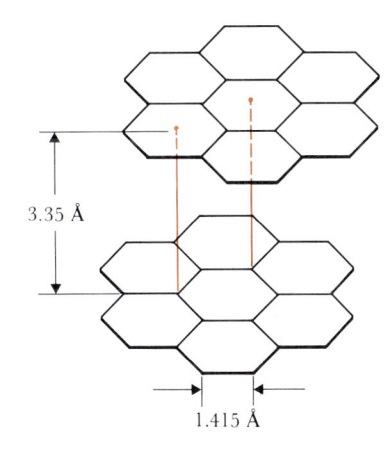

Fig. 16.1 (left). The crystalline structure of diamond. Each carbon atom is at the center of a regular tetrahedron with four other carbon atoms at the apexes.

Fig. 16.2 (right). The layered structure of graphite. The vertical lines connect points directly opposite each other in adjacent, parallel layers. Carbon atoms are located at the vertexes of the hexagons.

---

* Boron is also capable of forming a large variety of compounds and may possibly surpass carbon in this respect.

† Hardness may be expressed in a rough numerical fashion by the use of Mohs' hardness scale: 1, talc; 2, gypsum; 3, calcite; 4, fluorite; 5, apatite; 6, orthoclase; 7, quartz; 8, topaz; 9, corundum; and 10, diamond. If solid A will scratch solid B but solid B will not scratch solid A, solid A is said to be harder than solid B. A copper penny will barely scratch calcite but will not scratch fluorite; thus it has a hardness slightly above 3.

sembling chicken-wire fence or hexagonal bathroom tile. Atoms in the layer are located about 1.415 Å apart, and the layers themselves are about 3.35 Å apart. This rather large interlayer distance accounts for the low density of graphite relative to that of diamond. It may also account for some of the excellent lubricating properties of powdered graphite in that the layers can slip rather readily past each other since there are only weak electronic attractions between them. However, modern research suggests that gases and vapors adsorbed on the graphite surface and between layers account for most of the lubricating properties. The "lead" of a lead pencil (actually a rod of graphite mixed with various additives) owes its marking properties to these same weak interlayer attractions.

The bonding between carbon atoms within a single layer may be represented by three different Lewis structures as shown below. Carbon atoms are at hexagon vertexes and electron pairs are represented as dashes.

The fact that alternative Lewis structures can be written suggests that graphite C-to-C bonds are composites of single and double bonds. The fact that graphite is an excellent electrical conductor (it is the only nonmetal so endowed) is explained by assuming that the "mobile" bonding electrons occupy a partially filled conduction band (see Section 7.9). The bonding in diamond, on the other hand, is described by a single Lewis structure and is consistent with this material's behavior as an electrical insulator.

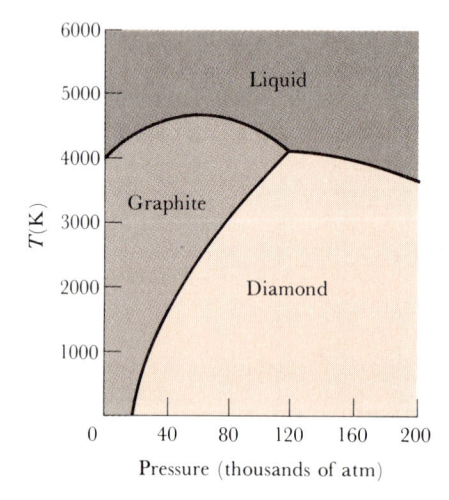

Fig. 16.3. A possible phase diagram for carbon. Experimental difficulties make precise location of the areas uncertain.

Graphite can be transformed to diamond at very high temperatures and very high pressures, for example, around 3000 K and 125,000 atm. Figure 16.3 shows what is believed to be an approximately correct portion of the phase diagram of carbon. Most synthetically produced diamonds are small—up to about 0.1 carat (0.02 g)—and are used only for industrial purposes, for example, as edges of cutting tools and tips of drills. In 1970 synthetic gem-quality diamonds several carats in weight were produced for the first time.

16.2 THE OXIDES OF CARBON

There are five known oxides of carbon: CO, CO_2, C_3O_2, C_5O_2, and $C_{12}O_9$; but only the first two, carbon monoxide and carbon dioxide, are of major importance.

Carbon monoxide, CO, is formed whenever carbon or carbon-containing compounds react with limited amounts of oxygen. Rather large amounts of CO are produced by internal combustion engines and by the burning of coal, petroleum, and wood. Surprisingly, over 90% of the 4 billion tons of CO entering the atmosphere each year arises from the oxidation of methane (CH_4) emitted by decaying organic matter. Many varieties of soil fungi, for example, common strains of *Penicillium* and *Aspergillus,* remove CO from the air very rapidly.*

Carbon monoxide is very toxic to humans and many other animals. Its toxicity results from the fact that it replaces O_2 in blood hemoglobin. Prolonged exposure to levels of 15 ppm can cause delayed reaction time and 40 ppm causes decreased intellectual functioning. Cigarette smoke has levels of up to 40 ppm of CO.†

Carbon dioxide, CO_2, is the usual product when carbon or carbon-containing compounds are burned in an adequate supply of oxygen. This substance dissolves readily in water and slowly reacts with it to form hydrogen carbonate, H_2CO_3. Hydrogen carbonate is a weak Brønsted-Lowry acid whose conjugate base, HCO_3^-, can also behave as an acid to produce the conjugate base CO_3^{2-}. These reactions are summarized below:

$$CO_2(g) \rightleftharpoons CO_2(aq) \rightleftharpoons H_2CO_3(aq) \underset{-H_2O}{\overset{+H_2O}{\rightleftharpoons}} H_3O^+ + HCO_3^- \underset{-H_2O}{\overset{+H_2O}{\rightleftharpoons}} H_3O^+ + CO_3^{2-}$$

Some experimental evidence for the above relationships is as follows: If CO_2-saturated water is added to a basic solution containing phenolphthalein indicator, the pink color fades very slowly, indicating that the reaction of CO_2 with water is rather slow. If a weak acid such as HOAc is added to the phenolphthalein solution, the pink color disappears almost instantly.

Carbon dioxide reacts with hydroxides to form hydrogen carbonate (HCO_3^-) and carbonate (CO_3^{2-}) ions. The reactions are

$$CO_2(aq) + OH^- \rightarrow HCO_3^-$$
$$HCO_3^- + OH^- \rightarrow CO_3^{2-} + H_2O$$

---

* One wonders how rapidly these organisms remove CO from the air of cities with wall-to-wall concrete.

† Evidently the smokers themselves are willing to tolerate this partial asphyxiation; their nonsmoking neighbors may not always feel accordingly.

The hydrogen carbonate ion has the resonance structures

$$\left[\text{H}-\ddot{\text{O}}-\underset{\underset{:\ddot{\text{O}}:}{\|}}{\text{C}}-\ddot{\text{O}}: \right]^{1-} \leftrightarrow \left[\text{H}-\ddot{\text{O}}-\underset{:\ddot{\text{O}}:}{\overset{}{\text{C}}}=\ddot{\text{O}} \right]^{1-}$$

The hydrogen atom, since it is attached to very electronegative oxygen, is easily removed by a base (as a proton H$^+$). If CO$_2$ gas is passed into a solution of NaOH and the solution evaporated to dryness, a mixture of solid NaHCO$_3$ (sodium hydrogen carbonate) and Na$_2$CO$_3$ (sodium carbonate) is obtained. Incidentally, carbon monoxide passed into NaOH solution produces sodium formate.* The fundamental reaction here is

$$\text{CO} + \text{OH}^- \rightarrow \underset{\underset{\text{O}}{\|}}{\text{HC}}\text{O}^-$$
formate ion

Carbon dioxide is readily produced in the laboratory by treating carbonates such as Na$_2$CO$_3$, NaHCO$_3$, or CaCO$_3$ with a strong acid. The reactions are

$$2\text{H}_3\text{O}^+ + \text{CO}_3^{2-} \rightarrow \text{CO}_2(g) + 3\text{H}_2\text{O}$$

$$\text{H}_3\text{O}^+ + \text{HCO}_3^- \rightarrow \text{CO}_2(g) + 2\text{H}_2\text{O}$$

Carbon dioxide is also produced by heating solid carbonates at very high temperatures. For example, when limestone, CaCO$_3$, is heated the reaction is

$$\text{CaCO}_3(s) \xrightarrow{\text{heat}} \text{CaO}(s) + \text{CO}_2(g)$$

John Tyndall (1820–1893), an Irish-born British scientist, did research on the nature of heat and was one of the first men ever to climb the Weisshorn. Tyndall also reached to within 800 feet of the summit of the Matterhorn a few years before its conquest by Whymper.

According to a theory first proposed by the British physicist John Tyndall in 1861, atmospheric CO$_2$ plays a major role in the regulation of the temperature of the earth via a mechanism called the **greenhouse effect**.† The effect occurs as follows:

---

* The CO$_2$ + OH$^-$ and CO + OH$^-$ reactions differ considerably in the way they actually take place, however. In the former, OH$^-$ simply bonds to carbon through oxygen and, in the latter, this is followed by migration of hydrogen from oxygen to carbon. The HCO$_3^-$ ion has an acidic hydrogen but the HCO$_2^-$ ion does not.

† Recent theoretical studies suggest that the analogy between the CO$_2$ of the atmosphere and the glass of a greenhouse is false. There is strong reason to believe that greenhouses function due to restriction of convection currents and not due to trapping of infrared radiation as was once supposed.

The gases which make up the earth's atmosphere allow the passage of solar radiation, mostly in the visible and ultraviolet ranges. This radiation is absorbed by the earth and some of it is reradiated as infrared radiation. However, CO_2 absorbs infrared radiation strongly and thus traps the outbound energy relatively close to the earth's surface. The trapped radiation energy becomes manifested as an increase in the temperature of the atmosphere.*

16.3 THE CARBON CYCLE IN NATURE

As is the case with oxygen (see Section 15.12), much of the carbon near the surface of the earth is in constant circulation and most of it has been cycled at one time or another through living organisms. Figure 16.4 is a simplified version of the main pathways along which carbon circulation occurs.

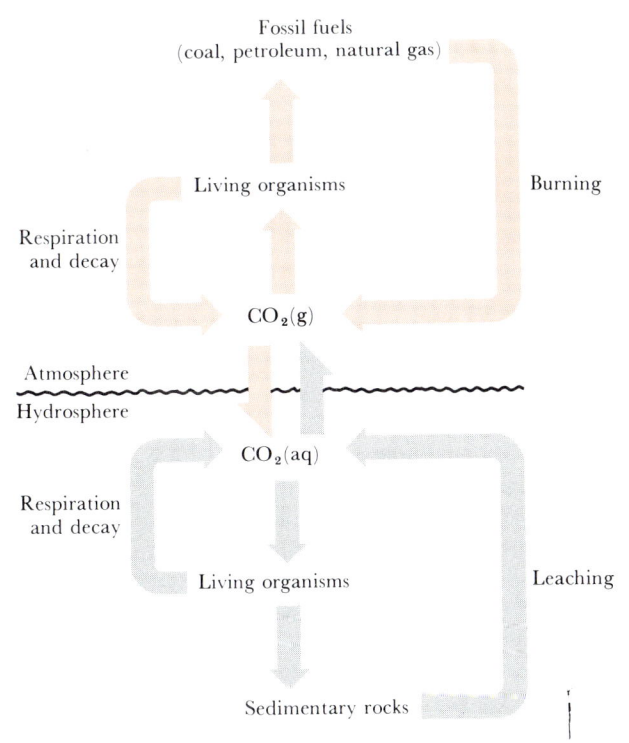

Fig. 16.4. A simplified illustration of the carbon cycle in nature.

---

* Climatologists do not agree as to the effect an increase of atmospheric temperature will have on the earth's climate. Some say this will help to melt the ice caps and thus raise ocean levels enough to flood coastal cities; others predict more snowfall and the advance of ice sheets. Also, dust and other particulate matter from industrial activity screens the earth from some solar radiation and tends to cancel out the greenhouse effect.

As Fig. 16.4 shows, there are actually two different main circulation pathways for carbon in nature. One of these is confined largely to the atmosphere and land areas, whereas the other is confined largely to the seas, oceans, and other large bodies of water. The two cycles are linked by a constant interchange brought about by wind and wave action at the atmosphere–hydrosphere interface. Perhaps about 100 billion tonnes of CO_2 is exchanged each year at this interface.

In the land–atmosphere cycle, CO_2 is used by plants in the photosynthetic production of organic compounds known as **carbohydrates**. Many carbohydrates have the empirical formula CH_2O (glucose, for example, is $C_6H_{12}O_6$) so that the general photosynthetic process is represented by

$$CO_2 + 2H_2A \xrightarrow{h\nu} CH_2O + H_2O + 2A$$

This is an endothermic process ($h\nu$ represents radiant energy—from the sun in this case) where A is usually oxygen (in which case 2A is O_2) but may be sulfur or an organic group. Some of the organic compounds produced by photosynthesis are used to provide energy for the plant's own needs so that some CO_2 becomes returned directly to the atmosphere as respiration products. The CO_2 which becomes incorporated into the basic structure of the plant is eaten by organisms or animals unable to fix CO_2 by photosynthesis or dependent on carbon in addition to CO_2 from the atmosphere. The rest of the photosynthesized matter slowly decays to CO_2 and to CO or becomes sidetracked in the formation of deposits of **fossil fuels**—coal, peat, natural gas, and oil. Although the formation of fossil fuels is exceedingly slow and involves but a minute fraction of the circulating carbon at any given time, eons and eons of such processes have resulted in the accumulation of roughly fifty times as much carbon in fossil fuel deposits as is found in living land organisms of all kinds.

Without the deliberate intervention of humans, the carbon locked in fossil fuel deposits can reenter the circulation path only by very slow and infrequently occurring processes such as seepage of natural gas and oil to the surface, weathering of exposed coal beds, and slow decay of surface peat deposits.

The carbon dioxide in the oceans is utilized mainly by organisms known as **phytoplankton**.* Such organisms depend on photosynthesis and live in those parts of the oceans where sunlight penetrates. Phytoplankton are eaten by other organisms and thereby begin the food chain for other aquatic life. Some of the organic matter accumulates as sediment on the ocean bottom and eventually forms sedimentary rocks such as limestone ($CaCO_3$) and dolomite ($MgCO_3$). Probably about 2000 times as much carbon is now tied up in sedimentary rocks as is present in fossil fuel deposits. The loss of carbon via the formation of sedimentary rocks is offset somewhat by the leaching action of surface water (rain, streams) on sedimentary rocks exposed on land by geological uplifting. This process returns small amounts of carbon to the oceans.

---

* There are two general types of plankton (small floating or weakly swimming bodies) present in water; **phytoplankton** are plants and **zooplankton** are animals.

16.4 THE CARBON-TO-CARBON BOND

Since carbon [(He)$2s^2 2p^2$] has four valence-shell electrons, it can form bonds with more electronegative elements such as chlorine (for example,

C : Ċl :

in CCl_4) and also with more electropositive elements such as sodium (for example, C : Na in CH_3Na). Furthermore, carbon can not only bond to itself but it can do so via extended rings and chains in an almost infinite variety of ways. No other element exhibits self-bonding ability to a degree approaching that of carbon. Although many other elements can bond to themselves—the diatomic molecules are examples—none is capable of forming extended arrays of such bonds.

TABLE 16.1 M—M AND M—O BOND ENERGIES FOR THE ELEMENTS M = C, Si, AND S

M	BOND ENERGY (eV) FOR[a] M—M	M—O
C	3.60	3.65
Si	2.26	3.84
S	2.22	3.5

[a] The bond energy represents an estimate of the energy required to break a given chemical bond as it exists in typical chemical compounds. The bond energy is defined in such a way that it is relatively constant from molecule to molecule and also that the sum of the bond energies over all the bonds in a molecule is a good approximation to the total dissociation energy of the molecule (see Section 18.6).

Since carbon is the smallest element in its group, it is capable of approaching another element or another carbon atom very closely to form a strong chemical bond. In Table 16.1 are listed the energies required to break the bonds between several pairs of like atoms. Note that the C-to-C bond energy is considerably higher than that of any other atom pair. Also of importance is the fact that C-to-O bonds require almost exactly the same amount of energy to rupture as does a C-to-C bond. This suggests that from a purely energetic standpoint there is no advantage to transforming a C-to-C bond to a C-to-O bond; hence C-to-C bonds should be relatively stable to oxidation. This is not true for oxygen bonds to the other elements listed in Table 16.1; in each case the M-to-O bond is energetically favored over the M-to-M bond. The fact that the structures of various living organisms are built around long chains and rings of C-to-C bonds partially explains the ability of life on earth to resist oxidation by the large quantities of oxygen in the atmosphere.*

---

* There are substances called **silicones** which exist as chains of silicon and oxygen atoms:

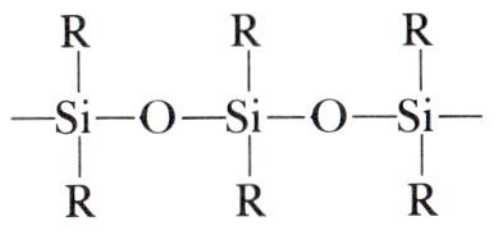

$$\begin{array}{ccccc} & R & & R & & R \\ & | & & | & & | \\ -\text{Si} & -\text{O} & -\text{Si} & -\text{O} & -\text{Si}- \\ & | & & | & & | \\ & R & & R & & R \end{array}$$

where R represents groups such as CH_3. Silicones have high heat stability and find use in lubricating oils. Perhaps life on some other planet has evolved on the basis of silicone chains.

Some of the more common ways of forming extended C-to-C linkages are as follows.

1. Unbranched chains:

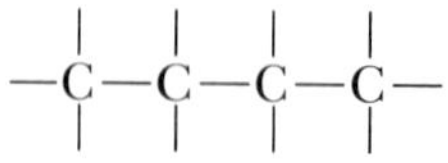

$$-\underset{|}{\overset{|}{C}}-\underset{|}{\overset{|}{C}}-\underset{|}{\overset{|}{C}}-\underset{|}{\overset{|}{C}}-$$

where each dash not linking two carbon atoms represents a two-electron bond to monovalent elements such as hydrogen, chlorine, or fluorine or to monovalent groups such as CN, NO_2, NH_2, and others to be introduced later.

2. Branched chains:

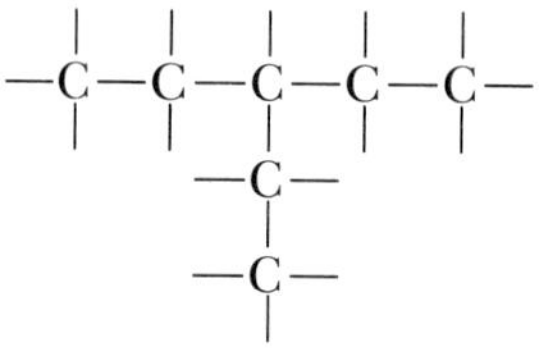

$$-\underset{|}{\overset{|}{C}}-\underset{|}{\overset{|}{C}}-\underset{|}{\overset{|}{C}}-\underset{|}{\overset{|}{C}}-\underset{|}{\overset{|}{C}}-$$

3. Cyclic or ring structures:

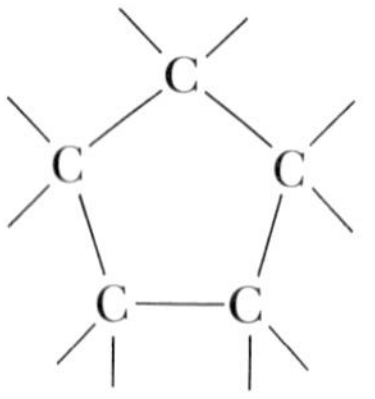

4. Combinations of 1, 2, and 3 (or only two of these):

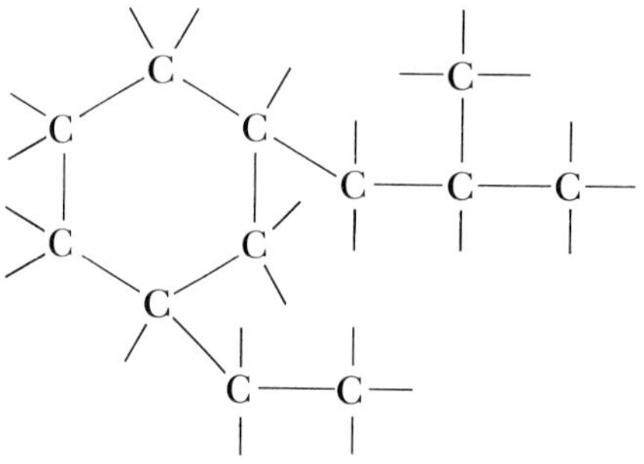

Variations involving double and triple bonds and other elements will be discussed later. Note that in each type of linkage there are four bonds to each and every carbon atom. For all practical purposes this is true for all organic compounds; carbon monoxide is the only important exception.

What is perhaps surprising is the high stability of some of these intricate linkages of carbon atoms. Carbon compounds with molecular weights in the millions are known, and some are exceedingly stable even at elevated temperatures.

16.5 ISOMERISM IN ORGANIC COMPOUNDS

Two or more different compounds with the same molecular formula are said to be **isomers** (Greek *isos,* equal and *meros,* share or part, that is, an *equal sharing* of formulas by two or more substances). Because of the variety of ways in which carbon can bond to itself and to other atoms and because of the three-dimensional nature of molecular structure, there are several fundamentally different ways in which isomerism can occur in organic compounds. Examples of each of these ways are given for those cases in which a given molecular formula leads to only two isomers or isomeric molecules.

There are two ways of linking up the four carbon atoms in the compound with the molecular formula C_4H_{10}. These are:

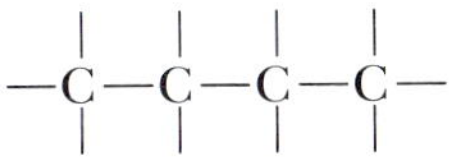

$$-\overset{|}{\underset{|}{C}}-\overset{|}{\underset{|}{C}}-\overset{|}{\underset{|}{C}}-\overset{|}{\underset{|}{C}}-$$

and

$$-\overset{|}{\underset{|}{C}}-\overset{|}{\underset{\overset{|}{\underset{|}{C}}}{C}}-\overset{|}{\underset{|}{C}}-$$

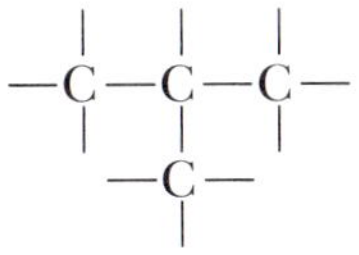

(Hydrogens have been omitted for clarity.) These are called **structural isomers** or, more descriptively, **linkage isomers**. Note that the C-to-C linkages are fundamentally different in the two cases; the first is an entirely unbranched chain and the second is a branched chain. No other different ways of linking the carbons are possible for the given molecular formula C_4H_{10}.

Isomerism can occur even though no differences in carbon-to-carbon linkages exist. To illustrate this type of isomerism requires the use of formulas which show the correct geometrical relationships of the atoms within the molecule. Consider the two planar molecules

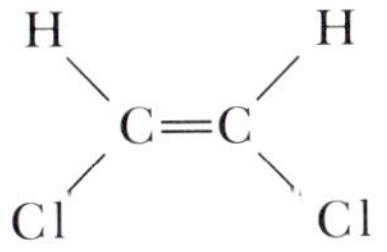

$$\begin{matrix} H & & H \\ \diagdown & & \diagup \\ & C=C & \\ \diagup & & \diagdown \\ Cl & & Cl \end{matrix}$$

and

$$\begin{matrix} H & & Cl \\ \diagdown & & \diagup \\ & C=C & \\ \diagup & & \diagdown \\ Cl & & H \end{matrix}$$

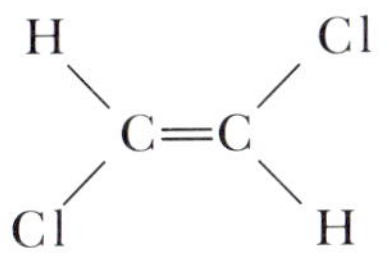

The presence of the double bond prevents rotation about the C-to-C axis; otherwise one isomer could readily convert to the other. Both molecules have the molecular formula $C_2H_2Cl_2$ and both have the same atom-to-atom linkages for all atoms, yet the two molecules have different three-dimensional spatial relationships among some of the atoms. In the first molecule the two hydrogen atoms are on the same side of the C=C bond and so are the two chlorine atoms. In the second molecule the two hydrogen atoms are on opposite sides of the C=C and so are the two chlorine atoms. The two molecules are said to be **stereoisomers** of a special type known as **diastereomers**. The two diastereomers have different physical and chemical properties.

There is one other type of stereoisomerism which involves considerably more subtlety. Consider the two molecules

where the large circle represents a carbon atom and a, b, c, and d are four *different* atoms or a group of atoms arranged tetrahedrally about it. Note first that the two molecules are drawn as mirror images, that is, when one of the molecules "looks" at the other it is as if it sees its own image in a mirror. Next, note that the two mirror images are not superimposable, that is, it is not possible to simply rotate one of the molecules so as to match it up, atom for atom, with the other. For example, if atoms a and b are matched up, atoms c and d are reversed. Molecules such as these are stereoisomers of a special type known as **enantiomers** or mirror-image isomers. Enantiomers always consist of a pair of molecules whose mirror images are nonsuperimposable. Right- and left-hand gloves bear this same relationship to each other; hence enantiomers are sometimes said to possess the property of **handedness** or **chirality**.

As discussed later, the members of an enantiomeric pair have almost—but not quite—the same physical and chemical properties.

16.6 HYDROCARBONS

Compounds containing carbon and hydrogen only are called **hydrocarbons**. These are perhaps the simplest of organic molecules. The chief source of hydrocarbons in nature is fossil fuel deposits, petroleum and natural gas in particular.

There are two different types of hydrocarbons: aliphatic and aromatic. **Aliphatic hydrocarbons** are characterized by unbranched-chain or branched-chain linkages of carbon atoms. Certain ring-type linkages whose chemical properties are similar to the unbranched-chain or branched-chain compounds are also usually considered as aliphatic compounds. These bear the specific name of **alicyclic hydrocarbons**. **Aromatic hydrocarbons** include benzene and compounds resembling benzene in chemical behavior. Since similarities between aliphatic and aromatic compounds are frequently more obvious than their differences, this classification scheme is not always a useful one. Aliphatic compounds are subdivided into four main groups: alkanes, alkenes, alkynes, and the alicyclic compounds just mentioned. The main characteristics of each of these groups are discussed in the next four sections.

There exist two or more separate sets of names for most hydrocarbons and related compounds. Many of the simpler, well-known compounds have so-called **common** names, names that became established before chemists realized the need for a logical and systematic approach to naming. Some of these names are so well-established—often widely used outside the chemical community—that it is unlikely that they will be totally replaced. The names of a second set are those recommended by the International Union of Pure and Applied Chemistry (**IUPAC**). These names are based on a systematic procedure which minimizes memorization and which emphasizes similarities among different compounds. This text will use mainly the IUPAC names except for those few cases where there is wide use of the common names.

16.7 ALKANES

Alkanes have the general molecular formula C_nH_{2n+2} ($n = 1, 2, 3, \ldots$). The simplest alkane, CH_4, is called **methane**. The formulas of all the higher alkanes are generated by adding one or more CH_2 units to the methane formula. Compounds related in this way are said to be **homologs**. The first homolog of methane is C_2H_6 and is called **ethane**. Both of the naming systems have as their foundation the alkane names listed in Table 16.2 for $n = 1$ through 12. Names for n beyond 12 are not often needed but may be found in textbooks of organic chemistry or other references.

TABLE 16.2 NAMES OF SOME UNBRANCHED-CHAIN ALKANES AND THEIR CORRESPONDING ALKYL GROUPS

n	ALKANE NAME	ALKYL RADICAL NAME
1	Methane	Methyl
2	Ethane	Ethyl
3	Propane	Propyl[a]
4	Butane	Butyl
5	Pentane	Pentyl
6	Hexane	Hexyl
7	Heptane	Heptyl
8	Octane	Octyl
9	Nonane	Nonyl
10	Decane	Decyl
11	Undecane	Undecyl
12	Dodecane	Dodecyl

[a] The group

$$\begin{array}{c} CH_3 \\ \diagdown \\ CH- \\ \diagup \\ CH_3 \end{array}$$

is called **isopropyl** in the common system even though a compound called isopropane does not exist.

The principal geometric characteristic of alkanes is that each carbon atom has its four bonds tetrahedrally oriented about itself; that is, each bond is described in terms of sp^3 hybrid orbitals.* As shown earlier in Fig. 6.15, this leads to bonds between atoms characterized by electron densities whose maximum values are along a line joining the two bonded atoms. Bonds having this characteristic are called **sigma bonds**. The C—H sigma bonds are formed by overlap of carbon sp^3 hybrid orbitals with hydrogen $1s$ orbitals; the C—C sigma bonds are formed by overlap of two carbon sp^3 hybrid orbitals. The geometric structure of ethane may then be illustrated as

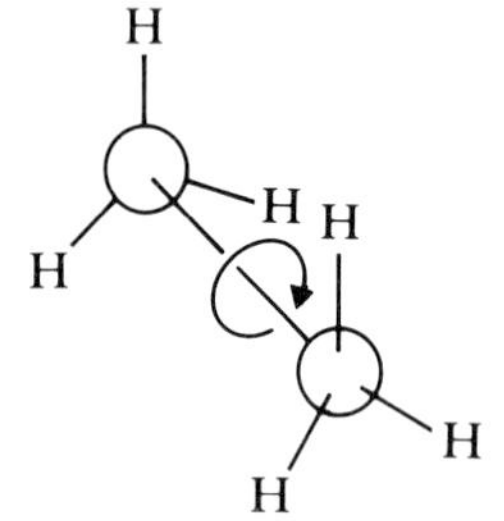

where the curved arrow indicates that there can be rotation about the C—C bond.

Two different geometrical arrangements of atoms such that one arrangement can be converted to the other by rotation about a single bond are said to represent different **conformations** of a given molecule. Although ethane has an infinite number of conformations, two are of special importance. Two alternative ways of illustrating these are shown below.

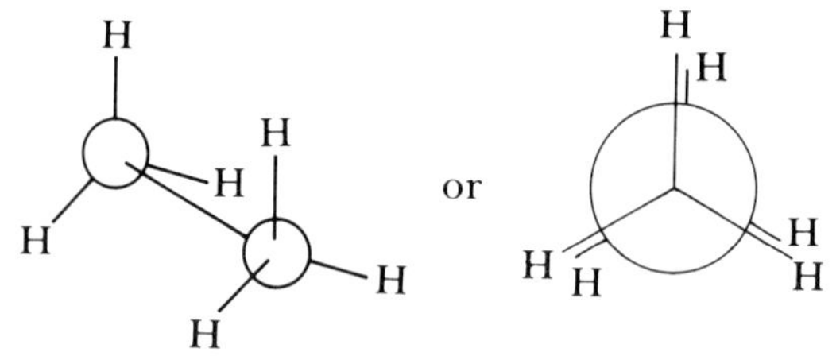

eclipsed conformation

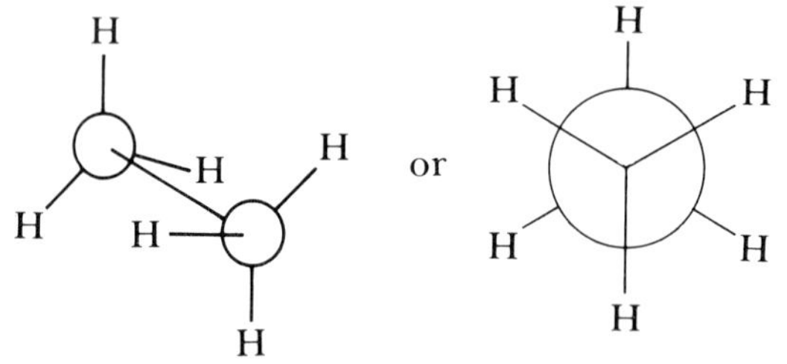

staggered conformation

---

* A typical piece of experimental evidence which supports the tetrahedral geometry of carbon is the fact that there is only one compound with the molecular formula CH_2Cl_2. If the two hydrogen atoms and two chlorine atoms formed a plane with carbon in the center, there would be *two* compounds with the molecular formula CH_2Cl_2; one compound would have hydrogen atoms adjacent and the other would have alternating hydrogen and chlorine atoms.

The Newman projection diagram is named after Melvin S. Newman (b. 1908), professor of chemistry at Ohio State University. Professor Newman has made numerous significant contributions to chemical science and is currently studying cancer-producing organic compounds, in addition to carrying on numerous other research activities.

The right-hand representations are called **Newman projection diagrams**. These diagrams show what the ethane molecules would look like if one sighted down the C—C bond axis; the large open circle represents the two in-line carbon atoms.

There is an energy difference of about 0.13 eV (3.0 kcal·mol^{-1} or 13 kJ·mol^{-1}) between the staggered and eclipsed conformations of ethane, with the staggered conformation being the lower. All other conformations have energies intermediate to these two. Consequently, even though a sample of ethane gas is really a mixture of all possible conformations—each rapidly changing to other conformations—it is customary to depict ethane as if it existed in the lower-energy staggered conformation alone.

One of the first things to get used to in organic chemistry is the practice of writing formulas in many different ways. For example, in addition to the geometric representations of ethane already presented and the molecular composition formula C$_2$H$_6$, the following types of formulas are also used:

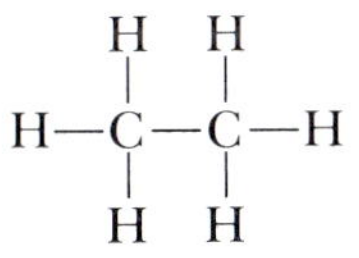

$$\begin{array}{c} H\ \ H \\ |\ \ \ | \\ H-C-C-H \\ |\ \ \ | \\ H\ \ H \end{array} \quad \text{and} \quad CH_3CH_3$$

The first formula, a **structural** or **linkage** formula, shows only in what sense the atoms are linked; this formula makes no attempt to illustrate the true three-dimensional geometry of the molecule. The second formula is a shorthand way of conveying the information supplied in the linkage formula; the fact that carbon has four bonds enables one to unambiguously link carbon and hydrogen atoms together in the proper manner. This latter type of formula is greatly favored by practicing chemists since it can be unambiguously interpreted and is quickly written down. It can also be quickly translated into a geometrical formula if there is a need to do so.

A simple and rapid way of illustrating the geometric structure of ethane is as follows:

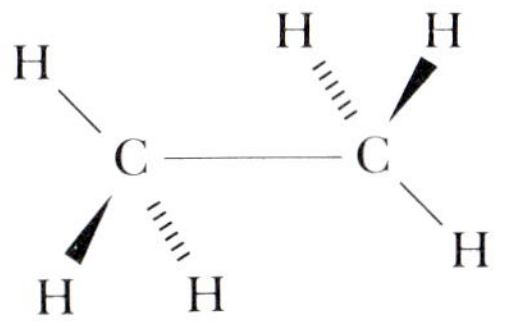

Solid bonds are understood to be in the plane of the paper, wedge bonds come out toward the reader, and dashed bonds incline toward the back side of the paper. That is, wedge and dashed bonds are assumed to be exactly opposite each other relative to the plane of the paper. Thus methane may be written

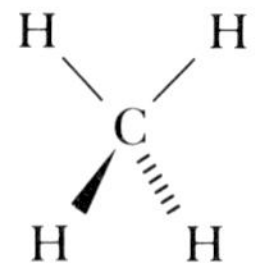

When two CH$_2$ units are added to CH$_4$ (or one CH$_2$ unit is added to CH$_3$CH$_3$) the molecule is called **propane**, denoted C$_3$H$_8$ or CH$_3$CH$_2$CH$_3$. Below are shown a variety of formulas for the lowest-energy conformation of the propane molecule:

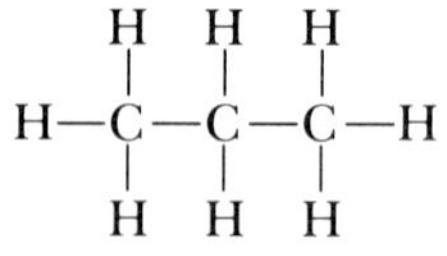

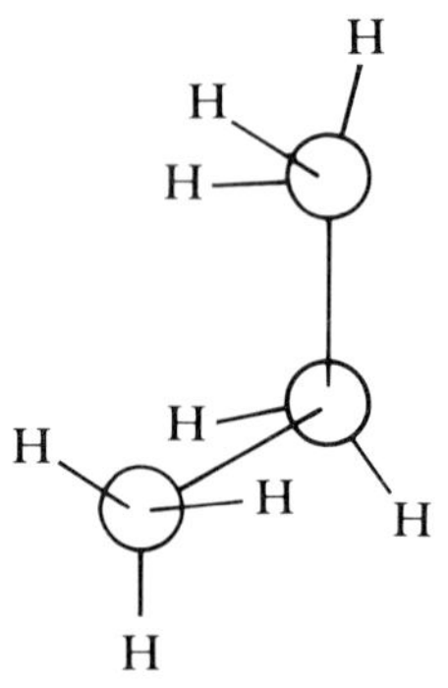

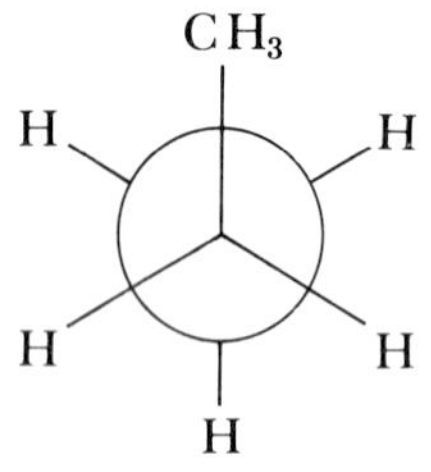

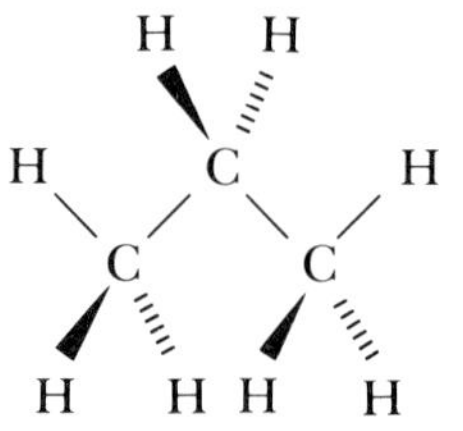

The next homolog of methane is C$_4$H$_{10}$. As discussed in Section 16.5, there are two linkage isomers with this molecular formula. The unbranched chain isomer is called *n*-butane in the common system (read "normal butane") and has the formulas

CH$_3$CH$_2$CH$_2$CH$_3$

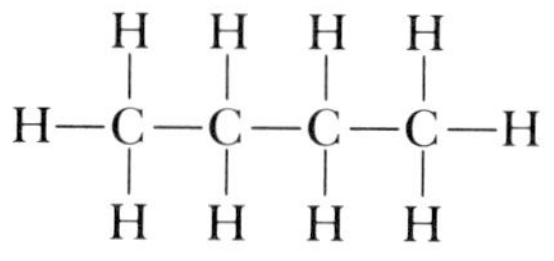

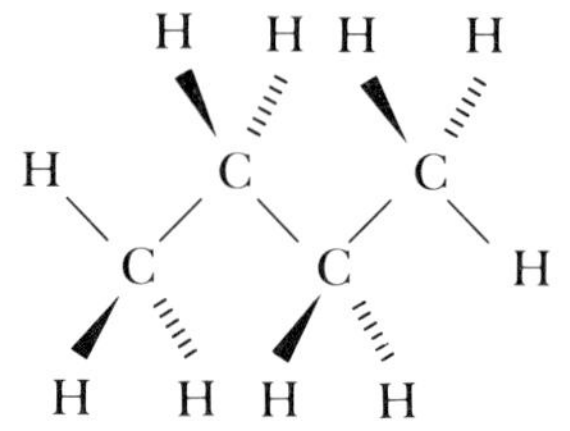

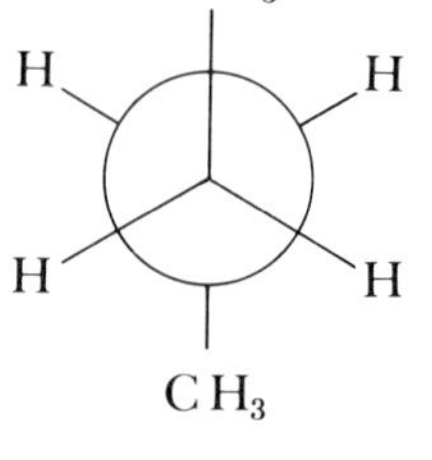

420   CARBON AND ITS COMPOUNDS

The IUPAC name is simply **butane**. The branched-chain isomer is called **isobutane** in the common system and **2-methylpropane** in the IUPAC system. 2-Methylpropane may be symbolized by any of the following formulas:

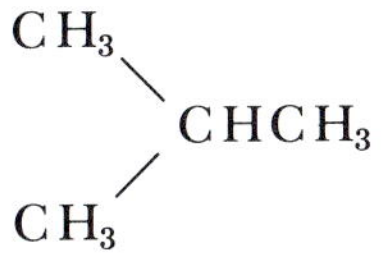

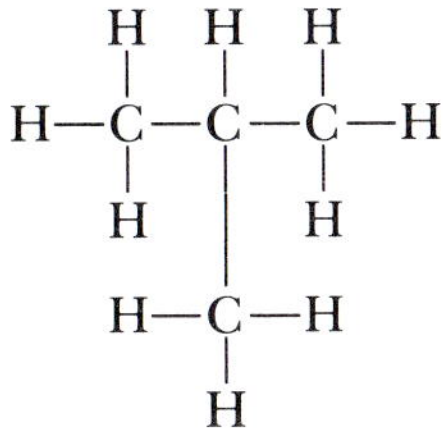

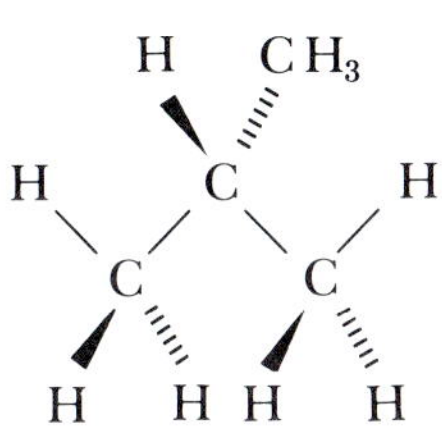

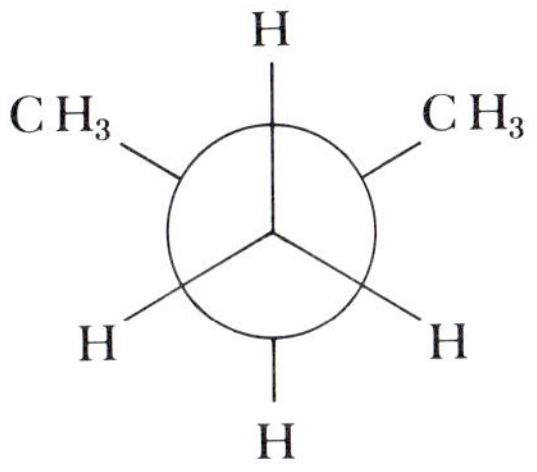

n-Butane and isobutane are different compounds with distinctly different chemical and physical properties. For example, the melting point and boiling point of n-butane are −138 °C and −0.5 °C, respectively, compared with −159 °C and −12 °C for isobutane.

When n in C_nH_{2n+2} is greater than 4, the number of possible isomers per given molecular formula eventually becomes very large (see Table 16.3). This is one of the reasons for the large number of carbon compounds; there are 366,319 isomers with the molecular formula $C_{20}H_{42}$ alone.

To see how IUPAC names are derived, it is necessary to look back to Table 16.2. An assembly of atoms corresponding to an alkane formula but missing a hydrogen atom is called an **alkyl group** and its name is formed by dropping the *-ane* ending of the corresponding alkane and replacing it with *-yl*. Thus —CH_3 is called a **methyl** group, —CH_2CH_3 is called an **ethyl** group, and similarly for other alkanes. The alkyl group names listed in Table 16.2 along with the alkane names are very important since they form the basis for naming many other organic molecules.

TABLE 16.3 NUMBER OF STRUCTURAL OR LINKAGE ISOMERS HAVING THE GENERAL FORMULA C_nH_{2n+2}

n	NUMBER OF LINKAGE ISOMERS
4	2
5	3
6	5
7	9
10	75
20	366,319

In the common naming system the linkage isomers with the unbranched-chain carbon skeleton are always named with the prefix *n-*, while those having the general form

$$\begin{array}{c} CH_3 \\ \diagdown \\ CH-(CH_2)_{n-4}CH_3 \\ \diagup \\ CH_3 \end{array}$$

are named with the prefix *iso-*. In both the common and IUPAC systems the other isomers are named on the basis of the longest continuous chain of carbon atoms, with side-chain names and positions used as prefixes. For example,

$$\overset{①}{C}H_3\overset{②}{C}H_2\overset{③}{C}H\overset{④}{C}H_2\overset{⑤}{C}H_3 \\ | \\ CH_3$$

is called **3-methylpentane**. The pentane portion of the name corresponds to the longest continuous chain ($n = 5$). The "3-methyl" indicates that a methyl group is attached to carbon atom number 3 as numbered from the end of the $n = 5$ chain. (To keep the numbers used as small as possible, the numbering starts from the end closest to the attached group if a choice exists.) Note that iso compounds are always 2-methylalkanes.

There are many organic compounds whose structures are basically those of hydrocarbons but which have one or more hydrogen atoms replaced by other atoms or groups of atoms. For example, the molecule

$$\overset{①}{C}H_3\overset{②}{C}HCl\overset{③}{C}H_3 \quad \text{or} \quad \overset{①}{C}H_3\overset{②}{C}H\overset{③}{C}H_3 \\ | \\ Cl$$

is named either 2-chloropropane (IUPAC) or isopropyl chloride (common). The molecules

$$Cl \\ | \\ \overset{①}{C}H_3\overset{②}{\underset{|}{C}}\overset{③}{C}H_2\overset{④}{C}H_3 \quad \text{and} \quad \overset{①}{C}H_3\overset{②}{C}H\overset{③}{C}H\overset{④}{C}H_3 \\ Cl ClCl$$

are called 2,2-dichlorobutane and 2,3-dichlorobutane, respectively. Note that the prefix 2 is repeated in the first case even though it may appear redundant to do so.

EXAMPLE 16.1 Name the alkane whose carbon skeleton is given below (hydrogen atoms are omitted for clarity):

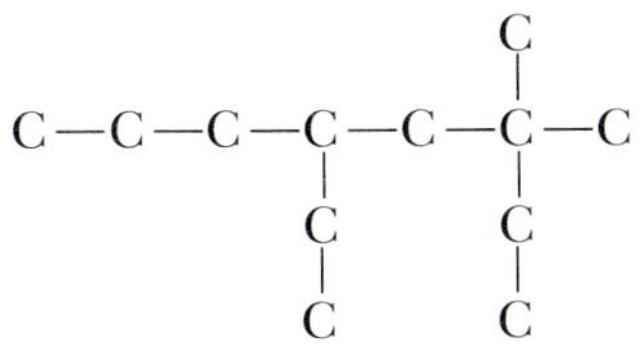

```
              C
              |
C—C—C—C—C—C—C
      |   |
      C   C
      |   |
      C   C
```

SOLUTION The longest sequence of carbon atoms is 8, thus the basic name is **octane**. To obtain the smallest numbers for indicating positions of attached groups, the chain is numbered as follows:

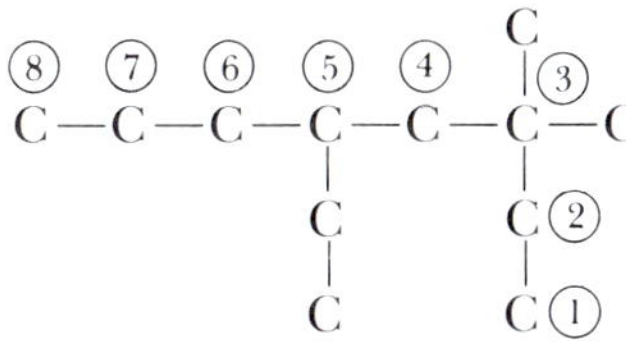

```
 ⑧  ⑦  ⑥  ⑤  ④   C
                  |③
C—C—C—C—C—C—C
         |    |
         C    C②
         |    |
         C    C①
```

There are two methyl groups on atom 3 and an ethyl group on atom 5. Thus the name is 3,3-dimethyl-5-ethyloctane.

EXAMPLE 16.2 Draw the carbon framework of the compound 2,2-dichloro-3-methylpentane.

SOLUTION Pentane has five carbon atoms. Thus the basic chain is

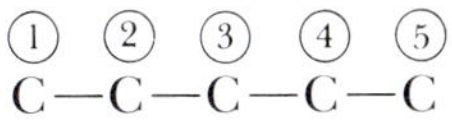

```
① ② ③ ④ ⑤
C—C—C—C—C
```

Putting two chlorines on atom 2 and a methyl group on atom 3 produces

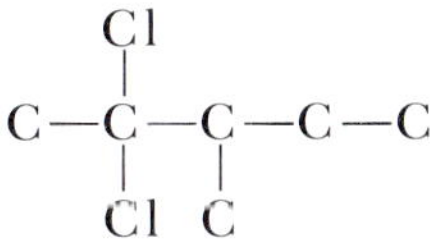

```
    Cl
    |
C—C—C—C—C
    |  |
    Cl C
```

16.8 ALKENES

Alkenes have the general molecular formula C_nH_{2n} ($n = 2, 3, 4, \ldots$). All alkenes contain the doubly bonded unit

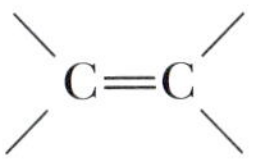

$$\diagdown_{\diagup}\!C\!=\!C\diagup_{\diagdown}$$

in which the carbon atoms are described in terms of sp^2 hybrid orbitals. This unit plus the four atoms attached to it lie in the same plane. The simplest alkene has the common name **ethylene** and the IUPAC name **ethene**. Ethene, C_2H_4 or $CH_2{=}CH_2$, has the planar structure

$$\begin{array}{c} H \\ \diagdown \\ C{=}C \\ \diagup \diagdown \\ H H \end{array} \begin{array}{c} H \\ \diagup \\ \\ \\ \end{array}$$

The electronic bonding in this molecule was illustrated earlier in Fig. 6.19. Each C—H bond is a sigma bond formed by overlap of a carbon sp^2 hybrid orbital with a hydrogen $1s$ orbital; the C=C double bond consists of a sigma bond formed by overlap of two carbon sp^2 hybrid orbitals and a pi bond formed by overlap of two carbon $2p_\perp$ orbitals. The electron density in such a pi bond is zero along the C—C bond axis and has its maximum on either side of the molecular plane some distance above and below the two carbon atoms. This bonding is often represented by a diagram such as Fig. 16.5.

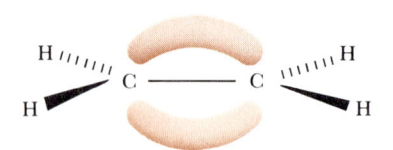

Fig. 16.5. Pi bonding in the ethene (ethylene) molecule. The electron density is greatest in regions above and below the plane of the six atoms.

The simplest homolog of ethylene (ethene) is C_3H_6, **propylene** (common name) or **propene** (IUPAC name).

$$CH_3CH{=}CH_2 \quad \text{or} \quad H-\underset{\underset{H}{|}}{\overset{\overset{H}{|}}{C}}-\overset{\overset{H}{|}}{C}{=}\overset{\overset{H}{|}}{C}-H \quad \text{or} \quad \begin{array}{c} CH_3 H \\ \diagdown \diagup \\ C{=}C \\ \diagup \diagdown \\ H H \end{array}$$

The CH_3 portion of the molecule exhibits the tetrahedral geometry of alkanes.

The next homolog of ethylene (ethene) is C_4H_8. There are four isomers having this molecular formula. These have the structures:

$$\begin{array}{c} CH_3CH_2 H \\ \diagdown \diagup \\ C{=}C \\ \diagup \diagdown \\ H H \\ \text{I} \end{array} \qquad \begin{array}{c} CH_3 H \\ \diagdown \diagup \\ C{=}C \\ \diagup \diagdown \\ CH_3 H \\ \text{II} \end{array}$$

$$\begin{array}{c} CH_3 CH_3 \\ \diagdown \diagup \\ C{=}C \\ \diagup \diagdown \\ H H \\ \text{III} \end{array} \qquad \begin{array}{c} CH_3 H \\ \diagdown \diagup \\ C{=}C \\ \diagup \diagdown \\ H CH_3 \\ \text{IV} \end{array}$$

These four different molecules represent three different atom linkages: I, II, and III or I, II, and IV (III and IV have the same linkages). The three different linkages represent three different linkage isomers. Compound I is called 1-butene, compound II is 2-methyl-1-propene, and compounds III and IV are both 2-butenes. Compounds III and IV constitute a pair of diastereomers which are distinguished as follows: Compound III is called *cis*-2-butene and compound IV is called *trans*-2-butene. The prefixes *cis*- and *trans*- refer to the relative locations of the two methyl groups (or the two hydrogen atoms) attached to the doubly-bonded carbon atoms; *cis*- means the like groups (methyl or hydrogen) are on the same side of the double bond and *trans*- means they are on opposite sides.

In Table 16.4 are listed the melting points and normal boiling points of the four isomers just discussed. Although these molecules have similar physical properties, they are by no means identical.

TABLE 16.4 THE MELTING POINTS AND BOILING POINTS OF THE ISOMERIC BUTENES

COMPOUND	MELTING POINT (°C)	NORMAL BOILING POINT (°C)
1-Butene	−185.4	−6.3
2-Methyl-1-propene	−141	−7
cis-2-Butene	−139	3.7
trans-2-Butene	−105.6	0.9

The basic alkene name is derived from the alkane name of the longest unbranched chain by changing *-ane* to *-ene*. Exceptions occur for $n = 2$ and 3; the common names, ethylene and propylene, are often used instead of the ethene and propene required by the IUPAC rules. The position of the double bond is indicated by the lower of the two numbers labeling the carbons in it—the numbering always beginning from that end of the longest carbon chain which is closer to the double bond. Side-chains and substituents replacing hydrogen are numbered and named as in alkanes. The two molecules

①　②　③　④　⑤　　　　①　②　③　④　⑤
$CH_3CH=CHCH_2CH_3$　　$CH_2=CHCH_2CH_2CH_3$

are named 2-pentene and 1-pentene, respectively. Although a chemist would understand if you called the first compound 3-pentene (this would happen if you number from the wrong end), chemical dictionaries list only the 2-pentene name.

Note that 2-pentene has two diastereomers. These are

$$\begin{array}{ccc} H \quad\quad H & \quad & CH_3 \quad\quad H \\ \diagdown\;\;\diagup & & \diagdown\;\;\diagup \\ C=C & & C=C \\ \diagup\;\;\diagdown & & \diagup\;\;\diagdown \\ CH_3 \quad\quad CH_2CH_3 & & H \quad\quad CH_2CH_3 \end{array}$$

Using the positions of the hydrogen atoms relative to the double bond as a reference, these are named *cis*-2-pentene and *trans*-2-pentene, respectively.

The isomers of $C_2H_2Cl_2$

$$\underset{Cl}{\overset{Cl}{\diagdown}}C=C\underset{H}{\overset{H}{\diagup}} \qquad \underset{Cl}{\overset{H}{\diagdown}}C=C\underset{Cl}{\overset{H}{\diagup}} \qquad \underset{Cl}{\overset{H}{\diagdown}}C=C\underset{H}{\overset{Cl}{\diagup}}$$

are named 1,1-dichloroethene, *cis*-1,2-dichloroethene, and *trans*-1,2-dichloroethene.

Names of **alkenyl groups** are somewhat harder to remember than names of alkyl groups. Below are just a few examples.

$CH_2=CH-$	vinyl or ethenyl
$CH_2=CH-CH_2-$	allyl or 2-propenyl
$CH_2=\underset{\vert}{C}-CH_3$	isopropenyl or 1-methylethenyl
$CH_3-CH=CH-$	1-propenyl

EXAMPLE 16.3 Name the following alkene using the IUPAC system:

$$\text{C}-\underset{\underset{Cl}{\vert}}{\text{C}}-\text{C}-\underset{\underset{\underset{C}{\vert}}{C}}{\overset{\Vert}{\text{C}}}-\text{C}$$

SOLUTION The longest chain containing C=C has six carbon atoms. Thus the basic name is **hexene**. The carbon atoms are numbered so that the double bond has the lowest possible number. This gives

$$\overset{⑥}{\text{C}}-\overset{⑤}{\underset{\underset{Cl}{\vert}}{\text{C}}}-\overset{④}{\text{C}}-\overset{③}{\underset{\underset{\underset{\text{C}①}{\vert}}{\text{C}②}}{\overset{\Vert}{\text{C}}}}-\text{C}$$

Since the double bond is between atoms 2 and 3, this is a **2-hexene**. Including a methyl group at carbon 3 and a chloro group at carbon 5 produces 3-methyl-5-chloro-2-hexene. This molecule may exist either as the *cis* or *trans* diastereomer.

EXAMPLE 16.4 Draw the framework of the compound 5-bromo-5-methyl-3-heptene (*cis* or *trans*).

SOLUTION The basic framework has seven carbon atoms with a double bond between atoms 3 and 4:

$$\overset{①}{\text{C}}-\overset{②}{\text{C}}-\overset{③}{\text{C}}=\overset{④}{\text{C}}-\overset{⑤}{\text{C}}-\overset{⑥}{\text{C}}-\overset{⑦}{\text{C}}$$

Adding Br and CH₃ to atom 5 produces

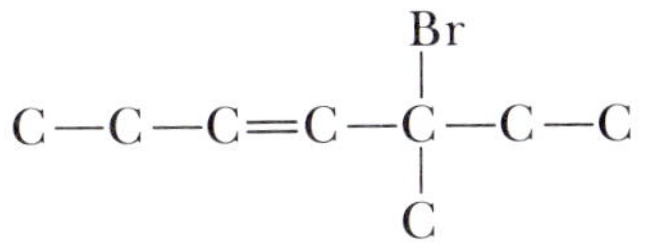

$$\text{C—C—C=C—}\underset{\underset{\text{C}}{|}}{\overset{\overset{\text{Br}}{|}}{\text{C}}}\text{—C—C}$$

16.9 ALKYNES AND DIENES

Alkynes have the general molecular formula C_nH_{2n-2} ($n = 2, 3, 4, \ldots$). All alkynes contain the triply bonded unit —C≡C— in which the carbon atoms are described in terms of *sp* hybrid orbitals. One of the C—C bonds is a sigma bond formed by overlap of two carbon *sp* hybrid orbitals. The other two bonds are pi bonds; each pi bond is the same as that of ethylene but due to absence of a molecular plane the electron density distribution maximum resembles a cylindrical cloud about the C—C axis.

The simplest alkyne has the formula C_2H_2 or CH≡CH. This is a linear molecule with the structure

H—C≡C—H

This molecule is properly called **ethyne** in the IUPAC system, but the common name, **acetylene**, is so well established that the former is seldom used. The pi bonding in acetylene is illustrated in Fig. 16.6.

Fig. 16.6. Pi bonding in the ethyne (acetylene) molecule. The electron density is as if the pi bond of ethene (Fig. 16.5) were spinning about the C—C axis. Thus it resembles a lemon with part of the inside scooped out.

The first homolog of acetylene is C_3H_4, **propyne**:

CH₃—C≡C—H

As the names ethyne and propyne suggest, the IUPAC alkyne names are formed from the alkane name by changing *-ane* to *-yne*. If necessary, the triple bond is given the number of the first triply bonded carbon encountered, starting from the end of the chain nearest the triple bond. Thus the molecule

⑤ ④ ③ ②①
CH₃CH₂C≡CCH₃

is named **2-pentyne**.

All alkynes with $n \geq 3$ are isomeric with a class of compounds called **dienes**. Dienes are characterized by two ethylene units per molecule. For example, another molecule with the same molecular formula as propyne, C_3H_4, is **propadiene** (common name **allene**)

$$CH_2=C=CH_2$$

The geometry of this molecule is

$$\begin{array}{c} H \\ \diagdown \\ \end{array} C=C=C \begin{array}{c} H \\ \diagup \\ \\ \diagdown \\ H \end{array}$$

where the two H—C—H units are at right angles to each other. The two pi bonds are also at right angles to each other.

Another important diene (isomeric with the butynes) is

$$CH_2=CH-CH=CH_2$$

and is called **1,3-butadiene**. The formula above provides a misleading description of the pi bonds in the molecule; the molecule does not have two isolated double bonds separated by a single bond but rather has three C—C bonds which are intermediate in nature to single and double bonds. Perhaps a better designation is

$$\overline{CH_2-CH-CH-CH_2}$$

where the dotted line symbolizes four electrons forming a bond encompassing all four carbon atoms. The molecular orbitals representing this bond are formed by overlap of the $2p^\perp$ atomic orbitals of the four carbon atoms. The *trans* isomer of 1,3-butadiene has the planar structure

The electron density of the pi bond electrons has its maximum values above and below the molecular plane. The *cis* isomer

does not exist at ordinary temperatures and pressures.

EXAMPLE 16.5 Name the alkyne

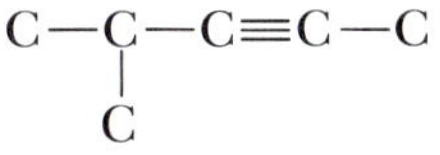

$$C-C-C\equiv C-C$$
$$\ \ \ \ \ |$$
$$\ \ \ \ \ C$$

SOLUTION The longest chain has five carbon atoms so the basic name is **pentyne**. The triple bond is between atoms 2 and 3 and there is a methyl group on atom 4. Thus the name is 4-methyl-2-pentyne.

EXAMPLE 16.6 Draw the framework of the alkyne 3-ethyl-1-hexyne.

SOLUTION The hexyne means there are six carbon atoms in the longest chain, with the triple bond between atoms 1 and 2. Putting an ethyl group on atom 3 leads to

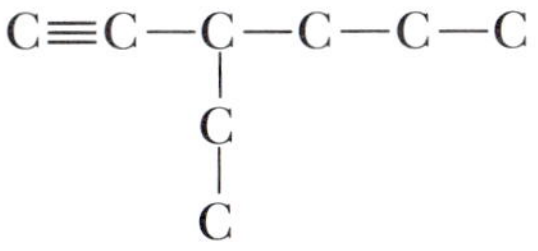

$$C\equiv C-C-C-C-C$$
$$\ \ \ \ \ \ \ \ \ |$$
$$\ \ \ \ \ \ \ \ \ C$$
$$\ \ \ \ \ \ \ \ \ |$$
$$\ \ \ \ \ \ \ \ \ C$$

EXAMPLE 16.7 What is the name of the compound

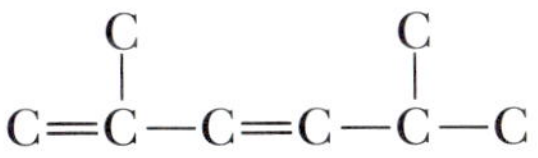

$$\ \ \ \ \ C\ \ \ \ \ \ \ \ \ C$$
$$\ \ \ \ \ |\ \ \ \ \ \ \ \ \ |$$
$$C=C-C=C-C-C$$

SOLUTION The longest chain containing both double bonds has six carbon atoms; hence the basic name is hexadiene. One double bond is between atoms 1 and 2 and the other is between 3 and 4 (using the lowest possible numbers); thus the molecule is a 1,3-hexadiene. There are methyl groups at atoms 2 and 5 so the final name is 2,5-dimethyl-1,3-hexadiene.

16.10 ALICYCLIC HYDROCARBONS

As the name implies, alicyclic compounds have cyclic or ring structures. Alicyclic compounds with the molecular formula C_nH_{2n} ($n \geq 3$) are called **cycloalkanes** but are isomeric with the alkenes. The alicyclic isomer of propylene is **cyclopropane**, with the structure

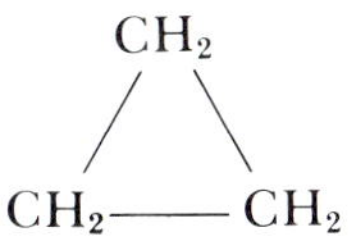

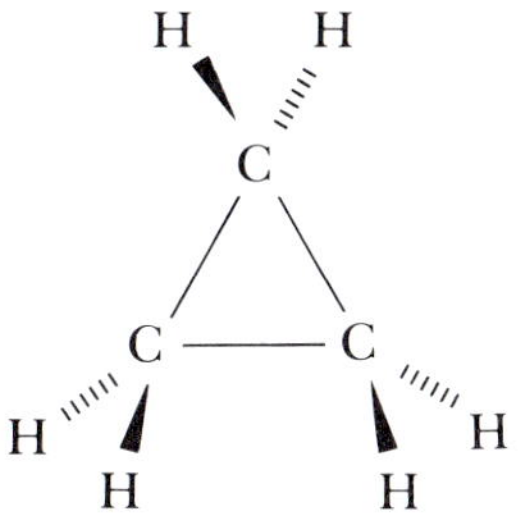

The next three homologs of cyclopropane are

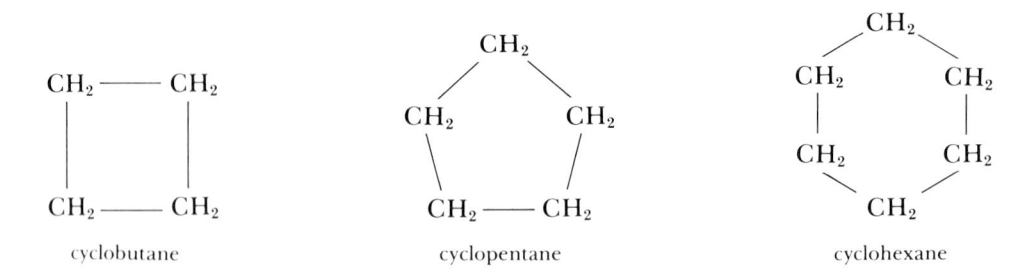

$$\begin{array}{cc} CH_2 \text{———} CH_2 \\ | \qquad\qquad | \\ CH_2 \text{———} CH_2 \end{array}$$
cyclobutane

$$\begin{array}{c} CH_2 \\ / \quad \backslash \\ CH_2 \qquad CH_2 \\ \backslash \qquad / \\ CH_2 \text{———} CH_2 \end{array}$$
cyclopentane

$$\begin{array}{c} CH_2 \\ / \quad \backslash \\ CH_2 \qquad CH_2 \\ | \qquad\qquad | \\ CH_2 \qquad CH_2 \\ \backslash \quad / \\ CH_2 \end{array}$$
cyclohexane

The cycloalkanes are usually symbolized by the simpler diagrams

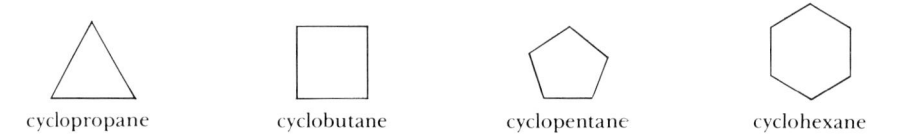

△   □   ⬠   ⬡

cyclopropane   cyclobutane   cyclopentane   cyclohexane

where each vertex represents a CH_2 unit ($-CH_2-$ is called a **methylene group**).

Cyclohexane is the most important alicyclic compound. The lowest-energy conformation of this molecule has a geometric shape indicated by the following:

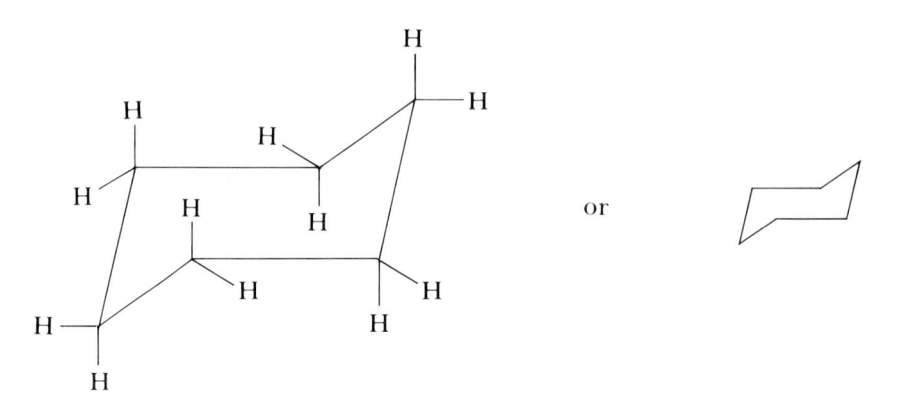

The carbon skeleton is a distorted hexagon; two opposite tips are bent out of the plane in opposite directions. This conformation is called the **chair conformation**. Another possible conformation is the **boat** form:

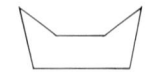

in which the opposite tips of the carbon hexagon are bent out of plane in the same direction.

EXAMPLE 16.8 Name the following alicyclic compound:

C—C—△

SOLUTION Two names are possible: The molecule can be considered either a substituted cyclopropane or a substituted ethane. Thus the possible names are ethylcyclopropane and cyclopropylethane.

EXAMPLE 16.9 Draw the framework of the compound 3-bromocyclobutene.

SOLUTION The basic structure has a ring of four carbon atoms and a double bond (cyclobutene). The double bond defines atoms 1 and 2 so that the framework is

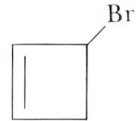

This is an example of a **cycloalkene**, a type of alicyclic compound containing a double bond and isomeric with alkynes.

16.11 AROMATIC HYDROCARBONS

The simplest aromatic hydrocarbon—and perhaps the most important—is **benzene**, C_6H_6. The structure of benzene was first proposed by Kekulé and Couper around 1866. Several Lewis resonance formulas may be written for benzene; the two most important are

A simpler representation, in which carbon atoms are implied by apexes of a hexagon and hydrogen atoms are not shown explicitly, is

Space-filling model of the benzene molecule.

These two formulas differ only in the interchange of single and double bonds. The fact that it is possible to write two such formulas suggests that rather than viewing the hexagon of carbon atoms as alternating single and double bonds, each carbon-to-carbon bond must be regarded as a composite of a single and a double bond. For convenience it is customary to represent benzene by the single symbol

where each vertex represents a CH unit and the circle represents the six electrons of three double bonds spread out, so to speak, over all six carbon atoms. Figure 16.7 illustrates how the symbolic representation of benzene has evolved since the last century.

The modern bonding description of benzene is as follows: Each carbon atom is described in terms of sp^2 hybrid orbitals, the C—H sigma bonds are formed by an overlap between a carbon sp^2 hybrid orbital and a hydrogen 1s orbital, and the C—C

As noted early in the chapter, the ring structure of benzene occurred to the German chemist Kekulé in a dream. Coincidentally, a Scottish-born chemist, A. S. Couper (1831–1892) proposed the same structure at almost the same time. Couper also invented the dash symbol (as in H—Cl) to represent a chemical bond.

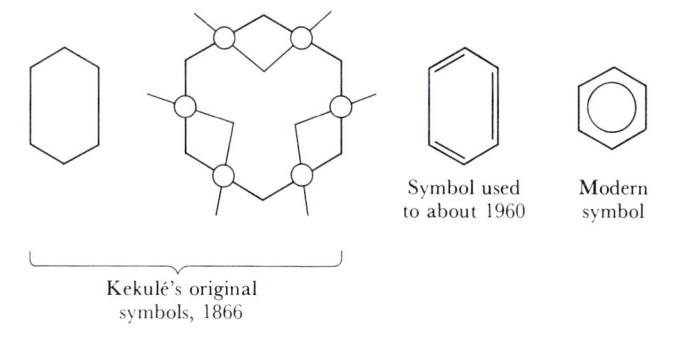

Fig. 16.7. The evolution of benzene symbols since 1866.

sigma bonds (represented by the sides of the hexagon) are formed by overlap of two carbon sp^2 hybrid orbitals. The molecular orbitals representing the six electrons of the "mobile" double bonds are formed by overlap of the $2p^{\perp}$ atomic orbitals of the six carbon atoms. These **pi electrons** exhibit maximum densities above and below the plane of the molecule and are usually called the **aromatic sextet**. A schematic representation of the pi electron density in benzene is shown in Fig. 16.8. It should be noted that the planar, regular hexagonal structure of benzene implies that the molecule has a zero dipole moment. This is in accord with experiment.

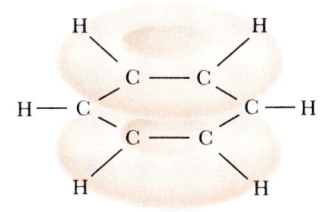

Fig. 16.8. An illustration of the charge cloud picture of the aromatic sextet of pi electrons in the benzene molecule.

Three other aromatic hydrocarbons of great importance are **naphthalene**, **anthracene**, and **phenanthrene**. Like benzene, these molecules have planar structures:

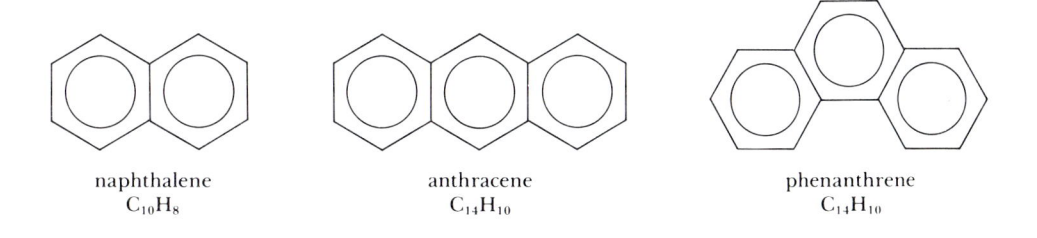

naphthalene
$C_{10}H_8$

anthracene
$C_{14}H_{10}$

phenanthrene
$C_{14}H_{10}$

Each circle represents pi electrons. However, these pi electrons are shared by the carbon atoms common to more than one ring, so that the total number of pi electrons is 10 in naphthalene and 14 in the isomers anthracene and phenanthrene. In general, the number of aromatic pi electrons is always $4n + 2$, where n is the number of rings in the compound.

The term **aryl group**, analogous to aliphatic alkyl and alkenyl groups, is used to denote an aromatic molecule which has a hydrogen atom removed. Below are the structures and names of a few common aryl groups:

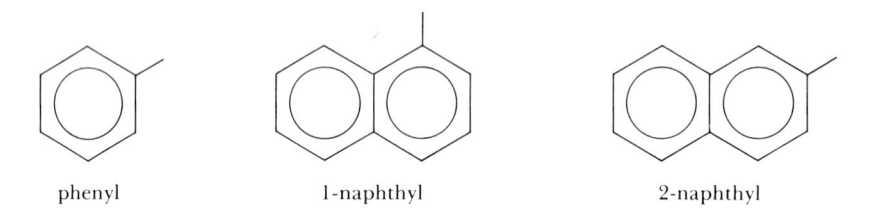

phenyl   1-naphthyl   2-naphthyl

A commonly occurring arylalkyl group is **benzyl**, with the structure

Compounds of benzene may be named in a manner similar in some respects to that of aliphatic compounds. For example, the compounds

are called **chlorobenzene** and **nitrobenzene**, respectively. Some compounds have special names, for example is called **toluene** rather than methylbenzene. When there are two groups attached to the benzene ring the names follow the pattern shown below:

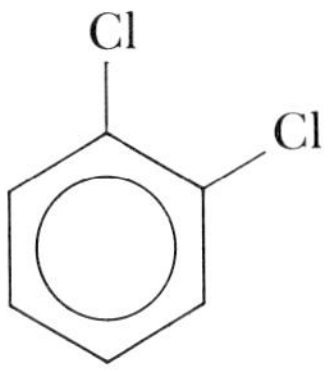

o-dichlorobenzene 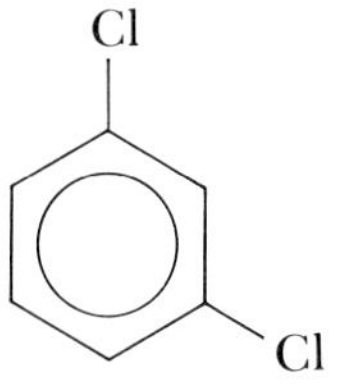   m-dichlorobenzene 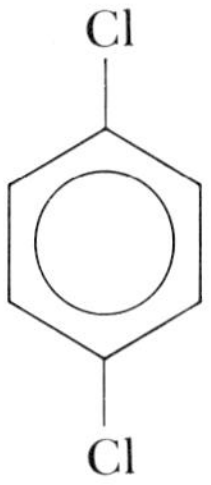   p-dichlorobenzene

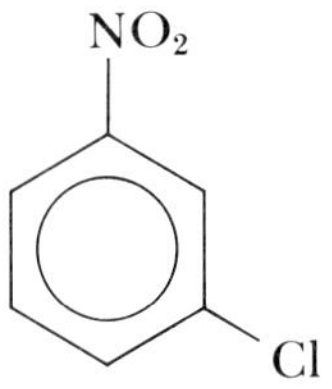

m-chloronitrobenzene 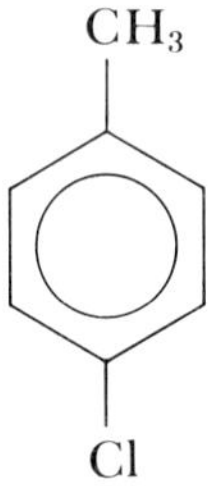   p-chlorotoluene

The prefixes o-, m-, and p- are read "**ortho**," "**meta**," and "**para**," respectively. The dimethylbenzenes are given the special name of **xylenes**.

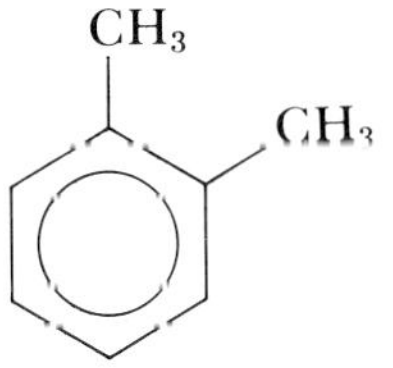

o-xylene 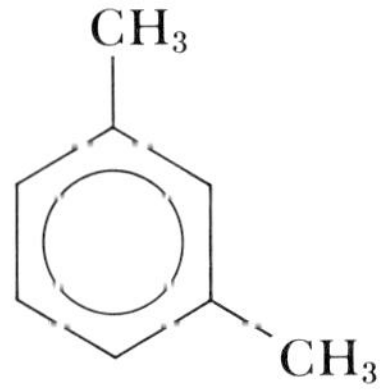   m-xylene 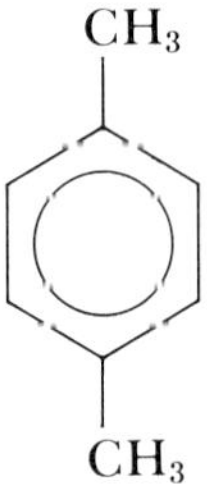   p-xylene

When there are three or more groups on the ring they are located by number. If all the groups are the same, they are given numbers such that the sequence gives the lowest possible combination of numbers. Thus

is called 1,2,4-trichlorobenzene and not 1,3,4-trichlorobenzene or some other combination. If the groups are different, the last named is assumed to be in position 1 and the other groups are numbered relative to this. For example

may be called 2-chloro-5-bromonitrobenzene. If one of the groups present leads to a special name, that group automatically becomes number 1 and the compound is named using the special name as follows: The compound

is 3-chloro-5-nitrotoluene.

Naphthalene compounds may be named in a manner similar to that of trisubstituted benzenes. The ring-numbering system is

436 CARBON AND ITS COMPOUNDS

Thus the compound

could be called 3-nitro-7-bromochloronaphthalene.

EXAMPLE 16.10 What are the names of the following compounds?

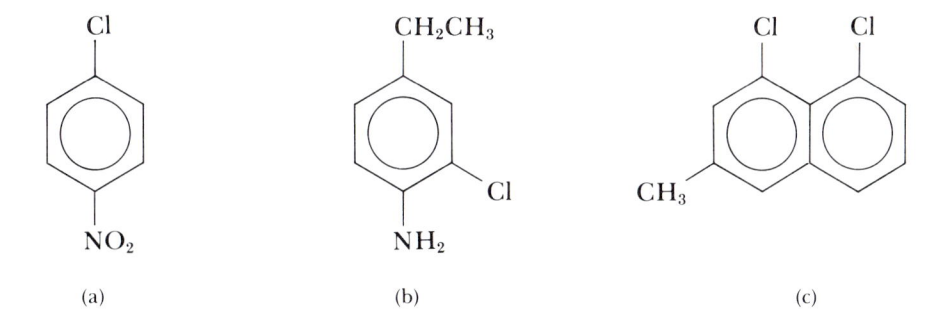

(a)   (b)   (c)

SOLUTION
a) There are only two groups on benzene so the *o-, m-, p-* nomenclature applies. The name is *p*-chloronitrobenzene. Another possibility is *p*-nitrochlorobenzene.
b) Benzene with one amino group (—NH₂) has the special name of **aniline**. Thus the basic structure is that of aniline and the NH₂ position is taken to be number 1. To use the smallest numbers, Cl is at 2 and ethyl at 4. The name is 2-chloro-4-ethylaniline.
c) Using the naphthalene numbering system given in the text leads to 1,8-dichloro-3-methylnaphthalene.

16.12 CLASSIFICATION OF ORGANIC COMPOUNDS AS TO FUNCTIONAL GROUPS

Although hydrocarbons are widely used as fuels, they are perhaps much more important as sources of other compounds. Of particular importance are the compounds that result when one or more hydrogen atoms of a hydrocarbon are replaced by another atom or, particularly, by groups of atoms. Such a group of atoms, called a **functional group**, may dominate the chemistry of the molecule in which it appears so that the hydrocarbon portion plays but a secondary or modifying role. Consequently, a classification of molecules according to the functional groups they contain is more useful to the chemist than is the aliphatic–aromatic classification. Eight different types of molecules, based on the functional groups they contain, are discussed in the following pages.

ALCOHOLS

The group of atoms —OH is called the **hydroxyl** group (not to be confused with the hydroxide ion OH⁻) and is characteristic of a class of compounds called **alcohols**. The most important type of alcohol has the general formula ROH, where R is an aliphatic group such as an alkyl or alkenyl group.

When R is a methyl group the alcohol CH_3OH is called either **methyl alcohol** (common name) or, preferably, **methanol** (IUPAC name). The latter name is formed from the name of the parent alkane by changing *-e* to *-ol*. Thus when R is an ethyl group the alcohol is CH_3CH_2OH and is named **ethanol** (IUPAC) or **ethyl alcohol** (common). Alcohols in which R contains more than two carbon atoms are named in an analogous fashion but it is necessary to indicate the location of the carbon atom to which the —OH is attached. For example,

$$\overset{③}{C}H_3\overset{②}{C}H_2\overset{①}{C}H_2OH \quad \text{and} \quad \overset{①}{C}H_3\overset{②}{C}HCH_3 \atop OH$$

are called **1-propanol** and **2-propanol**, respectively, in the IUPAC system. The common names, ***n*-propyl alcohol** and **isopropyl alcohol**, are also used.

Methanol, once called **wood alcohol** since it was commonly obtained from distillation of wood, is very toxic and can cause blindness and death if ingested. Part of its toxicity is due to the fact that it becomes metabolized in the liver to formaldehyde, a toxic compound which the liver is incapable of detoxifying by further metabolism. Ethanol is the alcohol used in intoxicating beverages. It is usually prepared by enzymic fermentation of natural sugars and starches. Pure ethanol is virtually tasteless; most of the flavor of whiskey comes from traces of other alcohols and other compounds. 2-Propanol is commonly available as **rubbing alcohol**.

When R in ROH is an aryl group, the compound is called a **phenol**. Three important phenols and their names are

phenol   1-naphthol   2-naphthol

CARBON AND ITS COMPOUNDS

Phenols are weak Brønsted-Lowry acids; the K_a's of the above phenols are all around 1×10^{-10} at 25 °C. Phenol itself is sometimes called **carbolic acid**; it is occasionally employed as an antiseptic.

If the carbon atom attached directly to OH is not part of an aromatic ring, the compound is an alcohol, for example:

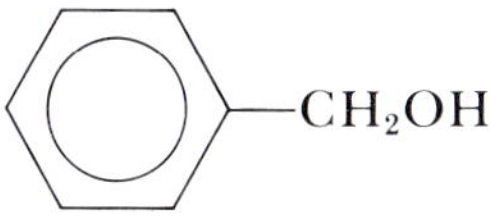

$$\text{C}_6\text{H}_5\text{—CH}_2\text{OH}$$

benzyl alcohol

EXAMPLE 16.11 Name the compound whose basic framework is

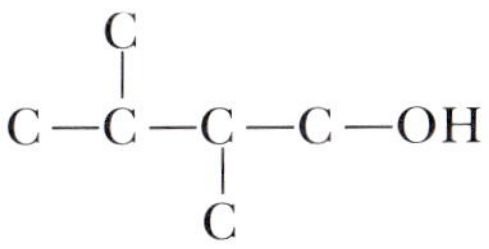

```
        C
        |
  C—C—C—C—OH
        |
        C
```

SOLUTION The longest chain has four carbon atoms; thus the compound is a **butanol**. Numbering begins at the carbon containing OH so that the full name is 2,3-dimethyl-1-butanol.

EXAMPLE 16.12 Write the structure of 2,3-dimethyl-2-butanol.

SOLUTION The basic 2-butanol structure is

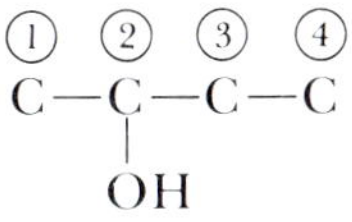

```
  ①   ②   ③   ④
  C—C—C—C
      |
      OH
```

The numbering begins with the left-hand carbon. Adding methyl groups at atoms 2 and 3 produces

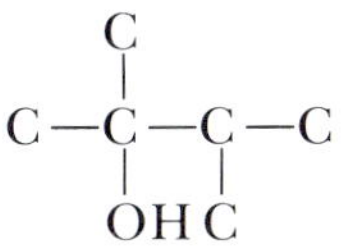

```
        C
        |
  C—C—C—C
      |   |
      OH  C
```

ETHERS

Ethers are characterized by the —C—O—C— group. The general formula of an ether is R—O—R′, where R and R′ may be the same or different. The naming of ethers follows the pattern shown below:

CH₃—O—CH₃   methyl ether

CH₃—O—CH₂CH₃   methyl ethyl ether

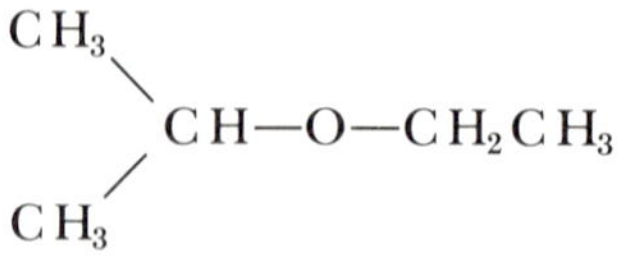

(CH₃)₂CH—O—CH₂CH₃   ethyl isopropyl ether

C₆H₅—O—CH₃   methyl phenyl ether (also called anisole)

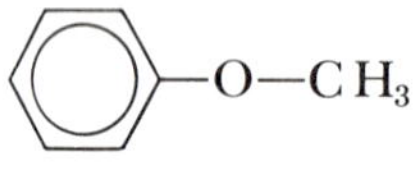

The ethers are isomeric with alcohols. For example, both methyl ether and ethanol have the molecular formula C_2H_6O. However, the two molecules have strikingly different physical and chemical properties. Since the oxygen atom in ethers is not directly bonded to hydrogen, ethers do not exhibit hydrogen bonding and, consequently, are very volatile. Whereas ethanol is a liquid at room temperature and 1 atm pressure, methyl ether is a gas. Liquid methyl ether has a normal boiling point of −23 °C.

Ethyl ether, $CH_3CH_2OCH_2CH_3$, has been used since 1846 as a general anesthetic and although it is being replaced with other substances, still remains the most widely used substance for that purpose. Many of the ethers find use as solvents for fats.

When R is an alkyl group, the group —OR, which is part of the ether structure, is called an **alkoxy group**; for example, —OCH₃ is a **methoxy** group and —OCH₂CH₃ is an **ethoxy** group.

CARBOXYLIC ACIDS

The functional group

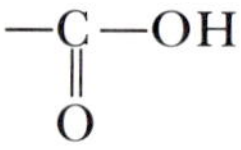

$$-\underset{\underset{O}{\|}}{C}-OH$$

is called a **carboxyl** group. Its presence identifies a **carboxylic acid** of the general form

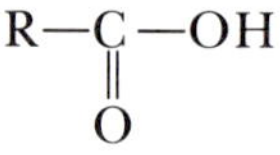

$$R-\underset{\underset{O}{\|}}{C}-OH$$

(also written RCO₂H or RCOOH), where R may be hydrogen. When R is hydrogen, the resulting acid formula is written

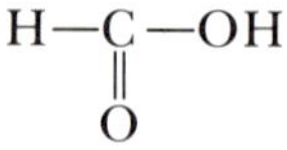

$$H-\underset{\underset{O}{\|}}{C}-OH \quad \text{or} \quad HCO_2H \quad \text{or} \quad HCOOH$$

This compound has the common name **formic acid**. The IUPAC name is **methanoic acid**. It is a weak Brønsted-Lowry acid in water solution with a K_a of 1.8×10^{-4} at

25 °C. In all of the carboxylic acids RCO_2H, it is the hydrogen bonded to oxygen which is donated as a proton to a base. The reaction with H_2O is

$$R-\underset{\underset{O}{\|}}{C}-OH + H_2O \rightleftharpoons R-\underset{\underset{O}{\|}}{C}-O^- + H_3O^+$$

When R is a methyl group, the acid formula is

$$CH_3\underset{\underset{O}{\|}}{C}OH \quad \text{or} \quad CH_3CO_2H \quad \text{or} \quad CH_3COOH$$

The common name is **acetic acid** and the IUPAC name is **ethanoic acid**. This is the acid symbolized as HOAc in Chapter 13. The symbol Ac represents the **acetyl** group

$$CH_3\underset{\underset{O}{\|}}{C}-$$

The general name of the group

$$R-\underset{\underset{O}{\|}}{C}-$$

is **acyl** group. The group

$$H-\underset{\underset{O}{\|}}{C}- \quad \text{(from the acid } H-\underset{\underset{O}{\|}}{C}-OH\text{)}$$

is called **formyl** (common name) or **methanoyl** (IUPAC name). Acyl group names are used in naming compounds such as the following:

$$CH_3\underset{\underset{O}{\|}}{C}-Cl \quad \text{or} \quad CH_3COCl$$

This molecule is known as **acetyl chloride** or **ethanoyl chloride**. The acyl name is formed by changing the *-ic acid* of the carboxylic acid name to *-yl* (common) or *-oyl* (IUPAC).

When R in RCO_2H is a phenyl group, the acid formula is written

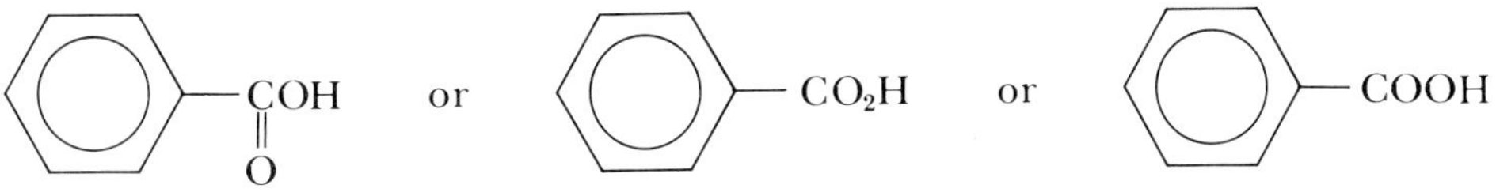

$$\text{C}_6\text{H}_5-\underset{\underset{O}{\|}}{C}OH \quad \text{or} \quad \text{C}_6\text{H}_5-CO_2H \quad \text{or} \quad \text{C}_6\text{H}_5-COOH$$

and is called **benzoic acid** in both the common and IUPAC systems.

The common names of carboxylic acids are based largely on their natural sources and are not easy to illustrate in a simple and definitive manner. For example, butyric

TABLE 16.5 NAMES OF SOME CARBOXYLIC ACIDS

FORMULA	COMMON NAME	IUPAC NAME
HCO_2H	Formic	Methanoic
CH_3CO_2H	Acetic	Ethanoic
$CH_3CH_2CO_2H$	Propionic	Propanoic
$CH_3CH_2CH_2CO_2H$ or $CH_3(CH_2)_2CO_2H$	Butyric	Butanoic
$CH_3(CH_2)_3CO_2H$	Valeric	Pentanoic
$CH_3(CH_2)_{10}CO_2H$	Lauric	Dodecanoic
$C_6H_5CO_2H$	Benzoic	Benzoic
$C_6H_5CH_2CO_2H$	Phenylacetic	Phenylethanoic
$CH_3(CH_2)_7CH=CH(CH_2)_7CO_2H$ (*cis* isomer)	Oleic	*cis*-9-Octadecenoic

acid, $CH_3CH_3CH_3CO_2H$, gets its name from its presence in rancid butter. Table 16.5 lists names for some commonly encountered carboxylic acids in both the common and IUPAC systems.

Formic acid may be obtained by the distillation of ants (the name is derived from the Latin *formica*, ant). It is a constituent of bee venom and the venom of other biting insects. Acetic acid is an important constituent of vinegar (*acetum* is the Latin name for vinegar). Vinegar is produced when the ethanol in apple cider oxidizes to acetic acid. Benzoic acid inhibits the growth of microorganisms in food. When the salt sodium benzoate is added to food which has a pH below 4.5, benzoic acid is produced by the following process:

$$\text{C}_6\text{H}_5\text{–CO}^- + \text{H}_3\text{O}^+ \rightleftharpoons \text{C}_6\text{H}_5\text{–COOH} + \text{H}_2\text{O}$$

      benzoate ion   benzoic acid

This is an example of a Brønsted-Lowry acid–base reaction.

EXAMPLE 16.13 Give the common names and the IUPAC names of the compounds

$$\text{C–C–CO}_2\text{H} \quad\quad \text{C–C–C–CO}_2\text{H}$$
$$\quad\quad |$$
$$\quad\quad \text{C}$$
        (a)   (b)

SOLUTION Given Table 16.5 as a guide, (a) and (b) have the common names isobutyric and butyric acid, respectively. The IUPAC names are (a) 2-methylpropanoic acid and (b) butanoic acid. The latter names are based on the number of carbon atoms in the longest chain, 3 in (a) and 4 in (b). Numbering begins at the —CO_2H. Actually, (a) could be called simply methylpropanoic acid; no position other than carbon 2 can occur (3-methylpropanoic acid is butanoic acid and would never be called by the former name).

EXAMPLE 16.14 Write the structure of phenylethanoic acid.

SOLUTION Ethanoic acid (acetic acid) has two carbon atoms. The phenyl group can be attached only to the carbon atom that is not part of CO_2H. Thus the structure is

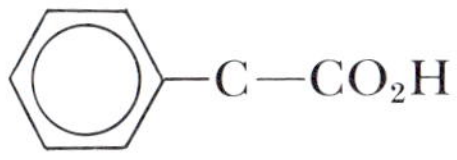

$$\text{C}_6\text{H}_5-\text{C}-\text{CO}_2\text{H}$$

This compound is also called phenylacetic acid.

ALDEHYDES AND KETONES

The functional group

$$-\underset{\overset{\|}{O}}{C}-$$

is called the **carbonyl** group. When the carbonyl group appears in the structure

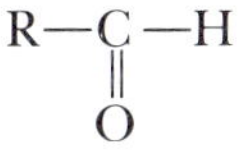

$$R-\underset{\overset{\|}{O}}{C}-H \quad \text{or} \quad RCHO \quad (R \text{ may be } H)$$

the compound is called an **aldehyde**. A commonly used way of naming aldehydes is based on the names of the acids which would arise from the oxidation reaction

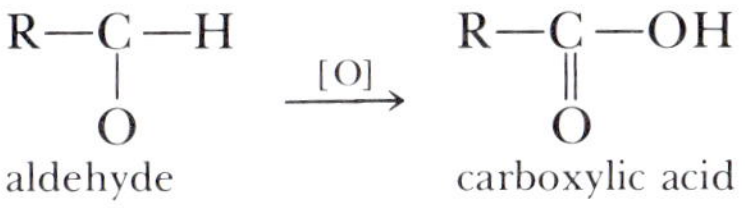

$$\underset{\text{aldehyde}}{R-\underset{\overset{|}{O}}{C}-H} \xrightarrow{[O]} \underset{\text{carboxylic acid}}{R-\underset{\overset{\|}{O}}{C}-OH}$$

where [O] denotes an unspecified oxidizing agent. In the common system, the aldehyde name is formed simply by replacing *-ic acid* or *-oic acid* of the acid name with *aldehyde*. Thus HCHO becomes **formaldehyde**, CH_3CHO becomes **acetaldehyde**, CH_3CH_2CHO becomes **propionaldehyde**, and C_6H_5CHO becomes **benzaldehyde**. Nonaromatic aldehydes are named in the IUPAC system by changing the *-e* of the corresponding alkane name to *-al*. Thus HCHO is **methanal**, CH_3CHO is **ethanal**, and CH_3CH_2CHO is **propanal**. Branched chains are located by number as in alkanes, using the carbon of the —CHO group as a reference. Thus the compound

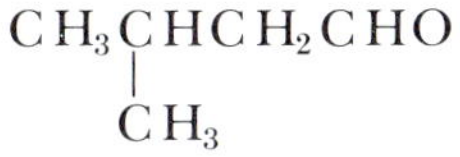

$$CH_3CHCH_2CHO$$
$$\;\;\;\;\;|$$
$$\;\;\;CH_3$$

may be called **3-methylbutanal**.

Water solutions of formaldehyde are called **formalin** and find use as a disinfectant and in the preservation of biological specimens.

When the carbonyl group appears in the structure

$$R-\underset{\underset{O}{\|}}{C}-R' \quad \text{(R, R' same or different but not H)}$$

the compound is called a **ketone**. The common system of naming ketones is similar to the naming of ethers, as the following examples show:

$CH_3\underset{\underset{O}{\|}}{C}CH_3$ methyl ketone (commonly called **acetone**)

$CH_3\underset{\underset{O}{\|}}{C}CH_2CH_3$ methyl ethyl ketone (commonly called **MEK**)

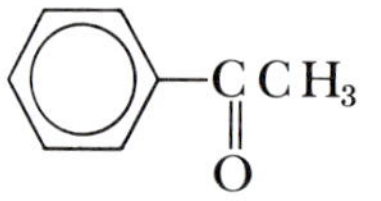

$C_6H_5-\underset{\underset{O}{\|}}{C}CH_3$ methyl phenyl ketone (commonly called **acetophenone**)

The IUPAC naming of nonaromatic ketones changes the -*e* of the longest unbranched chain alkane name to -*one* and locates the carbonyl group by number. For example, the first two ketones above become **propanone** and **butanone** (there is only one possible position for the carbonyl group, so its carbon-atom number is unnecessary). The ketones

$CH_3\underset{\underset{O}{\|}}{C}CH_2CH_2CH_3$ and $CH_3CH_2\underset{\underset{O}{\|}}{C}CH_2CH_3$

are called **2-pentanone** and **3-pentanone**, respectively.

Acetone is widely used as a commercial solvent and for the synthesis of many other chemicals. MEK finds similar uses.

EXAMPLE 16.15 Name these two compounds in the IUPAC system.

$$C-\underset{\underset{C}{|}}{C}-C-\underset{\underset{O}{\|}}{C}-C \qquad C-\underset{\underset{C}{|}}{C}-C-C-CHO$$

(a)   (b)

SOLUTION The longest chain in (a) has five carbon atoms. The carbonyl group is not at the end of a chain so the compound is a ketone, specifically a pentanone. The carbonyl carbon is numbered 2 (to give it the smaller of two possible numbers). The full name is 4-methyl-2-pentanone. Compound (b) is an aldehyde with five carbon atoms in the basic chain, hence a pentanal. The full name is 4-methylpentanal.

EXAMPLE 16.16 Write the structure of 2-methyl-3-pentanone.

444   CARBON AND ITS COMPOUNDS

SOLUTION The structure is like (a) in Example 16.15 but with the carbonyl on the third carbon atom (from either end). Note that this should not be called 4-methyl-3-pentanone since this gives the methyl group a higher number than necessary.

ESTERS

When a carboxylic acid reacts with an alcohol in the presence of a strong acid or base, a compound known as an **ester** is formed. The balanced equation for the reaction is

$$\underset{\text{carboxylic acid}}{R\underset{\underset{O}{\|}}{C}OH} + \underset{\text{alcohol}}{R'OH} \xrightarrow{H_3O^+ \text{ or } OH^-} \underset{\text{ester}}{R\underset{\underset{O}{\|}}{C}OR'} + H_2O$$

As a specific example, acetic acid reacts with ethanol to produce the ester called **ethyl acetate**:

$$\underset{\text{acetic acid}}{CH_3\underset{\underset{O}{\|}}{C}OH} + \underset{\text{ethanol}}{CH_3CH_2OH} \rightarrow \underset{\text{ethyl acetate}}{CH_3\underset{\underset{O}{\|}}{C}OCH_2CH_3} + H_2O$$

Esters are named as if they were salts of carboxylic acids; the alkyl name from the alcohol is followed by the name of the acid with *-ic acid* changed to *-ate*. Thus the ester

$$CH_3\underset{\underset{O}{\|}}{C}OCH_3$$

arises from methanol and acetic acid and is called methyl acetate (or methyl ethanoate). Care must be taken to distinguish between esters such as

$$CH_3CH_2\underset{\underset{O}{\|}}{C}OCH_3 \quad \text{and} \quad CH_3\underset{\underset{O}{\|}}{C}OCH_2CH_3$$

The first is methyl propanoate; the second is ethyl ethanoate. The noncarbonyl oxygen is always attached to the alcohol part of the ester.

EXAMPLE 16.17 Name the following ester:

$$C-C-C-O-\underset{\underset{O}{\|}}{C}-\underset{\underset{C}{|}}{C}-C$$

SOLUTION The alcohol part is to the left of the oxygen atom in the chain; the alcohol is 1-propanol. The acid part (to the right of the oxygen) is 2-methylpropanoic acid. Thus the ester is called propyl 2-methylpropanoate. A common name is *n*-propyl isobutyrate.

16.12 | CLASSIFICATION OF ORGANIC COMPOUNDS AS TO FUNCTIONAL GROUPS

EXAMPLE 16.18 Write the structure of the ester 2-propyl 3-methylbutanoate (isopropyl isovalerate).

SOLUTION The alcohol is 2-propanol (isopropyl alcohol) and the acid is 3-methylbutanoic acid (isovaleric acid). Thus the ester is

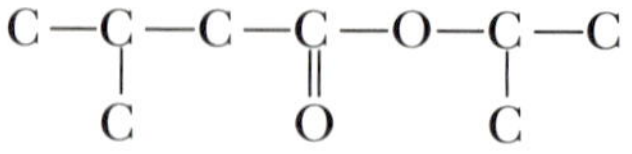

$$\text{C}-\underset{\underset{\text{C}}{|}}{\text{C}}-\text{C}-\underset{\underset{\text{O}}{||}}{\text{C}}-\text{O}-\underset{\underset{\text{C}}{|}}{\text{C}}-\text{C}$$

It is apparent that the H_2O produced as the result of an alcohol/acid reaction (called an **esterification** reaction) must come either from the alcohol H and the acid OH or from the alcohol OH and the acid H. By the use of ^{18}O as a tracer it is possible to demonstrate that the former is the predominant occurrence. The demonstration is carried out as follows (where O* represents the ^{18}O atom): Water containing 10% O* is allowed to react with Na to prepare O*-labeled hydroxide,

$$2H_2O^* + 2Na(s) \rightarrow 2Na^+(aq) + 2O^*H^-(aq) + H_2(g)$$

The O*H⁻ so formed is used to prepare labeled acid from an acyl chloride as follows:

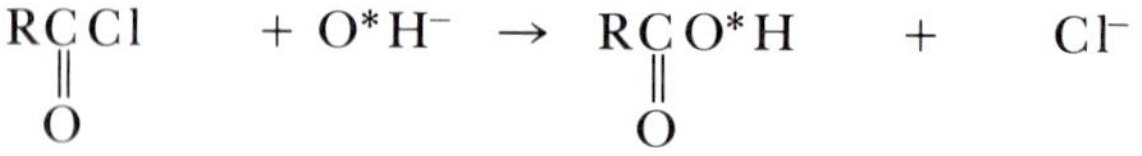

$$\underset{\underset{\text{O}}{||}}{\text{RCCl}} + \text{O*H}^- \rightarrow \underset{\underset{\text{O}}{||}}{\text{RCO*H}} + \text{Cl}^-$$

acyl chloride   labeled carboxylic acid

If the esterification reaction involves the H of the acid and the OH of the alcohol, the O* ends up in the ester, as the following reaction shows:

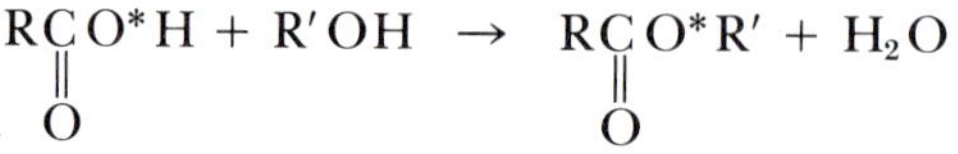

$$\underset{\underset{\text{O}}{||}}{\text{RCO*H}} + \text{R'OH} \rightarrow \underset{\underset{\text{O}}{||}}{\text{RCO*R'}} + H_2O$$

If the esterification reaction involves the OH of the acid and the H of the alcohol, the O* ends up in the H_2O:

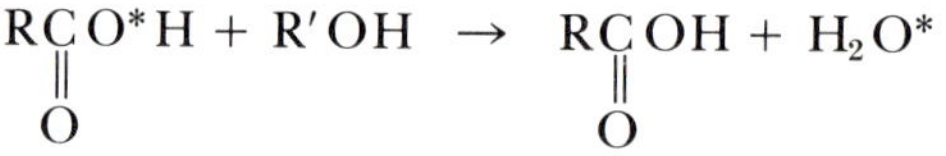

$$\underset{\underset{\text{O}}{||}}{\text{RCO*H}} + \text{R'OH} \rightarrow \underset{\underset{\text{O}}{||}}{\text{RCOH}} + H_2O^*$$

The point is settled by isolating the ester, purifying it, and determining whether or not it contains O*. One way of doing this is to heat the ester in a sealed quartz tube to produce CO. The CO is then analyzed in a mass spectrometer. The mass spectrometer converts CO to the ion CO⁺ and determines its mass. CO⁺ containing ^{16}O has a mass of 28 amu; CO⁺ containing ^{18}O has a mass of 30 amu. Experiment shows that all of the CO⁺ produced from the ester of an O*-labeled acid has a mass of 28 amu; therefore, the O* must end up in H_2O. Final confirming proof results when labeled alcohol R'O*H is esterified with unlabeled acid RCO_2H. In this case some of the CO⁺ from the ester is found to have a mass of 30 amu.

Many of the volatile esters have fruitlike odors; isobutyl propionate (2-methylpropyl propanoate) smells like rum and *n*-butyl butyrate (butyl butanoate) smells like pineapple. The bouquets of wines are attributed to mixtures of esters and other compounds.

AMINES

Amines are organic molecules containing the trivalent nitrogen linkage

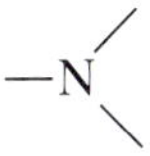

$$-N\diagup^{\diagdown}$$

in one of the following ways:

R—NH₂   primary amine

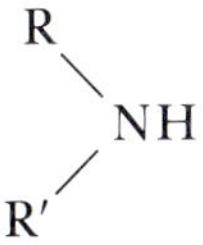

$$\begin{array}{c} R \\ \diagdown \\ NH \\ \diagup \\ R' \end{array}$$ secondary amine (R, R' same or different)

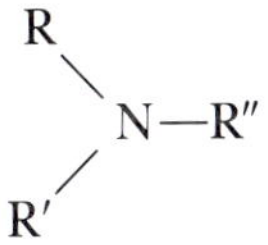

$$\begin{array}{c} R \\ \diagdown \\ N-R'' \\ \diagup \\ R' \end{array}$$ tertiary amine (R, R', R'' same or different)

The group —NH₂, found in primary amines, is called the **amino** group. Aliphatic amines are usually named by prefixing the name of the attached group to the word *amine*. For example:

CH₃NH₂   methylamine

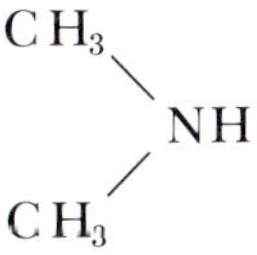

(CH₃)(CH₃)NH   dimethylamine

(CH₃CH₂)(CH₃)NH   ethylmethylamine

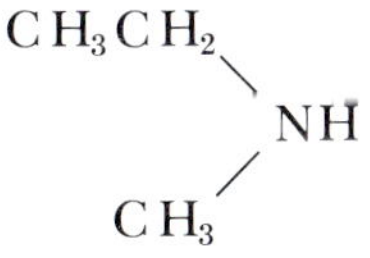

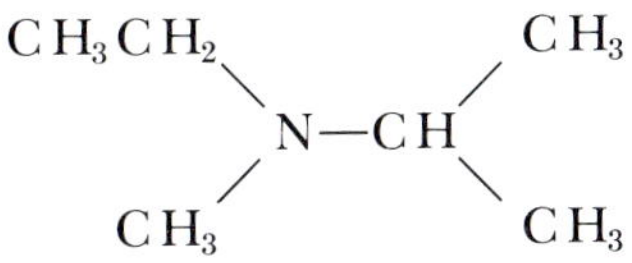

(CH₃CH₂)(CH₃)N—CH(CH₃)₂   ethylisopropylmethylamine

Amines may also be named as **amino hydrocarbons**. For example, ethylisopropylmethylamine can also be named as a 2-aminopropane, based on the longest alkane chain to which nitrogen is attached. To indicate that the nitrogen atom attached to the propane chain has one hydrogen replaced with ethyl and the other with methyl, the amino group is called (N-ethyl-N-methyl)amino-. Thus the full name becomes 2-(N-ethyl-N-methyl)aminopropane.

The simplest aromatic amine

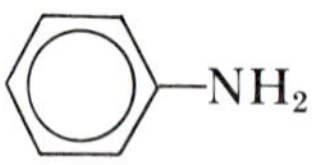

$$\text{C}_6\text{H}_5\text{—NH}_2$$

is called **aniline**. It could also be called **aminobenzene** or **phenylamine**, but these latter names are seldom, if ever, used.

Most aliphatic amines are weak Brønsted-Lowry bases with K_b values around 1×10^{-4} at 25 °C. Amines react with strong acids as follows:

$$RNH_2 + H_3O^+ \rightarrow RNH_3^+ + H_2O$$

The ion RNH_3^+ is analogous to the ammonium ion NH_4^+.

Amines are produced during the decomposition of living matter. Low-molecular-weight amines have odors resembling that of ammonia, NH_3, but as the molecular weight increases the odors range from "fishy" to putrid.

Aniline is an important starting material for many other important chemicals, including pharmaceuticals and dyes.

AMIDES

Amides contain the amino group bonded to an acyl group. The general formula is

$$R-\underset{\underset{O}{\|}}{C}-NH_2$$

The hydrogen atoms on the nitrogen may also be replaced by other groups to form so-called **N-substituted** amides. In the common system the names of amides are derived from the carboxylic acids having the same acyl group as the amide. The *-ic acid* or *-oic acid* part of the carboxylic acid name is replaced with *-amide*. Thus

$$H\underset{\underset{O}{\|}}{C}NH_2$$

is **formamide**,

$$CH_3\underset{\underset{O}{\|}}{C}NH_2$$

is **acetamide**, and

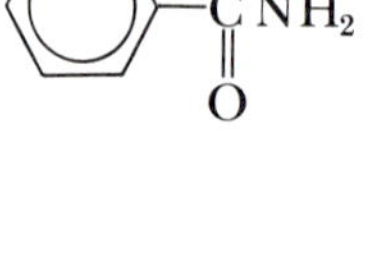

$$C_6H_5\underset{\underset{O}{\|}}{C}NH_2$$

is **benzamide**. The compound

$$\underset{O}{\overset{CH_3}{\underset{\|}{HCN}}}\diagdown_{CH_3} \quad \text{or} \quad \underset{O}{\overset{}{\underset{\|}{HCN(CH_3)_2}}}$$

is called N,N-dimethylformamide. Most chemists call it DMF. DMF is used as a solvent for some chemical reactions. In the IUPAC system amides are named after the corresponding parent hydrocarbon. Thus formamide and acetamide are called **methanamide** and **ethanamide**, respectively.

EXAMPLE 16.19 Name the amide whose framework is

$$\text{C}-\underset{\text{C}}{\overset{}{\underset{|}{\text{C}}}}-\text{C}-\underset{\text{O}}{\overset{}{\underset{\|}{\text{C}}}}-\text{NH}_2$$

SOLUTION The longest chain has four carbon atoms, so this is a butanamide. The methyl group on the third carbon atom makes it 3-methylbutanamide. The common name is isovaleramide, named after isovaleric acid.

AMINO ACIDS

Amino acids are carboxylic acids which also contain an amino group. Most of the important amino acids have the basic structure

$$\text{R}-\underset{\text{H}_2\text{N}}{\overset{}{\underset{|}{\text{CH}}}}-\underset{\text{O}}{\overset{}{\underset{\|}{\text{C}}}}\text{OH}$$

in which the amino group is attached to the carbon atom adjacent to the carboxyl group. However, there is strong evidence that amino acids actually have the dipolar ionic structure

$$\text{R}-\underset{\text{H}_3\text{N}^+}{\overset{}{\underset{|}{\text{CH}}}}-\underset{\text{O}}{\overset{}{\underset{\|}{\text{C}}}}-\text{O}^-$$

which is called a **zwitterion** (from the German *Zwitter*, hybrid). The zwitterion nature is consistent with the fact that amino acids are often crystalline solids at ordinary temperatures and are more soluble in polar solvents such as water than in nonpolar solvents. In addition, aqueous solutions of amino acids behave like aqueous solutions of substances with very high dipole moments. Also, the ionization constants K_a and K_b are much lower than those of comparable carboxylic acids and amines.

Amino acids are the basic units out of which proteins are constructed. So far, over twenty different amino acids have been found sufficient to account for the diversity of proteins found in materials such as feathers, muscle, and coverings of viruses.

Amino acids and proteins are discussed in more detail in Chapter 24. Amino acids are known almost exclusively by their common names. These are given in Chapter 24.

16.13 OPTICAL ACTIVITY OF ORGANIC MOLECULES

Optical activity may be characterized as follows: Ordinary light consists of electric and magnetic components whose magnitudes are directed at right angles to each other and to the direction of propagation and which constantly fluctuate with time. When light is passed through certain types of filtering devices (for example, Polaroid lenses), the fluctuations (or vibrations) in one direction can be weakened or eliminated and only those at right angles to this retained. The resultant beam of light, having vibrations confined to a single plane, is said to be **plane-polarized**. If a beam of plane-polarized light is passed through ordinary matter, the planes of the incident and the transmitted light are the same. However, if the plane-polarized beam is passed through a sample containing a single enantiomeric substance (see Section 16.5, on enantiomeric isomers), the plane of the transmitted light is generally different from that of the incident light (see Fig. 16.9). The angle between the two planes depends on the specific enantiomeric substance involved, its concentration, and the length of the optical path through the substance. The angle of rotation α may be expressed mathematically as

$$\alpha = [\alpha]_\lambda^T \, lc$$

where l is the length of the optical path in dm (1 dm is 10 cm), c is the concentration of enantiomer in $g \cdot cm^{-3}$, and $[\alpha]_\lambda^T$ is a constant characteristic of the enantiomer and called its **specific rotation**. The specific rotation $[\alpha]_\lambda^T$ also depends on the temperature T and the wavelength λ of the plane-polarized light used.

For example, the two enantiomers (mirror-image isomers) of lactic acid (2-hydroxypropanoic acid)

rotate plane-polarized light—that having the wavelength of the sodium D-line (5893 Å)—by 2.26° at 25 °C when in a tube 10 cm long and at a concentration of 1 $g \cdot cm^{-3}$. The two enantiomers differ in that one rotates the plane of plane-polarized light 2.26° to the *left* and the other rotates the plane of plane-polarized light 2.26° to the *right*. Thus the specific rotation $[\alpha]_D^{25}$ is $-2.26°$ $cm^3 \cdot dm^{-1} \cdot g^{-1}$ for the former and $+2.26°$ $cm^3 \cdot dm^{-1} \cdot g^{-1}$ for the latter. The two enantiomers are called $(-)$-lactic acid and $(+)$-lactic acid, respectively. Because of this difference in the sign of the rotation of light, enantiomers are often called **optical isomers**. Molecules which rotate the plane of polarized light are said to be **optically active**. Molecules with $(-)$ rotation are said to be **levorotatory** (left-rotating) and molecules with $(+)$ rotation are said to be **dextrorotatory** (right-rotating).

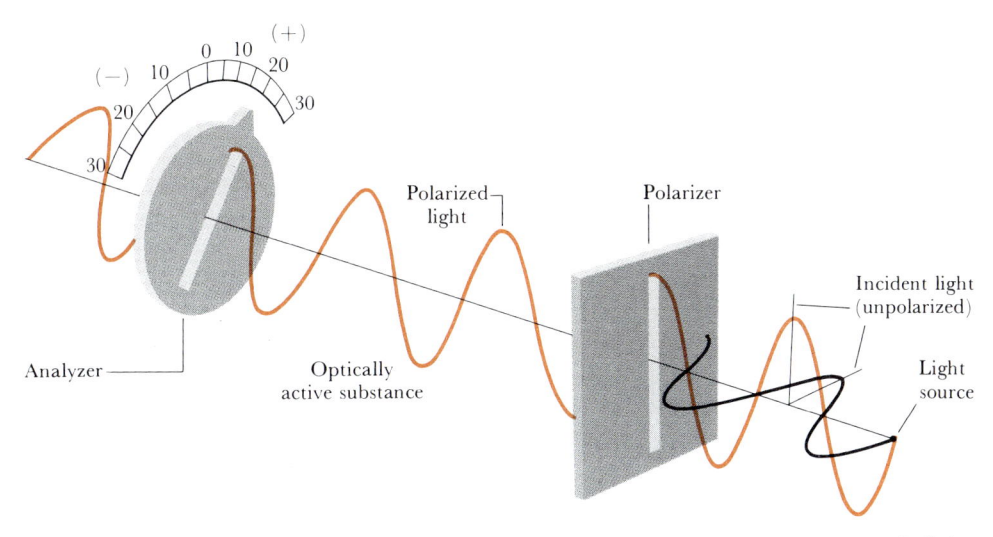

Fig. 16.9. Schematic illustration of the measurement of optical rotation. Monochromatic light enters the polarizer, is polarized, and passes through a solution of an optically active substance. The amount by which the plane of the polarized light is rotated is read off on the analyzer scale. For an optically inert material the polarizer and analyzer slots would be parallel and the analyzer scale would read 0°.

EXAMPLE 16.20 A solution of (−)-lactic acid has a concentration of 1.5 g·cm⁻³ and is placed into a 5-cm-long tube at 25 °C. By how much and in which direction is the plane of plane-polarized sodium D light rotated when passed through this solution?

SOLUTION

$$\alpha = [\alpha]_D^{25} lc = (-2.26° \text{ cm}^3 \cdot \text{dm}^{-1} \cdot \text{g}^{-1})(0.5 \text{ dm})(1.5 \text{ g} \cdot \text{cm}^{-3})$$

$$= -1.70° \quad \text{(rotation to the } \textit{left}\text{)}$$

It is not immediately obvious which of the enantiomers illustrated above is (+)-lactic acid and which is (−)-lactic acid. This fact can be ascertained only by determining the absolute configurations of the enantiomers. From X-ray studies by the Dutch chemist J. M. Bijvoet in 1949, it is now known that the left enantiomer is (+)-lactic acid and the right enantiomer is (−)-lactic acid.

When lactic acid is synthesized in the laboratory, a mixture containing equal amounts of the two enantiomers is obtained since one enantiomer is just as likely to be formed as the other under perfectly random synthesis conditions. Such a mixture is called a **racemic** mixture and is optically inactive. Separation of the racemic mixture into the optically active components can be carried out by physical or chemical means, depending on the specific substance involved. The first scientist to note that optical activity was related to structure was Louis Pasteur. Pasteur noted that crystals of sodium ammonium tartrate, an optically inactive substance, were of two different types which were mirror images of each other. Pasteur used tweezers to separate the two types of crystals and then showed each was optically active with specific rotations of equal magnitude but opposite signs.

Chemical separation of enantiomers is based on the principle that it takes a chiral molecule to distinguish between a pair of other chiral molecules. This is because only a chiral molecule is sensitive to the distinction between right and left. (In this sense Pasteur may be regarded as a chiral "molecule" since his separation was dependent on the ability to distinguish between the two types of crystals!) The chemical method

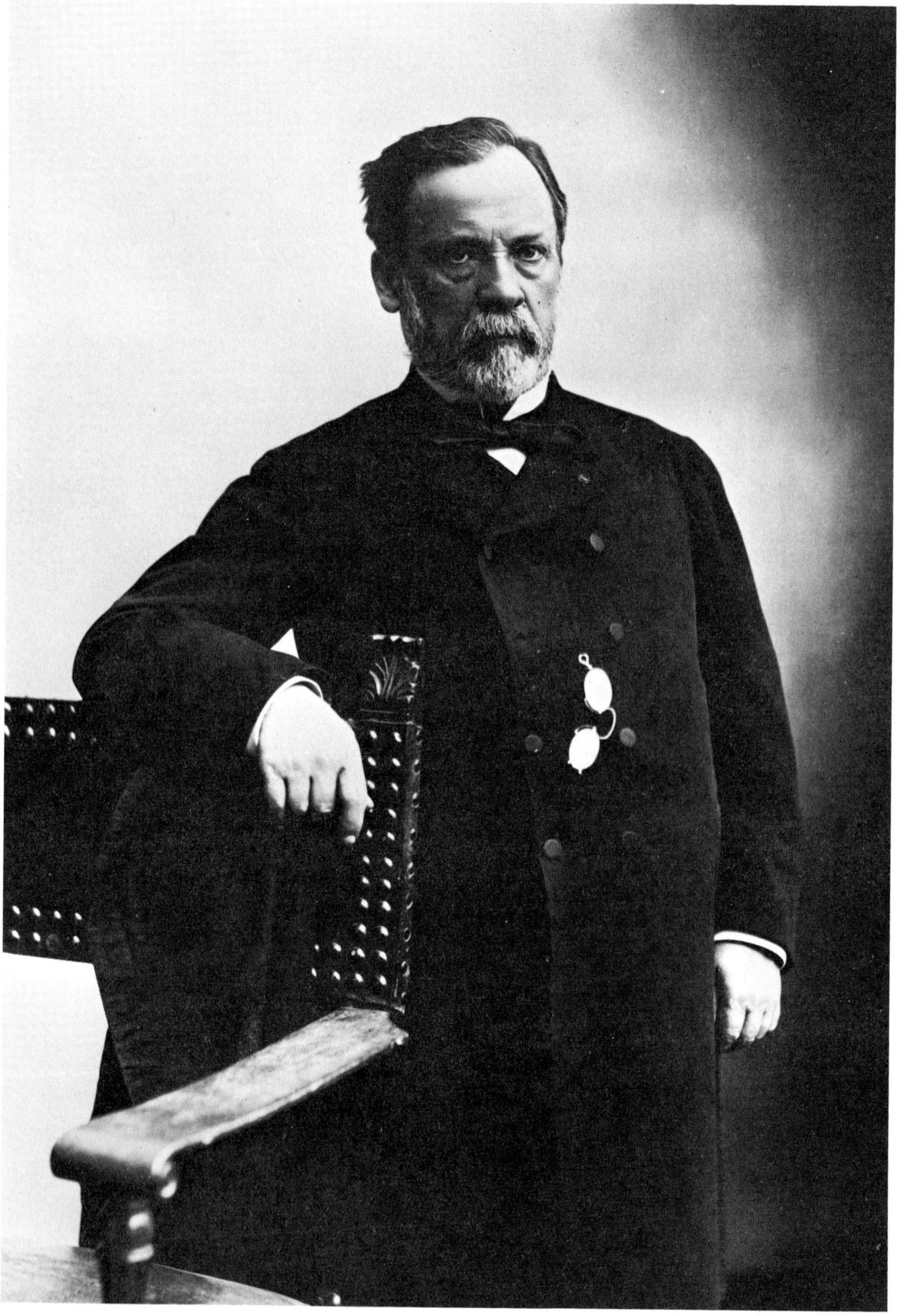

Louis Pasteur (1822–1895), a French chemist, is best known for laying the foundations of modern bacteriology, developing vaccines, and performing experiments concerning the spontaneous generation of life. An interesting and readable biography of Pasteur has been written by René Vallery-Radot (The Life of Pasteur, Garden City, N.Y.: Garden City Publishing, 1909).

involves finding a chiral molecule which will combine with each of the members of an enantiomeric pair to convert them to a pair of diastereomers. The diastereomers have different physical and chemical properties and, hence, are readily separated by ordinary means. Removal of the chiral agent then produces pure enantiomers.

Optically active molecules may be isolated from plant and animal substances. Most living organisms produce only one member of an enantiomeric pair. Thus the glucose produced in photosynthesis is (+)-glucose; the (−)-glucose enantiomer is not formed. (See Section 16.14 for structures.) Similarly, (+)-lactic acid is found in muscle and blood, and (−)-lactic acid is found in sour milk.

Both members of an enantiomeric pair have exactly the same melting and boiling points, the same density, and virtually all physical and chemical properties in common. The enantiomers will exhibit different properties only under conditions which are sensitive to chirality, that is, to a difference between "right-handedness" and "left-handedness." Some organisms can utilize only one member of an enantiomeric pair for food. The proteins found in nature are made up almost entirely of amino acids in which only one of the enantiomers is present. This enantiomer is almost always the one having the configuration

$$\begin{array}{c} CO_2H \\ | \\ C \\ / | \backslash \\ NH_2 \; R \; H \end{array}$$

The mirror image of this, the other enantiomer, is usually absent in proteins. Amino acids having the above configuration are called L-amino acids; the enantiomers are called D-amino acids. The two different enantiomers of an amino acid may differ in taste. For example, D-(+)-phenylalanine is sweet while L-(−)-phenylalanine is bitter. Similarly, D-(−)-valine is bitter and L-(+)-valine is sweet. Only L-(−)-phenylalanine and L-(+)-valine are found in natural proteins. In the case of sugars, D-(+)-glucose is sweet, but L-(−)-glucose tastes salty. The enantiomers of the drug amphetamine have the same physiological effect but the (+) enantiomer (trade name Dexedrine) is two to four times as potent as the (−) enantiomer. The racemic mixture of amphetamine is sold under the trade name Benzedrine.

Facing page: Louis Pasteur, French physical chemist who pioneered in molecular structure, germ theory of fermentation and disease, and the development of vaccines. (Courtesy of The Bettmann Archive.)

Why is the interaction of plane-polarized light with matter sensitive to chirality? Basically, what occurs when light passes through matter is an interaction between the electric and magnetic fields of the light and the electrical charges of which all matter is made. This interaction tends to distort the path of the light so that the plane of plane-polarized light becomes bent in one direction or another. However, if the substance is nonenantiomeric, that is, its own mirror image, any distortion of the plane of light by a molecule having a given orientation is very likely to eventually be cancelled out by an equal but opposite deflection brought about by another molecule having a mirror-image orientation to that of the first molecule. By the time the light has interacted with around one mole of molecules it is very likely that all deflections of the plane average out to zero. But if the substance consists of a single enantiomer, a deflection of the light's plane by a molecule in a given orientation can never be cancelled out by the mirror image of that molecule since no mirror image molecules are present. Thus no overall cancellation of deflections occurs and the net deflection will result in either a positive or negative rotation of the plane of the light.

16.14 HETEROCYCLIC MOLECULES

A **heterocyclic** molecule is a cyclic molecule in which one or more carbon atoms in the ring have been replaced by some other atom. Many of the molecules found in living organisms are heterocyclic. Atoms which commonly replace carbon in such molecules include nitrogen, oxygen, and sulfur.

Five heterocyclic molecules found in DNA (deoxyribonucleic acid) and RNA (ribonucleic acid) are:

adenine   guanine   thymine

cytosine   uracil

DNA and RNA play central roles in the chemical mechanisms of genetics.

A rather broad category of heterocyclic compounds known collectively as **sugars** plays such important roles in the structures and properties of living organisms that it merits separate attention. One of the most important sugars, D-(+)-glucose, has a structure resembling that of the chair form of cyclohexane:

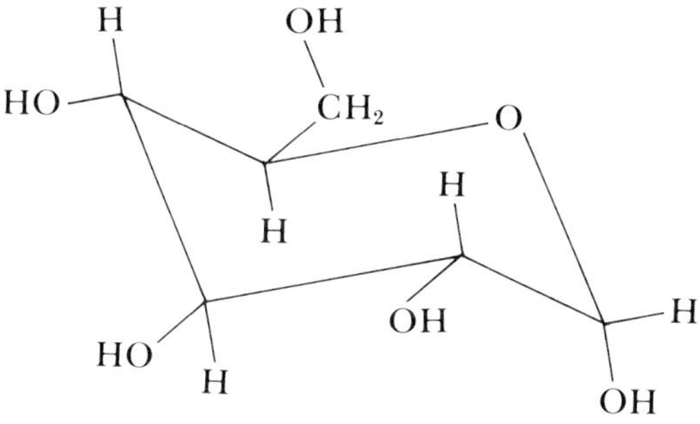

This is usually represented by the simpler diagram

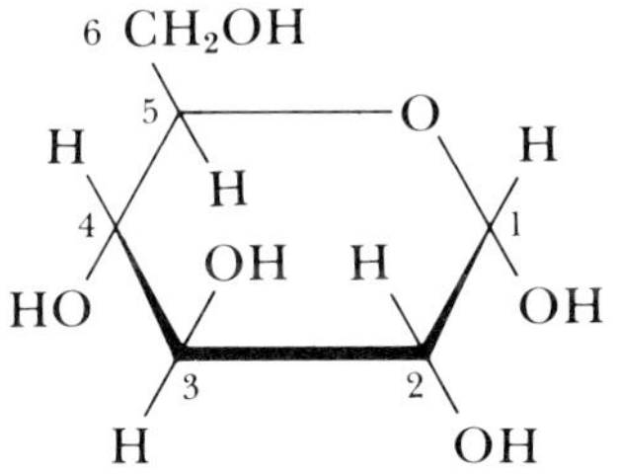

The shaded bottom edge of the hexagon indicates that one is looking at the ring from the side and from above. The puckering of the ring is not indicated in this diagram. The numbers indicate the system for numbering carbon atoms that is employed by sugar chemists. A diastereomeric variant of the above molecule has the H and OH positions about the first carbon atom reversed:

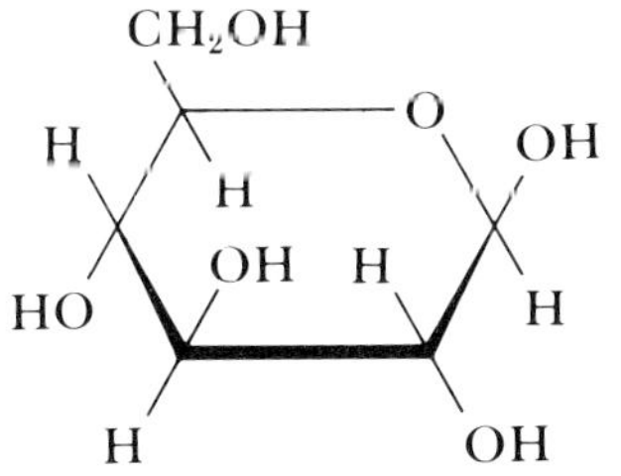

The two diastereomers are called **anomers**; the former is called α-D-(+)-glucose and the latter is called β-D-(+)-glucose. The α anomer has a specific optical rotation of +112° and a melting point of 146 °C, whereas the β anomer has a specific rotation of +19° and a melting point of 150 °C (the anomer with the larger rotation is usually called α).

Both of the above anomers have an enantiomeric counterpart or optical isomer

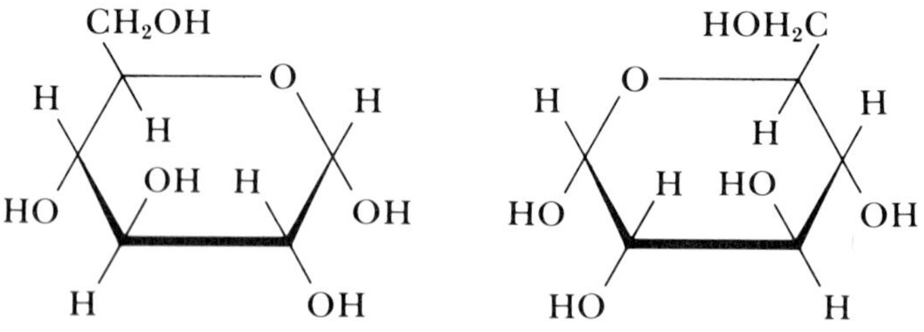

α-D-(+)-glucose   α-L-(−)-glucose

Note that α-L-(−)-glucose may also be written

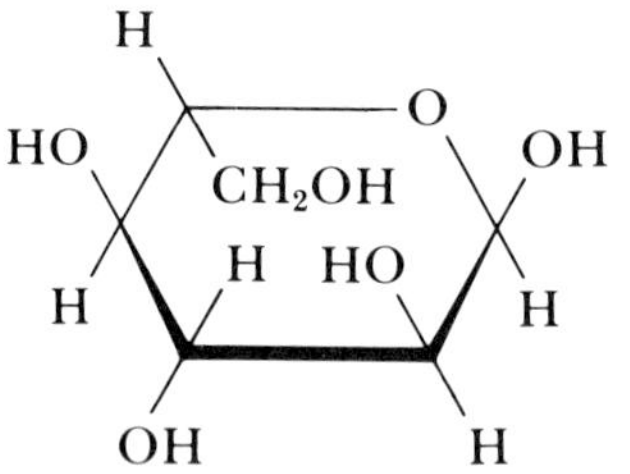

The symbols D and L refer to the absolute configuration of the sugar based on the known absolute geometries of the following reference enantiomeric pair of molecules:

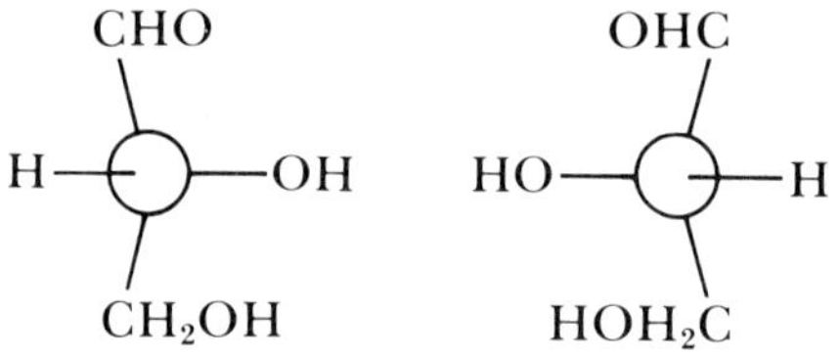

D-(+)-glyceraldehyde   L-(−)-glyceraldehyde

A molecule is called D if it can be shown to be synthesizable from D-(+)-glyceraldehyde by a sequence of reactions. Although in the examples shown above, the D compounds are dextrorotatory (+) and the L compounds are levorotatory (−), there is no necessary connection between the D, L designations and the ± designations. For example, the compounds D-(−)-glyceric acid and L-(+)-lactic acid are known to exist. The latter is found in muscle tissue.

Glucose often occurs in nature as long chains of molecules known as **polysaccharides**. The formation of a polysaccharide from glucose molecules may be represented stoichiometrically by

$$(n + 2)C_6H_{12}O_6 \rightarrow C_6H_{11}O_6(C_6H_{10}O_5)_nC_6H_{11}O_5 + (n + 1)H_2O$$

When n is very large, the empirical formula *appears* to be $(C_6H_{10}O_5)_n$. Such molecules are examples of carbohydrates.

Two of the most important naturally occurring polysaccharides are **starch** and **cellulose**. Plant starch consists of 20% by weight of a water-soluble fraction called **amylose** and the rest a water-insoluble residue called **amylopectin**. Amylose consists of straight chains of approximately 200 glucose units of the α anomer. Amylopectin is similar but the chain is highly branched—each branch containing 20–25 glucose units—and consists of about 1000 glucose units per chain. Animal starch (glycogen) resembles plant starch but the amylopectin portion tends to have slightly shorter branches. Cellulose consists largely of long chains of the β anomer (1000–5000 glucose units each) which are twisted together and held together in places by hydrogen bonds. These hydrogen-bonded areas are crystalline in nature.

The basic chain structure of starch may be symbolized by

while that of cellulose is

Different enzymes (see Section 20.9) are required to digest the two substances. Humans do not possess enzymes to digest cellulose but cows and other herbivores do. Thus the latter can eat and digest hay and other fodder crops while humans cannot.

Two other sugars, both of which play important roles in living organisms, are **ribose** and **deoxyribose**:

ribose   deoxyribose

The former is found in RNA and the latter in DNA.

16.14 | HETEROCYCLIC MOLECULES   **457**

SUMMARY

Elemental carbon occurs in nature as graphite and diamond. The latter is the hardest natural substance known. The two most important oxides of carbon are carbon monoxide, CO, and carbon dioxide, CO_2. CO is a toxic gas formed by incomplete combustion of carbon or carbon-containing substances; CO_2 is formed by complete combustion of carbon-containing substances. The latter is utilized by plants in photosynthesis to produce carbohydrates. Some of the carbon of the earth is in constant circulation between the atmosphere and the hydrosphere; large amounts have become stored in fossil fuel deposits and sedimentary rocks.

Carbon's unique ability to bond to itself and to many other elements leads to a diversity of carbon compounds, many of them constituting the basic substances of which plants and animals are built. Hydrocarbons are among the simplest of compounds in which large numbers of carbon atoms are linked to each other in long chains and rings. Important hydrocarbons include alkanes, alkenes, alkynes, and aromatic compounds. The structures of many of these constitute the basis for understanding much of organic chemistry.

The chemical properties of many organic compounds depend largely on small groups of atoms attached to the carbon skeleton. These so-called functional groups produce several fundamentally different types of compounds including alcohols, ethers, carboxylic acids, aldehydes, ketones, esters, amines, amides, and amino acids.

Many carbon compounds exist as isomers. The main types of isomers are called structural (linkage) isomers, diastereomers, and enantiomers. Two different isomers generally have different chemical and physical properties; enantiomers differ only in those properties sensitive to chirality—a sense of right- and left-handedness.

Many of the organic compounds found in living things are heterocyclic in nature and contain heteroatoms such as nitrogen, sulfur, and oxygen. Important heterocyclic compounds include adenine, cytosine, guanine, thymine, uracil, and the sugars. All of these are found in DNA and RNA—the chemical substances that control heredity—and the latter are found in starch and cellulose.

LEARNING GOALS

1. Be able to describe the lattice structures of graphite and diamond and know the principal properties of each.

2. Be familiar with the main properties of carbon monoxide and carbon dioxide. Know how each may be prepared.

3. Know the general features of the carbon cycle in nature.

4. Know in detail the general patterns whereby carbon atoms bind to each other in extended arrays.

5. Know the various ways isomerism can occur in carbon compounds.

6. Know the basic types of aliphatic hydrocarbons and be able to name unbranched-chain alkanes, alkenes, and alkynes with up to twelve carbon atoms.

7. Know how to name branched-chain hydrocarbons.

8. Know how alicyclic compounds are named.

9. Know the general structure of dienes and how they are named.

10. Know what is meant by a conformation and be able to sketch different conformations of molecules such as ethane, propane, and cyclohexane.

11. Be able to draw structures and be able to name aromatic hydrocarbons containing up to three rings.

12. Be able to describe the three-dimensional geometries of aliphatic and aromatic hydrocarbons in terms of the appropriate hybrid orbitals.

13. Be able to name aromatic compounds formed from benzene.

14. Know the general formulas of alcohols, phenols, ethers, carboxylic acids, aldehydes, ketones, esters, amines, amides, and amino acids.

15. Be able to name the simpler members of each of the types of compounds just listed.

16. Know the way in which a carboxylic acid and an alcohol combine to form an ester.

17. Know what an acyl group is and how to use it in naming compounds such as the acyl halides.

18. Know how optical activity arises and how it is measured experimentally.

19. Know what is meant by D, L, (+), and (−) and how these are used to name enantiomers.

20. Know the basic structure of glucose and how starch and cellulose are formed from it.

DEFINITIONS, TERMS, AND CONCEPTS TO KNOW

allotropes
carbohydrates
greenhouse effect
fossil fuels
phytoplankton
aliphatic compounds
alicyclic compounds
aromatic compounds
homolog
alkanes
alkenes
alkynes
dienes
cycloalkanes
Newman projection diagrams
structural (linkage) formula

stereoisomer
conformation
isomer
structural (linkage) isomer
diastereomer
enantiomer
anomer
alkyl group
alkenyl group
aryl group
ortho, meta, para positions
pi electrons
chair and boat conformations of cyclohexane
aromatic sextet
alcohol

phenol
ether
carboxylic acid
aldehyde
ketone
ester
amine
amide
amino acid
heterocyclic compound
chirality
optical activity
polarized light
specific rotation
racemic mixture
polysaccharide

QUESTIONS AND PROBLEMS

General

1. Data gathered by Apollo missions to the moon indicate that each square meter of the earth is hit by 1.0×10^8 carbon atoms per second from the solar wind. Assuming the earth is about 4.5×10^9 yr old and that the solar wind flux has been constant over this span of time, how many tonnes of carbon would the earth have accumulated from the sun during its lifetime? [*Note:* It is estimated that the earth now contains over 1×10^{11} tonnes of carbon in the form of living matter.]

2. Draw Lewis octet diagrams for H_2CO_3, HCO_3^-, and CO_3^{2-} and predict the geometry of each of these species.

3. Balance the oxidation–reduction reaction $CH_3OH + ClO_3^- \rightarrow HCO_2H + Cl^-$ in acid solution.

4. With respect to the molecule

$$\begin{array}{c} R \\ \diagdown \\ H \end{array} C{=}C \begin{array}{c} H \\ \diagup \\ R \end{array}$$

(where R≠H), would you expect zero dipole moment for all possible kinds of R? Explain.

5. Draw the structures of the carboxylic acids and alcohols of which each of the following esters is a combination.

a) C₆H₅—O—C(=O)—CH₃

b) C₆H₅—C(=O)—O—CH₃

6. The K_a of benzoic acid at 25 °C is 6.3×10^{-5}. What is the equilibrium constant of the reaction of sodium benzoate with H_3O^+ at 25 °C? This reaction accounts for the action of sodium benzoate as a food preservative. The product, benzoic acid, inhibits the growth of microorganisms.

7. The solubility of an amino acid in water is generally a minimum at a certain pH called the **isoelectric point**; from this point the solubility increases for either a decrease or an increase of pH. Using the fact that an amino acid can act as both a weak acid and a weak base, explain the above behavior.

8. Although the odors of amines tend to become more obnoxious with increasing molecular weight, very high-molecular-weight amines are practically odorless. Explain.

9. The normal boiling points of compounds of a group such as the alkanes generally increase as the molecular weight increases. Why should this be so? (See Section 6.4.)

10. Alkanes are often called **saturated** hydrocarbons whereas alkenes, alkynes, and aromatic compounds are called **unsaturated** hydrocarbons. What do you suppose is the reason for these terms?

11. Write molecular formulas for each of the following.
  a) An alkane with 35 carbon atoms.
  b) An alkene with 17 carbon atoms.
  c) An alkyne with 41 carbon atoms.
  d) A diene with 41 carbon atoms.
  e) A cycloalkane with 17 carbon atoms.
  f) An aromatic compound having 5 fused rings all in a line.

12. Compare the relative positions of the four carbon atoms in butane, 2-butene, and 2-butyne. Why is there such a difference?

13. Why does 2-butene exhibit *cis-trans* isomerization while 1-butene does not?

14. Suppose carbon atoms of alkanes were planar rather than tetrahedral. How many isomers of CHFClBr could there be? How many isomers does this molecule actually have?

15. Discuss the relative structures and properties of graphite and diamond. List as many contrasting pairs of properties as you can.

16. The alcohols $C_nH_{2n+1}OH$ become less and less water soluble as n increases. Why?

17. Which would you expect to be more water soluble (on a molar basis), $HO_2CCH_2CH_2CO_2H$ (succinic acid), or $CH_3CH_2CH_2CO_2H$ (butanoic acid)? Explain.

Isomers

18. How would you name the diastereomers of CHCl=CHBr?

19. Draw linkage (structural) formulas for all the isomers of the alcohol C_4H_9OH and name as many of these as you can.

20. Draw linkage (structural) formulas for all the isomers of pentene and give systematic names for each of these.

21. Sketch the carbon skeletons for all the isomers having the molecular formula C_4H_6. There are at least 10.

22. Draw all the isomers of C_6H_{12}.

23. Heptane has nine isomers. Draw the structure and give the name of each.

24. Name all the ways you can think of that prove humans are chiral. In what ways do humans appear to be achiral?

25. Sketch structures for all the isomers having the following molecular formulas.
  a) C_2H_5Cl b) $C_2H_4Cl_2$ c) $C_2H_3Cl_3$
  d) $C_2H_2Cl_4$ e) C_2HCl_5

Recognition of Functional Groups

26. Name the type of compound (alcohol, aldehyde, etc.) represented by each of the following formulas.

a) 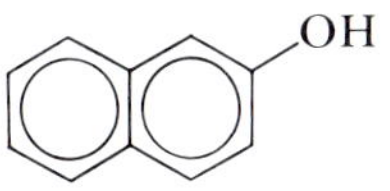 [naphthalen-2-ol structure with OH]

b) 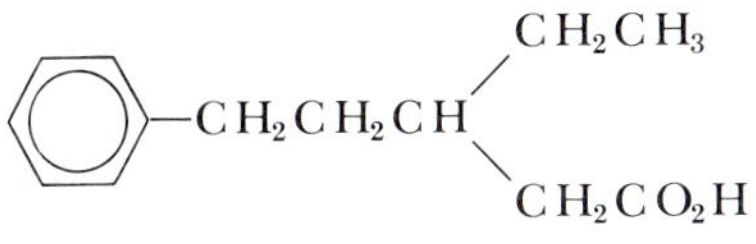 C₆H₅—CH₂CH₂CH(CH₂CH₃)(CH₂CO₂H)

c) 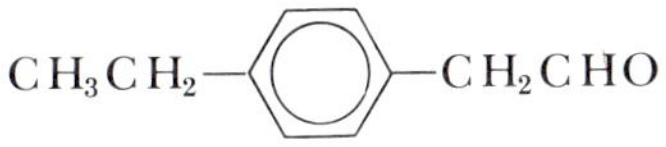 CH₃CH₂—C₆H₄—CH₂CHO

d) 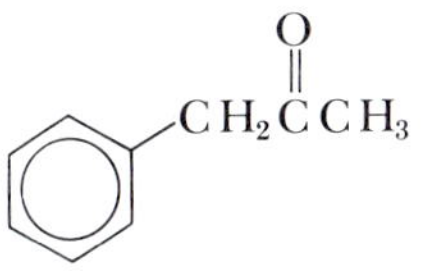 C₆H₅—CH₂C(=O)CH₃

e) CH₃—C₆H₄—CH₂OCH(CH₂—CH₂)(naphthalene)

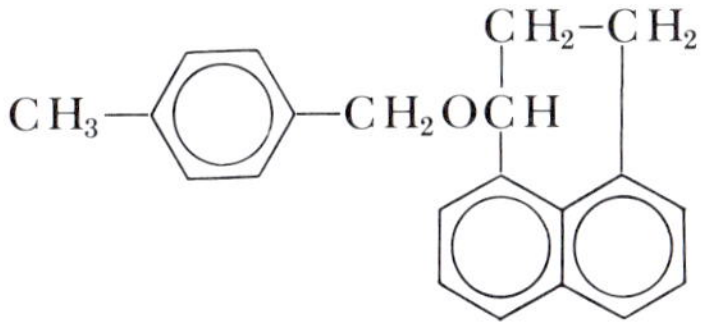

f) C₆H₁₁O₆—(C₆H₁₀O₅)₅₀₀—C₆H₁₁O₅

g) 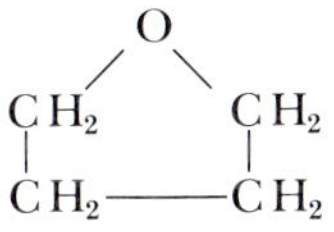 [oxetane-like ring: O, CH₂, CH₂, CH₂, CH₂]

h) 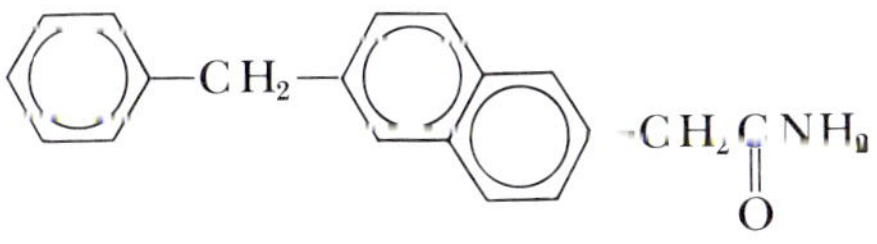 C₆H₅—CH₂—(naphthalene)—CH₂C(=O)NH₂

i) 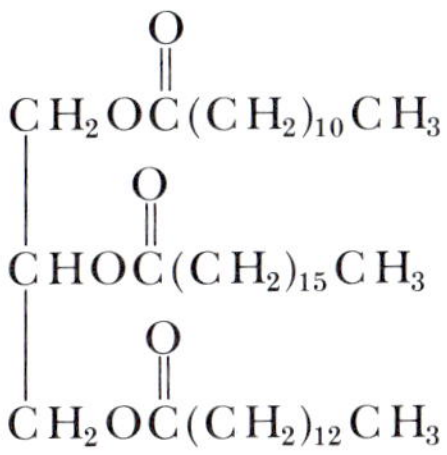
CH₂OC(=O)(CH₂)₁₀CH₃
|
CHOC(=O)(CH₂)₁₅CH₃
|
CH₂OC(=O)(CH₂)₁₂CH₃

j) 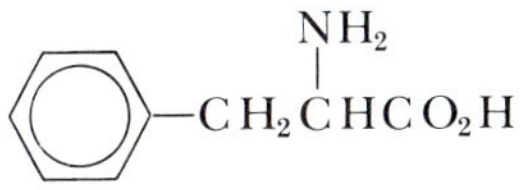 C₆H₅—CH₂CH(NH₂)CO₂H

k) 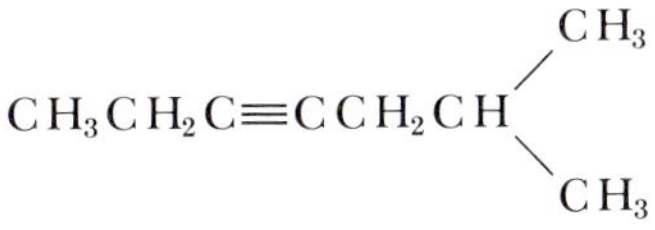 CH₃CH₂C≡CCH₂CH(CH₃)(CH₃)

l) 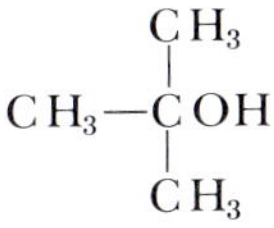 CH₃—C(CH₃)(CH₃)OH

27. Name the type of compound (alcohol, aldehyde, etc.) represented by each of the following formulas.

a) 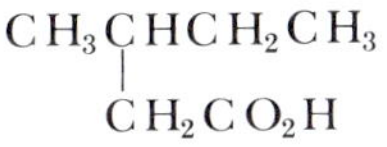 CH₃CH(CH₂CO₂H)CH₂CH₃

b) 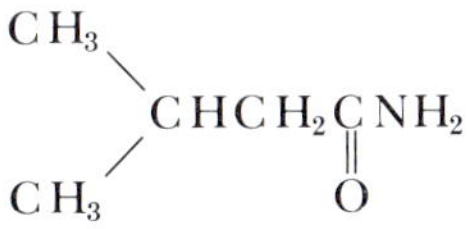 (CH₃)(CH₃)CHCH₂C(=O)NH₂

c) 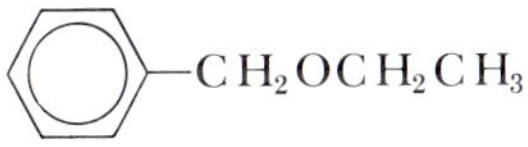 C₆H₅—CH₂OCH₂CH₃

d) 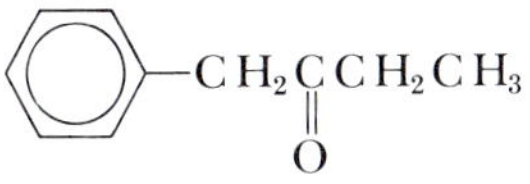 C₆H₅—CH₂C(=O)CH₂CH₃

e) 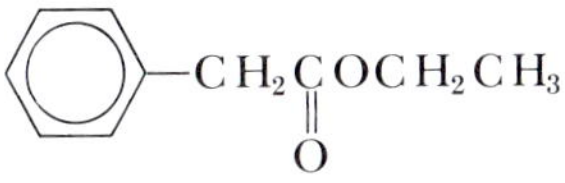 C₆H₅—CH₂C(=O)OCH₂CH₃

f) 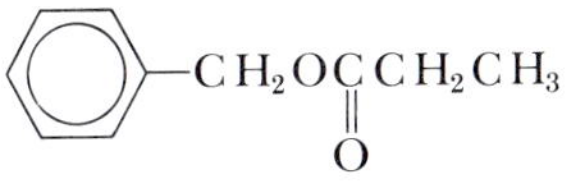 C₆H₅—CH₂OC(=O)CH₂CH₃

g) 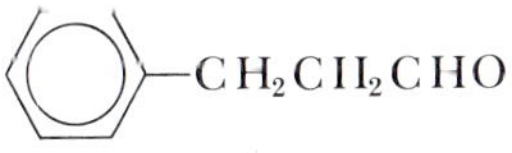 C₆H₅—CH₂CH₂CHO

h) HO₂CCH₂CH₂CH(CH₃)₂

i) 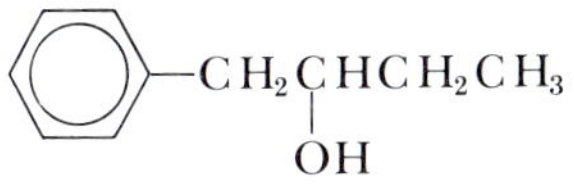 C₆H₅—CH₂CH(OH)CH₂CH₃

j) 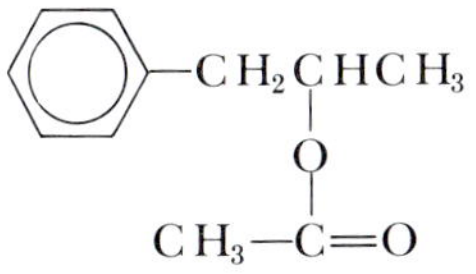 C₆H₅—CH₂CH(CH₃)—O—C(=O)CH₃

k) 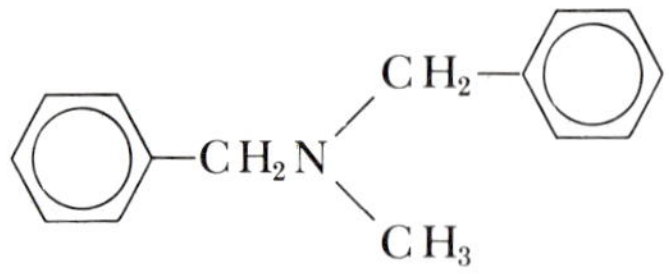 C₆H₅–CH₂N(CH₃)(CH₂C₆H₅)

l) 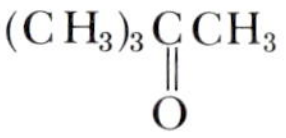 (CH₃)₃CCCH₃
           ‖
           O

Naming of Organic Compounds

28. Give specific names for each of the following compounds.

a) 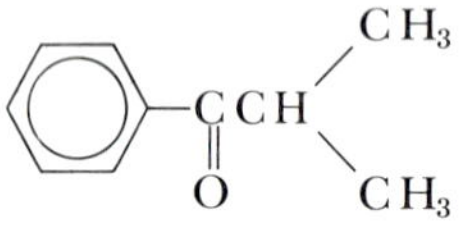 C₆H₅–C(O)–CH(CH₃)₂

b) CH₃CH₂CH₂OCCH₃
                ‖
                O

c) (CH₃)₂CHOH

d) 3-methylbenzoic acid structure (m-CH₃–C₆H₄–CO₂H)

e) CH₃CHCH₂C–CHCH₂CH₃
       | | |
       Cl   Br CH₃
            (with Br on central C)

f) CH₃CH₂CH₂CH₂CHO

g) 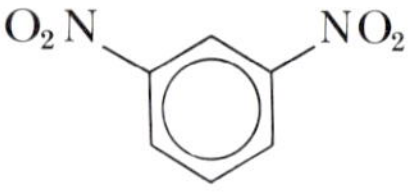 1,3-dinitrobenzene (O₂N–C₆H₄–NO₂, meta)

h)  (CH₃)₂CHOCH(CH₃)₂

i) CH₂ClCO₂H

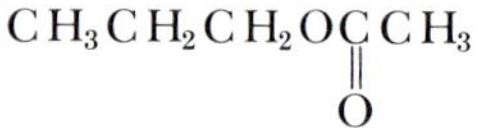

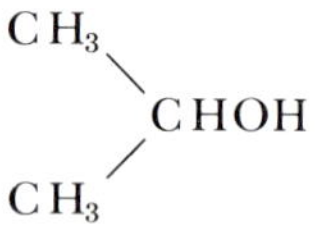

j) 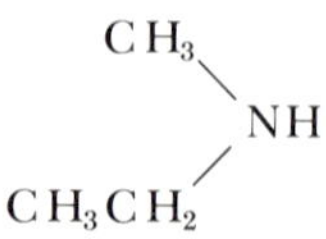 (CH₃)(CH₃CH₂)NH

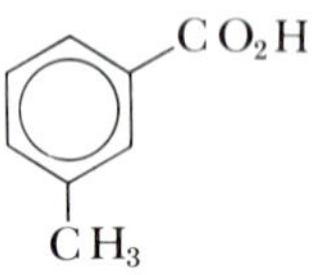

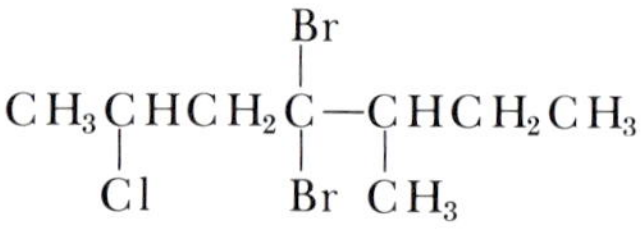

k) 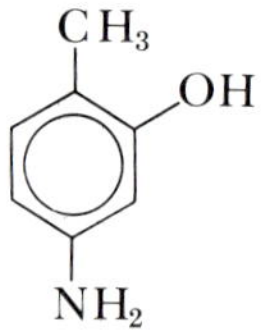 2-methyl-3-hydroxy... (CH₃, OH, NH₂ on benzene ring)

(name as a toluene, a phenol, or an aniline)

l) 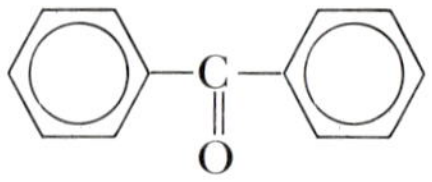 C₆H₅–C(O)–C₆H₅

m) 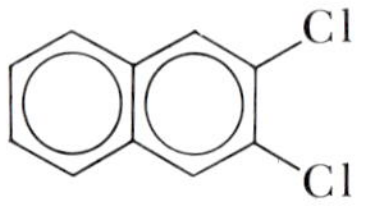 2,3-dichloronaphthalene

n) 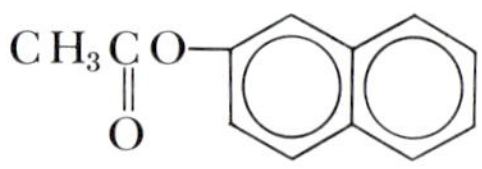 CH₃CO–O–(naphthyl)
      ‖
      O

o) CH₃CH₂CH₂CNH₂
              ‖
              O

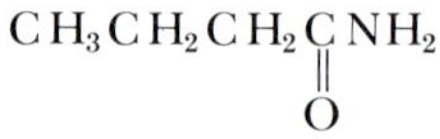

p) 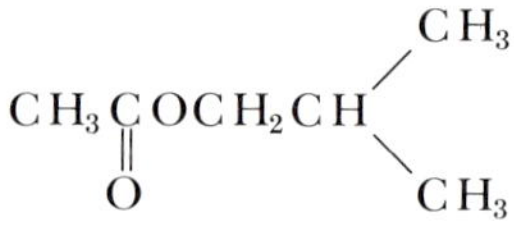 CH₃COCH₂CH(CH₃)₂
      ‖
      O

q) 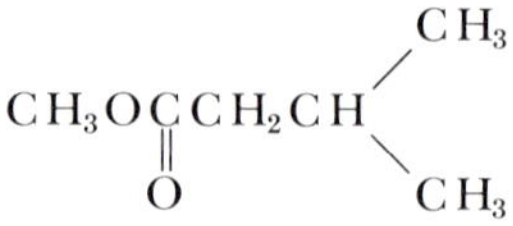 CH₃OCCH₂CH(CH₃)₂
       ‖
       O

r) 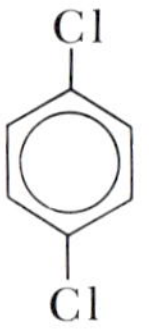 1,4-dichlorobenzene

s) 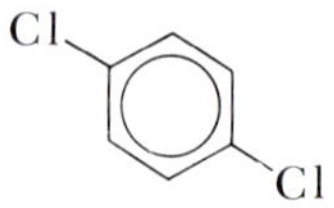 1,4-dichlorobenzene (Cl para Cl)

t) 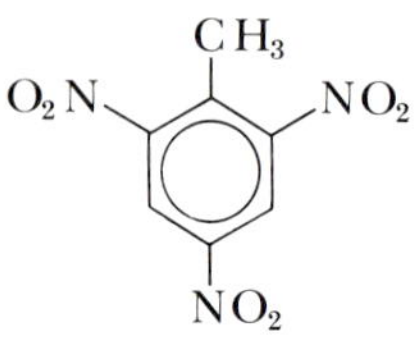 2-methyl-1,3,5-trinitrobenzene (O₂N, NO₂, NO₂ with CH₃ on ring)

29. Give specific names for each of the following compounds.

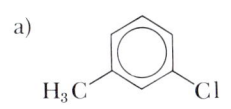

a) [benzene ring with H₃C and Cl substituents in meta positions]

b) $CH_3CHCH_2NH_2$
      |
      CH_3

c) $CH_3CHCH_2CO_2H$
      |
      CH_3

d) $CH_3CHCH_2C\underset{\underset{O}{\|}}{N}\overset{CH_3}{\underset{CH_3}{}}$
      |
      CH_3

e) $H_2N\underset{\underset{O}{\|}}{C}CH_2CH\overset{CH_3}{\underset{CH_3}{}}$

f) $CH_3\overset{CH_3}{\underset{|}{C}}HCH_2\underset{\underset{O}{\|}}{C}NH_2$

g) [benzene ring with NO₂, Cl, OH, Br substituents]

h) $CH_3CH_2CH_2\underset{\underset{O}{\|}}{C}CH_2CH(CH_3)_2$

i) [C=C with H, Cl on one carbon and CHCl₂, H on other]

j) [C=C with Cl, Cl on one carbon and CHCH₂Cl on other]

k) [C=C with CH₃, Br on one carbon and Br, CH₃ on other]

l) $CH_3CH_2CH_2CH_2O\underset{\underset{O}{\|}}{C}CH_2CH_2CH_3$

30. Draw structures for each of the following compounds.
- a) 1,1,3-trichloropropene
- b) 1,3,3-trichloropropene (*cis* isomer)
- c) 3-chloro-5-aminotoluene
- d) 3-methyl-5-chloroaniline
- e) 4-chloropentanoic acid
- f) 3-methylbutanenitrile
- g) 2-pentanone
- h) 3-fluorobutanal
- i) 2-methyl-6-chloro-3-heptyne
- j) 2-methylpropyl-4-chloropentanoate
- k) N-ethyl-N-methylbutanamide
- l) *m*-chloroaniline
- m) 1,5-dimethylnaphthalene
- n) benzyl ketone
- o) phenylmethanamide

31. Draw the structures of the esters formed from each of the following acids and alcohols and name each ester.

ACID	ALCOHOL
a) ethanoic (acetic)	2-butanol
b) ethanoic (acetic)	1-propanol
c) benzoic	2-propanol
d) propanoic	benzyl
e) methanoic (formic)	ethanol

Geometry of Molecules

32. Describe the bonding in propadiene (allene), $CH_2=C=CH_2$, in terms of appropriate hybrid orbitals, and sketch the expected geometry of this molecule.

33. Draw Newman projection diagrams for what you would expect to be the most stable conformation of each of the following molecules.
- a) $CH_3CH_2CH_2CH=CH_2$
- b) $HO_2CCH-CHCO_2H$
          | |
         CH_3 CH_3
- c) Cyclohexane
- d) [square]—CH_3 (methylcyclobutane)
- e) Cl_2CHCH_2Cl

34. Suggest modified Newman projection diagrams illustrating the geometries of acetaldehyde (ethanal) and acetic acid (ethanoic acid).

35. Carbon suboxide, C_3O_2, is an unpleasant-smelling gas formed by dehydrating malonic acid, $HO_2CCH_2CO_2H$, with P_4O_{10} in a vacuum at 140–150 °C. Draw the expected Lewis octet diagram of C_3O_2 and predict its geometry.

36. Acetic acid (ethanoic acid) can exist in a dimeric form (two molecules joined to form one molecule). Show a probable geometry for this molecule which involves two hydrogen bonds.

37. Explain why 2-butene can have *cis-trans* isomers while 2-butyne cannot.

38. Can the molecule 1,3-dichloropropadiene have *cis-trans* isomers? Explain.

39. The molecule CH_2=$CHCHO$ (commonly called **acrolein**; IUPAC name, **propenal**) is planar. Use appropriate hybrid orbitals to suggest why this is so. [*Hint:* Compare with 1,3-butadiene. Would you expect the *cis* or *trans* isomer to be the more stable?]

40. Discuss why benzene and cyclohexane have such different geometries.

41. Compare the geometries of ethanol and methyl ether.

42. Sketch a possible structure for the zwitterion of the amino acid glycine ($H_2NCH_2CO_2H$) in which two carbon atoms, one oxygen atom, one nitrogen atom, and one hydrogen atom form a five-membered planar ring involving hydrogen bonding.

Optical Isomerism

43. a) Which of the following molecules can have optical isomers?
b) Draw structural or linkage formulas to represent any possible enantiomers.

$NH_2CH_2CO_2H \quad\quad CH_3\underset{\underset{NH_2}{|}}{C}HCO_2H$

44. a) Are the following mirror-image molecules superimposable?
b) Under what conditions would you expect I and II to be optically active? [*Hint:* Consider the ease with which the different conformations can interconvert.]

(R ≠ H)

45. Explain the symbols D, L, and ± as used, for example, in the naming of sugars. Explain specifically what D-(−)-glyceric acid and L-(+)-lactic acid are and how they rotate the plane of polarized light. Glyceric acid is

$HOCH_2\underset{\underset{OH}{|}}{C}HCO_2H$

and lactic acid is

$CH_3\underset{\underset{OH}{|}}{C}HCO_2H$

46. When either anomer of D-(+)-glucose is dissolved in water, the specific optical rotation changes with time and eventually becomes constant at +52.7°. When the solution is evaporated to dryness, a mixture of both anomers is recovered. Explain. (The specific rotation of the α anomer is +112° and that of the β anomer is +19°). Calculate the composition of the dried mixture.

47. Pure samples of the enantiomers L-(+)-lactic acid and D-(−)-lactic acid melt at 53 °C, yet a racemic mixture of these enantiomers melts at 17 °C. Explain. [*Hint:* What happens to the freezing point of water when another substance is added to it?]

48. There are three different geometric isomers of 2,3-butanediol. Their Newman projection diagrams are:

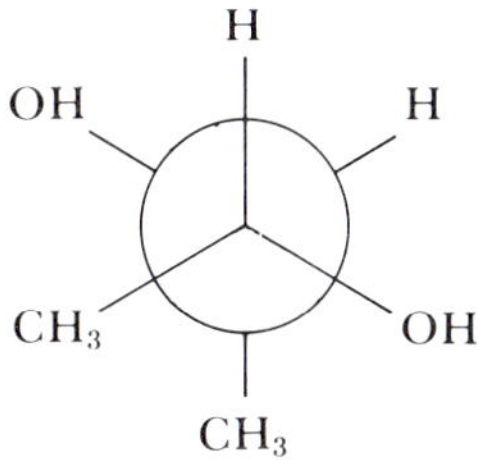

[Newman projection I: front carbon with H (top), CH₃, OH; back carbon with OH, H, CH₃]

[Newman projection II: front carbon with CH₃ (top), OH, H; back carbon with H, CH₃, HO]

[Newman projection III: front carbon with H (top), CH₃, HO; back carbon with H, CH₃, OH]

I 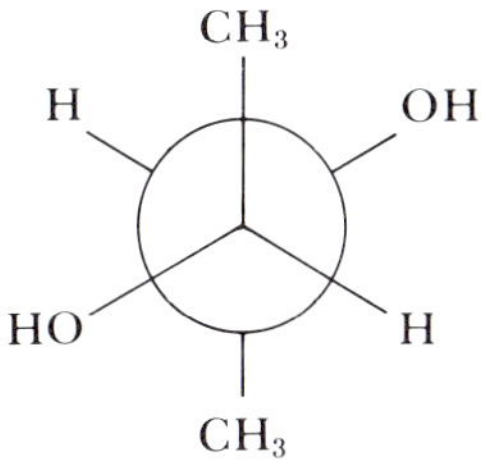   II 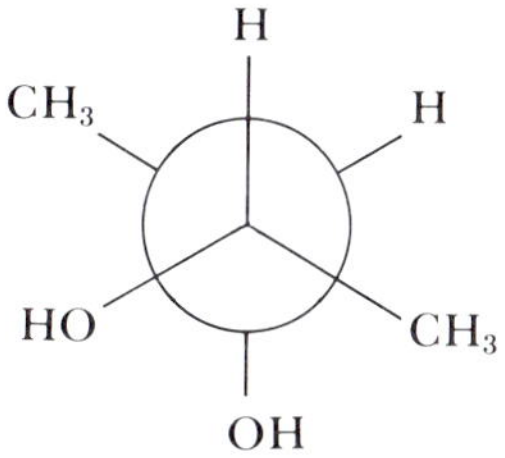   III

a) Identify which two of the above form a pair of enantiomers.

b) The remaining molecule (called a *meso* compound) is optically inactive. Why is this so?

49. Use Newman projection diagrams to show that the molecule 2-chloro-3-bromobutane has two pairs of enantiomers. Why does 2,3-dichlorobutane have only one pair of enantiomers?

50. Why is it impossible for a planar molecule to exhibit optical isomerism?

ADDITIONAL READINGS

Morrison, Robert T., and Robert N. Boyd, *Organic Chemistry*, 3rd edition. Boston: Allyn and Bacon, 1973. One of the best-organized and most readable textbooks of organic chemistry. Excellent for self-study.

Noller, Carl R., *Chemistry of Organic Compounds*. Philadelphia: Saunders, 1965. Excellent as a reference for the structures and properties of a wide variety of organic molecules, including many of commercial and biological significance.

Herz, Werner, *The Shape of Carbon Compounds*. New York: Benjamin, 1963. Provides a relatively complete introduction to organic chemistry with a great deal of emphasis on the geometry of molecules.

Wasserman, Edel, "Chemical Topology." *Scientific American*, November 1962. Discusses the structures and syntheses of some novel organic compounds existing as linked rings.

Hall, H. Tracy, "The Synthesis of Diamond." *J. Chem. Educ.* **38**, 484 (1961). One of the first scientists to synthesize diamonds discusses the principles and methods used.

Kaufman, George B., and Robin D. Myers, "The Resolution of Racemic Acid." *J. Chem. Educ.* **52**, 777 (1975). Discusses how to repeat Pasteur's separation of the optically active forms of sodium ammonium tartrate. The historical background of the original separation is treated in some detail.

17
THE SYNTHESIS OF ORGANIC COMPOUNDS

Friedrich Wöhler.
(Courtesy of Culver Pictures.)

Organic chemistry appears to me like a primeval forest . . . full of the most remarkable things . . . I must tell you that I can make urea without requiring a kidney or an animal, either man or dog.
FRIEDRICH WÖHLER, *in a letter to J. J. Berzelius, 1835*

Friedrich Wöhler (1800–1882) was a professor of chemistry at the University of Göttingen from 1836 to his death. Wöhler, although an M.D., never practiced medicine. He was the first to demonstrate that a product of living matter (urea,

$$H_2N-\underset{\underset{O}{\|}}{C}-NH_2$$

a major waste product in urine) could be made from a material originating in nonliving matter (ammonium cyanate, NH_4CNO*), in contradiction of the doctrine of the time. Wöhler also isolated aluminum in 1827 and beryllium in 1828.*

Well over a million organic compounds have been synthesized in the laboratory. Some of these are relatively abundant in nature but many either are very scarce or did not previously exist. Many of these compounds can be synthesized from very simple starting materials. Some of the simpler general methods used to synthesize representative molecules whose names and structures were presented in the preceding chapter are discussed in this chapter. The deliberate synthesis of molecules, first carried out hundreds of years ago by trial-and-error methods, now constitutes a highly developed, systematic activity of great depth and breadth. A large part of the industrial might of a developed country characteristically includes the production of chemicals ranging in diversity from simple hydrocarbon fuels to complex pharmaceuticals.

17.1 STRATEGY AND TACTICS IN ORGANIC SYNTHESES

Suppose a chemist wishes to use benzene as a starting material to prepare the compound whose formula is

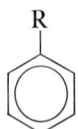

where R denotes a group such as methyl, chloro, nitro, or any one of a large number of possibilities. The choice of a method for doing this may be called **chemical tactics**; the specific tactic used will depend on the particular R group the chemist wishes to add to benzene. Suppose R is a nitro group; that is, the chemist wishes to prepare nitrobenzene. Laboratory studies have established that nitrobenzene is easily prepared by treating benzene with a nitric acid–sulfuric acid solution. The reaction is written

$$\text{benzene} \xrightarrow{HNO_3, H_2SO_4} \text{nitrobenzene}$$

If the R group is chloro, a different tactic or reaction must be used. It has been found that reaction of benzene with chlorine gas in the presence of Fe or FeCl$_3$ produces chlorobenzene. The reaction is

$$\text{benzene} \xrightarrow{Cl_2, Fe \text{ or } FeCl_3} \text{chlorobenzene}$$

Now suppose a chemist wishes to prepare *o*-chloronitrobenzene

Here is where **chemical strategy** enters the picture. On first thought, a reasonable strategy would be first to introduce the nitro group by nitration in a sulfuric acid solution, then to chlorinate the nitrobenzene with a chlorine/iron mixture. However, when this is done, most of the product turns out to be not the desired *o*-chlo-

Robert B. Woodward, American organic chemist noted for the synthesis of complex molecules. (Courtesy of Harvard University News Office.)

Some of the most impressive examples of the synthesis of complex organic molecules have been carried out by Harvard chemist Robert B. Woodward (b. 1917) and his colleagues. Woodward flunked out of M.I.T. at the end of his second year, was readmitted on probation, and earned his Ph.D. within the next two years. He has since received, in 1965, the Nobel prize in chemistry for the synthesis of substances such as chlorophyll, cholesterol, cortisone, quinine, vitamin B_{12}, and strychnine and for the determination of the molecular structures of many other substances (for example, some of the antibiotics). He and his colleagues have also made major contributions to the theoretical understanding of various types of organic reactions.

ronitrobenzene but the isomer *m*-chloronitrobenzene

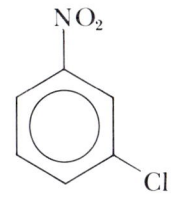

Obviously, a new strategy must be devised.

Extensive laboratory studies have shown that the nature of the first group on a benzene ring determines where a second group is most likely to be attached. For example, if the first group is nitro or carboxyl, subsequent groups tend to become attached in the *meta* position. However, if the first group is chloro, hydroxy, or methyl, subsequent groups tend to become attached in *ortho* or *para* positions. Thus the best strategy is to prepare chlorobenzene first and then to nitrate it. The result is a mixture of *o*-chloronitrobenzene and *p*-chloronitrobenzene from which the former may be isolated and purified.

By the clever use of experimental facts such as the above, organic chemists find it possible to put carbon skeletons and other groups together in a large number of controlled ways. Synthesizing a given organic molecule requires even more skill than playing a game of chess against an expert opponent and can be just as exciting or more so. Some strategies and tactics used in the preparations of a number of organic compounds are discussed in the following sections.

17.2 THIRTY-FOUR SYNTHESES BEGINNING WITH PROPANE

In Fig. 17.1 there is shown a block diagram for the synthesis of over thirty different compounds from the simple hydrocarbon propane. In an attempt to show something of the principles involved in transforming one molecule to another in a controlled fashion, each of the steps is discussed in some detail. The student should study each reaction step carefully and then use the block diagram as a guide for recall and study. It is not advisable to memorize each step, but the student should become familiar enough with each step to be able to say just what must be done to one molecule to transform it to the next molecule. For example, if given the two blocks

$$\boxed{\text{1-Chloropropane}} \rightarrow \boxed{\text{Propene}}$$

the student should know the structures of the reactant and product and know that a hydrogen atom and a chlorine atom must be removed from the reactant to form the product. At this stage of the study of chemistry, it is not necessary to know just what reagents are needed to effect the atom removal; such details are usually dealt with in a sophomore-level course in organic chemistry.

Reactions 1 and 2: Propane to 1-Chloropropane and 2-Bromopropane

When propane is treated with a halogen X_2 the general reaction may be written

$$\underset{\text{propane}}{CH_3CH_2CH_3} + \underset{\text{halogen}}{X_2} \rightarrow \underset{\text{1-halopropane}}{CH_3CH_2CH_2X} + \underset{\text{2-halopropane}}{CH_3\underset{\underset{X}{|}}{C}HCH_3}$$

Another product, HX, is also formed. In the reaction, called a **substitution reaction** (see Section 14.1), one atom of X_2 replaces H on the carbon skeleton and the other atom combines with the displaced H to form HX.

When X_2 is Cl_2 the products obtained at 25 °C are about 45% 1-chloropropane and 55% 2-chloropropane. Using Br_2 under the same conditions leads to 97% 2-bromopropane and only 3% 1-bromopropane. These facts must be taken into account by the chemist in order to obtain a maximum amount of the desired isomer.

The following shows two different but analogous sets of synthetic reactions; one set starts with 1-chloropropane and the other starts with 2-bromopropane. These analogous sets of synthetic pathways impart a high degree of symmetry to the block diagram of Fig. 17.1. For example, all syntheses beginning with 1-chloropropane are labeled with the numeral 1 and a letter; all syntheses beginning with 2-bromopropane are labeled with the numeral 2 and the same letter corresponding to the analogous 1-chloropropane reaction. Thus reactions 1C and 2C both refer to the transformation of an acid to an acyl chloride; reaction 1C begins with 1-chloro-

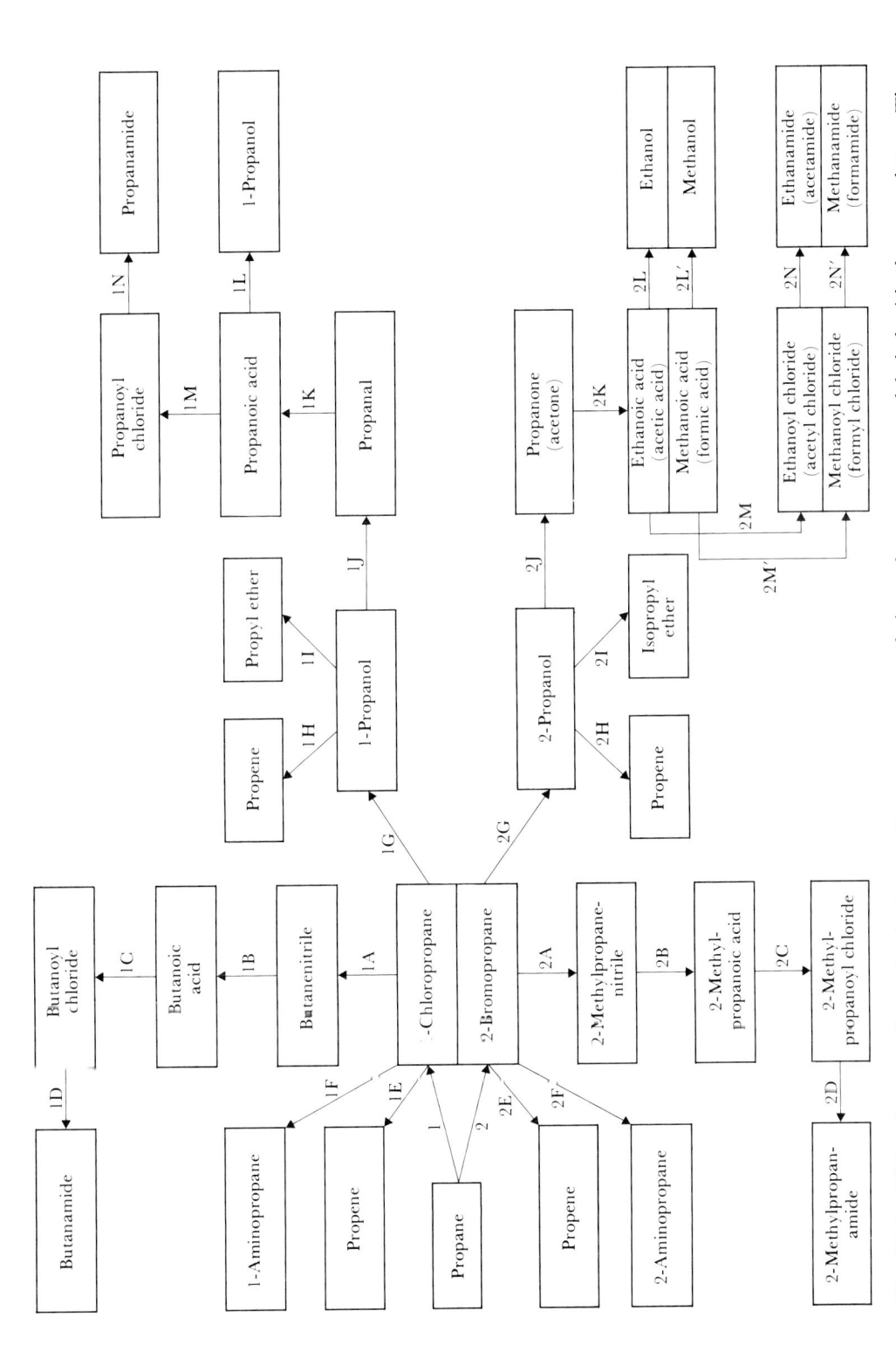

Fig. 17.1. Thirty-four syntheses beginning with propane. The diagram shows two different but analogous sets of synthetic pathways using propane, $CH_3CH_2CH_3$, as a starting material. The two different pathways are labeled 1 and 2 and their analogous steps are labeled with the same letter. Thus 1G and 2G are analogous reactions of pathways 1 and 2, respectively. Each labeled pathway is discussed in Section 17.2 of the text.

propane and reaction 2C begins with 2-bromopropane. This labeling makes it easy for the student to match the text description of a reaction with the overview provided by Fig. 17.1 and to compare analogous portions of different synthetic routes.

Reaction 1A: 1-Chloropropane to Butanenitrile

Treatment of 1-chloropropane with NaCN (sodium cyanide) at elevated temperatures leads to the reaction:

$$\underset{\text{1-chloropropane}}{CH_3CH_2CH_2Cl} + NaCN \xrightarrow{\text{heat}} \underset{\text{butanenitrile}}{CH_3CH_2CH_2C\equiv N} + NaCl$$

The product with the general formula RC≡N (or RCN) is called a **nitrile**. Nitriles may be named on the basis of the corresponding carboxylic acid by changing the *-ic acid* of the common name to *-onitrile*. Thus CH_3CN is acetonitrile and $CH_3CH_2CH_2CN$ is butyronitrile. The IUPAC system uses the name of the parent hydrocarbon followed by *nitrile* (all as one word). Thus CH_3CN is ethanenitrile. This reaction can be made a part of a general strategy which requires increasing the length of a carbon chain in a molecule.

Reaction 2A: 2-Bromopropane to 2-Methylpropanenitrile

Use of reaction 1A to convert 2-bromopropane to 2-methylpropanenitrile works but produces much propene as a by-product. The two reactions are

$$\underset{\text{2-bromopropane}}{\overset{CH_3}{\underset{CH_3}{>}}CHBr} + NaCN \longrightarrow \underset{\text{2-methylpropanenitrile}}{\overset{CH_3}{\underset{CH_3}{>}}CHC\equiv N} + NaBr$$

$$\underset{\text{propene}}{CH_2=CHCH_3} + NaBr + HCN$$

Reaction 1B: Butanenitrile to Butanoic Acid

When nitriles are treated with aqueous HCl, carboxylic acids are produced. The general reaction is

$$R-C\equiv N + HCl + 2H_2O \rightarrow RCO_2H + NH_4^+ + Cl^-$$

Thus treatment of butanenitrile with aqueous HCl produces butanoic acid.

Reaction 2B: 2-Methylpropanenitrile to 2-Methylpropanoic Acid

As indicated above, treatment of 2-methylpropanenitrile with aqueous HCl produces 2-methylpropanoic acid:

$$\underset{\text{2-methylpropanenitrile}}{\overset{CH_3}{\underset{CH_3}{>}}CHC\equiv N} + HCl + 2H_2O \rightarrow \underset{\text{2-methylpropanoic acid}}{\overset{CH_3}{\underset{CH_3}{>}}CHCO_2H} + NH_4^+ + Cl^-$$

In preparing 2-methylpropanoic acid from 2-bromopropane, a chemist would avoid going through reaction 2A since it produces an unwanted side product and thereby wastes starting material. An alternative strategy to prepare 2-methylpropanoic acid from 2-bromopropane is as follows: First treat 2-bromopropane with magnesium metal in ether solution to produce isopropylmagnesium bromide:

$$\begin{array}{c}CH_3\\ \diagdown \\ CHBr + Mg \xrightarrow{\text{ether}} \\ \diagup \\ CH_3\end{array} \quad \begin{array}{c}CH_3\\ \diagdown \\ CHMgBr\\ \diagup \\ CH_3\end{array}$$

Addition of CO_2 to the isopropylmagnesium bromide followed by acidification (with HCl, for example) produces 2-methylpropanoic acid:

$$\begin{array}{c}CH_3\\ \diagdown \\ CHMgBr\\ \diagup \\ CH_3\end{array} \xrightarrow{CO_2} \begin{array}{c}CH_3\\ \diagdown \\ CH-\underset{\underset{O}{\|}}{C}-OMgX\\ \diagup \\ CH_3\end{array} \xrightarrow{H_3O^+} \begin{array}{c}CH_3\\ \diagdown \\ CHCO_2H\\ \diagup \\ CH_3\end{array}$$

Reaction 1C: Butanoic Acid to Butanoyl Chloride

Carboxylic acids are converted to the corresponding acyl chlorides when treated with PCl_3, PCl_5, or $SOCl_2$ (sulfonyl chloride). Butanoyl chloride is produced from butanoic acid as follows:

$$CH_3CH_2CH_2CO_2H + PCl_3 \rightarrow CH_3CH_2CH_2\underset{\underset{O}{\|}}{C}-Cl$$

butanoic acid ⠀⠀⠀⠀⠀⠀⠀⠀⠀⠀⠀⠀ butanoyl chloride

Reaction 2C: 2-Methylpropanoic Acid to 2-Methylpropanoyl Chloride

The reaction is entirely analogous to reaction 1C:

$$\begin{array}{c}CH_3\\ \diagdown \\ CHCO_2H + PCl_3 \rightarrow\\ \diagup \\ CH_3\end{array} \quad \begin{array}{c}CH_3\\ \diagdown \\ CH\underset{\underset{O}{\|}}{C}-Cl\\ \diagup \\ CH_3\end{array}$$

2-methylpropanoic acid ⠀⠀⠀⠀⠀⠀ 2-methylpropanoyl chloride

Reaction 1D: Butanoyl Chloride to Butanamide

When acyl halides are treated with ammonia, the corresponding amides are produced. For example

$$CH_3CH_2CH_2\underset{\underset{O}{\|}}{C}Cl + NH_3 \rightarrow CH_3CH_2CH_2\underset{\underset{O}{\|}}{C}NH_2 + HCl$$

butanoyl chloride ⠀⠀⠀⠀⠀⠀⠀⠀⠀ butanamide

Reaction 2D: 2-Methylpropanoyl Chloride to 2-Methylpropanamide

The reaction is entirely analogous to reaction 1D:

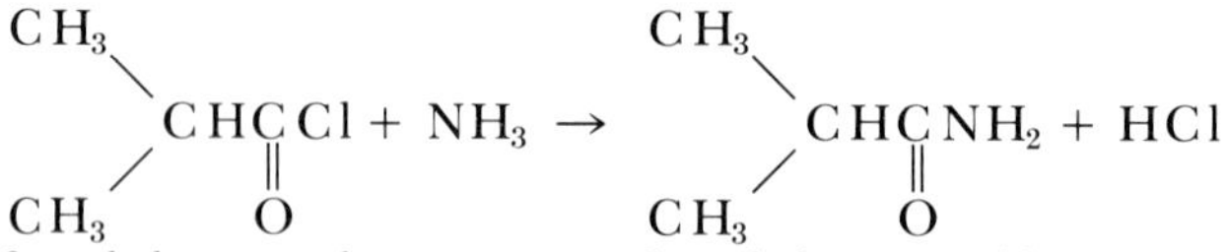

$$\begin{array}{c}CH_3\\ \diagdown\\ CHCCl\\ \diagup\|\\ CH_3O\end{array} + NH_3 \rightarrow \begin{array}{c}CH_3\\ \diagdown\\ CHCNH_2\\ \diagup\|\\ CH_3O\end{array} + HCl$$

2-methylpropanoyl chloride   2-methylpropanamide

Reaction 1E: 1-Chloropropane to Propene

When a haloalkane is treated with KOH in alcohol solution, the halogen atom is removed and so is a hydrogen atom on a carbon atom adjacent to the halogen atom. This reaction is a standard method for introducing the ethylenic linkage —C=C— into a molecule. The following reactions show how the position of the double bond may be controlled to some extent:

$$CH_3CH_2CH_2CH_2X \xrightarrow[\text{alcohol}]{\text{KOH}} CH_3CH_2CH{=}CH_2$$
1-halobutane   1-butene

$$CH_3CH_2\underset{\underset{X}{|}}{C}HCH_3 \xrightarrow[\text{alcohol}]{\text{KOH}} CH_3CH{=}CHCH_3 + CH_3CH_2CH{=}CH_2$$
2-halobutane   2-butene   1-butene
                 80%   20%

In the second case there are two different carbon atoms adjacent to the carbon atom to which the halogen atom is attached so that two different hydrogen atoms can be removed to form the double bond. The predominant product is the one in which the hydrogen atom attached to the carbon atom with the lesser number of hydrogen atoms has been removed.

When 1-chloropropane is treated with alcoholic KOH, only one reaction is possible and propene is produced:

$$CH_3CH_2CH_2Cl \xrightarrow[\text{alcohol}]{\text{KOH}} CH_3CH{=}CH_2$$
1-chloropropane   propene

Reaction 2E: 2-Bromopropane to Propene

Regardless of which hydrogen atom is removed when 2-bromopropane is treated with alcoholic KOH, only one product, propene, results.

$$CH_3\underset{\underset{Br}{|}}{C}HCH_3 \xrightarrow[\text{alcohol}]{\text{KOH}} CH_3CH{=}CH_2$$

Reaction 1F: 1-Chloropropane to 1-Aminopropane

Reaction of haloalkanes with ammonia produces amines. Thus 1-aminopropane is produced when 1-chloropropane is treated with ammonia. The reaction is

$$CH_3CH_2CH_2Cl + NH_3 \rightarrow CH_3CH_2CH_2NH_2 + HCl$$
1-chloropropane → 1-aminopropane

Reaction 2F: 2-Bromopropane to 2-Aminopropane

$$\underset{\underset{\text{2-bromopropane}}{Br}}{CH_3CHCH_3} + NH_3 \rightarrow \underset{\underset{\text{2-aminopropane}}{NH_2}}{CH_3CHCH_3} + HBr$$

This reaction is entirely analogous to reaction 1F. Reactions 1F and 2F are also similar to reactions 1C and 2C, in which acyl halides are converted to amides by the action of ammonia.

Reaction 1G: 1-Chloropropane to 1-Propanol

When a haloalkane is treated with an aqueous solution of hydroxide ion from a strong base, the halogen atom is replaced by OH to produce an alcohol. Thus 1-propanol is formed from 1-chloropropane as follows:

$$CH_3CH_2CH_2Cl + OH^- \xrightarrow{H_2O} CH_3CH_2CH_2OH + Cl^-$$
1-chloropropane → 1-propanol (*n*-propyl alcohol)

Reaction 2G: 2-Bromopropane to 2-Propanol

This reaction is entirely analogous to reaction 1G:

$$\underset{\text{2-bromopropane}}{\overset{CH_3}{\underset{CH_3}{>}}CHBr} + OH^- \xrightarrow{H_2O} \underset{\text{2-propanol (isopropyl alcohol)}}{\overset{CH_3}{\underset{CH_3}{>}}CHOH} + Br^-$$

Reaction 1H: 1-Propanol to Propene

When an alcohol is treated with a strong dehydrating agent such as H_2SO_4 at high temperatures, the OH group and a hydrogen atom from a carbon atom adjacent to the carbon atom to which the OH is attached is removed to produce an alkene:

$$CH_3CH_2CH_2OH \xrightarrow[\text{high } T]{H_2SO_4} CH_3CH=CH_2$$
1-propanol → propene

Reaction 2H: 2-Propanol to Propene

The reaction is entirely analogous to reaction 1H and produces the same product:

$$\underset{\underset{\text{2-propanol}}{OH}}{CH_3CHCH_3} \xrightarrow[\text{high } T]{H_2SO_4} CH_3CH=CH_2$$
propene

Reaction 1I: 1-Propanol to n-Propyl Ether

When an alcohol is treated with an excess of strong dehydrating agent such as H_2SO_4 at low temperatures, an OH is removed from one molecule of alcohol and a hydrogen atom from the OH of another molecule, and the two molecules are linked to form an ether. Thus 1-propanol forms *n*-propyl ether:

$$CH_3CH_2CH_2OH + HOCH_2CH_2CH_3 \xrightarrow[\text{low } T]{H_2SO_4} CH_3CH_2CH_2OCH_2CH_2CH_3$$
1-propanol ———— *n*-propyl ether

Reaction 2I: 2-Propanol to Isopropyl Ether

This reaction is exactly analogous to reaction 1I:

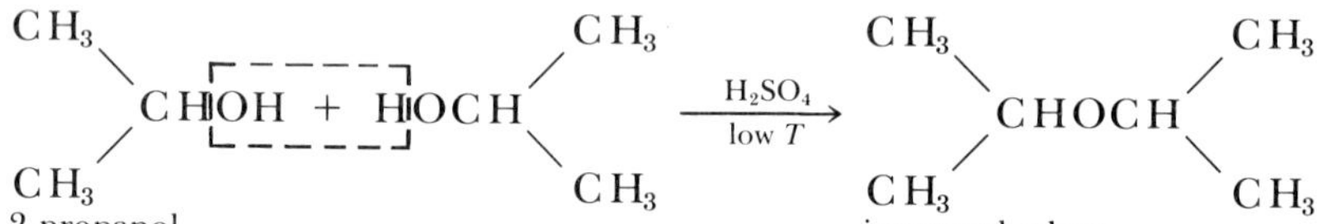

$$\begin{array}{c}CH_3\\ \diagdown\\ CHOH\\ \diagup\\ CH_3\end{array} + \begin{array}{c}CH_3\\ \diagup\\ HOCH\\ \diagdown\\ CH_3\end{array} \xrightarrow[\text{low } T]{H_2SO_4} \begin{array}{c}CH_3\\ \diagdown\\ CHOCH\\ \diagup\\ CH_3\end{array}\begin{array}{c}CH_3\\ \diagup\\ \\ \diagdown\\ CH_3\end{array}$$

2-propanol isopropyl ether

Reaction 1J: 1-Propanol to Propanal

When an alcohol having its OH group attached to the carbon atom at the end of an unbranched chain is treated with a relatively mild oxidizing agent such as $KMnO_4$ an aldehyde is produced. For example, 1-propanol is oxidized to propanal:

$$CH_3CH_2CH_2OH \xrightarrow{KMnO_4} CH_3CH_2CHO$$
1-propanol propanal

Since aldehydes do not have any hydrogen atoms attached to an oxygen atom, they do not exhibit hydrogen bonding and, consequently, are relatively volatile. Thus the above reaction is best carried out by removing the volatile aldehyde by distillation as it is formed. The alcohol, being strongly hydrogen bonded, does not distill over to any great extent.

Note that there are two ways by which a hydrocarbon can be oxidized by oxgyen. Either an oxygen atom can become incorporated between a carbon atom and a hydrogen atom, or hydrogen atoms can be removed from the molecule. An example of the first case is the oxidation of ethane to ethanol:

$$CH_3CH_3 \xrightarrow{[O]} CH_3CH_2OH$$

The second type of oxidation is exemplifed by the oxidation of ethanol to acetaldehyde (ethanal):

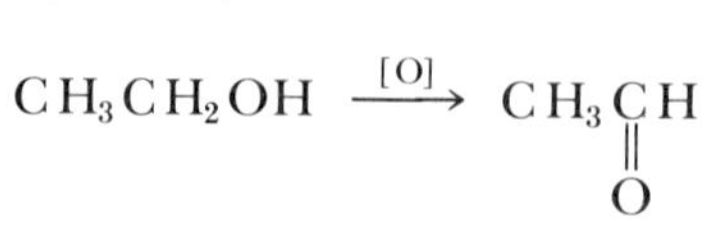

$$CH_3CH_2OH \xrightarrow{[O]} CH_3\underset{\underset{O}{\|}}{C}H$$

Note that two hydrogen atoms have been removed in the latter case.

476 THE SYNTHESIS OF ORGANIC COMPOUNDS

Reaction 2J: 2-Propanol to Acetone

When the above reaction is carried out on an alcohol having the OH group on a carbon atom which itself is attached to two other carbon atoms, a ketone rather than an aldehyde is formed. Thus 2-propanol is oxidized to acetone (propanone) via the reaction

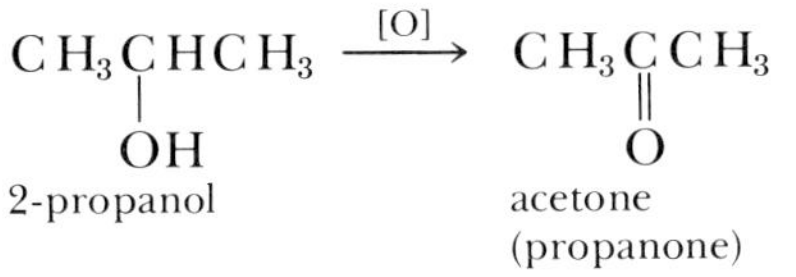

$$\underset{\text{2-propanol}}{\text{CH}_3\underset{\underset{\text{OH}}{|}}{\text{C}}\text{HCH}_3} \xrightarrow{[O]} \underset{\substack{\text{acetone} \\ \text{(propanone)}}}{\text{CH}_3\underset{\underset{\text{O}}{\|}}{\text{C}}\text{CH}_3}$$

Ketones may also be removed by distillation since they do not exhibit hydrogen bonding.

Reaction 1K: Propanal to Propanoic Acid

If reaction 1J is carried out with a strong oxidizing agent or if the reaction is allowed to run a long time without removal of the aldehyde by distillation, a carboxylic acid is formed. Thus propanal is oxidized to propanoic acid:

$$\underset{\text{propanal}}{\text{CH}_3\text{CH}_2\text{CHO}} \xrightarrow[\text{[O]}]{\text{strong}} \underset{\text{propanoic acid}}{\text{CH}_3\text{CH}_2\text{CO}_2\text{H}}$$

This is essentially what happens to the ethanol in apple cider when it is allowed to ferment; the ethanol is oxidized to acetaldehyde (ethanal) and ultimately to acetic acid (ethanoic acid) to produce vinegar.

Reaction 2K: Acetone to Acetic Acid and Formic Acid

When a ketone is oxidized it does not form a single product. Thus when acetone is oxidized as in reaction 1K, both formic acid and acetic acid are formed:

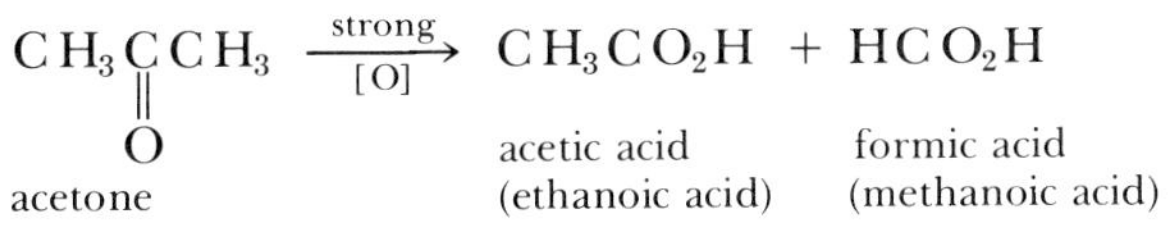

$$\underset{\text{acetone}}{\text{CH}_3\underset{\underset{\text{O}}{\|}}{\text{C}}\text{CH}_3} \xrightarrow[\text{[O]}]{\text{strong}} \underset{\substack{\text{acetic acid} \\ \text{(ethanoic acid)}}}{\text{CH}_3\text{CO}_2\text{H}} + \underset{\substack{\text{formic acid} \\ \text{(methanoic acid)}}}{\text{HCO}_2\text{H}}$$

Oxidation ruptures carbon-to-carbon bonds and then converts the fragments to the corresponding carboxylic acids. Thus the

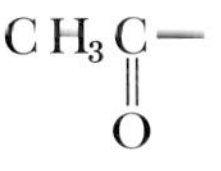

$$\text{CH}_3\underset{\underset{\text{O}}{\|}}{\text{C}}-$$

fragment becomes $\text{CH}_3\text{CO}_2\text{H}$ and the CH_3- fragment becomes HCO_2H.

Reaction 1L: Propanoic Acid to 1-Propanol

If carboxylic acids, aldehydes, or ketones are treated with a strong reducing agent such as LiAlH$_4$ (lithium aluminum hydride), alcohols are produced. Thus propanoic acid is reduced to 1-propanol via the reaction

$$\underset{\text{propanoic acid}}{CH_3CH_2CO_2H} \xrightarrow{LiAlH_4} \underset{\text{1-propanol}}{CH_3CH_2CH_2OH}$$

Note that the butanoic acid produced in reaction 1B would be reduced by LiAlH$_4$ to 1-butanol.

Reaction 2L and 2L': Acetic Acid to Ethanol and Formic Acid to Methanol

These reactions are exactly analogous to reaction 1L.

Reaction 1M: Propanoic Acid to Propanoyl Chloride

This reaction is exactly analogous to reaction 1C.

Reactions 2M and 2M': Acetic Acid to Acetyl Chloride (Ethanoyl Chloride) and Formic Acid to Formyl Chloride (Methanoyl Chloride)

These reactions are exactly analogous to reaction 2C and also to reaction 1M just preceding.

Reaction 1N: Propanoyl Chloride to Propanamide

This reaction is exactly analogous to reaction 1D.

Reactions 2N and 2N': Acetyl Chloride to Acetamide (Ethanamide) and Formyl Chloride to Formamide (Methanamide)

These reactions are exactly analogous to reaction 2D and also to reaction 1N above.

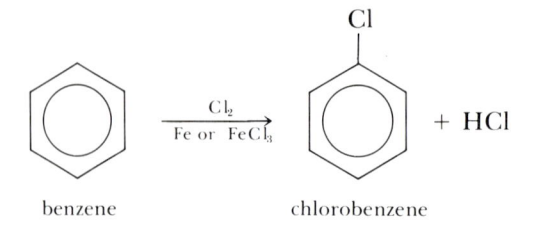

17.3 THIRTEEN SYNTHESES BEGINNING WITH BENZENE

In Fig. 17.2 is shown a block diagram for thirteen synthetic processes starting with the aromatic compound benzene. Although there are similarities between the chemistries of aliphatic compounds and aromatic compounds, there are many significant differences as well.

The thirteen synthetic steps in Fig. 17.2 are labeled A, B, . . . through M. Each step is discussed below.

Reaction A: Benzene to Chlorobenzene

Treatment of benzene with chlorine gas in the presence of Fe or FeCl$_3$ produces chlorobenzene:

benzene $\xrightarrow[\text{Fe or FeCl}_3]{Cl_2}$ chlorobenzene + HCl

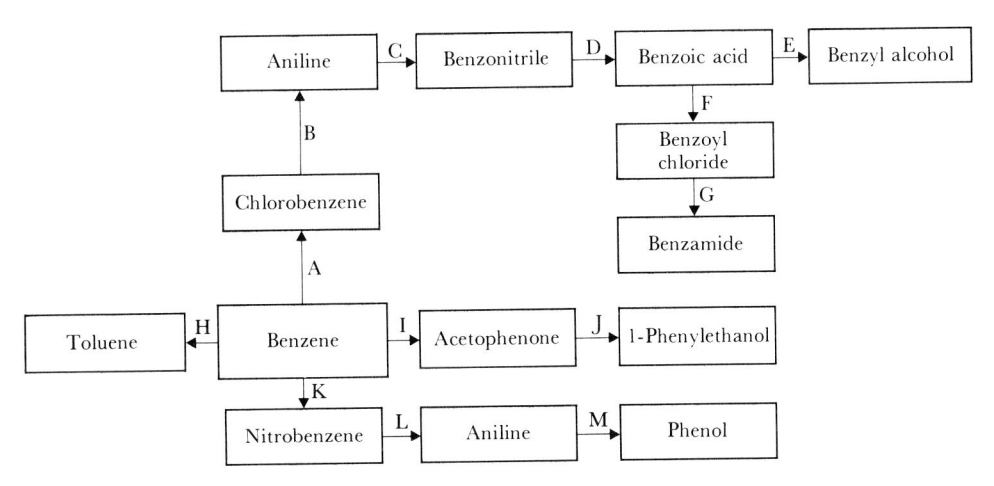

Fig. 17.2. Thirteen syntheses beginning with benzene. Each synthesis is labeled by the letters A through M and is discussed in Section 17.3 of the text.

When Cl_2 reacts with an alkane (see reaction 1 in Fig. 17.1), the attack involves a chlorine atom. In the case of the benzene reaction above, Cl_2 and Fe form $FeCl_4^-$ and Cl^+ and it is the latter which displaces H^+ on the benzene ring. Cl^+ is often called the **chloronium** ion.

Reaction B: Chlorobenzene to Aniline

The industrial production of aniline (aminobenzene) is carried out by treating chlorobenzene with ammonia at 200 °C and 60 atm in the presence of copper(I) oxide. The reaction is

$$C_6H_5Cl + NH_3 \xrightarrow[200\,°C\ 60\ atm]{Cu_2O} C_6H_5NH_2 + HCl$$

chlorobenzene → aniline

Synthesis of alkyl amines from alkyl chlorides requires much milder reaction conditions (see reactions 1F and 2F in Fig. 17.1).

Reaction C: Aniline to Benzonitrile

This process must be carried out in two steps. In the first step aniline is treated with sodium nitrite, NaNO$_2$, and HCl to produce benzenediazonium chloride:

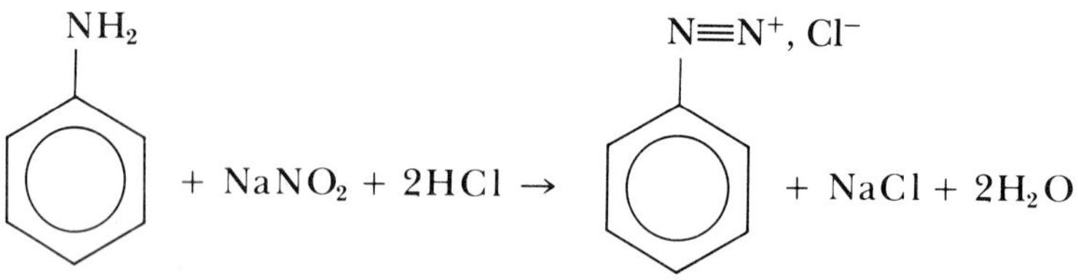

[Structure: aniline (C$_6$H$_5$NH$_2$) + NaNO$_2$ + 2HCl → benzenediazonium chloride (C$_6$H$_5$N≡N$^+$, Cl$^-$) + NaCl + 2H$_2$O]

Diazonium salts of aromatic compounds are very useful as intermediates in the preparation of other substances. When benzenediazonium chloride is treated with copper(I) cyanide at high temperature, benzonitrile is formed:

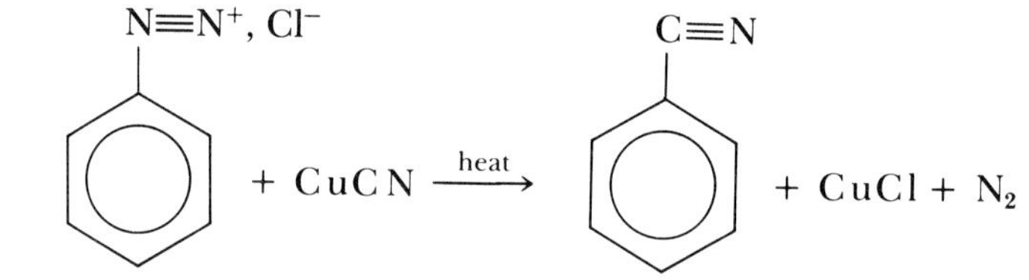

[Structure: benzenediazonium chloride + CuCN $\xrightarrow{\text{heat}}$ benzonitrile (C$_6$H$_5$C≡N) + CuCl + N$_2$]

Although aliphatic nitriles are made by treating the halides with NaCN, aromatic halides are too unreactive to be used in this way.

Reaction D: Benzonitrile to Benzoic Acid

Treatment of benzonitrile with aqueous acid produces benzoic acid. The reaction is

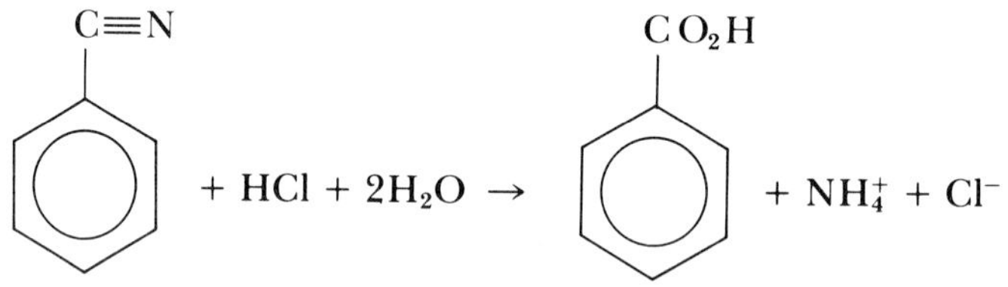

[Structure: benzonitrile (C$_6$H$_5$C≡N) + HCl + 2H$_2$O → benzoic acid (C$_6$H$_5$CO$_2$H) + NH$_4^+$ + Cl$^-$]

This reaction is analogous to reactions 1B and 2B in Fig. 17.1, in which an aliphatic nitrile is used to produce an aliphatic carboxylic acid.

Reaction E: Benzoic Acid to Benzyl Alcohol

When benzoic acid is treated with a strong reducing agent such as LiAlH$_4$, benzyl alcohol is produced:

$$\text{benzoic acid (C}_6\text{H}_5\text{CO}_2\text{H)} \xrightarrow{\text{LiAlH}_4} \text{benzyl alcohol (C}_6\text{H}_5\text{CH}_2\text{OH)}$$

This reaction is analogous to reactions 1L, 2L, and 2L' in Fig. 17.1, where aliphatic acids are reduced by LiAlH$_4$ to alcohols.

Reaction F: Benzoic Acid to Benzoyl Chloride

If benzoic acid is treated with PCl$_5$ at 100 °C, the acyl halide, benzoyl chloride, is formed:

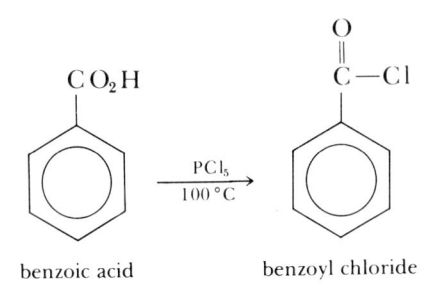

$$\text{benzoic acid (C}_6\text{H}_5\text{CO}_2\text{H)} \xrightarrow[\text{100 °C}]{\text{PCl}_5} \text{benzoyl chloride (C}_6\text{H}_5\text{COCl)}$$

This reaction is analogous to reactions 1M, 2M, and 2M' in Fig. 17.1, in which aliphatic acyl chlorides are synthesized from aliphatic carboxylic acids.

Reaction G: Benzoyl Chloride to Benzamide

Treatment of benzoyl chloride with ammonia produces benzamide via the reaction

$$\text{benzoyl chloride (C}_6\text{H}_5\text{COCl)} + \text{NH}_3 \rightarrow \text{benzamide (C}_6\text{H}_5\text{CONH}_2) + \text{HCl}$$

This reaction is analogous to reactions 1N, 2N, and 2N' in Fig. 17.1, where aliphatic acyl chlorides are treated with ammonia to produce aliphatic amides.

Reaction H: Benzene to Toluene

When benzene is treated with an alkyl halide in the presence of aluminum chloride, an alkyl benzene is produced:

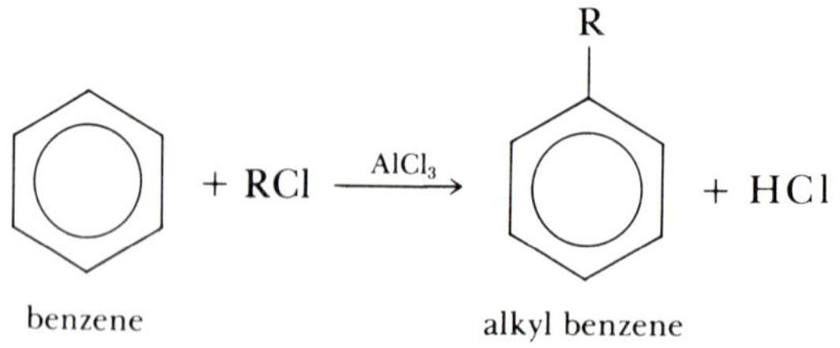

$$\text{benzene} + \text{RCl} \xrightarrow{\text{AlCl}_3} \text{alkyl benzene} + \text{HCl}$$

This is an example of an important general reaction known as a **Friedel-Crafts reaction**. When R is a methyl group, the product in the above reaction is toluene.

Reaction I: Benzene to Acetophenone

A variation of the Friedel-Crafts reaction replaces the alkyl chloride with an acyl chloride. The reaction product is then an acyl benzene. Thus when benzene is treated with acetyl chloride, acetophenone (phenyl methyl ketone) is produced as follows:

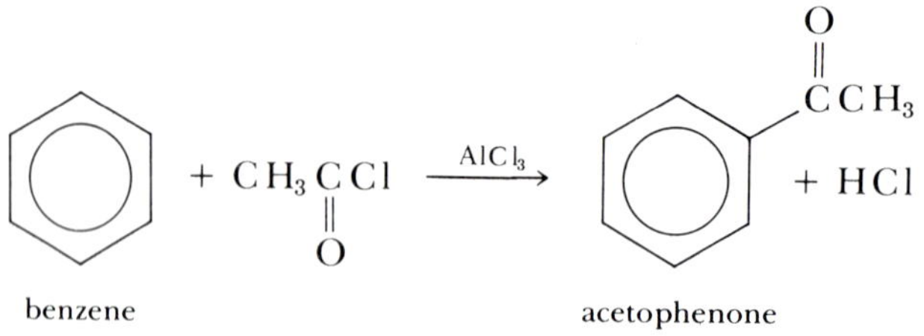

$$\text{benzene} + \text{CH}_3\text{CCl}=\text{O} \xrightarrow{\text{AlCl}_3} \text{acetophenone} + \text{HCl}$$

Reaction J: Acetophenone to 1-Phenylethanol

Treatment of an aromatic aldehyde, ketone, or carboxylic acid with a strong reducing agent such as LiAlH$_4$ reduces it to an alcohol. Thus acetophenone may be reduced to 1-phenylethanol:

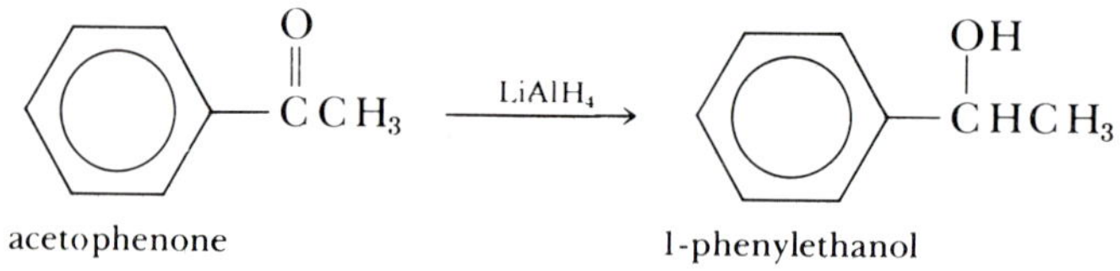

$$\text{acetophenone} \xrightarrow{\text{LiAlH}_4} \text{1-phenylethanol}$$

The Friedel-Crafts reaction is named after the French chemist Charles Friedel (1832–1899) and the American chemist James M. Crafts (1839–1917). Crafts worked for some time at M.I.T. and eventually joined forces with Friedel at the École des Mines in Paris. Ultimately Crafts returned to M.I.T. to become its president. Friedel also had strong interests in mineralogy.

---

482 THE SYNTHESIS OF ORGANIC COMPOUNDS

This reaction is analogous to reactions 1L, 2L, 2L' in Fig. 17.1 and reaction E in Fig. 17.2.

Reaction K: Benzene to Nitrobenzene

Nitrobenzene is produced when benzene is treated with a sulfuric acid solution of nitric acid. The reaction is

benzene + HONO$_2$ $\xrightarrow{H_2SO_4}$ nitrobenzene + H$_2$O

Sulfuric acid is a stronger acid than HNO$_3$ (written HONO$_2$ in the above) and forms **nitronium ion** NO$_2^+$ as follows:

$$H_2SO_4 + HONO_2 \rightleftharpoons HSO_4^- + NO_2^+ + H_2O$$

The nitronium ion displaces H$^+$ on the aromatic ring.

Reaction L: Nitrobenzene to Aniline

When nitrobenzene is treated with tin in hydrochloric acid solution, it is reduced to aniline:

nitrobenzene $\xrightarrow[\text{HCl}]{\text{Sn}}$ aniline

Reaction M: Aniline to Phenol

This reaction requires two steps. The first step is the formation of a diazonium salt as in reaction C:

aniline + NaNO$_2$ + 2H$_2$SO$_4$ → benzenediazonium hydrogen sulfate + NaHSO$_4$ + 2H$_2$O

17.3 | THIRTEEN SYNTHESES BEGINNING WITH BENZENE **483**

Treatment of the diazonium salt with aqueous acid produces phenol:

$$\underset{\substack{\text{benzenediazonium}\\\text{hydrogen sulfate}}}{C_6H_5-N\equiv N^+, HSO_4^-} + H_2O \xrightarrow{H_3O^+} \underset{\text{phenol}}{C_6H_5-OH} + N_2 + H_2SO_4$$

The synthesis of phenol cannot be carried out by treating chlorobenzene with strong base (as was done to produce an alcohol from an alkyl chloride) since chlorobenzene is too unreactive. In general, aromatic compounds tend to be more chemically inert than their aliphatic counterparts.

17.4 ADDITION REACTIONS

Many of the reactions we have discussed in the preceding two sections are of a type known as **substitution reactions** (see Section 14.1), that is, reactions in which an atom such as hydrogen on carbon is replaced by another atom or by a functional group. Another important class of reactions is known as **addition reactions** (see Section 14.1). In addition reactions, one or more atoms are added to a molecule without displacing any other atoms. An important type of addition reaction is the following: Under the proper conditions a general reagent AB capable of polarization A^+B^- can add to an alkene double bond:

$$-C=C- + AB \rightarrow -\underset{A}{\overset{|}{C}}-\underset{B}{\overset{|}{C}}-$$

Reactions are known for which AB is H_2, X_2, HX, RH, and H_2O (acting as H^+OH^-). Two specific examples of some of these are:

$$\underset{\text{ethylene}}{CH_2=CH_2} + H_2 \rightarrow \underset{\text{ethane}}{CH_3CH_3}$$

$$\underset{\text{propene}}{CH_3CH=CH_2} + Br_2 \rightarrow \underset{\text{1,2-dibromopropane}}{CH_3\underset{Br}{\overset{|}{C}}HCH_2Br}$$

When an unsymmetrical reagent such as HA adds to a double bond, the hydrogen adds predominantly to that carbon atom which already has the greater number of hydrogen atoms. For example,

$$\underset{\text{propene}}{CH_3CH=CH_2} + HBr \rightarrow \underset{\text{2-bromopropane}}{CH_3\underset{Br}{\overset{|}{C}}HCH_3}$$

THE SYNTHESIS OF ORGANIC COMPOUNDS

> *Markovnikov's rule takes its name from the Russian organic chemist Vladimir Vasil'evich Markovnikov (1838–1904). Vladimir Vasil'evich was largely responsible for instituting a new organic chemistry laboratory at Moscow University in 1887. Many of his students became famous chemists and made significant contributions to chemical science.*

Only a small amount of 1-bromopropane is formed. The above rule is known as **Markovnikov's rule**.

According to Markovnikov's rule, if propene is treated with H_2O in acid or basic solution, the main product should be 1-propanol.

One important distinguishing feature between aliphatic and aromatic compounds is that the latter seldom exhibit addition reactions. Side chains of aromatic compounds may undergo addition reactions but the rings themselves generally undergo substitution reactions only.

17.5 THE IDENTIFICATION OF SYNTHETIC PRODUCTS

After carrying out a chemical synthesis the chemist must separate the reaction products and identify them. Separation is usually carried out by fractional distillation, fractional crystallization, or on the basis of different solubility in one or more solvents. Identification of products is generally a more challenging task. This task assumes especially great importance when the chemist is doing exploratory research such as attempting to find a new pathway to an old compound or trying to make a new compound, one which no one has ever made before. Here it is vital that the chemist know precisely what compounds are present every step of the way. Some of the simpler methods used to help in the identification of reaction products are discussed in the following; a general treatment is beyond the scope of an elementary text.

There are many ways of verifying that a known compound is present in a reaction mixture. After it has been separated and is present in a pure form, a number of simple physical tests such as melting point, boiling point, color, and crystalline form can aid in its positive identification. Sometimes nonsolids may be converted to solid compounds by complexing them with other molecules. The physical properties of the complex can then be compared with known tabulated values. For example, aldehydes and ketones may be converted to **semicarbazones** by reaction with semicarbazide:

$$H_2N-\underset{\underset{O}{\|}}{C}-N\begin{smallmatrix}H\\ \\NH_2\end{smallmatrix} + O=C\begin{smallmatrix}R\\ \\R'\end{smallmatrix} \rightarrow H_2N-\underset{\underset{O}{\|}}{C}-N\begin{smallmatrix}H\\ \\N=C\end{smallmatrix}\begin{smallmatrix}R\\ \\R'\end{smallmatrix} + H_2O$$

semicarbazide   aldehyde or ketone   a semicarbazone

Semicarbazones of all the known aldehydes and ketones have been made and their melting points tabulated.

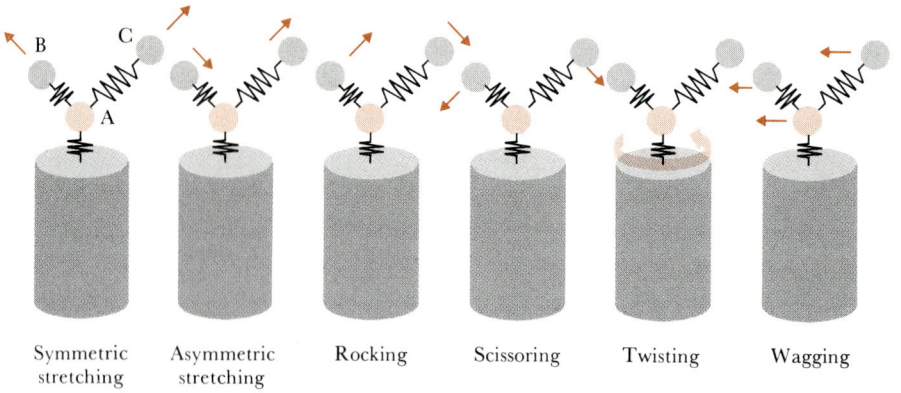

| Symmetric stretching | Asymmetric stretching | Rocking | Scissoring | Twisting | Wagging |

Fig. 17.3. Six different vibrational motions of the —ABC atom group. The cylinder represents the nonvibrating part of the rest of the molecule to which —ABC is attached. Each of these vibrations is excited by a different wavelength of infrared radiation. This wavelength is affected very little by the molecule to which —ABC— is attached. A, B, and C are hypothetical atoms.

If the compound is volatile, a sample may be introduced into a mass spectrometer and its molecular weight determined. Approximate molecular weights may also be estimated by freezing-point depression or other colligative property methods (see Section 10.8). The compound also may be analyzed for its constituent elements and the empirical formula determined.

One of the most useful methods used routinely by chemists for the identification of substances is **infrared spectroscopy**. This method can often not only identify previously existing compounds but also aid in establishing the nature of a totally new compound. This method is based on the facts that the atoms of a molecule are constantly vibrating and that the vibrational energy of a molecule is quantized just as is the electronic energy of an atom. The energy spacings between the vibrational energy levels are of such a magnitude that radiation in the infrared region (wavelengths above about 1×10^{-4} cm) is capable of inducing quantum transitions from one vibrational energy level to another. This is analogous to the electron in the hydrogen atom absorbing light to go to a higher quantum state (see Fig. 4.6). Both processes obey the same basic quantum mechanical relationship given by Eq. (4.14):

$$\Delta E = \frac{hc}{\lambda}$$

Many of the vibrational quantum states of a molecule act as if only a relatively small group of neighboring atoms are vibrating while the rest of the molecule behaves like a dead weight. Furthermore, each such group of atoms absorbs only one particular wavelength of infrared radiation. For example, the OH group in many molecules undergoes stretching vibrations which absorb infrared radiation of wavelengths in the narrow range of about 2.770×10^{-4} to 2.747×10^{-4} cm regardless of what molecule this group is attached to. Many other groups of atoms behave similarly; each has characteristic vibrational frequencies which are revealed by the absorption of a particular narrow range of infrared radiation. In Fig. 17.3 are shown some of the different ways a group such as —CH$_2$— can vibrate relative to the rest of the molecule. Each of these vibrations would tend to absorb its own characteristic infrared wavelength. Table 17.1 lists some often used infrared absorption ranges and the groups they represent. In keeping with the usual practice of infrared spectroscopists, the infrared ranges are given not in wavelength but in reciprocal wavelength (cm^{-1}).

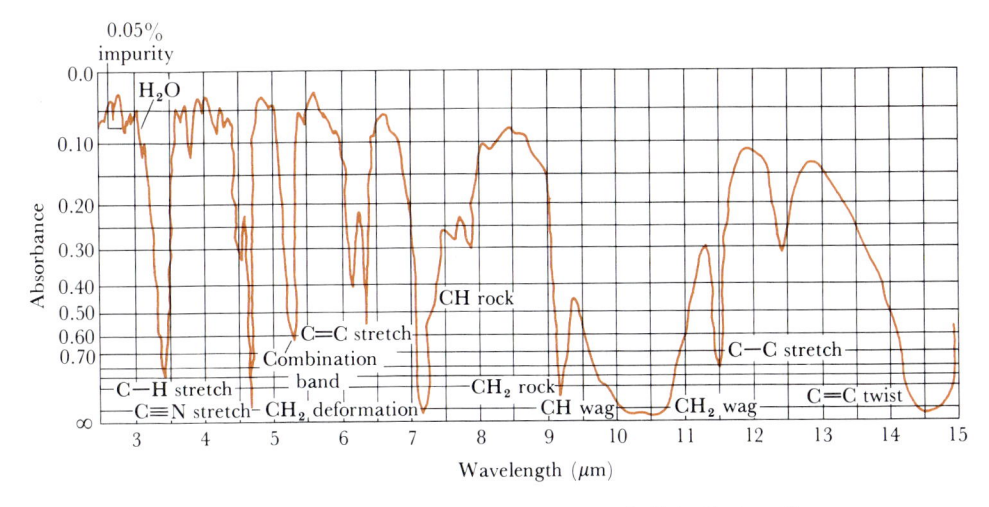

Fig. 17.4. An infrared spectrum of acrylonitrile (propenenitrile). Each valley represents a wavelength at which infrared radiation is strongly absorbed. Identification of the group of atoms whose vibrations are responsible for the absorptions is made by use of tables such as Table 17.1. (Courtesy of Perkin-Elmer Corporation.)

This quantity is generally given the symbol $\tilde{\nu}$ and is often referred to as a **frequency**. Strictly speaking, $\tilde{\nu}$ is equal to the frequency ν (in s^{-1}) divided by the velocity of light c (in $cm \cdot s^{-1}$).

Figure 17.4 is an infrared spectrum of acrylonitrile (propenenitrile). The valleys represent wavelengths of infrared radiation which propenenitrile absorbs strongly. The presence of a valley thus indicates that the molecule is absorbing infrared radiation of a particular wavelength to carry out a quantum transition from one vibrational energy level to another. The valleys are identified by use of Table 17.1 as to the

TABLE 17.1 CHARACTERISTIC GROUP ABSORPTION FREQUENCIES IN THE NEAR INFRARED[a]

GROUP	TYPE OF COMPOUND IN WHICH FOUND	NORMAL FREQUENCY RANGE (cm^{-1})
C—H	Alkanes	2850–2960
		1350–1470
	Alkenes	3020–3080 (m)
		675–1000
	Aromatic (ring)	3000–3100 (m)
		675–870
C—O	Alcohols, ethers, carboxylic acids, esters	1080–1300
C=O	Aldehydes, ketones, carboxylic acids, esters	1690–1760
O—H	Alcohols[b]	3610–3640 (v)
	Carboxylic acids	2500–3000[c]
N—H	Amines	3300–3500 (m)
C—N	Amines	1180–1360
C≡N	Nitriles	2210–2260 (v)
NO$_2$	Nitro compounds	1515–1560
		1345–1385

[a] All absorptions are strong unless marked as follows: m = moderate, v = variable. (Weak absorptions also occur but are not characteristic of the above examples.)
[b] When hydrogen bonding is present, a shift to lower $\tilde{\nu}$ (3200–3600 cm^{-1}) occurs and the absorption lines become broad.
[c] These absorption lines are broad.

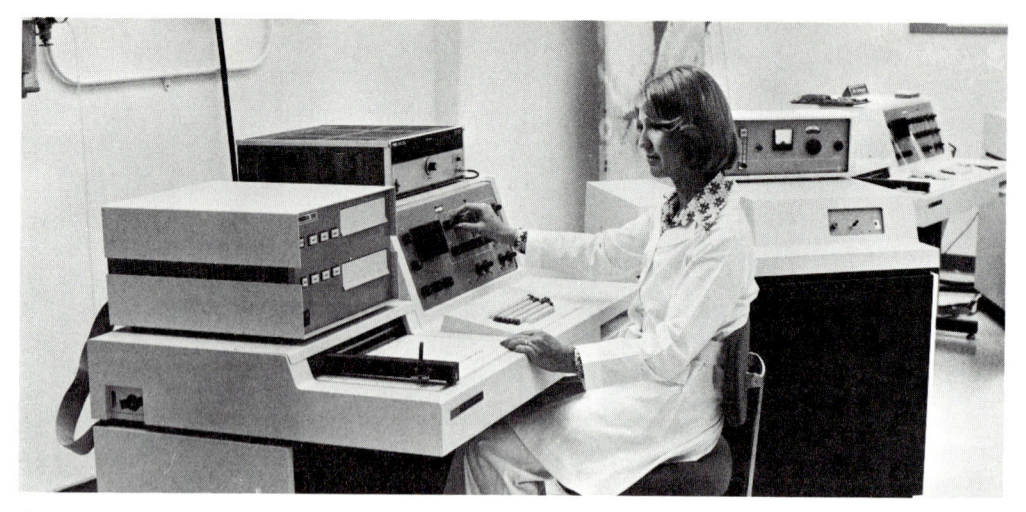

A nuclear magnetic resonance (NMR) spectrometer. This instrument measures magnetic interactions among nuclei which can be interpreted in terms of molecular structure. (Courtesy of Varian Associates, Inc., Palo Alto, Calif.)

specific groups vibrating in each quantum transition. Infrared spectroscopy can quickly indicate some of the groups present in a molecule. In some cases this may be enough to completely identify the molecule. This is true when the spectrum of the unknown molecule completely matches a spectrum of a known molecule in the library or files of the laboratory. If the molecule's infrared spectrum is not on file, additional information from other methods must be sought.

There are a variety of other methods used by chemists to aid in the identification of molecules and the determination of their structures. Many of these are analogous to infrared spectroscopy in that they are based on the quantized nature of molecular energies. Two of the most valuable methods are **ultraviolet spectroscopy**, based on quantization of molecular electronic energy, and **nuclear magnetic resonance spectroscopy**, depending on quantization of the interaction of nuclear magnetic moments with a magnetic field. Discussion of these is left to more advanced texts.

SUMMARY

The synthesis of organic compounds may be likened to a game such as chess in which varying challenges require the use of diverse strategies and tactics. Familiarity with the details of a large number of individual reactions enables the chemist to develop general tactics for the synthesis of various classes of compounds. The linking of these techniques to produce molecules of a precisely desired structure requires strategic considerations of the highest order.

For example, it is necessary to know how to introduce atoms at any desired position, how to put in double bonds where wanted, and how to avoid excessive side reactions.

It is possible to begin with a simple molecule such as propane or benzene and from it synthesize an almost endless variety of compounds. Both aliphatic and aromatic compounds undergo substitution reactions but only aliphatic compounds readily undergo addition reactions. Aromatic compounds are characterized by a relatively low level of chemical reactivity.

Reaction products may be identified by a variety of techniques, for example, by determination of melting points and by use of infrared spectroscopy. The latter technique is a very powerful aid in the determination of the structures of unknown compounds.

LEARNING GOALS

1. Given two adjacent blocks in Figs. 17.1 or 17.2, be able to do the following:
   a) Give structures of the main reactant and product.
   b) Using the text as a guide, be able to provide details of a synthesis using analogous reactants or products.

2. Be able to write the equations for the synthesis of each of the following from some other material: alcohol, ether, carboxylic acid, ketone, aldehyde, nitrile, amine, amide, and ester.

3. Know Markovnikov's rule and how to apply it.

4. Know the general features of some of the ways organic compounds are identified, particularly infrared spectroscopy.

5. Be able to name at least one important scientific contribution of each of the following men, as described in this chapter or Chapter 16: Kekulé, Wöhler, Couper, Pasteur, Friedel, and Crafts.

DEFINITIONS, TERMS, AND CONCEPTS TO KNOW

substitution reaction
addition reaction
diazonium salt
nitrile

Friedel-Crafts reaction
nitronium ion
chloronium ion
Markovnikov's rule

infrared spectroscopy
semicarbazone

QUESTIONS AND PROBLEMS

Syntheses and Reactions

1. Use Fig. 17.1 as the basis to review the following (see how many you can do without referring to the text).
   a) The formula for each molecule named.
   b) The method used in each individual synthesis.

2. Repeat Problem 1 for Fig. 17.2.

3. How would you synthesize the following? (Figures 17.1 and 17.2 should be used as a general guide.)
   a) 1-Butene from 1-bromobutane.
   b) Ethanamide from acetic acid.
   c) 1-Butene from 1-butanol.
   d) 2-Propanol from 1-propanol.
   e) o-Aminotoluene from o-nitrotoluene.
   f) Acetic acid from bromomethane.
   g) Methanol from formic acid.
   h) Formaldehyde from methanol.
   i) $\begin{array}{c}CH_3\\ \end{array}$CHCO$_2$H from propane.
   CH$_3$
   j) Toluene from benzene.
   k) Benzenediazonium chloride from aniline.

4. What would you expect to obtain as products if an alkyl halide **RX** is mixed with the sodium salt of an alcohol or phenol NaOR'? What would be the principal advantage of this reaction? [*Hint:* Look at the products of reactions 1I and 2I, Fig. 17.1.]

5. Suggest a synthetic route for the production of 2-methylpropanol from ethanol. [*Hint:* Consider the steps: alcohol to alkene to alkyl halide to nitrile to acid to alcohol, and repeat once.]

6. Complete each of the following reactions (balancing not required).

a) 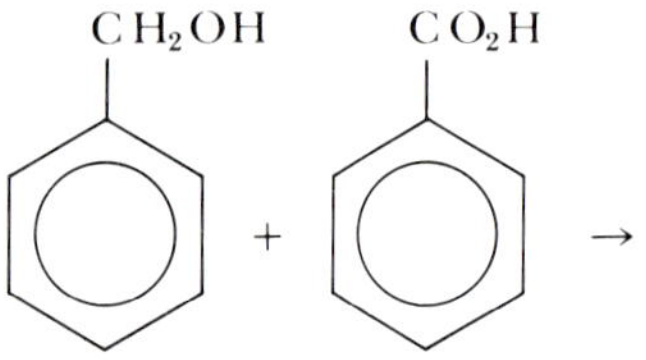 $C_6H_5CH_2OH + C_6H_5CO_2H \rightarrow$

b) $C_6H_5OH + C_6H_5CH_2CO_2H \rightarrow$

c) $CH_3CCl + NH_3 \rightarrow$
   $\|$
   O

d) $CH_3CH_2CN \xrightarrow[H_2O]{HCl}$

e) $\begin{array}{c}CH_3\\ \diagdown\\ CHCH_2OH\\ \diagup\\ CH_3\end{array} \xrightarrow{[O]}$

f) $CH_3CH=CH_2 + HBr \rightarrow$

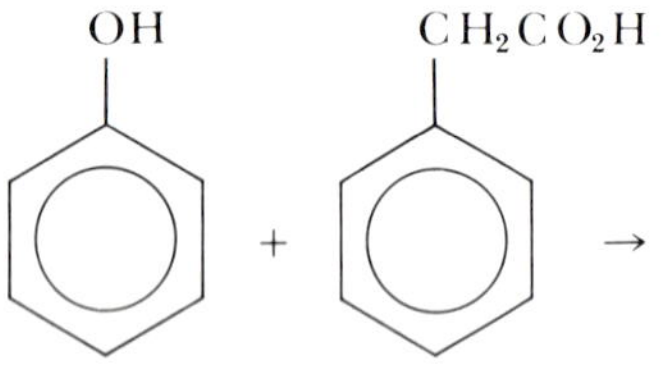

g) $CH_4 + Cl_2 \rightarrow$

h) 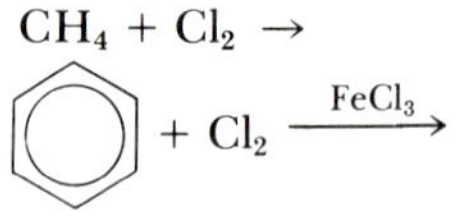 $C_6H_6 + Cl_2 \xrightarrow{FeCl_3}$

i) 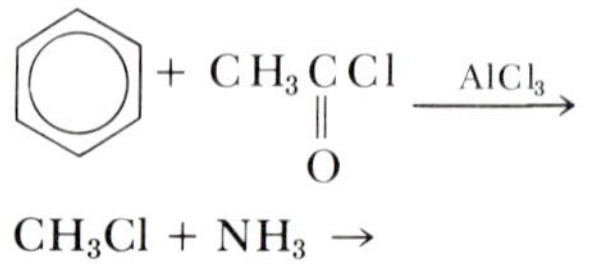 $C_6H_6 + CH_3CCl \xrightarrow{AlCl_3}$
   $\|$
   O

j) $CH_3Cl + NH_3 \rightarrow$

k) 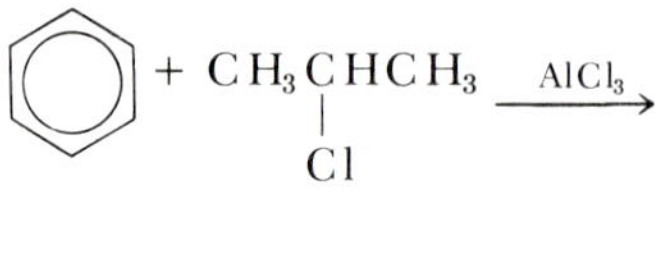 $C_6H_6 + CH_3CHCH_3 \xrightarrow{AlCl_3}$
   $|$
   Cl

l) 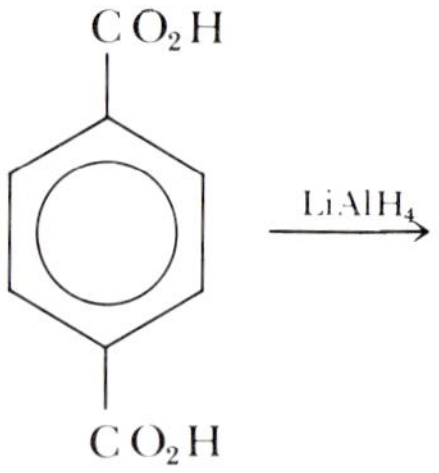 $p\text{-}C_6H_4(CO_2H)_2 \xrightarrow{LiAlH_4}$

m) $CH_3CH_2Br + NaCN \rightarrow$

n) $CH_3CH_2CCH_3 \xrightarrow{[O]}$
   $\|$
   O

o) naphthalene-COCl $+ NH_3 \rightarrow$

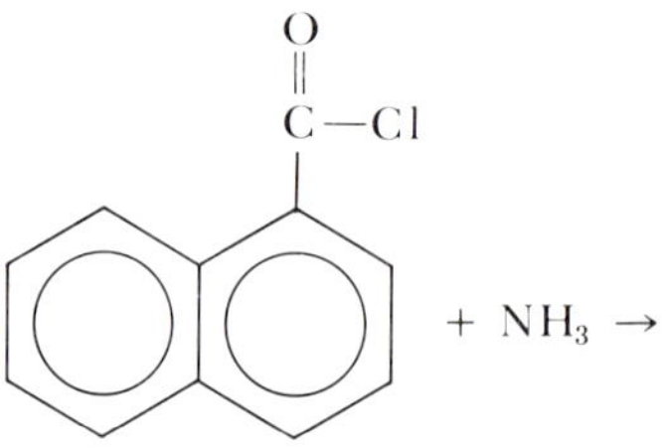

7. Use Markovnikov's rule to predict the main product in each of the following reactions.
   a) $CH_3CH_2CH=CH_2 + HI \rightarrow$
   b) $CH_3CH=CH_2 + H_2O \rightarrow$
   c) $CH_3CH=CHCH_3 + HBr \rightarrow$
   d) $CH_3C=CH_2 + HI \rightarrow$
      $|$
      CH_3
   e) $CH_3CH_2CH=C\begin{array}{c}CH_3\\ \diagup\\ \diagdown\\ CH_3\end{array} + H_2O \rightarrow$

8. The synthesis Wöhler was referring to in the quotation at the beginning of the chapter is

$$NH_4CNO \rightarrow H_2N-C-NH_2$$
$$\|$$
$$O$$

Basically, what is taking place in this reaction?

9. Show how to prepare each of the following.
   a) 2-Butanone from 1-bromobutane.
   b) Propanone from propane.
   c) 2-Propanol from 1-chloropropane.
   d) 1,2-Dichloroethane from ethene.
   e) Benzyl alcohol from benzene.

490   THE SYNTHESIS OF ORGANIC COMPOUNDS

f) 2-Propanol from propanoic acid.
h) Phenylethanoic acid from toluene.
i) 2-Chloro-2-methyl propane from 1-chloro-2-methyl propane.
j) Propanol from propanone.

10. The reaction

$$RX + R'ONa \rightarrow ROR' + NaX$$

where R is alkyl and R' is alkyl or aryl, is known as the **Williamson synthesis**. The compound R'ONa is formed from sodium metal and an alcohol or phenol. Show how you would use this method to prepare the following ethers.
   a) Phenyl methyl ether.
   b) Ethyl ether.
   c) Phenyl benzyl ether.

11. Suggest how the mild-oxidation products of 2-propanol and 2-methyl-2-propanol might differ.

12. How would you carry out the oxidation of 1-butanol in order to obtain (a) mainly butanal, (b) mainly butanoic acid?

13. When an ester is treated with excess acid or base, the following reaction (the reverse of esterification) occurs

$$\underset{\underset{O}{\|}}{R C O R'} + H_2O \rightarrow \underset{\underset{O}{\|}}{R C O H} + R'OH$$

This is called a **saponification** reaction. Suppose the reaction is carried out with water containing ^{18}O. Where would you expect the ^{18}O to appear, in the acid or in the alcohol? Justify your answer.

Identification of Compounds

14. How could you tell the difference between methyl ether and ethanol solely on the basis of infrared spectroscopy?

15. You are given four reagent bottles, each containing one of the compounds I–IV,

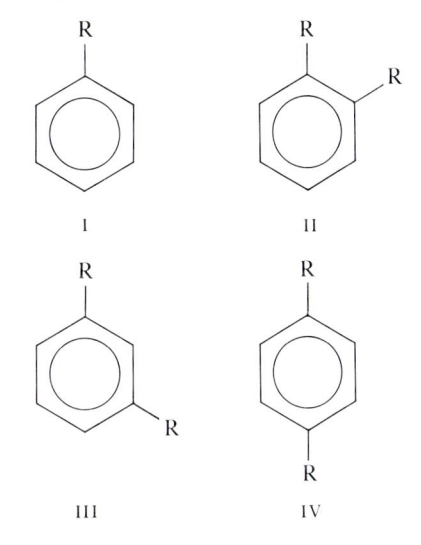

where R is some functional group. A chemical reaction is carried out on each sample in which one more R group is added to the ring. Upon analyzing the contents of each bottle *after* the addition of an R group, you obtain the following results:
   Bottle A: Three different compounds are present.
   Bottle B: Three different compounds are present.
   Bottle C: Two different compounds are present.
   Bottle D: One compound is present.
You also find that the compounds in bottle A have a higher molecular weight than those in bottle B. Which compound was originally in each bottle?

16. Suggest a way of distinguishing between propanone and propanal in the laboratory.

17. Suggest a way of distinguishing between 1-butanone and 2-butanone in the laboratory.

ADDITIONAL READINGS

Roberts, John D., "Organic Chemical Reactions." *Scientific American,* November 1957. A renowned California Institute of Technology professor tells how chemists study the behavior of complex compounds.

Wasserman, Edel, "Chemical Topology." *Scientific American,* November 1962. A Bell Telephone Laboratory scientist tells how to make compounds in which two independent rings are linked through each other.

18
THE ENERGETICS OF CHEMICAL CHANGE

Die Energie der Welt ist konstant; die Entropie der Welt strebt einem Maximum zu.
RUDOLF CLAUSIUS, *1865*

The quotation at the beginning of the chapter means, "The energy of the world is constant; the entropy of the world strives to attain a maximum value." This succinct statement of the first two laws of thermodynamics is due to Rudolf Julius Emmanuel Clausius (1822–1888), a German theoretical physicist who invented the term "entropy" and who made significant contributions to the kinetic theory of gases, thermodynamics, and electrolysis. Clausius held a number of university professorships: at the Royal Artillery and Engineering School in Berlin, Polytechnic School in Zurich, University of Zurich, University of Würzburg, and, finally, the University of Bonn.

All physical and chemical processes involve changes in energy. When a gram of ice at 0 °C is melted to liquid at 0 °C, about 331 J of energy must be supplied to carry out the endothermic process; the energy of the water sample is said to have increased by 331 J during the process. Similarly, when one mole of H_2 and one-half mole of O_2 are burned at 25 °C to produce one mole of liquid H_2O at 25 °C, about 286 kJ of energy is released by the exothermic process. Thus there is a decrease in energy of 286 kJ when the hydrogen and oxygen are transformed from the molecular elemental form to the combined chemical form of water at 25 °C. The 286 kJ of energy is not lost, however, but is taken up by the surroundings. The present chapter deals with some of the ways in which energy changes enter into chemical and physical processes, the laws the energy changes satisfy, and the roles they play in equilibrium and spontaneity.

18.1 THE CONCEPT OF ENERGY

Energy itself is an abstract concept, an invention of the human mind designed to facilitate the description of natural phenomena. Just as the French painters Manet,

◀ Steam tractor used in logging operations (1894). Steam was the main motive power of the Industrial Revolution. (Courtesy of Brown Brothers.)

Degas, and Renoir developed impressionist techniques to extend the range of artistic description and perception, physical scientists such as Clausius, Mayer, and Helmholtz developed the concept of energy to permit a deeper understanding of the subtleties of natural processes.

Energy is usually defined in terms of the ability to perform work. When a system has performed a given amount of work, it has done so at the expenditure of a given amount of its energy; its energy has changed by an amount commensurate with the work done. Strictly speaking, energy itself is never measured directly; rather it is *changes* in the property called energy that are measured. Measurement of the work done by a system provides a measure of the **energy change** of the system between the **initial state** (before the work was done) and the **final state** (after the work was done). If a system is able to do work upon its surroundings while passing from its initial state to its final state, there is a decrease in energy of the system. The process is said to be *exothermic* (see Chapter 10), and the energy change $\Delta E = E_{\text{final}} - E_{\text{initial}}$ is negative. Similarly, if work must be done by the surroundings upon the system in order to pass from the initial state to the final state, the energy of the system increases. Such a process is said to be *endothermic* and ΔE is positive.

Energy changes may be observed in a variety of ways depending on what particular process is involved. For example, a system may do mechanical work, moving a given mass with a given acceleration through a given distance. Measurement of the mass, acceleration, and displacement leads to a direct measure of the corresponding energy change in the system. The energy change could be measured more indirectly, for example, by employing the moving mass to turn a dynamo and thereby generate an electric current. In principle, ΔE could then be calculated from a measurement of the voltage and the amount of charge flowing. The above example is sometimes said to illustrate the changing of energy from one form to another. More accurately, it shows that the energy changes of one system may be used to bring about an equivalent energy change in another type of system.

18.2 ENERGY CHANGES IN PHYSICAL PROCESSES

The energy changes taking place in many physical processes are conveniently measured in terms of accompanying temperature changes. Such energy changes are referred to as **thermal energy**, or, popularly, as **heat**. A simple example of a physical process involving a change in temperature is the heating of a substance from some initial temperature T_i to some final temperature T_f. The actual energy change depends on the conditions under which the process occurs. If the conditions are such that the total volume of the system does not change, the energy change is given by

$$\Delta E = n\bar{C}_v \Delta T \tag{18.1}$$

where n is the number of moles of the substance, $\bar{C}_v$ is the **molar heat capacity** of the substance at constant volume, and ΔT is the temperature change $T_f - T_i$. The overbar on $\bar{C}_v$ denotes a **molar** quantity. As was stated in Section 9.4, the molar heat capacity is the specific heat times the molecular weight of the substance; it represents the energy required to raise the temperature of one mole of the substance by 1 °C (or 1 K) under conditions of constant volume. In actuality, the molar heat capacity at constant volume changes somewhat with temperature; this is often accounted for by using an *average* value over the temperature interval in question. For small temperature intervals, $\bar{C}_v$ often changes but slightly. In H_2O, $\bar{C}_v$ is 75.9 J·mol^{-1}·K^{-1} at 0 °C, 75.3 J·mol^{-1}·K^{-1} at 50 °C, and 75.9 J·mol^{-1}·K^{-1} at 100 °C. Consequently, the variation of $\bar{C}_v$ in this temperature interval is of little importance except for calculations requiring a very high degree of accuracy.

If the process is carried out under conditions of constant pressure, the energy is given the special name of **enthalpy** and the energy change is called an **enthalpy change** (symbol ΔH). The enthalpy change is given by

$$\Delta H = n\bar{C}_p \Delta T \tag{18.2}$$

where $\bar{C}_p$ is the molar heat capacity of the substance at constant pressure. In the case of solids and liquids, $\bar{C}_v$ and $\bar{C}_p$ are almost the same (since most solids and liquids do not show large volume changes on heating) and, consequently, ΔE and ΔH are virtually the same. However, $\bar{C}_p$ and $\bar{C}_v$ are quite different for gases since gases expand appreciably upon heating. If the gas is ideal, $\bar{C}_p$ and $\bar{C}_v$ are related by

$$\bar{C}_p = \bar{C}_v + R \tag{18.3}$$

where R is the ideal gas constant. The quantity R represents the work done on the surroundings by one mole of gas when expanding as a result of a one-degree change in temperature.

Some physical processes occur without an accompanying change of temperature, for example, the melting of a solid or the boiling of a liquid. Normally, such processes are carried out under constant-pressure conditions so that enthalpy changes are used to describe the process.

EXAMPLE 18.1 A sample of 2.5 moles of water at −10 °C and 1 atm pressure is heated until it melts and eventually ends up as liquid at 25 °C and 1 atm. Calculate the change in energy of the system (water sample).

SOLUTION Since H_2O does not change very much in volume when it is heated from −10 °C to 25 °C, the distinction between ΔE and ΔH is ignored even though it is the latter quantity actually involved in the constant-pressure process. It is convenient to divide the process into three separate steps, to calculate ΔH for each step, and then to add the ΔH's of the individual steps to obtain a total ΔH.

STEP 1: The heating of ice from −10 °C to 0 °C. The average specific heat of ice in this interval is about 2.1 J·g^{-1}·K^{-1}. Thus the molar heat capacity $\bar{C}_p$ is 2.1 J·g^{-1}·K^{-1} times 18 g·mol^{-1} or 38 J·K^{-1}·mol^{-1}. The enthalpy change for the step is

$$\Delta H_1 = n\bar{C}_p \Delta T = (2.5 \text{ mol})(38 \text{ J·K}^{-1}\text{·mol}^{-1})[0 - (-10)]\text{K} = 950 \text{ J}$$

STEP 2: The melting of ice at 0 °C. The heat of fusion of water is 331 J·g^{-1} or 5950 J·mol^{-1}. Thus

$$\Delta H_2 = (2.5 \text{ mol})(5950 \text{ J·mol}^{-1}) = 14{,}874 \text{ J}$$

STEP 3: The heating of liquid water from 0 °C to 25 °C. The average specific heat of liquid water is about 4.2 J·g^{-1}·K^{-1} in this interval. Thus the molar heat capacity $\bar{C}_p$ is 4.2 J·g^{-1}·K^{-1} times 18 g·mol^{-1} or 76 J·K^{-1}·mol^{-1}. The enthalpy change for step 3 is

$$\Delta H_3 = n\bar{C}_p \Delta T = (2.5 \text{ mol})(76 \text{ J}\cdot\text{K}^{-1}\cdot\text{mol}^{-1})(25 - 0)\text{K} = 4750 \text{ J}$$

The total enthalpy change is

$$\Delta H = \Delta H_1 + \Delta H_2 + \Delta H_3 = 950 + 14{,}874 + 4750 = 20{,}574 \text{ J}$$

Note that the process is endothermic; the water sample has increased its energy by 20,574 J.

18.3 ENERGY CHANGES IN CHEMICAL REACTIONS

The energy changes accompanying chemical reactions are measured experimentally by a device called a **calorimeter**. This device, which is typically used to study only exothermic processes, makes use of the fact that the energy change of the reaction may be used to change the temperature of an inert substance whose specific heat is known. The mass of the inert substance is chosen to be large enough so that the temperature change is kept small. Thus, since heat is constantly carried off from the reaction site and distributed to create only a small average rise in temperature, the chemical process may be regarded as approximately **isothermal**, that is, a constant-temperature process. Figure 18.1 illustrates one type of calorimeter used in measuring energy changes in combustion reactions. Combustion in the calorimeter is carried out in a heavy-walled steel bomb so that constant-volume conditions prevail. The energy change of an exothermic reaction taking place in the calorimeter is given by the expression

$$\Delta E = -C\Delta T \tag{18.4}$$

where C is the effective heat capacity of the calorimeter and ΔT is the change in temperature of the calorimeter bomb and surrounding inert material due to absorption of the exothermic heat of reaction. The effective heat capacity of the calorimeter is simply the amount of thermal energy needed to raise the temperature of the bomb and the surrounding inert substance by 1 °C (or 1 K). This quantity consists of three parts: (1) the thermal energy needed to raise the temperature of the bomb (usually stainless steel) by 1 °C; (2) the thermal energy needed to raise the temperature of the substance surrounding the bomb (usually water) by 1 °C; and (3) heat losses due to imperfect insulation. The second part is readily calculated from a measurement of the mass of water used; the first and third parts are usually determined as a single combined quantity by burning a weighed amount of substance whose ΔE value is accurately known. Using the relationship

$$C = C_{\text{bomb}} + C_{\text{H}_2\text{O}} + C_{\text{leakage}} \tag{18.5}$$

knowing what ΔE should be, and measuring $C_{\text{H}_2\text{O}}$ and ΔT allows the sum $C_{\text{bomb}} + C_{\text{leakage}}$ to be calculated for a given calorimeter.

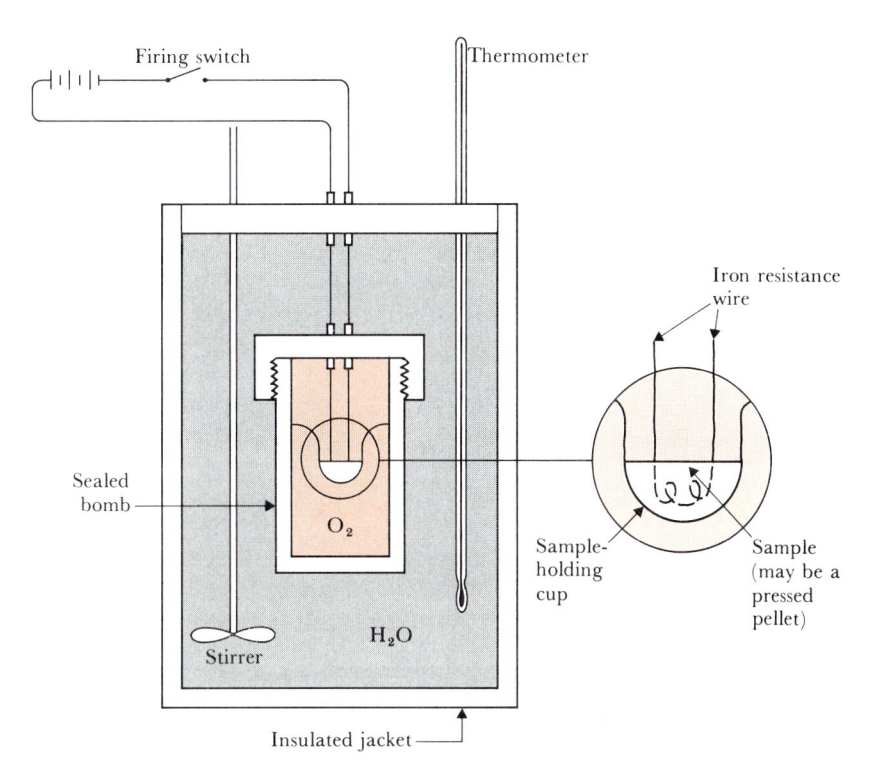

Fig. 18.1. Schematic diagram of a combustion calorimeter. When the firing switch is depressed, the iron resistance wire becomes red hot and ignites the sample. The heat released by the combustion is distributed between the bomb and the water. Solid samples are usually handled as pressed pellets molded about the resistance wire; gases and liquids must be handled by special techniques.

An oxygen-bomb calorimeter. A schematic illustration of the insides is found in Fig. 18.1. (Courtesy of Parr Instrument Company, Inc.)

18.3 | ENERGY CHANGES IN CHEMICAL REACTIONS **497**

EXAMPLE 18.2 A quantity of hydrogen gas weighing 1.0 g is burned in oxygen in a calorimeter to produce H$_2$O. The original temperature of the calorimeter was 24.192 °C; after the reaction this temperature rises to 25.514 °C. The calorimetric constant $C_{bomb} + C_{leakage}$ was previously determined to be 98,380 J·K^{-1}. The calorimeter contains 2 kg of water surrounding the bomb. Calculate the change in energy when one mole of H$_2$ reacts with O$_2$ to produce liquid H$_2$O at 25 °C.

SOLUTION The constant C in Eq. (18.4) is given by

$$C = C_{H_2O} + C_{bomb} + C_{leakage}$$
$$= (2000 \text{ g})(4.184 \text{ J}\cdot\text{g}^{-1}\cdot\text{K}^{-1}) + 98{,}380 \text{ J}\cdot\text{K}^{-1}$$
$$= 106{,}750 \text{ J}\cdot\text{K}^{-1}.$$

Thus
$$\Delta E = -(106{,}750 \text{ J}\cdot\text{K}^{-1})(25.514 - 24.192)\text{K} = -141{,}124 \text{ J or } -141.1 \text{ kJ}$$

This ΔE value is for 1 g or 0.5 mol of H$_2$. For one mole of H$_2$ burned, the value is twice this or -282.2 kJ. Thus ΔE of the reaction

$$H_2(g) + \tfrac{1}{2}O_2(g) \rightarrow H_2O(l)$$

is -282.2 kJ per mole of H$_2$O(l) formed. Since ΔE values for reactions such as this change very little with temperature, this ΔE value is assumed to be valid at 25 °C.

EXAMPLE 18.3 Calculate the enthalpy change for burning one mole of H$_2$ at 25 °C under conditions of constant pressure.

SOLUTION ΔH and ΔE are related by

$$\Delta H = \Delta E + \Delta(pV)$$

where $\Delta(pV)$ means $(pV)_{products} - (pV)_{reactants}$. As a good approximation, it is assumed that only gaseous products and reactants have significant changes in pV. If pV for each gaseous reactant or product is given by the ideal gas law, then it follows that

$$\Delta(pV) = \Delta nRT$$

where Δn = moles of gaseous products − moles of gaseous reactants. In the reaction of Example 18.2 there are zero moles of gaseous products and $1\tfrac{1}{2}$ moles of gaseous reactants so that $\Delta n = 0 - 1\tfrac{1}{2} = -1\tfrac{1}{2}$. Thus

$$\Delta H = \Delta E + \Delta nRT$$
$$= -282.2 \text{ kJ} + (-1\tfrac{1}{2})(8.314 \times 10^{-3} \text{ kJ}\cdot\text{K}^{-1}\cdot\text{mol}^{-1})(298 \text{ K})$$
$$= -282.2 - 3.7 \quad \text{or} \quad -285.9 \text{ kJ}\cdot\text{mol}^{-1}.$$

The -3.7-kJ contribution may be interpreted as work done on the system to decrease its volume.

Since most chemical processes are carried out under constant-pressure conditions, the enthalpy function may be regarded as an "invention" which simplifies discussion of energy changes in constant-pressure processes.

18.4 THE FIRST LAW OF THERMODYNAMICS

Many years of observing energy changes in diverse types of systems and processes have led scientists to the general observation that the energy of the universe appears to be constant. This means that energy cannot be created nor destroyed; for every energy change occurring in one part of the universe, there is a corresponding energy change of opposite sign someplace else in the universe so that the net change of energy is always zero. This fact means that the energy concept is a very useful invention for describing how nature operates. The principle of constancy of energy is generally referred to as the **first law of thermodynamics**, or, alternatively, the **law of conservation of energy**. This law, although supposedly valid on a cosmic scale, is actually based on observations on smaller systems which are isolated from their surroundings in such a way that no exchange of energy or matter occurs between system and surroundings. For example, a calorimeter may be assumed to be isolated from its surroundings by virtue of its thermally insulated outer jacket. The energy ΔE which is evolved by the combustion within the bomb is used to heat the bomb and its immediate surroundings. Thus the energy merely becomes redistributed and does not change in total amount.

Two of the foremost contributors to the law of conservation of energy were physicians: Julius Robert von Mayer (1814–1878) and Hermann Ludwig Ferdinand von Helmholtz (1821–1894). Both of these versatile Germans announced their formulations of the law in the early 1840s on the basis of physiological studies of animal heat. Helmholtz also made notable contributions to theoretical physics.

18.5 ENTHALPY CALCULATIONS FOR CHEMICAL REACTIONS

One of the most important consequences of the first law of thermodynamics is that the ΔE and ΔH of a process depend only upon the initial and final states of the system and are totally independent of how the process actually occurs; that is, it is immaterial what path is used in going from the initial state to the final state. This consequence of the first law makes it possible to calculate energy and enthalpy changes for many processes without actually carrying out the processes in the laboratory, provided that energy and enthalpy changes have been measured for a select number of other processes. Since most chemical reactions are carried out under constant-pressure conditions, the following discussion is limited to enthalpy changes. Nevertheless, the same principles apply to both energy and enthalpy changes.

Enthalpy changes depend on the temperatures and pressures of the substances involved and on their physical states. It is therefore convenient to base enthalpy values on what are known as **standard reference states** of substances. Enthalpy changes for other conditions can then be calculated from the standard-state values, if desired. The standard reference state of a substance is defined as the substance in its most stable form at the temperature in question when the pressure is 1 atm. Most standard-state enthalpy changes are tabulated for a temperature of 298 K but, in principle, other temperatures could have been chosen (this would lead to different numerical values for enthalpy changes, however). For example, a chemist specializing in high-temperature reactions might prefer to tabulate standard-state enthalpy values at 2000 K. Standard-state enthalpy changes are indicated by a superscript zero on the ΔH symbol: ΔH^0. Sometimes the particular reference temperature chosen is indicated via subscript, for example, ΔH^0_{298} or ΔH^0_{2000}. A reference temperature of 298 K will be assumed unless otherwise stated.

Enthalpy changes depend very strongly on the physical states of the substances involved. For example, although ΔH^0 of the reaction

$$H_2(g) + \tfrac{1}{2}O_2(g) \rightarrow H_2O(l)$$

is -286 kJ, the same process, when it produces gaseous water instead,

$$H_2(g) + \tfrac{1}{2}O_2(g) \rightarrow H_2O(g)$$

has $\Delta H^0 = -242$ kJ. The difference of 44 kJ is just the energy required to vaporize one mole of water at 298 K.

Let us consider a general chemical reaction

$$a\mathrm{A} + b\mathrm{B} \rightarrow l\mathrm{L} + m\mathrm{M}$$

where all reactants (A and B) and all products (L and M) are in their standard states at 298 K. The first law of thermodynamics states that the enthalpy change of the above reaction is given by

$$\Delta H^0_{rx} = l\bar{H}^0_L + m\bar{H}^0_M - a\bar{H}^0_A - b\bar{H}^0_B$$

where $\bar{H}^0_X$ (X = A, B, L, or M) is the **absolute** enthalpy of one mole of substance X in its standard state. However, since there is no unique way of defining a zero-value of enthalpy, it is not possible to obtain absolute values of enthalpies; experiments can only measure *changes* in enthalpy. By the same token, surveyors do not determine the absolute elevations of mountain peaks; they merely determine relative elevations above an arbitrarily selected point called sea level. In a somewhat analogous fashion, it is possible to choose an arbitrary reference point for measurement of $\bar{H}^0_X$ values on a *relative* basis. Since the same elements are present in the products L and M as in the reactants A and B, the following conventions establish a common reference point for all the relative $\bar{H}^0_X$ values in a given chemical reaction:

1. If X is an element in its standard state, then $\bar{H}^0_X = 0$.

2. If X is a compound in its standard state, then $\bar{H}_X^0$ is equal to $\Delta \bar{H}^0$ for the reaction:

   elements → compound

   The symbol $\bar{H}_X^0$ is then replaced with $\Delta \bar{H}_X^0$ and this is called the **standard-state enthalpy of formation** of compound X.

Note that the standard state of an element such as oxygen is the molecule O_2 and not the atom O. Thus $\bar{H}_{O_2}^0$ (or $\Delta \bar{H}_{O_2}^0$) is zero but $\bar{H}_O^0$ (or $\Delta \bar{H}_O^0$) is not zero. The latter quantity is the enthalpy change involved in dissociating molecular oxygen to atomic oxygen.

If absolute enthalpies were known, one could calculate ΔH_f^0 of AB as follows:
$$A + B \rightarrow AB$$
$$\Delta H_f^0(AB) = \bar{H}_{AB}^0 - \bar{H}_A^0 - \bar{H}_B^0$$
$$= 10 - 12 - 5 = -7$$

The relative scale sets $\bar{H}_A^0 = \bar{H}_B^0 = 0$ and $\bar{H}_{AB}^0 = \Delta H_f^0(AB)$; $\Delta H_f^0(AB)$ is measured experimentally.

$$AB + CD \rightarrow BC + AD$$
$$\Delta H_{rx}^0 = \bar{H}_{BC}^0 + \bar{H}_{AD}^0 - \bar{H}_{AB}^0 - \bar{H}_{CD}^0$$
$$= 1 + 5 - 10 - 5 = -9$$
$$= \Delta H_{BC}^0 + \Delta H_{AD}^0 - \Delta H_{AB}^0 - \Delta H_{CD}^0$$
$$= -8 - 10 - (-7) - (-2) = -9$$

∴ Relative enthalpies may be used to calculate ΔH^0 of reactions.

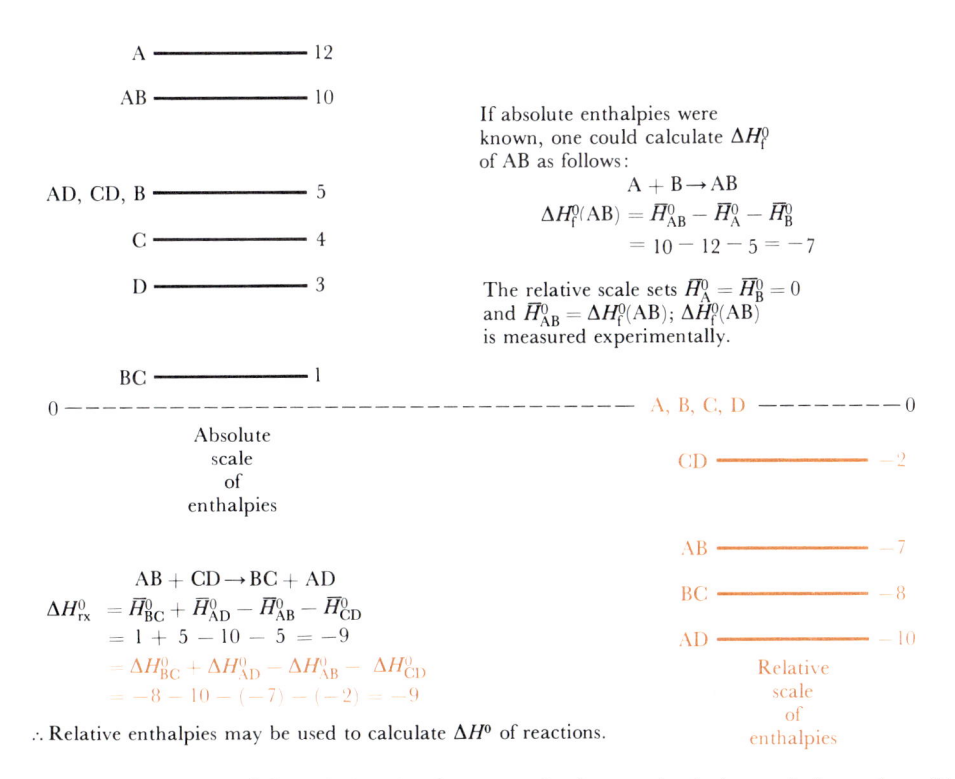

Fig. 18.2. Illustration of the relationship between absolute and relative enthalpy values. Note that both products and reactants contain the same atoms; thus one can use a common value (specifically, zero) for the enthalpies of all the elements in their standard states.

The above conventions are shown diagrammatically in Fig. 18.2. These conventions make it possible to set up tables of the enthalpy of formation of various molecules which can be used to calculate enthalpy changes for a great variety of chemical reactions, many of which might be difficult if not impossible to carry out in the laboratory. The calculation of ΔH_{rx}^0 always takes the simple form: the sum of the ΔH_f^0's of products minus the sum of the ΔH_f^0's of reactants. In Table 18.1 are listed the standard-state enthalpies of formation of several common molecules. The following examples illustrate some of the ways in which the tabulated enthalpies may be used to calculate enthalpy changes for various chemical reactions.

TABLE 18.1 STANDARD-STATE ENTHALPIES OF FORMATION OF SOME COMMON MOLECULES AT 298 K AND 1 ATM PRESSURE

		ΔH_f^0	
MOLECULE	STATE	kcal·mol^{-1}	kJ·mol^{-1}
HF	Gas	−64.80	−271.1
HCl	Gas	−22.06	−92.30
HBr	Gas	−8.66	−36.2
HI	Gas	+6.20	+25.9
H$_2$O	Liquid	−68.35	−286.0
H$_2$O	Gas	−57.8	−242
NH$_3$	Gas	−11.04	−46.19
CO	Gas	−26.41	−110.5
CO$_2$	Gas	−94.05	−393.5
CH$_4$	Gas	−17.89	−74.85
CH$_3$CH$_3$	Gas	−20.24	−84.68
CH$_2$=CH$_2$	Gas	+12.4	+51.9
CH≡CH	Gas	+54.2	+227
C$_6$H$_6$	Liquid	+27.78	+116.2
C$_6$H$_6$	Gas	+19.28	+80.67
NaCl	Solid	−98.23	−411.0
NO	Gas	+21.60	+90.37
N$_2$O	Gas	+19.49	+81.56
NO$_2$	Gas	+8.01	+33.5
C, graphite	Solid	0	0
C, diamond	Solid	+0.45	+1.9
O$_3$	Gas	+34.0	+140

EXAMPLE 18.4 Calculate the enthalpy change for the formation of ammonia from hydrogen and nitrogen at 298 K when all reactants and products are in their standard states.

SOLUTION The relevant chemical reaction is

$$N_2(g) + 3H_2(g) \rightarrow 2NH_3(g)$$

The standard-state enthalpy change of the reaction is given by

$$\Delta H_{rx}^0 = 2\Delta \bar{H}_{NH_3}^0 - \Delta \bar{H}_{N_2}^0 - 3\Delta \bar{H}_{H_2}^0$$

Table 18.1 shows that $\Delta \bar{H}_{NH_3}^0$ is $-46.19\,\text{kJ}\cdot\text{mol}^{-1}$. By convention 1, since N$_2$ and H$_2$ are elements in their standard states, $\bar{H}_{N_2}^0 = \Delta \bar{H}_{N_2}^0 = 0$ and $\bar{H}_{H_2}^0 = \Delta \bar{H}_{H_2}^0 = 0$ so that

$$\Delta H_{rx}^0 = 2(-46.19) - (0) - 3(0) = -92.38 \text{ kJ}$$

Note that ΔH_{rx}^0 is just twice $\Delta \bar{H}_{NH_3}^0$.

EXAMPLE 18.5 Calculate the standard-state enthalpy change for the reaction (at 298 K):

$$N_2O(g) + \tfrac{1}{2}O_2(g) \rightarrow 2NO(g)$$

SOLUTION

$$\Delta H^0_{rx} = 2\Delta \bar{H}^0_{NO} - \Delta \bar{H}^0_{N_2O} - \tfrac{1}{2}\Delta \bar{H}^0_{O_2} = 2(90.37) - (81.56) - \tfrac{1}{2}(0) = 99.18 \text{ kJ}$$

An alternative approach, which some people find easier to use, is to note that the following two reactions (both of the type elements → molecule, or the reverse)

$$N_2(g) + O_2(g) \rightarrow 2NO(g) \qquad \Delta H^0 = 2\Delta \bar{H}^0_{NO}$$

$$N_2O(g) \rightarrow N_2(g) + \tfrac{1}{2}O_2(g) \qquad \Delta H^0 = -\Delta \bar{H}^0_{N_2O}$$

add up to the desired reaction. Thus the enthalpy changes of the two reactions must also add up to the enthalpy change of the single reaction. This gives

$$\Delta H^0_{rx} = 2\Delta \bar{H}^0_{NO} - \Delta \bar{H}^0_{N_2O}$$
$$= 2(90.37) - (81.56) = 99.18 \text{ kJ}$$

As a general principle, any time a series of reactions (all carried out at the same temperature and pressure) can be added algebraically to produce a new reaction, the enthalpy change of the new reaction is the sum of the individual enthalpy changes of the reactions in the series. This principle is often called **Hess's law**.

The following example illustrates one way of determining the standard-state molar enthalpies of formation which appear in Table 18.1.

EXAMPLE 18.6 The standard-state enthalpy of combustion of glucose is determined in a calorimeter to be -2816 kJ·mol^{-1} at 25 °C. Determine the standard-state molar enthalpy of formation of glucose from the elements at 25 °C.

SOLUTION The combustion reaction occurring in the calorimeter is

$$C_6H_{12}O_6(s) + 6O_2(g) \rightarrow 6CO_2(g) + 6H_2O(l)$$

The calorimetric quantity of -2816 kJ·mol^{-1} is ΔE^0_{rx} for burning one mole of glucose. Since $\Delta n = 6$ moles of CO_2 minus 6 moles of $O_2 = 0$ (see Example 18.3), ΔE^0_{rx} is equal to ΔH^0_{rx}. This latter quantity is also given by

$$\Delta H^0_{rx} = 6\Delta \bar{H}^0_{CO_2} + 6\Delta \bar{H}^0_{H_2O(l)} - \Delta \bar{H}^0_{glucose} - 6\Delta \bar{H}^0_{O_2}$$

All of the above quantities are known except $\Delta \bar{H}^0_{glucose}$. Thus, solving for this quantity algebraically leads to

$$\Delta \bar{H}^0_{glucose} = 6\Delta \bar{H}^0_{CO_2} + 6\Delta \bar{H}^0_{H_2O(l)} - 6\Delta \bar{H}^0_{O_2} - \Delta H^0_{rx}$$
$$= 6(-393.5) + 6(-286.0) - 6(0) - (-2816)$$
$$= -1261 \text{ kJ·mol}^{-1}$$

Thermodynamic data other than enthalpies of formation of molecules may be utilized to determine enthalpy changes and other energy quantities. This is illustrated in the following two examples.

EXAMPLE 18.7 The standard-state enthalpy of formation of atomic oxygen is 247 kJ·mol⁻¹. Use this information along with information in Table 18.1 to calculate ΔH^0 for the reaction (at 298 K):

$$O_3(g) \rightarrow O_2(g) + O(g)$$

SOLUTION

$$\Delta H^0_{rx} = \Delta \bar{H}^0_{O_2} + \Delta \bar{H}^0_{O} - \Delta \bar{H}^0_{O_3}$$

By convention, $\Delta \bar{H}^0_{O_2}$ is zero. $\Delta \bar{H}^0_{O}$, which was given as 247 kJ·mol⁻¹, refers to the process

$$\tfrac{1}{2}O_2(g) \rightarrow O(g)$$

$\Delta \bar{H}^0_{O_3}$ is given in Table 18.1 as 140 kJ·mol⁻¹. Thus

$$\Delta H^0_{rx} = (0) + 247 - (140) = 107 \text{ kJ}$$

Note that ΔH^0 for the reverse reaction

$$O_2(g) + O(g) \rightarrow O_3(g)$$

is −107 kJ. This exothermic reaction requires the presence of a third body, otherwise the O_3 dissociates by recoil as fast as it is formed (see Section 15.11).

The electron affinity of an atom (see Section 5.1) may be calculated on the basis of the first law of thermodynamics. As an example, consider the reaction

$$M(s) + \tfrac{1}{2}X_2(g) \rightarrow M^+(s) + X^-(s)$$

where MX is a crystalline ionic solid formed from a metal M and a halogen X. One path from the initial state (reactants only) to the final state (product only) may be represented by the following scheme:

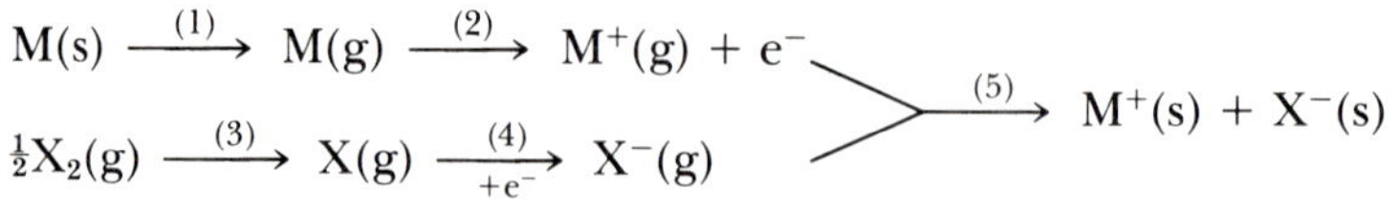

$$M(s) \xrightarrow{(1)} M(g) \xrightarrow{(2)} M^+(g) + e^-$$
$$\tfrac{1}{2}X_2(g) \xrightarrow{(3)} X(g) \xrightarrow[+e^-]{(4)} X^-(g)$$
$$\xrightarrow{(5)} M^+(s) + X^-(s)$$

This pathway is known as a **Born-Haber cycle**. If ΔH values for the steps are numbered as above, the total ΔH of the process is

$$\Delta H_{rx} = \Delta H_1 + \Delta H_2 + \Delta H_3 + \Delta H_4 + \Delta H_5$$

The Born-Haber cycle is named after the German physicist Max Born (1882–1970) and the German chemist Fritz Haber (1886–1934). Born is best known for his theoretical work on crystals and liquids and for the development of quantum mechanics. He received the Nobel prize in physics in 1954. Haber became famous for developing the first commercially feasible process for the synthesis of ammonia from the elements. He received the Nobel prize in chemistry in 1918. Although Haber was a fervently patriotic German, the Nazis forced him to resign from the Kaiser Wilhelm Institute in 1933 because of his Jewish ancestry. Haber moved to England and later to Switzerland where he died still incredulous that his own countrymen could have disowned him.

where

ΔH_1 = heat of vaporization (sublimation) of M(s)

ΔH_2 = ionization energy of M(g)

ΔH_3 = enthalpy of dissociation of X_2(g) to X(g)

ΔH_4 = electron affinity of X(g)

ΔH_5 = lattice energy of MX(s) (see Section 7.4)

All but the quantities ΔH_4 and ΔH_5 are readily determined by experimental methods and ΔH_5 may often be estimated quite accurately. Thus the remaining quantity, the electron affinity ΔH_4, can be calculated.

EXAMPLE 18.8 Calculate the electron affinity of the chlorine atom from the following data.

Enthalpy of formation of NaCl(s) = -411.0 kJ·mol^{-1} (Table 18.1)

Ionization energy of Na(s) = 5.1 eV (494.5 kJ·mol^{-1})

Heat of sublimation of Na(s) = 108 kJ·mol^{-1}

Enthalpy of formation of Cl(g) = 121 kJ mol^{-1}

Lattice energy of NaCl(s) = -770 kJ·mol^{-1} (Table 7.1)

SOLUTION

$$\Delta H_4 = \Delta \bar{H}^0_{\text{NaCl}} - \Delta H_1 - \Delta H_2 - \Delta H_3 - \Delta H_5$$
$$= -411.0 - 108 - 494.5 - 121 - (-770)$$
$$= -365 \text{ kJ·mol}^{-1} \quad \text{or} \quad -3.8 \text{ eV}$$

Recall that the conversion from kJ·mol⁻¹ to eV requires dividing by Avogadro's number followed by use of the equivalence of 1 eV and 1.602×10^{-19} J. Different compounds and different sources of data may lead to slightly different values of electron affinities. The values given in reference tables usually represent averages based on several different calculations. Also, inorganic chemists commonly change the sign of electron affinities to make them all positive. This obsolete and undesirable practice is apparently difficult to eliminate.

An examination of the terms making up the Born-Haber cycle energy expression shows that the enthalpy of formation of an ionic compound depends on the difference between the exothermic terms (electron affinity of the nonmetal and lattice energy of the solid) and the endothermic terms (vaporization and ionization of the metal, and dissociation of the nonmetal). The three terms which are of greatest magnitude, and hence contribute most to the overall energy, are the electron affinity of the nonmetal, the lattice energy of the solid, and the ionization energy of the metal. All other factors assumed equal, the stability of an ionic solid is favored by a large negative value for the electron affinity of the nonmetal, a large negative value for the lattice energy, and a low ionization energy for the metal.

18.6 BOND ENERGIES

The enthalpy change of the process

$$H_2(g) \rightarrow 2H(g)$$

is 435 kJ. This quantity is just twice $\Delta \overline{H}_H^0$, the standard-state enthalpy of formation of a mole of hydrogen atoms. The enthalpy change 435 kJ may also be interpreted as representing the **energy of the H—H bond**. The energies of other bonds involving diatomic molecules may be similarly defined. The average energy of a C—H bond may be defined as one-fourth the energy of the process

$$CH_4(g) \rightarrow C(g) + 4H(g) \qquad \Delta H^0 = 1657 \text{ kJ}$$

that is, $E(\text{C—H}) = 1657/4$ or 414 kJ·mol⁻¹ of bonds. The energy of a C—C bond may be calculated from the process

$$CH_3CH_3(g) \rightarrow 2C(g) + 6H(g) \qquad \Delta H^0 = 2833 \text{ kJ}$$

by noting that this involves breaking six C—H bonds and one C—C bond. It is assumed that

$$E(\text{C—C}) + 6E(\text{C—H}) = 2833 \text{ kJ}$$

If it is assumed that the energy of the C—H bond in ethane is the same as that of the C—H bond in methane, then the energy of the C—C bond is readily calculated as

$$E(\text{C—C}) = 2833 - 6E(\text{C—H})$$

$$= 2833 - 6(414) = 349 \text{ kJ·mol}^{-1} \text{ of bonds}$$

TABLE 18.2 THE ENERGIES OF SOME COMMON CHEMICAL BONDS AT 298 K[a]

BOND	ENERGY IN: kcal·mol⁻¹ of bonds	kJ·mol⁻¹ of bonds
H—H	104	435
O—O	33	14 × 10¹
O=O	118	494
O—H	111	464
C—H	99	41 × 10¹
C—C	83	35 × 10¹
C=C	148	619
C—O	82	34 × 10¹
C=O	169	707

[a] Some of the values given are averages computed on the basis of the enthalpies of dissociation of various molecules containing the bond being considered.

A continuation of this process, assuming that energies of comparable bonds in different molecules are equal, leads to a list of bond energies such as given in Table 18.2. Bond energies are sometimes used in the estimation of the dissociation energies of molecules for which insufficient calorimetric data are available.

EXAMPLE 18.9 Calculate the N—N bond energy in hydrazine (N_2H_4) using the following data: The enthalpy of dissociation of gaseous N_2H_4 and NH_3 to atoms is 1854 kJ·mol⁻¹ and 1172 kJ·mol⁻¹, respectively.

SOLUTION Hydrazine has the structure

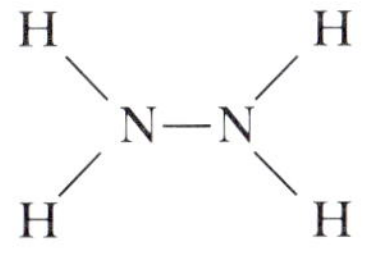

```
H           H
 \         /
  N — N
 /         \
H           H
```

Thus

$E(N—N) + 4E(N—H) = 1854$ kJ

Assuming $E(N—H)$ is the same in N_2H_4 as in NH_3 leads to

$3E(N—H) = 1172$ kJ

∴ $E(N—H) = 391$ kJ·mol⁻¹ of bonds

Then,

$E(N—N) = 1854 - 4E(N—H)$

$= 1854 - 4(391) = 290$ kJ·mol⁻¹ of bonds

Nicolas Léonard Sadi Carnot (1796–1832) was a French military engineer. Carnot set out to formulate the theory behind the operation of the steam engine when he realized its enormous potential for the development of industrial power.

18.7 THE SECOND LAW OF THERMODYNAMICS

The second law of thermodynamics—also based on many years of observations—is somewhat subtler and more difficult to express than is the relatively simple first law. One of the observations leading to this law arose during the era of steam engines; if a fuel is burned at one temperature T_2, then only part of the evolved thermal energy, q_2, can be converted to mechanical work. Part of the evolved energy, labeled q_1, gets tied up in the random molecular motions of the combustion products at a lower temperature T_1 and cannot be used to perform useful work. This situation is illustrated schematically in Fig. 18.3. If w, or $q_2 - q_1$, represents the work produced, then, as was shown by the French engineer Carnot, the *maximum* fraction of q_2 convertible to useful work is given by

$$\frac{w}{q_2} = \frac{T_2 - T_1}{T_2} \tag{18.6}$$

This relationship shows that 100% conversion of heat to work is possible only if $T_1 = 0$ K or $T_2 = \infty$. A typical automobile engine may have a combustion chamber temperature of around 2300 K. Assuming that exhaust products are at 900 K, the maximum efficiency of the engine is given by

$$\frac{2300 - 900}{2300} \times 100 = 61\%$$

This means that for every 100 J of energy produced by combustion of gasoline, a maximum of 61 J can be used to operate the automobile; 39 J is tied up in the exhaust gases. In practice, most of the 61 J becomes lost via thermal conduction, friction, and other processes so that, typically, only about 20–25 J ends up producing mechanical work.

The observations of Carnot and others led the German physicist Clausius to invent a new quantity called **entropy** which is extremely useful in acquiring an even deeper understanding of natural processes than that provided by the first law of thermodynamics alone. Clausius proposed a second law of thermodynamics as follows:

For every spontaneous process occurring in an isolated system there is always an increase, or at best, no change, in the entropy.

This is written in mathematical form as

$$\Delta S_{\text{isolated system}} \geq 0 \tag{18.7}$$

An **isolated system** is one that exchanges neither energy nor matter with its surroundings. Clausius and others were able to show that one of the consequences of

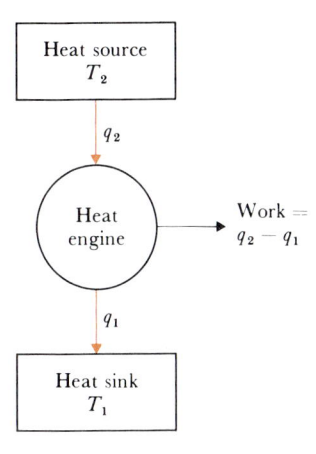

Fig. 18.3. One way of illustrating the second law of thermodynamics. T_2 is the temperature of the combustion chamber in which a chemical fuel is being burned, and T_1 is the temperature of the combustion products after the hot gases resulting from the combustion have operated the heat engine.

the second law is the Carnot relationship, Eq. (18.6). The equality holds whenever the process is carried out in a perfectly reversible manner; otherwise the inequality holds. A perfectly reversible process is one that is always in equilibrium, that is, it passes smoothly and continuously through a succession of equilibrium states; in practice, reversibility cannot be attained, although it can be approached quite closely. One of the interpretations of the second law is that all natural processes in which energy changes are involved tend to make the energy of the universe less available for performing useful work.

Entropy may be thought of as a measure of disorder, randomness, or chaos. The units of entropy are those of energy divided by temperature: $J \cdot K^{-1}$ in SI. The entropy function makes it possible to generalize the observation that natural systems, when left to themselves, tend to decay and degenerate; that which was once organized and ordered tends to become jumbled and disordered. A handful of alphabetized file cards tends to become randomized when dropped; the reverse is highly unlikely to occur. When a box containing 50 red balls and 50 black balls is shaken, the final arrangement is much more likely to be a relatively uniform mixture of red and black balls than a clustering of red balls in half of the box and black balls in the other half. Thus the second law is a statistical law in that it deals with the most probable course of natural events.

The entropy function is related to the impossibility of carrying out a total conversion of heat to work. Consider a steam engine as an example. Heat from burning fuel is used in the engine's cylinder. However, only those steam molecules moving in the direction the piston is to go can do useful work and push the piston. The entropy function, related to the disorder of the steam molecules, is also related to the energy tied up in motions which are in the wrong directions to push the piston. The second law of thermodynamics essentially says that no matter how clever we are, some energy will always be tied up in random motions that cannot be made to push the piston.

The tendency to disorder has its counterpart in all natural physical and chemical processes, although it is not always possible to describe it in easily visualized terms. A simple example is the melting of ice. The arrangement of molecules in the ice crystal represents a relatively ordered arrangement, whereas the less rigid arrangement in the liquid state is more disordered. When the ice melts, the entropy of water increases:

$$\Delta S_{H_2O} > 0 \quad \text{for} \quad H_2O(s) \rightarrow H_2O(l)$$

At the same time the surroundings change in entropy since the energy to melt the ice came from them.* The second law says that

$$\Delta S_{\text{surr}} + \Delta S_{\text{H}_2\text{O}} \geq 0$$

Thus, at best, $\Delta S_{\text{surr}} = -\Delta S_{\text{H}_2\text{O}}$. This occurs if the melting process is done in a perfectly reversible manner; otherwise ΔS_{surr} must always be larger than $-\Delta S_{\text{H}_2\text{O}}$. Since $\Delta S_{\text{H}_2\text{O}}$ is positive for the melting of ice, it follows that ΔS_{surr} must be negative for the reversible case. Thus an increase in the disorder of water can be matched by an equal increase of order in the surroundings; usually, however, there is a net increase in disorder due to lack of perfect reversibility.

As the foregoing shows, entropy changes can be less than zero (entropy decreases). However, some other part of the isolated system must then experience an increase in entropy so that the net entropy change is either zero or positive. Living organisms are outstanding examples of systems which are characterized by a decreasing entropy. While they are living they organize smaller molecules into larger molecules and thus create order out of chaos. However, they do this only at the expense of an entropy increase in their surroundings, an increase which more than compensates for the entropy decrease.

18.8 SPONTANEITY AND CHEMICAL EQUILIBRIUM

One of the most important deductions arising from the first and second laws of thermodynamics is the realization that the most probable course of events in a natural process occurring at a given pressure and temperature is dependent on two fundamental tendencies: the tendency to achieve a **minimum potential energy** and the tendency to achieve **maximum disorder**. These tendencies may be most simply described by inventing a new quantity G called the **Gibbs function** or **Gibbs free energy**. Changes in G (which has the same units as energy) occurring at constant temperature and pressure are given by the relationship

$$\Delta G = \Delta H - T\Delta S \qquad (18.8)$$

A process is said to be **intrinsically spontaneous** if it is characterized by a decrease in G, that is, if ΔG is negative. Physically, ΔG is interpreted as the negative of the useful work the process is intrinsically capable of doing if incorporated into a properly engineered apparatus. Thus an intrinsically spontaneous process is defined as one capable, at least in principle, of producing useful work. By useful work is meant work which could be harnessed to operate a machine, light a bulb, lift a weight, etc. The rate at which the useful work is produced is of no consequence here; some spontaneous processes produce useful work rapidly and others do so infinitesimally slowly.

Equation (18.8) shows that there are three different ways of having a negative ΔG: First, ΔH could be negative (implying a decrease in potential energy) and ΔS

---

* Each isolated system to which the second law is applied is viewed as consisting of two parts: the substance which is the focal point (the system) and everything else (the surroundings). The substance burned in a calorimeter is the system; the surrounding bomb and water are the surroundings.

Josiah Willard Gibbs, American mathematical physicist responsible for much of our basic understanding of phase transitions and other areas of thermodynamics. (Courtesy of Culver Pictures.)

The Gibbs function is named after Josiah Willard Gibbs (1839–1903), an American mathematical physicist who, while teaching in relative obscurity (and without salary!) at Yale University, established the theoretical foundations of thermodynamics, statistical mechanics, phase equilibria, and other areas. In one of his numerous papers, Gibbs made the very important comment that the purpose of theory (contrary to what most students may think) is to find that viewpoint from which experimental observations appear to fit the simplest *pattern. A penetrating biography of Gibbs* (Willard Gibbs) *has been written by Muriel Rukeyser, New York: E. P. Dutton & Co., N.Y., 1942 (also available in a 1964 paperback edition by the same publisher.)*

could be positive (implying an increase in disorder); second, ΔH could be positive (implying an increase in potential energy) but ΔS could be positive and $T\Delta S$ large enough in magnitude to still make the resultant $\Delta H - T\Delta S$ negative; and third, ΔS could be negative (implying an increase in order) but ΔH could be negative and large enough in magnitude to still make $\Delta H - T\Delta S$ negative. The latter two cases show that the two tendencies, to achieve minimum potential energy and to achieve maximum disorder, do not necessarily operate in unison.

Whenever ΔG is positive the process is intrinsically nonspontaneous and is not capable of producing useful work. In fact, work must be done on the process in order to carry it out.

Whenever ΔG is zero, the process is in a state of equilibrium; that is, there is no tendency for the process to go in either direction. The fact that all physical and chemical processes, if allowed to do so, proceed until equilibrium is reached, indicates that the state of equilibrium represents the best compromise between maximum

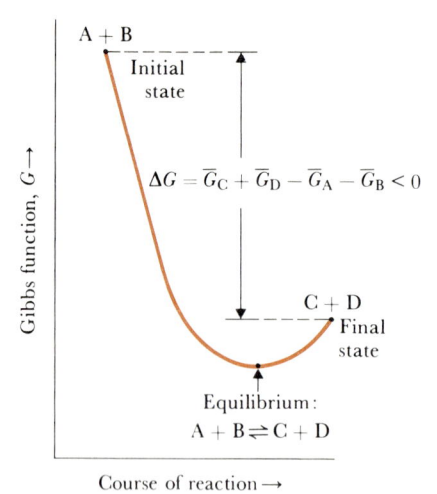

Fig. 18.4. Graphical illustration of the Gibbs function for reactants, products, and the equilibrium mixture.

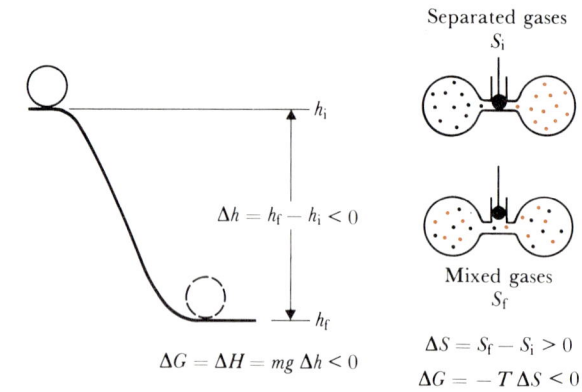

Fig. 18.5. Illustration of two spontaneous processes. The top process involves ΔH only; the bottom process involves ΔS only.

disorder and minimum potential energy. The position of equilibrium relative to pure products and pure reactants is illustrated schematically in Fig. 18.4. At equilibrium the total value of G for the reactants is equal to the total G of the products and there is no tendency for further reaction.

A simple illustration of the individual roles of ΔH and ΔS in determining intrinsic spontaneity is given in the following two examples. First, consider a boulder of mass m rolling down an incline from a higher point h_i to a lower point h_f. For this process, $\Delta H = mg\,(h_f - h_i) = mg\,\Delta h < 0$ since $\Delta h < 0$. If effects such as friction are absent, ΔS is not involved in this process and $\Delta G = \Delta H < 0$. Clearly, it would be possible to use the spontaneously rolling boulder to operate a machine or to perform some other kind of useful work (hydroelectric power plants use the energy of falling water in this way). Second, consider two flasks containing two different gases at the same temperature and pressure and connected by a tube with a closed stopcock. When the stopcock is opened the two gases will mix spontaneously and eventually come to equilibrium. Since ΔH is approximately zero for this process, $\Delta S > 0$ and $\Delta G < 0$. Since it would require the input of work to separate the mixed gases, it is clear that the mixing performed work. The two processes just described are illustrated in Fig. 18.5.

18.9 FREE ENERGY AND ENTROPY CALCULATIONS

The entropies of atoms and molecules may be measured experimentally by the use of various calorimetric techniques.* Values for the standard states of some representative substances at 298 K have been assembled in Table 18.3. These values can be used to calculate ΔS^0_{rx}, the standard-state entropy change of a chemical process. Similarly, ΔH^0_{rx} can be calculated by use of the data of Table 18.1. Finally, ΔS^0_{rx} and ΔH^0_{rx} may be used to calculate the standard-state free-energy change of the reaction using the relationship

$$\Delta G^0_{rx} = \Delta H^0_{rx} - T\Delta S^0_{rx} \tag{18.9}$$

TABLE 18.3 STANDARD-STATE ENTROPIES OF SOME COMMON ATOMS AND MOLECULES AT 298 K AND 1 ATM PRESSURE

ATOM OR MOLECULE	ENTROPY IN: cal·K^{-1}·mol^{-1}	J·K^{-1}·mol^{-1}
Br$_2$(g)	58.6	245
C(s) diamond	0.6	3
C(s) graphite	1.4	5.9
CH$_2$=CH$_2$(g)	52.4	219
CH$_3$CH$_3$(g)	54.9	230
CH$_4$(g)	44.5	186
C$_6$H$_6$(l)	40.3	169
C$_6$H$_6$(g)	64.3	269
CO(g)	47.2	197
CO$_2$(g)	51.1	214
Cl$_2$(g)	53.3	223
F$_2$(g)	48.6	203
H$_2$(g)	31.2	131
HBr(g)	47.4	198
HCl(g)	44.6	187
HF(g)	41.5	174
HI(g)	49.3	206
H$_2$O(l)	16.8	70.3
H$_2$O(g)	45.1	189
I$_2$(g)	62.3	261
Na(s)	12.3	51.5
NaCl(s)	17.3	72.4
N$_2$(g)	15.8	192
NH$_3$(g)	46.0	192
NO(g)	50.3	210
N$_2$O(g)	72.6	304
NO$_2$(g)	57.5	241
O$_2$(g)	49.0	205

---

* There is a **third law of thermodynamics** that states that a perfect crystal at absolute zero has perfect order and thus zero entropy. This law makes it possible to measure entropies on an *absolute* basis whereas most other thermodynamic quantities (E and H for example) can be measured only on a *relative* basis. There is no single unique starting point to use as a basis of assigning numerical values of E and H to substances in general.

EXAMPLE 18.10 Calculate the standard-state entropy change of the ammonia synthesis reaction given in Example 18.4.

SOLUTION The reaction is

$$N_2(g) + 3H_2(g) \rightleftharpoons 2NH_3(g)$$

The standard-state entropy of reaction is given by

$$\Delta S^0_{rx} = 2\bar{S}^0_{NH_3} - 2\bar{S}^0_{N_2} - 3\bar{S}^0_{H_2}$$

Given the molar entropies shown in Table 18.3:

$$\Delta S^0_{rx} = [2(192) - 192 - 3(131)] \text{ J} \cdot \text{K}^{-1} = -201 \text{ J} \cdot \text{K}^{-1}$$

The fact that ΔS^0_{rx} is negative implies that ΔS of the surroundings is $+201 \text{ J} \cdot \text{K}^{-1}$ or greater.

EXAMPLE 18.11 Use the results of Examples 18.4 and 18.10 to calculate ΔG^0_{rx} of the ammonia synthesis reaction.

SOLUTION $\Delta G^0_{rx} = \Delta H^0_{rx} - T\Delta S^0_{rx}$

From Example 18.4, $\Delta H^0_{rx} = -92.38$ kJ and from Example 18.10, $\Delta S^0_{rx} = -201 \text{ J} \cdot \text{K}^{-1}$. Since $T = 298$ K, ΔG^0_{rx} is calculated as follows:

$$\Delta G^0_{rx} = -92,380 \text{ J} - (298 \text{ K})(-201) \text{ J} \cdot \text{K}^{-1}$$
$$= -32,482 \text{ J} \quad \text{or} \quad -32.5 \text{ kJ}$$

Note that units of ΔH and ΔS must be compatible before combining to calculate ΔG. A common mistake is to use ΔH^0 in kJ and ΔS^0 in $\text{J} \cdot \text{K}^{-1}$.

The fact that ΔG^0_{rx} in Example 18.11 is negative means that the process, conversion of one mole of nitrogen in its standard state and three moles of hydrogen in its standard state to two moles of ammonia in its standard state, is intrinsically spontaneous at 298 K.

If it is assumed that ΔH^0_{rx} and ΔS^0_{rx} do not change with temperature (as is often approximately true), then it is clear that ΔG^0_{rx} of the ammonia reaction becomes more positive at higher temperatures, that is, the production of ammonia is favored by lower temperatures. This is the theoretical basis of Le Châtelier's principle, which was used earlier (Section 12.4) to make the same prediction.

Although Example 18.11 uses two tables of thermodynamic values to calculate ΔG^0_{rx}—one for ΔH^0 values and another for S^0 values—the calculation can be done on the basis of a single table of ΔG values for the formation of substances from the elements. Such tables, which are simply combinations of Tables 18.1 and 18.3, are found in handbooks and in some textbooks. Calculation of ΔG^0_{rx} from such tables is based on Hess's law just as are the calculations of ΔH^0_{rx} and ΔS^0_{rx}. For elements in their standard states, ΔG^0 of formation is taken to be zero, just like the corresponding ΔH^0 value.

EXAMPLE 18.12 The reaction $CH_4(g) \rightarrow C(s) + 2H_2(g)$ has $\Delta H^0_{rx} = 74.9$ kJ and $\Delta S^0_{rx} = 80.8$ J·K^{-1} at 298 K.

a) Is methane in its standard state stable with respect to decomposition to the elements in their standard states at 298 K?

b) If the answer to (a) is yes, at what temperature would methane become unstable with respect to decomposition to the elements (all products are reactants in their standard states)?

SOLUTION

a) $\Delta G^0_{rx} = \Delta H^0_{rx} - T\Delta S^0_{rx} = 74{,}900$ J $- (298$ K$)(80.8)$ J·K^{-1}

$$= 50{,}822 \text{ J}$$

Since $\Delta G^0_{rx} > 0$, decomposition of CH_4 to C and H_2 is not intrinsically spontaneous; therefore CH_4 is stable with respect to this decomposition.

b) Assuming that ΔH^0_{rx} and ΔS^0_{rx} do not change with temperature, ΔG^0_{rx} will become negative at all temperatures above that satisfied by

$$\Delta H^0_{rx} - T\Delta S^0_{rx} = 0$$

This gives

$$T = \frac{\Delta H^0_{rx}}{\Delta S^0_{rx}} = \frac{74{,}900 \text{ J}}{80.8 \text{ J·K}^{-1}} = 927 \text{ K}$$

Thus, above 927 K, methane in its standard state is intrinsically unstable with respect to decomposition to the elements in their standard states.

In Fig. 18.6 there is shown a graph of $-\Delta G/T$ as a function of T for the process

$$C_{\text{graphite}} \rightleftharpoons C_{\text{diamond}}$$

which was discussed earlier (Section 16.1). The data for this figure are based on experimental work and theoretical calculations by Frederick Rossini and co-workers at the U.S. Bureau of Standards. The shaded area represents the temperatures and pressures for which $\Delta G < 0$ and for which the synthesis of diamond from graphite

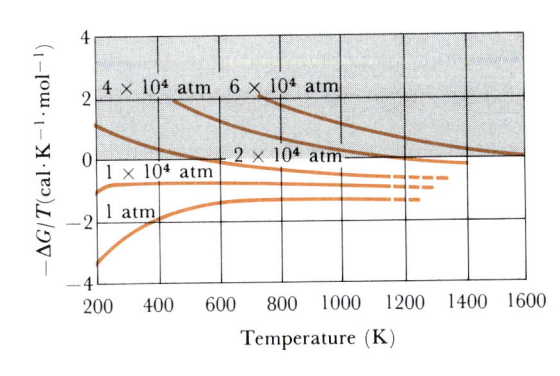

Fig. 18.6. Variation of $\Delta G/T$ with temperature at several pressures for the transformation of graphite into diamond. The shaded area represents combinations of pressure and temperature for which the conversion of graphite to diamond is thermodynamically feasible. (Redrawn from F. D. Rossini, *Chemical and Engineering News*, 5 April 1971; copyright © 1971 by the American Chemical Society. Used by permission of the copyright owner.)

should be intrinsically possible. It is seen that the process is favored by low temperatures and high pressures.

It is important to realize that the first and second laws of thermodynamics say nothing about how fast an intrinsically spontaneous process will proceed. Just because a reaction has a very negative ΔG value is no guarantee that the reaction will proceed fast enough to be of practical use; it may, in fact, proceed so slowly that to all appearances no reaction at all occurs! The factors which affect the rate at which a reaction occurs are discussed in Chapter 20.

18.10 FREE ENERGY AND EQUILIBRIUM

The quantity ΔG^0_{rx} refers to the intrinsic spontaneity of a very special situation: the transformation of reactants in their standard states to products in their standard states. Suppose a chemist is interested in transformations involving states other than standard states. It can be shown that nonstandard-state processes and standard-state processes have their free-energy changes related by the equation*

$$\Delta G = \Delta G^0 + 2.3\,RT \log Q_p \quad \text{or} \quad \Delta G^0 + RT \ln Q_p \tag{18.10}$$

where Q_p for the process $aA + bB \rightleftharpoons lL + mM$ is given by

$$Q_p = \frac{(p_L)^l (p_M)^m}{(p_A)^a (p_B)^b} \tag{18.11}$$

The partial pressures appearing in the Q_p expression are not necessarily equilibrium values. When $\Delta G = 0$, which occurs at equilibrium, Q_p becomes the equilibrium constant K_p given by

$$K_p = \frac{p_L^l p_M^m}{p_A^a p_B^b}$$

in which the partial pressures are now measured in the equilibrium mixture. This leads to the very important relationship

$$\Delta G^0 = -2.3\,RT \log K_p \quad \text{or} \quad -RT \ln K_p \tag{18.12}$$

For a given temperature, $\log K_p$ will be positive (which means $K_p > 1$) only if ΔG^0_{rx} is negative. When ΔG^0_{rx} is negative, the formation of products (in their standard states) from reactants (in their standard states) is intrinsically spontaneous. When $\log K_p$ is negative (which means $K_p < 1$), ΔG^0_{rx} is positive and the reaction is not intrinsically spontaneous for reactants and products in their standard states. Nevertheless, it may be possible to find nonstandard-state conditions such that $\Delta G_{rx} < 0$ even though $\Delta G^0_{rx} > 0$. Note that it is $\Delta G_{rx} = 0$ (*not* $\Delta G^0_{rx} = 0$) which characterizes equilibrium even though it is ΔG^0_{rx} which determines the numerical value of the equilibrium constant. The following example illustrates some of these important points.

---

* Students owning electronic calculators with a $\ln x$ key should use the second form. Any equation containing $2.3 \log x$ can be replaced by one containing $\ln x$ since $e^{2.3} = 10$, thus $2.3 \log x = \ln x$. For more accurate calculations, one should use the factor 2.303 rather than 2.3. The latter leads to small errors.

EXAMPLE 18.13 The reaction

$$N_2(g) + O_2(g) \rightleftharpoons 2NO(g)$$

has $\Delta G_{rx}^0 = +173$ kJ at 298 K.
a) Is this reaction intrinsically spontaneous for conversion of N_2 and O_2 in their standard states to NO in its standard state?
b) Under what general conditions will ΔG_{rx} be negative and what does this imply?
c) Calculate K_p of the reaction at 298 K.

SOLUTION
a) Since ΔG_{rx}^0 is positive, the conversion of N_2 and O_2 in their standard states to NO in its standard state is *not* intrinsically spontaneous.
b) Begin with Eq. (18.10)

$$\Delta G_{rx} = \Delta G_{rx}^0 + 2.3\, RT \log Q_p$$

and replace ΔG_{rx}^0 with $-2.3\, RT \log K_p$. This gives

$$\Delta G_{rx} = -2.3\, RT \log K_p + 2.3\, RT \log Q_p$$

This may be rearranged to

$$\Delta G_{rx} = 2.3\, RT \log \frac{Q_p}{K_p}$$

Since the logarithm of any number less than 1 is negative, ΔG_{rx} can be negative only if Q_p is made smaller than K_p. Write out the ratio Q_p/K_p in full to see what must be done to make $Q_p < K_p$:

$$\frac{Q_p}{K_p} = \frac{(p_{NO})^2/(p_{N_2})(p_{O_2})}{p_{NO}^2/p_{N_2}p_{O_2}}$$

Q_p can be made smaller by increasing the partial pressures of N_2 or O_2 or both. Or the partial pressure of NO could be decreased. In either case, it is possible to find nonstandard-state conditions such that the reaction is intrinsically spontaneous. There is no guarantee, however, that the required conditions are feasible to attain in practice.

Note that the same general results are obtained by applying Le Châtelier's principle to the balanced chemical equation.

c) It is convenient to first calculate $\log K_p$ and then to convert this to K_p:

$$\log K_p = \frac{-\Delta G_{rx}^0}{2.3\, RT} = \frac{-173{,}000\text{ J}}{(2.3)(8.314\text{ J}\cdot\text{K}^{-1})(298\text{ K})} = -30.4$$

Converting to the antilogarithm gives $K_p = 4.4 \times 10^{-31}$ at 298 K.
The same calculation in terms of natural logarithms is

$$\ln K_p = \frac{-\Delta G_{rx}^0}{RT}$$

$$= \frac{-173{,}000\text{ J}}{(8.314\text{ J}\cdot\text{K}^{-1})(298\text{ K})} = -69.8$$

Thus

$$K_p = e^{-69.8} = 4.7 \times 10^{-31}$$

The small discrepancy is due solely to the use of $e^{2.3} = 10$ rather than a more accurate relationship such as $e^{2.303} = 10$.

SUMMARY

Various changes which occur in nature are conveniently described in terms of accompanying changes in energy. Many years of observations indicate that nature obeys several laws based on energy changes; one of these deals with the conservation of energy and a second describes the tendency for energy to become less available for useful work in terms of a quantity, the entropy, which always increases. The intrinsic spontaneity of natural processes at constant temperature and pressure can be shown to depend on a composite of two tendencies: minimization of potential energy and maximization of entropy. These two tendencies permit the introduction of a new thermodynamic function called the Gibbs free energy. The change in the Gibbs free energy of a process in which all products and reactants are in their standard states provides a measure of the equilibrium constant of the process. Processes which are intrinsically spontaneous are characterized by a decrease in the Gibbs free energy; processes which are intrinsically nonspontaneous are characterized by an increase in the Gibbs free energy.

The free-energy change for a process involving reactants and products in their standard states is related to the free-energy change for a process in which reactants and products are in arbitrary states by

$$\Delta G = \Delta G^0 + RT \ln Q_p$$

At equilibrium (where $\Delta G = 0$), Q_p becomes equal to the equilibrium constant K_p and a very important relationship is obtained:

$$\Delta G^0 = -RT \ln K_p$$

This relationship permits calculation of equilibrium constants from free-energy changes or vice versa.

LEARNING GOALS

1. Know how energy is defined and how changes in energy are measured.

2. Be able to calculate the change in energy due to the heating of a substance from one temperature to another.

3. Know what an enthalpy change is and how it is measured for the heating of a substance from one temperature to another.

4. Know how energy changes are measured for chemical reactions; be able to use calorimetric data to calculate ΔE of a reaction.

5. Given the ΔE of a chemical reaction, be able to calculate the corresponding ΔH value.

6. Be able to calculate the ΔH of a reaction from the ΔH values of appropriate other reactions.

7. Know what is meant by the standard state of a substance.

8. Be able to use tables of ΔH^0 of formation values to calculate ΔH^0 values for chemical reactions.

9. Be able to use the Born-Haber cycle to calculate thermodynamic quantities such as electron affinities.

10. Know how bond energies are defined and how they may be calculated from thermodynamic data.

11. Know how to calculate the maximum possible conversion of heat to work using Carnot's equation.

12. Know the mathematical statement of the second law of thermodynamics and what it means.

13. Know how entropy is defined and how entropy changes of chemical reactions may be calculated from entropies of the reactants and products.

14. Know the equation $\Delta G = \Delta H - T\Delta S$ and be able to state to what situations it applies. Know the significance of zero, positive, and negative values of ΔG.

15. Be able to calculate ΔG^0 of a chemical reaction in various ways: from tables of ΔG^0 of formation and from tables of ΔH^0 of formation and S^0 values.

16. Be able to estimate the temperature above or below which a substance is unstable.

17. Know how ΔG is related to the equilibrium constant of a reaction. Be able to calculate K_p from ΔG^0 or vice versa.

18. Be able to name at least one scientific contribution made by each of the following: Mayer, Carnot, Born, Haber, Clausius, Helmholtz, and Gibbs.

DEFINITIONS, TERMS, AND CONCEPTS TO KNOW

energy
heat
molar heat capacity
calorimeter
isothermal process
enthalpy
standard state
standard-state enthalpy of formation (of a substance)
standard-state enthalpy of reaction
bond energy
first law of thermodynamics
law of conservation of energy
Hess's law
Born-Haber cycle
isolated system
second law of thermodynamics
efficiency of conversion of heat to work
entropy
Gibbs function (Gibbs free energy)
intrinsically spontaneous process

QUESTIONS AND PROBLEMS

Energy Changes in Physical Processes

1. Using data given in Example 18.1, calculate the enthalpy change of the transformation of 10 g of ice at $-5\,°C$ to steam at 110 °C. Assume the specific heat of steam is about $2.1\ \text{J}\cdot\text{K}^{-1}\cdot\text{g}^{-1}$. The heat of vaporization of water (an enthalpy term) is about $2260\ \text{J}\cdot\text{g}^{-1}$.

2. A substance absorbs 42 J of thermal energy from its surroundings.

a) If this energy causes the substance to shrink in volume, is E of the substance increased or decreased? Explain.

b) If the substance expands, what can you say about ΔE of the substance?

3. When a mass m falls a distance Δh, the work done is $mg\Delta h$, where g is the acceleration of gravity ($9.8\ \text{m}\cdot\text{s}^{-2}$). If $m = 1$ kg and $\Delta h = 1$ m, the work

done is 9.8 J. What is the minimum distance a 10-g weight would have to fall to provide enough energy to heat 1 L of water from 25 °C to 26 °C (say, by rotating a paddle wheel immersed in the water)?

4. Since the molecules of an ideal gas are assumed not to attract or repel each other, ideal gases have no potential energy. On that basis, what would you predict to be the value of ΔE (or ΔH) for the isothermal expansion or compression of a sample of ideal gas? Explain.

5. The sun radiates energy at the rate of about 1.4×10^2 W per square meter of the earth's surface. Assuming that reflective and other losses are about 75%, how long would it take to collect enough solar energy falling on a 1.0-m² collector panel to melt 1.0 L of ice at 0 °C and convert it to steam at 100 °C?

6. What happens in an energy sense when water freezes? Where does the energy go?

7. a) The first law of thermodynamics is often written in the form

$$\Delta E = q + w$$

where q is the heat involved in a process and w is the work involved. A positive value of q denotes that the system absorbs heat from the surroundings; a negative value denotes that heat is lost by the system to the surroundings. Similarly, a positive value of w means work is done by the surroundings on the system. When a sample of ideal gas is compressed or expanded isothermally, ΔE is found to be zero. Interpret this fact in terms of q and w.

b) When a liquid is vaporized at constant pressure and temperature, q is the heat of vaporization and w is the work done by expansion of liquid to vapor. This latter term is given by $p(V_1 - V_g)$, where p is the atmospheric pressure, V_1 is the volume of liquid, and V_g is the volume of the vapor. Calculate ΔE for the vaporization of one mole of water at 100 °C and 1 atm. Why is this not zero as in the expansion of an ideal gas?

8. What size light bulb uses up energy at the same rate as a man existing on a diet that provides 2500 kcal per day?

Calorimetric Calculations

9. A quantity of 1.8 g of glucose, $C_6H_{12}O_6$, is burned in a calorimeter which contains 750 cm³ of H₂O. The calorimeter temperature changes from 23.8 °C to 25.8 °C. $C_{bomb} + C_{leakage}$ is known to be 3100 J·K⁻¹. Calculate ΔE and ΔH of combustion for one mole of glucose at 25 °C.

10. When a sample of naphthalene, $C_{10}H_8$, weighing 0.258 g is burned in a calorimeter at 25 °C, the temperature change is 0.708 °C. The calorimeter contains 2500 cm³ of H₂O, and $C_{bomb} + C_{leakage}$ is 4184 J·K⁻¹.

a) Calculate ΔE and ΔH for the combustion of one mole of naphthalene at 25 °C.

b) Use appropriate data from Table 18.1 to calculate the standard enthalpy of formation of naphthalene at 25 °C.

11. Quantities of 50 cm³ of 1-N HCl and 50 cm³ of 1-N NaOH (both at 25 °C) are rapidly mixed. The final temperature after reaction is 31.8 °C. Assuming that the mixing container is a perfect insulator and that the heat capacity of dilute NaCl solution is the same as that of water, estimate the enthalpy of the reaction

$$H_3O^+ + OH^- \rightarrow 2H_2O(l)$$

12. Use data from Table 18.1 to calculate the enthalpy change at 25 °C for the reaction

$$3C(s) + 2O_3(g) \rightarrow 3CO_2(g)$$

13. Show how the enthalpy of combustion of methane, CH_4 (-890 kJ·mol⁻¹), can be used to calculate the enthalpy of formation of methane (see Example 18.6).

14. Some textbooks of inorganic chemistry state that a substance such as CaF (involving a monovalent group IIA metal) would be expected to be unstable since the process

$$2CaF(s) \rightarrow CaF_2(s) + Ca(s)$$

is exothermic. Use the Born-Haber cycle and the following data to estimate ΔH for the above reaction. [*Hint:* Calculate ΔH's for

$$Ca(s) + \tfrac{1}{2}F_2(g) \rightarrow CaF(s)$$

and

$$Ca(s) + F_2(g) \rightarrow CaF_2(s),$$

and combine.]

a) What one quantity plays the dominant role in the relative stability of CaF and CaF₂?

b) What must be true if it is possible to determine the stability of CaF on the basis of ΔH rather than ΔG?

REACTION	ΔH (kJ·mol^{-1})
Ca(s) → Ca(g)	167
Ca(g) → Ca$^+$(g) + e$^-$	586 (6.1 eV)
Ca(g) → Ca^{2+}(g) + 2e$^-$	1146 (11.9 eV)
$\frac{1}{2}$F$_2$(g) → F(g)	159 (1.65 eV)
F(g) + e$^-$ → F$^-$(g)	−337 (−3.5 eV)
CaF(s) → Ca$^+$(g) + F$^-$(g)	896 (assumed to be the same as for NaF; see Table 7.1)
CaF$_2$(s) → Ca^{2+}(g) + 2F$^-$(g)	4435

15. Use data given in Table 18.1 to calculate enthalpy changes for each of the following reactions at 298 K.

   a) $CH_3CH_3(g) \rightarrow CH_2{=}CH_2(g) + H_2(g)$
   b) $CO(g) + \frac{1}{2}O_2(g) \rightarrow CO_2(g)$
   c) $CH_4(g) + 2O_2(g) \rightarrow CO_2(g) + 2H_2O(l)$

16. The standard-state enthalpies of formation of SiH$_4$(g) and SiO$_2$(s) at 25 °C are 34 and −911 kJ·mol^{-1}, respectively. Calculate the enthalpy change of the reaction

$$SiH_4(g) + 2O_2(g) \rightarrow SiO_2(s) + 2H_2O(l)$$

at 25 °C.

17. The standard-state enthalpy change of the reaction

$$OF_2(g) + H_2O(g) \rightarrow 2HF(g) + O_2(g)$$

is −161.5 kJ at 25 °C. Calculate the standard-state enthalpy of formation of OF$_2$ at 25 °C.

18. The standard-state enthalpy of formation of Ca(OH)$_2$(s) at 25 °C is −896.6 kJ·mol^{-1}. Calculate ΔE^0 for the formation of Ca(OH)$_2$(s) from the elements.

19. Calculate the standard-state enthalpy change at 25 °C for the reaction

$$O_2(g) + O(g) \rightarrow O_3(g)$$

Why is a third body needed for this reaction?

20. The standard-state enthalpy of formation of O$_2$ is taken as zero. Why, then, are the standard state enthalpies of O and O$_3$ not zero?

21. The following reaction occurs when plaster of Paris is mixed with water.

$$CaSO_4 \cdot \tfrac{1}{2}H_2O(s) + 1\tfrac{1}{2}H_2O(l) \rightarrow CaSO_4 \cdot 2H_2O(s)$$

Given that a plaster cast newly placed on a broken arm or leg feels warm, what can you say about the relative values of the standard-state enthalpies of formation of CaSO$_4 \cdot \tfrac{1}{2}$H$_2$O(s) and CaSO$_4 \cdot 2$H$_2$O(s)?

22. The standard-state enthalpy of vaporization of liquid water at 25 °C is 43.9 kJ·mol^{-1}. Calculate ΔE^0 for this process.

23. Use the following data to calculate the electron affinity of bromine (in eV).

Enthalpy of formation of NaBr(s)	−360 kJ·mol^{-1}
Heat of vaporization of Br$_2$(l)	30.5 kJ·mol^{-1}
Dissociation energy of Br$_2$(g)	192 kJ·mol^{-1}
Lattice energy of NaBr(s)	734 kJ·mol^{-1}

Additional data are given in Example 18.8.

24. The enthalpy of combustion of propene is −2051 kJ·mol^{-1}. Calculate the standard-state enthalpy of formation of propene.

25. Given that the standard-state enthalpy of formation of Ca(OH)$_2$(s) is −896.6 kJ·mol^{-1} at 25 °C and that the standard state enthalpy change of the process

$$CaC_2(s) + 2H_2O(l) \rightarrow Ca(OH)_2(s) + HC{\equiv}CH(g)$$

is −125.5 kJ, calculate the standard-state enthalpy of formation of calcium carbide at 25 °C. You will also need data from Table 18.1.

Bond Energies

26. Using bond energies from Table 18.2, predict the enthalpy change of the process

$$CH_3CH{=}CH_2(g) \rightarrow 3C(g) + 6H(g)$$

27. Use bond energies from Table 18.2 to estimate the enthalpies of the following reactions (all at 25 °C).

   a) $H_2O(g) \rightarrow 2H(g) + O(g)$
   b) $HCOH(l) \rightarrow 2H(g) + 2O(g) + C(g)$
   ‖
   O

(The enthalpy of vaporization of formic acid is 41 kJ·mol^{-1}.)

28. The enthalpy of formation of hydrazine (H$_2$NNH$_2$) is 95.4 kJ·mol^{-1} at 25 °C. Use this value and data from Example 18.9 to estimate the enthalpy of formation of the amino radical, NH$_2$.

29. Estimate the standard-state enthalpy of formation of the methyl radical, CH$_3$, using data from Tables 18.1 and 18.2.

30. Use the bond energies given in Table 18.2 to

QUESTIONS AND PROBLEMS **521**

estimate the standard-state enthalpy of formation of propene at 25 °C. Compare with the calorimetric-derived value of 11 kJ·mol^{-1}.

31. Use the following quantities (plus those given in Tables 18.1 and 18.2) to estimate the enthalpy of sublimation of iodine.

H—I bond energy = 299 kJ·mol^{-1}

I—I bond energy = 151 kJ·mol^{-1}

[*Hint:* Express the reaction $H_2(g) + I_2(s) \rightarrow 2HI(g)$ in terms of H—H, I—I and H—I bond energies.]

Entropy

32. Predict whether you would expect $\Delta S > 0$ or $\Delta S < 0$ for each of the following processes.
    a) $NaCl(s) \rightarrow Na^+(g) + Cl^-(g)$
    b) $CH_2{=}CH_2(g) + H_2(g) \rightarrow CH_3CH_3(g)$
    c) Graphite → diamond
    d) $NaOH(aq) + HCl(aq) \rightarrow NaCl(aq) + H_2O$
    e) Crystalline sulfur → amorphous sulfur
    f) Water in a cloud → raindrops
    g) Liquid → solid
    h) Sperm + egg → embryo
    i) $CO_2 + H_2O \rightarrow CH_2O + O_2$
       (photosynthesis)
    j) $CH_2O + O_2 \rightarrow CO_2 + H_2O$
       (cell metabolism)

33. Calculate the standard-state entropy changes of the following reactions (all at 25 °C).
    a) $CO(g) + \frac{1}{2}O_2(g) \rightarrow CO_2(g)$
    b) $Cl_2(g) + H_2(g) \rightarrow 2HCl(g)$
    c) Graphite → diamond
    d) $H_2O(l) \rightarrow H_2O(g)$
    e) $CH_2{=}CH_2(g) + H_2(g) \rightarrow CH_3CH_3(g)$

34. Processes in which initial and final states are in equilibrium have $\Delta G = 0$. Thus when liquid water is vaporized at 100 °C and 1 atm, the entropy change for the process

$$H_2O(l) \rightleftharpoons H_2O(g)$$

is given by $\Delta H - T\Delta S = 0$ or $\Delta S = \Delta H/T$, where ΔH is the heat of vaporization of water (40.6 kJ·mol^{-1}). Compare the entropy change calculated in this way with that calculated from Table 18.3. Why is there a difference? Estimate the heat of vaporization of water at 25 °C.

35. Use the method suggested in Problem 34 to calculate the entropy change for the process

$$H_2O(s) \rightleftharpoons H_2O(l)$$

at 0 °C. The heat of fusion of water is 331 J·g^{-1}. Why is ΔS for vaporization larger than for fusion?

36. Predict the sign of ΔS (positive or negative) for each of the following.
    a) Drawing four aces in a row from a newly shuffled deck.
    b) Supersaturated solution → saturated solution.
    c) Freezing of water.
    d) A fire burning.
    e) Memorizing a set of rules.
    f) Forgetting how to calculate pH.
    g) The act of living.
    h) Straightening up your dorm room.
    i) Calculating ΔH_{rx} from Table 18.1.
    j) War.

Calculations and Questions Involving ΔG^0

37. The entropy change of the following reaction is -103 J·K^{-1} at 25 °C and the enthalpy change at the same temperature is -106 J:

$$H_2O_2(g) \rightarrow H_2O(g) + \tfrac{1}{2}O_2(g)$$

a) Use data from Table 18.3 to calculate the molar entropy of $H_2O_2(g)$ at 25 °C.
b) Calculate ΔG^0 of the reaction at 25 °C.
c) Would you expect $H_2O_2(g)$ to be highly stable at 25 °C? Explain.

38. Calculate ΔG^0 of formation of ethane at 298 K using the data of Tables 18.1 and 18.3.

39. A chemical reaction $A + B \rightleftharpoons C$ has $\Delta H = 42$ kJ and $\Delta S = 210$ J·K^{-1} at 300 K.
    a) What is ΔG of the reaction at 300 K?
    b) Is the reaction intrinsically spontaneous?
    c) Assuming ΔH and ΔS do not change with temperature, is there any temperature for which ΔG changes sign? What is the physical interpretation of this result?

40. Estimate the highest temperature at which H_2O in its standard state would be expected to be stable to decomposition to the elements in their standard states. What assumptions do you need to make in doing the calculation?

41. Use the data given in Tables 18.1 and 18.3 to calculate ΔG^0 and K_p for the formation of HI(g) from $H_2(g)$ and $I_2(g)$ at 298 K.

42. The standard-state enthalpy of formation of aqueous formic acid is 40 kJ·mol^{-1} at 25 °C; the molar entropy is 164 J·K^{-1}·mol^{-1}.

a) Calculate the enthalpy of the following reaction at 25 °C:

$$HCO_2H(aq) \rightarrow H_2O(l) + CO(g)$$

b) Calculate ΔS^0 of the above reaction.
c) Calculate ΔG^0 of the above reaction.

43. Consider an equilibrium process such as

solid or liquid $\rightleftharpoons$ vapor

for which $K_p = P_{vapor}$ and derive Eq. (9.1). Show that $A = \Delta H^0$ (vaporization or sublimation) and that $B = \Delta S^0/R$.

44. Explain how it is possible for a chemical reaction to be spontaneous even though its ΔG^0 value is positive.

45. Although ΔH and ΔS often vary but little as the temperature changes, this is definitely not true for ΔG. Explain why this is so.

46. Using data given in various tables in this chapter, calculate the following at 25 °C for the reaction

$$O_2(g) \rightarrow 2O(g)$$

a) ΔS^0
b) ΔG^0
c) K_p

At what temperature would K_p begin to exceed unity?

47. Calculate ΔG^0, ΔH^0, and ΔS^0 at 25 °C for the conversion of graphite to diamond. Discuss the results.

48. Calculate the equilibrium constant of the reaction

$$2SO_2(g) + O_2(g) \rightleftharpoons 2SO_3(g)$$

at 25 °C. The standard-state enthalpy of reaction is -198 kJ and the entropies of SO_2 and SO_3 at 25 °C are 248 and 256 $J \cdot K^{-1} \cdot mol^{-1}$, respectively.

49. The free energy of formation of ammonia in its standard state at 298 K is $-16{,}241$ $J \cdot mol^{-1}$. Calculate ΔG^0_{rx} for the reaction

$$N_2(g) + 3H_2(g) \rightarrow 2NH_3(g)$$

at 298 K. Compare with Example 18.11.

General

50. Keeping in mind the transformation of mass into energy (Eq. 11.6), what would you conclude about the strict validity of the laws of the conservation of mass and the conservation of energy? Suggest a combined mass–energy conservation law which appears to be more general in its application.

51. Nutritionists say that the average moderately active person requires an average daily food intake of about 2500 Calories. To melt a teaspoon (about 5 g) of ice consumes approximately 400 cal. Why is it not possible to design an effective reducing diet in which one eats as one wishes but sucks a few small ice cubes to offset the caloric intake? [*Hint:* What is meant by a "Calorie" as used by a nutritionist?]

52. If w_p, w_c, and w_f represent the weights (in grams) of protein, carbohydrate, and fat, respectively, in one ounce (28.4 g) of food, the number of kcal in an ounce of food is approximated by $4(w_g + w_c) + 9w_f$. Test this formula on cereals and other foods which list the needed data on the package.

ADDITIONAL READINGS

Mahan, Bruce H., *Elementary Chemical Thermodynamics.* New York: Benjamin, 1963. Contains extended discussions of all the thermodynamic subjects presented in this chapter and many more besides. Makes some use of the calculus.

Stevenson, Kenneth L., "Brief Introduction to the Three Laws of Thermodynamics," *J. Chem. Educ.* **52**, 330 (1975). A short nonmathematical summary of the laws of thermodynamics.

Hall, H. Tracy, "The Synthesis of Diamonds." *J. Chem. Educ.* **38**, 484 (1961). Discusses the theoretical and practical problems involved in the conversion of graphite to diamond.

Wilson, Mitchell, *Energy.* New York: Time-Life Books, 1963. Discusses calorimetry, entropy, conversion of energy, and many other topics found in Chapter 18, providing some historical perspective.

19
ELECTROCHEMISTRY

Electrochemistry at work on the Trans-Alaska pipeline. Twin ribbons of zinc are laid down along with the iron pipe to act as sacrificial anodes; the zinc corrodes rather than the iron. (Courtesy of Alyeska Pipeline Service Co.)

Galvanism I have found, by numerous experiments, to be a process purely chemical, *and to depend wholly on the oxydation of metallic surfaces, having different degrees of electric conducting power.*

SIR HUMPHRY DAVY, *in a letter to Davies Gilbert, October 1800.**

In 1786 Luigi Galvani, a professor of anatomy and gynecology at the University of Bologna, found that when the leg of a dissected frog was touched to two different metals such as zinc and silver, muscular contraction occurred (see Fig. 19.1). In 1799 the Italian physicist Alessandro Volta invented the first battery by separating discs of two different metals with sheets of paper or cloth soaked in acid or salt water. The phenomenon of producing electric current in this way was called **galvanism**; the device of Volta, when constructed as several alternating discs of the two different metals, was called a **voltaic pile** (see Fig. 19.2). It required many years of study by the best scientific minds of the time—and of succeeding generations—to firmly establish that the electric current arose from a chemical reaction between the two metals.

Today, there is a branch of physical chemistry known as **electrochemistry** which deals with the interconversion of chemical energy and electrical energy. This

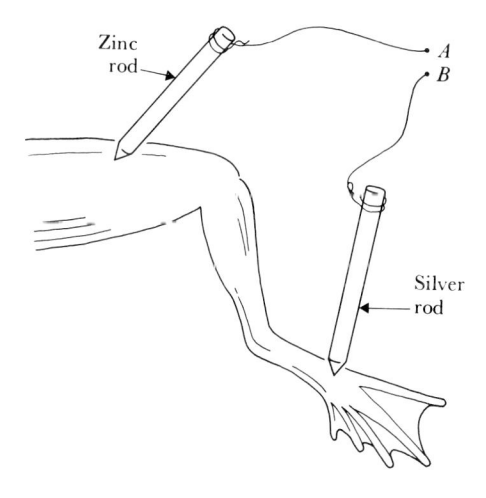

Fig. 19.1. Illustration of Galvani's frog leg experiment. The zinc rod touches a muscle and the silver rod touches a nerve. When the wire ends *A* and *B* are brought into contact, the frog leg contracts.

---

* Sir Humphry abandoned this idea in 1806 in favor of another theory. He was right the first time.

Metal A   Paper soaked   Metal B
in saline
solution

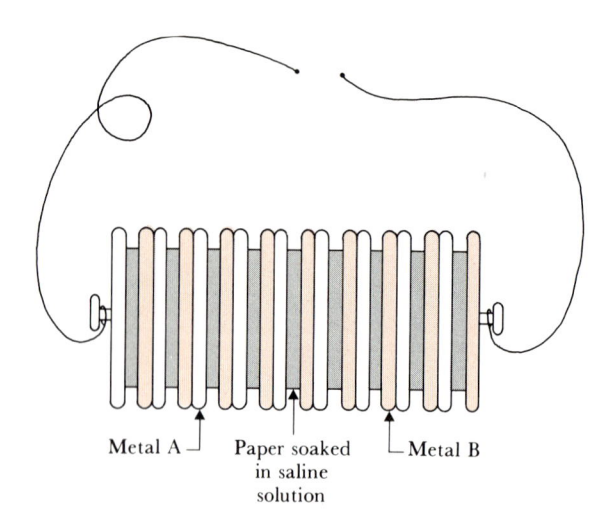

Fig. 19.2. A voltaic pile. A and B are discs of different metals; each pair is separated by paper discs which have been soaked in salt solution. Connecting the two ends of the pile with a wire causes a steady electric current to flow through the wire. A pile made up of about two dozen plates is capable of imparting an unpleasantly strong shock to anyone holding the ends of the wires with the hands. By using one of his piles as a source of current, Volta became the first to decompose water by electrolysis.

interconversion—which can be carried out in either direction—involves oxidation–reduction reactions. All oxidation–reduction reactions involve the transfer of electrons from one reactant to another, or a shift of electron density from one nucleus to another. An electric current results when the electron transfer or shift is forced to take place via an external pathway rather than directly between reactants. This flow of electrons by means of an indirect route constitutes an electric current. Just as a water wheel may be used to harness the flow of water over a falls or dam, so can the flow of electric current be harnessed by various devices to produce useful work.

19.1 VOLTAIC CELLS

The device in which chemical energy is made to produce an electric current is called a **voltaic cell**. Figure 19.3 shows a voltaic cell that uses the reaction

$$Zn(s) + Cu^{2+}(aq) \rightleftharpoons Zn^{2+}(aq) + Cu(s)$$

The electrode at which oxidation takes place is called the **anode**. The anode reaction is

$$Zn(s) \rightleftharpoons Zn^{2+}(aq) + 2e^-$$

Humphry Davy (1778–1829), an English scientist, was noted for his discovery of the elements sodium and potassium, the invention of a miner's safety lamp, establishment of the foundations of electrochemistry, and the "discovery" of Michael Faraday, who became his assistant and, later, successor at the Royal Institution.

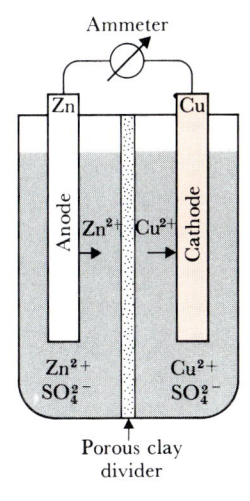

Fig. 19.3. A Daniell-type voltaic cell. The porous clay divider permits SO_4^{2-} to pass into the left-hand compartment so that charge neutrality is maintained throughout the cell, but direct migration of Cu^{2+} to Zn is minimized.

The electrode at which reduction takes place is called the **cathode**.* The cathode reaction is

$$Cu^{2+}(aq) + 2e^- \rightleftharpoons Cu(s)$$

Since the Zn(s) and Cu^{2+}(aq) are not in direct contact, the electron transfer from Zn(s) to Cu^{2+}(aq) can take place only through a wire connecting the zinc anode and the copper cathode. If an electrical device (for example, a small flashlight bulb) is placed into the external circuit, the electron flow can be made to produce work, that is, to light the bulb. The barrier between the $ZnSO_4$ and $CuSO_4$ solutions allows SO_4^{2-}(aq) to pass through to the $ZnSO_4$ solution so that electrical neutrality is maintained. This is necessary because as Zn^{2+}(aq) increases in the anode compartment, Cu^{2+}(aq) decreases in the cathode compartment. For each Cu^{2+}(aq) depleted, in the cathode compartment, an SO_4^{2-}(aq) ion flows through the barrier to the anode compartment.

The fact that the above reaction goes in the direction indicated—and not in the opposite direction—is easily verified experimentally. One way is to show that the zinc strip making up the anode decreases in mass as the cell operates. At the same time the copper cathode gains weight.

A voltaic cell provides a very direct way of measuring the change in the Gibbs free energy accompanying a chemical reaction. The Gibbs free-energy change is given by

$$\Delta G = -nF\mathscr{E} \tag{19.1}$$

where n is the moles of electrons transferred, F is the charge on a mole of electrons, and $\mathscr{E}$ is the **electromotive force** (abbreviated **emf**) of the voltaic cell.† The quantity F, which has the value of about 96,500 $C \cdot mol^{-1}$, is called the **faraday** or,

---

* The fact that *oxidation* occurs at the *anode* and *reduction* at the *cathode* is most easily remembered by noting that the letters *o* and *r* are in alphabetical order as are *a* and *c*.

† The cell voltage must be measured for the cell while it is operating perfectly reversibly. The cell is placed in an electrical circuit and the flow of current is reduced to zero by opposing it with another source of current of known voltage.

sometimes, the **Faraday constant**.* The emf of a voltaic cell is a measure of the difference in electrical potential between anode and cathode and is usually expressed in volts. The cell emf is somewhat analogous to the pressure gradient or hydrostatic head which determines the rate of flow of water through a pipe. Since ΔG and $\mathscr{E}$ have opposite signs, note that a voltaic cell which produces electric current spontaneously (that is, ΔG of the cell reaction is negative) will have a positive value of $\mathscr{E}$.

In the special case that all the substances participating in the cell reaction are in their standard states, Eq. (19.1) becomes

$$\Delta G^0 = -nF\mathscr{E}^0 \tag{19.2}$$

where $\mathscr{E}^0$ is called the **standard-state emf** of the cell. For ordinary calculations the standard states of ionic substances such as $Zn^{2+}(aq)$ and $Cu^{2+}(aq)$ are usually taken as a concentration of $1.0\ M$.

When Eqs. (19.1) and (19.2) are substituted into Eq. (18.10), there is obtained the general voltaic cell relationship

$$\mathscr{E} = \mathscr{E}^0 - \frac{2.3\ RT}{nF} \log Q_c \tag{19.3}$$

where Q_c corresponds to K_c as Q_p does to K_p. This relationship is known as the **Nernst equation**. At 25 °C, the quantity $2.3\ RT/F$ is equal to $0.059\ \text{V·mol}$ and the Nernst equation assumes the specific form in which it is most often used:

$$\mathscr{E} = \mathscr{E}^0 - \frac{0.059}{n} \log Q_c \tag{19.4}$$

It should be noted that because of Eqs. (19.2) and (18.12), the standard-state cell emf is related to the equilibrium constant of the cell reaction as follows:

$$\mathscr{E}^0 = \frac{2.3\ RT}{nF} \log K_c \tag{19.5}$$

19.2 STANDARD ELECTRODE POTENTIALS

It is convenient to regard cell emfs as consisting of two parts: a contribution from the anode and a contribution from the cathode. This is written as follows:

$$\begin{aligned}\mathscr{E} &= \mathscr{E}_{\text{anode}} + \mathscr{E}_{\text{cathode}} \\ \mathscr{E}^0 &= \mathscr{E}^0_{\text{anode}} + \mathscr{E}^0_{\text{cathode}}\end{aligned} \tag{19.6}$$

However, since it is impossible to carry out only the oxidation part or only the reduction part of a reaction, it is not possible to measure the anode and cathode contributions to the cell emf separately; measurement of the voltage of a voltaic cell

---

* Named after the English scientist Michael Faraday. For a biographical sketch of Faraday see Chapter 1.

Hermann Walther Nernst, German physical chemist who established much of the theoretical foundations of electrochemistry. (Courtesy of Culver Pictures.)

The Nernst equation is named after Hermann Walther Nernst (1864–1941), a German physical chemist noted for pioneering work in thermodynamics and electrochemistry. Nernst also developed the idea of the solubility product, developed methods of measuring pH with indicators, and suggested the use of buffer solutions. Nernst received the Nobel prize in chemistry in 1920. He also acquired some fame in certain circles by suggesting that the unit of beer drinkers' throughput (in litres of urine excreted per hour per litre of beer consumed) be called the **falstaff**. *History does not record Herr Doktor Professor Nernst's personal falstaffian index.*

gives only the sum of the two components. Nevertheless, it is possible to assign *relative* emf values to separate electrode reactions by using a standard reference electrode as a basis.

As a specific example of how this is done, consider the Zn/Cu^{2+} cell discussed previously in which the half-reaction

$$H_2(g) \rightleftharpoons 2H^+(aq) + 2e^-$$

is chosen as a reference electrode. Under standard-state conditions (H$_2$ pressure at 1 atm and [H$^+$] = 1M), $\mathscr{E}^0$ for this electrode is defined as exactly 0.00... volts at *all temperatures*. A voltaic cell (see Fig. 19.4) is now set up in which the anode reaction is Zn(s) $\rightleftharpoons$ Zn^{2+}(aq) + 2e$^-$ and the cathode reaction is 2H$^+$(aq) + 2e$^-$ $\rightleftharpoons$ H$_2$(g). The cell reaction is

$$Zn(s) + 2H^+(aq) \rightleftharpoons Zn^{2+}(aq) + H_2(g)$$

By use of an instrument called a galvanometer (actually a sensitive voltmeter) the cell

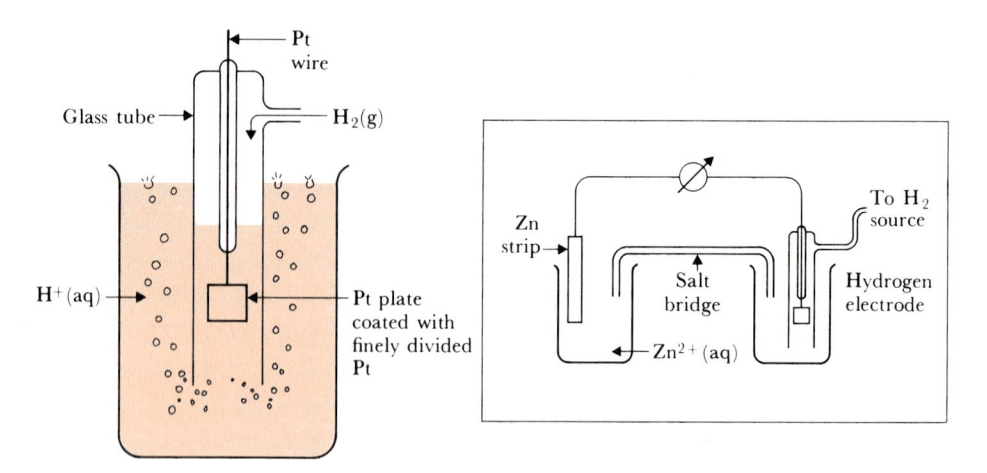

Fig. 19.4. Diagram of a hydrogen electrode. The inset shows a voltaic cell constructed from a hydrogen electrode and a zinc electrode. The salt bridge (a tube of gel containing KCl) allows electrical charges to move across the electrode boundary but prevents Zn and H⁺ from reacting directly.

emf at standard conditions is found to be 0.76 V at 25 °C. Since

$$\mathscr{E}^0_{cell} = \mathscr{E}^0_{Zn/Zn^{2+}} + \mathscr{E}^0_{H^+/H_2} = 0.76 \text{ V}$$

and since $\mathscr{E}^0_{H^+/H_2} = 0$ by convention, it follows that the value of $\mathscr{E}^0_{Zn/Zn^{2+}}$ *relative to the hydrogen reference electrode* is given by

$$\mathscr{E}^0_{Zn/Zn^{2+}} = 0.76 \text{ V}$$

The subscripts of electrode potentials indicate whether the electrode is functioning as an anode or cathode. To indicate the potential of the zinc electrode as a cathode, the order of the Zn and Zn²⁺ is reversed in the subscript and the sign of the potential changed. For example,

$$\mathscr{E}^0_{Zn^{2+}/Zn} = -0.76 \text{ V}$$

This refers to the reduction reaction $Zn^{2+}(aq) + 2e^- \rightleftharpoons Zn(s)$.
Similarly, a cell is set up with the chemical reaction

$$H_2(g) + Cu^{2+}(aq) \rightleftharpoons 2H^+(aq) + Cu(s)$$

in which the hydrogen electrode is now an anode and the copper electrode is the cathode. A galvanometer indicates that $\mathscr{E}^0_{cell}$ for this reaction is 0.34 V at 25 °C. Thus

$$\mathscr{E}^0_{Cu^{2+}/Cu} = 0.34 \text{ V} \quad \text{or} \quad \mathscr{E}^0_{Cu/Cu^{2+}} = -0.34 \text{ V}$$

The two relative values of the electrode potentials are now used to calculate an absolute value of the cell emf for the zinc–copper reaction. There is obtained

$$\mathscr{E}^0_{cell} = \mathscr{E}^0_{Zn/Zn^{2+}} + \mathscr{E}^0_{Cu^{2+}/Cu} = 0.76 + 0.34 = 1.10 \text{ V} \quad \text{(at 25 °C)}$$

This is precisely the value obtained by direct measurement.

It should be noted that the potential of an electrode reaction (and, consequently, the emf of a cell) does not depend on the actual number of moles of substances involved in the reaction. This is because voltage is a measure of potential *difference*. In just the same way, the gauge (width) of a railroad track is 4 ft. $8\frac{1}{2}$ in. regardless of how long the track is. Thus, electrode reactions such as $Zn(s) \rightarrow Zn^{2+}(aq) + 2e^-$ and $2Zn(s) \rightarrow 2Zn^{2+} + 4e^-$ have the same $\mathscr{E}^0$ value, namely, 0.76 V. However, the *total energy* produced by a cell does depend on the amounts of reactants. This dependence comes in via the value of n in Eq. (19.1).

By continuing this process for a variety of electrodes, extensive tables of **standard electrode potentials** are built up. Such data are very valuable in the calculation of ΔG, ΔG^0, and equilibrium constants. Values of some selected standard electrode potentials are given in Table 19.1. Care must be taken in using this table to see that the signs of the potentials correspond to use of the electrode as an anode or as a cathode. All of the potentials given in Table 19.1 are for use of the electrode as an anode; for use in cathode reactions all signs must be reversed.

The electrode potentials of Table 19.1 provide the theoretical basis of the activity series of the metals shown in Table 15.1. Any metal whose potential as an oxidation reaction is positive (compared with $\mathscr{E}^0_{H^+/H_2} = 0.000\ldots$) is able to react with acids or water to produce H_2; that is, all such metals are more easily oxidized than hydrogen.

The following examples illustrate some typical applications of standard electrode potentials.

TABLE 19.1 VALUES OF SELECTED STANDARD-STATE ELECTRODE POTENTIALS AT 25 °C[a]

ELECTRODE REACTION	$\mathscr{E}^0_{electrode}$ (V)
$Ag(s) \rightleftarrows Ag^+(aq) + e^-$	−0.7996
$Ag(s) + Cl^-(aq) \rightleftarrows AgCl(s) + e^-$	−0.2223
$Al(s) \rightleftarrows Al^{3+}(aq) + 3e^-$	1.706
$2Br^-(aq) \rightleftarrows Br_2(l) + 2e^-$	−1.065
$2Cl^-(aq) \rightleftarrows Cl_2(g) + 2e^-$	−1.3583
$Cu(s) \rightleftarrows Cu^{2+}(aq) + 2e^-$	−0.3402
$2F^-(aq) \rightleftarrows F_2(g) + 2e^-$	−2.85
$Fe(s) \rightleftarrows Fe^{2+}(aq) + 2e^-$	0.409
$Fe(s) \rightleftarrows Fe^{3+}(aq) + 3e^-$	0.036
$Fe^{2+}(aq) \rightleftarrows Fe^{3+} + e^-$	−0.770
$H_2(g) \rightleftarrows 2H^+(aq) + 2e^-$	0.0000
$2I^-(aq) \rightleftarrows I_2(s) + 2e^-$	−0.535
$Zn(s) \rightleftarrows Zn^{2+}(aq) + 2e^-$	0.7628
$2Hg(l) + 2Cl^-(aq) \rightleftarrows Hg_2Cl_2(s) + 2e^-$	−0.2682
$Hg(l) \rightleftarrows \frac{1}{2}Hg_2^{2+}(aq) + e^-$	−0.7986
$Pb(s) + SO_4^{2-}(aq) \rightleftarrows PbSO_4(s) + 2e^-$	0.36
$Pb(s) \rightleftarrows Pb^{2+}(aq) + 2e^-$	0.13
$Mg(s) \rightleftarrows Mg^{2+}(aq) + 2e^-$	2.36
$Co(s) \rightleftarrows Co^{2+}(aq) + 2e^-$	0.28
$Ni(s) \rightleftarrows Ni^{2+}(aq) + 2e^-$	0.25
$Sn(s) \rightleftarrows Sn^{2+}(aq) + 2e^-$	0.14

[a] Values taken from the extensive tables found in the 51st edition of the *Handbook of Chemistry and Physics* (Cleveland, Ohio: Chemical Rubber Co., 1970–1971).

EXAMPLE 19.1 For the special case of all products and reactants in their standard states, determine whether the reaction below is intrinsically spontaneous:

$$2Ag(s) + Pb^{2+}(aq) \rightleftharpoons 2Ag^{+}(aq) + Pb(s)$$

That is, is it possible to set up a voltaic cell in which the anode reaction is $Ag(s) \rightleftharpoons Ag^{+}(aq) + e^{-}$ and the cathode reaction is $Pb^{2+}(aq) + 2e^{-} \rightleftharpoons Pb(s)$ so that the cell reaction is $2Ag(s) + Pb^{2+}(aq) \rightleftharpoons 2Ag^{+}(aq) + Pb(s)$?

SOLUTION The standard cell emf is given by

$$\mathscr{E}^0_{cell} = \mathscr{E}^0_{Ag/Ag^+} + \mathscr{E}^0_{Pb^{2+}/Pb} = (-0.7996 - 0.13)V = -0.93 \text{ V}$$

Since $\mathscr{E}^0_{cell}$ is negative, the indicated cell reaction is not intrinsically spontaneous. The voltaic cell constructed from the two electrodes would operate but with $Pb(s) \rightleftharpoons Pb^{2+}(aq) + 2e^{-}$ as the anode reaction and $Ag^{+}(aq) + e^{-} \rightleftharpoons Ag(s)$ as the cathode reaction. The cell reaction would be

$$2Ag^{+}(aq) + Pb(s) \rightleftharpoons 2Ag(s) + Pb^{2+}(aq)$$

The standard cell emf would be $+0.93$ V.

EXAMPLE 19.2 Calculate the cell emf at 25 °C of the voltaic cell based on the reaction

$$2Fe(s) + 3Cu^{2+}(aq) \rightleftharpoons 2Fe^{3+}(aq) + 3Cu(s)$$

when $[Fe^{3+}] = 3.0 M$ and $[Cu^{2+}] = 1.5 M$.

SOLUTION Application of Eq. (19.4) to the cell leads to

$$\mathscr{E}_{cell} = \mathscr{E}^0_{cell} - \frac{0.059}{6} \log \frac{[Fe^{3+}]^2}{[Cu^{2+}]^3}$$

The electrode reactions and standard electrode potentials are:

anode: $2Fe(s) \rightleftharpoons 2Fe^{3+}(aq) + 6e^{-}$; $\quad \mathscr{E}^0_{Fe/Fe^{3+}} = 0.036$ V

cathode: $3Cu^{2+}(aq) + 6e^{-} \rightleftharpoons 3Cu(s)$; $\quad \mathscr{E}^0_{Cu^{2+}/Cu} = 0.3402$ V

Thus $\mathscr{E}^0_{cell}$ is $0.036 + 0.3402$ or 0.376 V, and

$$\mathscr{E}_{cell} = 0.376 - \frac{0.059}{6} \log \frac{(3.0)^2}{(1.5)^3} = 0.372 \text{ V}$$

Since $\mathscr{E}^0_{cell}$ is positive the voltaic cell would operate in the direction indicated. A crude verification of the spontaneity of this reaction may be made by dipping a clean steel knife blade into a solution of a Cu(II) salt; a copper coating forms almost instantly on the blade.

EXAMPLE 19.3 Calculate the solubility-product constant of silver chloride, AgCl, at 25 °C.

SOLUTION The solubility-product constant of AgCl refers to the reaction

$$AgCl(s) \rightleftharpoons Ag^+(aq) + Cl^-(aq)$$

If $\mathcal{E}^0_{rx}$ for the above is known, K_{sp} may be calculated from

$$\mathcal{E}^0_{rx} = \frac{2.3\,RT}{nF} \log K_{sp}$$

Examination of the electrode reactions of Table 19.1 shows that a combination of the following two electrodes leads to the desired reaction:

$$AgCl(s) + e^- \rightleftharpoons Ag(s) + Cl^-(aq)$$

$$Ag(s) \rightleftharpoons Ag^+(aq) + e^-$$

Thus $\mathcal{E}^0_{rx} = \mathcal{E}^0_{AgCl/Ag} + \mathcal{E}^0_{Ag/Ag^+} = (0.2223 - 0.7996)V = -0.5773$ V

and

$$\log K_{sp} = \frac{nF}{2.3\,RT}\,\mathcal{E}^0_{rx} = \frac{-0.5773}{0.059} = -9.8$$

Thus

$$K_{sp} = 10^{-9.8} \quad \text{or} \quad 1.6 \times 10^{-10}$$

This is one of the most accurate methods of calculating equilibrium constants.

19.3 ELECTROCHEMICAL MEASUREMENT OF pH

A convenient method of measuring pH very rapidly is based on an electrochemical cell in which one of the electrodes depends on hydrogen ion concentration. Such electrochemical cells are called **pH meters**; a typical pH meter may use the electrode reactions:

anode: $\frac{1}{2}H_2(g) \rightleftharpoons H^+(aq) + e^-$

cathode: $\frac{1}{2}Hg_2Cl_2(s) + e^- \rightleftharpoons Hg(l) + Cl^-(aq)$

cell reaction: $\frac{1}{2}H_2(g) + \frac{1}{2}Hg_2Cl_2(s) \rightleftharpoons H^+(aq) + Cl^-(aq) + Hg(l)$

If the $[Cl^-(aq)]$ and p_{H_2} are kept fixed, the emf of the cell may be written

$$\mathcal{E}_{cell} = \mathcal{E}' - 0.059 \log[H^+(aq)] \tag{19.7}$$

where $\mathcal{E}'$ contains the $\mathcal{E}^0$'s of the two electrodes plus contributions from $[Cl^-(aq)]$ and p_{H_2}. Since the pH is given as $-\log[H^+(aq)]$, the above equation becomes

$$\mathcal{E}_{cell} = \mathcal{E}' + 0.059\,\text{pH} \tag{19.8}$$

Solving for pH leads to

$$\text{pH} = \frac{\mathcal{E}_{cell} - \mathcal{E}'}{0.059} \tag{19.9}$$

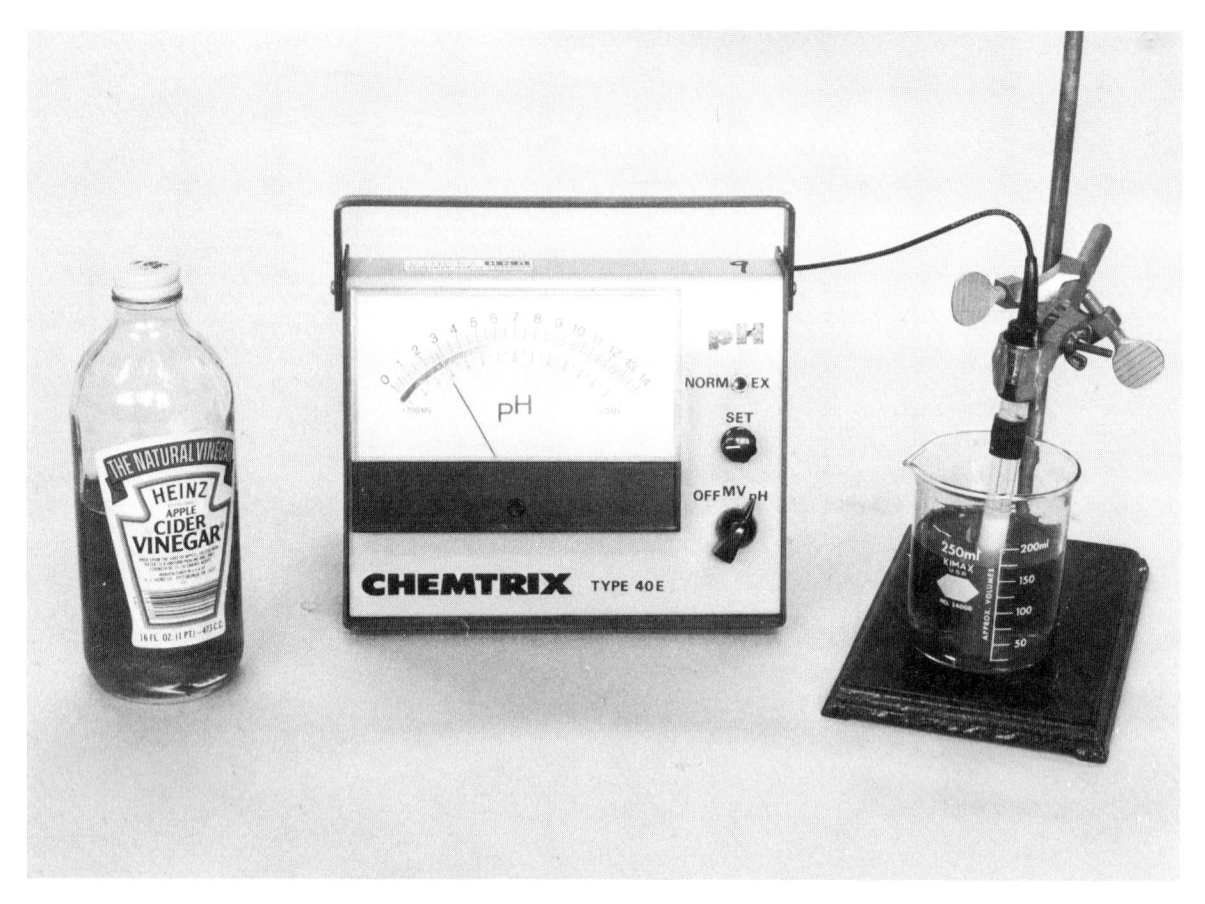

Fig. 19.5. A portable, battery-operated pH meter. The glass electrode of the meter has been immersed in a sample of household vinegar (approximately 5% acetic acid). The dial indicates a pH of about 3. (Photo by Charles W. Owens.)

The pH meter is calibrated as follows: The anode is made part of a solution whose pH is known exactly (thus [H$^+$(aq)] is known exactly). The voltmeter reading, $\mathscr{E}_{cell}$, is then labeled with the known pH. Repeating with several other known pH's produces a voltmeter scale that reads directly in pH. Once this calibration is completed, the anode can be used as part of a solution of unknown [H$^+$(aq)]; the corresponding pH is then read directly from the voltmeter scale.

It is not practical to use the hydrogen electrode as the H$^+$(aq)-sensitive electrode. Commercially available instruments use alternative H$^+$(aq)-sensitive electrodes that are simpler to set up and operate. Many different pH meters are now available on the market. Several are portable and may be used by scientists who wish to measure the pH of their favorite trout-fishing lake located at 9000 ft in the Sierra Nevada mountains, or some equally remote area. Figure 19.5 shows one type of commercially available pH meter.

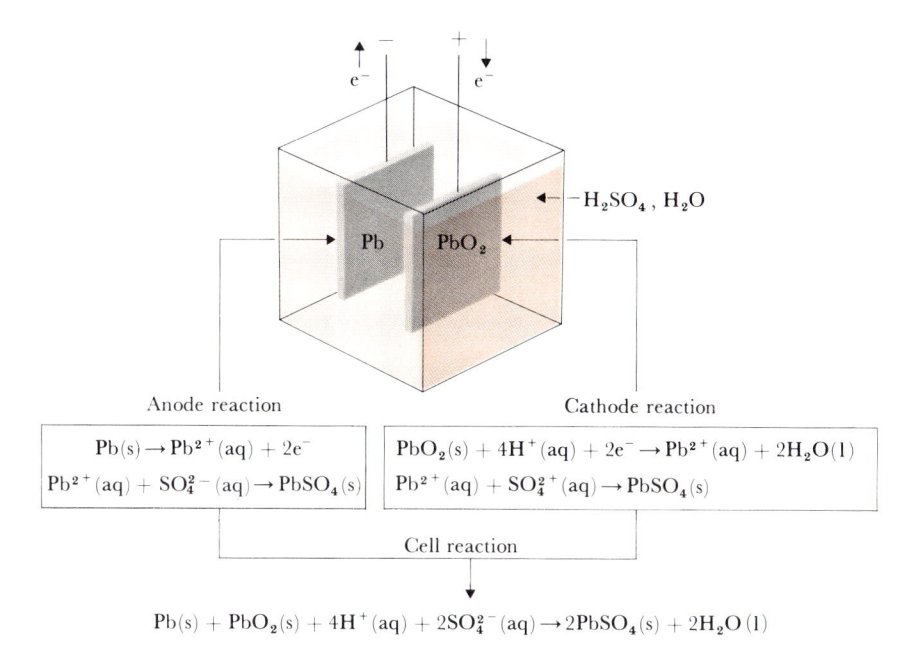

Fig. 19.6. Schematic diagram of one cell of a lead storage battery. Each such cell has a voltage of about 2.1 V. A 12-V car battery consists of six such cells connected in series. The electrode reactions and direction of electron flow shown are for spontaneous operation of the battery; in charging the reactions and direction of flow are reversed. The PbSO$_4$ formed in the reaction adheres to the plates.

19.4 STORAGE BATTERIES

Figure 19.6 is a schematic illustration of a lead storage battery,* the type used in most automobiles. The anode reactions are

$$Pb(s) \rightarrow Pb^{2+}(aq) + 2e^-$$

$$Pb^{2+}(aq) + SO_4^{2-}(aq) \rightarrow PbSO_4(s)$$

The cathode reactions are

$$PbO_2(s) + 4H^+(aq) + 2e^- \rightarrow Pb^{2+}(aq) + 2H_2O(l)$$

$$Pb^{2+}(aq) + SO_4^{2-}(aq) \rightarrow PbSO_4(s)$$

Adding all the anode and cathode reactions leads to the total cell reaction:

$$Pb(s) + PbO_2(s) + 4H^+(aq) + 2SO_4^{2-}(aq) \rightarrow 2PbSO_4(s) + 2H_2O(l)$$

---

* The term **battery** literally means two or more voltaic cells connected to form a single source of electrical power. However, in popular usage, even single voltaic cells (such as flashlight cells) are called batteries.

A 12-V automobile battery consists of six voltaic cells utilizing the above reactions. Each voltaic cell has a potential of about 2.1 V and when all six are connected in series the combined potential is over 12 V.

The battery is recharged by passing an electric current through it in the reverse direction; this reverses the cell reaction above and charges the battery. The reverse electric current is provided by the automobile's generator or alternator. Thus the automobile uses chemical energy to produce mechanical energy which, in turn, produces electrical energy to produce chemical energy. Service stations test the extent to which a car battery is charged by using a device called a **hydrometer**. A hydrometer measures the density of the battery solution and thus indicates its content of H_2SO_4. A battery has its maximum H_2SO_4 content when it is fully charged.

The batteries used in flashlights, portable radios, and similar devices are known as **dry cells**. A typical dry cell has a zinc outer cover as an anode and an inert carbon rod in the center connected to the cathode. The cell is not really dry since the electrodes are separated by paste which does contain some water. The anode reaction of the cell diagramed in Fig. 19.7 is

$$Zn + 2NH_4^+ + 2Cl^- + 2OH^- \rightarrow Zn(NH_3)_2Cl_2 + 2H_2O + 2e^-$$

The cathode reaction is

$$2MnO_2 + 2H_2O + 2e^- \rightarrow 2MnO(OH) + 2OH^-$$

Basically, Zn is oxidized to Zn(II) and Mn(IV) is reduced to Mn(III). Such a cell has a voltage of about 1.5 V.

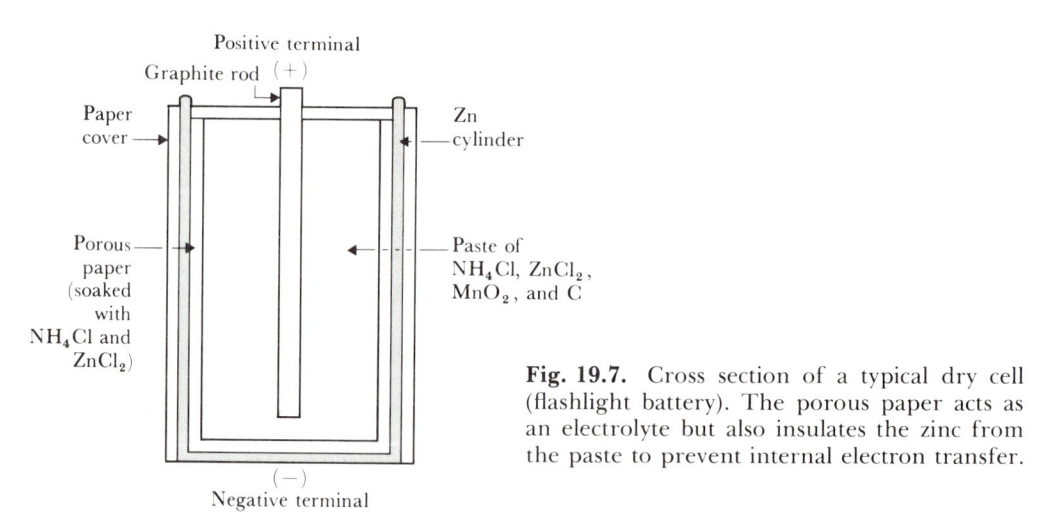

Fig. 19.7. Cross section of a typical dry cell (flashlight battery). The porous paper acts as an electrolyte but also insulates the zinc from the paste to prevent internal electron transfer.

19.5 FUEL CELLS

Of great interest to scientists and engineers is the development of a device which can produce a continuous electric current directly from the combustion of a substance normally used as a fuel. Such devices, termed **fuel cells**, have been used on a small scale (the H_2/O_2 fuel cells of the Apollo space missions are an example) but large-scale development has proven extremely difficult. The basic fuel cell reaction is

$$\text{fuel} + O_2 \rightarrow \text{oxidation products}$$

but, as in any voltaic cell, electrons must be made to flow from the fuel to oxygen via an indirect route involving an external circuit. In one type of fuel cell the fuel produces positive ions at a porous anode and electrons flow through the external circuit to a porous cathode where they are picked up by O_2. For example, if the combustion reaction is

$$CH_4(g) + 2O_2(g) \rightarrow CO_2(g) + H_2O(l)$$

the anode reaction is

$$CH_4(g) + 2H_2O(l) \rightarrow CO_2(g) + 8H^+ + 8e^-$$

and the cathode reaction is

$$2O_2(g) + 8e^- + 8H^+ \rightarrow 4H_2O(l)$$

The H^+ migrates to the cathode through the electrolyte. This fuel cell is diagramed in Fig. 19.8.

In the H_2/O_2 fuel cell, such as that used in Apollo Mission spacecraft, the cathode reaction is

$$\tfrac{1}{2}O_2(g) + H_2O(l) + 2e^- \rightarrow 2OH^-$$

The OH^- migrates through the electrolyte to the anode, where the reaction is

$$H_2(g) + 2OH^- \rightarrow 2H_2O(l) + 2e^-$$

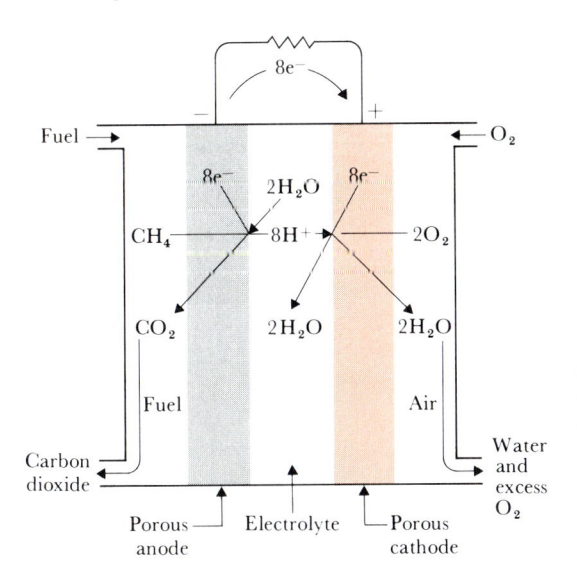

Fig. 19.8. Diagram of a typical fuel cell in which positive ions produced by the fuel migrate from anode to cathode. The cell reaction is the combustion of methane, CH_4.

This cell has an $\mathscr{E}^0$ of about 1.2 volt at 25 °C. Figure 19.9 illustrates the basic setup of this fuel cell.

One of the main advantages of the fuel cell, besides producing energy in a convenient form, is that its efficiency is very high. The efficiency of an engine or other energy-producing device which actually burns its fuel directly in oxygen is governed by the Carnot equation (18.6); this severely limits the amount of thermal energy which can be converted to useful work. The thermal efficiency of the fuel cell, since it does not operate as a heat engine, depends on a different relationship. Recall that $\Delta G = \Delta H - T\Delta S$; then the thermal efficiency of a fuel cell is given by

$$\frac{\Delta G}{\Delta H} = 1 - T\frac{\Delta S}{\Delta H} \tag{19.10}$$

Provided T is kept as low as possible and the ratio $\Delta S/\Delta H$ is small, theoretical efficiencies of over 80% can be obtained for a number of fuel cell reactions. Efficiencies of 50 to 60% have been obtained in actual practice. This should be compared with the 20–25% efficiencies typical of the ordinary gasoline-powered automobile.

The main reason fuel cells are not in common use today is that their electrodes must be made of expensive materials which are easily inactivated by impurities in the fuel. Future research may solve this problem.

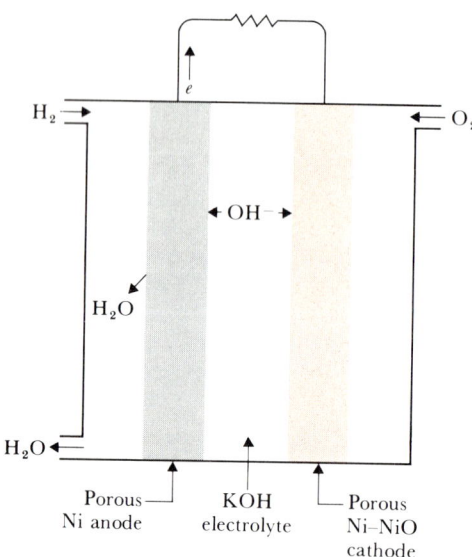

Fig. 19.9. Diagram of a typical hydrogen-oxygen fuel cell. The electrolyte (75% KOH in H_2O) must be kept at around 200 °C to remain liquid.

19.6 GALVANIC CORROSION

Examination of the plumbing of a typical home sometimes reveals a joint made by fastening a galvanized pipe (zinc-coated steel) to a copper pipe. Such a combination seldom lasts very long; eventually the galvanized pipe becomes badly corroded and a plumber must be called in to replace it. If the plumber knows any elementary chemistry, he will not simply replace the corroded galvanized pipe with a new galvanized pipe. Table 19.1 shows that the reaction

$$Zn(s) + Cu^{2+}(aq) \rightleftharpoons Zn^{2+}(aq) + Cu(s)$$

has an $\mathscr{E}^0$ of 1.10 V at 25 °C. In the presence of moisture the zinc–copper joint functions as an approximation to a zinc–copper voltaic cell; zinc spontaneously releases electrons to copper ions present in the oxide coating of the copper pipe. Thus, the plumber—unless he or she deliberately chooses to guarantee a return call—will replace the corroded galvanized pipe with a copper pipe.

Whenever two dissimilar metals are coupled in an electrolytic environment, galvanic corrosion may occur. The metal which has the larger electrode potential when acting as an anode will be oxidized (corroded) by the other metal. Since the rate of corrosion depends on the density of the corrosion current (due to flow of electrons from the more easily oxidized metal), corrosion can be minimized by making the mass of more easily oxidized metal large compared with that of the other metal. Thus it is better to use iron nails to fasten aluminum sheeting than it is to use aluminum nails to fasten iron sheeting. Corrosion occurs in both cases but the nails will last longer in the former case.

Iron tanks buried in the ground or iron shafts on boat propellers are often protected from corrosion by the use of so-called **sacrificial anodes**.* A quantity of material such as zinc, which oxidizes more readily than iron, is connected to the iron object via a small copper wire or, in the case of the boat propeller shaft, is made into a collar fitting around the shaft. As long as some zinc remains, it will oxidize rather than the iron. Oil pipelines are protected in this way by being connected to buried zinc anodes located at spaced intervals. This is a very important consideration in the design of the Alaska pipeline between Prudhoe Bay and Valdez since the oil in the line is hot, and thus corrosion would be very rapid.

---

*This is also called **cathodic protection**. The sacrificial anode protects the material acting as a cathode.

SUMMARY

Chemical reactions of the oxidation–reduction type can be made to produce electrical energy by forcing electron transfer between reactants to take place through an external circuit rather than directly. The electromotive force of a voltaic cell constitutes a direct measure of the Gibbs free energy change of the cell reaction and thus provides a valuable means of calculating the equilibrium constants of chemical reactions.

Voltaic cells in which one electrode depends on hydrogen ion concentration form the basis of pH meters, devices that permit the rapid measurement of pH.

Storage batteries consist of one or more voltaic cells whose cell reaction may be reversed by passing through an electric current in the reverse direction.

Voltaic cells utilizing fuel combustion reactions are known as fuel cells. Such cells show great intrinsic promise for the efficient production of electrical energy from fuels, but practical difficulties have hindered their widespread use.

Galvanic corrosion is the result of coupling two dissimilar metals in an electrolytic environment. Corrosion occurs when the metal with the higher anode potential is oxidized by the other metal.

LEARNING GOALS

1. Be able to diagram a voltaic cell such as that shown in Fig. 19.1 and to explain by means of electrode reactions how the cell produces an electric current.

2. Know how standard electrode potentials are determined and how to use them to calculate the standard voltage of a voltaic cell.

3. Know how to use the Nernst equation to calculate the voltage of a voltaic cell for other than standard states of the electrode substances.

4. Know the relationship between the Gibbs energy change of a reaction and the reaction emf.

5. Be able to use standard electrode potentials to calculate equilibrium constants of chemical reactions.

6. Be able to describe how a pH meter operates.

7. Be able to describe the basic construction of the lead storage battery and the chemical basis of its operation.

8. Be able to describe the basic construction of a typical dry cell.

9. Know what a fuel cell is and how it operates. Be able to discuss why fuel cells have high efficiency in the conversion of chemical energy to useful work.

10. Be able to explain what galvanic corrosion is and how it may be minimized. Know what is meant by a sacrificial anode and how it operates.

DEFINITIONS, TERMS, AND CONCEPTS TO KNOW

voltaic cell
electromotive force
faraday

standard electrode potential
anode
cathode

Nernst equation
galvanic corrosion
sacrificial anode

QUESTIONS AND PROBLEMS

Cell Reactions

1. Label each reaction below as spontaneous or nonspontaneous under standard conditions.
   a) $Zn(s) + Fe^{2+}(aq) \rightarrow Zn^{2+}(aq) + Fe(s)$
   b) $H_2(g) + 2Ag^+(aq) \rightarrow 2H^+(aq) + 2Ag(s)$
   c) $2H^+(aq) + Pb(s) \rightarrow H_2(g) + Pb^{2+}(aq)$
   d) $Zn^{2+}(aq) + 2Cl^-(aq) \rightarrow Zn(s) + Cl_2(g)$
   e) $2AgCl(s) \rightarrow 2Ag(s) + 2Cl_2(g)$

2. A voltaic cell is designed to operate on the basis of the reaction $3Fe(s) + 2Al^{3+}(aq) \rightleftharpoons 3Fe^{2+}(aq) + 2Al(s)$.
   a) In which direction is the cell reaction spontaneous under standard conditions?
   b) What is $\mathscr{E}^0$ of the spontaneous reaction?

3. What electrodes from Table 19.1 must be used to construct a voltaic cell with the cell reaction
   $$Ag(s) + \tfrac{1}{2}Cl_2(g) \rightleftharpoons AgCl(s)$$
   Is the cell reaction as written spontaneous under standard conditions?

4. Is the cell reaction
   $$FeBr_2(aq) \rightarrow Fe(s) + Br_2(l)$$
   spontaneous at 25 °C in a cell with $[Fe^{2+}] = [Br^-] = 1\ M$? Explain.

5. If necessary, rewrite each reaction below in the direction needed to function as a source of electric current, and calculate the standard potential.
   a) $Cu(s) + 2Ag^+(aq) \rightarrow Cu^{2+}(aq) + 2Ag(s)$
   b) $2AgCl(s) + H_2(g) \rightarrow 2Ag(s) + 2HCl(aq)$
   c) $H_2(g) + I_2(g) \rightarrow 2HI(aq)$
   d) $Mg(s) + ZnCl_2(aq) \rightarrow MgCl_2(aq) + Zn(s)$
   e) $2Ag(s) + Co(NO_3)_2(aq) \rightarrow 2AgNO_3(aq) + Co(s)$

6. What electrodes from Table 19.1 must be used to construct a voltaic cell with the cell reaction
   $$2Ag(s) + 2HCl(aq) \rightarrow 2AgCl(s) + H_2(g)$$
   What is $\mathscr{E}^0$ of this voltaic cell at 25 °C?

7. Compare the two different (but similar) voltaic cell reactions below
   $$Ag(s) + \tfrac{1}{2}Cl_2(g) \rightarrow AgCl(s)$$
   $$Ag(s) + HCl(aq) \rightarrow AgCl(s) + \tfrac{1}{2}H_2(g)$$
   Write anode and cathode reactions for each case and determine $\mathscr{E}^0$ values for the cells.

8. Will nickel metal dissolve in 1.0-M HCl? What about copper? Explain in terms of appropriate electrode reactions.

Voltaic Cell Calculations

9. A voltaic cell operates on the basis of the following reaction, for which $\mathscr{E}^0 = 0.6$ V at 25 °C:
   $$3A^+ + 2B(s) \rightarrow A_3(s) + B_2^{3+}$$
   a) Calculate Q_c for a cell containing 0.1-M A^+ and 0.01-M B_2^{3+}.
   b) Calculate the emf for the cell in part (a) at 25 °C.
   c) Write the anode and cathode reactions for the cell.
   d) Can you calculate $\mathscr{E}^0_{A^+/A_3}$ and $\mathscr{E}^0_{B/B_2^{3+}}$ from the above information? Explain.

10. An electrochemical cell operates on the basis of the reaction
    $$Cu^{2+} + Pb(s) + SO_4^{2-} \rightarrow Cu(s) + PbSO_4(s)$$
    a) Write down and identify the anode and cathode reaction when $[Cu^{2+}] = [SO_4^{2-}] = 1\ M$.
    b) Calculate $\mathscr{E}^0$ of the cell at 25 °C.

11. What is $\mathscr{E}^0_{Fe^{2+}/Fe^{3+}}$? What, in other words, is the value of the electrode potential (relative to the H^+/H_2 half-cell) of the oxidation $Fe^{2+} \rightarrow Fe^{3+} + e^-$?

12. A voltaic cell has the reaction
    $$Zn(s) + Hg_2Cl_2(s) \rightarrow 2Hg(l) + Zn^{2+}(aq) + 2Cl^-(aq)$$
    a) Write the anode reaction.
    b) Write the cathode reaction.
    c) Calculate $\mathscr{E}^0$ of the reaction at 25 °C.

13. The value of ΔG^0 of the reaction
    $$H_2(g) + Ni^{2+}(aq) \rightleftharpoons 2H^+(aq) + Ni(s)$$
    is 44.39 kJ at 25 °C.
    a) Write the cell reaction for a spontaneously operating cell at standard conditions.
    b) Calculate $\mathscr{E}^0_{Ni/Ni^{2+}}$ from the data given.

14. A voltaic cell has the reaction

$$2Fe^{2+}(aq) + Zn^{2+}(aq) \rightarrow 2Fe^{3+}(aq) + Zn(s)$$

a) Write the anode and cathode reactions and label which is which.
b) What is $\mathscr{E}^0$ of the cell?
c) Calculate ΔG^0 of the cell reaction at 25 °C.

15. A voltaic cell employs two electrodes whose half-reactions are:

$$H_2(g) \rightarrow 2H^+(aq) + 2e^-$$

$$Hg_2Cl_2(s) + 2e^- \rightarrow 2Hg(l) + 2Cl^-(aq)$$

a) What is the cell reaction?
b) Calculate $\mathscr{E}^0$ of the cell.
c) How will an increase in pH affect the cell emf? [Show that this can also be predicted from (a) using Le Châtelier's principle.]

16. Calculate $\mathscr{E}^0$ of each of the following reactions.
a) $2Al(s) + 3Zn^{2+} \rightarrow 2Al^{3+}(aq) + 3Zn(s)$
b) $Ag(s) + \frac{1}{2}Cl_2(g) \rightarrow AgCl(s)$
c) $2Ag(s) + Cl_2(g) \rightarrow 2AgCl(s)$
d) $Ag(s) + \frac{1}{2}Cl_2(g) \rightarrow Ag^+(aq) + Cl^-(aq)$
e) $2H^+(aq) + Pb(s) \rightarrow H_2(g) + Pb^{2+}(aq)$

17. A voltaic cell operates on the basis of the chemical reaction

$$Te(s) + 2Zn^{2+}(aq) \rightleftharpoons Te^{4+}(aq) + 2Zn(s)$$

a) Write down the half-reactions of the electrodes.
b) If $\mathscr{E}^0$ of the cell is 0.16 V, what is the value of $\mathscr{E}^0_{Te/Te^{4+}}$?
c) Calculate ΔG^0 of the cell reaction.

18. Consider the possible voltaic cell reaction

$$Fe^{3+}(aq) + Hg(l) \rightarrow Fe^{2+}(aq) + \frac{1}{2}Hg_2^{2+}(aq)$$

a) What is $\mathscr{E}^0$ of this cell?
b) Assuming $[Fe^{3+}] = [Fe^{2+}]$, what is the maximum concentration of Hg_2^{2+} that leads to the above as a spontaneous cell reaction?

19. What electrode potential must a metal have (as $M \rightarrow M^+ + e^-$) in order to liberate hydrogen gas from pure water? Assume that $[M^+] = 1.0\ M$.

20. The all-too-familiar rusting out of automobile bodies may be represented by the reaction

$$Fe(s) + 3H^+(aq) \rightarrow Fe^{3+}(aq) + 1\frac{1}{2}H_2(g)$$

which has $\mathscr{E}^0 = 0.036$ V at 25 °C. A major source of $H^+(aq)$ (in addition to dissolved CO_2) is hydrated metal ions $[M(H_2O)_n]^{x+}$ (see Section 13.10).

a) Use the Nernst equation to show how $[H^+]$ and temperature affect the emf of a voltaic cell having the above reaction.
b) Which would you expect to lead to more rusting, $CaCl_2$ or $NaCl$, when used to melt snow on the highways? (Ca^{2+} and Na^+ have almost the same ionic radii).

21. Calculate the equilibrium constant at 25 °C for the reaction

$$Cu(s) + Zn^{2+}(aq) \rightleftharpoons Cu^{2+}(aq) + Zn(s)$$

What does such a value mean?

22. Calculate an equilibrium constant at 25 °C for the reaction

$$Ni(s) + Sn^{2+}(aq) \rightleftharpoons Ni^{2+}(aq) + Sn(s)$$

23. Calculate ΔG^0 of the reaction given in Problem 20.

24. Calculate the solubility-product constant of Hg_2Cl_2 at 25 °C using data from Table 19.1.

25. The cell reaction

$$Pb(s) + Sn^{2+}(aq) \rightarrow Pb^{2+}(aq) + Sn(s)$$

is nonspontaneous at 25 °C with $[Pb^{2+}] = [Sn^{2+}] = 1.0\ M$ since $\mathscr{E}^0 = -0.01$ V. However, if the Pb^{2+} concentration is lowered by adding SO_4^{2-} to the lead electrode compartment [this removes Pb^{2+} as $PbSO_4(s)$], the cell reaction becomes spontaneous. If $[SO_4^{2-}]$ is adjusted to $1.0\ M$ the cell emf becomes 0.22 V.

a) Calculate the concentration of Pb^{2+} in the spontaneous cell having $[SO_4^{2-}] = 1.0\ M$.
b) Calculate the K_{sp} of $PbSO_4$. How does this compare with the experimental solubility of 0.0043 g per 100 cm³?

26. The cell reaction

$$H_2(g) + Hg_2Cl_2(s) \rightarrow 2Hg(l) + 2HCl(aq)$$

has an emf of 0.42 V when $[Cl^-] = 1.0\ M$ and $p_{H_2} = 1$ atm. What is the pH of the solution containing the hydrogen electrode?

27. Show how to use electrode potentials to answer the following questions
a) Which is the better reducing agent, aluminum or silver?
b) Which is the better oxidizing agent, chlorine or bromine?

28. Give an electrochemical explanation of the following facts.

a) If an iron nail and a copper nail are stuck into a lemon and a sensitive ammeter connected between them, a flow of electrical current is indicated.

b) If a penny and a dime are placed in opposite corners of a person's mouth and a sensitive ammeter connected between them, a flow of electric current is indicated.

What substance acts as the electrolyte in each case? Suggest possible electrode reactions in each case (formation of H_2 probably occurs at one of the electrodes in each case).

Galvanic Corrosion

29. When nickel and zinc are in contact, which is more likely to corrode? Justify your answer.

30. Which would probably be more successful from the standpoint of corrosion resistance, putting a thin plating of iron over copper, or vice versa? Explain.

31. Shortly after World War II several boat manufacturers used surplus aircraft aluminum to make boat hulls. Such hulls corroded very rapidly, especially when used in salt water. Chemical analysis showed the aluminum contained small amounts of iron as the major impurity. What caused the hulls to corrode? What do you suppose was done to make today's aluminum boats possible?

32. Why is galvanic corrosion more of a problem in salt water than in fresh water?

General

33. a) Will solid copper react with 1-M HCl to produce H_2? Prove your answer.
b) Repeat for solid aluminum in 1-M HCl.

34. It is not necessary to know the value of $\mathscr{E}'$ in Eq. (19.9) in order to calibrate a pH meter. Why is this so?

35. Use the data in Table 19.1 to calculate the solubility (in g·L^{-1}) of Hg_2Cl_2 in H_2O at 25 °C.

36. Tarnished silverware may be cleaned by placing it on a sheet of aluminum foil in a pan and covering it with $NaHCO_3$ solution. Suggest a chemical explanation of the cleaning process.

37. In what major ways do fuel cells differ from ordinary voltaic cells? Why wouldn't a large zinc-copper voltaic cell be just as feasible for running an automobile as a hydrogen-oxygen fuel cell?

38. What are the major advantages and disadvantages of fuel cells at their present level of development?

39. It is possible to purchase devices which recharge dry cells. How do you suppose such devices operate? Would these work if the dry cell ever became truly *dry*?

40. Explain why it is necessary to have electron transfer occur through an external circuit in a voltaic cell. Why can't the direct flow of electrons from the oxidized to the reduced substance be used as a source of electrical energy?

ADDITIONAL READINGS

Weissbart, J., "Fuel Cells: Electrochemical Converters of Chemical to Electrical Energy." *J. Chem. Educ.* **38**, 207 (1961).

Yeager, E., "Fuel Cells." *Science* **134**, 1178 (1961).

Benson, Richard, "Corrosion at Sea." *Sea Frontiers* **16**, 3 (May–June 1970): 172–181. A descriptive treatment of how corrosion occurs, the various types of corrosion, and how it can be minimized in marine applications.

Slaubaugh, W. H., "Corrosion." *J. Chem. Educ.* **51**, 218 (1974). A short, elementary article on some important aspects of corrosion and how it is controlled.

20
THE DYNAMICS OF CHEMICAL CHANGE

A rapid chemical reaction: A forest fire rages out of control. (Courtesy of Grant Heilman.)

If the number of molecules of A and B in unit volume be denoted by p *and* q, *the product* pq *will represent the frequency of the encounters of these molecules. If each motion of the various molecules be equally favourable to the formation of new substances, the velocity . . . may be made equal to* ϕpq, *the coefficient of velocity being supposed dependent on temperature.*
C. GULDBERG and P. WAAGE, *1879**

Knowing that a reaction is intrinsically spontaneous, that is, ΔG_{rx} is negative, tells nothing about how fast the reaction will go. For example, the decomposition of gaseous benzene to the elements

$$C_6H_6(g) \rightleftharpoons 6C(s) + 3H_2(g)$$

has a ΔG^0_{rx} of -128 kJ at 25 °C. This shows that benzene vapor in its standard state is thermodynamically unstable with respect to decomposition to carbon and hydrogen in their standard states. Yet, observation shows that benzene exhibits virtually no tendency to decompose at room temperature. Similarly, ΔG^0_{rx} of the ammonia synthesis reaction

$$N_2(g) + 3H_2(g) \rightleftharpoons 2NH_3(g)$$

is -33.3 kJ at 298 K and becomes even more negative at lower temperatures. Yet, ammonia synthesis is carried out only with great difficulty under rather special conditions. Furthermore, the rate of the synthesis reaction increases as the temperature increases, that is, as the intrinsic spontaneity of the reaction decreases.

The present chapter discusses how velocities of chemical reactions are expressed, what such velocities depend on, and what a study of reaction velocities reveals about the manner in which chemical substances react.

20.1 THE RATE OF A REACTION

Consider a general chemical reaction which may be expressed in the simple form

reactants $\rightleftharpoons$ products

Assume that initially no products are present in the reaction vessel as the reactants are mixed. The **rate** at which the reaction proceeds may be expressed mathematically as the moles of a given reactant disappearing to form products per liter of

---

* Biographical sketches of Guldberg and Waage are found on page 304.

volume per unit of time. Alternatively, the rate may be given in terms of the moles of a product appearing per liter of volume per unit of time. Consider the hydrogen–iodine reaction as an example,

$$H_2(g) + I_2(g) \rightarrow 2HI(g)$$

The rate of reaction can be given in the following three ways:

1. The rate of disappearance of hydrogen (in mol·L⁻¹ per unit of time). This may be given the symbol $-R_{H_2}$ (the minus sign means [H_2] decreases as time increases).
2. The rate of disappearance of iodine (in mol·L⁻¹ per unit of time). This is symbolized by $-R_{I_2}$.
3. The rate of appearance of hydrogen iodide (in mol·L⁻¹ per unit of time). This is symbolized by R_{HI}.

Since the balanced equation says that whenever one mole of H_2 disappears, two moles of HI appear, it follows that the rate of HI appearance is twice that of the rate of H_2 disappearance. Thus $R_{HI} = -2R_{H_2}$ and, by the same token, $R_{HI} = -2R_{I_2}$. The above rates are illustrated graphically in Fig. 20.1. The decrease of reaction rate as time goes on is a result of the law of mass action (Chapter 12); reactant molecules become progressively depleted and, consequently, there are fewer and fewer molecules to react as the reaction progresses. As the reaction proceeds in the forward direction (left to right), products accumulate and begin to participate in the reverse reaction (right to left). If the rates of the forward and reverse reactions are given the general symbols R_f and R_r, respectively, then the **net rate of reaction** at any instant is $R_f - R_r$. The rate of the reverse reaction is initially zero (since there are assumed to be no products originally present) but progressively increases as the reaction proceeds and more products are formed. Ultimately, as shown in Fig. 20.2, the forward rate and the reverse rate become equal; that is, reactants go to products as rapidly as products reform reactants. At this stage the net rate of reaction, $R_f - R_r$, becomes zero. This is the point of chemical equilibrium; the balance of opposing tendencies is simply the equality of the two opposing rates of reaction.

At the same time that R_f and R_r are approaching each other, the concentrations of reactants and products are changing as shown in Fig. 20.3. As the reaction proceeds, the concentration of products increases and the concentration of reactants decreases. Ultimately, both product and reactant concentrations reach limiting values as they begin to interconvert at equal rates at the point of chemical equilibrium.

There are four major factors that affect the rates of chemical reactions. These are:

1. the temperature at which the reaction is carried out,
2. the concentrations of reacting substances,
3. the intrinsic natures of the particular reactants,
4. external influences such as catalysts.

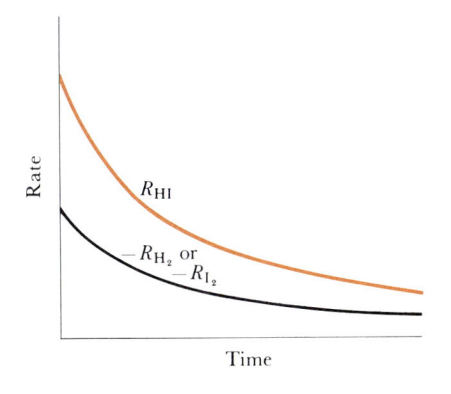

Fig. 20.1. Rates of hydrogen and iodine disappearance and hydrogen iodide appearance as a function of time in the reaction $H_2(g) + I_2(g) \rightarrow 2HI(g)$.

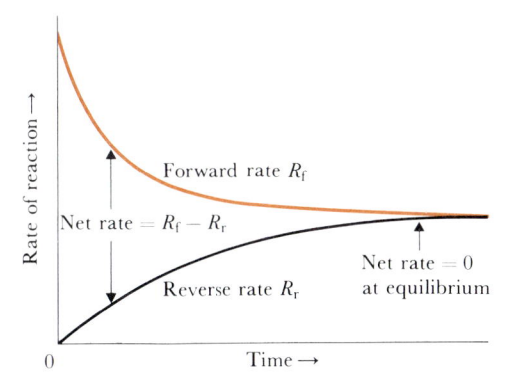

Fig. 20.2. The change of forward, reverse, and net reaction rates during the course of a chemical reaction. It is assumed that no products are initially present.

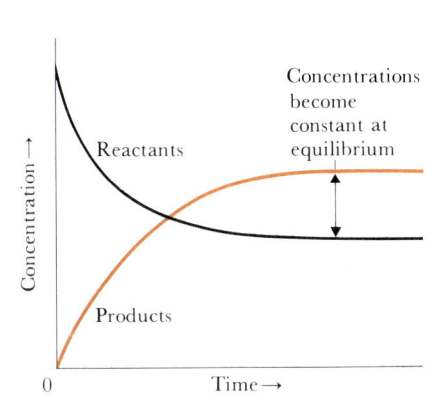

Fig. 20.3. Change in reactant and product concentrations during the course of a chemical reaction.

The effect of temperature is discussed in Sections 20.2 and 20.3, the effect of reactant concentrations is discussed in Sections 20.3 and 20.4, and catalysts are discussed in Sections 20.8 and 20.9. The third factor (not explicitly discussed in the present chapter) deals with the fact that different substances have different chemical reactivities. Take iron and sodium, for example; iron reacts rather slowly with water whereas sodium reacts with great vigor. Such differences are best appreciated by being familiar with the specific chemistries of a wide variety of substances.

The rate of a reaction is determined in the laboratory by measuring the concentration of a reactant or product as a function of time. To illustrate how this is done, consider the reaction

$$a\text{A} + b\text{B} \rightarrow \text{products}$$

Let C_{A_1} be the molar concentration of A at time t_1 and C_{A_2} the molar concentration at time t_2. If the time interval $\Delta t = t_2 - t_1$ is very small then the average rate of disappearance of A during this time interval is given approximately by

$$-R_A \simeq \frac{-\Delta C_A}{\Delta t}$$

where $\Delta C_A = C_{A_2} - C_{A_1}$ (see Fig. 20.4).* In a similar fashion, the concentration of B or the concentration of some product could be followed as a function of time. The choice of substance to monitor is based on how easy or convenient it is to identify; some substances are easier to do concentration measurements on than others.

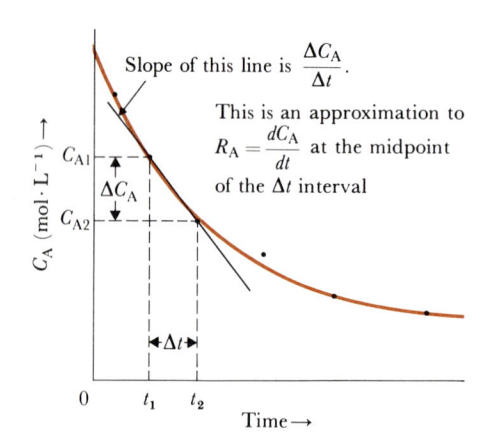

Fig. 20.4. Experimental determination of a rate of reaction. The substance A whose concentration is being followed as a function of time is one of the reactants, hence its concentration decreases as time progresses. Each circle represents a concentration of A measured at a given time.

EXAMPLE 20.1 One of the commercial preparations of chlorine utilizes the reaction

$$4\text{HCl}(g) + \text{O}_2(g) \rightarrow 2\text{H}_2\text{O}(g) + 2\text{Cl}_2(g)$$

At a given time the rate of appearance of chlorine (R_{Cl_2}) is equal to 0.01 mol·L^{-1}·min^{-1}. Calculate, for the same time, (a) the rate of disappearance of HCl ($-R_{HCl}$), (b) the rate of disappearance of O$_2$ ($-R_{O_2}$), (c) the rate of appearance of H$_2$O (R_{H_2O})

---

* Students familiar with the calculus will recognize that when Δt approaches zero, the quantity $-\Delta C_A/\Delta t$ becomes the first derivative, $-dC_A/dt$. This is just the slope of the plot of C_A vs. t.

SOLUTION

a) Each time two moles of Cl_2 appears, four moles (twice as much) HCl disappears (that's what the *balanced* equation says). Thus $-R_{HCl} = 2R_{Cl_2} = 0.02$ mol·L^{-1}·min^{-1}.

b) The balanced equation says O_2 disappears half as fast as Cl_2 appears. Thus $-R_{O_2} = \frac{1}{2}R_{Cl_2} = 0.005$ mol·L^{-1}·min^{-1}.

c) The balanced equation says H_2O and Cl_2 appear at the same rate. Thus $R_{H_2O} = R_{Cl_2} = 0.01$ mol·L^{-1}·min^{-1}.

20.2 THE EFFECT OF TEMPERATURE ON CHEMICAL REACTION RATES

Figure 20.5 illustrates several different ways in which temperature may affect the rate at which a chemical reaction occurs. In graph I the rate of reaction increases exponentially as the temperature increases. Such reactions comprise a large number of very important chemical reactions and are called **activated reactions**; the source of the name is discussed in Section 20.5. In graph II the rate of reaction increases with an increase in temperature—but only up to a point. Thereafter the rate of reaction drops and eventually becomes zero. A class of biological reactions called **enzymic reactions** behave in this fashion. Enzymes function sluggishly or not at all at very low temperatures, gradually become more active as the temperature increases, but ultimately become denatured and inactivated at high enough temperatures. The

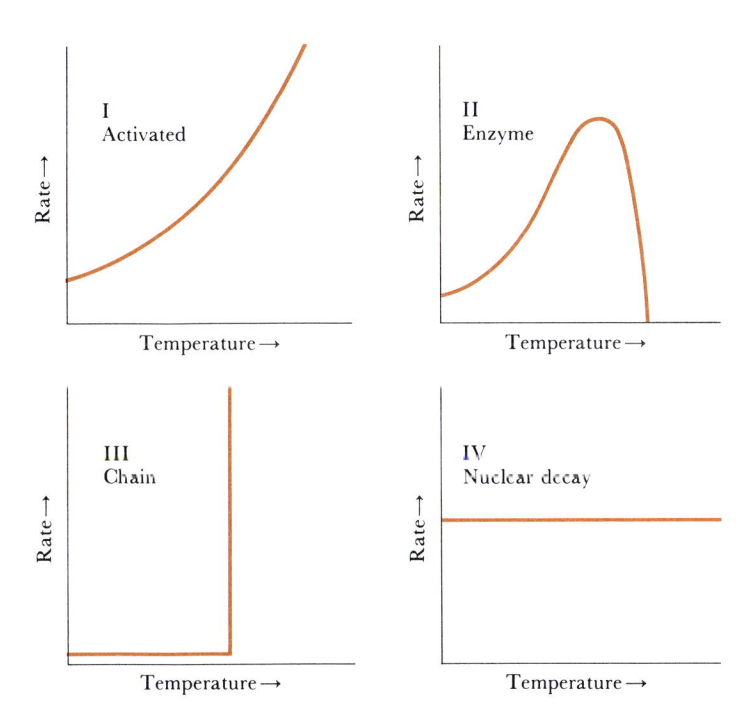

Fig. 20.5. The effect of temperature on the rates of four basic types of reactions.

rising of yeast dough is an example of an enzymic reaction. The dough must be kept in a warm place for rising to occur, but too much warmth will stop the rising. In graph III the rate of the reaction is essentially zero at low enough temperatures and then abruptly becomes very, very large at a given temperature. Reactions of this type are called **chain reactions**; an explosion of a fuel–air mixture is an example of a chain reaction. The temperature at which the explosion (a very rapid reaction) occurs is called the **ignition temperature**. The phenomenon in graph IV is not a chemical reaction at all but is included to show that nuclear decay processes (see Chapter 11) are unaffected by temperature changes which are normally encountered in practice.*

In many practical applications it becomes necessary to set up reaction conditions which are a compromise between minimum ΔG_{rx} (maximum intrinsic spontaneity) and maximum rate. For example, ΔG^0 of the ammonia synthesis reaction is favored by low temperatures, yet the reaction must be run at fairly high temperatures in order to obtain an acceptable rate of production of NH_3. The synthesis of diamond from graphite involves a similar problem. This reaction also has lower (more negative) ΔG at lower temperatures but the rate of reaction is too slow to be useful unless high temperatures are used.

20.3 ACTIVATED CHEMICAL REACTIONS

Consider a general type of chemical reaction symbolized by

$$a\text{A} + b\text{B} \rightleftharpoons \text{products}$$

If the equilibrium constant of the reaction is large, then the rate of the reverse reaction may be assumed to be zero during the early stages of the reaction. This means that $R_f - R_r \simeq R_f$; that is, the net rate of the reaction is approximately equal to the forward rate of the reaction. Thus the rate at which products are reforming reactants is negligible at the beginning of the reaction. Since the law of mass action (Section 12.1) states that the rate of a reaction is proportional to the concentrations of reactants, the **initial rate** of many reactions may be written in the general form†

$$R_f = k[\text{A}]^m[\text{B}]^n \tag{20.1}$$

where [A] is the molar concentration of reactant A at a given time, [B] is a similar quantity for reactant B, k is called the **rate constant**, and m and n are experimentally determined numbers. Equation (20.1) is called the **rate law** of the chemical reaction to which it refers.

The quantity m is called the **order** of the reaction with respect to reactant A, and n is called the order of the reaction with respect to reactant B. The **total order** of the reaction is defined as $m + n$. If $m + n$ has a value of 1, the reaction is said to be **first**

---

* Nuclear decay rates do depend somewhat on temperature, chemical form, pressure, and applied fields. See Wayne C. Wolsey, *J. Chem. Educ.* **55**, 303 (1978).

† The initial rate of a reaction is determined experimentally by measuring the change in concentration of a reactant or product over a very small time interval immediately after mixing the reactants, then dividing the concentration change by this time interval.

TABLE 20.1 EXPERIMENTAL RATE DATA FOR THE REACTION
$2NO(g) + 2H_2(g) \rightarrow N_2(g) + 2H_2O(g)$
AT A GIVEN TEMPERATURE

MIXTURE NUMBER	INITIAL RATE OF DISAPPEARANCE OF NO (mol·L^{-1}·min^{-1})	INITIAL REACTANT CONCENTRATIONS: [NO]	[H$_2$]
1	1	0.01	0.01
2	4	0.02	0.01
3	2	0.01	0.02

order; if $m + n$ is equal to 2, the reaction is said to be **second order**. It is possible for m and n or their sum to be nonintegral. Also, one or both may be zero or negative.

The quantities k, m, and n which make up the rate law can be determined only by experiment. The following shows one simple way in which this may be done. Consider the reaction

$$2NO(g) + 2H_2(g) \rightarrow N_2(g) + 2H_2O(g)$$

The rate law may be deduced by mixing several different amounts of reactants (NO and H$_2$) in separate reaction vessels and measuring the initial rate of reaction for each separate mixture. The temperature is kept the same for each separate mixture and throughout the reaction. Table 20.1 shows the results of measuring the rates of three different mixtures of NO and H$_2$. The rate law of the reaction is assumed to have the general form

$$-R_{NO} = k[NO]^m[H_2]^n$$

Examination of mixtures 1 and 2 shows that doubling [NO] while keeping [H$_2$] constant quadruples the rate of reaction. Thus an expression for the order m is as follows:

$$\frac{-R_{NO}(2)}{-R_{NO}(1)} = \frac{4}{1} = \frac{(0.02)^m(0.01)^n}{(0.01)^m(0.01)^n} = \frac{(0.02)^m}{(0.01)^m} = \left(\frac{0.02}{0.01}\right)^m = 2^m$$

Thus, since $2^m = 4$, it follows that m is 2; the reaction is second order with respect to NO. Similarly, the data for mixtures 1 and 3 show that doubling [H$_2$] while keeping [NO] constant doubles the reaction rate:

$$\frac{-R_{NO}(3)}{-R_{NO}(1)} = \frac{2}{1} = \frac{(0.01)^m(0.02)^n}{(0.01)^m(0.01)^n} = \frac{(0.02)^n}{(0.01)^n} = \left(\frac{0.02}{0.01}\right)^n = 2^n$$

Thus, since $2^n = 2$ it follows that n is 1; the reaction is first order with respect to H$_2$. The complete rate law is

$$-R_{NO} = k[NO]^2[H_2]$$

The total order is $m + n = 2 + 1 = 3$; the reaction is said to be **third order**. The rate law provides a mathematical description of just how the concentrations of reactants affect the rate of reaction. For example, it shows that tripling [NO] will increase the rate by a factor of 3^2 (or nine), whereas tripling [H$_2$] will increase the rate by a factor of 3^1 (or three).

The rate constant k may be calculated by using the data of any one of the mixtures given in Table 20.1 and substituting this into the rate law. For example, given the data of mixture 1:

$$k = \frac{-R_{NO}(1)}{[NO]^2[H_2]} = \frac{1 \text{ mol} \cdot L^{-1} \cdot \text{min}^{-1}}{(0.01 \text{ mol} \cdot L^{-1})^2(0.01 \text{ mol} \cdot L^{-1})}$$

$$= 1 \times 10^6 \text{ L}^2 \cdot \text{mol}^{-2} \cdot \text{min}^{-1}$$

This value of k is valid for any mixture of NO and H$_2$ at the same temperature as that for which the data of Table 20.1 were obtained. Thus this value may be used to predict a rate of reaction of any mixture of NO and H$_2$ at this given temperature. If the rate of reaction is desired at some other temperature, a new reaction mixture must be made up and the rate constant k must be calculated for this new temperature.

The radioactive decay process

nucleus A → decay products

is an example of a first-order process. Chapter 11 showed that if m is the mass of the decaying substance A, then rate of decay = km [see Eq. (11.1)]. Some chemical reactions of the type

A → products

have the rate law

$-R_A = k[A]$

Such a rate law is analogous to the rate of decay of a radioactive element. The concentration of substance A at any given time t is given by

$$[A] = [A]_0 \, 10^{-kt/2.3} \quad \text{or} \quad [A]_0 \, e^{-kt} \tag{20.2}$$

which is analogous to Eq. (11.2) if $[A]_0$ is the amount of A present at the start of the reaction. The main characteristic of any first-order process, chemical or nuclear, is that it takes the same amount of time for half of the original material to disappear regardless of its original mass. Thus it is also common to speak of the half-life of a first-order chemical reaction; this is the time required for one-half the reactant to react.

An example of a first-order chemical reaction is the thermal decomposition of azobisisobutyronitrile:

$$(CH_3)_2-\underset{CN}{CH}-N=N-\underset{CN}{CH}(CH_3)_2(aq) \xrightarrow{\text{heat}}$$

$$(CH_3)_2-\underset{CN}{CH}-\underset{CN}{CH}-(CH_3)_2(aq) + N_2(g)$$

The rate of reaction can easily be followed by measuring the amount of N$_2$ evolved per unit time. Since the N$_2$ is insoluble in the reaction solution, no complication due to the reverse reaction enters in.

EXAMPLE 20.2 The thermal decomposition of acetone (propanone) at 600 °C has a first-order rate constant of 8.63×10^{-3} s^{-1}.
   a) Calculate the half-life of this reaction.
   b) How long will it take for 80% of a sample of acetone to decompose at 600 °C?

SOLUTION
   a) The half-life is given by

$$t_{1/2} = \frac{0.69}{k} = \frac{0.69}{8.63 \times 10^{-3} \text{ s}^{-1}} = 80 \text{ s}$$

   b) Using

$$\log \frac{[A]_0}{[A]} = \frac{kt}{2.3}$$

   and solving for t produces

$$t = \frac{2.3}{k} \log \frac{[A]_0}{[A]} = \frac{2.3}{8.63 \times 10^{-3} \text{ s}^{-1}} \log \frac{1}{0.2} = 186 \text{ s}$$

20.4 MECHANISMS OF CHEMICAL REACTIONS

Although chemists often need to know just how fast a reaction proceeds under given conditions, they utilize rate studies for many other reasons as well. One of the most valuable uses of reaction rates arises from the information they are sometimes able to provide concerning *mechanisms* of chemical reaction. The **mechanism** of a chemical reaction may be defined as a detailed, step-by-step description of what the reactants are thought to do when they form products. Only when chemists know the mechanism of a reaction can they say they understand the reaction. Knowledge of the details of a chemical reaction makes it possible to:

1. design reaction conditions so that industrially important processes can be carried out under optimum conditions,
2. design reactions to produce new products not hitherto available,
3. design new drugs and medicines that are more effective in the control of disease.

Much modern medical research is being devoted to studies of the chemical mechanisms of cancers; in this way it may become possible to design ways of controlling or eliminating this dreaded condition.

The mechanism of a reaction is usually expressed as a series of what are known as **elementary steps**. An elementary step has the form

$$A + B + \cdots \rightarrow \text{products}$$

and has a rate law of the simple form

$$R_f = k[A][B]\cdots$$

in which each reactant appears to the first order. As a simple example, suppose an elementary step in a reaction sequence is second order, that is,

$$A + B \rightarrow C$$

$$R_f = k[A][B]$$

A simple model for such a step is one in which molecules A and B collide to form the product C. If there are a given number of molecules (say, 2A and 2B), then there are four collision possibilities which can lead to reaction:

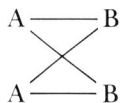

Now if the number of A molecules or the number of B molecules is doubled, the number of collision possibilities (and hence the rate of reaction) is also doubled:

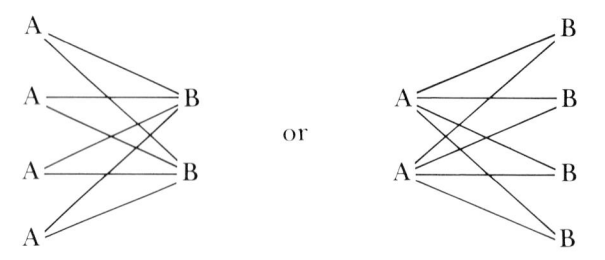

If both the number of A molecules and the number of B molecules are doubled, the number of collision possibilities will increase by a factor of 4 and the rate of reaction should increase by a factor of 4:

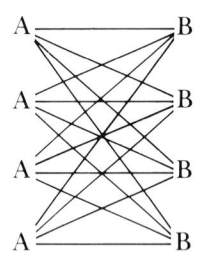

In some cases, one of the elementary steps in the mechanism of a reaction is so much slower than any of the other steps that it acts as a bottleneck and alone determines the overall rate of the reaction. Such a slow step is called a **rate-determining step** (often abbreviated as RDS). Some reactions may have more than one RDS, but a large number of important reactions appear to have but a single RDS.

The basic procedure in establishing the mechanism of a chemical reaction is to first determine the rate law and then to postulate a series of elementary steps which account for the experimental facts. The idea behind this is simple but its execution requires the highest skills of the art. The following examples illustrate the deduction of two rather simple mechanisms.

Alkyl halides of the type RCH_2X react with hydroxyl ion to form alcohols in a second-order reaction. The reaction and its rate law are

$$RCH_2X + OH^- \rightarrow RCH_2OH + X^-$$

$$R_f = k[RCH_2X][OH^-]$$

Running this reaction in a variety of solvents of varying polarity affects the rate but slightly. By contrast, the same reaction with an alkyl halide of the type R_3CX is first order with respect to halide and independent of hydroxyl ion concentration:

$$R_3CX + OH^- \rightarrow R_3COH + X^-$$

$$R_f = k[R_3CX]$$

The rate of this latter reaction increases significantly when run in more polar solvents.

A possible mechanism for the first reaction is

$$OH^- + \underset{\underset{H}{|}}{\overset{\overset{R}{|}}{C}}\diagdown X \rightleftharpoons \left[HO-\underset{\underset{H}{|}}{\overset{\overset{R}{|}}{C}}-X \right] \rightarrow \underset{\underset{H}{|}}{\overset{\overset{HO}{|}}{C}}\diagdown R + X^-$$

The reaction proceeds as follows: The carbon atom attached to the halogen atom is attacked on one side by OH^- to form an unstable intermediate which loses X^- to form the alcohol. Since the pathway involves collision between two species, OH^- and RCH_2X, the rate should have a first-order dependence on each of these. Additional evidence for this mechanism comes from the fact that if one of the hydrogen atoms on the alkyl halide is replaced with deuterium, then determination of the absolute configurations of halide and alcohol shows that the R group changes position as required by the mechanism.

The second reaction is postulated to consist of two elementary steps—a slow step followed by a fast step:

STEP 1: $R_3CX \rightleftharpoons R_3C^+ + X^-$ (slow)

STEP 2: $R_3C^+ + OH^- \rightarrow R_3COH$ (fast)

Just as a chain can be no stronger than its weakest link, many multistep reactions can be no faster than their slowest step. Thus the overall rate should depend only on the first reaction—the rate-determining step—and should be independent of $[OH^-]$. The observation that the reaction rate increases in more polar solvents is due to the fact that more polar solvents enhance the ionization of the alkyl halide.

The ozone decomposition reaction is an example of a somewhat more complicated mechanism:

$$2O_3(g) \rightarrow 3O_2(g)$$

Experiments show that the rate of disappearance of ozone obeys the rate law

$$-R_{O_3} = k[O_3]^2[O_2]^{-1}$$

Note that this is not of the general form (20.1) in that the product, O_2, enters in. The fact that $[O_2]$ comes in with a negative power means that the formation of product inhibits the decomposition of ozone. The following has been suggested as a possible mechanism:

STEP 1: $O_3(g) \rightleftharpoons O_2(g) + O(g)$ (fast equilibrium)

STEP 2: $O(g) + O_3(g) \xrightarrow{k'} 2O_2(g)$ (slow)

The rate of disappearance of O_3 in the RDS is given by

$$-R_{O_3} = k'[O][O_3] \tag{20.3}$$

Since [O] is not directly measurable, there is no way to verify the rate law as it stands above. However, [O] can be expressed in terms of measurable quantities as follows: If the equilibrium is established rapidly, then

$$K_c = \frac{[O_2][O]}{[O_3]} \quad \text{and} \quad [O] = K_c \frac{[O_3]}{[O_2]} \tag{20.4}$$

Substituting for [O] in Eq. (20.3) from Eq. (20.4) leads to

$$-R_{O_3} = k'K_c[O_3]^2[O_2]^{-1} \tag{20.5}$$

which is of the same form as the experimental rate law. Thus the postulated mechanism is consistent with the experimental facts.*

It is probably not possible to prove conclusively that a mechanism is absolutely correct. Sometimes two or more different mechanisms appear to fit the facts equally well. It can even happen that someone eventually comes up with yet another mechanism which seems more plausible than any of the existing mechanisms. Eventually, however, accumulating evidence usually focuses on a specific mechanism as the most likely of several alternatives.

---

* The NO, H_2 reaction discussed on page 551 is postulated to have the mechanism

Step 1. $NO + H_2 \underset{k_{-1}}{\overset{k_1}{\rightleftharpoons}} NOH_2$

Step 2. $NOH_2 + NO \xrightarrow{k_2} N_2 + H_2O_2$ (slow)

Step 3. $H_2O_2 + H_2 \rightarrow 2H_2O$ (fast)

The substance NOH_2 has never been isolated. It is postulated that it is so reactive that it is used up as fast as it is formed.

EXAMPLE 20.3 When a disaccharide such as (+)-maltase is dissolved in acidified water, D-(+)-glucose (a monosaccharide) is formed. The reaction is

$$C_{12}H_{22}O_{11} + H_2O \xrightarrow{H_3O^+} 2C_6H_{12}O_6$$

For dilute solutions the reaction is found to be first order, depending only on the (+)-maltase concentration. Explain.

SOLUTION The rate of reaction (a hydrolysis) may be written

$$-R_{\text{maltase}} = k[H_3O^+][H_2O][C_{12}H_{22}O_{11}]$$

which appears to be third order. However, H_3O^+ is a **catalyst** (see Section 20.8) so its concentration (for a given experiment) is constant. Similarly, if the solution is dilute, $[H_2O]$ is almost constant. If $k[H_3O^+][H_2O] = k'$, the rate law becomes

$$-R_{\text{maltase}} = k'[C_{12}H_{22}O_{11}]$$

The reaction is said to be a **pseudo** first-order reaction.

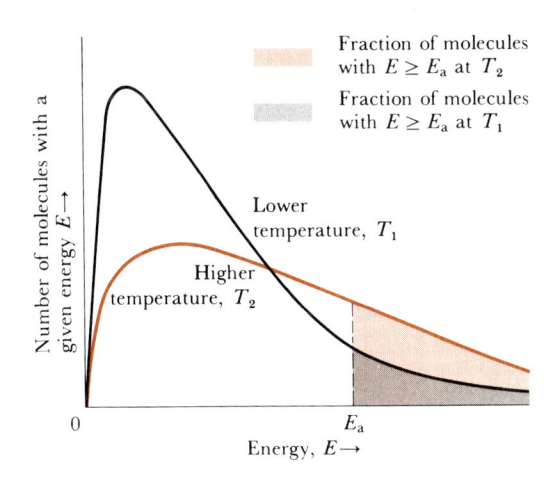

Fig. 20.6. The effect of temperature on the fraction of molecules having enough or more than enough energy to react.

20.5 THE ACTIVATION THEORY OF REACTION RATES

The fact that the initial rates of many reactions increase as the temperature increases (all other factors held constant) suggests the following model for the rate-determining step of chemical reactions: Let E_a represent the minimum energy necessary to bring about the RDS. To take an example, E_a may be the minimum energy needed to break the RCH_2—X bond to form the ion RCH_2^+. According to Fig. 20.6 (a Maxwell-Boltzmann distribution of molecular energies), the fraction of molecules having an energy E_a or higher should increase as the temperature increases; therefore, the rate of reaction should increase accordingly. This suggests that the rate constant k for a

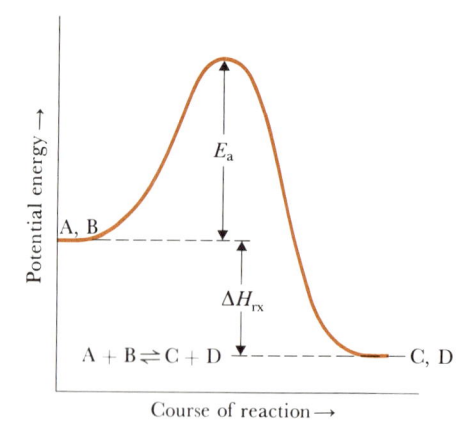

Fig. 20.7. The change in the potential energy of a chemically reacting system as reactants form products. Note that if the forward reaction is exothermic, the activation energy of the reverse reaction is $E_a + \Delta H_{rx}$.

reaction should be given by

$$k = \begin{pmatrix} \text{number of reactant colli-} \\ \text{sions per unit time} \end{pmatrix} \begin{pmatrix} \text{fraction of reactant molecules} \\ \text{having energy } E_a \text{ or higher} \end{pmatrix}$$

The fraction of reactant molecules having an energy E_a or higher is given approximately by

$$f_{E_a} = 10^{-E_a/2.3\,RT} \quad \text{or} \quad e^{-E_a/RT} \tag{20.6}$$

If A represents the number of reactant collisions per unit time (called the **frequency factor**), the rate constant becomes

$$k = A\,10^{-E_a/2.3\,RT} \quad \text{or} \quad Ae^{-E_a/RT} \tag{20.7}$$

This important relationship is known as the **Arrhenius equation**.*

The Arrhenius equation suggests a model for chemical reactions in which the reactants must become **activated** in order to pass over an energy barrier or hump of magnitude E_a before products can be formed. The energy E_a which reactants must acquire in order to transcend this barrier is called the **activation energy** of the reaction. The activation-energy model for a chemical reaction $A + B \rightleftharpoons C + D$ is shown schematically in Fig. 20.7. Note that if the forward reaction is exothermic, then the activation energy of the reverse reaction is $E_a + \Delta H_{rx}$.

Taking the logarithm of both sides of Eq. (20.7) leads to

$$\log k = \log A - \frac{E_a}{2.3\,RT} \tag{20.8}$$

or

$$\ln k = \ln A - \frac{E_a}{RT}$$

If k is measured experimentally at several different temperatures and $\log k$ is plotted versus $1/T$ (or $\ln k$ is plotted versus $1/T$), the result is a straight line of slope $-E_a/2.3R$ (or $-E_a/R$) from which E_a may be calculated (see Fig. 20.8). The point at which the line crosses the vertical axis (at $1/T = 0$) is the value of $\log A$ (or $\ln A$) from which the frequency factor A may be computed. Activation energies often aid in the estab-

---

* Named after the Swedish chemist Svante Arrhenius. For a short biographical sketch of Arrhenius see Chapter 1.

Svante A. Arrhenius, Swedish physicist whose doctoral thesis proposed the theory of electrolytic dissociation. (Courtesy of Brown Brothers.)

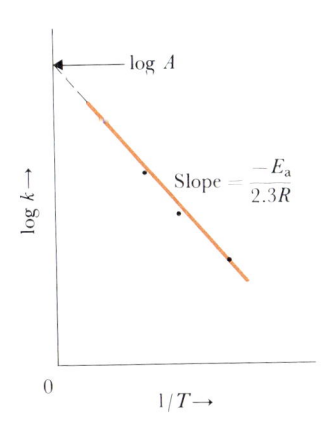

Fig. 20.8. Determination of the activation energy of a reaction. The reaction rates are measured at several different temperatures and the logarithms of the rate constants are plotted against the reciprocals of the reaction temperatures. The slope of the line obtained is related in a simple way to the activation energy of the reaction.

lishment or rejection of proposed mechanisms. Note that the greater the activation energy, the slower is the rate of reaction.

EXAMPLE 20.4 A first-order reaction has a rate constant of 1.2×10^{-1} s^{-1} at 15 °C and a rate constant of 2.5×10^{-1} s^{-1} at 25 °C. Calculate the activation energy and frequency factor of the reaction.

SOLUTION In practice three or more values of ln k at different temperatures would be plotted against $1/T$, and E_a and A would be determined graphically as in Fig. 20.8. However, when only two values of k are known, E_a and A may be calculated mathematically (assuming both values of k are accurate). The two values of k satisfy

$$k_1 = Ae^{-E_a/RT_1} \quad \text{and} \quad k_2 = Ae^{-E_a/RT_2}$$

Dividing one equation by the other leads to

$$\frac{k_2}{k_1} = \frac{Ae^{-E_a/RT_2}}{Ae^{-E_a/RT_1}} = \exp\left[\frac{-E_a}{R}\left(\frac{1}{T_2} - \frac{1}{T_1}\right)\right] = \exp\left[\frac{E_a}{R}\left(\frac{1}{T_1} - \frac{1}{T_2}\right)\right]$$

(The notation "exp" is used whenever e has a large, unwieldy exponent. Thus exp (x) means e^x).

Taking the natural logarithm of both sides gives

$$\ln \frac{k_2}{k_1} = \frac{E_a}{R}\left(\frac{1}{T_1} - \frac{1}{T_2}\right)$$

Solving for E_a gives

$$E_a = R\left(\frac{T_1 T_2}{T_2 - T_1}\right) \ln \frac{k_2}{k_1}$$

Letting $k_1 = 1.2 \times 10^{-1}$ s^{-1} when $T_1 = 288$ K and letting $k_2 = 2.5 \times 10^{-1}$ s^{-1} when $T_2 = 298$ K produces

$$E_a = 8.314 \text{ J} \cdot \text{K}^{-1} \cdot \text{mol}^{-1} \left(\frac{288 \times 298}{10}\right) \text{K} \ln \frac{2.5 \times 10^{-1} \text{ s}^{-1}}{1.2 \times 10^{-1} \text{ s}^{-1}}$$

$$= (8.314)(8582.4) \ln 2.08 = 52{,}257 \text{ J} \cdot \text{mol}^{-1} \quad \text{or} \quad 52.3 \text{ kJ} \cdot \text{mol}^{-1}$$

The frequency factor is calculated from

$$A = \frac{k}{e^{-E_a/RT}} = k e^{E_a/RT}$$

using either rate constant. From k_1,

$$A = (1.2 \times 10^{-1} \text{ s}^{-1}) e^{52{,}257 \text{ J}/8.314 \text{ J} \cdot \text{K}^{-1} \times 288 \text{ K}}$$
$$= 3.6 \times 10^8 \text{ s}^{-1}$$

Quite a few common chemical reactions double or triple their rates with a 10-degree increase in temperature. The above reaction has $k_2/k_1 = 2.08$ for a temperature change of 10 K.

THE DYNAMICS OF CHEMICAL CHANGE

20.6 TRANSITION-STATE THEORY OF REACTION RATES

Although the Arrhenius activation-energy model is still useful, most modern treatments of reaction rates are based on the following improved model of a chemical reaction

$$\text{reactants} \rightleftharpoons \begin{array}{c}\text{transition-state complex}\\ \text{or}\\ \text{activated complex}\end{array} \rightarrow \text{products}$$

The reactants are assumed to first form an intermediate molecule called a **transition-state complex** or **activated complex**. The transition-state complex, once formed, can do either of two things: decompose back to reactants or form products via an essentially irreversible path. The rate constant for the formation of products is given by

$$k = \frac{RT}{N_0 h} K\ddagger \tag{20.9}$$

where N_0 is Avogadro's number, h is Planck's constant, and $K\ddagger$ is the equilibrium constant for the reactants and the transition-state complex. With Eq. (18.12) as an analogy, it is possible to define the quantity

$$\Delta G\ddagger = -2.3\, RT \log K\ddagger \quad \text{or} \quad -RT \ln K\ddagger \tag{20.10}$$

called the **activation free energy** of the reaction.* As indicated in Fig. 20.9, $\Delta G\ddagger$ is the free-energy change in going from reactants to the transition-state complex. Given Eq. (18.8) as a model, $\Delta G\ddagger$ must also satisfy

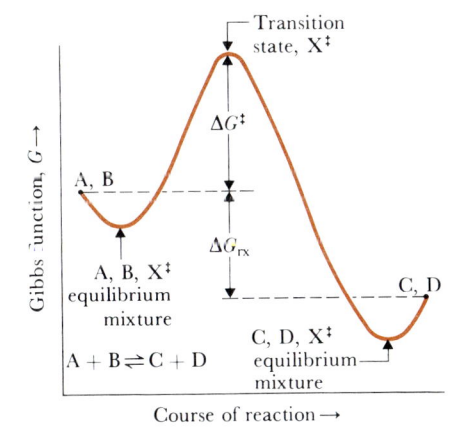

$$\Delta G\ddagger = \Delta H\ddagger - T\Delta S\ddagger \tag{20.11}$$

Fig. 20.9. The transition-state model of chemical reactions.

---

* $\Delta G\ddagger$ is read "delta gee double dagger" or "delta gee Christmas seal."

> *The Eyring rate equation is named after the American physical chemist Henry Eyring (b. 1901), who has made many fundamental contributions to an understanding of the rates of chemical and physical processes and the structures of liquids and solids. Up to his recent retirement, Professor Eyring was Dean of the Graduate School at the University of Utah. He is now Distinguished Professor of Chemistry at that university.*

where $\Delta H^\ddagger$ is the **activation enthalpy** and $\Delta S^\ddagger$ is the **activation entropy**. Equation (20.9) then becomes

$$k = \frac{RT}{N_0 h} 10^{\Delta S^\ddagger / 2.3 R} 10^{-\Delta H^\ddagger / 2.3 RT} \quad \text{or} \quad \frac{RT}{N_0 h} e^{-\Delta S^\ddagger / R} e^{-\Delta H^\ddagger / RT} \tag{20.12}$$

which is known as the **Eyring rate equation**. This equation permits a somewhat more sophisticated description of rate processes than is possible with the cruder Arrhenius equation. Note that the quantity

$$\frac{RT}{N_0 h} 10^{\Delta S^\ddagger / 2.3 R}$$

is analogous to the Arrhenius frequency factor A and that $10^{-\Delta H^\ddagger / 2.3 RT}$ is analogous to $10^{-E_a / 2.3 RT}$. The quantity $RT/N_0 h$ is a universal frequency factor which, unlike the Arrhenius A factor, depends on temperature; $10^{\Delta S^\ddagger / 2.3 R}$ represents the probability that the colliding reactants are properly oriented to form the transition-state complex; and $10^{-\Delta H^\ddagger / 2.3 RT}$ is the fraction of reactant molecules having sufficient energy to form the transition-state complex. Thus, in the one-step attack of RCH_2X by OH^-, the OH^- ion must be lined up with the oxygen atom pointing at the carbon of RCH_2X if the transition state is to form upon collision. In addition, the collision must have sufficient energy to form an HO bond to carbon.

Experiments show that as the R group on RCH_2X is made more bulky, then the reaction becomes slower. This occurs due to a smaller $\Delta S^\ddagger$; that is, it becomes less probable that a random collision with OH^- will involve the proper orientation for reaction even if the collision energy is adequate.

Thus it is easier to understand why it is possible for a reaction to be intrinsically spontaneous ($\Delta G < 0$) and yet proceed with a barely perceptible rate; the factors which govern intrinsic spontaneity (ΔG) and the factors which govern the rate of reaction ($\Delta S^\ddagger$ and $\Delta H^\ddagger$) have nothing to do with each other. The rate of a reaction is determined by how probable it is that reactants can attain the proper alignment to form the transition-state complex on collision (the $\Delta S^\ddagger$ factor) and by how much energy is required to form this complex (the $\Delta H^\ddagger$ factor). Fast reactions are characterized by high values of $\Delta S^\ddagger$ or low values of $\Delta H^\ddagger$, or both.

20.7 CHAIN REACTIONS

As indicated in Fig. 20.5, chain reactions exhibit what appears to be a very strange sort of temperature dependence. These reactions do not go at all as the temperature increases, until a particular temperature is reached. At this point the rate may abruptly become so large that the reaction proceeds with explosive force. One of the best-studied chain reactions is the reaction of hydrogen gas and bromine gas,

$$H_2(g) + Br_2(g) \rightarrow 2HBr(g)$$

A mixture of H_2 and Br_2 is quite stable unless heated, subjected to intense light, or sparked, whereupon an explosion occurs. The experimentally determined rate law for the reaction is

$$R_f = \frac{k_1[H_2][Br_2]^{1/2}}{k_2 + [HBr]/[Br_2]} \tag{20.13}$$

Clearly, this rate law does not fit the general form given by Eq. (20.1). Although the reaction is first order with respect to H_2, the orders with respect to HBr and Br_2 cannot be simply defined. An adequate discussion of this reaction is outside the scope of this book, but the mechanism is postulated to include elementary steps such as

1. $Br_2 \rightarrow 2Br\cdot$ chain initiation
2. $M + Br\cdot + Br\cdot \rightarrow Br_2 + M$ chain termination
3. $Br\cdot + H_2 \rightarrow HBr + H\cdot$ ⎫
   $H\cdot + Br_2 \rightarrow HBr + Br\cdot$ ⎬ propagation
4. $HBr + H\cdot \rightarrow H_2 + Br\cdot$ chain inhibition

where M represents a third body needed to carry away the energy of reaction to prevent decomposition of Br_2 by recoil (see Section 15.11 for a similar situation in the ozone reaction). By contrast with the activated reactions considered earlier, there is no single RDS in chain reactions. This is also true in many other types of complex reactions. The reason for the behavior of chain reactions is as follows: Only at a high enough temperature is it possible to break the bond whereby the initiating substance ($Br\cdot$ in the above) is formed. However, once a single $Br\cdot$ is formed, the propagation steps produce a constant supply of species such as $H\cdot$ and $Br\cdot$ which can continue the production of HBr. Once the reaction begins, the energy released raises the temperature of the reaction products to produce even more $Br\cdot$. Thus the production of $Br\cdot$ accelerates and the reaction rate very quickly reaches very high values.

The fission of a supercritical mass of plutonium or uranium (see Section 11.6) follows a similar mechanism.

20.8 CATALYSIS

Sometimes the rate of a chemical reaction can be increased by adding what is ostensibly a foreign substance to the reaction mixture. A simple example, easily carried out in the home or in a dormitory ashtray, is to take a cube of sugar and attempt to light it with a match; the sugar cube will melt and char but will not ignite and continue burning. Yet, if a small amount of cigarette ashes is placed on the edge of the cube, the cube can be lighted and will burn with a steady flame.

Another example involves the heating of potassium chlorate, KClO$_3$, to produce oxygen:

$$2KClO_3(s) \rightarrow 2KCl(s) + 3O_2(g)$$

Even at very high temperatures, very little O$_2$ is produced from pure KClO$_3$. However, if a small amount of powdered manganese(IV) oxide, MnO$_2$, is first mixed with the KClO$_3$, the reaction produces a steady stream of O$_2$ even upon gentle heating.

Reactions such as the above are said to be **catalyzed** and the foreign materials (ashes and MnO$_2$) are called **catalysts** (from the Greek *kata*, wholly + *lyein*, to loosen).* An important characteristic of a catalyst is that it does not become consumed during the course of the reaction; ideally, all of the catalyst can be recovered after the reaction is completed.

Catalysts have been recognized since early in the 19th century when chemists such as Berzelius and Mitscherlich noted that H$_2$O$_2$ decomposed more rapidly in the presence of silver or platinum, but only recently have scientists begun to acquire some insight as to how catalysts operate. According to the modern theory of catalysis, all catalysts belong to one of two basic types.

TYPE I: The catalyst provides a surface upon which one or more reactants become attached and thereby activated.

TYPE II: The catalyst participates chemically in one or more elementary steps but becomes regenerated in another elementary step.

Both types of catalyst systems are illustrated in the following reactions.

The reaction

$$2SO_2(g) + O_2(g) \rightarrow 2SO_3(g)$$

occurs extremely slowly in the absence of other substances. However, if the reactants SO$_2$ and O$_2$ are passed through finely divided platinum or over a platinum gauze, reaction occurs readily. The mechanism of the catalytic action of platinum is assumed to be as follows: Oxygen molecules are ordinarily highly unreactive due to the rather high value of the O=O bond energy (494 kJ), but oxygen atoms are extremely reactive. Thus oxygen molecules are adsorbed on the platinum surface as depicted in Fig. 20.10. Adsorption of O$_2$ on the platinum surface stretches the O=O bond and thereby weakens it. When an SO$_2$ molecule strikes the oxygen-covered platinum surface it is able to pick up an oxygen atom to form SO$_3$. The remaining oxygen atom may leave the platinum surface and eventually participate in a three-body process to produce a second SO$_3$:

$$SO_2(g) + O(g) + M \rightarrow SO_3(g) + M$$

Many reactions of the above type are known. The hydrogenation of alkenes to form

---

* It is possible that the ashes in the first example are not really a catalyst but rather act somewhat like a wick in an oil lamp. For example, touching a match to the surface of a cup of ordinary cooking oil will not ignite it, but a string or piece of cloth dipped in the oil ignites readily.

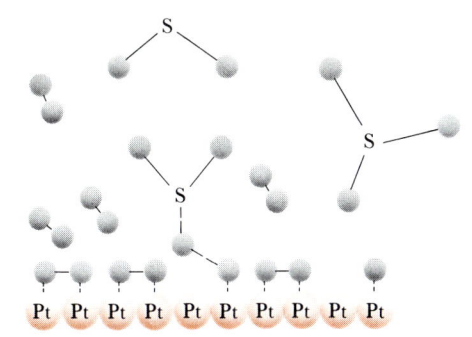

Fig. 20.10. Illustration of a possible mechanism for the platinum-catalyzed $2SO_2 + O_2 \rightarrow 2SO_3$ reaction. Note that the O—O bond becomes longer when O_2 is adsorbed on Pt.

alkanes

$$R-CH=CH-R' + H_2 \rightarrow RCH_2CH_2R'$$

is usually carried out in the presence of metals with catalytic activity. Vegetable oils, which are liquids containing many double bonds, are converted to solids by catalytic hydrogenation to produce shortenings such as Crisco and Spry. The ammonia synthesis reaction must be run in the presence of catalysts; otherwise inordinately high temperatures are required—temperatures at which ΔG_{rx} begins to become less negative. A similar situation obtains in the synthesis of diamond from graphite; ΔG_{rx} for this process becomes more negative at lower temperatures but the rate of the reaction at low temperatures is too low to be useful. The use of a catalyst makes it possible to strike an acceptable compromise between rate and intrinsic spontaneity.

The decomposition of formic acid in acid solution is catalyzed by hydrogen ions supplied by an acid HA. The overall reaction is

$$HCO_2H(aq) \xrightarrow{H^+} CO(g) + H_2O$$

Jöns Jacob Berzelius (1779–1848), a Swedish chemist, is known for the accurate determination of atomic weights, origin of the term "protein," development of the concept of isomerism, and pioneering work on catalysis. At age 56, Berzelius was made a baron by the King of Sweden—a royal gift to honor Berzelius's wedding. Berzelius was unquestionably one of the greatest chemists of the nineteenth century; his pioneering work determined the course of chemical science for over a hundred years.

Eilhard Mitscherlich (1794–1863) was a German chemist who left the study of oriental languages to take up medicine and ultimately began to carry out studies in chemistry. Mitscherlich worked for over a year with Berzelius in Stockholm and the two became lifelong friends. Mitscherlich coined the name benzene, discovered selenic acid, and studied the yeast-induced fermentation of sugar.

The mechanism is believed to be as follows. The first step is

$$H-\underset{\underset{O}{\|}}{C}-\ddot{O}-H + HA \rightleftharpoons \left[H-\underset{\underset{O}{\|}}{\overset{\overset{H}{\ddot{|}}}{C}}-\ddot{O}-H \right]^+ + A^-$$

This is followed by loss of water from the positive ion to form the formyl ion:

$$\left[H-\underset{\underset{O}{\|}}{\overset{\overset{H}{\ddot{|}}}{C}}-\ddot{O}-H \right]^+ \rightarrow H-\underset{\underset{O}{\|}}{C}\oplus + H_2O$$

The formyl ion now reacts with the conjugate base of HA to regenerate HA:

$$H-\underset{\underset{O}{\|}}{C}\oplus + A^- \rightarrow HA + CO(g)$$

Many biological reactions probably function in an analogous manner.

Whatever the particular mechanism of catalytic action, catalysts must decrease the $\Delta G^\ddagger$ of the reaction. This can be done either by decreasing the $\Delta H^\ddagger$ below that of the uncatalyzed reaction or by increasing the $\Delta S^\ddagger$ above that of the uncatalyzed reaction, or both. It is also clear that a catalyst should affect the rates of the forward and reverse reactions in the same way so that the point of equilibrium is not affected. This conclusion follows from the fact that a catalyst has no effect on ΔG_{rx} (see Fig. 20.11) and thus has no effect on ΔG_{rx}^0 or K_p. In short, catalysts act by providing more rapid routes between initial and final states.

Catalysts which slow down the rate of a reaction are also known. These are usually termed **negative catalysts** or **inhibitors**. Such substances are added to hydrogen peroxide solutions to increase their shelf life. Some food additives used to retard spoilage may function in a similar way.

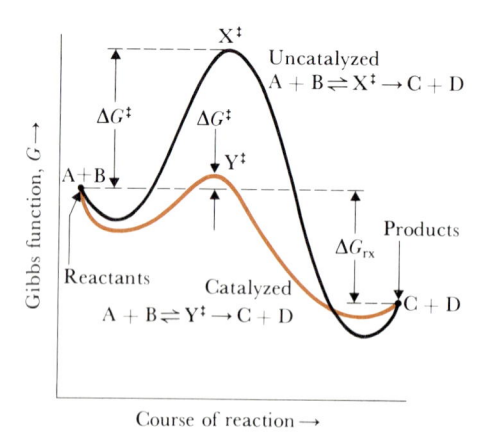

Fig. 20.11. The function of a catalyst according to the transition-state theory of reaction rates. Note that the uncatalyzed and catalyzed reactions go through different transition-state complexes.

EXAMPLE 20.5 The decomposition of hydrogen peroxide in aqueous solution is represented by the reaction

$$2H_2O_2(aq) \rightarrow 2H_2O(l) + O_2(g)$$

The reaction is slow unless catalyzed or under the action of light; iodide ion is one example of a catalyst. The rate law for the I^- catalyzed reaction is

$$-R_{H_2O_2} = k[I^-][H_2O_2] = k'[H_2O_2]$$

Explain.

SOLUTION Although the amount of catalyst affects the rate, its concentration is constant for a given experiment. Thus $k[I^-] = k'$ and the reaction is treated as if it were overall a first-order reaction. Example 20.3 involves a similar type of reaction. See Problem 44 for a mechanism of the hydrogen peroxide reaction.

20.9 ENZYME-CATALYZED REACTIONS

Many biological reactions are catalyzed by protein substances called **enzymes**. Enzymes are remarkably selective in their catalytic activity; normally a single enzyme catalyzes one and only one specific reaction. Compared with other types of catalysts, enzymes are remarkably efficient as well. For example, the following acid-catalyzed hydrolysis reaction may be carried out in the laboratory:

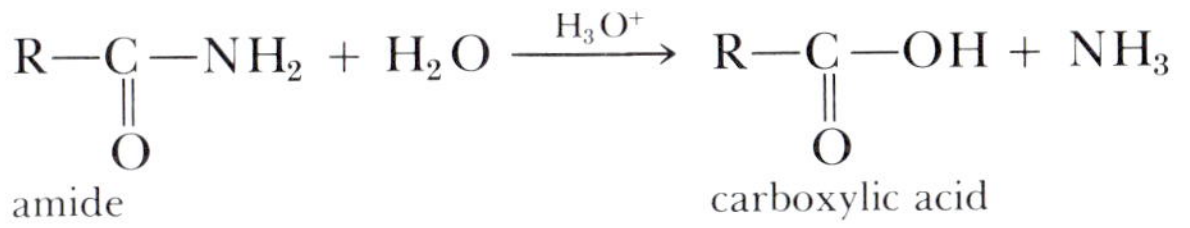

$$\underset{\text{amide}}{R-\underset{\underset{O}{\|}}{C}-NH_2} + H_2O \xrightarrow{H_3O^+} \underset{\text{carboxylic acid}}{R-\underset{\underset{O}{\|}}{C}-OH} + NH_3$$

When R is the group

$$CH_3\underset{\underset{O}{\|}}{C}NH\overset{|}{C}HCH_2-\!\!\!\left\langle\bigcirc\right\rangle\!\!\!-OH$$

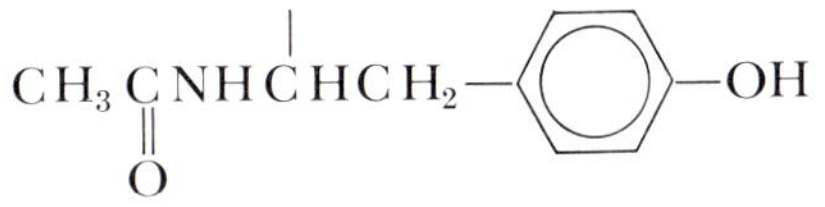

(the amide is then N-acetyl tyrosinamide) then the enzyme α-chymotrypsin carries out the hydrolysis about *one million times faster* than does H_3O^+.* This enzymic reaction is one of the processes occurring in food digestion.

Enzymes are complex protein structures (see Chapter 24) and the structures of only a few are known in any detail. Many appear to owe their catalytic abilities to a combination of shape and the presence of certain functional groups. The shape of the enzyme allows it to form weakly bonded complexes with specific reactants only, and the functional groups permit the enzyme to alter the reactant in a specific

---

* An enzymic rate enhancement of 10^6 is relatively modest; rate enhancements of 10^8 to 10^{12} are more common.

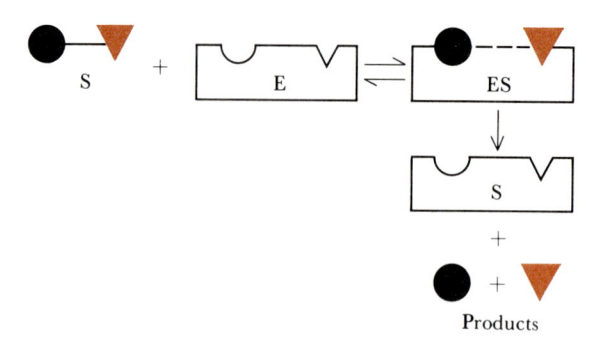

Fig. 20.12. Illustration of the "lock-and-key" model of enzyme catalysis. Decarboxylation reactions in which S = RCO$_2$H, O = RH, and △ = CO$_2$ are known.

chemical way so as to make it more reactive toward other reactants. Figure 20.12 depicts what is usually called the **lock-and-key** model of enzyme catalysis. The **substrate** (reactant molecule) fits into a niche of the right shape on the enzyme surface (called the **active site**) and becomes activated so that it decomposes to products. In some respects this process is a composite of the surface adsorption mechanism of SO$_2$ oxidation (Fig. 20.10) and the regeneration mechanism of formic acid decomposition (p. 566). The general mechanism of an enzyme-catalyzed process is sometimes written

$$E + S \rightleftharpoons ES \rightarrow E + \text{products}$$

where E represents the enzyme and S represents the molecule it activates (the substrate). ES is the intermediate complex formed by the enzyme and the substrate. Apparently S becomes more chemically reactive when bound to the enzyme than when it is unbound.

In Fig. 20.13 there is shown a simplified illustration of what is believed to be the mode of operation of the enzyme **lysozyme**. Lysozyme has a molecular weight of about 15,000 amu and contains about 130 amino acids joined in a long chain. The chain is coiled and folded back onto itself several times so that the enzyme has the shape of an egg with a cleft or open area in its side. This cleft is the active site of lysozyme. This enzyme is known to break down certain bacterial substances by cleaving long polysaccharide chains which are components of bacterial cell walls. Specifically, lysozyme acts as a pair of chemical scissors to snip a hexasaccharide chain into two shorter chains. View (a) shows the cleft in the lysozyme molecule which constitutes the "blades" of the chemical scissors and the bond in the hexasaccharide which these cut. The shape of the lysozyme molecule, although exaggerated, is actually quite similar to that illustrated. View (b) illustrates how the hexasaccharide fits into the active site of the enzyme, where it is held by electrostatic forces arising from the polarity of the bond to be broken and from polarities of portions of certain amino acids which occupy key positions in the cleft. As the

$$\diagdown_{\diagup}\!\!C\!-\!O\!-\!C\diagup^{\diagdown}$$

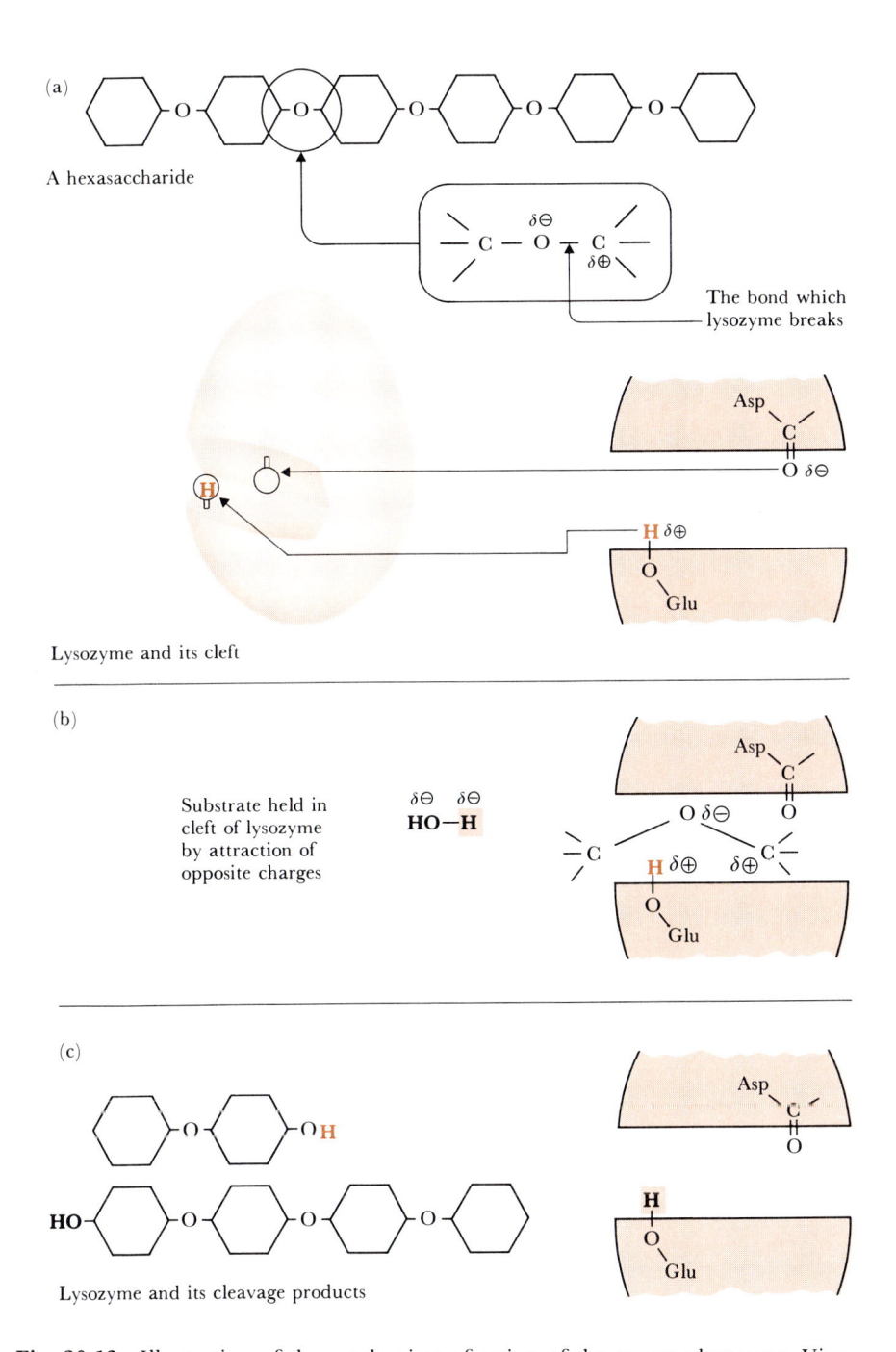

Fig. 20.13. Illustration of the mechanism of action of the enzyme lysozyme. View (a) shows the appearance of the substrate and the active site of the enzyme, view (b) shows how the substrate fits into the active site of the enzyme and is held there by electrostatic forces, and view (c) shows the products of the reaction and the enzyme ready for another substrate molecule.

20.9 | ENZYME-CATALYZED REACTIONS 569

bond breaks to

$$\begin{matrix}\diagdown\\ -\mathrm{C}-\mathrm{O}\ominus\\ \diagup\end{matrix} \quad \text{and} \quad \oplus\,\mathrm{C}\begin{matrix}\diagup\\ \\ \diagdown\end{matrix}$$

H$^+$ from one amino acid (labeled Glu for glutamine) adds to

$$\begin{matrix}\diagdown\\ -\mathrm{C}-\mathrm{O}\ominus\\ \diagup\end{matrix},$$

Next, OH$^-$ from H$_2$O adds to

$$\oplus\,\mathrm{C}\begin{matrix}\diagup\\ \\ \diagdown\end{matrix},$$

and H$^+$ from H$_2$O replaces the H$^+$ lost by glutamine. View (c) shows the disaccharide and tetrasaccharide products of the enzyme-catalyzed hydrolysis, along with the enzyme ready to cut another hexasaccharide chain.

It is interesting to note that living organisms consist of a great number of **metastable systems**—systems whose decomposition is intrinsically spontaneous but which can decompose (or react) only in the presence of specific enzymes. This very metastability is the key to life. Systems which are stable, that is, intrinsically unable to react, could not be activated by enzymes to produce work when the need arises. The high degree of selectivity of enzymes guarantees that reactions can be turned on and off as desired, rather than being allowed to occur at random.

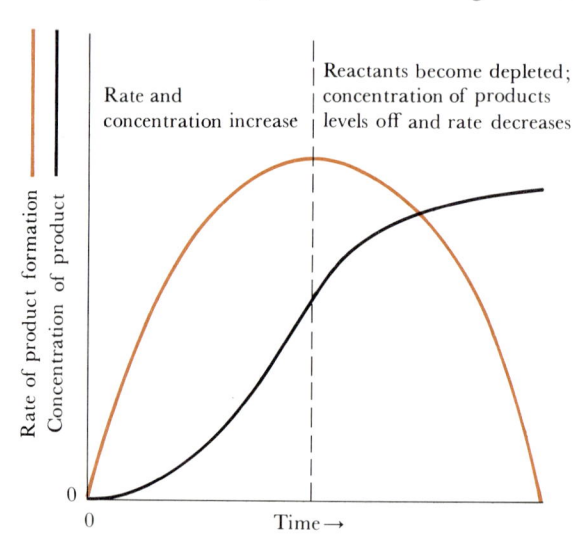

Fig. 20.14. Variation of rate of reaction with time for an autocatalytic process. The rate eventually goes to zero as all reactants are consumed.

20.10 AUTOCATALYSIS

There are some reactions in which one of the products of the reaction functions as a catalyst for that reaction. For example

$$2\mathrm{MnO_4^-(aq)} + 5\mathrm{C_2O_4^{2-}(aq)} + 16\mathrm{H^+(aq)} \rightarrow 2\mathrm{Mn^{2+}(aq)} + 10\mathrm{CO_2(g)} + 8\mathrm{H_2O(l)}$$

Initially, permanganate oxidizes the oxalate ion slowly; but as [Mn^{2+}] increases, the oxidation becomes progressively faster due to catalysis by Mn^{2+}. Figure 20.14 shows how the rate of a typical self-catalyzed or **autocatalyzed** reaction varies with time. Many scientists believe that autocatalysis may have played an important role in the chemical evolution of living systems. It has always been a puzzle as to how the complex molecules acting as precursors to life forms could have accumulated in large enough amounts to be available when needed. One simple possibility is that these complex molecules acted as autocatalysts in their own formation processes.

SUMMARY

The effect of temperature on the rates of chemical reactions indicates that there are at least three types of reactions: activated reactions, chain reactions, and enzymic reactions.

Activated reactions have rates which increase exponentially as the temperature increases. Experimentally determined rate laws for such reactions provide evidence for the deduction of reaction mechanisms. The rates of many activated reactions depend on a single rate-determining step. The rate constant of an activated reaction depends on the free energy of activation, the free-energy change involved in forming the transition-state complex. This free-energy change consists of an entropy term, which relates to the probability of correctly oriented colliding reactants, and an enthalpy term, which relates to the energy needed to form the transition-state complex.

Chain reactions have virtually zero rates of reaction until a certain threshold temperature is reached, whereupon the reaction rate abruptly becomes very large.

Catalysts are substances added to a reaction mixture which increase the rate of reaction but do not become used up in the reaction. Catalysts may function as surfaces upon which certain steps of the reaction take place or they may actually participate chemically in a reaction step only to become regenerated later. Catalysts act by increasing the activation entropy, decreasing the activation enthalpy, or both. They have no effect on the intrinsic spontaneity of the reaction.

Enzymes are biological catalysts and differ from other catalysts in that they are deactivated by either very low or very high temperatures. Enzymes are very specific in their actions and they are very efficacious in increasing reaction rates. It is thought that many enzymes operate via a lock-and-key mechanism in which the substrate fits into a definite space on the enzyme surface, where it becomes activated.

LEARNING GOALS

1. Know the four basic factors affecting the rates of chemical reactions.

2. Be able to determine the rate law of a simple chemical reaction using the appropriate experimental data.

3. Be able to distinguish among activated reactions, chain reactions, enzymic reactions, and nuclear reactions on the basis of temperature dependence.

4. Know what is meant by the mechanism of a reaction and what is meant by elementary steps and rate-determining steps.

5. Be able to explain in detail the two types of assumed mechanisms of the reaction of alkyl halides with OH^-.

6. Know in detail how the mechanism of ozone decomposition proceeds and how it correlates with the experimental rate law.

7. Know the Arrhenius equation and be able to identify all terms in it.

8. Know the Eyring rate equation and be able to identify all terms in it.

9. Be able to give the details of one example each of the two types of catalysis.

10. Know the basic mechanism thought to describe most enzyme reactions.

11. Be able to describe in fair detail how the enzyme lysozyme functions in the hydrolysis of a hexasaccharide.

12. Know the principal features of autocatalysis, including how the rate of reaction varies with time.

DEFINITIONS, TERMS, AND CONCEPTS TO KNOW

rate (velocity) of a chemical reaction
metastable system
initial rate of reaction
net rate of reaction
rate law
rate constant
order of a reaction

mechanism of a reaction
rate-determining step
activation energy
frequency factor
transition-state (or activated) complex
Eyring rate equation
meaning of $\Delta G^\ddagger$, $\Delta H^\ddagger$, and $\Delta S^\ddagger$

chain reaction
catalyst
enzyme
lock-and-key mechanism
active site
autocatalysis
Arrhenius rate equation

QUESTIONS AND PROBLEMS

Rate of Reaction

1. Consider the reaction $2NO(g) + Br_2(g) \rightarrow 2NOBr(g)$. At a given time the rate of appearance of NOBr is 0.03 mol·L^{-1}·min^{-1}.
  a) What is the rate of disappearance of NO at the same instant?
  b) What is the rate of disappearance of Br$_2$ at the same instant?

2. Consider the ammonia synthesis reaction
$$N_2(g) + 3H_2(g) \rightarrow 2NH_3(g)$$
At a given instant of time the rate of disappearance of nitrogen is measured to be 0.002 mol·L^{-1}·hr^{-1}.
  a) What is the rate of disappearance of hydrogen at the same instant?
  b) What is the rate of appearance of ammonia at the same instant?

3. Consider the reaction $2A + B \rightarrow C$.
  a) Show that the rate of disappearance of A is twice the rate of disappearance of B.
  b) Does the value of k in the rate law depend on how the rate of a reaction is expressed? Explain.

4. The reaction $A + B \rightarrow 2C$ has the rate law
$$R_C = k[A]^{1/2}[B]$$
with $k = 6 \times 10^{-2}$ L$^{1/2}$·mol$^{-1/2}$·min^{-1}. What is the rate of appearance of C when $[A] = 0.8\ M$ and $[B] = 0.25\ M$?

5. A reaction has the rate law $-R_B = k[A][B]^{1/2}$. At a given temperature and given concentrations of A and B the rate of disappearance of B is 0.01 mol·L^{-1}·min^{-1}.
  a) What is the rate of reaction if [B] is kept the same but [A] is doubled?
  b) What is the rate of reaction if [A] is kept the same but [B] is quadrupled?
  c) What is the rate of reaction if [A] is doubled and [B] is quadrupled at the same time?

6. The reaction $2A + 3B \rightarrow C + 4D$ is found to have the rate law

rate of disappearance of A (in mol·L^{-1}·min^{-1})
$$= k[A][B]^2$$
with $k = 10^{-3}$ L^2·mol^{-2}·min^{-1} at 25 °C.

572   THE DYNAMICS OF CHEMICAL CHANGE

a) What is the initial rate of reaction (in terms of A) when 1 mole of A and 2 moles of B are mixed in a 2-L container?
b) If the rate of disappearance of A is 2 mol·L⁻¹·hr⁻¹, what is the rate of disappearance of B at the same time?
c) How fast is D appearing at this time?

7. A 10-L flask at a given temperature contains 0.010 mol of ozone. After a certain amount of time the amount of ozone remaining is 0.008 mol.
a) What is the rate of disappearance of ozone (in units of k) at this time?
b) What is the rate of appearance of oxygen (in units of k) at this time?

8. What are the four major factors that affect the rates of chemical reactions? Give a specific example of each factor.

9. Consider the general reaction

$$aA + bB \rightarrow cC + dD$$

Show that the various ways of expressing the rate of the reaction are related as follows:

$$-\frac{1}{a}R_A = -\frac{1}{b}R_B = \frac{1}{c}R_C = \frac{1}{d}R_D$$

10. Use simple kinetic molecular theory (see Section 3.7) to explain why rates of reaction generally increase as temperature is increased.

11. The number of chirps a cricket makes per unit time increases uniformly as the temperature increases. Explain this phenomenon on the basis of chemical reaction rates.

12. Why is an automobile generally harder to start on a very cold morning than on a warm one?

13. Not all collisions between reactant molecules lead to chemical reaction. Suggest a simple way of proving this statement.

14. For each reaction below, list all the different ways of expressing the rate of reaction and show how the various ways are related.
a) $2NOCl(g) \rightarrow 2NO(g) + Cl_2(g)$
b) $3O_2(g) \rightarrow 2O_3(g)$
c) $H_2(g) + \frac{1}{2}O_2(g) \rightarrow H_2O(g)$

15. The decomposition of hydrazine follows the reaction

$$N_2H_4(g) \rightleftharpoons N_2(g) + 2H_2(g)$$

At a given time the rate of appearance of hydrogen is 0.001 mol·L⁻¹·hr⁻¹

a) What is the rate of disappearance of hydrazine at the same time?
b) What is the rate of appearance of nitrogen at the same time?
c) Sketch how $-R_{N_2H_4}$ and R_{N_2} would be expected to vary with time.

16. What is meant by the **net rate** of reaction? Under what conditions is the net rate of reaction equal to the rate of the forward reaction?

17. What is the effect of the doubling of the concentration of a reactant on the rate of reaction if that reactant has the following order?

a) 1   c) 2   e) 3
b) $\frac{1}{2}$ d) 0   f) $-\frac{1}{2}$

Rate Law Determination and Mechanisms

18. The following initial rates of reaction refer to the reaction $2A + B \rightarrow C$ at a given temperature.

EXPERIMENT NUMBER	INITIAL RATE OF PRODUCTION OF C (mol·L⁻¹·hr⁻¹)	INITIAL CONCENTRATIONS: [A]	[B]
1	1.0	0.01	0.04
2	2.0	0.04	0.04
3	0.5	0.01	0.02
4	2.8	0.02	0.08

a) What is the order of the reaction with respect to A?
b) What is the order of the reaction with respect to B?
c) What is the rate law of the reaction?
d) What is the total order of the reaction?
e) Calculate the rate constant k.
f) What is the rate of disappearance of A in experiment 4?

19. Initial rates of reaction were determined as follows for the reaction $A + B \rightarrow C$.

EXPERIMENT NUMBER	$-R_A$(mol·L⁻¹·s⁻¹)	[A]	[B]
1	0.02	0.001	0.002
2	0.18	0.003	0.002
3	0.02	0.001	0.003

a) What is the rate law of the reaction?
b) What is the value (in correct units) of the rate constant k?

20. A chemical reaction has the stoichiometry

$$A + 2B + C \rightarrow D + E$$

QUESTIONS AND PROBLEMS 573

The effect of varying reactant concentrations on initial rates is summarized below.

EXPERIMENT NUMBER	INITIAL RATE OF DISAPPEARANCE OF C (mol·L⁻¹·min⁻¹)	INITIAL CONCENTRATIONS: [A]	[B]	[C]
1	0.05	0.1	0.2	0.3
2	0.15	0.3	0.2	0.3
3	0.20	0.1	0.4	0.3
4	0.20	0.1	0.4	0.6

a) Write the rate law for the reaction.
b) What is the rate of the reaction when $[A] = 0.5\ M$, $[B] = 0.7\ M$, and $[C] = 2.0\ M$?

21. Use the following assumptions to demonstrate that the mechanism proposed for the NO + H_2 reaction (footnote on p. 556) leads to the experimentally observed rate law

Rate of formation of $NOH_2 = k_1[NO][H_2]$

Rate of disappearance of NOH_2
$= k_{-1}[NOH_2] + k_2[NOH_2][H_2] - k_1[NO][H_2]$

Net rate of formation of NOH_2
$= 0$ (used up as fast as it is formed)

Assume $k_{-1} \gg k_2[H_2]$.

22. The rate of growth of a sexually reproducing population would seem to be first order with respect to males and first order with respect to females. Yet evidence indicates that for humans it is more nearly first order with respect to total population (male plus female).
a) Show that the long pregnancy time and infertile periods of the female act like a rate-determining step with a first-order dependence on total population.
b) Would a high female/male ratio lead to a second-order process?

23. The following mechanism is proposed for a reaction:

Step 1 $2A \to B$
Step 2 $B + C \to E$
Step 3 $E + C \to D$

What is the net chemical reaction represented by the mechanism?

24. The formation of polyethylene from ethylene may be carried out by a chain reaction process thought to involve steps such as
a) $R\cdot + CH_2{=}CH_2 \to RCH_2CH_2\cdot$
b) $RCH_2CH_2\cdot + CH_2{=}CH_2 \to RCH_2CH_2CH_2CH_2\cdot$
c) $R_2 \to 2R\cdot$
d) $R(CH_2)_{100} + R(CH_2)_{50} \to R(CH_2)_{150}R$

Identify each step as initiation, propagation, or termination.

25. The data below were obtained at a given temperature for the reaction

$2NO(g) + Cl_2(g) \to 2NOCl(g)$

EXPERIMENT NUMBER	INITIAL RATE $-R_{NO}$(mol·L⁻¹·s⁻¹)	INITIAL CONCENTRATIONS: [NO]	[Cl₂]
1	2.78×10^{-6}	0.10	0.10
2	5.57×10^{-6}	0.10	0.20
3	11.1×10^{-6}	0.20	0.10
4	25.1×10^{-6}	0.30	0.10

a) What is the rate law of the reaction?
b) What is the rate constant of the reaction?
c) What is the initial rate of the reaction when 0.5 mol of NO and 0.3 mol of Cl_2 are mixed in a 1.0-L flask?

26. The rate of a reaction is second order in a given reactant.
a) What is the effect on the rate if the concentration of this reactant is increased from $0.021\ M$ to $0.056\ M$?
b) What is the effect on the rate if the concentration of this reactant is decreased from $0.045\ M$ to $0.011\ M$?

27. A reaction is postulated to have the following mechanism:

Step 1 $NO(g) + Br_2(g) \underset{k_{-1}}{\overset{k_1}{\rightleftarrows}} NOBr_2(g)$

Step 2 $NOBr_2(g) + NO(g) \xrightarrow{k_2} 2NOBr(g)$

a) What is the net reaction?
b) Write the rate expression for appearance of $NOBr_2$ in step 1.
c) Write the rate expression for the disappearance of $NOBr_2$ in steps 1 and 2.
d) What is the expression for the net rate of appearance of $NOBr_2$?
e) If step 2 is assumed to be the rate-determining step, what is the rate of appearance of NOBr in this step?
f) If $NOBr_2$ disappears as fast as it is formed, we can assume $R_{NOBr_2} = 0$. Show that this assumption

574   THE DYNAMICS OF CHEMICAL CHANGE

along with the additional assumption that k_{-1} is much larger than $k_2[Br_2]$ leads to the rate law

$$R_{NOBr} = k[NO]^2[Br_2]$$

Compare this situation with that discussed in Problem 21.

28. Discuss why it is seldom if ever possible to prove conclusively that a given mechanism is correct. Is it ever possible to prove conclusively that a given mechanism is wrong? Explain.

Calculations of the Rate Constant

29. The reaction $2A + B \rightarrow C$ has the rate law $-R_A = k[A][B]$. When $[A] = [B] = 0.1\ M$, the rate of disappearance of A is $0.001\ mol \cdot L^{-1} \cdot hr^{-1}$. Calculate the rate constant of the reaction.

30. The reaction $A + 2B \rightarrow C$ is first order with respect to A and is not affected by the concentration of B. At $t = 0$, with no C present, 0.4 mol of A is mixed with 1.5 mol B. Two hours later 0.3 mol of C is present.
  a) What is the half-life of A?
  b) What is the rate constant of the reaction?

31. A reaction of the type

$$2A + B \rightarrow products$$

has the rate law

$$-R_A = k[A]$$

It is found that it takes 96 min for a 0.010-M solution of A to decrease to 0.007 M.
  a) What is the value of the rate constant of the reaction?
  b) How long will it take for [A] to reach 0.005 M?

32. The reaction

$$2NO(g) + Br_2(g) \rightarrow 2NOBr(g)$$

has the rate law

$$-R_{NO} = k[NO]^2[Br_2]$$

When $[NO] = 0.10\ M$ and $[Br_2] = 0.05\ M$, the rate of disappearance of NO is found to be 3.12×10^{-5} $mol \cdot L^{-1} \cdot s^{-1}$.
  a) Calculate the rate constant k.
  b) What is the value of k if the rate law is written

$$-R_{Br_2} = k[NO]^2[Br_2]$$

  c) Explain why k is different in (a) and (b).

Activation Energy

33. What is the activation energy of a reaction whose rate constant at 35 °C is double that at 25 °C?

34. A reaction has an activation energy of 45 kJ·mol^{-1}. How much faster will this reaction proceed (for given concentrations of reactants) at 35 °C than at 25 °C?

35. A first-order reaction has a rate constant of 6.5×10^{-5} s^{-1} at 10 °C and 2.3×10^{-2} s^{-1} at 70 °C.
  a) Calculate the half-life of each reaction at each of the above two temperatures.
  b) What is the activation energy of the above reaction?
  c) What is the frequency factor (and its units) for the above reaction?

36. Chloroethane, CH_3CH_2Cl, decomposes by a first-order process. At 550 °C it takes 22 s for half of a given sample to decompose, and at 575 °C it takes 31 s for 75% of a given sample to decompose.
  a) What is the activation energy of the reaction?
  b) What is the half-life of chloroethane at 600 °C?

37. A given reaction has an activation energy of 120 kJ·mol^{-1}. At 200 °C the rate constant has a value of 3.5×10^{-3} L·mol^{-1}·s^{-1}.
  a) What is the order of the reaction?
  b) What is the value of the rate constant at 300 °C?

38. A given reaction has an activation energy of 80 kJ·mol^{-1} and a rate constant value of 3.2×10^{-3} s^{-1} at 25 °C. What is the value of the rate constant at 50 °C?

39. How fast would a reaction go if it had an activation energy of zero?

40. Some reactions (the burning of wood, for example) require a high temperature to get started but proceed spontaneously from then on. Explain.

Catalysis

41. Use Eq. (18.12) and Fig. 20.11 to explain why catalysts do not affect the point at which equilibrium is reached.

42. What effect would a catalyst have on a reaction such as $N_2(g) + 3H_2(g) \rightleftharpoons 2NH_3(g)$ if the catalyst is added *after* equilibrium has been reached?

43. The reaction $2SO_2(g) + O_2(g) \rightarrow 2SO_3(g)$ is catalyzed by platinum. However, the catalyst eventually becomes inactive and must be reactivated (for example, by heating it in an inert atmosphere). Suggest a reason for this behavior.

44. Hydrogen peroxide decomposes according to the reaction

$$2H_2O_2 \rightarrow 2H_2O + O_2$$

The reaction is catalyzed by iodide ion. A possible mechanism of the reaction is

$$H_2O_2 + I^- \rightarrow H_2O + IO^- \quad \text{(slow)}$$
$$IO^- + H_2O_2 \rightarrow H_2O + O_2 + I^- \quad \text{(fast)}$$

a) What is the rate-determining step?
b) Does the mechanism support the experimental fact that the rate law is $-R_{H_2O_2} = k[H_2O_2]$? Explain. (See Example 20.5.)

45. Discuss supercooling (see Section 9.3) in terms of thermodynamic metastability. Assuming $\Delta S\ddagger < 0$ and $\Delta H\ddagger \approx 0$ for the crystallization process, what is a thermodynamic explanation for the role of the seed crystal used to induce crystallization? Could you call the seed crystal a catalyst?

46. Explain why enzyme-catalyzed reactions have the unusual temperature dependence shown in Fig. 20.5.

47. Suppose you have calculated that a given reaction has a very large negative value of ΔG but yet you are unable to get the reactants to produce any products. Which one or more of the following approaches should you take? Justify your answer.
  a) Try the reaction at a lower temperature.
  b) Try the reaction at a higher temperature.
  c) Look for a catalyst for the reaction.
  d) Give up.
Would you change your approach if ΔG had turned out to have a very large positive value?

48. Many enzyme-catalyzed reactions have rate laws of the general form

$$\text{Rate} = a\left(\frac{1}{1 + \frac{b}{[S]}}\right)$$

where $[S]$ is the molar concentration of substrate and a and b are constants.
  a) Show that a is the maximum rate obtainable by using very large values of $[S]$, and that b is the value of $[S]$ needed in order to get a rate one-half the maximum possible value.
  b) Show that values for the constants a and b may be determined in either of the following two ways.
  i) Plot rate^{-1} against $[S]^{-1}$. The slope is b/a and the intercept is $1/a$.
  ii) Plot rate/$[S]$ against rate. The slope is $-1/b$ and the intercept is a/b.

49. Prior to 1888 one could go from Switzerland to Italy by crossing Saint Gotthard Pass in the Lepontine Alps. This pass has a minimum elevation of 6935 ft. Now there is a railway tunnel, 9.3 mi in length, at an elevation of 4000 ft. Compare this with the case of a catalyst making a chemical reaction go faster.

50. When vapors of formic acid, HCO_2H, are passed over certain solid catalysts, CO and H_2O are formed. However, some catalysts lead to CO_2 and H_2 instead. Suggest possible reasons for the different behaviors.

General

51. Both He_2 and C_6H_6 (benzene) are thermodynamically unstable at room temperature, yet He_2 cannot be isolated whereas benzene is produced industrially at a rate of millions of gallons per year. Discuss possible reasons for this difference in terms of transition-state theory.

52. What is the major consequence of each of the following for an activated reaction?
  a) A lowered activation energy.
  b) Having E_a of the forward reaction much greater than E_a of the reverse reaction.
  c) An increase in the temperature.

53. The reaction of iron filings with pure oxygen is much faster than is the reaction of a solid piece of iron with oxygen, all other factors (total mass of iron, oxygen pressure, etc.) held constant. Explain why this is so.

54. The oxygen content of our atmosphere is now about 20% by volume. Conjecture some possible consequences of changing this to (a) 30%, (b) 10%.

55. Heating an iron nail to redness in a flame blackens the nail somewhat but does not burn it. However, when the red-hot nail is inserted into a test tube of pure oxygen it burns brilliantly, throwing off large sparks. Explain.

57. Nitroglycerin,

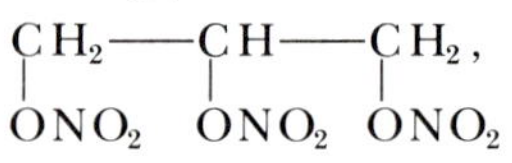

$$\begin{array}{ccc} CH_2 \!\!-\!\!\!\!&CH\!\!-\!\!\!\!&CH_2, \\ | & | & | \\ ONO_2 & ONO_2 & ONO_2 \end{array}$$

is an oily, unstable liquid which explodes when jarred slightly.

a) What general type of reaction is represented by the explosion of nitroglycerin?

b) When nitroglycerin is mixed with an inert solid material, the resulting mixture is much more stable to shock (this is called **dynamite**). Why is dynamite so much more stable than liquid nitroglycerin?

57. Zinc metal reacts with hydrogen ions according to the reaction

$$Zn(s) + 2H^+(aq) \rightarrow Zn^{2+}(aq) + H_2(g)$$

Why is this reaction so much faster when carried out in 1.0-M HCl than in pure water?

58. What experimental evidence is there that at equilibrium the rates of forward and reverse reactions are equal?

ADDITIONAL READINGS

Dence, Joseph B., Harry B. Gray, and George S. Hammond, *Chemical Dynamics.* New York: Benjamin, 1968. A rather comprehensive coverage of reaction rates and mechanisms designed to be taught at the end of a course in general chemistry. Considerable use of the calculus in certain sections.

King, Edward L., *How Chemical Reactions Occur.* New York: Benjamin, 1963. An elementary treatment of kinetics and mechanisms. Uses some calculus.

Haensel, V., and R. L. Burwell, Jr., "Catalysis." *Scientific American,* December 1971.

Phillips, David C., "The Three-Dimensional Structure of an Enzyme Molecule." *Scientific American,* November 1966. Describes the molecular structure of lysozyme and how this enzyme works. Contains beautiful color diagrams of the enzyme.

21
THE METALS OF GROUPS I AND II

Funeral mask of the Egyptian boy-king, Tutankhamun. The mask was hammered from a single piece of pure gold. (Photo by Harry Burton, courtesy of The Metropolitan Museum of Art.)

A small piece of potash [K_2CO_3], which had been exposed for a few seconds to the atmosphere so as to give conducting power to the surface, was placed upon an insulated disc of platina, connected with the negative side of a battery ... and a platina wire, communicating with the positive side, was brought into contact with the upper surface of the alkali. The potash began to fuse at both its points of electrization ... small globules having a high metallic luster, and being precisely similar in visible characters to quicksilver [mercury], appeared ...
SIR HUMPHRY DAVY, describing the first isolation of elemental potassium, 6 October 1807

The metals of groups I and II of the periodic table may be sub-grouped into four different categories:

Group IA: Li, Na, K, Rb, Cs, Fr
Group IIA: Be, Mg, Ca, Sr, Ba, Ra
Group IB: Cu, Ag, Au
Group IIB: Zn, Cd, Hg

This classification scheme is not always a good one; for example, there are greater similarities between group IA and IIA metals than there are between group IA and IB metals. The A subgroups differ from the B subgroups in that the latter have electron configurations containing ten electrons in d orbitals and, sometimes, fourteen electrons in f orbitals; this factor alone accounts for many of the differences between otherwise analogous groups.

This chapter discusses some of the basic properties, preparations, and uses of the most important members of the above four groups.

21.1 THE ELECTROLYTIC PRODUCTION OF METALS

Since the preparation and purification of many metals is carried out by the technique of electrolysis, a general discussion of this method is presented here.

As previously stated (Section 15.3), **electrolysis** implies decomposition by means of an electric current. When an electric current is passed through an **electrolyte***—a liquid which conducts electric current—chemical reactions occur at the two electrodes. The specific chemical reactions depend on the composition of the electrolyte but always comprise an oxidation reaction and a reduction reaction. Thus an electrolysis cell—as the apparatus is called—may be regarded as a voltaic

---

* An electrolyte is any liquid capable of conducting an electric current. Examples are a molten ionic salt or an aqueous solution containing ions.

cell forced to operate in an intrinsically nonspontaneous direction. Whereas a voltaic cell produces electrical energy from chemical energy, an electrolysis cell uses electrical energy to produce chemical reactions.

As a specific example consider passage of an electric current through an electrolyte consisting of an aqueous solution of $CuCl_2$. The chemical reactions occurring at the electrodes are

anode: $2Cl^-(aq) \rightarrow Cl_2(g) + 2e^-$ (oxidation)
cathode: $Cu^{2+}(aq) + 2e^- \rightarrow Cu(s)$ (reduction)

The result is production of metallic copper, which plates out on the cathode, and gaseous chlorine, which may be collected by suitable techniques. The cell reaction is

$Cu^{2+}(aq) + 2Cl^-(aq) \rightarrow Cu(s) + Cl_2(g)$

Electrode potential data from Table 19.1 show that $\mathscr{E}°$ of this reaction is -1.70 V at 25 °C. Clearly this reaction is not intrinsically spontaneous for reactants and products in their standard states. It is also clear that a voltaic cell utilizing the reverse reaction could be set up to produce electrical energy with an $\mathscr{E}^0$ of $+1.70$ V.

The English scientist Michael Faraday* was the first to show that definite numerical relationships exist between the amount of electrical current passed through an electrolysis cell and the amounts of substances produced at the electrodes. One of these important relationships, known as **Faraday's law**, may be stated as follows:

Chemical reactions occurring at the electrodes produce amounts of products proportional to the quantity of electric charge passed through the cell.

Specifically, for each mole of electrons passed through the cell, an equivalent weight (see Section 14.6) of product is formed at each electrode.

If the cell contains an aqueous solution of $CuCl_2$, copper metal is produced at the cathode via the reaction

$Cu^{2+}(aq) + 2e^- \rightarrow Cu(s)$

Each mole of electrons produces one equivalent weight or one-half mole of copper (31.77 g). The total charge on one mole of electrons is $(1.6021917 \times 10^{-19}$ C per electron$) \times (6.022169 \times 10^{23}$ electrons·mol^{-1}) or 96,486.70 C·mol^{-1}; this quantity of charge is called the **faraday** (or, sometimes, the Faraday constant) with the symbol F. For ordinary calculations, the approximate value of 96,500 C·mol^{-1} is often used for the faraday.

The same cell produces chlorine gas at the anode via the reaction

$2Cl^-(aq) \rightarrow Cl_2(g) + 2e^-$

According to Faraday's law, one faraday of charge will produce one equivalent weight or one-half mole of chlorine gas (35.457 g).

The total charge passed through an electrolysis cell is calculated from the rate of current flow (in amperes) and the time (in seconds) during which current flow

---

* A biographical sketch of Faraday is given in Chapter 1.

Michael Faraday, self-trained British scientist who made many basic discoveries in electricity and magnetism. (Courtesy of Culver Pictures.)

takes place. The relationship used is

charge passed (C) = [rate of current flow (A)] × [time of flow (s)]

It should be noted that an ampere (symbol A) is one $C \cdot s^{-1}$. Dividing the charge passed in C by 96,500 gives the charge passed in faradays.

EXAMPLE 21.1 A current of 5.00 mA is passed through a $CuCl_2$ solution for 10.0 hrs. Calculate the weight of copper formed at the cathode and the volume of chlorine gas produced at the anode (measured at 1 atm and 0 °C).

SOLUTION The quantity of charge passed through the solution is

$$\frac{(5.00 \times 10^{-3} \text{ A})(10 \text{ hr})(60 \text{ min} \cdot \text{hr}^{-1})(60 \text{ s} \cdot \text{min}^{-1})}{96,500 \text{ C} \cdot F^{-1}} = 1.87 \times 10^{-3} F$$

This means, according to Faraday's law, that $(1.87 \times 10^{-3})(\frac{1}{2})$ mol of copper is produced at the cathode and $(1.87 \times 10^{-3})(\frac{1}{2})$ mol of chlorine is produced at the anode. Thus the weight of copper produced at the cathode is

$$[(1.87 \times 10^{-3})(\tfrac{1}{2}) \text{ mol}](63.54 \text{ g} \cdot \text{mol}^{-1}) = 5.94 \times 10^{-2} \text{ g} \quad \text{or} \quad 59.4 \text{ mg}$$

The volume of chlorine produced may be estimated from the ideal gas law. There are $(1.87 \times 10^{-3})(\frac{1}{2})$ mol of chlorine produced at the anode. The volume at 1 atm and 0 °C is

$$V = \frac{nRT}{p} = \frac{[(1.87 \times 10^{-3})(\tfrac{1}{2}) \text{ mol}](0.082 \text{ L} \cdot \text{atm} \cdot \text{K}^{-1} \cdot \text{mol}^{-1})(273 \text{ K})}{1 \text{ atm}}$$

$$= 20.9 \times 10^{-3} \text{ L} \quad \text{or} \quad 20.9 \text{ cm}^3$$

EXAMPLE 21.2 With reference to the electrolysis of water discussed in Section 15.3, how many moles of H_2 and O_2 can be produced by passing a current of 5.0 A through an apparatus such as that shown in Fig. 15.1 for 15 min?

SOLUTION The quantity of charge passed through the electrolysis cell is given by

$$\frac{(5.0 \text{ A})(15 \text{ min})(60 \text{ s} \cdot \text{min}^{-1})}{96{,}500 \text{ C} \cdot F^{-1}} = 0.047 \, F$$

For hydrogen production $[2H^+(aq) + 2e^- \to H_2(g)]$, $0.047 \, F$ produces $(0.047/2)$ mol or 0.024 mol of H_2. For oxygen production two moles of O^{2-} (present in H_2O as OH^-) loses $4 \, F$ to produce one mole of O_2. Thus $0.047 \, F$ produces $(0.047/4)$ mol or 0.012 mol of O_2.

21.2 THE PREPARATION OF GROUP IA METALS

The group IA metals, lithium, sodium, potassium, cesium, and rubidium, may be prepared by electrolysis of their molten chlorides. Indeed, these metals are so chemically reactive that this is virtually the only practical way of preparing them. Francium has been prepared only in miniscule amounts and is not included in this discussion; presumably it could be prepared in the same manner as the other group IA metals. For NaCl as a specific example, the overall electrolysis process is

$$2Na^+Cl^-(l) \to 2Na(l) + Cl_2(g)$$

The separate electrode reactions are

cathode: $Na^+(l) + e^- \to Na(l)$ (reduction)
anode: $Cl^-(l) \to \frac{1}{2}Cl_2(g) + e^-$ (oxidation)

Figure 21.1 shows a schematic diagram of a typical electrolysis cell. To avoid having to operate at the high temperature of molten NaCl (above 800 °C), the metal may also be prepared by electrolysis of aqueous NaCl solutions, provided a mercury cathode is used. If a mercury cathode is not used, hydrogen gas is produced preferentially to sodium via the cathode reaction

$$2H_3O^+ + 2e^- \to H_2(g) + H_2O(l)$$

since the reaction of sodium with water

$$2Na(s) + 2H_2O(l) \to 2NaOH(aq) + H_2(g)$$

is so highly spontaneous. Normally, it is not possible to produce any metal by electrolysis of its aqueous solution unless that metal is found above hydrogen in the activity series (see Table 15.1). The role of the mercury in the cathode is to inactivate sodium by dissolving it to form an **amalgam**, that is, a mercury solution in which mercury is regarded as the solvent.

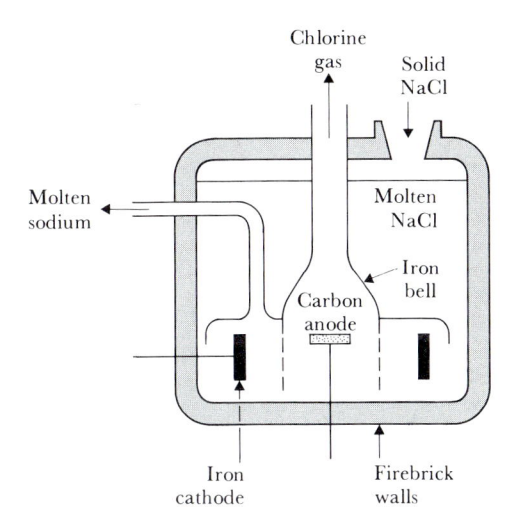

Fig. 21.1. Electrolysis cell for the production of elemental sodium and chlorine from molten sodium chloride. Electric current (an electron flow) enters the cathode (an iron ring) and reduces sodium ions to elemental sodium. At the graphite anode electrons leave chloride ions to produce elemental chlorine (an oxidation process). Molten sodium metal and chlorine gas are removed as they are formed and transferred to purification and storage containers.

EXAMPLE 21.3 How long would it take for a current of 1 A to produce 1 t of sodium metal?

SOLUTION Since the cathode reaction involved is $Na^+ + e^- \rightarrow Na$, it takes about 96,500 C to produce 1 mol (about 23 g) of sodium. Thus to produce 1 t (10^3 kg or 10^6 g) of sodium requires $(10^6 \text{ g}/23 \text{ g}\cdot\text{mol}^{-1})(96{,}500 \text{ C}\cdot\text{mol}^{-1}) = 4.2 \times 10^{10}$ C. A current of 1 A would have to operate 4.2×10^{10} s to transport this amount of charge. Since there are about 3.15×10^7 s in a year, this would take 1333 yr—a long time to wait for a tonne of sodium!

21.3 THE PROPERTIES OF THE GROUP IA METALS.

In Table 21.1 are summarized some properties of the group IA metals. Note that atomic and ionic radii increase and ionization energies decrease in going down the group, just as predicted on the basis of Section 4.5. Note also that the second ionization energies are considerably higher than the first; this shows why M^+ ions exist rather than M^{2+} ions.

Recall (Chapter 2) that group IA metals are known collectively as the **alkali metals**. Lithium, the first member of the group, has many properties which differ considerably from those of the remaining group IA metals. In fact, this is true for

TABLE 21.1 SOME PROPERTIES OF THE GROUP IA METALS

	Li	Na	K	Rb	Cs
Valence-shell electron configuration	$2s$	$3s$	$4s$	$5s$	$6s$
Atomic radius (Å)	1.23	1.57	2.03	2.16	2.35
Ionic radius (Å)	0.78	0.95	1.33	1.49	1.65
Ionization energy (eV): 1st	5.4	5.1	4.3	4.2	3.9
2nd	75.6	47.3	31.8	27.5	25.1
Melting point (°C)	179	97.8	63.7	39	29
Normal boiling point (°C)	1317	892	774	688	690
Electronegativity	1.0	0.9	0.8	0.8	0.7
$\mathscr{E}°$ for $M(s) \rightarrow M^+(aq) + e^-$ (volts)	3.04	2.71	2.92	2.99	3.02
Abundance on earth (atoms per 10,000)	—	48	10	16	—
Abundance rank[a]	—	9	12	—	—
Density (g·cm^{-3}) at 20 °C	0.53	0.97	0.86	1.53	1.88

[a] Other abundance ranks previously given are O (1) and C (14).

the lightest element in virtually each of the groups from IA to VIIA (excluding hydrogen from consideration). This has led to the half-serious observation that the periodic table may be divided into two categories, with the lightest elements of each group in the first category, and all other elements in the second.

With the exception of lithium, all of the group IA elements are soft at room temperature. Freshly exposed surfaces of the metals show a silvery luster but quickly react with O_2 and water vapor to become tarnished. Lithium alone reacts with atmospheric N_2 at room temperature to produce the nitride Li_3N.

Although the salts of the group IA metals are predominantly ionic, some lithium compounds are covalent in character. LiOH is the only hydroxide of group IA which is not 100% dissociated to hydrated ions in water solution, and it is less soluble than NaOH or KOH. LiOH also decomposes to the oxide when heated:

$$2\text{LiOH}(s) \xrightarrow{\text{heat}} \text{Li}_2\text{O}(s) + \text{H}_2\text{O}(g)$$

The other group IA hydroxides are stable to heating. Whereas sodium and potassium combine rather rapidly and rather violently with water, lithium does so only slowly.

In 1949 the Australian psychiatrist John F. J. Cade discovered that doses of lithium citrate reduced the episodes of excitability encountered frequently in manic patients and in some schizophrenics. Since then, other studies have shown that lithium salts (lithium carbonate is often used) can protect against depression, ameliorate unstable character disorders, and calm hyperactive children as well as violent and aggressive patients. Although research with lithium continues, the mechanism of its action is not understood.*

---

* Theoretical calculations predict that CH_3Li is planar (not tetrahedral!) and $CH_2=CLi$ is nonplanar (not planar like ethylene!). If the predictions are verified, it may be that this unusual effect on geometry plays some role in lithium's physiological properties.

Only lithium forms a single oxide, Li$_2$O. The other group IA metals also form peroxides, superoxides (see Section 15.10), and some unusual oxides.

Lithium has a very high specific heat (about 3.3 J·g^{-1}·K^{-1}) compared with those of the other metals, for example, 1.3 J·g^{-1}·K^{-1} for sodium and 0.8 J·g^{-1}·K^{-1} for potassium. This property has led to the use of liquid lithium as a coolant (in nuclear reactors, for example) but its highly corrosive nature is a serious disadvantage in this application.

All of the common salts (and most of the uncommon salts) of sodium and potassium are ionic and water soluble. The only truly covalent compounds are the diatomic molecules, Na$_2$ and K$_2$, which exist in small amounts in the vapor phase. The most commonly known salts of sodium and potassium are the halides, nitrites (NaNO$_2$, KNO$_2$), nitrates (NaNO$_3$, KNO$_3$), sulfates (Na$_2$SO$_4$, K$_2$SO$_4$), carbonates (Na$_2$CO$_3$, K$_2$CO$_3$), and hydrogen carbonates (NaHCO$_3$, KHCO$_3$). The latter compounds are commonly known by the misleading names of sodium bicarbonate and potassium bicarbonate; the correct names are sodium hydrogen carbonate and potassium hydrogen carbonate. The hydroxides of sodium and potassium, commonly known as caustic soda and caustic potash, respectively, are prepared by electrolysis of aqueous solutions of the chlorides. The cathode reaction is (see Problem 15.22)

$$2H_2O(l) + 2e^- \rightarrow H_2(g) + 2OH^-(aq)$$

and the anode reaction is

$$2Cl^-(aq) \rightarrow Cl_2(g) + 2e^-$$

The electrolyzed solution becomes enriched in OH$^-$ as H$^+$ is reduced so that, ultimately, Na$^+$(aq) and OH$^-$(aq) remain. Such a cell produces H$_2$ and Cl$_2$ at the electrodes and NaOH (or KOH) within the cell itself. The net cell reaction is

$$2H_2O(l) + 2Cl^-(aq) \rightarrow H_2(g) + Cl_2(g) + 2OH^-(aq)$$

Sodium, the most common and widely disseminated of the group IA metals, is perhaps the easiest to detect. Spectral analysis of almost any material found in nature inevitably reveals the bright yellow lines of wavelengths 5890 Å and 5896 Å called the sodium D-lines. Potassium is almost as widespread in nature but its D-lines have wavelengths above 7600 Å and are not as apparent as the sodium lines.

The ammonium ion, NH$_4^+$, resembles the potassium ion, K$^+$, in many ways. The two have similar ionic radii, 1.43 and 1.33 Å, respectively, and form compounds of similar structures and solubilities.

Sodium and potassium readily form salts with the carboxylic acids (Section 16.12). Some of these, such as the formates and acetates, have already been encountered. Of great practical importance are the salts of long-chain carboxylic acids containing 12 to 18 carbon atoms. Such salts are known as **soaps** and function as grease and soil removers. The commercial preparation of a soap involves the treatment of an ester of **glycerol** (an alcohol with three OH groups) with sodium hydroxide. The reaction is as follows.

$$\begin{array}{c}\text{CH}_2-\text{O}-\underset{\underset{\text{O}}{\|}}{\text{C}}-\text{R}\\ |\\ \text{CH}-\text{O}-\underset{\underset{\text{O}}{\|}}{\text{C}}-\text{R}'\\ |\\ \text{CH}_2-\text{O}-\underset{\underset{\text{O}}{\|}}{\text{C}}-\text{R}''\end{array} + 3\text{Na}^+\text{OH}^- \rightarrow \begin{array}{c}\text{CH}_2\text{OH}\\ |\\ \text{CHOH}\\ |\\ \text{CH}_2\text{OH}\end{array} + \begin{array}{c}\text{RCO}_2^-\text{Na}^+\\ \text{R}'\text{CO}_2^-\text{Na}^+\\ \text{R}''\text{CO}_2^-\text{Na}^+\end{array}$$

a triglyceride   glycerol   soaps
(ester of glycerol)

where R, R′, R″ typically contain 11 to 17 carbon atoms in an unbranched chain. **Triglycerides** are often obtained from animal fats. When the R groups are alkane the triglyceride tends to be a solid and is called a **fat**; when the R groups contain —C=C— units, the triglyceride tends to be a liquid and is called an **oil**. Our ancestors used a process similar to this to manufacture home-made soap. Caustic potash (crude KOH) was extracted from wood ashes and boiled with animal fats and oils to make a soft soap. Soaps made with NaOH tend to be much firmer.

Soap owes its cleaning action to the fact that the long carboxylic chain has a water-soluble end (the ionic —CO_2^- end) and an oil-soluble end (the hydrocarbon end, indicated by R, R′, or R″). As illustrated in Fig. 21.2 the oil-soluble end can become linked to a globule of grease, and the water-soluble end enables the globule to be borne along in suspension and ultimately rinsed away.

Soaps react with metal ions such as Mg^{2+} and Ca^{2+} to form sludges, the stuff that makes rings in bathtubs. The old solution to the sludge problem in laundering was to add Na_2CO_3 (washing soda) to the wash. The sodium carbonate forms a milky suspension of $MgCO_3$ and $CaCO_3$ which does not settle out. However, $CaCO_3$ and $MgCO_3$ eventually build up on clothing fibers to produce "tattle-tale gray." Also, Na_2CO_3 solutions have high pH and can cause severe skin burns.

It was to eliminate both of these disadvantages that modern detergents were developed. Detergents, having a water-soluble and an oil-soluble portion, function in the same way as soaps but many do not form sludges with Mg^{2+} and Ca^{2+}. In addition, they either are salts of strong acids (and thus very weak conjugate bases of those acids) or are otherwise neutral in pH. One type of detergent has the structure

which is fundamentally a sodium salt of benzene sulfonic acid. The R group may be attached at random positions of the benzene ring. In early detergents R was usually a highly branched hydrocarbon. Bacteria were unable to degrade the branched chains

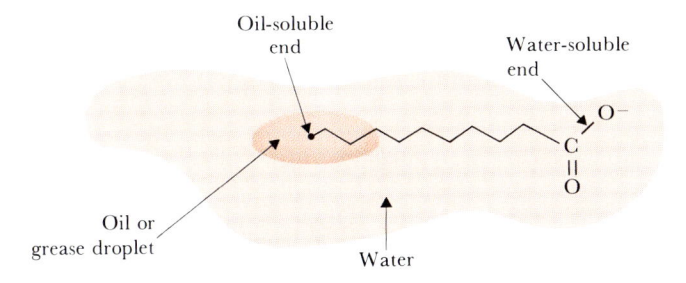

Fig. 21.2. How soap removes grease in water solution. The negatively charged soap-grease particles repel each other and remain in solution.

so that detergents of this type persisted in waterways and even entered public water supply systems. Their presence was unmistakable in glasses of tap water with heads of foam that would make a master brewer envious and in billowing suds surrounding waterfalls in natural streams. Later detergents were made at least partially biodegradable by using unbranched-chain R groups.

Rubidium, with a melting point of 39 °C, becomes a liquid at just 2° above normal body temperature. Rubidium is much more reactive than the lighter group IA metals and ignites spontaneously in air. Four oxides of rubidium, Rb_2O, Rb_2O_2, Rb_2O_3, and Rb_2O_4 are known. Cesium is the most electropositive of the known elements. With a melting point of 29 °C it may be liquefied at room temperature. The metal reacts explosively with cold water. Cesium hydroxide, CsOH, may be used to etch glass.

EXAMPLE 21.4 An aqueous solution of NaCl is electrolyzed for 10 hr using a 2-A current. How many moles of NaOH, H_2, and Cl_2 are produced?

SOLUTION The overall cell reaction may be written

$$2NaCl(aq) + 2H_2O(l) \rightarrow H_2(g) + Cl_2(g) + 2NaOH(aq)$$

Each mole of electron charge will produce 0.5 mol of H_2, 0.5 mol of Cl_2, and 1.0 mol of NaOH. The number of moles of electron charge passed is

$$\frac{(2A)(10 \times 60 \times 60)s}{96,500 \; C \cdot mol^{-1}} = 0.75 \; mol$$

Thus this cell will produce about 0.38 mol of H_2, 0.38 mol of Cl_2, and 0.75 mol of NaOH.

21.4 THE PREPARATION OF GROUP IIA METALS

Beryllium and its compounds have a sweet taste and are very poisonous. Beryllium metal is usually prepared by reducing beryllium fluoride with magnesium metal. The reaction is

$$BeF_2 + Mg \rightarrow Be + MgF_2$$

This reaction illustrates a very important general principle: A metal tends to reduce other metals of the same group which lie above it in the periodic table. Thus, sodium can be prepared by treating sodium chloride with potassium.

Magnesium metal is most often obtained by electrolysis of molten magnesium chloride, $MgCl_2$. Magnesium chloride is obtained from seawater and from certain wells. Magnesium may also be prepared from dolomite, a sedimentary rock containing magnesium and calcium carbonates. Heating of dolomite produces magnesium and calcium oxides which are then reduced to magnesium metal by ferrosilicon, FeSi.

Calcium may be prepared by electrolysis of molten calcium chloride. Some calcium fluoride is usually added to lower the melting point of the chloride.

Barium may be formed from electrolysis of the chloride. Strontium is similarly prepared. The latter is also obtained by reduction of the oxide with aluminum. The reaction is

$$3SrO + 2Al \rightarrow Al_2O_3 + 3Sr$$

The process is carried out in a vacuum at a high enough temperature to distill off the strontium.

Radium was first prepared in pure form by electrolysis of its chloride solution using a mercury cathode.

EXAMPLE 21.5 An electrolysis cell is used to produce magnesium metal from molten $MgCl_2$. What weight of pure magnesium is obtained in the cell after 1 week using a 250-mA current?

SOLUTION The cathode process is $Mg^{2+} + 2e^- \rightarrow Mg$. Thus one faraday produces 0.5 mol of Mg. The number of moles of electrons passed is

$$\frac{(0.250 \text{ A})(60 \times 60 \times 24 \times 7)\text{s}}{96{,}500 \text{ C} \cdot \text{mol}^{-1}} = 1.57 \text{ mol}$$

This produces $(1.57/2)$ mol $(24.3 \text{ g} \cdot \text{mol}^{-1})$ = 19 g of magnesium.

EXAMPLE 21.6 How much radium would be prepared from $RaCl_2$ using the same current and time as in Example 21.5?

SOLUTION The cathode process is $Ra^{2+} + 2e^- \rightarrow Ra$. Thus, just as for magnesium, one faraday produces 0.5 mol of radium. Thus the amount of radium produced is $(1.57/2)$ mol $(226 \text{ g} \cdot \text{mol}^{-1})$ = 177 g.

21.5 THE PROPERTIES OF THE GROUP IIA METALS.

Except for brief comments about other metals, the discussion here is confined primarily to magnesium and calcium, perhaps the most important group IIA metals.

Collectively, the group IIA metals are known as the **alkaline earth metals**. Table 21.2 lists some of the most important physical properties of the alkaline earth metals. Note that ionization energies and atomic and ionic radii generally follow the trends predicted in Section 4.5. It should be noted that the second ionization energies of the alkaline earth metals are rather low and not much larger than the first ionization energies. This is in sharp contrast to the analogous situation with the alkali metals and is in accord with the +2 oxidation state so characteristic of group IIA elements. Although there is no readily apparent reason why group IIA elements do not also exhibit the +1 oxidation state, no compounds of this oxidation state have ever been isolated. Theoretical calculations based on a Born-Haber cycle (see Section 18.5 and Problem 18.14) show that the process

$$2MX \rightarrow MX_2 + M$$

(where M is a group IIA metal and X is a typical univalent anion) should be highly exothermic. This suggests that +1 compounds, even if they had a transitory existence, could not be isolated.

TABLE 21.2 SOME PROPERTIES OF THE GROUP IIA METALS

	Be	Mg	Ca	Sr	Ba	Ra
Valence-shell electron configuration	$2s^2$	$3s^2$	$4s^2$	$5s^2$	$6s^2$	$7s^2$
Atomic radius (Å)	0.89	1.36	1.74	1.91	1.97	—
Ionic radius (Å)	0.35	0.78	1.06	1.12	1.34	1.43
Ionization energy (eV): 1st	9.3	7.6	6.1	5.7	5.2	8.3
2nd	18.2	15.0	11.9	11.0	10.0	10.1
Melting point (°C)	1278	176	141	769	725	700
Normal boiling point (°C)	2970	650	850	1384	1640	1140
Electronegativity	1.5	1.2	1.0	1.0	0.9	0.9
$\mathscr{E}°$ for $M(s) \rightarrow M^{2+}(aq) + 2e^-$ (volts)	1.70	2.375	2.76	2.89	2.90	—
Abundance on earth (atoms per 10,000)	—	1003	177	—	—	—
Abundance rank	—	4	6	—	—	—
Density (g·cm^{-3}) at 20 °C	1.848	1.74	1.55	2.54	3.5	5

Beryllium, the lightest member of the group, is, like lithium in the IA group, somewhat of a maverick. It has unusually high melting and boiling points and other properties somewhat out of line with its position in the group. Like lithium, beryllium also tends to form nonionic compounds, in contrast to other group members.

Magnesium resembles lithium in many of its properties. The salts of both elements have similar solubilities and their reactions have some striking parallels; for example, both react quite readily with N_2 to form nitrides. The two elements have fairly close melting points and apparently identical ionic radii. Sodium and calcium are similar in much the same way. It is not uncommon for two elements in adjacent groups, with the heavier one in a lower period, to exhibit such similarities.

All of the common Mg^{2+} salts are water soluble except MgF_2 and $MgCO_3$. Most Ca^{2+} salts are insoluble except $CaCl_2$, $CaBr_2$, CaI_2, $Ca(NO_3)_2$, and $Ca(OAc)_2$. In aqueous solutions the Mg^{2+} ion exists in the form of the complex ion $[Mg(H_2O)_6]^{2+}$.

MgO and CaO are basic oxides, reacting with water to form hydroxides:

$$MgO + H_2O \rightarrow Mg(OH)_2$$

$$CaO + H_2O \rightarrow Ca(OH)_2$$

In both of these reactions O^{2-} is acting as a Brønsted-Lowry base, with H_2O as an acid, to produce OH^-. $Mg(OH)_2$ is a weak base and is not very soluble in water. $Ca(OH)_2$ is a stronger base and somewhat more soluble in water.

Magnesium burns in air with an intense flame. The main reaction is

$$3Mg(s) + N_2(g) \rightarrow Mg_3N_2(s)$$

Some oxide is also formed. If Mg_3N_2 is wetted with H_2O, the odor of ammonia is readily discerned. The reaction is

$$Mg_3N_2(s) + 6H_2O(l) \rightarrow 3Mg(OH)_2(s) + 2NH_3(g)$$

This reaction shows that the nitride ion N^{3-} is a strong Brønsted-Lowry base relative to water. This should be compared to the reaction of O^{2-} and H_2O above.

Magnesium and calcium form carbonates when CO_2 gas is passed into solutions containing Mg^{2+} or Ca^{2+}. For example,

$$Ca^{2+}(aq) + CO_2(g) + 3H_2O(l) \rightarrow CaCO_3(s) + 2H_3O^+$$

Le Châtelier's principle suggests that $CaCO_3$ should dissolve in strong acids to produce CO_2. In the presence of excess CO_2 the $CaCO_3$ redissolves to form the hydrogen carbonate:

$$CaCO_3(s) + H_2O(l) + CO_2(g) \rightarrow Ca(HCO_3)_2(aq)$$

However, $Ca(HCO_3)_2$ cannot be isolated since it decomposes to $CaCO_3$ and CO_2 on removal of solvent.

When magnesium or calcium is heated with coke (carbon) in an electric furnace, the **carbides** are formed:

$$Mg(s) + 2C(s) \xrightarrow{heat} MgC_2(s)$$

$$Ca(s) + 2C(s) \xrightarrow{heat} CaC_2(s)$$

Calcium carbide finds use in the production of acetylene gas via the reaction

$$CaC_2(s) + 2H_2O(l) \rightarrow Ca(OH)_2(s) + HC \equiv CH(g)$$

Calcium carbide headlamps are now popular with spelunkers and Nordic skiers who like to be out at night. Solid calcium carbide probably consists of three-dimensional arrays of the following type:

$$Ca^{2+} \quad C \equiv C^{2-} \quad Ca^{2+} \quad C \equiv C^{2-}$$

$$C \equiv C^{2-} \quad Ca^{2+} \quad C \equiv C^{2-} \quad Ca^{2+}$$

$$Ca^{2+} \quad C \equiv C^{2-} \quad Ca^{2+} \quad C \equiv C^{2-}$$

X-ray studies show that calcium carbide forms tetragonal crystals (see the chart of crystal forms, Fig. 7.3).

Magnesium is found in the structures of many molecules of biological importance; the structure of chlorophyll, the green coloring matter of plants which plays an important role in photosynthesis, centers about an atom of magnesium as shown below.

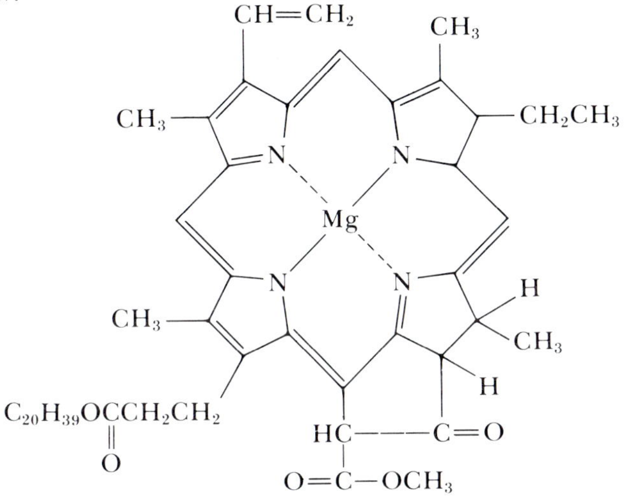

Magnesium is now being used in the manufacture of lightweight alloys which are replacing aluminum for some purposes, such as frames for backpacks. The density of magnesium is only $1.74 \text{ g} \cdot \text{cm}^{-3}$ as compared with $2.70 \text{ g} \cdot \text{cm}^{-3}$ for aluminum.

A constant supply of Ca^{2+} (and PO_4^{3-}) is needed in the blood for building bone and for the control of certain cell functions. The level of Ca^{2+} in the blood is controlled by vitamin D and by a parathyroid hormone. The former controls the uptake of Ca^{2+} from food, and the latter controls the release of Ca^{2+} from bone and also induces the kidney tubules to reabsorb Ca^{2+} about to be excreted. Calcium ions are known to play controlling roles in the normal contraction and relaxation of muscle tissue. Whenever the concentration of Ca^{2+} in the blood decreases to a certain level, nerve and muscle cells discharge spontaneously and voluntary muscles go into continuous contraction in a condition known as tetany.

There is considerable evidence that the $[Ca^{2+}]/[Na^+]$ ratio in the hypothalamus gland, a part of the forebrain containing vital autonomic nervous centers, plays a role in the regulation of body temperature in subhuman primates and, perhaps, in humans as well. An increase in this ratio causes body temperature to stabilize below the normal temperature while a decrease causes the opposite effect. This behavior is of interest as a possible means of protecting people from exposure to extremes of temperature.

21.6 THE GROUP IB METALS: COPPER, SILVER, AND GOLD

Copper, silver, and gold are known as the **coinage metals** although gold is now almost never used for coins and even silver is no longer used in great amounts. However, at one time gold and silver were almost the only metals used in coinage.

Copper and gold are unusual in that the elements do not possess a white or silver luster; copper is reddish in color and gold is yellow. Whereas copper and silver are moderately chemically reactive, gold is just about the most inert of all the metals. The only common reagent that reacts with gold is a mixture of one part HNO_3 and three parts HCl known as **aqua regia** (Latin for "royal water," so named because of its ability to dissolve the "king" of metals). Copper and silver will not dissolve in HCl solution since they are above hydrogen in the activity series (Table 15.1) and since Cl^-(aq) is not an oxidizing agent. However, both are oxidized by NO_3^-(aq) in nitric acid solution. The reactions in concentrated HNO_3 solution are

$$Cu(s) + 4H^+(aq) + 2NO_3^-(aq) \rightarrow Cu^{2+}(aq) + 2NO_2(g) + 2H_2O(l)$$

$$Ag(s) + 2H^+(aq) + NO_3^-(aq) \rightarrow Ag^+(aq) + NO_2(g) + H_2O(l)$$

Both metals are also oxidized by sulfuric acid solutions.

The electron configurations of the group IB metals are:

Cu:$(Ar)4s3d^{10}$

Ag:$(Kr)5s4d^{10}$

Au:$(Xe)6s4f^{14}5d^{10}$

Completely filled f levels (14 electrons) behave in many ways like completely filled d levels (10 electrons); both tend to act as part of an inert-gas core. Thus the above electron configurations suggest that the +1 oxidation state should be the most common. The +1 compounds are, indeed, among the best known, but +2 and +3 compounds of all are also known. Copper forms many Cu(II) compounds and several Au(III) compounds exist.

Table 21.3 lists some properties of the group IB metals. Note that all the group IB metals have first ionization energies significantly higher than do the group IA metals. This may be attributed to relative ineffectiveness of d electrons in shielding the nucleus. Groups IA and IB, in spite of some similarity in electron structure, differ considerably in their chemistries, largely due to the role of d electrons in the latter. Whereas group IA metals tend to be very soft and react readily with H_2O,

TABLE 21.3 SOME PROPERTIES OF THE GROUP IB METALS

	Cu	Ag	Au
Ionization energy (eV): 1st	7.7	7.6	9.2
2nd	20.3	21.5	20.5
3rd	36.8	34.8	—
Melting point (°C)	1083	961	1064
Normal boiling point (°C)	2595	2212	2940
Ionic radius (Å): +1 state	0.96	1.26	1.37
+2 state	0.72	0.89	—
+3 state	—	—	0.85
Atomic radius (Å)	1.17	1.44	1.44
Electronegativity	1.9	1.9	2.4
Density at 20 °C (g·cm^{-3})	9.0	10.5	19.3

group IB metals are hard and do not react with water. In contrast to the group IA metals, which are noted for a rather high degree of chemical reactivity, the group IB metals are relatively inert.

Copper has been mined for about 5000 years. The element is found free in nature and in combined form in many minerals, principally sulfides and carbonates. The ores are usually converted to oxides by **roasting** and then reduced with coke. Typical roasting reactions are

$$Cu_2S(s) + 2O_2(g) \xrightarrow{heat} 2CuO(s) + SO_2(g)$$

$$CuCO_3(s) \xrightarrow{heat} CuO(s) + CO_2(g)$$

The reduction process is

$$2CuO(s) + C(s) \xrightarrow{heat} 2Cu(s) + CO_2(g)$$

The element prepared in this way typically contains impurities which leave it brittle. It may be purified by dissolving in sulfuric acid and electrolyzing.

Copper has a very high electrical and thermal conductivity and is oxidized only superficially in air, eventually acquiring a greenish coating of carbonate and sulfate commonly called **patina**.* Copper is often used in **alloys**, mixtures of two or more metals. Two of the most important alloys are **brass** (copper and zinc) and **bronze** (copper and tin).

Aqueous solutions of copper salts have a characteristic blue or blue-green color due to the complex ion $[Cu(H_2O)_4]^{2+}$. Ammonia solutions contain the intensely blue complex $[Cu(NH_3)_4]^{2+}$. One of the most important salts of copper is $CuSO_4 \cdot 5H_2O$, copper(II) sulfate pentahydrate, known commercially as **blue vitriol**. It is used as an agricultural poison and as an algicide in water treatment. Cu(II) salts, having the electron configuration $(Ar)3d^9$, are paramagnetic. The existence of colored complexes, such as that exemplified by Cu(II), is often associated with ionic electronic configurations containing from one to nine d electrons (see Section 22.8).

A great variety of biologically important compounds contain copper. One is **hemocyanin**, a molecule which plays the role of hemoglobin in the blood of some mollusks and arthropods. The enzyme **phenolase**, a biological catalyst operative in the oxidation of phenolic compounds, contains copper.

Silver is a bright white metal whose attractive and distinctive luster has given rise to the adjective *silver*. Silver is found free in nature and as the mineral argentite, Ag_2S. The most important commercial use of silver is in photography. When silver salts on the surface of a film or plate are exposed to light, some finely divided silver is formed, blackening the exposed surface. The fact that the amount of silver produced is proportional to the duration and intensity of the light forms the basis of black-and-white photography. An important additional fact is that the plate coated with silver salt can be **fixed** after exposure to light. This means that if the unreduced silver salts are removed, the remaining reduced silver reproduces the pattern and

---

* You have probably noticed this patina on copper and bronze (an alloy of copper) statues in museums and some public buildings and parks. Since patina supposedly adds dignity and value to such objects, contemporary statuary (much of it of plastic composition) is often produced with artificial or imitation "patina."

intensity of the light to which the plate has been exposed to produce a permanent record of the light-exposure process.

Silver has the highest electrical and thermal conductivity of all the metals, although copper, being almost as good and much cheaper, is far more widely used as a conductor. Metallic silver rapidly loses its brilliant luster when exposed to air. Much of this is due to reaction with traces of ozone, H_2S, or other sulfur compounds normally present in air.

With the exception of silver nitrate, $AgNO_3$, silver salts are generally insoluble in water.

EXAMPLE 21.7 Use balanced chemical equations to illustrate how to obtain pure silver from an ore having the composition Ag_2S.

SOLUTION Roasting of Ag_2S produces Ag_2O:

$$Ag_2S(s) + 1\tfrac{1}{2}O_2(g) \xrightarrow{\text{heat}} Ag_2O(s) + SO_2(g)$$

Elemental silver is prepared by reducing Ag_2O with carbon:

$$2Ag_2O(s) + C(s) \xrightarrow{\text{heat}} 4Ag(s) + CO_2(g)$$

21.7 THE GROUP IIB METALS: ZINC, CADMIUM, AND MERCURY

The metals of group IIB have the electron configurations

Zn: $(Ar)4s^2 3d^{10}$

Cd: $(Kr)5s^2 4d^{10}$

Hg: $(Xe)6s^2 4f^{14} 5d^{10}$

All exhibit +2 oxidation states, and mercury exhibits the +1 oxidation state as well. Table 21.4 lists some of the properties of these metals.

Mercury is the most unusual of the three metals; zinc and cadmium are very

TABLE 21.4 SOME PROPERTIES OF THE GROUP IIB METALS

	Zn	Cd	Hg
Ionization energy (eV): 1st	9.4	9.0	10.4
2nd	18.0	16.9	18.8
3rd	40.0	37.5	34.2
Melting point (°C)	419	321	−39.9
Normal boiling point (°C)	907	767	357
Ionic radius (Å) +2 state	0.69	0.92	0.93
Atomic radius (Å)	1.25	1.41	1.44
Electronegativity	1.6	1.7	1.9
Density (g·cm^{-3}) at 20 °C	7.1	8.7	13.6

mercury, from the Roman god Mercury, symbol Hg (Latin *hydrargyrus*, "liquid silver"), chemical element, liquid metal of Group IIb or the zinc group of the periodic table. Mercury metal, or quicksilver, was known to the ancient Chinese and Hindus and has been found in an Egyptian tomb of c. 1500 BC. The only elemental metal that is liquid at ordinary temperatures (mercury melts at −39° C [−38° F], gallium melts at 30° C and rubidium at 39° C [102° F]) is very white, slowly tarnishes in moist air, solidifies into a soft metal like tin or lead at −39° C (−38° F), and alloys with most metals (iron is a notable exception) to form amalgams.

Mercury does not wet or cling to glass, and this property, coupled with its uniform volume expansion throughout its liquid range, makes it useful in thermometers and barometers. Many instruments utilize its high density and surface curvature. The good electrical conductivity of mercury makes it exceptionally useful in sealed electrical switches and relays. An electric discharge through mercury vapour produces a bluish glow rich in ultraviolet light, a phenomenon utilized in ultraviolet fluorescent and mercury

similar in many respects. Mercury is a liquid at room temperature whereas zinc and cadmium melt at quite high temperatures. The +1 oxidation state of mercury, not shared by zinc and cadmium, is represented by the dimeric ion Hg_2^{2+}, that is,

$^+$Hg—Hg$^+$

in which the two Hg$^+$ ions are covalently linked.

21.7 | THE GROUP IIB METALS: ZINC, CADMIUM, AND MERCURY

There is a greater resemblance between the group IIA and IIB metals than there is between the group IA and IB metals. Tables 21.2 and 21.4 show that groups IIA and IIB have similar first and second ionization energies, although those of the latter are generally higher. Salts of Zn^{2+} and Cd^{2+} resemble those of Ca^{2+} to some degree.

Zinc is found in nature in the form of ZnS, $ZnCO_3$, and ZnO. The ores are usually roasted to produce the oxide and then reduced with coke. Small amounts of cadmium are usually present in zinc and are removed by fractional distillation. Elemental zinc is of bluish-white luster and somewhat brittle at lower temperatures. Zinc metal burns readily in air and creates dense clouds of ZnO "smoke." Zinc is much used in alloys, particularly with copper to make brass. The oxide is used as a pigment in white paints.

Elemental zinc will react with either acids or bases. In acids the reaction is

$$Zn(s) + H_3O^+ \rightarrow Zn^{2+}(aq) + 2H_2O(l) + H_2(g)$$

The reaction with bases is

$$Zn(s) + 2OH^- + 2H_2O(l) \rightarrow [Zn(OH)_4]^{2-}(aq) + H_2(g)$$

where $[Zn(OH)_4]^{2-}$ (also written $ZnO_2^- \cdot 2H_2O$) is the hydrated **zincate ion**. ZnO is amphoteric and reacts similarly to elemental zinc with acids and bases.

Zinc is found in the enzyme **carbonic anhydrase**, which is present in red blood cells. This enzyme controls the formation and breakdown of HCO_3^- at rates fast enough for CO_2 transport; it is the enzyme which controls the conversion of HCO_3^- to CO_2 in the lungs just before the latter is exhaled. Carbonic anhydrase may also play a role in making CO_3^{2-} available for eggshell production in birds. There is some evidence that insecticides such as DDT may inhibit the action of carbonic anhydrase so that the eggshells produced are abnormally fragile, breaking before the egg is ready to hatch.

Zinc is needed in the normal growth process. Animals fed on zinc-deficient diets require up to 50% more food in order to gain as much weight as control animals on diets adequate in zinc.

Cadmium is a soft bluish-white metal which can be cut easily with a knife. The metal is much used to produce low-melting alloys. Practically all metallic cadmium is produced as a by-product of the treatment of ores of zinc and other metals.

Cadmium sulfide, CdS, is often used as a yellow paint pigment. Cadmium compounds are frequently used as phosphors for television picture tubes, black and white as well as color. Cadmium and its compounds are highly toxic and must be handled with care. The compounds cause emphysema and kidney damage, interfere with copper metabolism, and promote teratogenic effects in rats and mice.* Cadmium is also responsible for the disease known to the Japanese as "itai-itai byo" (ouch-ouch disease) in which the bones become so fragile that coughing can lead to broken ribs.

Unlike the amphoteric ZnO, CdO is basic. Both $Zn(OH)_2$ and $Cd(OH)_2$ are virtually insoluble in water. Both zinc and cadmium bond to carbon in organometallic

---

* **Teratogenic** means "monster forming." The birth of grossly deformed babies resulting from their mothers' having taken the drug thalidomide during pregnancy is an example of a teratogenic effect.

TABLE 21.5 SOME COMMON MERCURY COMPOUNDS

COMPOUND NAME	FORMULA	REMARKS
Dimethylmercury	$(CH_3)_2Hg$	Colorless liquid, soluble in alcohol, sweet odor
Methylmercury chloride	CH_3HgCl	White crystals, insoluble in H_2O, characteristic odor
Phenylmercury acetate	C_6H_5—$HgOCOCH_3$	Slightly soluble in H_2O, used as a fungicide and herbicide
Methoxyethylmercury acetate	$CH_3OCH_2CH_2HgOCOCH_3$	Soluble in H_2O, used as a fungicide and herbicide
Mercury(II) acetate	$Hg(O_2CCH_3)_2$	White crystalline powder, soluble in H_2O
Mercury(II) cyanide	$Hg(CN)_2$	White solid, slightly soluble in H_2O
Mercury(II) oxide	HgO	Red and yellow solid forms, insoluble in H_2O, decomposes to elements on heating to 500 °C
Mercury(II) sulfide	HgS	Red and black solid forms, insoluble in H_2O
Mercury(I) chloride (calomel)	Hg_2Cl_2	White solid, insoluble in H_2O, $K_{sp} = 2 \times 10^{-18}$ at 25 °C
Mercury(II) chloride (corrosive sublimate)	$HgCl_2$	White powder, solubility in H_2O is 6.9 g/100 cm³ at 20 °C
Mercury fulminate	$Hg(OCN)_2$	White crystals, slight solubility in H_2O, explodes on heating or sharp impact
Merthiolate	$C_6H_4(CO_2^-,Na^+)(SHgCH_2CH_3)$	Cream-colored crystals, very soluble in H_2O, used as an antiseptic

compounds such as RZnX (X is a halogen), R_2Zn, and R_2Cd. In this respect the two metals resemble magnesium, which forms compounds of the type RMgX.

Mercury is a bright, silvery liquid metal with a considerably higher density than either zinc or cadmium (see Table 21.4). The element is usually prepared from the ore **cinnabar**, HgS. Heating of cinnabar in air leads to the reaction

$$HgS(s) + O_2(g) \xrightarrow{heat} Hg(l) + SO_2(g)$$

Mercury is used in an incredibly large number of applications: mercury/vapor lamps, fungicides, pesticides, silent electrical switches and thermostats, thermometers, electrolytic-cell electrodes, dental preparations, and detonators for explosives. In Table 21.5 are listed some common mercury compounds. Mercury and most of its compounds are very toxic and act as cumulative poisons since they are eliminated very slowly by the human body; the average half-life of all forms of mercury in the body is estimated to be about two months. Typical symptoms of mercury poisoning include loss of muscular control, blurring of vision, loss of hair, loss of teeth, and progressive insanity. Coma and death may follow high levels of ingestion. Although it is not known with certainty just what the physiological basis of mercury toxicity is, it appears that mercury attacks nerve cells, especially those of the dorsal root ganglia and cerebellum; motor nerve cells appear to be only slightly affected. It may be that mercury denatures proteins by combining with the disulfide linkages and sulfhydryl side groups.

EXAMPLE 21.8 ZnO and PbO are both used as pigments in white paints. Which pigment would you use if the paint was expected to be used in atmospheres containing H_2S?

SOLUTION The two pigments react as follows with H_2S:

$$ZnO(s) + H_2S(g) \rightarrow ZnS(s) + H_2O(l)$$

$$PbO(s) + H_2S(g) \rightarrow PbO(s) + H_2O(l)$$

Since ZnS is white and PbS is black, the ZnO pigment is the more desirable.

21.8 MERCURY IN THE ENVIRONMENT

The widespread use of mercury in industry and commerce makes it inevitable that mercury and its compounds eventually become disseminated throughout the environment. According to standards set up by WHO (World Health Organization), the mercury content of food should not exceed 50 ppb (parts per billion). Yet, during the early 1950s more than a hundred Japanese died from eating mercury-contaminated fish from Minamata Bay, into which a plastics factory had been discharging inorganic mercury salts. Tests on the hair of some of the victims revealed mercury levels of 150,000 ppb. Analysis of fish caught in the bay showed mercury levels as high as 5000 ppb, one hundred times the WHO maximum. During the late 1960s it was discovered that fish caught in Lake St. Clair, which connects Lake Erie to Lake Huron, contained up to 7090 ppb of mercury.

Chemists and health authorities were surprised to find such high levels of mercury in fish even though it had long been known that large amounts of mercury compounds enter the world's waterways each year. Table 21.6 shows some of the sources of this mercury originating in the United States. The mercury which manages to find its way into waterways does so in the form of the liquid or as highly insoluble inorganic salts; presumably these should end up with bottom sludges and sediments to remain out of the way indefinitely. Unfortunately, that is not what happens; bottom-dwelling anaerobic bacteria, those same creatures that convert rotting vegetation to methane, protect themselves from mercury poisoning by converting mercury and its salts to volatile compounds such as dimethyl mercury, $(CH_3)_2Hg$, or methylmercury chloride, CH_3HgCl. These compounds become spread throughout the

TABLE 21.6 CONSUMPTION OF MERCURY IN THE UNITED STATES IN 1969 BY VARIOUS ACTIVITIES[a]

ACTIVITY OR SOURCE	TONS USED PER YEAR
Chlor-alkali production	800
Electrical apparatus	700
Paints	370
Industrial and control apparatus	265
Dental preparations	115
Catalysis	115
Agriculture[b]	100
General laboratory use	75
Pharmaceuticals	25
Pulp and paper industry	20
Other[c]	370
Total[d]	2955

[a] Data supplied by the U.S. Bureau of Mines.
[b] This includes fungicides and bactericides for industrial purposes.
[c] This includes university research and other research and government allocation for military and scientific purposes.
[d] An estimated 3000 tons of mercury is released to the air each year just from the burning of coal! This is in addition to the waste from the industrial mercury listed in this table, which is also dumped into the environment.

aquatic environment, are ingested by algae, plankton, and other organisms, and ultimately are spread upward throughout the food chain. As little fish eat mercury-contaminated plankton and are in turn eaten by bigger fish, the mercury compounds tend to become more and more concentrated as one goes up the predatory hierarchy. The highest mercury levels are therefore found in creatures such as tuna, swordfish, pickerel, and wild ducks. This phenomenon, the increase in the concentration of a substance as one goes up a food chain, is known as **biomagnification**. Biomagnification occurs not only because big fish eat littler fish but also because big fish eat so *many* littler fish.

SUMMARY

The metals of groups IA and IIA, the alkali metals and alkaline earth metals, respectively, are generally soft electropositive metals which are chemically reactive, combining readily with a variety of electronegative elements. The oxides are generally basic.

The metals of group IB bear little similarity to group IA metals in spite of similarities in electron configurations. The former are generally hard and are relatively inert chemically, especially gold, one of the most inert of all metals.

The group IIB metals zinc and cadmium do bear some similarity to group IIA metals in that they tend to be soft and fairly reactive. Mercury, however, is a liquid at room temperature with chemical properties unique to itself. Zinc is unusual in that its oxide is amphoteric.

Electrolytic cells, essentially voltaic cells run backwards, are useful in the production of many metals. Group IA and IIA metals are usually prepared by electrolysis of their molten salts. Copper and zinc are usually prepared by roasting of their ores followed by reduction with coke to the metal. Copper is often refined by dissolving in sulfuric acid and electrolyzing.

Mercury compounds become disseminated in nature in a process that begins with bacterial conversion of insoluble mercury salts in water sediments to water-soluble compounds. The soluble mercury is ingested by small organisms which serve as food for larger organisms. By the process of biomagnification, larger organisms in the predatory chain acquire larger amounts of mercury. Humans eventually acquire the mercury by using the mercury-contaminated organisms as food.

LEARNING GOALS

1. Be able to calculate the weight of a given element that can be produced by a given electric current operating for a specified amount of time; that is, be able to apply Faraday's law of electrolysis.

2. Be able to describe the preparation of group IA and IIA metals by electrolysis.

3. Be familiar with the principal physical properties of the group IA and IIA metals.

4. Know how substances such as NaOH and KOH are produced by electrolysis.

5. Be familiar with the principal chemical properties of sodium, potassium, magnesium, and calcium and their compounds.

6. Be able to describe how a soap acts to remove grease and how the bathtub "ring" is formed.

7. Be familiar with the main physical and chemical properties of the coinage metals.

8. Be able to describe how to obtain metallic copper from a sulfide ore.

9. Be familiar with the main physical and chemical properties of zinc, cadmium, and mercury.

10. Be able to discuss how mercury becomes an environmental hazard and how biomagnification occurs.

DEFINITIONS, TERMS, AND CONCEPTS TO KNOW

electrolysis
electrolyte
faraday
alkali metal
amalgam
soap

fat
oil
glycerol
triglyceride
ionic radius
alkaline earth metal

metal carbides
aqua regia
roasting of ores
biomagnification

QUESTIONS AND PROBLEMS

Electrolysis

1. The charge on a mole of electrons, the value of the faraday, has been accurately determined by the National Bureau of Standards as 96,489 C. If the charge on a single electron is 1.60210×10^{-19} C, what is the value of Avogadro's number? [*Note:* This is one of the better ways of determining Avogadro's number experimentally.]

2. Three containers are electrically connected in series: The first contains $AgNO_3$ solution, the second contains $CuSO_4$ solution, and the third contains $ZnCl_2$ solution. A current of 0.01 A is passed through the series circuit for two hours.
  a) What happens at each of the six electrodes (assume O_2 is produced in the first two cells at the anode)?
  b) How much of each substance (in moles) is produced at each electrode?

3. How long must one pass a current of 0.1 A through a $ZnSO_4$ solution to produce 0.5 mol of zinc?

4. A current of 25 A is passed through a sodium chloride solution for 2 hours. How many grams of Cl_2 are produced? How many moles of NaOH can be obtained from the solution?

5. It costs more to plate out one mole of zinc from Zn^{2+} than to plate out one mole of copper from Cu^{2+} even though the same number of faradays is involved in each case. Explain.

6. How many faradays would it take to produce a mole of N_2 from NH_3 in an electrolysis cell having the reaction $2NH_3 \rightarrow N_2 + 3H_2$?

7. A current of 50 mA is passed through two solutions in series; the first contains 0.1-M $AgNO_3$ and the second contains 0.5-M $CuSO_4$.
  a) Write balanced equations for the cathode reactions in each solution.
  b) How long will the current have to flow to produce 54 g of silver?
  c) What weight of copper is produced at the same time?

8. When fused NaH is electrolyzed, does hydrogen gas appear at the anode or cathode? Where does hydrogen appear when water is electrolyzed? Explain.

9. What current is required to prepare 5.0 g of Br_2 from a solution of KBr in 1 hr? What other products, and in what amounts, are produced at the same time?

10. A current of 0.50 A is passed through a 0.20-M solution of NaCl for 20 min. If the volume of the solution is 500 cm³, what is the pH of the solution after electrolysis is stopped?

11. How many seconds would it take to plate out 2.5 g of silver from a $AgNO_3$ solution using a current of 1.0 A?

12. How many faradays would be required to produce 10 L of O_2 gas (measured at 0 °C and 1 atm) from hydrogen peroxide?

13. Water is electrolyzed for 1 hr using a current of 2.0 A. What volumes of H_2 and O_2, measured at 0 °C and 1 atm, will be produced?

14. Two electrolysis cells are wired in series; one contains $CuSO_4$ solution, and the other contains a nitrate of an unknown metal. After electrolyzing for a given period of time, 0.98 g of copper is plated out at the cathode of one cell. At the other cathode 2.68 g of the unknown metal plates out.
  a) Assuming the electrolysis current was 1.0 A, how long was the apparatus operated?
  b) What is the equivalent weight of the unknown metal?
  c) What additional information is needed to calculate the atomic weight of the unknown metal?

Groups IA and IIA

15. The distance between adjacent Na^+ and Cl^- ions in NaCl is measured by X-ray diffraction to be 2.81 Å. The distance between K^+ and Cl^- in KCl is 3.14 Å. Use the Na^+ and K^+ ionic radii given in Table 21.1 to estimate r_{Cl^-} in the two compounds (See Section 7.5). Do the two values agree reasonably well? Compare your answer with the value given in Table 23.2.

16. Explain why water solutions of washing soda, Na_2CO_3, have a very high pH (see Sections 13.1 and 13.2).

17. A typical Ca^{2+} and Mg^{2+} analysis of natural water might be 40 and 22 ppm, respectively (ppm = parts per million by weight). What weight of soap ($C_{17}H_{35}CO_2Na$) would a gallon (about 4 L) of such water tie up as sludge?

18. a) How many liters of acetylene can be produced at $-10\,°C$ and 1 atm from 70 g of calcium carbide?

b) What volume of liquid water is needed to carry out the above reaction?

c) If the above reaction is to operate a headlamp for 2.5 hrs, how many $cm^3 \cdot min^{-1}$ must be burned?

19. The electrode reaction $M(s) \rightarrow M^+(aq) + e^-$, where M is an alkali metal, may be thought of in terms of three steps:

$M(s) \rightarrow M(g)$ (vaporization)

$M(g) \rightarrow M^+(g) + e^-$ (ionization)

$M^+(g) + xH_2O(l) \rightarrow M^+(aq)$ (hydration)

The first two steps normally have $\Delta G^0 > 0$ while the latter is negative. The lithium ion, Li^+, has an unusually large negative value of ΔG^0 for hydration due to its small ionic radius and consequent high positive charge density. Show how this is reflected in the unusually high value of $\mathscr{E}^0_{Li/Li^+}$ given in Table 21.1.

20. Molten $BeCl_2$ conducts electric current poorly, hence it is not practicable to produce beryllium by electrolysis of this salt (or any of its halides). What does this indicate about the nature of the Be—Cl bond? If some NaCl is added to the fused $BeCl_2$, beryllium metal can be produced by electrolysis. Explain.

21. When alkali metals are dissolved in liquid ammonia, the following process occurs

$Na(s) + (x+y)NH_3(l) \rightarrow Na^+ \cdot xNH_3 + e^- \cdot yNH_3$

where Na^+ and e^- are solvated by ammonia molecules. The solvated electron, $e^- \cdot yNH_3$, accounts for the deep blue colors of these solutions.

a) Would you expect solutions such as the above to be poor, fair, or excellent reducing agents? Explain.

b) Why doesn't an analogous reaction occur when water is the solvent?

22. Predict the products of the reaction

$Na(s) + Fe_2O_3(s) \xrightarrow{heat}$

23. Predict the products and balance the following reactions.
 a) $Na_2O(s) + H_2O(l) \rightarrow$
 b) $KOH(s) + heat \rightarrow$
 c) $KO_2(s) + H_2O(l) \rightarrow H_2O_2(aq) +$
 d) $Li(s) + N_2(g) \rightarrow$
 e) $NaHCO_3(s) + heat \rightarrow$
 f) $RaO(s) + H_2O(l) \rightarrow$
 g) $SrO(s) + Al(s) \rightarrow$
 h) $CaC_2(s) + H_2O(l) \rightarrow$
 i) $MgCO_3(s) + heat \rightarrow$
 j) $SrO(s) + SO_3(g) \rightarrow$
 k) $NaOH(s) + CO_2(g) \rightarrow$
 l) $Na_2O(s) + CO_2(s) \rightarrow$

24. Calcium oxide combines with sulfur dioxide to form calcium sulfite. This reaction has been used to remove sulfur dioxide from automobile exhaust fumes.

a) Write a balanced equation for the process above.

b) If 1 L of gasoline weighs about 900 g and contains 5 ppm of sulfur, all of which forms sulfur dioxide, what weight of calcium oxide is needed to treat the exhaust fumes arising from one tank full of gas (about 50 L)?

Groups IB and IIB

25. The first ionization energy of $K:(Ar)4s$ is 4.3 eV while that of $Cu:(Ar)4s\,3d^{10}$ is 7.7 eV. Recalling that the first ionization energy is roughly proportional to Z^2_{eff}/n^2 and that d electrons provide ineffective shielding of the nucleus, supply a rationale for this ionization energy difference.

26. a) Write balanced equations for the preparation of elemental copper from Cu_2S (chalcocite ore) and $Cu_2(OH)_2CO_3$ (malachite). Note the oxidation states of copper in the above ores.

b) A sample of malachite ore is 70% pure (30% is foreign material called **gangue**). What weight of copper could be obtained from 1 tonne of such ore? How much coke (assume 95% carbon) would be needed to produce this copper?

c) How many faradays are needed to refine each tonne of copper by electrolysis?

27. The electron configuration $(Ar)4s\,3d^{10}$ for copper is only slightly favored over $(Ar)4s^2\,3d^9$ (see Section 4.5). Show that this helps explain the existence of both $+1$ and $+2$ oxidation states in copper.

28. Write a balanced chemical equation for the oxidation of copper with SO_4^{2-} in acid solution. (SO_2 is a product.)

29. Use Le Châtelier's principle to explain why AgCl dissolves in NH_3 solution [note that $[Ag(NH_3)_2]^+$ salts are soluble in water]. The $[Ag(NH_3)_2]^+$ ion is colorless.

30. Balance the following equation and explain what happens in this reaction:

$$Hg_2Cl_2 + NH_3 \rightarrow HgNH_2Cl + Hg + NH_4Cl$$

The mercury is produced in finely divided form and appears black.

31. AgCl and Hg_2Cl_2 are both insoluble in water and both are white. Use Problems 29 and 30 to suggest a way of telling whether a solution contains Ag^+, Hg_2^{2+}, or both.

32. Complete and balance the following equations.
a) $Cu_2O(s) + H_2(g) \rightarrow$
b) $HgO(s) + heat \rightarrow$
c) $ZnO(s) + HCl(aq) \rightarrow$
d) $CuCl_2(s) + Zn(s) \rightarrow$
e) $Hg_2(NO_3)_2(aq) + HCl(aq) \rightarrow$
f) $ZnCl_2(aq) + H_2S(aq) \rightarrow$
g) $CdO(s) + H_2SO_4(aq) \rightarrow$
h) $(CH_3)_2Hg(l) + O_2(g) \rightarrow$
i) $Au(s) + HCl(aq) \rightarrow$
j) $Ag_2O(s) + C(s) \rightarrow$

33. It has been proposed to run the following sequence of reactions using solar energy as the source of heat:

$$CaBr_2 + 2H_2O \rightarrow Ca(OH)_2 + 2HBr$$

$$2HBr + Hg \xrightarrow{heat} HgBr_2 + H_2$$

$$HgBr_2 + Ca(OH)_2 \rightarrow HgO + CaBr_2 + H_2O$$

$$HgO \xrightarrow{heat} Hg + \tfrac{1}{2}O_2$$

a) What gets used up and what is produced in the above series of reactions; that is, what is the net reaction?
b) What are the products useful for?

34. Which would be a better oxidizing agent, Cu_2O or CuO? Explain.

35. The standard-state enthalpies of formation of the solids CuO, ZnO, and CaO are -155, -348, and -636 kJ·mol^{-1}, respectively.

a) Which oxide is most likely to undergo the reaction $2MO(s) + heat \rightarrow 2M(s) + O_2(g)$?
b) Which oxide is least likely to undergo the above reaction?

36. Write balanced equations for the preparation of zinc metal from zinc carbonate. Repeat for zinc sulfide.

37. How would you prepare cadmium metal from cadmium chloride? The value of $\mathscr{E}^0_{Cd/Cd^{2+}}$ is 0.40 V at 25 °C.

38. Ketones are often prepared by the general reaction

$$R_1CdR_1 + 2R_2\underset{\underset{O}{\|}}{C}Cl \rightarrow 2R_1\underset{\underset{O}{\|}}{C}R_2 + CdCl_2$$

where R_1 is aryl or primary alkyl. The compound R_1CdR_1 is called an organocadmium compound. Show how to use this method to prepare (a) 2-propanone, (b) acetophenone (phenyl methyl ketone).

39. Use the following data to calculate K_p for the reaction

$$CuO(s) \rightleftharpoons Cu(s) + \tfrac{1}{2}O_2(g)$$

at 25 °C.

$$\Delta \overline{H}^0_{CuO} = -155 \text{ kJ·mol}^{-1}$$
$$\overline{S}^0_{CuO} = 43.5 \text{ J·K}^{-1}\text{·mol}^{-1}$$
$$\overline{S}^0_{Cu} = 33.3 \text{ J·K}^{-1}\text{·mol}^{-1}$$
$$\overline{S}^0_{O_2} = 205 \text{ J·K}^{-1}\text{·mol}^{-1}$$

Mercury

40. Most Hg(I) salts are diamagnetic both as solids and in solution. Show that this lends support to the existence of the Hg_2^{2+} ion

41. The following reaction has been proposed as the basis of a process designed to recover small traces of mercury from industrial plant effluents:

$$4HgX_2 + NaBH_4 + 6OH^- \rightarrow$$
$$4Hg + BO_3^{3-} + 4HX + 4X^- + Na^+ + 3H_2O$$

If $NaBH_4$ (sodium borohydride) costs $13·lb^{-1} and mercury costs $5·lb^{-1} (1971 prices), what is the *minimum* cost of recovering a pound of mercury from its salts using the above process?

42. Each year about 5000–6000 metric tons (tonnes) of mercury enter the world's oceans. The average concentration of mercury in the world's oceans is now about 0.03 ppb. If the world's oceans contain 10^9 km³ of water, how long will it take to double the mercury concentration at the present rate of influx, assuming none is removed in the meantime?

43. The accompanying figure represents a map of a small lake located near a chlor-alkali plant and other sources of mercury. In 1970 the lake, which has an average depth of 25 ft, was found to contain 9 ppb of mercury. Estimate the total amount of mercury in the lake in terms of the volume the mercury would have as a liquid (density = 13.6 g·cm⁻³). In what form is the mercury probably present in the water?

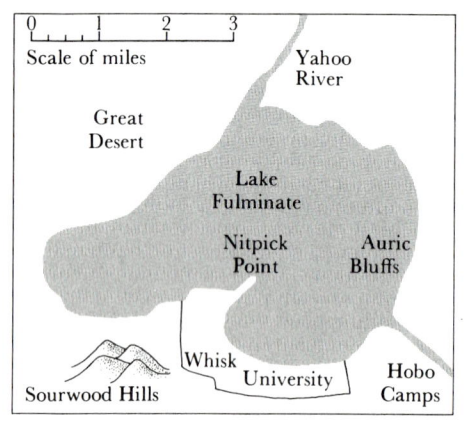

44. The reaction of aluminum with water

$$Al(s) + 3H_2O(l) \rightarrow Al^{3+}(aq) + 3OH^-(aq) + \tfrac{3}{2}H_2(g)$$

does not appear to take place at ordinary temperatures since the aluminum acquires an impermeable protective layer of $Al_2O_3 \cdot nH_2O$ (a hydrated oxide). However, if a drop of mercury is placed on a freshly scratched surface of aluminum and the metal is covered with water, steady evolution of hydrogen gas occurs. Suggest an explanation of this phenomenon.

45. The standard-state enthalpy of formation of solid HgO at 25 °C is -181.6 kJ·mol⁻¹. The molar entropies of Hg(l), O_2(g), and HgO(s) at 25 °C are 77.4, 205, and 72.0 J·K⁻¹, respectively. Calculate K_p for the decomposition of HgO(s) at 25 °C.

46. When mercury(II) compounds in solution are brought in contact with liquid mercury, the following reaction occurs

$$Hg^{2+}(aq) + Hg(l) \rightarrow Hg_2^{2+}(aq)$$

a) Explain what happens in the above reaction.
b) Suggest a simple way of proving that the above reaction actually takes place.

47. How much mercury could be prepared from a tonne of ore which contains 22% cinnabar?

48. A small lake has an area of 30 mi² and an average depth of 30 ft. Chemical analysis shows the water contains 11 ppb of mercury. What is the total volume of mercury (measured as a liquid with a density of 13.6 g·cm⁻³) in the lake?

General

49. Use relative electronegativities to explain why the following reactions occur:

$$Mg + BeF_2 \rightarrow MgF_2 + Be$$
$$K + NaCl \rightarrow KCl + Na$$

50. Which of the following reactions is more likely and why? See the previous problem.

$$Cl_2 + 2NaBr \rightarrow 2NaCl + Br_2$$
$$Br_2 + 2NaCl \rightarrow 2NaBr + Cl_2$$

ADDITIONAL READINGS

Cotton, F. A., and G. W. Wilkinson, *Advanced Inorganic Chemistry*, 3rd ed. New York: Wiley (Interscience), 1972. Although generally written at an advanced level, this very comprehensive text will be of some use to elementary students as a reference work.

Putnam, John J., and Robert W. Madden, "Mercury, Man's Deadly Servant." *National Geographic* (October 1972): 507–527. A well-illustrated overview of the hazards of mercury in our environment.

Rabenstein, Dallas L., "The Chemistry of Methylmercury Toxicology." *J. Chem. Educ.* **55**, 293 (1978). Tells how $(CH_3)_2Hg$ causes irreversible damage to the central nervous system.

Weeks, Mary Elvira, and Henry M. Leicester, *The Discovery of the Elements.* 7th ed. Cleveland: Chemical Education Publishing Co., 1968. Gives interesting historical accounts of how the elements were discovered and isolated. Sir Humphry Davy's isolation of group I and II metals is of special interest.

22
THE TRANSITION METALS

Iron, one of the commonest and least expensive of metals, is also one of the most beautiful. (Courtesy of Marshall Henrichs.)

Even when, to judge by the valence number, the combining power of certain atoms is exhausted, they still possess in most cases the power of participating further in the construction of complex molecules with the formation of very definite atomic linkages. The possibility of this action is to be traced back to the fact that, besides the affinity bonds designated as principal valencies, still other bonds on the atoms, called auxiliary valencies, may be called into action.
ALFRED WERNER, 1920

Alfred Werner (1866–1919), a professor of chemistry at the Zürich Polytechnic, is generally regarded as the founder of modern coordination chemistry. Werner was the first to show that an ion such as Co^{3+}, supposedly bereft of further bonding ability, could form bonds with substances containing lone pairs of electrons. In 1911 Werner demonstrated that coordination complexes could exhibit optical activity. A brilliant teacher, he received the Nobel prize in chemistry in 1913. The quotation at the head of the chapter appeared in a book published after Werner had died.

Although there is no universal agreement as to the definition of a **transition metal**, the following definitions are most commonly encountered:

1. Any element whose neutral atoms have electron configurations containing one to nine d electrons or one to thirteen f electrons. This definition includes groups IIIB through VIIIB of the periodic table.
2. Any element satisfying the above definition for either the neutral atom or any of the common oxidation states. This definition adds group IB to the above.

Since the group IB elements, copper, silver, and gold, resemble the other transition metals as a group, the latter definition is perhaps the preferred one. However, the group IB metals have already been discussed in Chapter 21 along with groups IA, IIA, and IIB so they will be largely ignored in the present chapter. The discussions will be limited to iron, cobalt, nickel, palladium, and platinum—five of the most important transition metals.

Table 22.1 lists some of the properties of the above transition metals along with those of copper, silver, and gold. Note that the eight elements have remarkably similar ionization energies for at least the first three ionizations. Also, the ionic radii are very similar for comparable oxidation states. The electronegativities of the elements are not given since they vary with oxidation state, but for the common oxidation states they are all around 1.8. It is typical of the transition elements that

TABLE 22.1 SOME PROPERTIES OF EIGHT TRANSITION METALS

METAL	IONIZATION ENERGIES (eV) 1	2	3	4	MELTING POINT (°C)	DENSITY AT 20 °C (g·cm^{-3})	IONIC RADII FOR VARIOUS OXIDATION STATES (Å)
Fe	7.9	16.2	30.6	56.8	1535	7.9	0.74(+2), 0.64(+3)
Co	7.9	17.1	33.5	83.1	1495	8.9	0.72(+2), 0.63(+3)
Ni	7.6	18.2	35.2	—	1453	8.9	0.69(+2)
Pd	8.3	19.4	32.9	—	1552	12.0	0.80(+2), 0.65(+4)
Pt	9.0	18.6	—	—	1772	21.5	1.01(+2), 0.90(+4)
Cu	7.7	20.3	36.8	—	1083	9.0	0.96(+1), 0.72(+2)
Ag	7.6	21.5	34.8	—	961	10.5	1.26(+1), 0.89(+2)
Au	9.2	20.5	—	—	1064	19.3	1.37(+1), 0.85(+3)

the similarities extend not only vertically, with respect to the arrangement of the periodic table, but horizontally as well. The horizontal similarities are due largely to the ns^2 electrons common to the valence shells of most of the transition elements.

As mentioned in Chapter 21, copper, silver, and gold are called the **coinage metals**. Iron, cobalt, and nickel are often called the **ferromagnetic metals**, and nickel, palladium, and platinum are called the **catalyst metals**. Note that nickel is both a ferromagnetic metal and a catalyst metal. The significance of these names will be discussed later.

22.1 COORDINATION COMPLEXES OF TRANSITION METALS

One of the outstanding characteristics of the transition metals is their ability to form **coordination complexes**. A coordination complex is a molecule or ion in which a transition metal ion is surrounded by a number of atoms or groups of atoms in a well-defined geometric arrangement. One simple type of coordination complex may be represented by the general formula ML_n, where the ion M is surrounded by n atoms or groups of atoms L called **ligands**. The ligands may be neutral or ionic. The number n is called the **coordination number** of the metal M. The bonds between the metal ion and ligand are generally of the Lewis acid–base type, that is, the ligand acts as an electron-pair donor and the metal ion as an electron-pair acceptor.

The reason that transition metal ions tend to form complexes with ligands, that is, to act as Lewis acids, is due to their electron configurations. These elements would have to gain or lose an inordinately large number of electrons in order to attain inert-gas configurations in the normal way; covalent bonding with Lewis bases makes it possible to attain inert-gas configurations (or configurations approaching these) much more easily.

As shown in Fig. 22.1, there are several different geometries commonly exhibited by coordination complexes. Among the most important are: linear, square planar, tetrahedral, and octahedral. Although there are many exceptions to the rule, the coordination number of a transition metal is often just twice its oxidation state in the complex. Thus complexes with Ag$^+$ usually exhibit a coordination number of 2, as exemplified by the complex [Ag(NH$_3$)$_2$]$^+$. This complex has a linear geometry; the

L—M—L

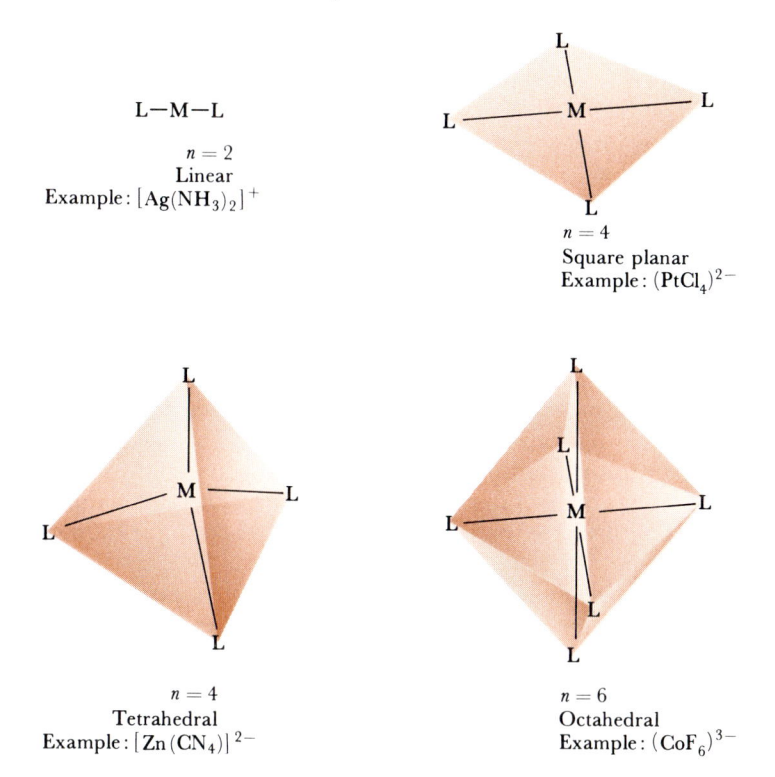

$n = 2$
Linear
Example: $[Ag(NH_3)_2]^+$

$n = 4$
Square planar
Example: $(PtCl_4)^{2-}$

$n = 4$
Tetrahedral
Example: $[Zn(CN_4)]^{2-}$

$n = 6$
Octahedral
Example: $(CoF_6)^{3-}$

Fig. 22.1. Four different geometries commonly exhibited by coordination complexes of transition metals. L is a monodentate ligand bound to the metal M by a lone pair of electrons.

two ammonia ligands are on opposite sides of the Ag^+ ion. Complexes with Co^{3+} tend to be octahedral, with coordination number 6. Thus the complex $(CoF_6)^{3-}$ consists of Co^{3+} surrounded by six fluoride-ion ligands distributed in an octahedral fashion. Copper complexes exhibit coordination numbers of 2 for Cu^+ and 4 for Cu^{2+} but the geometries are somewhat irregular.

Ligands which can occupy only one coordination site of a transition metal are said to be **monodentate** (having *one tooth*). Examples of monodentate ligands are molecules such as H_2O, NH_3, and CO and anions such as CN^-, Cl^-, F^-, SO_4^{2-}, NO_3^-, and

$$RC\underset{\underset{O}{\parallel}}{}O^-$$

Ligands which can occupy two coordination sites on a metal ion are called **bidentate**; 1,2-diaminoethylene (also called ethylenediamine), $H_2NCH_2CH_2NH_2$, is an example. This ligand is often abbreviated **en**. Thus the complex $[Co(en)_2NH_3Cl]^{2+}$ has two bidentate en ligands, a monodentate NH_3 ligand, and a Cl^- monodentate ligand. The central ion is Co(III). Tri-, tetra-, penta-, and even hexadentate ligands are also known.

The nomenclature of transition-metal complexes requires a special study. A brief summary is given in Appendix 4. Problems on nomenclature are given at the end of this chapter.

22.2 IRON AND ITS COMPOUNDS

In terms of mass, iron is the most abundant element on earth; 39.76% of the earth's mass is due to iron and 27.71% is due to oxygen. However, on an atom-number basis, oxygen is first with 48.63% and iron is a poor second with 19.99%. The high mass fraction of iron is due largely to the fact that the core of the earth, with a diameter of around 7000 km, is solid iron with up to 10% occluded hydrogen. As pointed out in Chapter 11, iron has the highest binding energy per nucleon of all the elements. This fact may play a role in its large natural abundance. The chemical symbol for iron, Fe, comes from the ancient Latin name *ferrum*.

The electron configuration of iron is $(Ar)4s^23d^6$. This suggests a maximum oxidation state of +8 but in fact no oxidation state beyond +6 is known. The +2 and +3 oxidation states are by far the most common. The +2 oxidation state is expected on the basis of the $4s^2$ portion of the electron configuration, whereas the +3 state is consistent with the expected high stability of a $3d^5$ electron configuration.

Elemental iron has a silvery luster but reacts rather rapidly with air, especially moist air, to form a brownish-red coating of hydrous oxide—mostly FeO(OH). On heating, this coating, which tends to flake away, is converted to iron(III) oxide, Fe_2O_3. One of the most outstanding characteristics of elemental iron and some of its compounds is its abnormally high paramagnetism. In fact, any material which has an unusually high paramagnetism is said to be **ferromagnetic**. The superparamagnetism of iron is usually explained on the basis of the **domain theory**, which views iron as consisting of a number of small regions or domains with different directions of magnetization. When a magnetic field is applied, the domains tend to become aligned and their magnetic natures become magnified.

Elemental iron exists in four allotropic forms designated α, β, γ, and δ. At 770 °C, the magnetic α form is converted to the nonmagnetic β form.

Metallic iron is usually prepared from one of its natural ores: Fe_2O_3 (hematite), Fe_3O_4 (magnetite, a magnetic oxide which is the **lodestone** of Sindbad the sailor tales), FeO(OH) (limonite), and $FeCO_3$ (siderite). Minerals of rather variable composition FeS_x (pyrites) also exist in abundance but are generally unsuitable as ores. The ores are usually treated with coke at high temperatures to produce elemental iron:

$$2Fe_2O_3(s) + 3C(s) \xrightarrow{heat} 4Fe(s) + 3CO_2(g)$$

Iron reacts readily with several other elements, particularly oxygen, the halogens, sulfur, silicon, carbon, and phosphorus. All of the oxides—FeO, Fe_2O_3, FeO(OH), and Fe_3O_4—exist in a variety of crystalline modifications depending on the conditions of formation. Iron reacts with sulfur to form nonstoichiometric compounds FeS_x, where x depends on the proportions of reactants and other conditions. The reaction must be started by heating but once begun proceeds rapidly with the liberation of large amounts of heat.

When elemental iron is dissolved in nonoxidizing acids such as HCl, and air is excluded, the Fe^{2+} salt is formed. The iron(II) ion exists in aqueous solution as the complex $[Fe(H_2O)_6]^{2+}$ and has a pale blue-green color. The coordination number of 6 for Fe(II) is an example of the failure of the simple rule for predicting the coordination number from the oxidation state. The complex is believed to be octahedral. In the presence of O_2 the Fe^{2+} ion oxidizes to Fe^{3+}:

$$2Fe^{2+}(aq) + \tfrac{1}{2}O_2(g) + 2H_3O^+ \rightarrow 2Fe^{3+}(aq) + 3H_2O(l)$$

Very strong oxidizing acids such as concentrated HNO_3 cause iron to assume a so-called **passive** state in which it resists further oxidation.

When a strong base is added to a Fe^{3+} solution, a voluminous red-brown precipitate is formed. This substance, sometimes written as iron(III) hydroxide, $Fe(OH)_3$, is more likely hydrated iron(III) oxide, $Fe_2O_3 \cdot nH_2O$.

The Fe^{3+} ion is believed to exist in aqueous solution as the complex

$$[Fe(H_2O)_{6-n}(OH)_n]^{(3-n)+}$$

where n depends on pH. At very low pH (acid solutions), the value of n is zero and the ion is

$$[Fe(H_2O)_6]^{3+} \quad [\text{hexaaquoiron(III) cation}]$$

which has an octahedral geometry. In solutions concentrated in Cl^- but dilute in Fe^{3+}, the ion is probably $(FeCl_4)^-$. When potassium thiocyanate, KSCN, is added to a solution of Fe^{3+}, a very deep-red complex, possibly $[Fe(SCN)_6]^{3-}$, is formed. This ion is called the hexathiocyanatoiron(III) cation. This reaction provides the basis for one of the most sensitive tests for the presence of Fe^{3+}.

22.3 THE BIOLOGICAL ROLE OF IRON

Two of the most remarkable molecules nature ever designed are **myoglobin** and **hemoglobin**, two proteins found in the blood of animals. Myoglobin functions as a storage receptacle for oxygen while hemoglobin transports oxygen from the lungs to muscles and tissues, carries carbon dioxide to the lungs from muscles and tissues, and helps to control blood pH.

Both of these molecules contain Fe(II); in each case the iron is incorporated in a very unusual structure called the **heme** group. As shown below, the iron-heme group consists of a single Fe^{2+} surrounded by four linked pyrrole molecules.

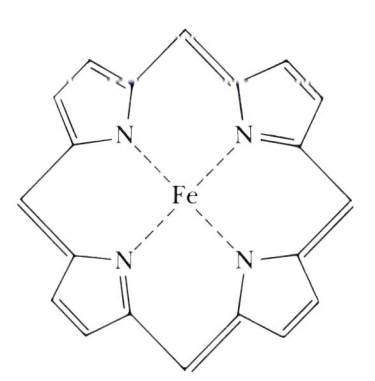

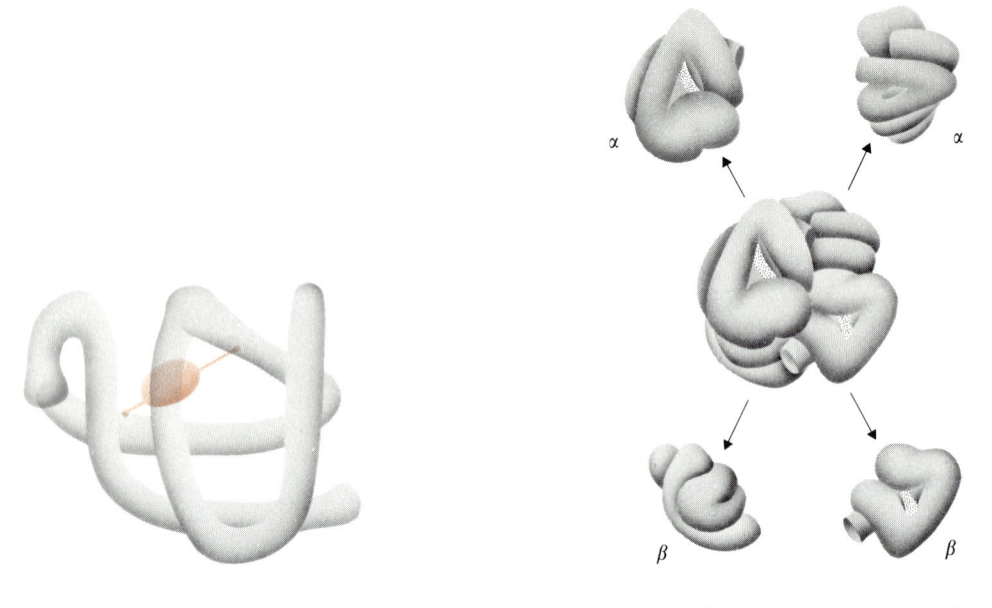

Fig. 22.2 (left). The three-dimensional shape of the myoglobin molecule. The shaded disk indicates the location of the iron–heme group. The viscera-like tubes indicate the areas occupied by coiled chains of proteins.

Fig. 22.3 (right). The structure of the hemoglobin molecule. (Reproduced by permission from *Journal of Chemical Education* **52**, 760 (1975). Copyright © by the American Chemical Society.)

Except for some side chains attached to the 3 and 4 positions of the pyrroles (not shown above), the iron-heme group is planar.

The structure of myoglobin is known in considerable detail. As shown in Fig. 22.2, this molecule consists of a single **polypeptide chain** (see Section 24.9) enclosing an iron-heme group. Myoglobin contains 153 amino acids and has a molecular weight of around 17,500 amu.

The structure of hemoglobin is less well known but it consists of four polypeptide units, each unit containing one iron-heme group. Two of the polypeptide units contain 141 amino acids each, and the other two contain 146 amino acids each. The molecular weight of hemoglobin is 64,500 amu, roughly four times that of myoglobin. The entire molecule appears spheroidal with approximate dimensions of 64 Å × 55 Å × 50 Å. Figure 22.3 shows a schematic illustration of how the four polypeptide units of hemoglobin appear in gross overall form and how they probably fit together to form a molecule. The four units are probably held together by hydrogen-bonding and various intermolecular forces.

The iron-heme group is incorporated into a polypeptide unit as follows: The nitrogen atoms of the four pyrroles occupy four coordinate sites of iron; a nitrogen atom from an amino acid (histidine) of the enclosing protein occupies another site, and a bound oxygen molecule (if present) occupies the sixth site.

The detailed roles of myoglobin and hemoglobin are best understood by examining Fig. 22.4, which shows how these molecules absorb oxygen at various partial

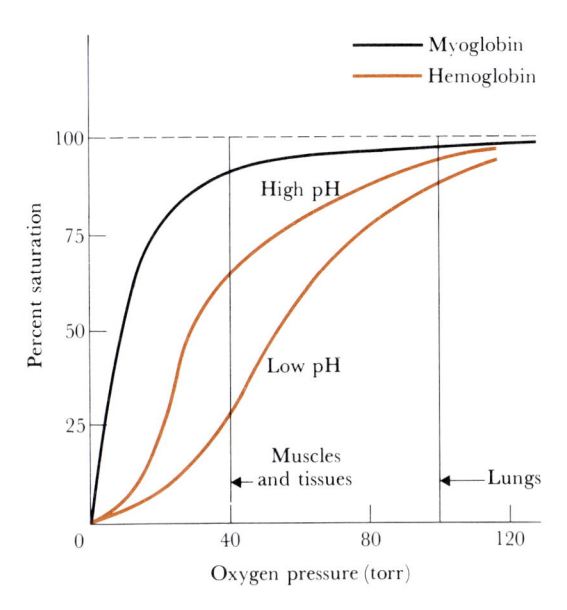

Fig. 22.4. Oxygen binding curves for myoglobin and hemoglobin. Note that myoglobin absorbs (binds) oxygen better at low oxygen pressures than does hemoglobin. Also, hemoglobin releases oxygen better at low pH than at high pH.

pressures of oxygen. It is seen that the two molecules behave quite differently in the absorption of oxygen. Whereas myoglobin absorbs oxygen with increasing ability as p_{O_2} increases and is unaffected by pH, hemoglobin absorbs oxygen rather poorly at low p_{O_2} and low pH, and begins to absorb better as both p_{O_2} and pH increase. These facts explain much of the physiological roles of the two molecules.

When hemoglobin (present in the erythrocytes or red blood cells) is in the lungs, a region of high p_{O_2}, it absorbs oxygen quite readily. The reaction may be symbolized by

$$Hb + nO_2 \rightarrow Hb \cdot nO_2 + H_3O^+$$

where $n = 4$ at 100% saturation. This corresponds to one O_2 per iron-heme group. The molecule $Hb \cdot nO_2$ is called **oxyhemoglobin**; it is bright red in color. Hb is called **deoxyhemoglobin**; it is purple in color. The H_3O^+ in the reaction is probably released by amino acid side chains such as those of histidine

The oxyhemoglobin is transported from the lungs via the arteries and eventually reaches the muscles and various tissues. These are areas of low p$_{O_2}$ and low pH; the latter is a result of waste products of metabolism, mainly CO_2 and lactic acid. Carbon dioxide leads to H_3O^+ production via the equilibria

$$CO_2(aq) + H_2O \rightleftharpoons H_2CO_3(aq) \rightleftharpoons H_3O^+ + HCO_3^-(aq)$$

As shown in Fig. 22.4, these conditions promote release of oxygen. The increase in H_3O^+ promotes the reverse of the lung reaction

$$Hb \cdot nO_2 + H_3O^+ \rightarrow Hb + nO_2$$

This is an example of Le Châtelier's principle. It appears as if the Fe^{2+} moves out of the plane slightly as oxygen is released; at the same time the four polypeptides also shift position somewhat. For this reason hemoglobin is sometimes said to be a "breathing" molecule. Some of the released oxygen is used at once for cellular metabolism; the rest is transferred to myoglobin where it remains until needed. The red color of steak is that of myoglobin in the tissues.

The deoxyhemoglobin formed in muscle and tissue now returns to the lungs by way of the veins. The venous blood also contains the acid waste products of the muscles and tissues; these would lower the pH of the blood were it not for the buffers in blood. Also, deoxyhemoglobin functions as an additional buffer, removing H_3O^+ via the reaction

Since each liter of blood contains 120 to 180 g of deoxyhemoglobin, this buffering action is considerable.

Deoxyhemoglobin also removes CO_2 from the muscles and tissues by two different mechanisms. Some CO_2 is bound to amino acid side chains and the rest is converted to HCO_3^- by removal of H_3O^+ (Le Châtelier's principle again!). This HCO_3^- is transported by the blood to the lungs.

Release of H_3O^+ by oxygenation at the lungs now converts HCO_3^- to CO_2 and the latter is exhaled. The entire cycle now begins anew.

Hemoglobin also combines with cyanide ion (CN^-), carbon monoxide (CO), and some other molecules. Carbon monoxide is bound much more readily by hemoglobin than is oxygen; consequently very small amounts of carbon monoxide can seriously lower the oxygen capacity of hemoglobin and lead to asphyxiation.

22.4 COBALT AND NICKEL

Cobalt, $(Ar)4s^23d^7$, and nickel, $(Ar)4s^23d^8$, are usually found together in nature. Note their similar melting points, densities, and ionic radii in Table 22.1. Both are ferromagnetic and not especially reactive chemically. Nickel, being somewhat more resistant to attack by O_2 and H_2O, once found wide use as a coating for more oxidizable metals. At present it finds wide use in various alloys, including stainless steel and Monel. Although cobalt dissolves in mineral acids, nickel, like iron, becomes passive in concentrated HNO_3 and does not dissolve. Both cobalt and nickel exhibit the +2 oxidation state in their compounds, but only cobalt forms +3 compounds to any extent. Cobalt(II) and nickel(II) form complexes with geometries such as tetrahedral, square planar, and octahedral, but cobalt(III) forms octahedral complexes almost exclusively. Unless complexed, cobalt(III) compounds are relatively unstable in aqueous solutions, tending to be reduced by water to cobalt(II). The cobalt(II) complex $[Co(H_2O)_6]^{2+}$ is stable.

Cobalt and nickel—as well as iron—react with carbon monoxide to form **carbonyls**. Iron and nickel form carbonyls by direct interaction:

$$Ni(s) + 4CO(g) \rightarrow Ni(CO)_4(l)$$

$$Fe(s) + 5CO(g) \rightarrow Fe(CO)_5(l)$$

The nickel tetracarbonyl forms at ordinary temperatures and pressures, but the iron pentacarbonyl requires elevated temperatures and pressures. Ultrapure nickel metal is sometimes prepared by thermal decomposition of nickel tetracarbonyl. Cobalt carbonyl is prepared at high pressure (around 200 atm) and moderately high temperature (around 150 °C) from the carbonate by the reaction

$$2CoCO_3(s) + 2H_2(g) + 8CO(g) \rightarrow Co_2(CO)_8(g) + 2CO_2(g) + 2H_2O(g)$$

$Co_2(CO)_8$ is called dicobalt octacarbonyl. The iron and nickel carbonyls are flammable liquids with highly toxic vapors; dicobalt octacarbonyl, unlike the nickel and iron carbonyls, is unstable in the presence of air.

22.5 THE BIOLOGICAL ROLE OF COBALT

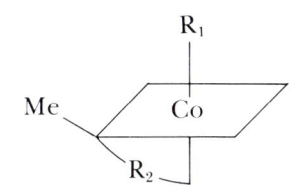

Many enzymes owe their activities only to their protein structures, for example, lysozyme (discussed in Section 20.9). Other enzymes require in addition nonprotein structures for activity. These nonprotein structures, which may be a metal ion, a complex organic molecule, or both, are called **coenzymes**. Cobalt(III) is found in a compound known as **vitamin B**$_{12}$ or **cobalamin**, a derivative of which acts as a coenzyme in several biological processes. Vitamin B_{12} has a rather complicated structure which is best illustrated a step at a time. The basic geometry about cobalt(III) is octahedral and may be represented by where Me stands for a methyl group and R_1 and R_2 are the following groups:

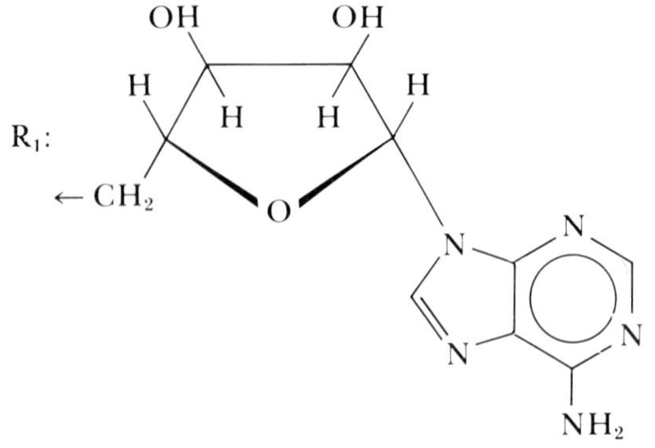

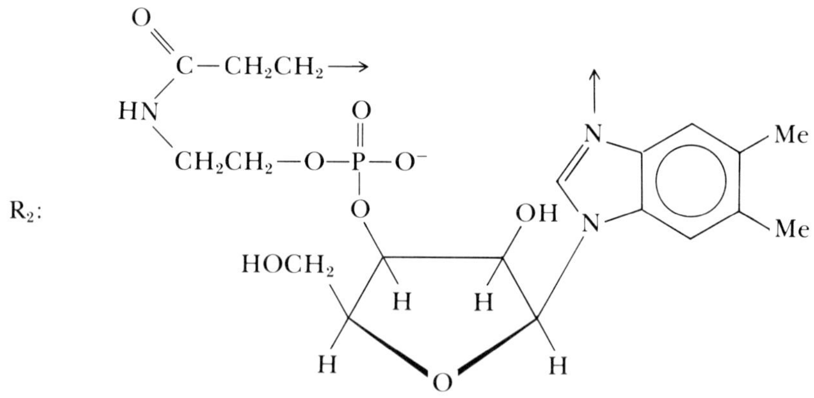

The arrows indicate the points of attachment to the basic structure. The plane in the basic structure perpendicular to the plane of this paper consists of a structure similar to that of heme and of chlorophyll:

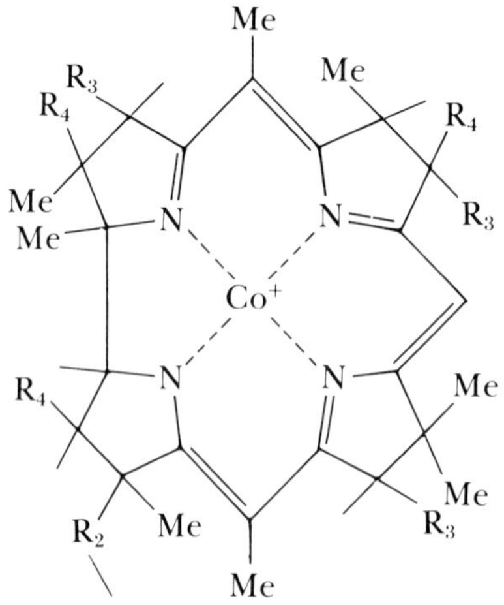

where R$_3$ and R$_4$ are

R$_3$: —CH$_2$CH$_2$CNH$_2$ R$_4$: —CH$_2$CNH$_2$
               ‖   ‖
               O   O

Vitamin B$_{12}$ is involved in the production of mature red blood cells. In a condition known as **pernicious anemia**, the gastric juices are lacking in a low-molecular-weight mucoprotein necessary for the absorption of vitamin B$_{12}$.

There is a vitamin B$_{12}$-like substance called **methylcobalamin** which may play a role in the formation of dimethyl mercury from inorganic mercury in stream and lake sediments (see Section 21.8). When this substance is treated with CO$_2$ in a reducing atmosphere, acetic acid is produced. The essential features of the reaction may be represented by

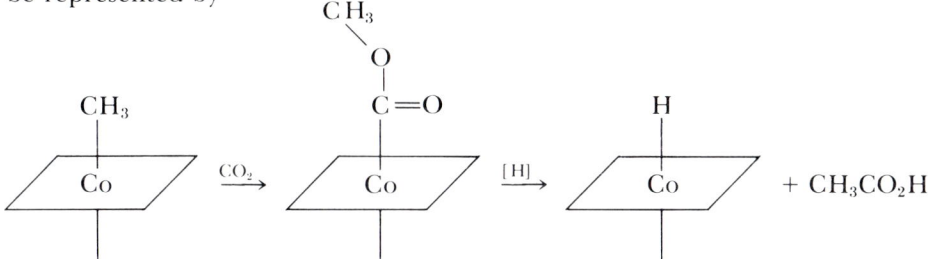

$$\text{CH}_3\text{-Co} \xrightarrow{\text{CO}_2} \text{CH}_3\text{OC(=O)-Co} \xrightarrow{[\text{H}]} \text{H-Co} + \text{CH}_3\text{CO}_2\text{H}$$

The methylation of mercury is believed to be carried out by anaerobic bacteria which contain methylcobalamin. Somehow the methyl group of cobalamin becomes transferred to mercury ions as part of the bacterium's attempt to protect itself against the toxicity of ingested mercury salts.

22.6 THE CATALYST METALS: NICKEL, PALLADIUM, AND PLATINUM

Nickel, palladium, and platinum have the electron configurations

Ni: (Ar)$4s^2 3d^8$

Pd: (Kr)$4d^{10}$

Pt: (Xe)$6s\,4f^{14}5d^9$

All three elements form mainly +2 oxidation state compounds. Whereas nickel(II) complexes are tetrahedral, square planar, or octahedral, palladium(II) and platinum(II) complexes tend to be square planar only. This is true, for example, of (PdCl$_4$)$^{2-}$ and (PtCl$_4$)$^{2-}$ [tetrachloropalladiate(II) anion and tetrachloroplatinate(II) anion].

Palladium and platinum are grayish-white lustrous metals and, like nickel, are ductile and malleable. All three metals are much used as hydrogenation catalysts, usually in the form of a finely divided powder. Finely divided nickel is used as a catalyst in hydrogenation of vegetable oils to make solid shortenings and margarines. Platinum is finding wide use in catalytic converters on automobiles. These converters consume unburned fuel and convert nitrogen oxides to ammonia.

Palladium and platinum have the ability to absorb large volumes of hydrogen gas. At 80 °C and 1 atm, palladium absorbs about 900 times its own volume of hydrogen gas. The resulting structure corresponds to an empirical formula of $Pd_{10}H_7$ and is an example of a so-called interstitial hydride (Section 15.4).

Solid platinum has an interesting catalytic property. When a platinum wire is heated to redness and held just above the surface of liquid methanol, a catalytic oxidation of methanol vapor to formaldehyde occurs. The reaction is

$$2CH_3OH(g) + O_2(g) \xrightarrow{Pt} 2HCHO(g) + H_2O(g)$$

Since the process is exothermic and takes place on the surface of the platinum wire, the wire stays red-hot and the reaction continues as long as methanol vapor and air are available. Some commercial applications of this and similar processes include catalytic hand warmers and catalytic heaters for campers.

Palladium and platinum are relatively rare elements and are usually regarded as among the most inert of the metals. However, both metals are attacked by molten alkali metal oxides and peroxides and by chlorine at red-hot temperatures. Palladium is also attacked by nitric and sulfuric acids. The reaction is accelerated by the presence of O_2 and oxides of nitrogen. Platinum and nickel, like iron, are rendered passive by concentrated nitric acid. Nitrates of passive metals are made by treating other salts of the metals with nitric acid.

Like nickel, palladium and platinum find wide use in alloys, particularly the gold alloys used in jewelry.

22.7 CHELATES

Multidentate ligands which can bond to a metal ion in such a way as to enclose it in a pincerlike manner are called **chelating agents** and the complexes formed are called **chelates** (after the Greek *chele*, claw). The chelating agent dimethylglyoxime

$$\begin{array}{cc} H_3C & CH_3 \\ | & | \\ HO-N=C-C=N-OH \end{array}$$

forms a bright-red complex with Ni^{2+} in basic solution. The structure of the complex is

Two OH^- are involved in the formation reaction. The formation of this complex is the basis of a very sensitive test for the presence of Ni^{2+}; concentrations as low as 2.5 ppm can be detected in this way.

Another common chelating agent is ethylenediaminetetraacetic acid (EDTA), known commercially as Versene or Sequestrene. The sodium salt has the formula

$$\begin{matrix} ^-O_2CCH_2 & & CH_2CO_2^- \\ & \diagdown \diagup & \\ & NCH_2CH_2N & \\ & \diagup \diagdown & \\ ^-O_2CCH_2 & & CH_2CO_2^- \end{matrix} \quad , 4Na^+$$

The Fe(III) complex of EDTA is

Note that EDTA is a hexadentate ligand in the above complex. In some EDTA complexes EDTA acts as a pentadentate ligand; one of the —CO_2^- groups remains uncoordinated and the position it would have had on the metal ion is occupied by a water molecule.

EDTA is one of the very few chelating agents capable of complexing with group IA and IIA metals. EDTA is frequently added to detergent formulations, where it acts to complex ions such as Ca^{2+} and Mg^{2+} found in hard water. The complexed ions are incapable of inactivating the detergent and are, in effect, removed by the EDTA. Chelating agents used in this way are called **sequestering agents**. EDTA is also used to treat heavy metal poisoning; the toxicity of many metals is eliminated or reduced when their ions are in chelated form.

Chlorophyll, heme, and cobalamin are chelates of Mg^{2+}, Fe^{2+}, and Co^{3+}, respectively.

22.8 OPTICAL AND MAGNETIC PROPERTIES OF TRANSITION-METAL COMPLEXES

One of the most interesting features of substances containing transition-metal ions is their often attractive colors.* In the case of the transition-metal complexes, chang-

---

* Many precious and semiprecious gems owe their characteristic colors to trace amounts of transition-metal ions. For example, the mineral corundum (Al_2O_3) is colorless when pure but small amounts of Cr^{3+} in place of Al^{3+} produce the blood-red color of ruby. Other impurities in corundum lead to other gems; for example, Fe^{3+} produces topaz (usually yellow) and manganese produces amethyst (purple or violet).

ing from one ligand to another generally brings out dramatic changes in color. For example, [Co(NH$_3$)$_6$]Cl$_3$ is orange whereas [Co(NH$_3$)$_5$Cl]Cl$_2$ is purple, yet the two complexes differ by only one ligand; one NH$_3$ in the former is replaced by Cl$^-$ in the latter.

The color of a substance is determined by the wavelengths of light it will absorb and reflect. A substance appears black if it absorbs all wavelengths of visible light; it appears white if it reflects all wavelengths of visible light. In general, a substance has the color of the light it does not absorb; that is, it has the color of the light it reflects. If a substance absorbs blue light (and no other color), it will reflect all light of other colors. In this case these other colors will look orange, the complement of blue on the primary color scale. The three primary colors are red, blue, and yellow; the complement of each primary color is a mixture of the other two primary colors. Thus the complement of blue is red + yellow or orange, the complement of red is yellow + blue or green, and the complement of yellow is red + blue or violet. Consequently, a substance that looks orange ([Co(NH$_3$)$_6$]Cl$_3$ for example) absorbs blue light. This implies that [Co(NH$_3$)$_6$]Cl$_3$ has two electronic states whose energy difference corresponds to a photon of blue light. Similarly, a substance that looks green must have two electronic states whose energy difference corresponds to a photon of red light.

EXAMPLE 22.1 Solutions of the hexathiocyanatoferrate(III) anion are red. What color of light does this complex absorb?

SOLUTION The anion must absorb light complementary to red, that is, blue + yellow or green.

EXAMPLE 22.2 Chlorophyll, the pigment in the leaves of most plants, is green. What color of light does chlorophyll absorb?

SOLUTION Chlorophyll must absorb light complementary to green, that is, red.

Transition-metal complexes, just like organic compounds, display geometric isomerism. Figure 22.5 shows a general example of *cis-trans* diastereomers in an octahedral complex. The ligand L$_1$ is bidentate and L$_2$ is monodentate. Figure 22.6 shows an example of optical isomerism. Here L is a bidentate ligand. Such isomers will rotate the plane of polarized light in opposite directions but by equal magnitudes (see Section 16.13).

Transition-metal complexes are either diamagnetic or paramagnetic (see Section 4.6) according to the absence or presence of unpaired electrons. These complexes are often classified as **high-spin** complexes or **low-spin** complexes. High-spin complexes are those in which the complex has the same number of unpaired electrons as the free central ion M^{x+}; low-spin complexes have fewer unpaired electrons than M^{x+}. For example, Co^{3+} has the electron configuration (Ar)3d^6 and hence has four unpaired electrons as shown by the orbital box diagram:

Co^{3+}: (Ar) [↑↓|↑|↑|↑|↑]
                    3d

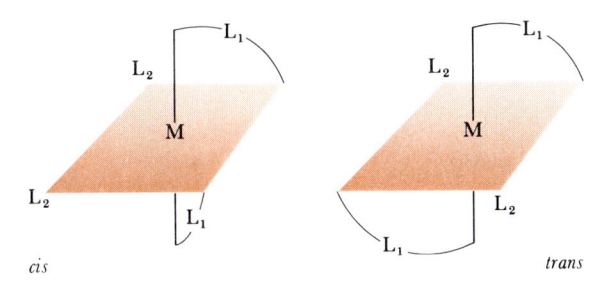

Fig. 22.5. *Cis-trans* isomerism in an octahedral transition-metal complex. Ligand L_1 is bidentate and ligand L_2 is monodentate.

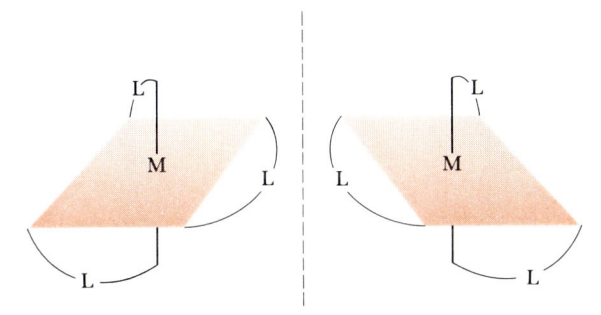

Fig. 22.6. Optical isomerism in an octahedral transition-metal complex. L is a bidentate ligand.

The complex $(CoF_6)^{3-}$ also has four unpaired electrons and is a high-spin complex. On the other hand, $[Co(NH_3)_6]^{3+}$ is diamagnetic and hence is a low-spin complex.

The number of unpaired electrons is usually deduced by the experimental value of the compound's **magnetic moment**.* This quantity is related approximately to the number of unpaired electrons by the formula

$$\mu = \sqrt{n(n+2)}\ \beta \qquad (22.1)$$

where n is the number of unpaired electrons and β is a unit of magnetic moment called the **bohr magneton**. Usually, experimental magnetic moments are slightly larger than those predicted by Eq. (22.1). This is due to a small contribution from the angular momentum of the electron. The Co(II) ion has the electron configuration

Co^{2+}: (Ar) [↑↓|↑↓|↑|↑|↑] $3d$

Thus $n = 3$ and $\mu = 3.87\ \beta$. The experimental value is 4 to 5 β depending on the compound involved. Similarly, Co^{3+}: $(Ar)3d^6$ has $n = 4$ and $\mu = 4.90\ \beta$; the experimental values range from 5 to about 5.4.

---

* The magnetic moment of a magnet is a measure of the torque experienced by the magnet when it is at right angles to a uniform magnetic field. An unpaired electron behaves in some respects like a tiny magnet.

22.9 QUANTUM MECHANICAL MODELS OF BONDING IN TRANSITION-METAL COMPLEXES

Quantum mechanics has proved to be highly successful in accounting for and providing insight into the structures of transition-metal complexes. Among the properties accounted for are the following:

1. The geometric arrangement of ligands about the central ion.
2. Color differences in various complexes.
3. The nature of high-spin and low-spin complexes.

The geometric arrangement of ligands about the central ion is accounted for qualitatively by the use of suitable hybrid orbitals of the central ion. The tetrahedral geometry of $[Zn(NH_3)_4]^{2+}$ is accounted for as follows: The ion Zn^{2+} has the electron configuration $(Ar)3d^{10}$. If vacant $4s$ and $4p$ atomic orbitals are mixed to form sp^3 hybrid orbitals, it is possible to account for each zinc–ligand bond as an overlap of an sp^3 hybrid orbital of zinc with an sp^3 hybrid orbital of $:NH_3$. Here both pairs of electrons in the bond orbital σ_{ZnN} come from the ammonia lone pair. This shows that Zn^{2+} is acting as a Lewis acid and NH_3 is acting as a Lewis base.

The square planar geometry of $[Ni(CN)_4]^{2-}$ is accounted for similarly. Ni^{2+} has the electron configuration $(Ar)\ 3d^8$. The orbital box diagram is

Ni^{2+}: (Ar) [↑↓|↑↓|↑↓|↑|↑]
                    $3d$

Experiment shows $[Ni(CN)_4]^{2-}$ is a low-spin complex (diamagnetic). Consequently, it is assumed that, first, maximum pairing of electrons occurs in the $3d$ atomic orbitals, leading to

Ni^{2+}: (Ar) [↑↓|↑↓|↑↓|↑↓|]
                    $3d$

Now the unused $3d$ atomic orbital along with vacant $4s$ and $4p$ atomic orbitals can be mixed to form **dsp^2** hybrid orbitals. These four hybrids lie in a single plane and form angles of 90° with each other. Ions $[:C:::N:]^{1-}$ can bond to Ni^{2+} via an overlap of sp_C and dsp^2 of Ni^{2+}. This leaves all electron spins paired.

The complex $[Co(NH_3)_6]^{3+}$ is octahedral and diamagnetic (low spin). The ion Co^{3+} has the orbital box diagram

Co^{3+}: (Ar) [↑↓|↑|↑|↑|↑]
                    $3d$

Again, maximum pairing leads to

Co^{3+}: (Ar) [↑↓|↑↓|↑↓| |]
                    $3d$

Now the two unused 3d atomic orbitals can mix with the vacant 4s and 4p to form d^2sp^3 hybrids which are octahedral. Each cobalt–ligand bond is an overlap of d^2sp^3 of Co^{3+} with lone pair sp^3 of NH_3. Since all electrons are paired, the complex is diamagnetic.

However, in order to explain why $(CoF_6)^{3-}$ is a high-spin complex, it is necessary to make the assumption that maximum pairing does not occur in Co^{3+}. In this case **d^2sp^3** hybrids are formed by mixing 4s, 4p, and 4d (not 3d) atomic orbitals of Co^{3+} (these are often called **sp^3d^2** hybrids).* Why is it necessary to make such different assumptions in what appears to be virtually analogous situations?

The answer to the above is that the hybrid orbital model is inadequate to handle details of the magnetic properties of transition-metal complexes. A much more satisfactory model known as the **crystal-field** model has been proposed by Bethe and Van Vleck and greatly extended by others. The crystal-field model assumes that the ligands can be treated as point charges surrounding the central ion. Figure 22.7 illustrates this model for an octahedral complex. The heart of the model is its consideration of how the surrounding negative charges interact with electrons in d orbitals of the central ion. Because the different d orbitals are oriented differently with respect to the ligand point charges, different interactions result. With the 3d orbital boundary surface diagrams of Fig. 22.8 as a guide, it is seen that the $3d_{z^2}$ and

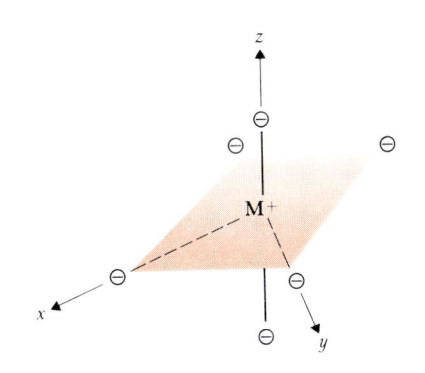

Fig. 22.7. Crystal-field model of an octahedral transition-metal complex. The central ion M^+ and the surrounding ligands are treated as point charges.

Hans A. Bethe (b. 1906) is a professor of physics at Cornell University. Bethe was one of the pioneers in the theory of thermonuclear fusion and was among the first to suggest nuclear fusion as the source of the sun's energy. He received the Nobel prize in physics in 1967.

J. H. Van Vleck (b. 1899), a physics professor at Harvard University, shared in the 1977 Nobel prize in physics for fundamental theoretical investigations on the electronic structure of magnetic and disordered solids. Van Vleck has been called the "father of modern magnetism."

---

* Compounds using 3d with 4s and 4p are also called *inner-orbital complexes*; those using 4d with 4s and 4p are called *outer-orbital complexes*.

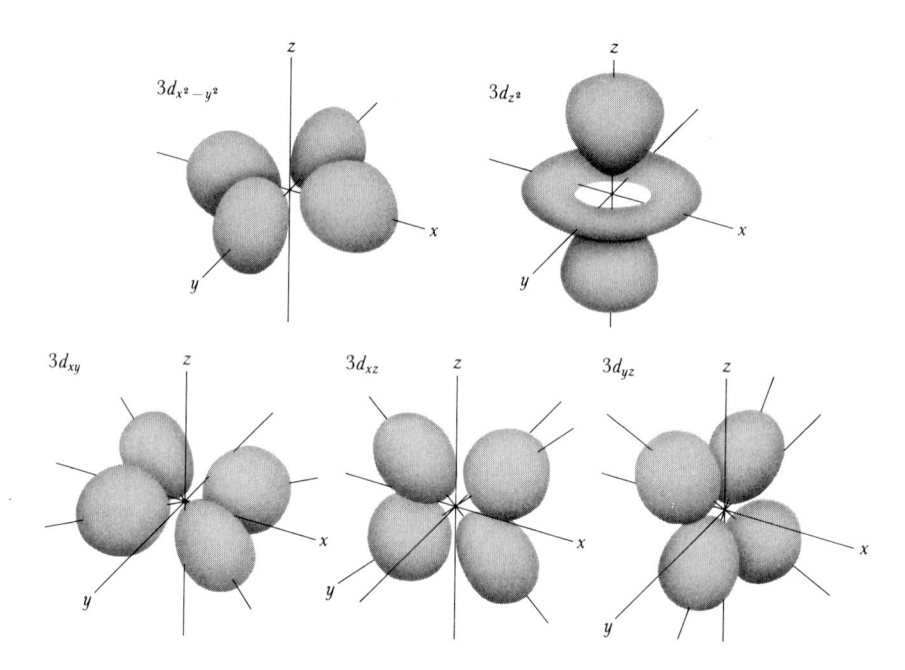

Fig. 22.8. Boundary surface diagrams of $3d$ orbitals.

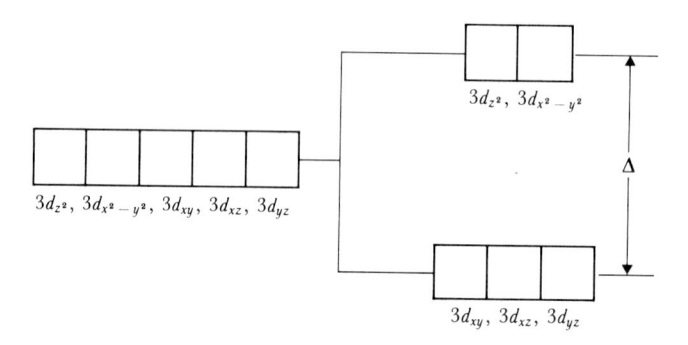

Fig. 22.9. Splitting of $3d$ orbitals by a crystal field of octahedral symmetry. The quantity Δ is the energy difference between the two sets of $3d$ orbitals.

$3d_{x^2-y^2}$ atomic orbitals point directly at ligand point charges whereas the remaining atomic orbitals ($3d_{xy}$, $3d_{xz}$, and $3d_{yz}$) point between two ligand charges. This means that there is a greater electron–electron repulsion in the former so that the five $3d$ atomic orbitals, instead of having the same energy, have two different energies. Specifically, the energy of the $3d_{z^2}$ and $3d_{x^2-y^2}$ atomic orbitals is higher than that of $3d_{xy}$, $3d_{xz}$, and $3d_{yz}$ by an amount represented by Δ (see Fig. 22.9). The quantity Δ, called the **crystal-field splitting parameter**, depends on the central ion and on the nature of the ligands. For a given central ion, the following ligands lead to increasingly larger values of Δ:

$$Cl^- < F^- < H_2O < NH_3 < NO_2^- < CN^-$$

This is an abbreviated sample of a list called the **spectrochemical series**.

The crystal-field model provides a very simple qualitative explanation for the colors of transition-metal complexes. First of all, Δ turns out to have just the correct magnitude so that the wavelengths calculated from

$$\Delta = \frac{hc}{\lambda} \tag{22.2}$$

correspond to the visible portion of the electromagnetic spectrum (about 400 to 700 nm). This suggests that light absorption by transition-metal complexes involves a transition between two electronic states representable by exciting an electron from a lower $3d$ orbital to an upper $3d$ orbital. Furthermore, this model shows why colors of complexes change when ligands are changed. Since $[Co(NH_3)_6]Cl_3$ is orange, the transition from a lower $3d$ orbital to an upper $3d$ orbital must require absorption of a photon of blue light. Replacement of one NH_3 by Cl^- lowers Δ slightly (since $Cl^- < NH_3$) and thus a photon of lower energy (higher wavelength) is capable of causing the electronic transition. This means that $[Co(NH_3)_5Cl]Cl_2$ will reflect light of higher energy (lower wavelength) than $[Co(NH_3)_6]Cl_3$. The purple color of $[Co(NH_3)_5Cl]Cl_2$ is in accord with this model.

EXAMPLE 22.3 Salts of $[Ti(H_2O)_6]^{3+}$ [hexaaquotitanium(III)] are violet and absorb light of wavelength 492 nm. What is the value of Δ for this complex?

SOLUTION

$$\Delta = \frac{hc}{\lambda} = \frac{(6.63 \times 10^{-34}\ J \cdot s)(3.00 \times 10^8\ m \cdot s^{-1})}{492 \times 10^{-9}\ m}$$

$$= 4.04 \times 10^{-19}\ J \quad \text{or about 2.5 eV}$$

Low-spin and high-spin complexes are readily accounted for by the crystal-field model. The fact that $[Co(NH_3)_6]^{3+}$ is low-spin and $(CoF_6)^{3-}$ is high-spin follows very simply from the assumption that NH_3 has a larger Δ value than does F^-. Suppose six electrons are to be placed into the split $3d$ orbitals below:

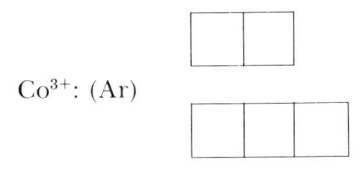

Co^{3+}: (Ar)

When the two sets of $3d$ orbitals have the same energy, Hund's rule applies and the electrons go in as

But when the two sets of orbitals have different energies, Hund's rule applies only if the energy difference Δ is very small, that is, if the energy stabilization due to maximum spin is greater than the energy needed to carry out the promotion

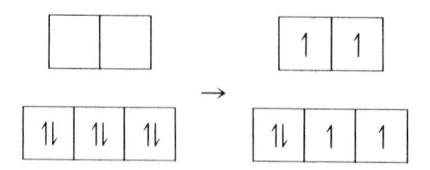

Assuming that Δ due to F⁻ is small enough for Hund's rule to apply leads to

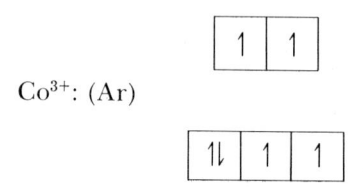

Co³⁺: (Ar)

Ligand bonds now use sp^3d^2 hybrids formed from vacant 4s, 4p, and 4d atomic orbitals of Co³⁺, leading to a high-spin complex. In the case of NH₃ ligands, it is assumed that Δ is large enough that it is energetically lower to have maximum spin pairing in the lower 3d atomic orbitals as follows:

Co³⁺: (Ar)

Ligand bonds now use d^2sp^3 hybrids formed from 3d (not 4d), 4s, and 4p atomic orbitals of Co³⁺, leading to the low-spin complex.

Square planar complexes [such as (PtF₄)²⁻] are treated by removing two charges from the octahedral model which are opposite each other. As the charges are removed the $3d_{z^2}$ and $3d_{x^2-y^2}$ atomic orbitals split, the former going down in energy and the latter going up. The remaining atomic orbitals also split, $3d_{xy}$ going up and the other two going down together to almost the same level. This is diagrammed in Fig. 22.10.

In the case of tetrahedral complexes, the crystal-field splitting is just the opposite of octahedral; that is, the $3d_{z^2}$ and $3d_{x^2-y^2}$ atomic orbitals go down in energy and the other three atomic orbitals go up. However, since none of the d atomic orbitals point exactly at any of the ligand charges, the splitting Δ is small. In fact, all other factors being equal, Δ for tetrahedral splitting is about half (actually 4/9) that for octahedral splitting. As a consequence, Hund's rule of maximum unpaired spins always applies and all tetrahedral complexes are high-spin.

The crystal-field model also fails in some areas. The best models in use today are those based on a very general molecular orbital theory. Since these models make

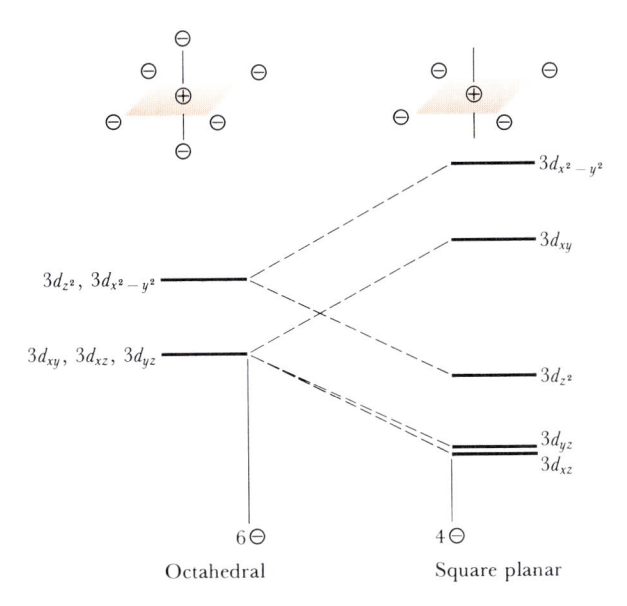

Fig. 22.10. Splitting of $3d$ orbitals by a crystal field of square planar symmetry as deduced from the octahedral case. As a pair of opposite ligands is removed from the octahedral complex, the splitting of $3d$ orbitals changes as shown in the diagram. Note that the $3d_{xz}$ and $3d_{yz}$ orbitals are separated only slightly in energy.

use of some rather sophisticated mathematics, they lie beyond the scope of a general chemistry text.

EXAMPLE 22.4 Describe the bonding in the hexaamminechromium(III) cation and predict whether it is a low-spin or high-spin complex.

SOLUTION The cation is $[Cr(NH_3)_6]^{3+}$. The electron configurations of Cr and Cr^{3+} are

Cr: $(Ar)4s\,3d^5$

Cr^{3+}: $(Ar)3d^3$ (in an octahedral field)

The magnitude of Δ is immaterial in this case since all electrons are in the lower $3d$ atomic orbitals. The six NH_3 are bonded to Cr^{3+} via d^2sp^3 octahedral hybrids (formed from $3d$, $4s$, and $4p$). The complex has three unpaired electrons (just like Cr^{3+}) and is thus high-spin.

SUMMARY

Transition metals are characterized by one to nine d electrons or one to thirteen f electrons. Compounds in which the transition-metal ion functions as an electron acceptor and other molecules or anions function as electron donors are called coordination complexes. Many transition-metal ions have coordination numbers which are twice their oxidation numbers, but exceptions to this rule occur. Iron, cobalt, and nickel are ferromagnetic. Iron and cobalt play important biological roles when incorporated in heme structures and complexed with proteins. Iron is found in hemoglobin, a molecule which transports oxygen in the blood, and in myoglobin, a molecule which stores oxygen for use in metabolism in muscle and tissues. Cobalt is found in vitamin B_{12}, a derivative of which is a coenzyme for the production of mature red blood cells. Nickel, palladium, and platinum are relatively inert metals with pronounced catalytic properties, particularly in hydrogenation reactions.

Chelates are complexes of multidentate ligands and metal ions. Of special importance are chelating agents such as EDTA which can complex with group IA and IIA ions. Chelates may be used to sequester metal ions in solution, thus, in effect, removing these ions.

Transition-metal complexes, like organic compounds, exhibit geometrical isomerism. This includes *cis-trans* diastereomers and optical isomers.

One of the most characteristic features of transition-metal compounds is their varied colors. In the case of coordination complexes, color changes accompany changes in ligands about a given central ion.

Transition-metal complexes are diamagnetic or paramagnetic; complexes having the same number of unpaired electrons as the isolated central ion are called high-spin complexes and those having a smaller number of unpaired electrons are called low-spin complexes.

Quantum mechanics provides various models that account for the properties of transition-metal complexes, properties such as color, geometry, and magnetic behavior. One of the simplest models, the crystal-field model, treats ligands as point charges which split the energies of the $3d$ orbitals of the central ion into two or more groups. The types and magnitudes of these splittings can be used to account for color and magnetic properties in a rather simple manner.

LEARNING GOALS

1. Be able to provide a definition of a transition element.

2. Be able to sketch the geometries of the most common types of transition-metal complexes and to illustrate *cis-trans* and optical isomerism.

3. Given the formula of a transition-metal complex, be able to determine the oxidation state of the central ion and its coordination number.

4. Know the common oxidation states of iron, cobalt, nickel, palladium, and platinum.

5. Be familiar with the properties of each of the above pure elements and know their principal chemical properties.

6. Be able to write the electron configurations of iron, cobalt, nickel, palladium, and platinum and their important ions.

7. Be able to explain, using balanced equations, how iron is prepared from its ores.

8. Know the general features of the biological roles of hemoglobin and myoglobin and be able to describe how oxygen is transported in the blood.

9. Be familiar with chelates, their structures, and their uses.

10. Be able to explain what produces colors in substances.

11. Given the formula of a transition-metal complex, be able to describe the bonding in terms of the appropriate hybrid orbitals and to use the crystal-field model to account for high- and low-spin complexes.

12. *Optional:* If your teacher so requires, be able to give the name of a transition-metal complex from its formula and its formula from its name. You will need Appendix 4 for the nomenclature rules.

DEFINITIONS, TERMS, AND CONCEPTS TO KNOW

transition metal
coordination complex
coordination number
ligand
monodentate
bidentate
ferromagnetism
domain theory
passive state

oxyhemoglobin
deoxyhemoglobin
myoglobin
heme structure
metal carbonyl
coenzyme
cobalamin
chelate
dsp^2

d^2sp^3
sp^3d^2
high-spin complex
low-spin complex
magnetic moment
crystal-field model
crystal-field splitting parameter
spectrochemical series

QUESTIONS AND PROBLEMS

General

1. Predict the coordination numbers of the following.

a) Cr^{3+}
b) Au^{3+}
c) V^{3+}
d) Au^+
e) W^{4+}
f) Cd^{2+}

2. What is the maximum number of unpaired electrons a transition-metal atom can have?

3. a) What is the oxidation number of iron in magnetite, Fe_3O_4?
b) Suppose one Fe in Fe_3O_4 is +2 and two Fe's are +3, what is the average oxidation number?

4. A sulfide ore of nickel is roasted to NiO and reduced with coke.
a) Write a balanced equation for the reduction reaction.
b) How many tonnes of coke (90% carbon by weight) would be needed to produce each tonne of nickel?
c) Suggest an easy way of obtaining a small amount of ultrapure nickel.

5. One cm³ of human blood can absorb a maximum of about 0.20 cm³ of O_2 (measured at standard temperature and pressure). One cm³ of blood contains about 5×10^9 blood cells and each blood cell contains about 2.5×10^8 hemoglobin molecules.
a) How many O_2 molecules can a single blood cell hold?
b) How many O_2 molecules can one molecule of hemoglobin carry?
c) Analysis shows hemoglobin is 0.335% iron. What is the minimum molecular weight of hemoglobin?
d) If each Fe in hemoglobin can hold one O_2 molecule, what is the molecular weight of hemoglobin?

6. A solution of $Ni(NO_3)_2$ is used in a nickel-plating bath. How much nickel can be plated out in 2 hr with a current of 0.01 A?

7. Write electron configurations for the following ions.

a) Pt^{2+} c) Fe^{2+}
b) Cu^+ d) V^{3+}

8. A compound contains 25.33% nickel, 36.26% nitrogen, 30.59% chlorine, and 7.82% hydrogen. What is the empirical formula and name of the compound?

9. The ion MnO_4^- is commonly called *permanganate*. What is the IUPAC name of this ion?

10. Silver chloride is insoluble in water but readily dissolves in ammonia solution. What reaction accounts for this?

11. Oxalic acid (see Table A4.1 for the conjugate base, oxalate ion) can remove rust stains from clothing. Use balanced equations to show how this works.

12. Determine the coordination number of the central ion in each of the following.

a) $(AlCl_4)^-$
b) $[Co(NH_3)_5Cl]Cl_2$
c) $[Fe(EDTA)]^{2-}$
d) $[Co(en)_2(NH_3)_2]^{3+}$
e) $[Zn(NH_3)_2]Cl_2$
f) $K[Pt(NH_3)Cl_3]$
g) $[Pt(NH_3)_4Cl_2]Cl$
h) $[Cr(NH_3)_5Cl]NO_3$

13. The relationship between chromate ion (CrO_4^{2-}) and dichromate ion $(Cr_2O_7^{2-})$ is given by the equation

$$2CrO_4^{2-}(aq) + H_2O(l) \rightleftharpoons Cr_2O_7^{2-}(aq) + 2OH^-(aq)$$

Would you predict dichromate to predominate at low pH or high pH? Explain.

Color in Transition-Metal Complexes

14. Within what energy range (in eV) must Δ lie if a transition-metal complex is to absorb visible light?

15. Which of the following would you expect to be colored?

a) $[Ag(NH_3)_2]^+$ c) VO^{2+} e) $[Zn(NH_3)_4]^{2+}$
b) $[Cu(CN)_2]^-$ d) $TiCl_4$

16. Explain why Al^{3+} and Zn^{2+} compounds are generally colorless, e.g., Al_2O_3, ZnO, ZnS, and various complexes.

17. One of these complexes has a blue color and the other is buff (yellowish): $[Ni(NH_3)_4]^{2+}$, $[Ni(CN)_4]^{2-}$. Decide which has which color and state what color of light each will absorb.

18. All other factors being equal, which should get warmer faster when exposed to sunlight, a white panel or a black panel? Explain.

19. Show that the colors of deoxyhemoglobin and oxyhemoglobin (Section 22.3) can be explained on the basis of high-spin and low-spin complexes, respectively.

Which compound in each of the following would absorb light of the longer wavelength?

a) $[Cr(NH_3)_6]^{3+}$ or $(CrCl_6)^{3-}$
b) $[Ni(H_2O)_6]^{2+}$ or $[Ni(NH_3)_6]^{2+}$
c) $[Co(NH_3)_4Cl_2]^+$ or $[Co(NH_3)_5Cl]^{2+}$

21. The dimethylglyoxime complex of nickel (see Section 22.7) is bright red in color. What color of light does this complex absorb?

Bonding and Geometry

22. Discuss the possibility of *cis-trans* isomerism in square planar complexes such as $[Pt(NH_3)_2]Cl_2$.

23. Illustrate by means of sketches the molecular geometries of the isomers of $[Co(en)_2NH_3Cl]^{2+}$.

24. Use appropriate hybrid orbitals to describe the structure of $[Mn(H_2O)_6]^{2+}$. Would you expect this to be a high-spin or a low-spin complex? Repeat for $[Fe(CN)_6]^{3-}$.

25. Can $[Pt(en)_2]^{2+}$ have optical isomers? Explain.

26. What is the oxidation state of the central ion in $[Pt(NH_3)_3Cl_3]^+$? Predict the geometry of the complex and indicate any possible isomers.

27. The sodium salt of nitrilotriacetic acid (NTA) is

$$:N\begin{matrix}-CH_2CO_2^-\\-CH_2CO_2^-\\-CH_2CO_2^-\end{matrix}\ ,\ 3Na^+$$

The anion can act as a tetradentate chelating agent with an ion M^{2+}. Sketch the geometric shape of the chelate formed. [*Hint:* The nitrogen atom and three oxygen atoms form a tetrahedron about M^{2+}.]

28. Compare the geometry of CrO_4^{2-} with that of SO_4^{2-}. Do chromium and sulfur have similar electron configurations?

29. Predict the number of unpaired electrons in each of the following complexes.
    a) $[Cr(CN)_6]^{4-}$
    b) $[Cr(H_2O)_6]^{2+}$

30. Would you expect the square planar complex $[Cu(NH_3)_4]^{2+}$ to be low-spin or high-spin?

31. Square planar complexes of Ni(II) are diamagnetic while tetrahedral complexes are paramagnetic. Explain.

32. Show that $[Pt(NH_3)_2]Br_2$ has two isomers (cis and trans).

33. Consider the complexes $[Co(en)_2ClNO_2]Cl$ and $[Co(en)_2Cl_2]NO_2$. What happens to each of these if some $AgNO_3$ is added to their water solution? Show that each of the above exists in several isomeric forms.

Naming of Transition-Metal Complexes

34. Give the IUPAC name of each of the following compounds or ions.
    a) $[Co(H_2O)_4F_2]F_2$
    b) $[Pt(NH_3)_2]Cl_2$
    c) $[Au(CN)_4]^-$
    d) $[Co(en)_2(CN)_2]ClO_4$
    e) $[Mn(H_2O)_6]^{2+}$
    f) $[Pt(NH_3)_3Cl_3]Cl$
    g) $[Co(en)_2NH_3Cl]CO_3$
    h) $[Zn(NH_3)_4]^{2+}$

35. Write formulas for each of the following.
    a) chloronitrobis(ethylenediamine)cobalt(III)
    b) dioxovanadium(V)hexafluoroantimonate(V) (first part is VO_2)
    c) potassium bromochlorodicyanonickelate(II)
    d) tris(ethylenediamine)chromium(III) cyanide.
    e) diamminesilver chloride

36. Write chemical formulas for each of the following.
    a) Tetraoxochromate(VI) ion
    b) Potassium hexacyanomanganate(III)
    c) Diiodoargentate(I) ion
    d) Hexachlorovanadate(III) ion
    e) Calcium tetracyanoaurate(III)

37. Give the IUPAC names for the following ions or compounds.
    a) $K_2Cr_2O_7$
    b) $[Co(H_2O)_4(NH_3)_2]Cl_2$
    c) $[Mn(C_2O_4)_3]^{3-}$
    d) $[Co(NO_2)_6]^{3-}$
    e) $[Fe(SCN)_6]^{3-}$

ADDITIONAL READINGS

Perutz, M. F., "The Hemoglobin Molecule." *Scientific American*, November 1964. A Nobel-prize-winning chemist explains how hemoglobin functions and discusses what is known of its structure.

Kendrew, J. C., "The Three-Dimensional Structure of a Protein Molecule." *Scientific American*, December 1961. The author, who shared the 1962 Nobel prize in chemistry with M. F. Perutz for work on globular proteins, discusses how the structure of myoglobin was worked out.

Russo, Salvatore F., and Ralph B. Sorstokke, "Hemoglobin: Isolation and Chemical Properties." *J. Chem. Educ.* **50**, 347 (1973).

Keffer, Frederic, "The Magnetic Properties of Materials." *Scientific American*, September 1967. Discusses how magnetism in atoms leads to magnetism in bulk materials.

Cotton, F. A., and G. Wilkinson, *Advanced Inorganic Chemistry*. New York: Interscience, 1972. Useful even to the beginning student as a reference.

Larsen, Edwin M., *Transitional Elements*. New York: Benjamin, 1965. An excellent treatment of the general types of reactions typical of the transition elements.

Basolo, F., and R. C. Johnson, *Coordination Chemistry*. New York: Benjamin, 1964. A classic work on coordination compounds.

Companion, Audrey L., *Chemical Bonding*. New York, McGraw-Hill, 1964. Chapter 6 is a very clear account of the structure of transition-metal complexes.

23
NITROGEN, PHOSPHORUS, SULFUR, AND THE HALOGENS

The nodules attached to the roots of leguminous plants such as clover and alfalfa contain rhizobium bacteria which fix atmospheric nitrogen to fertilize the soil. (Courtesy of Hugh Spencer, Photo Researchers, Inc.)

... If one measure of nitrous air [NO, now called nitric oxide] be put to two measures of common air, in a few minutes ... there will want about one-ninth of the original two measures; and if both the kinds of air be very pure, the diminution will still go on very slowly, till in a day or two there will remain only one-fifth of the original quantity of common air. ... I hardly know any experiment that is more adapted to amaze and surprise than this is, which exhibits a quantity of air, which, as it were, devours a quantity of another kind of air half as large as itself, and yet is so far from gaining any addition to its bulk, that it is considerably diminished by it.
JOSEPH PRIESTLEY, *Experiments and Observations on Different Kinds of Air*, 1774

Joseph Priestley (1733–1804), an English-born theologian and chemist, was the writer of about 150 books (mostly on theology and education) and discoverer of NH_3, SO_2, CO, HCl, NO, and H_2S. Priestley also shared the independent discovery of oxygen with Carl Wilhelm Scheele of Sweden. In 1794, after his religious and political views made him increasingly unpopular in England, Priestley emigrated to the United States. His last ten years were spent in Northumberland, Pennsylvania.

Next to carbon, hydrogen, and oxygen, the most important nonmetals are perhaps nitrogen, phosphorus, sulfur, and the halogens. Although nitrogen and phosphorus are both group VA elements, they are different enough physically and chemically to warrant separate treatment. Sulfur, a member of the same group as oxygen, group VIA, also differs greatly from its nearest neighbor in the periodic table. Nitrogen, phosphorus, sulfur, and one of the halogens (chlorine) play very important roles in the chemistry of biological processes.

23.1 NITROGEN

The electronic configuration of nitrogen, $(He)2s^22p^3$, suggests that possible oxidation states range from a minimum of -3 to a maximum of $+5$. One of the most striking properties of nitrogen is that it actually exhibits all of its possible oxidation states. Furthermore, unlike most other elements with variable oxidation states, all of the oxidation states are exemplified by important compounds. In Table 23.1 are listed examples of compounds of each of the nine oxidation states of nitrogen.

Even though it forms almost 80% of the earth's atmosphere, nitrogen is not a very abundant element. It ranks far below carbon, which is fourteenth in terms of atom number.

Elemental nitrogen, N_2, is a very unreactive substance. The high dissociation

TABLE 23.1 EXAMPLES OF THE NINE OXIDATION STATES OF NITROGEN

OXIDATION STATES	COMPOUND
−3	Ammonia, NH_3
−2	Hydrazine, H_2N-NH_2
−1	Hydroxylamine, H_2NOH
0	Elemental nitrogen, N_2
+1	Nitrous oxide, N_2O
+2	Nitric oxide, NO
+3	Nitrous acid, HNO_2
+4	Nitrogen dioxide, NO_2
+5	Nitric acid, HNO_3

energy, probably about 9.8 eV, is consonant with the triply bonded Lewis structure :N≡N: and the molecular orbital electron configuration (ISNB) $(1\pi)^4(3\sigma)^2$. One of the most important nitrogen reactions

$$N_2(g) + 3H_2(g) \rightleftharpoons 2NH_3(g)$$

can be made to yield substantial amounts of products only at high pressures, relatively high temperatures, and in the presence of special catalysts (see Section 20.8). By contrast, certain lowly bacteria living in the roots of leguminous plants, as well as some water-dwelling blue-green algae, can carry out an equivalent process at room temperature and atmospheric pressure. Eugene van Tamelen and co-workers at Stanford University have been examining several laboratory processes to convert atmospheric N_2 to ammonia or amines. Several processes utilizing organotitanium compounds as catalysts have shown promise. Many of the reactions are of the general type

$$N_2(g) + \text{reducing agent} \xrightarrow{\text{Ti(IV) complex}} NH_3 \text{ or amines}$$

Much of the ammonia produced industrially is used directly as fertilizer. Large amounts are also converted to nitric acid by the **Ostwald process**:

$$4NH_3(g) + 5O_2(g) \xrightarrow{\text{catalyst}} 4NO(g) + 6H_2O(g)$$

$$2NO(g) + O_2(g) \rightarrow 2NO_2(g)$$

$$3NO_2(g) + H_2O(l) \rightarrow 2HNO_3(aq) + NO(g)$$

The NO produced in the third step is recycled via the second step. The nitric acid so produced is used for a variety of purposes, including the manufacture of nitrates, which are used as fertilizers. There are probably very few, if any, large manufacturing activities which do not use nitric acid in one way or another.

Nitric acid also may be prepared in the laboratory by the action of sulfuric acid on nitrate salts, for example

$$NaNO_3(s) + H_2SO_4(aq) \rightarrow NaHSO_4(aq) + HNO_3(aq)$$

The Ostwald process is named after the German chemist Friedrich Wilhelm Ostwald (1853–1932), who pioneered in the areas of reaction rates and catalysis. Ostwald was largely a publicizer of the theories of others, notably van't Hoff and Gibbs. He received the Nobel prize in chemistry in 1909 for work on chemical equilibrium and reaction rates. Ostwald was one of the last great scientists to accept Dalton's atomic theory, and he did so only after Einstein's explanation of Brownian motion (1905) left him no alternative.

The HNO_3 is volatile and may be removed by heating the reaction mixture. This process is an ancient one, having been used over 1000 years ago. Then, earthenware vessels were used in place of today's Pyrex glassware.

The nitrate ion, NO_3^-, is a very strong oxidizing agent. The relatively unreactive metal copper is attacked by even dilute nitric acid solutions as follows:

$$3Cu(s) + 2NO_3^-(aq) + 8H_3O^+ \rightarrow 3Cu^{2+}(aq) + 12H_2O(l) + 2NO(g)$$

The type of nitrogen oxide actually formed as a product depends on the HNO_3 concentration; concentrated HNO_3 produces NO_2 instead. One unusual property of the nitrate ion is that all of its metal salts are water soluble. This fact makes it possible to prepare water solutions of all the metallic ions.

Nitric oxide, NO, is a colorless, paramagnetic gas. It is so reactive toward O_2 that it rapidly oxidizes in air to form reddish-brown nitrogen dioxide, NO_2. The NO_2 then forms *nitrogen tetroxide*, N_2O_4, and participates in the equilibrium

$$2NO_2(g) \rightleftharpoons N_2O_4(g)$$

The compound N_2O_4 is colorless. At 25 °C about 20% of the equilibrium mixture is NO_2; at 135 °C the proportion of NO_2 becomes 99%. Thus, as a mixture of NO_2 and N_2O_4 is heated, it turns darker in color as the relative amount of NO_2 increases.

Nitrogen atoms occur in a wide variety of linkages in organic compounds, especially those of biological origin. Besides occurring in amines and amides, nitrogen forms the following types of linkages:

pyridine   pyrrole   nitrile   nitro group   diazonium salt   peptide

The dotted lines indicate partial double-bond character due to pi electrons. The electron bonding in pyridine is described in terms of sp^2 hybrids about nitrogen. Two of the sp^2 hybrids form bonds to adjacent carbon atoms and the remaining sp^2 hybrid contains a lone pair of electrons. The remaining valence-shell electron is in a $2p$ orbital of nitrogen, an orbital perpendicular to the plane of the molecule. It is this electron which forms part of an aromatic sextet. Thus pyridine has the structure

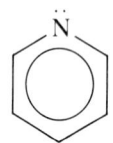

The lone-pair electrons on nitrogen enable pyridine to act as a Lewis base. In pyrrole there are two electrons in a $2p$ orbital which can form part of an aromatic sextet. Thus pyrrole has the structure

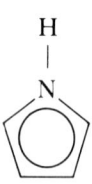

in which four pi electrons come from four ring carbons and the other two pi electrons come from nitrogen. In both pyridine and pyrrole, the nitrogen atom is sp^2 hybridized. The nitrogen in the peptide linkage is also sp^2 hybridized and two electrons in a $2p$ orbital of nitrogen are shared with two pi electrons of the carbonyl group so that the six atoms shown in the diagram all lie in the same plane. The peptide linkage is one of the main bonding features of proteins (see Section 24.9).

The isotope ^{15}N forms 0.37% of naturally occurring nitrogen; the rest is ^{14}N. Although not radioactive, the heavier isotope is useful as a tracer; mass spectrographic techniques are used to identify it. Nitrogen which becomes involved in biological pathways tends to become depleted in ^{15}N since this heavier isotope is more sluggish and tends to remain behind in various evaporation and diffusion processes of nature. Hence, measurement of $^{15}N/^{14}N$ ratios of the nitrogen content of a stream can sometimes be used to distinguish between pollution by natural sources or by fertilizer run-off. The $^{15}N/^{14}N$ ratio in artificial fertilizers is just that of atmospheric N_2 since it is not subject to ^{15}N depletion.

Nitrogen gas is normally prepared by fractional distillation of liquid air. Since the normal boiling points of N_2 and O_2 are 77 and 90 K, respectively, nitrogen distills off in greater proportion than oxygen. Many chemical reactions sensitive to oxidation are carried out in nitrogen atmospheres.

When deep-sea divers are subjected to high pressures of air, their bloodstreams accumulate increased amounts of dissolved nitrogen gas due to Henry's law (see Section 10.3). If the diver is returned suddenly to lower pressures, the dissolved nitrogen comes out of the bloodstream, collects in the joints, and causes a painful and often fatal condition called the **bends**. The bends are avoided by allowing the diver to reach normal atmospheric pressure slowly, in stages, or by using atmospheres not containing nitrogen. For example, replacement of nitrogen gas with helium minimizes the problem of the bends.*

---

* But helium atmospheres make divers sound like Donald Duck when they talk.

23.2 NITROGEN OXIDES AND SMOG PRODUCTION

Unburned hydrocarbons arising from incomplete combustion of gasoline and diesel fuel interact with oxides of nitrogen present in exhaust gases to form the eye and membrane irritants associated with **photochemical smog**. The source of the problem is the high temperature at which modern high-compression-ratio internal combustion engines operate.* These temperatures favor the oxidation of atmospheric nitrogen to various nitrogen oxides, mainly NO. An average automobile exhaust may contain on the order of 4000 ppm of NO and smaller amounts of NO_2. The chemical reactions which occur are many, but the following appear to be the most significant: The majority of exhaust NO reacts with atmospheric oxygen to produce NO_2:

$$2NO(g) + O_2(g) \rightarrow 2NO_2(g)$$

The NO_2 absorbs sunlight to carry out the photochemical reaction

$$NO_2(g) \xrightarrow{light} NO(g) + O(g)$$

This reaction occurs with almost 100% efficiency with ultraviolet radiation of wavelengths of 3600 Å and less. The highly reactive oxygen atoms produced by the reaction may participate in a variety of reactions, among which are

$$O(g) + O_2(g) + M \rightarrow O_3(g) + M$$

$$O(g) + NO_2(g) \rightarrow NO(g) + O_2(g)$$

$$O(g) + \text{hydrocarbons} \rightarrow \text{aldehydes, alkyl radicals}$$

The hydrocarbons originate from unburned fuel present in the exhaust gases along with the nitrogen oxides. The eye irritants characteristic of smog arise from processes such as the following:

$$O_3(g) + \text{hydrocarbons} \rightarrow R\underset{\underset{O}{\|}}{C}O \cdot \text{(oxyacyl radicals)}$$

$$O_2(g) + R \cdot \text{(alkyl radicals)} \rightarrow RO_2 \cdot \text{(peroxyalkyl radicals)}$$

$$NO(g) + RO_2 \cdot \rightarrow RO \cdot \text{(oxyalkyl radicals)} + NO_2(g)$$

One of the most common irritants, the chief visible component of smog, is peroxyacetylnitrate

$$CH_3 \underset{\underset{O}{\|}}{C} - O - O - NO_2$$

better known as **PAN**. The components of smog generally become dispersed in the atmosphere or are swept away by winds. However, a combination of weather conditions and geograhical features can cause a concentration of these components such

---

* The higher the compression ratios the higher the operating temperature of the engine. This higher temperature is desirable since, by Eq. (18.6), the efficiency of the transformation of heat into work is thereby increased.

that they constitute serious hazards to health. For example, if a large city is located in a bowl-like depression and this depression acquires a stagnant air mass due to a thermal inversion—a mass of cold air trapped beneath a layer of warmer air—the smog components can build up to high levels. Los Angeles is perhaps the best known example of this type of smog.

23.3 THE NITROGEN CYCLE IN NATURE

The circulation of nitrogen may be viewed as taking place among four different zones or spheres of activity as represented by the diagram below

$$\text{lithosphere} \rightleftarrows \text{biosphere} \rightleftarrows \text{hydrosphere} \rightleftarrows \text{atmosphere} \rightleftarrows \text{lithosphere}$$

The **biosphere** refers to that part of the cycle in which living organisms (marine and terrestrial) participate, the **lithosphere** refers to that part of the cycle taking place under the earth's surface, and the **hydrosphere** refers to that part of the cycle taking place in aqueous solution, mostly in rivers, lakes, and oceans. In Fig. 23.1 is shown a more detailed diagram of that portion of the nitrogen cycle taking place in the atmosphere, terrestrial biosphere, and lithosphere. The portion of the cycle confined to the marine biosphere and hydrosphere is not understood in very much detail.

The process whereby elemental nitrogen is converted to compounds is referred to as **fixation**. An important type of nitrogen fixation is carried out by *Rhizobium* bacteria in symbiotic existence with leguminous plants, and by certain blue-green algae* to produce ammonia and amino compounds. Some nitrogen-utilizing bacteria do not fix nitrogen but rather oxidize ammonium compounds to nitrite and nitrate. This process, known as **nitrification**, provides nutrients for many plants. Other bacteria reduce inorganic nitrogen back to N_2 or to N_2O and divert these products to the atmosphere:

$$NO_2^-, NO_3^- \xrightarrow[\text{reduction}]{\text{enzymic}} N_2, N_2O$$

This process is termed **denitrification**. In addition to the above, some bacteria act on decaying organic matter and convert it to ammonium salts and amines such as cadaverine, $H_2N(CH_2)_5NH_2$, and putrescine, $H_2N(CH_2)_4NH_2$.

Plants, which incorporate nitrogen into their cellular structures, are eaten by animals who convert the nitrogen into protein and other nitrogen compounds. When the animal dies, bacteria convert the protein nitrogen to NH_4^+, some of which becomes denitrified to N_2 and N_2O and returned to the atmosphere. Some nitrogen is introduced into the lithosphere in the form of animal waste products. Additional nitrogen is added to the soil as commercial fertilizer, either as NH_3 or as nitrate. A small amount of nitrogen enters the lithsophere and the hydrosphere

---

* Legumes such as alfalfa and beans have nodules on their roots where nitrogen-fixing bacteria reside. See the photo on page 632.

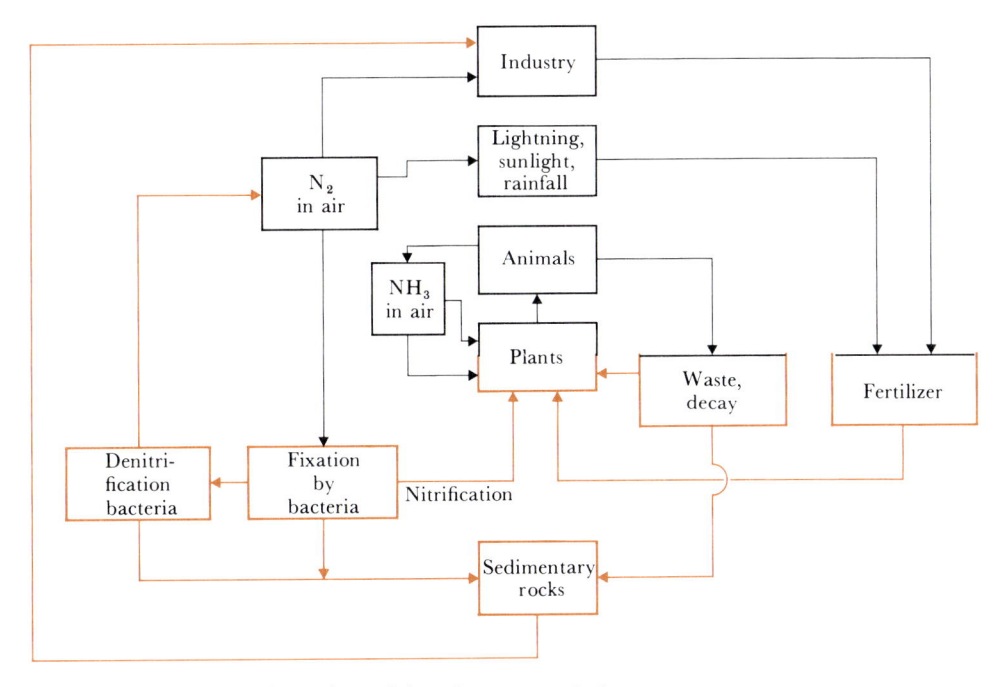

Fig. 23.1. A simplified version of the nitrogen cycle in nature.

from the atmosphere with each rainfall. Nitrogen oxides produced either by lightning discharges or by ultraviolet radiation from the sun are dissolved by falling rain and brought to earth. Small lakes occasionally become visibly greener soon after a lightning storm and rainfall, a phenomenon known as **algal bloom**. Presumably, nitrates added by the rainfall lead to a sudden spurt in the growth of algae.

The atmosphere normally contains between 20 and 200 $\mu g \cdot m^{-3}$ of ammonia which arises from the combustion of coal and natural gas, auto exhausts, animal feeding yards, agricultural fertilizers, municipal sewage, and chemical industries. Most of this is absorbed by the soil and some is absorbed directly through the leaves of plants.

Although nitrogen is not an abundant element, its large number of relatively stable oxidation states make it a versatile and efficient circulator. The high solubility of most nitrogen compounds, especially the nitrates, makes it unlikely that much nitrogen ever becomes sidetracked in sedimentary deposits.

23.4 PHOSPHORUS

Although phosphorus belongs to the same group of the periodic table as nitrogen, the two elements have very few similarities. Whereas nitrogen is ordinarily a diatomic gas, phosphorus is a tetraatomic solid. Nitrogen exhibits eight relatively stable oxidation states, but phosphorus has only two important ones: +3 and +5.

Phosphorus is the thirteenth most abundant element in terms of atom number; eleven atoms out of every 10,000 atoms are phosphorus atoms. Naturally occurring phosphorus is essentially 100% ^{31}P. The longest-lived artificially produced isotope is ^{33}P, a β-emitter with a half-life of 25 days. The most common source of elemental phosphorus is the mineral **apatite**, an ore with the formula $Ca_5(PO_4)_3X$, where X^- is OH^-, F^-, or Cl^-. The production of phosphorus from this ore may be represented by the equation

$$2Ca_3(PO_4)_2(s) + 5C(s) + 6SiO_2(s) \xrightarrow{\text{high } T} P_4(g) + 6CaSiO_3(l) + 5CO_2(g)$$

Elemental phosphorus exists in several allotropic forms. The white crystalline modification (also called yellow phosphorus), is very reactive and spontaneously ignites in air. It must be stored under water. The other common form, red phosphorus, is stable in air. Red phosphorus is prepared from white phosphorus by heating at 400 °C for several hours. Black phosphorus (also called violet phosphorus) also exists.

The most important phosphorus compounds are the halides PX_3 and PX_5 and the oxides P_4O_6 and P_4O_{10}.* The oxides P_4O_6 and P_4O_{10} are the anhydrides of phosphorous acid, H_3PO_3, and phosphoric acid, H_3PO_4, respectively. These acids may also be prepared by the hydrolysis of the halides:

$$PX_3(g) + 3H_2O(l) \rightarrow H_3PO_3(aq) + 3HX(aq)$$

$$PX_5(g) + 4H_2O(l) \rightarrow H_3PO_4(aq) + 5HX(aq)$$

Phosphoric acid (also called orthophosphoric acid) has three replaceable hydrogens as indicated by the Lewis formula

$$\begin{array}{c} :\!\ddot{O}\!: \\ \| \\ H\ddot{O}\!-\!P\!-\!\ddot{O}H \\ | \\ :\!\ddot{O}\!: \\ | \\ H \end{array}$$

However, phosphorous acid has only two replaceable hydrogens

$$\begin{array}{c} :\!\ddot{O}\!: \\ \| \\ H\ddot{O}\!-\!P\!-\!\ddot{O}H \\ | \\ H \end{array}$$

since one of the hydrogens is bonded directly to phosphorus. Generally, a hydrogen atom must be bonded to an electronegative atom such as oxygen if it is to be capable of participating in a proton donor reaction. The structures of the two acids are confirmed by the observation that although the salt Na_3PO_4 is known, the salt Na_3PO_3 is not. Phosphorous acid produces only the salts NaH_2PO_3 and Na_2HPO_3 whereas phosphoric acid produces NaH_2PO_4, Na_2HPO_4, and Na_3PO_4.

An important phosphorus compound is **sodium tripolyphosphate**, with the structure

$$\begin{array}{c} O^- \quad\; O^- \quad\; O^- \\ |\qquad |\qquad | \\ ^-O\!-\!P\!-\!O\!-\!P\!-\!O\!-\!P\!-\!O^-, \; 5\,Na^+ \\ |\qquad |\qquad | \\ O \quad\;\; O \quad\;\; O \end{array}$$

---

* Figure 2.3 illustrates the geometric structure of P_4O_{10}. That of P_4O_6 is similar but lacks the four outer oxygen atoms.

This compound is frequently added to detergent formulations. Its function is to tie up Mg^{2+} and Ca^{2+}, two ions commonly present in hard water, so that these ions do not react with the detergent and thus render it ineffective. The phosphate also controls pH, suspends soils, and emulsifies oils and grease. Unfortunately, such large amounts of phosphates have entered some of our natural waters, presumably via detergents, that aquatic growth has become overstimulated to produce a condition known as **eutrophication**. In eutrophication, aquatic plant growth becomes excessive due to abundance of critical nutrients such as phosphorus and nitrogen. Consequently, waterways become choked with vegetation; when the vegetation dies its decay depletes the water of available oxygen, and a stagnant, prematurely aged water system results.

A biologically important molecule, **adenosine triphosphate** (ATP), has a structure similar to that of sodium tripolyphosphate:

$$R-O-\underset{O}{\underset{|}{\overset{OH}{\overset{|}{P}}}}-O-\underset{O}{\underset{|}{\overset{OH}{\overset{|}{P}}}}-O-\underset{O}{\underset{|}{\overset{OH}{\overset{|}{P}}}}-O-R$$

where R is the **adenosine group**

This molecule is produced from ADP (**adenosine diphosphate**) by the enzyme-controlled reaction

$$ADP + H_3PO_4 \rightarrow ATP + H_2O$$

which has an enthalpy of reaction of about 33 kJ·mol^{-1}. The energy for this reaction is obtained from the metabolism of foodstuff or (in the case of plants) from sunlight. When water is added to ATP, other enzymes convert it back to ADP—and release the 33 kJ of stored energy. This energy is used by the organism for various processes: respiration, protein synthesis, and other biological reactions. Thus ATP stores chemical energy in much the same way as a compressed steel-spring stores mechanical energy. ATP is the universal energy currency of living cells; other sources of energy are also used but only ATP is common to all forms of life.

Phosphorus analogs of NH_3 and NH_4^+ exist in the form of PH_3, phosphine, and PH_4^+, phosphonium ion, but they are relatively unimportant.

Phosphorus compounds play important roles in photosynthesis, in food metabolism, and in protein synthesis.

23.5 SULFUR

Sulfur has the electron configuration $(Ne)3s^23p^4$, analogous to $(He)2s^22p^4$ of oxygen. However, the two elements resemble each other but slightly. Whereas the common oxidation states of oxygen are -1 and -2, sulfur exhibits -2, $+4$, and $+6$ in its important compounds.

Sulfur is eighth in abundance both in terms of weight and atom number; about 56 atoms out of every 10,000 are sulfur. The element occurs in nature as sulfides, sulfates, and also in the elemental form in large underground deposits in the Gulf States. The latter deposits are mined by forcing superheated water into underground pipes, melting the sulfur, and forcing it to the surface with compressed air. This process, known as the **Frasch process**, is illustrated in Fig. 23.2.

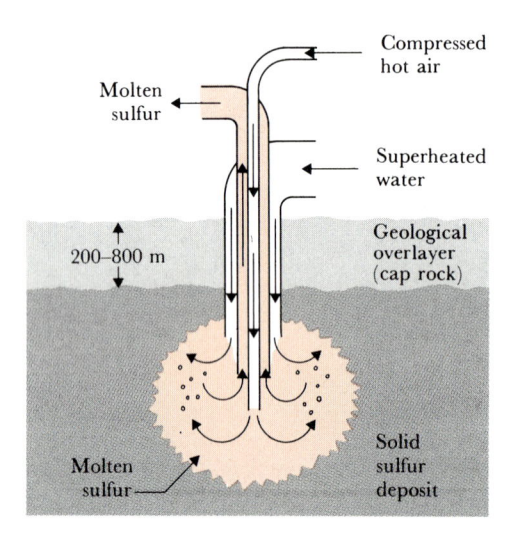

Fig. 23.2. The Frasch process for the mining of sulfur. Vast deposits of almost pure sulfur are found in the Gulf States in association with geological structures known as salt domes.

The element sulfur is a pale yellow, odorless, brittle solid. It was known to the ancients as **brimstone**. The element exists in a variety of allotropes but the two most important are the **rhombic form** (stable above 95.5 °C) and the **monoclinic form** (stable below 95.5 °C). See Fig. 23.3. Both contain the molecule S_8, a puckered octagon. Molecules such as S_2, S_4, S_6, and S_8 exist in the vapor phase.

Hydrogen sulfide, H_2S, is superficially similar to H_2O. One major difference is that H_2S exhibits very little hydrogen bonding and consequently, unlike H_2O, is a gas at room temperature. The HSH bond angle is only 92° as compared with 105° for HOH. Hydrogen sulfide may be produced by heating metal sulfides with mineral acids:

$$FeS(s) + 2HCl(aq) \rightarrow FeCl_2(aq) + H_2S(g)$$

Rhombic

Monoclinic

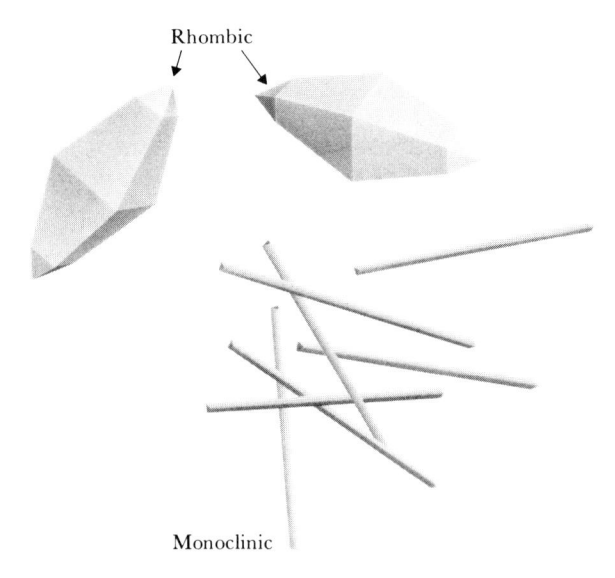

Fig. 23.3. Rhombic and monoclinic crystals of sulfur as they would appear under a magnifying lens. The long, needlelike monoclinic crystals are the ones commonly observed when molten sulfur is cooled slowly to room temperature.

The gas is very poisonous, in fact just as poisonous as hydrogen cyanide, HCN, but is readily detectable in very low concentrations—well below the lethal limit—by virtue of its strong rotten-egg odor.

Hydrogen sulfide is produced by certain anaerobic bacteria living in mud. Some bacteria utilize H_2S instead of H_2O in photosynthesis in the following manner:

$$CO_2 + 2H_2S \xrightarrow{light} CH_2O + H_2O + 2S$$

Some more typical examples of H_2S as a reducing agent are the following:

$$H_2S + H_2SO_4 \rightarrow SO_2 + 2H_2O + S$$

$$H_2S + \tfrac{1}{2}O_2 \rightarrow H_2O + S \qquad \text{(in limited } O_2 \text{ supply)}$$

$$H_2S + 1\tfrac{1}{2}O_2 \rightarrow H_2O + SO_2 \qquad \text{(in excess } O_2 \text{ supply)}$$

$$H_2S + I_2 \rightarrow 2HI + S$$

The ionic radius of the sulfide ion S^{2-} is 1.84 Å as compared with 1.32 Å for O^{2-}. Consequently, solid sulfides, unlike oxides, have rather low lattice energies.

The main oxides of sulfur are SO_2 (sulfur dioxide) and SO_3 (sulfur trioxide). When elemental sulfur is burned in air, only the SO_2 compound is formed. SO_3 may be prepared from SO_2 and O_2 by passing a mixture of these gases over a finely divided platinum catalyst (see Section 20.8):

$$2SO_2(g) + O_2(g) \xrightarrow{Pt} 2SO_3(g)$$

SO_2 and SO_3 are the anhydrides of H_2SO_3 (sulfurous acid) and H_2SO_4 (sulfuric acid), respectively. The latter substance is of great importance in many industrial processes and chemical syntheses. In fact, it may well be the most important single chemical of the modern technological world. Sulfuric acid has the Lewis formula

$$\text{HO}-\overset{\overset{\displaystyle :\ddot{O}:}{|}}{\underset{\underset{\displaystyle :\ddot{O}:}{|}}{S}}-\ddot{O}\text{H}$$

Since the hydrogen atoms are attached to highly electronegative oxygen which in turn is attached to another relatively electronegative atom, loss of H^+ occurs readily. Hence, H_2SO_4 is a strong Brønsted-Lowry acid in water solution.

Concentrated sulfuric acid is also used as a dehydrating agent and as a desiccant. When concentrated H_2SO_4 is placed in contact with a carbohydrate such as sugar (sucrose, $C_{12}H_{22}O_{11}$) a black mass of carbon results. The reaction is

$$C_{12}H_{22}O_{11}(s) \rightarrow 12C(s) + 11H_2O(l)$$

In one of the commercial preparations of sulfuric acid—the **contact process**—SO_3 is absorbed in previously existing H_2SO_4 to produce $H_2S_2O_7$, a substance known as **oleum**. The oleum is diluted with water to make 98% sulfuric acid. This acid is a colorless, oily liquid boiling at 340 °C. The steps in the process are

$$SO_3(g) + H_2SO_4(l) \rightarrow H_2S_2O_7(l)$$

$$H_2S_2O_7(l) + H_2O(l) \rightarrow 2H_2SO_4(l)$$

Direct preparation of H_2SO_4 by dissolving SO_3 in H_2O is too slow a process to be practicable.

There is evidence that the clouds of the planet Venus contain considerable amounts of sulfuric acid. Venus is known to absorb infrared radiation of just those wavelengths characteristic of vibrations associated with H_3O^+ and HSO_4^-.

Sulfurous acid, H_2SO_3, is produced by the reaction

$$SO_2(g) + H_2O(l) \rightarrow H_2SO_3(aq)$$

This acid has the Lewis structure

$$\text{HO}-\overset{\ddot{}}{\underset{\underset{\displaystyle :\ddot{O}:}{|}}{S}}-\ddot{O}\text{H}$$

Sulfurous acid is a weak acid in water solution.

An important compound of sulfur, thionyl chloride ($SOCl_2$), is useful in preparative chemistry (the prefix *thio* comes from Greek *theion*, brimstone). This molecule can be used to prepare acyl chlorides from carboxylic acids (see Section 17.2). Thionyl chloride is prepared by the reaction

$$SO_2 + PCl_5 \rightarrow SOCl_2 + POCl_3$$

Sulfur may replace oxygen in many organic molecules, as in the thioalcohols RSH and the thioethers RSR'. Sulfur may replace oxygen in a carbonyl group as in thioacetamide (thioethanamide):

$$\text{CH}_3\underset{\underset{S}{\|}}{C}\text{NH}_2$$

This molecule produces H$_2$S and ethanamide when heated in aqueous solution. The ethanamide reacts with water to form acetic acid and ammonia.

Sulfur is found in many compounds of biological significance, for example, in vitamins and proteins.

23.6 THE AQUEOUS CHEMISTRY OF METAL SULFIDES

With the exception of the group IA and IIA metals, all metal sulfides are sparsely soluble in water. Because of this, a useful analysis scheme can be set up for separating and identifying many metallic ions through their sulfides. The separation is done on the basis of the varying solubilities of the sulfides as a function of pH. Using CdS as an example, the dissolution of a sulfide in water is given by

$$\text{CdS(s)} \rightleftharpoons \text{Cd}^{2+}\text{(aq)} + \text{S}^{2-}\text{(aq)}$$

The solubility-product constant is given by

$$K_{sp} = [\text{Cd}^{2+}][\text{S}^{2-}]$$
$$= 3.6 \times 10^{-29} \quad \text{at } 25\,°\text{C} \tag{23.1}$$

In analyzing a solution for Cd^{2+}, the sulfide is formed by passing H$_2$S into the solution. The reaction is

$$\text{H}_2\text{S(aq)} + \text{Cd}^{2+}\text{(aq)} + 2\text{H}_2\text{O(l)} \rightarrow \text{CdS(s)} + 2\text{H}_3\text{O}^+$$

Most metal sulfides are black, but CdS is yellow-orange. In a few cases color can be used as a qualitative indication of the presence of an ion in solution. The maximum amount of Cd^{2+} which can exist in equilibrium with S^{2-} is given by

$$[\text{Cd}^{2+}] = \frac{K_{sp}}{[\text{S}^{2-}]} \tag{23.2}$$

It is clear that the greater the value of [S^{2-}], the smaller is the amount of Cd^{2+} that can exist in the solution. The following shows that [S^{2-}] depends on pH and, consequently, the amount of Cd^{2+} also depends on pH. When H$_2$S dissolves in water one of the reactions is

$$\text{H}_2\text{S(aq)} + 2\text{H}_2\text{O(l)} \rightleftharpoons 2\text{H}_3\text{O}^+ + \text{S}^{2-}\text{(aq)}$$

The acid equilibrium constant is given by

$$K_a = K_{a_1}K_{a_2} = \frac{[\text{H}_3\text{O}^+]^2[\text{S}^{2-}]}{[\text{H}_2\text{S}]} = 6.7 \times 10^{-23} \tag{23.3}$$

where K_{a_1} and K_{a_2} are given by

$$H_2S(aq) + H_2O(l) \rightleftharpoons H_3O^+ + HS^-(aq)$$

$$K_{a_1} = \frac{[H_3O^+][HS^-]}{[H_2S]} = 5.6 \times 10^{-8} \tag{23.4}$$

$$HS^-(aq) + H_2O(l) \rightleftharpoons H_3O^+ + S^{2-}(aq)$$

$$K_{a_2} = \frac{[H_3O^+][S^{2-}]}{[HS^-]} = 1.2 \times 10^{-15} \tag{23.5}$$

Equation (23.3) shows that $[S^{2-}]$ depends on pH, that is, on $[H_3O^+]$, via the relationship

$$[S^{2-}] = \frac{[H_2S]}{[H_3O^+]^2} K_a \tag{23.6}$$

This means that $[Cd^{2+}]$ can be controlled by adjusting the pH of the solution. At 1 atm and 25 °C a saturated solution of H_2S in water is about $0.1\ M$ in H_2S. Thus, as a good approximation,

$$[S^{2-}] = \frac{0.1}{[H_3O^+]^2}(6.7 \times 10^{-23}) = \frac{6.7 \times 10^{-24}}{[H_3O^+]^2} \tag{23.7}$$

Substituting Eq. (23.7) into Eq. (23.2) leads to

$$[Cd^{2+}] = \frac{[H_3O^+]^2}{6.7 \times 10^{-24}} K_{sp} = 5.4 \times 10^{-6}[H_3O^+]^2 \tag{23.8}$$

As a specific example, the maximum amount of Cd^{2+} that can exist in a solution of pH = 1 ($[H_3O^+] = 0.1\ M$) which is saturated in H_2S is given by

$$[Cd^{2+}] = 5.4 \times 10^{-6}(0.1)^2 = 5.4 \times 10^{-8}\ M$$

It should be noted that as the pH decreases, the amount of Cd^{2+} increases; in other words, CdS can be dissolved in an acid solution. This same conclusion follows qualitatively from Le Châtelier's principle. A look at the reactions

$$CdS(s) \rightleftharpoons Cd^{2+}(aq) + S^{2-}(aq)$$

$$H_3O^+ + S^{2-}(aq) \rightleftharpoons HS^-(aq) + H_2O(l)$$

shows that removal of S^{2-} by reaction with H_3O^+ will cause more CdS to dissolve.

The preceding discussion shows that if a solution contains two different ions whose sulfides have different K_{sp} values, then it is possible to adjust the pH so that saturation with H_2S precipitates only one of the sulfides.

23.7 THE HALOGENS

The halogens are among the most electronegative elements. Like their electropositive counterparts, the alkali metals, the halogens are far too reactive to occur in nature in the elemental form. With the exception of salts such as CaF_2, and the halides of lead(II), mercury(I), and silver(I), most halides are very soluble in water and, consequently, are distributed widely throughout the earth's crust. In Table 23.2 are listed some basic properties of the halogens F_2, Cl_2, Br_2, and I_2. The heaviest halogen, astatine At_2, is too rare to have been studied in detail.*

Fluorine and chlorine are gases at room temperature and 1 atm pressure. The former is pale yellow and the latter is greenish yellow. Both are extremely hazardous to handle and cause frightful injury to the lungs and membranes when breathed. Fluorine is so reactive that many organic compounds spontaneously ignite in fluorine atmospheres. Bromine is a reddish-brown liquid with a high vapor pressure. The vapors are also very irritating to the lungs and should be avoided. Iodine is an almost black (somewhat bluish) lustrous solid which readily sublimes to form a violet vapor. Iodine vapors are also irritating to breathe.

The halogen elements have the electron configurations

F: (He)$2s^22p^5$
Cl: (Ne)$3s^23p^5$
Br: (Ar)$3d^{10}4s^24p^5$
I: (Kr)$4d^{10}5s^25p^5$

The elements exist as the diatomic molecules X_2 with the Lewis formula

$$:\ddot{\ddot{X}}:\ddot{\ddot{X}}:$$

The relatively high ionization energies and the large negative electron affinities of the halogens are in accord with their pronounced tendency to form ionic compounds with the -1 oxidation state. However, fluorine compounds are often anomalous in being—or appearing to be—less ionic than might be expected. Also, the higher general reactivity of fluorine appears inconsistent with the fact that its electron affinity is less negative than that of chlorine. Only by examining the detailed energetics of the reaction steps of F_2 and Cl_2 in comparable reactions is it possible to rationalize their relative chemistries. When a halogen reacts, it must first dissociate to atoms:

$$\tfrac{1}{2}X_2(g) \rightarrow X(g)$$

This process is endothermic and involves 0.83 eV for F_2 and 1.24 eV for Cl_2. The lower energy for F_2 dissociation clearly enhances its reactivity. Adding the above to the energy of the process

$$X(g) + e^- \rightarrow X^-(g)$$

gives $0.83 - 3.5$ or -2.7 eV for F_2 and $1.24 - 3.6$ or -2.4 eV for Cl_2. These latter energies represent the overall process

$$\tfrac{1}{2}X_2(g) + e^- \rightarrow X^-(g)$$

---

* The most stable isotope, ^{210}At, has a half-life of only 8.3 hrs.

Thus the F_2 reaction is favored over that of Cl_2 by 0.3 eV per atom (30 kJ·mol^{-1}). In addition, fluoride ions are smaller than chloride ions so that the former generally have larger lattice energies.

Chlorine, bromine, and iodine can be prepared by the oxidation of halide compounds. Chlorine is commonly prepared by the electrolysis of molten chlorides or aqueous chloride solutions. When alkali metal halides are employed, these processes also produce alkali metals and alkali metal hydroxides, respectively (see Section 21.2). Chemical oxidation reactions may also be employed. Some of those used to produce chlorine are

$$4HCl(g) + O_2(g) \xrightarrow{high\ T} 2H_2O(g) + 2Cl_2(g)$$

$$4HCl(aq) + MnO_2(s) \rightarrow Cl_2(g) + MnCl_2(aq) + 2H_2O(l)$$

$$6Cl^-(aq) + Cr_2O_7^{2-}(aq) + 14H_3O^+ \rightarrow 3Cl_2(g) + 2Cr^{3+}(aq) + 21H_2O(l)$$

Bromine and iodine can be prepared by similar reactions.

Bromine is often obtained commercially from seawater by the process

$$2Br^-(aq) + Cl_2(g) \rightarrow Br_2(aq) + 2Cl^-(aq)$$

When air is blown through the solution, Br_2 is removed as the vapor. The basis of the process is that Cl_2 is a stronger oxidizing agent than is Br_2 (note the relative $\mathscr{E}^0_{X_2/X^-}$ values given in Table 19.1). Most of the bromine made by this process is used in the manufacture of 1,2-dibromoethane, a gasoline additive which acts as a scavenger for lead.

Iodine is often prepared by the reduction of sodium iodate, $NaIO_3$, a compound found in deposits of Chile saltpeter, $NaNO_3$; hydrogen sulfite ion, HSO_3^- is used as a reducing agent:

$$2IO_3^-(aq) + 5HSO_3^-(aq) + 2H_3O^+ \rightarrow I_2(s) + 5HSO_4^-(aq) + 3H_2O(l)$$

Note that the oxidation state of iodine changes from +5 to 0 in the above.

Fluorine, since it has the highest electronegativity of all the elements, except for some of the inert gases, cannot be prepared by chemical oxidation of fluorides. Rather, it is prepared by the electrolysis of a fused mixture of KF and HF, sometimes written KHF_2. The process is carried out at 250 °C using a graphite anode. The electrode reactions are

cathode: $2HF_2^-(l) + 2e^- \rightarrow H_2(g) + 4F^-(l)$

anode: $\ \ \ 2HF_2^-(l) \rightarrow F_2(g) + 2HF(g) + 2e^-$

The overall reaction is

$$4HF_2^-(l) \rightarrow H_2(g) + F_2(g) + 2HF(g) + 4F^-(l)$$

Note that the reaction mixture does not become depleted in F^-. The electrolysis cell can be made to operate continuously by constantly adding HF.

TABLE 23.2 SOME PROPERTIES OF THE HALOGENS

	F_2	Cl_2	Br_2	I_2
Atomic radius (Å)	0.72	0.99	1.14	1.33
Ionic radius (Å)	1.36	1.81	1.95	2.16
Ionization energy (eV)	17.4	13	11.8	10.5
Electron affinity (eV)	−3.5	−3.6	−3.3	−3.1
Electronegativity	4.0	3.0	2.8	2.5
Normal boiling point of X_2 (°C)	−188	−34	59	183
Dissociation energy of X_2 (eV)	1.65	2.48	1.97	1.54
Normal boiling point of HX (°C)	19.4	−84.9	−67	−35
Enthalpy of formation of HX (kJ·mol^{-1})	−272	−92	−50	+25
Dissociation energy of HX (eV)	5.8	4.43	3.75	3.06

23.8 THE HYDROGEN HALIDES

All of the hydrogen halides, HX, are strong acids in aqueous solution except for HF, which has a K_a of only 6.7×10^{-4} at 25 °C. Table 23.2 shows that hydrogen fluoride has an unusually high boiling point as compared with the other hydrogen halides; this is an excellent example of hydrogen bonding. The boiling points of HCl, HBr, and HI increase regularly with increasing molecular weight, as expected on the basis of ordinary van der Waals interactions (see Section 6.4). In fact, hydrogen bonding is so strong in HF that even the vapor is largely (HF)$_n$ with n from 2 to 6, as opposed to $n = 1$ for the other hydrogen halides. Hydrogen fluoride is also unusual in that it has such a strong dissociation energy, 5.8 eV—a fact which appears to be out of line with the weak F—F bond energy of 1.65 eV. No simple, truly satisfactory rationalization of this anomaly has been given.

Hydrogen fluoride is usually prepared from the mineral **fluorite**, CaF_2, by the reaction

$CaF_2(s) + H_2SO_4(aq) \rightarrow 2HF(g) + CaSO_4(s)$

Hydrogen fluoride has the ability to etch glass and quartz. The reaction with quartz is

$SiO_2(s) + 4HF(g) \rightarrow SiF_4(g) + 2H_2O(l)$

Hydrogen chloride is most conveniently prepared by treating group IA and IIA chlorides with sulfuric acid:

$NaCl(s) + H_2SO_4(aq) \rightarrow NaHSO_4(aq) + HCl(g)$

Hydrogen bromide and hydrogen iodide cannot be prepared by the same process since SO_4^{2-} oxidizes Br^- and I^- to Br_2 and I_2, respectively. For example

$2Br^-(s) + SO_4^{2-}(aq) + 4H_3O^+ \rightarrow Br_2(g) + SO_2(g) + 6H_2O(l)$

However, H_3PO_4 may be used instead of H_2SO_4:

$$NaBr(s) + H_3PO_4(aq) \rightarrow HBr(g) + NaH_2PO_4(aq)$$

An alternative preparation method is to drop liquid Br_2 onto wet phosphorus. The reaction is the net result of the steps

$$P_4(s) + 6Br_2(l) \rightarrow 4PBr_3(l)$$
$$PBr_3(l) + 3H_2O(l) \rightarrow H_3PO_3(aq) + 3HBr(g)$$

Hydrogen iodide may be prepared analogously by dropping H_2O onto a mixture of I_2 and P_4.

The hydrogen halides may also be prepared by direct union of the elements:

$$H_2(g) + X_2(g) \rightarrow 2HX(g)$$

The reaction is explosive and instantaneous with F_2 and explosive with Cl_2 and Br_2 once initiated; the reaction with I_2 proceeds at a moderate rate. The reactions with F_2, Cl_2, and Br_2 are most likely chain reactions (see Section 20.7).

The fluoride ion is known to play a vital role in the prevention of tooth decay. If young people use drinking water containing small amounts of F^-, their tooth enamel develops compounds such as $CaF_2 \cdot Ca_3(PO_4)_2$ which are highly resistant to chemical attack.* Tin(II) fluoride (stannous fluoride) added to toothpaste appears to owe its anticaries activity to an ability to react with incipient cavities.

23.9 THE OXY ACIDS OF THE HALOGENS

When Cl_2, Br_2, or I_2 reacts with water, the following reaction (a disproportionation) can occur:

$$X_2 + H_2O \rightarrow HOX + HX$$

Here, the HOX compound contains the +1 oxidation state of the halogen. Fluorine generally reacts in a different manner with water.

$$2F_2(g) + 2H_2O(l) \rightarrow 4HF(aq) + O_2(g)$$

However, it is claimed that HOF and HF are formed when F_2 is passed over wet surfaces at 0 °C.

The halogens (except for fluorine) also form various oxygen-containing compounds in which the +3, +5, and +7 oxidation states occur. These compounds (and those of the type HOX) are called the **oxy acids** of the halogens. The known oxy acids are:

F: HOF (uncertain)
Cl: HOCl, $HClO_2$, $HClO_3$, $HClO_4$
Br: HOBr, $HBrO_3$, $HBrO_4$
I: HOI, HIO_3, HIO_4

---

* Too much F^- in drinking water leads to mottled teeth but this effect, although cosmetically undesirable, is harmless.

TABLE 23.3 NOMENCLATURE OF THE CHLORINE OXY ACIDS AND THEIR SALTS

ACIDS	SALTS
HOCl, hypochlorous acid	NaOCl, sodium hypochlorite
HClO$_2$, chlorous acid	NaClO$_2$, sodium chlorite
HClO$_3$, chloric acid	NaClO$_3$, sodium chlorate
HClO$_4$, perchloric acid	NaClO$_4$, sodium perchlorate

Corresponding salts also exist. All of the above acids are strong oxidizing agents. For the chlorine compounds as examples, the acids and their sodium salts are named as shown in Table 23.3. As would be expected on the basis of the number of oxygen atoms, the acid strengths of the oxy acids of chlorine increase in the order HOCl < HClO$_2$ < HClO$_3$ < HClO$_4$.

Hypochlorites can be made by dissolving Cl$_2$ in base:

$$Cl_2 + 2OH^- \rightarrow OCl^- + H_2O + Cl^-$$

When the hypochlorite is heated, the chlorate is formed:*

$$3OCl^- \xrightarrow{\text{heat}} ClO_3^- + 2Cl^-$$

Chlorites are not easy to prepare. One approach is to first prepare ClO$_2$, chlorine (IV) oxide, as follows:

$$3KClO_3 + 3H_2SO_4 \rightarrow HClO_4 + 3KHSO_4 + 2ClO_2 + H_2O$$

The ClO$_2$ reacts with base to produce a mixture of chlorite and chlorate,

$$2ClO_2 + 2OH^- \rightarrow ClO_2^- + ClO_3^- + H_2O$$

Perchlorates are prepared by electrolytic oxidation of chlorates in basic solution:

$$ClO_3^- + 2OH^- \rightarrow ClO_4^- + H_2O + 2e^-$$

Perchloric acid, HClO$_4$, can be prepared pure by fractional distillation from aqueous solution. It is a colorless liquid which reacts explosively with even small traces of oxidizable material.

The halogens form some oxides, for example, OF$_2$, O$_2$F$_2$, ClO$_2$, Cl$_2$O$_7$, Br$_2$O, BrO$_2$, and I$_2$O$_5$. The halogens also form some compounds by interacting among themselves, the simple diatomic molecules BrCl and ClF and more complex ones such as ICl$_3$, BrF$_3$, IF$_7$, and ClF$_3$ are examples.

23.10 ORGANIC HALOGEN COMPOUNDS

Some of the reactions of Br$_2$ and Cl$_2$ with organic molecules were discussed in Chapters 16 and 17. Fluorine is very reactive toward hydrocarbons and is used to

---

* In Section 15.9 it was shown that heating of KClO$_3$ with a MnO$_2$ catalyst was a good way to prepare O$_2$ in the laboratory.

produce many useful substances. One way of obtaining **fluorinated hydrocarbons** is by use of metallic fluorides, for example,

$$C_7H_{16} + 32CoF_3 \rightarrow C_7F_{16} + 16HF + 32CoF_2$$
heptane   perfluoroheptane

The prefix *per-* (from the Latin, "thoroughly") is used to indicate that *all* of the hydrogen atoms in the alkane have been replaced by fluorine. The well-known material Teflon is perfluoropolyethylene and has a chain structure of perfluoromethylene units,

$$\begin{array}{c} \text{F} \quad \text{F} \quad \text{F} \quad \text{F} \\ | \quad | \quad | \quad | \\ -\text{C}-\text{C}-\text{C}-\text{C}- \\ | \quad | \quad | \quad | \\ \text{F} \quad \text{F} \quad \text{F} \quad \text{F} \end{array}$$

Reaction of carbon tetrachloride with a metallic fluoride produces Freon-12, a substance used as a refrigerant.

$$3CCl_4 + 2SbF_3 \rightarrow 3CCl_2F_2 + 2SbCl_3$$
Freon-12

Other Freon-type compounds find wide use as propellents in aerosol sprays. The suspected role of fluorochlorohydrocarbons in the depletion of stratospheric ozone was discussed in Section 15.11.

Fluorinated hydrocarbons are very stable at high temperatures. The C—F bond energy is about 5 eV, as compared with 4.3 eV for C—H and 3.4 eV for C—Cl.

The replacement of hydrogen atoms by halogen atoms affects both the chemical and physical properties of organic molecules. In Table 23.4 are listed the ionization constants of chlorine-substituted acetic acids.

TABLE 23.4 THE IONIZATION CONSTANTS OF THE CHLOROACETIC ACIDS

ACID	K_a AT 25 °C
Acetic	1.8×10^{-5}
Monochloroacetic	1.4×10^{-3}
Dichloroacetic	5.5×10^{-2}
Trichloroacetic	2.3×10^{-1}

The structure of monochloroacetic acid (chloroethanoic acid) is

$$\begin{array}{c} \text{H} \\ | \\ \text{Cl}-\text{C}-\text{C}-\text{OH} \\ | \quad \| \\ \text{H} \quad \text{O} \end{array}$$

The effect of the substituted chlorine is called an **inductive effect** and is characteristic of highly electronegative atoms. The electron density tends to be skewed toward the chlorine end of the molecule at the expense of the carboxylic end; that is, the electronegative chlorine *induces* a positive charge or electron deficit at its neighbor carbon atom. Thus the O—H bond is weakened and it is easier to remove a proton. This means that the K_a of the acid increases. By judicious use of halogen substitution, organic acids of differing K_a values may be synthesized.

SUMMARY

Nitrogen's outstanding characteristic is the large number of stable oxidation states represented by its compounds. Elemental nitrogen is very unreactive, but, nevertheless, nitrogen compounds are widespread in nature and circulate throughout the biosphere, atmosphere, lithosphere, and hydrosphere. Phosphorus, although in the same period as nitrogen, resembles it but slightly in chemical nature. Phosphorus has only two important oxidation states, +3 and +5, and its compounds circulate much less extensively in nature than do those of nitrogen. Both nitrogen and phosphorus play important roles in biological compounds. Phosphates are widely used in detergents to remove ions which cause hardness in water. Sulfur is also widely distributed in nature, not only as metal sulfide ores but also in biological compounds. One of the most important sulfur compounds is sulfuric acid, a compound of great use in industrial processes. Many metal ions can be identified and separated on the basis of the varying solubilities of their sparsely soluble sulfides.

The halogens are electronegative elements of high chemical reactivity. Fluorine, the most electronegative of all the chemical elements, differs from the other halogens in many of its properties. The halogens form binary hydrogen compounds which, with the exception of HF, are strong Brønsted-Lowry acids in water solution. All the halogens also form oxy acids of various types. The most common oxidation state of the halogens is −1, but +1, +3, +5, and +7 are also represented by important compounds.

LEARNING GOALS

1. Be able to give an example of at least one compound for each of the nine oxidation states of nitrogen.

2. Be familiar with the optimum conditions necessary for the synthesis of ammonia from nitrogen.

3. Be able to write balanced chemical equations for the Ostwald process for producing nitric acid.

4. Be able to write a balanced chemical equation for the production of nitric acid from nitrates.

5. Be able to discuss the bonding of nitrogen in pyridine, pyrrole, nitriles, nitro groups, and the peptide linkage.

6. Know how nitrogen is prepared by fractional distillation of air.

7. Know and be able to discuss the main steps in the production of photochemical smog.

8. Know the overall features of the nitrogen cycle in nature.

9. Know how elemental phosphorus is prepared from the mineral apatite.

10. Be able to write balanced equations for the preparation of the halides and oxides of phosphorus and for the preparation of H_3PO_3 and H_3PO_4 from these.

11. Be able to write Lewis formulas for H_3PO_3 and H_3PO_4 and to predict their acid properties from these.

12. Know the role played by tripolyphosphates in detergent formulations.

13. Be able to explain what eutrophication is and how it arises.

14. Be able to explain the Frasch process for the production of sulfur.

15. Be able to write balanced chemical equations for the production of hydrogen sulfide from metal sulfides.

16. Know how the two main oxides of sulfur are formed and be able to write balanced equations for these.

17. Be able to describe, using balanced chemical equations, the contact process for the production of sulfuric acid.

18. Be able to describe the basic features of how sulfides are used to identify and separate metallic cations.

19. Know how the elemental halogens may be prepared and be able to write balanced equations for the processes.

20. Know how the hydrogen halides may be prepared and be able to write balanced equations for the processes.

21. Know how the oxy acids of the halogens and their salts are prepared.

22. Know how perfluorohydrocarbons are formed and what some of these (for example, Freon and Teflon) are used for.

23. Be able to describe how the inductive effect of chlorine makes the chloroacetic acids stronger acids than acetic acid.

DEFINITIONS, TERMS, AND CONCEPTS TO KNOW

photochemical smog
PAN
fixation of nitrogen
nitrification
denitrification
algal bloom
eutrophication
oxy acid
hypochlorous acid
Ostwald process

chlorous acid
chloric acid
perchloric acid
hypochlorite
chlorite
chlorate
perchlorate
perfluorohydrocarbons
inductive effect

Frasch process
nitrite
nitrate
phosphorous acid
phosphoric acid
sulfite
sulfate
sulfurous acid
sulfuric acid

QUESTIONS AND PROBLEMS

Nitrogen

1. Concentrated nitric acid reacts with metallic copper to produce NO_2 as one of the products. Write the balanced equation for this process.

2. Write a balanced chemical equation for the reaction of nitric acid with a soluble metal sulfide (such as Na_2S) if the products are NO and sulfate. What substance is oxidized and what substance is reduced in this reaction?

3. What (in addition to NO or NO_2) are the products of a reaction between nitric acid and copper(I) sulfide?

4. Predict two possible geometrical structures for the molecule N_2F_2 (order of atoms FNNF). Both isomers are known experimentally.

5. What weight (in kg) of ammonia is needed to produce 1 metric ton (tonne) of nitric acid by the Ostwald process?

6. Explain why HNO_3 reacts with benzene in the following manner:

$$\text{C}_6\text{H}_6 + HNO_3 \xrightarrow{H_2SO_4} \text{C}_6\text{H}_5NO_2 + H_2O$$

That is, why does $-NO_2$ and not $-NO_3$ replace hydrogen? What is the role of the sulfuric acid? See Section 17.3

7. The chemical reactions

$$2NO(g) \rightarrow N_2(g) + O_2(g)$$
$$2NO(g) + O_2(g) \rightarrow 2NO_2(g)$$

have ΔG^0 values at 298 K of -173 and -69.5 kJ, respectively. The first reaction is very slow and the second is very fast.
  a) What is the consequence of the relative rates with respect to air pollution?
  b) What would be the consequences if the relative rates were reversed?
  c) Can you think of a practical use for a catalyst which would increase the rate of the first reaction only?

8. Classify the following processes as fixation, nitrification, or denitrification.

  a) $3Ca(s) + N_2(g) \rightarrow Ca_3N_2(s)$
  b) $2CH_3NH_2(g) + 5\tfrac{1}{2}O_2(g) \rightarrow 2CO_2(g) + 5H_2O(l) + 2NO(g)$
  c) $2NH_3(g) + 2\tfrac{1}{2}O_2(g) \rightarrow 2NO(g) + 3H_2O(l)$
  d) $2NO_2(g) + 4H_2(g) \rightarrow N_2(g) + 2H_2O(g)$
  e) $2NH_3(g) \rightarrow N_2(g) + 3H_2(g)$

9. Will NO_2 react better with acidic or basic water? Explain.

10. What volume of air (at 25 °C and 1 atm) is needed to produce 1 t of the fertilizer ammonium nitrate, assuming all of the nitrogen comes from air? Assume air is 20% nitrogen and that the processes have an average efficiency of 60%.

11. Solid ammonium nitrate, although used as a commercial fertilizer, is also a powerful explosive. Taking into account that NH_4^+ is a reducing agent and NO_3^- is an oxidizing agent, write the equation for one of the reactions probably occurring in the explosion.

12. Assuming that the formation of NO by lightning is a chain reaction, suggest some possible steps in the chain.

13. Sodium reacts with water to form sodium hydroxide and hydrogen. Using this as an analogy, predict what happens when sodium reacts with ammonia. Show that this reaction is a sum of the two steps

$$Na(s) + (x+y)NH_3(l) \rightarrow Na^+ \cdot xNH_3 + e^- \cdot yNH_3$$
$$e^- \cdot yNH_3 \rightarrow \tfrac{1}{2}H_2(g) + NH_2^- \cdot (y-1)NH_3$$

The second reaction occurs on long standing. The substance $Na^+ \cdot xNH_3$ is a *solvated* sodium ion; the solvent (ammonia) has formed a weak association with the ion, much as water molecules do in hydration.

14. Pure ammonia undergoes the self-ionization

$$2NH_3 \rightleftharpoons NH_4^+ + NH_2^-$$

At 25 °C, $[NH_4^+] = [NH_2^-] = 3.2 \times 10^{-16}$ M.
  a) What is the equilibrium constant (analogous to K_w of water) for the self-ionization?
  b) Calculate the concentration of ammonium ion in a 0.01-M solution of $NaNH_2$ in ammonia.

c) Given that pA = $-\log$ [NH$_4^+$], what is the pA of a neutral ammonia solution?

d) What is the pA of a 0.001-M solution of NH$_4$Cl in ammonia?

Phosphorus

15. Draw a Lewis structure for PO$_4^{3-}$ and indicate its expected geometry.

16. A typical phosphate-containing detergent may contain 10% by weight of phosphorus. The average family, approximately four people, may use a total of 50 lb of detergent (laundry, dishes, car washing, etc.) per year. In a city of 200,000 people, how many moles of phosphorus will be put into the drains each day on the average?

17. The liquid effluent from a municipal sewage treatment plant may have 5 mg of phosphorus per liter. Express this as molar concentration of PO$_4^{3-}$.

18. Floating lake algae contain about 0.5% by weight of phosphorus (dry weight). If a certain lake contains 0.1 mg of PO$_4^{3-}$ per liter, how many liters of water are needed to produce 1 g of dry algae? Assume all the PO$_4^{3-}$ can be utilized for this purpose.

19. An average cow produces 2 kg of manure (dry weight) per day. Chemical analysis shows this is about 0.8% by weight in phosphorus (some chemists will analyze anything!). Only about 10% of this, called the **available** phosphorus, is water soluble. Estimate the maximum amount of phosphorus (measured as kg of PO$_4^{3-}$) that a herd of 100 cows could add to runoff water during the course of a year.

20. The accompanying graph depicts some data based on chemical analyses of bottom sediments of a midwestern lake. The year of deposition of each sediment sample was estimated from the annular pattern revealed in test borings and from the depths of each sample from the present bottom level. Use this graph to suggest dates or time spans characteristic of the following events and developments.

a) Before this date, the land surrounding the lake was occupied by Indians who regularly burned off the grasslands and forests.

b) During this time span settlers began to displace the Indians.

c) The settlers begin to raise increasing amounts of livestock.

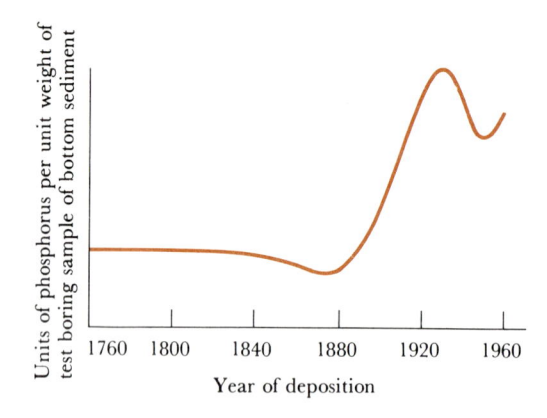

d) Economic depression and drought strike the area.

e) Economic recovery begins but mechanization begins to displace horses. Also, cattle raising moves to other areas of the midwest.

f) Detergents begin to appear on the market.

21. The undiscovered planet Pyggi has a population of 500,000 intelligent creatures whom we shall call Pyggians. The planet's entire water supply consists of one lake containing approximately 5×10^8 m^3 of water. The Pyggians obtain all of their oxygen from the algae living in the lake. The algae, in turn, depend on the phosphate-containing waste material that the Pyggians dump into the lake. Over the course of past centuries the Pyggians have managed to keep the lake phosphate level at about 1 ppm (as phosphorus) and this has maintained a stable algae population of about 3 tonnes. Theoretical calculations by Pyggian scientists indicate that a tenfold increase of algae would raise the atmospheric oxygen concentration to the point where spontaneous combustion of Pyggians would become a serious hazard. One Pyggian month ago (10 Pyggian days = 1 Pyggian month and 1 Pyggian day = 20 earth days), as a result of a high-intensity advertising campaign that convinced them that it was socially desirable to have the whitest tennis shoes on the block, Pyggians began using a new miracle detergent containing 20% (by weight) of sodium tripolyphosphate. Previous to this time Pyggians never washed and used no soaps or detergents. Assuming that there are four Pyggians in an average family and that each family does ten loads of washing per Pyggian day to keep their tennis shoes sparkling white, estimate how long it will take (in Pyggian days) before Pyggians begin to self-destruct. Assume an average wash uses 25 gal of water and 1 lb of detergent and that 75% of the phosphate reaches the water supply. You may also assume that the lake's algal concentration is directly proportional to the phosphate level.

22. A solution of Na_3PO_4 in water is basic. Show by means of an appropriate reaction why this is so.

23. Although nitrogen is more electronegative than phosphorus, the latter is more reactive at room temperature. Explain.

24. a) Which is the better oxidizing agent and why: HNO_3 or H_3PO_4?
b) Which is the better reducing agent and why: NH_3 or PH_3?

25. Each of the following solutions was evaporated and heated to dryness. What compound(s) is present in each?
  a) A solution containing 5.0 g H_3PO_4 and 3.5 g KOH.
  b) A solution containing 5.0 g H_3PO_4 and 7.0 g KOH.
  c) A solution containing 5.0 g H_3PO_4 and 7.5 g NaOH.

26. What product is formed when H_3PO_3 is treated with excess NaOH?

Sulfur

27. What are the expected geometrical shapes of the SO_3^{2-} and SO_4^{2-} ions?

28. The ion $S_2O_3^{2-}$ is called the thiosulfate ion.
  a) How are the ions SO_4^{2-} and $S_2O_3^{2-}$ related?
  b) What are the oxidation numbers of sulfur in the two ions? Explain.

29. Draw Lewis diagrams for SO_2 and SO_3 and predict the geometries of these molecules. Show that SO_2 resembles O_3 in structure.

30. Which would you expect to be a stronger Brønsted-Lowry base, S^{2-} or O^{2-}? Explain.

31. When thioacetamide,

$$CH_3\underset{\underset{S}{\|}}{C}NH_2$$

is heated in aqueous solution, acetic acid, NH_3, and H_2S are produced. Write a balanced chemical equation for this process. What volumes of NH_3 and H_2S (measured at 25 °C and 1 atm pressure) could one prepare from 1 L of 0.5-M thioacetamide?

32. A solution is 0.001 M in Cd^{2+} and 0.001 M in Ni^{2+}. If the K_{sp}'s of CdS and NiS are 3.6×10^{-29} and 1.4×10^{-24} at 25 °C, respectively, at what pH can one precipitate one of the sulfides and not the other by the addition of H_2S? Which sulfide will precipitate first?

33. The amount of SO_2 in air may be measured by scrubbing the air with a solution containing $FeNH_4(SO_4)_2$ and ferrozine, an organic compound which forms a magenta-colored complex with Fe^{2+}. The reaction (not balanced) is

$$SO_2(g) + Fe^{3+} \rightarrow SO_4^{2-} + Fe^{2+}$$

The Fe(II) ferrozine complex absorbs light of wavelength 5620 Å very strongly and thus its concentration may be measured spectrophotometrically. When 100 L of air is so treated with a 100-cm³ solution, the concentration of Fe(II) ferrozine complex produced is 1.6×10^{-6} M. Calculate the concentration of SO_2 in the air sample in ppm.

34. The Frasch process requires water superheated to about 180 °C (sulfur melts at about 119 °C). How can water be heated to such a temperature without being vaporized?

35. Explain why it is possible for sulfurous acid to act as either an oxidizing agent or as a reducing agent, whereas sulfuric acid is an oxidizing agent only.

36. Write expected chemical formulas for a compound containing sulfur and each of the following elements.
  a) Cu(II)   c) Ca   e) O
  b) Na   d) Al   f) C

37. Predict the products of the reaction between sulfide ion and nitrate ion in acid solution and write the balanced chemical equation for the process.

38. Since SO_2 has a critical temperature of 157.2 °C and a critical pressure of 77.7 atm, it was once widely used as a refrigerant gas. Today most home refrigerators use Freon (a much more expensive gas) instead. Give one practical reason for this change.

39. 1000 L of SO_2-contaminated air is passed through 150 cm³ of a $FeNH_4(SO_4)_2$ and ferrozine solution (see Problem 33). The concentration of Fe(II) ferrozine complex is 2.1×10^{-5} M. Calculate the concentration of SO_2 in the air sample in ppm.

40. What weight of FeS$_2$ [iron(II) disulfide] is required to produce 1 t of sulfuric acid, assuming 90% efficiency for the process?

41. The mineral chalcopyrite contains 34.62% copper, 30.43% iron, and 34.95% sulfur. What is the simplest formula of this mineral?

The Halogens

42. A current of 2 A is passed through a molten KHF$_2$ solution for 10 hrs. What products, and in what amounts, are formed at each electrode?

43. Four grams of hydrogen chloride is placed into a 1-L flask and heated to 50 °C. The pressure of the gas is 2 atm. Explain.

44. Discuss the preparation of HBr and HI using elemental phosphorus and show why the two compounds require slightly different procedures.

45. a) How much sodium hydrogen sulfite (in grams) is needed to obtain the maximum amount of elemental iodine from 1 tonne of Chile saltpeter, assuming a sodium iodate content of 1%?
b) What weight of iodine would be produced in the above?

46. How many moles of potassium chlorate can be made from one mole of chlorine gas which is added to KOH solution and then heated?

47. Draw Lewis diagrams and predict the geometries of HOCl, HOClO, HOClO$_2$, HOClO$_3$, OCl$^-$, ClO$_2^-$, ClO$_3^-$, and ClO$_4^-$.

48. The reaction

$$2HBr(g) + Cl_2(g) \rightarrow 2HCl(g) + Br_2(l)$$

proceeds quite readily but the reaction

$$2HCl(g) + I_2(g) \rightarrow 2HI(g) + Cl_2(g)$$

does not. Explain. Would the following reaction go?

$$2HI(g) + Br_2(l) \rightarrow 2HBr(g) + I_2(s)$$

49. Arrange the halogens in order of (a) decreasing electronegativity, (b) increasing ionization energy, (c) increasing oxidizing ability, (d) decreasing ease of preparation from the halide ion.

50. Chlorine can be prepared by oxidizing Cl$^-$ (from NaCl, for example) with potassium dichromate (K$_2$Cr$_2$O$_7$) in acid solution. What is the equivalent weight of potassium dichromate in this reaction? One product contains Cr(III).

51. From the standpoint of ease of reaction, which would be better to use in the following reaction, bromoethane or chloroethane?

$$CH_3CH_2X + H_2O \xrightarrow{H_3O^+} CH_3CH_2OH + HX$$

52. Give as many examples as you can think of in which halogen-substituted organic compounds are used in organic syntheses. See Chapter 17 for a review of such uses.

53. Many interhalogen compounds are known in which chlorine, bromine, and iodine have positive oxidation numbers, for example, ClF$_3$, BrF$_3$, IF$_5$, IF$_7$, etc. However, there are no fluorine compounds in which fluorine has a positive oxidation number. Explain.

General

54. Discuss the relationship of carbon to silicon as being analogous to that of nitrogen to phosphorus. You will need to look up some basic facts of silicon chemistry in some reference text, for example, Cotton and Wilkinson.

55. Write balanced chemical equations for preparing each of the following.
  a) HNO$_3$ from NH$_3$
  b) H$_3$PO$_4$ from P$_4$
  c) H$_2$SO$_4$ from S
  d) Cl$_2$ from NaCl
  e) I$_2$ from KI
  f) HCl from NaCl
  g) Br$_2$ from KBr using Cl$_2$

56. In Section 4.5 it was pointed out that atomic radii vary roughly as n^2/Z_{eff}, where Z_{eff} is the number of valence electrons. How would you expect ionic radii of positive and negative ions to compare with the corresponding atomic radii? Do the data in Tables 21.1 and 23.2 support these predictions?

57. Complete and balance the following reactions (indicate if no reaction occurs).
  a) Mg(s) + Br$_2$(l) $\rightarrow$
  b) Cl$_2$(g) + NaF(s) $\rightarrow$
  c) H$_2$O(l) + F$_2$(g) $\rightarrow$
  d) H$_2$S(aq) + HNO$_3$(aq) $\rightarrow$
  e) PH$_3$(g) + H$_2$SO$_4$(aq) $\rightarrow$
  f) NaBr(aq) + KMnO$_4$(aq) + H$_3$O$^+$ $\rightarrow$
  g) NO(g) + H$_2$O$_2$(l) $\rightarrow$
  h) H$_2$SO$_4$(aq) + HNO$_3$(aq) $\rightarrow$
  i) H$_3$PO$_4$(aq) + KClO$_3$(aq) $\rightarrow$
  j) N$_2$H$_4$(g) + O$_2$(g) $\rightarrow$

58. Give the names of each of the following compounds.
- a) Ca(ClO$_3$)$_2$
- b) KHSO$_3$
- c) IF$_5$
- d) LiOI
- e) HNO$_2$
- f) Al$_2$S$_3$
- g) MnS
- h) N$_2$F$_2$
- i) (NH$_4$)$_2$HPO$_3$
- j) RbBrO$_4$

59. Write formulas for each of the following compounds.
- a) Magnesium iodite
- b) Lithium nitrite
- c) Copper(I) sulfide
- d) Sodium phosphite
- e) Chlorine trifluoride
- f) Nickel(II) phosphate
- g) Ammonium sulfide
- h) Perchloric acid
- i) Silver sulfite
- j) Iron(III) nitrate

ADDITIONAL READINGS

Delwiche, C. C., "The Nitrogen Cycle." *Scientific American*, September 1970.

Deevey, Edward S., Jr., "Mineral Cycles." *Scientific American*. September 1970. Discusses the roles of sulfur, phosphorus, and other elements in the biosphere.

Schneller, Stewart W., "Nitrogen Fixation." *J. Chem. Educ.* **49**, 786 (1972). A short discussion of the biological, chemical, and nonenzymic fixation of nitrogen as an interdisciplinary research frontier.

Cotton, F. A., and G. Wilkinson, *Advanced Inorganic Chemistry.* New York: Interscience, 1972. Useful as a reference.

Skinner, Karen Joy, "Nitrogen Fixation." *Chemical & Engineering News,* 4 Oct. 1976, pp. 22–35. Discusses recent work in unraveling the complexities of nitrogen fixation.

24
GIANT MOLECULES

Herman Staudinger, German organic chemist who was the first to propose that rubber consisted of large molecules whose atoms were bound by the same forces found in small molecules. (Courtesy of Wide World Photos.)

One of the problems which I am going to start work on has to do with substances of high molecular weight. I want to attack the problem from the synthetic side. One part would be to synthesize compounds of high molecular weight and known constitution . . . another phase of the problem will be to study the action of substances xAx on yBy where A and B are divalent radicals and x and y are functional groups capable of reacting with each other.
WALLACE HUME CAROTHERS, *in a letter to Dr. John R. Johnson of Cornell University, 14 February 1928*

Wallace Hume Carothers (1896–1937) was an Iowa-born organic chemist. As Head of Organic Research at DuPont, he developed the theory of step-growth polymerization, directed the synthesis of nylon and many other synthetic polymers, and laid the foundations for the development of neoprene. While still a graduate student at the University of Illinois, Carothers wrote a definitive paper on the electronic structure of the double bond. Subject to periods of deep depression, he committed suicide fourteen months after his marriage and seven months before the birth of a daughter.

Hermann Staudinger (1881–1965), a German organic chemist, is generally regarded as the founder of polymer chemistry. After working on problems such as the chemical nature of the aroma of coffee, Staudinger began a study of the structure of natural rubbers which ultimately led to his concept of macromolecules. Staudinger was awarded the Nobel prize in chemistry in 1953; seldom has this honor been more deserved.

Prior to about 1920 few chemists believed that a stable molecule containing thousands of atoms could exist. The insoluble substances occasionally produced in their reaction vessels and the substances isolated from animal and vegetable sources were believed to be aggregates of small molecules held together by special, as yet unknown, types of forces. Then in 1922 the German chemist Hermann Staudinger made the bold proposal that substances such as rubber were simply very large molecules in which the atoms were bound by the same chemical valence bonds as characterized small molecules. Staudinger proposed the name **macromolecule** (after the Greek *makros*, long) for such large molecules. Staudinger's assertion, held in disbelief by most of his contemporaries, was followed by an even bolder one; Staudinger said that it was possible that life itself somehow emerges when molecules of sufficient size and complexity are assembled. Today it is realized that we live in a world dominated by macromolecules: synthetic plastics and fibers, the cellulose of plants, and the proteins of living things. Indeed, small molecules are the real anomalies, existing mainly under the special protective conditions of the laboratory.

24.1 THE NATURE OF GIANT MOLECULES

It would be a mistake to assume that a large molecule differs from a small molecule only by virtue of its mass.* As Staudinger's prescient statement suggests, size and complexity themselves open up a new dimension in chemical behavior.

Except for isotopic variation, one small molecule is just like another insofar as molecular weight is concerned. In large molecules, however, there may not exist a unique entity whose replicas constitute the substance; each molecule may differ from the others in molecular weight. This is generally true in the case of synthetically produced macromolecules but may not be true for naturally occurring macromolecules. Thus, it is customary to speak of *average* molecular weights when dealing with macromolecules. In general, then, macromolecular substances are heterogeneous in mass.

Two macromolecules of identical mass may also differ greatly as to conformation and linkage of atoms. Such variations are also possible in small molecules but not to a pronounced degree. A very long molecule, for example, a hydrocarbon such as $C_{500}H_{1002}$, may be twisted, bent, and looped in a large number of different ways to produce a large number of different conformations. In addition, there may be variations in linkages such that branched chains of various types exist. Strictly speaking, any linkage variations should be treated as isomers but this concept is not practicable to apply when the number of closely related isomers becomes very, very large.

Most small molecules are stable as gases, liquids, or solids and are characterized by definite melting points and boiling points. In large molecules, disorder in the chains often prevents complete crystallinity from developing so that true solids may not exist. If an attempt is made to vaporize a large molecule, the temperatures high enough to overcome intermolecular attractive forces are also high enough to break chemical bonds, and the molecule degrades. In small molecules the area of contact between molecules is generally so small that intermolecular attractions are relatively weak; in macromolecules, however, the area of contact may be very large, leading to large total interaction forces.

*Intra*molecular interactions can also play important roles in large molecules, whereas these are usually absent or insignificant in small molecules. For example, the coiling and folding of a single large molecule can lead to interactions between different portions of the same molecule—a possibility which is ordinarily absent in small molecules due to their less flexible structures.

In small molecules, purely chemical properties—for example, the natures of functional groups—determine the behavior almost entirely; molecular geometry

---

* There is no sharp dividing line between small and large molecules. Quite arbitrarily we shall call *small* those molecules with molecular weights below 1000 amu and *large* those molecules with molecular weights above 5000 amu. Most of the important macromolecules have molecular weights ranging from tens of thousands to several million amu.

plays but a secondary role. In large molecules, size permits complexity of organization which, in turn, may permit highly specialized and sophisticated behavior going beyond that inherent in any functional groups present. For example, the enzyme lysozyme (see Section 20.9) owes its remarkable catalytic ability not only to the chemical natures of the functional groups of two amino acids but also to the precise geometrical location of those functional groups and the size and shape of the cleft in which they are located. Only a very large molecule possesses the potential of meeting such demanding requirements.

24.2 POLYMERS AND MONOMERS

Most synthetic and natural macromolecules can be viewed as composed of a large number of simpler molecules or the major parts of such molecules. Such macromolecules are called **polymers** (after the Greek *polys*, many, and *meros*, part). Each of the subunits making up a polymer is called a **monomer**. For example, polyethylene is a polymer made up of linked ethylene monomers, cellulose is a polymer made up of linked glucose monomers, and proteins are polymers made up of linked amino acid monomers. Nylon generally consists of two monomers, a diamine and a dicarboxylic acid linked in alternating fashion.

The first synthetic polymer was invented by an American chemist of Belgian origin, Leo Baekeland, in 1907. While searching for a way to make an artificial shellac (natural shellac is made from resin-secreting insects called lac bugs), Baekeland combined phenol and formaldehyde to make a gummy substance which he heated under pressure. Upon cooling of the reaction mixture a clear, hard, amber-like material molded in the shape of the "cooking" apparatus was formed. This material, now known as **Bakelite**, is strong, light, resistant to heat, acid, and weather, and can be dyed or colored. It also has excellent properties as an electrical insulator. The basic chemical reaction is now known to be

Leo H. Baekeland (1863–1944), a Belgian-born American chemist, devised the first commercially successful photographic contact developing paper and also invented Bakelite, the first major industrial synthetic plastic. Baekeland's original corporation for the manufacture of Bakelite has been absorbed by the Union Carbide and Carbon Company.

followed by processes such as

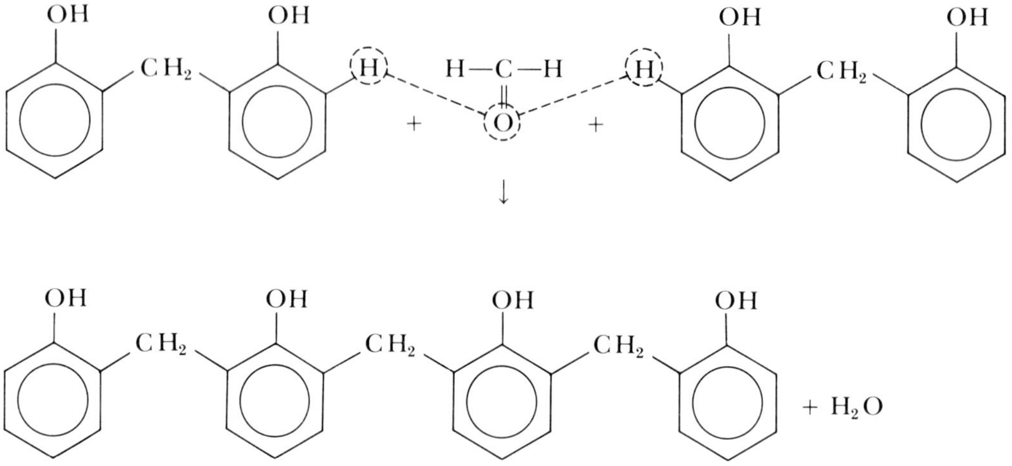

[Chemical reaction scheme showing two phenol-CH₂-phenol dimers reacting with formaldehyde (H-C(=O)-H) to form a tetrameric chain of phenol rings linked by CH₂ groups, plus H₂O]

Heating induces cross-linking between chains such as shown below:

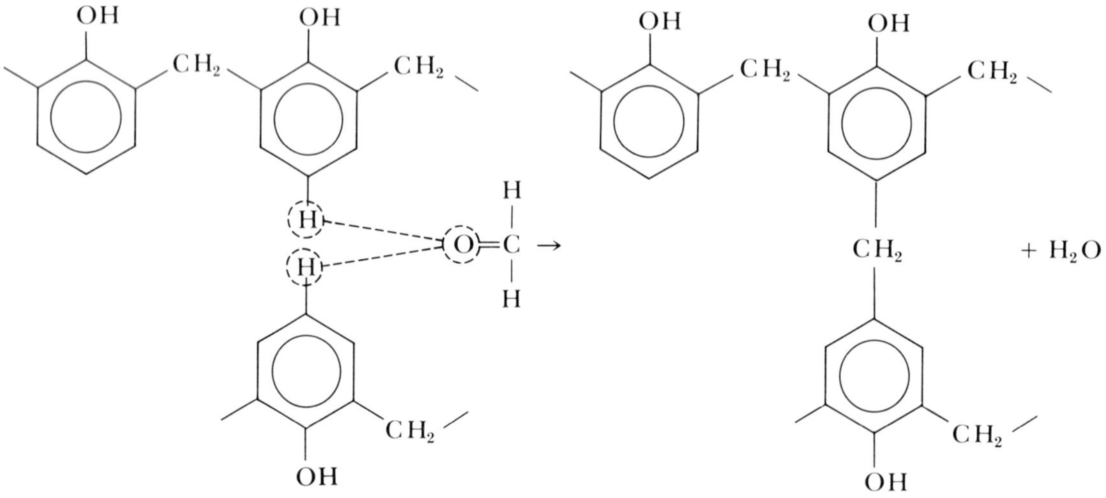

[Chemical reaction scheme showing two phenol-formaldehyde chains cross-linking via a formaldehyde molecule (O=CH₂) to produce a cross-linked network connected by a CH₂ bridge, plus H₂O]

The resulting polymer is a complex three-dimensional network of phenol residues linked by methylene groups.

Polymers such as Bakelite are called **thermosetting**. The basic resin (here the phenol-formaldehyde gummy mass) is put into a mold and heated to induce cross-linking. A thermosetting polymer, once formed, cannot be melted and molded into a new object.* Polymers which can be melted, poured into molds, and cooled are

---

* Thermosetting polymers decompose if heated to high temperatures.

called **thermoplastic**. Ideally, thermoplastic polymers can be remelted and remolded indefinitely. Polyethylene is an example of a thermoplastic polymer.

The process whereby one or more monomers react to form a high-molecular-weight polymer is called **polymerization**. Polymerization reactions are of two distinct types: **chain polymerization** and **step-growth polymerization**. The essential features of both of these are discussed in the following sections.

24.3 CHAIN POLYMERIZATION

Chain polymerizations are chain reactions (see Section 20.7). The process is most easily described by using the simple example of a substituted ethylene molecule (called a vinyl-type monomer) adding to itself. The general formula of a vinyl-type monomer is

$$\begin{array}{c} H \\ \diagdown \\ C=C \\ \diagup \\ H \end{array} \begin{array}{c} H \\ \diagup \\ \\ \diagdown \\ Y \end{array} \quad \text{or} \quad CH_2{=}CHY$$

where Y represents a univalent chemical group such as H, Cl, or CH_3. Chain polymerizations may be initiated by free radicals, cations, or anions. An example of a free radical producer is benzoyl peroxide. Upon heating, benzoyl peroxide produces **free-radical initiators** as follows:

$$\text{benzoyl peroxide} \xrightarrow{\text{heat}} 2 \text{ free-radical initiator}$$

The first step of the polymerization, the **chain initiation** step, proceeds as follows:

$$R\cdot + CH_2{=}CHY \rightarrow RCH_2{-}\dot{C}HY$$

where R· stands for the free-radical initiator. The new free radical, $RCH_2\dot{C}HY$, interacts with another monomer molecule in the basic **propagation step**:

$$RCH_2\dot{C}HY + CH_2{=}CHY \rightarrow RCH_2CHCH_2\dot{C}HY \\ \phantom{RCH_2CHCH_2\dot{C}HY mmm} | \\ \phantom{RCH_2CHCH_2\dot{C}HY mmm} Y$$

The chain continues to grow rapidly in this fashion until, eventually, two growing chains happen to combine in a **termination** step such as

$$R{-}(CH_2CHY)_x{-}CH_2\dot{C}HY + \\ Y\dot{C}HCH_2{-}(CHYCH_2)_y{-}R \rightarrow R{-}(CH_2CHY)_{\overline{x+y+2}}R$$

A process known as **chain branching** can also occur. For example, a growing chain, $R_1CH_2\dot{C}HY$, can abstract a hydrogen atom from a completed polymer chain as follows:

$$R_1CH_2\dot{C}HY + R_2CH_2CHYR_3 \rightarrow R_1CH_2CH_2Y + R_2\dot{C}HCHYR_3$$

where R_1, R_2, and R_3 are polymer chain segments. The new free radical, $R_2\dot{C}HCHYR_3$, can continue to grow by adding monomer units as side chains:

$$R_2\dot{C}HCHYR_3 + CH_2{=}CHY \rightarrow \begin{array}{c} R_2CHCHYR_3 \\ | \\ CH_2 \\ | \\ \cdot CHY \end{array}$$

Polymers made by free-radical initiated reactions are characterized by a high degree of branching.

In Table 24.1 are listed some of the more common vinyl-type monomers and the polymers which they produce. In Table 24.2 are listed a few vinyl-type polymers whose monomers are variations of the $CH_2{=}CHY$ form.

Several very important polymers are obtained from monomers of the 2-substituted 1,3-butadiene type, such as

$$CH_2{=}\underset{\underset{Y}{|}}{C}{-}CH{=}CH_2$$

where Y is usually H, Cl, or CH_3. When this monomer polymerizes, only one of the double bonds is utilized so that four different types of repeating units can arise. Two of these are

Polymers containing either of these repeating units are said to be of 1,2-type. Since the three carbon-to-carbon bonds of 1,3-butadiene are each composites of sigma and pi bonds (see Section 16.9), there is another way in which just one of the double bonds can participate in the polymerization reaction. For 1,3-butadiene itself as an example, the previous two repeating units may be viewed as arising as follows: Two of the electrons in the 1,2 double bond are shifted to convert the intramolecular pi bond into two intermolecular sigma bonds to other monomers. This may be

TABLE 24.1 VINYL-TYPE POLYMERS OF THE MONOMER CH$_2$=CHY

Y	MONOMER NAME	FORMULA	REPEATING POLYMERIC UNIT	POLYMER NAME AND USES
H	Ethylene	CH$_2$=CH$_2$	—CH$_2$—CH$_2$—	Polyethylene or polythene, used in films, fibers, molded objects, kitchenware, tubing, electrical insulation
CH$_3$	Propylene	CH$_2$=CHCH$_3$	—CH$_2$—CH(CH$_3$)—	Polypropylene, used in fibers, molded objects, kitchenware
Cl	Vinyl chloride	CH$_2$=CHCl	—CH$_2$—CH(Cl)—	Polyvinylchloride, used in phonograph records, floor coverings
OAc	Vinyl acetate	CH$_2$=CHOAc	CH—CH(OAc)	Polyvinylacetate, used in floor coverings, chewing gum, textile coatings, adhesives
CN	Acrylonitrile	CH$_2$=CHCN	CH$_2$—CH(CN)—	Polyacrylonitrile, used in Acrilan and Orlon fibers
C$_6$H$_5$ (phenyl)	Styrene	CH$_2$=CH(C$_6$H$_5$)	—CH$_2$—CH(C$_6$H$_5$)—	Polystyrene, used in molded objects, kitchenware, electrical insulation

TABLE 24.2 SOME ADDITIONAL VINYL-TYPE POLYMERS

MONOMER NAME	FORMULA	REPEATING UNIT	POLYMER NAME AND USES
Perfluoroethylene	CF$_2$=CF$_2$	—CF$_2$—CF$_2$—	Teflon, used in heat-resistant nonstick coatings of cooking utensils
1,1-Dichloroethylene or vinylidene chloride	CH$_2$=CCl$_2$	—CH$_2$—CCl$_2$—	Saran, used in films noted for their "clingy" nature. Usually made with some added vinyl chloride monomer.
Methylmethacrylate	CH$_2$=C(CH$_3$)(C(=O)OCH$_3$)	—CH$_2$—C(CH$_3$)(C(=O)OCH$_3$)—	Plexiglas, Lucite, used to make clear glass-like objects and as shatterproof replacements for glass panes.

symbolized as shown below:

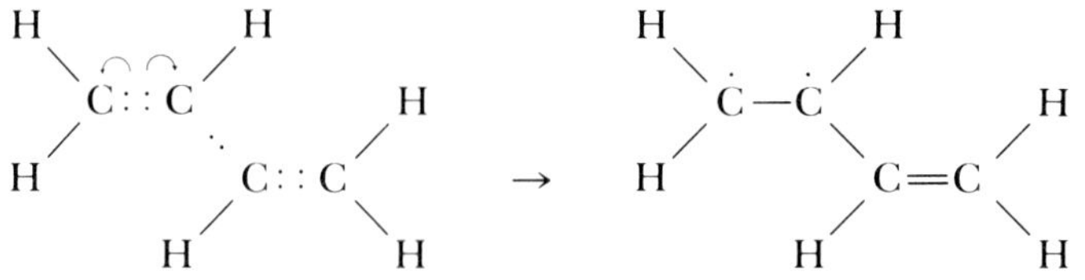

$$\begin{matrix} H & & H \\ \diagdown & & \diagup \\ & C::C & \\ \diagup & & \diagdown \\ H & & C::C \\ & \diagup & \diagdown \\ & H & H \end{matrix} \quad \rightarrow \quad \begin{matrix} H & & H \\ \diagdown & & \diagup \\ & \dot{C}-\dot{C} & \\ \diagup & & \diagdown \\ H & & C=C \\ & \diagup & \diagdown \\ & H & H \end{matrix}$$

However, it can also happen that the two electrons needed for the intermolecular sigma bonds come from different ends of the molecule, for example,

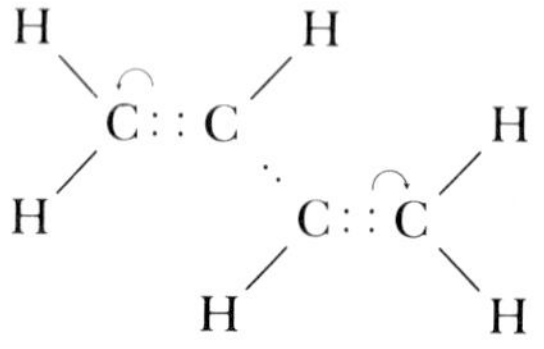

$$\begin{matrix} H & & H \\ \diagdown & & \diagup \\ & C::C & \\ \diagup & & \diagdown \\ H & & C::C \\ & \diagup & \diagdown \\ & H & H \end{matrix}$$

But, in order to satisfy the octet rule, electrons on atoms 2 and 3 must shift to form a pi bond between atoms 2 and 3, that is,

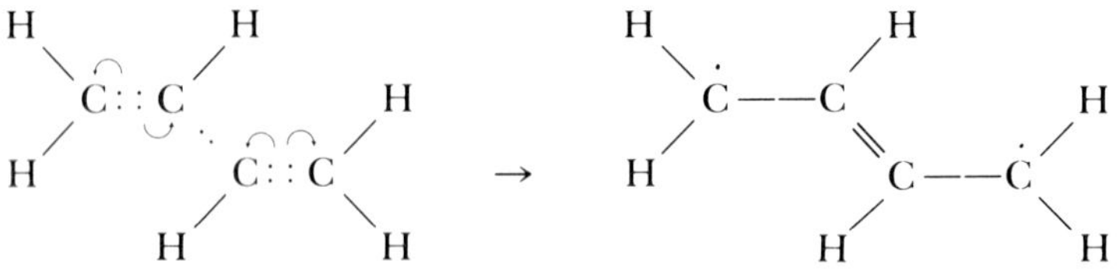

$$\begin{matrix} H & & H \\ \diagdown & & \diagup \\ & C::C & \\ \diagup & & \diagdown \\ H & & C::C \\ & \diagup & \diagdown \\ & H & H \end{matrix} \quad \rightarrow \quad \begin{matrix} H & & H \\ \diagdown & & \diagup \\ & \dot{C}--C & \\ \diagup & & \diagdown \\ H & & C--\dot{C} \\ & \diagup & \diagdown \\ & H & H \end{matrix}$$

Thus two other repeating units are

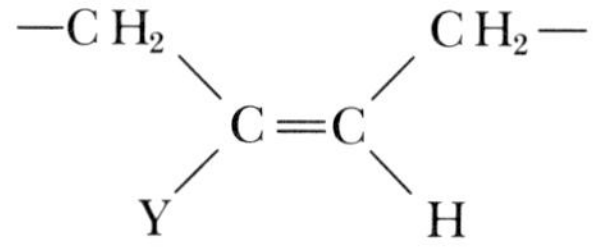

$$\begin{matrix} -CH_2 & & CH_2- \\ \diagdown & & \diagup \\ & C=C & \\ \diagup & & \diagdown \\ Y & & H \end{matrix}$$

and

$$\begin{matrix} -CH_2 & & H \\ \diagdown & & \diagup \\ & C=C & \\ \diagup & & \diagdown \\ Y & & CH_2- \end{matrix}$$

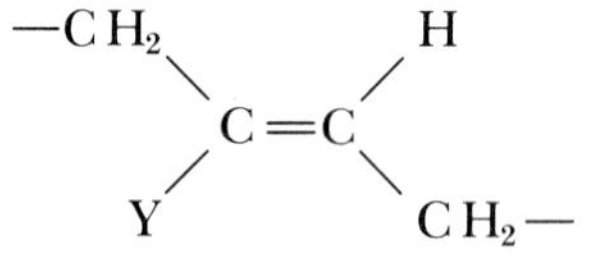

These are called *cis* and *trans*, respectively, and are said to be of 1,4-type.

Natural rubber is a polymer of isoprene

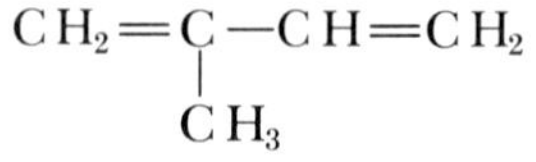

$$CH_2=C-CH=CH_2$$
$$\quad\quad |$$
$$\quad\quad CH_3$$

The repeating unit is virtually all of *cis* geometry. Thus natural rubber, or *cis*-polyisoprene, has the structure

Trans-polyisoprene also occurs in nature as a tough substance known as **gutta-percha**. This substance is used to make coverings for golf balls. Gutta-percha has the structure

Cis-polybutadiene resembles natural rubber somewhat but has very poor skid resistance. *Trans*-polybutadiene resembles gutta-percha.

When Y is Cl, the monomer is called **chloroprene**. *Trans*-polychloroprene is known as **neoprene**; it is an oil-resistant synthetic rubber.

When any of the 2-substituted 1,3-butadienes are polymerized in the laboratory using a free-radical initiator, most of the product is a mixture of 1,2-type and a small amount of *trans*-1,4. No *cis* polymer is formed.

24.4 STEP-GROWTH POLYMERIZATION

One type of step-growth polymerization involves a single monomer with two different functional groups which are designated by A and B. The general polymerization reaction may be written

$$n\text{A}-\text{R}-\text{B} \rightarrow \text{A}-(\text{R})_n-\text{B} + (n-1)\text{AB}$$

where AB is generally a small molecule obtained along with the polymer A—(R)$_n$—B. The first step of the polymerization may be written

$$\text{A}-\text{R}-\text{B} + \text{A}-\text{R}-\text{B} \rightarrow \text{A}-\text{R}-\text{R}-\text{B} + \text{AB}$$

This is followed by a similar step

$$A-R-R-B + A-R-B \rightarrow A-R-R-R-B + AB$$

In this fashion chains grow and combine as long as there are any A and B functional groups present. Cellulose and starch are natural polymers of this type. The monomers are

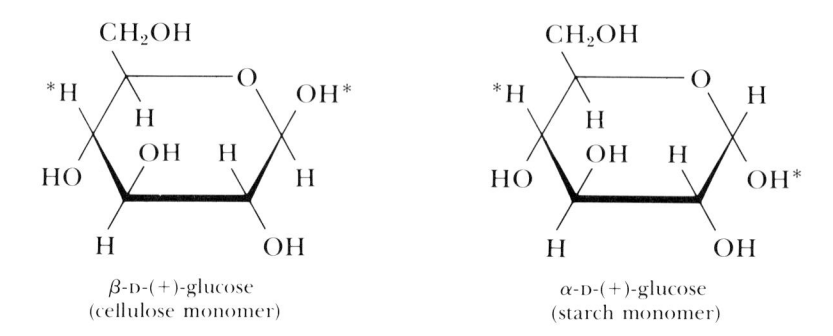

β-D-(+)-glucose
(cellulose monomer)

α-D-(+)-glucose
(starch monomer)

Asterisks indicate the functional groups H and OH which combine to form H_2O. The structures of these polymers are shown in Section 16.14 as examples of polysaccharides.

A synthetic polymer of the above type is made from the monomer dimethylsilanediol

$$HO-\underset{\underset{CH_3}{|}}{\overset{\overset{CH_3}{|}}{Si}}-OH$$

The polymeric structure is

$$H-\left[O-\underset{\underset{CH_3}{|}}{\overset{\overset{CH_3}{|}}{Si}}\right]_n-OH$$

This polymer is of a general type called a **silicone**. Silicones are used as temperature-insensitive lubricants and rubbers (the oils maintain a constant viscosity over a wide range of temperature) and for water-repellent coatings.

Most synthetic step-growth polymers are made from two different monomers, each having a pair of identical functional groups. This type of polymerization has the general form

$$nA-R_1-A + nB-R_2-B \rightarrow A-(R_1-R_2)_n B + (2n-1)AB$$

The polymerization of ethylene glycol and dimethyl terephthalate is an example of

670 GIANT MOLECULES

the above type:

$$n\,HOCH_2CH_2OH + n\,CH_3OC\text{—}\underset{O}{\overset{\|}{C}}\text{—}\bigcirc\text{—}\underset{O}{\overset{\|}{C}}OCH_3 \rightarrow$$

ethylene glycol   dimethyl terephthalate

$$H\text{—}[OCH_2CH_2O\underset{O}{\overset{\|}{C}}\text{—}\bigcirc\text{—}\underset{O}{\overset{\|}{C}}]_n OCH_3 + (2n-1)CH_3OH$$

polyethyleneterephthalate   methanol

This polymer contains repeated ester linkages and, consequently, is a **polyester**. Fibers made from this polymer are marketed as **Dacron**; films are marketed as **Mylar**. Both products are discussed further in Section 24.8.

When hexamethylenediamine is polymerized with adipic acid

$$n\,H_2N(CH_2)_6NH_2 + n\,HO_2C(CH_2)_4CO_2H \rightarrow$$
hexamethylenediamine   adipic acid

$$H\text{—}[HN(CH_2)_6NH\text{—}\underset{O}{\overset{\|}{C}}\text{—}(CH_2)_4\underset{O}{\overset{\|}{C}}\text{—}]_n OH + (2n-1)H_2O$$

nylon 66

the resulting polymer is known as **nylon 66** (the 6's signify that each monomer has six carbon atoms). Nylon contains repeated amide linkages,

$$\text{—}NH\text{—}\underset{O}{\overset{\|}{C}}\text{—}$$

and, hence, is a **polyamide**. Proteins are also polyamides.

Bakelite, discussed in Section 24.2, is also an example of a step-growth polymer.

24.5 COPOLYMERIZATION

Copolymerization is usually used to designate the polymerization of two or more different vinyl-type polymers with each other. For example, when butadiene and styrene are mixed in a 3:1 molar ratio and polymerized, the result is a copolymer whose "average" repeating unit is

$$\text{—}(CH_2\text{—}CH\text{=}CH\text{—}CH_2)_{\overline{3}}\,CH_2\text{—}CH\text{—}CH_2\text{—}CH\text{=}CH\text{—}CH_2\text{—}$$
$$\qquad\qquad\qquad\qquad\qquad\qquad\quad |$$
$$\qquad\qquad\qquad\qquad\qquad\qquad\;\;\bigcirc$$

This material is a synthetic rubber known as Buna S or GR-S. Structurally, this polymer is a cross between polybutadiene and polystyrene and, correspondingly, its properties are to some extent a composite of those of both polymers. Whereas polybutadiene alone produces an inferior synthetic rubber and polystyrene is rigid and brittle, the copolymer GR-S is a very good substitute for natural rubber and can be used in the manufacture of tires. In a similar manner a copolymer of vinyl chloride and vinyl acetate provides a better material for floor tiles than does a polymer of either material alone.

Copolymerization is a convenient way of tailoring the properties of a polymeric material. Many years of experience have taught polymer technologists what monomers to copolymerize and in what proportions in order to obtain copolymers with a wide variety of properties. Copolymerization is also used in those instances where a monomer does not readily polymerize with itself (homopolymerize) but does polymerize with another monomer. For example, maleic anhydride

$$\begin{array}{c} CH=CH \\ /\quad\quad\backslash \\ C\quad\quad C \\ /\!/\;\backslash\;/\;\backslash\!\backslash \\ O\quad\;O\quad\;O \end{array}$$

copolymerizes better with styrene or vinyl chloride than it homopolymerizes.

Copolymers may be classified as to different types based on the manner in which the monomers are linked. Some of the more common variations are illustrated in Fig. 24.1.

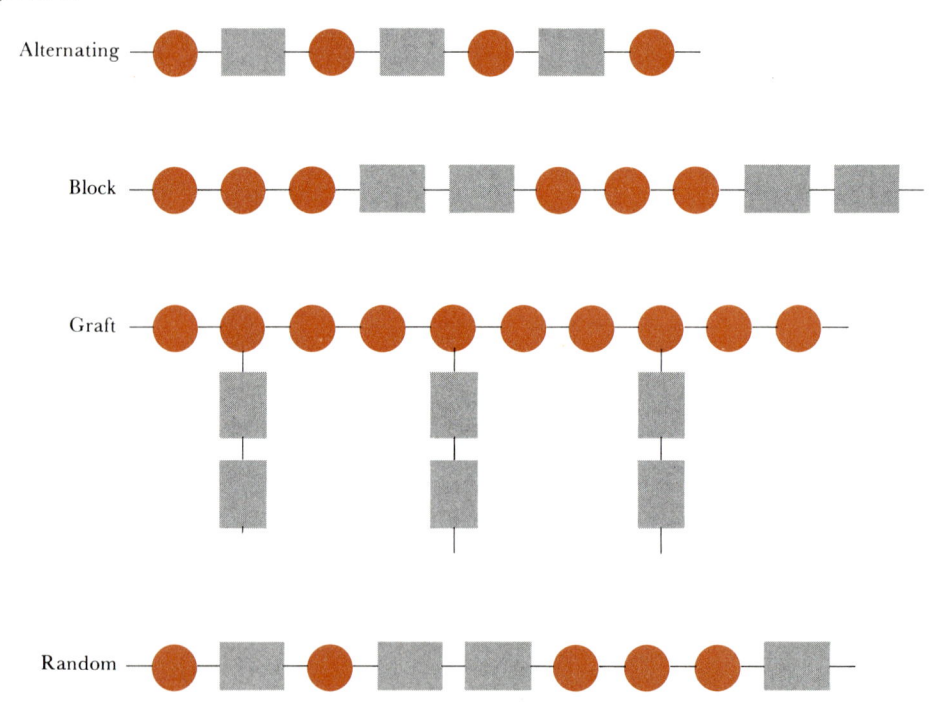

Fig. 24.1. Some common types of copolymers resulting from combination of two different monomers.

24.6 STERIC CONTROL OF POLYMERIZATION REACTIONS

In 1963 Karl Ziegler and Giulio Natta shared the Nobel prize in chemistry for the development of techniques for **stereospecific polymerization**—the synthesis of polymers with controlled geometries. For example, the monomer propylene can be incorporated in polypropylene in two different stereoregular ways. One possibility is

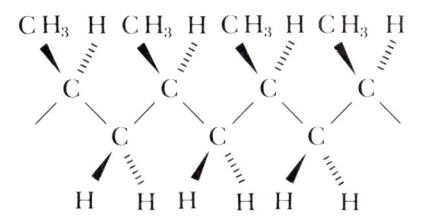

where all methyl groups lie on the same side of the chain. Such a polymer chain is said to be **isotactic** (from the Greek *isos*, equal, and *taxis*, arrangement). A second regular possibility is

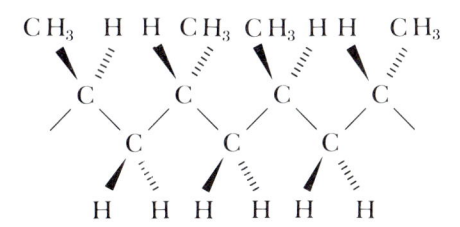

where successive methyl groups are staggered on opposite sides of the chain. Such a polymer chain is said to be **syndiotactic** (from the Greek *syndios*, alternating). A third possibility, but not a regular one, is a random distribution of methyl groups; such a chain is said to be **atactic**.

The Ziegler-Natta stereospecific polymerization method employs an aluminum compound AlR_3 (R is typically alkyl or alkoxy) and a salt MX_n where M is usually

---

Karl Ziegler (1898–1974), a German chemist, shared the 1963 Nobel prize in chemistry with Italy's Giulio Natta for the development of methods of stereoregular polymerizations. In 1943 Ziegler became director of the Kaiser Wilhelm Institute (now the Max Planck Institute) for Coal Research.

Giulio Natta (b. 1903) is a professor of industrial chemistry at the Polytechnic Institute of Milan in Italy. Natta was the first chemist to synthesize methanol; later he invented a method of converting certain alcohols to butadiene, a feat which led to the founding of Italy's synthetic rubber industry.

titanium, zirconium, vanadium, or cobalt and X is chlorine, bromine, or iodine. The mechanism of the polymerization is believed to involve an organometallic complex formed between the components AlR_3 and MX_n. The complex may have the general form

$$\begin{array}{c} \diagup A:\ominus \\ R \\ \diagdown M \oplus \end{array}$$

and this is believed to be bound to a solid surface. The anion $-A:\ominus$ acts as an initiator, and the metal cation $-M\oplus$ guides the monomer so that it becomes involved in the propagation step in a specific geometric way. If propylene is used as a monomer, the first step of the process may be

$$\begin{array}{c}\diagup A:\ominus \\ R \\ \diagdown M\oplus\end{array} \;+\; \begin{array}{c}\delta\oplus \\ CHCH_3 \\ \| \\ CH_2 \\ \delta\ominus\end{array} \;\rightarrow\; \begin{array}{c}\diagup A\text{----}\overset{\ominus}{C}HCH_3 \\ R | \\ \diagdown \underset{\oplus}{M}\text{----}CH_2\end{array}$$

where $\delta\oplus$ and $\delta\ominus$ represent the dipole charges on the vinyl carbon atoms. A second monomer may now slip in between the metal ion and the first monomer as follows:

$$\begin{array}{c}\diagup A\text{----}\overset{\ominus}{C}HCH_3 \\ R | \\ \diagdown \underset{\oplus}{M}\text{----}CH_2\end{array} \;+\; \begin{array}{c}\delta\oplus \\ CHCH_3 \\ \| \\ CH_2 \\ \delta\ominus\end{array} \;\rightarrow\; \begin{array}{c} AH \\ | \\ CHCH_3 \\ | \\ \diagup \ominus CH\text{---}CHCH_3 \\ R | \\ \diagdown \underset{\oplus}{M}\text{----}CH_2\end{array}$$

Continuation of the process leads to a growing chain of the form

$$\begin{array}{c} CH_3 \\ | \\ (CHCH)_n\text{—}AH \\ \diagdown \\ CHCH_3 \\ \diagdown \\ CH\ominus \\ \diagup \\ R \\ \diagdown M\oplus \end{array}$$

674 GIANT MOLECULES

Note that the above is an anion which is constantly propagating; the role of the cation is to see to it that the monomer is inserted with correctly oriented polarity. If the complex is attached to a solid surface, it is likely that the monomer can be added to the growing chain only if the CH_3 group is always oriented in the same way, that is, so as to point away from the solid surface. This would account for the stereoregularity of the product. The mechanism also accounts for the fact that the resulting polymers are highly **linear**; very little, if any, branching appears to occur.

A specific example of a Ziegler-Natta polymerization complex is that formed from triethylaluminum, $Al(CH_2CH_3)_3$, and titanium(IV) chloride, $TiCl_4$. When these two components are mixed in a hydrocarbon solvent, there is formed a flocculent mass, dirty brown to black in color, which is believed to be

$$\begin{array}{c} CH_3 \\ CH_3CH_2 \diagdown \diagup CH Cl \\ \diagup \ominus \oplus \diagup \\ Al Ti \\ CH_3CH_2 \diagup \diagdown Cl \diagdown Cl \end{array}$$

When ethylene gas is bubbled into the flocculent mass, highly linear polyethylene is produced very rapidly. By contrast, the polyethylene produced by free-radical initiation is highly branched.

Isotactic polypropylene is obtained when titanium(III) chloride is used with AlR_3. If the polymerization is carried out at a very low temperature, say 203 K, syndiotactic polypropylene is obtained. In general, the type of product obtained depends on the specific AlR_3/MX_n systems employed, the ratio of AlR_3 to MX_n, and the polymerization temperature.

Pure *cis*-polybutadiene may be prepared from 1,3-butadiene by using titanium(IV) bromide or iodide with AlR_3 in a hydrocarbon solvent. The $AlClR_2/CoCl_2$ system also produces *cis*-polybutadiene. *Cis*-polyisoprene, whose structure duplicates that of natural rubber, is produced with the same systems. *Trans*-polybutadiene results when $Al(CH_2CH_3)_3$ and vanadium(III) chloride are used. Ziegler-Natta polymerization systems are also known for the production of 1,2-polybutadienes.

The development of stereospecific polymerization by Ziegler and Natta has been hailed as representing a "revolution" in polymer chemistry. The method has not only led to the production of new types of polymers, but has also contributed to our understanding of how geometry may be chemically controlled. There is quite a strong resemblance between the Ziegler-Natta processes and the way in which enzymes control the stereospecific synthesis of many biopolymers.

24.7 DETERMINATION OF THE MOLECULAR WEIGHTS OF GIANT MOLECULES

As mentioned before, most polymers, especially synthetic polymers, are mixtures of chains of different length and often—as in the case of branched polyethylene—

different shapes as well. Thus the term **average molecular weight** is used in describing such polymers. In some polymers the spread in mass between the smallest chain and the largest chain may be very, very large.

In practice it is convenient to define two distinctly different types of average molecular weights. To illustrate these, consider a polymer sample containing N_1 chains of mass M_1 each and N_2 chains of mass M_2 each. The usual average, called the **number-average molecular weight**, is defined in the standard way as

$$\bar{M}_N = \frac{N_1 M_1 + N_2 M_2}{N_1 + N_2}$$

A totally different, less familiar average is the **weight-average molecular weight**, defined by

$$\bar{M}_w = \frac{m_1 M_1 + m_2 M_2}{m_1 + m_2}$$

where $m_1 = N_1 M_1$ (the total mass of chains with mass M_1 each) and $m_2 = N_2 M_2$ (the total mass of chains with mass M_2 each).

The following example illustrates the calculation of the two types of average molecular weights.

EXAMPLE 24.1 A polymer contains equal numbers of two different chains, one with a mass of 90 amu and the other with a mass of 9910 amu. Calculate the average molecular weights $\bar{M}_N$ and $\bar{M}_w$.

SOLUTION Let $M_1 = 90$ amu and $M_2 = 9910$ amu. Of every 100 chains, 50 have masses of 90 amu each and 50 have masses of 9910 each. Thus $N_1 = N_2 = 50$ and the number-average molecular weight is

$$\bar{M}_N = \frac{(50)(90 \text{ amu}) + (50)(9910 \text{ amu})}{100} = 5000 \text{ amu}$$

Of every 100 amu of polymer, the lighter chain contributes 0.90 amu and the heavier chain contributes 99.10 amu. Thus $m_1 = 0.90$ amu and $m_2 = 99.10$ amu, and the weight-average molecular weight is

$$\bar{M}_w = \frac{(0.90 \text{ amu})(90 \text{ amu}) + (99.10 \text{ amu})(9910 \text{ amu})}{100 \text{ amu}} = 9822 \text{ amu}$$

$\bar{M}_w$ is always greater than or equal to $\bar{M}_N$. The ratio of the two types of averages, $\bar{M}_w/\bar{M}_N$, is called the **polydispersity** of the sample. The polydispersity is a measure of the heterogeneity of the polymer sample; when $\bar{M}_w/\bar{M}_N = 1$ the sample is said to be **monodisperse**; that is, it contains only one size of chain. The sample in Example 24.1 has a polydispersity of 9822/5000 or about 1.96. Biopolymers such as hemoglobin and the enzyme lysozyme are monodisperse; synthetic polymers generally have high values of the polydispersity, indicating a large amount of heterogeneity in chain size.

An experimental way of measuring the number-average molecular weight is based on the **osmotic pressure** (see Section 10.8). If the concentration of a polymer solution is expressed in terms of the weight of polymer per volume of solution (symbolized by c), then an approximate relationship for the osmotic pressure of the solution is given by

$$\pi = \frac{cRT}{\bar{M}_N} + Ac^2$$

where A is a constant. This equation may be rearranged to

$$\frac{\pi}{c} = \frac{RT}{\bar{M}_N} + Ac$$

In practice this equation is used to obtain fairly accurate values of $\bar{M}_N$ by the following procedure: Values of π are measured at several c values and π/c is plotted against c. Extrapolation of the data to $c = 0$ leads to $RT/\bar{M}_N$ as the point where the line crosses the $c = 0$ axis. $\bar{M}_N$ can then be calculated from this intercept value. Due to the difficulty of measuring small osmotic pressures reliably, this method is generally limited to molecular weights of one million amu or less.

EXAMPLE 24.2 The following osmotic pressures were obtained for a cyclohexane solution of polyisobutylene at 25 °C. Calculate the number-average molecular weight of the polyisobutylene sample.

c (g·L^{-1})	π (N·m^{-2}) × 10^{-2}
20	16.09
15	9.92
10	5.29
5	2.03

SOLUTION The above data are used to calculate π/c for each value for c. The results are

c (g·L^{-1})	π/c (N·m^{-2}·L·g^{-1}) × 10^{-2}
20	0.80
15	0.66
10	0.53
5	0.41

$(\pi/c) \times 10^{-2}$ is plotted against c as shown in Fig. 24.2 and the $c = 0$ intercept estimated as 0.3×10^2 N·m^{-2}·L·g^{-1}. The number-average molecular weight is then calculated as

$$\bar{M}_N = \frac{RT}{\text{intercept}} = \frac{(8.314 \text{ J·K}^{-1}\text{·mol}^{-1})(298 \text{ K})}{(0.3 \times 10^2 \text{ N·m}^{-2}\text{·L·g}^{-1})(10^{-3} \text{ m}^3\text{·L}^{-1})}$$

$$= 8 \times 10^4 \text{ g·mol}^{-1} \quad \text{or} \quad 8 \times 10^4 \text{ amu}$$

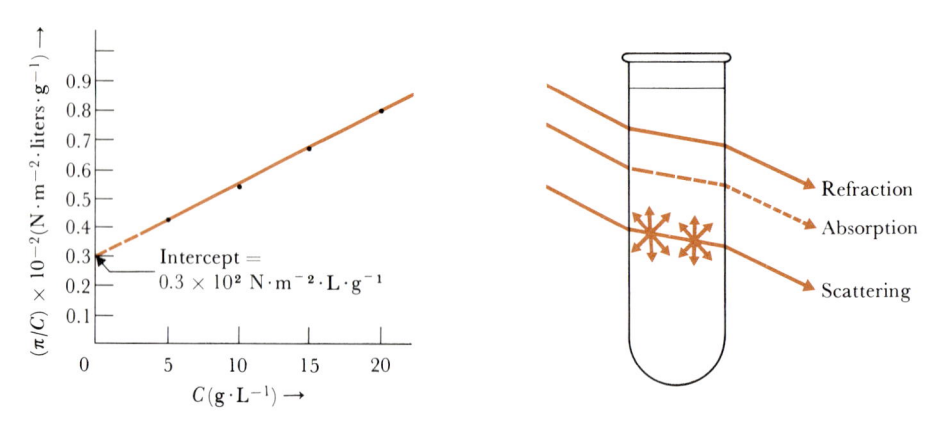

Fig. 24.2 (left). Osmotic pressure method for the determination of the number-average molecular weight of polyisobutylene in cyclohexane solution at 25 °C. The molecular-weight calculation is illustrated in Example 24.2.

Fig. 24.3 (right). Refraction, absorption, and scattering of light by a polymer solution. The scattering occurs along the entire path length and in all directions.

Molecular weights above about 10^6 amu are best measured by a method known as **light scattering**. This method measures the weight-average molecular weight $\overline{M}_w$. When a beam of light passes through a polymer solution, part of the light is reflected, some is refracted, some is absorbed and converted to heat, and some is **scattered**, that is, absorbed and reemitted in random directions (see Fig. 24.3). The latter occurs because each molecule behaves as a small antenna capable of receiving radiation and rebroadcasting it.* The amount of the scattering (which is a measure of the turbidity, or cloudiness, of the solution) is directly proportional to the number of polymer molecules in the solution times the square of their **polarizability**, a measure of the ease of distortion of the molecule's electron distribution. At the same time, the amount by which the transmitted light is refracted (bent) is directly proportional to the number of polymer molecules in the solution times the polarizability. Thus a measure of the turbidity and of the refraction makes it possible to calculate the number of molecules in the solution. Knowing the total mass of the polymer then leads to the molecular weight.

Since the polarizability of a molecule is proportional to its size, the increase in turbidity for a molecule twice the size of another is double that of the smaller molecule. For example, in a sample with equal numbers of two different-size molecules with masses 10,000 amu and 40,000 amu, each larger molecule would scatter four times as much light as would each smaller molecule. The solution would behave as if 20% of the molecules had masses of 10,000 amu and 80% had masses of 40,000 amu. This would lead to an average molecular weight of (0.20)(10,000) + (0.80)(40,000) or 34,000 amu. This is, of course, the weight-average molecular weight of the sample; the number-average molecular weight is 25,000 amu. Thus the polydispersity of the sample is 34,000/25,000 or 1.36.

---

* The intensity of scattered radiation is inversely proportional to the fourth power of the wavelength of the absorbed radiation. The blue color of the sky results when the atmosphere scatters short-wavelength radiation (blue end of the spectrum) more than long-wavelength radiation (red end of the spectrum).

678 GIANT MOLECULES

24.8 THE PHYSICAL PROPERTIES OF SYNTHETIC POLYMERS

The physical properties of a polymer at a given temperature depend on such factors as the chemical nature and bulkiness of side groups, the number and kind of cross-links, the nature of the repeating units, and mechanical working of the bulk material.

Natural rubber, largely *cis*-polyisoprene, is a rather sticky, tacky material at room temperature unless the chains are artificially cross-linked. This fact was discovered by Charles Goodyear in 1839 when he accidentally spilled a mixture of raw rubber (latex) and sulfur on a hot stove. To Goodyear's surprise the "cooked" mixture was no longer sticky but was elastic when stretched and twisted; it was also more thermally and chemically stable than the original latex. It is now known that the action of sulfur is to form cross-links between the polyisoprene chains and thereby to increase the structural rigidity of the polymer. This process, called **vulcanization**, probably connects double bonds in adjacent polyisoprene chains as follows:

$$\begin{array}{c} -\overset{|}{C}=\overset{|}{C}- \\ -\overset{|}{C}=\overset{|}{C}- \end{array} + 4\,S \rightarrow \begin{array}{c} -\overset{|}{C}-\overset{|}{C}- \\ \overset{|}{S}\quad \overset{|}{S} \\ \overset{|}{S}\quad \overset{|}{S} \\ -\overset{|}{C}-\overset{|}{C}- \end{array}$$

The less sulfur used, the softer the rubber. Hard rubber may contain up to 35% sulfur by weight, with very few double bonds remaining.

Polyethyleneterephthalate (Section 24.4) may be used as an example of how mechanical working of a bulk polymer can affect its physical properties. When the molten, viscous polymer is pulled into long threads which are allowed to cool, strong fibers useful for the weaving of fabrics are produced. This substance is known by the trade name of Dacron in the U.S. and Terylene in Great Britain. When the same polymer is rolled into sheets, a tough, transparent film known as Mylar is produced.

A simplified explanation of what happens in the two processes is as follows: When a molten polymer is pulled out like taffy, there is a tendency for some of the adjacent chains to align themselves in a parallel fashion and thus create regions of crystallinity. This process, called **drawing**, is said to **orient** the polymer. An increase in the degree of crystallinity leads to a stronger polymer and produces very strong fibers. This is what happens when Dacron fibers are made. Nylon fibers are produced in a similar fashion (see Fig. 24.4). When the molten polymer is rolled into sheets, orientation occurs in two dimensions and a very tough film is produced by the induced crystallization. This is what happens when Mylar is produced.

Charles Goodyear (1800–1860) was the American inventor who accidentally discovered the process of vulcanization. Goodyear received the Medal of the Legion of Honor from Napoleon III but never succeeded in obtaining financial success with his invention. He died heavily in debt and a pauper.

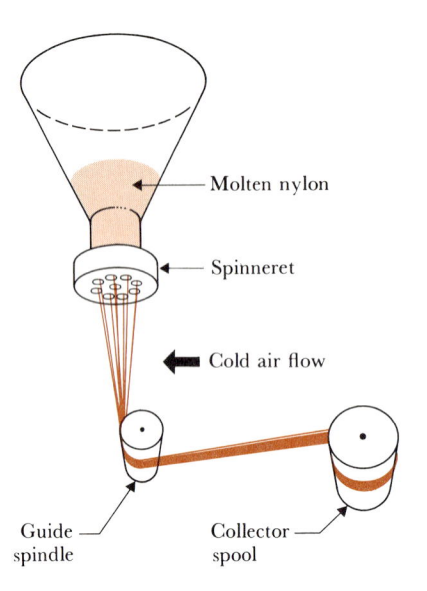

Fig. 24.4. Schematic illustration of the production of nylon fibers.

On the basis of the way long chains are oriented with respect to each other, polymers may be classified into two general types, **completely amorphous** polymers and **partly crystalline** polymers. The chains in an amorphous polymer may be likened to the randomly oriented strands in a batch of cooked spaghetti (see Fig. 24.5). At sufficiently low temperatures such a polymer exists in what is termed the **glassy state** and can be viewed as a supercooled liquid. A polymer in the glassy state has a relatively high degree of rigidity. Upon heating the polymer gradually loses rigidity, becoming leathery or rubbery. This occurs because the molecules begin to loosen up and shorter chains acquire some freedom of movement. Eventually, at a high enough temperature, provided the polymer does not first decompose, it loses the last traces of rigidity and begins to behave like a very viscous liquid. In this state, the **liquid-flow state**, all the chains are able to slide past each other. Partly crystalline polymers may be compared to bundles of uncooked spaghetti. The crystalline regions consist of short sections of chains aligned in a parallel fashion and separated by amorphous regions (see Fig. 24.6). The actual degree of crystallinity may be as high as 95% in linear polyethylene. The crystalline regions can be detected by X-rays and are even visible to the naked eye under illumination with polarized light.

To more clearly show the difference in behaviors of completely amorphous and partly crystalline polymers, it is useful to examine how some experimentally measurable property—for example, the **modulus of elasticity**—varies with temperature. The modulus of elasticity, a measure of the strength or toughness of a polymer, is defined as the tensile force (force tending to stretch) divided by the fractional elongation it produces. This quantity varies with temperature in a way which accurately reflects some of the internal structure of the polymer. Figure 24.7 shows how the modulus of elasticity varies with temperature for a typical amorphous polymer. Note that there is a region where the modulus of elasticity drops rather suddenly, levels off briefly, and then drops abruptly again. The region where the first drop occurs is called the **glass-transition region**. As its name suggests, this region represents the

Fig. 24.5. Schematic illustration of a completely amorphous polymer. The randomly oriented strands resemble a batch of cooked spaghetti.

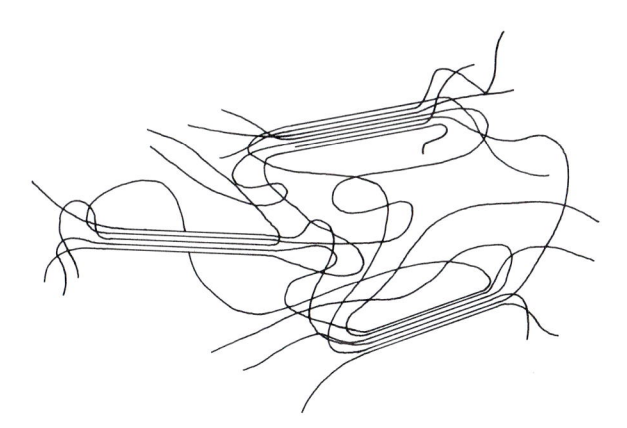

Fig. 24.6. Schematic illustration of a partly crystalline polymer. The crystalline regions resemble bundles of uncooked spaghetti surrounded by randomly oriented cooked strands.

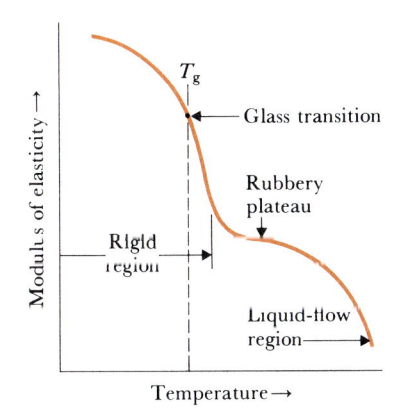

Fig. 24.7. The modulus of elasticity of a typical amorphous polymer as a function of temperature.

transition from the rigid glassy state to the rubbery state. The midpoint of the glass-transition region is called the **glass-transition temperature**, T_g. Every polymer has its characteristic glass-transition temperature. Polymers with glass-transition temperatures below room temperature tend to be soft and flexible while those with glass-transition temperatures above room temperature tend to be hard and brittle. The region beyond T_g is called the **rubbery plateau**. Within this region the polymer is relatively soft and may be stretched like rubber. The rubbery state is analogous in many respects to the gaseous state. For example, the stretching of rubber is similar to the compression of a gas in that both are processes which involve a decrease in entropy (increase in order) and both are exothermic.* As the temperature is increased beyond the rubbery region the polymer abruptly loses all of its structural strength as it enters the **liquid-flow region**.

Substances which have glass-transition temperatures well below room temperature are called **elastomers** and are used as rubbers. Many elastomers have some degree of cross-linking. When a highly cross-linked polymer is heated, it generally begins to degrade due to breaking of cross-links before going into the liquid-flow state.

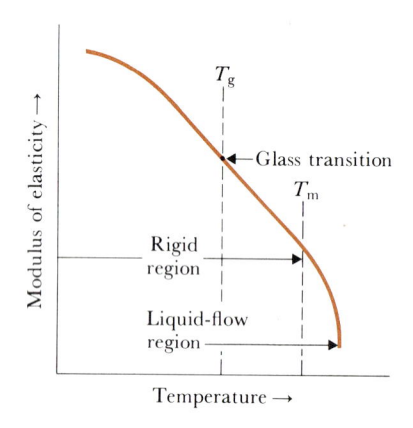

Fig. 24.8. The modulus of elasticity of a typical partly crystalline polymer as a function of temperature. Compare with Fig. 24.7 for an amorphous polymer, and note the absence of a rubbery plateau and the persistence of relative rigidity up to the liquid-flow region.

Figure 24.8 illustrates how the modulus of elasticity of a partly crystalline polymer varies with temperature. The chief difference between the crystalline polymer and the amorphous polymer is that the former does not pass through a rubbery state. The glass-transition temperature is characterized by a decrease in brittleness—as in amorphous polymers—but rigidity is maintained until the crystalline regions melt and the liquid-flow state is entered. Partly crystalline polymers enter the liquid-flow state at a fairly definite temperature called the **crystalline melting point**, T_m. Table 24.3 lists the glass-transition temperatures and crystalline melting points of some common polymers.

The glass-transition temperature of a polymer can be lowered by mixing it with a substance called a **plasticizer**. For example, the polymer polyvinylchloride ($T_g = 70\ °C$) is too brittle at room temperature for many purposes (it could not be

---

* The exothermic nature of stretching a rubber band is easily demonstrated. Touch an unstretched rubber band to your lips, then stretch it and again touch it to your lips. The lips, which are very sensitive heat sensors, clearly detect that heat is produced by the stretching process. An elongation of three to four times can produce a rise in temperature of 10 °C.

TABLE 24.3 GLASS-TRANSITION TEMPERATURES AND CRYSTALLINE MELTING POINTS OF SOME COMMON POLYMERS

POLYMER	T_g (°C)	T_m (°C)
Nylon 66	47	235
Polystyrene (isotactic)	80	220
Polyethylene (linear)	30	110
Polyvinyl chloride	70	140

used for shower curtains, for instance). However, the addition of substances such as di-2-ethylhexylphthalate, known commercially as DOP, will produce flexible films of polyvinylchloride.

The glass-transition temperature can also be controlled by copolymerization. When random copolymers are made from two different monomers, the copolymer T_g usually lies between the T_g's of the homopolymers.

24.9 THE STRUCTURE OF PROTEINS

Proteins are biopolymers of L-amino acids (see Section 16.13). They are polyamides with some similarity to nylon (see Section 24.4). About twenty different amino acids are found in proteins; these are listed in Table 24.4 along with their linkage formulas and abbreviated names. The formation of a protein is a step-growth process whose basic step may be written

$$H_2N-CH(R_1)-C(=O)-OH + H-N(H)-CH(R_2)-CO_2H \rightarrow H_2N-CH(R_1)-C(=O)-NH-CH(R_2)-CO_2H + H_2O$$

The two amino acids are bonded through the peptide (or amide) linkage

$$\begin{array}{c} H \\ | \\ N \\ / \quad \backslash \\ C \\ \| \\ O \end{array}$$

The four atoms in the peptide linkage plus the carbon atoms to which the link is joined all lie in a plane. Polymers of amino acids are generally called **polypeptides**. Proteins are naturally occurring polypeptides, generally with molecular weights above 5000 amu.

It is customary to symbolize polypeptides by links of the amino-acid abbreviated names. For example, the symbol Arg–Gly–Ala–Gly represents a tetrapeptide of the amino acids arginine, glycine, alanine, and glycine—linked in that order. It is under-

TABLE 24.4 NAMES, STRUCTURAL FORMULAS, AND ABBREVIATIONS OF TWENTY L-AMINO ACIDS FOUND IN NATURE

$$\begin{array}{c} R-CH-C-OH \\ | \quad \| \\ NH_2 \quad O \end{array}$$

NAME	ABBREVIATION	R[c]
L-(+)-Alanine	Ala	—CH$_3$
L-(+)-Arginine[a]	Arg	—(CH$_2$)$_3$—NH—C(=NH)—NH$_2$
L-(−)-Asparagine	AspN	—CH$_2$—C(=O)—NH$_2$
L-(+)-Aspartic acid	Asp	—CH$_2$—C(=O)—OH
L-(−)-Cysteine	Cys	—CH$_2$ SH
L-(+)-Glutamic acid	Glu	—(CH$_2$)$_2$—C(=O)—OH
L-(+)-Glutamine	GluN	—(CH$_2$)$_2$—C(=O)—NH$_2$
L-Glycine	Gly	—H
L-(−)-Histidine[a]	His	—CH$_2$—(imidazole)
L-(+)-Isoleucine[a]	Ileu	—CH(CH$_3$)(CH$_2$CH$_3$)
L-(−)-Leucine[a]	Leu	—CH$_2$CH(CH$_3$)$_2$

stood that the left-hand amino acid, arginine in the above, has a free —NH$_2$ group and that the right-hand amino acid, glycine in the above, has a free —CO$_2$H group.

The structural formula of this tetrapeptide is

$$\begin{array}{c} H_2N-CH-C-NH-CH-C-NH-CH-C-NH-CH-C-OH \\ \quad | \quad \| \quad\quad | \quad \| \quad\quad | \quad \| \quad\quad | \quad \| \\ \quad R_1 \quad O \quad\quad R_2 \quad O \quad\quad R_3 \quad O \quad\quad R_2 \quad O \end{array}$$

684 GIANT MOLECULES

NAME	ABBREVIATION	R[c]
L-(+)-Lysine[a]	Lys	—(CH$_2$)$_4$NH$_2$
L-(−)-Methionine[a]	Met	—(CH$_2$)$_2$SCH$_3$
L-(−)-Phenylalanine[a]	Phe	—CH$_2$—C$_6$H$_5$
L-(−)-Proline[b]	Pro	—CH$_2$CH$_2$CH$_2$—
L-(−)-Serine	Ser	—CH$_2$OH
L-(−)-Threonine[a]	Thr	—CH(OH)CH$_3$
L-(−)-Tryptophane[a]	Trp or Try	—CH$_2$—(indole)
L-(−)-Tyrosine	Tyr	—CH$_2$—C$_6$H$_4$—OH
L-(+)-Valine[a]	Val	—CH(CH$_3$)$_2$

[a] One of the so-called essential amino acids, which cannot be synthesized in the human body and must be supplied in the diet.

[b] The structure of proline deviates from the general pattern and is

$$\text{HN—CH—CO}_2\text{H with CH}_2\text{—CH}_2\text{—CH}_2 \text{ ring}$$

The presence of this amino acid in an α-helix causes a sharp bend in the helix.

[c] The R groups of amino acids play vital roles in determining the chemical and physical properties of proteins. For example, they determine polarity, water solubility, sizes and shapes of chains, and what inter- and intramolecular links are formed.

where R$_1$, R$_2$, and R$_3$ represent the R groups of Arg, Gly, and Ala, respectively, as indicated in Table 24.4.

Two general types of proteins are the **fibrous proteins** found in hair, wool, hooves, connective tissues, and silks and **globular proteins** such as hemoglobin and myoglobin (see Section 22.3). Fibrous proteins are made up of three basic types of structures: the **α-helix**, the **β-sheet**, and the **triple helix**; globular proteins consist largely of α-helixes. Let us consider each of these three structural types separately.

In 1951 Pauling and Corey* proposed that some proteins existed as right-handed helical chains like those shown in Fig. 24.9. This structure is now called the α-helix. The rigidity of the helix is maintained by hydrogen bonds between amino nitrogens and carbonyl oxygens.† Although the general shape of the helix is determined by intramolecular hydrogen bonding, the overall three-dimensional shape of many proteins depends strongly on —S—S— linkages between cysteine amino acids. Such linkages can occur both intramolecularly (between cysteine residues in the same α-helix) and intermolecularly (between cysteine residues in different α-helixes). Both

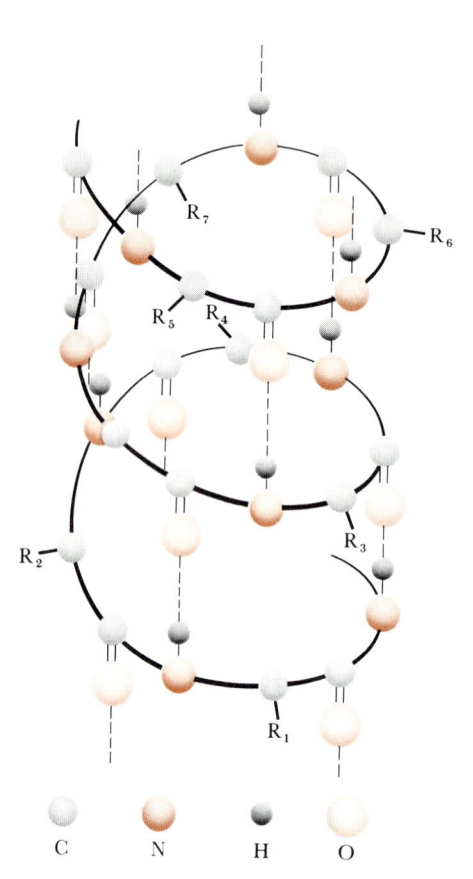

Fig. 24.9. The α-helical structure of a protein. Hydrogen bonds between loops are shown by dashed lines. The pitch is about 5.4 Å with 3.6 to 3.7 amino acids per turn.

---

* A biographical sketch of Linus Pauling is given in Chapter 4. Robert B. Corey (b. 1897) is a specialist in X-ray crystallography and electron diffraction at the California Institute of Technology.

† Only α-amino acids can produce the geometries characteristic of proteins. For example, the planar peptide region is not formed with β-amino acids. Consequently, β-amino acids cannot produce the α-helix, β-sheet, or other typical protein structures.

types of disulfide linkages are illustrated in the following:

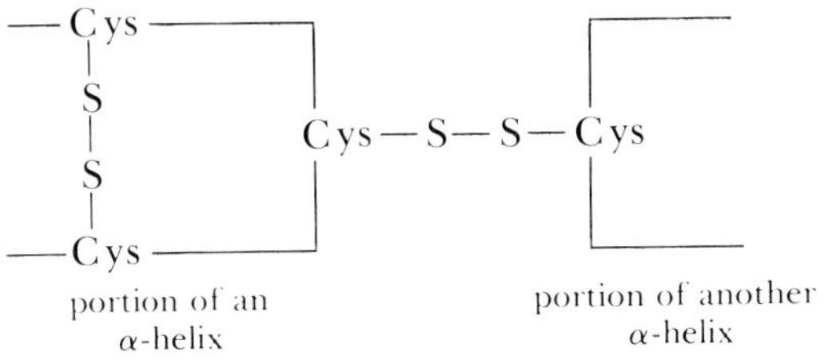

```
—Cys——————┐
   |       |
   S       |    Cys—S—S—Cys
   |       |
   S       |
   |       |
—Cys——————┘
portion of an      portion of another
  α-helix              α-helix
```

Flexible proteins such as skin have a lower sulfur content (fewer —S—S— linkages) than do stiff proteins such as nails and hair. In wool and hair, three α-helixes are twisted around each other in a left-handed sense and eleven of these (two in the middle and nine grouped around them) form a bundle known as a **microfibril**. The fact that wool and hair are very flexible and extensible indicates that the amount of interchain hydrogen bonding is very small.

The proteins called **collagens**, which are used in ligaments and connective tissues, have the triple helix structure. The triple helix consists of three α-helixes twisted around each other in a right-handed sense—just opposite to that in wool and hair. An abundance of interchain hydrogen bonding makes collagens rather rigid and incapable of stretching very much.

The structure of silk is that of the β-sheet. As shown in Fig. 24.10, a typical silk consists of long sheets of peptide chains cross-linked with interchain hydrogen bonds.

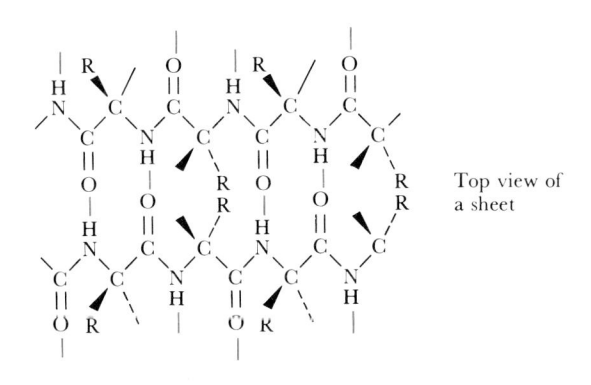

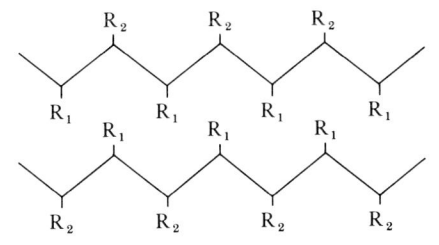

Top view of a sheet

Fig. 24.10 The β-sheet structure of silklike proteins. The sheet consists of peptide chains cross-linked with hydrogen bonds. Note that alternate peptide chains run in opposite directions. The sheets themselves are pleated and stacked in layers held together by weak van der Waals forces.

Side view of two parallel sheets. Typically R_1 = H (from glycine) and R_2 = CH_3 (from alanine).

Adjacent peptide chains run in opposite directions. The sheets themselves are held together by relatively weak van der Waals forces. Thus each of the three dimensions of silk is characterized by a different type of bonding: covalent bonding along each peptide chain, hydrogen bonding between peptide chains, and van der Waals bonding between essentially parallel sheets. These features make silk a strong substance with little stretch but great flexibility. The peptide chains of silk consist almost entirely of alternating glycine and alanine, with serine replacing approximately every third alanine. Thus the basic repeating unit appears to be –Gly–Ser–Gly–Ala–Gly–Ala–. Without the occasional serine and its bulky size, silk would be a highly crystalline and stiff polymer. The bulky serine reduces the crystallinity. This is nature's way of using copolymerization to control properties.

Many natural proteins are found associated with water. The role of the water is analogous to that of a plasticizer; that is, the water lowers the glass-transition temperature of the protein and allows it to maintain a high degree of flexibility. When a protein is dehydrated, loss of water results in an irreversible degradation of the structure.

Globular proteins such as myoglobin and hemoglobin contain several sections of α-helixes. Some of the α-helixes are connected by sharp bends and fold back onto each other, and others are connected by nonhelical sections.

24.10 DETERMINATION OF PROTEIN STRUCTURES

The detailed three-dimensional structure of a protein can be determined only by rather long, exhaustive studies of many kinds. The various steps include determining the amino acid content, the amino acid sequence, the structure of individual chains, the number of chains and how they are joined, and the precise manner in which all chains are folded.

When a protein is treated with 6-M HCl and the solution allowed to "digest" for an extended period of time, a mixture of the amino acids which constitute the protein is eventually obtained. Individual amino acids may be separated by various techniques and identified by appropriate tests. One of the most useful general tests for the presence of an amino acid is based on the **ninhydrin reaction**,

The ionic substance (occurring as an ammonium salt) is a blue-colored complex known as **Ruhemann's purple**. All α-amino acids, protein residues containing α-NH$_2$, and amines containing —CHNH$_2$ react in this way. Proline and hydroxyproline, which have no α-NH$_2$, give a different product and color but also produce CO$_2$. In all cases the reaction goes virtually to completion so that the amount of CO$_2$ produced can be used to estimate the amount of amino acid present. Very small amounts of amino acids may be detected by the ninhydrin reaction; a fingerprint on the page of a book may be revealed by spraying with ninhydrin solution and heating.

The order in which the amino acids are linked in the protein may be determined by a variety of techniques. One of the most useful employs enzymes which cleave only certain types of peptide bonds and those which remove amino acids from a peptide chain one at a time beginning at one end. For example, the enzyme **pepsin** hydrolyzes most rapidly those peptide bonds in which the amino group comes from an amino acid containing an aromatic R group. By contrast, the enzyme **chymotrypsin** hydrolyzes most rapidly those peptide bonds in which the carboxyl group comes from an amino acid containing an aromatic R group. Thus, in the polypeptide chain segment

$$-HN-CH-C+NH-CH-C+NH-CH-C-$$

(with phenyl, CH$_3$, O, phenyl, O substituents)

chymotrypsin would tend to break the chain at the point indicated at the left and pepsin would tend to break the chain at the point indicated at the right. A zinc-containing enzyme, **carboxypeptidase** (found in pancreatic juice), cleaves only that end of a peptide chain having a —CO$_2$H group. Similarly, the enzyme **aminopeptidase** cleaves amino acids from the —NH$_2$ end of a polypeptide. These latter two enzymes are extremely useful in determining amino acid sequences.

In addition to amino acid analyses, chemists make use of X-ray diffraction techniques to determine detailed geometries of proteins. In some cases the structure of a protein is proved by synthesizing it in the laboratory from amino acids, using reaction steps whose detailed pathways and geometrical consequences are known. The synthesis of a protein whose properties match those of a natural protein provides invaluable insight into the structure of the latter.

24.11 LABORATORY SYNTHESIS OF PROTEINS

It is not possible to synthesize proteins by simply mixing amino acids and allowing them to react. For one thing, such a procedure won't link the amino acids in the correct order. Furthermore, the amino acids aren't reactive enough to form peptide linkages spontaneously. Consequently, laboratory methods must provide some means of chemically activating the amino acids so that peptide linkages form, and they must also provide a way of controlling the order in which the amino acids link up. Nature does all of these things by means of enzymes (Section 20.9) and the chemical code recorded in DNA (see Section 25.2).

One of the most commonly used methods of synthesizing proteins in the laboratory is the **solid-support method** developed by R. B. Merrifield,[*] in which long polypeptide chains are grown while anchored to the surface of beads made of a copolymer of styrene and divinylbenzene. The bead surface may be depicted as a plane with phenyl groups sticking out of it:

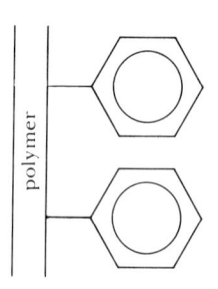

The first step in the synthesis is to activate the attached phenyl groups so that amino acids may be anchored to them. One approach utilizes a Friedel-Crafts chloromethylation reaction (see Section 17.3):

$$\text{polymer-C}_6\text{H}_5 + \text{ClCH}_2\text{OCH}_3 \xrightarrow{\text{SnCl}_4} \text{polymer-C}_6\text{H}_4-\text{CH}_2\text{Cl} + \text{CH}_3\text{OH}$$

chloromethyl ether

The next step is to anchor an amino acid to the chloromethyl group by coupling to the acid's $-\text{CO}_2\text{H}$ group. Before doing this one must block or protect the amino group of the amino acid so that it does not couple to the chloromethyl group of the solid support. The protected amino acid may be symbolized by

$$\text{HOC}\underset{\underset{O}{\parallel}}{\text{C}}\text{HN}\underset{\boxed{P}}{\overset{R_1 \quad H}{\diagup}}$$

where $\boxed{P}$ represents the group which protects $-\text{NH}_2$ by blocking it. One specific example of an effective amino-protecting group is the tertiary butoxy carbonyl group

$$-\underset{\underset{O}{\parallel}}{\text{C}}-\text{O}-\underset{\underset{\text{CH}_3}{|}}{\overset{\overset{\text{CH}_3}{|}}{\text{C}}}-\text{CH}_3$$

---

[*] R. Bruce Merrifield (b. 1921) is professor of biochemistry at Rockefeller University in New York City.

Addition of the amino-protected amino acid to the activated solid support produces a benzyl ester linkage

$$\text{—}\bigcirc\text{—CH}_2\text{O}\underset{\text{O}}{\overset{\|}{\text{C}}}\underset{R_1}{\overset{|}{\text{C}}}\text{HN}\diagdown_{\boxed{P}}^{H}$$

benzyl ester linkage

Next, the amino-protecting group is removed (deprotection) by using a reagent which breaks the N—$\boxed{P}$ bond but does not affect the benzyl ester bond. Hydrogen chloride in absolutely dry nonaqueous solution functions in this way. Thus there is obtained

$$\text{—}\bigcirc\text{—CH}_2\text{O}\underset{\text{O}}{\overset{\|}{\text{C}}}\underset{R_1}{\overset{|}{\text{C}}}\text{HNH}_2$$

The amino acid added next must have an activated carboxylic acid group and a protected amino group. This amino acid may be symbolized by

$$\boxed{A}\text{—O}\underset{\text{O}}{\overset{\|}{\text{C}}}\underset{R_2}{\overset{|}{\text{C}}}\text{HN}\diagdown_{\boxed{P}}^{H}$$

where $\boxed{A}$ represents a CO_2H-activating group. A specific example of such an activating group is the 4-nitrophenyl group

$$O_2N\text{—}\bigcirc\text{—}$$

Addition of the activated-protected amino acid to the solid support leads to the dipeptide

$$\text{—}\bigcirc\text{—CH}_2\text{O}\underset{\text{O}}{\overset{\|}{\text{C}}}\underset{R_1}{\overset{|}{\text{C}}}\text{HNH}\underset{\text{O}}{\overset{\|}{\text{C}}}\underset{R_2}{\overset{|}{\text{C}}}\text{HN}\diagdown_{\boxed{P}}^{H} + O_2N\text{—}\bigcirc\text{—OH}$$

The amino-protecting group is now removed and another activated-protected amino acid is ready to be added. Continuation of the process leads to polypetides of any desired length. The finished polypeptide is removed from the support by treatment with a mixture of HBr and trifluoroacetic acid. In practice, the entire process may be automated so that reagents are added and reactions timed by mechanical and electronic means. Merrifield's apparatus has been dubbed the "protein-making" machine.

Several naturally occurring proteins have been synthesized by the solid-support method. One of these is the hormone **insulin**, which plays a role in the proper metabolism of glucose in the body. Insulin is made up of two polypeptide chains joined by two —S—S— linkages. One of the other polypeptide chains contains an intramolecular —S—S— linkage. The molecule has a total of 17 different amino acids. The synthesis was achieved by making the two polypeptide chains separately and then linking them under the proper conditions. In 1969 Merrifield used the solid-support method to carry out the first synthesis of an enzyme, **ribonuclease**. This enzyme plays a role in the biological synthesis of proteins. Ribonuclease contains 124 amino acids; the synthesis involved 369 chemical reactions and 11,931 steps—all carried out by the automated "protein-making" machine.* An independent synthesis of ribonuclease was accomplished almost simultaneously by Denkewalter and Hirschmann of the Merck, Sharpe, and Dohme Research Laboratories using a somewhat different method.

SUMMARY

Giant molecules are molecules of large numbers of atoms bound by ordinary chemical valence forces. By virtue of their large size, giant molecules can attain a complexity of structure which enables them to exhibit properties much more sophisticated and subtle than those of small molecules.

Many of the synthetic and natural giant molecules are polymers, that is, molecules composed of large numbers of smaller molecules bonded to each other. Two general methods of synthesizing polymers are by chain reactions and by step-growth reactions. Sometimes two or more different monomers are combined or copolymerized. By suitable choice of monomers and the order in which they are joined, some control of polymer properties may be achieved.

Ziegler-Natta polymerization systems allow the synthesis of stereoregular polymers, for example, linear polyethylene and isotactic and syndiotactic polypropylene. Natural rubber, cis-polyisoprene, may be duplicated in this way.

Two different types of average molecular weights are used to characterize polymers: number-average and weight-average molecular weights. The ratio of the two types of averages, the polydispersity, is a measure of the heterogeneity of the polymer.

A typical polymer is composed of amorphous regions, where polymer chains are randomly coiled as in cooked spaghetti, and crystalline regions, where the chains are aligned in parallel. The physical properties of a polymer depend on the ratio of its amorphous to its crystalline regions.

Proteins are biopolymers of amino acids joined in long polypeptide chains. Three im-

---

* The synthetic proteins produced so far have only partial biological activity, indicating that not all of the synthesized molecules are precise duplicates of the natural ones.

portant structural forms of proteins are the α-helix, the β-sheet, and the triple helix. Hydrogen bonding and disulfide linkages play important roles in determining and maintaining the three-dimensional shapes of proteins.

Techniques exist whereby the amino acid composition of proteins may be determined; in many cases even the sequence in which the amino acids are linked can be deduced. Proteins such as the hormone insulin and the enzyme ribonuclease have been synthesized in the laboratory.

LEARNING GOALS

1. Be able to give a fairly detailed discussion of just how giant molecules differ from small molecules.

2. Know the difference between a thermosetting and a thermoplastic polymer.

3. Be able to describe chain polymerization in terms of its four basic types of reactions.

4. Be able to illustrate the four types of polymers that can arise from 1,3-butadiene.

5. Be able to write the equations for the two types of step-growth polymerizations.

6. Know what is meant by copolymerization and why it is important.

7. Know what is meant by isotactic and syndiotactic polymers and be able to give an example of each.

8. Be able to calculate number-average and weight-average molecular weights and know what is meant by polydispersity.

9. Know how to calculate the number-average molecular weight of a polymer using the osmotic pressure method.

10. Be able to describe how Dacron and Mylar differ and how they are prepared.

11. Know what is meant by the glass-transition temperature and how it is related to the physical properties of polymers.

12. Be able to describe how an amorphous polymer and a partly crystalline polymer differ over a range of temperature for a property such as the modulus of elasticity.

13. Know what is meant by a polypeptide and be able to sketch its basic structure.

14. Know what is meant by an α-helix and how it maintains its shape.

15. Be able to describe the basic structure of the β-sheet and to explain the different types of forces contributing to its three-dimensional structure.

16. Know the basic principles of how the sequence of amino acids in a polypeptide can be determined experimentally.

17. Know the basic steps of the solid-support method for the synthesis of proteins.

18. Be able to name at least one important scientific accomplishment of each of the following: Baekeland, Staudinger, Ziegler, Natta, Goodyear, Pauling and Corey, and Merrifield.

DEFINITIONS, TERMS, AND CONCEPTS TO KNOW

polymer
polymerization
chain and step-growth polymerization
monomer
initiator
chain initiation
chain propagation
chain termination
chain branching
silicones
polyester
polyamide
copolymerization
graft copolymer
block copolymer

alternating copolymer
random copolymer
stereospecific polymerization
isotactic polymer
syndiotactic polymer
linear and nonlinear polyethylene
vulcanization
amorphous polymer
crystalline polymer
glassy state
modulus of elasticity
glass-transition temperature
rubbery plateau
elastomer
crystalline melting point

drawing
orientation
plasticizer
number-average and weight-average molecular weight
polydispersity
light scattering
thermosetting
thermoplastic
peptide linkage
polypeptide
protein
α-helix
β-sheet
triple helix

QUESTIONS AND PROBLEMS

General

1. Illustrate some of the specific reactions that can take place if some glycerol, $HOCH_2CHOHCH_2OH$, is added to the polyethyleneterephthalate condensation reaction.

2. Name the substance you would expect to obtain when polyvinyl acetate is treated with a strong NaOH solution.

3. Caprolactam has the molecular formula indicated below. In the presence of water the molecule behaves like $NH_2-(CH_2)_5CO_2H$. What is the structure of the nylon formed when caprolactam is heated in the presence of a trace of water?

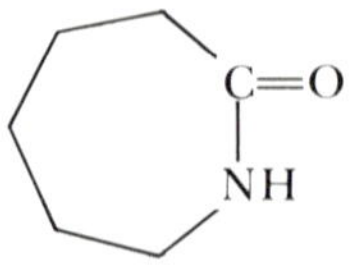

4. What stereoregular polymers can arise from the monomer

$$\underset{R_1}{CH}=\underset{R_2}{CH} \quad (R_1 \neq R_2 \neq H)$$

Repeat for $CH_2=CR_1R_2$

5. Using the data below, choose the best polymer for each of the following applications.
a) To make flexible hinges for operation at room temperature.
b) To make Christmas toys guaranteed not to survive to 26 December.

POLYMER	T_g (°C)	T_m (°C)
A	−20	35
B	−10	75
C	55	120
D	100	200

6. Two different amorphous polymers have T_g's of 0 °C and 50 °C. Contrast the behaviors of the two polymers at 25 °C.

7. When polystyrene is treated with H_2SO_4, the phenyl rings become substituted with $-SO_3H$ groups. The polymer repeating units are

$$-CH_2-CH(C_6H_3(SO_3H)_2)-$$
(with HO_3S and SO_3H substituents on the phenyl ring)

Predict what will happen when water containing Na^+ is passed over a porous resin made of this substance. [*Note:* This is an example of an ion-exchange resin used in the purification of water.]

8. Enzymes are rather large molecules, having molecular weights of several tens of thousands amu. Suggest an explanation for this fact.

9. Identify each amino acid below as a D-form or an L-form (see Section 16.13).

a) b) c)

10. The monomer

$$CH_2=C(R)-CH=CH_2$$

(a 2-substituted-1,3-butadiene) can polymerize in a number of ways. The possible repeating units are:

$$-CH_2-C(R)-CH=CH_2 \qquad CH_2=C(R)-CH-CH_2-$$

$$-CH_2-C(R)=CH-CH_2-$$

Discuss the stereoregular possibilities for each repeating unit.

11. Why do large-molecular-weight compounds not exist as gases?

12. When a strong acid such as HCl is spilled on nylon, the nylon disintegrates rapidly. Suggest a chemical explanation for this fact.

13. The IUPAC name for ethylene glycol is 1,2-ethanediol. What would be the IUPAC name for glycerol, $CH_2OHCHOHCH_2OH$?

14. Water is often said to be a plasticizer for proteins. Explain. What effect does water have on the glass transition temperature of proteins?

15. When NH_4Cl and PCl_5 are heated, a polymer called a **polyphosphazene** is formed. This polymer has the repeating unit

$$-N=P(Cl)(Cl)-$$

a) This polymer degrades readily in the presence of water. Why?
b) Suggest a way of replacing the Cl groups with other groups, say alkyl groups. Would this stabilize the polymer with respect to hydrolysis? Explain.

16. Explain why most proteins are copolymers.

Molecular Weight Determination

17. The following osmotic pressure data were obtained for a solution of nitrocellulose in acetone at 20 °C. Calculate the molecular weight of nitrocellulose.

c (g·cm^{-3}) × 10^3	π (mm of H_2O)
1.16	6.2
3.66	25.6
8.38	80.0
19.0	254

18. Ceruloplasmin, a blood protein, contains 0.33% (by weight) of copper.
a) What is the minimum molecular weight of ceruloplasmin?
b) If the actual molecular weight of ceruloplasmin is 150,000 amu, how many copper atoms are there per molecule?

19. A 1.0-g sample of polyester is titrated to an endpoint (using a suitable indicator) with 5.0 cm^3 of standard 0.01-M NaOH. Estimate $\bar{M}_N$, assuming each chain contains only one —CO_2H group.

20. An electrophoretic study of an aqueous solution of protein reveals two species with $M_1 = 60{,}000$ amu and $M_2 = 120{,}000$ amu. A solution of the protein contains 1.76% of protein by weight. The fraction of the heavier protein is found to be 61%. Calculate $\bar{M}_w$, $\bar{M}_N$, and the polydispersity.

21. Fractionation of a polymer sample gave the results shown in the following table. Calculate $\bar{M}_w$, $\bar{M}_N$, and the polydispersity.

MOLECULAR WEIGHT (AMU)	% BY WEIGHT
10,000	10
25,000	20
50,000	60
100,000	10

22. A polymer sample was fractionated and found to have the following composition:

MOLECULAR WEIGHT (AMU)	% BY WEIGHT
9,000	5
11,000	10
15,000	20
17,000	10
20,000	25
22,000	15
25,000	10
30,000	5

Calculate $\bar{M}_w$, $\bar{M}_N$, and the polydispersity of the polymer.

23. The oxygen-storing protein, myoglobin (see Section 22.3) contains 0.319% iron by weight. If each molecule contains one iron atom, what is the molecular weight of myoglobin?

24. The weight-average and number-average molecular weights of many proteins, for example, myoglobin, hemoglobin, enzymes, etc., are the same. Why is this so?

25. Why is the osmotic pressure method not useful for molecular weights over 10^6 amu?

26. If osmotic pressure is to be in pascals, what units must the polymer concentration c have in the osmotic pressure equation? What are the units of the constant A?

Proteins and Polypeptides

27. a) How many different polypeptides could one make from four different amino acids? Would all of these have different chemical and physical properties?
b) How many different dipeptides can one make from 20 different amino acids?
c) How many different kilopeptides (peptides with 1000 amino acid residues) can one make from 20 different amino acids?

28. An excess of ninhydrin is added to a mixture of amino acids and heated. The solution turns blue and evolves CO_2. When the CO_2 is passed through a tube containing Ascarite (NaOH on asbestos), the tube increases in weight by 5 g. How many moles of amino acids were present in the mixture?

29. Outline all the steps needed to synthesize the tripeptide Try–Ala–Gly using Merrifield's solid-support method.

30. Compare the structures of proteins and nylon, pointing out how they are similar and how they are different.

31. A time-course analysis of the action of carboxypeptidase and aminopeptidase on a hexapeptide yielded the following results.

ENZYME	ORDER OF APPEARANCE OF AMINO ACIDS, FIRST TO LAST
Carboxypeptidase	Val, Arg, Gly, Ala; Cys and Arg at same time
Aminopeptidase	Cis, Arg, Ala, Gly; Val and Arg at same time

a) Why do two amino acids appear simultaneously in each of the above?
b) What is the amino acid sequence in the hexapeptide?

32. The average turn length of an α-helix in human hair is about 5.4 Å. Assuming hair grows about 0.5 in. per month, how many turns of the α-helix are produced each second?

33. A tetrapeptide contains phenylalanine, aspartic acid, glutamic acid, and tryptophane. Treatment with chymotrypsin produces phenylalanine and a tripeptide.
a) What can you say definitely about the location of phenylalanine?
b) What are all the possible structures of the tetrapeptide?
c) Suppose you knew that aspartic acid was at the amino end. What is the structure of the tetrapeptide?

34. Draw the structure of the tripeptide Ala–Gly–His.

35. The presence of proline (Pro) in a peptide leads to a sharp bend in the α-helix. Suggest a reason for this.

36. A polypeptide contains alanine, aspartic acid, glycine, phenylaline, and tryptophane. The amino end is glycine. Cleavage with chymotrypsin yields a tripeptide and the individual amino acids tryptophane and aspartic acid. What is the structure of the peptide?

ADDITIONAL READINGS

O'Driscoll, Kenneth F., *The Nature and Chemistry of High Polymers.* New York: Reinhold, 1964. An elementary and readable account of many of the most important aspects of polymer chemistry.

Cowie, J. M. C., *Polymers: Chemistry & Physics of Modern Materials.* New York: Intext Educational Publishers, 1973. An excellent introduction at a somewhat more advanced level than O'Driscoll.

Mark, Herman F., *Giant Molecules.* New York: Time-Life Books, 1966. An excellent descriptive account of synthetic polymers and their uses. The history of the development of polymer science is very well done.

"Giant Molecules." The September 1957 issue of *Scientific American* is devoted to this subject. Some of the articles are now dated but many are still worth reading.

Natta, Giulio, "Precisely Constructed Polymers." *Scientific American,* August 1961. The Nobel-prize-winning chemist describes the technique he and Karl Ziegler developed for the stereospecific synthesis of polymers.

Doty, Paul, "Proteins." *Scientific American,* September 1957. An elementary discussion of some early work on the structure of proteins involving the α-helix.

Pauling, L., Robert B. Corey, and Roger Hayward, "The Structure of Proteins." *Scientific American,* July 1954. Discusses the structures of fibrous proteins.

Merrifield, R. Bruce, "The Automatic Synthesis of Proteins." *Scientific American,* March 1968. Discusses how the "protein machine" operates. Written by one of the first scientists ever to synthesize an enzyme.

Denkewalter, Robert G., and Ralph Hirschmann, "The Synthesis of an Enzyme." *American Scientist* **57**, 389 (1969). Describes the Merck, Sharpe, and Dohme synthesis of ribonuclease.

Allcock, Harry R., "Polyphosphazenes: New Polymers with Inorganic Backbone Atoms." *Science* **193**, 1214 (1976). Discusses a new class of macromolecules which promise to solve many of the problems associated with conventional organic polymers.

Meloon, Clifton E., "Fibers: Natural and Synthetic," *Chemistry* **51**, No. 3 (April 1978), p. 8. Lots of interesting material on permanent-press materials, soil repellants, etc.

Chemistry **51**, No. 5 (June 1978 issue). Totally devoted to articles on polymers, all at a very readable level.

25
THE CHEMISTRY OF LIFE: A BRIEF OVERVIEW

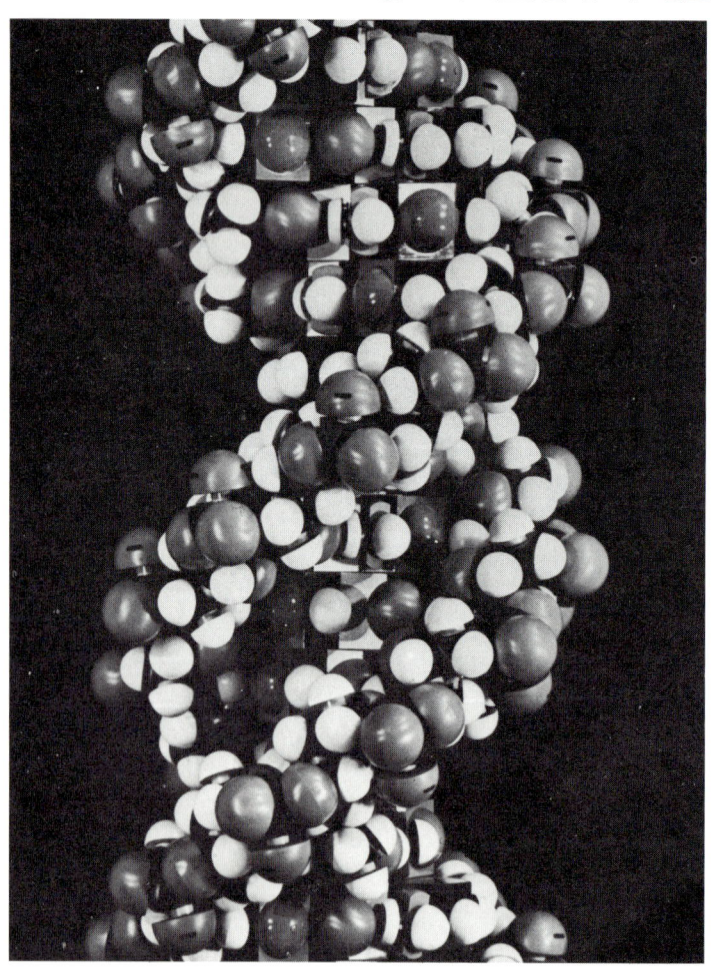

Space-filling model of the double helix of DNA. (Courtesy of Professor M. H. J. Wilkins, The Biophysics Department, King's College, London.)

What drives life is . . . a little electric current, kept up by the sunshine.
ALBERT SZENT-GYÖRGYI

Albert Szent-Györgyi (b. 1893), a Hungarian-born American biochemist, received the 1937 Nobel prize in physiology and medicine for his work on the explanation of cell metabolism. Szent-Györgyi also explained the role of vitamins in metabolism and the chemistry and structure of muscle tissue. When last seen by the author (about 1970) Szent-Györgyi was doing a very athletic Swedish dance in the company of several attractive lady scientists.

As recently as 20 years ago, if asked to discuss the role of chemistry in life processes, a biologist would have answered only in the most general of terms. Today, a biologist can tell you just what molecules play roles in a large number of complex biological processes and what the probable mechanisms of many of these processes are. Some scientists now devote entire careers to studying the possibility that all life began as a random combination of atoms that somehow found ways to reproduce and to evolve to increasingly sophisticated forms. Little by little, the molecular bases of various diseases—both bacterial and viral—are being unraveled. This final chapter summarizes five broad areas of biological nature in which the role of chemistry is now being studied on a molecular basis.

25.1 PHOTOSYNTHESIS AND CELL METABOLISM

A cell, either plant or animal, is the world's most extraordinary chemical factory. Of the hundreds—and perhaps thousands—of chemical reactions carried out by living cells, two of the best understood are **photosynthesis**, the conversion of carbon dioxide and water to carbohydrate, and **cell metabolism**, the breakdown of carbohydrate and similar substances to carbon dioxide and water. Photosynthesis, drawing its energy from the sun, produces the fuel to power the earth's living organisms.

Many of the basic steps of photosynthesis and cell metabolism are very similar—they merely take place in opposite sequences and directions. Photosynthesis takes place in parts of cells (those of plants and of some algae) called **chloroplasts** where a magnesium-containing compound called **chlorophyll** (Section 21.5) plays a role in converting solar energy to chemical energy. Cell metabolism—the burning of the fuel produced by photosynthesis and other processes—takes place in cell regions called **mitochondria**. Basically, the mitochondria transform carbohydrate to CO_2 and H_2O (an exothermic process) and make the energy available for the needs of the entire organism. In summary, photosynthesis is analogous to the charging of a battery using an external source of energy (the sun); discharge of the battery produces energy just as cell metabolism produces energy.

The first step of photosynthesis is the light-induced breakdown of water. This process may be symbolized by the reaction

$$2H_2O \xrightarrow{\text{light}} 4H + O_2$$

The four hydrogen atoms do not exist in free form but are passed on from compound to compound by various enzyme-controlled reactions until they have acquired enough energy to reduce carbon dioxide to carbohydrate. This process is summarized by the reaction

$$4H + CO_2 \rightarrow CH_2O + H_2O$$

where CH_2O is the empirical formula of most carbohydrates. Glucose, for example, is a carbohydrate with the molecular formula $(CH_2O)_6$ or $C_6H_{12}O_6$. The above reaction does not take place in a single step. When glucose, $C_6H_{12}O_6$, is the product, the reaction first produces a three-carbon intermediate and, via a sequence of steps, eventually leads to glucose. Figure 25.1 symbolizes the entire process in simplified form.

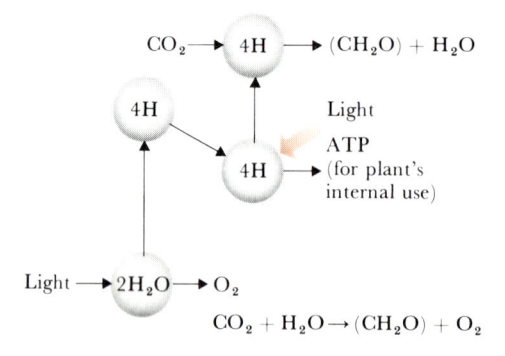

Fig. 25.1. Schematic illustration of the photosynthesis process. Sunlight splits water into hydrogen atoms and O_2 and activates the hydrogen atoms to combine with CO_2 to form carbohydrate.

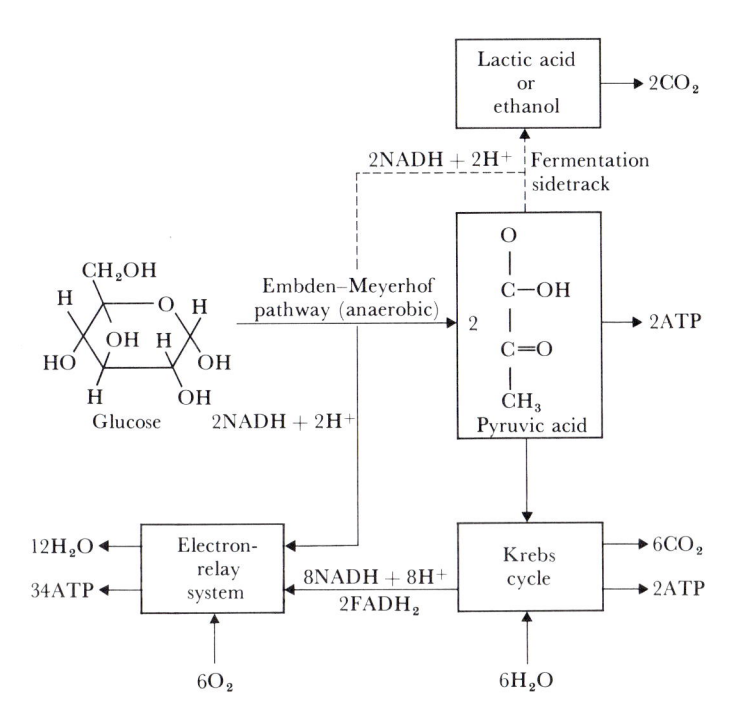

Fig. 25.2. The combustion of glucose in the mitochondrion of a living cell. NADH and FADH$_2$ are enzyme systems regulating the chemical reactions.

Cell metabolism of glucose to CO_2 and H_2O, although producing the same energy as burning glucose in a calorimeter, is a long-drawn-out, enzyme-controlled process involving many, many steps. First, the six-carbon molecule is broken down to two three-carbon molecules and these are then degraded to CO_2 and H_2O. The entire process is shown in broad outline in Fig. 25.2.

The energy produced in many biological reactions is often stored for future use with the aid of a remarkable substance called **adenosine triphosphate** or **ATP**. (See Section 23.4.) Each mole of ATP is capable of storing about 33 kJ of energy. When this energy is needed by the organism (either in a synthesis or in physical activity) it is released by the reaction

$$ATP + H_2O \rightarrow ADP + phosphate$$

where ADP stands for adenosine diphosphate.

25.2 THE CHEMISTRY OF REPRODUCTION AND SURVIVAL

Ever since Gregor Mendel discovered the general pattern by which characteristics of one generation are passed on to succeeding generations, scientists have attempted to find out just how genetics operates on the molecular basis. Today, thanks to the work of Watson, Crick, Franklin, Wilkins, Chargaff, and many others, it is known that genetic information is stored in a giant molecule known as deoxyribonucleic acid (DNA). This information—called the **genetic code**—is built into DNA by the sequence of adenine–thymine and guanine–cytosine linkages which join the double-helical strands of DNA; each sequence of three such linkages is a chemical code symbol for a specific amino acid (see Fig. 25.3). In this way DNA can specify the amino acid sequence of proteins—proteins which are part of the enzymes that control all life processes.

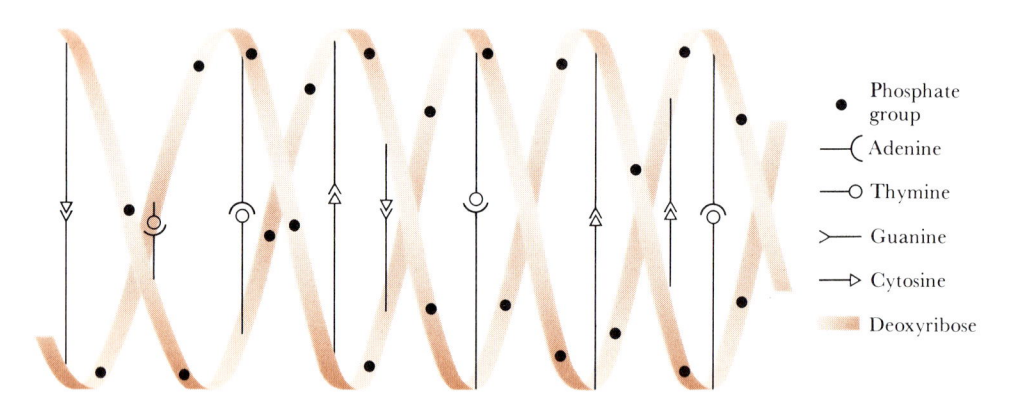

Fig. 25.3. The DNA double helix. The two intertwined ribbons are chains of alternating deoxyribose (a sugar) and phosphate groups. The double helix configuration is maintained by the adenine–thymine and guanine–cytosine "rungs" between the spiral chains.

Unraveling of the mechanism of genetics also led to a solution of the nature of virus-caused diseases. Viruses are bits of DNA or a closely related substance, RNA (ribonucleic acid), enclosed in a protective sheath of protein. A virus infects a cell by injecting its DNA or RNA portion through the cell membrane. Thereupon, the DNA or RNA—by virtue of the protein-synthesis code it carries—takes over the cell and uses its materials to construct replicas of the invading virus. When the cell's materials are exhausted, its membrane ruptures and hordes of newly made viruses are released to invade new cells to repeat the process.

Figure 25.4 shows a virus known as the T4 bacteriophage, a virus which infects cells of the colon bacterium, *Escherichia coli*. The "head" and "body" portion of the virus is a hollow protein membrane containing DNA. The insert shows how the six "leg" fibers bind to the cell membrane while the DNA is injected into the cell.

Chemistry plays biological roles other than that of genetic transformation of information. Many forms of life, particularly primitive forms, communicate by means of chemical interactions. Chemical substances that serve as messages between members of a single species are known as **pheromones**. For example, when the female tiger moth is ready to mate, she releases the hydrocarbon 2-methylheptadecane from her abdomen as a tiny puff of vapor. A male tiger moth needs to contact no more than about 100 of these molecules in order to become inspired to seek out the emitter posthaste. Similarly, aphids warn each other of the approach of enemies by emitting droplets of a compound called *trans-β*-farnesene. Similar examples abound throughout the plant and animal kingdoms.

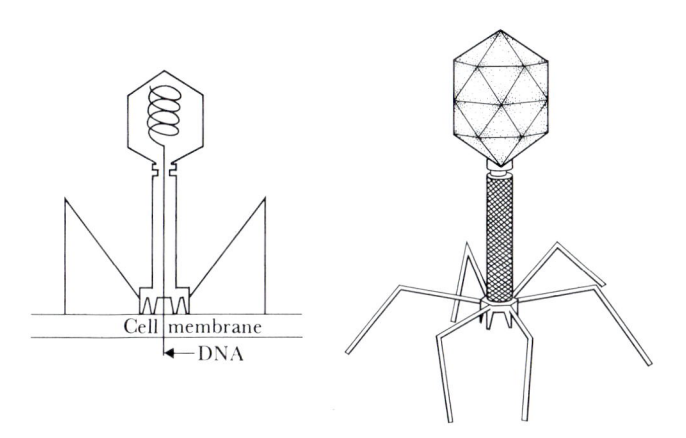

Fig. 25.4 Structure of the T4 bacteriophage. The left-hand drawing illustrates how the virus injects its DNA into a cell.

25.3 CHEMICAL EVOLUTION AND THE ORIGIN OF LIFE

Consider the following scenario: A gaseous mixture of water, ammonia, hydrogen, and methane envelops the primeval earth. As giant storms rage in the skies, huge electrical discharges rend the atmosphere and activate its components. The activated components react to form various short-lived compounds but occasionally a few molecules of amino acids, such as glycine and alanine, and other compounds are formed and survive long enough to be washed down into the primeval seas. Here, the amino acids occasionally link to form polypeptides, most of which do not long survive. However, one day, purely by chance, a polypeptide that just happens to be catalytically active finds itself in the proper milieu to catalyze the synthesis of a molecule large enough to possess some degree of self-organization. Eventually, by a continuation of such random processes, very few of which lead to anything notable, there are enough giant molecules of the right kind in one place to create the first primitive organism. From this point, under the influence of evolution, the life forms of today ultimately develop.

Fantastic? Unbelievable? Science fiction? Perhaps—and then, perhaps not. Ever since the Russian biochemist Oparin first proposed essentially such a chemical ori-

Aleksandr I. Oparin, Russian biochemist who made the first complete proposal of how life may have originated and evolved by purely chemical means (Courtesy of Wide World Photos.)

gin of life, more and more reputable scientists have given considerable thought to its very real possibility. Today, the scenario given above has been greatly expanded, its details filled in, and experiments carried out to test its plausibility. Amazingly, the more work that is done, the more probable the whole idea seems to become! Beginning with the fossil record, which shows life began with extremely primitive forms and slowly evolved to increasingly complex forms, the accumulating evidence supports a chemical origin of life more and more strongly.

25.4 CHEMICAL MESSENGERS AND REGULATORS

Although the basic chemical reactions of life are controlled by enzymes, something must in turn control the enzymes; that is, there must be some mechanism for turning enzymes on and shutting them off. Such a role is played by chemical regulators known as **hormones**. Hormones are chemical substances produced by the ductless glands in response to a chemical or physical stimulus (injury, hunger, danger, etc.) and carried via the bloodstream to various parts of the body. Each cell contains proteins capable of recognizing a variety of specific hormones. When a cell protein recognizes a certain hormone, this represents a message sent to that cell. The protein responds to the message by producing a compound known as **cyclic AMP** (adenosine-3′,5′-monophosphate) which acts as a second messenger by controlling the enzymes that govern the cell's chemistry.

Vitamins are chemical substances that participate in various enzyme systems. Some vitamins are synthesized by the organism utilizing it; others must be supplied in the diet. Vitamin A plays an important role in the photochemical reactions involved in vision. Vitamins such as niacin and riboflavin form parts of the enzyme systems operative in photosynthesis and in cell metabolism.

25.5 MOLECULAR PHARMACOLOGY: THE CHEMISTRY OF DRUGS

Human use and misuse of drugs predates written history. Today, drugs are used to fight disease, control pain, alter moods, escape from reality, and, sometimes, just because they are illegal.

Among the most beneficial of drugs are **antibiotics** and **antivirals**. Penicillin, one of the best known antibiotics, has played a major role in reducing death from many types of infection. On a molecular basis, penicillin appears to kill bacteria by inactivating the enzymes that control the synthesis of bacterial cell membranes. Viruses are much harder to control chemically than are bacteria. Presumably, this is because bacteria carry out many more reactions than do viruses and thus are more vulnerable to chemical control. Viruses carry out relatively few chemical reactions, many of these in concert with the host, so that any chemical that interferes with a virus's chemical reactions may also injure the host.

Use of **depressants** and **stimulants** has increased dramatically during the last two decades. The most popular depressants, tranquilizers, are used to calm violence, relieve anxiety, and end withdrawal. Detailed molecular bases of depressant activity are not known, but it appears likely that such molecules somehow alter chemical processes taking place in the brain itself. Chemical stimulants, such as amphetamine, act in a fashion opposite to that of depressants and are often called **psychic energizers**. Amphetamine enables a fatigued person to perk up and perform at a high level of activity. However, the effect wears off and the individual ultimately succumbs to the fatigue.

Hallucinogenic drugs produce hallucinations, distortion of perceptions, and, occasionally, states of mind resembling psychoses. Alcohol may also act in these ways. Among the better-known hallucinogens are lysergic acid diethylamide (LSD), and tetrahydrocannabinol (found in marijuana).

ADDITIONAL READINGS

Cheldelin, Vernon H., and R. W. Newburgh, *The Chemistry of Some Life Processes*. New York: Reinhold, 1964. Elementary account of how chemistry is involved in life processes. An inexpensive paperback of slightly over 100 pages.

Pfeiffer, John, *The Cell*. New York: Time-Life Books, 1964. Superbly illustrated semipopular account of the structure, properties, and roles of cells in living systems.

APPENDIX 1
THE METRIC SYSTEM

A1.1 THE INTERNATIONAL SYSTEM OF UNITS

In December 1975, President Ford signed the Metric Conversion Act of 1975, which provides for a United States Metric Board of 17 members charged with coordinating a voluntary nationwide conversion to the metric system. Specifically, the Metric Board is expected to recommend the adoption of the Système International d'Unites (abbreviated SI) which has been used mainly by physicists for a number of years. The SI has seven base units to be used with sixteen prefixes to designate multiples and submultiples. In addition, there are a number of special derived units, many having names of reknowned scientists. The six base units of greatest interest to chemists are listed below. A complete discussion of SI units has been given by Paul.*

MEASUREMENT	BASIC UNIT	SYMBOL
Length	meter	m
Mass	kilogram	kg
Time	second	s or sec
Electric current	ampere	A
Temperature	kelvin	K
Amount of a substance	mole	mol

The United States National Bureau of Standards and the National Physical Laboratory of the United Kingdom have agreed to replace the American spelling of "meter" with "metre" and the British spelling of "gramme" with "gram." The form "metre" is common to most of the world's languages and thus promotes international unity in the spelling of SI names. Another important rationale behind this spelling change is that the spelling "meter" will always signify a measuring instrument (barometer, thermometer, aenomometer, etc.) while "metre" will always signify a unit of length. Thus the word "micrometer" (the machinist's gauge) will not be confused with "micrometre" (one-millionth of a metre). However, not all federal agencies have adopted this change, and the American Metric Council prefers -er while the American Society

---

* Martin A. Paul, "International System of Units (SI)," *Chemistry* **45** (October 1972), p. 14.

for Testing and Materials (ASTM) uses -re. Consequently this text has continued to use the traditional American form.

Fractions and multiples of the base units are to be indicated via the prefixes shown below (the name of each prefix is followed by its abbreviation):

SUBMULTIPLES		MULTIPLES	
10^{-1}	deci (d)	10^{1}	deka (da)
10^{-2}	centi (c)	10^{2}	hecto (h)
10^{-3}	milli (m)	10^{3}	kilo (k)
10^{-6}	micro (μ)	10^{6}	mega (M)
10^{-9}	nano (n)	10^{9}	giga (G)
10^{-12}	pico (p)	10^{12}	tera (T)
10^{-15}	femto (f)	10^{15}	peta (P)
10^{-18}	atto (a)	10^{18}	exa (E)

Thus 10^{-12} m becomes one pm (picometer) and 10^6 mol becomes one Mmol (megamole). The use of compound prefixes is to be avoided. For example, 10^{-5} m may be written as 10^{-2} mm or 10^{-3} cm but not one cmm or one mcm. Thus deka (da) cannot be mistaken for deciatto since the latter should never be used.

Strict adherence to the SI conventions is unlikely to be the rule—at least in the immediate future. There exist vast libraries of older literature in which the older cgs (centimeter-gram-second) system of units and other systems are used. Also, since conversion is voluntary, many people will continue to use non-SI units. Accordingly, until such time as uniformity of practice occurs, there is no alternative but to become familiar with both SI and non-SI units. This text uses SI units except in areas where non-SI units are still the rule. In such cases, both SI and non-SI units are often indicated.

A1.2 LENGTH, AREA, AND VOLUME

The basic unit of length in the metric system is the **meter (m)**. The meter is slightly longer than 3 ft (1 m is equivalent to approximately 39.37 inches).* The most commonly used submultiples of the meter are the millimeter (mm), the centimeter (cm), and the nanometer (nm). Use of the angstrom unit (1 Å = 10^{-10} m) is to be allowed for a limited time, but it is preferable to express a quantity such as 5000 Å as 500 nm instead. The micrometer (μm) is used to express wavelengths of infrared radiation. An older unit, the micron (μ), was used for what is now the μm. The use of the micron should be abandoned and μ used only as a prefix. The kilometer (km) is the only widely used multiple of the meter insofar as chemists are concerned.

Some relationships among the various metric units of length and their relationship to some common English units are given below:

1 mm = 0.001 m	or	1 m = 1000 mm
1 cm = 0.01 m	or	1 m = 100 cm
1 km = 1000 m	or	1 m = 0.001 km

---

* The meter is officially defined as exactly 1,650,763.73 wavelengths of the radiation involved between two specific quantum states of ^{86}Kr (the orange-red line).

Also

1 mm = 0.1 cm	or	1 cm = 10 mm
1 cm = 0.3937 in.	or	1 in. = 2.540 cm
1 km = 0.6214 statute mile	or	1 statute mile = 1609 km

The symbol = is to be read as "is equivalent to" in the above context.

The fundamental unit of area in SI is the **square meter** (**m^2**), but the square of any legitimate SI length unit may be used, for example, mm^2, cm^2, nm^2. Note that nm^2 means $(10^{-9}$ m$)^2 = 10^{-18}$ m^2. A non-SI unit, the **are** (**a**), is often used in land measurement. This unit is equivalent to a square dekameter (dam^2).*

The SI unit for volume is the **cubic meter** (**m^3**) but it is permissible to employ a non-SI unit, the liter (L). Note that the abbreviation has been changed from a lowercase to a capital letter. The liter has now been redefined as exactly 10^{-3} m^3 so that 1 L is equivalent to 1000 cm^3. The unit milliliter (mL) may also be used and is synonymous with cm^3.† In spite of the SI allowance of the unit liter, it is recommended that the SI unit dm^3 (cubic decimeter) be used instead. For example, molar concentration as $mol \cdot dm^{-3}$ is preferred over the older $mol \cdot L^{-1}$.

The unit cm^3 is sometimes written **cc** but this usage definitely is discouraged. Besides not conforming to any systematic rules, "cc" lacks mnemonic dimensional suggestion; for example, it is instantly obvious that $cm^3/cm = cm^2$ but much less apparent when written cc/cm.

A1.3 MASS

The SI base mass, the kilogram (kg), is officially defined as the mass of a particular cylinder of platinum-iridium alloy (called the International Prototype Kilogram) preserved in a vault at Sèvres, France, by the International Bureau of Weights and Measures.

Since the base unit of mass already has a prefix and since compound prefixes are to be avoided, the term **gram** (**g**) has been retained in SI (otherwise one would need to use millikilogram, mkg). The units **milligram** (**mg**) and **microgram** (**μg**) are also retained in SI. A metric ton (1000 kg) is now called a **tonne** (**t**) in lieu of a megagram (Mg).

Although biochemists frequently use the **dalton** (**d**) in place of the **atomic mass unit** (**amu**) for atomic and molecular masses, SI has made no provision for such usage.

Some useful interrelationships among commonly used mass units are given below:

1 kilogram (kg)	= 1000 g	or	1 g = 0.001 kg
1 milligram (mg)	= 0.001 g	or	1 g = 1000 mg
1 tonne (t)	= 1000 kg	or	1 kg = 0.001 t

---

* Frankly, some people won't give a decameter for this information!

† Formerly, a liter was defined as the volume of 1 kg of water at 1 atm and 3.98 °C; this led to 1 L = 1000.028 cm^3.

Also

$1 \text{ kg} = 2.204 \text{ lb}$

$1 \text{ lb} = 453.6 \text{ g}$

When multiple units such as grams per cubic centimeter are used, the individual units must be separated by a space, or, if there is a chance of confusion, by a dot. Thus either g cm^{-3} or g·cm^{-3}. The latter form should always be used in cases such as m·L^{-1} (meter per liter) to distinguish it from mL^{-1} (reciprocal milliliter). Alternatively, forms such as g/cm^3 may be used for g·cm^{-3}. However, the latter is preferred in more complicated cases such as L·atm·K^{-1}·mol^{-1}.

A1.4 FORCE, PRESSURE, ENERGY, AND RELATED QUANTITIES

Many quantities whose dimensions are composites of those of several SI units have special names. In most cases these names are those of scientists whose major contributions were related to the unit in question. For example, the SI unit of force (kg·m·s^{-2}) is called the **newton** (**N**) in honor of Sir Isaac Newton, who introduced the concept of force in its modern form. The older cgs unit still used is the **dyne** (g·cm·s^{-2}); according to SI this unit is to be abandoned. It is readily deduced that 1 N is equivalent to 10^5 dyne.

Pressure (force per unit area) has the units of newton per square meter (N·m^{-2}) and is called the **pascal** (**Pa**). One **atmosphere** (**atm**) is equivalent to approximately 101.3 kilopascal (kPa) or, roughly, 10^5 Pa. A **torr** (1/760 of an atm) is about 133.3 Pa or 1.333 hPa (hectopascal). SI recommends that atm and torr be dropped.

The SI unit of energy (kg·m^2·s^{-2}) is called the **joule** (**J**).* The older cgs unit, the **erg** (g·cm^2·s^{-2}), is to be abandoned. It is readily verified that 1 J is equivalent to 10^7 erg.

The **electron volt** (**eV**) is to be retained. Note that 1 eV is equivalent to 1.602×10^{-19} J or 0.1602 aJ.

A practical unit of energy much used when dealing with thermal energy is the **calorie** (**cal**). The calorie was originally defined as the amount of energy required to change the temperature of 1 g of air-free water from 15 °C to 16 °C. The modern definition of a calorie is in terms of the joule; 1 cal = 4.184 J. However, SI recommends that the calorie be abandoned, and it is becoming increasingly common to see thermal energies reported in joules and kilojoules (kJ) rather than cal and kcal.

The English unit of energy comparable to the calorie is the **British Thermal Unit** (**Btu**). The Btu is the energy required to change the temperature of one pound of air-free water from 39 °F to 40 °F. One Btu is equivalent to 1055.06 J.

The SI unit for absorbed dose of radiation is the **gray** (**Gy**) or J·kg^{-1}. An older unit is the **rad** (100 erg·g^{-1}). Thus 1 Gy is equivalent to 100 rad.

A1.5 ELECTRICAL QUANTITIES

The SI units for electrical potential, charge, and power are the **volt** (**V**), the **coulomb** (**C**), and the **watt** (**W**), respectively. Note that a coulomb is an ampere second (A·s)

---

* The word *joule* (after British scientist James Prescott Joule) is usually rhymed with *fool* or *foul*; historical evidence suggests the Joule family pronounced it to rhyme with *foal*.

and a watt is a joule per second (J·s^{-1}). A volt coulomb (V·C) is equivalent to a joule.

The SI unit of frequency is the **hertz (Hz)** (cycles·s^{-1} or s^{-1}). Note that a **cycle** is a dimensionless quantity.

A1.6 TEMPERATURE

SI temperatures are given in **kelvin (K)**. A kelvin corresponds to what was previously called a *degree Kelvin,* now written, for example, 298 K rather than 298°K.

The Celsius or centigrade temperature scale is retained but a space is now left between the numerical value of the temperature and the symbol for degree Celsius (°C). Thus the older 25°C is now written 25 °C. The size of a kelvin and a degree Celsius is the same.

Since the United States (and a few other nations) use the awkward **Fahrenheit** scale, in which 32 °F and 212 °F correspond to 0 °C and 100 °C, respectively, it is occasionally necessary to convert temperatures between the Celsius and Fahrenheit scales. *The student should never, never waste time and mental energy memorizing formulas for doing this!* Conversions can be carried out *rapidly* and *accurately* by remembering only the four numbers 0 °C, 100 °C, 32 °F, and 212 °F (and how they correspond). Most people know these numbers already. To convert 86 °F to °C note that there are 180 Fahrenheit degrees (212 − 32 = 180) to every 100 Celsius degrees (100 − 0 = 100) so that the range 86 °F to 32 °F (54 Fahrenheit degrees) corresponds to

$$54 \times \frac{100}{180} = 54 \times \frac{5}{9} = 30 \text{ °C}$$

To convert 60 °C to °F note that 60 Celsius degrees corresponds to

$$60 \times \frac{180}{100} = 60 \times \frac{9}{5} = 108 \text{ Fahrenheit degrees}$$

which must be added to 32 °F. Thus 60 °C corresponds to 108 °F + 32 °F = 140 °F. A little practice in doing such conversions will make you an expert and will free your mind to study some of the important aspects of science.

APPENDIX 2
EXPONENTIAL NOTATION AND SIGNIFICANT FIGURES

A2.1 EXPONENTIAL NOTATION

Scientists and engineers frequently find occasion to make measurements which must be expressed in either very small numbers or very large numbers. For example, the mass of an electron is found to be 0.00000000000000000000000000009110 g, and the number of carbon atoms in 12.000... g of ^{12}C is found to be 602,300,000,000,000,000,000,000.* Such unwieldy numbers are much more convenient to handle when written in the compact form 9.110×10^{-28} g and 6.023×10^{23}, respectively. In the former the decimal point has been moved *28 places to the right* and the result (9.110) multiplied by 10^{-28}; in the latter the decimal point has been moved *23 places to the left* and the result (6.023) multiplied by 10^{23}. By the same approach, 0.0008 becomes 8×10^{-4} and 8000 becomes 8×10^{3}. The number 0.0008 could also be written as 80×10^{-5}, 0.8×10^{-3}, or in a number of equivalent ways. Similarly, 8000 could be written 80×10^{2}, 0.8×10^{4} or in several other ways. Usually, however, it is preferable to write the nonexponential (or preexponential) part of the number as one digit followed by the decimal point. Just how many digits (or zeros) are retained after the decimal point will be discussed later.

The **exponential notation** just illustrated is of immense help in simplifying many types of arithmetical computations and aids in the minimization of errors such as misplaced decimal points. For example, suppose you wish to divide 0.00018 by 90,000. This may be written

$$\frac{0.00018}{90,000} = \frac{18 \times 10^{-5}}{9 \times 10^{4}} = \frac{18}{9} \times \frac{10^{-5}}{10^{4}} = 2 \times 10^{-9}$$

---

* The values of these quantities are known accurately to more figures than those quoted.

(Note that $10^a/10^b = 10^{a-b}$; also $10^a \times 10^b = 10^{a+b}$.) As another example, let us evaluate the square root of 0.000036. Rewriting this as 36×10^{-6} shows at once that the square root is 6×10^{-3}. Similarly, the square root of 0.00036 could be rewritten $(3.6 \times 10^{-4})^{1/2}$. From tables, a slide rule, an electronic calculator, or the square-root-extraction procedure we find that $(3.6)^{1/2} = 1.9$, and the answer is 1.9×10^{-2}. Of course, 0.00036 can be entered directly into an electronic calculator and a key pressed to obtain 0.019 as the answer.

The following, more involved calculation shows more vividly the usefulness of exponential notation:

$$\frac{(0.00031)(760)}{(93,000)(0.005)} = \frac{(3.1 \times 10^{-4})(7.6 \times 10^2)}{(9.3 \times 10^4)(5 \times 10^{-3})} = \frac{(3.1)(7.6)}{(9.3)(5)} \times \frac{(10^{-4})(10^2)}{(10^4)(10^{-3})}$$

$$= 0.51 \times \frac{10^{-2}}{10^1} = 0.51 \times 10^{-3} = 5.1 \times 10^{-4}$$

Setting up calculations in this manner is very convenient when an electronic calculator is not available or cannot handle the exponents involved. The preexponential part is divided and multiplied separately and the exponentials are reduced by addition and subtraction to a single power of 10.

A2.2 SIGNIFICANT FIGURES

All measuring instruments such as meter sticks, clocks, and balances have scales which are subdivided into various units and subunits. The smallest subunit into which the scale is marked places a limit upon the accuracy of the measurement that can be made with that instrument. For example, if a meter stick is marked with mm as the smallest units, it is not feasible to measure the length of an object with complete confidence any closer than the nearest mm; at best it is possible to *estimate* to the nearest 0.1 mm but this last figure is uncertain. Suppose, for example, the object is between 13 and 14 mm long; that is, it is certain that the object is at least 13 mm long and shorter than 14 mm. Suppose the length is estimated as 13.7 mm. This measurement is said to have three **significant figures**; the general rule is that the number of significant figures in a measured quantity is always all the digits that are not in doubt plus one additional one which has been estimated. Therefore, the results of measurements are always expressed in such a way that *only the last figure* is in doubt.

Note that 13.7 mm, 1.37 cm, and 0.0137 m all have the same number of significant figures; the number of significant figures in a measurement cannot be changed by switching units! The only way to increase the number of significant figures in the above measurement is to use a meter stick accurately marked off in 0.1-mm subunits (or smaller subunits). Thus if a mass is reported as 1.3945 g, it is implied that the balance used in the measurement had a scale marked off in units of 0.001 g (1 mg) and that tenths of a milligram were estimated. Strict adherence to the proper use of significant figures makes it possible to report results of measurements in such a way that their reliability may be judged accurately by others.

When combining numbers which arise from measurements, it is necessary to observe certain rules in order to avoid obtaining a result which implies a higher degree of reliability than is warranted. Only two rules are generally needed: one rule when numbers are being multiplied (or divided) and another rule when numbers are being added (or subtracted). These rules are:

MULTIPLICATION RULE: The number obtained as a result of the multiplication or division (or a combination) of two or more numbers obtained by measurement must have no more significant figures than that number used in the multiplication (or division) which has the *least* number of significant figures. For example,

$$(1.39)(3.456) = 4.80$$

$$932/56 = 17$$

$$\frac{(31.23)(4.56)}{9.41395} = 15.1$$

The numbers which determine the number of significant figures in the answers are 1.39 (three significant figures), 56 (two significant figures) and 4.56 (three significant figures), respectively.

ADDITION RULE: The number obtained as a result of adding or subtracting (or a combination) of two or more numbers obtained by measurement must have no more significant figures *to the right of the decimal point* than that number used in the addition (or subtraction) which has the least number of significant figures to the right of its decimal point. For example,

$$1.31 + 2.1 = 3.4$$

$$9.631 - 4.21 = 5.42$$

In a sense, the above rules mean that the reliability of results calculated from a series of measurements depends on the least reliable measurements.

It should be noted that the number 0.000370 has three significant figures. Zeroes count as significant figures only if they occur to the right of at least one nonzero digit.

When using exponential numbers, the preexponential part must be written to the correct number of significant figures. For example, 9.3×10^4 and 9.30×10^4 have two and three significant figures, respectively. In both cases it would be misleading to write 93,000 since the latter has five significant figures. Similarly, 3×10^{-3} means 0.003 whereas 3.0×10^{-3} means 0.0030.*

Some numbers have an infinite number of significant figures, e.g., certain defined numbers and numbers obtained by counting. For example, one-half the length of an object measured to be 90.4 cm long is

$$\frac{90.4 \text{ cm}}{2} = 45.2 \text{ cm}$$

---

* Some texts state that when a number ends in zeroes that are not to the right of a decimal point (for example, 93,000), the zeroes are not necessarily significant and the actual number of significant figures is ambiguous. This text uses the more sensible convention that a number such as 93,000 definitely has *five* significant figures. Otherwise, it would be written as 9.3×10^4 (for 2 significant figures), 9.30×10^4 (for 3 significant figures), etc.

In this case the number 2 must be treated as having an infinite number of significant figures. Twelve times the mass of a marble of mass 51.392 g is expressed as

$$12 \times 51.392 \text{ g} = 616.70 \text{ g}$$

where 12 is treated as having an infinite number of significant figures. Note, however, that the total mass of a dozen marbles, each of measured mass 51.392 g, would be 616.704 g, a result obtained by adding 51.392 to itself twelve times.

A2.3 PRECISION AND ACCURACY

Accuracy refers to the closeness of a measured quantity to the true value of that quantity. **Precision** refers to the reproducibility of the repeated measurement of a single quantity. It is possible to have measurements which are very precise but yet highly inaccurate.

Suppose a block of metal weighs 1.000 g and three successive weighings produce

0.993 g
0.992 g
0.993 g

Another set of measurements may give

0.989 g
0.995 g
0.990 g

Neither set of measurements is very accurate but the first set is the more precise (each number deviates less from the average). A set of measurements such as

1.001 g
1.000 g
1.001 g

is both precise and accurate (relative to the other measurements). Both high precision and high accuracy are desirable.

APPENDIX 3
LOGARITHMS AND THEIR USE

Since most students now own electronic calculators, it is possible for them to simply press a button to obtain numerical values of logarithms and other mathematical functions *without having the foggiest notion what such functions mean.* Nevertheless, such mindless button-pushing is to be strongly discouraged. This appendix is written for those students who do not know what logarithms are (or have forgotten) and who are apt to need to push the button marked "ln x" on their pocket calculators.

A3.1 DECIMAL LOGARITHMS AND THEIR DETERMINATION

Let N be any positive number.* The number N may always be expressed as 10 raised to some power. If x symbolizes the power to which 10 is taken to produce N, this fact may be written

$$N = 10^x \tag{A3.1}$$

For example, if $N = 1000$, $x = 3$, i.e., $1000 = 10^3$. The power (or exponent) x is called the **logarithm of the number N**. This definition may be written in the algebraic form

$$\log N = x \tag{A3.2}$$

This is read: "the logarithm of N is x." The logarithm (often abbreviated in spoken as well as written language to **log**) of N is simply the power to which 10 must be taken to produce N. Equations (A3.1) and (A3.2) are simply two different but equivalent ways of saying the same thing, e.g., $1000 = 10^3$ is the same as $\log 1000 = 3$.

The logarithm of a number N can always be expressed as the *sum* of two separate numbers called the **characteristic** C_N and the **mantissa** M_N. This may be written as

$$\log N = C_N + M_N \tag{A3.3}$$

Values of the characteristic are determined by the order of magnitude of the number, i.e., that integral power of 10 which is smaller than (or equal to) the number in question. Values of mantissas are recorded in logarithm tables (see Table A3.1). All numbers having the same digits (e.g., 0.0123, 0.123, 1.23, 12.3, 12,300, etc.) have the same mantissa but different characteristics.

---

* The concept of logarithms can be applied to negative numbers but this is of no practical importance in the present context.

TABLE A3.1 Four-Place Logarithms of Numbers

N	0	1	2	3	4	5	6	7	8	9
10	0000	0043	0086	0128	0170	0212	0253	0294	0334	0374
11	0414	0453	0492	0531	0569	0607	0645	0682	0719	0755
12	0792	0828	0864	0899	0934	0969	1004	1038	1072	1106
13	1139	1173	1206	1239	1271	1303	1335	1367	1399	1430
14	1461	1492	1523	1553	1584	1614	1644	1673	1703	1732
15	1761	1790	1818	1847	1875	1903	1931	1959	1987	2014
16	2041	2068	2095	2122	2148	2175	2201	2227	2253	2279
17	2304	2330	2355	2380	2405	2430	2455	2480	2504	2529
18	2553	2577	2601	2625	2648	2672	2695	2718	2742	2765
19	2788	2810	2833	2856	2878	2900	2923	2945	2967	2989
20	3010	3032	3054	3075	3096	3118	3139	3160	3181	3201
21	3222	3243	3263	3284	3304	3324	3345	3365	3385	3404
22	3424	3444	3464	3483	3502	3522	3541	3560	3579	3598
23	3617	3636	3655	3674	3692	3711	3729	3747	3766	3784
24	3802	3820	3838	3856	3874	3892	3909	3927	3945	3962
25	3979	3997	4014	4031	4048	4065	4082	4099	4116	4133
26	4150	4166	4183	4200	4216	4232	4249	4265	4281	4298
27	4314	4330	4346	4362	4378	4393	4409	4425	4440	4456
28	4472	4487	4502	4518	4533	4548	4564	4579	4594	4609
29	4624	4639	4654	4669	4683	4698	4713	4728	4742	4757
30	4771	4786	4800	4814	4829	4843	4857	4871	4886	4900
31	4914	4928	4942	4955	4969	4983	4997	5011	5024	5038
32	5051	5065	5079	5092	5105	5119	5132	5145	5159	5172
33	5185	5198	5211	5224	5237	5250	5263	5276	5289	5302
34	5315	5328	5340	5353	5366	5378	5391	5403	5416	5428
35	5441	5453	5465	5478	5490	5502	5514	5527	5539	5551
36	5563	5575	5587	5599	5611	5623	5635	5647	5658	5670
37	5682	5694	5705	5717	5729	5740	5752	5763	5775	5786
38	5798	5809	5821	5832	5843	5855	5866	5877	5888	5899
39	5911	5922	5933	5944	5955	5966	5977	5988	5999	6010
40	6021	6031	6042	6053	6064	6075	6085	6096	6107	6117
41	6128	6138	6149	6160	6170	6180	6191	6201	6212	6222
42	6232	6243	6253	6263	6274	6284	6294	6304	6314	6325
43	6335	6345	6355	6365	6375	6385	6395	6405	6415	6425
44	6435	6444	6454	6464	6474	6484	6493	6503	6513	6522
45	6532	6542	6551	6561	6571	6580	6590	6599	6609	6618
46	6628	6637	6646	6656	6665	6675	6684	6693	6702	6712
47	6721	6730	6739	6749	6758	6767	6776	6785	6794	6803
48	6812	6821	6830	6839	6848	6857	6866	6875	6884	6893
49	6902	6911	6920	6928	6937	6946	6955	6964	6972	6981
50	6990	6998	7007	7016	7024	7033	7042	7050	7059	7067
51	7076	7084	7093	7101	7110	7118	7126	7135	7143	7152
52	7160	7168	7177	7185	7193	7202	7210	7218	7226	7235
53	7243	7251	7259	7267	7275	7284	7292	7300	7308	7316
54	7324	7332	7340	7348	7356	7364	7372	7380	7388	7396

TABLE A3.1 Four-Place Logarithms of Numbers (cont.)

N	0	1	2	3	4	5	6	7	8	9
55	7404	7412	7419	7427	7435	7443	7451	7459	7466	7474
56	7482	7490	7497	7505	7513	7520	7528	7536	7543	7551
57	7559	7566	7574	7582	7589	7597	7604	7612	7619	7627
58	7634	7642	7649	7657	7664	7672	7679	7686	7694	7701
59	7709	7716	7723	7731	7738	7745	7752	7760	7767	7774
60	7782	7789	7796	7803	7810	7818	7825	7832	7839	7846
61	7853	7860	7868	7875	7882	7889	7896	7903	7910	7917
62	7924	7931	7938	7945	7952	7959	7966	7973	7980	7987
63	7993	8000	8007	8014	8021	8028	8035	8041	8048	8055
64	8062	8069	8075	8082	8089	8096	8102	8109	8116	8122
65	8129	8136	8142	8149	8156	8162	8169	8176	8182	8189
66	8195	8202	8209	8215	8222	8228	8235	8241	8248	8254
67	8261	8267	8274	8280	8287	8293	8299	8306	8312	8319
68	8325	8331	8338	8344	8351	8357	8363	8370	8376	8382
69	8388	8395	8401	8407	8414	8420	8426	8432	8439	8445
70	8451	8457	8463	8470	8476	8482	8488	8494	8500	8506
71	8513	8519	8525	8531	8537	8543	8549	8555	8561	8567
72	8573	8579	8585	8591	8597	8603	8609	8615	8621	8627
73	8633	8639	8645	8651	8657	8663	8669	8675	8681	8686
74	8692	8698	8704	8710	8716	8722	8727	8733	8739	8745
75	8751	8756	8762	8768	8774	8779	8785	8791	8797	8802
76	8808	8814	8820	8825	8831	8837	8842	8848	8854	8859
77	8865	8871	8876	8882	8887	8893	8899	8904	8910	8915
78	8921	8927	8932	8938	8943	8949	8954	8960	8965	8971
79	8976	8982	8987	8993	8998	9004	9009	9015	9020	9025
80	9031	9036	9042	9047	9053	9058	9063	9069	9074	9079
81	9085	9090	9096	9101	9106	9112	9117	9122	9128	9133
82	9138	9143	9149	9154	9159	9165	9170	9175	9180	9186
83	9191	9196	9201	9206	9212	9217	9222	9227	9232	9238
84	9243	9248	9253	9258	9263	9269	9274	9279	9284	9289
85	9294	9299	9304	9309	9315	9320	9325	9330	9335	9340
86	9345	9350	9355	9360	9365	9370	9375	9380	9385	9390
87	9395	9400	9405	9410	9415	9420	9425	9430	9435	9440
88	9445	9450	9455	9460	9465	9469	9474	9479	9484	9489
89	9494	9499	9504	9509	9513	9518	9523	9528	9533	9538
90	9542	9547	9552	9557	9652	9566	9571	9576	9581	9586
91	9590	9595	9600	9605	9609	9614	9619	9624	9628	9633
92	9638	9643	9647	9652	9657	9661	9666	9671	9675	9680
93	9685	9689	9694	9699	9703	9708	9713	9717	9722	9727
94	9731	9736	9741	9745	9750	9754	9759	9763	9768	9773
95	9777	9782	9786	9791	9795	9800	9805	9809	9814	9818
96	9823	9827	9832	9836	9841	9845	9850	9854	9859	9863
97	9868	9872	9877	9881	9886	9890	9894	9899	9903	9908
98	9912	9917	9921	9926	9930	9934	9939	9943	9948	9952
99	9956	9961	9965	9969	9974	9978	9983	9987	9991	9996

To find the mantissa of, say, 546 (or 54.6 or 0.546, etc.) in Table A3.1, start by locating the first two digits (54) in the column headed N, then read across to the entry under the heading for the third digit (6). The table gives the value 7372, which is interpreted 0.7372.

The number of places to which the mantissa is carried depends only upon the size of the logarithm table. The log table in this book lists mantissas for numbers having from one to three digits (1 to 999) and is called a **four-place table**. The Handbook of Chemistry and Physics (Chemical Rubber Publishing Co.) contains a **five-place table** for numbers having from one to four digits (1 to 9999).

To illustrate how to find the characteristic of a number it is simpler to first consider only numbers which are multiples and submultiples of 10, e.g.,

N	C_N
1000 (10^3)	3
100 (10^2)	2
10 (10^1)	1
1 (10^0)	0
0.1 (10^{-1})	-1
0.01 (10^{-2})	-2
0.001 (10^{-3})	-3

Note that the characteristic is simply the exponent of the number written as a power of ten. Each of the above numbers has the mantissa 0.0000 $\cdots$. This is indicated in the logarithm table as 0000, i.e., without any decimal point. Thus

$$\log 1000 = 3.0000$$
$$\log 100 = 2.0000$$
$$\log 10 = 1.0000$$
$$\log 1 = 0.0000$$
$$\log 0.1 = -1.0000$$
$$\log 0.01 = -2.0000$$
$$\log 0.001 = -3.0000$$

Now consider a number such as 23 which is not a multiple or submultiple of 10. To get C_N note that 23 lies between 10^1 and 10^2. The characteristic of 23 is just the power of the 10 which is *smaller* than 23, i.e., $C_{23} = 1$. This means that 10 must be raised to some power greater than 1 but less than 2 to produce 23. The mantissa of 23 (found by the table to be 3617) indicates just what power between 1 and 2 must be used. Thus

$$\log 23 = C_{23} + M_{23}$$
$$= 1 + 0.3617 = 1.3617$$

Alternatively, this may be written in the exponential form

$$10^{1.3617} = 23$$

Note that 0.3617 is also the mantissa of 230, 0.23, etc.; only the characteristic for each of these will be different. Log 230 is found as follows: Since 230 lies between 10^2 and 10^3, its characteristic is 2 and log 230 = 2.3617.

To find log 0.23 note that 0.23 lies between 10^{-1} and 10^0. Since 10^{-1} is smaller than 10^0, the characteristic of 0.23 is -1. Thus

$$\log 0.23 = -1 + 0.3617 = -0.6383$$

or

$$10^{-0.6383} = \frac{1}{10^{0.6383}} = 0.23$$

However, when working back from logarithms to numbers, the notation -0.6383 is not convenient, because it suppresses the mantissa as found in tables. To find what number has the logarithm -0.6383, this logarithm must first be rewritten as $-1 + 0.3617$. To avoid this problem, it is customary to write $\bar{1}.3617$ or $9.3617 - 10$ instead of -0.6383. The bar over the 1 indicates that the characteristic (*only*) is negative; the same thing is accomplished by writing $9 - 10$. Note that the form $\bar{1}.3617$ (or $9.3617 - 10$) gives the mantissa at once and the $\bar{1}$ (or $9 - 10$) indicates where the decimal point belongs.

Extension of the above procedure to log 0.0023 yields $C_{0.0023} = -3$, $M_{0.0023} = 0.3617$ so that log 0.0023 = $\bar{3}.3617$ or $7.3617 - 10$.

A rapid way of determining characteristics is as follows: For numbers greater than 1, the characteristic is the number of digits to the left of the decimal point minus one; for numbers less than 1, the characteristic is the negative of one more than the number of zeroes immediately to the right of the decimal point. For example, $C_{1972.45} = 3$ and $C_{0.00014} = -4$ or $6 - 10$. Alternatively, rewrite the number in exponential form using only one digit to the left of the decimal point in the preexponential part. The exponent of 10 is the desired characteristic. For example, if $N = 156 = 1.56 \times 10^2$, $C_{156} = 2$ and if $N = 0.00156 = 1.56 \times 10^{-3}$, $C_{0.00156} = -3$ or $7 - 10$.

A3.2 THE USE OF LOGARITHMS IN COMPUTATIONS

There are three fundamental relationships involving logarithms which provide the keys to their most important practical uses. If A and B are any two positive numbers and n is any positive or negative number, these may be written

$$\log AB = \log A + \log B \tag{A3.4}$$

$$\log \frac{A}{B} = \log A - \log B \tag{A3.5}$$

$$\log A^n = n \log A \tag{A3.6}$$

The first relationship makes it possible to use logs for multiplication, the second for division, and the third for obtaining roots and powers.

EXAMPLE A3.1 Multiply 294 by 0.0045.

SOLUTION By Eq. (A3.4) and Table A3.1,

$$\log (294)(0.0045) = \log 294 + \log 0.0045$$
$$= 2.4683 + 7.6532 - 10$$
$$= 10.1215 - 10$$
$$= 0.1215$$

The log table shows that 1215 is the mantissa of 132. Since $C_N = 0$, it follows that N is 1.32. Thus

$$(294)(0.0045) = 1.32$$

Of course, as noted in Appendix 2, if the numbers 294 and 0.0045 arose from measurement, their product should have only two significant figures.

EXAMPLE A3.2 Divide 0.15 by 91.

SOLUTION By Eq. (A3.5) and Table A3.1,

$$\log \frac{0.15}{91} = \log 0.15 - \log 91$$
$$= 9.1761 - 10 - 1.9590$$
$$= 7.2171 - 10$$

The log table shows that 2171 is the mantissa of 165. Since $C_N = -3$, $N = 0.00165$ or 1.65×10^{-3}. Thus

$$\frac{0.15}{91} = 1.65 \times 10^{-3}$$

Again, only two significant figures should appear in the answer if 0.15 and 91 arise from measurement.

EXAMPLE A3.3 Find the seventh root of 31, i.e., evaluate $\sqrt[7]{31}$ or $(31)^{1/7}$.

SOLUTION By Eq. (A3.6) and Table A3.1,

$$\log (31)^{1/7} = \tfrac{1}{7} \log 31$$
$$= \tfrac{1}{7}(1.4914) = 0.2131$$

Since 2131 is the mantissa of 163 and $C_N = 0$, the number N is 1.63. Thus

$$(31)^{1/7} = 1.63 \text{ or } (1.63)^7 = 31$$

This calculation would be very difficult to do without the use of logarithms. Those with electronic calculators would simply use the $\sqrt[x]{y}$ key.

EXAMPLE A3.4 Raise 0.31 to the 15th power, i.e., evaluate $(0.31)^{15}$.

SOLUTION It is left to the reader to show that $\log (0.31)^{15} = 15 \log 0.31$, which leads to $(0.31)^{15} = 2.35 \times 10^{-8}$.

EXAMPLE A3.5 Evaluate

$$\frac{(72)(0.61)^{0.2}}{35} = N$$

SOLUTION Using a combination of Eqs. (A3.4), (A3.5), and (A3.6) leads to

$$\log N = \log 72 + 0.2 \log 0.61 - \log 35$$

$$= 1.8573 + 0.2(9.7853 - 10) - 1.5441$$

$$= 1.8573 + 9.9571 - 10 - 1.5441$$

$$= 0.2703$$

Thus $N = 1.86$

Note that $0.2(9.7853 - 10)$ comes out as $1.9571 - 2$, which is equivalent to $9.9571 - 10$, since $1 - 2 = 9 - 10 = -1$.

Negative numbers are treated as positive numbers and the correct sign is appended to the answer, e.g., to multiply 5 by -2, the result is just the negative of multiplying 5 by 2.

A3.3 NATURAL LOGARITHMS

Logarithms need not be restricted to numbers expressed as powers of 10; in theory any base other than 10 can be used.* For example, since $3^2 = 9$, 3 could be used as a base instead of 10. Consequently, the logarithm of 9 (to the base 3) is 2. This is written $\log_3 9 = 2$. By contrast, $\log_{10} 9 = 0.9542$. By convention, when the base is 10, the subscript 10 in $\log_{10}$ is omitted; otherwise the base must be indicated.

Many mathematical problems involving logarithms are most simply expressed when the base $2.7183\cdots$ (a number which, like pi, does not terminate) is used. A logarithm for this rather unusual base is written **ln** (for *l*ogarithm, *n*atural). Tables of natural logarithms are found in many mathematical handbooks, for example, *Handbook of Chemistry and Physics*. From such a table it can be found for example, that

---

* When Baron John Napier (1550–1617) invented logarithms (and coined the name) he made no use of exponents and used a base less than unity. A fellow countryman, Henry Briggs of London, pointed out the convenience of 10 as a base. Briggs deserves much of the credit for constructing the first reliable log tables.

ln 9 = 2.19722. The base 2.7183 is given the symbol e; thus $e^{2.19722} = 9$. Tables of e for a range of positive and negative powers are also given in handbooks.

Electronic calculators almost always have a key for ln x and may also have a separate key for log x (base 10). They also have keys for e^x and 10^x.* However, conversion between the two bases is very simple and is based on the relationship

$$e^{2.303} = 10 \quad \text{or} \quad 2.303 \log x = \ln x$$

Alternatively,

$$10^{1/2.303} = e \quad \text{and} \quad \log x = \frac{1}{2.303} \ln x$$

Many equations in the text have been set up for direct use with base-10 logarithms, for example, the radioactive decay equations

$$m = m_0 \, 10^{-kt/2.303} \quad \text{and} \quad \log \frac{m}{m_0} = \frac{-kt}{2.303}$$

If your electronic calculator has ln x and e^x, but not log x and 10^x, simply replace the above equations with

$$m = m_0 e^{-kt} \quad \text{and} \quad \ln \frac{m}{m_0} = -kt$$

that is, ignore the 2.303 (or its approximation, 2.3) and replace log with ln.

Warning: Don't foget that pH is normally defined in terms of base-10 logarithms! In terms of natural logarithms, the definition of pH is

$$\text{pH} = -\frac{1}{2.303} \ln [H_3O^+]$$

---

* Most calculators handle 10^x with a y^x key, where y may be any base.

APPENDIX 4
THE NAMING OF SIMPLE INORGANIC COMPOUNDS

A4.1 BINARY COMPOUNDS

A binary compound is a compound which consists of only two different elements, e.g., HCl, H_2O, NH_3, and CaC_2. The name of a binary compound is usually given in the form: name of the less electronegative element followed by the name of the more electronegative element with its last syllable (or, sometimes, the last two syllables) changed to *-ide*. Thus

HCl	hydrogen chloride
H_2O	hydrogen oxide (water)
NH_3	hydrogen nitride (ammonia)
CaC_2	calcium carbide
CaO	calcium oxide
K_2S	potassium sulfide
Al_2Se_3	aluminum selenide

Note that H_2O and NH_3 are usually not referred to by their correct chemical names as given above. This often happens in the case of compounds which have been known for a long time and which are quite common. The names *water* and *ammonia* are often said to be **trivial** names since they convey no information as to the composition of the molecule. Also, NH_3 should really be written H_3N (the less electronegative element first), but long usage has sanctioned the former.

There are some exceptions to the general rule. For example, the ion NH_4^+ is called ammonium and its salts named as binary compounds. Thus NH_4Cl is called ammonium chloride. Also, some negative ions (especially OH^- and CN^-) are named as if they derived from single elements, viz., hydroxide and cyanide, respectively. Thus $Ba(OH)_2$ is barium hydroxide and KCN is potassium cyanide.

If the less electronegative element is a metal and exists in two or more oxidation states, the oxidation state is included in the compound name as a roman numeral following the metal.

$FeCl_2$	iron(II) chloride
$FeCl_3$	iron(III) chloride
$MnCl_2$	manganese(II) chloride
MnO_2	manganese(IV) oxide

The same system may be used when the less electronegative element is a nonmetal, but this is not commonly done. For example, CO and CO_2 could be called carbon(II) oxide and carbon(IV) oxide, respectively, but the names carbon monoxide and carbon dioxide are preferred. The naming of the oxides of nitrogen and phosphorus is especially confusing since little uniformity exists in practice. NO is usually called nitric oxide although nitrogen(II) oxide or nitrogen monoxide are clearer. NO_2 is called nitrogen dioxide—a better choice than nitrogen(IV) oxide since the latter does not distinguish NO_2 from N_2O_4 (dinitrogen tetroxide).* N_2O is often called nitrous oxide but nitrogen(I) oxide or dinitrogen monoxide are clearer. P_4O_6 and P_4O_{10} are often called phosphorus trioxide and phosphorus pentoxide, respectively, but these names are terribly misleading. The names tetraphosphorus hexoxide and tetraphosphorus decoxide, although a mouthful to handle, are certainly unambiguous. Phosphorus(III) oxide and phosphorus(V) oxide are also used but these could mean P_2O_3 and P_2O_5, respectively. Some other binary compounds and their usual names are

PCl_3	phosphorus trichloride
PCl_5	phosphorus pentachloride
SO_2	sulfur dioxide
SO_3	sulfur trioxide

A4.2 ACIDS DERIVED FROM BINARY HYDROGEN COMPOUNDS

When a binary compound HA dissolves in water, a Brønsted-Lowry acid–base reaction occurs (see Section 13.1), viz.,

$$HA + H_2O \rightleftharpoons H_3O^+ + A^-$$

The solution is named by changing the *hydrogen* of the HA name to *hydro-* and the *-ide* ending of the A^- name to *-ic*, and putting the word *acid* after it all. Thus HCl in water becomes *hydrochloric acid*. Some other acid names are

HBr in water:	hydrobromic acid
HCN in water:	hydrocyanic acid
H_2S in water:	hydrosulfuric acid

---

* Unfortunately, most chemists call N_2O_4 *nitrogen* tetroxide!

When the acid is weak, the solution may be named as the binary hydrogen compound, e.g., H_2S in water may be referred to as a hydrogen sulfide solution.

A solution of ammonia in water is sometimes referred to as ammonium hydroxide since it contains some NH_4^+ and OH^-. The formula NH_4OH should never be used to represent this solution.

A4.3 ACIDS CONTAINING OXYGEN (THE OXY ACIDS)

The most important oxy acids and their names are:

- H_2SO_4 sulfuric acid
- HNO_3 nitric acid
- H_3PO_4 phosphoric or orthophosphoric acid
- $HClO_3$ chloric acid
- H_2CO_3 carbonic acid

There are some important acids which differ from the above by having *one less* oxygen atom. A few of these are:

- H_2SO_3 sulfurous acid
- HNO_2 nitrous acid
- H_3PO_3 phosphorous acid
- $HClO_2$ chlorous acid

Note that the *-ic* ending of the first set changes to *-ous* in the second set. There are also a few acids containing *two less* oxygen atoms than the *-ic* acids. Some of these are:

- $HClO$ hypochlorous acid
- H_3PO_2 hypophosphorous acid
- $(HNO)_2$ hyponitrous acid

Note that the names are obtained by using the prefix *hypo-* with the *-ous* name.

Occasionally acids are encountered which have *one more* oxygen atom than the *-ic* acid. These are named by using the prefix *per-* with the *-ic* name, e.g.,

- $HClO_4$ perchloric acid

A4.4 SALTS OF THE OXY ACIDS

The salts of the oxy acids are named as follows: The metal is named as in binary compounds (indicating the oxidation state, if necessary) and the rest is named by changing -ic endings to -ate, and -ous endings to -ite. For example,

Na_2SO_4	sodium sulfate
$KClO$	potassium hypochlorite
$NaClO_4$	sodium perchlorate
$Cu(NO_3)_2$	copper(II) nitrate
$NaHCO_3$	sodium hydrogen carbonate*
K_2SO_3	potassium sulfite
$LiNO_2$	lithium nitrite
$KHSO_4$	potassium hydrogen sulfate
NaH_2PO_4	sodium dihydrogen phosphate
$NaClO_2$	sodium chlorite

A4.5 TRANSITION-METAL COMPLEXES

Some of the rules used to name transition-metal complexes are the same as for ordinary inorganic compounds, but some additional complexity has to be introduced. According to the IUPAC system the following rules apply to cations such as $[Co(NH_3)_4Cl_2]^+$ and anions such as $(PtCl_6)^{2-}$.

1. As in ordinary inorganic salts, cations are named first and anions second. Thus $[Co(NH_3)_4Cl_2]Cl$ is named as a chloride and $K_2(PtCl_6)$ is named as a potassium salt.

2. Ligands are named in the following order and preceding the metal ion: anionic ligands first and then neutral ones. Thus in $[Co(NH_3)_4Cl_2]^+$ the order of names is: chloride, ammonia, cobalt. Note that the opposite order is used in the formula itself.

3. With two common exceptions, anionic ligands bear the ending -o, whereas neutral ones bear the name of the molecule. The two common exceptions are *aquo* for H_2O and *ammine* for NH_3. Thus $[Co(NH_3)_4Cl_2]^+$ has *chloro* and *ammine*, and $(PtCl_6)^{2-}$ has *chloro*. Names of the more common ligands are given in the table at the end of this appendix.

4. When more than one ligand of a given type is present, the Greek prefixes *di-*, *tri-*, *tetra-*, *penta-*, and *hexa-* are used. Thus $[Co(NH_3)_4Cl_2]^+$ has *dichloro* and *tetraammine* and $(PtCl_6)^{2-}$ has *hexachloro*. An exception occurs whenever the ligand name itself contains a Greek prefix *di-*, *tri-*, etc., such as in ethylenediamine (en). In this case alternate prefixes, *bis-*, *tris-*, *tetrakis-*, *pentakis-*, and *hexakis-* are used in the following manner: $[Co(en)_3]^{3+}$ would contain tris(ethylenediamine) with the ligand name in parentheses.

---

* This compound is also commonly called sodium *bicarbonate* and sodium *acid carbonate*. The former name, which is especially misleading, originated back in the days when chemical compositions and structures were not accurately known and before systematic approaches to naming were developed.

TABLE A4.1 NAMES OF SOME IMPORTANT LIGANDS

FORMULA OR SYMBOL	USUAL NAME	LIGAND NAME	DENTATE NUMBER
H_2O	Water	Aquo	1
NH_3	Ammonia	Ammine	1
Cl^-	Chloride	Chloro	1
en	Ethylenediamine or 1,2-diaminoethane	Ethylenediamine	2
CN^-	Cyanide	Cyano	1
CNO^-	Cyanate	Cyanato	1
CNS^-	Thiocyanate	Thiocyanato	1
CO_3^{2-}	Carbonate	Carbonato	1
NO	Nitric oxide	Nitrosyl	1
CO	Carbon monoxide	Carbonyl	1
OH^-	Hydroxide	Hydroxo	1
S^{2-}	Sulfide	Sulfo or thio	1
$^-O-C(=O)-C(=O)-O^-$	Oxalate	Oxalato	2
NO_2^-	Nitrite	Nitro[a]	1
		Nitrito[b]	1

[a] If coordinated through nitrogen.
[b] If coordinated through oxygen.

5. The metal ion is named with its elemental name followed by its oxidation state in roman numerals. Thus the full name of $[Co(NH_3)_4Cl_2]^+$ is dichlorotetraamminecobalt(III) cation. The compound $[Co(NH_3)_4Cl_2]Cl$ is called dichlorotetraamminecobalt(III) chloride.

6. Anionic complexes have names ending in -*ate*. Thus $(PtCl_6)^{2-}$ is called hexachloroplatinate(IV). Sometimes the Latin name of the metal is used; for example, $[Fe(CN)_6]^{4-}$ is called hexacyanoferrate(II). The compound $K_2(PtCl_6)$ is called potassium hexachloroplatinate(IV).

Those interested in the naming of more complex compounds should consult a textbook of advanced inorganic chemistry or the *Handbook of Chemistry and Physics* (Cleveland: Chemical Rubber Co.).

APPENDIX 5
THE CONSTRUCTION AND USE OF CONTOUR DIAGRAMS

In order to understand how contour diagrams are used to represent electron charge distributions in atoms and molecules, it is helpful to first understand something about their use in a much simpler situation, i.e., the representation of a three-dimensional surface by means of a two-dimensional diagram. Topographic maps which represent the hills and valleys of the earth's surface are used in this sense. Those of you who have been in Scouts or who have done serious hiking in the White Mountains, the High Sierras, the Rockies, the Great Smokies, or other wilderness areas will already know how to obtain an accurate idea of terrain by the use of a topographic map. However, this does not mean that you can interpret an electron density diagram in a conceptually correct manner since some very important mental adjustments must be made in going from solid surfaces to ethereal quantities such as electron "clouds."

Figure A5.1 shows the construction of a contour diagram of a simple surface—that of a cone. Note that each contour line represents a "path" circumscribing the cone at a given distance above the base line. Furthermore, successive contour lines represent "paths" which are separated by a *vertical* distance of 20 ft. This latter quantity is called the **contour interval**. The point at the center is the highest contour line and represents the summit (if the summit does not coincide with the contour interval it is indicated by a dot marked with the summit elevation, e.g., 95 ft). Note that the contour diagram represents what would actually be seen by looking down on the cone from immediately above the summit and if the contour "paths" were marked on the cone surface itself.

Figure A5.2 shows a contour diagram representation of a hill shaped like a cone whose sides become increasingly steeper in going from the base to the summit. The contour diagram is characterized by contour lines which become closer and closer together in going from the outer contour to the inner contours. Again, the contour diagram represents what would actually be seen by looking down on the cone from immediately above the summit and if the contour "paths" were marked on the cone surface itself.

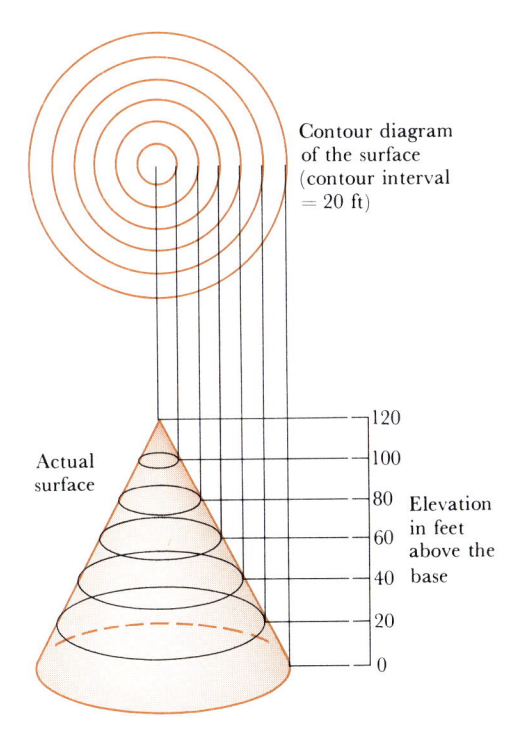

Fig. A5.1. Illustration showing the construction of a contour diagram of the surface of a simple cone. A trained observer looking at the contour diagram imagines he or she is looking down on the cone from a point immediately above its tip. The fact that the contours are uniformly spaced indicates that the sides of the cone have a constant steepness.

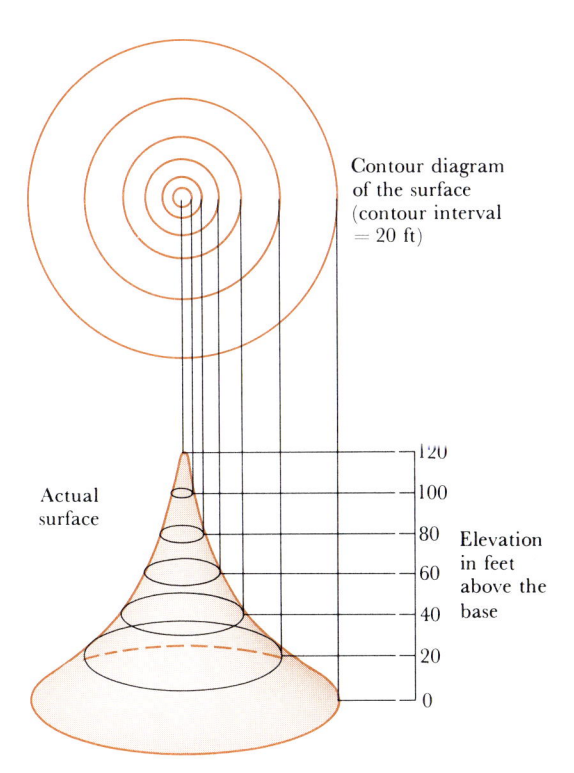

Fig. A5.2. Contour diagram of a "cone" whose sides become increasingly steeper from base to summit. Note that the contour lines become closer and closer together in going from the outer contour inwards.

APPENDIX 5: THE CONSTRUCTION AND USE OF CONTOUR DIAGRAMS   **729**

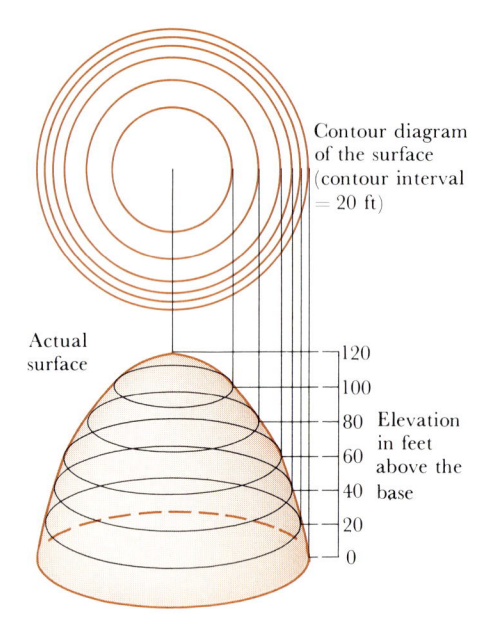

Fig. A5.3. Contour diagram of a "cone" whose sides become decreasingly steeper from base to summit. Note that the contour lines become farther and farther apart in going from the outer contour inwards.

Fig. A5.4. Contour diagram of a hill with two peaks—an aiguille (left) and a dome (right). The contour diagram shows that the aiguille is about 20 ft shorter than the dome and that the col between the two peaks dips down to between 40 and 60 ft. With more closely spaced contours, details of the hill could be represented more accurately. The line AB represents a traverse which passes to the right of the dome.

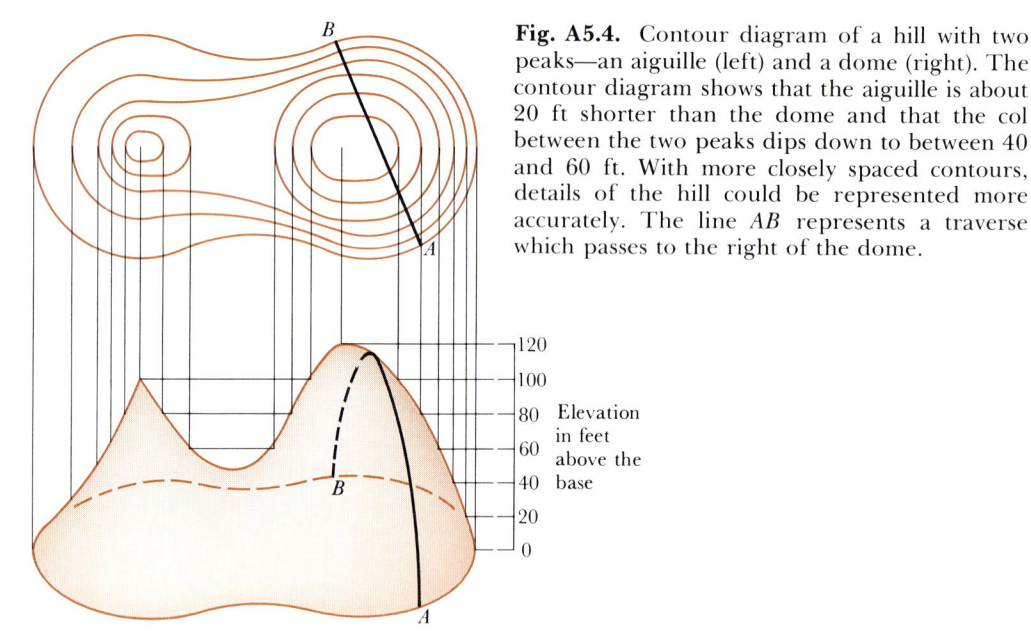

Figure A5.3 shows a contour diagram representation of a hill shaped like a cone whose sides become increasingly less steep in going from the base to the summit (an inverted bowl). The contour diagram is characterized by contour lines which become farther and farther apart in going from the outer contour to the inner contour.

Figure A5.4 shows a contour diagram of a surface which is a composite of those shown in Figs. A5.2 and A5.3. The diagram has been drawn to indicate that the left peak is 20 ft lower than the right peak. The col between the two peaks has a maximum elevation between the 40-ft and 60-ft contour lines.

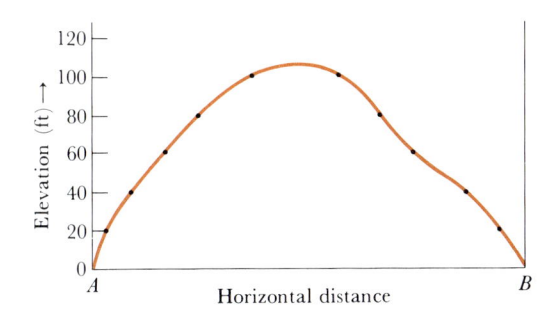

Fig. A5.5. A profile of the traverse shown in Fig. A5.4.

The line labeled A to B in Fig. A5.4 represents a traverse which passes near the higher peak of the hill. Figure A5.5 shows what the profile of this traverse would look like. This also shows the shape of a cross-section of the hill such as would result from slicing the hill with a huge knife along the line A to B. Mountain climbers use such profiles to enable them to see the ups and downs of a particular trail and their steepnesses. The profile need not be restricted to a straight-line path as shown. It is easy to deduce that the profile in Fig. A5.5 is constructed from the contour diagram in Fig. A5.4 by plotting the horizontal location of the point where the traverse intersects a contour line against the vertical distance represented by that contour.

In using contour diagrams to represent electron densities certain mental readjustments must be made. Most important, it must be remembered that it is the *magnitude* of the electron density at a point *in an arbitrary plane* which is analogous to the *elevation* of a point on a surface, and that this electron density varies within a plane as do the elevations of the points on the surface. Many students fall into the trap of visualizing electron density contours as shapes of electron clouds, and thereby they are led to conceptually incorrect interpretations. The contours do *not* attempt to show what the electron "cloud" would look like when viewed from afar but rather how its "thickness" or opacity changes from point to point in an arbitrary plane as viewed from within the atom or molecule. For example, the electron density represented by a 1s orbital changes from point to point *in a plane containing the nucleus* in a way represented by the contour diagram of Fig. A5.2. To correctly show how the 1s electron density varies in the three-dimensional space about the nucleus would require contour diagrams for a series of planes parallel to that represented in Fig. A5.2. For example, a contour diagram of the electron density in a plane whose closest distance from the nucleus is 1 Å would resemble that of Fig. A5.2 but the maximum density (the elevation of the summit) would be less than in the nuclear plane. A profile constructed along a line passing through the nucleus would look as in Fig. A5.6. A profile for the 1-Å plane could have the same general shape but would have a lower peak in the center.

Electron density in an atom or molecule may be crudely compared with the distribution of sugar grains in the air resulting when an astronaut in a weightless environment breaks a bag of sugar and the grains begin to slowly float about to fill the space capsule. The astronaut could record the distribution of sugar grains in space (at any given instant) by mentally subdividing the capsule compartment into a series of arbitrary planes and counting the number of sugar grains per unit area over the entire surface of each plane. The number of sugar grains per cm^2 of the plane's

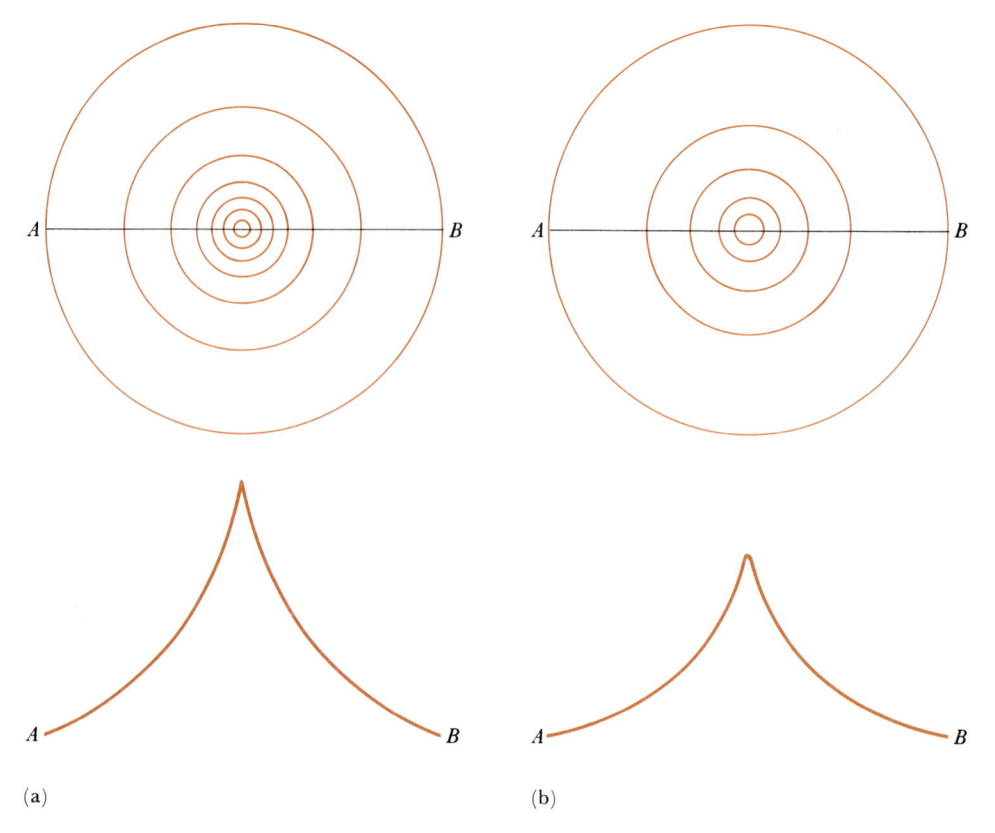

Fig. A5.6. Contour diagrams and profiles of the electron density represented by a 1s orbital. In (a) the density illustrated is in a plane containing the nucleus; in (b) the density illustrated is in a plane whose closest approach to the nucleus is 1 Å.

surface at every point of the surface would be analogous to the elevations of various points on the surface of the earth and could be illustrated by a contour diagram. Each plane, however, would have to be represented by a new contour diagram—a situation which has no analogy in the example of the earth's surface. Similarly, each different contour diagram representing the sugar grain density distribution of a given plane would have its own set of profiles. In the case of the earth's surface there is only one such contour diagram and its accompanying profiles.

Most of the contour diagrams used in this text (Chapters 4 and 5) represent the electron densities in planes containing the nucleus, or—in the case of molecules—two or more nuclei. The profiles constructed from these are usually adequate for acquiring an accurate idea of how the electronic density in an atom or molecule varies in space. It is not too difficult to guess from these what contours and profiles in other planes would look like, but the need for this seldom arises at elementary levels of discussion.

ANSWERS TO SELECTED PROBLEMS

CHAPTER 1

1. O/N weight ratios = A/B/C = 1.141/0.572/2.289 = 2/1/4
2. a) 83, 126, 83 b) 30, 36, 28 c) 27, 32, 27 d) 36, 48, 36 e) 9, 10, 10
3. A: 45, 58, 45; B: 27, 32, 27; C: 75, 33, 33; D: 27, 27, 27; E: 53, 103, 53
4. 17, 8, 10 **5.** 35.46 amu **6.** 74.49% ^{35}Cl, 25.51% ^{37}Cl **13.** 6.1×10^{24} kg **14.** 4 g·cm^{-3}
15. a) 1.738×10^3 kg·m^{-3} b) 13.99 cm^3·mol^{-1} c) 2.323×10^{-23} cm^3·atom^{-1}
16. 210 g **17.** 10.5 g·cm^{-3} **18.** 6.8 kg
25. 1.91×10^{16} yr **26.** 200 km **27.** 9.03×10^{23} **31.** 0.35 J·g^{-1}·K^{-1} **32.** 2×10^2 J·kg^{-1}·K^{-1} **33.** 9.649×10^4 J **34.** 10,460 kJ **35.** 1.31×10 kJ
36. a) 0.98 J b) 0.98 J c) 0.98 J d) Sum of KE and PE is constant.
44. 57×10^4 W·m^{-2} **45.** N·m^2·C^{-2} or J·m·C^{-2} **46.** 1.2 g·cm^{-3}
47. a) Yes b) 2.6 m·s^{-1}

CHAPTER 2

1. A: NO; B: N$_2$O; C: NO$_2$
2. a) 82% C, 18% H b) C$_3$H$_8$ c) No (need mol wt)
3. a) C$_3$H$_4$N b) C$_6$H$_8$N$_2$ c) 108.1456 amu
4. C$_{23}$H$_{26}$O$_2$ **5.** CHCl$_3$ **6.** C$_2$HCl
7. C$_{10}$H$_{14}$N$_2$ **8.** a) CF$_3$ b) C$_2$F$_6$ **9.** FeSO$_4$
20. a) 1, 1, 2 b) 1, 2, 1, 2 c) 2, 2 d) 1, 3, 2 e) 1, 2, 2, 1 f) 3, 1, 1 g) 1, 5/4, 1, 3/2 h) 1, 12, 12, 11 i) 1, 4, 6, 6 j) 1, 7, 6, 2, 2
21. a) 3Fe + 2O$_2$ → Fe$_3$O$_4$ b) 4NH$_3$ + 5O$_2$ → 6H$_2$O + 4NO c) NH$_3$ + $\frac{7}{4}$O$_2$ → $\frac{3}{2}$H$_2$O + NO$_2$ d) C$_6$H$_6$ + Cl$_2$ → C$_6$H$_5$Cl + HCl e) CuSO$_4$ + 5H$_2$O → CuSO$_4$·5H$_2$O f) B$_2$H$_6$ + 3O$_2$ → 3H$_2$O + B$_2$O$_3$
24. a) H$_2$CO$_3$ → H$_2$O + CO$_2$ b) 0.25 mol c) 4.5 g
25. a) 9.5 lb b) 128.6 klk
26. a) 1, 3, 2 and 2, 2, 2, 1 b) 46 g c) 14 g Li left
27. 153 amu
28. a) 1028 g b) 253.5 g
29. a) 280 g b) 400 g c) 2.5 mol NO, 3.75 mol H$_2$O produced, 3.38 mol NH$_3$ unreacted
30. a) 39.8 g b) 0.25 mol
31. a) 5.94 mol b) 154.4 g
40. RbClO$_3$: +1, +5, −2; SnO$_2$: +4, −2; K$_2$Cr$_2$O$_7$: +1, +6, −2; NH$_4$Br: −3, +1, −1; H$_3$PO$_3$: +1, +3, −2; H$_3$PO$_4$: +1, +5, −2; S$_2$O$_3^{2-}$: +2, −2; Fe$_2$O$_3$: +3, −2; FeO: +2, −2; Fe$_3$O$_4$: +$\frac{8}{3}$, −2; S$_4$O$_6^{2-}$: +2.5, −2; O$_3$: 0; HN$_3$: +1, −$\frac{1}{3}$
41. a) N$_2$O, NO, N$_2$O$_3$, NO$_2$, N$_2$O$_5$ b) NO$_2^-$, NO$_3^-$
43. a) I b) D
44. 2.46×10^{-23} cm^3 **45.** 4
46. a) 342 b) 42.1% C, 6.4% H, 51.5% O
47. a) 1.25×10^{23} b) 3.75×10^{23} c) 0.21 d) 3.09×10^{21}.

CHAPTER 3

1. a) 9 L b) 20 torr c) 20 lb·in.$^{-2}$ d) 10 M
2. 1 k
6. a) 200 L b) 600 K (327 °C) c) 20 cm^3 d) 15 bbl
7. 20 G **11.** 6.4 L **12.** 488 K, 215 °C **13.** 0.44 mol **14.** 690 **15.** 1196 K (923 °C)
16. a) 600 torr b) 644 torr
17. a) 140 torr b) 800 torr c) 940 torr
18. a) 0.0041 mol b) 0.0020 mol c) 0.25 atm
19. 0.023 g·L^{-1} **20.** 32.8 amu **21.** 1.93 g·L^{-1}
22. 1.5 to 1.9 mol CO per yr
24. a) 415 kPa b) 161 L

25. a) 20.5 atm   b) 79.5 atm   c) 19.4 mol   d) 89.8 atm (69.3 atm O_2, 20.5 atm H_2O)
26. 456 torr
27. 4366 m (14,321 ft). Official elevation: 14,410 ft
29. 6.9×10^2 m·s^{-1} **30.** 2, T has no effect
31. 0.4%
35. a) 7.688 atm   b) 7.632 atm
36. (c) is best
37. Multiply by V^2, multiply out, regroup in powers of V
38. a) 224 K   b) Yes, above $T = 2a/Rb$
40. a) 328 K   b) 4.6×10^4 g
41. Buoyancy depends on different masses of equal volumes, thus different gravitational forces.
42. 1.1×10^{18} kg   **43.** 1.9×10^{11}
44. a) 41 °C   b) -62.5 °M   c) -87.5 °M, 162.5 °M
45. a) 150 ft^3 b) 1190 ft^3
46. 1.67 atm   **47.** 8X

CHAPTER 4

1. 24 m·s^{-1} **2.** 5×10^6 m
3. a) 50 s^{-1} b) 2×10^6 nm, 2×10^7 Å
6. No   **7.** 1.03 eV
8. a) 326 nm   b) 442 nm
12. 2.6×10^{-4} m·s^{-1}
13. Man: 7.9×10^{-36} m; no; electron: 1.2 Å; yes
16. 0.544 eV
17. a) 4.8×10^{-19} J, 3.0 eV   b) 7.3×10^{14} s^{-1} c) 411 nm, 4110 Å
18. a) 0.136 eV   b) 122.4 eV   c) 489.6 eV
19. 138 nm, emitted
21. a) 19×10^{-19} J   b) 198 nm   c) Emit photon of energy 14×10^{-19} J (wavelength 141 nm)   d) 2.13×10^{15} s^{-1}
22. b) 700 nm, 2333 nm   c) 1.13×10^{-19} J   d) 1.3×10^{14} s^{-1}
23. a) 2×10^{-16} J, 1240 eV   b) 3×10^{16}, 3×10^{19} s^{-1}
24. a) 4.0×10^{-19} J   b) 2.5 eV
35. a) (Ar)$4s^2 3d^{10} 4p^6$ b) (Ar)$3d^2$ c) (Ar)   d) (He)$2s^2 2p^6$ e) (He)$2s^2 2p^6$ f) (He)$2s^2$
36. l = 3, 2, 1, 0; orbitals are 4f, 4d, 4p, 4s
37. a) 6   b) 9   c) 2
38. All but zinc are paramagnetic.
46. Atoms are quanta (smallest units) of matter.
47. a) Z   b) X   c) Z
48. 2s density is maximum at nucleus, drops to zero, increases and then decreases again; 2p density zero at nucleus, increases to a maximum, then decreases again.
49. Quantum jumps in nature are usually small. Terms abrupt, sudden, or discontinuous should be used.

CHAPTER 5

1. F: 3.9; Br: 2.8; I: 2.5; Li: 1.0; Cs: 0.7
2. $\chi = 0$ implies $I = E$.   **3.** a) D   b) M
7. 6.1 eV   **8.** Na_2S **9.** C—F   **10.** Nonpolar
11. a) D   b) covalent   c) A   d) BE_2 e) D
17. a) H:C::N: b) H:N::N:H   c) H:C̈:C̈:H with H H above and below
d) :C::C: with H H   e) H:C::C:H   f) [:Ö:Cl̈:]$^{1-}$ g) [:C::N:]$^{1-}$
h) :Cl̈:P̈:Cl̈: with :Cl̈: below   i) H:Ö:Ö:H   j) [:Ö:P:Ö: with :Ö: above and :Ö: below]$^{3-}$
18. :Ö:Ö::Ö ↔ Ö::Ö:Ö:
19. :N̈:Ö::N̈, N̈::N̈::Ö ↔ :N̈::N:Ö: ↔ :N̈:N:::Ö:
20. [Ö::N:Ö:]$^{1-}$ ↔ [:Ö:N::Ö]$^{1-}$
29. a) N_2 b) CO^+
30. [:C::N:]$^{1-}$, $(1\sigma)^2 (1\sigma^*)^2 (2\sigma)^2 (2\sigma^*)^2 (3\sigma)^2 (1\pi)^4$
31. All have nonzero bond orders.
32. Shortest to longest: O_2^+, O_2, O_2^-, O_2^{2-}. All are stable.
33. Longest to shortest: OF^-, OF, OF^+ (only OF is paramagnetic).
40. a) 4.61×10^{-19} J   b) 1 Å   c) 3.16 Å
41. 2.85 nm   **42.** 96.3 kJ, 23 kcal
43. Ionization energies too large.

CHAPTER 6

1. Zero
2. a) II   b) Either I or III   c) I (↔), II (↕)
3. Bond moments occur in equal and opposite pairs.
4. III(0), II(1.72 D), I(2.50 D)
5. Calculated value 11.5 D. Charge separation really not complete.
6. N—F and lone-pair dipoles almost cancel. In NH_3 these act in same direction.
7. Expect large dipole (exp. 2.33 D).
15. a) Linear ($\mu = 0$)   b) Linear ($\mu \neq 0$)   c) Bent ($\mu \neq 0$)   d) Distorted tetrahedron ($\mu \neq 0$)
e) regular tetrahedron ($\mu = 0$)   f) Distorted trigonal pyramid ($\mu \neq 0$)   g) Bent ($\mu \neq 0$)   h) Two tetrahedra with central O in common ($\mu \neq 0$)
i) Zigzag ($\mu \neq 0$ but small)   j) Linear ($\mu = 0$)
k) Linear ($\mu = 0$)   l) Trigonal bipyramid ($\mu = 0$)
m) Trigonal pyramid with dog-leg ($\mu \neq 0$)
n) Bent ($\mu \neq 0$)   o) Triangle about S ($\mu = 0$)

16. a)

[structure: H–C(=O)–O–H with angle arrow]

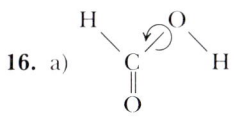

b) iii, since OH is bulkier than H and squeezes the HC=O angle
18. a) Linear b) Linear c) H₂NN as trigonal pyramid with two H as dog-legs d) Bent e) Trigonal pyramid f) Tetrahedral g) Bent h) Planar (O's form triangle about C) i) Trigonal pyramid j) Trigonal pyramid k) Linear l) Square planar (Xe in center of square with F's at corners) m) Planar n) Planar except for atoms on N and S
19. 1.75 Å 21. a) I with Cl's (structure) b) Br–F (structure)
c) Octahedral about Sn
29. Lone pairs of F best described by sp^3.
30. Expect lone-pair repulsions of O's to make it nonplanar.
31. H—C is $1s(H) + sp^2(C)$, C=O is $sp^2(C) + sp^2(O)$ for σ and $2p(C) + 2p(O)$ for π, C—O is $sp^2(C) + sp^3(O)$, O—H is $sp^3(O) + 1s(H)$, lone pairs on C=O are $sp^2(O)$, lone pairs on C—O are $sp^3(O)$
32. C—O is $sp(C) + sp^2(O)$ for σ, $2p(C) + 2p(O)$ for π, lone pairs on O are $sp^2(O)$
33. Use $sp^2(O)$ for σ, $2p(O)$ for π, $sp^2(O)$ for lone pairs; use $sp(N)$ for σ, $2p(N)$ for π's, $sp(N)$ for lone pairs.
34. Cannot get good overlap, geometry more likely :N≡N—N̈—H (bent about NNH).
35. $2p(B) + 2p(F)$ in which *both* electrons come from the more electronegative element is unlikely.
42. Ethanol (due to H-bonding)
43. CH₄, SiH₄, GeH₄, PbH₄
44. a) Dispersion b) and c) Dispersion, dipole–dipole, and H-bonding d) Dispersion e) Dispersion and dipole-induced dipole f) and g) Dispersion and dipole–dipole h) Dispersion
45. a), d), and h) Dispersion b) and c) H-bonding e) Dipole-induced dipole f) and g) Dipole–dipole

CHAPTER 7

1. a) 3.36 Å b) 29.6°
2. 2.81 Å
3. Atoms are mostly empty space.
6. 63.56 g·mol⁻¹ 7. 0.03 cal·K⁻¹·g⁻¹
10. Cl₂ has more electrons, hence greater van der Waals attractions.
11. 6.8×10^2 kJ·mol⁻¹ 14. 1:3 15. 1.2×10^{19}

16. a) 3.14 Å b) 1.28 Å
17. 6.26×10^{23}

CHAPTER 8

1. 28.8 mm 2. 29.1 mm 3. 24.9 mN·m⁻¹
4. 1486 mm, 14.86 mm 8. 18.4 mm, 29.1 mm
9. B 10. A
13. Orderly arrangement due to intermolecular interactions is optimized at some concentration ratio
14. a) Large contact areas b) Intermolecular attractions greater than molecular kinetic energies c) Above a given T, kinetic energies prevent ordered alignment.
15. Smaller average intermolecular distances
16. NH₃, due to H-bonding
17. a) 3×10^{-23} cm³ b) 2.5×10^{-23} cm³

CHAPTER 9

1. At least 34 torr b) Wait for pressure to reach a limiting value.
2. 59% 3. 80% 4. 19.3 torr 5. 23.8 torr
6. 1.28×10^3 mol
7. B.p. of H₂O too low (not enough heat to cook an egg in 3 min)
8. Use Eq. (9.1) for each temperature, eliminate B, solve for A; ΔH_{vap} is about 60 kJ·mol⁻¹.
17. 30.7 kJ 18. 44.4 J 19. 27 °C 20. 8.16 g
25. a) Increases m.p. b) 34 atm·K⁻¹ c) gas
26. Steep line means large pressure change needed to cause small change in m.p.
27. No matter how large p is, when $T > T_c$, point always falls in gas area.
33. Shrinks when warmed.
34. Some ice melts. 37. 1.6×10^{10} kg
38. a) SO₂ (has lower v.p.) b) SO₂ (higher T_c)
39. Joule-Thomson effect, T drops as gas expands, some liquid freezes.
40. Evaporation is endothermic, heat removed from hands.
41. Ionic bonds are very strong.
42. All involve changing relative positions of adjacent atoms or molecules.

CHAPTER 10

1. 39.4 g
2. a) 0.31 b) 25.1 m c) 34.7 M
3. Benzene (0.70)
4. a) 349 cm³ b) 1.43 M c) 1.39 M d) 0.976
5. 0.87 g·cm⁻³ 6. 39 cm³

ANSWERS TO SELECTED PROBLEMS **735**

7. a) 200 cm³ b) 0.20 mol c) 0.5 M
8. Use $\frac{1}{3}$L of 6-M HCl and add H$_2$O to total volume of 1 L.
9. 18.2 M 10. Dilute 0.060 L H$_2$SO$_4$ to 1 L.
18. a) Unsaturated b) Unsaturated c) Supersaturated
19. Exothermic
20. a) 0.33 b) 0.29
21. a) -10 kJ·mol^{-1} b) Less soluble at 30 °C
22. Rigid structure and order disrupted in both cases.
23. NaCl(s) → NaCl(aq) is endothermic, cools sidewalk.
24. Each layer is a dilute solution of one component in the other.
30. a) 2×10^{-5} M·torr^{-1} b) 3×10^{-3} M, c) k decreases.
31. 5.3×10^{-4} M·atm^{-1}; yes
32. m, M almost equal for dilute solutions.
35. a) 46.6 torr b) Benzene: 35.1 torr; toluene: 11.5 torr c) 0.75
36. a) 74.5 torr b) 0.19
40. a) 320 g·mol^{-1} b) 2.73 m^{-1}
41. a) 52 g·mol^{-1} b) 50 amu c) A$_4$B$_2$
42. a) 2.5 m^{-1} b) 0.12 m c) 18 g·mol^{-1}
43. 100 g·mol^{-1} 44. S$_8$ 45. 102.1 torr, 1389 mm 46. 1.23 L
47. Can't distinguish between AB$_2$ and A$^+$, AB$_4^-$.
48. Pure H$_2$O, 0.5-m alcohol, 0.5-m NaCl, 2-m KCl, 1-m CaCl$_2$
49. b) $i = 3$

CHAPTER 11

1. Test BaSO$_4$ ppt with a Geiger counter.
2. a) $^1_1$H b) $^1_0$n c)$^{21}_{10}$Ne d) $^{62}_{28}$Ni e) $^{229}_{90}$Th
3. a) $^{101}_{43}$Tc → $^{0}_{-1}$e + $^{101}_{44}$Ru b) $^{126}_{57}$La → $^{0}_{1}$e + $^{126}_{56}$Ba c) $^{148}_{64}$Gd → $^{4}_{2}$He + $^{144}_{62}$Sm
4. Many of U-235's fission products emit β.
9. a) $1\frac{2}{3}$ yr b) 0.41 yr^{-1} c) 5.1 yr
10. a) 69 min b) 1.0×10^{-2} min^{-1}
11. $t_{1/2}$ (yr): 2, 12, 33.3, 66.7, 80, 84, 1000, 1600
18. a) 0.008485 amu b) 2.6 MeV
19. 7.7×10^8 K   20. 19 MeV   21. 3.7 MeV
26. a) About 3760 B.C. b) Unlikely but possible
27. 1928   28. 2529 yr ago
29. Need to know natural level of C-14.
33. 1.0045, slow but it works.

CHAPTER 12

1. $K_c = 10$ 2. 0.0022 M 3. 1.44×10^{-2}

4. a) [A] = 0.008 M, [B] = 0.019 M, [D] = 0.001 M b) 0.82
5. a) [H$_2$] = 0.047 M (orig. amt. NO should be 0.01 mol), [N$_2$] = 0.0015 M, [H$_2$O] = 0.103 M b) 147
6. a) 0.015 M b) 50 c) $p_{H_2} = 0.74$ atm, $p_{I_2} = 0.02$ atm, $p_{HI} = 0.89$ atm d) 1.65 atm
7. 1.6×10^{-3} 8. $K_x = K_p p^{-\Delta n} = K_c (n/V)^{-\Delta n}$
9. 0.065 M 10. N$_2$: 27 g; H$_2$: 3.86 g
11. [AC][B]/[A][BC]
16. a) [O$_3$]²/[O$_2$]³ b) [DCl][HBr]/[HCl][DBr] c) [Ba^{2+}][F$^-$]² d) [H$_2$O]³/[H$_2$]⁶ e) [Cl$_2$]²/[HCl]⁴[O$_2$]
17. a) $K_p = 4 \times 10^{-11}$, $K_c = 1.6 \times 10^{-12}$ b) 6.3×10^{11}, reaction virtually complete
20. a) [C]/[A]²[B] b) 50 c) →, ←, ←
21. a) Decrease, increase b) Decrease, decrease c) Decrease, increase d) In each case, product added is increased, other product is decreased.
22. a) No effect in either b) ←, no effect
23. a) 0.63 b) A(1.15 atm), B(0.85 atm) c) At 1 atm, X$_A$ = 0.46; at 2 atm, X$_A$ = 0.56
24. Endothermic
31. a) PbI$_2$(s) ⇌ Pb^{2+}(aq) + 2I$^-$(aq) b) [Pb^{2+}][I$^-$]² c) 0.69 g·L^{-1} d) 0.13 g e) Endothermic
32. 1.95×10^{-4}
33. AgBr ppts first, takes 7.7×10^{-6} drops!
34. a) 16.8 mg·L^{-1} b) 0.03 mg·L^{-1} ([KF] should be 0.01 M) c) 4.91 mg·L^{-1} ([CaCl$_2$] should be 0.01 M)

CHAPTER 13

1. ClO$_2^-$, H$_3$O$^+$, NH$_2^-$, CO$_3^{2-}$, H$_2$CO$_3$, (ZnO$_2$H)$^-$
2. Products, direction (L or R): a) HNO$_3$ + OAc$^-$, L b) HOAc + CN$^-$, L c) SO$_4^{2-}$ + H$_2$S, R d) HS$^-$ + H$_2$S, neither e) H$_2$CO$_3$ + CO$_3^{2-}$, L
3. Poisonous HCN produced, $K_c = 1.7 \times 10^7$, unsafe.
4. a) Base b) Oxyhemoglobin/hemoglobin
5. a) H$_3$O$^+$ b) S^{2-}
6. K_a(NH$_3$) = $10^{-14}/K_b$(NH$_2^-$), K_b(Cl$^-$) = $10^{-14}/K_a$(HCl), NH$_2^-$ is a strong base, HCl is a strong acid.
7. H$_3$O$^+$ + H$_2$O ⇌ H$_2$O + H$_3$O$^+$, OH$^-$ + H$_2$O ⇌ H$_2$O + OH$^-$
8. 4×10^{-23} 9. 3.6×10^{-5} M 10. 4×10^{-5}
17. 0.042   18. 0.02 M 19. 0.95
20. a) i = moles solute/c = $[c(1 - \alpha) + \alpha c + \alpha c]/c = 1 + \alpha$
b) ΔT_f gives i, i gives α, α gives K_a (or K_b)
23. a) 7 b) 3 c) 11 d) 4.9 e) 9.4 f) 12
24. There is a limit to the number of H$_3$O$^+$ in a fixed volume.

25. 0.073 M
26. a) 11 b) 2
27. $[H_3O^+]$ from self-ionization cannot be neglected.
28. a) 10^{-12} M b) 2.4×10^{-11} M
29. a) 3.9×10^{-9} b) 6.6
30. a) 4.5×10^{-11} b) 8.8
34. $HCN + H_2O \rightleftharpoons H_3O^+ + CN^-$; HCN resupplies H_3O^+; CN^- uses up extra H_3O^+
35. $C_2H_5NH_2 + H_2O \rightleftharpoons C_2H_5NH_3^+ + OH^-$; $C_2H_5NH_2$ resupplies OH^-; $C_2H_5NH_3^+$ uses up extra OH^-
36. 5.6 37. 1.76
38. Small $[A^-]$ provides little capacity to remove H_3O^+.
40. 9 41. b) 7×10^{-10} 42. 1667 cm³
43. a) 20 g b) 1 N
44. 0.011
48. Lewis acid/Lewis base: a) BF_3/F^- b) CO_2/OH^- c) Ag^+/NH_3^+

49. H:B̈:C::Ö:, CO is a Lewis base.
    H   H

CHAPTER 14

1. a) Ion-transfer or acid–base, no redox b) Addition, redox c) Decomposition, redox d) Addition, no redox e) Disproportionation, redox f) Displacement, redox g) Substitution, no redox h) Acid–base, redox i) Elimination, redox
2. Substitution and acid–base
3. H^- is a proton acceptor, H changes oxidation number from -1 to 0.
7. a) $5C_7H_8 + 18H^+ + 6MnO_4^- \rightarrow 5C_7H_6O_2 + 6Mn^{2+} + 14H_2O$ (Mn red, C ox)
b) $2FeS_2 + 5\frac{1}{2}O_2 \rightarrow Fe_2O_3 + 4SO_2$ (Fe, S ox, O red)
c) $5H_2O_2 + 2MnO_4^- + 16H^+ \rightarrow 5O_2 + 2Mn^{2+} + 8H_2O$ (O ox, Mn red)
d) $2I_3^- + 2NO_2 \rightarrow 3I_2 + 2NO + 2H_2O$ (I ox, N red)
e) $2ClO_2^- + O_2^{2-} + 2H_2O \rightarrow 2ClO_2 + 4OH^-$ (Cl ox, O red)
f) $P_4 + 3H_2O + 3OH^- \rightarrow PH_3 + 3H_2PO_2^-$ (P ox, red)
8. $6NO_2 + 2H_2S \rightarrow 2H_2O + 2SO_2 + 6NO$
9. 162.4 g 10. 98.07 g
17. 139 cm³ $AgNO_3$ and 46.3 cm³ NaCl
18. a) $Hg_2^{2+}(aq) + 2Cl^-(aq) \rightarrow Hg_2Cl_2(s)$ b) KNO_3
19. Oxidizing agent CO_2, reducing agent H_2O
20. Reducing agent when $O^- \rightarrow O_2$, oxidizing agent when $O^- \rightarrow O^{2-}$.

CHAPTER 15

1. a) $H_2 + Br_2 \rightarrow 2HBr$ b) $SnO_2 + 2H_2 \rightarrow Sn + 2H_2O$ c) $Zn + 2HOAc \rightarrow Zn(OAc)_2 + H_2$ d) $2Al + 3H_2SO_4 \rightarrow Al_2(SO_4)_3 + 6H_2$ e) $NaH + H_2O \rightarrow NaOH + H_2$
2. a) 0.33 mol b) 21.6 g c) 736.2 torr d) 0.079 mol e) 0.079 mol
3. Reducing agent/oxidizing agent: a) Na/H b) H/Cl
4. Neutron flux of upper atmosphere may have varied in the past.
9. Process is sum of 3 steps; lattice energy is only endothermic step.
10. 1257 g KO_2, 576 L O_2
11. a) $-\frac{1}{2}$ b) Paramagnetic c) $(ISNB)(1\pi)^4(3\sigma)^2(1\pi^*)^3$ d) Longest to shortest: O_2^{2-}, O_2^-, O_2
12. 1.95 eV
13. 397 kJ·mol⁻¹, O_3 filters out light capable of breaking chemical bonds.
14. a) None b) More O_3 formed c) Less O_3 formed
15. 4.6% 21. H_2O is polar 22. Cathode: $2H_2O + 2e^- \rightarrow H_2 + OH^-$, anode: $2OH^- \rightarrow \frac{1}{2}O_2 + H_2O + 2e^-$, net: $H_2O \rightarrow H_2 + \frac{1}{2}O_2$
23. a) 18 b) 30 c) 9 d) 57 at least
24. 0.4 eV 25. 13.6 yr 26. 8 cm
27. Without H-bonding H_2O might boil at T close to that of CH_4 (-164 °C).
28. Co^{3+} accepts lone pair of H_2O, ∴ Co^{3+} is a Lewis acid, H_2O is a Lewis base.
31. $CaCl_2·2H_2O$ 32. 56%
33. a) $K_p = p_{H_2O}$ b) 1.03×10^{-2} c) 4.2×10^{-4} d) Endothermic
38. Products are: a) Sn, H_2O b) Pb, HCl c) Ca^{2+}, Cl^-, H_2 d) NaH e) SnH_2 f) Mg^{2+}, OH^-, H_2 g) $Ca(OH)_2$ h) H_3PO_4 i) H_2O, Fe^{3+} j) Fe_2O_3 k) $CuSO_4·5H_2O$ l) Cs^+, OH^-, H_2 m) Sn^{2+}, Cl^-, H_2 n) CO_2, H_2O o) CO_2, H_2O p) K_2CO_3, O_2 q) Fe, H_2O r) H_2O, O_2 s) $B(OH)_3$, Fe^{2+}, H^+ t) Ag^+, NO_3^-, NO_2, H_2O
39. $24NO_3^- + 5C_6H_{12}O_6 + 24H^+ \rightarrow 12N_2 + 30CO_2 + 42H_2O$
40. 1.31×10^8 yr
41. Light bottle shows larger Δm_{O_2} due to photosynthetic production of O_2 by algae; negative Δm_{O_2} in dark bottles means O_2 used up by oxidizable material; the deeper the sample, the

less O₂ produced—no algae on bottom to produce O₂ so both bottles same.

CHAPTER 16

1. 1.4×10^{11} t

2. H:Ö:C:Ö:H , [H:Ö:C:Ö:]¹⁻ ,
   :Ö: :Ö:

   [:Ö:C:Ö:]²⁻ , all have planar CO_3 part.
   :Ö:

3. $3CH_3OH + 2ClO_3^- \rightarrow 3HCO_2H + 2Cl^- + 3H_2O$
4. No, only for R's that are properly symmetric.
5. a) Phenol, acetic acid b) Methanol, benzoic acid
6. 1.6×10^4
7. At isoelectric point, concentration of zwitterion is a maximum. This behaves like a neutral molecule. Increase of pH increases
   RCHCO₂⁻,
   |
   NH₂

   decrease of pH increases
   RCHCO₂H
   |
   NH₃⁺

8. Vapor pressures decrease as molecular weights increase.
9. Dispersion forces increase as number of electrons increases.
18. *cis* and *trans* 1-chloro-2-bromoethene.
19. CCCCOH (1-butanol), CCCC (2-butanol),
                              |
                              OH
    CCCOH (2-methyl-1-propanol),
    |
    C
    C
    |
    CCOH (dimethylethanol)
    |
    C

20. CCCC=C (1-pentene), CCC=CC (2-pentene),
    CCC=C (3-methyl-1-butene),
    |
    C
    CC=CC (2-methyl-2-butene),
    |
    C
    C=CCC (2-methyl-1-butene)
    |
    C

26. a) Phenol b) Carboxylic acid c) Aldehyde
    d) Ketone e) Ether f) Polysaccharide g) Ether
    h) Amide i) Ester j) Amino acid k) Alkyne
    l) Alcohol

28. a) 2-Propyl phenyl ketone b) Propyl ethanoate or propyl acetate c) 2-Propanol or isopropyl alcohol d) *m*-Methylbenzoic acid e) 2-Chloro-4,4-dibromo-5-methylheptane f) Pentanal g) *m*-Dinitrobenzene h) Isopropyl ether i) Chloroethanoic or chloroacetic acid j) Methyl ethyl amine k) 2-Hydroxy-4-aminotoluene, 2-methyl-5-aminophenol, 3-hydroxy-4-methylaniline l) Phenyl ketone m) 2,3-Dichloronaphthalene n) β-Naphthyl ethanoate or β-naphthyl acetate o) Butanamide p) 2-Methylpropyl ethanoate or 2-methylpropyl acetate q) Methyl 3-methylbutanoate r) *p*-Dichlorobenzene s) *p*-Dichlorobenzene t) 2,4,6-Trinitrotoluene (TNT)

32. End C's use sp^2, middle uses sp. Each σ is $sp^2 + sp$, each π is $p + p$. Two CH₂'s are 90° to each other C=C=C

33. a) Newman projection with CH₃, H, H on front and CH=CH₂, H, H on back.
    b) Newman projection with HO₂C, CH₃, H on front and CH₃, CO₂H, H on back.
    c) Two Newman projections (cyclohexane-like).

d) [Newman projection with H, H, H, CH₃, H, H]

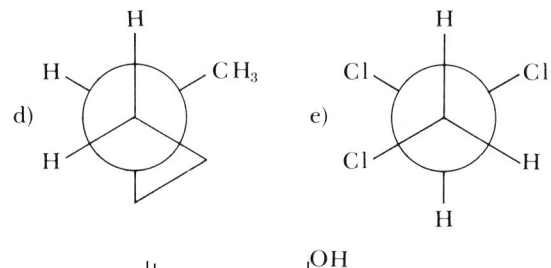

e) [Newman projection with H, Cl, Cl, H, Cl, H]

34. [Two Newman projections with =O and OH groups]

35. $\ddot{C}::O::\ddot{C}::O::\ddot{C}$, linear

36.
$$CH_3-C\begin{smallmatrix}O-H\cdots O\\ \\O\cdots H-O\end{smallmatrix}C-CH_3,$$

43. a) Second one

b) L-form CO_2H — C with NH₂, H, CH₃   D-form CO_2H — C with H, NH₂, CH₃

44. a) No b) Rotation about C—C must be hindered.

45. D-(−)-glyceric acid rotates light to left and L-(+)-glyceric acid rotates light to right. The two are mirror images. The D-isomer is

[Structure: CO₂H, H, OH, CH₂OH]

46. 36.3% α, 63.7% β

47. Each depresses the other's freezing point.

CHAPTER 17

3. a) $CH_3(CH_2)_3Br + KOH \xrightarrow{alc} CH_3CH_2CH=CH_2$

b) $CH_3CO_2H + PCl_3 \rightarrow CH_3COCl \xrightarrow{NH_3} CH_3CONH_2$

c) $CH_3(CH_2)_3OH \xrightarrow{Al_2O_3} CH_3CH_2CH=CH_2$
or use H_2SO_4 at high temp.

d) $CH_3CH_2CH_2OH \xrightarrow[\text{high T}]{H_2SO_4} CH_3CH=CH_2 \xrightarrow{H_2O} CH_3CHOHCH_3$

e) [Benzene-NO₂] $\xrightarrow[HCl]{Sn}$ [Benzene-NH₂] (with CH₃ groups)

f) $CH_3Br \xrightarrow{NaCN} CH_3C\equiv N \xrightarrow[H_2O]{HCl} CH_3CO_2H$

g) $HCO_2H \xrightarrow{LiAlH_4} CH_3OH$

h) $CH_3OH \xrightarrow[\text{distill}]{[O]} HCHO$

i) $CH_3CH_2CH_3 \xrightarrow{Br_2} CH_3CHBrCH_3 \xrightarrow{NaCN} (CH_3)_2CHC\equiv N \xrightarrow[H_2O]{HCl} (CH_3)_2CHCO_2H$

j) [Benzene] + $CH_3Cl \xrightarrow{AlCl_3}$ [Toluene]

k) [Benzene-NH₂] $\xrightarrow[HCl]{NaNO_2}$ [Benzene-N₂⁺Cl⁻]

4. ROR′. Used for ethers when R and R′ are different.

5. $CH_3CH_2OH \xrightarrow[\text{High T}]{H_2SO_4} CH_2=CH_2 \xrightarrow{HBr}$
$CH_3CH_2Br \xrightarrow{NaCN} CH_3CH_2C\equiv N \xrightarrow[H_2O]{HCl}$
$CH_3CH_2CO_2H \xrightarrow{LiAlH_4} CH_3CH_2CH_2OH \xrightarrow[\text{High T}]{H_2SO_4}$
$CH_3CH=CH_2 \xrightarrow{HBr} CH_3CHBrCH_3 \xrightarrow{NaCN}$
$(CH_3)_2CH\equiv N \xrightarrow[HCl]{H_2O}$
$(CH_3)_2CHCO_2H \xrightarrow{LiAlH_4} (CH_3)_2CHCH_2OH$

6. Products (φ = phenyl): a) $\phi CO_2CH_2\phi$
b) $\phi CH_2CO_2\phi$ c) CH_3CONH_2 d) $CH_3CH_2CO_2H$
e) $(CH_3)_2CHCHO$ and $(CH_3)_2CHCO_2H$
f) $CH_3CHBrCH_3$ g) $CH_3Cl, CH_2Cl_2, CHCl_3, CCl_4$ are possible h) ϕCl i) $\phi COCH_3$ j) CH_3NH_2

k) $\phi\text{CH}(\text{CH}_3)_2$ l) $\phi(\text{CH}_2\text{OH})_2$ (para)
m) $\text{CH}_3\text{CH}_2\text{CN}$ n) $\text{CH}_3\text{CH}_2\text{CO}_2\text{H}$, $\text{CH}_3\text{CO}_2\text{H}$, HCO_2H

o) [naphthalene with C(=O)—NH₂ substituent]

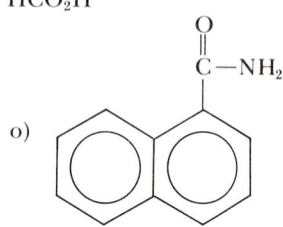

14. Latter will show O—H stretch, former will not, etc.
15. A(III), B(I), C(II), D(IV)

CHAPTER 18

1. 30,425 J
2. a) Increased   b) Increases, decreases, or stays the same
3. 42.9 km   **9.** $\Delta \bar{E} = \Delta \bar{H} = -1{,}250$ kJ·mol^{-1}
10. a) -5165 kJ·mol^{-1} b) 86 kJ·mol^{-1}
11. -57 kJ·mol^{-1} **12.** -1460.5 kJ   **13.** -75.1 kJ·mol^{-1} **14.** a) Lattice energy of CaF_2 b) ΔS is either negligible or positive.
15. a) 311.38 kJ   b) -504 kJ   c) -890.65 kJ
26. 3429 kJ
27. a) 928 kJ   b) 1962 kJ
28. 192.7 kJ·mol^{-1}
32. > 0 for (a), (e), (j), < 0 for others
37. a) 394.5 J·K^{-1}·mol^{-1} b) -75.3 kJ   c) No, ΔG^0 for decomposition is negative.
38. -32.59 kJ·mol^{-1}
39. a) -21 kJ   b) Yes   c) 200 K (Below this, reaction not spontaneous.)
40. 1752 K, assume ΔH^0, ΔS^0 do not change with T.
41. 22.92 kJ·mol^{-1} of HI, 9.6×10^{-5}
42. a) -436.5 kJ   b) 103.3 J·K^{-1} c) -467.3 kJ
50. Mass and energy conserved together.
51. Nutritionist's Calorie is a kilocalorie. Need to suck 5 L of ice to offset 400 kcal.
52. Typical sample: Cheerios breakfast cereal, $w_p = 4$ g, $w_c = 20$ g, $w_f = 2$ g. This gives 114 kcal·oz^{-1} (package says 110).

CHAPTER 19

1. (a), (b), (c) are spontaneous.
2. R→L is spontaneous direction.   b) 1.297 V
3. $\text{Ag} + \text{Cl}^- \rightarrow \text{AgCl} + e^-$ and $\text{Cl}_2 + 2e^- \rightarrow 2\text{Cl}^-$, spontaneous
4. Not spontaneous since $\mathscr{E}^0_{\text{cell}} = -1.474$ V

9. a) 10   b) 0.58 V   c) No; either $\mathscr{E}^0_{\text{A}^+/\text{A}_3}$ or $\mathscr{E}^0_{\text{B}/\text{B}^{3+}}$ must be known (relative to a standard) in order to determine the other.
10. a) Anode: $\text{Pb}(s) + \text{SO}_4^{2-}(aq) \rightarrow \text{PbSO}_4(s) + 2e^-$; cathode: $\text{Cu}^{2+}(aq) + 2e^- \rightarrow \text{Cu}(s)$ b) 0.7002 V
11. -0.770 V
12. a) $\text{Zn}(s) \rightarrow \text{Zn}^{2+}(aq) + 2e^-$ b) $\text{Hg}_2\text{Cl}_2(s) + 2e^- \rightarrow 2\text{Hg}(l) + 2\text{Cl}^-(aq)$ c) 1.03 V
13. a) $2\text{H}^+(aq) + \text{Ni}(s) \rightarrow \text{H}_2(g) + \text{Ni}^{2+}(aq)$ b) 0.23 V
14. a) Anode: $\text{Zn}(s) \rightarrow \text{Zn}^{2+}(aq) + 2e^-$; cathode: $\text{Fe}^{3+}(aq) + e^- \rightarrow \text{Fe}^{2+}(aq)$ b) 1.5328 V   c) -295.8 kJ
15. a) $\text{Hg}_2\text{Cl}_2(s) + \text{H}_2(g) \rightarrow 2\text{Hg}(l) + 2\text{H}^+(aq) + 2\text{Cl}^-(aq)$ b) 0.2682 V   c) $\mathscr{E}_{\text{cell}}$ increases. Removal of H$^+$ shifts equilibrium to right, makes ΔG more negative and $\mathscr{E}$ more positive.
16. a) 0.9432 V   b) 1.136 V   c) Same as (b)   d) 0.5587 V   e) 0.13 V
17. a) Anode: $\text{Te}(s) \rightarrow \text{Te}^{4+}(aq) + 4e^-$; cathode: $\text{Zn}^{2+}(aq) + 2e^- \rightarrow \text{Zn}(s)$ b) 0.92 V   c) $-61{,}760$ J
29. Zn since $\mathscr{E}^0_{\text{Zn}/\text{Zn}^{2+}} > \mathscr{E}^0_{\text{Ni}/\text{Ni}^{2+}}$
30. Thin coating of Cu over Fe better
33. a) No, $\text{Cu}(s) + 2\text{H}^+(aq) \rightarrow \text{Cu}^{2+}(aq) + \text{H}_2(g)$ has $\mathscr{E}^0 = -0.3402$ V   b) Yes, $\text{Al}(s) + 3\text{H}^+(aq) \rightarrow \text{Al}^{3+}(aq) + 1\frac{1}{2}\text{H}_2(g)$ has $\mathscr{E}^0 = 1.706$ V
34. $\mathscr{E}'$ is a constant.
35. 3.1×10^{-4} g·L^{-1}
36. May have reaction such as $\text{Al} + 3\text{Ag}^+ \rightarrow \text{Al}^{3+} + 3\text{Ag}$.

CHAPTER 20

1. a) 0.03 mol·L^{-1}·min^{-1} b) 0.015 mol·L^{-1}·min^{-1}
2. a) 0.006 mol·L^{-1}·hr^{-1} b) 0.004 mol·L^{-1}·hr^{-1}
3. b) Yes. If $-R_A = k[A]^m[B]^n$, then $-R_B = k'[A]^m[B]^n$ where $k' = \frac{1}{2}k$.
4. 1.34×10^{-2} mol·L^{-1}·min^{-1}
5. In terms of $-R_B$: a) 0.02 mol·L^{-1}·min^{-1} b) 0.02 mol·L^{-1}·min^{-1} c) 0.04 mol·L^{-1}·min^{-1}
6. a) 5×10^{-4} mol·L^{-1}·min^{-1} b) 3 mol·L^{-1}·hr^{-1} c) 4 mol·L^{-1}·hr^{-1}
7. a) 4k   b) 6k
18. a) $\frac{1}{2}$ b) 1   c) $R_C = k[A]^{1/2}[B]$ d) $1\frac{1}{2}$ e) 250 L$^{1/2}$·mol$^{-1/2}$·hr^{-1} f) 5.6 mol·L^{-1}·hr^{-1}
19. a) $-R_A = k[A]^2$ b) 2×10^4 L·mol^{-1}·s^{-1}
20. a) $-R_C = k[A][B]^2$ b) 3.06 mol·L^{-1}·min^{-1}
21. RDS is $-R_{\text{NO}} = k_2[\text{NOH}_2][\text{NO}]$. Net rate of formation of $\text{NOH}_2 = 0 = k_1[\text{NO}][\text{H}_2] - k_{-1}[\text{NOH}_2] - k_2[\text{NOH}_2][\text{NO}]$. Solve for $[\text{NOH}_2]$,

substitute into RDS to get

$$-R_{NO} = \frac{k_1 k_2 [NO]^2 [H_2]}{k_{-1} + k_2 [NO]}$$

If $k_{-1} \gg k_2[NO]$, get

$$-R_{NO} = \frac{k_1 k_2}{k_{-1}} [NO]^2 [H_2] = k[NO]^2 [H_2]$$

22. If M = male, F = female, and B = baby, reproduction proceeds as $MF \to B$ (slow) since B usually results from married pairs MF. Thus $R_B = k[MF]$ and $[MF]\alpha$ total population. If $[F] \gg [M]$ and reproduction not confined to married pairs, $M + F = B$ and $R_B = k[M][F]$.
29. $10 \; L \cdot mol^{-1} \cdot hr^{-1}$
30. a) 1 hr b) 0.693 hr^{-1}
33. 52.9 $kJ \cdot mol^{-1}$
34. 1.8 faster
35. a) 1.06×10^4 s at 10 °C and 30 s at 70 °C b) 76 $kJ \cdot mol^{-1}$ c) $8.63 \times 10^9 \; s^{-1}$
41. K_p depends only on ΔG^0; catalyst changes only $\Delta G\ddagger$.
42. None whatsoever
43. Pt becomes inactivated by absorbed impurities.
44. a) First step b) Yes, since $k[I^-]$ is constant.
45. Seed crystal increases $\Delta S\ddagger$ and lowers $\Delta G\ddagger$
51. $\Delta G\ddagger$ for He reaction is very small; that for benzene is very high
52. a) Rate increases. b) Reaction has large K_c. c) Rate increases.
53. Iron filings have greater surface area. This acts like an increased effective concentration.
54. a) Fires could rage out of control. b) Hard to start fires and to exercise vigorously.

CHAPTER 21

1. 6.0230×10^{23}
2. a) $AgNO_3$ solution; cathode, $Ag^+(aq) + e^- \to Ag(s)$; anode, $4OH^- \to O_2 + 2H_2O + 4e^-$; $CuSO_4$ solution: cathode, $Cu^{2+}(aq) + 2e^- \to Cu(s)$; anode, $4OH^- \to O_2 + 2H_2O + 4e^-$; HCl solution: cathode, $2H^+(aq) + 2e^- \to H_2(g)$; anode, $2Cl^-(aq) \to Cl_2(g) + 2e^-$ b) $AgNO_3$ solution: 7.5×10^{-4} mol Ag, 1.88×10^{-4} mol O_2; $CuSO_4$ solution: 3.8×10^{-4} mol Cu, 1.88×10^{-4} mol O_2; HCl solution: 3.8×10^{-4} mol H_2, 3.8×10^{-4} mol Cl_2
3. 11.2 da   4. 66.2 g Cl_2, 1.87 mol NaOH
5. Minimum energy needed is $nF\mathscr{E}$, $\mathscr{E}_{Zn/Zn^{2+}} > \mathscr{E}_{Cu/Cu^{2+}}$
6. 6F
7. a) $Ag^+(aq) + e^- \to Ag(s)$, $Cu^{2+}(aq) + 2e^- \to Cu(s)$ b) 11.2 da c) 31.77 g
15. $r_{Cl^-} = 1.86$ Å (NaCl), 1.81 Å (KCl)

16. CO_3^{2-} is a fairly strong proton acceptor.
17. 4.9 g
18. a) 23.6 L b) 39.4 cm^3 c) 0.26 $cm^3 \cdot min^{-1}$
19. Without third step, would have $\mathscr{E}^0_{Li/Li^+} < \mathscr{E}^0_{Na/Na^+} < \mathscr{E}^0_{K/K^+}$, etc. Large negative ΔG^0 of third step changes position of Li.
25. $Z^{Cu}_{eff} > Z^{K}_{eff}$ due to ineffective d-shielding, $\therefore I_{Cu} > I_K$.
26. a) $Cu_2S + 2O_2 \to 2CuO + SO_2$, $2CuO + C \to 2Cu + CO_2$, $Cu_2(OH)_2CO_3 \to 2CuO + H_2O + CO_2$ b) 0.402 t Cu, 0.04 t coke c) 3.15×10^{10} F
27. +1 arises by $4s3d^{10} \to 3d^{10} + e^-$, +2 by $4s3d^{10} \to 4s^23d^9 \to 3d^9 + 2e^-$. $4s3d^{10}$ and $4s^23d^9$ have almost the same energy.
28. $Cu(s) + SO_4^{2-}(aq) + 4H^+(aq) \to Cu^{2+}(aq) + SO_2(g) + 2H_2O(l)$
29. $AgCl(s) + 2NH_3(aq) \rightleftharpoons [Ag(NH_3)_2]^+(aq) + Cl^-(aq)$. Adding NH_3 shifts equilibrium to right.
30. One Hg(I) oxidized to Hg^{2+}, other reduced to Hg.
31. Add HCl to ppt Ag^+ and Hg_2^{2+} as chlorides. Separate solids by filtration. Add $NH_3(aq)$ to solid(s). If only AgCl, all solid dissolves; if only Hg_2Cl_2 or both Hg_2Cl_2 and AgCl, get black color. Add HCl to filtered-off liquid; get AgCl if Ag^+ also present.
40. Hg is $6s^25d^{10}4f^{14}$, $\therefore Hg^+$ would be paramagnetic.
41. \$0.16 lb^{-1}
42. 5×10^3 yr
43. 2.6×10^2 L if lake area estimated as 50 km^2.

CHAPTER 22

1. a) 6 b) 6 c) 6 d) 2 e) (but 6 is more common) f) 4
2. Normally 7 (half-filled f orbitals)
3. a) $2\frac{2}{3}$ b) $2\frac{2}{3}$
4. a) $2NiO(s) + C(s) \to 2Ni(s) + CO_2(g)$ b) 0.114 t c) Make $Ni(CO)_4$, purify, decompose thermally.
5. a) 1.1×10^9 b) 4 c) 16,000 amu d) 64,000 amu
6. 0.022 g   14. 1.8 to 3.1 eV
15. Only those with 1 to 9 d electrons are colored: (b) and (c).
16. Al^{3+} has no d electrons, Zn^{2+} has 10. Can't excite electrons between d orbitals; this is what produces color.
17. $[Ni(NH_3)_4]^{2+}$ is blue.
18. Black panel reflects less and absorbs more.
19. Deoxyhemoglobin is paramagnetic (thus low Δ); oxyhemoglobin is diamagnetic (thus high Δ). Thus former absorbs lower energy.

22.

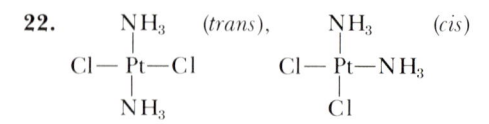

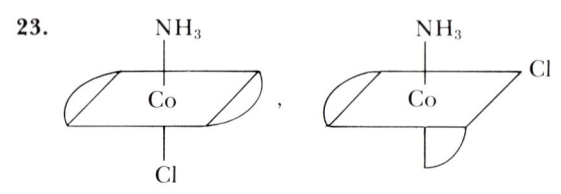

NH₃ (trans)   NH₃ (cis)
Cl—Pt—Cl   Cl—Pt—NH₃
NH₃   Cl

23.

[octahedral Co complex with NH₃ axial, Cl axial], [octahedral Co complex with NH₃ and Cl],

latter has an optical isomer.

24. Simple hybrid orbital model predicts low spin for both; crystal field model shows [(Mn(H₂O)₆]²⁺ is borderline but [Fe(CN)₆]³⁻ should be high spin.
25. No   26. +4, octahedral, 2 isomers
34. a) Difluorotetraaquocobalt(III) fluoride
b) Diammineplatinum(II) chloride
c) Tetracyanoaurate(III) anion
d) Dicyanobis(ethylenediamine)cobalt(III) perchlorate   e) Hexaaquomanganese(II) cation
f) Trichlorotriammineplatinum(IV) chloride
g) chloroamminebis(ethylenediamine)cobalt(II) carbonate   h) Tetraamminezinc(II) cation
35. a) [Co(en)₂NO₃Cl]⁺   b) VO₂[SbF₆]
c) K₂[Ni(CN)₂BrCl]   d) [Cr(en)₃](CN)₃
e) [Ag(NH₃)₂]Cl

CHAPTER 23

1. Cu + 4HNO₃ → Cu(NO₃)₂ + 2NO₂ + 2H₂O
2. 3S²⁻ + 8NO₃⁻ + 8H⁺ → 3SO₄²⁻ + 8NO + 4H₂O, S is oxidized (−2 to +6) and N is reduced (+5 to +2).
3. Cu²⁺, SO₄²⁻, and H₂O
4.

F   F
 \   /
  N=N   and   N=N
 /   \
F   F

5. 0.33 t
6. H₂SO₄ is a much stronger acid than HNO₃, thus HONO₂ + 2H₂SO₄ ⇌ H₃O⁺ + 2HSO₄⁻ + NO₂⁺. NO₂⁺ attacks the benzene ring to displace H⁺.
7. a) Reaction 2 makes NO₂ available for photochemical smog production.   b) Auto exhausts would produce N₂ rather than NO₂.
c) Catalytic exhaust systems.
15. Tetrahedral   16. 3.6×10^6 mol·yr⁻¹
17. 1.6×10^{-4} M   18. 153 L   19. 179 kg·yr⁻¹
20. a) 1820   b) 1820–1880   c) 1880–1930
d) 1930–1940   e) 1940–1950   f) 1950
21. 2.1×10^3 Pyggian days (115 earth yr)

27. SO₃²⁻ is a trigonal pyramid, SO₄²⁻ is a tetrahedron.
28. a) S₂O₃²⁻ is SO₄²⁻ with one O replaced with S.
b) SO₄²⁻ (+6), S₂O₃²⁻ (+2); latter is average of +6 and −2.
29. SO₂ is bent (like O₃) and SO₃ is triangular planar.
30. O²⁻ (OH⁻ is a weaker acid than HS⁻.)
31. CH₃CSNH₂ + 2H₂O → CH₃CO₂H + H₂S + NH₃, 12.2 L of each
32. CdS ppts first; NiS ppts when pH ≤ 1.2.
33. 0.043 ppm   42. Cathode: 0.37 mol H₂, 0.75 mol K, 1.50 mol F⁻; anode: 0.37 mol F₂, 0.75 mol HF
43. Ideal gas equation predicts 2.9 atm. Intermolecular attractions decrease this.
44. Since I₂ is a solid, best to mix it with P₄ and drop H₂O on both.
45. a) 1.26×10^2 mol   b) 6.5×10^3 g
46. 0.33 mol   54. Gaseous compounds of C (CO, CO₂, CH₄) analogous to NO, NO₂, NH₃ of N; solid compounds of Si (SiO₂, silicates) analogous to P₄O₆, P₄O₁₀, phosphates of P + many others
55. a) 4NH₃ + 5O₂ → 4NO + 6H₂O, 2NO + O₂ → 2NO₂, 3NO₂ + H₂O → 2HNO₃ + NO   b) P₄ + 5O₂ → P₄O₁₀, P₄O₁₀ + 6H₂O → 4H₃PO₄   c) S + O₂ → SO₂, 2SO₂ + O₂ → 2SO₃, SO₃ + H₂SO₄ → H₂S₂O₇, H₂S₂O₇ + H₂O → 2H₂SO₄   d) 10Cl⁻ + 2MnO₄⁻ + 16H⁺ → 5Cl₂ + 2Mn²⁺ + 8H₂O   e) 2KI + Cl₂ → 2KCl + I₂   f) 2NaCl + H₂SO₄ → Na₂SO₄ + 2HCl   g) 2KBr + Cl₂ → 2KCl + Br₂
56. Na → Na⁺, Z_{eff} goes up, r goes down; Cl → Cl⁻, Z_{eff} goes down, r goes up. Supported by data in Tables 21.1 and 23.2.

CHAPTER 24

1. OH in 2-position can start a side chain and produce cross-linking.
2. Polyvinylalcohol
3. ⁺NH(CH₂)₅C(=O)⁺ₙ, called nylon 6
4. CHR₁=CHR₂ has 5 variations: R₁, R₂ both isotactic and on same side; R₁, R₂ both isotactic but on opposite sides; R₁ isotactic and R₂ syndiotactic; R₁ syndiotactic and R₂ isotactic; R₁, R₂ both syndiotactic. CH₂=CR₁R₂ has two variations: both isotactic or both syndiotactic.
5. a) B   b) D
6. Polymer with T_g = 0 °C will be flexible; other will be brittle.

7. Na$^+$ and H$^+$ exchange, resin used to purify water.
17. 4.9×10^4 g·mol^{-1}
18. a) 19,250 amu b) 8
19. 2×10^4 g·mol^{-1}
20. $\overline{M}_w = 96{,}000$ amu, $\overline{M}_N = 86{,}333$ amu
21. $\overline{M}_w = 46{,}000$ amu, $\overline{M}_N = 32{,}258$ amu
27. a) 24 b) 400 c) 10^{1300}
28. 0.114 mol
29. (T = tertiary butoxy carbonyl group, Q = $O_2N-\langle\bigcirc\rangle-OH$). Protect amino end of Gly with T, react with activated support, remove T, add Ala which has NH$_2$ protected with T and CO$_2$H activated with Q, remove T and repeat sequence with activated and protected Trp. Finally remove tripeptide from resin with HBr/CF$_3$CO$_2$H.
30. Proteins have peptide bonds linked through α-carbons; nylon has these separated by hydrocarbon chains. Protein monomer is of type x-A-y; nylon has two monomers x-A-x and y-B-y.
31. a) Last segment cleaved is a dipeptide; its two amino acids must appear simultaneously. b) Cys-Arg-Ala-Gly-Arg-Val
32. 9 turns per second.

GLOSSARY

This glossary provides short definitions and descriptions of most of the items appearing in the lists of terms, definitions, and concepts at the ends of chapters. If you cannot find an item listed here, consult the index.

Absolute temperature A temperature expressed with respect to the lowest possible temperature.

Acid–base titration Addition of a standard acid (base) to a solution of base (acid) of unknown concentration in order to determine the concentration of the latter.

Acid ionization constant Defined for the reaction of an acid HA with water,

$$HA + H_2O \rightleftharpoons H_3O^+ + A^-$$

as

$$K_a = K_c[H_2O] = \frac{[H_3O^+][A^-]}{[HA]}$$

Acidic oxide An oxide, generally of a nonmetal, which forms an acid when dissolved in water. Also called an acid anhydride.

Activated complex *See* Transition state complex.

Activation energy The minimum energy reactants must have in order to form the activated complex of a reaction.

Active site The region on an enzyme which is assumed to bind and activate the substrate.

Activity series A listing of metals in the order of their relative ability to displace hydrogen from water or acid solutions.

Addition reaction A chemical reaction in which two or more substances combine to form a single substance. Usually refers to the case where one substance is a large molecule and the substance added is a small molecule or an atom.

Alcohol A general type of organic compound with the formula ROH, where R is a carbon-containing group. The carbon of R which is attached to OH must belong to an aliphatic group.

Aldehyde A general type of organic compound with the formula RCHO, where R is a carbon-containing group or hydrogen.

Algal bloom A sudden spurt in the growth of algae in a body of water. Usually brought about by a sudden increase in nutrients such as that accompanying a rainstorm which removes nitrogen oxides from the air.

Alicyclic compounds Organic compounds in which the carbon atoms are linked in rings but which have properties similar to those of aliphatic compounds.

Aliphatic compounds Organic compounds in which the carbon atoms are linked in open chains rather than rings.

Alkali metals The elements of group IA of the periodic table, from lithium down to francium. The alkali metals are characterized by ns valence-shell electron configurations.

Alkaline earth metals The elements of group IIA of the periodic table, from beryllium down to radium. The alkaline earth metals are characterized by ns^2 valence-shell electron configurations.

Alkanes Aliphatic hydrocarbons having the general formula C_nH_{2n+2}.

Alkenes Aliphatic hydrocarbons having the general formula C_nH_{2n} and containing one C=C bond.

Alkenyl group A group of atoms equivalent to an alkene molecule less one of its hydrogen atoms.

Alkyl group A group of atoms equivalent to an alkane molecule less one of its hydrogen atoms.

Allotropes Different crystalline forms of the same substance.

Alpha (α) helix The geometrical configuration of a protein chain characterized by a spiral analogous to the threads of a right-handed screw. The shape is maintained by hydrogen bonding between parallel "threads."

Alpha (α) particle Name for a nucleus of doubly ionized helium when it occurs as a radioactive-decay product.

Alternating copolymer A copolymer in which the different monomers appear with a constant frequency throughout a chain.

Amalgam A solution of a metal in liquid mercury.

Amide A general type of organic compound with the formula

$$R\underset{\underset{O}{\|}}{C}NH_2$$

where R is a carbon-containing group or hydrogen.

Amine A general type of organic compound with the formula

$$R'N\begin{matrix}R'\\ \\R''\end{matrix}$$

where R, R', and R" (same or different) are carbon-containing groups. One or two of these groups may be hydrogen.

Amino acid A general type of organic compound containing both an amino group and a carboxylic acid group.

Amorphous polymer A polymer having no regularity in alignment of chains, that is, no crystalline nature.

Anaerobic process A process occurring in the absence of oxygen.

Angstrom unit (Å) 1×10^{-10} m, 1×10^{-8} cm, or 0.1 nm.

Anode That electrode of a voltaic cell or electrolysis cell at which oxidation occurs.

Antibonding molecular orbital *See* Pi* (π*) molecular orbital, Sigma* (σ*) molecular orbital.

Antimatter A form of matter that becomes spontaneously annihilated when placed in contact with ordinary matter. Ordinary matter and antimatter particles generally have the same mass but charges of opposite sign.

Aqua regia A mixture of one part HNO_3 to three parts HCl, capable of reacting chemically with gold.

Aromatic compounds Organic compounds in which the carbon atoms are linked in one or more fused six-membered rings, each ring resembling that of the simplest member, benzene.

Aromatic sextet A name often applied to the six pi electrons of the benzene ring.

Arrhenius rate equation An equation relating the rate constant of a reaction to the activation energy of formation of the activated complex.

Aryl group A group of atoms equivalent to an aromatic molecule less one of its hydrogen atoms.

Atmosphere (atm) The pressure exerted by a column of mercury 76 cm high and 1 cm² in cross section at a temperature of 0 °C at sea level and 045° latitude.

Atom The smallest possible subunit of an element.

Atomic number The total number of protons in the nucleus of an atom.

Atomic pile A device for harnessing the energy of nuclear fission.

Atomic volume The apparent volume occupied by a single atom. Usually given as the molar volume divided by Avogadro's number.

Atomic weight The average mass of an atom of an element as it exists in nature.

Atomic solid A solid whose repeating units are individual atoms.

Atomic symbol A one- or two-letter symbol used to represent an element.

Atomic theory of matter The assumption that all matter consists of tiny, indivisible particles called atoms.

Aufbau principle The set of rules (Pauli exclusion principle, energy minimization, and Hund's rule) used to predict electron configurations.

Autocatalysis The catalysis of a reaction by one of its own products.

Avogadro's law At a given pressure and temperature, equal volumes of different gases contain the same number of molecules.

Azeotrope That composition of a liquid solution for which the composition of the vapor equals the composition of the liquid.

Background radiation The radiation occurring in the natural materials about us, or that due to cosmic-ray bombardment from outer space.

Band theory A theory of solids in which electrons are thought to occupy closely spaced molecular orbitals called bands.

Base ionization constant Defined for the reaction of a base B with water,

$$B + H_2O \rightleftharpoons BH^+ + OH^-$$

as

$$K_b = K_c[H_2O] = \frac{[BH^+][OH^-]}{[B]}$$

Basic oxide An oxide, generally of a metal, which forms a base (hydroxide) when dissolved in water. Also called a basic anhydride.

Beta (β) particle Name given to an electron when it occurs as a radioactive-decay product.

Beta (β) sheet A planar arrangement of antiparallel protein chains held together by hydrogen bonding between adjacent chains.

Bidentate ligand A ligand capable of furnishing two pairs of electrons to a metal ion in the formation of a coordination complex.

Binary solution A solution containing two different substances.

Binding energy The difference between the energies of the atoms of a molecule and the energy of the molecule (where it is a minimum). This is the energy needed to dissociate the molecule to atoms.

Biomagnification The increase in the concentration of a substance in the body of a predatory animal as one moves higher and higher up a food chain.

Block copolymer A type of copolymer in which each chain consists of segments of one type of monomer alternating with segments of another type of monomer.

Boat conformation of cyclohexane A conformation of cyclohexane in which carbon atoms 1 and 4 are located on the same side of a plane defined by the other four carbon atoms.

Boiling point The temperature at which the vapor pressure of a liquid equals the external pressure on the liquid.

Bond energy The energy assumed to be required to break a given chemical bond in a substance without causing any other changes in that substance.

Bonding molecular orbital *See* Sigma (σ) molecular orbital, Pi (π) molecular orbital.

Bond moment A dipole moment attributed to individual bonds of a polyatomic molecule such that the vector sum of all such bond moments equals the total dipole moment of the molecule.

Bond orbital A bonding molecular orbital formed by overlapping the orbitals (hybrid or otherwise) of two adjacent atoms.

Bond order The number of pairs of electrons constituting covalent bonding between a pair of atoms; one-half the net number of electrons in bonding molecular orbitals of a pair of atoms.

Born-Haber cycle A specific application of Hess's law in which the formation of an ionic solid is viewed in terms of components such as sublimation, dissociation, ionization, electron affinity, lattice energy, etc.

Boyle's law At a given temperature for a fixed amount of gas, pressure and volume are inversely proportional to each other; that is, the greater the pressure, the less the volume—and vice versa.

Breeder reactor A nuclear reactor which is capable of producing more fissionable material than it consumes.

Brønsted-Lowry acid A substance capable of donating a proton to another substance.

Brønsted-Lowry base A substance capable of accepting a proton from another substance.

Buffer solution A mixture of a conjugate acid–base pair having the ability to maintain an almost constant pH when small amounts of strong acids or bases are added to it.

Calorie A unit of thermal energy defined as 4.184 joules. Nutritionists sometimes use Calorie (capital C) to represent kcal.

Calorimeter A device used to measure energy changes in chemical and physical processes. Usually used for exothermic chemical processes.

Capacitance A measure of the ability of a substance to store electrical energy.

Capillary A channel having a circular cross section, usually of very small diameter.

Capillary flow The spontaneous flow of a liquid through a capillary.

Carbohydrate Any polyhydroxy aldehyde or ketone, or any compound that can be hydrolyzed to one of them.

Carboxylic acid A general type of organic compound with the formula RCO_2H, where R is a carbon-containing group or hydrogen.

Carnot efficiency *See* Efficiency of conversion of heat to work.

Catalyst A substance capable of altering the rate of a reaction and which can be recovered at the end of the reaction.

Cathode That electrode of a voltaic cell or electrolysis cell at which reduction occurs.

Cathode rays Name given to the electrons emitted by the cathode of a cathode ray tube.

Chain branching A process occurring in chain growth which results in a nonlinear growth of the chain; that is, side chains or branches develop and chain growth continues along these.

Chain initiation The first step of a chain reaction, for example, the reaction of a free radical R· and ethene, $CH_2\!=\!CH_2$ to form $RCH_2CH_2\cdot$, a new free radical which is then capable of continuing the reaction.

Chain polymerization A type of polymerization in which one monomer adds to another to form a long chain. Also called *addition* polymerization.

Chain propagation That step of a chain reaction which continues the growth of the chain, once initiation has occurred.

Chain reaction A reaction in which the products of one or more of the steps are capable of initiating the reaction, thereby allowing the reaction to proceed at a faster and faster rate.

Chain termination Any process during a chain reaction which leads to cessation of chain growth.

Chair conformation of cyclohexane A conformation of cyclohexane in which carbon atoms 1 and 4 are located on opposite sides of a plane defined by the other four carbon atoms.

Charles's law At a given pressure for a fixed amount of gas, the volume of the gas is directly proportional to its absolute temperature; that is, the greater the absolute temperature, the greater the volume—and vice versa.

Chelate A ligand capable of totally surrounding a metal ion and thus effectively removing it or sequestering it from the solution in which it originally occurred.

Chemical analysis The process of determining the identity or elemental composition of a sample of matter.

Chemical equation A symbolic representation in terms of chemical symbols indicating specific reactants forming specific products.

Chemical equilibrium A state of a chemical reaction characterized by constant composition; at this point the rate of product formation equals the rate of reactant disappearance.

Chemical formula A combination of atomic symbols used to represent the composition of chemical substances.

Chemical synthesis The process of forming one substance from other substances.

Chemical transformation A transformation in which one or more substances are changed to one or more different substances.

Chirality The property of "handedness," that is, of existing in forms analogous to right- and left-hand gloves.

Chlorate A salt of chloric acid, containing the chlorate ion ClO_3^-.

Chloric acid A water solution of hydrogen chlorate, $HClO_3$.

Chlorite A salt of chlorous acid, containing the chlorite ion ClO_2^-.

Chloronium ion The ion Cl^+, thought to be the substituting species in the chlorination of aromatic hydrocarbons.

Chlorous acid A water solution of hydrogen chlorite, $HClO_2$.

Cobalamin Another name for vitamin B_{12}, a derivative of which occurs as a coenzyme in several biological processes.

Coenzyme The nonprotein part of an enzyme, usually an organic group or a metal ion.

Colligative property A property of a solution which depends only upon the relative number of solute particles and not upon the nature of the solute itself.

Combination reaction Another name for an addition reaction; a reaction in which two or more substances combine to form a new substance.

Common-ion effect The decrease in the solubility of a sparsely soluble salt in the presence of an ion common to that salt.

Compound A substance that is a chemical union of two or more different elements.

Condensation The transformation of a vapor to a liquid.

Conductor A substance which readily allows the passage of electrical current.

Conformation A particular arrangement of atoms in a given molecule such that all such arrangements can be converted into each other without breaking and reforming chemical bonds.

Conjugate acid–base pair A pair of substances related as follows:

acid − proton = base

or, equivalently,

base + proton = acid

Control substance A substance having the ability to absorb neutrons and thus to control the neutron flux in a nuclear reactor.

Coordination complex A molecule or ion formed by a Lewis acid–base reaction in which a transition metal ion is the electron acceptor.

Coordination number The number of pairs of electrons a metal ion is capable of accepting in forming a complex by a Lewis acid–base reaction.

Copolymerization Polymerization involving two or more different monomers which are linked to form a polymer.

Coulomb's law The force acting between a pair of charged particles is directly proportional to the product of their charges and inversely proportional to the square of the distance between them.

Covalent bond A chemical bond involving the sharing of electrons by two atoms.

Covalent hydride A binary compound of hydrogen in which hydrogen is covalently bonded; the oxidation number of hydrogen in such a hydride may be either +1 or −1, depending on the relative electronegativity of the other atom.

Critical mass The smallest mass of a pure fissionable material which will lead to a spontaneous, self-sustaining chain reaction in which each fission produces one new fission.

Critical temperature of a gas The highest temperature at which a gas can be liquefied by increasing the pressure.

Crystal defect An imperfection in the structure of a crystal, for example, a missing ion, a foreign atom, or some other structural irregularity.

Crystal-field model A model which explains certain properties of transition metal complexes in terms of how ligands (acting as point charges) split the d orbitals of the metal ion.

Crystal-field splitting parameter The energy difference between the d orbitals of a transition metal ion when split by an electrostatic field provided by the surrounding ligands.

Crystalline melting point The temperature at which a mostly crystalline polymer enters the liquid-flow state.

Crystalline polymer A polymer with a high degree of regularity in the alignment of neighboring chains.

Crystallization The transformation of a liquid to a solid (also called solidification or freezing).

Cycloalkanes Alicyclic hydrocarbons isomeric with alkenes but containing only C—C bonds.

Cyclotron A device for accelerating charged particles to very high velocities by means of a cyclic path maintained by a magnetic field.

Decomposition reaction A chemical reaction in which a given substance forms two or more substances of lower molecular weight.

Deductive reasoning Determination of the specific consequences of a general theory or model.

Deliquescence The ability of a hydrate to absorb water from the atmosphere.

Denitrification Reduction of nitrites and nitrates to elemental nitrogen or to nitrous oxide.

Density Mass per unit volume.

Deoxyhemoglobin Name used to designate hemoglobin when it contains no oxygen.

Desiccant A substance capable of absorbing water vapor from the atmosphere.

Desiccator A closed container into which a desiccant is placed, used to remove moisture from other substances.

Deuterium The isotope of hydrogen with mass number 2; symbol ^2_1H.

Diamagnetic substance A substance which is repelled by a magnetic field. Such substances have equal numbers of electrons of both spins.

Diastereomers Stereoisomers which have superimposable mirror images.

Diazonium salt An ionic compound having the general formula R—N≡N⁺, X⁻, where X⁻ is an anion such as Cl⁻, Br⁻, NO₃⁻, etc., and R is an aryl group.

Dielectric breakdown The onset of electrical conductivity when an insulator is subjected to very high voltages.

Dielectric constant The ratio of the capacitance of a substance to the capacitance of a vacuum, both capacitances being measured between the same two parallel plates.

Dienes Aliphatic hydrocarbons isomeric with alkynes but containing two C=C bonds.

Differential manometer A device used to measure the difference of two gas pressures.

Diffusion The random motion of one substance through another.

Dipole–dipole forces An intermolecular force due to interaction of the dipole of one molecule with the dipole of another molecule.

Dipole-induced dipole forces An intermolecular interaction between the permanent dipole of one molecule and the dipole this molecule induces in an ordinarily nonpolar molecule.

Dipole moment A vector quantity which is a measure of the polarity of a molecule. Defined as the magnitude of the charge separation times the distance between centers of unlike charges.

Disproportionation reaction An oxidation–reduction reaction in which only one element is oxidized and reduced.

Dissociation energy *See* Binding energy.

Domain theory A theory which attempts to explain why some metals such as iron have unusually large paramagnetic properties.

Doping Addition of controlled amounts of impurities to a substance such as silicon or germanium to make it a semiconductor.

Drawing A mechanical process in which a molten crystalline polymer is rolled out and cooled so as to induce crystallization in one or more directions.

Dynamic scattering The clouding up of a nematic liquid crystal containing a small amount of dissolved ionic material when an electric field is turned on.

Efficiency of conversion of heat to work (also called the Carnot efficiency) The maximum fraction of heat energy convertible to work by a heat engine operating between the temperatures T_2 and T_1, given by $(T_2 - T_1)/T_2$.

Efflorescence The spontaneous loss of water by a hydrate.

Elastomers Polymers with glass-transition temperatures well below room temperature and which can be used as rubbers.

Electrolysis Chemical reactions brought about by passing an electrical current through a substance called an electrolyte.

Electrolyte A liquid capable of conducting an electric current.

Electromotive force The energy per unit charge that a voltaic cell (or dynamo) is capable of converting reversibly from a chemical (or mechanical) process.

Electron A subatomic particle having a mass of 9.1×10^{-31} kg and an electrical charge of -1.6×10^{-19} C. Electrons constitute the negatively charged components of all matter.

Electron affinity The energy involved when a neutral atom acquires an extra electron.

Electron configuration Designation of the orbitals that are being used to approximate the total electron distribution of an atom or molecule.

Electron density The probability of finding an electron in a given volume of space, usually represented as the "thickness" of an electron charge cloud.

Electron distribution A description of the relative probabilities of finding an electron in various regions of space in an atom or molecule.

Electronegativity A measure of the electron-attracting ability of an element in a compound.

Electropositive element A term used to designate those elements toward the left-hand side of the periodic table, that is, those of relatively low electronegativity. Such elements tend to lose rather than gain electrons in chemical reactions.

Element Matter in its simplest form—a chemical substance that consists of one type of atom only.

Elimination reaction A chemical reaction in which a relatively large molecule decomposes to form two products, one of them a small molecule or atom.

Empirical (simplest) formula That formula of a substance which expresses the relative amounts of the constituent atoms in terms of the smallest whole-number ratio.

Enantiomers Stereoisomers which have nonsuperimposable mirror images.

Endpoint of a titration The apparent equivalence point of an acid–base titration as indicated by the titration method used, for example, by the color change of an indicator.

Energy The capacity or ability to do work; work done on an object increases its energy and work done by an object decreases its energy.

Energy state A state of an atom or molecule characterized by a certain definite value of its total energy. The atom or molecule can acquire a different energy (go to a different energy state) only by the emission or absorption of a quantum of radiant energy ($h\nu$) which satisfies $\Delta E = h\nu$.

Energy transition The process whereby an atom or molecule changes from one quantized energy state to another.

Enthalpy A quantity used to describe energy changes taking place at constant pressure.

Entropy A thermodynamic quantity used as a measure of disorder, randomness, or chaos.

Enzyme A protein substance functioning as a catalyst in a biological reaction.

Equilibrium bond distance The internuclear bond distance of a molecule at which the molecular energy is a minimum.

Equilibrium constant A quantitative index of the relative amounts of products and reactants existing at the point of chemical equilibrium.

Equivalent weight For an acid or base, that weight capable of donating or accepting one mole of protons; for an oxidizing or reducing agent, that weight capable of accepting or donating one mole of electrons.

Escape velocity The minimum velocity an object must have if it is to be able to escape the gravitational field of the earth.

Ether A general type of organic compound with the formula ROR′, where R and R′ (which may be the same or different) are carbon-containing groups. Each group is bonded to the ether oxygen through a carbon atom.

Eutrophication A process whereby an increase in nutrients stimulates the growth of aquatic vegetation, which in turn depletes the water of oxygen, thereby producing an environment favoring plant life over animal life.

Excess reagent A reactant in a chemical reaction whose amount is in excess of that able to react with another reactant.

Exponential decay A decay process in which the amount of material remaining at any time t is proportional to e^{-kt}, where k is a constant.

Extended lattice crystals Solids whose repeating units are all mutually bonded throughout the entire crystal lattice.

Eyring rate equation An equation relating the rate constant of a reaction to the Gibbs free energy of formation of the transition state complex.

Faraday (or Faraday constant) The total charge on a mole of electrons.

Fat A naturally occurring ester of a carboxylic acid and glycerol which is a solid at room temperature.

Ferromagnetism The unusually large paramagnetism associated with a metal such as iron.

Filled shell A set of orbitals of given $n\ell$ occurring in an electron configuration and containing the maximum possible number of electrons, for example, $2p^6$, $2s^2$, $3d^{10}$.

First law of thermodynamics Two out of many alternative statements are: (1) the energy of the universe (or of any isolated system) is constant; (2) the energy change occurring in any substance is equal to the heat absorbed from (or lost to) the surroundings plus work done by (or done on) the surroundings.

Fixation of nitrogen Any process whereby elemental nitrogen is converted to a compound.

Formula weight The sum of weights of the ions indicated in the simplest formula of an ionic compound (usually in amu). Also used synonymously with the weight of one mole of the ions in the simplest formula (usually in grams). The latter is more correctly called the gram-formula-weight.

Fossil fuel Any fuel, such as peat, coal, petroleum, or natural gas, which originated from plant and animal forms that lived long, long ago and have been buried below the earth's surface.

Fractional crystallization The purification of a solid by crystallization from solution followed by redissolution and recrystallization, the cycle being repeated until the desired degree of purity is achieved.

Fractional distillation A process for the separation of the components of a liquid solution whereby the vapor becomes enriched in the more volatile component and the residue becomes enriched in the less volatile component.

Frasch process A process for the mining of underground sulfur deposits. By use of concentric pipes, superheated water is forced down to melt the sulfur, which is then forced to the surface by compressed air.

Frequency The number of wavelengths passing a given point per unit time; the velocity of wave motion in wavelengths per unit time.

Frequency factor The number of collisions occurring between reactants per unit of time to form the activated complex.

Friedel-Crafts reaction A reaction between an aromatic hydrocarbon and an alkyl halide or acyl halide in which a hydrogen atom on the aromatic hydrocarbon is replaced by an alkyl or acyl group.

Galvanic corrosion A spontaneous oxidation–reduction reaction occurring when two dissimilar metals are placed in contact in an electrolytic environment. The corrosion generally involves the oxidation of one metal by the other.

Gamma-(γ) ray Name given to the very-high-energy electromagnetic radiation accompanying some radioactive decay processes.

Geiger counter A device used to measure the number of radioactive decay particles emitted per unit of time by a radioactive sample.

General gas law $pV = nRT$ (also called the ideal gas law).

Geometric formula A structural molecular formula which represents the three-dimensional shape of a molecule or ion.

Gibbs function A thermodynamic quantity whose changes at constant temperature constitute a measure of useful work.

Glasses Substances which have the rigidity and appearance of solids but which lack crystalline structure.

Glass-transition temperature That temperature at which an amorphous polymer passes from the rigid glassy state to the rubbery state.

Glassy state That physical state of an amorphous polymer below its glass transition temperature characterized by brittleness and lack of flexibility.

Glycerol A common name for the trihydroxy alcohol 1,2,3-propanetriol.

Graft copolymer A type of copolymer in which chains of one type of monomer are attached as side chains to chains of another type of monomer.

Graham's law The root-mean-square velocity of a gas molecule at a given temperature is inversely proportional to the square root of its molecular weight.

Gray A quantitative measure of radiation exposure, the amount of radiation that subjects one kilogram of matter to one joule of energy.

Greenhouse effect The trapping of infrared radiation emitted by the earth's surface by the carbon dioxide of the atmosphere.

Groups Subdivisions of the periodic table exhibited as vertical columns of elements with similar chemical and periodic properties.

Half-filled shell A set of orbitals of given $n\ell$ occurring in an electron configuration and containing exactly one-half the maximum possible number of electrons, for example, $2p^3$, $2s$, $3d^5$.

Half-life The time required for one-half of a radioactive material to decay.

Halogens The elements of group VIIA of the periodic table, from fluorine to astatine. The halogens are characterized by ns^2np^5 valence-shell electron configurations.

Heat A term used to denote an energy change accompanying a change in temperature.

Heat of fusion The energy needed to transform a solid to a liquid.

Heat of solution The energy change occurring when a given amount of solute is dissolved in a given amount of solvent at a given temperature.

Heat of vaporization The energy needed to transform a liquid to a vapor.

Heme structure Name of the planar structure consisting of four pyrrole rings and functioning as a tetradentate ligand in molecules such as chlorophyll, hemoglobin, and vitamin B_{12}.

Henry's law The solubility of a gas in a liquid or solid is directly proportional to the partial pressure of the gas in equilibrium with the liquid or solid at the given temperature.

Hess's law Whenever the equations of two or more processes, all carried out at the same temperature and pressure, can be added to form the equation of a resultant process, the enthalpy change of the resultant process is the sum of the enthalpy changes of the component processes.

Heterocyclic compound A ring compound in which one or more carbon atoms of the ring are replaced by another atom, for example, O, S, or N.

High-spin complexes Coordination complexes in which the complex has the same number of unpaired electrons as the free metal ion.

Homologs Two or more different compounds differing only in the number of methylene groups (—CH_2—) they contain.

Hund's rule Electrons are placed into orbitals of the same $n\ell$ value in such a way that their spins are as much as possible the same.

Hybrid orbitals A mixture or combination of two or more different atomic orbitals of the same atom. The most important examples are as follows. (1) sp, a combination of one s orbital with one p orbital, both on the same atom (also called digonal orbitals). The two digonal orbitals are oriented at 180° to each other. (2) sp^2, a combination of one s orbital with two p orbitals, all on the same atom (also called trigonal orbitals). The three trigonal orbitals lie in a plane and are oriented 120° to each other. (3) dsp^2, a combination of one part d, one part s, and two parts p. Such hybrid orbitals are also called square planar since they are oriented at 90° to each other in a single plane. (4) d^2sp^3, a combination of two parts d, one part s, and three parts p, in which the principal quantum number of d is one less than that of the s and p. Also called octahedral orbitals since they are oriented at 90° to each other as in a regular octahedron. These hybrid orbitals are used to describe so-called inner-orbital complexes. (5) sp^3d^2, a combination of one part s, three parts p, and two parts d, in which the s, p, d have the same principal quantum number. Also called octahedral orbitals since they are oriented at 90° to each other as in a regular octahedron. These hybrid orbitals are used to describe so-called outer-orbital complexes.

Hydrate A crystalline substance containing stoichiometric amounts of water in the crystal lattice.

Hydration energy The energy change involved when the ions of a solid (freed from the crystal lattice) interact with water.

Hydrogen bonding A specific type of intermolecular attraction in which a hydrogen atom bonded to N, O, or F in one molecule interacts strongly with N, O, or F atoms of another molecule.

Hydrologic cycle The pathways water follows in nature.

Hypochlorite A salt of hypochlorous acid, containing the hypochlorite ion OCl^- (also written ClO^-).

Hypochlorous acid A water solution of hydrogen hypochlorite, HClO (also written HOCl).

Ideal or perfect gas A hypothetical gas, serving as a model for real gases, which obeys the laws of Boyle, Charles, and Avogadro.

Ideal gas constant The constant R appearing in the ideal gas equation and equal to pV/nT when $p = 1$ atm, $V = 22.4$ L, $T = 273$ K, and $n = 1$ mol.

Indicator A substance used to estimate the point at which acid and base are present in equivalent amounts during a titration.

Inductive effect The withdrawal of electron charge from an atom in a molecule by another, more electronegative atom.

Inductive reasoning The assumption of a general consequence on the basis of a limited number of individual observations.

Inert-gas core That part of the electron configuration of an atom that corresponds to the total electron configuration of a group VIIIA atom.

Inert gases The elements of group VIIIA of the periodic table, from helium down to radon. The inert gases are relatively inert chemically and are characterized by ns^2np^6 electron configurations.

Infrared spectroscopy An area of spectroscopy which measures the absorption of infrared radiation by molecules. Used to identify molecules.

Initial rate of a reaction The rate of a reaction at the instant the reactants are mixed and begin to react.

Initiator A substance that initiates or begins a chain reaction.

Inner-shell nonbonding electrons That part of the electron configuration of a diatomic molecule characterized by filled σ and σ^* molecular orbitals.

Insulator A substance which does not readily allow the passage of electrical current.

Interionic attraction theory A model which accounts for the apparently anomalous colligative properties of concentrated ionic solutions by means of mutual interferences among the ions.

Interstitial defect A type of crystal defect in which a foreign substance occupies otherwise empty space in a lattice.

Interstitial hydride A combination of hydrogen and a metal in which hydrogen is bound by filling vacant sites in the crystal lattice. Such compounds are usually nonstoichiometric in composition.

Intrinsically spontaneous process Any process which, in principle, is capable of producing useful work.

Ion A single atom or chemically bound aggregate of atoms, having a net positive or negative charge.

Ionic bond The electrostatic attraction between positive and negative ions in an ionic compound; a chemical bond involving the actual transfer of one or more electrons from one atom to another.

Ionic compound A compound composed of positive and negative ions in such proportions that the net electrical charge is zero.

Ionic crystals Solids whose repeating units are ions.

Ionic hydride A binary compound of hydrogen in which hydrogen exists as the hydride ion, H^-.

Ionic radius A hypothetical radius of an ion as it appears in the crystalline solid.

Ionization energy The energy needed to remove an electron from an atom or molecule.

Ion-product constant of water An equilibrium constant for the self-ionization reaction of water,

$$2H_2O \rightleftharpoons H_3O^+ + OH^-$$

given by

$$K_w = K_c[H_2O]^2 = [H_3O^+][OH^-]$$

This has the value 1.0×10^{-14} at 25 °C.

Ion-transfer reaction A chemical reaction in which an ion is transferred from one substance to another.

Isoelectronic ions Ions having the same total number of electrons.

Isolated system Any part of the universe which does not exchange energy or matter with its surroundings. The universe as a whole is taken to be an isolated system.

Isomers Two or more different molecules having the same molecular formula.

Isotactic polymer A stereospecific polymer in which each monomer in a chain has the same geometric configuration.

Isothermal process A process occurring at a constant temperature.

Isotope effect The difference in the rate of reaction of a substance when one or more of its atoms are replaced by another isotope of that atom.

Isotopes A general term for the atoms of a given element that have different mass numbers.

Joule-Thomson effect The change in temperature accompanying a change in the pressure of a gas.

Ketone A general type of organic compound with the formula

$$\begin{matrix} R\underset{\underset{O}{\parallel}}{C}R' \end{matrix}$$

where R and R' (same or different) are carbon-containing groups.

Kinetic energy The energy of motion, always given as $\frac{1}{2}mv^2$.

Kinetic molecular equation The relationship between the kinetic energy of an ideal gas and the absolute temperature of the gas: $\frac{1}{2}Mu^2 = \frac{3}{2}RT$.

Kinetic molecular theory of gases A model of gases which assumes randomly moving particles of zero volume and no intermolecular interactions.

Lattice energy The energy needed to transform a solid to gaseous atoms, molecules, or ions.

Law A statement of the way in which a natural process is found to operate.

Law of combining volumes When gases combine chemically they do so in volume ratios which are simple whole numbers, provided all volumes are measured at the same pressure and temperature.

Law of conservation of energy *See* First law of thermodynamics.

Law of conservation of matter The total amount of matter existing before an ordinary chemical or physical process always equals the total amount of matter existing after the process.

Law of definite composition Different samples of a given chemical substance always contain the same relative amounts of the constituent elements.

Law of mass action The rate of a chemical reaction is proportional to the concentrations of the reactants.

Law of multiple proportions When two or more different compounds contain the same two elements, the ratio of the weights of one of the elements—for a fixed reference weight of the other element—is always that of small, whole numbers.

Le Châtelier's principle Whenever a system in equilibrium is subjected to a stress (disturbance), the system responds in such a way as to undo or lessen the effect of this stress insofar as possible.

Lewis acid A substance capable of accepting a pair of electrons to form a covalent bond.

Lewis base A substance capable of donating a pair of electrons to form a covalent bond.

Ligand A general name for a molecule or ion acting as a Lewis base in the formation of a coordination complex with a metal ion.

Light scattering A method of measuring the weight-average molecular weight of a polymer.

Limiting reagent A reactant in a chemical reaction whose amount is insufficient to completely use up another reactant.

Linear accelerator A device which is capable of accelerating charged particles to very high velocities by means of a linear succession of electrostatic fields.

Linear polyethylene Polyethylene in which all monomers are linked to each other with no side chains or branching.

Liquid crystal A substance having the structural properties of a liquid and the optical properties of a solid.

Lock-and-key mechanism A simple mechanism for enzymic reactions in which the substrate is thought to become activated by occupying a cleft in the enzyme complementary to its own size and shape.

London dispersion forces An intermolecular interaction between atoms or molecules due to the simultaneous attractions among all their nuclei and electrons.

Lone-pair electrons *See* Nonbonding electrons.

Low-spin complexes Coordination complexes in which the complex has fewer unpaired electrons than the free metal ion.

Lyotropic liquid crystal A liquid crystal in which the mesophase depends on the relative amounts of two or more components.

Magnetic moment A measure of the torque experienced by a magnet when it is at right angles to a uniform magnetic field.

Markovnikov's rule When an unsymmetrical reagent HA adds to a double bond, the H atom goes to that end of the double bond already having the greater number of hydrogen atoms.

Mass An attribute of matter characterized by a resistance to a change in the state of its motion (inertia).

Mass defect The difference between the actual mass of an atom and the sum of the masses of its constituent electrons, protons, and neutrons.

Mass number The total number of nucleons (protons and neutrons) in the nucleus of an atom.

Mass spectrograph A device used to make very accurate measurements of the masses of atoms and small molecules.

Mechanism of a reaction A detailed, step-by-step description of how a chemical reaction is thought to proceed.

Melting point The temperature at which the vapor pressure of the solid equals the vapor pressure of the liquid.

Meniscus The curved surface of a liquid as it appears at the air–liquid interface in a capillary tube.

Mesophase That state of a substance in which it exhibits the properties of a liquid crystal.

Meta position Two groups two carbon-carbon bonds apart on the benzene ring are said to have positions *meta* to each other.

Metal carbides A compound of a metal (usually Ca or Mg) and carbon having the empirical formula MC_2 (M^{2+}, C_2^{2-}) in which C_2^{2-} has the structure $[:C\equiv C:]^{2-}$.

Metal carbonyl A compound formed from combination of carbon monoxide and a transition metal.

Metals Electropositive elements found toward the left-hand side of the periodic table and generally possessing electrical conductivity, malleability, and shiny luster.

Metastable system A system which is intrinsically unstable but appears to be stable relative to small disturbances, for example, a long, thin rod standing on its end (touch it and it falls over).

Model *See* Theory.

Moderator A substance used to slow down neutrons and thence to provide more neutrons capable of splitting fissionable material used in a nuclear reactor.

Modulus of elasticity The tensile force (force tending to stretch) divided by the elongation it produces.

Molal concentration The number of moles of a substance per kilogram of another substance in a solution.

Molar concentration The number of moles of one substance per liter of solution.

Molar heat capacity The amount of energy needed to change the temperature of one mole of a substance by 1 K (1 °C).

Mole A number equal to the number of atoms in exactly 12.000 . . . grams of ^{12}C.

Molecular crystals Solids whose repeating units are individual molecules.

Molecular formula That formula of a substance which expresses the type and number of atoms of each type contained in each molecule of the substance.

Molecular orbital A mathematical function which describes the probability of finding an electron in the space about the nuclei of a molecule (abbreviation: MO). *See* Sigma, Sigma*, Pi, and Pi* molecular orbitals.

Molecular weight The weight of a single molecule of a substance (usually in amu). Also used synonymously with the weight of a mole of molecules of a substance (usually in grams). The latter is more correctly called the gram-molecular-weight or weight of one mole.

Molecule The smallest possible subunit of a compound—an aggregate of two or more different atoms, or a union of two or more atoms of the same element.

Mole fraction concentration The number of moles of a given substance divided by the total number of moles of components in the solution.

Mole-ratio factor (abbreviated MRF) The ratio of the number of moles of one product or reactant to the number of moles of another product or reactant, both numbers being those occurring in the balanced equation for the chemical reaction in question.

Monodentate ligand A ligand capable of furnishing one pair of electrons to a metal ion in the formation of a coordination complex.

Monomer A general name for the relatively light and simple molecules which link together to form a polymer.

Myoglobin A protein-iron complex capable of storing oxygen.

Negatron Another name for the electron.

Nernst equation The general equation relating the emf of a voltaic cell to its standard cell potential and the concentrations of reactants and products.

Net rate of a reaction The difference between the rate of the forward reaction and the reverse reaction at any given time. The net rate of reaction becomes zero at chemical equilibrium.

Neutron A subatomic particle found in the nuclei of all elements (except the ^1_1H isotope of hydrogen) and having zero electrical charge and a mass of about 1.7×10^{-27} kg (almost the same as a proton).

Newman projection diagrams Diagrams used to represent various conformations of ethane and related compounds.

Nitrate A salt of nitric acid, containing the ion NO_3^-.

Nitrification Oxidation of ammonia and ammonium compounds to nitrites and nitrates.

Nitrile A general type of organic compound having the formula R—C≡N, where R is a group containing carbon.

Nitrite A salt of nitrous acid, containing the ion NO_2^-.

Nitronium ion The ion NO_2^+, formed when nitric acid, HNO_3, reacts with a stronger acid such as H_2SO_4.

Nonbonding electrons The electrons of a Lewis structure which belong to only one atom and thus do not represent bonding between a pair of atoms (also called *lone-pair* electrons).

Nonlinear polyethylene Polyethylene in which some side-chains or branching exists.

Nonmetals The elements found toward the right-hand side of the periodic table and generally possessing poor electrical conductivity, brittleness, and (if solid) dull luster.

Normality The number of equivalent weights of a substance per liter of solution (also called *normal concentration*).

n-type semiconductor A semiconductor characterized by a small number of electrons in its otherwise empty conduction band.

Nuclear binding energy The energy equivalent of the mass defect of an atom.

Nuclear fission The splitting of a single heavier nucleus into two smaller nuclei, generally of roughly equal masses.

Nuclear fusion The combination of two or more lighter nuclei to form a single heavier nucleus.

Number-average molecular weight The sum of the products of the molecular weights of each type of molecule in a sample and the number of such molecules, divided by the total number of molecules in the sample.

Octahedron A geometric figure having eight sides, each of which is a triangle. If each side is an equilateral triangle, the octahedron is called *regular*.

Octet rule Lewis formulas of stable compounds usually require that each atom be surrounded by eight electrons (an octet).

Oil A naturally occurring ester of a carboxylic acid and glycerol which is a liquid at room temperature.

Optical activity The ability of a substance to rotate the plane of plane-polarized light.

Optical birefringence The ability of a substance to split ordinary unpolarized light into two polarized components.

Orbital A mathematical function which describes the probability (cloud density) of an electron in the space about an atom or molecule.

Orbital energy The negative of the energy required to remove an electron from a given orbital.

Order of a reaction The order of a reaction having a rate law of the general form

Rate = $k[A]^m[B]^n$

is defined as $m + n$. Reactions not having a rate law of this form cannot be classified on the basis of order.

Orientation The result of drawing a polymer. This induces crystallization in one or two directions and increases

GLOSSARY **753**

the strength of the polymer. Used to make fibers and films from polymers.

Ortho position Two groups on adjacent carbon atoms on the benzene ring are said to have positions *ortho* to each other.

Osmotic pressure The pressure needed to just prevent the flow of solvent across a semipermeable membrane separating two solutions of the same components but with different concentrations.

Ostwald process A commercial process for the production of nitric acid from ammonia; ammonia is oxidized to nitric oxide, this is oxidized further to nitrogen dioxide, and the nitrogen dioxide is dissolved in water to produce nitric acid and nitric oxide.

Oxidation A process in which an element increases its oxidation number, that is, is oxidized. The oxidized substance also loses electrons.

Oxidation number A signed number (+ or −) assigned to an element in a compound on the basis of its relative electronegativity and electron configuration.

Oxidizing agent A substance capable of oxidizing another substance. The oxidizing agent contains an element whose oxidation number decreases during the reaction; that is, this element is reduced.

Oxy acid Any Brønsted-Lowry acid whose molecular formula contains oxygen in addition to hydrogen and at least one other element. Examples are HNO_3, $HClO_3$, etc.

Oxyhemoglobin Name used to designate hemoglobin when it is carrying oxygen.

PAN Abbreviation for peroxyacetylnitrate, one of the eye irritants present in photochemical smog.

Paramagnetic substance A substance which is attracted by a magnetic field. Such substances have more electrons of one spin than of the other.

Para position Two groups three carbon-carbon bonds apart on the benzene ring are said to have positions *para* to each other.

Partial pressure The pressure a gas in a mixture of gases would exert if it alone occupied the container at the given temperature.

Passive state A state in which iron and some other metals become resistant to oxidation even under conditions when they are normally quite reactive.

Pauli exclusion principle No two electrons in a single atom or molecule can have the same total set of quantum numbers. (In an atom the total set has *four* quantum numbers.)

Peptide linkage A special name for the amide linkage characteristic of polymers of α-amino acids.

Perchlorate A salt of perchloric acid containing the perchlorate ion ClO_4^-.

Perchloric acid A water solution of hydrogen perchlorate, $HClO_4$.

Perfluorohydrocarbons Compounds formed by replacing all the hydrogen atoms in a hydrocarbon by fluorine atoms. For example, CF_4 is perfluoromethane (also carbon tetrafluoride).

Periodic law The chemical and physical properties of the elements are periodic functions of their atomic numbers.

Periods Sequences exhibited in the periodic table as horizontal rows of elements whose chemical and physical properties change progressively.

Peroxide A binary compound of oxygen in which oxygen has an oxidation number of −1. The peroxide ion is O_2^{2-}.

pH The negative of the decimal logarithm of the total hydronium ion concentration of an aqueous solution; $pH = -\log[H_3O^+]_{tot}$.

Phase A homogeneous, mechanically separable, and physically distinct portion of a heterogeneous system.

Phase diagram A diagram showing under what conditions various phases of a substance can exist or coexist.

Phenol A general type of organic compound with the formula ROH, where R is a carbon-containing group. The carbon of R which is attached to OH must belong to an aromatic group; otherwise the compound is called an *alcohol*.

Phosphoric acid The acid formed by hydrolysis of phosphorus(V) oxide and having the formula H_3PO_4. This acid has three acidic protons.

Phosphorous acid The acid formed by hydrolysis of phosphorus(III) oxide and having the formula H_3PO_3. This acid has only two acidic protons.

Photochemical smog Name applied to the atmospheric conditions characterized by the presence of eye irritants and haze, conditions caused by the action of sunlight on nitrogen oxides and unburned hydrocarbons.

Photon Name given to a quantum ($h\nu$) of electromagnetic radiation. Also refers to the "particles" of which light appears to consist in phenomena such as the photoelectric effect.

Photosynthesis The process whereby chlorophyll-containing plants and some bacteria utilize the energy of sunlight to convert carbon dioxide and water to carbohydrate and oxygen.

Photovoltaic cell A semiconductor-based device which produces a flow of electric current in response to illumination with light.

Physical transformation A transformation in which one form of a substance is changed to another form of the same substance.

Phytoplankton Tiny, free-floating plants found in the upper layers of ocean water.

Pi bond A chemical bond between a pair of atoms characterized by a buildup of electron density on both sides of the bond axis.

Pi electrons Those electrons assigned to pi molecular orbitals of a molecule.

Pi (π) molecular orbital A molecular orbital characterized by a buildup of internuclear electron density on both sides of the bond axis. Such an orbital is said to be *bonding* since it implies an increase or buildup of electron density between a pair of atoms.

Pi* (π*) molecular orbital A molecular orbital characterized by a buildup of electron density on both sides of the bond axis and *outside* the internuclear region. Such an orbital is said to be *antibonding* since it implies a decrease of electron density between a pair of atoms.

Plasticizer A substance added to a polymer to decrease its effective glass-transition temperature, therefore making it more flexible at a given temperature.

Polar bond A chemical bond characterized by a noncoincidence of positive and negative charge centers.

Polarity The quality of possessing poles, for example, the north and south poles of a magnet or the positive and negative poles of an electrostatically charged object.

Polarized light Light in which the vibrational characteristics have been uniformly altered, for example, by being restricted to a single plane.

Polar molecule A molecule that possesses a dipole moment since its centers of positive and negative charges do not coincide.

Polyamide A polymer whose repeating units are joined by amide linkages. Examples: nylons and proteins.

Polydispersity The ratio of the weight-average molecular weight to the number-average molecular weight. A measure of the molecular homogeneity of a polymer; a polymer all of whose molecules have the same mass has a polydispersity of unity.

Polyester A polymer whose repeating units are joined by ester linkages. Example: Dacron (polyethyleneterephthalate).

Polymer A compound consisting of a large number of repeated linked units, each a relatively light and simple molecule or a major part of such a molecule.

Polymerization The process whereby relatively light and simple molecules are linked to form a polymer.

Polypeptide A polymer of α-amino acids. The term *protein* is generally used to designate polypeptides having more than 100 α-amino acids, or above 5000 amu in molecular weight.

Polyprotic acid An acid capable of donating more than one mole of protons per mole of acid.

Polysaccharide A polymer of a saccharide such as glucose.

Positron The antimatter counterpart of the electron, having the same mass as the electron but a charge of opposite sign.

Potential energy The energy a substance can have by virtue of its position or state.

Pressure Force per unit area.

Protein A polypeptide, or combination of polypeptides, each generally containing 100 α-amino acids or more. A protein may also contain metal atoms and other molecules.

Protium The isotope of hydrogen with mass number 1; symbol ^1_1H.

Proton A subatomic particle found in the nuclei of all elements and having a mass of 1.7×10^{-27} kg and an electrical charge of $+1.6 \times 10^{-19}$ C. Protons account for the positively charged portions of all matter.

p-type semiconductor A semiconductor characterized by a small number of missing electrons (holes) in its otherwise filled valence band.

Quantization The occurrence of energy (or other quantity) in only certain amounts.

Quantum numbers Numbers (usually integers or half-integers) which govern the sizes of quantized properties such as energies.

Racemic mixture A 50-50 mixture of an enantiomeric pair of substances.

Rad A unit used to designate amount of radiation exposure. It is the amount of radiation that subjects one kg of matter to 0.1 J of energy.

Radioactive decay constant A constant of proportionality between the decay rate of a radioactive substance and the instantaneous mass of that substance.

Radioactivity The emission of particles and/or radiation by the nucleus of an element as it spontaneously transforms to an isotope or to another element.

Radioisotope dating A method used to estimate the age of a sample of matter by measuring its content of a given radioactive isotope and comparing this with the original content of this isotope.

Random copolymer A copolymer in which the different monomers appear with random frequency throughout a chain.

Raoult's law The partial pressure of a component of a liquid solution in the vapor which is in equilibrium with that solution is equal to the mole fraction of that component in the solution times the vapor pressure of that component in its pure state.

Rate constant The constant of proportionality relating the rate of a reaction to the concentrations of products and reactants of the reaction.

Rate-determining step The slowest step in the sequence of steps characterizing the assumed mechanism of a chemical reaction.

Rate law A mathematical expression relating the rate of a reaction to the concentrations of products and reactants of the reaction.

Rate (velocity) of a chemical reaction The change in reactant or product concentration per unit time at any point of a chemical reaction. *See also* Net rate of a reaction, Initial rate of a reaction.

RBE Radiation biological effectiveness, a measure of the relative ability of different types of radiation to cause damage to biological tissue.

Reducing agent A substance capable of reducing another substance. The reducing agent contains an element whose oxidation number increases during the reaction; that is, this element is oxidized.

Reduction A process in which an element decreases its oxidation number, that is, is reduced. The reduced substance also gains electrons.

Relative humidity The pressure of water vapor in the atmosphere divided by the equilibrium vapor pressure of water at that same temperature.

Rem A measure of radiation exposure defined as 100 times the dosage in grays times the radiation biological effectiveness (RBE) of the radiation.

Resonance structures Two or more Lewis formulas of a given molecule which differ only in the specific assignment of electrons to atoms.

Reversible reaction A reaction in which reactants and products are in equilibrium.

Roasting of ores Conversion of sulfide and carbonate ores to oxides by heating.

Root-mean-square velocity The square root of the average of the square of the velocity ($u = \sqrt{(v^2)_{av}}$). This is *not* equal to $\sqrt{(v_{av})^2}$.

Rubbery plateau That region above the glass-transition temperature of an amorphous polymer in which the polymer is relatively soft and may be stretched like rubber.

Sacrificial anode A strip of easily oxidized metal placed in electrical contact with a less easily oxidized metal so as to protect the latter from corrosion.

Saline hydride *See* Ionic hydride.

Saturated solution A solution which contains the maximum amount of solute that a given amount of solvent can dissolve at the given temperature; a solution in which the dissolved solute and undissolved solute are in equilibrium.

Scientific method The collection of attitudes and processes by which scientific discoveries are critically analyzed and eventually incorporated into models or theories.

Screw dislocation A type of crystal defect in which atomic planes, instead of being parallel, are offset at an angle.

Second law of thermodynamics Two out of many alternative statements are: (1) the entropy of any spontaneous process occurring in an isolated system always increases or, at best, stays the same; (2) when heat flows spontaneously, it always does so from a hotter body to a colder body.

Seed crystal A small solute crystal used to initiate crystallization in a supersaturated solution.

Semicarbazone An addition product of a ketone or aldehyde with semicarbazide used in the separation and identification of ketones and aldehydes.

Semiconductor A substance having an electrical conductivity intermediate to that of conductors and insulators.

Sigma bond A chemical bond between a pair of atoms characterized by a buildup of electron density along the bond axis.

Sigma (σ) molecular orbital A molecular orbital characterized by a cylindrically symmetric electron distribution about the bond axis and having the maximum electron density along this bond axis and *between* the bonded atoms. Such an orbital is said to be *bonding* since it implies a buildup of electron charge between a pair of atoms.

Sigma* (σ^*) molecular orbital A molecular orbital characterized by a cylindrically symmetric electron distribution about the bond axis and having the maximum electron density along this bond axis and *outside* the internuclear region. Such an orbital is said to be *antibonding* since it implies a decrease of electron density between a pair of atoms.

Silicones Polymers containing chains of alternating silicon and oxygen atoms.

Soap A salt of an alkali metal (usually sodium or potassium) and a carboxylic acid having 11 to 17 carbon atoms.

Solubility-product constant An equilibrium constant for the solubility of a sparsely soluble salt.

Solute One of the components of a solution, usually that component of greater interest or that present in lesser amount.

Solution A homogeneous mixture of two or more substances.

Solvent One of the components of a solution, usually that component of lesser interest (perhaps serving only as an inert medium) or that present in greater amount.

Solvent leveling effect The inability of a solvent to distinguish relative strengths of acids much stronger than the conjugate acid of that solvent.

Specific heat The amount of energy required to change the temperature of one gram of a substance by 1 K (or 1 °C).

Specific rotation The number of degrees by which a substance of concentration 1 g·cm^{-3} rotates the plane of polarized light when the optical path is 1 dm. The specific rotation also depends on the wavelength of light used and on the temperature.

Spectral lines Those specific wavelengths of electromagnetic radiation that an atom or molecule is capable of emitting or absorbing.

Spectrochemical series A list of various ligands in their order of ability to split the *d* orbitals of a transition metal ion in a coordination complex.

Standard electrode potential The emf of a voltaic cell constructed from the electrode in question and the hydrogen electrode, both electrodes being in their standard states.

Standard state A reference state for thermodynamic measurements; usually defined as the most stable form of a substance at 1 atm at the temperature in question, or, in the case of solutes, as a 1-M solution.

Standard-state enthalpy of formation (of a substance) The enthalpy change involved when a substance in its standard state is formed from its constituent elements in their standard states.

Standard-state enthalpy of reaction The enthalpy change involved when reactants in their standard states form products in their standard states.

Step-growth polymerization A type of polymerization in which the monomers link by eliminating a small molecule such as H_2O between them. Also called *condensation* polymerization.

Stereoisomers Two or more isomers having the same structural formulas but differing in the spatial locations of one or more atoms.

Stereospecific polymerization A polymerization process in which the monomers are incorporated in the chain according to specific geometric configurations.

Strong acid An acid which is capable of giving up most of its protons to a reference substance, usually the solvent.

Strong base A base which is capable of accepting a large fraction of the protons it has a capacity for, from a reference substance, usually the solvent.

Structural (linkage) formula A molecular formula showing the atoms contained in a molecule and how these are linked to each other.

Structural (linkage) isomers Two or more isomers having one or more of their atoms linked differently.

Sublimation Transformation of a solid directly to the vapor without forming the liquid state.

Substitutional defect A crystal defect in which a foreign substance occupies a lattice site.

Substitution reaction A chemical reaction in which an atom or a group of atoms of a substance is replaced by another atom or group of atoms.

Sulfate A salt of sulfuric acid containing the sulfate ion SO_4^{2-}.

Sulfite A salt of sulfurous acid, containing the sulfite ion SO_3^{2-}.

Sulfuric acid The acid formed by hydrolysis of sulfur trioxide (which must be carried out indirectly) and having the formula H_2SO_4.

Sulfurous acid The acid formed by hydrolysis of sulfur dioxide and having the formula H_2SO_3.

Supercooling The cooling of a liquid below its melting point (freezing point) without the onset of crystallization.

Superheating The heating of a liquid above its boiling point without the onset of boiling.

Superoxide A binary compound of oxygen in which oxygen has an oxidation number of $-\frac{1}{2}$. The superoxide ion is O_2^-.

Supersaturated solution An unstable solution which contains more than the maximum amount of solute that a given amount of solvent can normally dissolve at the given temperature.

Surface tension The force required to extend the perimeter of the surface of a fluid by unit length or, equivalently, the energy required to increase the surface area by unit amount.

Syndiotactic polymer A stereospecific polymer in which alternating monomers in a chain have the same geometric configurations; that is, adjacent monomers have different geometric configurations.

Tetrahedron A geometric figure having four sides, each of which is a triangle. If each side is an equilateral triangle, the tetrahedron is called *regular*.

Theory An assumed model or structure of nature which incorporates known natural laws and attempts to predict hitherto undiscovered laws.

Thermoplastic polymer A polymer which can be melted and solidified over and over again.

Thermosetting polymer A polymer, such as bakelite, that forms upon the application of heat and becomes permanently hardened.

Thermotropic liquid crystal A type of liquid crystal in which the mesophase depends on the temperature.

Three-body reaction A gas-phase exothermic reaction between two substances in which a third substance collides with the product soon after its formation to carry away excess energy before this energy causes the product to dissociate back to reactants.

Torr A pressure unit equal to exactly 1/760 of an atmosphere.

Torricelli barometer A mercury manometer used to measure absolute pressures of gases.

Tracer A radioactive isotope used to trace the path of a substance in a chemical process.

Transition element An element containing 1 to 9 d electrons or 1 to 13 f electrons.

Transition metal Any element whose neutral atom or ions have electron configurations containing 1 to 9 d electrons or 1 to 13 f electrons.

Transition-state complex A metastable aggregate of reactant molecules formed by reactants prior to the formation of products. Also called an activated complex.

Triglyceride A general name for an ester of glycerol and carboxylic acids.

Trigonal bipyramid A geometric figure having six sides and formed by placing two triangular-base pyramids base to base.

Trigonal geometry A single atom surrounded by three other atoms in the same plane, the latter atoms defining the corners of a triangle.

Triple helix A geometrical configuration of protein chains in which three α-helixes are twisted together in a left-handed sense.

Triple point The temperature and pressure at which the liquid, vapor, and solid forms of a pure substance can coexist.

Tritium The isotope of hydrogen with mass number 3; symbol ^3_1H.

Uncertainty principle It is impossible to measure simultaneously the exact position and exact velocity of a moving particle.

Unsaturated solution A solution which contains less than the maximum amount of solute that a given amount of solvent can dissolve at the given temperature.

Valence-shell electrons Those electrons in an atom's electron configuration in addition to the inert-gas core and which account for the chemical properties of the atom.

Van der Waals equation An extension of the ideal gas equation which takes into account the actual volumes of gas molecules and their intermolecular interactions.

Van der Waals forces A general term encompassing dipole–dipole, dipole-induced dipole, and London dispersion forces.

Vapor pressure The pressure the vapor form of a solid or liquid substance has when it is in equilibrium with the solid or liquid.

Viscosity A measure of a fluid's resistance to flow.

Voltaic cell A device in which the energy of an oxidation–reduction reaction is manifested as an electrical current.

VSEPR model A set of rules which allows predictions of molecular geometries from Lewis formulas.

Vulcanization The cross-linking of the chains of rubber accomplished by the addition of sulfur and heating.

Water table The level at which underground water exists and from which it may be recovered by means of wells. Lakes, ponds, and streams may form when the water table intersects the surface of the earth.

Wavelength The distance between two successive comparable points on a periodic disturbance such as a wave.

Wave-particle duality The manifestation of wave properties by a particle, and particle properties by a wave; that is, both waves and particles have some attributes of the other.

Weak acid An acid which is capable of donating but a small fraction of its protons to a reference substance, usually the solvent.

Weak base A base which is capable of accepting but a small fraction of the protons it has a capacity for, from a reference substance, usually the solvent.

Weight A force equal to the mass of an object times the acceleration of gravity acting on the object.

Weight-average molecular weight The sum of the products of the molecular weights of each type of molecule in a sample and the mass of such molecules, divided by the total mass of molecules in the sample.

Weight of a mole The weight of one mole of molecules of a substance (usually in grams).

Wetting The ability of a substance to attract the molecules of another substance better than it attracts its own molecules.

Work A product of a force times the distance through which it acts.

X-ray scattering Absorption of X-rays by a crystal, followed by reemission of the X-rays.

Zeolite A complex silicate which retains its crystalline form when it loses water of hydration, thereby acquiring a porous structure.

Zone refining A method of obtaining ultrapurified substances by repeated meltings of a narrow zone which is moved from one end of a solid sample to another. Impurities generally become concentrated in the molten zone and are thus swept to the ends of the sample.

INDEX

Absolute alcohol, meaning of, 260
Absolute zero, 66, 77
Accuracy, meaning of, 714
Acetaldehyde, see Ethanal
Acetamide, see Ethanamide
Acetic acid, see Ethanoic acid
Acetone, see Propanone
Acetophenone, see Methyl phenyl ketone
Acetyl chloride, see Ethanoyl chloride
Acetyl group, see Ethanoyl group
Acetylene, see Ethyne
Acid anhydrides, 385
Acids, Brønsted-Lowry, 328
    carboxylic, 440 ff
    conjugate, 328
    degree of dissociation of, 334 ff
    ionization constants of, 331 ff
    Lewis, 350
    naming of, 724 ff
    neutralization of, 345
    polyprotic, 344
    strengths of, 329 ff
Acrylonitrile, see Propanenitrile
Activated complex, 561
Activated reactions, temperature effect on, 549
Activation energy, calculation of, 558 ff
    meaning of, 558
Activation enthalpy, 562
Activation entropy, 562
Activation free energy, 561
Activation theory, 557 ff
Activity, meaning of, 306
Activity series of metals, 379
    table, 378
Acyl group, definition of, 441
Addition reactions, 358
    nature of, 484
Adenine, structure of, 454
Adenosine diphosphate, see ADP
Adenosine group, structure of, 641
Adenosine triphosphate, see ATP
ADP, 641

Alchemy, practice of, 3
Alcohols, naming of, 438 ff
    structure of, 438
Aldehydes, naming of, 443
    structure of, 443
Algal bloom, causes of, 639
Alkali metals, 52
Alkaline earth metals, 53
Alkanes, general formula of, 417
    names of (table), 417
    naming of, 422 ff
Alkenes, general formula of, 423
    naming of, 425 ff
Alkenyl groups, names of, 426
Alkoxy group, definition of, 440
Alkyl group, definition of, 421
    names of (table), 417
Alkyl halides, hydrolysis, mechanism of, 555
Alkynes, general formula of, 427
    naming of, 427
Allene, see Propadiene
Allotropes, definition of, 201
Alpha-helix, structure of, 686
Alpha-particle, charge on, 19
    in radioactive decay, 274
Alum, definition of, 256
Aluminum, corrosion of, 206, 543
Aluminum hydride, preparation of, 380
Amalgam, meaning of, 582
Amides, naming of, 448
    structure of, 448
Amines, as Brønsted-Lowry bases, 448
    odors of, 448
    primary, 447
    secondary, 447
    structure of, 447
    tertiary, 447
Amino acids, D and L isomers of, 453
    structure of, 449
    table of, 684
    zwitterion form of, 449
Amino group, definition of, 447

Aminobenzene, see Aniline
Aminopeptidase, use in protein analysis, 689
Aminopropanes, synthesis of, 474 ff
Ammonia, electron configuration of, 179
    geometry of, 170 ff
    synthesis of, 634
Ammonium ion, geometry of, 172
Amorphous solids, 195
Ampere, definition of, 14
Amphetamines, 705
Amphoteric oxides, 385
amu, 22
Analysis, chemical, 8
Anderson, Carl D., 277
Angstrom unit, definition of, 16
Angular momentum, 100
Anhydrides, acid and base, 385
Aniline, structure of, 437, 448
    synthesis of, 479, 483
Anion, definition of, 138
Anisole, see Methyl phenyl ether
Annealing, meaning of, 206
Anode, definition of, 376, 526
Anomers, meaning of, 455
Anthracene, structure of, 433
Antibiotics, 705
Antibonding orbital, 150
Antimatter, characteristics of, 277
Antineutrino, nature of, 277
Antiproton, nature of, 277
Antivirals, 705
Apatite, composition of, 639
Aphids, chemical warning system of, 703
Aqua regia, nature of, 592
Area, SI units of, 708
Argon-40, use in dating, 295
Aromatic compounds, rules for naming, 434 ff
Aromatic hydrocarbons, 431
Aromatic sextet, 433
Arrhenius, Svante A., 13
    acid–base theory of, 327

759

picture of, 559
Arrhenius equation, 558
Aryl groups, names of, 434
Ashoka pillar, 206
Aston, F. W., 20
Atmosphere (pressure unit), definition of, 64
Atomic bomb, 287
Atomic hydrogen, liquid, nature of, 374
　production of, 381
Atomic hydrogen torch, 383
Atomic mass unit, definition of, 22
Atomic number, definition of, 21
Atomic pile, nature of, 285
Atomic theory, 7, 10
Atomic volume, periodic variation of, 123
Atomic weight, definition of, 22
Atoms, immutability of, 10
　meaning of, 10
ATP, function of, 701
　properties of, 641
　structure of, 641
Aufbau principle, 120
Autocatalysis, 570
Average atomic weight, 22
Avogadro, Amadeo, 69
Avogadro's law, 69
Avogadro's number, definition of, 22
　determination by X-ray diffraction, 198
Azeotrope, definition of, 260

Background radiation, meaning of, 281
Bacon, Francis, 4
Baekeland, Leo, 663
Bakelite, production of, 663 ff
　structure of, 664
Balances, chemical, 24, 25
Balancing equations, general rules, 42 ff
　oxidation-reduction reactions, 361 ff
Balmer-Rydberg-Ritz formula, 103
Band gaps in solids, 208
Band theory of solids, 208
Bands in solids, 208
Barium, preparation of, 588
Bases, conjugate, 328
　degree of dissociation of, 334 ff
　ionization constants of, 331 ff
　neutralization of, 345
　strengths of, 329 ff
Basic anhydrides, 385
Battery, meaning of, 535
　storage, operation of, 535
Becquerel, Antoine Henri, 273, 274
Bends, role of nitrogen in, 636

Benzamide, structure of, 448
　synthesis of, 481
Benzene, pi electrons in, 432
　resonance formulas of, 431
　resonance structures in, 147, 148
Benzoic acid, structure of, 441
　synthesis of, 480
　use of, 442
Benzonitrile, synthesis of, 479 ff
Benzoyl chloride, synthesis of, 481
Benzyl alcohol, structure of, 439
　synthesis of, 480 ff
Benzyl group, structure of, 434
Berthollet, Claude Louis, 303
Beryllium, preparation of, 587
　properties of, 589
Berzelius, Jöns Jacob, 565
Beta-emission, nature of, 277
Beta-particles, in radioactive decay, 274
Beta sheet, structure of, 687
Bethe, Hans A., 623
Bidentate, meaning of, 609
Binary compounds, naming of, 723 ff
Binding energy, 139
Biomagnification, meaning of, 599
Bohr, Niels, 92, 99
Bohr magneton, 621
Bohr radius, 101
Boiling, nature of, 228
Boiling chips, use of, 234
Boiling point, definition of, 233
Boiling-point elevation, equation for, 262
Boltwood, Bertram, 294
Bond distance, equilibrium, 140
Bond energies, for carbon and other atoms, 413
　table of, 507
Bond energy, meaning of, 506
Bond moment, definition of, 167
Bond order, correlation with bond length, 154
　definition of, 152
Bonding orbital, 150
Born, Max, 505
Born-Haber cycle, 504
Boron trifluoride, electron configuration of, 181
　geometry of, 174
　Lewis acid nature, 350
　resonance structures, 147
Bosch process, problem on, 400
Boundary-surface diagrams of orbitals, 111, 624
Boyle, Robert, 4, 63
Boyle's law, 63
　graph of, 64
Bragg, Sir William, 7, 197
Bragg equation, 196

Branched carbon chains, 414
Brass, composition of, 593
Breeder reactor, nature of, 286
Bristlecone pine, use in radiocarbon dating, 294
British thermal unit, definition of, 709
Bromine, disproportionation in, 366
　preparation of, 648
2-Bromopropane, synthesis of, 470
Bronowski, J., 6
Brønsted, Johannes N., 13, 328
Brønsted-Lowry, acid-base theory of, 328
Bronze, composition of, 593
Brown, Robert, 11
Brownian motion, 11
Buffer solutions, nature of, 340
Butadiene, polymers of, 669
1,3-Butadiene, cis and trans isomers of, 428
　structure of, 428
Butanamide, synthesis of, 473
Butanenitrile, synthesis of, 472
Butanes, structure of, 420, 421
Butanoic acid, formula of, 442
　synthesis of, 472
Butanols, structure of, 439
Butanone, structure of, 444
Butanoyl chloride, synthesis of, 473
1-Butene, structure of, 424
2-Butene, diastereomers of, 424
Butenes, synthesis of, 474
Butyric acid, see Butanoic acid

Cadaverine, 638
Cadmium, properties of, 596
Calcium, compounds of, 589 ff
　preparation of, 588
Calcium carbide, preparation of, 590
　structure of, 590
　uses of, 590
Calcium ions, role in blood, 591
Calorie, definition of, 27, 235
Calorimeter, description of, 496
　diagram of, 497
Capacitance, definition of, 165
Capillary, 216
Capillary flow, 216
Carbides, preparation of, 590
Carbohydrates, nature and origin of, 412
Carbolic acid, see Phenol
Carbon, abundance of, 407
　allotropes of, 407
　oxides of, 409
Carbon chains, branched, 414
　unbranched, 414
Carbon cycle, 411

Carbon dioxide, formation from
    limestone, 410
  geometry of, 174
  laboratory production of, 410
  occurrence and properties of,
    409
  reaction with OH$^-$, 409
  reaction with water, 409
Carbon-14, use in dating, 294
Carbon monoxide, in cigarette
    smoke, 409
  natural removal from air, 409
  occurrence of, 409
  toxicity of, 409, 614
Carbon-to-carbon bond, stability of,
    413
Carbon-to-carbon linkages, types
    of, 414
Carbonic anhydrase, role in
    respiration, 596
Carbonyl compounds, preparation
    of, 615
Carbonyl group, structure of, 443
Carboxyl group, structure of, 440
Carboxylic acids, naming of, 440
  structure of, 440
  table of names, 442
Carboxypeptidase, use in protein
    analysis, 689
Carnot, Nicolas L. S., 508
Carnot relationship, 508
Carothers, Wallace Hume, 661
Catalysis, 563 ff
  in sulfur trioxide formation, 564
Catalyst, meaning of, 564
  negative, 566
Catalyst metals, properties of, 617
    ff
Cathode, definition of, 376, 527
Cathode ray tube, description of,
    15
Cathodic protection, 539
Cation, definition of, 138
Cell emf, and free energy, 527
Cell metabolism, general nature of,
    699
Cellulose, structure of, 457
Celsius, Anders, 66
Celsius temperature scale, 710
Cementite, role in structure of
    steel, 206, 255
Centigrade temperature scale, 710
Cesium, properties of, 587
Chadwick, James, 19, 20
Chain branching, 665
Chain initiation, 665
Chain polymerization, nature of,
    665
Chain reaction, nuclear, 285
  description of, 562 ff
  temperature effect on, 549

Changes of state, meaning of, 227
Chargaff, Edwin, 702
Charles, Jacques, 67
Charles's law, 66, 68
Chelates, definition of, 618
Chemical changes, nature of, 227
Chemical equilibrium, 305
  use of Le Châtelier's principle in,
    314
Chemical formulas, 37
Chemical transformations, nature
    of, 303
Chile saltpeter, 648
Chirality, meaning of, 416
Chlorates, preparation of, 651
Chlorine, oxy compounds,
    nomenclature of, 651
  preparation of, 648
Chlorine trifluoride, geometry of,
    176
Chlorites, preparation of, 651
Chlorobenzene, structure of, 434
  synthesis of, 478
Chloroethanoic acids, ionization
    constants of, 652
  structure of, 652
Chloronium ion, 479
Chloroplasts, role in
    photosynthesis, 700
1-Chloropropane, synthesis of, 470
Cholesteric liquid crystals, 221
Chymotrypsin, use in protein
    analysis, 689
Clausius, Rudolf J. E., 493
Cobalamin, 615 ff
Cobalt, biological role of, 615 ff
  properties of, 614 ff
Coenzymes, nature of, 615
Coinage metals, 591
Collagens, nature of, 687
Colligative property, 261
Combination reactions, 358
Combustion, enthalpy of, 496 ff
Common-ion effect, 320
Compound, meaning of, 7
Compressibility factor of gases, 82
Concentration units, use of, 248 ff
Condensation, meaning of, 236
  nature of, 229
Conduction band, 208
Conductors, nature of, 210
Conformation, meaning of, 418
Conjugate acid-base pairs,
    definition of, 328
Conservation of energy, law of, 27,
    499
Conservation of matter, law of, 8
Constant-boiling mixtures, 260
Contact process, 644
Continuum, in hydrogen atom
    spectrum, 104

Contour diagrams, interpretation
    of, 728 ff
Control substances, in nuclear
    reactors, 285
Coordination complexes, nature of,
    608
Coordination number, meaning of,
    608
Copolymerization, uses of, 672
Copolymers, general nature of,
    671 ff
  general types of, 672
Copper, alloys of, 593
  oxidation of, 592
  production of, 593
Copper salts, nature of, 593
Corey, Robert B., and protein
    structure, 686
Corrosion, galvanic, 538 ff
Coulomb, definition of, 14
Coulomb's law, 18, 135, 136
Coulson, Charles A., 165
Couper, A. S., 432
Covalent bond, nature of, 139 ff
Covalent bonding, 138 ff
Cracking of hydrocarbons, 379
Crafts, James M., 482
Crick, Francis H. C., 702
Critical mass, definition of, 287
Critical temperature, 80
Cryoscopic constant, 263
Crystal, extended lattice, 201
Crystal-field model, 623
Crystal-field splitting parameter,
    624
Crystal lattice, 197
Crystalline melting point of a
    polymer, 682
Crystallization, fractional, 254
  meaning of, 236
Crystallographic systems, 199
Crystals, electron-density maps of,
    198
  growing of, 256
  ionic, 200
  molecular, 201
Curie, Marie Sklodovska, 273, 274
Curie, Pierre, 273, 274
Cyclic AMP, role of, 705
Cyclic carbon chains, illustration of,
    414
Cycloalkanes, general formula of,
    429
Cycloalkenes, example of, 431
Cyclobutane, structure of, 430
Cyclohexane, conformations of, 430
  structure of, 430
Cyclopentane, structure of, 430
Cyclopropane, structure of, 429
Cyclotron, diagram of, 279
Cytosine, structure of, 454

Dacron, composition of, 671
 production of, 679
Dalton, John, 9, 10, 74
Dalton, unit of mass, 708
Dalton's atomic theory, 10
Dalton's law of partial pressures, 74
Dating, radioisotope, 293 ff
Davy, Sir Humphry, 13
 quotations from, 525, 579
De Broglie, Louis Victor, 101
Debye, Peter, 167
Debye unit, definition of, 166
Decay, exponential, 280
Decomposition reactions, 358
Deductive reasoning, 5
Defect, interstitial, 205
 substitutional, 205
Defects in solids, 205 ff
Definite composition, law of, 8
Deliquescence, definition of, 393
Democritus, atomic theory of, 10, 11
Denitrification, 638
Denkewalter, Robert G., 692
Density, definition of, 24
 of a gas, calculation from ideal gas law, 72
Deoxyhemoglobin, 612 ff
Deoxyribose, structure of, 457
Depressants, nature of, 705
Desiccants, function of, 393
Detergents, nature of, 586
Diamagnetic substance, definition of, 127
Diamond, free energy of formation of, 515
 photograph of, 406
 properties and structure of, 407
Diastereomers, definition of, 416
 cis and trans structures of, 424
Diazonium salts, 480
Dielectric breakdown, 208
Dielectric constant, definition of, 165
Dielectrics, description of, 166
Dienes, general formula of, 427
 naming of, 428
 polymers of, 666 ff
Differential manometer, 65
Diffraction grating, description of, 196
Diffusion, definition of, 220
 rate of, in a gas, 78
 role in chemical reactions, 220
 variation in rate of, 220
Dimethyl mercury, nature of, 598
Dimethylamine, structure of, 447
Dimethylbenzene, see Xylenes
N,N-Dimethylformamide, structure of, 449
Dipole moment, definition of, 166
 table of, 169

Dipole moments and boiling points (table), 188
Dirac, Paul A. M., 107
Displacement reaction, 359
Disproportionation reactions, 366
Dissociation energy, 139
Distillation, fractional, 259
Distillation column, 260
Disulfide linkages, in proteins, 686
DMF, see N-N-Dimethylformamide
Domain theory, 610
Doping, 255
Dot formula, Lewis, 144 ff
Double displacement reactions, 359
Drawing of polymers, 679
Dry cell, diagram of, 536
Duality, wave–particle, 96
Ducks, floating of, 218
Dulong, Pierre Louis, 207
Dynamic scattering, 223

Ebulliometric constant, 263
Efflorescence, description of, 394
Einstein, Albert, 97
 and Brownian motion, 11
 mass-energy equivalence relationship of, 283
 and photoelectric effect, 96 ff
 and rate of diffusion, 220
 and specific heats of solids, 207 ff
Elasticity, modulus of, 680
Elastomers, nature of, 682
Electric eye, 96
Electrical quantities, SI units of, 709 ff
Electrochemistry, meaning of, 525
Electrode potentials, table of, 531
Electrolysis, definition of, 376, 579
Electrolysis cell, diagram of, 583
Electrolyte, definition of, 13, 53, 579
Electromagnetic spectrum, 95
Electromotive force, meaning of, 527
Electron, basic nature of, 13 ff
 charge-to-mass ratio of, 14
 mass of, 14
 origin of name, 13
Electron affinity, calculation of, 505
 meaning of, 136
Electron configuration, 113
Electron density, meaning of, 106
Electron distribution, meaning of, 106
Electron dot formulas, 144 ff
Electron volt, definition of, 709
Electronegativity, calculation of, 136 ff
 meaning of, 136
 scales of, 137
Electrons, inner-shell nonbonding, 153
 lone pair, 144

 nonbonding, 144
 valence-shell, 121
Electrovalent bonding, 138 ff
Element, meaning of, 7
Elementary step, 553
Elimination reactions, 358
emf, 527
Empedocles, theory of, 7
Enantiomers, definition of, 416
Endothermic, meaning of, 494
Endothermic processes, 251
Endpoint of a titration, 346
Energy, activation, 558
 binding, 139
 chemical, 496 ff
 conservation of, 27, 499
 definition of, 26, 493 ff
 dissociation, 139
 electrical, 527, 710
 hydration, 253
 interconversion of, 494
 kinetic, 26, 76
 lattice, 202
 law of conservation of, 499
 mass equivalence of, 283
 nuclear, 283 ff
 orbital, 112
 potential, 26, 76
 quantization of, 96
 radiation, 103
 solution, 251
 of spectral transitions, 103
 thermal, 494
 units of, 27, 709
 vibrational, 204, 486
 zero-point, 205
Energy-level diagrams, 102, 105, 155
Energy minimization principle, 112
Energy states, of hydrogen atom, 101
Enthalpies of formation, table, 502
Enthalpy, definition of, 495
 of formation, 501
Entropies, table of, 513
Entropy, meaning of, 508
Enzyme reactions, lock-and-key model, 568
 temperature effect on, 549
Enzymes, function of, 567
Equations, balancing of, 42 ff
 chemical, 42
Equilibria, in solutions of sparsely soluble salts, 318
Equilibrium, and free energy, 516
 chemical, nature of, 305
 nature of, 229
Equilibrium constant, calculations with, 309 ff
 definition of, 306, 307
 in terms of partial pressures, 312
Equivalence point, 346

Equivalent weight, definition of, 367
Esterification, tracer studies of, 446
Esterification reactions, 446
Esters, formation of, 445
 odors of, 447
 structure of, 445
Ethanal, formula of, 443
Ethanamide, structure of, 448
 synthesis of, 478
Ethane, representation of structure, 419
 staggered and eclipsed conformations of, 419
Ethanoic acid, formation from propanone, 477
 ionization of, 331
 source of, 442
 structure of, 441
Ethanol, formula of, 458
 synthesis of, 478
Ethanoyl chloride, structure of, 441
 synthesis of, 478
Ethanoyl group, structure of, 441
Ethene, electron configuration of, 182
 structure of, 424
Ethers, naming of, 440
 structure of, 439
 uses of, 440
Ethoxy group, structure of, 440
Ethyl acetate, see Ethyl ethanoate
Ethyl alcohol, see Ethanol
Ethyl ethanoate, structure of, 445
Ethyl ether, nature of, 440
Ethyl isopropyl ether, structure of, 440
Ethylene, see Ethene
Ethylmethylamine, structure of, 447
Ethyne, electron configuration of, 182
 structure of, 427
Eutrophication, 641
Evaporation, nature of, 229
Evolution, chemical, 703 ff
Excess reagent, 46
Excited states, meaning of, 104
Exciton, 210
Exclusion principle, 112
Exothermic, meaning of, 494
Exothermic processes, 251
Explosions, 550, 563
Exponential notation, 711
Eyring, Henry, 562
Eyring rate equation, 562

Fahrenheit temperature scale, 710
Falstaff, unit of beer drinker's throughput, 529
Faraday, Michael, 12, 526, 580–581
Faraday, definition of, 527 ff
 value and meaning of, 580

Faraday constant, 527 ff, 580
Faraday's law, 580
Fast breeder, nature of, 286
Fats, definition of, 586
Xeldspars, nature of, 202
Fermi, Enrico, 284, 285
Ferromagnetism, domain theory of, 610
Filled shells, role in electron properties, 125
Fisher spider, 218
Fission, nuclear, 284
Fixation of nitrogen, 638
Fluids, definition of, 215
Fluorescence, definition of, 273
Fluoride ion, role in dental caries prevention, 650
Fluorine, preparation of, 648
Fluorite, production of F_2 from, 649
Force, 23, 709
 SI unit of, 65, 709
Force constant, meaning of, 375
Forces, dipole–dipole, 186
 dipole-induced dipole, 186
 hydrogen-bonding, 186
 intermolecular, 185 ff
 London dispersion, 186
 van der Waals, 186
Formaldehyde, see Methanal
Formalin, 443
Formamide, see Methanamide
Formate ion, formation from CO, 410
Formic acid, see Methanoic acid
Formula weight, 39
Formulas, chemical, 11
 Dalton's, 38
 empirical, 37
 geometric, 40
 linkage, 40
 molecular, 38
 network, 40
 simplest, 37
 structural, 40
Formyl group, see Methanoyl group
Fossil fuels, nature and origin of, 412
Fractional distillation, 259
Frank, F. C., crystal growth theory of, 255
Franklin, Rosalind, 702
Frasch process, 642
Fraunhofer, Josef, 99
Fraunhofer lines, 99
Free energy, and cell emf, 527
 and equilibrium, 516
Freezing-point depression, equation for, 262
Freons, structure and properties, 652
Frequency, definition of, 94

Frequency factor, 558
Friedel, Charles, 482
Friedel-Crafts reaction, 482
Fuel cells, description of, 536 ff
 diagrams of, 537, 538
 efficiency of, 538
Functional groups, 437
Fusion, nuclear, 284

Galvani, Luigi, 524
Galvanic corrosion, 538 ff
Galvanism, discovery of, 525
Gamma rays, 274
Gangue, meaning of, 602
Gases, liquefaction of, 82
Gay-Lussac, Joseph Louis, 67
Gay-Lussac's law, 66, 68
Geiger counter, diagram of, 281
General gas law, 68 ff
Genetic code, nature of, 702
Gibbs, J. Willard, 511
Gibbs free energy, 510
Gibbs function, 510
Glass, definition of, 195
Glass-transition region of polymers, 680
Glass-transition temperature in polymers, 681
Glucose, anomers of, 455
 determination of enthalpy of formation, 503
 structure of, 455
D-(+)-Glucose, see Glucose
Glyceraldehyde, D and L forms of, 456
Glycerol, in soaps, 586
Goodyear, Charles, 679
Graham, Thomas, 78
Graham's law, 78
Gram formula weight, 40
Gram molecular weight, 40
Graphite, conversion to diamond, 409, 515
 photograph of, 406
 structure and properties of, 407
Gravitational constant, meaning of, 23
Gray, definition of, 290, 709
Greenhouse effect, 410
Ground state, definition of, 104
Group (of periodic table), 49
Group IA metals, preparation of, 582
 properties of, 583 ff
 table of properties, 584
Group IIA metals, preparation of, 587 ff
 properties of, 588 ff
 table of properties, 589
Group IB metals, properties of, 591 ff
 table of properties, 592

INDEX 763

Group IIB metals, properties of, 594
   table of properties, 594
Guanine, structure of, 454
Guericke, Otto von, 62
Guldberg, Cato, 304
   quotation from, 545
Gutta-percha, structure of, 669

Haber, Fritz, 505
Hahn, Otto, 284, 285
   and Strassmann experiment, 284
Haldane, J. B. S., quotation by, 93
Half-filled shells, role in electron properties, 125
Half-life, calculations involving, 282
   of carbon-14, 293
   of a chemical reaction, 552
   radioactive decay, 280
   of tritium, 294
Half-reactions, 529
Hallucinogenic drugs, nature of, 705
Halogens, 53
   diatomic molecules of, 647
   disproportionation in, 650
   general properties, 647
   in organic compounds, 651 ff
   oxides of, 651
   oxy acids of, 650
   preparation of, 648
   table of properties, 649
Hardness, scale of, 407
Heat, flow of, 77
   meaning of, 494
Heat capacity, 85
Heat of solution, meaning of, 251
Heavy water, production of, 376
   properties of, 395
Heisenberg, Werner, 104, 105
Heisenberg uncertainty principle, 96
Helium, discovery of, 99
Helmholtz, Hermann L. F., 499
Helmont, Joannes van, 63
Heme group, structure of, 611
Hemocyanin, nature of, 593
Hemoglobin, properties of, 613 ff
   structure of, 612
Henderson-Hasselbalch equation, 342
Henry, William, 253
Henry's law, 253
Hertz, definition of, 95
Hess's law, 503
Heterocyclic molecules, 454
2-Hexene, structure of, 426
High-spin complexes, 620
Hirschmann, Ralph, 692
Hole, octahedral, in crystals, 206
Holes, in semiconductors, 210
Hormones, role of, 704

Humidity, relative, definition of, 232
Hund, Friedrich, 114
Hund's rule, 114
Hybrid orbital, definition of, 177
   dsp^2, 622
   dsp^3, 185
   d^2sp^2, 623
   d^2sp^3, 185
   sp, 182
   sp^2, 180
   sp^3, 177
   sp^3d^2, 623
Hydrates, 392 ff
Hydration, meaning of, 251
Hydrides, classification of, 379
   covalent, 380
   interstitial, 381
   ionic, 379
   saline, 379
Hydrocarbons, alicyclic, 417, 429
   aliphatic, 416 ff
   aromatic, 416, 431
   nature of, 416
   types of, 416
Hydrogen, abundance of, 373, 374
   atomic, 381
   basic physical properties of, 373
   industrial preparation of, 379
   isotope effect in, 375
   isotopes of, 21 ff, 374 ff
   laboratory preparation of, 376
   presence on Jupiter, 374
   reduction of metal oxides and chlorides with, 381
   role in photosynthesis, 374
Hydrogen atom, energy-level diagram of, 102
   spectrum of, 102
Hydrogen bomb, 289
Hydrogen bond, general nature of, 187
Hydrogen bromide, preparation of, 650
Hydrogen carbonate ion, structure of, 410
Hydrogen chloride, preparation of, 649
Hydrogen electrode, 529 ff
Hydrogen fluoride, preparation of, 649
Hydrogen halides, preparation of, 380, 649 ff
Hydrogen iodide, preparation of, 650
Hydrogen sulfide, aqueous equilibria of, 646
   preparation of, 642
   properties of, 642 ff
Hydrologic cycle, 395 ff
Hydrometer, use of, 536
Hydronium ion, nature of, 329

Hydroxides, characteristics of, 52
Hydroxyl (OH) radicals, role in radiation damage, 290
Hygroscopic substances, 393
Hypochlorites, preparation of, 651

Ice, melting point of, 233
Ice point, definition of, 258
Ideal gas, definition of, 70
Ideal gas constant, 70
Ideal gas equation, 70
Ignition temperature, meaning of, 550
Impurities, role of in solids, 205 ff
Indicator, choice of, 347 ff
   acid–base, use of, 347 ff
Inductive effect, 652
Inductive reasoning, 4
Inert-gas core, meaning of, 121
Inert gases, 53 ff
Inertia, relationship to mass, 23
Infrared spectroscopy, uses of, 486
Inhibitors, 566
Inner-orbital complexes, 623
Inorganic compounds, naming of, 723 ff
Insulators, description of, 166
Insulin, description of, 692
Interhalogen compounds, 651
Interionic interaction theory, 266
Iodine, preparation of, 648
Ion, definition of, 12
Ion-transfer reaction, 357
Ionic bond, rule for predicting, 139
Ionic bonding, 138 ff
Ionic radius, 203
Ionic sizes, periodic trends in, 203
Ionization, definition of, 123
Ionization constant, acids, 331
   bases, 332
Ionization energy, definition of, 124
   periodic nature of, 125
Ionizing powers, relative, of various types of radiation, 290
Ions, electron configurations of, 126
   role in electrolytes, 13
Iron, biological role of, 611 ff
   compounds of, 610
   passive state of, 611
   preparation of, 610
   properties of, 610
Irreversible reactions, nature of, 317
Isoalkanes, general structure of, 422
Isobutane, structure of, *see* 2-Methylpropane
Isoelectronic ions, sizes of, 203, 204
   meaning of, 203
Isolated system, meaning of, 508

Isomers, definition of, 415
  linkage, 415
  optical, 450
  structural, 415
  types of, 415
Isoprene, polymers of, 668, 669
Isopropane, polymers of, 668, 669
Isopropyl alcohol, *see* 2-Propanol
Isopropyl ether, synthesis of, 476
Isotactic polymers, 673
Isothermal, meaning of, 496
Isotope, definition of, 20
Isotope effect, meaning of, 375

Joule, definition of, 27, 709
Joule-Thomson coefficient, 85
Joule-Thomson effect, 82

K-capture, nature of, 295
K_{sp}, definition of, 318
  electrochemical calculation of, 532 ff
Kekulé, Friedrich August, 405
Kelvin, Lord, *see* Thomson, William
Kelvin temperature scale, 710
Ketones, naming of, 444
  structure of, 444
Kinetic energy, 26
Kinetic molecular equation, 77
Kinetic molecular theory of gases, 74 ff

Langmuir, Irving, 240
Laplace, Pierre Simon, 216
Lattice energy, 202
Lavoisier, Antoine, 8, 9
  acid–base duality theory of, 327
Law of atomic heats, 207
Law of combining volumes, 73
Law of partial pressures, Dalton's, 73 ff
Laws, scientific, 4
Le Châtelier, Henri Louis, 240
  principle of, 240 ff
Length, SI units of, 707
Lethal dose (radiation), meaning of, 291
Leukippos, atomic theory of, 10, 11
Leveling effect, of solvent, 331
Lewis, Gilbert N., 143
Lewis acid–base, definition of, 350
Lewis diagrams, 144 ff
Libby, Willard, 293
Life, chemical origin of, 703 ff
Ligand, meaning of, 608
  table of names, 727
Light, plane-polarized, 450
  velocity of, 94
Light-scattering, use in molecular weight determination, 678
Limestone caverns, formation of, 397

Limiting reagent, 46
Linear accelerator, diagram of, 278
Liquid crystals, definition of, 221
  types of, 221
Liquid-flow region, 682
Liquid-flow state, in polymers, 680
Liter, definition of, 708
Lithium, properties of, 584 ff
Lithium aluminum hydride, preparation of, 380
  use in organic syntheses, 477 ff
Lockeyer, Sir Joseph, 99
Lodestone, nature of, 610
Logarithm, definition of, 715
Logarithms, basic relationships of, 719
  characteristics and mantissas, 715
  examples of use, 719 ff
  natural, 721
  table, 716–717
Lone-pair electrons, 144
Lord Kelvin, *see* Thomson, William
Low-spin complexes, 620
Lowry, Thomas M., 328
LSD, 705
Lysozyme, action of, 568 ff

Macromolecules, meaning of, 202
  nature of, 661 ff
Magdeburg experiment, 62
Magnesium, preparation of, 588
  properties of, 589
Magnetic moment, meaning of, 621
Magnetic properties, of atoms, 127
  of molecules, 158, 619 ff
Magnetism, 127
Manometer, principle of operation, 65
Marijuana, 705
Markovnikov, Vladimir V., 485
Markovnikov's rule, 485
Mass, nature of, 23
  SI unit of, 708
Mass action, law of, 304
Mass defect, 22
  definition of, 283
Mass number, definition of, 20
Mass spectrograph, 20
Mass spectrometer, 20
  diagram of, 21
Matter, forms of, 23
  nature of, 23 ff
Mayer, Julius R., 499
Maxwell-Boltzmann distribution, 75, 557
Mechanism, lock-and-key, 568
Mechanism of a reaction, determination of, 555 ff
  meaning of, 553
MEK, *see* Butanone
Melting, nature of, 228
Melting point, definition of, 233

Mendel, Gregor, 701
Mendeleev, Dmitri, 48, 49
Meniscus, 216
Mercury, in amalgams, 582
  compounds (table), 597
  environmental problems with, 598
  physiological action of, 598
  preparation of, 597
  properties of, 594 ff
  uses of, 597
Merrifield, R. B., 690
Mesons, nature of, 277
Mesophase, meaning of, 221
Meta, use in naming, 435
Metal sulfides, qualitative tests for, 645
  solubility of, 645
Metals, general characteristics of, 49; *see also* Group IA metals, Group IIA metals, Group IB metals, Group IIB metals
Metastable systems, 570
Metathesis, 359
Methanal, electronic structure of, 184
  formula of, 443
  geometry of, 173
  Lewis formula of, 147
Methanamide, structure of, 448
  synthesis of, 478
Methane, electron configuration of, 178
  geometry of, 171
Methanoic acid, catalyzed decomposition of, 565 ff
  formation from propanone, 477
  K_a of, 440
  source of, 442
  structure of, 440
Methanol, formula of, 438
  synthesis of, 478
Methanoyl chloride, synthesis of, 478
Methanoyl group, structure of, 441
Methoxy group, structure of, 440
Methyl acetate, *see* Methyl ethanoate
Methyl alcohol, *see* Methanol
Methyl ethanoate, structure of, 445
Methyl ether, structure of, 440
Methyl ethyl ether, structure of, 440
Methyl ethyl ketone, *see* Butanone
Methyl ketone, *see* Propanone
Methyl phenyl ether, 440
Methyl phenyl ketone, structure of, 444
Methyl propanoate, structure of, 445
Methylamine, structure of, 447
Methylbenzene, *see* Toluene

3-Methylbutanal, structure of, 443
Methylcobalamin, role in mercury dissemination, 617
Methylene group, definition of, 430
3-Methylpentane, structure of, 422
2-Methylpropanamide, synthesis of, 474
2-Methylpropane, structure of, 421
2-Methylpropanenitrile, synthesis of, 472
2-Methylpropanoic acid, synthesis of, 472
2-Methylpropanoyl chloride, synthesis of, 473
Meyer, Julius Lothar, 48, 49
Microfibril, 687
Millikan, Robert A., 14, 97, 98
Mitochondria, function of, 700
Mitscherlich, Eilhard, 565
Mixture, meaning of, 8
Model, meaning of, 5
Moderator, use in nuclear fission, 285
Molal concentration, definition of, 248
Molality, 248
Molar concentration, definition of, 248
Molar heat capacity, 85, 235, 494
Molar heat of fusion, 236
Molar heat of vaporization, 236
Molarity, 248
Mole, definition of, 11, 22
  weight of, 40
Mole-fraction composition, definition, 248
Mole ratio, actual, 46
  theoretical, 46
Mole-ratio factor, definition of, 45
  use of in problems, 44 ff
Molecular orbital, density diagrams, 156, 157
  for heteratomic diatomic molecules, 158
  meaning of, 148
  pi type, 150
  sigma type, 148
Molecular orbital method, 148 ff
Molecular pharmacology, 705
Molecular weight, 37
  determination from colligative properties, 264
  of a gas, calculation from ideal gas law, 72
  number-average, 676
  weight-average, 676
Molecule, meaning of, 10
Monochloroacetic acid, 652
Monodentate, meaning of, 609
Monomer, definition of, 663
Moseley, H. G. J., 21, 126

Moseley's law, explanation of, 126 ff
Mulliken, Robert S., 137
Multiphase reactions, equilibria in, 316
Multiple proportions, law of, 9
Mylar, composition of, 671
  production of, 679
Myoglobin, 612 ff

Nagaoka, atomic model of, 16
Naphthalene, ring numbering system of, 436
  structure of, 433
1-Naphthol, structure of, 438
2-Naphthol, structure of, 438
1-Naphthyl group, structure of, 434
2-Naphthyl group, structure of, 434
Natta, Giulio, 673
Nematic liquid crystals, 221
Neon, isotopes of, 21 ff
Neoprene, structure of, 669
Nernst, Hermann Walther, 529
Nernst equation, 528
Neutralization, meaning of, 345
Neutron, discovery of, 19, 20
Neutrons, production of, 278
Newman, Melvin S., 419
Newman projection diagrams, 418 ff
Newton, definition of, 709
Nickel, properties of, 614 ff
Ninhydrin reaction, 688
Nitrate ion, resonance structures in, 146
  preparation of, 634
Nitric acid, preparation of, 634
  reaction with copper, 635
Nitric oxide, nature of, 635
Nitrification, meaning of, 638
Nitriles, general formula of, 472
  naming of, 472
Nitrobenzene, structure of, 434
  synthesis of, 483
Nitrogen, compounds of, 634 ff
  denitrification, 638
  disproportionation in, 367
  fixation of, 638
  isotopic composition of, 636
  molecular, electronic structure of, 634
  nitrification, 638
  occurrence in organic compounds, 635
  properties of, 633 ff
Nitrogen cycle in nature, 638
Nitrogen dioxide, nature of, 635
Nitrogen gas, preparation of, 636
Nitrogen tetroxide, nature of, 635

Nitronium ion, nature of, 483
Noble gases, *see* Inert gases
Nonmetals, general characteristics of, 49
Normality, definition of, 367
  definition for acids and bases, 344
Nuclear binding energy, definition of, 283
Nuclear bomb, principle of, 287
Nuclear fusion, controlled, 289
Nuclear magnetic resonance spectroscopy, 488
Nuclear reactor, diagram of, 286
Nucleon, definition of, 22
Nylon, structure of, 671
  synthesis of, 671
Nylon fibers, production of, 680

Octahedral hole, in crystals, 206
Octahedron, description of, 174
Octet, Lewis, 144
Octet rule, 144
Oil-drop experiment, 15
Oils, definition of, 586
Oleum, nature of, 644
Oparin, Alexsandr I., 704
Optical activity, 450
Optical birefringence, nature of, 222
Orbital, antibonding, 150
  bond, 177
  bonding, 150
  boundary-surface diagrams of, 111, 624
  box diagrams of atoms, 113 ff
  hybrid, *see* Hybrid orbitals
  meaning of, 108
  molecular, *see* Molecular orbital
  order of filling of, 120
  3d, splitting in an octahedral field, 624
Order of a reaction, meaning of, 550
Organic chemistry, definition of, 405
Ortho, use in naming, 435
Osmosis, description of, 264 ff
Osmotic pressure, definition of, 264
  use in molecular weight measurements, 677
Ostwald, Friedrich Wilhelm, 635
Ostwald process, 634
Outer-orbital complexes, 623
Oxidation, definition of, 360
Oxidation numbers, 54
  and periodic table, 122
  table of, 55
Oxidation–reduction reactions, 360 ff

balancing by half-reaction method, 364 ff
balancing by oxidation number method, 361 ff
Oxidation state, 54
Oxides, acidic, 385
  amphoteric, 385
  basic, 384
  types of, 384, 385
Oxidizing agent, definition of, 360
Oxonium ion, *see* Hydronium ion
Oxygen, abundance of, 383
  calculation of atomic weight of, 22
  diamagnetic, production of, 383
  isotopes of, 383
  laboratory preparation of, 383
  occurrence in photosynthesis, 384
  origin of, 383
  preparation of, 384
  properties of, 383
  uses as a tracer, 384
Oxygen cycle, 389
Oxyhemoglobin, 612 ff
Ozone, Lewis formulas for, 387
  preparation of, 387
  properties of, 387
  structure of, 387
Ozone decomposition, mechanism of, 556
Ozone layer, depletion of, 388, 389
  formation of, 387
  role of, 387 ff

Palladium, properties of, 617 ff
PAN, role in photochemical smog, 637
Para, use in naming, 435
Paramagnetic substance, definition of, 127
Partial pressure, definition of, 73
Pascal, Blaise, 215
Pascal, definition of, 65, 709
Pasteur, Louis, 452, 453
  and separation of enantiomers, 451
Pauli, Wolfgang, 113
Pauli exclusion principle, 112
Pauling, Linus, 134, 137
  and protein structure, 686
Penetrating powers, relative, of various types of radiation, 290
Penicillin, action of, 705
2-Pentanone, structure of, 444
3-Pentanone, structure of, 444
1-Pentene, structure of, 425
2-Pentene, diastereomers of, 425
Pepsin, use in protein analysis, 689
Percentage composition, definition of, 248
Perchlorates, preparation of, 651

Period (of periodic table), 49
Periodic law, 49
Periodic property, definition of, 47
Periodic table, general nature of, 49
  and law of definite proportions, 123
  and quantum rules, 118 ff
  table, 50–51
Pernicious anemia, role of cobalt in, 617
Peroxyacetylnitrate, *see* PAN
Peroxides, 385
Perrin, Jean-Baptiste, 14
Petit, Alex-Thérèse, 207
pH, calculation of, 336 ff
  definition of, 336
  electrochemical measurement of, 533
  of world's oceans, 343
pH meter, operation of, 533 ff
  photograph of, 534
Phase, definition of, 236
Phase diagram, description of, 236
Phenanthrene, structure of, 433
Phenolase, nature of, 593
Phenols, K_a's of, 439
  structure of, 438
  synthesis of, 483
Phenyl group, structure of, 434
Phenyl methyl ketone, synthesis of, 482
Phenylacetic acid, *see* Phenylethanoic acid
Phenylamine, *see* Aniline
Phenylethanoic acid, structure of, 443
1-Phenylethanol, synthesis of, 482
Pheromones, general nature of, 702
Philosopher's stone, 7
Phlogiston, 14, 357
Phosphoric acid, preparation of, 640
  structure of, 640
Phosphorous acid, preparation of, 640
  structure of, 640
Phosphorus, compounds of, 640 ff
  preparation of, 639 ff
  properties of, 639 ff
Phosphorus(III) oxide, reaction of, 640
Phosphorus(V) oxide, geometric structure of, 40, 41
  reaction of, 640
Photochemical smog, role of nitrogen oxides in, 637 ff
Photoelectric cell, diagram of, 97
Photoelectric effect, 96
Photography, use of silver in, 593

Photon, definition of, 96
Photosynthesis, general nature of, 699 ff
Photovoltaic cell, 211
Physical changes, nature of, 227
Phytoplankton, 412
Pi bond, 154
Planck, Max, 97
Planck's constant, 96
Plasticizers, roles of, 682
Platinum, properties of, 617 ff
p-n junction, 211
Polar bonds, 141
Polarizability, 678
Polonium, discovery of, 273
Polyamides, structure of, 671
Polydispersity, definition of, 676
Polyesters, structure of, 671
Polymer, amorphous, 680
  crystalline melting points of (table), 683
  crystallinity in, 680
  definition of, 663
  diene-type, 666 ff
  glass-transition temperatures of (table), 683
  isotactic, 673
  molecular weight of, 675 ff
  syndiotactic, 673
  vinyl-type, tables of, 667
Polymerization, step-growth, 669
  stereospecific, 673
  Ziegler-Natta mechanism of, 674
Polypeptides, 683
Polyphosphazenes, 695
Polypropylene, 673 ff
Polyprotic acids, equilibria in, 344
Polysaccharides, general nature of, 456
Positron, nature of, 277
Positron emission, nature of, 277
Potassium, properties of, 585
Potential energy, 26
Power, definition of, 27
  SI unit of, 709
Precision, meaning of, 714
Pressure, definition of, 64
  SI units of, 65
  variation with altitude, 78 ff
Priestley, Joseph, 633
Prism, operation of, 99
Propadiene, structure of, 428
Propagation of chain reactions, 665
Propanal, formula of, 443
  synthesis of, 476
Propanamide, synthesis of, 478
Propane, structure of, 420
Propanenitrile, infrared spectrum of, 487
Propanoic acid, synthesis of, 477
1-Propanol, structure of, 438

INDEX **767**

2-Propanol, structure of, 438
Propanols, synthesis of, 475
Propanone, structure of, 444
   synthesis of, 477
   uses of, 444
Propanoyl chloride, synthesis of, 478
Propene, structure of, 424
   synthesis of, 474
Propionaldehyde, see Propanal
n-Propyl alcohol, see 1-Propanol
n-Propyl ether, synthesis of, 476
Propylene, see Propene
Propyne, structure of, 427
Protein-making machine, 692
Proteins, fibrous, 685
   laboratory synthesis of, 689 ff
   structural determination of, 688
   structures of, 683 ff
Proton, charge of, 19
   mass of, 19
Pseudo-order reactions, 557
Psychic energizers, 705
Putrescine, 638

Quantum, definition of, 95 ff
Quantum number, azimuthal, 107
   in Bohr theory, 101
   magnetic, 107
   orbital angular momentum, 107
   principal, 106
   spin, 107
Quantum theory, main features of, 96
   origin of concepts, 95 ff
   of transition-metal complexes, 622 ff

Racemic mixture, 451
Rad, definition of, 709
Radiation, effect on living organisms, 291
   electromagnetic, 94 ff
Radio frequency heating, 255
Radioactive decay, rate of, 279
Radioactivity, nature of, 273
Radium, decay series of, 275
   discovery of, 273
   preparation of, 588
Raoult, François Marie, 257
Raoult's law, 257
Rate constant, meaning of, 550
Rate-determining step, 554
Rate law, meaning of, 550
Rate of reaction, definition of, 545 ff
   factors affecting, 540
   laboratory measurement of, 548
RBE, definition of, 291
Redox reactions, see Oxidation-reduction reactions
Reducing agent, definition of, 360

Reduction, definition of, 360
Refrigerators, basis of operation, 82
Rem, definition of, 290
Reproduction, chemistry of, 702 ff
Resonance structures, 146, 147
Reversible process, meaning of, 509
Reversible reaction, 305
Ribonuclease, description of, 692
Ribose, structure of, 457
Rock candy, production of, 256
Rubber, structure of, 668, 669
Rubbery plateau in polymers, 682
Rubbing alcohol, see 2-Propanol
Rubidium, properties of, 587
Ruhemann's purple, 688
Rule of eight, 144
Rutherford, Ernest, 16, 17, 273, 274
   atomic model of, 17 ff
   experiment of, 17

Sacrificial anode, 539
Salt, general meaning of, 345
   naming of, 725 ff
Saponification, problem on, 491
Scattering, meaning of, 196
Schrödinger, Erwin, 106, 107
Scientific method, steps in, 4
Screw dislocation, 205
Seed crystal, use of, 255 ff
Semicarbazones, structure of, 485
Semiconductor, doped, 210
   intrinsic, 210
   n-type, 210
   p-type, 212
Semimetals, 52
Sequestering agents, 619
SI units, 706
Sigma bond, 154
   in alkanes, 418
Significant figures, definition, 712
   use of, 712
Silicones, structure of, 413, 670
Silk, structure of, 687
Silver, properties of, 593
Smectic liquid crystals, 221
Soap, cleaning action of, 586
   preparation of, 585 ff
Soddy, Frederick, 274, 276
Sodium, properties of, 585
   spectral identification of, 585
Solid-support method, in protein synthesis, 690
Solids, amorphous, definition of, 195
   atomic, 200
   crystalline, definition of, 195
   metallic, 200
   nonmetallic, 200
Solubility, definition of, 250
   factors affecting, 250
   temperature effect on, 252

Solubility-product constant, definition of, 318
   electrochemical determination of, 532
Solute, meaning of, 247
Solution, binary, meaning of, 247
   definition of, 247
   saturated, equilibrium in, 252
   saturated, meaning of, 250
Solution process, 251 ff
Solvation, meaning of, 251
Solvent, meaning of, 247
Solvent leveling effect, 331
Sørensen, Søren Peter Lauritz, 336
Sparsely soluble salts, equilibria in, 318
Specific gravity, definition of, 26
Specific heat, definition of, 27, 235
   of fusion, 235
   of vaporization, 235
Specific rotation, meaning of, 450
Spectrochemical series, for transition-metal complex ligands, 625
Spectrum, nature of, 98
Standard electrode potentials, 531
Standard state, meaning of, 500
Standard-state emf, definition of, 528
Standard substance, meaning of, 347
Standard temperature and pressure, 70
Stannous fluoride, use in toothpaste, 650
Starch, structure of, 457
Staudinger, Hermann, 661
Steel, 206, 255
   tempering and annealing of, 255
Stereoisomers, types of, 416
Stimulants, nature of, 705
Stoney, George J., 13
STP, definition of, 70
Strategy, in chemical synthesis, 468
Strontium, preparation of, 588
Structural isomers, table of, 421
Sublimation, meaning of, 240
   molar heat of, 232
Substitution reactions, 359, 470
Sulfate ion, SO_4^{2-}, geometry of, 175
Sulfur, allotropic forms of, 642
   compounds of, 642 ff
   in organic compounds, 644
   properties of, 642
Sulfur dioxide, oxidation of, 564
   preparation of, 643
   resonance structures in, 146
Sulfur hexafluoride, geometry of, 174
Sulfur monochloride, S_2Cl_2, geometry of, 175

Sulfur trioxide, catalyzed formation of, 564
　preparation of, 643
Sulfuric acid, occurrence on Venus, 644
　preparation of, 644
　structure of, 644
Sulfurous acid, preparation of, 644
　structure of, 644
Sun, nuclear fusion processes in, 288
Supercooling, nature of, 235
Superheating, nature of, 234
Superoxides, 386
Supersaturation, meaning of, 255
Surface tension, definition of, 217
　table of, 219
Survival, chemistry of, 702 ff
Symbols, chemical, 11
Syndiotactic polymers, 673
Synthesis, chemical, 8
Synthetic products, identification of, 485 ff
Szent-Györgi, Albert, 25, 699

Tactics, in chemical synthesis, 468
Temperature, absolute, 66
　Celsius scale, 66
　Kelvin scale, 66
Temperature conversions, 710
Temperature scales, 710
Tempering, meaning of, 206
Termination of a chain reaction, 665
Tetany, causes of, 591
Tetrahedron, description of, 171
　drawings of, 41
Tetrahydrocannabinol, 705
T4 bacteriophage, 702
Theban papyrus, 3
Theory, meaning of, 5
Thermodynamics, first law of, 499
　second law of, 508
　third law of, 513
Thermoplastic, definition of, 665
Thermosetting, definition of, 664
Thermotropic liquid crystals, types of, 221
Thioacetamide, see Thioethanamide
Thioethanamide, structure of, 645
　uses of, 645
Thionylchloride, preparation of, 644
Thompson, Francis, quotation from, 135
Thompson, atomic model of, 16
Thomson, Sir J. J., 14
Thomson, William (Lord Kelvin), 67
Three-body reaction, description of, 388
Threshold frequency, 97
Thymine, structure of, 454

Tiger moths, mating of, 703
Titration, meaning of, 344
Toluene, structure of, 434 ff
　synthesis of, 482
Torr, definition of, 65
Torricelli, Evangelista, 65
Torricelli barometer, 65
Tracers, meaning of, 296
　uses of, 296
Tranquilizers, 705
Transition, meaning of, 103
Transition elements, electron configurations of, 122
Transition metal, definition of, 607
Transition-metal complexes, color of, 620
　crystal-field model of, 623
　magnetic properties of, 620 ff
　naming of, 726
　optical properties of, 619 ff
　structural models of, 622 ff
Transition-metal ions, hybrid orbitals of, 622
Transition metals, table of properties, 608
Transition-state complex, 561
Transition-state theory, 561
Triglycerides, soap from, 586
Trigonal pyramid, description of, 174
Triple helix, structure of, 687
Triple point, definition of, 237
Tripolyphosphate, sodium, properties of, 641
　structure of, 640
Tritium, use in dating, 294
Tutankamun, gold mask of, 578
Tyndall, John, 410

Ultraviolet spectroscopy, 488
Unbranched carbon chains, 414
Uncertainty principle, 96
　illustration of, 104 ff
Unit conversions, 28 ff
Uracil, structure of, 454
Uranium, fission of, 284
Urea, structure of, 467
　synthesis of, 467
Ussher, James, Archbishop, 295

Valence-shell electron pair, repulsion model, 170
Van der Waals, Johannes D., 187
Van der Waals constants of gases, table, 85
Van der Waals equation, 84
Van Vleck, J. H., 623
Van't Hoff factor, 271, 354
Vapor, definition of, 228
Vapor pressure, equation for calculation of, 231
　meaning of, 229 ff
　measurement of, 230, 231
Vaporization, molar heat of, 232
Vectors, addition of, 168
　definition of, 167
Velocity, root-mean-square, of a gas, 75
Vibrational motions in solids, 204
Virus, action of, 702
Viscosity, definition of, 219
Vitamin B_{12}, structure of, 615 ff
Vitamins, roles of, 705
Volta, Alessandro, 525
Voltaic cell, definition of, 526
　diagram of, 526
Voltaic pile, diagram of, 526
Volume, SI unit of, 708
VSEPR model, 170
Vulcanization, 679

Waage, Peter, 304
　quotation from, 545
Wastes, radioactive, 293
Water, abnormal boiling point of, 187
　electron configuration of, 179
　geometry of, 172, 173
　heavy, 395
　ion-product constant of, 332
　rainfall, uses of, 397
　self-ionization of, 332
　unusual properties of, 390 ff
Water cycle, 395
Water gas, problem involving, 400
Water strider, 218
Water table, meaning of, 396
Watson, James D., 702
Watt, definition of, 27, 709
Wave function, meaning of, 108
Wave mechanics, 106
Wave–particle duality, 96
Wavelength, definition of, 94
Weight, definition of, 23
Werner, Alfred, 607
Wetting, nature of, 216
Wilkins, Maurice, 702
Williamson synthesis of ethers, problem on, 491
Wöhler, Friedrich, 466, 467
Wood alcohol, see Methanol
Woodward, Robert B., 469
Work, meaning of, 26
Work function, 98

Xylenes, structure of, 435

Zeolites, nature of, 393
Ziegler, Karl, 673
Zinc, properties of, 596
Zone refining, 255
Zwitterions, 449

SOME COMMON ENERGY UNITS AND THEIR INTERCONVERSIONS*

	cal	joule	eV
1 calorie (cal)	1	4.184	2.612×10^{19}
1 joule (J)	0.2390	1	6.242×10^{18}
1 electron volt (eV)†	3.829×10^{-20}	1.602×10^{-19}	1

* Example: To convert 1 joule to calories, find 1 joule in the left-hand column and look in this row under the cal column to find the number 0.2390. Thus, 1 joule is equivalent to 0.2390 cal.

† To convert 1 eV per molecule to $kJ \cdot mol^{-1}$, multiply the quantity in the joule column by 6.023×10^{23} and divide by 10^3. Thus 1 eV per molecule is equivalent to 96.49 $kJ \cdot mol^{-1}$. This is also equivalent to 23.06 $kcal \cdot mol^{-1}$.